Springer Series in Statistics

Series editors
Peter Bickel, CA, USA
Peter Diggle, Lancaster, UK
Stephen E. Fienberg, Pittsburgh, PA, USA
Ursula Gather, Dortmund, Germany
Ingram Olkin, Stanford, CA, USA
Scott Zeger, Baltimore, MD, USA

More information about this series at http://www.springer.com/series/692

Jan G. De Gooijer

Elements of Nonlinear Time Series Analysis and Forecasting

Springer

Jan G. De Gooijer
University of Amsterdam
Amsterdam, The Netherlands

ISSN 0172-7397   ISSN 2197-568X (electronic)
Springer Series in Statistics
ISBN 978-3-319-43251-9   ISBN 978-3-319-43252-6 (ebook)
DOI 10.1007/978-3-319-43252-6

Library of Congress Control Number: 2017935720

© Springer International Publishing Switzerland 2017
This work is subject to copyright. All rights are reserved by the Publisher, whether the whole or part of the material is concerned, specifically the rights of translation, reprinting, reuse of illustrations, recitation, broadcasting, reproduction on microfilms or in any other physical way, and transmission or information storage and retrieval, electronic adaptation, computer software, or by similar or dissimilar methodology now known or hereafter developed.
The use of general descriptive names, registered names, trademarks, service marks, etc. in this publication does not imply, even in the absence of a specific statement, that such names are exempt from the relevant protective laws and regulations and therefore free for general use.
The publisher, the authors and the editors are safe to assume that the advice and information in this book are believed to be true and accurate at the date of publication. Neither the publisher nor the authors or the editors give a warranty, express or implied, with respect to the material contained herein or for any errors or omissions that may have been made. The publisher remains neutral with regard to jurisdictional claims in published maps and institutional affiliations.

Printed on acid-free paper

This Springer imprint is published by Springer Nature
The registered company is Springer International Publishing AG
The registered company address is: Gewerbestrasse 11, 6330 Cham, Switzerland

To Jeanne

Preface

Empirical time series analysis and modeling has been deviating, over the last 40 years or so, from the linear paradigm with the aim of incorporating nonlinear features. Indeed, there are various occasions when subject-matter, theory or data suggests that a time series is generated by a nonlinear stochastic process. If theory could provide some understanding of the nonlinear phenomena underlying the data, the modeling process would be relatively easy, with estimation of the model parameters being all that is required. However, this option is rarely available in practice. Alternatively, a particular nonlinear model may be selected, fitted to the data and subjected to a battery of diagnostic tests to check for features that the model has failed adequately to approximate. Although this approach corresponds to the usual model selection strategy in linear time series analysis, it may involve rather more problems than in the linear case.

One immediate problem is the selection of an appropriate nonlinear model or method. However, given the wealth of nonlinear time series models now available, this is a far from easy task. For practical use a good nonlinear model should at least fulfill the requirement that it is general enough to capture some of the nonlinear phenomena in the data and, moreover, should have some intuitive appeal. This implies a systematic account of various aspects of these models and methods.

The Hungarian mathematician John von Neumann once said that the study of nonlinear functions is akin to the study of non-elephants.[1] This remark illustrates a common problem with nonlinear theory, which in our case is equivalent to nonlinear models/methods: the subject is so vast that it is difficult to develop general approaches and theories similar to those existing for linear functions/models. Fortunately, over the last two to three decades, the theory and practice of "non-elephants" has made enormous progress. Indeed, several advancements have taken place in the nonlinear model development process in order to capture specific nonlinear features of the underlying data generating process. These features include symptoms such as

---

[1] A similar remark is credited to the Polish mathematician Stanislaw M. Ulam saying that using a term like nonlinear science is like referring to the bulk of zoology as the study of non-elephant animals; Campbell, Farmer, Crutchfield, and Jen (1985), "Experimental mathematics: The role of computation in nonlinear science". *Communications of the ACM*, 28(4), 374–384.

non-Gaussianity, aperiodicity, asymmetric cycles, multi-modality, nonlinear causal relationships, nonstationarity, and time-irreversibility, among others. Additionally, considerable progress has been made in the development of methods for real, out-of-sample, nonlinear time series forecasting.[2]

Unsurprisingly, the mass of research and applications of nonlinear time series analysis and forecasting methods is scattered over a wide range of scientific disciplines and numerous journal articles. This does not ensure easy access to the subject. Moreover, different papers tend to use different notations making it difficult to conceptualize, compare, and contrast new ideas and developments across different scientific fields. This book is my attempt to bring together, organize, extend many of the important ideas and works in nonlinear time series analysis and forecasting, and explain them in a comprehensive and systematic statistical framework.

While some mathematical details are needed, the main intent of the book is to provide an overview of the current state-of-the-art of the subject, focusing on practical issues rather than discussing technical details. To reach this goal, the text offers a large number of examples, pseudo-algorithms, empirical exercises, and real-world illustrations, as well as other supporting additions and features. In this respect, I hope that the many empirical examples will testify to the breadth of the subject matter that the book addresses. Some of the material presented in the book is my own or developed with co-authors, but a very large part is based on the contributions made by others. Extensive credit for such previously published work is given throughout the book, and additional bibliographic notes are given at the end of every chapter.

Who is this book for?

The text is designed to be used with a course in *Nonlinear Time Series Analysis*, *Statistical System Processing* or with a course in *Nonlinear Model Identification* that would typically be offered to graduate students in system engineering, mathematics, statistics, and econometrics. At the same time, the book will appeal to researchers, postgraduates, and practitioners in a wide range of other fields. Finally, the book should be of interest to more advanced readers who would like to brush up on their present knowledge of the subject. Thus, the book is not written toward a single prototypical reader with a specific background, and it is largely self-contained. Nevertheless, it is assumed that the reader has some familiarity with basic linear time series ideas. Also, a bit of knowledge about Markov chains and Monte Carlo simulation methods is more than welcome.

The book is selective in its coverage of subjects, although this does not imply that a particular topic is unimportant if it is not included. For instance, Bayesian approaches – that can relax many assumptions commonly made on the type and nature of nonlinearity – can be applied to all models. Of course, the extensive list of

---

[2]Throughout the book, I will use the terms forecast and prediction interchangeably, although not quite precisely. That is, prediction concerns statements about the likely outcome of unobserved events, not necessarily those in the future.

references allows readers to follow up on original sources for more technical details on different methods. As a further help to facilitate reading, each chapter concludes with a set of key terms and concepts, and a summary of the main findings.

What are the main features?
Here are some main features of the book.

- The book shows concrete applications of "modern" nonlinear time series analysis on a variety of empirical time series. It avoids a "theorem-proof" format.
- The book presents a toolbox of discrete-time nonlinear models, methods, tests, and concepts. There is usually, but not in all cases, a direct focus on the "best" available procedure. Alternative procedures that boast sufficient theoretical and practical underpinning are introduced as well.
- The book uses graphs to explore and summarize real-world data, analyze the validity of the nonlinear models fitted and present the forecasting results.
- The book covers time-domain and frequency-domain methods both for the analysis of univariate and multivariate (vector) time series. In addition, the book makes a clear distinction between parametric models on the one hand, and semi- and nonparametric models/methods on the other. This offers the reader the possibility to concentrate exclusively on one of these ways of time series analysis.
- One additional feature of the book are the numerous algorithms in pseudo code form which streamline many ideas and material in a systematic way. Thus readers can rapidly obtain the general gist of a method or technique. Moreover, it is relatively easy to convert a pseudocode to programming language.

Real data
It is well known that real data analysis can reduce the gap between theory and practice. Hence, throughout the book a broad set of empirical time series, originating from many different scientific fields, will be used to illustrate the main points of the text. This already starts off in Chapter 1 where I introduce five empirical time series which will be used as "running" examples throughout the book. In later chapters, other concrete examples of nonlinear time series analysis will appear. In each case, I provide some background information about the data so that the general context becomes clear. It may also help the reader to get a better understanding of specific nonlinear features in the underlying data generating mechanism.

About the chapters
The text is organized as follows. Chapter 1 introduces some important terms and concepts from linear and nonlinear time series analysis. In addition, this chapter offers some basic tools for initial data analysis and visualization. Next, the book is structured into two tracks.

The first track (Chapters 2, 3, 5 – 8, and 10) mainly includes parametric nonlinear models and techniques for univariate time series analysis. Here, the overall outline basically follows the iterative cycle of model identification, parameter estimation, and model verification by diagnostic checking. In particular, Chapter 2 concentrates on some important nonlinear model classes. Chapter 3 introduces the concepts of stationarity and invertibility. The material on time-domain linearity testing (Chapter 5), model estimation and selection (Chapter 6), tests for serial dependence (Chapter 7), and time-reversibility (Chapter 8) relates to Chapter 2. Although Chapter 7 is clearly based on nonparametric methods, the proposed test statistics try to detect structure in "residuals" obtained from fitted parametric models, and hence its inclusion in this track. If forecasting from parametric univariate time series models is the objective, Chapter 10 provides a host of methods. As a part of the entire forecasting process, the chapter also includes methods for the construction of forecast intervals/regions, and methods for the evaluation and combination of forecasts.

When sufficient data is available, the flexibility offered by many of the semi- and nonparametric techniques in the second track may be preferred over parametric models/methods. A possible starting point of this track is to test for linearity and Gaussianity through spectral density estimation methods first (Chapter 4). In some situations, however, a reader can jump directly to specific sections in Chapter 9 which contain extensive material on analyzing nonlinear time series by semi- and nonparametric methods. Also some sections in Chapter 9 discuss forecasting in a semi- and nonparametric setting. Finally, both tracks contain chapters on multivariate nonlinear time series analysis (Chapters 11 and 12). The following exhibit gives a rough depiction of how the two tracks are interrelated.

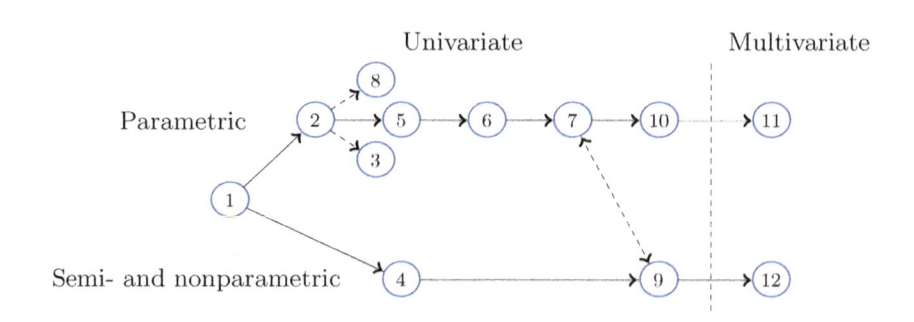

Each solid directed line, denoted by $a \to b$, represents a suggestion that Chapter a be read before Chapter b. The medium-dashed lines indicate that some specific chapters can be read independently. Chapters 2, 7, and 9 are somewhat lengthy, but the dependence among sections is not very strong.

At the end of each chapter, the book contains two types of exercises. Theory exercises illustrate and reinforce the theory at a more advanced level, and provide results that are not available in the main text. The chapter also includes empir-

Preface

ical and simulation exercises. The simulation questions are designed to provide the reader with first-hand information on the behavior and performance of some of the theoretical results. The empirical exercises are designed to obtain a good understanding of the difficulties involved in the process of modeling and forecasting nonlinear time series using real-world data.

The book includes an extensive list of references. The many historical references should be of interest to those wishing to trace the early developments of nonlinear time series analysis. Also, the list contains references to more recent papers and books in the hope that it will help the reader find a way through the bursting literature on the subject.

Reading roadmaps

I do not anticipate that the book will be read cover to cover. Instead, I hope that the extensive indexing, ample cross-referencing, and worked examples will make it possible for readers to directly find and then implement what they need. Nevertheless, those who wish to obtain an overall impression of the book, I suggest reading Chapters 1 and 2, Sections 5.1 – 5.5, Sections 6.1 – 6.2, Sections 7.2 – 7.3, and Chapters 9 and 10. Chapter 3 is more advanced, and can be omitted on a first reading. Similarly, Chapter 8 can be read at a later stage because it is not an essential part of the main text. In fact this chapter is somewhat peripheral.

Readers who wish to use the book to find out how to obtain forecasts of a data generating process maybe "expected" to have nonlinear features, may find the following reading suggestions useful.

- Start with Chapter 1 to get a good understanding of the central concepts such as linearity, Gaussianity, and stationarity. For instance, by exploring a recurrence plot (Section 1.3.4) one may detect particular deviations from the assumption of strict stationarity. This information, added to the many stationarity tests available in the literature, may provide a starting point for selecting and understanding different nonlinear (forecasting) models.
- To further support the above objectives, Sections 2.1 – 2.10 are worth reading next. It is also recommended to read Section 6.1 on model estimation.
- Section 3.5 introduces the concept of invertibility, which is directly linked to the concept of forecastability. So this section should be a part of the reading-list.
- Continue by reading Sections 5.1 on Lagrange multiplier type tests. These tests are relatively easy to carry out in practice, provided the type of nonlinearity is known in advance. The diagnostic tests of Sections 5.4, and the tests of Section 5.5, may provide additional information about potential model inadequacies.
- Next, continue reading Section 6.2.2 on model selection criteria.
- Finally, reading all or parts of the material in Chapter 10 is a prerequisite for model-based forecasting and forecast evaluation. Alternatively, readers with an interest in semi- and nonparametric models/methods may want to consult (parts of) Chapter 12.

Do it yourself ... with a little help from software code

It is likely that the reader is tempted to reproduce the presented results, and also apply some of the nonlinear methods described here to other time series data. This suggest the need of writing ones own programming code. Fortunately, many researchers and specialists have already carried out this task, and results are freely available through the Internet. In addition, there are many user-friendly software packages, often with a graphical interface, that fit the need of a nonlinear time series analyst and, moreover, are easy to use by non-specialists and students. Hence, I decided not to integrate any software package in the text. Rather, at the end of each chapter I provide references to websites where relevant, sometimes even complete programs and/or toolboxes are available for downloading. In doing so, I am certainly taking a risk; Internet is a dynamic environment and sites may change, move, or even disappear. Despite this potential risk, I believe that the benefits of providing links outweighs the aforementioned drawbacks. After all, scientific knowledge is only advancing by making data, software and other material publicly accessible.

Some software programs written for MATLAB and the R system have been kindly made available by researchers working in the field. If appropriate, the Solutions Manual contains the whole source-code of many of the examples and the empirical/simulation exercises. In some cases, however, I have simplified the code and added explanatory text. It goes without saying that the available code and functions are to be used at one's own risk.

The data sets are stored at the website http://extras.springer.com/. My personal web page http://www.jandegooijer.nl contains computer codes, data sets, and other information about the book; see also the link on the book's website.

Acknowledgments

The first step in writing a book on nonlinear time series analysis dates back to the year 1999. Given the growing interest in the field, both Bonnie K. Ray and I felt that there was a need for a book of this nature. However, our joint efforts on the book ended at an early stage because of a change of job (BKR) and various working commitments (JDG). Hence, it is appropriate to begin the acknowledgement section by thanking Bonnie for writing parts of a former version of the text. I also thank her for valuable feedback, comments and suggestions on earlier drafts of chapters.

Many of the topics described in the book are outgrowths of co-authored research papers and publications. These collaborations have greatly added to the depth and breadth of the book. In particular, I would like to acknowledge Kurt Brännäs, Paul De Bruin, Ali Gannoun, Kuldeep Kumar, Eric Matzner–Løber, Martin Knotters, Selliah Sivarajasingham, Antoni Vidiella–i–Anguera, Ao Yuan, and Dawit Zerom. In addition, I am very grateful to Roberto Baragona, Cees Diks, and Mike Clements who read selective parts of the manuscript and offered helpful suggestions for improvement. Thanks also go to the many individuals who have been willing to share their computer code and/or data with me. They are: Tess Astatkie, Luca Bagnato, Francesco Battaglia, Brendan Beare, Arthur Berg, Yuzhi Cai, Kung-Sik Chan, Yi-Ting Chen, Daren Cline, Kilani Ghoudi, Jane L. Harvill, Yongmia Hong, Rob

Hyndman, Nusrat Jahan, Leena Kalliovirta, Dao Li, Dong Li, Guodong Li, Jing Li, Shiqing Ling, Sebastiano Manzan, Marcelo Medeiros, Marcella Niglio, Tohru Ozaki, Li Pan, Dimitris N. Politis, Nikolay Robinzonov, Elena Rusticelli, Hans J. Skaug, Chan Wai Sum, Gyorgy Terdik, Howell Tong, Ruey S. Tsay, David Ubilava, Yingcun Xia, and Peter C. Young (with apologies to anyone unintentionally left out). Finally, I would like to thank all the publishers for permission to use materials from papers that have appeared in their journals.

Amsterdam *Jan G. De Gooijer*

Contents

	Preface	vii
1	**INTRODUCTION AND SOME BASIC CONCEPTS**	1
1.1	Linearity and Gaussianity	2
1.2	Examples of Nonlinear Time Series	4
1.3	Initial Data Analysis	9
	1.3.1 Skewness, kurtosis, and normality	10
	1.3.2 Kendall's (partial) tau	14
	1.3.3 Mutual information coefficient	18
	1.3.4 Recurrence plot	19
	1.3.5 Directed scatter plot	21
1.4	Summary, Terms and Concepts	22
1.5	Additional Bibliographical Notes	22
1.6	Data and Software References	23
	Exercises	25
2	**CLASSIC NONLINEAR MODELS**	29
2.1	The General Univariate Nonlinear Model	30
	2.1.1 Volterra series expansions	30
	2.1.2 State-dependent model formulation	32
2.2	Bilinear Models	33
2.3	Exponential ARMA Model	36
2.4	Random Coefficient AR Model	39
2.5	Nonlinear MA Model	39
2.6	Threshold Models	41
	2.6.1 General threshold ARMA (TARMA) model	41
	2.6.2 Self-exciting threshold ARMA model	42
	2.6.3 Continuous SETAR model	44
	2.6.4 Multivariate thresholds	45
	2.6.5 Asymmetric ARMA model	47
	2.6.6 Nested SETARMA model	49
2.7	Smooth Transition Models	51
2.8	Nonlinear non-Gaussian Models	53
	2.8.1 Newer exponential autoregressive models	53

		2.8.2 Product autoregressive model	54
2.9	Artificial Neural Network Models		56
	2.9.1	AR neural network model	58
	2.9.2	ARMA neural network model	61
	2.9.3	Local global neural network model	62
	2.9.4	Neuro-coefficient STAR model	65
2.10	Markov Switching Models		66
2.11	Application: An AR–NN model for EEG Recordings		69
2.12	Summary, Terms and Concepts		72
2.13	Additional Bibliographical Notes		72
2.14	Data and Software References		75
Appendix			76
	2.A	Impulse Response Functions	76
	2.B	Acronyms in Threshold Modeling	78
Exercises			81

3 PROBABILISTIC PROPERTIES 87
3.1	Strict Stationarity		88
3.2	Second-order Stationarity		90
3.3	Application: Nonlinear AR–GARCH model		91
3.4	Dependence and Geometric Ergodicity		95
	3.4.1	Mixing coefficients	95
	3.4.2	Geometric ergodicity	96
3.5	Invertibility		101
	3.5.1	Global	101
	3.5.2	Local	107
3.6	Summary, Terms and Concepts		110
3.7	Additional Bibliographical Notes		110
3.8	Data and Software References		111
Appendix			112
	3.A	Vector and Matrix Norms	112
	3.B	Spectral Radius of a Matrix	114
Exercises			115

4 FREQUENCY-DOMAIN TESTS 119
4.1	Bispectrum		120
4.2	The Subba Rao–Gabr Tests		126
	4.2.1	Testing for Gaussianity	126
	4.2.2	Testing for linearity	128
	4.2.3	Discussion	129
4.3	Hinich's Tests		130
	4.3.1	Testing for linearity	131
	4.3.2	Testing for Gaussianity	132
	4.3.3	Discussion	133

	4.4	Related Tests		133
		4.4.1	Goodness-of-fit tests	133
		4.4.2	Maximal test statistics for linearity	136
		4.4.3	Bootstrapped-based tests	136
		4.4.4	Discussion	139
	4.5	A MSFE-Based Linearity Test		140
	4.6	Which Test to Use?		146
	4.7	Application: A Comparison of Linearity Tests		148
	4.8	Summary, Terms and Concepts		149
	4.9	Additional Bibliographical Notes		149
	4.10	Software References		151
	Exercises			151
5	**TIME-DOMAIN LINEARITY TESTS**			**155**
	5.1	Lagrange Multiplier Tests		156
	5.2	Likelihood Ratio Tests		168
	5.3	Wald Test		178
	5.4	Tests Based on a Second-order Volterra Expansion		179
	5.5	Tests Based on Arranged Autoregressions		182
	5.6	Nonlinearity vs. Specific Nonlinear Alternatives		186
	5.7	Summary, Terms and Concepts		187
	5.8	Additional Bibliographical Notes		188
	5.9	Software References		190
	Appendix			191
		5.A	Percentiles of LR–SETAR Test Statistic	191
		5.B	Summary of Size and Power Studies	191
	Exercises			194
6	**MODEL ESTIMATION, SELECTION, AND CHECKING**			**197**
	6.1	Model Estimation		198
		6.1.1	Quasi maximum likelihood estimator	198
		6.1.2	Conditional least squares estimator	202
		6.1.3	Iteratively weighted least squares	223
	6.2	Model Selection Tools		227
		6.2.1	Kullback–Leibler information	227
		6.2.2	The AIC, AIC_c, and AIC_u rules	228
		6.2.3	Generalized information criterion: The GIC rule	230
		6.2.4	Bayesian approach: The BIC rule	231
		6.2.5	Minimum descriptive length principle	232
		6.2.6	Model selection in threshold models	233
	6.3	Diagnostic Checking		236
		6.3.1	Pearson residuals	236
		6.3.2	Quantile residuals	240
	6.4	Application: TARSO Model of a Water Table		242

6.5		Summary, Terms and Concepts	246
6.6		Additional Bibliographical Notes	247
6.7		Data and Software References	250
	Exercises		251

7 TESTS FOR SERIAL INDEPENDENCE — 257

7.1	Null Hypothesis	258
7.2	Distance Measures and Dependence Functionals	260
	7.2.1 Correlation integral	260
	7.2.2 Quadratic distance	260
	7.2.3 Density-based measures	263
	7.2.4 Distribution-based measures	265
	7.2.5 Copula-based measures	266
7.3	Kernel-Based Tests	267
	7.3.1 Density estimators	268
	7.3.2 Copula estimators	269
	7.3.3 Single-lag test statistics	270
	7.3.4 Multiple-lag test statistics	272
	7.3.5 Generalized spectral tests	273
	7.3.6 Computing p-values	276
7.4	High-Dimensional Tests	278
	7.4.1 BDS test statistic	278
	7.4.2 Rank-based BDS test statistics	282
	7.4.3 Distribution-based test statistics	284
	7.4.4 Copula-based test statistics	286
	7.4.5 A test statistic based on quadratic forms	290
7.5	Application: Canadian Lynx Data	291
7.6	Summary, Terms and Concepts	294
7.7	Additional Bibliographical Notes	295
7.8	Data and Software References	297
	Appendix	298
	7.A Kernel-based Density and Regression Estimation	298
	7.B Copula Theory	305
	7.C U- and V-statistics	308
	Exercises	310

8 TIME-REVERSIBILITY — 315

8.1	Preliminaries	316
8.2	Time-Domain Tests	317
	8.2.1 A bicovariance-based test	317
	8.2.2 A test based on the characteristic function	319
8.3	Frequency-Domain Tests	322
	8.3.1 A bispectrum-based test	322
	8.3.2 A trispectrum-based test	323

	8.4	Other Nonparametric Tests	324
		8.4.1 A copula-based test for Markov chains	325
		8.4.2 A kernel-based test	327
		8.4.3 A sign test	328
	8.5	Application: A Comparison of TR Tests	330
	8.6	Summary, Terms and Concepts	332
	8.7	Additional Bibliographical Notes	332
	8.8	Software References	333
		Exercises	334
9	**SEMI- AND NONPARAMETRIC FORECASTING**		**337**
	9.1	Kernel-based Nonparametric Methods	338
		9.1.1 Conditional mean, median, and mode	338
		9.1.2 Single- and multi-stage quantile prediction	341
		9.1.3 Conditional densities	347
		9.1.4 Locally weighted regression	352
		9.1.5 Conditional mean and variance	355
		9.1.6 Model assessment and lag selection	358
	9.2	Semiparametric Methods	360
		9.2.1 ACE and AVAS	360
		9.2.2 Projection pursuit regression	363
		9.2.3 Multivariate adaptive regression splines (MARS)	365
		9.2.4 Boosting	369
		9.2.5 Functional-coefficient AR models	374
		9.2.6 Single-index coefficient model	378
	9.3	Summary, Terms and Concepts	380
	9.4	Additional Bibliographical Notes	382
	9.5	Data and Software References	384
		Exercises	387
10	**FORECASTING**		**391**
	10.1	Exact Least Squares Forecasting Methods	392
		10.1.1 Nonlinear AR model	392
		10.1.2 Self-exciting threshold ARMA model	394
	10.2	Approximate Forecasting Methods	398
		10.2.1 Monte Carlo	398
		10.2.2 Bootstrap	399
		10.2.3 Deterministic, naive, or skeleton	399
		10.2.4 Empirical least squares	400
		10.2.5 Normal forecasting error	401
		10.2.6 Linearization	404
		10.2.7 Dynamic estimation	406
	10.3	Forecast Intervals and Regions	408
		10.3.1 Preliminaries	408

		10.3.2	Conditional percentiles	408
		10.3.3	Conditional densities	413
	10.4	Forecast Evaluation		415
		10.4.1	Point forecast	415
		10.4.2	Interval evaluation	419
		10.4.3	Density evaluation	422
	10.5	Forecast Combination		425
	10.6	Summary, Terms and Concepts		426
	10.7	Additional Bibliographical Notes		428
	Exercises			431

11 VECTOR PARAMETRIC MODELS AND METHODS — 439

	11.1	General Multivariate Nonlinear Model		440
	11.2	Vector Models		441
		11.2.1	Bilinear models	441
		11.2.2	General threshold ARMA (TARMA) model	446
		11.2.3	VSETAR with multivariate thresholds	449
		11.2.4	Threshold vector error correction	452
		11.2.5	Vector smooth transition AR	453
		11.2.6	Vector smooth transition error correction	455
		11.2.7	Other vector nonlinear models	455
	11.3	Time-Domain Linearity Tests		458
	11.4	Testing Linearity vs. Specific Nonlinear Alternatives		464
	11.5	Model Selection Tools		471
	11.6	Diagnostic Checking		472
		11.6.1	Quantile residuals	474
	11.7	Forecasting		476
		11.7.1	Point forecasts	476
		11.7.2	Forecast evaluation	478
	11.8	Application: Analysis of Icelandic River Flow Data		481
	11.9	Summary, Terms and Concepts		484
	11.10	Additional Bibliographical Notes		485
	11.11	Data and Software References		488
	Appendix			489
		11.A	Percentiles of the LR–VTAR Test Statistic	489
		11.B	Computing GIRFs	489
	Exercises			490

12 VECTOR SEMI- AND NONPARAMETRIC METHODS — 495

	12.1	Nonparametric Methods		496
		12.1.1	Conditional quantiles	496
		12.1.2	Kernel-based forecasting	498
		12.1.3	K-nearest neighbors	501
	12.2	Semiparametric methods		502

		12.2.1	PolyMARS	502
		12.2.2	Projection pursuit regression	504
		12.2.3	Vector functional-coefficient AR model	506
	12.3	Frequency-Domain Tests		510
	12.4	Lag Selection		512
	12.5	Nonparametric Causality Testing		514
		12.5.1	Preamble	514
		12.5.2	A bivariate nonlinear causality test statistic	515
		12.5.3	A modified bivariate causality test statistic	516
		12.5.4	A multivariate causality test statistic	518
	12.6	Summary, Terms and Concepts		521
	12.7	Additional Bibliographical Notes		521
	12.8	Data and Software References		523
	Appendix			523
		12.A	Computing Multivariate Conditional Quantiles	523
		12.B	Percentiles of the $\widehat{R}(\ell)$ Test Statistic	525
	Exercises			526

References 529

Books about Nonlinear Time Series Analysis 597

Notation and Abbreviations 599

List of Pseudocode Algorithms 607

List of Examples 609

Subject index 613

Chapter 1

INTRODUCTION AND SOME BASIC CONCEPTS

Informally, a time series is a record of a fluctuating quantity observed over time that has resulted from some underlying phenomenon. The set of times at which observations are measured can be equally spaced. In that case, the resulting series is called discrete. Continuous time series, on the other hand, are obtained when observations are taken continuously over a fixed time interval. The statistical analysis can take many forms. For instance, modeling the dynamic relationship of a time series, obtaining its characteristic features, forecasting future occurrences, and hypothesizing marginal statistics. Our concern is with time series that occur in discrete time and are realizations of a stochastic/random process.

The foundations of classical time series analysis, as collected in books such as Box et al. (2008), Priestley (1981), and Brockwell and Davis (1991), to name just a few, is based on two underlying assumptions, stating that:

- The time series process is stationary, commonly referred to as weak or second-order stationarity, or can be reduced to stationarity by applying an appropriate transformation;

- The time series process is an output from a *linear filter* whose input is a purely random process, known as *white noise* (WN), usually following a Gaussian, or normal, distribution. A typical example of a stationary linear Gaussian process is the well-known class of autoregressive moving average (ARMA) processes.

Although these twin assumptions are reasonable, there remains the rather problematic fact that in reality many time series are neither stationary, nor can be described by linear processes. Indeed, there are many more occasions when subject-matter, theory or data suggests that a stationarity-transformed time series is generated by a nonlinear process. In addition, a large fraction of time series cannot be easily transformed to a stationary process. Examples of nonstationary and/or nonlinear time series abound in the fields of radio engineering, marine engineering, servo-systems, oceanography, population biology, economics, hydrology, medical engineering, etc.; see, e.g., the various contributions in the books by Galka (2000), Small (2005), and Donner and Barbosa (2008).

Before focusing on particular models and methods, we deem it useful to introduce some of the basic concepts and notions from linear and nonlinear time series analysis. Specifically, in Section 1.1 we start off by discussing the notion of linearity, and thus nonlinearity, to attempt to reduce potential misunderstandings or disagreements. In Section 1.2, as a prelude to a more detailed analysis in later sections, we discuss five real data sets taken from different subject areas. These series illustrate some of the common features of nonlinear time series data. Each data set is accompanied with some background information. Next, in Section 1.3, we introduce some techniques for initial data analysis. These techniques are complemented with tests for exploratory data analysis.

1.1 Linearity and Gaussianity

There are various definitions of a linear process in the literature. Often it is said that $\{Y_t, t \in \mathbb{Z}\}$ is a *linear process* with mean zero if for all $t \in \mathbb{Z}$

$$Y_t = \sum_{i=-\infty}^{\infty} \psi_i \varepsilon_{t-i}, \text{ where } \sum_{i=-\infty}^{\infty} \psi_i^2 < \infty, \; \{\varepsilon_t\} \stackrel{\text{i.i.d.}}{\sim} (0, \sigma_\varepsilon^2), \tag{1.1}$$

i.e., $\{\varepsilon_t\}$ is a sequence of independent and identically (i.i.d.) random variables with mean zero and finite variance σ_ε^2. Such a sequence is also referred to as *strict white noise* as opposed to *weak white noise*, which is a stationary sequence of uncorrelated random variables. Obviously the requirement that $\{\varepsilon_t\}$ is i.i.d. is more restrictive than that this sequence is serially uncorrelated. Independence implies that third and higher-order non-contemporaneous moments of $\{\varepsilon_t\}$ are zero, i.e., $\mathbb{E}(\varepsilon_t \varepsilon_{t-i} \varepsilon_{t-j}) = 0$ $\forall i, j \neq 0$, and similarly for fourth and higher-order moments. When $\{\varepsilon_t\}$ is assumed to be Gaussian distributed, the two concepts of white noise coincide.

More generally, the above concepts of white noise are in increasing degree of "whiteness" part of the following classification system:

(i) *Weak white noise:*

$$\{\varepsilon_t\} \sim \text{WN}(0, \sigma_\varepsilon^2),$$

i.e., $\mathbb{E}(\varepsilon_t) = 0$, $\gamma_\varepsilon(\ell) = \mathbb{E}(\varepsilon_t \varepsilon_{t+\ell}) = \sigma_\varepsilon^2$ if $\ell = 0$ and 0 otherwise ($\ell \in \mathbb{Z}$).

(ii) *Stationary martingale difference:*

$$\mathbb{E}(\varepsilon_t | \mathcal{F}_{t-1}) = 0, \text{ and } \mathbb{E}(\varepsilon_t^2) = \sigma_\varepsilon^2, \quad \forall t \in \mathbb{Z},$$

where $\mathcal{F}_t$ is the σ-algebra (information set) generated by $\{\varepsilon_s, s \leq t\}$.

1.1 LINEARITY AND GAUSSIANITY

(iii) *Conditional white noise:*

$$\mathbb{E}(\varepsilon_t|\mathcal{F}_{t-1}) = 0, \text{ and } \mathbb{E}(\varepsilon_t^2|\mathcal{F}_{t-1}) = \sigma_\varepsilon^2, \quad \forall t \in \mathbb{Z}.$$

(iv) *Strict white noise:*

$$\{\varepsilon_t\} \stackrel{\text{i.i.d.}}{\sim} (0, \sigma_\varepsilon^2).$$

(v) *Gaussian white noise:*

$$\{\varepsilon_t\} \stackrel{\text{i.i.d.}}{\sim} \mathcal{N}(0, \sigma_\varepsilon^2).$$

The process $\{Y_t, t \in \mathbb{Z}\}$ is said to be *linear causal* if $\psi_i = 0$ for $i < 0$, i.e., if

$$Y_t = \varepsilon_t + \sum_{i=1}^{\infty} \psi_i \varepsilon_{t-i}, \text{ where } \sum_{i=1}^{\infty} \psi_i^2 < \infty, \; \{\varepsilon_t\} \stackrel{\text{i.i.d.}}{\sim} (0, \sigma_\varepsilon^2). \tag{1.2}$$

This infinite moving average (MA) representation should not be confused with the Wold decomposition theorem for *purely nondeterministic* time series processes. In (1.2) the process $\{\varepsilon_t\}$ is only assumed to be i.i.d. and not weakly WN as in the Wold representation. The linear representation (1.2) can also be derived under the assumption that the spectral density function of $\{Y_t, t \in \mathbb{Z}\}$ is positive almost everywhere, except in the Gaussian case when all spectra of order higher than two are identically zero; see Chapter 4 for details. Note that a slightly weaker form of (1.2) follows by assuming that the process $\{\varepsilon_t\}$ fulfills the conditions in (iii).

Time series processes such as (1.2) have the convenient mathematical property that the best H-step ahead ($H \geq 1$) mean squared predictor, or forecast, of Y_{t+H}, denoted by $\mathbb{E}(Y_{t+H}|Y_s, -\infty < s \leq t)$, is identical to the best linear predictor; see, e.g., Brockwell and Davis (1991, Chapter 5). This result has been the basis of an alternative definition of linearity. Specifically, a time series is said to be *essentially linear*, if for a given infinite past set of observations the linear least squares predictor is also the least squares predictor. In Chapter 4, we will return to this definition of linearity.

Now suppose that $\{\varepsilon_t\} \sim \text{WN}(0, \sigma_\varepsilon^2)$ in (1.2). In that case the best mean square predictor may not coincide with the best linear predictor. Moreover, under this assumption, the complete probabilistic structure of $\{\varepsilon_t\}$ is not specified: thus, nor is the full probabilistic structure of $\{Y_t\}$. Also, by virtue of $\{\varepsilon_t\}$ being uncorrelated, there is still information left in it. A partial remedy is to impose the assumption that $\{Y_t, t \in \mathbb{Z}\}$ is a Gaussian process, which implies that the process $\{\varepsilon_t\}$ is also Gaussian. Hence, (1.2) becomes

$$Y_t = \varepsilon_t + \sum_{i=1}^{\infty} \psi_i \varepsilon_{t-i}, \text{ where } \sum_{i=1}^{\infty} \psi_i^2 < \infty, \; \{\varepsilon_t\} \stackrel{\text{i.i.d.}}{\sim} \mathcal{N}(0, \sigma_\varepsilon^2). \tag{1.3}$$

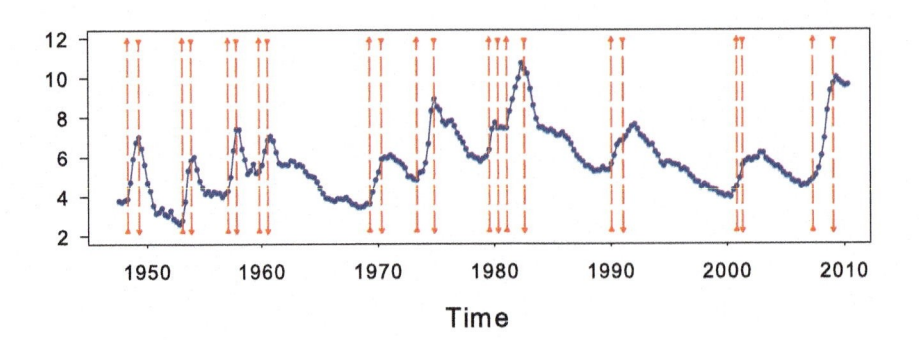

Figure 1.1: *Quarterly U.S. unemployment rate (in %) (252 observations); red triangle up = business cycle peak, red triangle down = business cycle trough.*

Then, the best mean square predictor of $\{Y_t, t \in \mathbb{Z}\}$ equals the best linear predictor. So, in summary, we classify a process $\{Y_t, t \in \mathbb{Z}\}$ as *nonlinear* if neither (1.1) nor (1.2) hold.

Finally, we mention that it is common to label a combined stochastic process, such as (1.1) or (1.2), as the *data generating process* (DGP). A model should be distinguished from a DGP. A DGP is a complete characterization of the statistical properties of $\{Y_t, t \in \mathbb{Z}\}$. On the other hand, a model aims to provide a concise and reasonably accurate reflection of the DGP.

1.2 Examples of Nonlinear Time Series

Example 1.1: U.S. Unemployment Rate

It has long been argued that recessions in economic activity tend to be steeper and more short-lived than recoveries. This implies a *cyclical asymmetry* between the two main phases, expansion and contraction, of the business cycle. A typical example is the quarterly U.S. civilian unemployment rate, seasonally adjusted, covering the time period 1948(i) – 2010(iv) (252 observations) shown in Figure 1.1.[1] The series displays steep increases that end in sharp peaks and alternate with much more gradual and longer declines that end in mild troughs. Time series that exhibit such strong asymmetric behavior cannot be adequately modeled by linear time series models with normally distributed innovations. Such models are characterized by symmetric joint conditional density functions and that rules out asymmetric sample realizations. The vertical (short dashed) red lines in Figure 1.1 denote the business cycle contractions that run from peak to trough as dated by the U.S. National Bureau of Economic Research (NBER).

---

[1] Most of the figures in this book are obtained using **Sigmaplot**, a scientific data analysis and graphing software package. **Sigmaplot**® is a registered trademark of Systat Software, Inc.

1.2 EXAMPLES OF NONLINEAR TIME SERIES

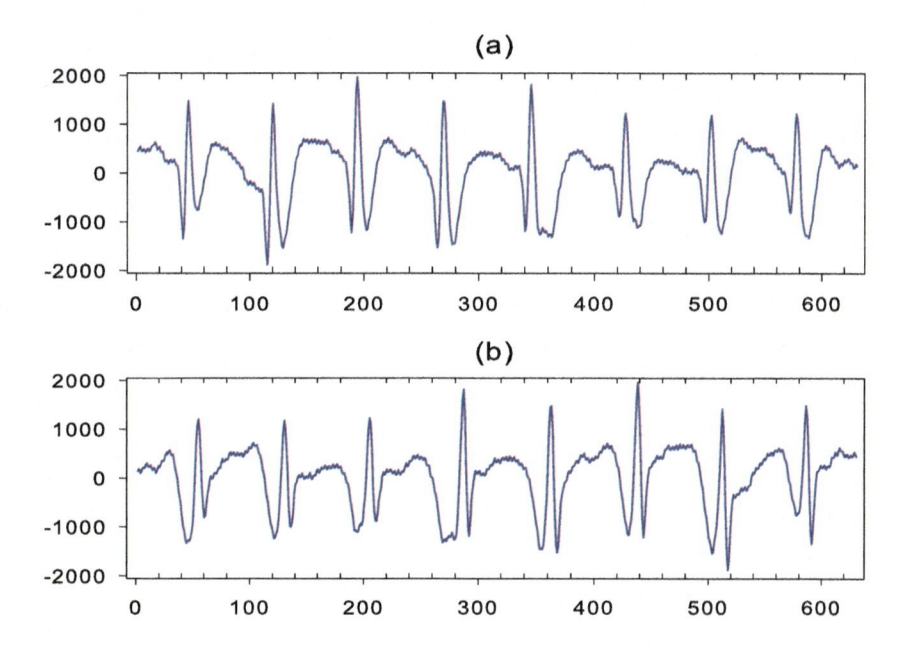

Figure 1.2: *(a) EEG recordings in voltage (μV) for a data segment of 631 observations (just over 3 seconds of signal), and (b) the reversed data plot.*

The NBER uses many sources of information to determine business cycles, including the U.S. unemployment rate. To know the duration and turning points of these cycles it is important to accurately forecast unemployment rates. This applies particularly during contractionary periods.

Example 1.2: EEG Recordings

An electroencephalogram (EEG) is the recording of electrical potentials (activity) of the brain. Special sensors (electrodes) are uniformly distributed over the scalp and linked by wires to a computer. EEG signals are analyzed extensively for diagnosing conditions like epilepsy, memory impairments, and sleep disorder. In particular, a certain type of epileptic EEG, called spike and wave activity, has attracted the attention of many researchers due to its highly nonlinear dynamics.

Figure 1.2(a) shows a short approximately stationary, segment of only 631 observations of an EEG series from an 11-year-old female patient suffering from generalized epilepsy, with absence of seizures. Scalp recordings were obtained at the F3 derivation (F means frontal, and 3 is the location of a surface electrode). The sampling frequency was 200 hertz (Hz), or 5–msec epoch. This is common in EEG data analysis. Further a low-pass filter from 0.3 to 30 Hz was used, which removes high frequency fluctuations from the time series. Most of the cerebral activity oscillation observed in the scalp EEG falls in the range 1 – 20 Hz. Activity below or above this range is likely to be an artifact of non-cerebral origin under standard normal recording techniques.

The spike and wave activity is clearly visible with periodic spikes separated by slow waves. Note that there are differences in the rate at which the EEG series rises to a maximum, and the rate at which it falls away from it. This is an indication that the DGP underlying the series is not *time-reversible*.

A strictly stationary process $\{Y_t, t \in \mathbb{Z}\}$ is said to be time-reversible if its probability structure is invariant with respect to the reversal of time indices; see Chapter 8 for a more formal definition. If such invariance does not hold, the process is said to be *time-irreversible*. All stationary Gaussian processes are time-reversible. The lack of time-reversibility is either an indication to consider a linear stationary process with non-Gaussian (non-normal) innovations or a nonlinear process. No point transformation, like the Box–Cox method, can transform a time-irreversible process into a Gaussian process because such a transformation only involves the marginal distribution of the series and ignores dependence.

One simple way to detect departures from time-reversibility is to plot the time series with the time axis reversed. Figure 1.2(b) provides an example. Clearly, the mirror image of the series is not similar to the original plot. Thus, there is evidence against reversibility. In general, looking at a reverse time series plot can reinforce the visual detection of seasonal patterns, trends, and changes in mean and variance that might not be obvious from the original time plot.

Example 1.3: Magnetic Field Data

The Sun is a source of continuous flows of charged particles, ions and electrons called the solar wind. The terrestrial magnetic field shields the Earth from the solar wind. Changes in the magnetic field induce considerable currents in long conductors on the Earth's surface such as power lines and pipelines. Other undesirable effects include power blackouts, increased radiation to crew and passengers on long flights, and effects on communications and radio-wave propagation.

The primary scientific objectives of the NASA satellite Ulysses are to investigate, as a function of solar latitude, the properties of the solar wind and the interplanetary magnetic field, of galactic cosmic rays and neutral interstellar gas, and to study energetic particle composition and acceleration. Onboard data processing yields hourly time series measurements of the magnetic field. Field vector components are given in units of nanoteslas (nT) and in RTN coordinates, where the R axis is directed radially way from the Sun through the spacecraft (or planet). The T (tangential) axis is the cross product of the solar rotation axis and the R axis. The N (north) axis is the cross product of R and T. Figure 1.3 shows the daily averages of the T component, covering the time period February 17, 1992 – June 30, 1997.

1.2 EXAMPLES OF NONLINEAR TIME SERIES   7

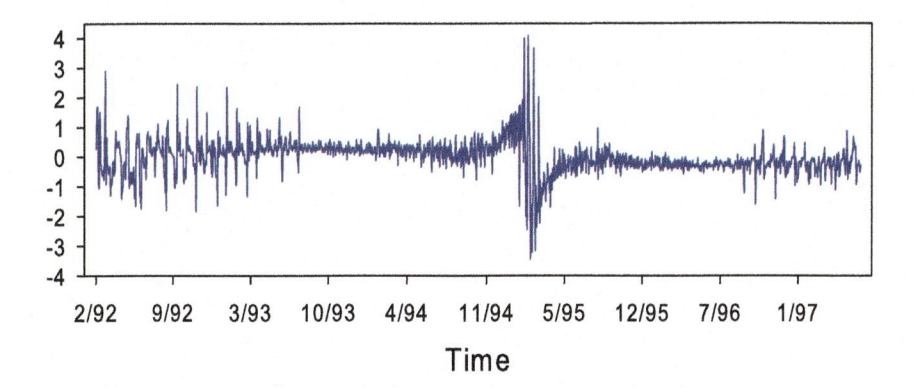

Figure 1.3: Magnetic field *data set,* T *component (in* nT *units) in* RTN *coordinate system. Time period:* February 17, 1992 – June 30, 1997 *(1,962 observations).*

We see relatively large interplanetary shock waves at the beginning of the series followed by a relatively stable period. Then, a considerable increase in wave activity occurs on and around January 11, 1995. In general there is a great variability in the strength of the magnetic field at irregular time intervals. No linear model can account for these effects in the data.

Example 1.4: ENSO Phenomenon

The El Niño–Southern Oscillation phenomenon (ENSO) is the most important source of interannual climate variability. Studies have shown that ENSO events have a tendency to amplify weather conditions such as droughts or excess precipitation in equatorial and subequatorial regions of the globe. Figure 1.4(a) shows the Niño 3.4 index for the time period January 1950 – March 2012 (748 observations) which is the departure in sea surface temperature (SST) from its long-term mean, averaged over the area of the Pacific Ocean between $5°N - 5°S$ and $170°W - 120°W$. Based on this index ENSO events are commonly defined as 5 consecutive months at or above the $+0.5°C$ anomaly for warm (El Niño) events and at or below the $-0.5°C$ anomaly for cold (La Niña) events. Figure 1.4(b) shows the 5-month running average of the Niño 3.4 index with the ENSO events identified by this method.

There is no indication of nonstationarity in the time series plot of the index. However, we see from Figure 1.4(b) that there is a pronounced asymmetry between El Niño and La Niña, the former being very strong. There is obviously a time of year effect, i.e. El Niño and La Niña events typically develop around spring (autumn) in the Northern (Southern) Hemisphere and these events occur every three to five years. These observations suggest that the DGP underlying ENSO dynamics may well be represented by a nonlinear time series

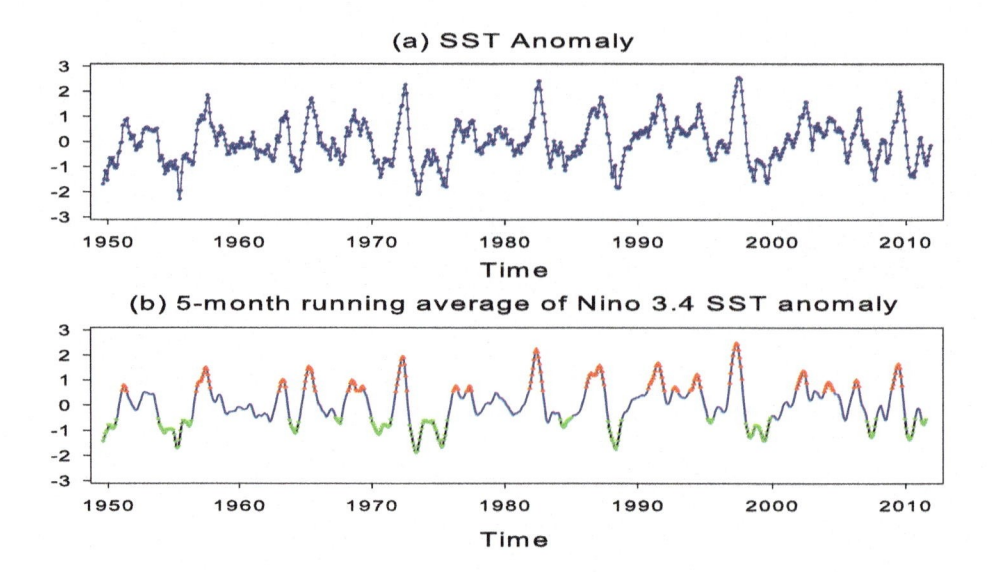

Figure 1.4: *(a) Plot of the* Niño 3.4 *index for the time period* January 1950 – March 2012 *(748 observations); (b) 5-month running average of the* Niño 3.4 *index with El Niño events (red triangle up) and La Niña events (green triangle down).*

model that allows for a smooth transition from an El Niño to a La Niña event, and vice versa.

Example 1.5: Climate Change

One of the major uncertainties associated with the "greenhouse effect" and the possibility of global warming lies within the ocean. To gain a better understanding of how the ocean responds to climate change, it is important to explore and quantify patterns of deep ocean circulation between 3 and 2 million years ago, the interval when significant northern hemisphere glaciation began. To this end the oxygen isotope $\delta^{18}O$ is often used as an indicator of global ice volume. Another important climate variable is the carbon isotope $\delta^{13}C$ which mainly reflects the strength of North Atlantic Deep Water formation.

One of the longest and most reliable data records comes from the Ocean Drilling Program (ODP) site 659, located on the Cape Verde Plateau west of Africa. The sample period corresponds to the past 5,000 ka (1 ka = 1,000 years). The available data set is divided into four distinctive climatic periods: with some climate variability in the oldest period (5,000 – 3,585 ka), but not as strong as the glaciation of the Northern Hemisphere which came in the late Pliocene between 3,885 and 2,625 ka. Then the early Pleistocene started (2,470 – 937 ka) with a time of gradual cooling and additional build-up of ice. Subsequently, after a relatively abrupt increase of global ice volume (the mid-Pleistocene Climatic Transition), the late Pleistocene ice ages started (since 894 ka). Below, and in forthcoming examples, we focus on climatological variables observed during the youngest period.

1.3 INITIAL DATA ANALYSIS

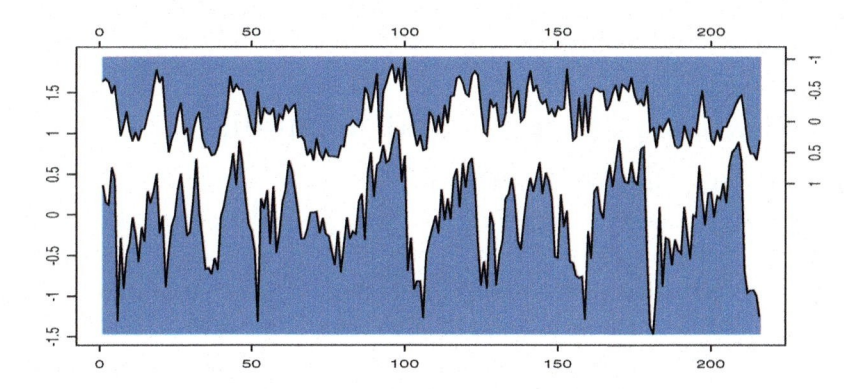

Figure 1.5: *Cave plot of the δ^{13}C (top, axis on the right) and δ^{18}O (bottom, axis on the left) time series. Time interval covers 896 – 2 ka (1 ka = 1,000 years); $T = 216$.*

Figure 1.5 shows two plots of the univariate time series δ^{13}C (denoted by $\{Y_{1,t}\}$) and δ^{18}O (denoted by $\{Y_{2,t}\}$), both of length $T = 216$, for the late Pleistocene ice ages.[2] The graph is called a *cave plot* since the visual distance between the two curves resembles the inside of a cave. The cave plot is constructed so that if the dependence of $\{Y_{1,t}\}$ on $\{Y_{2,t}\}$ is linear and constant over time then the visual distance between the curves is constant. In the present case, this is accomplished by a linear regression of the series $\{Y_{2,t}\}$ on $\{Y_{1,t}\}$ and obtaining the "transformed" series $\{Y_{1,t}\}$ as the fitted values.[3]

From the plot we see that the difference between the curves is not constant during this particular climatic period. This feature makes the data suitable for nonlinear modeling. In addition, we notice a clear correlation between series, with values of δ^{13}C increasing when δ^{18}O decreases, and vice versa. This suggests some nonlinear causality between the two series. In general, these graphs can give a useful visual indication of joint (non)linear short- and long-term periodic fluctuations, even if the two series are observed at irregular times as in the present case.

1.3 Initial Data Analysis

In any data analysis, it is good practice to start with some fairly simple descriptive techniques which will often detect the main features of a given series. For analysis of nonlinear time series, a host of formal and informal statistical methods and visu-

---

[2] The delta (δ) notation refers to the relative deviation of isotope ratios in a sample from a reference (ref) standard. For example, δ^{18}O (‰ vs. ref) = $\{\{(^{18}\text{O}/^{16}\text{O})_{\text{sample}} - (^{18}\text{O}/^{16}\text{O})_{\text{ref}}\}/(^{18}\text{O}/^{16}\text{O})_{\text{ref}}\} \times 1{,}000$. An analogous definition gives δ^{13}C in terms of ^{13}C and ^{12}C.

[3] Transformation used: -0.1136 (intercept), and -0.7628 (slope).

1.3.1 Skewness, kurtosis, and normality

Independent data: Jarque–Bera test

Departures from normality often take the form of asymmetry, or *skewness*. Let $\mu_{r,X} = \mathbb{E}[(X - \mu_X)^r]$ be the rth ($r \in \mathbb{N}$) central moment of a continuous random variable X with mean μ_X and standard deviation σ_X. Assume that the first four moments exist. Then a measure (one of many) of symmetry is given by the third central moment $\mu_{3,X}$. The fourth central moment, $\mu_{4,X}$, measures the tail behavior of X. Normalizing $\mu_{3,X}$ by σ_X^3, and $\mu_{4,X}$ by σ_X^4 gives rise to the *skewness* and *kurtosis* of X, defined as

$$\tau_X = \frac{\mu_{3,X}}{\sigma_X^3} = \frac{\mathbb{E}[(X - \mu_X)^3]}{[\mathbb{E}(X - \mu_X)^2]^{3/2}}, \qquad \kappa_X = \frac{\mu_{4,X}}{\sigma_X^4} = \frac{\mathbb{E}[(X - \mu_X)^4]}{[\mathbb{E}(X - \mu_X)^2]^2}.$$

For a symmetric distribution $\mu_{3,X} = 0$, and thus τ_X will be zero. The kurtosis for the normal distribution is equal to 3. When $\kappa_X > 3$, the distribution of X is said to have fat tails.

Let $\{X_i\}_{i=1}^n$ denote an i.i.d. random sample of X of size n. Then $\mu_{r,X}$ can be consistently estimated by the sample moments $\widehat{\mu}_{r,X} = n^{-1} \sum_{i=1}^n (X_i - \overline{X})^r$, where $\overline{X} = n^{-1} \sum_{i=1}^n X_i$. Sample analogues of τ_X and κ_X are given by

$$\widehat{\tau}_X = \frac{1}{n\widehat{\sigma}_X^3} \sum_{i=1}^n (X_i - \overline{X})^3, \quad \widehat{\kappa}_X = \frac{1}{n\widehat{\sigma}_X^4} \sum_{i=1}^n (X_i - \overline{X})^4, \tag{1.4}$$

where

$$\widehat{\sigma}_X^2 \equiv \widehat{\mu}_{2,X} = \frac{1}{n} \sum_{i=1}^n (X_i - \overline{X})^2.$$

If $\{X_i\} \stackrel{\text{i.i.d.}}{\sim} \mathcal{N}(0, \sigma_X^2)$ then, as $n \to \infty$,

$$\sqrt{n} \begin{pmatrix} \widehat{\tau}_X \\ \widehat{\kappa}_X \end{pmatrix} \xrightarrow{D} \mathcal{N}\left(\begin{pmatrix} 0 \\ 3 \end{pmatrix}, \begin{pmatrix} 6 & 0 \\ 0 & 24 \end{pmatrix}\right). \tag{1.5}$$

Using this asymptotic property, we can perform a Student t-test for testing the null hypothesis $\mathbb{H}_0 : \tau_X = 0$, or testing $\mathbb{H}_0 : \kappa_X - 3 = 0$, separately. A joint test of the null hypothesis $\mathbb{H}_0 : \tau_X = 0$ and $\kappa_X - 3 = 0$, is often used as a test statistic for normality. This leads to the so-called JB (Jarque and Bera, 1987) test statistic, i.e.,

$$\text{JB} = n\left(\frac{\widehat{\tau}_X^2}{6} + \frac{(\widehat{\kappa}_X - 3)^2}{24}\right), \tag{1.6}$$

which has an asymptotic χ_2^2 distribution under $\mathbb{H}_0$, as $n \to \infty$.

1.3 INITIAL DATA ANALYSIS

Independent data: Lin–Mudholkar test

The Lin–Mudholkar test statistic is based on the well-known fact that the sample mean $\overline{X}$ and sample variance $S_X^2 = n\widehat{\sigma}_X^2/(n-1)$ of a random sample $\{X_i\}_{i=1}^n$ are independent if and only if the parent distribution is normal. The practical computation involves three steps. First, obtain the n pairs of *leave-one-out* estimates $(\overline{X}^{-i}, (S_X^{-i})^2)$, where

$$\overline{X}^{-i} = \frac{1}{n-1}\sum_{j\neq i} X_j, \quad S_X^{-i} = \left[\frac{1}{n-2}\sum_{j\neq i}(X_j - \overline{X}^{-i})^2\right]^{1/2}, \quad (i=1,\ldots,n).$$

Next, apply the approximately normalizing cube-root transformation $Y_i = (S_X^{-i})^{2/3}$, and compute the sample correlation coefficient

$$r_{XY} = \frac{\sum_{i=1}^n (X_i - \overline{X})(Y_i - \overline{Y})}{\sqrt{\sum_{i=1}^n (X_i - \overline{X})^2 \sum_{i=1}^n (Y_i - \overline{Y})^2}}$$

as a measure of dependence between $\overline{X}$ and S_X^2. Finally, in view of the robustness and skewness reducing character of the Fisher z-transform, obtain the test statistic

$$Z_2 = \frac{1}{2}\log\left(\frac{1+r_{XY}}{1-r_{X,Y}}\right). \tag{1.7}$$

If the series $\{X_i\}_{i=1}^n$ consists of i.i.d. normal variables, then it can be shown (Lin and Mudholkar, 1980) that Z_2 is asymptotically normally distributed with mean 0 and variance $3/n$.

Within a time series framework, the JB and Z_2 test statistics are typically applied to the residuals, usually written simply as $\widehat{\varepsilon}_t$, of a fitted univariate (non)linear time series model as a final diagnostic step in the modeling process. A drawback of the JB test is that the finite-sample tail quantiles are quite different from their asymptotic counterparts. Alternatively, p-values of the JB test can be determined by means of bootstrapping (BS) or Monte Carlo (MC) simulation. A better-behaved JB test statistic can be obtained using exact means and variances instead of the asymptotic mean and variance of the standardized third- and fourth moments (cf. Exercise 1.5). Nevertheless, the JB and Z_2 tests only rely on the departure of the symmetry of possible alternatives to the normal distribution. However, the question whether for instance a positive skewness in the original series is reproduced by the fitted nonlinear model cannot be answered by analyzing the residuals alone.

Example 1.6: Summary Statistics

Table 1.1 reports summary statistic for the series introduced in Section 1.2. Except for the U.S. unemployment rate, for which we take the first differences, we consider the original data. Note from the last column that the sample kurtosis of the U.S. unemployment rate and the magnetic field data are much

Table 1.1: *Summary statistics for the time series introduced in* Section 1.2.

Series	T	Mean	Med.	Min.	Max.	Std. Dev.	Skewness	Kurtosis
U.S. unemployment rate [1]	252	0.023	-0.033	-0.967	1.667	0.399	1.113	5.741
EEG recordings	631	28.003	194	-1890	1955	630	-0.617	3.233
Magnetic field data	1,962	-0.004	-0.003	-3.448	4.094	0.572	0.337	10.226
ENSO phenomenon	748	-0.024	-0.090	-2.320	2.520	0.845	0.264	3.045
Climate change δ^{13}C	216	-0.103	-0.105	-1.020	0.630	0.392	-0.095	2.115
δ^{18}O	216	-0.035	0.005	-1.470	1.050	0.538	-0.342	2.571

[1] First differences of original data.

larger than the kurtosis for a normal distribution, indicating that both series have heavy tails. Further, the sample skewness of the series indicates no evidence of asymmetry. Below we search for more evidence to support these observations, using a skewness-kurtosis test statistic that is able to account for serial correlation.

Weakly dependent data: A generalized JB test

For testing normality in time series data, we need to introduce some additional notation similar to that given above. In particular, let $\{Y_t, t \in \mathbb{Z}\}$ be an ergodic strictly stationary process (see Chapter 3 for a formal definition of ergodicity) with mean μ_Y, rth central moment $\mu_{r,Y} = \mathbb{E}[(Y_t - \mu_Y)^r]$, and lag ℓ ($\ell \in \mathbb{Z}$) autocovariance function (ACVF) $\gamma_Y(\ell) = \mathbb{E}[(Y_t - \mu_Y)(Y_{t+\ell} - \mu_Y)]$. Given a set of T observations the corresponding sample statistics are $\overline{Y} = T^{-1}\sum_{t=1}^{T} Y_t$, $\widehat{\mu}_{r,Y} = T^{-1}\sum_{t=1}^{T}(Y_t - \overline{Y})^r$, and $\widehat{\gamma}_Y(\ell) = T^{-1}\sum_{t=1}^{T-\ell}(Y_t - \overline{Y})(Y_{t+\ell} - \overline{Y})$, respectively.

Assume that $\{Y_t, t \in \mathbb{Z}\}$ is a Gaussian short memory or weakly dependent process, i.e. $\sum_{j=0}^{\infty}|\gamma_Y(\ell)| < \infty$. Then it can be shown (Lomnicki, 1961; Gasser, 1975) that, as $T \to \infty$,

$$\sqrt{T}\begin{pmatrix} \widehat{\mu}_{3,Y} \\ \widehat{\mu}_{4,Y} - 3\widehat{\mu}_{2,Y}^2 \end{pmatrix} \xrightarrow{D} \mathcal{N}\left(\begin{pmatrix} 0 \\ 0 \end{pmatrix}, \begin{pmatrix} 6F_{3,Y} & 0 \\ 0 & 24F_{4,Y} \end{pmatrix}\right), \quad (1.8)$$

where

$$F_{r,Y} = \sum_{\ell=-\infty}^{\infty}\bigl(\gamma_Y(\ell)\bigr)^r, \quad (r = 3, 4).$$

A consistent estimator of $F_{r,Y}$ is given by $\widehat{F}_{r,Y} = \sum_{|\ell|<T}\bigl(\widehat{\gamma}_Y(\ell)\bigr)^r$, and hence a generalized JB (GJB) statistic for testing normality in weakly dependent data is given by

$$\text{GJB} = \frac{T\widehat{\mu}_{3,Y}^2}{6\widehat{F}_{3,Y}} + \frac{T(\widehat{\mu}_{4,Y} - 3\widehat{\mu}_{2,Y}^2)^2}{24\widehat{F}_{4,Y}}, \quad (1.9)$$

which has an asymptotic χ_2^2 distribution under the null hypothesis (Lobato and Velasco, 2004). Moreover, the test statistic is consistent under the alternative hypothesis.

Comparing (1.6) and (1.9), we see that asymptotically the GJB test statistic reduces to the JB test statistic if the DGP is i.i.d., since $\widehat{\gamma}_Y(\ell) \to 0$, $\forall \ell \neq 0$, and $\widehat{\gamma}_Y(0) = \widehat{\mu}_{2,Y} \neq 0$. Also observe that with positive serial correlation in the first few lags, the denominator in (1.9) will be larger than in JB. Consequently, the chance of rejecting normality will decrease when using the GJB test statistic.

Weakly dependent data: A robust JB test

Consider the coefficient of skewness and its sample analogue, respectively defined as

$$\tau_Y = \mu_{3,Y}/\mu_{2,Y}^{3/2}, \qquad \widehat{\tau}_Y = \widehat{\mu}_{3,Y}/\widehat{\mu}_{2,Y}^{3/2}.$$

Let $\mathbf{Z}_t = \left((Y_t - \mu_Y)^3 - \mu_{3,Y}, (Y_t - \mu_Y), (Y_t - \mu_Y)^2 - \sigma_Y^2\right)'$ be a 3×1 vector. Then, under the null hypothesis that $\tau_Y = 0$ (or, equivalently, $\mu_{3,Y} = 0$), it can be shown (Bai and Ng, 2005) that, as $T \to \infty$,

$$\sqrt{T}\widehat{\tau}_Y \xrightarrow{D} \mathcal{N}\left(0, \frac{\boldsymbol{\alpha}'\boldsymbol{\Gamma}_{22}\boldsymbol{\alpha}}{\sigma_Y^6}\right),$$

where $\boldsymbol{\alpha} = (1, -3\sigma_Y^2)'$ is a 2×1 vector, and $\boldsymbol{\Gamma}_{22}$ is the first 2×2 block matrix of $\boldsymbol{\Gamma} = \lim_{T \to \infty} T\mathbb{E}(\widetilde{\mathbf{Z}}\widetilde{\mathbf{Z}}')$ with $\widetilde{\mathbf{Z}}$ the sample mean of $\{\mathbf{Z}_t\}$.

In applications, $\boldsymbol{\alpha}$ can be consistently estimated by its sample counterpart $\widehat{\boldsymbol{\alpha}} = (1, -3\widehat{\sigma}_Y^2)'$. A consistent and robust estimate, say $\widehat{\boldsymbol{\Gamma}}_{22}$, of the long-run covariance matrix $\boldsymbol{\Gamma}_{22}$ can be obtained by kernel-based estimation. Let $s(\widehat{\tau}_Y) = (\widehat{\boldsymbol{\alpha}}'\widehat{\boldsymbol{\Gamma}}_{22}\widehat{\boldsymbol{\alpha}}/\widehat{\sigma}_Y^6)^{1/2}$. Then, under the null hypothesis $\tau_Y = 0$, the limiting distribution of the estimated coefficient of skewness is given by

$$\widehat{\pi}_{3,Y} = \frac{\sqrt{T}\widehat{\tau}_Y}{s(\widehat{\tau}_Y)} \xrightarrow{D} \mathcal{N}(0,1), \tag{1.10}$$

where it is assumed that $\mathbb{E}(Y_t^6) < \infty$.

Also, Bai and Ng (2005) develop a statistic for testing kurtosis. Similar to the i.i.d. case, the coefficient of kurtosis and its sample analogue are defined as

$$\kappa_Y = \mu_{4,Y}/\mu_{2,Y}^2, \qquad \widehat{\kappa}_Y = \widehat{\mu}_{4,Y}/\widehat{\mu}_{2,Y}^2.$$

Suppose that $\mathbb{E}(Y_t^8) < \infty$. Let $\mathbf{W}_t = \left((Y_t - \mu_Y)^4 - \mu_{4,Y}, (Y_t - \mu_Y), (Y_t - \mu_Y)^2 - \sigma_Y^2\right)'$ be a 3×1 vector. Then, under the null hypothesis $\kappa_Y = 3$, and as $T \to \infty$, it can be shown that

$$\sqrt{T}(\widehat{\kappa}_Y - 3) \xrightarrow{D} \mathcal{N}\left(0, \frac{\boldsymbol{\beta}'\boldsymbol{\Omega}\boldsymbol{\beta}}{\sigma_Y^8}\right),$$

where $\boldsymbol{\beta} = (1, -4\mu_{3,Y}, -6\sigma_Y^2)'$ is a 3×1 vector, and $\boldsymbol{\Omega} = \lim_{T \to \infty} T\mathbb{E}(\widetilde{\mathbf{W}}\widetilde{\mathbf{W}}')$ with $\widetilde{\mathbf{W}}$ the sample mean of $\{\mathbf{W}_t\}$.

In practice, β can be consistently estimated by $\widehat{\beta} = (1, -4\widehat{\mu}_{3,Y}, -6\widehat{\sigma}_Y^2)'$. Let $s(\widehat{\kappa}_Y) = (\widehat{\beta}'\widehat{\Omega}\widehat{\beta}/\widehat{\sigma}_Y^8)^{1/2}$ where $\widehat{\Omega}$ denotes a consistent estimate, using kernel-based estimation of Ω. This result implies that, as $T \to \infty$, under the null hypothesis $\kappa_Y = 3$,

$$\widehat{\pi}_{4,Y} = \frac{\sqrt{T}(\widehat{\kappa}_Y - 3)}{s(\widehat{\kappa}_Y)} \xrightarrow{D} \mathcal{N}(0,1). \tag{1.11}$$

Moreover, it can be shown that $\widehat{\pi}_{3,Y}$ and $\widehat{\pi}_{4,Y}$ are asymptotically independent under normality. Thus, combining both test statistics, a robust generalization of the JB test statistic (1.6) to dependent data is

$$\widehat{\pi}_{34,Y} = \widehat{\pi}_{3,Y}^2 + \widehat{\pi}_{4,Y}^2, \tag{1.12}$$

which is asymptotically distributed as χ_2^2.

Note that the first component of $\{\mathbf{W}_t\}$ depends on the fourth moment of $(Y_t - \mu_Y)^4$, which is a highly skewed random variable even if $\{Y_t, t \in \mathbb{Z}\}$ is not skewed. This will have a considerable impact on the finite-sample properties of both test statistics $\widehat{\pi}_{4,Y}$ and $\widehat{\pi}_{34,Y}$, even with fairly large samples ($T > 1{,}000$), and may lead to incorrect decisions in applied work. Another limitation of both test statistics is that asymptotic theory assumes the existence of moments up to order eight. However, it is a stylized fact that many financial time series are leptokurtic and have heavy-tailed marginal distributions. Thus, the existence of high-order moments cannot taken for granted and should generally be verified.

Example 1.7: Summary Statistics (Cont'd)

Table 1.2 reports values for the sample skewness $\widehat{\pi}_{3,Y}$, the sample kurtosis $\widehat{\pi}_{4,Y}$, the normality tests $\widehat{\pi}_{34,Y}$, and the GJB test statistic for the series introduced in Section 1.2. At the 5% nominal significance level, we find no evidence of skewness in the magnetic field series, the ENSO data, and the two series δ^{13}C and δ^{18}O. We fail to reject the null hypothesis of kurtosis in the EEG recordings, the ENSO data, and the δ^{18}O time series. Interestingly, with $\widehat{\pi}_{34,Y}$ only three time series (U.S. unemployment rate, EEG recordings, and magnetic field data) reject very strongly the null hypothesis of normality (symmetry) with a critical value of $\chi_2^2 = 5.991$ at the 5% nominal significance level. The GJB test statistic confirms these results.

1.3.2 Kendall's (partial) tau

For linear time series processes, the sample autocorrelation function (ACF) and sample partial autocorrelation function (PACF) are useful tools to determine a value for the time lag, or delay, ℓ ($\ell \in \mathbb{Z}$). Often these statistics are used in conjunction with the asymptotic Bartlett 95% confidence band, which for a time series of length

1.3 INITIAL DATA ANALYSIS

Table 1.2: *Test statistics for serially correlated data. The long-run covariance matrices of the test statistics $\hat{\pi}_{3,Y}$, $\hat{\pi}_{4,Y}$, and $\hat{\pi}_{34,Y}$ are estimated by the kernel method with Parzen's lag window; see (4.18).*

Series	Skewness $(\hat{\pi}_{3,Y})$	Kurtosis $(\hat{\pi}_{4,Y})$	Normality $(\hat{\pi}_{34,Y})$	GJB
U.S. unemployment rate[1]	2.602	2.032	6.943	89.400
EEG recordings	-2.805	0.337	8.873	5.731
Magnetic field data	0.927	2.630	7.267	2127
ENSO phenomenon	1.212	0.070	1.488	1.547
Climate change $\delta^{13}C$	-0.508	-2.005	5.280	4.150
$\delta^{18}O$	-1.805	-0.794	3.609	3.720

[1] First differences of original data.

T is given by $\pm 1.96/\sqrt{T}$. However, using Bartlett's formula can lead to spurious results (Berlinet and Francq, 1997) as it is derived under the precise assumptions of linearity of the underlying DGP and vanishing of its fourth-order cumulants (cf. Exercise 1.3).

Kendall's tau test statistic

One simple nonparametric measure for capturing the complete dependence, including nonlinear dependence if present, is Kendall's τ test statistic. It is defined as follows. For pairs of observations $\{(X_i, Y_i)\}_{i=1}^n$ ($n \geq 3$), define the second-order symmetric kernel function $h(i,j)$ to be

$$h(i,j) = h(j,i) = \text{sign}[(X_j - X_i)(Y_j - Y_i)],$$

where $\text{sign}(u) = 1\,(-1, 0)$ if and only if $u > (<, =) \, 0$. Then Kendall's τ test statistic is defined as

$$\hat{\tau} = \binom{n}{2}^{-1} \sum_{i<j}^n h(i,j) = \frac{N_c - N_p}{\frac{1}{2}n(n-1)}. \tag{1.13}$$

Here N_c (c for *concordant*) is the number of pairs for which $h(i,j)$ is positive, and N_d (d for *disconcordant*) is the number of pairs for which $h(i,j)$ is negative.

It is immediately verifiable that (1.13) always lies in the range $-1 \leq \hat{\tau} \leq 1$, where values 1, -1, and 0 signify a perfect positive relationship, a perfect negative relationship, and no relationship at all, respectively. The null hypothesis, $\mathbb{H}_0$, is that the random variables X and Y are independent while the alternative hypothesis, $\mathbb{H}_1$, is they are not independent. For large samples, the asymptotic null distribution of $\hat{\tau}$ is normal with mean zero and variance $2(2n+5)/9n(n-1) \approx 4/9n$. Note that one of the properties of $\hat{\tau}$ is that one of its variables of (X_i, Y_i) can be replaced by its associated ranks. The resulting test statistic is commonly known as the Mann–Kendall test statistic, which has been used as a nonparametric test for trend detection and seasonality within the context of linear time series analysis.

To obtain a version of Kendall's τ test statistic suitable for testing against serial dependence in a time series $\{Y_t\}_{t=1}^T$, simply replace $\{(X_i, Y_i)\}_{i=1}^n$ by $\{(R_i, R_{i+\ell})\}_{i=1}^{T-\ell}$ where $\{R_i\}$ are the ranks of $\{Y_t\}$. Then Kendall's τ test statistic may be defined as

$$\widehat{\tau}(\ell) = 1 - 2N_d(\ell) / \binom{T-\ell}{2} = 1 - \frac{4N_d(\ell)}{(T-\ell)(T-\ell-1)}, \tag{1.14}$$

with

$$N_d(\ell) = \sum_{i=1}^{T-\ell} \sum_{j=1}^{T-\ell} I(R_i < R_j, R_{i+\ell} > R_{j+\ell}).$$

Using the theory of U-statistics for weakly dependent stationary processes (see Appendix 7.C), it can be shown (Ferguson et al., 2000) that under the null hypothesis of serial independence $\sqrt{T}\widehat{\tau}(1)$ is asymptotically distributed as a normal random variable with mean zero and variance 4/9 for $T \geq 4$. For $\ell > 1$, explicit expressions for $\text{Var}(\widehat{\tau}(\ell))$ are rather cumbersome to obtain. However, under the null hypothesis of randomness, any K-tuple of the form $3\sqrt{T}(\widehat{\tau}(1), \ldots, \widehat{\tau}(K))'/2$ is asymptotically multinormal, with mean vector zero and unit covariance matrix.

Table 1.3: *Indicator patterns of the sample ACF and values of Kendall's τ test statistic.*

Series		Lag ℓ									
		1	2	3	4	5	6	7	8	9	10
U.S. unemployment rate	ACF (1)	+*	+*	−	−*	−*	−*	−	−*	−	−
	$\widehat{\tau}(\ell)$ (2)	+•	+•	+	−	−•	−•	−	−•	−	−
EEG recordings	ACF	+*	+*	+*	+*	+*	+*	−	−	−	−
	$\widehat{\tau}(\ell)$	+•	+•	+•	+•	+•	+•	+•	+•	+•	+•
Magnetic field data	ACF	+*	+*	+*	+*	+*	+*	+*	+*	+*	+*
	$\widehat{\tau}(\ell)$	+•	+•	+•	+•	+•	+•	+•	+•	+•	+•
ENSO phenomenon	ACF	+*	+*	+*	+*	+*	+*	+*	+*	+*	+
	$\widehat{\tau}(\ell)$	+•	+•	+•	+•	+•	+•	+•	+•	+•	+•
Climate change δ^{13}C	ACF	+*	+*	+*	+*	+*	+*	+	+	+	+
	$\widehat{\tau}(\ell)$	+•	+•	+•	+•	+•	+•	+•	+	+	+
δ^{18}O	ACF	+*	+*	+*	+	+	−	−	−*	−*	−*
	$\widehat{\tau}(\ell)$	+•	+•	+•	+•	+	−	−	−•	−•	−•

(1) +* indicates a sample ACF value greater than $1.96T^{-1/2}$, −* indicates a value less than $-1.96T^{-1/2}$, and + (−) indicates a positive (negative) value between $-1.96T^{-1/2}$ and $1.96T^{-1/2}$.

(2) • marks a p-value smaller than 5%, and + (−) marks a positive (negative) value of the test statistic with a p-value larger than 5%.

1.3 INITIAL DATA ANALYSIS   17

Example 1.8: Sample ACF and Kendall's tau test statistic

Table 1.3 contains indicator patterns of the sample ACFs and Kendall's τ test statistic for the time series introduced in Section 1.2. A number of observations are in order.

- For the U.S. unemployment series the sample ACF suggests, as a first guess, a linear AR(8) model with significant parameter values at lags 1, 2, 4 – 6, and 8. The results for $\hat{\tau}(\ell)$ match those of the sample ACF.

- The sample ACF of the EEG recordings suggests a linear AR(6) model. On the other hand, Kendall's $\hat{\tau}(\ell)$ test statistics are all significant up to and including lag $\ell = 10$. So it is hard to describe the series by a particular (non)linear model.

- Both the sample ACF and $\hat{\tau}(\ell)$ are not very helpful in identifying preliminary models for the magnetic field data and the monthly ENSO time series. Clearly, the fact that normality is strongly rejected for the magnetic field data has an impact on the significance of the series' test results. The sample ACF of the ENSO series has a significant negative peak (5% level) at lag 21 and a positive (insignificant) peak at lag 56. This reflects the fact that ENSO periods lasted between two and five years in the last century.

- The sample ACFs of the δ^{13}C and δ^{18}O series indicate that both series can be represented by a low order AR process, but there are also some significant values at lags 8 – 10. The test results for $\hat{\tau}(\ell)$ match those of the sample ACFs.

Kendall's partial tau test statistic

A variation on Kendall's τ test statistic (1.13), commonly referred to as Kendall's partial tau (Quade, 1967), is a nonparametric measure of the association between two random variables X and Y while controlling for a third variable Z. Given a time series sequence $\{Y_t\}_{t=1}^T$ and its associated ranks $\{R_i\}_{i=1}^T$, Kendall's partial τ test statistic is the correlation obtained after regressing R_i and $R_{i+\ell}$ on the intermediate observations $R_{i+1}, \ldots, R_{i+\ell-1}$. By analogy with (1.14), it may be defined as

$$\hat{\tau}_p(\ell) = 1 - \frac{4N_p(\ell)}{(T-\ell)(T-\ell-1)}. \tag{1.15}$$

Here $N_p(\ell)$ is the number of pairs $\{(R_i, R_{i+\ell})\}_{i=1}^{T-\ell}$ such that $\|\mathbf{Z}_i - \mathbf{Z}_j\| \leq T_Z$, for T_Z a predefined "tolerance" (e.g. $T_Z = 0.2T$), with $\mathbf{Z}_i = (R_{i+1}, \ldots, R_{i+\ell-1})'$ ($i = 1, \ldots, T - \ell$), and $\|\cdot\|$ is a norm. The statistic $\hat{\tau}_p(\ell)$ has similar properties as $\hat{\tau}(\ell)$. Moreover, it can be shown that $\hat{\tau}_p(\ell)$ has an asymptotically normal distribution under the null hypothesis of no serial dependence.

1.3.3 Mutual information coefficient

Granger and Lin (1994) develop a nonparametric statistic for measuring the complete dependence, including nonlinear dependence if present, based on the *mutual information* coefficient. Let X be a continuous random variable with probability density function (pdf) $f_X(x)$. Mutual information is directly related to the *Shannon entropy*, defined as

$$H(X) = -\int \log\{f_X(x)\} f_X(x)\, \mathrm{d}x, \tag{1.16}$$

which is just the mathematical expectation of $-\log f_X(x)$, i.e., $-\mathbb{E}\bigl(\log f_X(x)\bigr)$. Similarly, for a pair of random variables (X, Y) with joint pdf $f_{XY}(x, y)$ the *joint entropy* is defined as

$$H(X, Y) = -\iint f_{XY}(x, y) \log f_{XY}(x, y)\, \mathrm{d}x \mathrm{d}y. \tag{1.17}$$

The mutual information, also called *Kullback–Leibler* (KL) *divergence* or *relative entropy*, is defined as

$$I^{\mathrm{KL}}(X, Y) = \iint \log\left(\frac{f_{XY}(x, y)}{f_X(x) f_Y(y)}\right) f_{XY}(x, y)\, \mathrm{d}x \mathrm{d}y. \tag{1.18}$$

The mutual information measures the average information contained in one of the random variables about the other. It is a symmetric measure of dependence between X and Y as becomes obvious after expressing (1.18) in terms of entropies:

$$I^{\mathrm{KL}}(X, Y) = H(X) + H(Y) - H(X, Y). \tag{1.19}$$

The mutual information is invariant not only under scale transformations of X and Y, but more generally, under all continuous one-to-one transformations. It is also non-negative, $I^{\mathrm{KL}}(X, Y) \geq 0$, with equality if and only if $f_{XY}(x, y) = f_X(x) f_Y(y)$ (cf. Exercise 1.4).

If there exists perfect dependence between X and Y, $I^{\mathrm{KL}}(X, Y) \to \infty$. However, this property is not very attractive for developing a test statistic. Indeed, an ideal measure for testing (serial) dependence should take values in the range $[0, 1]$ or $[-1, 1]$. Moreover, for interpretation purposes it is useful to relate the measure to the correlation coefficient $\rho_{XY} = \mathbb{E}(XY)/\sqrt{\mathbb{E}(X^2)\mathbb{E}(Y^2)}$ when (X, Y) has a standard bivariate normal distribution. One way to establish these objectives, is to transform $I^{\mathrm{KL}}(X, Y)$ as follows

$$R(X, Y) = [1 - \exp\{-2I^{\mathrm{KL}}(X, Y)\}]^{1/2}, \tag{1.20}$$

which takes values in the range $[0, 1]$, with values increasing with $I^{\mathrm{KL}}(\cdot)$; $R(\cdot) = 0$ if and only if X and Y are independent, and $R(\cdot) = 1$ if X and Y are exact functionally related. Further, it can be shown (Pinsker, 1964, p. 123) that

$$I^{\mathrm{KL}}(X, Y) = \log\sqrt{\frac{1}{1 - \rho_{XY}}},$$

1.3 INITIAL DATA ANALYSIS

so that $R(X, Y) = |\rho_{XY}|$.

In a time series framework, $R(\cdot)$ can be used to measure the strength of association between lagged values of an observed time series $\{Y_t\}_{t=1}^T$. More specifically, the analogue to (1.20) at lag ℓ is given by

$$R(Y_t, Y_{t+\ell}) \equiv R_Y(\ell) = [1 - \exp\{-2I^{\text{KL}}(Y_t, Y_{t+\ell})\}]^{1/2}. \tag{1.21}$$

The corresponding sample estimate, say $\widehat{R}_Y(\ell)$, follows from estimating functionals of density functions. No distributional theory is currently available for $\widehat{R}_Y(\cdot)$, but empirical critical values may be computed for specific choices of T and ℓ; see, e.g., Granger and Lin (1994, Table III). Simulations show that $\widehat{R}_Y(\ell)$ has a positive bias. One way to avoid such a bias is to redefine (1.21) as $R_Y^*(\ell) = 1 - \exp\{-2I^{\text{KL}}(Y_t, Y_{t+\ell})\}$.

1.3.4 Recurrence plot

An appealing and simple graphical tool that enables the assessment of stationarity in an observed time series is the *recurrence plot* due to Beckman et al. (1987). The recurrence plot is a two-dimensional scatter diagram where a dot is placed at the point (t_1, t_2) whenever Y_{t_1} is "close" to Y_{t_2}, given some pre-specified threshold h, usually not larger than 1/10 of the standard deviation. It can be mathematically expressed as

$$R_{t_1, t_2} = I(\|\mathbf{Y}_{t_1}^{(\ell)} - \mathbf{Y}_{t_2}^{(\ell)}\| < h), \quad (t_1, t_2 = 1, \ldots, T),$$

where $\mathbf{Y}_t^{(\ell)}$ is an m-dimensional ($m \in \mathbb{Z}^+$) lag ℓ ($\ell \in \mathbb{Z}$) delay vector,[4,5] also called a *state* or *reconstruction* vector, given by

$$\mathbf{Y}_t^{(\ell)} = (Y_t, Y_{t-\ell}, \ldots, Y_{t-(m-1)\ell})',$$

and $\|\cdot\|$ is a norm.[6]

If $\{Y_t, t \in \mathbb{Z}\}$ is strictly stationary, the recurrence plot will show an approximately uniform density of recurrences as a function of the time difference $t_1 - t_2$. However, if $\{Y_t, t \in \mathbb{Z}\}$ has a trend or another type of nonstationarity, with a behavior that is changing over time, the regions of $\mathbf{Y}_t^{(\ell)}$ visited will change over time. The result will be that there are relatively few recurrences far from the main diagonal in the recurrence plot, that is for large values of $|t_1 - t_2|$. Also, if there are only recurrences

---

[4]In the analysis of deterministic chaos, i.e. irregular oscillations that are not influenced by random inputs, m is often called the *embedding dimension*. Within that context, it is important to choose m sufficiently large, such that the so-called m-dimensional *phase space* enables for a "proper" representation of the dynamical system.

[5]In economics and finance, but not in other fields, it is common to fix ℓ at one. So m takes over the role of ℓ. In that case we write $\mathbf{Y}_t$, suppressing the dependence on ℓ.

[6]In fact, the supremum norm is very popular for recurrence plots; see Appendix 3.A for more information on vector and matrix norms.

near $t_1 = t_2$ and for values of $|t_1 - t_2|$ that are of the order of the total length T, $\{Y_t, t \in \mathbb{Z}\}$ can be considered nonstationary. Obviously, in alliance with the choice of ℓ and m, visual interpretation of recurrence plots requires some experience.

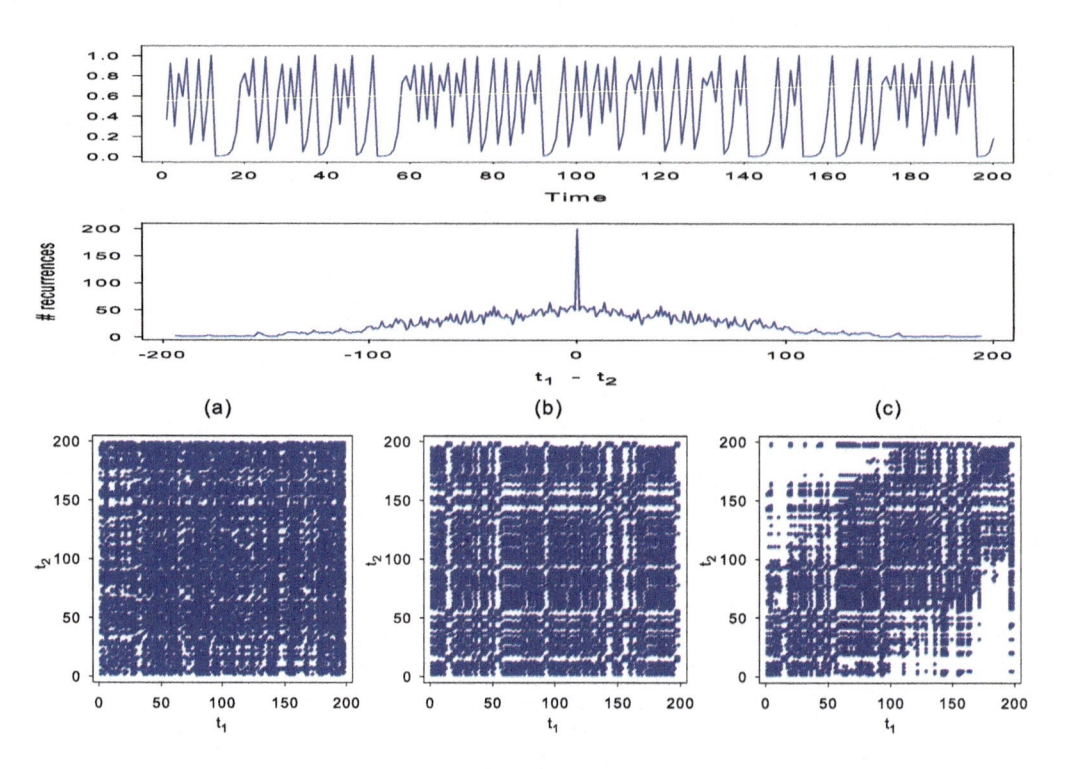

Figure 1.6: *Upper panel: a time series $\{Y_t\}_{t=1}^{200}$ generated by (1.22) with $a = 4$. Middle panel: number of recurrences for the recurrence plot in (b) of the lower panel. Lower panel: (a) a plot of R_{t_1,t_2} for a time series following an i.i.d. $U(0,1)$ distribution, (b) a plot of R_{t_1,t_2} for $\{Y_t\}$, and (c) a recurrence plot for the time series $Y_t + 0.005t$; $m = 3$ and $\ell = 1$.*

Example 1.9: The Logistic Map

The *logistic map* may be interpreted as a simple biological, completely deterministic, model for the evolution of a population size Y of some species over time. Due to limited natural resources there is a maximum population size which in suitable units is equal to unity. The population size must be larger than or equal to zero. The evolution rule is

$$Y_t = aY_{t-1}(1 - Y_{t-1}), \quad (t = 1, 2, \ldots), \qquad (1.22)$$

where $a > 1$ denotes the growth rate at time t of the species in the case of unlimited natural sources. The factor $(1 - Y_{t-1})$ describes the effect of over-population. In some cases, a particular solution of (1.22) can be found, depending on the value of a and the starting value Y_0.

1.3 INITIAL DATA ANALYSIS

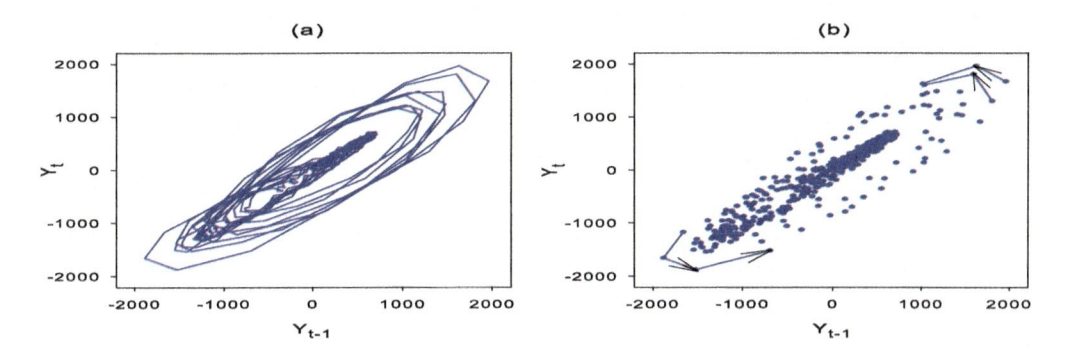

Figure 1.7: *(a) Directed scatter plot at lag 1 for the EEG recordings, and (b) a scatter plot with the two largest and two smallest values connected with the preceding and the following observations.*

Figure 1.6, top panel, shows the first 200 observations of a time series $\{Y_t\}$ generated with (1.22) for $a = 4$. The plot shows an erratic pattern, akin to that of a realization from some stochastic process. Still, the evolution of $\{Y_t\}$ is an example of chaos. The recurrence plot for $\{Y_t\}_{t=1}^{200}$ is shown in the bottom panel of Figure 1.6(b).

It is interesting to contrast the main features of graph (b) with the characteristic features of graph (a), showing a recurrence plot of an i.i.d. $U(0,1)$ distributed time series, and with the patterns in graph (c), showing a recurrence plot of the time series $Y_t + 0.005t$. Graph (a) has a homogeneous typology or pattern, which is an indicator that the series originated from a stationary DGP. In contrast, a non-homogeneous or disrupting typology, as with the recurrence plot in graph (c), indicates a nonstationary DGP. Finally, graph (b) shows a recurrence plot with a diagonal oriented periodic structure due to the oscillating patterns of $\{Y_t\}$. This is supported by the plot in the middle panel. The white areas of bands in the recurrence plots indicate changes in the behavior of a time series, perhaps due to outliers or structural shifts. As an exercise the reader is recommended to obtain recurrence plots for higher values of the embedding dimension m, and see whether or not the overall observations made above remain unchanged.

1.3.5 Directed scatter plot

This is a scatter diagram, at lag ℓ ($\ell \in \mathbb{Z}$), of an observed time series $\{Y_t\}_{t=1}^{T}$ (vertical axis) against $Y_{t-\ell}$ (horizontal axis) with straight lines connecting the adjacent observations, such as $(Y_{t-\ell}, Y_t)$ and $(Y_{t-\ell+1}, Y_{t+1})$. The plot can reveal clustering and/or cyclical phenomena. Also, any asymmetries around the diagonal are an indication of time-irreversibility.[7]

---

[7]An obvious three-dimensional extension is to plot $(Y_t, Y_{t-\ell}, Y_{t-\ell'})$ ($\ell \neq \ell'; \ell = \ell' = 1, 2, \ldots$). For this purpose the function **autotriples** in the R-tsDyn package can be used. Alternatively, the function **autotriples.rgl** displays an interactive trivariate plot of (Y_{t-1}, Y_{t-2}) against Y_t.

Example 1.10:   EEG Recordings (Cont'd)
Figure 1.7(a) provides a directed scatter plot of the EEG recordings, denoted by $\{Y_t\}_{t=1}^{631}$, of Example 1.2. The spirals indicate some cyclical pattern within the series. This becomes more apparent in Figure 1.7(b) where the observations for the two largest negative and two largest positive values of $\{Y_t\}$ are connected with the preceding and the following observations. The anti-clockwise route indicated by the arrows suggests a stochastically perturbed cycle.

1.4   Summary, Terms and Concepts

Summary

In this chapter we described some nonlinear characteristics of times series, arising from a variety of real-life problems. Using graphical tools for explanatory data analysis one can recognize a nonlinear feature of a particular data set. Generally, we noticed that a nonlinear time stationary series has a more complex behavior than a linear series. Further we introduced some terms and statistical concepts that are needed later in the book. Finally, we provided a brief treatment of test statistics for skewness, kurtosis and normality for initial data analysis, both for independent and weakly dependent data.

Terms and Concepts

cave plot, 9	logistic map, 20
(dis)concordant, 15	mutual information, 18
cyclical asymmetry, 4	phase space, 19
data generating process, 4	recurrence plot, 19
directed scatter plot, 21	Shannon entropy, 18
essentially linear, 3	skewness, 10
Gaussian white noise, 1	time-reversible, 6
Kendall's tau, 14	weak white noise, 2
kurtosis, 10	

1.5   Additional Bibliographical Notes

Section 1.1: The definition that a time series process is linear if the linear predictor is optimal is due to Hannan (1979); see also Hannan and Deistler (2012). It is considered to be the minimum requirement. The definition has been used in the analysis of time series neural networks; see, e.g., Lee et al. (1993).

Section 1.3.1: The univariate JB normality test of residuals, has been known among statisticians since the work by Bowman and Shenton (1975). Doornik and Hansen (2008) transform the coefficients of skewness and kurtosis such that they are much closer to the standard normal distribution, and thus obtain a refinement of the JB test (see, e.g., the R-normwhn.test package). Brys et al. (2004) and Gel and Gastwirth (2008) suggest some robust versions of the JB-test in the i.i.d. case. Koizumi et al. (2009) derive some multivariate JB tests. Fiorentini et al. (2004) show that the JB test can be applied to a broad class of GARCH-M processes. Boutahar (2010) establishes the limiting distributions for the JB test statistic for long memory processes. Kilian and Demiroglu (2000) find that the JB test statistic applied to the residuals of linear AR processes is too conservative in the sense that it hardly will reject the null hypothesis of normality in the residuals. Using the same setup as with the Lin–Mudholkar test statistic, Mudholkar et al. (2002) construct a test statistic based on the correlation between the sample mean and the third central sample moment.

Section 1.3.2: Nielsen and Madsen (2001) propose generalizations of the sample ACF and sample PACF for checking nonlinear lag dependence founded on the local polynomial regression method (Appendix 7.A). Some of the methodology discussed in that paper is implemented in the MATLAB and R source codes contained in the zip-file comp_ex_1_scrips_2011.zip, which can be downloaded from http://www2.imm.dtu.dk/courses/02427/.

If $\{Y_t\}_{t=1}^T$ follows a linear causal process, as defined by (1.2), but now the ε_t's are i.i.d. with mean zero and *infinite* variance rather than i.i.d. with finite variance, then the sample ACF for heavy tailed data, defined as $\widehat{\rho}_Y(\ell) = \sum_{t=1}^{T-\ell} Y_t Y_{t+\ell} / \sum_{t=1}^{T} Y_t^2$, still converges to a constant $\rho_Y(\ell) = \sum_{i=0}^{\infty} \psi_i \psi_{i+\ell} / \sum_{i=0}^{\infty} \psi_i^2$ ($\ell \in \mathbb{Z}$). However, for many nonlinear models $\widehat{\rho}_Y(\ell)$ converges to a nondegenerate random variable. Resnick and Van den Berg (2000a,b) use this fact to construct a test statistic for (non)linearity based on subsample stability of $\widehat{\rho}_Y(\ell)$; see the S-Plus code at the website of this book.[8]

Section 1.3.3: Several methods have been proposed for the estimation of the mutual information (Kullback–Leibler divergence) such as kernel density estimators, nearest neighbor estimators and partitioning (or binning) the XY plane. This latter approach, albeit in a time series context, is available through the function mutual in the R-tseriesChaos package. Khan et al. (2007) compare the relative performance of four mutual information estimation methods. Wu et al. (2009) discuss the estimation of mutual information in higher dimensions and modest samples ($500 \leq T \leq 1{,}000$).

1.6 Data and Software References

Data

Example 1.1: The quarterly U.S. unemployment rate can be downloaded from various websites, including U.S. Bureau of Labor Statistics (http://data.bls.gov/timeseries/LNS14000000), the website of the Federal Reserve Bank of St. Louis (http://research.stlouisfed.org/fred2/release?rid=202&soid=22), or from the website of this book. The series has been widely used in the literature to exhibit certain nonlinear characteristics, however, often covering a much shorter time-period; see, e.g., Montgomery et al. (1998).

Example 1.2: The EEG recordings have been analyzed by Tohru Ozaki and his co-workers in a number of papers; see, e.g., Miwakeichi et al. (2001) and the references therein. The data set can be downloaded from the website of this book. A link to other EEG time series is: http://epileptologie-bonn.de/cms/front_content.php?idcat=193&lang=3; see Stam (2005) for a review.

Example 1.3: The daily averages of the T component of the interplanetary magnetic field have been analyzed by Terdik (1999). The complete data set (24 hourly basis) can be

---
[8]S-Plus is a registered trademark of Insightful Corp.

downloaded from `http://nssdc.gsfc.nasa.gov/` along with further information on the magnetic field measurements. Also, the data set is available at the website of this book.

Example 1.4: The ENSO anomaly, Niño 3.4 index, is derived from the index tabulated by the Climate Prediction Center at the National Oceanic and Atmospheric Administration (NOAA);`http://www.cpc.ncep.noaa.gov/data/indices/ersst3b.nino.mth.ascii`.The series is available at the website of this book. The complete data set has been analyzed by Ubilava and Helmers (2013). Ubilava (2012) investigates a slightly different version of the ENSO data set. To replicate the main results of that study, R code is available at `http://onlinelibrary.wiley.com/doi/10.1111/j.1574-0862.2011.00562.x/suppinfo`. The 5-month running average in Figure 1.4(b) is used to smooth out variations in SSTs. Unfortunately, there is no single definition of an El Niño or La Niña event.

Example 1.5: Extensive information about the Ocean Drilling Program, including books, reports, and journal papers, can be found at `http://www-odp.tamu.edu/publications/citations/cite108.html`. The δ^{13}C and δ^{18}O time series plotted in this example were made available by Cees Diks; see also Diks and Mudelsee (2000). The data for all four climatic periods can be downloaded from the website of this book.

Software References

Section 1.2: Becker et al. (1994) introduce the cave plot for comparing multiple time series. The plot in Figure 1.5 is produced with an S-Plus function written by Henrik Aalborg Nielsen; see the website of this book. Alternatively, cave plots can be obtained using the R-grid package. Note, McLeod et al. (2012) provide an excellent overview of many R packages for plotting and analyzing, primarily linear, time series.

Section 1.3.1: The Jarque–Bera test statistic is a standard routine in many software packages. The generalized JB test statistic can be easily obtained from a simple modification of the code for the JB test. GAUSS[9] code for the Bai–Ng tests for skewness, kurtosis, and normality is available at `http://www.columbia.edu/~sn2294/research.html`. A MATLAB[10] function for computation of theses test statistics can be downloaded from the website of this book.

Section 1.3.2: FORTRAN77 subroutines for calculating Kendall's (partial) tau for univariate and multivariate (vector) time series, created by Jane L. Harvill and Bonnie K. Ray, are available at the website of this book.

Section 1.3.4: The results in Figures 1.6(a) – (c) can be reproduced with the function recurr in the R-tseriesChaos package. Alternatively, one can analyze the data with the function recurrencePlot in the R-fNonlinear package. The R-tsDyn package contains functions for explorative data analysis (e.g. recurrence plots, and sample (P)ACFs), and nonlinear AR estimation.

User-friendly programs for delay coordinate embedding, nonlinear noise reduction, mutual information, false-nearest neighbor, maximal Lyapunov exponent, recurrence plot, determinism test, and stationarity test can be downloaded from `http://www.matjazperc.com/ejp/time.html`. Alternatively, `http://staffhome.ecm.uwa.edu.au/~00027830/` contains MATLAB functions to accompany the book by Small (2005). Another option for applying nonlinear dynamic methods is the TISEAN package. The package is publicly available from

---

[9]GAUSS is a registered trademark of Aptech Systems, Inc.
[10]MATLAB is a registered trademark of MathWorks, Inc.

`http://www.mpipks-dresden.mpg.de/~tisean/`. The book by Kantz and Schreiber (2004) provides theoretical background material. Similar methods are available in the comprehensive MATLAB package TSTOOL: `http://www.physik3.gwdg.de/tstool/`. The package comes with a complete user manual including a large set of bibliographic references, which makes it useful for those researchers interested in getting started with nonlinear time series analysis methods from a dynamic system perspective.

Exercises

Theory Questions

1.1 Let the ARCH(1) process $\{Y_t, t \in \mathbb{Z}\}$ be defined by $Y_t|(Y_{t-1}, Y_{t-2}, \ldots) = \sigma_t \varepsilon_t$ where $\sigma_t^2 = \alpha_0 + \alpha_1 Y_{t-1}^2$, and $\{\varepsilon_t\} \stackrel{\text{i.i.d.}}{\sim} \mathcal{N}(0,1)$.[11] Assume $\alpha_0 > 0$ and $0 < \alpha_1 < 1$. Rewrite $\{Y_t^2, t \in \mathbb{Z}\}$ in the form of an AR(1) process. Then show that the error process of the resulting model does *not* have a constant conditional variance, i.e. $\{Y_t^2, t \in \mathbb{Z}\}$ is not a weakly linear time process.

1.2 Consider the process $Y_t = \beta Y_{t-2} \varepsilon_{t-1} + \varepsilon_t$, where $\{\varepsilon_t\}$ is an i.i.d. sequence such that $\mathbb{E}(\varepsilon_t) = \mathbb{E}(\varepsilon_t^3) = 0$, $\mathbb{E}(\varepsilon_t^2) = \sigma_\varepsilon^2$, and $\mathbb{E}(\varepsilon_t^4) < \infty$, and where β is a real constant such that $\beta^4 < 1$. Let $\varepsilon_0 = 0$ and $Y_{-1} = Y_0 = 0$ be the starting conditions of the process.

  (a) Show that $\{Y_t, t \in \mathbb{Z}\}$ is an uncorrelated process. Is it also a weak WN process?

  (b) Show that $\{Y_t^2, t \in \mathbb{Z}\}$ is an uncorrelated process.

1.3 Consider the estimator $\widehat{\gamma}_Y(1) = T^{-1} \sum_{t=1}^T Y_t Y_{t+1}$ of $\gamma_Y(1) = \mathbb{E}(Y_t Y_{t+1})$. If $\{\varepsilon_t\} \sim \text{WN}(0, \sigma_\varepsilon^2)$, the theoretical ACF is zero for all lags $\ell \geq 1$. Then Bartlett's formula for the asymptotic covariance between sample autocovariances implies that $\gamma_\varepsilon^{-2}(0) \text{Var}(\sqrt{T} \widehat{\gamma}_\varepsilon(1)) \to 1$, as $T \to \infty$.

Show that the ARCH process in Exercise 1.1 does *not* satisfy the white noise condition, i.e. $\lim_{T \to \infty} \gamma_Y^{-2}(0) \text{Var}(\sqrt{T} \widehat{\gamma}_Y(1))$ increases monotonically from 1 to ∞, as α_1 increases from 0 to $1/\sqrt{3}$.

1.4 Consider the divergence measure $I^{\text{KL}}(X, Y)$ as defined by (1.18).

  (a) Show that $I^{\text{KL}}(X, Y)$ is non-negative, and 0 if and only if X and Y are independent.

  (b) Suppose there exists a functional $h(\cdot)$ such that $X = h(Y)$. Show that $I^{\text{KL}}(X, Y) = \infty$.

1.5 Suppose $\{Y_i\}_{i=1}^n$ is a sequence of i.i.d. random variables of Y with mean zero. If the rth moment of Y exists, then the *semi-invariants* or *cumulants* are defined by the identity in t $\exp\{\sum_{p=1}^\infty k_p (it)^p/p!\} = \phi(t)$ with $\phi(t)$ the characteristic function.

---

[11] Throughout the book, we assume that the reader is familiar with the class of so-called (generalized) autoregressive conditional heteroskedastic (abbreviated as (G)ARCH) models; see, e.g., the excellent, and up-to-date, book by Francq and Zakoïan (2010).

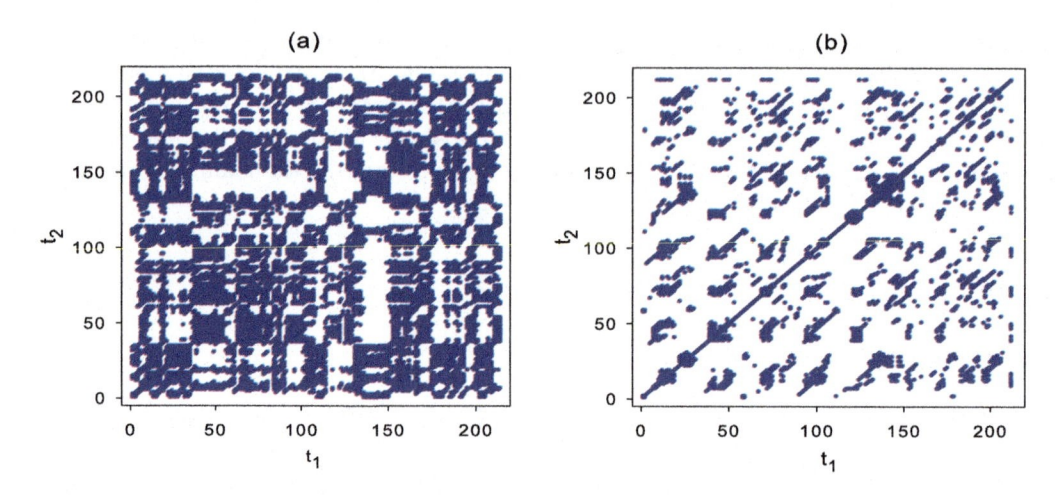

Figure 1.8: Climate change data set. (a) Recurrence plot of the δ^{13}C time series, and (b) recurrence plot of the δ^{18}O time series. Embedding dimension $m = 3$, and $\ell = 1$.

Subject to conditions of existence of moments, k_p can be expressed in terms of the central sample moments as

$$k_2 = \frac{n}{n-1}\widehat{\mu}_{2,Y}, \quad k_3 = \frac{n^2}{(n-1)(n-2)}\widehat{\mu}_{3,Y}, \quad k_4 = \frac{n^2[(n+1)\widehat{\mu}_{4,Y} - 3(n-1)\widehat{\mu}_{2,Y}^2]}{(n-1)(n-2)(n-3)}.$$

In normal samples it can be shown that $\overline{Y}$, $\widehat{\mu}_{2,Y}$ and $\widehat{\mu}_{\nu,Y}\widehat{\mu}_{2,Y}^{-3/2}$ ($\nu = 3, 4, \ldots$) are independent, and hence that

$$\text{Var}\left(\frac{k_3}{k_2^{3/2}}\right) = \frac{6n(n-1)}{(n-2)(n+1)(n+3)}, \quad \text{Var}\left(\frac{k_4}{k_2^2}\right) = \frac{24n(n-1)^2}{(n-3)(n-2)(n+3)(n+5)}.$$

(a) Using the above results, show that the *exact* mean and variance of the sample coefficient of skewness $\widehat{\tau}_Y$ and the sample coefficient of kurtosis $\widehat{\kappa}_Y$ are, respectively, given by

$$\mathbb{E}(\widehat{\tau}_Y) = 0, \quad \text{Var}(\widehat{\tau}_Y) = \frac{6(n-2)}{(n+1)(n+3)},$$

$$\mathbb{E}(\widehat{\kappa}_Y) = \frac{3(n-1)}{n+1}, \quad \text{Var}(\widehat{\kappa}_Y) = \frac{24n(n-2)(n-3)}{(n+1)^2(n+3)(n+5)}.$$

(b) Given the results in part (a) define an alternative for the JB test statistic (1.6).

Empirical and Simulation Questions

1.6 Figure 1.8(a) displays the recurrence plots of the δ^{13}C and δ^{18}O time series, respectively; see Example 1.5. Provide a global characterization of each plot, in terms of homogeneity, periodicity, and trend or drift.

EXERCISES

1.7 Figure 1.9 shows raw data plots of length $T = 100$, together with corresponding directed scatter plots, for three simulated time series processes:

    i) $Y_t = \varepsilon_t$,   (Gaussian white noise),
    ii) $Y_t = 0.6 Y_{t-1}\varepsilon_{t-1} + \varepsilon_t$,   (a stationary BL process; see Section 2.2),
    iii) $Y_t = \sigma_t \varepsilon_t$, $\sigma_t^2 = 1 + 1.2 Y_{t-1}^2$, (a nonstationary ARCH(1) process),

where in all cases $\{\varepsilon_t\} \overset{\text{i.i.d.}}{\sim} \mathcal{N}(0,1)$. The graphs are listed in random order. Which set of graphs corresponds to the listed processes?

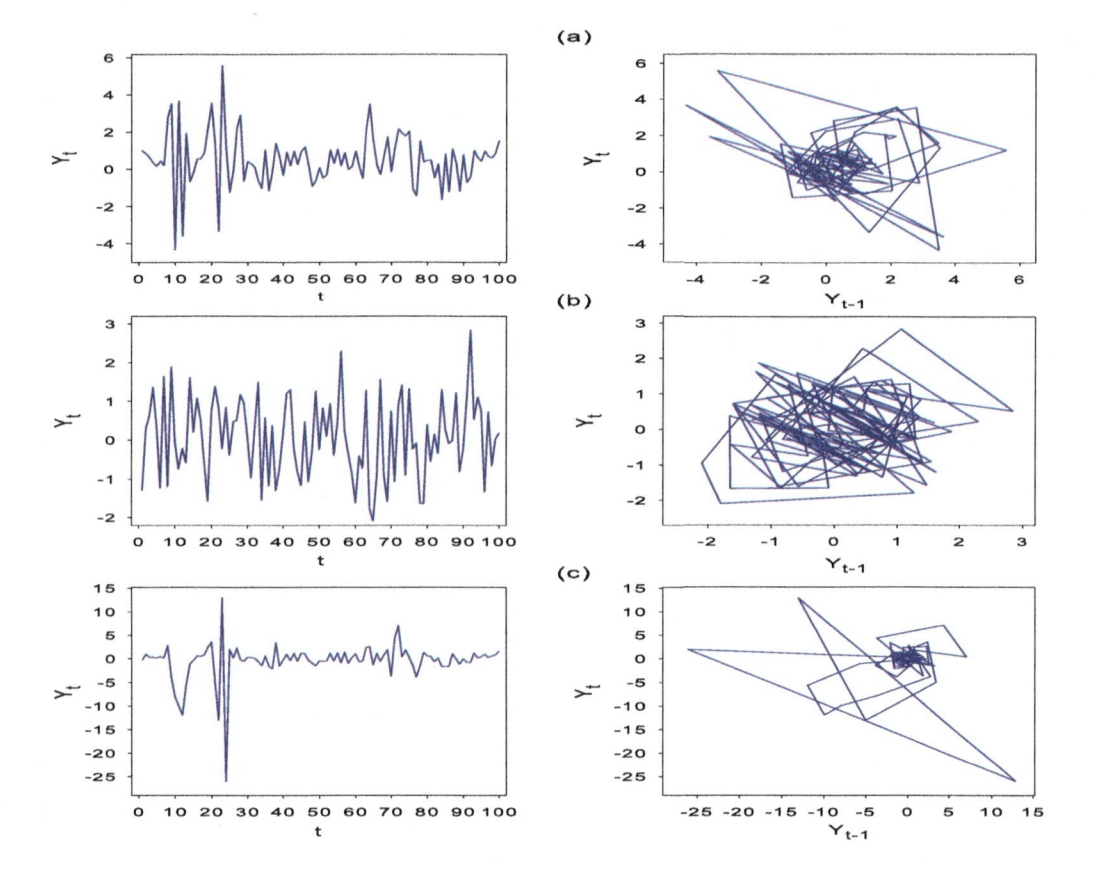

Figure 1.9: *Three time series plots and associated directed scatter plots.*

1.8 Consider the δ^{13}C time series, denoted by $\{Y_t\}_{t=1}^{216}$ and introduced in Example 1.5. Download the data from the website of this book.

    (a) Obtain the reversed time series, say $\{Y_t^{\text{R}}\}_{t=1}^{216}$. Plot both time series, i.e. $\{Y_t\}$ and $\{Y_t^{\text{R}}\}$. Is the process $\{Y_t, t \in \mathbb{Z}\}$ time-reversible?

    (b) Obtain the series $X_t(\ell) = Y_t - Y_{t-\ell}$ for $\ell = 1$ and 2. Draw histograms of $\{X_t(\ell)\}$ with superimposed Gaussian distributions using sample means and standard deviations of the two series. Is the process $\{Y_t, t \in \mathbb{Z}\}$ time-reversible?

    (c) Compute the JB and GJB test statistics and compare the results with the graphs plotted in part (b).

Chapter 2

CLASSIC NONLINEAR MODELS

In Section 1.1, we discussed in some detail the distinction between linear and nonlinear time series processes. In order to make this distinction as clear as possible, we introduce in this chapter a number of classic parametric univariate nonlinear models. By "classic" we mean that during the relatively brief history of nonlinear time series analysis, these models have proved to be useful in handling many nonlinear phenomena in terms of both tractability and interpretability. The chapter also includes some of their generalizations. However, we restrict attention to univariate nonlinear models. By "univariate", we mean that there is one output time series and, if appropriate, a related unidirectional input (exogenous) time series. In Chapter 11, we deal with vector (multivariate) parametric models in which there are several jointly dependent time series variables. Nonparametric univariate and multivariate methods will be the focus of Chapters 4, 9 and 12.

The chapter is organized as follows. In Section 2.1, we introduce a general nonlinear time series model followed by a representation as a so-called state-dependent model (SDM). The SDM builds upon the basic structure of the linear ARMA model. In particular, it generalizes the ARMA model to the nonlinear version by allowing the coefficients to take on more complex, and hence, flexible forms. As we will see in Sections 2.2 – 2.5, by imposing appropriate restrictions on the parameters of the SDM several important classes of nonlinear models emerge. In Section 2.6, we introduce the class of regime switching threshold models. Basically, these models can be regarded as piecewise linear approximations to the general nonlinear time series model of Section 2.1. Next, to allow for slow changes between various states of the DGP, we discuss smooth transition models in Section 2.7. In Section 2.8, we introduce some nonlinear non-Gaussian models. Section 2.9 deals with artificial neural networks (ANNs) which are useful for DGPs that have an unknown functional form. In Section 2.10, we focus on Markov switching models where the regimes are determined by an unobservable process. In the final section, we illustrate a number of practical issues of ANN modeling via a case study.

In addition, the chapter contains two appendices. In Appendix 2.A, we briefly introduce the concept of (non)linear impulse response functions. We will see that these response functions are a convenient tool for illustrating the dynamics of (non)linear time series models. Appendix 2.B provides a list of abbreviations for threshold-type nonlinear models which have been introduced in the literature since the early 1970s.

2.1 The General Univariate Nonlinear Model

2.1.1 Volterra series expansions

One of the purposes of univariate time series analysis is to study the dependence structure of a given sample realization. This is usually done by considering some functional form that describes the relationship between past and present values, say $(\ldots, Y_{t-2}, Y_{t-1}, Y_t)$, of a time series process in such a way that an observed time series $\{Y_t\}$ is filtered into a strict WN process $\{\varepsilon_t\}$. Let $h(\cdot)$ denote a suitably smooth (usually analytic) real-valued function. Then a general form for modeling $\{Y_t, t \in \mathbb{Z}\}$ can be expressed as

$$h(Y_t, Y_{t-1}, Y_{t-2}, \ldots) = \varepsilon_t, \tag{2.1}$$

which is independent of future observations and due to its generality may be considered as a nonlinear model. Model (2.1) is also referred to as *causal* or *non-anticipative* in the sense that future values, which typically are not available, do not participate in the functional form of the model.

Now we face the problem of finding $h(\cdot)$ such that (2.1) is *causally invertible*, i.e. it can be "solved" for Y_t as a function of $\{\ldots, \varepsilon_{t-2}, \varepsilon_{t-1}, \varepsilon_t\}$,

$$Y_t = \widetilde{h}(\varepsilon_t, \varepsilon_{t-1}, \varepsilon_{t-2}, \ldots). \tag{2.2}$$

In addition, while maintaining their generality, the functions $h(\cdot)$ and $\widetilde{h}(\cdot)$ must be tractable for the purpose of statistical analysis. However, as (2.2) stands not much can be said or done as far as analysis of a given time series is concerned. Therefore, we assume that $\widetilde{h}(\cdot)$ is a sufficiently well-behaved function so that we can expand (2.2) in a Taylor series about some fixed time point – say $\mathbf{0} = (0, 0, \ldots)'$. Then we can write

$$Y_t = \mu + \sum_{u=0}^{\infty} g_u \varepsilon_{t-u} + \sum_{u,v=0}^{\infty} g_{uv} \varepsilon_{t-u} \varepsilon_{t-v} + \sum_{u,v,w=0}^{\infty} g_{uvw} \varepsilon_{t-u} \varepsilon_{t-v} \varepsilon_{t-w} + \cdots, \tag{2.3}$$

where

$$\mu = g(\mathbf{0}), \quad g_{u_1} = \left(\frac{\partial \widetilde{h}}{\partial \varepsilon_{t-u_1}}\right)_\mathbf{0}, \ldots, g_{u_1,\ldots,u_n} = \left(\frac{\partial^n \widetilde{h}}{\partial \varepsilon_{t-u_1} \cdots \partial \varepsilon_{t-u_n}}\right)_\mathbf{0}.$$

This expansion is known as the *discrete-time Volterra series*, a nonparametric representation, where the sequences $\{g_u\}, \{g_{uv}\}, \{g_{uvw}\}, \ldots$ are called the *Volterra*

2.1 THE GENERAL UNIVARIATE NONLINEAR MODEL

kernels.[1] The first two terms in (2.3) correspond to a linear causally invertible model.

One may also consider the *dual Volterra series*, which is obtained by a Taylor series expansion applied to (2.1) – assuming invertibility of $\widetilde{h}(\cdot)$ and smoothness of $h(\cdot)$ – to obtain

$$\varepsilon_t = \mu' + \sum_{u=0}^{\infty} g'_u Y_{t-u} + \sum_{u,v=0}^{\infty} g'_{uv} Y_{t-u} Y_{t-v} + \sum_{u,v,w=0}^{\infty} g'_{uvw} Y_{t-u} Y_{t-v} Y_{t-w} + \cdots, \quad (2.4)$$

where the sequences $\{g'_u\}, \{g'_{uv}\}, \{g'_{uvw}\}, \ldots$ are defined in a similar way as above. Next, to obtain a more parsimonious representation, we truncate the sequences of Volterra kernels in (2.3) and (2.4) at the fixed points q and p, respectively. Then, by combining (2.3) and (2.4), we get

$$\mu' + \sum_{u=0}^{p} g'_u Y_{t-u} + \sum_{u,v=0}^{p} g'_{uv} Y_{t-u} Y_{t-v} + \sum_{u,v,w=0}^{p} g'_{uvw} Y_{t-u} Y_{t-v} Y_{t-w} + \cdots =$$

$$\mu + \sum_{u=0}^{q} g_u \varepsilon_{t-u} + \sum_{u,v=0}^{q} g_{uv} \varepsilon_{t-u} \varepsilon_{t-v} + \sum_{u,v,w=0}^{q} g_{uvw} \varepsilon_{t-u} \varepsilon_{t-v} \varepsilon_{t-w} + \cdots, \quad (2.5)$$

which can be expressed more generally as,

$$h^*(Y_t, \ldots, Y_{t-p}) = g^*(\varepsilon_t, \ldots, \varepsilon_{t-q}). \quad (2.6)$$

A further generalization, assuming $h^*(\cdot)$ is invertible, is given by

$$Y_t = G(Y_{t-1}, \ldots, Y_{t-p}, \varepsilon_t, \ldots, \varepsilon_{t-q}). \quad (2.7)$$

Note that (2.7) treats $\{\varepsilon_t\}$ as an observable input; therefore, the input-output relationships are expressed in terms of a finite number of past inputs and outputs.[2]

When $\{\varepsilon_t\}$ is unobservable and instead is taken as a random variable, we may reduce the observed time series $\{Y_t\}$ into a strict WN series by redefining $G(\cdot)$ as

$$Y_t = \widetilde{G}(Y_{t-1}, \ldots, Y_{t-p}, \varepsilon_{t-1}, \ldots, \varepsilon_{t-q}) + \varepsilon_t. \quad (2.8)$$

With $\widetilde{G}(\cdot)$ so defined, $\{\varepsilon_t\}$ is considered as the *innovation process* for $\{Y_t\}$, while $\widetilde{G}(\cdot)$ defines the relevant information on Y_t which is contained in past values of $\{Y_t\}$ and its innovation process $\{\varepsilon_t\}$. Observe that $\mathbb{E}(Y_t|\mathcal{F}_{t-1}) = \widetilde{G}(Y_{t-1}, \ldots, Y_{t-p}, \varepsilon_{t-1}, \ldots, \varepsilon_{t-q})$. Clearly, the above formulation is not restricted to the case where $\{\varepsilon_t\}$ is unobservable. It can also be adopted to the case where $\{\varepsilon_t\}$ is a controlled input variable which may enter the model linearly as a factor influencing current output $\{Y_t\}$.

---

[1] Named in honor of Vito Volterra, who studied integral equations involving kernels of this form in the first half of the 20th century.

[2] In neural network studies the Volterra expansion with finite sums is often called the Kolmogorov–Gabor polynomial, or alternatively the Ivakhnenko polynomial.

2.1.2 State-dependent model formulation

Let (2.8) serve as the basis for the general nonlinear finite-dimensional model, and assume that $\widetilde{G}(\cdot)$ is a sufficiently well-behaved function; then, we may proceed by expanding the right-hand side of (2.8) in a Taylor series about the fixed time point $(0, 0, \ldots, 0)'$. For simplicity we shall retain only the first term in the series expansion, i.e.

$$Y_t = \mu(\mathbf{S}_{t-1}) + \sum_{i=1}^{p} f_i(\mathbf{S}_{t-1}) Y_{t-i} + \sum_{j=1}^{q} g_j(\mathbf{S}_{t-1}) \varepsilon_{t-j} + \varepsilon_t, \qquad (2.9)$$

where

$$\mathbf{S}_t = (Y_t, \ldots, Y_{t-p+1}, \varepsilon_t, \ldots, \varepsilon_{t-q+1})',$$
$$\mu(\mathbf{S}_{t-1}) = \widetilde{G}(Y_{t-1}, \ldots, Y_{t-p}, \varepsilon_{t-1}, \ldots, \varepsilon_{t-q}),$$
$$f_i(\mathbf{S}_{t-1}) = \left(\frac{\partial \widetilde{G}}{\partial Y_{t-i}}\right)_{\mathbf{S}_{t-1}}, \qquad g_j(\mathbf{S}_{t-1}) = \left(\frac{\partial \widetilde{G}}{\partial \varepsilon_{t-j}}\right)_{\mathbf{S}_{t-1}}.$$

Rewriting (2.9) in ARMA-like notation gives,

$$Y_t = \mu(\mathbf{S}_{t-1}) + \sum_{i=1}^{p} \phi_i(\mathbf{S}_{t-1}) Y_{t-i} + \varepsilon_t + \sum_{j=1}^{q} \theta_j(\mathbf{S}_{t-1}) \varepsilon_{t-j}. \qquad (2.10)$$

Model (2.10) has been introduced by Priestley (1980). It is called the *state-dependent model* (SDM) of order (p, q) and may be regarded as a local linearization of the general nonlinear model (2.9). The unknown parameters of the model are $\phi_i(\cdot)$ $(i = 1, \ldots, p)$, $\theta_j(\cdot)$ $(j = 1, \ldots, q)$, the "local mean" $\mu(\cdot)$, all of which depend on the *state* $\mathbf{S}$ of the process at time $t - 1$, and σ_ε^2.[3]

Due to the characterization of the SDM as a locally linear ARMA model we impose a pair of 'identifiability' like conditions of the following form.

(i) The polynomials $\{1 - \sum_{i=1}^{p} \phi_i(\mathbf{x}) z^i\}$ and $\{1 + \sum_{j=1}^{q} \theta_j(\mathbf{x}) z^j\}$ have no common factors for all fixed vectors $\mathbf{x}$, and all their roots lie outside the unit circle.

(ii) $\phi_p(\mathbf{x}) \neq 0$ and $\theta_q(\mathbf{x}) \neq 0$ $\forall \mathbf{x}$.

The generality of (2.10) becomes more apparent as one imposes certain restrictions on $\mu(\cdot)$, $\phi_i(\cdot)$, and $\theta_j(\cdot)$. One simple case is to take all these parameters as constants, i.e. independent of $\mathbf{S}_{t-1}$. Then (2.10) becomes the well-known linear ARMA(p, q) model. Some more elaborate characterizations of (2.10) are introduced in the following Sections.

---

[3] In fact, an equivalent vector state space representation of (2.10) is easily written down.

2.2 BILINEAR MODELS

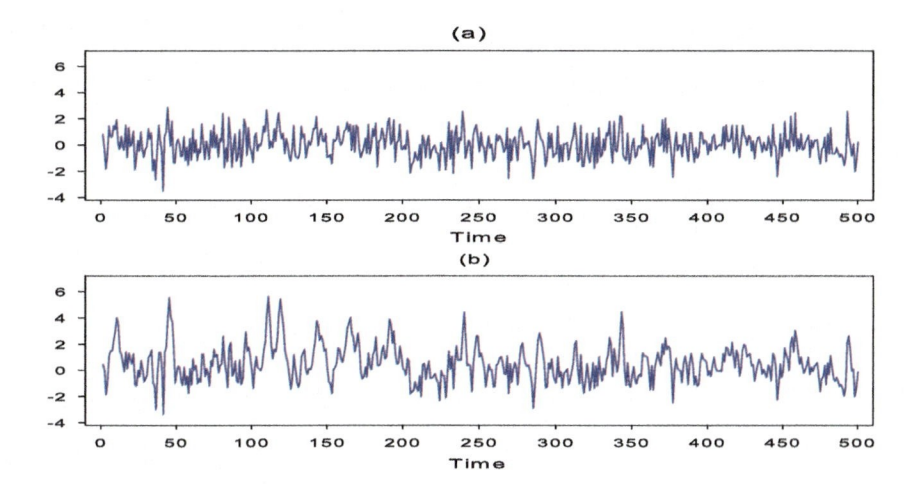

Figure 2.1: *(a) A realization of $\{\varepsilon_t\}_{t=1}^{500}$ with $\{\varepsilon_t\} \stackrel{i.i.d.}{\sim} \mathcal{N}(0,1)$, and (b) a realization of the BL(1,0,1,1) model (2.14), for parameter combination ($\phi = 0.5, \psi = 0.2$), with the generated WN series in panel (a) as input.*

2.2 Bilinear Models

Let $\mu(\mathbf{S}_{t-1}) = \phi_0$, $\phi_i(\mathbf{S}_{t-1}) = \phi_i$ ($i = 1, \ldots, p$), i.e. a sequence of constants, and let $\theta_j(\mathbf{S}_{t-1}) = \theta_j + \sum_{v=1}^{Q} \psi_{jv} Y_{t-v}$ ($j = 1, \ldots, q$), i.e. a linear combination of $Y_{t-1}, Y_{t-2}, \ldots, Y_{t-Q}$ ($Q \geq 1$). Then (2.10) becomes

$$Y_t = \phi_0 + \sum_{i=1}^{p} \phi_i Y_{t-i} + \varepsilon_t + \sum_{j=1}^{q} \theta_j \varepsilon_{t-j} + \sum_{j=1}^{q} \sum_{v=1}^{Q} \psi_{jv} Y_{t-j} \varepsilon_{t-v}. \qquad (2.11)$$

This is a special case of a general bilinear (BL) model of order (p, q, P, Q) where P is constrained to be equal q. The *general* BL model[4] is defined as

$$Y_t = \phi_0 + \sum_{i=1}^{p} \phi_i Y_{t-i} + \varepsilon_t + \sum_{j=1}^{q} \theta_j \varepsilon_{t-j} + \sum_{u=1}^{P} \sum_{v=1}^{Q} \psi_{uv} Y_{t-u} \varepsilon_{t-v}. \qquad (2.12)$$

This model is linear in the Y_t's and also in the ε_t's separately but not in both. In other words, provided $\psi_{uv} \neq 0$, the ARMA(p, q) model is nested within (2.12). The following example illustrates this feature.

Example 2.1: A BL Time Series

Consider the BL(1, 0, 1, 1) model

$$Y_t = \phi Y_{t-1} + \varepsilon_t + \psi Y_{t-1} \varepsilon_{t-1}, \qquad (2.13)$$

---
[4]There are several alternative ways to define a BL model. Since we are concerned with input-output model representations, we adopt definition (2.12) throughout this book unless it is explicitly noted otherwise.

where $\psi = \psi_{11}$. This process is stationary and ergodic if $\phi^2 + \psi^2\sigma_\varepsilon^2 < 1$; see Chapter 3. Its mean is $\mathbb{E}(Y_t) = \psi\sigma_\varepsilon^2$. Notice that (2.13) can be rewritten as

$$Y_t = (\phi + \psi\varepsilon_{t-1})Y_{t-1} + \varepsilon_t. \tag{2.14}$$

Equation (2.14) looks like a linear AR(1) process except that the AR parameter $\phi + \psi\varepsilon_{t-1}$ is now time dependent, i.e. it may be viewed as a random variable with mean ϕ. If ψ is positive, the AR parameter will increase with positive values of ε_{t-1} and decrease with negative values of ε_{t-1}. However, positive shocks will be more persistent than negative shocks in the sense that they have a more sizeable effect on the conditional variability of $\{Y_t, t \in \mathbb{Z}\}$.

To illustrate this point, we simulate (2.14) with parameter combinations ($\phi = 0.5, \psi = 0.2$) and ($\phi = 0.5, \psi = 0$), with the second process nested within the BL process. For both processes, we generate an identical set of i.i.d. $\mathcal{N}(0,1)$ random numbers. Figures 2.1(a) – (b) show $T = 500$ realizations of, respectively, $\{\varepsilon_t\}$ and the BL process $\{Y_t, t \in \mathbb{Z}\}$. Since ψ is positive, it can be seen that the value of $\{\varepsilon_{t-1}\}$ has a direct effect on the value of $\{Y_t\}$ but that this effect is larger for positive than for negative shocks, with values of $\{Y_t\}$ in the range $[-3.45, 5.59]$. In contrast, the AR(1) process is having values in the range $[-3.70, 3.45]$.

By focusing completely on the nonlinear structure, i.e. setting $p = q = \phi_0 = 0$, (2.12) becomes the *complete* BL model:

$$Y_t = \varepsilon_t + \sum_{u=1}^{P}\sum_{v=1}^{Q} \psi_{uv} Y_{t-u}\varepsilon_{t-v}. \tag{2.15}$$

Three special cases are of interest:

- If $\psi_{uv} = 0\ \forall u \neq v$, model (2.15) is called *diagonal*.

- If $\psi_{uv} = 0\ \forall u > v$, (2.15) is called *superdiagonal*. Here the multiplicative terms with non-zero coefficients are such that the input variable ε_{t-v} occurs after Y_{t-u} so that these terms are independent. This fact makes analysis somewhat easier.

- Model (2.15) is said to be *subdiagonal* if $\psi_{uv} = 0\ \forall u < v$. In this case the variable Y_{t-u} occurs strictly after ε_{t-v}, making analysis more difficult.[5]

---

[5]The terms *super* and *sub* are not quite natural, because it is purely by convention if lags in $\{Y_t, t \in \mathbb{Z}\}$ correspond to the first index (u) and lags in $\{\varepsilon_t\}$ correspond to the second index (v).

2.2 BILINEAR MODELS

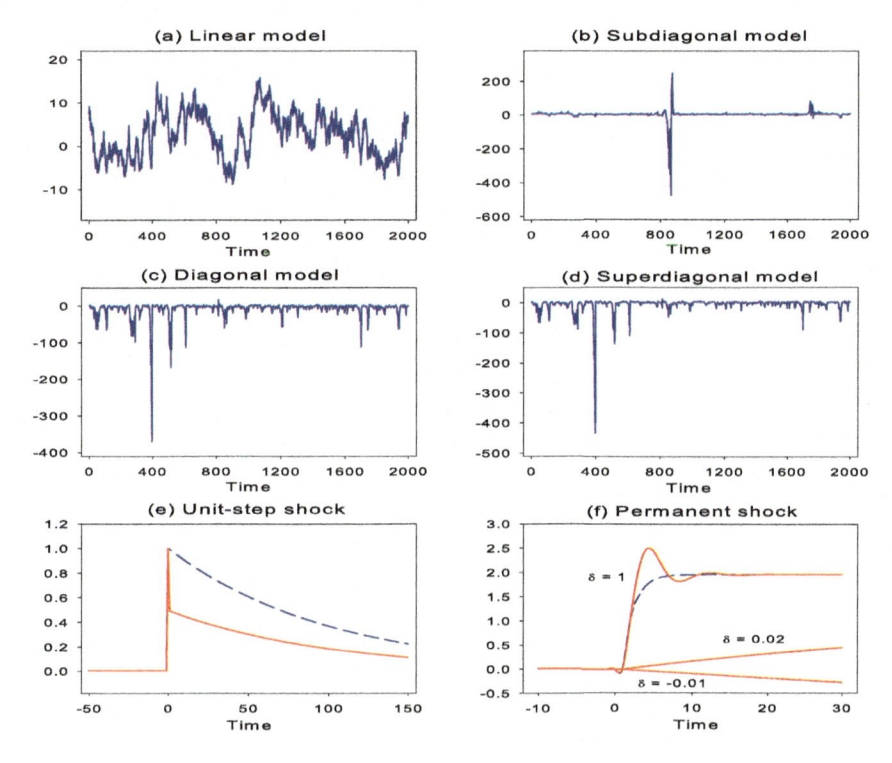

Figure 2.2: *(a) – (d) Realizations of the processes (2.16) – (2.19), respectively; (e) Generalized impulse response functions (GIRFs) for both diagonal and subdiagonal models (blue medium dashed line), and superdiagonal model (red solid line) for a unit-shock at $t = 1$; (f) GIRFs for both diagonal and superdiagonal models (blue medium dashed line) and subdiagonal model (red solid lines) for a permanent shock δ of magnitude -0.01, 0.02, and 1 at time $t = 1$.*

Example 2.2: Comparing BL Time Series

Some of the differences between the three special cases of the BL model can be seen by considering the following specifications:

$$Y_t = \phi Y_{t-1} + \varepsilon_t \qquad \text{(linear AR(1))} \qquad (2.16)$$
$$Y_t = \phi Y_{t-1} + \varepsilon_t + \psi Y_{t-2}\varepsilon_{t-1} \qquad \text{(subdiagonal)} \qquad (2.17)$$
$$Y_t = \phi Y_{t-1} + \varepsilon_t + \psi Y_{t-1}\varepsilon_{t-1} \qquad \text{(diagonal)} \qquad (2.18)$$
$$Y_t = \phi Y_{t-1} + \varepsilon_t + \psi Y_{t-1}\varepsilon_{t-2} \qquad \text{(superdiagonal)} \qquad (2.19)$$

with $\phi = 0.99$ and $\psi = -0.5$, and where $\{\varepsilon_t\} \stackrel{\text{i.i.d.}}{\sim} \mathcal{N}(0,1)$.

Figures 2.2(a) – (d) show plots of the time series. The linear AR(1) model, as a simple "baseline" specification, exhibits some evidence of long-term drift-like behavior, consistent with the fact that this model is close to a random walk. In marked contrast, model (2.17) exhibits two large, highly localized bursts; similar to the extreme peaks in Figure 1.3. Also, note that the series seems to have a sample mean zero, which is consistent with the result $\mathbb{E}(Y_t) = 0$

established in Exercise 1.2. The series generated by the diagonal model also exhibits a sample mean zero, but here the general character of the series is quite different from the subdiagonal case. In particular, we see many isolated negative bursts, occurring frequently enough to achieve a non-zero (specifically, negative) sample mean, which is agreement with the fact that $\mathbb{E}(Y_t) = -0.5$.

Example 2.3: Dynamic Effects of a BL Model

Consider the BL time series models (2.17) – (2.19) with $Y_0 = 0$. It is useful to compare these models through the effect of a one-unit shock on Y_t at time $t = 1$, i.e. $\varepsilon_1 = 1$, and $\varepsilon_2 = \varepsilon_3 = \ldots = 0$, given the history ω_{t-1}. As discussed in Appendix 2.A, this can be measured by the difference between the conditional expectation with and without the shock (called *generalized impulse response function* (GIRF)) and in this case given by

$$\text{GIRF}_Y(t, 1, \omega_{t-1}) = \mathbb{E}[Y_t|\varepsilon_1 = 1, \varepsilon_2 = 0, \varepsilon_3 = 0, \ldots] - \mathbb{E}[Y_t|\varepsilon_1 = 0, \varepsilon_2 = 0, \ldots].$$

Iterating each BL model, we get the following response functions for the three models:

$$\text{GIRF}^{(\text{sub})} = \phi^{t-1}, \quad \text{GIRF}^{(\text{diag})} = \phi^{t-1}, \quad \text{GIRF}^{(\text{super})} = \phi^{t-2}(\phi + \psi), \quad (t \geq 2).$$

Figure 2.2(e) shows these responses for the case $\phi = 0.99$ and $\psi = -0.5$. Note, the series generated by the superdiagonal model appears to exhibit somewhat similar behavior to the diagonal model. In contrast, the GIRF of the superdiagonal model defined by equation (2.19) is different from the other two models. In fact, the response functions of models (2.16) – (2.18) are identical (blue medium dashed line). For the superdiagonal model the term $-0.5Y_{t-1}\varepsilon_{t-2}$ is non-zero for $t = 2$, and hence has a direct effect on the impulse response function for $t > 2$ (red solid line).

Figure 2.2(f) presents a global picture of what happens when each of the three BL models are hit by a permanent shock δ at time $t = 1$. The step responses for $\delta = -0.01, 0.02$, and 1 for the diagonal and superdiagonal models are identical (blue medium dashed line). In fact, both step responses are described by an equivalent AR(1) process with parameter $\phi + \psi\delta$. The subdiagonal model (2.17), on the other hand, exhibits much faster step responses (red solid lines). There is a slight overshoot for this model, reflecting the fact that its equivalent linear model is an AR(2) process, i.e. $Y_t = 0.99Y_{t-1} - 0.5\delta Y_{t-2} + \varepsilon_t$.

2.3 Exponential ARMA Model

Let $\mu(\mathbf{S}_{t-1}) = \phi_0$, $\theta_j(\mathbf{S}_{t-1}) = \theta_j + \tau_j \exp(-\gamma Y_{t-d}^2)$ $(j = 1, \ldots, q)$, and $\phi_i(\mathbf{S}_{t-1}) = \phi_i + \xi_i \exp(-\gamma Y_{t-d}^2)$, $(i = 1, \ldots, p)$. Then (2.10) yields the *exponential autoregressive*

2.3 EXPONENTIAL ARMA MODEL

moving average (ExpARMA) model of order (p,q) and delay d $(d \leq p)$:

$$Y_t = \phi_0 + \sum_{i=1}^{p}\{\phi_i + \xi_i \exp(-\gamma Y_{t-d}^2)\}Y_{t-i} + \sum_{j=1}^{q}\{\theta_j + \tau_j \exp(-\gamma Y_{t-d}^2)\}\varepsilon_{t-j} + \varepsilon_t, \quad (2.20)$$

where the parameter $\gamma > 0$ denotes a scaling factor.

Essentially this model changes smoothly between two extreme linear models, since for large $|Y_{t-d}|$, the coefficients of (2.20) are almost ϕ_i's and θ_j's. For small values of $|Y_{t-d}|$, they are $\phi_i + \xi_i$ and $\theta_j + \tau_j$ and the exponential function changes smoothly between these two extreme values. A sufficient condition for strict stationarity for the ExpARMA process (2.20) is that all the roots of the associated characteristic equation

$$z^p - c_1 z^{p-1} - \cdots - c_p = 0 \quad (2.21)$$

are inside the unit circle, where $c_i = \max\{|\phi_i|, |\phi_i + \xi_i|\}$ $(i = 1, \ldots, p)$. Hence, the characteristic roots of (2.20) are amplitude-dependent, instead of constant. Consequently, $\{Y_t\}$ can be locally small or large. For this reason, (2.20) is also referred to as amplitude-dependent ExpARMA process.

One of the purposes of proposing (2.20) is to reproduce certain features of nonlinear random vibrations through a nonlinear time series model. Originally (2.20), with μ fixed at zero, $p = 2$, and $q = 0$, was derived from the stochastic second-order differential equation $\ddot{X}(t) + f(\dot{X}(t)) + g(X(t)) = \eta(t)$, where $f(\cdot)$ (the "*damping force*") and $g(\cdot)$ (the "*restoring force*") are nonlinear functions, and $\dot{X}(t)$ and $\ddot{X}(t)$ denote the first and second derivatives of the stochastic response $X(t)$ respectively. The function $\eta(t)$ is an external random input, or external force, representing nonlinear random vibrations.

The asymptotic solution of the nonlinear homogeneous differential equation $\ddot{X}(t) + f(\dot{X}(t)) + g(X(t)) = 0$ is a *periodic function* called *limit cycle*. A limit cycle refers to the phenomenon that the trajectories of $X(t)$ do not wind into a singular point, but they eventually go round on closed loops, leaving an interior region untraversed if they wind from outside, or leaving an exterior region untraversed if they wind from inside. Sometimes a limit cycle is *self-excited*, i.e., it remains "active" under zero input. Some nonlinear time series models with this property can produce useful long-term forecasts, as opposed to stationary linear models that have an "eventual forecast function" which gradually approaches a constant for increasing forecast horizons. In other cases a limit cycle requires a certain input to "excite" it. A formal definition of a limit cycle is as follows.

Let $\{\mathbf{Y}_t, t \in \mathbb{Z}\}$ denote an m-dimensional state vector satisfying the equation

$$\mathbf{Y}_t = f(\mathbf{Y}_{t-1}), \quad \mathbf{Y}_0 \in \mathbb{R}^m.$$

A set $\mathbf{\Lambda} = (c_1, \ldots, c_N)$ is called a limit cycle of period $N \in \mathbb{Z}^+$ if (i) $\exists \mathbf{Y}_0 \notin \mathbf{\Lambda}$, $\{\mathbf{Y}_t\}$ will ultimately fall into $\mathbf{\Lambda}$ as t increases, and (ii)

$$c_i = f(c_{i-1}) \quad (i = 1, \ldots, N+1),$$
$$f(c_N) = c_1, \text{ and } f(c_i) \neq c_1 \ (i = 2, \ldots, N).$$

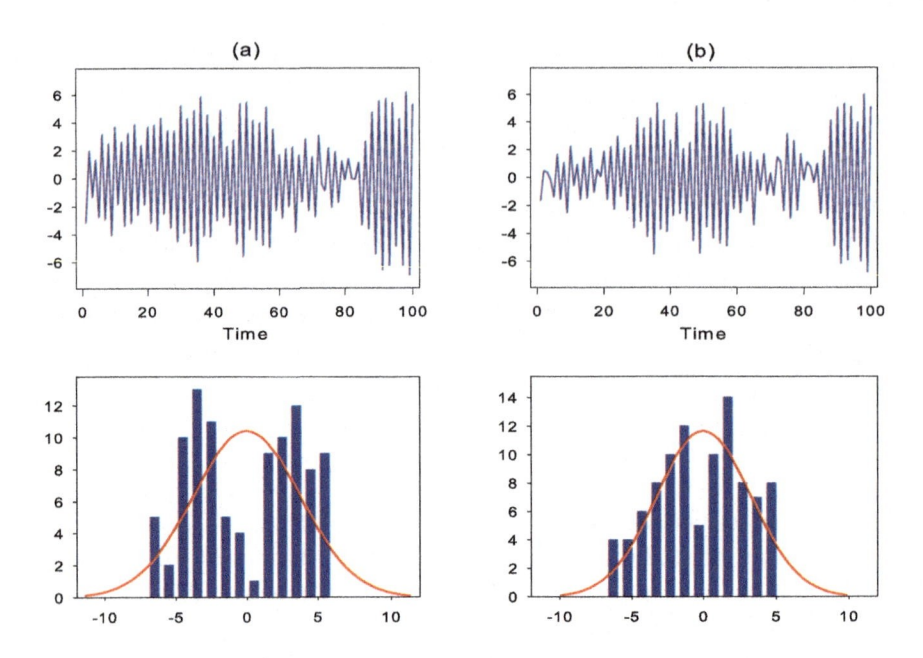

Figure 2.3: *(a) A realization of the* ExpAR(1) *model* (2.23) *with* $\xi = -0.95$ *and corresponding histogram; (b) A realization of the* ExpAR *model* (2.23) *with* $\xi = 0.95$ *and corresponding histogram;* $T = 100$.

In addition to (2.21), a necessary (but not sufficient) condition for the existence of a limit cycle of the ExpAR(p) process is that at least one of the roots of

$$z^p - (\phi_1 + \xi_1)z^{p-1} - \cdots - (\phi_p + \xi_p) = 0 \qquad (2.22)$$

lies outside the unit circle. Example 2.4 illustrates this feature of the ExpAR process via MC simulation.

Example 2.4: ExpAR Time Series

Consider the ExpAR(1) model

$$Y_t = \{-0.9 + \xi \exp(-Y_{t-1}^2)\}Y_{t-1} + \varepsilon_t, \quad \{\varepsilon_t\} \stackrel{\text{i.i.d.}}{\sim} \mathcal{N}(0,1). \qquad (2.23)$$

Figure 2.3 shows $T = 100$ observations from (2.23) with $\xi = -0.95$ and $\xi = 0.95$, respectively, with corresponding histograms below each graph. Both time plots demonstrate the two types of amplitude-dependent frequency, i.e. increasing and decreasing frequency. For both values of ξ condition (2.21) is satisfied. However, only in the case $\xi = -0.95$, a limit cycle exists. Indeed, it follows directly from the above definition that the skeleton of (2.23), i.e., its noise-free ($\varepsilon_t \equiv 0$) representation, has a limit cycle $(\tau_1, \tau_2) = (-1.50043, 1.50043)$. Still the up- and down patterns in both time series plots are very similar. Both histograms show a bimodal distribution with light and short tails, which is a characteristic of some distributions in the ExpAR family. The second histogram is slightly more peaked than the first histogram.

Note that if $|Y_{t-1}| \to 0$, the exponential term in (2.23) approaches 1. So for $\xi = -0.95$ behaves increasingly like an explosive (nonstationary) process, and for $\xi = 0.95$ as a stationary linear AR(1) process. In the latter case the impulse response of the ExpAR model will be approximated by the impulse response function of this linear process, which for a shock $\varepsilon_t = \delta$ is readily determined to be $(0.05)^{t/2}\delta$ if t is even, and 0 otherwise. Conversely, if $|Y_{t-1}|$ is sufficiently large, the exponential term is small, so the process behaves like a stationary AR(1) process for both values of ξ. Its impulse response function is $(-0.9)^{t/2}\delta$ if t is even, and 0 otherwise.

2.4 Random Coefficient AR Model

Let $\mu(\mathbf{S}_{t-1}) = \mu$ as constant, $\theta_j(\mathbf{S}_{t-1}) = 0 \ \forall j$, and $\phi_i(\mathbf{S}_{t-1}) = \{\phi_i + \beta_{i,t}\}$. Then (2.10) reduces to,

$$Y_t = \mu + \sum_{i=1}^{p}\{\phi_i + \beta_{i,t}\}Y_{t-i} + \varepsilon_t, \qquad (2.24)$$

where $\{\mathbf{B}_t = (\beta_{1,t}, \ldots, \beta_{p,t})'\}$ is a sequence of i.i.d. random vectors with zero mean $\mathbb{E}(\mathbf{B}_t) = \mathbf{0}$ and $\text{Cov}(\mathbf{B}_t) = \mathbf{\Sigma}_\beta$, and $\{\mathbf{B}_t\}$ is independent of $\{\varepsilon_t\}$.

Model (2.24) is termed a *random coefficient* AR (RCAR) model of order p. If $p = 1$, a necessary and sufficient condition for second-order stationarity is that $\phi^2 + \sigma_\beta^2 < 1$; see Anděl (1976, 1984) for more complicated stationarity conditions when $p > 1$. Note, by introducing random coefficients to an ARMA model, we can generalize the RCAR model. Alternatively, by assuming the coefficients $\beta_{i,t}$ are not independent but follow an arbitrary strictly stationary stochastic process (say an MA process) defined on the same probability space as $\{\varepsilon_t\}$, one obtains the so-called *doubly stochastic* model (Tjøstheim, 1986a,b).

2.5 Nonlinear MA Model

Let $\mu(\mathbf{S}_{t-1}) = 0$, $\phi_i(\mathbf{S}_{t-1}) = 0 \ \forall i$, and with a slight change of notation we define $\{\theta_j(\cdot)\}$ as,

$$\theta_{j,i_1}(\mathbf{S}_{t-1}) = \begin{cases} \sum_{i_1=0}^{Q} \beta_{i_1} & j = 1, \\ \sum_{i_1=0}^{Q} \sum_{i_2=0}^{Q} \beta_{i_1,i_2} \varepsilon_{t-i_2} & j = 2, \\ \vdots \\ \sum_{i_1=0}^{Q} \sum_{i_2=0}^{Q} \cdots \sum_{i_q=0}^{Q} \beta_{i_1,i_2,\ldots,i_q} \varepsilon_{t-i_2} \cdots \varepsilon_{t-i_q} & j = q. \end{cases}$$

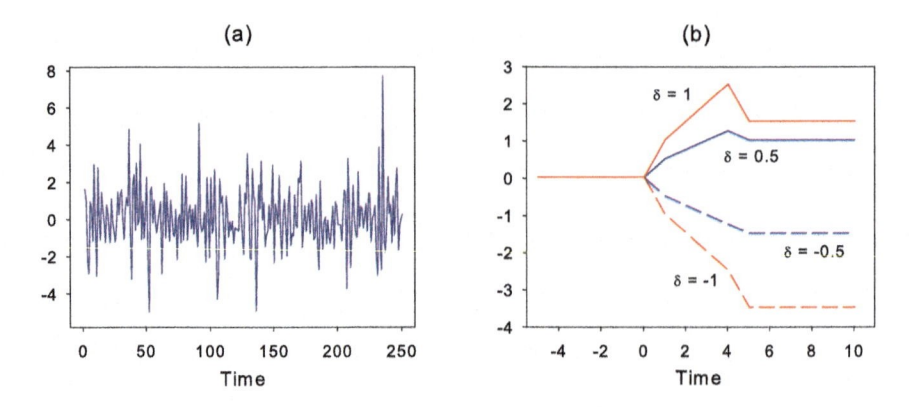

Figure 2.4: *(a) A realization of the NLMA model (2.26) with $\{\varepsilon_t\} \stackrel{i.i.d.}{\sim} \mathcal{N}(0,1)$, $\beta = 0.5$ and $T = 250$; (b) Four permanent step response functions.*

With these restrictions, (2.10) becomes

$$Y_t = \varepsilon_t + \sum_{j=1}^{q} \theta_{j,i_1}(\mathbf{S}_{t-1})\varepsilon_{t-ji_1}$$

$$= \varepsilon_t + \sum_{i_1=0}^{Q} \beta_{i_1}\varepsilon_{t-i_1} + \sum_{i_1=0}^{Q}\sum_{i_2=0}^{Q} \beta_{i_1,i_2}\varepsilon_{t-i_2}\varepsilon_{t-2i_1} + \cdots$$

$$+ \sum_{i_1=0}^{Q}\sum_{i_2=1}^{Q}\cdots\sum_{i_q=0}^{Q} \beta_{i_1,i_2,\ldots,i_q}\varepsilon_{t-i_2}\varepsilon_{t-i_3}\cdots\varepsilon_{t-\eta i_q}, \quad (2.25)$$

where η is the highest order of summations. The model is termed *nonlinear moving average* (NLMA) of order (Q, q).

Note, a similar NLMA representation follows from restricting the Volterra expansion (2.5).

Example 2.5: Dynamic Effects of an NLMA Model

To illustrate the general range of qualitative behavior seen in an NLMA model, consider the following model

$$Y_t = \varepsilon_t + \beta(\varepsilon_{t-1} + \varepsilon_{t-2} + \varepsilon_{t-3}) - \varepsilon_t \varepsilon_{t-4}. \quad (2.26)$$

The response of (2.26) to a one-unit shock at $t = 1$ is easily seen to be β for $t = 2$ and 3, and 0 otherwise. For a sequence of permanent shocks of size δ, starting at $t = 1$, we get the following response function: $\delta(1 + (t-1)\beta)$ for $t = 2, 3, 4$, and $\delta(1 + 3\beta) - \delta^2$ for $t \geq 5$.

Figure 2.4(a) shows a typical realization of model (2.26) with $\beta = 0.5$. The interesting feature of this model lies in its potential to produce large values of $\{Y_t\}$ given large values of ε_t and ε_{t-4}. Figure 2.4(b) shows the response function of (2.26) to a sequence of permanent shocks of magnitude ± 0.5 and ± 1. As with a single-step impulse response, these step responses all reach their steady-state values in finite time (here, 5 time steps). In contrast to the impulse response, however, these step responses give clear evidence of the asymmetric nature of the nonlinearity.

2.6 Threshold Models

Threshold models are a very general class of models, which can capture certain nonlinear features, such as limit cycles, asymmetries, and jump phenomena. The essential idea underlying this class of models is the piecewise linear approximation of the general nonlinear model (2.8) by the introduction of thresholds. Thresholds follow from partitioning the real line $\mathbb{R}$ into $k \geq 1$ non-overlapping intervals, or *regimes*, $\mathbb{R}^{(i)}$ such that $\cup_{i=1}^{k} \mathbb{R}^{(i)} = \mathbb{R}$ and $\mathbb{R}^{(i)} \cap \mathbb{R}^{(i')} = \emptyset$ if $i \neq i'$. Each interval $\mathbb{R}^{(i)}$ is given by $\mathbb{R}^{(i)} = (r_{i-1}, r_i]$, where $r_0 = -\infty$, $r_1, \ldots, r_{i-1} \in \mathbb{R}$, and $r_k = \infty$. The values $r_0 < r_1 < \cdots < r_{k-1} < r_k$ are called *thresholds*. These values determine the actual regimes, or mix of regimes. The ordering of the thresholds guarantees the identifiability of the model. The regime-switching dynamics can be driven by the observed time series $\{Y_t\}$ itself, the model is said to be *self-exciting*. Alternatively, the transition from one member of the set of thresholds to another can be driven by an external (exogenous) time series variable. Further, the transition can be abrupt or follow some smooth function over time. These observations have resulted in several versions of threshold models, some of which we discuss below.

2.6.1 General threshold ARMA (TARMA) model

Let $\{Y_t, t \in \mathbb{Z}\}$ be a strictly stationary time series process, and $\{J_t\}$ be a random (indicator) variable taking values in $\{1, 2, \ldots, k\}$. Given this setup, there are various equivalent ways to write down a threshold model each having its advantages, depending on the context and purpose. One general definition, due to Tong and Lim (1980), of a TARMA(p, q) model for the process $\{(Y_t, J_t), t \in \mathbb{Z}\}$ is given by

$$Y_t = \phi_0^{(J_t)} + \sum_{u=1}^{p} \phi_u^{(J_t)} Y_{t-u} + \varepsilon_t + \sum_{v=1}^{q} \theta_v^{(J_t)} \varepsilon_{t-v}, \tag{2.27}$$

where $\{\varepsilon_t\} \stackrel{\text{i.i.d.}}{\sim} (0, \sigma_\varepsilon^2)$, and the coefficients $\phi_u^{(J_t)}$ ($u = 1, \ldots, p$), $\theta_v^{(J_t)}$ ($v = 1, \ldots, q$) are constants. For each t, the process $\{J_t\}$ acts as the *switching* mechanism between the k regimes. The process can be observable, hidden, or a combination of both.

Writing $\mathbf{Y}_t = (Y_t, \ldots, Y_{t-p+1})'$, a canonical (vector) form of (2.27) is given by

$$\mathbf{Y}_t = \mathbf{C}^{(J_t)} + \mathbf{\Phi}^{(J_t)} \mathbf{Y}_{t-1} + \mathbf{\Theta}^{(J_t)} \boldsymbol{\varepsilon}_t, \tag{2.28}$$

where, for $J_t = i$,

$$\mathbf{C}^{(i)} = (\phi_0^{(i)}, 0, \ldots, 0)', \quad \mathbf{\Phi}^{(i)} = \begin{pmatrix} \phi_1^{(i)} & \cdots & \phi_{p-1}^{(i)} & \phi_p^{(i)} \\ \mathbf{I}_{p-1} & & \mathbf{0}_{(p-1)\times 1} \end{pmatrix} \quad \text{(a companion matrix)}$$

$$\mathbf{\Theta}^{(i)} = \begin{pmatrix} \theta_1^{(i)} & \cdots & \theta_q^{(i)} \\ & \mathbf{0}_{(p-1)\times q} & \end{pmatrix}, \quad \boldsymbol{\varepsilon}_t = (\varepsilon_t, \ldots, \varepsilon_{t-q+1})',$$

and $\boldsymbol{\varepsilon}_t$ is independent of $\{\mathbf{Y}_s\}$ ($s < t$).

2.6.2 Self-exciting threshold ARMA model

The general setting (2.28) includes as a special case the so-called *self-exciting threshold ARMA* (SETARMA) model of order $(k; p_1, \ldots, p_k, q_1, \ldots, q_k)$ and *delay parameter* $d \in \mathbb{Z}^+$. Taking $\mathbf{\Phi}^{(i)}$, $\mathbf{\Theta}^{(i)}$, $\mathbf{C}^{(i)}$ as above, with the additional conditions that, for $i = 1, \ldots, k$,

$$\phi_u^{(i)} = 0 \text{ for } u = p_i + 1, p_i + 2, \ldots, p, \text{ and } p = \max(p_1, \ldots, p_k, d),$$
$$\theta_v^{(i)} = 0 \text{ for } v = q_i + 1, q_i + 2, \ldots, q, \text{ and } q = \max(q_1, \ldots, q_k).$$

Assume that the indicator variable J_t takes the value i if $Y_{t-d} \in \mathbb{R}^{(i)}$.[6] Then the general SETARMA is defined as

$$Y_t = \sum_{i=1}^{k} \left(\phi_0^{(i)} + \sum_{u=1}^{p_i} \phi_u^{(i)} Y_{t-u} + \varepsilon_t^{(i)} + \sum_{v=1}^{q_i} \theta_v^{(i)} \varepsilon_{t-v}^{(i)} \right) I(Y_{t-d} \in \mathbb{R}^{(i)}), \quad (2.29)$$

where $\varepsilon_t^{(i)} = \sigma_i^2 \varepsilon_t$, and $\{\varepsilon_t\} \overset{\text{i.i.d.}}{\sim} (0, 1)$. Note that (2.29) may be viewed as a generalization of a nonhomogeneous linear ARMA model since the noise variances $\text{Var}(\varepsilon_t^{(i)})$ are different for different i.

Example 2.6: Dynamic Effects of a SETAR Model

To illustrate the effect of a one-unit shock or a permanent shock on $\{Y_t, t \in \mathbb{Z}\}$, it is instructive to consider the SETAR(2; 1, 0) model with threshold parameter r and delay $d = 1$, i.e.

$$Y_t = \begin{cases} 2Y_{t-1} + \varepsilon_t & \text{if } |Y_{t-1}| \leq r, \\ \varepsilon_t & \text{if } |Y_{t-1}| > r, \end{cases} \quad (2.30)$$

where $\{\varepsilon_t\} \overset{\text{i.i.d.}}{\sim} (0, \sigma_\varepsilon^2)$. We see that the model switches between a locally nonstationary process and a locally stationary process. Globally, however, the process is stationary, as may be deduced from Figure 2.5(a).

---

[6]There is no loss of generality in assuming $d \leq p$, since if $d > p$ we can introduce additional coefficients $\phi_u^{(i)} = 0$ for $u = p+1, \ldots, d$.

2.6 THRESHOLD MODELS

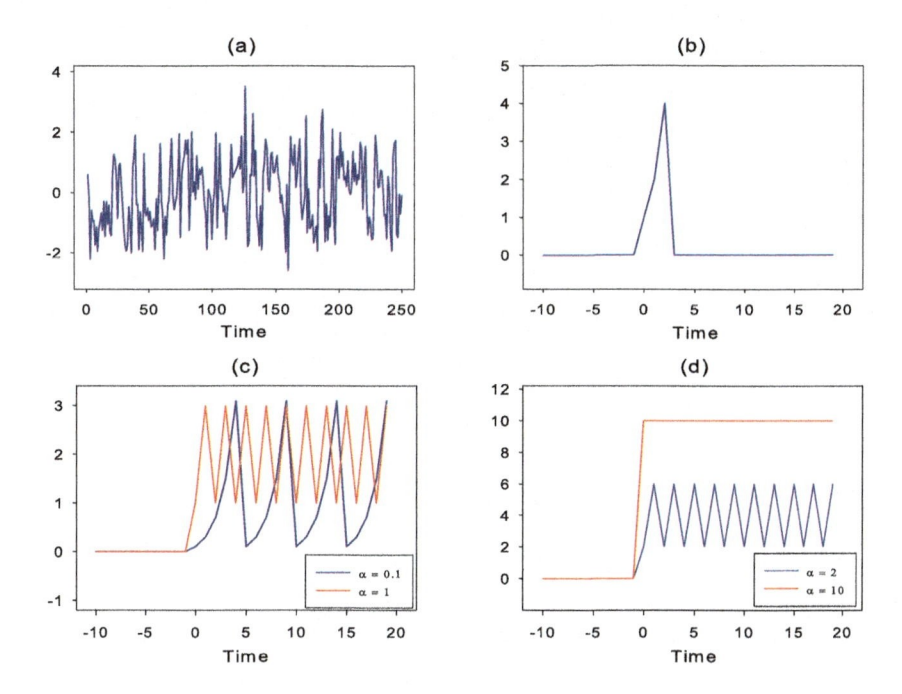

Figure 2.5: *(a) A realization of model (2.30) with $r = 2$, $T = 250$, and $\{\varepsilon_t\} \overset{\text{i.i.d.}}{\sim} (0, 1)$; (b) Impulse response function for a one-unit shock at time $t = 1$; (c) Permanent step responses for $\delta = 0.1$ and $\delta = 1$; (d) Permanent step responses for $\delta = 2$ and $\delta = 10$.*

Figure 2.5(b) shows the impulse response function of (2.30) for a one-unit shock at time $t = 1$ when $r = 2$, and $Y_0 = 0$. More generally, for an impulse response of magnitude δ, initially $Y_t = 0$ for $t \leq 0$, while $Y_1 = \delta$. Next, for $0 < \delta \leq r$, the resulting responses are $\{2\delta, 2^2\delta, \ldots, 2^n\delta, 0, \ldots, 0\}$, where n is the largest integer satisfying $2^n\delta \leq r$. If $\delta > r$, it follows that $Y_1 = \delta$ and $Y_t = 2Y_{t-1} + \varepsilon_t = 0$ for $t \geq 2$. Consequently, the impulse response function exhibits a one sample duration for $\delta > r$.

Given a threshold value $r = 2$, Figures 2.5(c) – (d) show responses to steps of four different magnitudes δ. Since $\varepsilon_t = \delta \ \forall t \geq 1$, the process does not remain in the domain of the unstable first-order linear model $Y_t = 2Y_{t-1} + \varepsilon_t$ but is periodically driven into the domain of $Y_t = \varepsilon_t$, where it "switches back" to the initial unstable model. So, for $|\delta| \leq 2$ the step response function oscillates with a period determined by the magnitude of the step input, between the two regimes. Note that the time required to "escape" from the lower regime depends on the input value δ. If $|\delta| > 2$ the step response function is simply the input step $\varepsilon_t = \delta \ \forall t \geq 1$.

2.6.3 Continuous SETAR model

Clearly, the SDM formulation (2.10) does not contain (2.29), because the passage from one regime to the another is not smooth, the conditional distribution of the process is discontinuous. More formally, consider a two-regime SETAR model of order (p,p). Let $\boldsymbol{\phi}_i = (\phi_0^{(i)}, \phi_1^{(i)}, \ldots, \phi_p^{(i)})'$ be the corresponding coefficient vector $(i = 1, 2)$. Then the model is said to have a *discontinuous* AR function if there exists $\mathbf{Z}_* = (1, Z_{p-1}, \ldots, Z_0)'$, where $Z_{p-d} = r$, such that $(\boldsymbol{\phi}_1 - \boldsymbol{\phi}_2)'\mathbf{Z}_* \neq 0$. In this case, the threshold parameter r constitutes the jump point of the AR function. Otherwise, that is, if $(\boldsymbol{\phi}_1 - \boldsymbol{\phi}_2)'\mathbf{Z}_* = 0$ for all $\mathbf{Z}_*$ satisfying the above condition, the model has a *continuous* AR function.

It is easy to see that the latter case is equivalent to the requirement that $\phi_u^{(1)} = \phi_u^{(2)}$ $(1 \leq u \neq d \leq p)$, and that $\phi_0^{(1)} + r\phi_d^{(1)} = \phi_0^{(2)} + r\phi_d^{(2)}$. Therefore, in the continuous case, the SETAR model can be written as

$$Y_t = \phi_0 + \sum_{u=1, u \neq d}^{p} \phi_u Y_{t-u} + \begin{cases} \phi_d^-(Y_{t-d} - r) + \sigma_1 \varepsilon_t & \text{if } Y_{t-d} \leq r, \\ \phi_d^+(Y_{t-d} - r) + \sigma_2 \varepsilon_t & \text{if } Y_{t-d} > r, \end{cases} \quad (2.31)$$

where

$$\phi_0 = \phi_0^{(1)} + r\phi_d^{(1)}, \quad \phi_d^- = \phi_d^{(1)}, \quad \phi_d^+ = \phi_d^{(2)}, \text{ and } \phi_u = \phi_u^{(1)} \text{ for } u \neq d.$$

We use the acronym CSETAR to distinguish (2.31) from discontinuous SETAR models. This distinction is important because the asymptotics of the conditional least squares (CLS) estimator of the parameter $\boldsymbol{\theta} = (\boldsymbol{\phi}_1', \boldsymbol{\phi}_2', r, d)'$ is different in both cases.[7] While, for a time series of length T, the CLS estimator $\widehat{\boldsymbol{\phi}}_{i,T}$ of $\boldsymbol{\phi}_i$ always converges to a normal distribution with mean zero at rate $\sqrt{T}$, the asymptotic covariance matrix depends upon whether the model is continuous or not. In fact, we shall see in Section 6.1.2 that in the discontinuous case the CLS estimator $\widehat{r}_T$ of r converges to a nonstandard distribution at a rate T (super-consistent), and is asymptotically independent of $\widehat{\boldsymbol{\phi}}_{i,T}$. For CSETAR models, $\widehat{r}_T$ converges to a normal distribution at the usual rate $\sqrt{T}$ and is asymptotically correlated with $\widehat{\boldsymbol{\phi}}_{i,T}$; see Chan and Tsay (1998).

The conditional expectation of model (2.31) is given by

$$\mathbb{E}(Y_t; \boldsymbol{\theta} | \mathcal{F}_{t-1}) = \phi_0 + \sum_{u=1, u \neq d}^{p} \phi_u Y_{t-u} + \phi_d^-(Y_{t-d} - r)^- + \phi_d^+(Y_{t-d} - r)^+, \quad (2.32)$$

where $\mathcal{F}_t$ is the σ-algebra generated by $\{Y_s, s \leq t\}$, and where $(y)^- = \min(0, y)$ and $(y)^+ = \max(0, y)$. Observe that the right-hand side of (2.32) can be written as $\sum_{u=1}^{p} g_u(Y_u)$ where $g_u(\cdot)$ $(u \neq d)$ are linear functions and $g_d(\cdot)$ is piecewise linear.

---

[7] The class of CSETAR(MA) models should not be confused with the class of continuous-time threshold ARMA models which may be viewed as a continuous-time analogue of (2.29); see, e.g., Brockwell (1994).

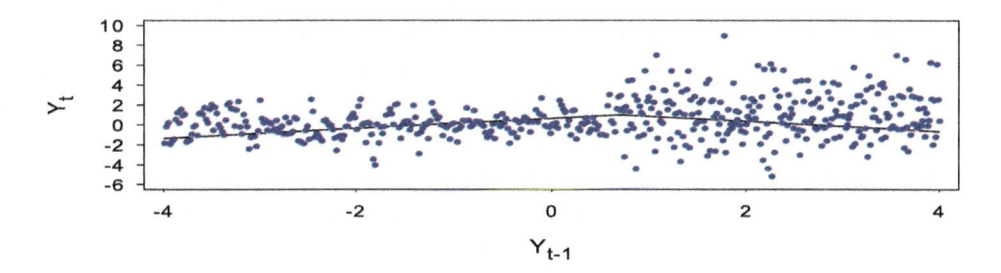

Figure 2.6: *Scatter plot of a typical realization of the* CSETAR *model (2.33) with the true* AR *functions overlaid (black solid lines);* $T = 500$.

Thus, the CSETAR model is additive. In fact, it is a subclass of the nonlinear additive functional-coefficient models to be discussed in Section 9.2.5, and a special case of the multivariate adaptive regression splines model of Section 9.2.3.

Example 2.7: A Simulated CSETAR Process

Consider the CSETAR(2; 1, 1) model

$$Y_t = 1 + \begin{cases} 0.5(Y_{t-1} - 0.7) + \varepsilon_t & \text{if } Y_{t-1} \leq 0.7, \\ -0.5(Y_{t-1} - 0.7) + 2\varepsilon_t & \text{if } Y_{t-1} > 0.7, \end{cases} \quad (2.33)$$

where $\{\varepsilon_t\} \stackrel{\text{i.i.d.}}{\sim} \mathcal{N}(0, 1)$. Figure 2.6 shows a scatter plot of Y_t versus Y_{t-1} for a typical simulated time series of length $T = 500$, and the true AR functions are overlaid. Given (2.32), the CLS parameter estimates follow from minimizing the sum of squared residuals following similar steps as in Algorithm 6.2; see also Chan and Tsay (1998). For the simulated series, we obtain the fitted model

$$\widehat{Y}_t = 1.02_{(0.11)} + \begin{cases} 0.56_{(0.06)}(Y_{t-1} - 0.72_{(0.21)}) & \text{if } Y_{t-1} \leq 0.72_{(0.21)}, \\ -0.48_{(0.12)}(Y_{t-1} - 0.72_{(0.21)}) & \text{if } Y_{t-1} > 0.72_{(0.21)}, \end{cases} \quad (2.34)$$

where the asymptotic standard errors of the parameter estimates are in parentheses. The standard errors of the residuals are $\widehat{\sigma}_1 = 1.08$ and $\widehat{\sigma}_2 = 3.98$. The sample sizes for the two regimes are 303 and 196, respectively. Comparing (2.33) and (2.34), we see that the two models are similar. The closeness in absolute value of the two lag-one coefficients in (2.34) is indicative of using a CSETAR model; see Gonzalo and Wolf (2005) for a formal test statistic.

2.6.4 Multivariate thresholds

The dynamics of the SETARMA model (2.29) are controlled by the single threshold variable Y_{t-d} with $d > 0$. A more flexible self-exciting threshold model can be obtained by introducing multivariate thresholds, assuming the relationships between the threshold variables is linear, but *unknown*. For ease of explanation we formulate the resulting model in terms of a SETAR specification. First, we introduce a general framework.

Consider an m-dimensional Euclidean space $\mathbb{R}^m$ and a point $\mathbf{x}$ in that space. Let $\boldsymbol{\omega} = (\omega_1, \ldots, \omega_m)'$ denote an m-dimensional unknown parameter vector. These parameters define a *hyperplane* as follows $\mathbb{H} = \{\mathbf{x} \in \mathbb{R}^m | \boldsymbol{\omega}'\mathbf{x} = r\}$, where r is a scalar. The direction of $\boldsymbol{\omega}$ determines the orientation of the hyperplane whereas r represents the position of the hyperplane in terms of its distance from the origin. The hyperplane $\mathbb{H}$ induces a partition of the space into two regions defined by the half spaces $\mathbb{H}^- = \{\mathbf{x} \in \mathbb{R}^m | \boldsymbol{\omega}'\mathbf{x} \leq r\}$ and $\mathbb{H}^+ = \{\mathbf{x} \in \mathbb{R}^m | \boldsymbol{\omega}'\mathbf{x} > r\}$. In terms of the indicator function $I(\cdot)$, this partition is given by $I(\mathbf{x}) = 1$ if $\mathbf{x} \in \mathbb{H}^-$ and 0 otherwise.

Now, assume that an m-dimensional space is spanned by the vector of time series values $\widetilde{\mathbf{X}}_{t-1} = (Y_{t-1}, \ldots, Y_{t-m})'$. Further, suppose that there are k functions $I(\boldsymbol{\omega}_i'\widetilde{\mathbf{X}}_{t-1} \leq r_i)$ $(i = 1, \ldots, k)$ where $\boldsymbol{\omega}_i = (\omega_1^{(i)}, \ldots, \omega_m^{(i)})'$ and r_i are real parameters. Thus, each of these functions defines a threshold. Then a SETAR model with m ($1 \leq m \leq p$) thresholds and order $(k; p, \ldots, p)$, denoted by SETAR$(k; p, \ldots, p)_m$, is defined as

$$Y_t = \phi_0 + \sum_{u=1}^{p} \phi_u Y_{t-u} + \sum_{i=1}^{k} \left\{ \xi_0^{(i)} + \sum_{u=1}^{p} \xi_u^{(i)} Y_{t-u} \right\} I(\boldsymbol{\omega}_i'\widetilde{\mathbf{X}}_{t-1} \leq r_i) + \varepsilon_t$$

$$= \boldsymbol{\phi}'\mathbf{X}_{t-1} + \sum_{i=1}^{k} \boldsymbol{\xi}_i'\mathbf{X}_{t-1} I(\boldsymbol{\omega}_i'\widetilde{\mathbf{X}}_{t-1} \leq r_i) + \varepsilon_t, \qquad (2.35)$$

where

$$\boldsymbol{\phi} = (\phi_0, \ldots, \phi_p)', \; \boldsymbol{\xi}_i = (\xi_0^{(i)}, \ldots, \xi_p^{(i)})', \text{ and } \mathbf{X}_{t-1} = (1, Y_{t-1}, \ldots, Y_{t-p})'.$$

Note that (2.35) is not identified. For identification purpose, we impose the restriction $r_1 \leq \cdots \leq r_k$. Further, due to the fact that $I(x) = 1 - I(-x)$, a convenient normalization condition is to set one element of $\boldsymbol{\omega}_i$ equal to unity.

Example 2.8: A Simulated SETAR$(2; 1, 1)_2$ Model

Consider the SETAR$(2; 1, 1)_2$ model

$$Y_t = 0.5 + 0.9 Y_{t-1} - 1.8 Y_{t-1} I(\boldsymbol{\omega}_1'\widetilde{\mathbf{X}}_{t-1} \leq 0) - I(\boldsymbol{\omega}_2'\widetilde{\mathbf{X}}_{t-1} \leq 0) + \varepsilon_t, \quad (2.36)$$

where $\boldsymbol{\omega}_1 = (1, -1)'$, $\boldsymbol{\omega}_2 = (0, 1)'$, and $\widetilde{\mathbf{X}}_{t-1} = (Y_{t-1}, Y_{t-2})'$. Thus the dynamics of (2.36) is controlled by two threshold functions. The first one is a bi-dimensional threshold when $Y_{t-1} - Y_{t-2} = 0$. The second one is a single threshold when $Y_{t-2} = 0$. Figure 2.7(a) shows the threshold boundaries.[8]

---

[8]Tiao and Tsay (1994) generalize the single threshold SETAR to a similar model as in (2.36) with *known* parameters $\boldsymbol{\omega}_i$ $(i = 1, 2)$.

2.6 THRESHOLD MODELS

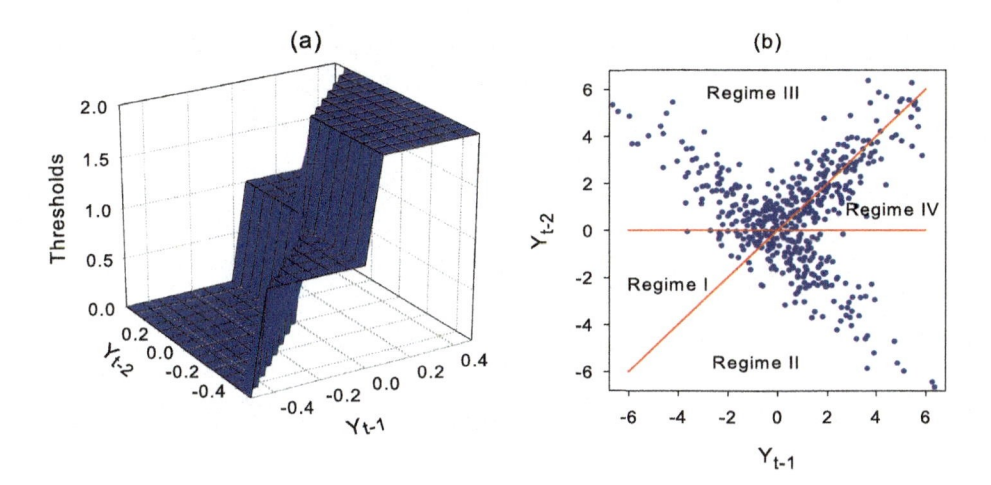

Figure 2.7: *(a) Threshold boundaries of model (2.36); (b) Scatter plot of Y_{t-2} versus Y_{t-1} with two separating hyperplanes (red solid lines); $T = 500$, $\{\varepsilon_t\} \stackrel{i.i.d.}{\sim} \mathcal{N}(0,1)$.*

Rewriting (2.36) in four separate regimes gives

$$Y_t = \begin{cases} -0.5 - 0.9Y_{t-1} + \varepsilon_t, & \text{I:} \quad \text{if } Y_{t-1} - Y_{t-2} \leq 0 \text{ and } Y_{t-2} \leq 0, \\ -0.5 + 0.9Y_{t-1} + \varepsilon_t, & \text{II:} \quad \text{if } Y_{t-1} - Y_{t-2} > 0 \text{ and } Y_{t-2} \leq 0, \\ 0.5 - 0.9Y_{t-1} + \varepsilon_t, & \text{III:} \quad \text{if } Y_{t-1} - Y_{t-2} \leq 0 \text{ and } Y_{t-2} > 0, \\ 0.5 + 0.9Y_{t-1} + \varepsilon_t, & \text{IV:} \quad \text{if } Y_{t-1} - Y_{t-2} > 0 \text{ and } Y_{t-2} > 0. \end{cases}$$

If we reconsider the U.S. unemployment series of Example 1.1 in terms of the above model specification the four regimes (I – IV) have a direct meaning. Regime I indicates that the economy changed from a contraction period ($Y_{t-2} \leq 0$) to an even worse one ($Y_{t-1} \leq Y_{t-2}$). In Regime II, the economy is still in recession ($Y_{t-2} \leq 0$), but improving ($Y_{t-1} > Y_{t-2}$). Regime III can be viewed as a contraction period with negative growth. Finally, Regime IV is an expansion period with positive growth. Figure 2.7(b) shows a scatter plot of Y_{t-2} versus Y_{t-1} based on one realization of (2.36) with $\{\varepsilon_t\} \stackrel{i.i.d.}{\sim} \mathcal{N}(0,1)$, and $T = 500$. The solid lines denote the two separating hyperplanes.

2.6.5 Asymmetric ARMA model

A strictly stationary time series $\{Y_t, t \in \mathbb{Z}\}$ is said to follow an *asymmetric autoregressive moving average* model of order (p, q), or for short asARMA(p, q), if it takes the form

$$Y_t = \phi_0 + \sum_{i=1}^{p} \phi_i^+ Y_{t-i}^+ + \sum_{i=1}^{p} \phi_i^- Y_{t-i}^- + \varepsilon_t + \sum_{j=1}^{q} \theta_j^+ \varepsilon_{t-j}^+ + \sum_{j=1}^{q} \theta_j^- \varepsilon_{t-j}^-. \tag{2.37}$$

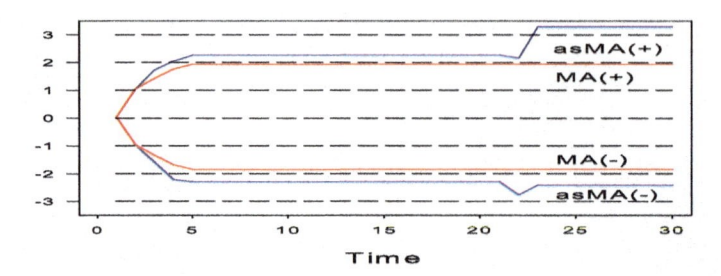

Figure 2.8: *Impact of a maintained unit shock from zero to one onwards from t = 10 (MA(+), asMA(+), blue solid lines) and a corresponding negative unit shock (MA(−), asMA(−), red solid lines) on the series* $\{Y_t\}$. *From* Brännäs and De Gooijer (1994).

Here $Y_t^{\pm}$ and $\varepsilon_t^{\pm}$ denote the asymmetric component processes, defined as

$$Y_t^- = Y_t I(\varepsilon_t \leq 0), \quad Y_t^+ = Y_t I(\varepsilon_t > 0), \quad \varepsilon_t^- = \varepsilon_t I(\varepsilon_t \leq 0), \quad \varepsilon_t^+ = \varepsilon_t I(\varepsilon_t > 0),$$

with $\{\varepsilon_t\} \sim \text{WN}(0, \sigma_\varepsilon^2)$. If $p \neq 0$ and $q = 0$, (2.37) reduces to an asymmetric AR(p) (asAR) model. It is called an asymmetric MA(q) (asMA) model for $p = 0$ and $q \neq 0$. Note that (2.37) has four filters, two for positive innovations and two for negative innovations.

An alternative way to write (2.37) is

$$Y_t = \sum_{i=1}^{p} \left(\phi_i^- + \alpha_i I(\varepsilon_{t-i} > 0) \right) Y_{t-i} + \varepsilon_t + \sum_{j=1}^{q} \left(\theta_j^- + \beta_j I(\varepsilon_{t-j} > 0) \right) \varepsilon_{t-j}, \quad (2.38)$$

where $\alpha_i = \phi_i^+ - \phi_i^-$ ($i = 1, \ldots, p$), $\beta_j = \theta_j^+ - \theta_j^-$ ($j = 1, \ldots, q$).[9] We see that the asAR and asMA parts add two weighted sums of positive innovations to a conventional ARMA model. In addition, we see that (2.38) belongs to the class of threshold models with $I(\varepsilon_{t-i} > 0)$ ($i = 1, \ldots, \max(p,q)$) controlling the transition between the two regimes.

Example 2.9: Dynamic Effects of an asMA Model

Consider the asMA model

$$Y_t = 0.01 + \varepsilon_t + 0.69\varepsilon_{t-1}^+ + 0.34\varepsilon_{t-2}^+ + 0.22\varepsilon_{t-3}^+ - 0.11\varepsilon_{t-21}^+ + 1.12\varepsilon_{t-22}^+ \\ + 0.61\varepsilon_{t-1}^- + 0.64\varepsilon_{t-2}^- - 0.07\varepsilon_{t-3}^- + 0.48\varepsilon_{t-21}^- - 0.35\varepsilon_{t-22}^-. \quad (2.39)$$

Brännäs and De Gooijer (1994) fitted the above model successfully to quarterly growth rates in U.S. real GNP, using first differences of logged values of the original series. Evidence of asymmetry may be noted from the sign and magnitude of the parameter values. For instance, at lag 22 the response to a

---

[9] If there is a threshold value $r \neq 0$ in the $\varepsilon_t^{\pm}$ functions, it can be accounted for by including a constant term in (2.38) and retaining $r = 0$ as a threshold value.

2.6 THRESHOLD MODELS

positive innovation is stronger than to a negative shock. In addition, the responses are of the same sign. Figure 2.8 shows this phenomenon in a slightly different way. The accumulated effect of a permanent positive or negative unit change from $t = 10$ onwards from a value zero in $\{\varepsilon_t\}$ is displayed for model (2.39) and a best fitted MA(3) model which is given by

$$Y_t = 0.01 + \widehat{\varepsilon}_t + 0.38\widehat{\varepsilon}_{t-1} + 0.34\widehat{\varepsilon}_{t-2} + 0.17\widehat{\varepsilon}_{t-3},$$

where $\widehat{\varepsilon}_t$ denotes the tth residual. For the MA(3) model a positive or negative shock has, apart from a change in sign, a similar effect on $\{Y_t\}$. On the other hand, for model (2.39), asymmetry is clearly present in the resulting series. There is a more rapid decline to a lower level for a negative shock than there is an increase to a higher level for a positive shock.

Note that the graph only gives the two most extreme outcomes out of $5^2 = 25$ possible parameter combinations. Each combination corresponds to a particular sequence of positive and negative innovations. There is equal probability for each combination when the innovations are i.i.d. from a symmetric distribution. Each combination of an asMA model can be given a corresponding AR representation. With 25 combinations, equally many AR representations will arise. These can be seen as a reasonable approximation to, for instance, a STAR model, discussed in Section 2.7.

2.6.6 Nested SETARMA model

The general setting (2.27) can be extended to allow for regime-switches controlled by multiple observable input variables. One general class of models suitable for this purpose is the so-called nested SETARMA (NeSETARMA) model of Astatkie et al. (1997). Suppose, without loss of generality, that a strictly stationary process $\{Y_t, t \in \mathbb{Z}\}$ (output) has two input variables $\{X_t, t \in \mathbb{Z}\}$ and $\{Z_t, t \in \mathbb{Z}\}$. Moreover, assume that the regime-switching is conditional on the values of the delayed observable variables Y_t and X_t. Using these variables the complete dynamic system is divided in two subsystems, or *stages*. Each stage consists of regimes, with the second stage regimes *nested* within those of the first stage. The regimes are formed in such a way that there is a linear relationship between Y_t and its lagged values, and a linear relationship between Y_t and lagged values of X_t. If Y_t is used as regime-switching variable in the first stage, then X_t will be used in the second stage and the resulting model is called an *output-input* NeSETARMA model. On the other hand, if X_t is used in the first stage and Y_t in the second, then the model is called an *input-output* NeSETARMA model. The (possibly lagged) relationship between Y_t and Z_t may be linear or quadratic.

Below we focus on an output-input NeSETARMA model. Before defining its structure, we introduce some notation.

- Let $k_1 \geq 1$ be the number of first-stage regimes formed by partitioning the values of Y_{t-d_1} into non-overlapping intervals with $d_1 \in \mathbb{Z}^+$ the *first-stage delay*.

- Let $\mathbb{R}^{(i)} = (r_{i-1}, r_i]$ denote the ith ($i = 1, \ldots, k_1$) interval with $r_0 = -\infty$ and $r_{k_1} = \infty$. The parameters $r_1, \ldots, r_{k_1-1}$ are the *first-stage thresholds*.

- Let $\ell_{i,2} \geq 1$ ($i = 1, \ldots, k_1$) be the number of second-stage regimes formed by using X_{t-d_2} as a threshold variable with $d_2 \in \mathbb{Z}$ the *second-stage delay*.

- Let $\mathbb{R}^{(i,j)} = (r_{i,j-1}, r_{i,j}]$ ($i = 1, \ldots, k_1; j = 1, \ldots, \ell_{i,2}$) denote the jth second-stage regime within the ith first-stage regime with $r_{i,0} = -\infty$ and $r_{i,\ell_{i,2}} = \infty$. The set $\{r_{i,1}, \ldots, r_{i,\ell_{i,2}-1}\}$ represents the *second-stage thresholds*.

Given the above setup, a general NeSETARMA model is defined as

$$Y_t = \sum_{i=1}^{k_1} \Big\{ \sum_{j=1}^{\ell_{i,2}} \Big(\phi_0^{(i,j)} + \sum_s \phi_s^{(i,j)} Y_{t-s} + \sum_u \xi_u^{(i,j)} X_{t-u} \\ + \sum_v \eta_v^{(i,j)} Z_{t-v} + \varepsilon_t^{(i,j)} + \sum_w \theta_w^{(i,j)} \varepsilon_{t-w} \Big) I(X_{t-d_2} \in \mathbb{R}^{(i,j)}) \Big\} I(Y_{t-d_1} \in \mathbb{R}^{(i)}), \tag{2.40}$$

where $\{\varepsilon_t^{(i,j)}\} \overset{\text{i.i.d.}}{\sim} (0, 1)$. Clearly, (2.40) consists of $\sum_{i=1}^{k_1} \ell_{i,2}$ regimes.

Several (non)linear models emerge as special cases of (2.40):

- If $k_1 = \ell_{1,2} = 1$, $\phi_s \neq 0$, $\xi_u = \eta_v = 0$, and $\theta_w \neq 0$ $\forall s, u, v, w$, then the NeSETARMA model reduces to an ARMA model.

- If $k_1 = \ell_{1,2} = 1$, $\phi_s \neq 0$, $\xi_u \neq 0$, $\eta_v = 0$, and $\theta_w \neq 0$ $\forall s, u, v, w$, then the NeSETARMA reduces to an ARMAX (loosely speaking a *transfer function*) model.

- If $k_1 > 1$, $\ell_{i,2} = 1$ $\forall i$, and $\xi_u = \eta_v = 0$, and $\theta_w \neq 0$ $\forall s, u, v, w$, then the NeSETARMA becomes a SETARMA model.

- If $k_1 = 1$, $\phi_s \neq 0$, $\xi_u \neq 0$, and $\eta_v = \theta_w = 0$ $\forall s, u, v, w$, then (2.40) reduces to the so-called *open-loop* SETAR (or TARSO) model of Tong (1990). This model is defined as

$$Y_t = \phi_0^{(j)} + \sum_{s=1}^{m_j} \phi_s^{(j)} Y_{t-s} + \sum_{u=0}^{m'_j} \xi_u^{(j)} X_{t-u} + \varepsilon_t^{(j)} \tag{2.41}$$

conditional on $X_{t-d} \in \mathbb{R}^{(j)}$ ($j = 1, \ldots, \ell$). We fit a (subset-)TARSO model to an empirical time series in Section 6.4. Exercise 2.10 shows estimation results for a NeSETAR model.

2.7 Smooth Transition Models

For some time series processes, it may not seem reasonable to assume an abrupt change in the regimes. Instead the speed of transition may be smooth over time. Let $G(\cdot)$ denote a smooth continuous function, the so-called *transition function*. Then a (two-regime) *smooth transition autoregressive* (STAR) model of order $(2;p,p)$ is defined as

$$Y_t = \left\{\phi_0^{(1)} + \sum_{u=1}^{p} \phi_u^{(1)} Y_{t-u}\right\}(1 - G(z_t)) + \left\{\phi_0^{(2)} + \sum_{u=1}^{p} \phi_u^{(2)} Y_{t-u}\right\} G(z_t) + \varepsilon_t,$$

$$= \phi_0 + \sum_{u=1}^{p} \phi_u Y_{t-u} + \left\{\xi_0 + \sum_{u=1}^{p} \xi_u Y_{t-u}\right\} G(z_t) + \varepsilon_t, \quad (2.42)$$

where $\phi_u = \phi_u^{(1)}$ and $\xi_u = \phi_u^{(2)} - \phi_u^{(1)}$ $(u = 0, 1, \ldots, p)$. The transition function $G(\cdot)$ allows the conditional expectation of the model to change smoothly from $\mathbb{E}(Y_t|Y_s; s \leq t) = \phi_0 + \sum_{u=1}^{p} \phi_u Y_{t-u}$ to $\mathbb{E}(Y_t|Y_s; s \leq t) = \phi_0 + \sum_{u=1}^{p} \phi_u Y_{t-u} + \{\xi_0 + \sum_{u=1}^{p} \xi_u Y_{t-u}\}$ with Y_t.

Various formulations for $G(\cdot)$ have been proposed in the literature. For example, one may use $G(z_t) \equiv G(Y_{t-d}; \gamma, c) = \Phi(\gamma\{Y_{t-d} - c\})$, where $\Phi(\cdot)$ is the cumulative distribution function (CDF) of the standard normal distribution. Here, $d \geq 1$ is again the delay parameter, c is a location value, indicating when the transition is occurring, whereas $\gamma > 0$ is a slope parameter. The role played by γ in $\Phi(\cdot)$ is that of smoothing. When the value of γ increases, the transition is completed in a short period of time, and $\Phi(\gamma\{Y_{t-d} - c\})$ approaches the indicator function $I(Y_{t-d} - c)$. In that case (2.42) reduces to a SETAR$(2; p, p)$ model. On the other hand, when γ is sufficiently close to zero (2.42) may be well approximated by a linear AR(p) model.

Two plausible alternative transition functions are the *logistic function* and the *exponential function*. The logistic function is defined as

$$G(Y_{t-d}; \gamma, c) = \frac{1}{1 + \exp\{-\gamma(Y_{t-d} - c)\}}, \quad \gamma > 0, \quad (2.43)$$

and the resulting model is then called *logistic smooth transition autoregressive* (LSTAR). The exponential function is specified as

$$G(Y_{t-d}; \gamma, c) = 1 - \exp\{-\gamma(Y_{t-d} - c)^2\}, \quad \gamma > 0, \quad (2.44)$$

and the resulting model is referred to as *exponential smooth transition autoregressive* (ESTAR) model. If $c = 0$ and $d = 1$, then the ESTAR(p) becomes identical to the ExpAR(p) model.

Figure 2.9 shows some examples of the relationship between γ, Y_{t-d} for (a) the logistic transition function (2.43), and for (b) the exponential transition function (2.44) where, for ease of interpretation, we set $c = 0$ and $d = 1$. Some observations are in order:

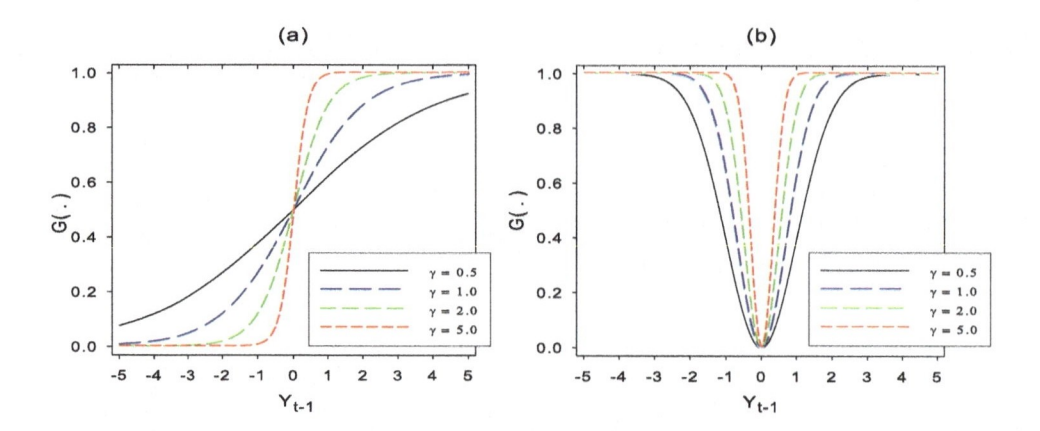

Figure 2.9: *Effects of various values of the smoothness parameter γ on (a) the logistic transition function (2.43), and (b) the exponential transition function (2.44). Both functions with $c = 0$ and $d = 1$.*

- In the limit, as $\gamma \to 0$, both transition functions switch between 0 and 1 very smoothly and slowly. Both models reduce to an AR(p) model as γ becomes small, with $G(\cdot) \to 0.5$ for the LSTAR(p) model, and with $G(\cdot) \to 0$ for the ESTAR(p) model.

- For the LSTAR(p) model, as $\gamma \to \infty$, $G(Y_{t-1}; \gamma, c) \to I(Y_{t-1} > c)$. Hence, the LSTAR($p$) model approaches a SETAR(2; p, p) model. In contrast, as $\gamma \to \infty$, (2.44) approaches the indicator function $I(Y_{t-1} = c)$, and consequently the ESTAR model does not nest the SETAR model as a special case.

- The ESTAR transition function is symmetric about c in the sense that the local dynamics are the same for high as for low values of Y_{t-1}, whereas the mid-range behavior, for values close to c, is different. Thus, the distance between Y_{t-1} and c matters, but not the sign. For the LSTAR model, the local dynamics depends on the distance between Y_{t-1} and c, as well as the sign.

Note that an asMA model of Section 2.6.5, contains 2^q separate MA(q) regimes. In some cases, it may also seem plausible to think of a continuum of MA regimes and that the transition from one extreme regime to the other is smooth. This requires modifying the transition function $I(\varepsilon_{t-j} \geq 0)$ into a smooth function $G_j(\gamma \varepsilon_{t-j})$ ($\gamma > 0; j = 1, \ldots, q$). Since the transition function multiplying ε_{t-j} has ε_{t-j} as its argument $\forall j$, the resulting nonlinear model is additive in structure. For instance, setting $p = 0$, an *additive smooth transition moving average* (ASTMA) model of order q is given by

$$Y_t = \varepsilon_t + \sum_{j=1}^{q} \Big(\theta_j + \delta_j G_j(\gamma \varepsilon_{t-j})\Big) \varepsilon_{t-j}. \tag{2.45}$$

In Example 3.7, we discuss the invertibility of this process.

2.8 Nonlinear non-Gaussian Models

In an attempt to capture the behavior of, possibly observed, nonlinear time series processes with explicit non-Gaussian marginal distributions a number of nonlinear non-Gaussian models have been introduced. In the following subsections we shall briefly discuss two models which seem to be promising to use in practice and have known statistical properties.

2.8.1 Newer exponential autoregressive models

To introduce this class of models, let $\{J_t, t \in \mathbb{Z}\}$, and $\{\varepsilon_t, t \in \mathbb{Z}\}$, be two independent sequences of i.i.d. discrete random variables. Consider the SDM (2.10) with $\mu(\mathbf{S}_{t-1}) = 0$, $\theta_j(\mathbf{S}_{t-1}) = 0$ $\forall j$, and $\phi_i(\mathbf{S}_{t-1}) = \beta^{(J_t)}$ $(i = 1, \ldots, p)$ where $\{J_t\}$ has the following distribution

$$J_t = \begin{cases} 0 & \text{with prob.} \quad \alpha_0, \\ 1 & \text{with prob.} \quad \alpha_1, \\ \vdots & \quad \vdots \\ p & \text{with prob.} \quad \alpha_p. \end{cases}$$

Here $\{\alpha_i\}_{i=0}^{p}$ is a non-negative sequence whose elements sum up to one. Let $\beta^{(0)}(\equiv 0)$, $\beta^{(1)}, \ldots, \beta^{(p)}$ be $p+1$ constants, satisfying $0 \leq \beta^{(j)} \leq 1$ $(1 \leq j \leq p)$. Under the above restrictions the SDM reduces to

$$Y_t = \beta^{(J_t)} Y_{t-J_t} + \varepsilon_t. \qquad (2.46)$$

If the $\{Y_t, t \in \mathbb{Z}\}$ process is assumed to have an exponential marginal distribution function then (2.46) is known as *newer exponential* AR (NEAR) model of order p, NEAR(p). Note that the NEAR(p) model is a special case (sub-class) of the RCAR model (2.24). It is obvious how the concept of "switching" comes into play in (2.46). The degree of AR dependence structure may switch among several, p, possibilities which are controlled by an external (unobserved) random variable J_t, which is independent of past values of the process $\{Y_t, t \in \mathbb{Z}\}$.

Example 2.10: NEAR(1) Model

The NEAR(1) is defined as,

$$Y_t = \varepsilon_t + \begin{cases} \beta Y_{t-1} & \text{with prob.} \quad \alpha, \\ 0 & \text{with prob.} \quad 1-\alpha, \end{cases}$$
$$= \beta J_t Y_{t-1} + \varepsilon_t, \qquad (2.47)$$

where

$$\varepsilon_t = \begin{cases} E_t & \text{with prob.} \quad p_1 = (1-\beta)/(1-(1-\alpha)\beta) \\ (1-\alpha)\beta E_t & \text{with prob.} \quad 1-p_1 = \alpha\beta/(1-(1-\alpha)\beta) \end{cases} \qquad (2.48)$$

$$J_t = \begin{cases} 0 & \text{with prob. } (1-\alpha) \\ 1 & \text{with prob. } \alpha, \end{cases} \quad (2.49)$$

where $\{E_t, t \in \mathbb{Z}\}$ is a sequence of i.i.d. unit mean exponential random variables. The form of the ε_t's is chosen to ensure that the marginal distribution of $\{Y_t, t \in \mathbb{Z}\}$ is exponential with mean unity, i.e. $f_Y(y) = \exp(-y)$ $(0 \leq y < \infty)$. The parameters α and β are allowed to take values over the domain defined by $0 \leq \alpha, \beta \leq 1$ with $\alpha = \beta \neq 1$. We note that due to the distributional assumption underlying $\{E_t\}$, the innovation process is not allowed to take on negative values, i.e. $\mathbb{P}(E_t \leq 0) = 0$. Again, the "switching" characteristic of (2.47) is evident. Due to the AR(1) setup of the model, (2.47), and the restricted domain of the parameters, it follows that for $Y_0 \sim \text{Exp}(1)$ and being independent of $\{E_t, t > 0\}$, the process $\{Y_t, t \in \mathbb{Z}\}$ is stationary – by construction.

Setting $\alpha = 1$, $0 \leq \beta \leq 1$ in (2.47) yields the so-called *exponential* AR model of order 1, or EAR(1) (Lawrence and Lewis, 1980),[10] where fixing $\beta = 1$, $0 \leq \alpha < 1$ give rise to the so-called *transposed* EAR (TEAR) model of order 1 (Lawrance and Lewis, 1981).[11] Both are extreme cases of a NEAR(1) process.[12] The main properties are: the ACF at lag $\ell \in \mathbb{Z}$ is given by $\rho_Y(\ell) = (\alpha\beta)^\ell$, and the regression curve $\mathbb{E}(Y_{t+1}|Y_t = y) = \alpha\beta y$, which is thus linear. This makes maximum likelihood (ML) estimation of α and β possible by numerical optimization. Another interesting feature, is that the NEAR(1) process is not time-reversible (cf. Exercise 2.5).

2.8.2 Product autoregressive model

As a natural extension of the linear AR(1) model, McKenzie (1982) proposes the so-called *product* AR model of order 1, or PAR(1). It consists of an exponentiation of a strictly stationary AR(1) process $\{Y_t, t \in \mathbb{Z}\}$ such that the additive form is being transformed into a linear form. Specifically,

$$Y_t = Y_{t-1}^\alpha V_t, \quad (0 \leq \alpha < 1), \quad (2.50)$$

where the log-transform is given by

$$\log Y_t = \alpha \log Y_{t-1} + \log V_t,$$

---

[10]This acronym should not be confused with the ExpAR model defined in Section 2.3.

[11]Corresponding to the EAR(1) model is the EMA(1), which takes the form $Y_t = \gamma E_t$ with probability γ, and $Y_t = \gamma E_t + E_{t-1}$ with probability $1 - \gamma$ $(0 \leq \gamma \leq 1)$. By bringing together the EAR(1) and EMA(1) processes, the EARMA(1,1) process can be defined.

[12]Both the EAR(1) and TEAR(1) models are somewhat limited in scope for practical application due to the sample paths these models generate. In particular, for the EAR(1) model large values arise when E_t is included (i.e. $J_t = 1$), which are followed by runs of decreasing value, with the runs having geometrically distributed lengths. For the TEAR(1) model the behavior of the sample paths, for a large α, shows geometrically distributed runs of rising values (i.e. $J_t = 1$) followed by sharp declines when the selection $J_t = 0$ is made. One can overcome these shortcoming by using high-order models.

2.8 NONLINEAR NON-GAUSSIAN MODELS

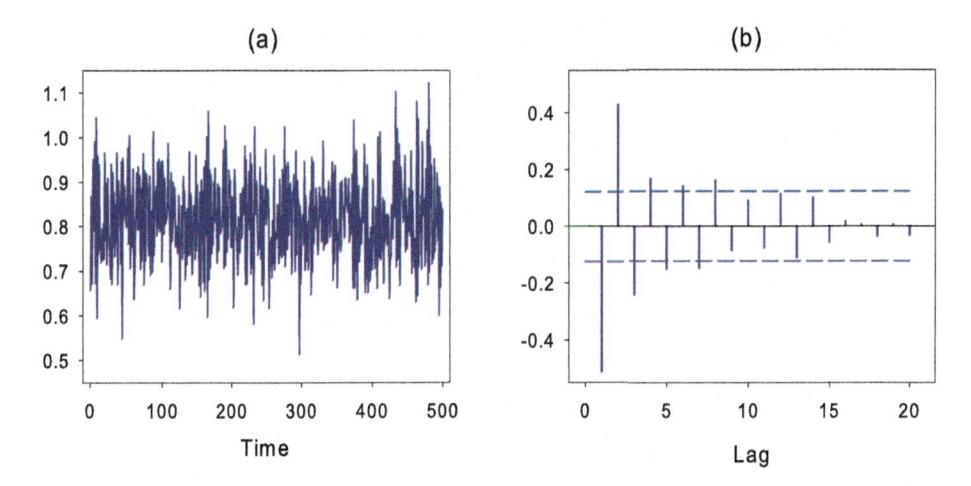

Figure 2.10: *(a) A realization of the* PAR(2) *model* $Y_t = (0.3Y_{t-1}^{-0.9} + 0.5Y_{t-2}^{0.4})\varepsilon_t$, *with* $\{\varepsilon_t\} \stackrel{i.i.d.}{\sim} \mathcal{N}(1, 0.1)$, *and* $T = 500$; *(b) Sample ACF of the time series in (a) with 95% asymptotic confidence limits (blue medium dashed lines).*

with $\{V_t\}$ a sequence of i.i.d. nonnegative random variables, and Y_0 is independent of V_1. We may classify the PAR(1) model as an *intrinsically* linear model, i.e. a nonlinear model which can be linearized. It differs from the NEAR models which cannot be linearized due to their switching nature.

Writing $Y_t = \{\prod_{i=0}^{\ell-1} V_{t-i}^{\alpha^i}\} Y_{t-\ell}^{\alpha^\ell}$. Then, dropping unnecessary subscripts, we have $\mathbb{E}(Y_t Y_{t-\ell}) = \{\prod_{i=0}^{\ell-1} \mathbb{E}(V^{\alpha^i})\} \mathbb{E}(Y^{\alpha^\ell+1})$. From (2.50), $\mathbb{E}(Y^s) = \mathbb{E}(Y^{\alpha s}) \mathbb{E}(V^s)$, and therefore

$$\mathbb{E}(Y_t Y_{t-\ell}) = \prod_{i=0}^{\ell-1} \left\{ \frac{\mathbb{E}(Y^{\alpha^i})}{\mathbb{E}(Y^{\alpha^{i+1}})} \right\} \mathbb{E}(Y^{\alpha^\ell+1}) = \frac{\mathbb{E}(Y)\mathbb{E}(Y^{\alpha^\ell+1})}{\mathbb{E}(Y^{\alpha^\ell})}. \qquad (2.51)$$

Hence, the ACF at lag ℓ is given by

$$\rho_Y(\ell) = \frac{\mathbb{E}(Y_t)\{\mathbb{E}(Y_{t-\ell}^{\alpha^\ell+1}) - \mathbb{E}(Y_{t-\ell}^{\alpha^\ell})\mathbb{E}(Y_{t-\ell})\}}{\mathbb{E}(Y_{t-\ell}^{\alpha^\ell}) \mathrm{Var}(Y_t)}.$$

Note, the ACF depends only on the moments of the stationary marginal distribution. In the particular case of the gamma distribution such moments exist, and this distribution is the only one for which the PAR(1) model has the same ACF structure as an AR(1) process (McKenzie, 1982), hence its name.

More generally, the PAR(p) ($p \geq 2$) model with non-additive noise is defined as

$$Y_t = V_t \left(\sum_{i=1}^{p} \phi_i Y_{t-i}^{\alpha_i} \right). \qquad (2.52)$$

Figure 2.10(a) shows a realization of a PAR(2) process, and 2.10(b) its corresponding sample ACF. We see that the pattern of the sample ACF is compatible with the sample ACF of an AR(2) model.

2.9 Artificial Neural Network Models

The *artificial neural network* (ANN) has been widely used for nonlinear processes with unknown functional form. Probably the most commonly used ANN architecture is the *multi-layer perceptron* (MLP), also known as *feed-forward* network. MLPs receive a vector of inputs x, the explanatory variables, and compute a response or output $y(x)$ by propagating x through the interconnected processing elements, called *neurons* or *nodes*. The processing elements are arranged in layers and the data, x, flows from each layer to the successive one. Within each layer or *"hidden unit"* (processing element), x is nonlinearly transformed by so-called nonlinear *activation-level* functions and propagated to the next layer. Finally, at the output layer $y(x)$, which can be scalar – or vector-valued, is computed. Thus, information flows only in one direction (feed-forward) from input to output units. Without loss of generality we focus here on single layer ANNs.

Figure 2.11 shows the basic architecture of a single hidden layer perceptron with two input units, three hidden units, and one output unit, called a 2-3-1 feed-forward network. The hidden (middle) layer performs a weighted summation of the input units. In fact, the jth node in the hidden layer is defined as

$$h_j = G_j\Big(\alpha_{0j} + \sum_{i \to j} \omega_{ij} x_i\Big), \tag{2.53}$$

where x_i is the value of the ith input node, α_{0j} is a constant (the "bias"), the summation $\sum_{i \to j}$ means summing over all input nodes feeding to j, and ω_{ij} are the connecting weights. The nonlinearity enters the model through the activation-level function $G_j(\cdot)$, usually a "smooth" transition function such as the logistic function in (2.43).

For the output layer, the node is defined as

$$o = \psi\Big(\alpha_{0o} + \sum_{j \to o} \omega_{jo} h_j\Big), \tag{2.54}$$

where $\psi(\cdot)$ is another activation-level function, which is almost always taken to be either linear or an indicator function. Combining (2.53) and (2.54), the output of a single-layer feed-forward ANN can be written as

$$o = \psi\Big[\alpha_{0o} + \sum_{j \to o} \omega_{jo} G_j\Big(\alpha_{0j} + \sum_{i \to j} \omega_{ij} x_i\Big)\Big]. \tag{2.55}$$

Let m be the number of input units, and k the number of nodes in the hidden layer. Then, the network weight vector, say $\boldsymbol{\theta}$, consists of a $(k+1) \times 1$ vector of biases $(\alpha_{0o}, \alpha_{0j})'$, an $mk \times 1$ vector of input layer to hidden layer weights $(\boldsymbol{\omega}'_1, \ldots, \boldsymbol{\omega}'_k)'$ with $\boldsymbol{\omega}_j = (\omega_{1j}, \ldots, \omega_{mj})'$ $(j = 1, \ldots, k)$, and a $k \times 1$ vector of hidden layer to output layer weights $(\omega_{1o}, \ldots, \omega_{ko})'$. Thus, for an m–k–1 network the total number of weights, or dimension of $\boldsymbol{\theta}$, is equal to $r = (m+1)k + (k+1)$. Usually the weight vector $\boldsymbol{\theta}$

2.9 ARTIFICIAL NEURAL NETWORK MODELS

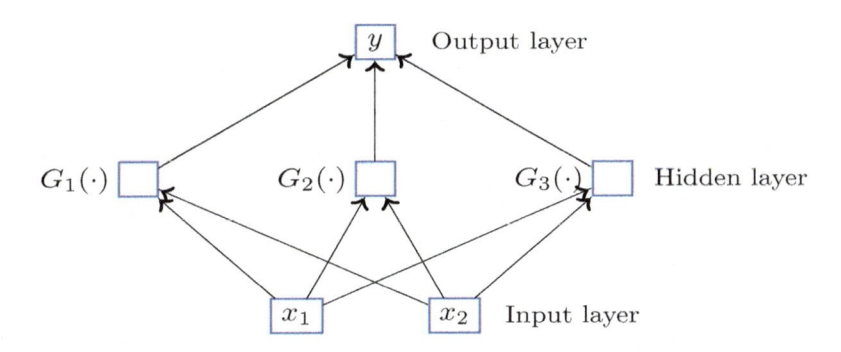

Figure 2.11: *The architecture of a single hidden layer ANN with two input units, three hidden units, and one output unit, a so-called $m - k - 1 = 2 - 3 - 1$ feed-forward network with 13 weights.*

is assumed to take values in the weight space Θ, a subset of the finite-dimensional space $\mathbb{R}^r$. That means, the ANN considered has bounded model complexity and contains a finite number of hidden units k and a finite number of input units m.

In time series applications one also allows an ANN to have so-called *skip-layer*, or direct, connections from inputs to outputs. Then, the output of a feed-forward ANN becomes

$$o = \psi\left[\alpha_{0o} + \sum_{i \to o} \alpha_{io} x_i + \sum_{j \to o} \omega_{jo} G_j\left(\alpha_{0j} + \sum_{i \to j} \omega_{ij} x_i\right)\right]. \tag{2.56}$$

Thus, when $\psi(\cdot)$ is a linear activation-level function, there are direct linear connections from the input to the output nodes.

The weights $\boldsymbol{\theta}$ are the adjustable parameters of the network, and they are obtained through a process called *training*. Let $\{(\boldsymbol{x}_i, y_i)\}_{i=1}^{N}$ denote the training set, where $\boldsymbol{x}_i$ denotes a vector of inputs, and y_i is the variable of interest. The objective of training is to determine a mapping from the training set to a set of possible weights so that the network will produce predictions $\widehat{y}_i$, which in some sense are "close" to the y_i's. For a given network, let $o(\boldsymbol{x}_i; \boldsymbol{\theta})$ be the output for a given $\boldsymbol{x}_i$. Then by far the most common measure of closeness is the ordinary least squares function, i.e.

$$L_N(\boldsymbol{\theta}) = \sum_{i=1}^{N} \{y_i - o(\boldsymbol{x}_i; \boldsymbol{\theta})\}^2.$$

Assume that the network weight space Θ is a compact subset of the r-dimensional Euclidean space $\mathbb{R}^r$, which ensures that the true ANN model is locally unique with regard to the objective function used for training. Then the weights are found as:

$$\widehat{\boldsymbol{\theta}} = \arg\min_{\boldsymbol{\theta}\in\Theta}\{L_N(\boldsymbol{\theta})\},$$

using some kind of iterative minimization scheme. A popular method is the *backpropagation* algorithm, i.e. a gradient descent algorithm where the computations are ordered in a simple fashion by taking advantage of the special structure of an ANN.

2.9.1 AR neural network model

The *autoregressive neural network* (AR–NN) of order p with k regimes and a single output, denoted by AR–NN$(k;p,\ldots,p)$,[13] is defined as

$$Y_t = h(\mathbf{X}_{t-1};\boldsymbol{\theta}) + \varepsilon_t,$$
$$= \phi_0 + \boldsymbol{\phi}'\mathbf{X}_{t-1} + \sum_{j=1}^{k}\xi_j G(\boldsymbol{\omega}_j'\mathbf{X}_{t-1} - c_j) + \varepsilon_t, \quad (2.57)$$

where $h(\cdot)$ denotes a hidden layer containing k nodes, with no activation-level function at the output unit, with hidden activation-level function $G(\cdot):\mathbb{R}\to\mathbb{R}$, a Borel-measurable function of the input vector $\mathbf{X}_{t-1} = (Y_{t-1},\ldots,Y_{t-p})'$, and with the network weight vector $\boldsymbol{\theta}\in\mathbb{R}^{(p+2)k+p+1}$ defined as

$$\boldsymbol{\theta} = (\boldsymbol{\phi}',\boldsymbol{\xi}',\boldsymbol{\omega}',\mathbf{c}',\phi_0)',$$

where

$$\boldsymbol{\phi} = (\phi_1,\ldots,\phi_p)', \quad \boldsymbol{\xi} = (\xi_1,\ldots,\xi_k)', \quad \mathbf{c} = (c_1,\ldots,c_k)',$$
$$\boldsymbol{\omega} = (\boldsymbol{\omega}_1',\ldots,\boldsymbol{\omega}_k')', \text{ with } \boldsymbol{\omega}_j = (\omega_{1j},\ldots,\omega_{pj})', \ (j=1,\ldots,k).$$

In ANN terminology the elements of the $p\times 1$ vector $\boldsymbol{\phi}$ are called the *shortcut connections*, the $k\times 1$ vector $\boldsymbol{\xi}$ consists of the *hidden unit to output connections*, the elements of the $k\times 1$ vector $\mathbf{c}$ are called the hidden unit "*bias*" weights, and the elements of the $pk\times 1$ vector $\boldsymbol{\omega}$ are the so-called *input unit to hidden unit connections*. Thus, jointly with the intercept ϕ_0, the dimension r of the network weight vector $\boldsymbol{\theta}$ is equal to $(p+2)k+p+1$. Note, (2.57) does not include lags of $\{\varepsilon_t\}$ in the set of input variables, and therefore is a feed-forward ANN.

Now, assume that the activation-level function is bounded, i.e. is $|G(x)| < \delta < \infty$ $\forall x\in\mathbb{R}$. Let $\phi(z)$ be the characteristic function associated with the shortcut connections. Then it can be shown (Trapletti et al., 2000) that the condition $\phi(z)\neq 0$ $\forall z$, $|z|\leq 1$ is sufficient, but not necessary for the ergodicity of the Markov chain $\{Y_t\}$. Furthermore, if this condition holds, then $\{Y_t, t\in\mathbb{Z}\}$ is geometrically ergodic (see

---

[13] Analogue to the notation introduced for SETAR models, we refer to the number of regimes k first, and to the order $p,\ldots,p$ of the AR–NN model second. In contrast, some books use the notation AR–NN(p,k).

2.9 ARTIFICIAL NEURAL NETWORK MODELS

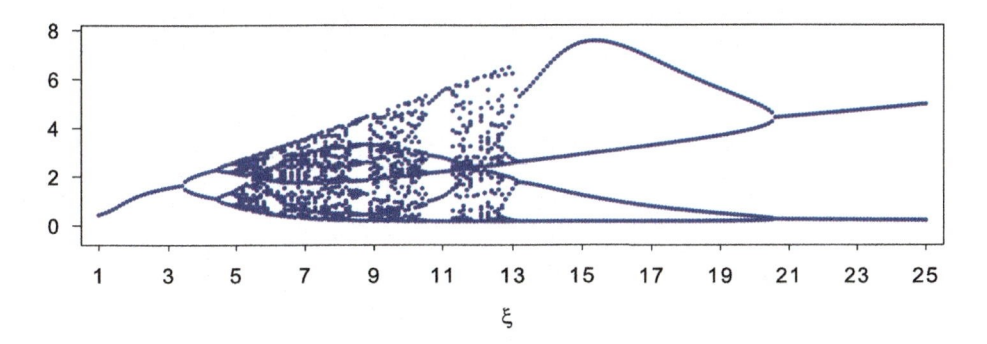

Figure 2.12: *Skeleton $h(\mathbf{X}_{t-1}; \boldsymbol{\theta})$ of the AR–NN(2; 0, 1) model (2.58) for 25 iterations of $\{Y_t\}$ for each value of $\xi = 1, 1.1, \ldots, 24.9, 25$.*[15]

Section 3.4.2) and the associated AR–NN process is called *asymptotically stationary*. Typical choices for $G(\cdot)$ are the hyperbolic tangent (tanh) function and the logistic function.

Certain special cases of the AR–NN model are of interest. If the sum in (2.57) vanishes, then the model reduces to a linear AR(p) model. For $k > 0$, this can be achieved by either setting $\xi_j = 0$ or $\boldsymbol{\omega}_j = \mathbf{0}$ $\forall j$. For the latter case, the sum is a constant, independent of $\mathbf{X}_{t-1}$, and can be absorbed in the intercept ϕ_0.

Example 2.11: Skeleton of an AR–NN(2; 0, 1) Model

Consider the single hidden layer feed-forward AR–NN(2; 0, 1) model

$$Y_t = 0.15 + \xi \tanh(Y_{t-1} - 1) - \xi \tanh(Y_{t-1} - 1.5) + \varepsilon_t, \qquad (2.58)$$

where $\tanh(x) = (\exp(2x) - 1)/(\exp(2x) + 1)$, and with initial condition $Y_0 = 0.1$. Thus, in terms of model specification (2.57), we have $\boldsymbol{\phi} = \mathbf{0}$, and $\boldsymbol{\xi} = (\xi, -\xi)'$, $\mathbf{c} = (1, 1.5)'$, and $\omega = (1, 1)'$.

To illustrate that a relative simple AR–NN model can generate complex dynamical patterns, we consider the *skeleton* $h(\mathbf{X}_{t-1}; \boldsymbol{\theta})$, i.e. the noise-free ($\varepsilon_t \equiv 0$) representation of (2.58) with $\xi = 1, 1.1, \ldots, 24.9, 25$. For each ξ, we perform 2,000 iterations of (2.58). Figure 2.12 shows a scatter plot of the values of $\{Y_t\}$ versus ξ after discarding the first 1,975 iterations. For approximately $1 \leq \xi \leq 3.4$ the model converges to a stable fixed point. Then, for approximately $3.4 < \xi < 4.5$ we see a local stable oscillation of period 2. The oscillation period is doubled for $4.5 < \xi < 5.8$. At about $\xi = 5.8$, the plot hints at deterministic chaos, i.e. the model looses predictability.

---

[15]This type of graph is commonly referred to as a bifurcation diagram in the chaos literature. The skeleton is the underlying dynamical system, i.e. the process without noise.

Example 2.12: Skeleton of an AR–NN$(3; 1, 1, 1)$ Model

Consider the single hidden layer feed-forward AR–NN$(3; 1, 1, 1)$ model composed of one linear and three logistic activation-level functions

$$h(\mathbf{X}_{t-1}; \boldsymbol{\theta}) = 1 - 0.5 Y_{t-1} + \sum_{j=1}^{3} G(Y_{t-1}; \omega_{1j}), \tag{2.59}$$

where

$$G(Y_{t-1}; \omega_{11}) = (1 + \exp(-10[Y_{t-1} - 2]))^{-1},$$
$$G(Y_{t-1}; \omega_{12}) = (1 + \exp(-2Y_{t-1}))^{-1},$$
$$G(Y_{t-1}; \omega_{13}) = (1 + \exp(-20[Y_{t-1} - 1]))^{-1}.$$

Figure 2.13 shows (2.59) as a function of the input series $\{Y_{t-1}\}$, with Y_{t-1} taking values in the set $\{-3, -2.9, \ldots, 2.9, 3\}$ (blue solid line). The values of the activation-level functions $G(Y_{t-1}; \omega_{1j})$ ($j = 1, 2, 3$) are displayed as blue dashed-dotted, dashed-doted-doted, and dotted lines, respectively.

For $Y_{t-1} < -1$ all three logistic activation-level functions are approximately equal to zero in value, so the behavior of (2.59) is determined largely by the slope of the linear activation-level function. For approximately $-1 \leq Y_{t-1} \leq 0.7$ the function $G(Y_{t-1}; \omega_{12})$ slowly starts increasing, but the values of the functions $G(Y_{t-1}; \omega_{11})$ and $G(Y_{t-1}; \omega_{13})$ remain approximately equal zero. As a result, the downward trend of $h(\mathbf{X}_{t-1}; \boldsymbol{\theta})$ levels off. At about $Y_{t-1} = 0.8$, the function $G(Y_{t-1}; \omega_{13})$ changes from 0 to 1 fairly rapidly, and the value of the skeleton increases. Next, for approximately $1.2 < Y_{t-1} \leq 1.7$, the skeleton resumes its gradual declining, owing to the fact that $G(Y_{t-1}; \omega_{12})$ and $G(Y_{t-1}; \omega_{13})$ essentially achieve their maximum values while the function $G(Y_{t-1}; \omega_{11})$ is still not very active. Then, at about $Y_{t-1} = 1.8$, the function $G(Y_{t-1}; \omega_{11})$ begins to activate, resulting in a slow increase of $h(\mathbf{X}_{t-1}; \boldsymbol{\theta})$ up till about the point $Y_{t-1} = 2.3$. Finally, for $Y_{t-1} \geq 2.4$ all three logistic functions are approximately equal unity. So, once again, the linear activation-level function causes the gradual decline of the function $h(\mathbf{X}_{t-1}; \boldsymbol{\theta})$.

In general, the AR–NN model can be either interpreted as a semi-parametric approximation to any Borel-measurable function, or as an extension of the threshold class of models (SETAR and LSTAR) where the transition variable can be a linear combination of stochastic variables. For instance, assume that the variable controlling the switching is composed of a particular subset, say $\widetilde{\mathbf{X}}_{t-1} = (Y_{t-1}, \ldots, Y_{t-q})'$ ($1 \leq q \leq p$) of the elements of $\mathbf{X}_{t-1}$. Then, using the indicator function as activation-level function, i.e. $G(\cdot) = I(\cdot)$, it is easy to see that (2.55) reduces to (2.35) with $k = m$.

Note that the AR–NN model (2.57) is, in principle, neither globally nor locally identified. Three characteristics of the model cause non-identifiability. First, due

2.9 ARTIFICIAL NEURAL NETWORK MODELS

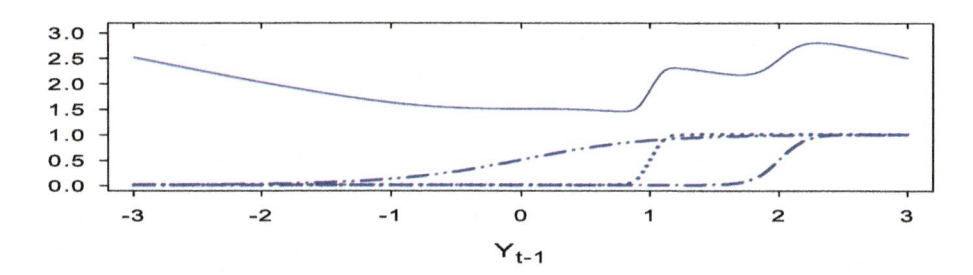

Figure 2.13: *Skeleton $h(\mathbf{X}_{t-1}; \boldsymbol{\theta})$ of an AR–NN(3; 1, 1, 1) model (2.59) (blue solid line). The values of the logistic functions $G(Y_{t-1}; \omega_{1j})$ ($j = 1, 2, 3$) are shown as blue dashed-dotted, dashed-dotted-dotted, and dotted lines, respectively.*

to the symmetries in the ANN architecture the value of the likelihood function remains unchanged if the hidden units are permuted, resulting in $k!$ possibilities for each one of the coefficients of the model. This problem is resolved by imposing the restrictions $c_1 \leq \cdots \leq c_k$ or $\xi_1 \geq \cdots \geq \xi_k$. The second characteristic is caused by the fact that $G(x) = 1 - G(-x)$, where $G(\cdot)$ is the logistic function. This problem can be circumvented, for instance, by imposing the restriction $\omega_{1j} > 0$ ($j = 1, \ldots, k$). Finally, the presence of irrelevant hidden units in the nonlinear part of the AR–NN model can be eliminated by assuming that each hidden unit makes a unique non-trivial contribution to the overall AR–NN process, i.e. $\xi_j \neq 0$, $\boldsymbol{\omega}_j \neq 0$ $\forall j$ ($j = 1, \ldots, k$), and $(\boldsymbol{\omega}'_i, c_i) \neq \pm(\boldsymbol{\omega}'_j, c_j)$ $\forall i \neq j$ ($i, j = 1, \ldots, k$). In practice, these latter assumptions are a part of the model specification stage, applying statistical inference techniques.

2.9.2 ARMA neural network model

The *autoregressive moving average network* ARMA–NN of order $(k; p, q)$ is defined as

$$Y_t = h(\mathbf{X}_{t-1}, \mathbf{e}_{t-1}; \boldsymbol{\theta}) + \varepsilon_t, \tag{2.60}$$

where

$$h(\mathbf{X}_{t-1}, \mathbf{e}_{t-1}; \boldsymbol{\theta}) = \phi_0 + \boldsymbol{\phi}'\mathbf{X}_{t-1} + \boldsymbol{\psi}'\mathbf{e}_{t-1} + \sum_{j=1}^{k} \xi_j G(\boldsymbol{\omega}'_j \mathbf{X}_{t-1} + \boldsymbol{\vartheta}'_j \mathbf{e}_{t-1} - c_j)$$

with the activation-level function $G(\cdot)$ as introduced in Section 2.9.1, an observed input vector $\mathbf{X}_{t-1} = (Y_{t-1}, \ldots, Y_{t-p})'$, and a $q \times 1$ input vector $\mathbf{e}_{t-1} = (e_{t-1}, \ldots, e_{t-q})'$ with a feedback through a linear MA-polynomial $\boldsymbol{\vartheta}_j$ ($j = 1, \ldots, k$) for filtering past residuals. In ANN terminology this feature means that the ARMA–NN network is *recurrent*: future network inputs depend on present and past network outputs.

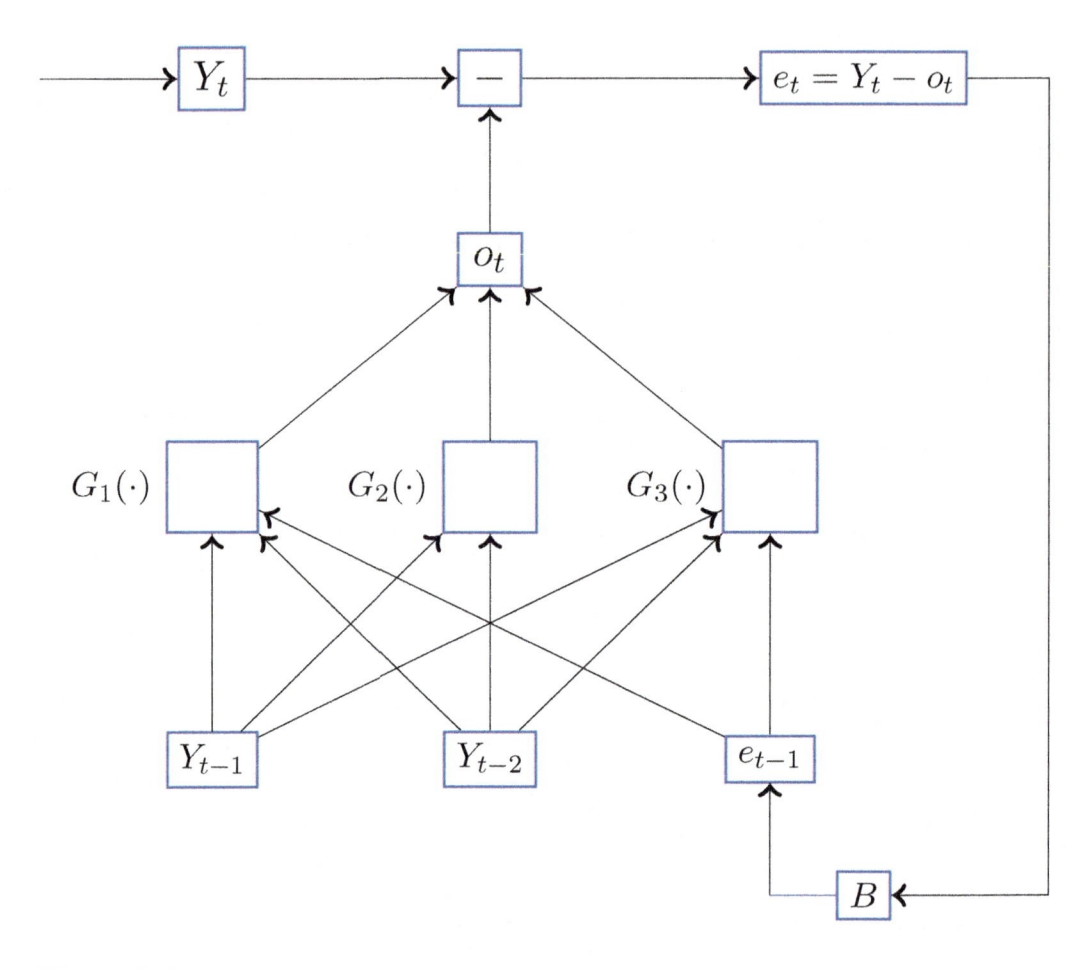

Figure 2.14: *A typical recurrent ARMA–NN(3; 2, 1) model with two lagged variables Y_{t-1} and Y_{t-2} and one recurrent variable e_{t-1} in the set of inputs; o_t denotes the network output at time t, and B is the backward shift operator.*

The network weight vector $\boldsymbol{\theta} \in \mathbb{R}^{(p+q+2)k+p+q+1}$ is composed of various subvectors in an analogous way as given in Section 2.9.1 for an AR–NN($k; p$) model. Indeed, for $p \leq 1$ and $q = 0$, the ARMA–NN($k; p, q$) model reduces to (2.57). Figure 2.14 displays the architecture of a single hidden recurrent layer feed-forward ARMA–NN(3; 2, 1) model.

2.9.3 Local global neural network model

Another member of the regime switching family, derived from ANNs, is the *local global neural network* (LGNN) model. The central idea of LGNN is to express the input-output mapping of a single hidden layer feed-forward ANN, containing k nodes, by a piecewise structure. In particular, the LGNN output describes a combination of pairs of smooth continuous functions, each composed of a p-dimensional

2.9 ARTIFICIAL NEURAL NETWORK MODELS

nonlinear approximation function $L\colon \mathbb{R}^p \to \mathbb{R}$ of $\mathbf{X}_{t-1} = (Y_{t-1}, \ldots, Y_{t-p})'$, and a q-dimensional activation-level function $B\colon \mathbb{R}^q \to \mathbb{R}$ of $\widetilde{\mathbf{X}}_{t-1} = (Y_{t-1}, \ldots, Y_{t-q})'$ ($1 \leq q \leq p$). The resulting model, denoted by LGNN$(k;p)_q$, is defined as

$$Y_t = \sum_{j=1}^{k} L(\mathbf{X}_{t-1}; \boldsymbol{\theta}_{L_j}) B(\widetilde{\mathbf{X}}_{t-1}; \widetilde{\boldsymbol{\theta}}_{B_j}) + \varepsilon_t, \qquad (2.61)$$

where $B(\widetilde{\mathbf{X}}_{t-1}; \widetilde{\boldsymbol{\theta}}_{B_j})$ is defined as the difference between two opposed logistic functions, i.e.

$$B(\widetilde{\mathbf{X}}_{t-1}; \widetilde{\boldsymbol{\theta}}_{B_j}) = -\Big(\frac{1}{1+\exp(-\gamma_j[\widetilde{\boldsymbol{\omega}}'_j\widetilde{\mathbf{X}}_{t-1} - c_{1j}])} \\ - \frac{1}{1+\exp(-\gamma_j[\widetilde{\boldsymbol{\omega}}'_j\widetilde{\mathbf{X}}_{t-1} - c_{2j}])}\Big), \qquad (2.62)$$

and where $\boldsymbol{\theta}_{L_j} = (\boldsymbol{\omega}'_j, \gamma_j, c_{1j}, c_{2j})'$ with $\boldsymbol{\omega}_j = (\omega_{1j}, \ldots, \omega_{pj})'$, γ_j the slope parameter, and (c_{1j}, c_{2j}) ($j = 1, \ldots, k$) the location parameters. Similarly, $\widetilde{\boldsymbol{\theta}}_{B_j} = (\widetilde{\boldsymbol{\omega}}'_j, \gamma_j, c_{1j}, c_{2j})'$ with $\widetilde{\boldsymbol{\omega}}_j = (\omega_{1j}, \ldots, \omega_{qj})'$.

Let $q = p$. Then a special case of (2.61) is the *local linear global neural network* of order p, or L^{2}GNN$(k;p)$ model, where the approximation functions are assumed to be linear, that is, $L(\mathbf{X}_{t-1}; \boldsymbol{\theta}_{L_j}) = \xi_{0j} + \boldsymbol{\xi}'_j \mathbf{X}_{t-1}$ with $\boldsymbol{\xi}_j = (\xi_{1j}, \ldots, \xi_{pj})'$. The L^{2}GNN$(k;p)$ model resembles the structure of the AR–NN$(k;p)$ model (2.57), and is defined as

$$Y_t = \sum_{j=1}^{k} (\xi_{0j} + \boldsymbol{\xi}'_j \mathbf{X}_{t-1}) B(\mathbf{X}_{t-1}; \boldsymbol{\theta}_{B_j}) + \varepsilon_t, \qquad (2.63)$$

where, similar to the AR–NN of Section 2.9.1, restrictions on the parameters need to be imposed to ensure identifiability. Further, it is easy to verify that (2.61) is related to the SETAR$(k;p,\ldots,p)_m$ model of Section 2.6.4, with a similar geometric interpretation.

Example 2.13: A Simulated L^{2}GNN$(2;1,1)$ Time Series

Consider the single hidden layer feed-forward L^{2}GNN$(2;1,1)$ model

$$Y_t = L(Y_{t-1}; \boldsymbol{\theta}_{L_1}) B(Y_{t-1}; \boldsymbol{\theta}_{B_1}) + L(Y_{t-1}; \boldsymbol{\theta}_{L_2}) B(Y_{t-1}; \boldsymbol{\theta}_{B_2}) + \varepsilon_t, \qquad (2.64)$$

where

$$L(Y_{t-1}; \boldsymbol{\theta}_{L_1}) = 1 - 1.2 Y_{t-1}, \quad L(Y_{t-1}; \boldsymbol{\theta}_{L_2}) = 1 - 0.5 Y_{t-1},$$

$$B(Y_{t-1}; \boldsymbol{\theta}_{B_1}) = -\Big(\frac{1}{1+\exp(10(Y_{t-1}+6))} - \frac{1}{1+\exp(10(Y_{t-1}-1))}\Big),$$

$$B(Y_{t-1}; \boldsymbol{\theta}_{B_2}) = -\Big(\frac{1}{1+\exp(5(Y_{t-1}+2))} - \frac{1}{1+\exp(5(Y_{t-1}-2))}\Big),$$

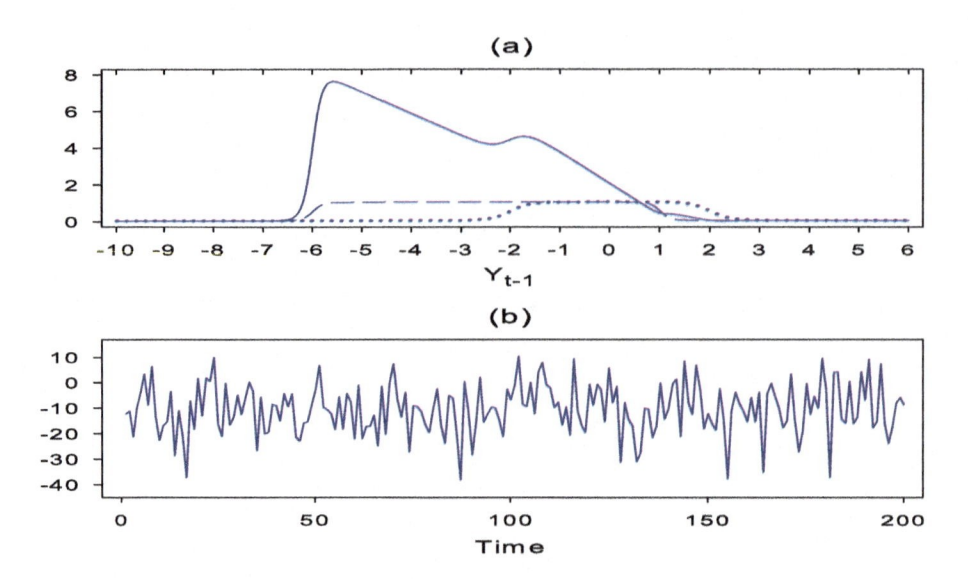

Figure 2.15: (a) Skeleton (the combined approximation and activation-level function) of the L^{2}GNN(2; 1, 1) model (2.64) (blue solid line) with activation-level functions $B(Y_{t-1}; \boldsymbol{\theta}_{B_1})$ (blue medium dashed line) and $B(Y_{t-1}; \boldsymbol{\theta}_{B_2})$ (blue dotted line); (b) A typical realization of the L^{2}GNN(2; 1, 1) model (2.64); $T = 200$.

and $\{\varepsilon_t\} \stackrel{\text{i.i.d.}}{\sim} \mathcal{N}(0, 1)$. Note that (2.64) is composed of a nonstationary AR(1) process, given by the linear approximation function $L(Y_{t-1}; \boldsymbol{\theta}_{L_1})$, and a stationary AR(1) process.

Figure 2.15(a) shows the skeleton of (2.64), i.e. the values of the combined approximation and activation-level function as a function of the input series $\{Y_{t-1}\}$ (blue solid line). The values of $B(Y_{t-1}; \boldsymbol{\theta}_{B_j})$ ($j = 1, 2$) are displayed near the bottom of Figure 2.15(a). For approximately $Y_{t-1} < -6.5$ both activation-level functions are almost equal to zero. Around the point $Y_{t-1} = -6.5$, the function $B(Y_{t-1}; \boldsymbol{\theta}_{B_1})$ changes rapidly from 0 to 1, causing a steep increase in $L(Y_{t-1}; \boldsymbol{\theta}_{L_1})B(Y_{t-1}; \boldsymbol{\theta}_{B_1})$ when $-6.5 < Y_{t-1} < -5.6$. Then, when $-5.6 < Y_{t-1} < -2.2$, the values of the skeleton drop, due to $L(Y_{t-1}; \boldsymbol{\theta}_{L_1})$. At $Y_{t-1} = -2.2$, there is a slight increase in the values of the skeleton when the function $B(Y_{t-1}; \boldsymbol{\theta}_{B_2})$ begins to activate. Next, at $Y_{t-1} = -1.7$ a further decline sets in, with a small increase in the values of the skeleton when the function $B(Y_{t-1}; \boldsymbol{\theta}_{B_1})$ begins to deactivate. Finally, the skeleton goes to zero at about $Y_{t-1} = 2$.

In general, as $\{Y_t\}$ grows in absolute value, the functions $B(Y_{t-1}; \boldsymbol{\theta}_{B_i}) \to 0$ ($i = 1, \ldots, k$), and thus $\{Y_t\}$ is driven back to 0. By imposing some weak conditions on the parameters $\boldsymbol{\omega}_i$, and using the above result, it can be proved (Suárez–Fariñas et al., 2004) that the L^{2}GNN model is asymptotically stationary with probability one, even if the model is a mixture of one or two explosive AR processes.

2.9 ARTIFICIAL NEURAL NETWORK MODELS

Figures 2.15(b) shows a $T = 200$ realization from the L^{2}GNN model (2.64). We observe that the series is fluctuating around a fixed sample mean of -10.780, with a standard deviation of 9.978, suggesting that the process is asymptotically stationary. There are, however, occasional large negative values ($\max\{Y_t\} = 10.109$; $\min\{Y_t\} = -38.428$), indicating local nonstationarity.

```
┌─────────────────────────────────────────────────────────────────┐
│ NCTAR(k; p, ..., p)_q:                                          │
│ Y_t = φ_0 + φ'X_{t-1} + Σ_{j=1}^k (ξ_{0j} + ξ'_j X_{t-1})G(X̃_{t-1}; ω̃_j, c_j) + ε_t │
└─────────────────────────────────────────────────────────────────┘
```

$G(\cdot) = I(\cdot)$ $\phi_0 = 0$, $\boldsymbol{\phi} = \mathbf{0}$, $G(\cdot) = B(\cdot)$

SETAR$(k; p, \ldots, p)_q$ LGNN$(k; p, \ldots, p)_q$

$q = p$ $q = p$ $q = p$, $\xi_{0j} = 0$, $\widetilde{\mathbf{X}}_{t-1} = \mathbf{Y}_{t-d}$ $q = 1$

SETAR$(k; p, \ldots, p)$ L^{2}GNN$(k; p)$ AR-NN$(k; p)$ LSTAR$(k; p)$

$\boldsymbol{\xi}_j = \mathbf{0}$ $\boldsymbol{\xi}_j = \mathbf{0}$ (or $\widetilde{\boldsymbol{\omega}}_j = \mathbf{0}$) $\xi_{0j} = 0$, $\boldsymbol{\xi}_j = \mathbf{0}$

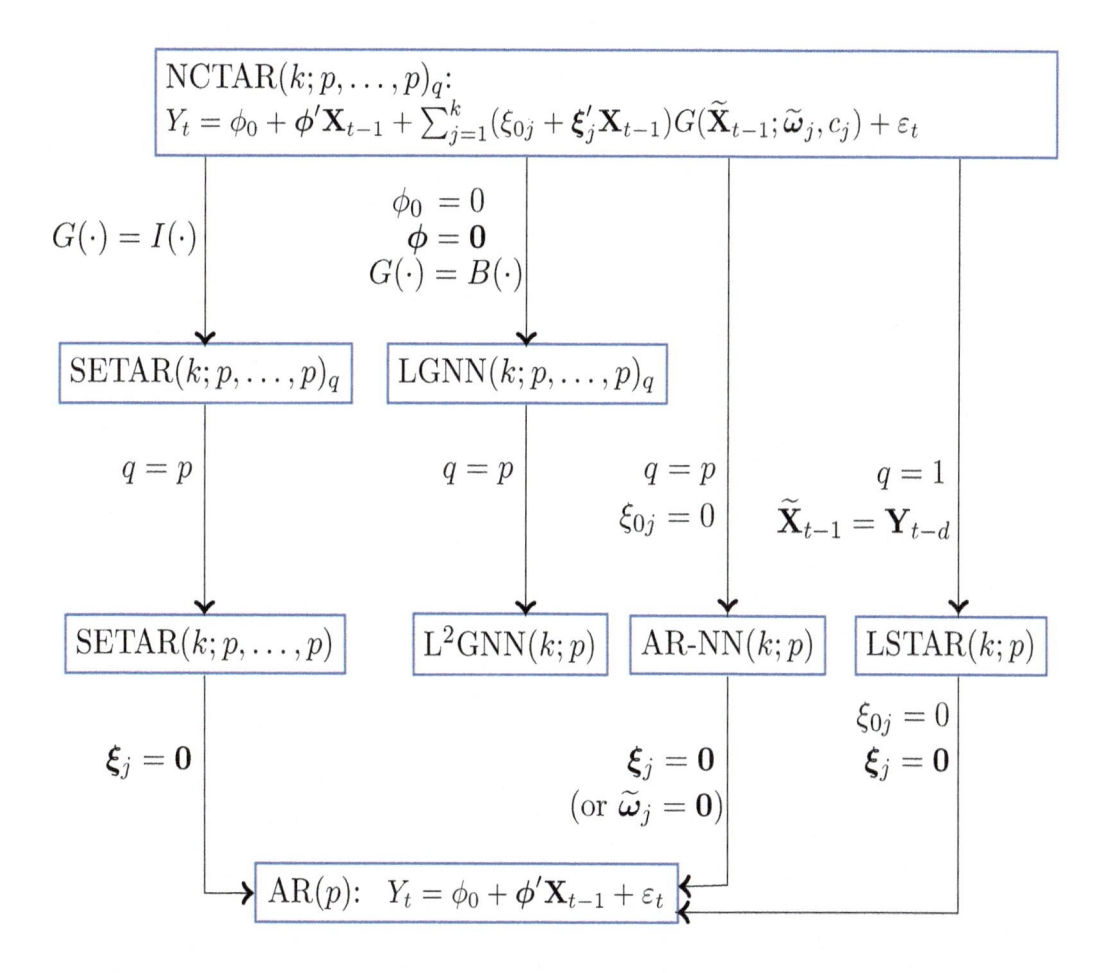

AR(p): $Y_t = \phi_0 + \boldsymbol{\phi}'\mathbf{X}_{t-1} + \varepsilon_t$

Figure 2.16: *Flow diagram of various relationships between (non)linear AR models.*

2.9.4 Neuro-coefficient STAR model

The *neuro-coefficient smooth transition autoregressive* (NCSTAR) model is a generalization of some of the previously described models and can handle multiple regimes and multiple smooth transition functions, using a logistic q-dimensional activation-level function $G(\cdot)$. In particular, the NCTAR model of order p with q

activation-level functions, denoted by NCTAR$(k;p)_q$, is defined as

$$Y_t = \phi_0 + \boldsymbol{\phi}'\mathbf{X}_{t-1} + \sum_{j=1}^{k}(\xi_{0j} + \boldsymbol{\xi}_j'\mathbf{X}_{t-1})G(\widetilde{\mathbf{X}}_{t-1}; \widetilde{\boldsymbol{\omega}}_j, c_j) + \varepsilon_t, \qquad (2.65)$$

where

$$G(\widetilde{\mathbf{X}}_{t-1}; \widetilde{\boldsymbol{\omega}}_j, c_j) = (1 + \exp(-[\widetilde{\boldsymbol{\omega}}_j'\widetilde{\mathbf{X}}_{t-1} - c_j]))^{-1},$$

with

$$\mathbf{X}_{t-1} = (Y_{t-1}, \ldots, Y_{t-p})', \ \widetilde{\mathbf{X}}_{t-1} = (Y_{t-1}, \ldots, Y_{t-q})'$$
$$\widetilde{\boldsymbol{\omega}}_j = (\widetilde{\omega}_{1j}, \ldots, \widetilde{\omega}_{qj})', \ \boldsymbol{\xi}_j = (\xi_{1j}, \ldots, \xi_{pj})', \ (j=1,\ldots,k).$$

Imposing the same parameter restrictions for the AR–NN model given in Section 2.9.1 guarantees identifiability of the NCTAR model. Figure 2.16 shows a flow diagram of various relationships between the (non)linear AR models.

2.10 Markov Switching Models

Markov chains have received wide attention in many areas of science. Before discussing Markov switching models, we introduce some basic notions. As is well known, a Markov chain $\{S_t\}$ is a discrete stochastic process $S_t \in \{1, \ldots, k\}$, satisfying

$$\mathbb{P}(S_t = j | S_{t-1} = i, S_{t-2} = r, \ldots) = \mathbb{P}(S_t = j | S_{t-1} = i) = p_{ij},$$

$$\sum_{j=1}^{k} p_{ij} = 1, \ p_{ij} \geq 0, \quad \forall i, j \in \{1, \ldots, k\}.$$

Loosely speaking, a Markov process is called *irreducible* if any state j can be reached from state i in a few steps, and it is termed *aperiodic* if the number of steps it needs to return to a state has no period. Furthermore, a Markov chain is *ergodic* if it is irreducible and aperiodic.

Any Markov chain has a stationary distribution $\{\pi_j = \mathbb{P}(S_t = j)\}_{j=1}^{k}$ satisfying

$$\pi_j = \sum_{j=1}^{k} \pi_j p_{ij}, \qquad (2.66)$$

or in matrix form $\boldsymbol{\pi} = \mathbf{P}'\boldsymbol{\pi}$ where $\boldsymbol{\pi} = (\pi_1, \ldots, \pi_k)'$ is the $k \times 1$ vector of *steady-state probabilities*, and $\mathbf{P} = (p_{ij})$ is the $k \times k$ *transition probability matrix*. For an ergodic Markov chain, $\pi_j = \lim_{n \to \infty} \mathbb{P}(S_n = j | S_1 = i)$ (independent of i).

Markov switching ARMA model
Consider a univariate time series process $\{Y_t, t \in \mathbb{Z}\}$ that is influenced by a hidden

2.10 MARKOV SWITCHING MODELS

discrete stochastic Markov process $\{S_t\}$. Then a *Markov-switching* ARMA (MS–ARMA) is defined as

$$Y_t = \sum_{i=1}^{k} \delta_{ti}\left(\phi_0^{(i)} + \sum_{u=1}^{p_i} \phi_u^{(i)} Y_{t-u} + \varepsilon_t^{(i)} + \sum_{v=1}^{q_i} \theta_v^{(i)} \varepsilon_{t-v}^{(i)}\right), \qquad (2.67)$$

where

$$\delta_{ti} = \begin{cases} 1 & \text{if } S_t = i, \\ 0 & \text{otherwise,} \end{cases}$$

with $\varepsilon_t^{(i)} = \sigma_i^2 \varepsilon_t$, and $\{\varepsilon_t\} \stackrel{\text{i.i.d.}}{\sim} (0,1)$, independent of $\{S_t\}$. So, S_t denotes the regime or state prevailing at time t, one of k possible cases, i.e. it plays the role of $\{J_t\}$ in (2.27). In the case $k = 1$ there is only one state and $\{Y_t, t \in \mathbb{Z}\}$ degenerates to an ordinary ARMA process. Adding exogenous variables, such as trends, is a straightforward extension of (2.67). Another extension of the model is to allow for generalized autoregressive conditional heteroskedastic (GARCH) errors. Multivariate modeling, including modeling cointegrated processes, is also an option.

Emphasis has been on two-state ($k = 2$) Markov switching AR (MSA or MSAR) models with $q_i = 0$ ($i = 1, \ldots, k$) and $w_1 = p_{12}$, $w_2 = p_{21}$. The resulting process is ergodic, with no absorbing states, if $0 < w_1 < 1$ and $0 < w_2 < 1$. The stationary probabilities are $\pi_1 = w_2/(w_1 + w_2)$ and $\pi_2 = w_1/(w_1 + w_2)$ (cf. Exercise 2.7). Moreover, the system stays in regime i for geometrically distributed time with mean $1/w_i$.

Example 2.14: A Two-regime Simulated MS–AR(1) Time Series

Consider a two-regime ($k = 2$) MS–AR(1) process given by

$$Y_t = \begin{cases} \phi_1^{(1)} Y_{t-1} + \sigma_1 \varepsilon_t & \text{if } S_t = 1, \\ \phi_1^{(2)} Y_{t-1} + \sigma_2 \varepsilon_t & \text{if } S_t = 2, \end{cases} \qquad (2.68)$$

where

$\phi_1^{(1)} = -\phi_1^{(2)} = 0.9$, $\sigma_1^2 = 1$, $\sigma_2^2 = 0.25$, $p_{11} = 0.8$, and $p_{22} = 0.9$.

Figure 2.17(a) shows a realization of (2.68) with $\{\varepsilon_t\} \stackrel{\text{i.i.d.}}{\sim} \mathcal{N}(0,1)$. A scatter plot of Y_t versus Y_{t-1} (not shown here) depicts two linear relationships: one showing a positive relationship and one with a negative linear relationship between the two variables.

There are various ways to estimate the MS–AR model. Because $\{S_t\}$ is not observed, the model does not directly give a likelihood function. Let $\boldsymbol{\theta} = (\phi_1^{(1)}, \phi_1^{(2)}, \sigma_1^2, \sigma_2^2, p_{11}, p_{22})'$ be the vector of parameters, and $\mathcal{F}_t$ the σ-algebra generated by $\{Y_s, s \leq t\}$. Maximum likelihood (ML) estimation requires

$$f(Y_t|\mathcal{F}_{t-1}; \boldsymbol{\theta}) = \sum_{j=1}^{2} f(Y_t|\mathcal{F}_{t-1}, S_t = j; \boldsymbol{\theta}) \mathbb{P}(S_t = j|\mathcal{F}_{t-1}; \boldsymbol{\theta}), \qquad (2.69)$$

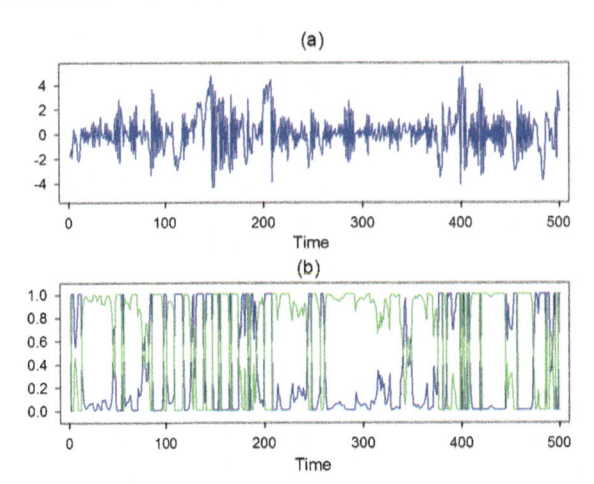

Figure 2.17: *(a) A realization of the MS–AR(1) model (2.67), $T = 500$; (b) Estimated smoothed probabilities in state 1 and 2 are plotted as blue and green solid lines, respectively.*

where $f(Y_t|\mathcal{F}_{t-1}, S_t = j; \boldsymbol{\theta})$ follows directly from the model, and $\mathbb{P}(S_t = j|\mathcal{F}_{t-1}; \boldsymbol{\theta})$ can be obtained recursively from Bayes' rule:

$$\mathbb{P}(S_t = j|\mathcal{F}_{t-1}; \boldsymbol{\theta}) = \sum_{i=1}^{2} \mathbb{P}(S_{t-1} = i|\mathcal{F}_{t-1}; \boldsymbol{\theta}) p_{ij}, \tag{2.70}$$

$$\mathbb{P}(S_t = i|\mathcal{F}_t; \boldsymbol{\theta}) = \frac{f(Y_t, S_t = i|\mathcal{F}_{t-1}; \boldsymbol{\theta})}{f(Y_t|\mathcal{F}_{t-1}; \boldsymbol{\theta})}$$

$$= \frac{f(Y_t|\mathcal{F}_{t-1}, S_t = i; \boldsymbol{\theta})\mathbb{P}(S_t = i|\mathcal{F}_{t-1}; \boldsymbol{\theta})}{\sum_{i=1}^{2} f(Y_t|\mathcal{F}_{t-1}, S_t = i; \boldsymbol{\theta})\mathbb{P}(S_t = i|\mathcal{F}_{t-1}; \boldsymbol{\theta})}. \tag{2.71}$$

Starting from the initial stationary probability

$$\mathbb{P}(S_1 = 1|\mathcal{F}_1) = \pi_1 = \frac{w_2}{w_1 + w_2} = 1 - \mathbb{P}(S_t = 2|\mathcal{F}_2),$$

we can construct the quasi log-likelihood function by evaluating (2.70), (2.69) and (2.71) iteratively for $t = 2, \ldots, T$. This is known as the *Hamilton* (Hamilton, 1994, Chapter 22) filter (closely related to the Kalman filter). Under stationarity conditions, the quasi maximum likelihood (QML) estimator $\widehat{\boldsymbol{\theta}}$ of $\boldsymbol{\theta}$ has the usual asymptotic properties. After maximizing the likelihood function, a similar Bayesian argument can be used to produce estimated smoothed probabilities

$$\mathbb{P}(S_t = 1|\mathcal{F}_T; \widehat{\boldsymbol{\theta}}) = 1 - \mathbb{P}(S_t = 2|\mathcal{F}_T; \widehat{\boldsymbol{\theta}}), \quad t = 1, \ldots, T.$$

For the simulated data of Figure 2.17(a), we obtain the parameter estimates

$$\widehat{\phi}_1^{(1)} = 0.93_{(0.02)}, \quad \widehat{\phi}_1^{(2)} = -0.88_{(0.02)}, \quad \widehat{\sigma}_1^2 = 0.94_{(0.12)}, \quad \widehat{\sigma}_2^2 = 0.28_{(0.02)},$$

$$\widehat{p}_{11} = 0.78, \quad \widehat{p}_{22} = 0.89,$$

with asymptotic standard errors of the parameter estimates in parentheses. The expected duration (length of stay) in the first regime is $1/(1-\widehat{p}_{11}) \approx 4.56$ time periods, and in the second regime $1/(1-\widehat{p}_{22}) \approx 9.33$ time periods. In conjunction with this result, Figure 2.17(b) shows the estimated smoothed state probabilities.

2.11 Application: An AR–NN model for EEG Recordings

To illustrate the application of a single hidden layer feed-forward AR–NN model, we reconsider the EEG recordings (epilepsy data). Let $\{Y_t\}_{t=1}^{631}$ denote the time series under study. The aim will be to reconstruct the dynamics underlying $\{Y_t\}$ and to predict future values. From the discussion in Example 1.2 it is reasonable to treat $\{Y_t\}$ as a realization of a stationary process. If, however, this is not the case we recommend to transform the series to a stationary series if possible (e.g. by differencing) before training an ANN on it.

Implementation
Implementing an AR–NN model requires several decisions to be made. First, we need to decide whether the data need scaling. Rescaling the data is linked to initial values of the weights $\boldsymbol{\omega}_j$ ($j = 1, \ldots, k$). These weights must vary over a reasonable range, neither too wide nor too narrow, compared with the range of the data. If this is not the case, the criterion function will have a number of local minima. Although, it is difficult to offer a general advice on the choice of scaling, the data in the training set is often standardized to have zero mean and variance one. Still it is recommended to train an AR–NN a couple of times, using different initial weights. For the EEG recordings we decided to use the original data. Since the values of the inputs are large, but centered around zero, we followed a recommendation in the R documentation of the nnet package to take the initial values of the weights randomly from a uniform $[-1/\max\{|Y_t|\}, 1/\max\{|Y_t|\}]$ ($t = 1, \ldots, N$) distribution with N the size of the training data set, also called the total number of in-sample observations.

The next issue is the choice of $G(\cdot)$. A commonly used activation function is the logistic function, which we adopt here. Furthermore, we need to choose the number p of input (lagged) variables, and the number of hidden units k. Various strategies have been proposed for this purpose. One strategy is to perform a grid search over a pre-specified range of pairs (p, k) and select the AR–NN on the basis of minimizing a model selection criterion. Recall, $r = (p + 2)k + p + 1$ denotes the number of parameters fitted in the model. Then Akaike's information criterion (AIC) and the Bayesian information criterion (BIC) are, respectively, given by

$$\text{AIC} = N\log(\widehat{\sigma}_\varepsilon^2) + 2r, \quad \text{BIC} = N\log(\widehat{\sigma}_\varepsilon^2) + r\ln(N),$$

where $\widehat{\sigma}_\varepsilon^2$ denotes the residual variance.

Table 2.1: *Comparison of various AR–NN models applied to the EEG recordings; $T = 631$. Blue-typed numbers indicate minimum values of a number of "key" statistics.*

			Measures of fit			Forecast accuracy	
k	p	r	$\widehat{\sigma}_\varepsilon^2$	AIC	BIC	RMSFE	MAFE
0	7	8	3875.15	7937.34	7971.73	65.76	51.29
1	7	17	3833.37	7949.44	8022.53	65.76	51.91
2	7	26	3852.46	7970.15	8081.92	65.27	51.20
3	7	35	3807.98	7981.83	8132.29	65.24	51.60
4	7	44	3744.84	7990.73	8179.89	63.76	49.87
5	7	53	3490.68	7970.50	8198.34	63.80	50.17
0	8	9	3146.67	7810.71	7849.38	51.99	40.43
1	8	19	3091.29	7821.07	7902.71	52.76	40.33
2	8	29	3041.18	7832.19	7956.81	52.23	39.75
3	8	39	3118.29	7865.79	8033.38	51.77	39.95
4	8	49	2702.61	7808.10	8018.66	51.25	39.12
5	8	59	2653.02	7818.05	8071.58	53.04	43.26

An alternative strategy is to select a linear AR(p) model first, using AIC or BIC. In the second stage hidden units are added to the model. Then, the improvement in fit is measured again by the AIC and BIC. In practice, we recommend the use of both order selection criteria. The reason is that the number of parameters in an AR–NN model is typically much larger than in traditional time series models, the ordinary AIC does not penalize the addition of extra parameters enough in contrast to the BIC. Section 6.2.2 contains some alternative versions of AIC which, for large values of p, penalize extra parameters (much) more severely than AIC.

Subsamples

Since the time-interval between oscillations in the original time series of EEG recordings is about 80, we divide the data into two subsamples. The first subsample, used for modeling, consists of a total of 551 observations. The remaining 80 observations are used in the second sample for out-of-sample forecasting.

Table 2.1, columns 4 – 6, contains values of $\widehat{\sigma}_\varepsilon^2$, AIC, and BIC for subselection of AR–NN models fitted to the data in the first subsample. Blue-typed numbers denote minimum values of these statistics. BIC selects an AR–NN(0; 8) model. This result is in line with the linear AR(8) model preferred by AIC on the basis of the complete data set of 631 observations. In particular, the resulting estimated model is given by

$$Y_t = 16.96_{(98.42)} + 2.71_{(0.06)} Y_{t-1} - 3.21_{(0.11)} Y_{t-2} + 2.52_{(0.16)} Y_{t-3} - 1.89_{(0.19)} Y_{t-4}$$
$$+ 0.84_{(0.19)} Y_{t-5} + 0.68_{(0.16)} Y_{t-6} - 1.14_{(0.11)} Y_{t-7} + 0.46_{(0.04)} Y_{t-8} + \widehat{\varepsilon}_t,$$

where asymptotic standard errors of the parameters are in parentheses, and where the residual variance is given by $\widehat{\sigma}_\varepsilon^2 = 3080.48$. In contrast, AIC picks the AR–

2.11 APPLICATION: AN AR–NN MODEL FOR EEG RECORDINGS

Table 2.2: EEG recordings. *Biases and weights of the best fitted* AR–NN(4; 8, ..., 8) *model.*

		Hidden layer h_1	h_2	h_3	h_4	Output layer o
Bias	$\alpha_0 \to$	-0.19	0.00	1.03	-0.01	-78.85
Input layer	$i_1 \to$	-16.57	19.59	-4.32	3.19	2.70
	$i_2 \to$	-1.74	10.80	-3.88	2.43	-3.25
	$i_3 \to$	-10.14	5.88	0.63	2.51	2.63
	$i_4 \to$	-6.17	3.40	2.97	1.69	-2.03
	$i_5 \to$	2.42	-2.65	4.96	0.61	0.96
	$i_6 \to$	-10.64	-4.51	-0.74	1.22	0.56
	$i_7 \to$	-10.87	-1.57	-7.31	1.62	-1.05
	$i_8 \to$	7.66	-4.56	-17.27	1.69	0.46
Hidden layer	$h_1 \to$					25.84
	$h_2 \to$					50.01
	$h_3 \to$					49.15
	$h_4 \to$					29.31

NN(4; 8, ..., 8) model and gives much results in terms of residual variance than BIC.

Table 2.2 shows the biases and weights of the single-layer AR–NN(4; 8, ..., 8) model. Evidently, the weights correspond to the coefficients in the logistic activation-level functions $G_j(\cdot)$ ($j = 1, ..., 4$). As can be seen from the values of ω_{jo} ($j = 1, 2$), the first two neurons h_1 and h_2 have much more effect on the output than the third and fourth neurons. The inputs at lags 1, 2, 3, 6, 7 and 8 have the largest effect, in absolute value, on the first hidden layer h_1, whereas all inputs contribute less to the second hidden layer h_2. Clearly, all inputs have an effect on h_3, but less on h_4. The signs tell us the nature of the correlation between the inputs to a neuron and the output from a neuron. The negative values of w_{ij} at lags $i = 2$ ($j = 1, 2, 4$), $i = 4$ ($j = 1, 2, 3$), and $i = 7$ ($j = 1, 2, 3$) match the signs of the parameter estimates in the fitted linear AR(8) model. This is about all that can be said about the weights here. Indeed, it is unwise to try to interpret the weights any further, unless we reduce the influence of local minima by using different initial weights.

Forecasting

We consider the forecast performance of the AR–NN($k; p, p$) models in a *"rolling"* forecasting framework with parameter estimates based on a $(551 - p) \times p$ matrix consisting of the in-sample observations: $\{Y_t\}_{t=p}^{550}, \{Y_t\}_{t=p-1}^{550-1}, ..., \{Y_t\}_{t=1}^{550-(p-1)}$ (here, $p = 7$ and $p = 8$); see Section 10.4.1 for details on various forecasting schemes. We evaluate the fitted model on the basis of $H = 1$ to $H = H_{\max} = 80$-steps ahead forecasts. So, we use an $80 \times p$ matrix consisting of the out-of-sample observations: $\{Y_t\}_{t=551}^{630}, \{Y_t\}_{t=551-1}^{630-1}, ..., \{Y_t\}_{t=551-(p-1)}^{630-(p-1)}$. Finally, the 80 forecast errors are summarized in two accuracy measures: the sample *root mean squared forecast error*

(RMSFE) and the sample *mean absolute forecast error* (MAFE); see the last two columns of Table 2.1. Note that the difference between the AR–NN(5; 8, ..., 8) and AR–NN(0; 8) models is minimal, in terms of RMSFE and MAFE.

2.12 Summary, Terms and Concepts

Summary
In this chapter we summarized the main features of various classic and popular nonlinear model classes introduced in the literature and some of the generalizations/extensions of these models. Much of the material should be familiar to researchers and practitioners already working in the field, but it is worth reviewing. Specifically, the chapter may be viewed as a useful basis for discussing the statistical properties of a number of these models in later chapters. One important practical point about these nonlinear models is that many model classes relate to one another, either through the Volterra representation or via the SDM. In addition, we have seen that some simple specializations of these models can produce interesting qualitative nonlinear behavior. More specializations will be examined throughout the rest of this book.

Terms and Concepts

activation-level, 56
aperiodic, 66
asymptotically stationary, 59
back-propagation, 58
doubly stochastic, 39
exponential function, 51
feed-forward, 56
hidden unit, 56
hyperplane, 46
impulse response function, 36
innovation process, 31
irreducible, 66
limit cycle, 37
logistic function, 51
multi-layer perceptron (MLP), 56

neurons (nodes), 56
periodic function, 37
random coefficient, 39
recurrent, 61
regimes, 41
self-exciting, 41
shortcut connections, 58
skip-layer, 57
state-dependent model (SDM), 32
super (sub) diagonal, 34
threshold, 41
training, 57
transition probability matrix, 66
Volterra, 30

2.13 Additional Bibliographical Notes

Section 2.1: The beginning of nonlinear time series has been attributed to Volterra (1930); see, e.g., Brockett (1976). Wiener (1958) suggests a linear combination of nonlinear functions using high order moments and high order polynomial models. The use of Wiener's approach died out in the 1960s largely due to the complexity of the proposed model and associated problems of parameter estimation.

2.13 ADDITIONAL BIBLIOGRAPHICAL NOTES

Section 2.2: D'Alessandro et al. (1974) provide a set of necessary and sufficient conditions for a Volterra series to admit a BL realization and showed there is a clear-cut method for determining the Volterra series for a BL system. Brockett (1977) links Volterra series and geometric control theory by proving that over a finite time interval, a BL model, which is itself a special case of Wiener's model, can approximate any "nice" Volterra series with an arbitrary degree of accuracy. Priestley (1988) discusses how BL models may be regarded as the natural nonlinear extension of the ARMA model. A considerable amount of research deals with various properties of BL models; see, e.g., the monographs by Granger and Andersen (1978a), and Subba Rao and Gabr (1984).

Section 2.3: Haggan and Ozaki (1980, 1981) propose the ExpAR model when $p = 2$, $d = 1$, and $\phi_0 = 0$. Earlier, Ozaki and Oda (1978) investigate the ExpAR(1) model with $\phi_0 = 0$ and $d = 1$. Jones (1978) considers methods for approximating the stationary distribution of nonlinear AR(1) processes, including ExpAR(1) processes.

Section 2.4: The monograph by Nicholls and Quinn (1982) provides a good source of the early works on RCAR models. These authors also generalize Andel's (1976) results to multivariate RCAR models. Amano (2009) proposes a G-estimator (named after Godambe) for RCAR models. Aue et al. (2006) deal with QML estimation of an RCAR(1) model. Pourahmadi (1986) presents sufficient conditions for stationarity and derives explicit results for double stochastic AR(1) processes with $\log(\beta_{1,t}^2)$ in (2.24) following a stationary Gaussian process, an AR(1) process, and an MA(q) process.

Section 2.5: Robinson (1977) and Lentz and Mélard (1981) consider estimation of simple nonlinear MA models using moment methods and ML, respectively. Ashley and Patterson (2002) use GMM to obtain estimates of the coefficients of a quadratic MA model. Ventosa–Santaulària and Mendoza–Velázquez (2005) propose a nonlinear MA conditional heteroskedastic (NLMACH) model with similar properties as the ARCH-class specifications.

Sections 2.6.1 – 2.6.2: Tong (1977, 1980, 1983, 1990) explores (self-exciting) TAR models in a number of papers, and two subsequent books; see also Tong (2007). Other influential publications are: Petruccelli (1992), who shows that threshold ARMA (TARMA) models, with and without conditional heteroskedastic (ARCH) errors, can approximate SDMs almost surely; Tong and Lim (1980), who demonstrate the versatility of SETAR models in capturing nonlinear phenomena; and K.S. Chan and Tong (1986), who discuss the problem of estimating the threshold parameter. Nevertheless, as noted by Tong (2011, 2015), these early publications did not attract many followers. Indeed, the real exponential growth of the threshold approach, and its extensions took off only in the late 1990s. The impact of Tong's SETAR models is enormous across many scientific fields. For instance, Hansen (2011) provides an extensive list of 75 papers published in the economics and econometrics literatures, which contribute to both the theory and application of the SETAR model. Similarly, Chen et al. (2011b) review the vast and important developments of the threshold model in financial applications.

Section 2.6.3: Gonzalo and Wolf (2005) propose a subsampling method for constructing asymptotically valid confidence intervals for the threshold parameter in (dis)continuous SETAR models. Stenseth et al. (2004) consider an extension of the CSETAR model, which they call functional coefficient threshold AR model, that specifies some coefficients of the SETAR model to be functions of some covariates.

Section 2.6.4: Medeiros et al. (2002b) propose SETAR models with unknown multivariate thresholds. For most practical problems a search over all possible threshold combinations is infeasible. Therefore these authors propose a procedure based on a greedy randomized adaptive search procedure (GRASP) which solves optimization problems which have a high number, but not infinite, of possible solutions; see, e.g., Feo and Resende (1995).

Section 2.6.5: Wecker (1981) introduces the class of asMA models, and Brännäs and De Gooijer (1994) extend this class to ARasMA models combining a linear AR with an asMA part. Further extensions include asMA models with an analogously defined asymmetric parameterization of the conditional variance (Brännäs and De Gooijer, 2004), and vector ARasMA models with asymmetric quadratic ARCH errors (Brännäs et al., 2011). Guay and Scaillet (2003) introduce a TMA model, as an asMA model which allows for contemporaneous asymmetry, and which does not restrict the threshold to be equal to zero.

Section 2.6.6: Astatkie et al. (1996) and Astatkie (2006) apply NeSETAR to time series data of daily streamflow. Hubrich and Teräsvirta (2013) discuss a vector nested SETAR (VNSETAR or VNTAR) version of (2.40) with only two regimes in each stage, and implicitly assuming that $\mathbb{R}^{(i,j)} \equiv \mathbb{R}^{(j,i)}$ ($i,j = 1,2$). An application of a special type of vector NeSTAR (called structural break TVAR) is in Galvão (2006).

Section 2.7: An early reference to the term smooth transition is Bacon and Watts (1971), which deals with the problem of two-phase regressions. K.S. Chan and Tong (1986) introduce STAR models into the nonlinear time series literature. The STAR family of models are popularized by, for instance, Granger and Teräsvirta (1992a) and Teräsvirta (1994). Van Dijk et al. (2002) provide a survey of various extensions and modifications of STAR models. Lopes and Salazar (2006) discuss Bayesian STAR models. The ASTMA model was introduced in Brännäs et al. (1998). Aznarte et al. (2007) establish the functional equivalence between STAR models and fuzzy rule-based systems.

Chini (2013) proposes a generalized STAR (GSTAR) model which allows the STAR family to capture the dynamic asymmetry in the conditional mean of a time series process, by using a particular generalization of the logistic smooth transition function.

Section 2.8.1: Raftery (1980) and Lawrance and Lewis (1985) derive properties and limit theorems of the NEAR(p) ($p = 1,2$) model. Chan (1988) obtains a necessary and sufficient condition for the existence of an "innovation" process and a stationary ergodic process satisfying a NEAR(p) model ($p \geq 1$). Smith (1986), Karlsen and Tjøstheim (1988), and Perera (2002, 2004) consider the problem of estimating the NEAR(1) and NEAR(2) models. Raftery (1982) proposes various modifications of the NEAR(1) model. He also introduces three nonstationary generalizations of the NEAR(1) model, including one which is appropriate when a seasonal effect is present. Moreover, he points out how the NEAR(1) model can be extended into a multivariate specification. Lawrence and Lewis (1977) develop the EMA(1) model, and Jacobs and Lewis (1977) introduce the EARMA(1,1) model.

Section 2.8.2: The PAR(1) may be viewed as a special case of the multiplicative error model for modeling non-negative processes of Engle (2002). Both McKenzie (1982) and Abraham and Balakrishna (2012) provide an algorithm for the simulation of PAR(1) models in the case of a gamma marginal distribution. Jose and Thomas (2012) study the properties of a PAR(1) model with a log-Laplace marginal distribution. Further, they consider multivariate extensions.

Section 2.9: A good understanding of neural networks can be obtained from, for instance, the (text)books of Hertz et al. (1992) and Nørgaard et al. (2000). Recurrent neural network models were introduced by Elman (1990). The motivation to consider a single hidden layer feed-forward ANNs with $\Psi(\cdot)$ a linear activation-level function stems from the fact that, under certain regularity conditions, it can provide arbitrarily accurate approximations to any measurable function in a variety of normed function spaces, given sufficiently many hidden units; see, e.g., Hornik et al. (1989). This also unveils the main weakness of the ANNs since they may end up fitting the noise in the data rather than the underlying DGP.

Sections 2.9.1 – 2.9.3: Lapedes and Farber (1987) propose an AR–NN model for time series prediction. Recurrent ARMA–NNs are defined by Connor et al. (1994). Aznarte and Benítez (2010) establish the functional equivalence between AR-NN time series models and fuzzy rule-based systems. Suárez–Fariñas et al. (2004) present the LGNN and L^{2}GNN models of Section 2.9.3. They consider parameter estimation by concentrated ML, and introduce a model building strategy. Furthermore, they address the fundamental differences between their model and the stochastic neural network model of Lai and Wong (2001) and the NCTAR model of Section 2.9.4.

Section 2.9.4: Medeiros and Veiga (2002a, 2005) propose the NCSTAR model. The model is related to the functional-coefficient AR model of Section 9.2.5, and to the single-index coefficient regression model of Section 9.2.6. Medeiros and Veiga (2003) address the issue of NCSTAR model evaluation by presenting a number of diagnostic (LM-type) test statistics.

Section 2.10: Kim and Nelson (1999) and Frühwirth–Schnatter (2006) provide an extensive introduction and discussion of MS models. Ephraim and Merhav (2002) present a detailed overview of many statistical and information-theoretic aspects of hidden Markov chains, including switching AR processes with Markov regime. Franke (2012) reviews the latest developments, and discusses various estimation methods, including Gibbs sampling. Bayesian estimation of MS–ARMA–GARCH models is the subject of a number of papers; see, e.g., Henneke et al. (2011). Davidson (2004) gives recursive formulae for multi-step point forecasts of MS models with ARMA(∞, q) dynamics and ARCH(∞) errors. Both Timmermann (2000) and Zhang and Stine (2001) derive the autocovariance structure of MS processes. The assumption of fixed transition probabilities have been relaxed by a number of authors; see, e.g., Bazzi et al. (2014) and the references therein.

2.14 Data and Software References

Exercise 2.10: The Jökulsá Eystri riverflow data set were made available by Tess Astatkie. The flow series is also listed in Tong (1990, Appendix 3). The complete data set can be downloaded from the website of this book. Related to this data set, and also available for downloading, is a set with three years series of daily data (January 1988 – December 1990) on flow, precipitation, and temperature of the Oldman River near Brocket in Alberta, Canada. In analogy with the results in Exercise 2.10, Astatkie et al. (1996) fit a NeSETAR to this data set.

Section 2.6: The R-tsDyn package contains a host of functions for testing and modeling univariate and multivariate threshold- and smooth transition type models. An R function programmed by K.S. Chan was used to obtain the fitted CSETAR model in (2.34). The code is available at the website of this book. Marcelo Medeiros contributed MATLAB code for estimating SETARs with multivariate thresholds using GRASP; see the website of this book.

Section 2.7: Chapter 18 in the book by Zivot and Wang (2006) covers some popular non-linear time series models and methods. Examples include SETAR, STAR, Markov-switching

(MS–)AR, and MS-state space models. S-Plus script files, using the S-Plus FinMetrics module, are available at http://faculty.washington.edu/ezivot/MFTS2ndEditionScripts.htm. R scripts are available at http://faculty.washington.edu/ezivot/MFTSR.htm. The R-MSwM package deals with univariate MS–AR models for linear and generalized models using the EM algorithm.

The website https://sites.google.com/site/marcelocmedeiros/Home/codes offers a set of MATLAB codes to estimate logistic smooth transition regression models with and without long memory; see McAleer and Medeiros (2008).

Section 2.9: MATLAB offers a toolbox for the analysis of ANNs. The toolbox NNSYSID contains a number of m-files for training and evaluation of multi-layer perceptron type neural networks; see http://www.iau.dtu.dk/research/control/nnsysid.html. There are functions for working ordinary feed-forward networks as well as for identification of nonlinear dynamic systems and time series analysis. Various ANN packages are available in R. For instance, nnet, neuralnet, RSNNS, and darch.

Section 2.10: MS_Regress is a MATLAB package for estimating Markov regime switching models written by Marcelo Perlin and available at https://sites.google.com/site/marceloperlin/. He also wrote a lighter version of the package in R which, however, is no longer being maintained; search for FMarkovSwitching on R-forge. The MATLAB code MS_Regress_tvtp is for estimating Markov-switching (MS) models with time varying transition probabilities. Its implementation is based on the code written by Perlin.

Data and software (mainly GAUSS code) for estimating MS models is available from James D. Hamilton's website at http://econweb.ucsd.edu/~jhamilton/software.htm. The site also offers links to software code written by third parties. The R-MSBVAR package includes methods for estimating MS Bayesian VARs.

Appendix

2.A Impulse Response Functions

Impulse response analysis consists in evaluating and examining the time evolution of the output sequence of a model when a particular input sequence changes in a very short time. Using the Wold decomposition, the dynamic behavior of a *linear* strictly stationary time series process $\{Y_t, t \in \mathbb{Z}\}$ is commonly described by an impulse response function defined as the difference between two realizations of Y_{t+H} ($H \geq 1$). Both realizations start from the same history ω_{t-1}, but one realization assumes that between t and $t+H$ the process is hit by a shock of size δ at time t (i.e. $\varepsilon_t = \delta$), while in the other realization (called *benchmark profile*) no shock occurs at time t. Furthermore, all shocks in intermediate time periods between t and $t+H$ are set equal to zero in both realizations, such that the *"traditional" impulse* (TI) *response function* is defined by

$$\text{TI}_Y(H, \delta, \omega_{t-1}) = \mathbb{E}[Y_{t+H} | \varepsilon_t = \delta, \varepsilon_{t+1} = \cdots = \varepsilon_{t+H} = 0, \omega_{t-1}]$$
$$- \mathbb{E}[Y_{t+H} | \varepsilon_t = 0, \varepsilon_{t+1} = \cdots = \varepsilon_{t+H} = 0, \omega_{t-1}], \quad (H \geq 1). \quad \text{(A.1)}$$

Nonlinear time series models do not have a Wold representation, however. In these models, the impact at time $t+H$ of a shock that occurs at time t typically depends on the history of the process up to the time the shock occurs, on the sign and the size of the shock, and on the shocks that occur in intermediate periods $t+1, \ldots, t+H$. This may, for instance, be deduced from the discrete-time Volterra series expansion (2.3). To avoid these problems, a natural thing to do is to use the expectation operator conditioned on only the history and/or shock. Given this choice, the benchmark profile for the impulse response function is then defined as the conditional expectation given only the history of the process ω_{t-1}. This approach leads to the GIRF, originally developed by Potter (1995, 2000) in a univariate framework and by Koop et al. (1996) in the multiple time series case. For a specific current shock, $\varepsilon_t = \delta$, and history ω_{t-1}, the GIRF is defined as

$$\text{GIRF}_Y(H, \delta, \omega_{t-1}) = \mathbb{E}[Y_{t+H}|\varepsilon_t = \delta, \omega_{t-1}] - \mathbb{E}[Y_{t+H}|\omega_{t-1}], \quad (H \geq 1). \quad (A.2)$$

It is easily seen that for linear models (A.2) is equivalent to (A.1).

Clearly, the GIRF in (A.2) depends on δ and ω_{t-1}, which are realizations of the random variables ε_t and $\mathcal{F}_{t-1}$ the σ-field generated by $\{Y_s, s \leq t-1\}$. Hence, $\text{GIRF}_Y(T, \delta, \omega_{t-1})$ itself is a realization of the random variable given by

$$\text{GIRF}_Y(H, \varepsilon_t, \mathcal{F}_{t-1}) = \mathbb{E}[Y_{t+H}|\varepsilon_t, \mathcal{F}_{t-1}] - \mathbb{E}[Y_{t+H}|\mathcal{F}_{t-1}], \quad (H \geq 1). \quad (A.3)$$

In general, the GIRF can be defined as a random variable conditional on particular subsets of shocks (e.g. only negative shocks) and histories (e.g. $Y_{t-1} \leq 0$).[16]

Note, the above impulse response analysis concerns a single, transitory, shock δ at time t. An alternative scenario is to measure the effect of a sequence of deterministic shocks $\{\delta_1, \delta_2, \ldots, \delta_t, \ldots\}$ on $\{\varepsilon_1, \varepsilon_2, \ldots, \varepsilon_t, \ldots\}$. Recall that a strictly stationary nonlinear time series process $\{Y_t, t \in \mathbb{Z}\}$ may be plausibly described by a discrete-time Volterra expansion, which can be expressed as

$$Y_t = G(\varepsilon_t, \varepsilon_{t-1}, \ldots, \varepsilon_1, \boldsymbol{\varepsilon}_0),$$

where $\{\varepsilon_t\} \stackrel{\text{i.i.d.}}{\sim} \mathcal{N}(0, 1)$, $\boldsymbol{\varepsilon}_0 = (\varepsilon_0, \varepsilon_{-1}, \ldots)$, and $G(\cdot)$ is a suitably smooth real-valued function. Again, the goal is to summarize the effect of the shocks on the time evolution of Y_t by a single measure. Since, however, future innovations are unknown, both the benchmark profile and the profile after the arrival of a shock are random variables. Let $\{\varepsilon_1^s, \varepsilon_2^s, \ldots, \varepsilon_t^s, \ldots\}$ denote a future path for the innovations, where $\varepsilon_1^s, \varepsilon_2^s, \ldots, \varepsilon_t^s, \ldots$ are i.i.d. $\mathcal{N}(0, 1)$ conditional on $\boldsymbol{\varepsilon}_0$. The random benchmark profile, or benchmark path, is equal to

$$Y_t^s(\boldsymbol{\varepsilon}_0) = G(\varepsilon_t^s, \varepsilon_{t-1}^s, \ldots, \varepsilon_1^s, \ldots, \boldsymbol{\varepsilon}_0),$$

whereas the time path after the shock arrival is given by

$$Y_t^s(\boldsymbol{\delta}, \boldsymbol{\varepsilon}_0) = G(\varepsilon_t^s + \delta_t, \varepsilon_{t-1}^s + \delta_{t-1}, \ldots, \varepsilon_1^s + \delta_1, \ldots, \boldsymbol{\varepsilon}_0),$$

where $\boldsymbol{\delta} = (\delta_1, \delta_2, \ldots, \delta_t, \ldots)$. Then the difference of expectations, conditional on $\boldsymbol{\varepsilon}_0 = \mathbf{0}$, of the two time paths of the responses is given by

$$\mathbb{E}[Y_t^s(\boldsymbol{\delta}, \boldsymbol{\varepsilon}_0)|\boldsymbol{\varepsilon}_0 = \mathbf{0}] - \mathbb{E}[Y_t^s(\boldsymbol{\varepsilon}_0)|\boldsymbol{\varepsilon}_0 = \mathbf{0}]. \quad (A.4)$$

---

[16]Unlike the linear case there are no general analytic expressions for the conditional expectations in the GIRF for nonlinear models. However, assuming the nonlinear model is completely known, MC simulation or BS can be used to obtain estimates of the impulse response measures; see, e.g., Exercise 2.11. Appendix 11.B describes the procedure to estimate the GIRF from multivariate nonlinear time series models along the lines of Koop et al. (1996).

Observe that this approach ignores the dependence between the benchmark and perturbed paths, accounted by the joint distribution of $(Y_t^s(\varepsilon_0), Y_t^s(\boldsymbol{\delta}, \boldsymbol{\varepsilon}_0), t \geq 1)$. Moreover, since the distribution of $\{\varepsilon_t\}$ is symmetric, positive and negative shocks will have the same infinitesimal occurrence. We refer to Gouriéroux and Jasiak (2005) for an alternative impulse response analysis, using the concept of nonlinear innovations, which eliminates these problems and provides straightforward interpretation of transitory or symmetric shocks.

Example A.1: Impulse Response Analysis

As a simple example, consider the BL model $Y_t = (\phi + \psi\varepsilon_t)Y_{t-1} + \varepsilon_t$ where $\{\varepsilon_t\} \stackrel{\text{i.i.d.}}{\sim} \mathcal{N}(0,1)$. The effect of a shock δ that occurs at time $t = 1$ is given by the perturbed path $Y_t(\delta) = (\phi + \psi\varepsilon_t)Y_{t-1}(\delta) + \varepsilon_t$ ($t \geq 2$). The difference (D) between the benchmark path and the perturbed path is equal to

$$Y_t^{\text{D}}(\delta) = Y_t(\delta) - Y_t = (\phi + \psi\varepsilon_t)Y_{t-1}^{\text{D}}(\delta)$$
$$= \prod_{\tau=2}^{t}(\phi + \psi\varepsilon_\tau)(1 + \psi Y_0)(\delta\varepsilon_1).$$

So that, for all $t \geq 2$, the effect of a shock as measured by the conditional expectation of the process $\{Y_t^{\text{D}}(\delta), t \in \mathbb{Z}\}$ is given by

$$\mathbb{E}[Y_t^{\text{D}}(\delta)|Y_0] = \phi^{t-1}(1 + \psi Y_0)(\delta\varepsilon_1).$$

Clearly, this effect converges toward zero if $|\phi| < 1$, which is a more stringent condition than the necessary and sufficient condition for stationarity of this model, i.e. $\mathbb{E}[\log(\phi + \psi\varepsilon_t)] < 0$; see Chapter 3.

2.B   Acronyms in Threshold Modeling

The TAR model has become a standard in nonlinear time series analysis. Many elaborate extensions/generalization of this model have been introduced since Tong (1977). Broadly these offsprings can be classified in two groups: TAR-related models with nonlinearities in the conditional mean, and models which extend the threshold idea to include both conditional mean and conditional heteroskedastic effects in a time series.[17] Against this background there is a growing use of acronyms and catchy abbreviations. Below, we provide a short list of abbreviations, including some key-references, without pretending to be complete. In case a model is introduced for the first time in the book, we include a reference to the appropriate section. For compactness, we exclude STAR-type models and Markov regime switching models from the list.

Conditional mean models

(AR)asMA   (Autoregressive) Asymmetric MA model. When the switching dynamics in a threshold MA model depends on lagged values of the noise process; Brännäs and De Gooijer (1994) and Section 2.6.5.

---

[17] Tong (1990) refers to a *second-generation* model when nonlinear features in both the conditional mean and the conditional variance are combined, as opposed to a *first-generation* model which concentrates on the conditional mean.

BAND–TAR	A TAR model with the characteristic feature that the time series process returns to an equilibrium band rather than an equilibrium point; Balke and Fomby (1997).
C–(M)STAR	Contemporaneous (multivariate) STAR model. When the mixing weights are determined by the probability that contemporaneous latent variables exceed certain threshold variables; Dueker et al. (2011).
CSETAR	Continuous SETAR; Section 2.6.3.
EDTAR	Endogenous delay TAR model. The model differs from the standard TAR implementation by using previously unexploited information about the length of time spent in regimes. This allows the construction of "sub-regimes" with "major" regimes. Parsimony is maintained by tightly restricting parameters across the sub-regimes; Pesaran and Potter (1997), Koop and Potter (2003), and Koop et al. (1996).
EQ–TAR	Equilibrium TAR. When the process tends towards an equilibrium value when it moves outside the threshold bounds; Balke and Fomby (1997).
GTM	Generalized threshold mixed model. A generalization of the TARX model to take account of non-Gaussian errors; Samia et al. (2007).
LTVEC	Level TVEC model. When the equilibrium error process is different in each regime; De Gooijer and Vidiella-i-Anguera (2003b).
M–TAR	Momentum TAR, with the thresholding based on the differences of the time series; Enders and Granger (1998).
MSETAR	Multivariate SETAR model. The model allows the threshold space to be equal to the dimension of the multivariate process using lagged values of the vector input series; Arnold and Günther (2001).
MUTARE	Multiple SETAR model. The threshold variable is applied to all the historical observations with a hierarchical substructure imposed upon the submodels; Hung (2012).
NeTARMA	Nested SETARMA model. The model defines primary level separated regimes using a threshold function which depends on one source and within each regime of the first stage, two more regimes are nested that are defined by a threshold function which depends on another source; Section 2.6.6.
PLTAR	Piecewise linear threshold AR model. When the coefficients of the SETAR model are linear functions of the state vector $\mathbf{Y}_{t-d}$ for some delay d; Baragona et al. (2004a).
Q–SETAR	Quantile SETAR model. When the existence of different regimes depends on the quantile of the series to be modeled. By estimating a sequence of conditional quantiles, the model describes the dynamics of the conditional distribution of a time series, not just the conditional mean; Cai and Stander (2008).

RD–TAR	Returning drift TAR model. Where a unit root is present in every regime, but the drift parameters move the process back to the equilibrium band when the process is outside the threshold; Balke and Fomby (1997).
RETAR	REduced-rank TAR model whose principal component process is a piecewise linear vector-valued function of past lags of the panel of time series variables; Li and Chan (2007).
SBTVAR	Structural break threshold VAR model. A special case of a two-regime VNTAR model; Galvão (2006).
SEASETAR	Seasonal SETAR model (both multiplicative and additive); De Gooijer and Vidiella-i-Anguera (2003a).
SEMTAR	SETAR model with multivariate thresholds: Section 2.6.4.
SEMI–TAR	Semiparametric TAR; Gao (2007) and Gao et al. (2013).
SETARMA	Self-exciting threshold ARMA. When parameter values depend on lagged values of series being explained; Section 2.6.2.
SSETARMA	Subset SETARMA model; Baragona et al. (2004b).
(SS)TARSO	(Subset) open-loop threshold AR (TAR) system with observable (O) input; Section 2.6.6 and Knotters and De Gooijer (1999).
TARMA(X)	Threshold ARMA (eXogenous) model. ARMA model with a step function having time-varying parameters; Section 2.6.1.
TARSV	Threshold AR stochastic volatility. When the leverage effect in a financial time series is described by an open-loop TAR(1) process; Breidt (1996), and Diop and Guégan (2004).
TVEC	Threshold vector error correction. When the cointegrating relationship is inactive inside a given range and then becomes active once the process gets too far from the equilibrium relationship; Balke and Fomby (1997) and Section 11.2.4.
VASTAR(X)	Vector adaptive spline threshold AR (eXogenous) model; Section 12.2.1.
VNTAR	Vector nested TAR model; Hubrich and Teräsvirta (2013).
VSETAR	Vector SETAR model with a single component series or exogenous variable to determine the different regimes (also called multivariate SETAR (MSETAR) model); Section 11.2.2.
VTARMA	Vector threshold ARMA; Section 11.2.2.

Conditional mean and variance models

ANST–GARCH	Asymmetric smooth transition–GARCH model; Anderson et al. (1999).
asMA–asQGARCH	Asymmetric MA – asymmetric quadratic GARCH model; Brännäs and De Gooijer (2004).
DT(G)ARCH	Double threshold (generalized) AR(MA) conditionally heteroskedastic (also abbreviated as SETAR-(G)ARCH). When the conditional mean is specified as a linear AR(MA) process and the driving random component in the (G)ARCH part is not observable, but rather linked to the innovations of the TAR(MA) model; Li and Li (1996) and Section 6.1.3.
(G)SSAR(I)–ARCH	(Generalized) simultaneous switching (integrated) AR models with ARCH errors. When the switching dynamics depends on lag-one values of the time series; Kunitomo and Sato (2002).

H(G)AR(CH)	Hysteretic (or buffered) GARCH model (also called buffered AR (BAR)). When the switching back and forth between two regimes depends on two different thresholds; Zhu et al. (2014).
SETAR–(G)ARCH	SETAR with (generalized) ARCH structure for conditional heteroskedasticity; Section 3.3.
SETAR–THSV	SETAR with threshold stochastic volatility; So et al. (2002).
TCAV$_{(X)}$	Threshold conditional autoregressive Value-at-Risk (CAViaR) with two regimes, and if appropriate an exogenous (X) threshold variable; Gerlach et al. (2011).
T–CAViaR–IG	A two-regime TCAV with an indirect GARCH$(1,1)$ model; Gerlach et al. (2011).
TDAR	Threshold double AR model. When both the conditional mean and the conditional variance specifications are piecewise linear AR processes but with the conditional variance specified as a function of the observations, rather than the innovations; Li et al. (2016).
T(G)ARCH	Threshold (G)ARCH; Rabemananjara and Zakoïan (1993), Zakoïan (1994), and Exercise 2.8.
TIG	Threshold indirect GARCH$(1,1)$ model; Yu et al. (2010).
TRIG	Threshold range indirect GARCH$(1,1)$ model: A two-regime TCAV model which replaces return data with range data; Chen et al. (2012a).
TRV	Threshold range value. A two-regime TCAV model which allows for different responses to high and low ranges in return data; Chen et al. (2012a).

Exercises

Theory Questions

2.1 Show that any BL(p,q,P,Q) model may be "converted" into a superdiagonal BL model by replacing ε_t with $\omega_t = \varepsilon_{t+L}$ for some $L \in \mathbb{N}$. Take as examples models (2.17) and (2.18).

2.2 Consider the ExpARMA(p,q) model in (2.20) with $d = 1$. Let $\{\varepsilon_t\} \overset{\text{i.i.d.}}{\sim} (0, \sigma_\varepsilon^2)$ with a density function which is strictly positive on $\mathbb{R}^{p+q}$. Assuming that the DGP is completely known, express $\{Y_t, t \in \mathbb{Z}\}$ as a convergent series via repeated substitution. Discuss briefly how this representation can be used to prove that the process is invertible if $\max_{1 \leq j \leq q}(|\theta_j| + |\tau_j|) < 1$.

2.3 A Markov process $\{Y_t\}$ is said to be *ergodic* if starting at any point $Y_1 = y$, the distribution of Y_T converges to a stationary distribution $\pi(x) = \lim_{T \to \infty} \mathbb{P}(Y_T < x | Y_1 = y)$, independent of y. It is called *geometrically ergodic* if this convergence occurs at an exponential rate. Geometric ergodicity is a concept of stability of the process; it excludes explosive or trending behavior; see Chapter 3. For the SETAR$(2; 1, 1)$ process

$$Y_t = \begin{cases} \phi_1 Y_{t-1} + \varepsilon_t & \text{if } Y_{t-1} \leq 0, \\ \phi_2 Y_{t-1} + \varepsilon_t & \text{if } Y_{t-1} > 0, \end{cases}$$

necessary and sufficient conditions for geometric ergodicity are $\phi_1 < 1$, $\phi_2 < 1$ and $\phi_1\phi_2 < 1$. These conditions imply the following three possible cases:

(i) $|\phi_1| < 1$ and $|\phi_2| < 1$;
(ii) $\phi_2 \leq -1$ and $-1 \leq \frac{1}{\phi_2} < \phi_1 < 1$;
(iii) $\phi_1 \leq -1$ and $-1 \leq \frac{1}{\phi_1} < \phi_2 < 1$.

Note that in each case, at least one of the two regimes is stationary ($|\phi_i| < 1$).

(a) Suppose that, in cases (ii) or (iii), the system starts in a nonstationary regime (i.e., $\phi_i < -1$). Explain (intuitively) why the system will always move to the other (stationary) regime in a few steps, i.e., the probability that it will stay in the nonstationary regime for the next T periods goes to zero as $T \to \infty$. Assume $\{\varepsilon_t\} \stackrel{i.i.d.}{\sim} \mathcal{N}(0, \sigma_\varepsilon^2)$.

(b) Explain why the system will not be stable if $\phi_1 = -1.25$ and $\phi_2 = -0.8$ (even though the second regime is stationary).

(c) Consider a SETAR$(k;1,1)$ process. It has been proved that the conditions for geometric ergodicity are $\phi_1 \leq 1$, $\phi_k < 1$ and $\phi_1\phi_k < 1$. Explain, using the appropriate versions of (i) – (iii), why the values of the AR parameters in the intermediate regimes ($\phi_2, \ldots, \phi_{k-1}$) are irrelevant for the stability of the process.

2.4 Consider the SETAR$(2;1,1)$ model

$$Y_t = \begin{cases} \phi Y_{t-1} + \varepsilon_t & \text{if } Y_{t-1} \leq 0, \\ -\phi Y_{t-1} + \varepsilon_t & \text{if } Y_{t-1} > 0, \end{cases}$$

where $0 < \phi < 1$, and $\{\varepsilon_t\} \stackrel{i.i.d.}{\sim} \mathcal{N}(0,1)$. The stationary marginal pdf of $\{Y_t, t \in \mathbb{Z}\}$ is given by

$$f(y) = 2\left(\frac{1-\phi^2}{2\pi}\right)^{1/2} \exp\left\{-\frac{1}{2}(1-\phi^2)y^2\right\}\Phi(-\phi y),$$

with $\Phi(\cdot)$ the standard normal distribution function.

(a) Prove that $f(y)$ is a solution of the equation

$$f(y) = \frac{1}{\sqrt{2\pi}} \int_{-\infty}^{0} \exp\left\{-\frac{1}{2}(y-\phi x)^2\right\} f(x) dx$$

$$+ \frac{1}{\sqrt{2\pi}} \int_{0}^{\infty} \exp\left\{-\frac{1}{2}(y+\phi x)^2\right\} f(x) dx.$$

(b) Prove that the mean and variance of $\{Y_t, t \in \mathbb{Z}\}$ are respectively given by

$$\mathbb{E}(Y_t) = -(2/\pi)^{1/2}\phi(1-\phi^2)^{-1/2}, \quad \text{Var}(Y_t) = (1-\phi^2)^{-1}\left(1 - \frac{2\phi^2}{\pi}\right).$$

[Hint:

$$\int_{-\infty}^{\infty} u\Phi(au+b)\varphi(u)du = \frac{a}{\sqrt{1+a^2}}\varphi\left(\frac{b}{\sqrt{1+a^2}}\right),$$

$$\int_{-\infty}^{\infty} u^2\Phi(au+b)\varphi(u)du = \Phi\left(\frac{b}{\sqrt{1+a^2}}\right) - \frac{a^2 b}{\sqrt{1+a^2}}\varphi\left(\frac{b}{\sqrt{1+a^2}}\right)$$

with the standard normal pdf $\varphi(u) = (2\pi)^{-1/2}\exp(-u^2/2)$.]

EXERCISES

2.5 Consider the asMA(1) model

$$Y_t = \begin{cases} \varepsilon_t + \theta^+ \varepsilon_{t-1} & \text{if } \varepsilon_{t-1} \geq 0, \\ \varepsilon_t + \theta^- \varepsilon_{t-1} & \text{if } \varepsilon_{t-1} < 0, \end{cases}$$

where $\{\varepsilon_t\} \overset{\text{i.i.d.}}{\sim} \mathcal{N}(0,1)$.

(a) Prove that the mean and variance are respectively given by

$$\mu_Y = \mathbb{E}(Y_t) = \frac{\theta^+ - \theta^-}{\sqrt{2\pi}}, \quad \text{Var}(Y_t) = 1 + \left((\theta^+)^2 + (\theta^-)^2\right)\frac{1}{2} - \mu_Y^2.$$

(b) Assuming stationarity, it is easy to see that the conditional pdf of $\{Y_t, t \in \mathbb{Z}\}$, given $\varepsilon_{t-1} = u \geq 0$, is normally distributed with mean $\mu_+ = \mathbb{E}(Y_t|u) = \theta^+ u$ and variance unity. Similarly, the conditional pdf of $\{Y_t\}$, given $\varepsilon_{t-1} = u < 0$, is normally distributed with mean $\mu_- = -\theta^- u$ and variance unity. Given these results, prove that the marginal pdf of $\{Y_t, t \in \mathbb{Z}\}$ is given by

$$f(y) = \frac{1}{\{1+(\theta^+)^2\}^{1/2}\sqrt{2\pi}} \exp\left\{\frac{-y^2}{2\{1+(\theta^+)^2\}}\right\} \Phi\left(\frac{\theta^+ y}{\{1+(\theta^+)^2\}}\right)$$
$$+ \frac{1}{\{1+(\theta^-)^2\}^{1/2}\sqrt{2\pi}} \exp\left\{\frac{-y^2}{2\{1+(\theta^-)^2\}}\right\} \Phi\left(\frac{-\theta^- y}{\{1+(\theta^-)^2\}}\right).$$

(c) Consider the case $\theta^+ = -\theta^- \equiv \theta$. Using part (b), prove that the marginal pdf of $\{Y_t, t \in \mathbb{Z}\}$ is identical to the marginal pdf of the SETAR(2;1,1) model in Exercise 2.3 with $\phi = \theta/(1+\theta^2)^{1/2}$.

2.6 (a) Verify the statement in Section 2.8.1 that the NEAR(1) process is not time-reversible using the third order cumulants of the process; see for cumulants (4.2).

(b) Consider the PAR(1) process (2.50) with an exponential marginal distribution of unit mean. Similar as in part (a), show that the process $\{Y_t, t \in \mathbb{Z}\}$ is not time-reversible.

2.7 Let $S_t \in \{1, 2\}$ follow a two-state Markov chain with switching probabilities $0 < w_1 < 1$ and $0 < w_2 < 1$.

(a) Show that the stationary probabilities are $\pi_1 = w_2/(w_1+w_2)$ and $\pi_2 = w_1/(w_1+w_2)$, so that $\mu = \mathbb{E}(S_t) = 1 + \pi_2 = (2w_1+w_2)/(w_1+w_2)$.

(b) Show that the process $\{S_t - 1\}$ is an i.i.d. Bernoulli sequence if $w_1 + w_2 = 1$.

(c) Show that $\mathbb{E}(S_t|S_{t-1}, S_{t-2}, \ldots) = \mu(1-\phi) + \phi S_{t-1}$, with $\phi = 1 - w_1 - w_2$, so that $\{S_t\}$ follows an AR(1) process.

2.8 Let $\{P_t\}$ denote the price of an asset at time t (not paying dividend), then the continuously compound return, or *log-return* (often called *return*), is defined as

$$r_t = \log(1 + R_t) = \log\frac{P_t}{P_{t-1}} = p_t - p_{t-1},$$

where $R_t = (P_t - P_{t-1})/P_{t-1}$ is the one-period *simple return*, and $p_t = \log P_t$. The k-period return is the sum of the one-period log-returns: $r_t[k] = p_t - p_{t-k} = \sum_{j=0}^{k-1} r_{t-j}$ ($k = 1, 2, \ldots$). Now, assume that $\{r_t, t \in \mathbb{Z}\}$ follows the TGARCH(1,1) model $r_t = Y_t = \sigma_t \varepsilon_t$, with $\sigma_t^2 = \alpha_0 + (\alpha_1 + \gamma_1 I(Y_{t-1} < 0))Y_{t-1}^2 + \beta_1 \sigma_{t-1}^2$ and $\{\varepsilon_t\} \stackrel{\text{i.i.d.}}{\sim} (0, 1)$, independent of σ_t, with $\mathbb{E}(\varepsilon_t^3) = 0$. The parameters satisfy $\alpha_0 > 0$, $\alpha_1 \geq 0$, $\beta_1 \geq 0$ and $\gamma_1 > 0$. Assume that the parameters also satisfy conditions such as $\sigma_Y^2 = \text{Var}(Y_t)$ and $\mathbb{E}(|Y_t|^3) < \infty$.

(a) Show that the (one-period) returns $r_t[1] = r_t = Y_t$ have skewness zero, i.e.

$$\tau_Y = \frac{\mathbb{E}(Y_t^3)}{\sigma_Y^3} = 0.$$

(b) Obtain an expression for the skewness of the two-period returns $r_t[2] = Y_t + Y_{t-1}$, and show that it is negative if $\gamma_1 > 0$.

Empirical and Simulation Questions

2.9 The file eeg.dat contains the EEG recordings used to estimate the AR–NN models in Section 2.11. Use the data to replicate the results reported in Tables 2.1 and 2.2.

[*Note*: The results need not be exactly as shown in both tables since they depend heavily on the initial weights chosen by random in the R-function nnet, unless set.seed(1).]

2.10 Consider the quarterly U.S. unemployment rate in Example 1.1, which we denote by $\{U_t\}_{t=1}^{252}$. If we were to work directly with this series, the assumption of a symmetric error process would be inappropriate. Various instantaneous data transformations have been employed in the analysis of $\{U_t\}$. These include the logistic transformation, first differences, the logarithmic transformation, and log-linear detrended. Because $\{U_t\}$ takes values between 0 and 1, we adopt the logistic transformation, i.e., $\{Y_t = \log(U_t/(1 - U_t))\}_{t=1}^{252}$. The transformed series (see Figure 6.2(a)) is now unbounded, and it is reasonable to assume that the error process $\{\varepsilon_t, t \in \mathbb{Z}\}$ of the nonlinear DGPs considered below is conditionally Gaussian distributed. The data are in the file USunemplmnt_logistic.dat.

(a) Estimate a SETAR(2; 2, 2) model with delay $d = 2$.
[*Hint*: Use the R-tsDyn-package.]

(b) Estimate a CSETAR(2; 2, 2) model with delay $d = 2$ and compare the results with the SETAR results obtained in part (a).

(c) The 250×3 matrix USunemplmnt_matrix.dat contains the transformed (logistic transform) U.S. unemployment data in the first column. The first- and second lags of the data are in columns 2 and 3. Estimate a two-state MS–AR model, and compare the estimation results with the SETAR results obtained in part (a).
[*Hint*: Use the R-MSwM-package.]

2.11 Astatkie et al. (1997) develop a NeSETAR model for an Icelandic streamflow system for the years 1972 – 1974, i.e. the Jökulsá Eystri in north-west Iceland. The dynamic system consists of daily data on flow (Q_t), precipitation (P_t), and temperature (T_t).

After some experimentation, it was found that the best-fitting NeSETAR model for Q_t is

$$Q_t = \begin{cases} 4.82_{(0.68)} + 0.82_{(0.03)}Q_{t-1} & \text{if } Q_{t-2} \leq 92\,\text{m}^3/\text{s and } \overline{T}_t \leq -2°\text{C}, \\ 1.32_{0.06)}Q_{t-1} - 0.32_{(0.06)}Q_{t-2} \\ \quad +0.20_{(0.03)}P_{t-1} + 0.52_{(0.10)}T_t & \text{if } Q_{t-2} \leq 92\,\text{m}^3/\text{s and } -2°\text{C} < \overline{T}_t \leq 1.8°\text{C}, \\ 1.15_{(0.04)}Q_{t-1} - 0.18_{0.04)}Q_{t-2} + 0.01_{(0.00)}P_{t-1}^2 \\ \quad +1.22_{(0.13)}T_t - 0.89_{(0.17)}T_{t-3} & \text{if } Q_{t-2} \leq 92\,\text{m}^3/\text{s and } \overline{T}_t > 1.8°\text{C}, \\ 49_{(13.6)} + 0.45_{(0.12)}Q_{t-1} \\ \quad +3.47_{(1.55)}T_t + 3.75_{(1.71)}T_{t-1} - 6.08_{(1.43)}T_{t-3} & \text{if } Q_{t-2} > 92\,\text{m}^3/\text{s}, \end{cases}$$
(2.72)

where $\overline{T}_t = (T_{t-1} + T_{t-2} + T_{t-3})/3$, and with asymptotic standard errors of the parameter estimates in parentheses. The model includes 16 parameters and produces a pooled residual variance of $27.4[\text{m}^3/\text{s}]^2$. As a comparison, Tong et al. (1985) and Tong (1990, Section 7.4.4) use a TARSO model with 42 parameters to the describe the streamflow data, resulting in a residual variance of $31.8[\text{m}^3/\text{s}]^2$.

The file jokulsa.dat contains the series stored in a $1{,}086 \times 32$ matrix with variables $(Q_t, Q_{t-1}, \ldots, Q_{t-10}, P_t, P_{t-1}, \ldots, P_{t-10}, T_t, T_{t-1}, \ldots, T_{t-9})$.

(a) Using the notation introduced in Section 2.6.6, specify the structure of the NeSETAR model (2.72). Interpret the fitted relationship.

(b) Using the supersmoother (function R-supsmu) proposed by Friedman (1984), regression estimates of Q_t on Q_{t-1} and Q_{t-2} reveals that there are two linear pieces in the data, with a threshold estimate $\widehat{r}_1 = 92$ m³/s.

Using the same method as above, verify the estimated second-stage threshold $\widehat{r}_{2,1} = -2°\text{C}$.

(c) Form subset data sets for each regime, and estimate the final model by least squares. Plot the sample ACF and sample PACF of the normalized residuals and comment.

2.12 Consider the simple SETAR(2; 1, 1) model

$$Y_t = \phi_1 Y_{t-1} + \phi_2 I(Y_{t-1} \leq 0) + \varepsilon_t, \quad \{\varepsilon_t\} \stackrel{\text{i.i.d.}}{\sim} \mathcal{N}(0,1).$$

(a) Derive an explicit expression for the one-period TI response function (A.1). Comment on the resulting time path.

(b) Use bootstrapping to compute the GIRF in (A.3) for horizons $H = 1, \ldots, 10$, and $\delta = \{1, -1\}$. Set $\phi_1 = 0.9$, $\phi_2 = -0.5$, and $B = 1{,}000$ replicates. Assume the model is completely known.

Comment on the resulting time path. Also compare the GIRF with the analytic expression for the TI response function of the AR(1) process $Y_t = \phi Y_{t-1} + \varepsilon_t$ with parameter $\phi = (0.9 - 0.5) = 0.4$.

[*Hint*: The total number of draws for an initial history is $(B-1)(H+1)$. The relevant computer code should include a loop through the data to change the initial condition, and a loop through each horizon of impulses: one with the initial condition based on a bootstrap draw, and one based on $\varepsilon_t + \delta$. Next, average over each horizon, for each initial condition. Finally, average over histories.]

Chapter 3

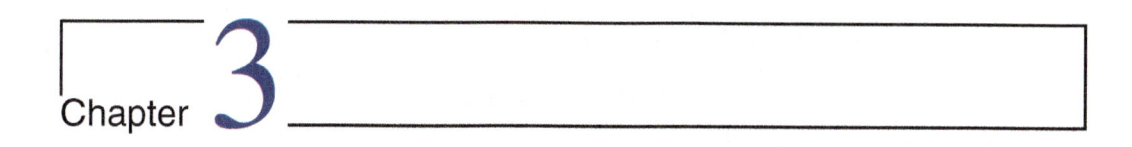

PROBABILISTIC PROPERTIES

From the previous two chapters we have seen that the richness of nonlinear models is fascinating: they can handle various nonlinear phenomena met in practice. However, before selecting a particular nonlinear model we need tools to fully understand the probabilistic and statistical characteristics of the underlying DGP. For instance, precise information on the stationarity (ergodicity) conditions of a nonlinear DGP is important to circumscribe a model's parameter space or, at the very least, to verify whether a given set of parameters lies within a permissible parameter space. Conditions for invertibility are of equal interest. Indeed, we would like to check whether present events of a time series are associated with the past in a sensible manner using an NLMA specification. Moreover, verifying (geometric) ergodicity is required for statistical inference.

In this chapter, we address the above topics. To find a balance between the many works on stationarity and ergodicity of nonlinear DGPs and yet to achieve results of general practical interest, we first discuss in Section 3.1 the existence of strict stationarity of processes embedded within the class of stochastic recurrence equations (SREs). Associated with the SRE, we define the notion of a Lyapunov exponent which measures the "geometric drift" of a process. This notion plays a central role throughout the rest of this chapter. In Section 3.2, we briefly mention a criterion for checking second-order stationarity. Next, in Section 3.3, we focus on the stationarity (ergodicity) of the class of nonlinear AR-(G)ARCH models as a special case, and application of the class of SREs. In Section 3.4, we collect some Markov chain terminologies and relevant results ensuring not only ergodicity, but also geometric ergodicity of a DGP. In Section 3.5, we discuss ergodicity, global and local invertibility of NLMA models with special emphasis on the SETMA model. This section also contains an empirical method to assess the notion of invertibility in practice.

Two appendices are added to the chapter. Appendix 3.A reviews some basic properties of vector and matrix norms, while Appendix 3.B discusses the spectral radius of a matrix.

3.1 Strict Stationarity

Suppose $\{Y_t, t \in \mathbb{Z}\}$ is a stochastic process. Then, in a multivariate setting, a *stochastic recurrence equation* (SRE) is defined as

$$\mathbf{Y}_t = \mathbf{A}_t \mathbf{Y}_{t-1} + \mathbf{B}_t, \quad t \in \mathbb{Z}, \tag{3.1}$$

where $\mathbf{Y}_t = (Y_t, \ldots, Y_{t-m+1})'$ and $\mathbf{B}_t$ are random vectors in $\mathbb{R}^m$, $\mathbf{A}_t$ are random $m \times m$ matrices, and $\{(\mathbf{A}_t, \mathbf{B}_t), t \in \mathbb{Z}\}$ is an i.i.d. sequence. Clearly, (3.1) is the defining equation of a vector AR(1) process with random coefficient matrix $\mathbf{A}_t$. Hence, it is also called a *generalized (multivariate) random coefficient* AR process or RCA for short. The process (3.1) is Markovian with transition probability $\mathcal{P}(\mathbf{y}, \cdot)$ ($\mathbf{y} \in \mathbb{R}^m$) equal to the distribution of $\mathbf{A}_t \mathbf{y} + \mathbf{B}_t$. The SRE embeds many of the nonlinear DGPs introduced in Chapter 2.

Now a sequence $\{\mathbf{Y}_t, t \in \mathbb{Z}\}$ of random vectors in $\mathbb{R}^m$ is said to be strictly (or strongly) stationary if the joint distributions of $(Y_{t_1}, \ldots, Y_{t_n})'$ and $(Y_{t_1+h}, \ldots, Y_{t_n+h})'$ are the same for all $n, h \in \mathbb{N}$, $t_1, \ldots, t_n \in \mathbb{Z}$. Of course, it is not *a priori* clear for which distributions of $\{(\mathbf{A}_t, \mathbf{B}_t)\}$ a strictly stationary solution to (3.1) exists. Below we give a sufficient condition in terms of the so-called top (or upper, or max-plus) Lyapunov exponent. However, first we introduce some additional notation: Let $\|\cdot\|$ be any vector norm in $\mathbb{R}^m$; see also Appendix 3.A. For a matrix $\mathbf{A} \in \mathbb{R}^{m \times m}$, the corresponding matrix norm $\|\mathbf{A}\|_s$ ($s \in [1, \infty)$) is defined as

$$\|\mathbf{A}\|_s = \sup_{\mathbf{y} \in \mathbb{R}^m, \mathbf{y} \neq 0} \frac{\|\mathbf{A}\mathbf{y}\|_s}{\|\mathbf{y}\|_s}. \tag{3.2}$$

Then, for an i.i.d. sequence of $m \times m$ matrices $\{\mathbf{A}_n, n \in \mathbb{Z}\}$ with $\mathbb{E}(\log^+ \|\mathbf{A}_1\|) < \infty$, we define the associated *top Lyapunov exponent* $\gamma(\cdot)$ by

$$\gamma(\mathbf{A}) = \inf_{n \in \mathbb{N}} \frac{1}{n} \mathbb{E}(\log\|\mathbf{A}_1 \mathbf{A}_2 \cdots \mathbf{A}_n\|) \stackrel{\text{a.s.}}{=} \lim_{n \to \infty} \frac{1}{n} \log\|\mathbf{A}_1 \mathbf{A}_2 \cdots \mathbf{A}_n\|, \tag{3.3}$$

where the last equality (Furstenberg and Kesten, 1960) shows that $\gamma(\cdot)$ is independent of the chosen norm.

By recursive substitution of the lagged values of $\mathbf{Y}_t$, (3.1) can be rewritten as

$$\mathbf{Y}_t = \Big(\prod_{i=0}^{s} \mathbf{A}_{t-i}\Big) \mathbf{Y}_{t-s-1} + \sum_{i=0}^{s} \Big(\prod_{j=0}^{i-1} \mathbf{A}_{t-j}\Big) \mathbf{B}_{t-i}, \quad \forall s \in \mathbb{N}, \tag{3.4}$$

with the usual convention $\prod_{j=0}^{-1} \mathbf{A}_{t-j} = \mathbf{I}_m$. If $\lim_{s \to \infty} \big(\prod_{i=0}^{s} \mathbf{A}_{t-i}\big) \mathbf{Y}_{t-s-1} \stackrel{\text{a.s.}}{=} \mathbf{0}_m$ holds, then it is reasonable to hope that (3.4) has a solution process $\{\mathbf{Y}_t, t \in \mathbb{Z}\}$ that is stationary. Indeed, suppose that $\gamma(\mathbf{A}) < 0$. Then, under some mild conditions, the series

$$\mathbf{Y}_t = \mathbf{B}_t + \sum_{s=1}^{\infty} \mathbf{A}_t \mathbf{A}_{t-1} \cdots \mathbf{A}_{t-s+1} \mathbf{B}_{t-s}, \tag{3.5}$$

3.1 STRICT STATIONARITY

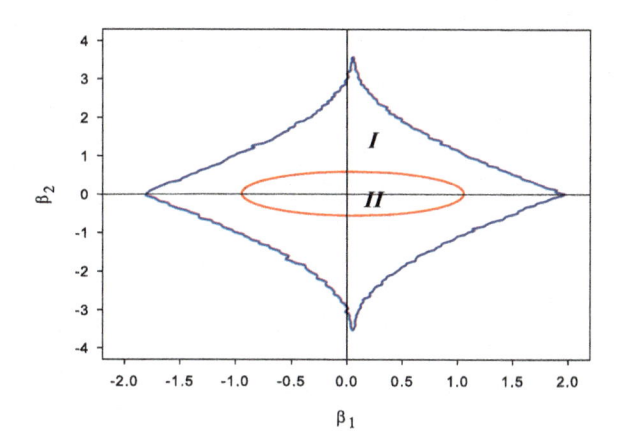

Figure 3.1: *Strict stationarity parameter region* $(I \cup II)$ *based on estimates of the top Lyapunov exponent, and second-order stationarity parameter region* (II) *for model* (3.6) *with* $\{\varepsilon_t\} \stackrel{i.i.d.}{\sim} \mathcal{N}(0,1)$.

converges a.s., and the process $\{\mathbf{Y}_t, t \in \mathbb{Z}\}$ is a non-anticipative stationary solution to (3.4); Brandt (1986). Here, *non-anticipative* (or causal) means that $\{\mathbf{Y}_t, t \in \mathbb{Z}\}$ is independent of $\{(\mathbf{A}_{t+h}, \mathbf{B}_{t+h}), h \in \mathbb{N}\}$ for each t. Further, the condition $\gamma(\mathbf{A}) < 0$ is sufficient when $\{(\mathbf{A}_t, \mathbf{B}_t)\}$ is strictly stationary and ergodic (Bougerol and Picard, 1992).

Note that $\gamma(\mathbf{A}) < 0$ holds if $\mathbb{E}(\log\|\mathbf{A}_1\|) < 0$ (take $n = 1$ in the definition of $\gamma(\cdot)$). Now assume $m = 1$. Then, $\{Y_t, t \in \mathbb{Z}\}$ as in (3.5) is the unique strictly stationary solution of (3.1) provided $-\infty \leq \mathbb{E}(\log|A_1|) < 0$ and $\mathbb{E}(\log^+|B_1|) < \infty$. These two conditions are easy to check, and $\gamma(A) = \mathbb{E}(\log|A_1|)$ can be obtained explicitly.

Example 3.1: Evaluating the Top Lyapunov Exponent

Consider the stochastic process

$$Y_t = \varepsilon_t + \beta_1 Y_{t-1}\varepsilon_{t-1} + \beta_2 Y_{t-2}\varepsilon_{t-2}^2, \quad \{\varepsilon_t\} \stackrel{i.i.d.}{\sim} (0, \sigma_\varepsilon^2). \tag{3.6}$$

Then (3.6) can be written in the form of the SRE (3.1) with

$$\mathbf{Y}_t = \begin{pmatrix} Y_t \\ Y_{t-1} \end{pmatrix}, \quad \mathbf{A}_t = \begin{pmatrix} \beta_1\varepsilon_{t-1} & \beta_2\varepsilon_{t-2}^2 \\ 1 & 0 \end{pmatrix}, \quad \mathbf{B}_t = \begin{pmatrix} \varepsilon_t \\ 0 \end{pmatrix}.$$

When $\beta_2 = 0$ (i.e., $m = 1$), the strict stationarity condition based on the top Lyapunov exponent takes the simple form $\gamma(A) = \mathbb{E}(\log|\beta_1\varepsilon_t|) = \log|\beta_1| + \mathbb{E}(\log|\varepsilon_t|) < 0$. If $\{\varepsilon_t\} \stackrel{i.i.d.}{\sim} \mathcal{N}(0, \sigma_\varepsilon^2)$, the condition reduces to $\sigma_\varepsilon|\beta_1| < \sqrt{2}\exp(C/2) = 1.8874\cdots$, where C is Euler's constant.

When $m > 1$, closed form expressions for $\gamma(\mathbf{A})$ are hard to obtain, and one has to resort to MC simulations. Figure 3.1 shows parameter regions for strict stationarity $(I \cup II)$, based on estimates of $\gamma(\mathbf{A})$ (using sequences of length 10,000), and for second-order stationarity (II), based on the constraint $\beta_1^2 \mathbb{E}(\varepsilon_t^2) + \beta_2^2 \mathbb{E}(\varepsilon_t^4) < 1$, for model (3.6) with $\{\varepsilon_t\} \overset{i.i.d.}{\sim} \mathcal{N}(0,1)$. Note, the parameter region II is much smaller than the region for strict stationarity. In the case of strict-stationarity the curve for $\gamma(\mathbf{A}) = 0$ passes through the points $(\beta_1, \beta_2) = (0, \pm 3.7748)$ and $(\beta_1, \beta_2) = (\pm 1.8874, 0)$.

3.2 Second-order Stationarity

A sequence $\{\mathbf{Y}_t, t \in \mathbb{Z}\}$ of random vectors in $\mathbb{R}^m$ is called *second-order stationary*, or *weakly stationary*, if $\mathbb{E}\|\mathbf{Y}_t\|^2 < \infty$ for all $t \in \mathbb{Z}$, $\mathbb{E}(\mathbf{Y}_t) \in \mathbb{R}^m$ is independent of $t \in \mathbb{Z}$, and the covariance matrices satisfy

$$\mathrm{Cov}(\mathbf{Y}_{t_1+h}, \mathbf{Y}_{t_2+h}) = \mathrm{Cov}(\mathbf{Y}_{t_1}, \mathbf{Y}_{t_2}), \quad \forall t_1, t_2, h \in \mathbb{Z}.$$

Clearly, every strictly stationary process which satisfies $\mathbb{E}\|\mathbf{Y}_t\| < \infty$ is also second-order stationary. In the sequel, we focus on the m-vector time series $\{\mathbf{Y}_t, t \in \mathbb{Z}\}$ generated by (3.1).

Given the strict stationary solution in (3.5), the vector process $\{\mathbf{Y}_t, t \in \mathbb{Z}\}$ is a Cauchy sequence in L_2 if and only if $\|(\prod_{j=0}^{s-1} \mathbf{A}_{t-j})\mathbf{B}_{t-s}\|_2$ exists and converges to 0 at an exponential rate as $s \to \infty$. Using the i.i.d. property of $\{(\mathbf{A}_t, \mathbf{B}_t), t \in \mathbb{Z}\}$ and Kronecker product notation, we have

$$\mathbb{E}\|\mathbf{A}_t \cdots \mathbf{A}_{t-s+1}\mathbf{B}_{t-s}\|^2 = \mathbb{E}(\mathbf{B}'_{t-s}\mathbf{A}'_{t-s+1} \cdots \mathbf{A}'_t \mathbf{A}_t \cdots \mathbf{A}_{t-s+1}\mathbf{B}_{t-s})$$
$$= \mathbb{E}\{\mathbf{B}'_{t-s} \otimes \mathbf{B}'_{t-s}\}\{\mathbb{E}(\mathbf{A}'_t \otimes \mathbf{A}')\}^s \mathrm{vec}\,\mathbf{I}_m.$$

Now, the *spectral radius* $\rho(\mathbf{M})$ of a square matrix $\mathbf{M}$ (see Appendix 3.B) is defined as

$$\rho(\mathbf{M}) = \sup\{|\lambda| : \lambda \text{ is eigenvalue of } \mathbf{M}\}.$$

Then, provided $\mathbb{E}\|\mathbf{B}_t\|^2 < \infty$, it can be deduced (see, e.g., Nicholls and Quinn, 1982; Tjøstheim, 1990) that

$$\rho\big(\mathbb{E}(\mathbf{A}_t \otimes \mathbf{A}_t)\big) < 1 \tag{3.7}$$

is a necessary and sufficient condition for the moments of order two to exist. This condition has a similar implication as that the characteristic polynomial associated with a linear AR process has no roots on and within the unit circle. If, in addition $\mathbf{A}_t$ has finite moments of order $2m$ ($m > 1$), then a necessary and sufficient condition ensuring finiteness of higher-order moments is $\rho[\mathbb{E}\{(\mathbf{A}_t)^{\otimes 2m}\}] < 1$, where $\mathbf{M}^{\otimes m} = \mathbf{M} \otimes \cdots \otimes \mathbf{M}$ (m factors); see, e.g., Pham (1986, Lemma 2). Finally, if $\{\mathbf{A} = \mathbf{A}_t\}$ is a deterministic process, then from (3.9) it follows that $\gamma(\mathbf{A}) = \log \rho(\mathbf{A})$.

3.3 Application: Nonlinear AR–GARCH model

Stability and stationarity of the class of conditionally heteroskedastic nonlinear AR models have been the focus of many papers; see, e.g., Meitz and Saikkonen (2010) and the references therein. These works often establish geometric ergodicity using conditions which overly restrict the parameter space. Unfortunately, the SRE framework does not allow for nonlinear AR models with (G)ARCH-type conditional heteroskedasticity. In fact, the random coefficients embedding of these models in (3.1) leads to "coefficients" that are no longer independent nor can one assume *a priori* that the process $\{(\mathbf{A}_t, \mathbf{B}_t), t \in \mathbb{Z}\}$ is stationary. This requires a more subtle approach than evaluating the asymptotic behavior of random matrices as in (3.3); see Cline and Pu (1999a,b, 2004).

The m-dimensional Markov (state space) representation of a nonlinear AR–GARCH time series model is of the form

$$\mathbf{Y}_t = B\left(\frac{\mathbf{Y}_{t-1}}{\|\mathbf{Y}_{t-1}\|}, \varepsilon_t\right) \|\mathbf{Y}_{t-1}\| + C(\mathbf{Y}_{t-1}, \varepsilon_t), \tag{3.8}$$

where $0 < \|B(\mathbf{y}/\|\mathbf{y}\|, u)\| \leq \bar{b}(1 + |u|)$ and $\|C(\mathbf{y}, u)\| \leq \bar{c}(y)(1 + |u|)$ for finite $\bar{b}$ and $\bar{c}(x) = o(\|\mathbf{y}\|)$, and where $\{\varepsilon_t\}$ are i.i.d. random variables with a density symmetric about 0 and positive on the real line. We also presume that $\mathbb{E}(|\varepsilon_t|^r) < \infty$ for some $r > 0$. Note, (3.8) includes the SRE in (3.1). Cline (2007c) provides explicit expressions for $B(\mathbf{y}/\|\mathbf{y}\|, u)\|\mathbf{y}\|$ in the case of a SETAR model with GARCH errors depending on past squared values of $\{Y_t\}$, a nonlinear AR–GARCH model, and a nonlinear AR model with (possibly nonlinear) GARCH errors.

For stability of (3.8) we need a tool which measures the geometric "drift" of the process when $\|\mathbf{Y}_{t-1}\|$ is large (and $C(\mathbf{Y}_{t-1}, \varepsilon_t)$ is negligible). To this end, we define the top Lyapunov exponent of the process $\{\mathbf{Y}_t, t \in \mathbb{Z}\}$ as

$$\gamma = \liminf_{n \to \infty} \limsup_{\|\mathbf{y}\| \to \infty} \frac{1}{n} \mathbb{E}\left(\log\left(\frac{1 + \|\mathbf{Y}_n\|}{1 + \|\mathbf{Y}_0\|}\right) \Big| \mathbf{Y}_0 = \mathbf{y}\right). \tag{3.9}$$

Under some regularity conditions $\gamma < 0$ implies geometric ergodicity while the converse $\gamma > 0$ ensures that $\{\mathbf{Y}_t, t \in \mathbb{Z}\}$ is transient (explosive); Cline and Pu (1999a, 2001).

Evaluating the double limit in (3.9) by MC simulation is difficult. However, by establishing ergodicity for a process associated with $\{\mathbf{Y}_t, t \in \mathbb{Z}\}$, one can express γ in terms that are more easy to compute. In particular, observe that only the first term on the right in (3.8) is *homogeneous* in $\mathbf{Y}_{t-1}$, and it dominates the behavior of $\mathbf{Y}_t$ when $\|\mathbf{Y}_{t-1}\|$ is very large. To exploit this characteristic, and following Cline (2007c), we consider the homogeneous version of (3.8). That is

$$\mathbf{Y}_t^* = B\left(\frac{\mathbf{Y}_{t-1}^*}{\|\mathbf{Y}_{t-1}^*\|}, \varepsilon_t\right) \|\mathbf{Y}_{t-1}^*\|, \tag{3.10}$$

where $\mathbf{Y}_t^* = (Y_t^*, \ldots, Y_{t-m+1}^*)'$. Let $\boldsymbol{\Theta} = \{\|\mathbf{y}\| \in \mathbb{R}^m : \|\mathbf{y}\| = 1\}$ be the unit sphere in $\mathbb{R}^m$. Furthermore, define

$$w(\boldsymbol{\theta}, u) = \|B(\boldsymbol{\theta}, u)\|, \quad \eta(\boldsymbol{\theta}, u) = \frac{B(\boldsymbol{\theta}, u)}{\|B(\boldsymbol{\theta}, u)\|}, \quad \text{for } \boldsymbol{\theta} \in \boldsymbol{\Theta}, \; u \in \mathbb{R}.$$

The homogeneous process can be collapsed to $\boldsymbol{\Theta}$:

$$\boldsymbol{\theta}_t^* = \frac{\mathbf{Y}_t^*}{\|\mathbf{Y}_t^*\|} = \eta(\boldsymbol{\theta}_{t-1}^*, \varepsilon_t). \tag{3.11}$$

Also, let

$$W_t^* = w(\boldsymbol{\theta}_{t-1}^*, \varepsilon_t).$$

Evidently the *collapsed* process $\{\boldsymbol{\theta}_t^*\}$ is Markovian. More importantly, $\{\boldsymbol{\theta}_t^*\}$ is uniformly ergodic (Cline, 2007c) with some stationary distribution, say π. Then the Lyapunov exponent for $\{\mathbf{Y}_t, t \in \mathbb{Z}\}$

$$\gamma = \int_{\boldsymbol{\Theta}} \mathbb{E}\big(\log W_1^* | \boldsymbol{\theta}_0^* = \boldsymbol{\theta}\big) \pi(\mathrm{d}\boldsymbol{\theta}) = \int_{\boldsymbol{\Theta}} \mathbb{E}\big(\log w(\boldsymbol{\theta}, \varepsilon_t)\big) \pi(\mathrm{d}\boldsymbol{\theta}) \tag{3.12}$$

is finite. Specifically,

$$\gamma \stackrel{\text{a.s.}}{=} \lim_{n \to \infty} \frac{1}{n} \sum_{t=1}^n \log W_t^*.$$

Thus, we can estimate γ simply by simulating the collapsed process and obtaining the sample average of $\{\log W_t^*\}$. Alternatively, γ may be determined numerically through an iterative procedure; see, e.g., Example 3.3.

Example 3.2: An Explicit Expression for γ (Cline, 2007b)

As a special case of (3.8), consider the Markov chain on $\mathbb{R}$ given by

$$Y_t = A(Y_{t-1}, \varepsilon_t) \stackrel{\text{def}}{=} B\Big(\frac{Y_{t-1}}{|Y_{t-1}|}, \varepsilon_t\Big) |Y_{t-1}| + C(Y_{t-1}, \varepsilon_t), \tag{3.13}$$

where the process $\{\varepsilon_t\} \stackrel{\text{i.i.d.}}{\sim} (0,1)$, $|B(y/|y|, u)| \leq \overline{b}(1 + |u|)$ and $C(y, u) \leq \overline{c}(1 + |u|)$ for finite $\overline{b}, \overline{c}$. Furthermore, we have the two-regime SETAR–ARCH model of order 1 and delay 1:

$$Y_t = A(Y_{t-1}, \varepsilon_1) = \begin{cases} \phi_0^{(1)} + \phi_1^{(1)} Y_{t-1} + (\alpha_0^{(1)} + \alpha_1^{(1)} Y_{t-1}^2)^{1/2} \varepsilon_t & \text{if } Y_{t-1} \leq 0, \\ \phi_0^{(2)} + \phi_1^{(2)} Y_{t-1} + (\alpha_0^{(2)} + \alpha_1^{(2)} Y_{t-1}^2)^{1/2} \varepsilon_t & \text{if } Y_{t-1} > 0, \end{cases} \tag{3.14}$$

with each $\alpha_j^{(i)} \geq 0$ ($i = 1, 2; j = 0, 1$). Then, by setting

$$B(-1, u) = -\phi_1^{(1)} + (\alpha_1^{(1)})^{1/2} u, \quad B(1, u) = \phi_1^{(2)} + (\alpha_1^{(2)})^{1/2} u,$$

3.3 APPLICATION: NONLINEAR AR–GARCH MODEL

and

$$C(y, u) = A(y, u) - B(y/|y|, u)|y|,$$

we can decompose (3.14) in the form (3.13), where $B(\cdot)$ and $C(\cdot)$ are respectively a homogeneous and a locally bounded function in Y_{t-1}. Now, analogous to (3.11), the homogeneous form of (3.13) can be collapsed to the process $\{\theta_t^* = \eta(\theta_{t-1}^*, \varepsilon_t)\}$ which is a two-state Markov chain on $[-1, 1]$. Let

$$p_{ij} = \mathbb{P}(\theta_1^* = j | \theta_0^* = i) = \mathbb{P}(\eta(i, \varepsilon_1) = j), \quad i, j \in \{-1, 1\}.$$

Then, the stationary distribution of $\{\theta_t^*\}$ is given by $\pi_1 = 1 - \pi_{-1} = p_{-1,1}/(p_{1,-1} + p_{-1,1})$ (cf. Exercise 2.7). To establish the uniform ergodicity of $\{\theta_t^*\}$, Cline (2007a) shows that there exists a function $\nu \colon \{-1, 1\} \to \mathbb{R}$ and a constant γ which solve the following identity, also known as the *Poisson equation*,

$$\mathbb{E}\big(\nu(\theta_1^*) - \nu(\theta_0^*) + \log W_1^* | \theta_0^* = i\big) = \gamma, \quad i = \pm 1.$$

The solution is given by

$$v(\pm 1) = \pm \frac{\mathbb{E}(\log W_1^* | \theta_0^* = 1) - \mathbb{E}(\log W_1^* | \theta_0^* = -1)}{2(p_{1,-1} + p_{-1,1})},$$

with Lyapunov exponent

$$\gamma = \pi_{-1}\mathbb{E}\big(\log|B(-1, e_1)|\big) + \pi_1\mathbb{E}\big(\log|B(1, e_1)|\big). \quad (3.15)$$

Example 3.3: Numerical Evaluation of γ (Cline, 2007c)

Consider the two-regime SETAR–ARCH model of order 2 and delay 1:

$$Y_t = \begin{cases} \phi_0^{(1)} + \sum_{i=1}^{2} \phi_i^{(1)} Y_{t-i} + (\alpha_0^{(1)} + \sum_{i=1}^{2} \alpha_i^{(1)} Y_{t-i}^2)^{1/2} \varepsilon_t & \text{if } Y_{t-1} \leq 0, \\ \phi_0^{(2)} + \sum_{i=1}^{2} \phi_i^{(2)} Y_{t-i} + (\alpha_0^{(2)} + \sum_{i=1}^{2} \alpha_i^{(2)} Y_{t-i}^2)^{1/2} \varepsilon_t & \text{if } Y_{t-1} > 0, \end{cases}$$
(3.16)

where $\{\varepsilon_t\} \overset{\text{i.i.d.}}{\sim} (0, 1)$, and each $\alpha_j^{(i)} \geq 0$ ($i = 1, 2; j = 0, 1, 2$). In this case we have the state vector $\mathbf{Y}_t = (Y_t, Y_{t-1})'$ and the collapsed process $\{\boldsymbol{\theta}_t^*\}$ takes values on the unit circle in $\mathbb{R}^2$. In addition, there are thresholds located at $\text{arc}(\theta) = \pm \pi/2$ on the unit circle. Since $m > 1$, one can only evaluate the Lyapunov exponent either by direct MC simulation or by numerically analyzing a uniformly ergodic process. Below we show results for γ obtained by solving numerically an equilibrium equation given by

$$\nu(\boldsymbol{\theta}) = \mathbb{E}\big(\nu(\boldsymbol{\theta}_1^*) + \log\|w(\boldsymbol{\theta}, \varepsilon_1)\| \big| \boldsymbol{\theta}_0^* = \boldsymbol{\theta}\big) - \gamma, \quad \text{s.t.} \int_\Theta \nu(\boldsymbol{\theta}) d\boldsymbol{\theta} = 0. \quad (3.17)$$

Simply stated, the solution follows from a one-dimensional numerical integration method combined with an iteration step for linear interpolation of a piecewise continuous function with linear extensions beyond the knots near a discontinuity point and at the extremes.

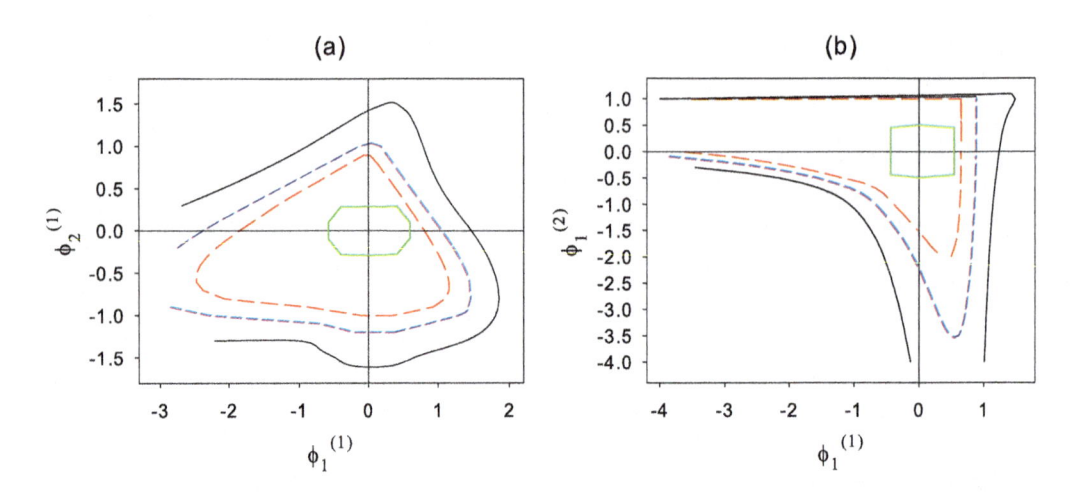

Figure 3.2: *Strict stationarity parameter regions (black solid line) for a SETAR–ARCH model, parameter regions for checking the existence of the first moment (blue medium dashed lines) and second moment (red medium dashed lines), and parameter regions for second-order stationarity (green solid lines) of* $\{\mathbf{Y}_t = (Y_t, Y_{t-1})', t \in \mathbb{Z}\}$.

Suppose $\gamma < 0$, then it is often useful to determine which moments are finite for the stationary distribution of $\{\mathbf{Y}_t, t \in \mathbb{Z}\}$. For general nonlinear AR–GARCH processes it can be shown (Cline, 2007a) that the rth moment exists when there is a bounded, positive function $\lambda(\boldsymbol{\theta})$ such that

$$\sup_{\boldsymbol{\theta} \in \Theta} \mathbb{E}\left(\frac{\lambda(\boldsymbol{\theta}^*)}{\lambda(\boldsymbol{\theta})} (W_t^*)^r \Big| \boldsymbol{\theta}_0^* = \boldsymbol{\theta}\right) < 1 \quad \text{for } r > 0. \tag{3.18}$$

A solution of (3.18) may be obtained by a numerical procedure analogous to evaluating γ through (3.17). For the quadrature (numerical integration) the results presented below are based on 100 evenly spaced points in $(-5, 5)$, and 200 points are used for interpolating $\nu(\cdot)$ and $\lambda(\cdot)$. Only eight parameters are critical for the stability of $\{\mathbf{Y}_t, t \in \mathbb{Z}\}$. Their values are:

$$\phi_1^{(1)} = 0.3, \ \phi_2^{(1)} = 0.2, \ \phi_1^{(2)} = -0.4, \ \phi_2^{(2)} = -0.1,$$
$$\alpha_1^{(1)} = (0.7)^2, \ \alpha_2^{(1)} = (0.2)^2, \ \alpha_1^{(2)} = (0.3)^2, \ \text{and } \alpha_2^{(2)} = (0.1)^2.$$

Figures 3.2(a) and (b) show parameter regions for strict stationarity (black solid lines) of the SETAR–ARCH model in (3.16) with in each case six parameters fixed and the remaining two parameters varying over a range of values. The figures also contain parameter regions for checking the existence of the first- and second moments of $\{\mathbf{Y}_t, t \in \mathbb{Z}\}$. Obviously, both regions are contained within the strict-stationarity region though covering a more restrictive set of parameter values. Indeed, we observe that for strict-stationarity the leading coefficient $\phi_1^{(1)}$ can be quite negative provided the other leading coefficient is not too big. Note that the stability region in Figure 3.2(b) closely resembles the stability region of a SETAR(2; 1, 1) model given in Figure

3.3(a). Presumably the values of $\phi_1^{(1)}$ and $\phi_1^{(2)}$ dominate the general pattern of the stability region while the other parameters have hardly any effect.

Figures 3.2(a) and (b) also show the parameter regions for second-order stationarity (green solid lines). The corresponding condition follows from (3.7) in Section 3.2, and is given by

$$\big(\max(|\phi_1^{(1)}|,|\phi_1^{(2)}|) + \max(|\phi_2^{(1)}|,|\phi_2^{(2)}|)\big)^2 + \max\{\alpha_1^{(1)}, \alpha_1^{(2)}\} \\ + \max\{\alpha_2^{(1)}, \alpha_2^{(2)}\} < 1. \tag{3.19}$$

We see that (3.19) is far too restrictive compared to the strict stationarity condition. Imposing them would unduly limit the dynamics permitted by the SETAR–ARCH model. In fact, as we see from the shape of the region enclosed by the red medium dashed lines, some parameters may have values much bigger than one, while the second moment still is finite.

3.4 Dependence and Geometric Ergodicity

3.4.1 Mixing coefficients

For i.i.d. sequences, the laws of large numbers and the central limit theorem are the cornerstone for making statistical inferences. In the context of analyzing time series, the i.i.d. assumption is practically always violated. Therefore, there is a continuous search for conditions weaker than independence for proving the above limit theorems, or variants thereof. Weak dependence is often quantified in terms of mixing conditions. Roughly speaking, mixing means that the future behavior of a time series becomes "almost independent" of the past, as time goes by. There exist several notions of mixing; see, e.g., Doukhan (1994). Here we concentrate on two standard dependence structures.

Let $\{\mathbf{Y}_t, t \in \mathbb{Z}\}$ be a strictly stationary time series in $\mathbb{R}^m$ defined on the probability space $(\Omega, \mathcal{F}, \mathbb{P})$. Denote by $\mathcal{F}_{-\infty}^0$ and $\mathcal{F}_t^\infty$ the σ-algebras generated by $\{\mathbf{Y}_s, s \leq 0\}$ and $\{\mathbf{Y}_s, s \geq t\}$ respectively. For each $k \geq 1$, define the following dependence coefficients

$$\alpha(k) = \sup_{A \in \mathcal{F}_{-\infty}^0, B \in \mathcal{F}_k^\infty} |\mathbb{P}(A \cap B) - \mathbb{P}(A)\mathbb{P}(B)|, \tag{3.20}$$

$$\beta(k) = \frac{1}{2} \sup_{A_i \in \mathcal{F}_{-\infty}^0, B_j \in \mathcal{F}_k^\infty} \sum_{i=1}^{I} \sum_{j=1}^{J} |\mathbb{P}(A_i \cap B_j) - \mathbb{P}(A_i)\mathbb{P}(B_j)|, \tag{3.21}$$

where in the definition of $\beta(k)$ the supremum is taken over all pairs of finite partitions $\{A_1, \ldots, A_I\}$ and $\{B_1, \ldots, B_J\}$ of Ω such that $A_i \in \mathcal{F}_{-\infty}^0$ for each i and $B_j \in \mathcal{F}_k^\infty$ for each j.

The quantities $\alpha(k)$ and $\beta(k)$ are called *mixing coefficients*. The process $\{\mathbf{Y}_t, t \in \mathbb{Z}\}$ is called *strongly mixing* (or α-*mixing*) if $\lim_{k \to \infty} \alpha(k) = 0$, and β-*mixing* (or *absolutely regular mixing*) if $\lim_{k\to\infty} \beta(k) = 0$. Additionally, the process is said to be strongly mixing with geometric rate if $\{\mathbf{Y}_t, t \in \mathbb{Z}\}$ is α-mixing (or β-mixing) with exponentially decaying coefficients. Since $\alpha(k) \leq (1/2)\beta(k)$, β-mixing implies α-mixing. The α-mixing is the weakest condition among all currently available mixing conditions. One way of checking mixing or stationarity conditions is to express (or approximate) the nonlinear model as a suitably chosen Markov chain and use Markov chain theory. This will be the focus of Section 3.4.2.

Mixing conditions are helpful in proving limit theorems. For instance, for the special case of strongly mixing sequences, these conditions imply the following central limit theorem (CLT) (Herrndorf, 1984, Corollary 1). Let $\{Y_t\}_{t=1}^{\infty}$ be a zero-mean univariate stochastic process, where

$$\sup_t \|Y_t\|_{2+a} < \infty \quad \text{and} \quad \sum_{k=1}^{\infty} \{\alpha(k)\}^{a/(2+a)} < \infty \quad \text{for some } a \in (0, \infty).$$

Assume that $\sigma^2 = \lim_{T\to\infty} \text{Var}(T^{-1/2} \sum_{t=1}^{T} Y_t) > 0$. Then, $T^{-1/2} \sum_{t=1}^{T} Y_t \xrightarrow{D} \mathcal{N}(0, \sigma^2)$, as $T \to \infty$; see also Rio (1993). The generalization of this CLT to a centered vector-valued stochastic process $\{\mathbf{Y}_t, t \in \mathbb{Z}\}$ is obvious.

3.4.2 Geometric ergodicity

Feigin and Tweedie (1985) develop a way of checking sufficient conditions for strong mixing. We adopt their notation and terminology. So we let $\{\mathbf{Y}_t, t \in \mathbb{N}\}$ be a temporarily homogeneous Markov chain taking values in $(\mathbb{E}, \mathcal{E})$, where $\mathbb{E} \subset \mathbb{R}^m$ and $\mathcal{E}$ is the Borel σ-algebra on $\mathbb{E}$. We denote its tth step transition probability by $\mathcal{P}^t(\mathbf{y}, C)$, i.e.

$$\mathcal{P}^t(\mathbf{y}, C) = \mathcal{P}(\mathbf{Y}_t \in C | \mathbf{Y}_0 = \mathbf{y}), \quad \mathbf{y} \in \mathbb{R}^m, C \in \mathcal{E},$$

with $\mathcal{P}(\mathbf{y}, C) = \mathcal{P}(\mathbf{Y}_1 \in C | \mathbf{Y}_0 = \mathbf{y}) = \mathcal{P}^1(\mathbf{y}, C)$, and where $\mathcal{P}$ is the probability measure on the underlying probability space on which $\mathbf{Y}_0$ is defined. A measure π is an *invariant* measure for the Markov chain $\{\mathbf{Y}_t, t \in \mathbb{Z}\}$ if

$$\pi(A) = \int_{\mathbb{E}} \mathcal{P}(\mathbf{x}, A)(\mathrm{d}\mathbf{y}). \tag{3.22}$$

Assume $\pi(\mathbb{E}) = 1$. If there exists a finite measure with property (3.22) and we run a Markov chain with initial probability distribution π, then the resulting process is stationary and its marginal distribution is π at any time point t.

It is of course not yet clear whether the distribution of $\{\mathbf{Y}_t, t \in \mathbb{Z}\}$ converges towards an invariant distribution π. If such a convergence happens with respect to the total variation norm $\|\cdot\|_V$, and with a fixed geometric rate, the Markov chain $\{\mathbf{Y}_t, t \in \mathbb{Z}\}$ is called *geometrically ergodic*. This means that there exists a constant $0 < \rho < 1$ such that $\forall \mathbf{y} \in \mathbb{R}^m$,

$$\lim_{t\to\infty} \rho^{-t} \|\mathcal{P}^t(\mathbf{y}, \cdot) - \pi(\cdot)\|_V = 0 \tag{3.23}$$

3.4 DEPENDENCE AND GEOMETRIC ERGODICITY

for almost all initial states $\mathbf{y} \in \mathbb{R}^m$ provided $\pi(\cdot) < \infty$. Thus, a geometrically ergodic stationary Markov chain is also strongly mixing with geometric rate. More precisely, for $\alpha(k)$ as defined by (3.20), we have $\alpha(k) \leq K\rho^k$ for some constants $K > 0$ and $\rho \in (0, 1)$. If (3.23) holds when $\rho = 1$, then $\{\mathbf{Y}_t, t \in \mathbb{Z}\}$ is said to be *Harris ergodic*.

As usual in the theory of Markov chains, we restrict attention to the case of irreducible Markov chains. Let φ be a non-trivial (i.e. $\varphi(\mathbb{R}^m) > 0$) σ-finite measure on $(\mathbb{R}^m, \mathcal{E})$. Then the Markov process defined above is called φ-*irreducible* if $\forall C \in \mathcal{E}$ with $\varphi(C) > 0$, $\forall \mathbf{y} \in \mathbb{R}^m$,

$$\sum_{t=1}^{\infty} \mathcal{P}^t(\mathbf{y}, C) > 0.$$

This simply states that almost all parts of the state space are accessible from all points $\mathbf{y}$ of $\mathbb{R}^m$. Further, a Markov chain is a (weak) *Feller chain* if for every bounded continuous function $g(\cdot)$ on $\mathbb{E} = \mathbb{R}$ the function

$$\mathbb{E}\{g(\mathbf{Y}_t)|\mathbf{Y}_{t-1} = \mathbf{y}\} \qquad (3.24)$$

is also continuous in $\mathbf{y} \in \mathbb{E}$.

Next, we state a result due to Feigin and Tweedie (1985, Thm. 1) which ensures geometric ergodicity. Suppose that (i) $\{\mathbf{Y}_t, t \in \mathbb{Z}\}$ is a Feller chain, and there exist a measure φ and a compact set C with $\varphi(C) > 0$ such that

(ii) $\{\mathbf{Y}_t, t \geq 0\}$ is φ-irreducible;

(iii) There exists a non-negative continuous function $V \colon \mathbb{E} \to \mathbb{R}$ satisfying $V(\mathbf{y}) \geq 1$ $\forall \mathbf{y} \in C$ and for some $\delta > 0$,

$$\mathbb{E}\{V(\mathbf{Y}_t)|\mathbf{Y}_{t-1} = \mathbf{y}\} \leq (1-\delta)V(\mathbf{y}), \quad \mathbf{y} \notin C.$$

Then $\{\mathbf{Y}_t, t \in \mathbb{Z}\}$ is geometrically ergodic.

As already mentioned, for a φ-irreducible Markov chain geometric ergodicity and strict stationarity are equivalent. Thus, verification of the conditions of the above result will not only ensure the existence of a unique strictly stationary solution of $\{\mathbf{Y}_t, t \in \mathbb{Z}\}$ but also the geometric rate of convergence of the marginals to the stationary distribution if the chain is not initially in its stationary regime. The function $V(\cdot)$ is the so-called *test* (or Lyapunov) function which is set in advance. In the vector case, a fashionable choice is $V(\mathbf{y}) = 1 + \mathbf{y}'\mathbf{Q}\mathbf{y}$, where $\mathbf{Q}$ is a suitably positive definite matrix. Condition (iii) is a drift condition for non-explositivity.

Example 3.4: Geometric Ergodicity of the SRE (Basrak et al., 2002)
Consider the SRE in (3.1) with either $\mathbf{A}_1$ or $\mathbf{B}_1$ having a strictly positive density over $\mathbb{R}^m$. Moreover, suppose there exists an $\epsilon > 0$ such that $\mathbb{E}\|\mathbf{A}_1\|^{\epsilon} < 1$ and $\mathbb{E}|\mathbf{B}_1|^{\epsilon} < \infty$. It is clear that $\{\mathbf{Y}_t, t \in \mathbb{N}\}$ is a Markov chain. We will show that the process $\{\mathbf{Y}_t, t \in \mathbb{Z}\}$ is geometrically ergodic by checking the conditions (i) – (iii) above.

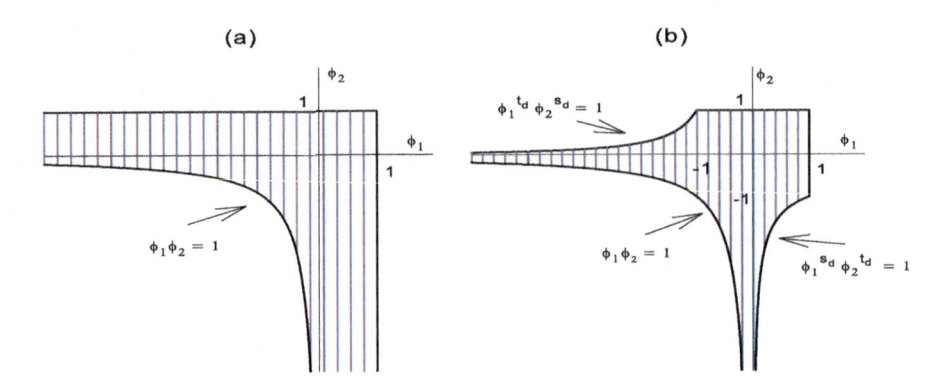

Figure 3.3: *Stationarity region of a* SETAR(2; 1, 1) *model; (a) d = 1, and (b) general d.*

(i) Lebesgue's dominated convergence theorem ensures that for any bounded continuous function $V(\cdot)$, $\mathbb{E}\{V(\mathbf{Y}_t)|\mathbf{Y}_{t-1} = \mathbf{y}\}$ is continuous in $\mathbf{y}$, and hence the Markov chain is Feller.

(ii) Given $\mathbf{Y}_0 = \mathbf{y}$, the law of $\mathbf{Y}_1 = \mathbf{A}_1\mathbf{y} + \mathbf{B}_1$ admits a strictly positive density with respect to Lebesgue measure μ^{Leb}, and so the chain is ϕ-irreducible with $\phi = \mu^{\text{Leb}}$.[1]

(iii) The condition $\mathbb{E}\|\mathbf{A}_1\|^\epsilon < 1$ for some $\epsilon > 0$ implies $\mathbb{E}(\log\|\mathbf{A}_1\|) < 0$, using Jensen's inequality. Now, without loss of generality, let $\epsilon \in (0, 1]$ and

$$V(\mathbf{y}) = 1 + |\mathbf{y}|^\epsilon, \quad \mathbf{y} \in \mathbb{R}^m.$$

Obviously,

$$\mathbb{E}\{V(\mathbf{Y}_t)|\mathbf{Y}_{t-1} = \mathbf{y}\} \leq 1 + \mathbb{E}|\mathbf{A}_1\mathbf{y}|^\epsilon + \mathbb{E}|\mathbf{B}_1|^\epsilon$$
$$\leq 1 + \mathbb{E}\|\mathbf{A}_1\|^\epsilon |\mathbf{y}|^\epsilon + \mathbb{E}|\mathbf{B}_1|^\epsilon$$
$$= \mathbb{E}\|\mathbf{A}_1\|^\epsilon V(\mathbf{y}) + (1 + \mathbb{E}|\mathbf{B}_1|^\epsilon - \mathbb{E}\|\mathbf{A}_1\|^\epsilon).$$

Choose C as the closed ball in $\mathbb{R}^m$ with center $\mathbf{0}$ and radius $M > 0$ so large that $\varphi(C) > 0$ and

$$\mathbb{E}\{V(\mathbf{Y}_t)|\mathbf{Y}_{t-1} = \mathbf{y}\} \leq (1 - \delta)V(\mathbf{y}), \quad |\mathbf{y}| > M$$

for some constant $1 - \delta > \mathbb{E}\|\mathbf{A}_1\|^\epsilon$. This proves the so-called drift condition and completes the argument.

Thus, the stationary solution (3.5) of the SRE is geometrically ergodic, and hence strongly mixing with geometric rate.

---

[1] Lesbesgue measure μ^{Leb} is a unique positive measure on the class $\mathbb{R}$ of linear Borel sets. It is specified by the requirement: $\mu^{\text{Leb}}(a, b] = b - a \ \forall a, b \in \mathbb{R}$ ($a \leq b$). Lebesgue measure on the class $\mathbb{R}^m$ of m-dimensional Borel sets is constructed similarly using the area of bounded rectangles as a basic definition; see, e.g., Billingsley (1995, Chapter 2).

3.4 DEPENDENCE AND GEOMETRIC ERGODICITY

Example 3.5: SETAR Geometric Ergodicity

Figure 3.3(a) shows the geometric ergodicity (strict stationarity) region for SETAR$(2;1,1)$ models with $d = 1$; see Table 3.1. Note that in contrast with the stationarity of linear AR models, the region is unbounded. Moreover, we see a much larger region of stationarity than the region $|\phi_1| < 1$ and $|\phi_2| < 1$ which would result if only sufficient conditions for stationarity were applied. Figure 3.3(b) shows the stationarity region in the parameter space implied by SETAR$(2;1,1)$ models with $d \geq 2$. Comparing these two plots, we see clearly the effect of the delay parameter d.

In Markov chain terminology, it can be proved (Guo and Petruccelli, 1991) that the SETAR$(2;1,1)$ model with $d \geq 1$ is positive Harris recurrent in the blue-striped "interior" and "boundary" areas; and it is transient (explosive) in the "exterior" of the parameter space. The SETAR$(2;1,1)$ model is null recurrent on the boundaries, and regular in the strict interior parameter space which in this case implies that the process $\{Y_t, t \in \mathbb{Z}\}$ is geometrically ergodic. In other words, the limit cycle behavior of the SETAR model arises from the alternation of explosive, dormant, and rising regimes.

Table 3.1 gives an overview of necessary and sufficient conditions for geometric ergodicity of some threshold models. The proofs are given under the assumption that $\{\varepsilon_t\}$ is i.i.d. with positive pdf over the real line $\mathbb{R}$ and $\mathbb{E}|\varepsilon_t| < \infty$. If appropriate, it is also assumed that for each i $\{\varepsilon_t^{(i)}\}$ are i.i.d. and $\{\varepsilon_t^{(i)}, i = 1, \ldots, k\}$ are independent. Finally, note that for the general SETARMA model $d \leq p$, since if $d > p$ one can introduce additional coefficients $\phi_j^{(i)} = 0$ for $i > p$.

Observe that for the SETARMA model, stationarity is completely determined by the linear AR pieces defined on the two boundary threshold regimes. That is, the MA part of the model does not affect stationarity. In fact, a pure SETMA model is always stationary and ergodic as is the linear MA model. Another interesting feature of SETARMA models is that overall (global) stationarity does not require the model to be stationary in each regime. The ergodicity conditions given by Liu and Susko (1992) and Lee and Shin (2001) illustrate this remark; see, also, Exercise 2.2. In general, distinguishing between local and global stationarity and between local and global invertibility (see Section 3.5) is important for physical motivation and for application of nonlinear time series models. However, it is quite complicated to derive explicit (analytical) conditions for local stationarity and local invertibility.

Table 3.1: *Necessary and sufficient conditions for geometric ergodicity of SETAR(MA) models.*

Reference	Model	Ergodicity conditions
Petruccelli and Woolford (1984)	SETAR(2;1,1): $Y_t = \phi_1 I(Y_{t-1} \leq 0)$ $+ \phi_2 I(Y_{t-1} > 0) + \varepsilon_t$	$\phi_1 < 1, \phi_2 < 1, \phi_1\phi_2 < 1$ (necessary and sufficient)
Chan et al. (1985)	SETAR(k;1,...,1): $Y_t = \sum_{i=1}^{k}\{\phi_0^{(i)} + \phi_1^{(i)}Y_{t-1} + \varepsilon_t^{(i)}\}I(Y_{t-1} \in \mathbb{R}^{(i)})$	$\phi_1^{(1)} < 1, \phi_1^{(k)} < 1$, and $\phi_1^{(1)}\phi_1^{(k)} < 1$ (sufficient)
Chen and Tsay (1991) [1]	SETAR(2;1,1): $Y_t = \phi_1 I(Y_{t-d} \leq 0)$ $+ \phi_2 I(Y_{t-d} > 0) + \varepsilon_t$ $(d \geq 2)$	$\phi_1 < 1, \phi_1\phi_2 < 1, \phi_1^{s_d}\phi_2^{t_d} < 1,$ $\phi_1^{t_d}\phi_2^{s_d} < 1$ where $t_d, s_d \in \mathbb{N}$, $t_d = s_d + 1$, and $s_d = 1_2, 3_3, 7_4, 1_5,$ $3 1_6, 63_7, 1_8, 33_9, 3 10$ (necessary and sufficient)
Brockwell et al. (1992) [2]	SETAR(k;p,...,p) - MA(q): $Y_t = \sum_{i=1}^{k}\{\phi_0^{(i)} + \phi_1^{(i)}Y_{t-d} + \varepsilon_t$ $+ \sum_{j=1}^{q}\psi_j\varepsilon_{t-j}\}I(Y_{t-d} \in \mathbb{R}^{(i)})$	$\rho(\max_i\{|\mathbf{A}^{(i)}|\}) < 1$ $(i = 1,\ldots,k)$ with $\mathbf{A}^{(i)} =$ $\begin{pmatrix} \phi_1^{(i)} & \cdots & \phi_p^{(i)} \\ \mathbf{I}_{p-1} & \mathbf{0}_{(p-1)\times 1} \end{pmatrix}$ (sufficient)
Liu and Susko (1992)	$Y_t = \sum_{i=1}^{k}\{\phi_0^{(i)} + \phi_1^{(i)}Y_{t-d} + \varepsilon_t^{(i)}$ $+ \sum_{j=1}^{q}\psi_j^{(i)}\varepsilon_{t-j}^{(i)}\}I(Y_{t-d} \in \mathbb{R}^{(i)})$	$\phi_1^{(1)} < 1, \phi_1^{(k)} < 1, \phi_1^{(1)}\phi_1^{(k)} < 1$ (sufficient) $\phi_1^{(1)} \leq 1$ and $\phi_1^{(k)} \leq 1$ (necessary)
Amendola et al. (2009a)	SETAR(2;p,q,p,q): $Y_t = \sum_{i=1}^{2}\{\phi_0^{(i)} + \sum_{j=1}^{p}\phi_j^{(i)}Y_{t-j} + \varepsilon_t$ $+ \sum_{j=1}^{q}\psi_j^{(i)}\varepsilon_{t-j}\}I(Y_{t-d} \in \mathbb{R}^{(i)})$	$\max_i\{\rho(\mathbf{A}^{(i)})\} < 1$ $(i = 1, 2)$ (sufficient, but weaker than $\rho(\max_i\{|\mathbf{A}^{(i)}|\}) < 1$)
Niglio and Vitale (2010a)	SETARMA(k;1,q,...,1,q): $Y_t = \sum_{i=1}^{k}\{\phi_1^{(i)}Y_{t-d} + \varepsilon_t^{(i)}$ $+ \sum_{j=1}^{q}\psi_j^{(i)}\varepsilon_{t-j}^{(i)}\}I(Y_{t-d} \in \mathbb{R}^{(i)})$	$\prod_{i=1}^{k}|\phi_1^{(i)}|^{p_i} < 1$ $(i = 1,\ldots,k)$, where $p_i = \mathbb{E}[I(Y_{t-d} \in \mathbb{R}^{(i)})]$, with $0 < p_i < 1$ [3] and $\sum_{i=1}^{k}p_i = 1$ (sufficient)
Lee and Shin (2000)	MTAR(2;1,1): $Y_t = \phi_1 Y_{t-1} I(Y_{t-1} \geq Y_{t-2}) + \phi_2 Y_{t-1} I(Y_{t-1} < Y_{t-2})$ $+ \varepsilon_t$	$\phi_1 < 1, \phi_2 < 1, \phi_1\phi_2 < 1, \phi_1\phi_2^2 < 1,$ and $\phi_1^2\phi_2 < 1$ (sufficient)
Lee and Shin (2001)	MTAR(2;1,1) with partial unit roots	$(\phi_1 = 1, |\phi_2| < 1)$ or $(|\phi_1| < 1, \phi_2 = 1)$ (necessary and sufficient)

[1] Lim (1992) derives necessary and sufficient conditions for stability of the deterministic SETAR(2;1,1) model with general d.

[2] Ling (1999) shows that a sufficient condition for strict stationarity of the SETARMA(k;$p,q,\ldots,p,q$) model is given by $\sum_{j=1}^{p}\max_i|\phi_j^{(i)}| < 1$ $(i = 1,\ldots,k)$ which is equivalent to the condition given by Brockwell et al. (1992).

[3] The k-regime SETARMA model becomes a linear ARMA model when $p_i = 1$, and a k^*-regime SETAR model ($k^* < k$) when $p_i = 0$.

3.5 Invertibility

The *classical invertibility* concept for univariate linear time series processes loosely says that a time series process is invertible when we are able to express the noise process $\{\varepsilon_t\}$ as a convergent series of the observations $\{Y_t\}$, given that the DGP is completely known. From the theory of linear time series it is well known that the invertibility concept is pivotal when one tries to recover the innovations from the observations of a DGP. Indeed, invertibility assures that there is a unique representation of the model which can be used for forecasting. In this section, we discuss conditions for the global and local invertibility of nonlinear DGPs, where in the latter case the boundary region is a part of the possible parameter space.

3.5.1 Global

To begin with, suppose $\{Y_t, t \in \mathbb{Z}\}$ is generated by the stationary and ergodic NLARMA(p,q) model

$$Y_t = g(Y_{t-1}, \ldots, Y_{t-p}, \varepsilon_{t-1}, \ldots, \varepsilon_{t-q}; \boldsymbol{\theta}) + \varepsilon_t, \tag{3.25}$$

where $\{\varepsilon_t\} \stackrel{\text{i.i.d.}}{\sim} (0, \sigma_\varepsilon^2)$, and $g(\cdot; \boldsymbol{\theta})$ is a known real-valued function for a known parameter vector $\boldsymbol{\theta}$. For nonlinear time series there exist (at least) three concepts of invertibility.

(i) *Granger–Andersen invertibility* (Granger and Andersen, 1978a,b)
  Suppose that q initial values, say $\overline{\varepsilon}_j$ ($j = -q+1, \ldots, 0$), of the process in (3.25) are given and that all Y_t are known. Let $\{\widehat{\varepsilon}_t, t \in \mathbb{Z}\}$ be a sequence of *innovations* (or residuals) generated by

$$\widehat{\varepsilon}_t = Y_t - g(Y_{t-1}, \ldots, Y_{t-p}, \widehat{\varepsilon}_{t-1}, \ldots, \widehat{\varepsilon}_{t-q}; \boldsymbol{\theta}), \tag{3.26}$$

where $\widehat{\varepsilon}_i = \overline{\varepsilon}_i$ for $i \leq 0$. Define the *reconstruction errors* as

$$e_t = \varepsilon_t - \widehat{\varepsilon}_t. \tag{3.27}$$

Then the model (3.25) is said to be invertible, if

$$\mathbb{E}[e_t^2] \to 0 \quad \text{as} \quad t \to \infty. \tag{3.28}$$

A more general form of (3.28) requires that

$$\mathbb{E}|e_t|^r \to 0 \quad \text{as} \quad t \to \infty, \quad (r = 1, 2, \ldots), \tag{3.29}$$

provided the q initial values $\overline{\varepsilon}_j$ ($j = -q+1, \ldots, 0$) are arbitrarily chosen. If (3.25) involves estimated parameters, which are obtained from an earlier finite length of data and not updated, condition (3.29) becomes

$$\mathbb{E}|e_t|^r \to c \quad \text{as} \quad t \to \infty, \tag{3.30}$$

Table 3.2: *Necessary and sufficient conditions for invertibility of* NLMA-*type models* [1].

Reference	Model	Condition
Ling and Tong (2005)	SETMA$(2; p, q)$: $Y_t = \sum_{i=1}^{p} \phi_i \varepsilon_{t-i}$ $+ \sum_{i=1}^{q} \psi_i I(Y_{t-d} \leq r) \varepsilon_{t-i} + \varepsilon_t$	$\sum_{i=1}^{p} \lvert \phi_i \rvert < 1$, and $\sum_{i=1}^{p} \lvert \phi_i + \psi_i \rvert < 1$ where $\psi_i = 0$ for $i > q$ (sufficient)
Ling et al. (2007)	SETMA$(k; 1, \ldots, 1)$: $Y_t = \{\psi_0$ $+ \sum_{i=1}^{k} \psi_i I(r_{i-1} < Y_{t-1} \leq r_i)\} \varepsilon_{t-1}$ $+ \varepsilon_t$	$\prod_{i=1}^{k} \{\lvert \psi_0 + \psi_i \rvert^{F_Y(r_i) - F_Y(r_{i-1})}\} < 1$ [2] and *not* invertible if $\prod_{i=1}^{k} \{\lvert \psi_0 + \psi_i \rvert^{F_Y(r_i) - F_Y(r_{i-1})}\} > 1$, where $F_Y(\cdot)$ is the CDF of $\{Y_t, t \in \mathbb{Z}\}$ (necessary and sufficient) [3]
Niglio and Vitale (2010b)	SETMA$(k; q, \ldots, q)$: $Y_t = \varepsilon_t$ $+ \sum_{i=1}^{k} \left(\sum_{j=1}^{q} \psi_j^{(i)} \varepsilon_{t-j} \right) I(Y_{t-d} \in \mathbb{R}^{(i)})$	$\prod_{i=1}^{k} \rho(\boldsymbol{\Psi}^{(i)})^{p_i} < 1$ with $\boldsymbol{\Psi}^{(i)} = \begin{pmatrix} \psi_1^{(i)} & \cdots & \psi_q^{(i)} \\ \mathbf{I}_{q-1} & \mathbf{0}_{(q-1) \times 1} \end{pmatrix}$ and $p_i = \mathbb{E}[I(Y_{t-d} \in \mathbb{R}^{(i)})]$ ($0 < p_i < 1$) (sufficient)
Marek (2005)	RCMA(1)): $Y_t = A_{t,0} \varepsilon_t + A_{t,1} \varepsilon_{t-1}$ where $\{A_{t-i,k}\}_{i=0}^{\infty}$ and $\{\varepsilon_{t-k+j}\}_{j=0}^{\infty}$ are independent ($k = 0, 1$)	$\mathbb{E} \log \lvert A_{t,1} \rvert < \mathbb{E} \log \lvert A_{t,0} \rvert$ where $\{A_{t,k}\}$ ($k = 0, 1$) is a stationary and ergodic process (sufficient)

[1] Assuming $\{Y_t, t \in \mathbb{Z}\}$ is strictly stationary and ergodic, and $\{\varepsilon_t\} \overset{\text{i.i.d.}}{\sim} (0, \sigma_\varepsilon^2)$.
[2] This condition is much weaker than the one of Ling and Tong (2005). A similar result can be found in Ling (1999).
[3] It remains to prove that the model is *not* invertible when $\prod_{i=1}^{k} \{\lvert \phi_0 + \psi_i \rvert^{F_Y(r_i) - F_Y(r_{i-1})}\} = 1$.

where $c < \infty$ is some constant. Clearly, the concept of invertibility is intimately related to the estimation of parameters. If some least squares method is used for this purpose, it is appropriate to set $r = 2$, i.e. consider the mean-square error convergence of the reconstruction errors. Most studies focus on this case, and we refer to $\{Y_t, t \in \mathbb{Z}\}$ as *invertible* if and only if

$$\mathbb{E}[e_t^2] \to 0 \quad \text{as} \quad t \to \infty, \tag{3.31}$$

for any initial $\bar{\varepsilon}_j$ ($j = -q + 1, \ldots, 0$).

(ii) *Generalized invertibility* (Hallin, 1980)

Suppose that a realization of the process has been observed from time $a - p$, and the innovations $\widehat{\varepsilon}_t$ are generated by (3.26) with $\widehat{\varepsilon}_{a-j} = \bar{\varepsilon}_{a-j}$, and $\bar{\varepsilon}_{a-j}$ ($j = 1, \ldots, q$) are arbitrarily chosen initial values. Define the reconstruction errors as in (3.27). Then (3.25) is said to be invertible, if

$$\mathbb{E}[e_t^2] \to 0 \quad \text{as } a \to -\infty, \quad \forall t \in \mathbb{Z}. \tag{3.32}$$

Hallin (1980) shows that in nonlinear models with constant coefficients definitions (i) and (ii) are equivalent. When the coefficients are not time dependent and the DGP is linear, (3.32) coincides with the classical invertibility condition.

(iii) *Pham–Tran invertibility* (Pham and Tran, 1981)

Suppose that $\{Y_t, t \in \mathbb{Z}\}$ in (3.25) admits an equivalent first-order Markovian representation $\{Z_t\}$. Let $\widehat{\boldsymbol{\theta}}$ be some guess or estimate of the true parameter vector $\boldsymbol{\theta}$. In that case the innovations can be computed recursively from the Markovian representation of the NLARMA model with $\boldsymbol{\theta}$ replaced by $\widehat{\boldsymbol{\theta}}$. Conditional on a chosen initial value z_0 for Z_0, we denote the resulting value by $\varepsilon_t(\widehat{\boldsymbol{\theta}}|z_0)$, to indicate its dependence on $\widehat{\boldsymbol{\theta}}$. Then the process (3.25) is said to be invertible at $\widehat{\boldsymbol{\theta}}$ relative to $\{Y_t, t \in \mathbb{Z}\}$ if there exists a stationary process, say $\{\varepsilon_t(\widehat{\boldsymbol{\theta}})\}$, such that $\varepsilon_t(\widehat{\boldsymbol{\theta}}|z_0) - \varepsilon_t(\widehat{\boldsymbol{\theta}})$ converges to 0 in some sense as $t \to \infty$. Thus, this invertibility concept is "open", as we may choose an appropriate measure of convergence. In contrast, the Granger–Andersen invertibility concept requires only that the second moment of $\varepsilon_t - \widehat{\varepsilon}_t$ tends to a limit.

Table 3.3: *Necessary and sufficient conditions for stationarity and invertibility of* BL *models. In all cases* $\{\varepsilon_t\} \overset{\text{i.i.d.}}{\sim} (0, \sigma_\varepsilon^2)$ *unless otherwise specified.*

Reference	Model	Condition
Quinn (1982)	$Y_t = \varepsilon_t + \psi Y_{t-u}\varepsilon_{t-v}$ $(u, v > 0)$, with $\mathbb{E}\log|\varepsilon_t| < \infty$	$\log|\psi| + \mathbb{E}\log|Y_t| < 0$ (necessary and sufficient)
	$Y_t = \varepsilon_t + \psi Y_{t-u}\varepsilon_{t-v}$ $(u, v > 0, u > v)$	$|\psi|\sigma_\varepsilon < 1/\sqrt{2} = 0.7071$
	$Y_t = \varepsilon_t + \psi Y_{t-u}\varepsilon_{t-v}$ $(u, v > 0, u > v)$, with $\{\varepsilon_t\} \overset{\text{i.i.d.}}{\sim} \mathcal{N}(0, \sigma_\varepsilon^2)$	$|\psi|\sigma_\varepsilon < \{2\exp C/(1+2\exp C)\}^{1/2} = 0.8836$
Liu (1985)	$Y_t = \sum_{i=1}^p \phi_i Y_{t-i} + \varepsilon_t + \theta\varepsilon_{t-1}$ $+ \sum_{u=1}^Q \psi_{1u} Y_t \varepsilon_{t-1}$	$\mathbb{E}|\log\theta + \mathbf{C}'\mathbf{B}\mathbf{Y}_t| < 0$ [1] with $\mathbf{B} = \begin{pmatrix} \psi_{11} \cdots \psi_{1Q} \, 0 \cdots 0 \\ \mathbf{0}_{(s-1)\times s} \end{pmatrix}$, $\mathbf{Y}_t = (Y_t, \ldots, Y_{t-s+1})'$, $\mathbf{C} = (1, 0, \cdots, 0)'$, and $s = \max(p, Q)$ (sufficient)
Liu (1990)	$Y_t = \sum_{i=1}^p \phi_i Y_{t-i} + \varepsilon_t + \theta\varepsilon_{t-1}$ $+ \sum_{u=1}^p \sum_{v=1}^Q \psi_{uv} Y_{t-u}\varepsilon_{t-v}$ with $\mathbb{E}\{\log^+|\varepsilon_1|\} < \infty$	$\mathbb{E}\{\log\|\prod_{j=1}^p \mathbf{B}(t-j)\|\} < 0$ with $\mathbf{B}(t) = \begin{pmatrix} \phi_1 + \sum_{v=1}^Q \psi_{1v}\varepsilon_{t-v} \cdots \phi_p + \sum_{v=1}^Q \psi_{pv}\varepsilon_{t-v} \\ \mathbf{I}_{p-1} \qquad\qquad \mathbf{0}_{(p-1)\times 1} \end{pmatrix}$ (sufficient)
Marek (2005)	$Y_t = \varepsilon_t + (a + \beta Y_{t-2})\varepsilon_{t-1}$, $Y_t = (a + \beta\varepsilon_{t-1})\varepsilon_t + \alpha\varepsilon_t$, $(a \neq 0, \alpha \neq 0, \beta > 0)$, $|\varepsilon_t| < 1$	$\beta^2\sigma_\varepsilon^2 < (1-a^2)/2$ $|\alpha| < |a|$ and $\beta < (|a| - |\alpha|)/3$ (sufficient)

[1] The condition reduces to the sufficient condition of Subba Rao (1981) for a BL$(p, 0, p, 1)$ model. In the case $p = Q = 1$ the condition becomes $|\psi| < \exp(-\mathbb{E}\log|\mathbf{Y}_t|)$, earlier obtained by Pham and Tran (1981).

Assuming that $\{Y_t, t \in \mathbb{Z}\}$ is an ergodic strictly stationary process, together with some additional assumptions on $\{\varepsilon_t\}$, it is possible to find sufficient conditions for invertibility for various NLMA- and BL-type models. Tables 3.2 and 3.3 summarize some of the theoretical works for these models. Note, most invertibility conditions are only sufficient and are written in general terms. Indeed, apart from a few simple cases, explicit conditions for the invertibility of nonlinear models are sparse. From Table 3.2 we see that, in contrast with the stationarity of SETAR models, all regimes

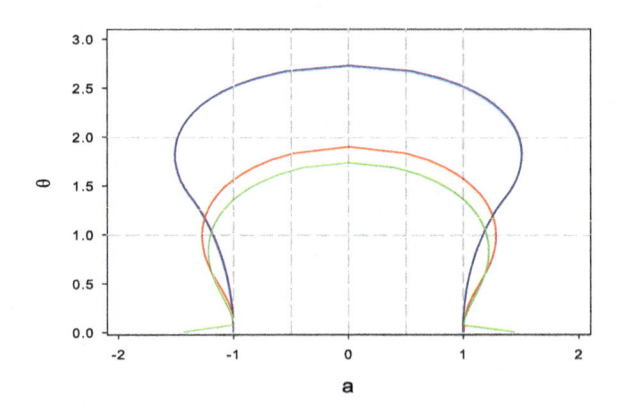

Figure 3.4: *Invertibility regions of the* RCMA(1) *model with* $A_{t,1}$ *following respectively a* $U(a-\theta, a+\theta)$ *distribution (blue solid curve), a* $\mathcal{N}(a, \theta^2)$ *distribution (red solid curve), and a Student* $t_6(a, \theta)$ *distribution (green solid curve).*

play a role to ensure invertibility of the SETMA model. For the SETMA model there is no difficulty in extending the results to the case where the data are generated by a SETARMA model.

Example 3.6: Invertibility of an RCMA(1) Model

Consider the RCMA(1) model of the form

$$Y_t = \varepsilon_t + (a + \theta Y_{t-2})\varepsilon_{t-1}, \quad \{\varepsilon_t\} \stackrel{\text{i.i.d.}}{\sim} (0, \sigma_\varepsilon^2), \qquad (3.33)$$

where a, and $\theta > 0$ are real-valued parameters, $\{Y_t, t \in \mathbb{Z}\}$ is a stationary and ergodic process. Thus, in the general notation of the RCMA model (see Table 3.2), $A_{t,0} = 1$ and $A_{t,1} = a + \theta Y_{t-2}$. Assume that $\{A_{t,1}\} \stackrel{\text{i.i.d.}}{\sim} U(a - \theta, a + \theta)$. Then it is easy to see that

$$\mathbb{E}(\log|A_{t,1}|) = \frac{1}{2\theta}\big[(a+\theta)\log|a+\theta| - (a-\theta)\log|a-\theta| - 2\theta\big]. \qquad (3.34)$$

If $\{A_{t,1}\} \stackrel{\text{i.i.d.}}{\sim} \mathcal{N}(a, \theta^2)$, we have

$$\mathbb{E}(\log|A_{t,1}|) = \log\theta + \int_{-\infty}^{\infty} \frac{1}{\sqrt{2\pi}} \exp\Big\{\frac{-y^2}{2}\Big\} \log\Big|y + \frac{a}{\theta}\Big| dy. \qquad (3.35)$$

Figure 3.4 shows the parameter regions for both sequences $\{A_{t,1}\}$ using the invertibility condition $\mathbb{E}(\log|A_{t,1}|) < 0$. Note that in the case of (3.34) the blue solid curve passes through the point $(a, \theta) = (0, e)$, while in the case of (3.35) the red solid curve goes through the point $(0, 1.8874\cdots)$.

Figure 3.4 also includes the parameter region for invertibility of the RCMA(1) model when $\{A_{t,1}\} \stackrel{\text{i.i.d.}}{\sim} t_6(a, \theta)$ distributed (green solid curve), which is a

3.5 INVERTIBILITY

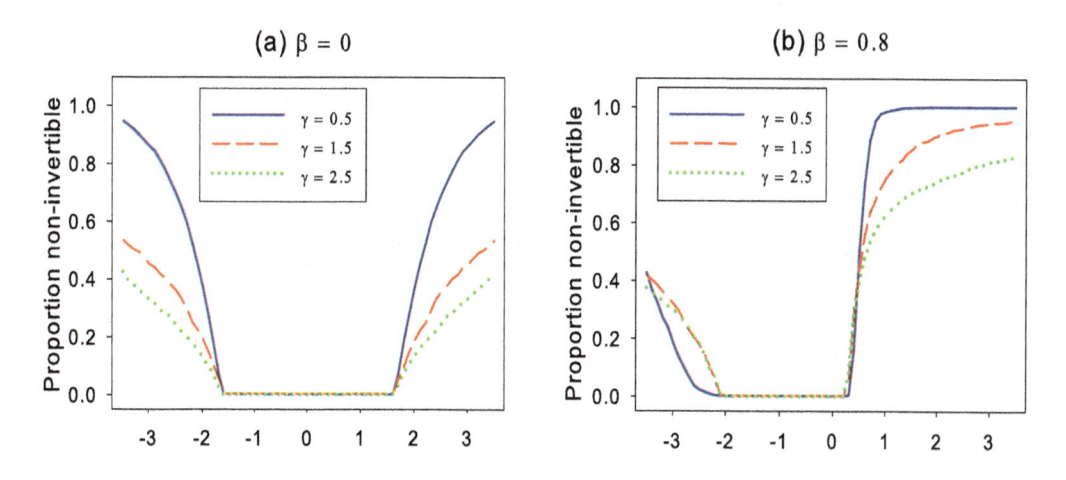

Figure 3.5: *Proportion of* ASTMA(1) *models classified as non-invertible as a function of* ψ *(horizontal axis);* $T = 100$, *1,000 MC replications.*

location-scale transformation of a standard Student t distribution with 6 degrees of freedom. Clearly, this invertibility region is smaller than the ones enclosed by the previous two distributions with a notable part indicating the heavy tails of the t_6 distribution when $\theta \downarrow 0$ and $|a| > 1$.

As a practical and operational alternative to the conditions in Tables 3.2 and 3.3, good sufficient conditions for invertibility can be obtained by MC simulation. Indeed, given definition (3.31), De Gooijer and Brännäs (1995) propose the following ready-to-use method.

Algorithm 3.1: Empirical invertibility of an NLARMA(p,q) model

(i) Generate a random sample of i.i.d. innovations $\{\widetilde{\varepsilon}_t\}_{t=T+1}^{N}$ from the known distribution function (e.g., normal) of the residual series $\{\widehat{\varepsilon}_t\}_{t=1}^{T}$, where N is some large value, say $N = 1,000$.

(ii) Replace ε_t by $\widetilde{\varepsilon}_t$ for $t = T+1, \ldots, N$ and use past values Y_{t-k} ($k = 0, \ldots, p$), and $\widehat{\varepsilon}_{t-k}$ ($k = 0, \ldots, q$), to generate a new set of observations $\{\widetilde{Y}_t\}_{t=T+1}^{N}$.

(iii) Calculate $\{\widehat{e}_t = \widetilde{Y}_t - \widehat{Y}_t\}_{t=T+1}^{N}$, where $\widehat{Y}_t$ are the out-of-sample fitted values. Estimate $\mathbb{E}(e_t^2)$ by $(\tau - T)^{-1} \sum_{t=T+1}^{\tau} \widehat{e}_t^2$. If for all values of $\tau = T+1, \ldots, N$, this sequence does not exceed a pre-fixed value the process $\{Y_t, t \in \mathbb{Z}\}$ is said to be *empirically invertible*, otherwise it suggests non-invertibility.

Example 3.7: Invertibility of an ASTMA(1) Model

Consider an additive smooth transition MA(1), or ASTMA(1), model of the

form
$$Y_t = \varepsilon_t + \beta\varepsilon_{t-1} + \psi F(\varepsilon_{t-1})\varepsilon_{t-1}, \quad \{\varepsilon_t\} \stackrel{\text{i.i.d.}}{\sim} \mathcal{N}(0,1), \qquad (3.36)$$
where $F(\varepsilon_{t-1}) = [1 + \exp(-\gamma\varepsilon_{t-1})]^{-1}$, and $\gamma > 0$. No explicit invertibility conditions have yet been derived for this model.

For $T = 100$, we generated 1,000 time series $\{Y_t\}_{t=1}^{100}$. Dropping the first 150 observations to avoid start-up effects and using Algorithm 3.1 with $N = 1,000$, we computed a sequence of estimates of $\mathbb{E}(e_t^2)$. Next, the process was classified as empirically invertible if for all values $\tau = T+1, \ldots, N$ the values of the sequence did not exceed 10^{-10}.

Figures 3.5(a) and (b) show curves of the proportion of non-invertible models as a function of the parameter ψ for three different values of γ. Note that the empirical invertibility region remains the same as γ increases when $\beta = 0$, while the region reduces when $\beta = 0.8$. For $\gamma = 0.5$ the width of the empirical region is about the same in both figures. For larger values of γ the size of the invertibility region becomes smaller when $\beta = 0.8$. Moreover, the curves show a clear difference in the proportion of non-invertible models for $\psi > 0$ as opposed to $\psi < -2$.

Throughout the previous part, we assumed that (3.25) is an ergodic strictly stationary process. Within a Markov chain framework this requires verifying the irreducibility condition as a part of the Feigin–Tweedie result to establish geometric ergodicity. For general nonlinear MA models this is a non-trivial problem. Interestingly, Li (2012) derives an explicit/closed form of the unique strictly stationary and ergodic solution to the multiple-regime SETMA model without resorting to Markov chain theory. Using a different approach, his work generalizes results of Li, Ling, and Tong (2012) for two-regime SETMA models. The main idea is to re-formulate the model as a SRE and adopt the notion of the top Lyapunov exponent as we discussed in Section 3.1.

Consider a k-regime SETMA model of order q which we write in the form
$$Y_t = a_t^{(k)} + \sum_{i=1}^{k-1}(a_t^{(i)} - a_t^{(k)})I(Y_{t-d} \in \mathbb{R}^{(i)}), \qquad (3.37)$$
where
$$a_t^{(i)} = \psi_0^{(i)} + \varepsilon_t + \sum_{j=1}^{q}\psi_j^{(i)}\varepsilon_{t-j}, \quad (i = 1, \ldots, k).$$

Here, $\{\varepsilon_t\}$ is assumed to be a strictly stationary and ergodic process rather than the usual and more restrictive assumption that $\{\varepsilon_t\}$ is i.i.d. It follows from (3.37) that
$$I(Y_t \in \mathbb{R}^{(i)}) = I(a_t^{(k)} \in \mathbb{R}^{(i)}) + \sum_{j=1}^{k-1}\{I(a_t^{(j)} \in \mathbb{R}^{(i)}) - I(a_t^{(k)} \in \mathbb{R}^{(i)})\}I(Y_{t-d} \in \mathbb{R}^{(j)}),$$
$$(i = 1, \ldots, k-1). \qquad (3.38)$$

3.5 INVERTIBILITY

To represent (3.38) as a SRE, we define

$$\mathbf{I}_t = \big(I(Y_t \in \mathbb{R}^{(1)}), \ldots, I(Y_t \in \mathbb{R}^{(k-1)})\big)', \quad \mathbf{a}_t = \big(I(a_t^{(k)} \in \mathbb{R}^{(1)}), \ldots, I(a_t^{(k)} \in \mathbb{R}^{(k-1)})\big)',$$

and

$$\mathbf{A}_t = (a_{ij,t}) \text{ with } a_{ij,t} = I(a_t^{(j)} \in \mathbb{R}^{(i)}) - I(a_t^{(k)} \in \mathbb{R}^{(i)}) \ (i, j = 1, \ldots, k-1).$$

Then

$$\mathbf{I}_t = \mathbf{A}_t \mathbf{I}_{t-d} + \mathbf{a}_t. \tag{3.39}$$

Observing that $\|\mathbf{A}_t\|$ takes values 0, 1, or 2, we have $\mathbb{E}(\log^+(\|\mathbf{A}_t\|) \leq 2 < \infty$. Moreover, it is easy to see that $\mathbb{P}(\|\mathbf{A}_t\| = 0) > 0$. Thus, the associated top Lyapunov exponent $\gamma(\mathbf{A})$ defined by (3.9) is $-\infty$ since $\mathbb{E}(\log\|\mathbf{A}_t\|) = \sum_{i=0}^{2}(\log i)\mathbb{P}(\|\mathbf{A}_t\| = i) = -\infty$. Then, following similar arguments as in Section 3.1, $\gamma(\mathbf{A}) < 0$ is a sufficient condition for equation (3.39) to have a unique strictly stationary and ergodic solution given by

$$\mathbf{I}_t = \sum_{s=1}^{\infty} \Big(\prod_{i=0}^{s-1} \mathbf{A}_{t-id}\Big) \mathbf{a}_{t-sd}, \quad \text{a.s.}, \tag{3.40}$$

which is of the form (3.5). So, a unique strictly stationary and ergodic solution of $\{Y_t, t \in \mathbb{Z}\}$ is given by

$$Y_t = a_t^{(k)} + (a_t^{(1)} - a_t^{(k)}, \ldots, a_t^{(k-1)} - a_t^{(k)}) \mathbf{I}_{t-d}, \quad \text{a.s.}, \tag{3.41}$$

where $\mathbf{I}_{t-d} = \sum_{s=1}^{\infty}\big(\prod_{i=1}^{s-1} \mathbf{A}_{t-id}\big)\mathbf{a}_{t-sd}$. It is immediate that (3.41) does not require any restriction on the coefficients of the process, which is different from SETAR models.

3.5.2 Local

Within the setting of a nonlinear stochastic difference equation, it is possible (Chan and Tong, 2010) to link local invertibility with the stability (in a suitable sense) of an attractor in a dynamical system. Let $\mathbf{e}_t = (e_t, \ldots, e_{t-q+1})'$ be the vector of reconstruction errors, and $\boldsymbol{\varepsilon}_t = (\varepsilon_t, \ldots, \varepsilon_{t-q+1})'$ ($q > 1$). Then (3.25) can be rewritten as a homogeneous (deterministic) equation associated with the SRE (3.1) in which $\mathbf{B}_t$ is replaced by the zero vector, i.e.

$$\begin{aligned}\mathbf{e}_t &= F(\mathbf{e}_{t-1}, \boldsymbol{\varepsilon}_{t-1}; \boldsymbol{\theta}) \\ &= \big(g(\varepsilon_{t-1}, \ldots, \varepsilon_{t-q}; \boldsymbol{\theta}) - g(e_{t-1} + \varepsilon_{t-1}, \ldots, e_{t-q} + \varepsilon_{t-q}; \boldsymbol{\theta}), e_{t-1}, \ldots, e_{t-q+1}\big)', \end{aligned} \tag{3.42}$$

where $F: \mathbb{R}^q \to \mathbb{R}^q$ is a vector function. Since $\mathbf{0} = F(\mathbf{0}, \boldsymbol{\varepsilon}; \boldsymbol{\theta})$ for all $\boldsymbol{\varepsilon}$ and with $\mathbf{0} \in \mathbb{R}^q$, it is clear that the origin is an equilibrium (limit) point. Then invertibility implies that the origin is an asymptotically globally attractor, in probability. Local invertibility can be established by a linear approximation of $\{\mathbf{e}_t\}$ around $\mathbf{e}_t = \mathbf{0}$, i.e.

$$\mathbf{e}_t = \mathbf{0} + \Big(\prod_{s=1}^{t} \dot{\mathbf{F}}_s\Big)\mathbf{e}_0, \qquad (3.43)$$

where $\dot{\mathbf{F}}_s = \partial F(\mathbf{e}_s, \boldsymbol{\varepsilon}_s; \boldsymbol{\theta})/\partial \mathbf{e}_s$ evaluated at $\mathbf{e}_s = \mathbf{0}$.

Note that (3.43) is the deterministic counterpart of the product of random matrices in the case of the SRE. Stability of (3.43) implies the existence of a suitable Lyapunov exponent $\gamma(\cdot)$. Hence, in analogy with the preceding results, if $\mathbb{E}(\log^+ \|\dot{\mathbf{F}}_1\|) < \infty$, a necessary condition for non-explosiveness (invertibility) is given by

$$\lim_{t \uparrow \infty} \frac{1}{t} \log \Big\| \prod_{s=1}^{t} \dot{\mathbf{F}}_s \Big\| = \gamma(\dot{\mathbf{F}}). \qquad (3.44)$$

When $q = 1$, $\gamma(\dot{\mathbf{F}}) = \mathbb{E}(\log\|\dot{\mathbf{F}}_1\|)$, by the independence of the $\dot{\mathbf{F}}_s$'s. For $q > 1$ a sufficient local invertibility condition can be obtained using the following property of a matrix norm: $\|\prod_s \mathbf{A}_s\| \leq \prod_s \|\mathbf{A}_s\|$ for a sequence of regular matrices $\mathbf{A}_s$ in $\mathbb{R}^{q \times q}$. Then, assuming that $\{\dot{\mathbf{F}}_s\}$ is a function of a stationary and ergodic process, we have

$$t^{-1}\mathbb{E}\Big(\log(\|\prod_{s=1}^{t} \dot{\mathbf{F}}_s\|)\Big) \leq t^{-1} p\, \mathbb{E}(\log(\|\prod_{j=1}^{m} \dot{\mathbf{F}}_j\|)) + t^{-1}\mathbb{E}^r(\log(\|\dot{\mathbf{F}}_1\|)),$$

where $t = mp + r$, and p and r are integers with $0 \leq r < m$.

Thus, $t^{-1}\mathbb{E}(\log(\|\prod_{s=1}^{t} \dot{\mathbf{F}}_s\|)) \to 0$ as $t \uparrow \infty$. So, by the independence of the $\dot{\mathbf{F}}_s$'s, the NLMA(q) model (3.25) is *locally invertible* if $\mathbb{E}(\log(\|\dot{\mathbf{F}}_1\|)) < 0$, and *locally non-invertible* if $\mathbb{E}(\log(\|\dot{\mathbf{F}}_1\|)) > 0$. More generally, these results apply to stationary NLARMA(p,q) processes, for which $F(\cdot)$ is a function of $(\mathbf{e}_t, Y_{t-1}, \ldots, Y_{t-p}, \boldsymbol{\varepsilon}_{t-1}; \boldsymbol{\theta})$. For typical SETARMA models where $h(\cdot)$ is conditionally linear in the innovations given Y_t's, local invertibility analysis is equivalent to global invertibility analysis.

Example 3.8: Invertibility of a SETMA Model

Consider a SETMA($2; q, \ldots, q$) model of the form

$$Y_t = \varepsilon_t + \Big(\sum_{j=1}^{q} \psi_j^{(1)} \varepsilon_{t-j}\Big) I(Y_{t-d} \leq r) + \Big(\sum_{j=1}^{q} \psi_j^{(2)} \varepsilon_{t-j}\Big)\big(1 - I(Y_{t-d} \leq r)\big), \qquad (3.45)$$

where $\{\varepsilon_t\} \stackrel{\text{i.i.d.}}{\sim} (0, \sigma_\varepsilon^2)$. From (3.41), we know that $\{Y_t, t \in \mathbb{Z}\}$ is strictly stationary. The reconstruction errors satisfy the stochastic difference equation $\mathbf{e}_t = \dot{\mathbf{F}}_t \mathbf{e}_{t-1}$, where $\dot{\mathbf{F}}_t$ is a *companion matrix* with its first row equal to

$$\psi_1^{(2)} + (\psi_1^{(1)} - \psi_1^{(2)})I(Y_{t-d} \leq r), \ldots, \psi_q^{(2)} + (\psi_q^{(1)} - \psi_q^{(2)})I(Y_{t-d} \leq r).$$

3.5 INVERTIBILITY

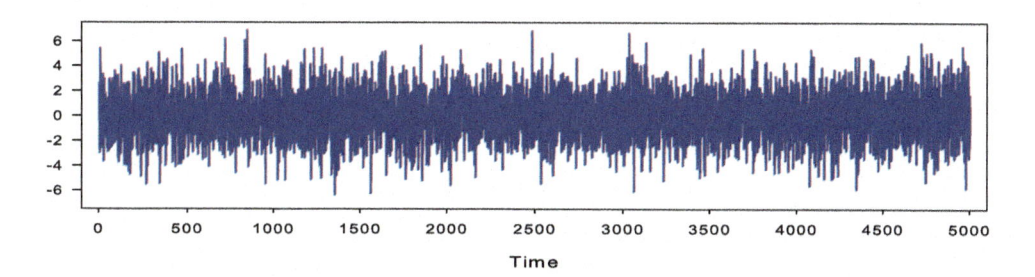

Figure 3.6: *Plot of a strictly stationary and ergodic time series generated by a globally invertible, but locally non-invertible SETMA(2; 2, 2) model; T = 5,000.*

Ling et al. (2007) show that for the SETMA(k; 1, ..., 1) model $Y_t = \{\psi_0 + \sum_{i=1}^{k} \psi_i I(r_{i-1} < Y_{t-1} \leq r_i)\}\varepsilon_{t-1} + \varepsilon_t$ the spectral radius $\rho(\dot{\mathbf{F}})$ is given by

$$\rho(\dot{\mathbf{F}}) = \exp\left(\gamma(\dot{\mathbf{F}})\right) = \prod_{i=1}^{k}\{|\psi_0 + \psi_i|^{F_Y(r_i) - F_Y(r_{i-1})}\},$$

where $0 \leq F_Y(r_i) = \mathbb{P}(Y_t \leq r_i) \equiv p_i \leq 1$. The process $\{Y_t, t \in \mathbb{Z}\}$ is (locally) invertible if $\rho(\dot{\mathbf{F}}) < 1$, and is not invertible if $\rho(\dot{\mathbf{F}}) > 1$. The case $\rho(\dot{\mathbf{F}}) = 1$ is undecided, but Ling et al. (2007) conjectured non-invertibility.

When $q > 1$, a strictly stationary and ergodic SETMA(k; q, ..., q) process is invertible if the spectral radius of each sub-MA(q) processes is less than one (see, e.g., Amendola et al., 2009b). Verifying this condition is rather straightforward. Consider, for instance, the SETMA(2; 2, 2) process

$$\begin{aligned}Y_t =\,& \varepsilon_t + (\psi_1^{(1)}\varepsilon_{t-1} + \psi_2^{(1)}\varepsilon_{t-2})I(Y_{t-1} \leq 0) \\ & + (\psi_1^{(2)}\varepsilon_{t-1} + \psi_2^{(2)}\varepsilon_{t-2})\big(1 - I(Y_{t-1} \leq 0)\big),\end{aligned}$$

where $\psi_1^{(1)} = 1.4$, $\psi_2^{(1)} = -0.7$, $\psi_1^{(2)} = 1.5$, $\psi_2^{(2)} = -0.5$, and $\{\varepsilon_t\} \stackrel{\text{i.i.d.}}{\sim} \mathcal{N}(0,1)$; see Figure 3.6 for a typical realization. The corresponding 2×2 companion matrices $\mathbf{\Psi}^{(i)}$ ($i = 1, 2$) (see Table 3.2) have eigenvalues $\lambda_{1,2}^{(1)} = 0.7 \pm 0.4583i$ and $\lambda_{1,2}^{(2)} = 0.75 \pm 0.25$, respectively. So, the MA process in the first ($Y_{t-1} \leq 0$) regime is invertible. When $Y_{t-1} > 0$, the MA process is not invertible with one root on the unit circle and one root less than one. However, the process $\{Y_t, t \in \mathbb{Z}\}$ is globally invertible even though it is locally non-invertible in the upper regime. Indeed, with $\rho(\mathbf{\Psi}^{(1)}) = |0.7 \pm 0.4583i| = 0.8367$ and $\rho(\mathbf{\Psi}^{(2)}) = |0.75 + 0.25| = 1$, we have

$$\rho(\mathbf{\Psi}^{(1)})^{1-p_1} \times \rho(\mathbf{\Psi}^{(2)})^{p_1} = (0.8367)^{0.4984} \times (1)^{0.5016} < 1,$$

where $\widehat{p}_1 = 0.5016$ is an estimate of $p_1 = \mathbb{E}(Y_{t-1} < 0)$. If the stationary probability p_1 of the lower regime approaches 0, as $r \to \infty$, the SETMA(2; 2, 2) process degenerates to a linear MA(2) process with the well-known invertibility condition $\rho(\mathbf{\Psi}^{(1)}) < 1$.

3.6 Summary, Terms and Concepts

Summary

We reviewed some of the important probabilistic properties of a Markov chain on a general state space. Necessary and sufficient conditions for stationarity and invertibility were also mentioned. The link between stability and ergodicity was investigated for the deterministic skeleton of the SRE. Furthermore, we discussed the use of the associated Lyapunov exponent in inferring stationarity and stability. Conditions for local and global invertibility were achieved. Verifying the invertibility requirement is essential when an NLMA model is used to forecast. Consequently, we provided a practical procedure for this purpose. Unfortunately, explicit/closed form expressions for the stationarity and invertibility of nonlinear models have been found only in a few simple cases.

Terms and Concepts

collapsed Markov chain, 92	mixing coefficients, 95
empirically invertible, 105	non-anticipative, 89
Feller chain, 97	Poisson equation, 93
globally (non-)invertible, 101	reconstruction errors, 101
generalized random coefficient AR, 88	stochastic recurrence equation, 88
geometric ergodic, 96	strong mixing, 95
Harris ergodic, 97	top Lyapunov exponent, 88
locally (non-)invertible, 108	

3.7 Additional Bibliographical Notes

Section 3.1: Most of the properties of a SRE are well known, including conditions for the existence and uniqueness of a stationary solution, or for the existence of moments for a stationary distribution, cf. Pourahmadi (1988). In the context of SREs, Kristensen (2009) gives necessary and sufficient conditions for stationarity of two broad classes of (non)linear GARCH models in terms of $\gamma(\cdot)$. Ispány (1997) does the same for an additive BL state space model.

Akamanam et al. (1986) show the existence of strict stationarity and ergodicity of BL time series models of the form (2.12) with $u \geq v$. Bhattacharaya and Lee (1995) and An and Chen (1997) consider (geometric) ergodicity of a general NLAR model.

Section 3.2: As a special case of the MS-ARMA model (2.67), Holst et al. (1994) give a sufficient condition for the switching AR with Markov regime to be second-order stationary. Francq and Zakoïan (2005) derive necessary and sufficient conditions for existence of moments of any order of GARCH models with Markov regime switching. For these models, the regime switching depends directly on a hidden Markov chain and only indirectly on the current state of the process itself, i.e. the process $\{(\mathbf{A}_t, \mathbf{B}_t), t \in \mathbb{Z}\}$ in (3.1) is no longer i.i.d.

Section 3.3: Goldsheid (1991) provides a CLT which may be used to construct asymptotic confidence bands for estimators of the top Lyapunov exponent, while Gharavi and Anantharan (2005) derive an upper bound for $\gamma(\cdot)$. In a review paper, Lindner (2009) addresses the question of strictly stationary and weakly stationary solutions for pure GARCH processes.

Section 3.4: In the early 80s the most part of the literature consider sufficient, and rarely necessary, conditions for stationarity and ergodicity for nonlinearities in the conditional mean; see, e.g., Chan and Tong (1985), Liu (1989a, 1995), Pham (1986), Pham and Tran (1985), Liu and Brockwell (1988) and the references therein. During the last two decades the focus is mainly on studying conditions for combined models with nonlinearities in both the conditional mean and the conditional variance; see, e.g., Fonseca (2004) and Chen et al. (2011b) for references to the main contributions. More recent developments are by Chen and Chen (2000), Ferrante et al. (2003), Fonseca (2005), Liebscher (2005) and Meitz and Saikkonen (2008, 2010), among others.

Section 3.4.2: Meyn and Tweedie (1993, Appendix B) propose a four-step procedure to classify a SETAR model as being ergodic, transient, and null recurrent. This procedure may also serve as a template for analyzing other nonlinear time series models.

Section 3.5: In the case when (3.25) has time dependent coefficients, Hallin (1980) generalizes the notion of invertibility in (3.31). Using the solution to the SETMA process (3.41), Li (2012) and Li, Ling, and Tong (2012) derive explicit expressions for the moments and ACF of some special TMA models. Amendola et al. (2006a, 2007) give examples of moment and ACF expressions of SETARMA models. Chen and Wang (2011) investigate some probabilistic properties of a combined linear–nonlinear ARMA model with time dependent MA coefficients.

3.8 Data and Software References

Section 3.3: R code (ctarch.eigen.r) for evaluating the Lyapunov exponent γ in the case of SETAR-ARCH models (Example 3.3) is available at the website of this book.

Section 3.5.1: MATLAB code for checking the empirical invertibility (Algorithm 3.1) of a BL model is available at the website of this book. The code can be quite easily modified to assess invertibility of other nonlinear models.

Exercise 3.8: Initially the West German data set was downloaded from datamarket. Description: Monthly unemployment figures in West Germany 1948 – 1980. DataMarket became a part of Qlik® in the year 2014; http://www.qlik.com/us/products/qlik-data-market.

Appendix

3.A Vector and Matrix Norms

Vector norms:
At various places in this book we require some method to measure the size of a vector or a matrix. We refer to these measures collectively as *norms*. Given a vector/linear space V, then a *vector norm*, denoted by $\|\mathbf{x}\|$ is a function $\mathbf{x} \to \|\mathbf{x}\|$ that assigns a nonnegative real number $\|\mathbf{x}\|$ to every vector $\mathbf{x} \in V$ with the following properties.

$$\|\mathbf{x}\| > 0, \ \forall \mathbf{x} \neq \mathbf{0}, \ (\|\mathbf{0}\| = 0) \tag{A.1}$$

$$\|\alpha \mathbf{x}\| = |\alpha| \, \|\mathbf{x}\|, \ \alpha \in \mathbb{R} \tag{A.2}$$

$$\|\mathbf{x} + \mathbf{y}\| \leq \|\mathbf{x}\| + \|\mathbf{y}\|. \tag{A.3}$$

The inequality (A.1) requires the size to be positive, and property (A.2) requires the size to be scaled as the vector $\mathbf{x}$ is scaled. Property (A.3) is known as the *triangle inequality*.

Any mapping of an n-dimensional vector space onto a subset of $\mathbb{R}$ that satisfies (A.1) – (A.3) is a norm. The following are some basic examples of norms.

(i) *The normed linear space:*
Let $\mathbf{x} = (x_1, \ldots, x_n)'$ be a vector in $V \equiv \mathbb{R}^n$ (Euclidean space). Then an obvious definition of a norm is

$$\|\mathbf{x}\|_p = \Big(\sum_{i=1}^n |x_i|^p \Big)^{1/p}, \ p \geq 1. \tag{A.4}$$

The function $\mathbf{x} \to \|\mathbf{x}\|_p$ is known as the L_p-normed linear space. The most common linear spaces are the *one-norm*, L_1, and the *two-norm*, L_2, where $p = 1$ and $p = 2$, respectively.

(ii) *The infinity-norm:*
Let $\mathbf{x} = (x_1, \ldots, x_n)'$ be a vector in $\mathbb{R}^n$. Another standard norm is the *infinity*, or *maximum*, or *supremum*, norm given by the function

$$\|\mathbf{x}\|_\infty = \max_{1 \leq i \leq n} (|x_i|). \tag{A.5}$$

The vector space $\mathbb{R}^n$ equipped with the infinity norm is commonly denoted L_∞.

(iii) *Continuous linear functionals:*
Let $V = C[a,b]$ be the space of all continuous functionals $f(\cdot)$ on the finite interval $[a,b]$. Then a natural norm is

$$\|f\|_p = \Big(\int_a^b |f(x)|^p \mathrm{d}x \Big)^{1/p}, \ p \geq 1, \tag{A.6}$$

with $p = 1$ and $p = 2$ the usual cases, and $\|f\|_\infty = \max_{a \leq x \leq b} |f(x)|$.

Matrix norms:
Suppose $\{\mathbb{R}^n, \|\mathbf{x}\|_p\}$ is a normed linear space with $\|\mathbf{x}\|_p$ some norm. Let $\mathbf{A} = (a_{ij})_{m \times n}$ be a real matrix. Then the norm of $\mathbf{A}$, subordinate to the vector norm $\|\mathbf{x}\|_p$, is defined as

$$\|\mathbf{A}\|_p = \sup_{\mathbf{x} \neq \mathbf{0}} \frac{\|\mathbf{A}\mathbf{x}\|_p}{\|\mathbf{x}\|_p} = \sup_{\|\mathbf{x}\|_p = 1} \|\mathbf{A}\mathbf{x}\|_p, \quad \mathbf{x} \in \mathbb{R}^n, \ \mathbf{A}\mathbf{x} \in \mathbb{R}^m. \tag{A.7}$$

APPENDIX 3.A

So, $\|\mathbf{A}\|_p$ is the largest value of the vector norm of $\mathbf{A}\mathbf{x}$ in the space $V = \mathbb{R}^n$ normalized over all non-zero vectors $\mathbf{x}$. In particular,

$$\|\mathbf{A}\|_1 = \max_j \sum_i |a_{ij}|, \quad \|\mathbf{A}\|_2 = \left(\text{maximum eigenvalue of } (\mathbf{A}'\mathbf{A})\right)^{1/2}.$$

The norm $\|\mathbf{A}\|_2$ is often called the *spectral norm*. When $p = 1$ and 2, the matrix norm satisfies the following four properties:

$$\begin{array}{lll} \text{Positivity:} & \infty > \|\mathbf{A}\|_p > 0,\ \forall \mathbf{A} \neq \mathbf{0},\ \text{except } \|\mathbf{0}\|_p = 0, & \text{(A.8)} \\ \text{Homogeneity:} & \|\alpha \mathbf{A}\|_p = |\alpha|\, \|\mathbf{A}\|_p,\ \alpha \in \mathbb{R}, & \text{(A.9)} \\ \text{Triangle inequality:} & \|\mathbf{A} + \mathbf{B}\|_p \leq \|\mathbf{A}\|_p + \|\mathbf{B}\|_p, & \text{(A.10)} \\ \text{Compatibility:} & \|\mathbf{A}\mathbf{x}\|_p \leq \|\mathbf{A}\|_p\, \|\mathbf{x}\|_p. & \text{(A.11)} \end{array}$$

Here, (A.8) – (A.10) are generalizations of the three properties (A.1) – (A.3). Property (A.11) is a direct consequence of the definition (A.4). A special case of (A.11) is

$$\|\mathbf{A}\mathbf{B}\|_p \leq \|\mathbf{A}\|_p\, \|\mathbf{B}\|_p, \tag{A.12}$$

which is a simple but often useful property. Another special case of (A.11) is

$$|a_{ij}| \leq \|\mathbf{A}\|_p, \quad \forall i, j. \tag{A.13}$$

An important use of matrix norms is in proving convergence of powers of matrices. Suppose $\mathbf{A}_1, \mathbf{A}_2, \ldots$ is a sequence of square matrices. Then,

$$\lim_{i \to \infty} \|\mathbf{A}_i\|_p = 0 \iff \lim_{i \to \infty} \mathbf{A}_i \to \mathbf{0}, \tag{A.14}$$

where $\mathbf{0}$ is a square matrix consisting of zeros. Now, suppose $\mathbf{A}_i$ is given as a product of another sequence of matrices $\mathbf{B}_1, \mathbf{B}_2, \ldots$, so that $\mathbf{A}_i = \prod_{j=1}^{i} \mathbf{B}_j$. In that case the desired conclusion of (A.14) will follow if there exists a ρ such that for all j, $\|\mathbf{B}_j\| < \rho < 1$. However, within the context of formulating conditions for (multivariate) stationarity and invertibility, we will encounter the case where the $\mathbf{B}_j$ are block matrices. In particular, for $n \times n$ matrices $\mathbf{C}_{u,j}$ ($u = 1, \ldots, p$) and $\mathbf{D}_v$ ($v = 1, \ldots, p-1$), we will see the block structure

$$\mathbf{B}_j = \begin{pmatrix} \mathbf{C}_{1,j} & \mathbf{C}_{2,j} & \cdots & \mathbf{C}_{p-1,j} & \mathbf{C}_{p,j} \\ \mathbf{D}_1 & \mathbf{0}_{n \times n} & \cdots & \mathbf{0}_{n \times n} & \mathbf{0}_{n \times n} \\ \mathbf{0}_{n \times n} & \mathbf{D}_2 & & & \vdots \\ \vdots & & \ddots & & \vdots \\ \mathbf{0}_{n \times n} & \cdots & \cdots & \mathbf{D}_{p-1} & \mathbf{0}_{n \times n} \end{pmatrix}.$$

If some or all of the matrices $\mathbf{D}_v = \mathbf{I}_n$, as with the so-called *companion matrix*, then by (A.13), $\|\mathbf{B}_j\| \geq 1$. So, the condition leading to (A.13) is not fulfilled. One can get around this problem by multiplying together sufficiently many $\mathbf{B}_j$'s before taking the norm.

3.B Spectral Radius of a Matrix

A quantity associated with matrices is the *spectral radius* of a matrix. A square matrix $\mathbf{A} = (a_{ij})_{n \times n}$ has n eigenvalues λ_i ($i = 1, \ldots, n$). The spectral radius of $\mathbf{A}$, which we denote by $\rho(\mathbf{A})$, is defined as

$$\rho(\mathbf{A}) = \max_{1 \leq i \leq n} (|\lambda_i|). \tag{B.1}$$

Note that $\rho(\mathbf{A}) \geq 0$ for all $\mathbf{A} \neq \mathbf{0}$. Furthermore,

$$\rho(\mathbf{A}) \leq \|\mathbf{A}\|_p, \tag{B.2}$$

for all subordinate matrix norms. This property can be easily proved. Note that $\rho(\mathbf{A})$ is *not* a norm since it can be shown that $\rho(\mathbf{A} + \mathbf{B}) \not\leq \rho(\mathbf{A}) + \rho(\mathbf{B})$.

The following properties are often useful. For any positive integer m, and a constant $c > 0$, we have

$$|(\mathbf{A}^m)_{ij}| \leq c\big(\rho(\mathbf{A})\big)^m, \quad \forall i, j \tag{B.3}$$

$$\rho(\mathbf{A}) \leq \max_{1 \leq i \leq n} \sum_{j=1}^{n} |a_{ij}| \leq n \max_{1 \leq i,j \leq n} |a_{ij}|, \tag{B.4}$$

$$\rho(\mathbf{A} \otimes \mathbf{A}) < 1 \quad \text{if and only if} \quad \rho(\mathbf{A}) < 1. \tag{B.5}$$

Also, it is easy to prove that

$$\|\mathbf{A}\|_2^2 = \rho(\mathbf{A}'\mathbf{A}), \tag{B.6}$$

i.e. the maximum eigenvalue of the symmetric matrix $\mathbf{A}'\mathbf{A}$.

In Chapter 11, we mention briefly the concept of *joint spectral radius* which is a generalization of the notion of spectral radius of a matrix, to sets of matrices. Consider a set of bounded square matrices $\mathcal{A} \subset \mathbb{R}^{n \times n}$. The joint spectral radius is defined by

$$\rho(\mathcal{A}) = \limsup_{p \to \infty} \Big(\sup_{A \in \mathcal{A}^{(p)}} \|\mathcal{A}\| \Big)^{1/p}, \tag{B.7}$$

where $\mathcal{A}^{(p)} = \{\mathbf{A}_1 \mathbf{A}_2 \cdots \mathbf{A}_p : \mathbf{A}_i \in \mathcal{A}, i = 1, \ldots, p\}$ and $\|\cdot\|$ can be any matrix norm; see, e.g., Liebscher (2005) for more results about the joint spectral radius.

Exercises

Theory Questions

3.1 Consider an EXPAR(1) model of the form

$$Y_t = \{\phi + \xi \exp(-\gamma Y_{t-1}^2)\} Y_{t-1} + \varepsilon_t, \quad (|\phi| < 1 <, \gamma > 0),$$

where $\{\varepsilon_t\}$ are i.i.d. random variables, each having a strictly positive and continuous density $f(x) = (1/2) \exp(-|x|)$. Prove that $\{Y_t, t \in \mathbb{Z}\}$ is geometrically ergodic and $\mathbb{E}|Y_t^m| < \infty \ \forall m \in \mathbb{Z}^+$.

3.2 Consider the k-regime asymmetric MA(1) model

$$Y_t = \varepsilon_t + \psi(\varepsilon_{t-1})\,\varepsilon_{t-1},$$

where $\psi(\varepsilon) = \sum_{i=1}^k \beta^{(i)} F_{\mathbb{R}^{(i)}}(\varepsilon)$ with $F_{\mathbb{R}^{(i)}}(\cdot)$ the characteristic function of set $\mathbb{R}^{(i)}$ ($i = 1, \ldots, k$). Assume $|\beta^{(i)}| \leq \gamma < 1$ and $\mathbb{E}|\varepsilon_t|^m \leq c < \infty$ ($m \in \mathbb{Z}^+$), where γ and c are real positive constants. Furthermore, assume that the residual $\widehat{\varepsilon}_0 = 0$.

Show that the process $\{Y_t, t \in \mathbb{Z}\}$ is invertible in the sense that $\limsup_{t \to \infty} \mathbb{E}|e_t|^m \leq c^*$, where $\{e_t\}$ are the reconstruction errors, and $c^* < \infty$ is some constant.

3.3 Consider the quadratic MA(1) model

$$Y_t = \varepsilon_t - \beta \varepsilon_{t-1}^2, \quad \{\varepsilon_t\} \stackrel{\text{i.i.d.}}{\sim} \mathcal{N}(0,1),$$

where $\beta \neq 0$. Granger and Andersen (1978a, p. 28) claim that this model is never invertible with respect to the non-zero value of the parameter β.

(a) Show that under the condition $|\beta| < (C + \log 2)/4$ the model is locally invertible where C is Euler's constant.

(b) Consider Algorithm 3.1 with $N = 1{,}000$. Set $T = 50$ and $T = 100$. Then, using 1,000 MC replications, show that the model is empirically invertible for $|\beta|$ values smaller than approximately 0.85.

3.4 Consider the first-order BL(1,0,1,1) model

$$Y_t = \phi Y_{t-1} + \psi Y_{t-1}\varepsilon_{t-1} + \varepsilon_t, \quad \{\varepsilon_t\} \stackrel{\text{i.i.d.}}{\sim} \mathcal{N}(0, \sigma_\varepsilon^2). \tag{3.46}$$

Using the above model, Terdik (1999, p. 207) obtains the following estimation results for the magnetic field data (Example 1.3):

$$Y_t = 0.5421 Y_{t-1} + 0.0541 Y_{t-1}\widehat{\varepsilon}_{t-1} + \widehat{\varepsilon}_t, \quad \widehat{\sigma}_\varepsilon^2 = 0.2765. \tag{3.47}$$

(a) Verify that the fitted BL model is a weakly (second-order) stationary process, assuming it is first-order stationary.

(b) Show that (3.46) is invertible if ϕ and ψ satisfy the condition

$$2(1+\phi)\lambda^4 + 2(1-\phi)\lambda^2 - (1-\phi)^2(1+\phi) < 0, \quad \lambda = \psi \sigma_\varepsilon. \tag{3.48}$$

(c) Using (3.48), verify that the fitted model is invertible.

3.5 Consider the BL model

$$Y_t = \phi_0 + \sum_{i=1}^{p} \phi_i Y_{t-i} + \sum_{j=1}^{q} \theta_j \varepsilon_{t-j} + \sum_{i=1}^{Q}\sum_{j=0}^{P} \psi_{ij} Y_{t-i-j}\varepsilon_{t-i}, \quad \{\varepsilon_t\} \stackrel{\text{i.i.d.}}{\sim} (0, \sigma_\varepsilon^2).$$

Show that the model can be represented as

$$Y_t = Z_{1,t-1} + \theta_0 \varepsilon_t,$$

where the process $\mathbf{Z}_t = (Z_{1,t}, \ldots, Z_{n,t})' \in \mathbb{R}^n$, with $n = \max(p, P+q, P+Q)$, solves the SRE representation $\mathbf{Z}_t = \mathbf{A}_t \mathbf{Z}_{t-1} + \mathbf{B}_t$ and where the $\mathbf{A}_t \in \mathbb{R}^{n \times n}$ and $\mathbf{B}_t \in \mathbb{R}^n$ is a random matrix, and a random vector of polynomials in $\{\varepsilon_t\}$ of degree 1 and 2 respectively.

(Kristensen, 2009)

Empirical and Simulation Questions

3.6 Consider the asMA(1) model

$$Y_t = \varepsilon_t + \beta^+ \varepsilon_{t-1}^+ + \beta^- \varepsilon_{t-1}^-, \quad \{\varepsilon_t\} \stackrel{\text{i.i.d.}}{\sim} \mathcal{N}(0, \sigma_\varepsilon^2),$$

where $\varepsilon_t^+ = I(\varepsilon_t \geq 0)\varepsilon_t$ and $\varepsilon_t^- = I(\varepsilon_t < 0)\varepsilon_t$.

(a) Using Algorithm 3.1 with $N = 1{,}000$, obtain a graphical representation of the empirical invertibility region for a simulated time series of size $T = 100$, using 1,000 MC replications.

(b) Wecker (1981) derives the following sufficient invertibility conditions: $|\beta^+| < 1$ and $|\beta^-| < 1$. Compare and contrast the resulting invertibility region with the one obtained in part (a). Suggest a necessary and sufficient condition for invertibility.

3.7 (a) Consider the asMA(1) model in Exercise 3.6. Rewrite the model in the form

$$Y_t = \beta(t-1)Y_{t-1} - \beta(t-1)\beta(t-2)Y_{t-2} + \cdots - \beta(t-1)\cdots\beta(1)Y_1 + \varepsilon_t,$$

where $\beta(T-1) \cdots \beta(1) = (\beta^+)^j (\beta^-)^{(T-1-j)}$ $(j = 0, \ldots, T-1)$.

(b) Using the specification in part (a), suggest an alternative notion of invertibility for the asMA(1) model. Give a graphical representation of the resulting invertibility region.

(c) Now, rewrite the asMA(1) model as follows:

$$Y_t = \varepsilon_t + \beta(\varepsilon_{t-1}),$$

where $\beta(\varepsilon_{t-1}) = \sum_{i=1}^{2} \beta_i I(\varepsilon_{t-1} \in S_i)\varepsilon_{t-1}$ with $\beta_1 = \beta^+$, $\beta_2 = \beta^-$, $S_1 = [0, \infty)$ and $S_2 = (-\infty, 0)$. Verify the invertibility condition $\mathbb{E}|e_t| \to 0$ as $t \to \infty$. Show that the corresponding invertibility region is given by

$$|\beta_1| < 1, \quad |\beta_2| < 1, \quad \text{and} \quad |\beta_1| + |\beta_2| < 1.$$

EXERCISES

3.8 Subba Rao and Gabr (1984, pp. 211 – 212) consider the monthly West German unemployment data (X_t) for the time period January 1948 – May 1980 (389 observations). They use the first 365 observations of the series $Y_t = (1 - B)(1 - B^{12})X_t$ for fitting a subset BL model, and the last 24 observations for out-of-sample forecasting. It is therefore vital that the fitted model is invertible. The best fitted subset BL model is given by

$$\begin{aligned}
Y_t &- 0.0874 Y_{t-1} + 0.1261 Y_{t-2} - 0.0426 Y_{t-9} - 0.2556 Y_{t-11} + 0.5067 Y_{t-12} \\
&= -4598.325 - 0.1315 \times 10^{-4} Y_{t-1}\widehat{\varepsilon}_{t-10} - 0.1279 \times 10^{-5} Y_{t-2}\widehat{\varepsilon}_{t-5} \\
&\quad - 0.3790 \times 10^{-6} Y_{t-5}\widehat{\varepsilon}_{t-4} + 0.1902 \times 10^{-5} Y_{t-11}\widehat{\varepsilon}_{t-7} \\
&\quad + 0.1513 \times 10^{-5} Y_{t-12}\widehat{\varepsilon}_{t-4} - 0.2267 \times 10^{-5} Y_{t-12}\widehat{\varepsilon}_{t-2} \\
&\quad - 0.9507 \times 10^{-6} Y_{t-4}\widehat{\varepsilon}_{t-10} - 0.1948 \times 10^{-5} Y_{t-10}\widehat{\varepsilon}_{t-8} \\
&\quad + 0.2715 \times 10^{-5} Y_{t-1}\widehat{\varepsilon}_{t-9}, \quad \widehat{\sigma}_\varepsilon^2 = 0.36665 \times 10^{10}.
\end{aligned}$$

Assuming the above model is correctly specified, check the empirical invertibility of the fitted BL model using Algorithm 3.1 with $N = 1{,}000$. The complete (undifferenced) data set (German_unemplmnt.dat) is available at the website of this book.

Chapter 4

FREQUENCY-DOMAIN TESTS

The specification and estimation of a nonlinear model may be difficult in practice and sometimes no substantial improvements in forecasting accuracy can be achieved by using a nonlinear model instead of a familiar ARMA model. Therefore, one may wish to start the model building from a linear model and abandon it only if sufficiently strong evidence for a nonlinear alternative can be found. This approach can be applied using a linearity test, often in combination with a test for Gaussianity. Several test statistics, both in the time domain and frequency domain, have been proposed for this purpose.

In this chapter, we will restrict attention to frequency-domain linearity and Gaussianity test statistics. These tests are nonparametric, or model-free, having an alternative hypothesis that only states that the DGP is nonlinear, and not specifying the type of nonlinearity. Within the frequency domain the simplest higher-order spectrum is the second-order spectrum, or bispectrum. Based on the asymptotic properties of the estimated normalized bispectrum, we introduce various test statistics. Most tests follow a two-stage approach. The first stage tests if a time series process has a zero third-order cumulant function, but is often interpreted as a test of white noise. If a process is WN then the second-order covariances and second-order spectra will contain all the useful information. In that case all its higher-order moments, or higher-order spectra, are identically zero. If on the other hand the null hypothesis of zero third-order cumulant function is rejected in stage one, then the second stage is to test for linearity.

The outline of the rest of this chapter is as follows. In Section 4.1 we define the normalized bispectrum and indicate how it motivates tests of Gaussianity and linearity. Next, in Sections 4.2 and 4.3, we introduce two "classical" methods, the Subba Rao and Gabr (1980) and the Hinich (1982) test statistics, and discuss their major shortcomings. In fact, the Hinich and the Subba Rao–Gabr tests for Gaussianity and linearity are only useful when large amounts of data are available, and rely on the asymptotic normality of the estimator of the bispectrum which may be a poor approximation for small sample sizes. Between the two, Hinich's test statistics have long been preferred in applications. However, these test statistics tend to have low power and require the specification of a smoothing or window-width parameter. Consequently, various improvements and modifications of the Hinich bispectral test statistics have been proposed; see Section 4.4 for a brief overview. First, in Section 4.4.1, we apply goodness-of-fit techniques to the asymptotic properties of the estimated bispectrum, resulting in new test statistics with increased power. In the following subsection, we describe a method to eliminate the arbitrariness concerning the selection of the smoothing parameter. In Section 4.4.3, we discuss another improvement based on a bootstrap algorithm, which approximates the finite-sample null distribution of Hinich's test statistics.

As we saw in Section 1.1, the differences between linear and nonlinear DGPs can also be defined in terms of mean squared forecast errors (MSFEs). In Section 4.5, we discuss a frequency domain linearity test statistic based on an additivity property of the bispectrum of the innovation process of a stationary linear Gaussian process. The bispectrum is used to check if the best predictor of an observed time series is linear, and the series is deemed to be linear if this null hypothesis is not rejected against the alternative hypothesis that the best forecast is quadratic. Section 4.6 contains a summary of numerical studies related to the size and power of most of the test statistics discussed in this chapter. Finally, in Section 4.7, we apply a number of test statistics to the six time series introduced in Chapter 1.

4.1 Bispectrum

Apart from Section 4.5, throughout this chapter we assume that $\{Y_t\}_{t=1}^{T}$ is a time series arising from a real-valued third-order strictly stationary stochastic process $\{Y_t, t \in \mathbb{Z}\}$ that – for ease of notation – is assumed to have mean zero. One basic tool for quantifying the inherent strength of dependence is the ACVF given by $\gamma_Y(\ell) = \mathbb{E}(Y_t Y_{t+\ell})$ ($\ell \in \mathbb{Z}$). For testing nonlinearity and non-Gaussianity, another useful function is the *third-order cumulant*, defined as $\gamma_Y(\ell_1, \ell_2) = \mathbb{E}(Y_t Y_{t+\ell_1} Y_{t+\ell_2})$, $(\ell_1, \ell_2 \in \mathbb{Z})$. Both functions are time invariant and unaffected by permutations in their arguments, which creates the symmetries

$$\gamma_Y(\ell) = \gamma_Y(-\ell), \tag{4.1}$$

$$\gamma_Y(\ell_1, \ell_2) = \gamma_Y(\ell_2, \ell_1) = \gamma_Y(-\ell_1, \ell_2 - \ell_1) = \gamma_Y(\ell_1 - \ell_2, -\ell_2). \tag{4.2}$$

The *spectral density* function, or *spectrum*, of $\{Y_t, t \in \mathbb{Z}\}$ is formally defined as the discrete-time *Fourier transform* (FT) of the ACVF, i.e.,

$$f_Y(\omega) = \sum_{\ell=-\infty}^{\infty} \gamma_Y(\ell) \exp(-2\pi i \omega \ell), \quad \omega \in [0, 1], \tag{4.3}$$

where ω denotes the frequency. A sufficient, but not necessary, condition for the existence of the spectrum is that $\sum_{\ell=-\infty}^{\infty} |\gamma_Y(\ell)| < \infty$.

4.1 BISPECTRUM

If, in addition, $\sum_{\ell_1,\ell_2=-\infty}^{\infty} |\gamma_Y(\ell_1, \ell_2)| < \infty$, then the *bispectral density* function, or *bispectrum*, exists and is defined as the bivariate, or double, FT of the third-order cumulant function,

$$f_Y(\omega_1, \omega_2) = \sum_{\ell_1,\ell_2=-\infty}^{\infty} \gamma_Y(\ell_1, \ell_2) \exp\{-2\pi i(\omega_1 \ell_1 + \omega_2 \ell_2)\}, \quad (\omega_1, \omega_2) \in [0,1]^2. \tag{4.4}$$

Note that in a similar fashion higher-order spectral functions can be defined whose corresponding multi-dimensional FTs are termed *polyspectra*. The spectrum is real-valued and nonnegative. In contrast, the bispectrum and higher-order spectra are complex-valued.

In view of (4.1) – (4.4), we have the relations,

$$f_Y(\omega) = f_Y(-\omega), \tag{4.5}$$
$$f_Y(\omega_1, \omega_2) = f_Y(\omega_2, \omega_1) = f_Y(\omega_1, -\omega_1 - \omega_2) = f_Y(-\omega_1 - \omega_2, \omega_2). \tag{4.6}$$

The third-order cumulant and the bispectrum are mathematically equivalent, as are the spectrum and the ACVF. Clearly $f_Y(\omega)$ is symmetric about 0.5. From (4.4), and due to the periodicity of the FT (4.3), the bispectrum in the entire plane can be determined from the values inside one of the twelve sectors shown in Figure 4.1. Therefore, it is sufficient to consider only frequencies in the first triangular region (cf. Exercise 4.1), which we define as the *principal domain*

$$\mathcal{D} = \{(\omega_1, \omega_2) : \omega_1 = \omega_2, \ \omega_1 = 0, \ \omega_1 = (1-\omega_2)/2\}; \tag{4.7}$$

recall that we have assumed a normalized sampling frequency of 1 Hz.

If $\{X_t, t \in \mathbb{Z}\}$ and $\{Y_t, t \in \mathbb{Z}\}$ are two statistically independent processes and $Z_t = X_t + Y_t$, then $\gamma_Z(\ell_1, \ell_2) = \gamma_X(\ell_1, \ell_2) + \gamma_Y(\ell_1, \ell_2)$, and hence $f_Z(\omega_1, \omega_2) = f_X(\omega_1, \omega_2) + f_Y(\omega_1, \omega_2)$. If $\{X_t, t \in \mathbb{Z}\}$ is Gaussian and i.i.d., then $\gamma_X(\ell_1, \ell_2) = 0$, $\forall (\ell_1, \ell_2)$, and $f_X(\omega_1, \omega_2) = 0$, $\forall (\omega_1, \omega_2)$, so $f_Z(\omega_1, \omega_2) = f_Y(\omega_1, \omega_2)$, in other words symmetric noise is suppressed in the bispectrum.

Another useful property of the bispectrum is that its imaginary part (denoted by $\Im(\cdot)$), should be zero for a time-reversible process. In that case, the third-order cumulant function of $\{Y_t, t \in \mathbb{Z}\}$ has the additional symmetry property that $\gamma_Y(\ell_1, \ell_2) = \gamma_Y(-\ell_1, -\ell_2)$, and hence

$$\Im\{f_Y(\omega_1, \omega_2)\} = \sum_{\ell_1,\ell_2=-\infty}^{\infty} \gamma_Y(\ell_1 \ell_2,) \sin 2\pi(\omega_1 \ell_1 + \omega_2 \ell_2)$$
$$= \sum_{\ell_1,\ell_2=0}^{\infty} \gamma_Y(\ell_1, \ell_2)\{\sin 2\pi(\omega_1 \ell_1 + \omega_2 \ell_2) + \sin 2\pi(-\omega_1 \ell_1 - \omega_2 \ell_2)\}$$
$$= 0 \tag{4.8}$$

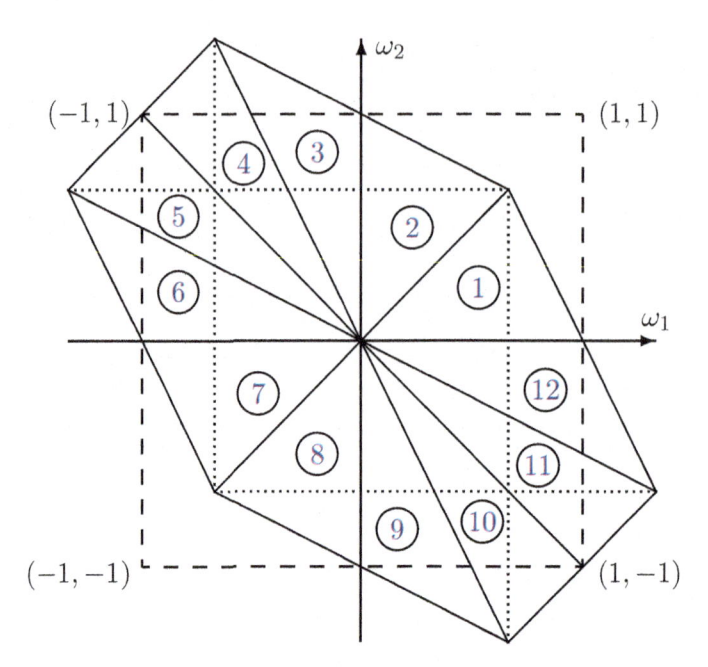

Figure 4.1: *Values of $f_Y(\omega_1, \omega_2)$ defined over the entire plane, as completely specified by the values over any one of the twelve labeled sectors.*

using the identity $\sin A + \sin B = 2 \sin \frac{A+B}{2} \cos \frac{A-B}{2}$.

For reasons to be apparent soon, a convenient normalization for the bispectrum is obtained by simply dividing the modulus of $f_Y(\omega_1, \omega_2)$ by the appropriate spectra, giving the *normalized bispectrum*, defined by

$$B_Y(\omega_1, \omega_2) = \frac{f_Y(\omega_1, \omega_2)}{\sqrt{f_Y(\omega_1) f_Y(\omega_2) f_Y(\omega_1 + \omega_2)}}, \quad (\omega_1, \omega_2) \in \mathcal{D}. \tag{4.9}$$

The third-order cumulant of the general linear causal process (1.2) is given by

$$\gamma_Y(\ell_1, \ell_2) = \mathbb{E}\Big(\sum_{j=0}^{\infty} \sum_{j'=0}^{\infty} \sum_{j''=0}^{\infty} \psi_j \varepsilon_{t-j} \psi_{j'} \varepsilon_{t+\ell_1-j'} \psi_{j''} \varepsilon_{t+\ell_2-j''} \Big)$$

$$= \mathbb{E}(\varepsilon_t^3) \sum_{j=0}^{\infty} \psi_j \psi_{j+\ell_1} \psi_{j+\ell_2}.$$

Hence, the bispectrum becomes

$$f_Y(\omega_1, \omega_2) = \mathbb{E}(\varepsilon_t^3) \sum_{\ell_1, \ell_2 = -\infty}^{\infty} \sum_{\ell=0}^{\infty} \psi_\ell \psi_{\ell+\ell_1} \psi_{\ell+\ell_2} \exp\{-2\pi i(\omega_1 \ell_1 + \omega_2 \ell_2)\}$$

$$= \mathbb{E}(\varepsilon_t^3) \sum_{\ell_1,\ell_2=-\infty}^{\infty} \sum_{\ell=0}^{\infty} \psi_{\ell_1}\psi_{\ell_2}\psi_\ell \exp\{-2\pi i\big(\omega_1(\ell_1-\ell)+\omega_2(\ell_2-\ell)\big)\}$$

$$= \mathbb{E}(\varepsilon_t^3) \sum_{\ell_1=0}^{\infty} \psi_{\ell_1} \exp\{-2\pi i \omega_1 \ell_1\} \sum_{\ell_2=0}^{\infty} \psi_{\ell_2} \exp\{-2\pi i \omega_2 \ell_2\}$$

$$\times \sum_{\ell=0}^{\infty} \psi_\ell \exp\{2\pi i (\omega_1+\omega_2)\ell\}$$

$$= \mathbb{E}(\varepsilon_t^3) H(\omega_1) H(\omega_2) H^*(\omega_1+\omega_2), \qquad (4.10)$$

where $H(\omega) = \sum_{j=0}^{\infty} \psi_j \exp(-2\pi i \omega_j)$ is known as the *transfer function*, and $H^*(\omega) = H(-\omega)$ its complex conjugate. Furthermore, it is well known that if $\{Y_t, t \in \mathbb{Z}\}$ is linear, then the spectral density function in (4.3) reduces to

$$f_Y(\omega) = \sigma_\varepsilon^2 |H(\omega)|^2. \qquad (4.11)$$

Combining (4.10) and (4.11), the square modulus of the normalized bispectrum, called *frequency bicoherence*, is simply

$$|B_Y(\omega_1,\omega_2)|^2 = \frac{\{\mathbb{E}(\varepsilon_t^3)\}^2}{\sigma_\varepsilon^6} \equiv \frac{\mu_{3,\varepsilon}^2}{\sigma_\varepsilon^6}, \quad (\omega_1,\omega_2) \in \mathcal{D}, \qquad (4.12)$$

where $\mu_{3,\varepsilon} = \mathbb{E}(\varepsilon_t^3)$. This fundamental property is the basis of frequency-domain tests for Gaussianity and linearity which we detail in the next sections.

Note that the right-hand side of (4.12) is the squared skewness of the process $\{\varepsilon_t, t \in \mathbb{Z}\}$. If $\{Y_t, t \in \mathbb{Z}\}$ is linear, and the distribution of $\{\varepsilon_t\}$ is symmetric, then $\mu_{3,\varepsilon} = 0$ and so $|B_Y(\omega_1,\omega_2)|^2 \equiv 0$, $\forall (\omega_1,\omega_2) \in \mathcal{D}$. However, this is also true for linear Gaussian time series processes. Thus the skewness function is a constant if $\{Y_t, t \in \mathbb{Z}\}$ is linear and that constant is zero if $\{Y_t, t \in \mathbb{Z}\}$ is Gaussian. Consequently, the null hypotheses of interest are, respectively,

$$\mathbb{H}_0^{(1)}: \quad f_Y(\omega_1,\omega_2) = 0, \quad \forall (\omega_1,\omega_2) \in \mathcal{D}; \quad \text{and} \qquad (4.13)$$

$$\mathbb{H}_0^{(2)}: \quad |B_Y(\omega_1,\omega_2)|^2 = \text{constant}, \quad \forall (\omega_1,\omega_2) \in \mathcal{D}. \qquad (4.14)$$

Given actual data of size T, consistent estimates of the spectrum and bispectrum can be obtained through various techniques. Broadly these techniques can be classified into three categories: nonparametric or conventional methods, parametric or model-based methods (e.g. AR modeling), and criterion-based methods (e.g. Burg's (1967) maximum entropy algorithm). The first category includes two classes: the *direct method* which is based on computing the third-order extension of the sample periodogram, known as the *third-order periodogram*, and the *indirect method*, which is the extension of the FT of the sample ACVF to the third-order cumulant. Both methods are easy to understand and easy to implement, but are limited by their resolving power when T is small, i.e., the ability to separate two closely spaced harmonics. Nevertheless, conventional methods dominate the literature.

The (sample) periodogram, as a natural estimator of the spectrum, is defined as the discrete FT of the sample ACVF, i.e.

$$I_T(\omega) = \sum_{\ell=-(T-1)}^{T-1} \widehat{\gamma}_Y(\ell) \exp\{-2\pi i \omega \ell\}, \quad \omega \in [0, \tfrac{1}{2}], \tag{4.15}$$

where $\widehat{\gamma}_Y(\ell) = T^{-1} \sum_{t=1}^{T-\ell} Y_t Y_{t+\ell}$. The periodogram, however, is not a consistent estimator of $f_Y(\omega)$. Similarly, the third-order periodogram, is an inconsistent estimator of $f_Y(\omega_1, \omega_2)$. Consistent estimators of $f_Y(\omega)$ and $f_Y(\omega_1, \omega_2)$ are obtained by "smoothing" the periodogram and third-order periodogram, and the resulting estimators are defined as

$$\widehat{f}_Y(\omega) = \sum_{\ell=-M}^{M} \lambda\Big(\frac{\ell}{M}\Big) \widehat{\gamma}_Y(\ell) \exp(-2\pi i \omega \ell), \quad \omega \in [0, \tfrac{1}{2}], \tag{4.16}$$

$$\widehat{f}_Y(\omega_1, \omega_2) = \sum_{\ell_1, \ell_2 = -M}^{M} \lambda\Big(\frac{\ell_1}{M}, \frac{\ell_2}{M}\Big) \widehat{\gamma}_Y(\ell_1, \ell_2) \exp\{-2\pi i(\omega_1 \ell_1 + \omega_2 \ell_2)\},$$

$$(\omega_1, \omega_2) \in \mathcal{D}, \tag{4.17}$$

where $\widehat{\gamma}_Y(\ell_1, \ell_2) = T^{-1} \sum_{t=1}^{T-\beta} Y_t Y_{t+\ell_1} Y_{t+\ell_2}$, with $\beta = \max\{0, \ell_1, \ell_2\}$, $(\ell_1, \ell_2 = 0, 1, \ldots, T-1)$ and $1 \leq M \ll T$ (*truncation point*).

The function $\lambda(\cdot)$ is a *lag window*, satisfying $\lambda(0) = 1$ and the symmetry condition (4.1). Furthermore, $\lambda(\cdot, \cdot)$ is a two-dimensional lag window satisfying the same symmetries as the third-order moment, and is real-valued and finite. A standard window is *Parzen's lag window*, which is defined as

$$\lambda(u) = \begin{cases} 1 - 6u^2 + 6|u|^3, & |u| \leq \tfrac{1}{2}, \\ 2(1 - |u|)^3, & \tfrac{1}{2} |u| \leq 1, \\ 0, & |u| > 1. \end{cases} \tag{4.18}$$

A two-dimensional lag window can be constructed from any one-dimensional window, and is given by $\lambda(\ell_1, \ell_2) = \lambda(\ell_1) \lambda(\ell_2) \lambda(\ell_1 - \ell_2)$. In general, $M \equiv M(T)$ is chosen such that as $T \to \infty$ then $M \to \infty$, but the ratio $M^2/T \to 0$. A large value of M will increase the variance and decrease the bias of the estimates of the spectrum and bispectrum.

Example 4.1: Third-order Cumulant and Bispectrum

Suppose the series $\{Y_t\}_{t=1}^T$ is generated by a diagonal BL(0, 0, 1, 1) process of the form

$$Y_t = \beta Y_{t-1} \varepsilon_{t-1} + \varepsilon_t, \tag{4.19}$$

4.1 BISPECTRUM

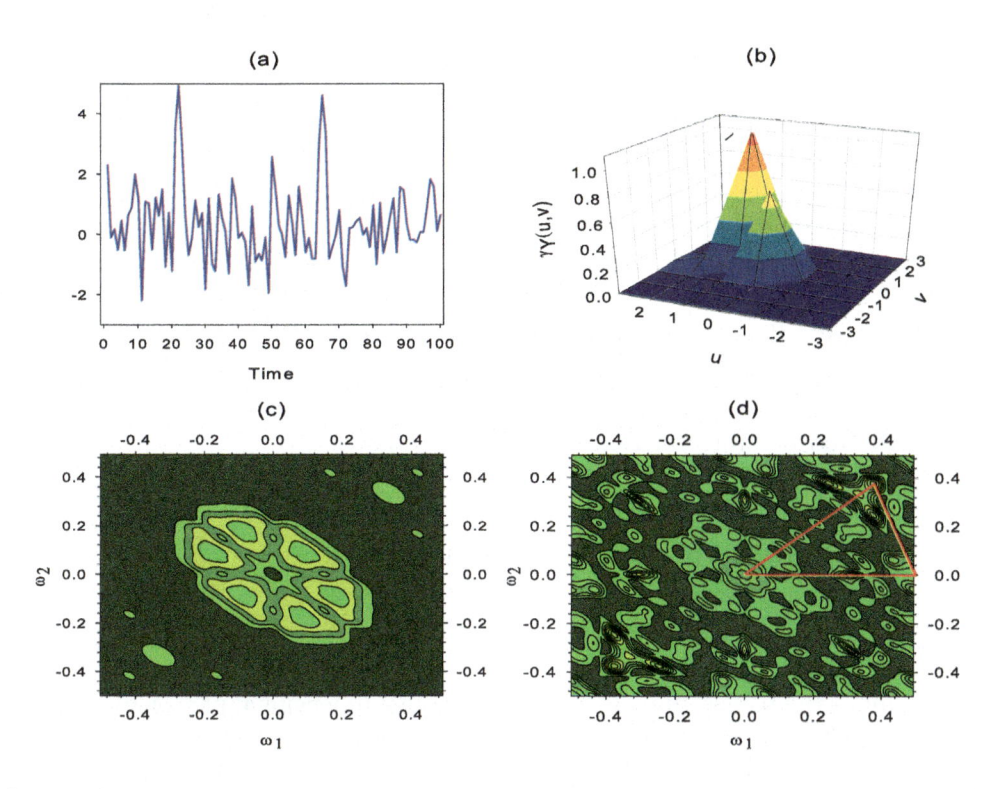

Figure 4.2: *(a) A realization of the diagonal* BL(0, 0, 1, 1) *process* $Y_t = 0.4Y_{t-1}\varepsilon_{t-1} + \varepsilon_t$ *with* $\{\varepsilon_t\} \stackrel{i.i.d.}{\sim} \mathcal{N}(0,1)$; *(b) Three-dimensional plot of* $\gamma_Y(u,v)$; *(c) Contour plot of the frequency bicoherence estimates of the* BL *process in (a); (d) Contour plot of the bicoherence of a series generated by the* AR(1) *process* $Y_t = 0.4Y_{t-1} + \varepsilon_t$ *with* $\{\varepsilon_t\} \stackrel{i.i.d.}{\sim} \mathcal{N}(0,1)$. *Superimposed is a plot of the principal domain (4.7);* $T = 100$.

where $\{\varepsilon_t\} \stackrel{i.i.d.}{\sim} \mathcal{N}(0,\sigma_\varepsilon^2)$. To ease notation, it is convenient to define $\lambda = \beta\sigma_\varepsilon$. The process is stationary and ergodic if $|\lambda| < 1$. According to Kumar (1986), the third-order cumulant is given by

$$\gamma_Y(\ell_1,\ell_2) = \begin{cases} 2\lambda^3\sigma_\varepsilon^3\left(\frac{4+5\lambda^2}{1-\lambda^2}\right), & (\ell_1,\ell_2) = (0,0), \\ \frac{2\beta\sigma_\varepsilon^4(1+\lambda^2+\lambda^4)}{1-\lambda^2}, & (\ell_1,\ell_2) = (1,1), \\ \frac{4\beta^3\sigma_\varepsilon^6(1+2\lambda^2\sigma_\varepsilon^2+3\lambda^4\sigma_\varepsilon^4)}{1-\lambda^2}, & (\ell_1,\ell_2) = (1,0), \\ \beta^3\sigma_\varepsilon^6, & (\ell_1,\ell_2) = (2,1), \\ \frac{6\beta^{2\ell_2+1}\sigma_\varepsilon^{2\ell_2+4}(1+\lambda^2+2\lambda^4)}{1-\lambda^2}, & (\ell_1=0,\ell_2=2,3,\ldots), \\ 0, & \text{otherwise.} \end{cases} \quad (4.20)$$

Figure 4.2(a) shows a plot of a realization of the BL(0, 0, 1, 1) process with $\beta = 0.4$. The plot gives an indication of the series periodicity, stationarity, and also whether there are any intermittent periods. Figure 4.2(b) shows a plot of $\gamma_Y(\ell_1,\ell_2)$ for $(\ell_1,\ell_2) = -3,\ldots,3$, with $\sigma_\varepsilon^2 = 1$. Note the peak in the third-order cumulant at $(\ell_1,\ell_2) = (1,1)$. For a diagonal BL(0, 0, p, p) $(p > 0)$

zero-mean process, $\gamma_Y(\ell_1, \ell_2)$ will have a peak in the third-order cumulant at $(\ell_1, \ell_2) = (p, p)$. Then the modulus of the bispectrum will be periodic on the manifolds $\omega_1 = 0$ and $\omega_2 = 0$ with frequency inversely proportional to p.

Figure 4.2(c) shows a contour plot of the bicoherence using the direct fast FT based estimation method. We see peaks at $(\omega_1, \omega_2) = (0, 0)$ and the 11 other symmetric locations indicative of nonlinear phenomena. Figure 4.2(d) gives the bicoherence for a realization of a stationary AR(1) process with the same parameter value as the simulated BL process. The plot also includes the first triangular region, i.e., the principal domain (4.7). We see, that in contrast to the BL process, the bicoherence is constant, indicating that the process is linear, and possibly Gaussian, or normally, distributed.

4.2 The Subba Rao–Gabr Tests

A first heuristic step of assessing non-Gaussianity (or more broadly asymmetry), and nonlinearity is to examine the real and imaginary parts of the bispectrum, as well the modulus of the bispectrum estimates by a three-dimensional plot or by a contour plot. This can be a useful exercise, but like interpreting a plot of the sample ACF it is an inexact art. A number of formal frequency domain tests for non-Gaussianity and nonlinearity have been based on the frequency bicoherence result (4.12). In this section, we discuss two test statistics proposed by Subba Rao and Gabr (1980, 1984).

4.2.1 Testing for Gaussianity

Subba Rao and Gabr (1980) suggest testing for Gaussianity first by forming an estimate of $f_Y(\omega_1, \omega_2)$ on a set of lattice frequencies in the principle domain $\mathcal{D}$, and then testing those quantities for constancy, by estimating $|B_Y(\omega_1, \omega_2)|^2$. The procedure for computing the Gaussianity test statistic consists of the following steps.

Algorithm 4.1: The Subba Rao–Gabr Gaussianity test

(i) Choose M, and estimate $f_Y(\omega)$ by (4.16).

(ii) Construct a set of estimators $\widehat{f}_Y(\omega_j, \omega_k)$ at a "coarse" grid of *designated frequencies* $(\omega_j, \omega_k) \in \mathcal{D}$, with $\omega_j = j/K$, $(j = 1, \ldots, \lfloor 2K/3 \rfloor)$, $\omega_k = k/K$, $(k = j+1, \ldots, K - \lfloor j/2 \rfloor - 1)$. Here, K must be chosen such that $K \ll T$ and its value lies inside $\mathcal{D}$. This is accomplished by defining a "fine" grid of $N = 4r + 1$ frequencies $\omega_{j_p} = \omega_j + \frac{pd}{2T}$, $(p = -r, -r+1, \ldots, -1, 0, 1, \ldots, r-1, r)$, $\omega_{k_q} = \omega_k + \frac{qd}{2T}$, $(q = -r, -r+1, \ldots, -1, 1, \ldots, r-1, r)$, which extend vertically and horizontally from each of the (ω_j, ω_k).

4.2 THE SUBBA RAO–GABR TESTS

Algorithm 4.1: The Subba Rao–Gabr Gaussianity test (Cont'd)

(ii) (Cont'd)
The distance d between the new frequencies is such that the bispectral estimates at neighboring points on this fine grid are approximately uncorrelated.

(iii) Use (4.17) at each of the $(\omega_{j_p}, \omega_{k_q})$ in the finer grid, to obtain $\widehat{f}_Y(\omega_{j_p}, \omega_{k_q})$, as N unbiased, approximately uncorrelated, estimates of $f_Y(\omega_j, \omega_k)$.

(iv) Place each of the $\widehat{f}_Y(\omega_{j_p}, \omega_{k_q})$ in a $P \times N$ matrix $\mathbf{D} = (\boldsymbol{\xi}_1, \ldots, \boldsymbol{\xi}_N)$ where $\boldsymbol{\xi}_i = (\xi_{1i}, \ldots, \xi_{Pi})'$ $(i = 1, \ldots, N)$ with $\xi_i = \widehat{f}_Y(\omega_{j_p}, \omega_{k_q})$, suitably relabeled, and where $P = \sum_{i=1}^{[2K/3]}(K - \lfloor i/2 \rfloor - 1 - i)$. The P row vectors of this matrix are asymptotically complex Gaussian with mean $\boldsymbol{\eta}$, a vector of length N, and variance-covariance matrix $\boldsymbol{\Sigma}_f$, say. Under $\mathbb{H}_0^{(1)}$, $\boldsymbol{\eta} = \mathbf{0}$.

(v) The test statistic for Gaussianity is developed as a complex analogue of Hotelling's T^2 test statistic. Specifically, calculate the statistic $T_1^2 = N\widehat{\boldsymbol{\eta}}^* \mathbf{A}^{-1} \widehat{\boldsymbol{\eta}}$, where $\mathbf{A} = N\widehat{\boldsymbol{\Sigma}}_f$ and $*$ denotes complex conjugate. For practical application, it is recommended to use the test statistic

$$F_1 = \frac{2(N-P)}{2P} T_1^2. \qquad (4.21)$$

Under $\mathbb{H}_0^{(1)}$, and as $T \to \infty$,

$$F_1 \xrightarrow{D} F_{\nu_1, \nu_2} \qquad (4.22)$$

with degrees of freedom $\nu_1 = 2P$ and $\nu_2 = 2(N-P)$.

Example 4.2: Principal Domain of the Subba Rao–Gabr Gaussianity Test

The choice of K has a direct effect on the selected frequencies in the principal domain. Suppose $T = 250$, $K = 6$, $d = 8$, and $r = 2$.[1] Then, $N = 4r+1 = 9$, and $P = (6-2) + (6-4) + (6-5) = 7$, resulting in 63 frequency pairs (ω_1, ω_2) from the total of approximately $(1/3)\{(T/2)+1\}^2 = 5,292$ in $\mathcal{D}$. Figure 4.3(a) shows a plot of the corresponding principal domain. Figure 4.3(b) displays similar results for $K = 7$ ($P = 10$). Observe that there is a lack of selected frequencies near the left and bottom edges of $\mathcal{D}$ in both figures. So, in practice, the Subba Rao–Gabr Gaussianity test statistic can be sensitive to small, or missing, values of the estimates of $f_Y(\omega_1, \omega_2)$ in certain areas of $\mathcal{D}$.

---

[1] Choosing K as a multiple of T results in ordinates that directly match the Fourier frequencies.

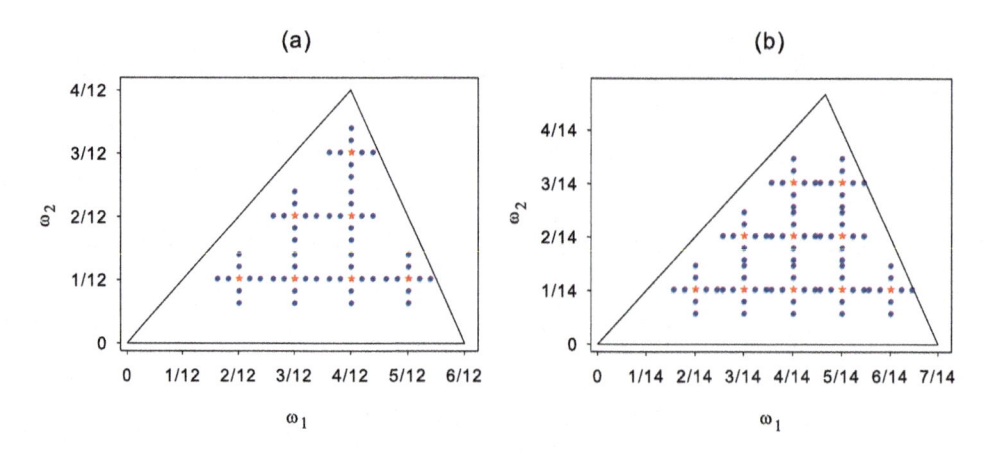

Figure 4.3: *(a) Principal domain for the bispectrum with frequency pairs $(\omega_{j_p}, \omega_{k_q})$ (blue dots) $(p = -2, -1, 0, 1, \ldots, 2; q = -2, -1, 1, 2)$ and designated frequency pairs (red stars) for $d = 8$, $T = 250$; (a) $K = 6$, and (b) $K = 7$.*

4.2.2 Testing for linearity

If the symmetry null hypothesis $\mathbb{H}_0^{(1)}$ is rejected, Subba Rao and Gabr (1980) consider testing $\mathbb{H}_0^{(2)}$. As in the Gaussianity test, estimates of $|B_Y(\omega_{j_p}, \omega_{k_q})|$ are constructed at the N points in the fine grid $(\omega_{j_p}, \omega_{k_q})$. Place these NP estimates in a $P \times N$ matrix. Average the values in the columns of this matrix to obtain a random sample of N estimates of the $P \times 1$ mean vector $\mathbf{Z} = (Z_1, \ldots, Z_P)'$, suitably relabeled. These estimates, denoted by $\mathbf{Z}_1^*, \ldots, \mathbf{Z}_N^*$, are asymptotically normally distributed (Brillinger, 1965). If $\mathbb{H}_0^{(2)}$ is "true" then all the elements of the mean vector $\mathbf{Z}$ are identical. Equality of the means under the null hypothesis can be expressed as $P-1$ comparisons, i.e. $Z_i - Z_{i-1} = 0$ $(i = 1, \ldots, P - 1)$. This expression can be written in matrix form. To this end, define a $(P-1) \times 1$ column vector $\boldsymbol{\beta}$ such that $\boldsymbol{\beta} = \mathbf{BZ}$, where $\mathbf{B}$ is the $(P-1) \times P$ matrix:

$$\mathbf{B} = \begin{pmatrix} 1 & -1 & 0 & \cdots & 0 & 0 \\ 0 & 1 & -1 & \cdots & 0 & 0 \\ \vdots & \vdots & \vdots & \vdots & \vdots & \vdots \\ 0 & 0 & 0 & \cdots & 1 & -1 \end{pmatrix}.$$

Under the null hypothesis $\mathbb{H}_0^{(2)}$, $\boldsymbol{\beta}$ is asymptotically jointly normally distributed with mean $\mathbf{0}$, and variance-covariance matrix $\mathbf{B}\boldsymbol{\Sigma}_Z\mathbf{B}'$.

Given the above results, the remaining part of the procedure to compute the test statistic goes as follows.

> **Algorithm 4.2: The Subba Rao–Gabr linearity test**
>
> (i) Compute
> $$\widehat{\boldsymbol{\beta}} = \mathbf{B}\overline{\mathbf{Z}}, \quad \text{and} \quad \widehat{\mathbf{S}} = \mathbf{B}\widehat{\mathbf{S}}_Z\mathbf{B}',$$
> where
> $$\overline{\mathbf{Z}} = N^{-1}\sum_{i=1}^{N}\mathbf{Z}_i^*, \quad \text{and} \quad \widehat{\mathbf{S}}_Z = N^{-1}\sum_{i=1}^{N}(\mathbf{Z}_i^* - \overline{\mathbf{Z}})(\mathbf{Z}_i^* - \overline{\mathbf{Z}})'$$
> are the ML estimates of the mean and variance-covariance matrix, respectively.
>
> (ii) Compute the likelihood ratio test statistic
> $$F_2 = \frac{N-P+1}{P-1}T_2^2, \qquad (4.23)$$
> where $T_2^2 = N\widehat{\boldsymbol{\beta}}'\widehat{\mathbf{S}}^{-1}\widehat{\boldsymbol{\beta}}$. Under $\mathbb{H}_0^{(2)}$, and as $T \to \infty$,
> $$F_2 \xrightarrow{D} F_{\nu_1,\nu_2} \qquad (4.24)$$
> with degrees of freedom $\nu_1 = P-1$ and $\nu_2 = N-P+1$.

4.2.3 Discussion

There are some drawbacks to the test statistics (4.21) and (4.23). Typically the user has to decide on the choice of the lag window, the truncation point M, and the placing of the grids, i.e., the parameters d, K, and r. Based on 500 generated BL(2,1,1,1) time series W.S. Chan and Tong (1986) note that the results of the Subba Rao–Gabr linearity test statistic is sensitive to the choice of the lag window. The choice of the truncation point M is another delicate issue; see, e.g., Subba Rao and Gabr (1984, Section 3.1) for various suggestions. One recommendation is that $M < T^{1/2}$. A more formal approach is to minimize the mean squared error (MSE) of the bispectral estimate, which is a function of $f_Y(\omega_1)$, $f_Y(\omega_2)$ and $f_Y(\omega_1, \omega_2)$, with respect to M.

The parameters d, K, and r should be chosen as follows. First, it is required that $N \times [2K/3] < T$, where $[\cdot]$ denotes the integer part; see step (iv) of Algorithm 4.1. Next, to ensure that the spectral and bispectral estimates at different points of the grid are effectively uncorrelated, it is necessary to choose d such that d/T is larger than the spectral window corresponding to the lag window $\lambda(s)$. Similarly, r should be chosen such that r/T is less than the lag window. Finally, to ensure that points in different fine grids do not overlap, it is essential that $d \leq T/\{K(r+1)\}$. In summary, great skill is necessary in applying both test statistics (4.21) and (4.23) because of the large number of parameters involved.

4.3 Hinich's Tests

Hinich (1982) modifies the Subba Rao–Gabr tests to use all the bispectrum Fourier frequency gridpoints. However, rather than using the windowed sample ACVF method, or indirect method, the test statistics are based on a consistent estimator of the bispectrum at frequency pair (ω_m, ω_n) obtained by smoothing the third-order periodogram over adjacent frequency pairs.

The general framework can be summarized as follows. Let $\omega_j = (j-1)/T$ $(j = 1, \ldots, \lfloor T/2 \rfloor + 1)$. For each pair (j,k) $(j,k \in \mathbb{Z})$, define the complex random variable

$$F_Y(\omega_j, \omega_k) = Y(\omega_j)Y(\omega_k)Y^*(\omega_{j+k})/T, \qquad (4.25)$$

where

$$Y(\omega_j) = \sum_{t=1}^{T} Y_t \exp\{-2\pi i \omega_j (t-1)\}.$$

Since $Y(\omega_{j+T}) = Y(\omega_j)$ and $Y(\omega_{T-j}) = Y^*(\omega_j)$, the principal domain of $F_Y(\omega_j, \omega_k)$ is the triangular set

$$\triangle = \{(j,k) : 0 < j \leq T/2,\ 0 < k \leq j,\ 2j + k \leq T\}, \qquad (4.26)$$

assuming T is even. A straightforward approach to obtain a consistent estimate of the bispectrum is to average the $F_Y(\omega_j, \omega_k)$ in a square of M^2 points, where the centers of the squares are defined by a lattice $\mathcal{L}$ of points such that $\mathcal{L} \in \triangle$; see Figure 4.4 for two examples. Then the resulting direct estimator of $f_Y(\omega_1, \omega_2)$ is given by

$$\widehat{f}_Y(\omega_m, \omega_n) = \frac{1}{M^2} \sum_{j,k=(m-1)M}^{mM-1} F_Y(\omega_j, \omega_k), \qquad (4.27)$$

with $M = \lfloor T^c \rfloor$ ($\frac{1}{2} < c < 1$). The complex variance of this estimator, assuming the terms in the summations are restricted to $\triangle$, excluding the manifolds $\omega_m = 0$, $\omega_m = \omega_n$, is given by

$$\operatorname{Var}\{\widehat{f}_Y(\omega_m, \omega_n)\} = \frac{T}{M^4} Q_{m,n} f_Y(\delta_m) f_Y(\delta_n) f_Y(\delta_{m+n}) + \mathcal{O}(M/T),$$

where $\delta_x = (2x-1)M/(2T)$ and $Q_{m,n}$ is the number of (j,k) in the squares that are in $\triangle$, but not on the boundaries $j = k$ or $(2j+k) = T$, plus *twice* the number on these boundaries. Note, $TM^{-4}Q_{m,n} \leq TM^{-2} = T^{1-2c} \to 0$ if $T \to \infty$, since $Q_{m,n} \leq M^2$.

4.3 HINICH'S TESTS

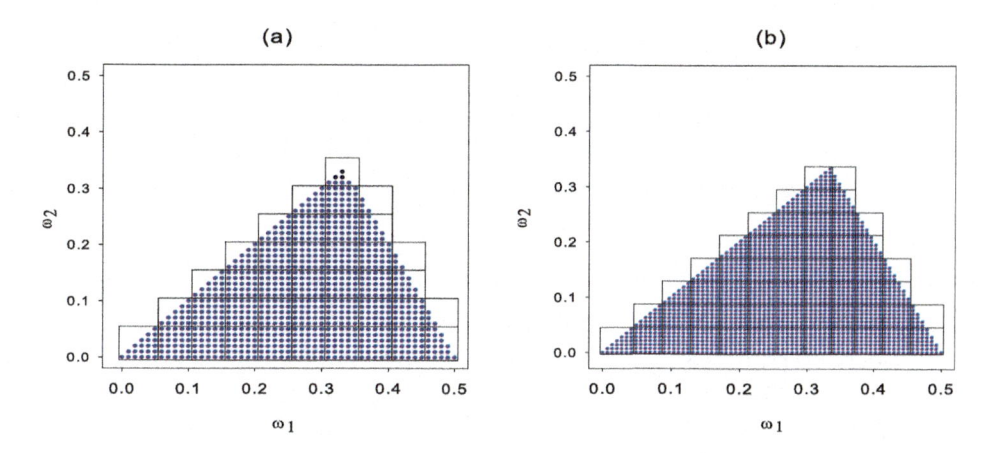

Figure 4.4: *(a) Lattice in the principal domain for the bispectrum with $K = 10$, and $r = 5$; (b) Lattice $\mathcal{L}$ in the principal domain of the bispectrum for estimating* Hinich's test statistics; $T = 144$ and $c = 1/2$.

It can be shown (Hinich, 1982) that the asymptotic distribution of each estimator is complex normal, and that the estimators are asymptotically independent inside the principal domain. Therefore, the distribution of the statistic

$$\widehat{B}_Y(\omega_m, \omega_n) = \frac{\widehat{f}_Y(\omega_m, \omega_n)}{\{T^{1-4c}Q_{m,n}\widehat{f}_Y(\delta_m)\widehat{f}_Y(\delta_n)\widehat{f}_Y(\delta_{m+n})\}^{1/2}} \quad (4.28)$$

is complex normal with unit variance, with $\widehat{f}_Y(\cdot)$ the estimator of the spectral density function constructed by averaging M adjacent periodogram ordinates. Now $2|\widehat{B}_Y(\omega_m, \omega_n)|^2$ is approximately distributed as $\chi_2^2(\lambda_{m,n})$, i.e. a noncentral chi-square distribution with two degrees of freedom and noncentrality parameter

$$\lambda_{m,n} = 2(T^{1-4c}Q_{m,n})^{-1}|B_Y(\omega_m, \omega_n)|^2 \geq 2T^{2c-1}|B_Y(\omega_m, \omega_n)|^2. \quad (4.29)$$

Thus, the value of (4.29) increases when a smaller set of frequency pairs (ω_m, ω_n) is considered.

The choice of the parameter c controls the trade-off between the bias and variance of $\widehat{B}_Y(\cdot, \cdot)$. The smallest bias is obtained for $c = 1/2$, whereas the smallest variance is for $c = 1$. The power of the test for a zero bispectrum depends on $T^{1/2}$ when T^{1-c} is large, c should be slightly larger than $1/2$ to give a consistent estimate.

4.3.1 Testing for linearity

Assume $\{Y_t, t \in \mathbb{Z}\}$ follows the zero-mean stationary linear (L) process (1.2). Then, for all squares in $\triangle$, so that $Q_{m,n} = M^2$, the noncentrality parameter reduces to

$$\lambda_{m,n} = 2T^{2c-1}\frac{\mu_{3,\varepsilon}^2}{\sigma_\varepsilon^6} \equiv \lambda_0.$$

Thus, the noncentrality parameter becomes a constant. Since $\mathbb{E}(|\widehat{B}_Y(\omega_m, \omega_n)|^2) = 1 + \lambda_{m,n}/2$, it follows from (4.29) and the asymptotic properties of $\widehat{B}_Y(\omega_m, \omega_n)$ that the parameter λ_0 can be consistently estimated by

$$\widehat{\lambda}_0 = \frac{2}{PM^2} \sum_{(m,n)\in\mathcal{L}} Q_{m,n}\left(|\widehat{B}_Y(\omega_m, \omega_n)|^2 - 1\right), \tag{4.30}$$

where P, the number of (m,n) in $\mathcal{L}$, is approximately $T^2/(12M^2)$. Consequently, the distribution $\chi_2^2(\widehat{\lambda}_0)$ converges to a $\chi_2^2(\lambda_0)$ variate, as $T \to \infty$.

If $\mathbb{H}_0^{(2)}$ is true, expression (4.30) shows that the noncentrality parameter of the asymptotic distribution of the statistic $2|\widehat{B}_Y(\omega_m, \omega_n)|^2$ is constant $\forall (m,n) \in \mathcal{L}$, and squares wholly in $\triangle$. If the null hypothesis is false, the noncentrality parameter will be different for different values of m and n. As a result, the sample dispersion of $2|\widehat{B}_Y(\omega_m, \omega_n)|^2$ will be larger than expected under the null hypothesis. This dispersion can be measured in many ways.

One way to proceed is to use the asymptotic normality of the *interquartile range*, say IQR$_M$, of the $2|\widehat{B}_Y(\omega_m, \omega_n)|$'s entirely within the principle domain. Let $q_{0.25}$ and $q_{0.75}$ denote respectively the first and third quartile of a $\chi_2^2(\lambda_0)$ random variable, and let $q_{0.75} - q_{0.25}$ be the IQR from this distribution. Then, under $\mathbb{H}_0^{(2)}$, the approximate distribution of IQR$_M$, as deduced from the theory of order statistics, is given by

$$Z_{\text{IQR}}^{\text{L}} = \frac{\text{IQR}_M - (q_{0.75} - q_{0.25})}{\sigma_0} \xrightarrow{D} \mathcal{N}(0,1), \text{ as } T \to \infty, \tag{4.31}$$

where

$$\sigma_0^2 = \frac{3[f_{\chi_2^2(\lambda_0)}(q_{0.25})]^{-2} - 2[f_{\chi_2^2(\lambda_0)}(q_{0.25})f_{\chi_2^2(\lambda_0)}(q_{0.75})]^{-1} + 3[f_{\chi_2^2(\lambda_0)}(q_{0.75})]^{-2}}{16P}, \tag{4.32}$$

and $f_{\chi_2^2(\lambda_0)}(\cdot)$ is the density function of a $\chi_2^2(\lambda_0)$ random variable. It is not difficult to estimate $q_{0.25}$, $q_{0.75}$, and (4.32) for a given value of λ_0. In practice, the estimator (4.30) is used in the computations of these values.

4.3.2 Testing for Gaussianity

If the error process $\{\varepsilon_t, t \in \mathbb{Z}\}$ in the linear DGP (1.2) is Gaussian (G), then $\lambda_0 \equiv 0$. In that case the following test statistic may be used

$$T^{\text{G}} = 2 \sum_{(m,n)\in\mathcal{L}} |\widehat{B}_Y(\omega_m, \omega_n)|^2, \tag{4.33}$$

which is asymptotically distributed as a central χ_{2P}^2 variate under $\mathbb{H}_0^{(2)}$, with $P \approx T^2/(12M^2)$; see (4.30). Note that (4.33) is essentially the Subba Rao–Gabr test statistic T_1^2, i.e., instead of using an estimate of the bispectral density in the sum of squares (4.33) uses an estimate of the normalized bispectrum.

4.3.3 Discussion

For relatively large sample sizes Ashley et al. (1986) examine in an MC simulation study the size and power of Hinich's linearity and Gaussianity test statistics. Overall, the sizes of these test statistics are satisfactory. What seems more important, however, is that the power of the linearity test statistic is disturbingly low in distinguishing between linear and nonlinear time series processes. In particular, this seems to be the case for ExpAR and SETAR behavior. Furthermore, Harvill and Newton (1995) show that uncommonly large time series sample sizes are necessary before the normal distribution in (4.32) is reliable for calculating p-values. Additionally, these authors point out that the asymptotics of this problem are present in three interwoven forms: the length T of the observed time series, the number of points M used to estimate the normalized bispectrum, and the number P of normalized bispectral estimates used in calculating the IQR. For instance, to have $P = 100$ requires a series of length $T = 1,200$ when using $M = \lfloor T^{1/2} \rfloor$.

Although Hinich's approach is robust to outliers in the case of linearity, a disadvantage of using the IQR is that if the null hypothesis is false and the process is of a type of nonlinearity which would result in a peak in $|B_Y(\omega_m, \omega_n)|^2$, the range effectively ignores that distinguishing feature. So the test statistic may differentiate between linear and nonlinear processes but provides no clue as to the form of nonlinearity. To some extent this may be overcome by visually assessing plots of the frequency bicoherence.

More importantly, Garth and Bresler (1996) raise some concerns with the assumptions required to form the linearity test statistic. As the number of discrete FT values of $\{Y_t\}_{t=1}^T$ increase as $T \to \infty$, the assumption that $|\widehat{B}_Y(\omega_m, \omega_n)|^2$ will converge to the proposed noncentral $\chi_2^2(\lambda_0)$ distribution is violated, as this requires a finite number of bispectral estimates. Ignoring the finite-dimensionality constraint leads to a different asymptotic distribution; it can also lead to dependence between two estimates, smoothed over distinct frequency regions. The dependence is eliminated by summing the discrete FT over a finite subset of points, which is true for the indirect estimate of the bispectrum. This approach, however, introduces the additional problem of carefully choosing the spectral bandwidth M, as with the Subba Rao–Gabr test statistics.

4.4 Related Tests

4.4.1 Goodness-of-fit tests

Recall that under Gaussianity, the noncentrality parameter of the test statistic $2|B_Y(\omega_m, \omega_n)|^2$ is identically zero $\forall (\omega_m, \omega_n) \in \mathcal{L}$. So the noncentral chi-square distribution with two degrees of freedom and noncentrality parameter $\lambda_0 = 0$ reduces to a central χ_2^2 distribution, i.e., an exponential distribution with mean 2. This suggests that a *goodness-of-fit* (GOF) test statistic might be effective in measuring the difference between the empirical distribution function (EDF) of $2|\widehat{B}_Y(\omega_m, \omega_n)|^2$ and the noncentral $\chi_2^2(\lambda_{m,n})$ as the null distribution.

Unfortunately, finding the null distribution of the resulting EDF-based test statistic is intractable. Jahan and Harvill (2008) overcome this problem by approximating the noncentral $\chi_2^2(\cdot)$ distribution by a normal distribution in the following way. Let $X \sim \chi_\nu^2(\lambda)$. Then a remarkably accurate approximation (Sankaran, 1959) for the tails of the $\chi_\nu^2(\lambda)$ distribution consists of replacing X by $Y = (X/(\nu + \lambda))^h$, where the exponent h is given by

$$h = 1 - \frac{2(\nu + \lambda)(\nu + 3\lambda)}{3(\nu + 2\lambda)^2}. \tag{4.34}$$

Specifically, Y has an approximate normal distribution with mean and variance given respectively by

$$\mu_Y = 1 + h(h-1)\frac{\nu + 2\lambda}{(\nu + \lambda)^2} - h(h-1)(2-h)(1-3h)\frac{(\nu + 2\lambda)^2}{2(\nu + \lambda)^4}, \tag{4.35}$$

$$\sigma_Y^2 = h^2 \frac{2(\nu + 2\lambda)}{(\nu + \lambda)^2}\left[1 - (1-h)(1-3h)\frac{\nu + 2\lambda}{(\nu + \lambda)^2}\right]. \tag{4.36}$$

If λ is unknown, it is recommended to replace λ by the method of moment based estimator

$$\widehat{\lambda} = \begin{cases} \overline{Y} - \nu & \text{if } \overline{Y} > \nu, \\ 0 & \text{otherwise}, \end{cases} \tag{4.37}$$

where $\overline{Y}$ is the sample mean. Under the null hypothesis of Gaussianity, $\widehat{\lambda}$ is a consistent estimator for λ.

Stephens (1974) shows that in a wide variety of situations the Anderson–Darling (AD) GOF test statistic is the most powerful EDF-based test followed by the (one-sample) Cramér–von Mises (CvM) test statistic. In the case of testing for Gaussianity and linearity, using the bispectrum, these test statistics can be computed as follows. Let $\{Q_{(i)}\}_{i=1}^P$ denote the quantiles computed from the ordered values $2|\widehat{B}_Y(\omega_i^{(1)}, \omega_i^{(2)})|^2$ ($i = 1, \ldots, P$). Note, that for testing Gaussianity, the data are assumed to come from a fully specified normal distribution. Then a modified form of the CvM-type test statistics is given by

$$\text{CvM}^* = (\text{CvM} - 0.4/P + 0.6/P^2)(1 + 1/P), \tag{4.38}$$

where

$$\text{CvM} = \frac{1}{12P} + \sum_{i=1}^P \left(Q_{(i)} - \frac{(2i-1)}{2P}\right)^2.$$

However, for all $P \geq 5$, the AD-type test statistic for testing Gaussianity needs no modification, i.e., its calculation can be based on the formula

$$\text{AD} = -P - \frac{1}{P}\sum_{i=1}^P (2i-1)\left[\log Q_{(i)} + \log(1 - Q_{((P+1)-i)})\right], \tag{4.39}$$

4.4 RELATED TESTS

assuming $Q_{(i)} \neq 0$ or 1.

For testing linearity both mean and variance of the transformed random variables are unknown. In that case these quantities are estimated by $\overline{B}_Y$, the sample mean of the $\widehat{B}_Y(\omega_i^{(1)}, \omega_i^{(2)})$ ($i = 1, \ldots, P$), and the sample standard variance $(P-1)^{-1} \sum_{i=1}^{P} (\widehat{B}_Y(\omega_i^{(1)}, \omega_i^{(2)}) - \overline{B}_Y)^2$. Then, according to Stephens (1986, Table 4.9), the asymptotic upper-tail p-value can be computed from first transforming CvM to the modified (m) statistic $\text{CvM}_m = \text{CvM}(1 + 0.5/P)$ and next calculating a parabolic approximation, i.e.,

$$p = \begin{cases} \exp\left(0.886 - 31.62\,\text{CvM}_m + 10.897\,(\text{CvM}_m)^2\right), & 0.051 < \text{CvM}_m < 0.092, \\ \exp\left(1.111 - 34.242\,\text{CvM}_m + 12.832\,(\text{CvM}_m)^2\right), & \text{CvM}_m \geq 0.092. \end{cases}$$

For the modified statistic $\text{AD}_m = \text{AD}(1 + 0.75/P + 2.25/P^2)$ $(P \geq 8)$, the formula for the asymptotic upper tail p-value is given by

$$p = \begin{cases} \exp(0.9177 - 4.279\,\text{AD}_m - 1.38\,(\text{AD}_m)^2), & 0.340 \leq \text{AD}_m < 0.600, \\ \exp(1.2937 - 5.709\,\text{AD}_m + 0.0186\,(\text{AD}_m)^2), & 0.600 \leq \text{AD}_m \leq 13. \end{cases}$$

Below we summarize the two-stage procedure for testing for Gaussianity and linearity.

Algorithm 4.3: Goodness-of-fit test statistics

(i) **Testing for Gaussianity (G):**

   (a) Compute the quantiles $Q_{(i)}$ ($i = 1, \ldots, P$) of the ordered $2|\widehat{B}_Y(\omega_m, \omega_n)|^2$ values, using the exponential(2) CDF. That is, $Q_{(i)} = 1 - \exp(-\widehat{B}_{(i)}/2)$, where $\widehat{B}_{(i)}$ are the arranged (ascending order) values of the $2|\widehat{B}_Y(\omega_i^{(1)}, \omega_i^{(2)})|^2$'s.

   (b) Apply these quantiles to the expressions in (4.38) or (4.39) to compute the value of, say, CvM_m^G or AD_m^G.

   (c) Compare the value of the test statistic with the appropriate critical value.

(ii) **Testing for linearity (L):**

   (a) For each i transform the random variable $\widehat{B}_{(i)}$ into $Y_i = \left(B_{(i)}/(2+\widehat{\lambda})\right)^{\widehat{h}}$, where $\widehat{h}$ is as in (4.34) with $\nu = 2$, and replacing λ with (4.37).

   (b) Standardize the P random variables Y_i, using (4.35) and (4.36) with $\nu = 2$ and λ given by (4.37).

   (c) Compute the quantiles $Q_{(i)}$ ($i = 1, \ldots, P$) of these variates, using the standard normal CDF.

   (d) Compute the values of, say, CvM_m^L or AD_m^L.

   (e) Compare the value of the test statistic with the appropriate critical value.

4.4.2 Maximal test statistics for linearity

As noted in Section 4.3.2, Hinich's Gaussianity and linearity tests involve the selection of the number of points M. The larger (smaller) M, the smaller (larger) the finite-sample variance of (4.27) and the larger (smaller) the sample bias. Because of this trade-off, Rusticelli et al. (2009) compute the maximal values of Hinich's bispectral test statistic for linearity $2|\widehat{B}_Y(\omega_m, \omega_n)|^2$ over the computationally feasible range of values for M. The upper bound (M^H) of this range is set at the total number of frequency pairs $(\omega_m, \omega_n) \in \mathcal{D}$ that at least exceeds one. The lower bound (M^L) is determined by the requirement that $\widehat{\lambda}_0$ in (4.30) should be positive. Then a well-sized test, giving the highest power against a wide set of nonlinear DGPs, is the maximal standardized *interdecile* (IDR) *fractile statistic*, MD_{IDR}^L, defined as

$$\text{MD}_{\text{IDR}}^L = \max_{M^L \leq M \leq M^H}\{\text{IDR}_M\}, \tag{4.40}$$

where

$$\text{IDR}_M = \frac{\{f_{\chi_2^2(\lambda_{m,n})}(q_{0.9}) - f_{\chi_2^2(\lambda_{m,n})}(q_{0.1})\} - \{f_{\chi_2^2(\widehat{\lambda}_0)}(q_{0.9}) - f_{\chi_2^2(\widehat{\lambda}_0)}(q_{0.1})\}}{\widehat{\sigma}_0} \tag{4.41}$$

is the standardized IDR fractile. The estimate $\widehat{\sigma}_0^2$ of σ_0^2 follows from (4.32) with $f_{\chi_2^2(\lambda_0)}(\cdot)$ replaced by $f_{\chi_2^2(\widehat{\lambda}_0)}(\cdot)$. The use of the IDR rather than the IQR in (4.41) is in line with Hinich et al. (2005) who, from numerous real and artificial applications, notice that the IDR gives more robust test results.

In an analogous way, maximal test statistics can be defined on the basis of the IQR, and 80% fractiles of $\widehat{B}_Y(\omega_m, \omega_n)$. Following the same arguments as in Hinich (1982), it can be shown that all these maxi-minimal test statistics are asymptotically distributed as $\mathcal{N}(0,1)$ under the null hypothesis that $\{Y_t, t \in \mathbb{Z}\}$ is a linear DGP, as defined by (1.2).

4.4.3 Bootstrapped-based tests

In finite samples, one cannot assess the validity of Hinich's linearity test statistic on the basis of critical values determined from the two asymptotic distributions – the noncentral $\chi_2^2(\lambda_0)$ distribution and the normal distribution (4.31). Data-dependent *bootstrapping* (resampling) the distributions of the linearity test is a way out, and several approaches have been proposed for this purpose. Often these bootstrap approaches involve, as a first step, prewhitening the time series by fitting an AR(p) model to the data, and separating out the residuals of the fit. A more appropriate approach is to allow the order p to be an increasing function of the sample size T, thereby creating an approximating *sieve* of AR models. This is the essence of the AR-sieve, or AR(∞) bootstrap, adopted by Berg et al. (2010) to formulate a bootstrap procedure for Hinich's linearity and Gaussianity test statistics.

4.4 RELATED TESTS

The proposed bootstrap algorithm is based on a 'kernelized' form of Hinich's test using the indirect bispectral estimation method. Specifically, asymptotically unbiased and consistent estimators of $f_Y(\omega)$ and $f_Y(\omega_1, \omega_2)$ are defined respectively by (4.16) and (4.17). where $\lambda(\cdot)$ and $\lambda(\cdot, \cdot)$ are non-negative one- and two dimensional lag windows (continuous weight functions), respectively, with compact support. This latter assumption can be relaxed with a trade-off of a more involved asymptotic theory. Very often $\lambda(\cdot)$ and $\lambda(\cdot, \cdot)$ are chosen such that they satisfy the symmetry conditions

$$\lambda(\omega) = \lambda(-\omega),$$
$$\lambda(\omega_1, \omega_2) = \lambda(\omega_2, \omega_1) = \lambda(-\omega_1, \omega_2 - \omega_1). \quad (4.42)$$

Clearly, both conditions mimic (4.1) and (4.2), or (4.5) and (4.6). But condition (4.42) is not required for proving consistency or asymptotic normality of (4.17).

Let $\omega_j = (\omega_j^{(1)}, \omega_j^{(2)})$ ($j = 1, \ldots, P$) denote the jth frequency pair in the lattice $\mathcal{L}$. Then, as already noted in Section 4.2, the kernel estimators $\widehat{f}_Y(\omega_j^{(1)}, \omega_j^{(2)})$ as in (4.17) are approximately complex Gaussian with variance

$$\text{Var}\{\widehat{f}_Y(\omega_j^{(1)}, \omega_j^{(2)})\} = \frac{M^2}{T} W_2 f_Y(\omega_j^{(1)}) f_Y(\omega_j^{(2)}) f_Y(\omega_j^{(1)} + \omega_j^{(2)}), \quad (4.43)$$

where

$$W_2 = \int_{-\infty}^{\infty} \int_{-\infty}^{\infty} \lambda^2(\omega_j^{(1)}, \omega_j^{(2)}) d\omega_j^{(1)} d\omega_j^{(2)}. \quad (4.44)$$

Then define the statistics

$$\widehat{Z}_Y(\omega_j^{(1)}, \omega_j^{(2)}) = \frac{\widehat{f}_Y(\omega_j^{(1)}, \omega_j^{(2)})}{\{M^2 W_2/T\}^{1/2} \{\widehat{f}_Y(\omega_j^{(1)}) \widehat{f}_Y(\omega_j^{(2)}) \widehat{f}_Y(\omega_j^{(1)} + \omega_j^{(2)})\}^{1/2}}. \quad (4.45)$$

Hence, the statistics $2|\widehat{Z}_Y(\omega_j^{(1)}, \omega_j^{(2)})|^2$ ($j = 1, \ldots, P$) are asymptotically distributed as independent noncentral χ_2^2 variates, with noncentrality parameter $|f_Y(\omega_j^{(1)}, \omega_j^{(2)})|^2/(M^2 W_2/T) f_Y(\omega_j^{(1)}) f_Y(\omega_j^{(2)}) f_Y(\omega_j^{(1)} + \omega_j^{(2)})$. For the purpose of testing linearity and Gaussianity, the set of random variables $2|\widehat{Z}_Y(\omega_j^{(1)}, \omega_j^{(2)})|^2$ for all $(\omega_j^{(1)}, \omega_j^{(2)})$ is considered to be a random sample from a continuous distribution with CDF $F(\cdot)$.

Before detailing the steps involved in the AR(∞)-sieve bootstrap procedure, we collect the spectral and bispectral density estimators into one long vector, i.e.,

$$V_T = (\widehat{f}_Y(\omega_1^{(1)}), \ldots, \widehat{f}_Y(\omega_P^{(1)}), \widehat{f}_Y(\omega_1^{(2)}), \ldots, \widehat{f}_Y(\omega_P^{(2)}), \widehat{f}_Y(\omega_1^{(1)} + \omega_1^{(2)}), \ldots,$$
$$\widehat{f}_Y(\omega_P^{(1)} + \omega_P^{(2)}), \widehat{f}_Y(\omega_1^{(1)}, \omega_1^{(2)}), \ldots, \widehat{f}_Y(\omega_P^{(1)}, \omega_P^{(2)})).$$

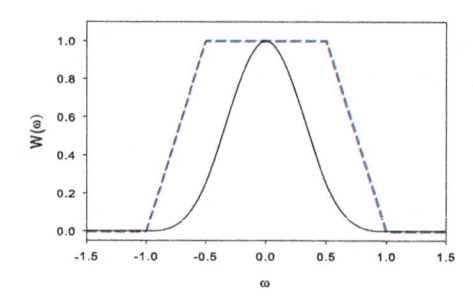

Figure 4.5: *Profiles of the Parzen lag window (black solid line) given by (4.18), and the trapezoid-shaped lag window (blue medium dashed line) as given by (4.50).*

The hypotheses of interest are:

$$\mathbb{H}_0^{(3)}: \quad \text{Linear but non-Gaussian (L+nG)}, \tag{4.46}$$

$$\mathbb{H}_0^{(4)}: \quad \text{Linear and symmetric (L+S), and} \tag{4.47}$$

$$\mathbb{H}_0^{(5)}: \quad \text{Gaussian (G)}. \tag{4.48}$$

Depending on the purpose of the analysis, one of the above three hypotheses are considered in the following bootstrap algorithm.

Algorithm 4.4: Bootstrap-based tests

(i) According to some order selection criterion choose p, fit (e.g., via the Yule-Walker equations) a strictly stationary AR(p) model $Y_t = \sum_{k=1}^{p} \phi_k Y_{t-k} + \varepsilon_t$ to $\{Y_t\}_{t=1}^T$, and separate out the residuals of the fit $\{\widehat{\varepsilon}_t\}_{t=p+1}^T$.

(ii) • When testing for $\mathbb{H}_0^{(3)}$:

   (a) Center the residuals, to obtain $\widetilde{\varepsilon}_t = \widehat{\varepsilon}_t - \overline{\varepsilon}$, where $\overline{\varepsilon} = (T-p)^{-1}\sum_t \widehat{\varepsilon}_t$.

   (b) Draw $T + b^*$ independent bootstrap residuals ε_t^* from the EDF F_T of $\{\widetilde{\varepsilon}_t\}$, where $b^* > 0$ denotes the so-called "burn-in" period to ensure the approximate stationarity of the bootstrap.

   (c) Generate, with the AR model found in (i) a series $\{Y_t^*\}_{t=1}^T$ of pseudo-observations, and obtain the corresponding EDF $F_T^{(3)}$.

• When testing for $\mathbb{H}_0^{(4)}$:

   (a) Draw $T - p$ independent bootstrap residuals ε_t^+ from $F_T^{(3)}$.

   (b) Transform the ε_t^+'s into pseudo-observations $\varepsilon_t^* = S_t \varepsilon_t^+$ with $\{S_t\} \overset{\text{i.i.d.}}{\sim} U[-1, 1]$, where U denotes the discrete uniform distribution on -1 and 1.

   (c) Obtain the corresponding EDF $F_T^{(4)}$.

4.4 RELATED TESTS

> **Algorithm 4.4: Bootstrap-based tests (Cont'd)**
>
> (ii) • When testing for $\mathbb{H}_0^{(5)}$:
>
> (a) Compute the residual variance $\hat{\sigma}_\varepsilon^2 = (T-p)^{-1} \sum_t (\hat{\varepsilon}_t - \bar{\varepsilon})^2$.
>
> (b) Draw $T-p$ independent bootstrap residuals ε_t^* from $\mathcal{N}(0, \hat{\sigma}_\varepsilon^2)$, and obtain the corresponding EDF $F_T^{(5)}$.
>
> (iii) Compute the vector of pseudo-statistics $V_T^{(i)}(Y_t^{(b)})$ ($i = 3, 4, 5$) analogous to V_T, but with the series $\{Y_t^{(b)}\}$ generated from the fitted AR(p) model with error process $\{\varepsilon_t^{(b)}\} \overset{\text{i.i.d.}}{\sim} F_T^{(i)}$.
>
> (iv) Repeat steps (ii) – (iii) B times, to obtain $\{V_T^{(i)}(Y_t^{(b)})\}_{b=1}^B$ ($i = 3, 4, 5$). The EDF of these bootstrap statistics can then be used to approximate the distribution of V_T under $\mathbb{H}_0^{(i)}$ ($i = 3, 4, 5$). In Table 4.1 we label the corresponding test statistics, based on the IQR, as: $Z_{\text{IQR}}^{\text{L+nG}}$, $Z_{\text{IQR}}^{\text{L+S}}$, and $T_{\text{IQR}}^{\text{G}}$.
>
> (v) Reject $\mathbb{H}_0^{(i)}$ ($i = 3, 4, 5$) when the p-value is less than a pre-specified significance level.

Suppose, in addition to the assumptions imposed on $\gamma_Y(\cdot)$ and $\gamma_Y(\cdot, \cdot)$, that

$$\sum_{\ell=-\infty}^{\infty} \ell^2 |\gamma_Y(\ell)| < \infty, \text{ and } \sum_{\ell_1, \ell_2 = -\infty}^{\infty} (1 + \ell_j^2)\gamma_Y(\ell_1, \ell_2) < \infty \quad (j = 1, 2). \quad (4.49)$$

Then Berg et al. (2010) prove the asymptotic consistency of the bootstrap test procedure under both the null hypothesis and the alternative hypothesis. They estimate the spectrum by a *trapezoid-shaped lag window* function (see Figure 4.5), and the bispectrum with a *right-pyramidal frustum-shaped* lag function (see Figure 4.6(a)). These functions are, respectively, defined by

$$\lambda(s) = 2(1 - |s|)^+ - (1 - 2|s|)^+, \quad (4.50)$$
$$\lambda(u, v) = 2\lambda_0(u, v) - \lambda_0(2u, 2v), \quad (4.51)$$

where

$$\lambda_0(x, y) = \begin{cases} (1 - \max(|x|, |y|))^+, & -1 \leq x, y \leq 0 \text{ or } 0 \leq x, y \leq 1, \\ (1 - \max(|x+y|, |x-y|))^+, & \text{otherwise}, \end{cases}$$

with $(x)^+ = \max(0, x)$. Both infinite-order functions can produce higher-order accurate estimators of the spectral and bispectral densities.

4.4.4 Discussion

Similar to the original Hinich's test statistics, the user of the AD- and CvM-type test statistics has to select M (the bispectral bandwidth), and P (the number of gridpoints). Consequently, the test statistics may still be sensitive to these user-specified parameters within the EDF framework. The automatic choice of M in the maximal test (4.40) reduces the bias-variance trade-off associated with the Hinich linearity test statistic. However, the resulting MD_{IDR}^L test statistic still relies on the asymptotic normality of the bispectrum.

On the other hand, no asymptotic distributions are utilized with the bootstrap based tests which may be viewed as a great advantage over the above test statistics. The disadvantage of this method is that one has to choose M and P. In addition, the order p of the AR approximation needs to be selected. One approach is to adopt order selection criteria as AIC or BIC. Alternatively, a bootstrap method for AR order selection may be included into the bootstrap algorithm; see, e.g., Zoubir (1999). Berg et al. (2010) report that, in general, there is not much sensitivity of the obtained test results due to the selection of the above parameters.

Furthermore, with the bootstrapped-based tests a decision needs to be made about the number of resamples B. Fortunately with greater computing power, one can often be very conservative and choose a much larger B than needed without any statistical consequences. As the number of resamples increases so does the accuracy of the test results. One simple diagnostic is to run the bootstrap algorithm twice with the same size B. If the results are adjudged to be similar, and the conclusions drawn remain the same, then the resample size can be considered to be adequate.

Finally, the bootstrap algorithm uses the direct estimation method of the bispectrum, similar to the Subba Rao–Gabr test statistics. However, a problem with both the direct and indirect estimate is that *leakage* may occur when a real frequency is not matched by a Fourier frequency in the observed data. The effect of this frequency is then leaked into the closest Fourier frequencies. With the indirect estimate, which uses a truncated estimate of the third-order cumulant, the influence of $\widehat{\gamma}_Y(0,0)$ on estimated values of the bispectrum at locations other than $(0,0)$ is potentially greater at lower frequencies. As the estimated value of $\gamma_Y(0,0)$ reflects the skewness of the series $\{Y_t\}_{t=1}^T$ this is more likely to be an issue for non-symmetric time series, especially when T is relatively small.

4.5 A MSFE-Based Linearity Test

In Section 1.1, we introduced a second notion of linearity of a time series process, following the simple definition that a process is linear if the *linear forecast* is optimal in the MSE sense. Terdik and Máth (1998) and Terdik (1999) use this notion to propose a linearity test statistic based on one-step ahead forecast errors. Suppose we are to make a prediction of Y_{t+1}, at origin t. If $\{Y_t, t \in \mathbb{Z}\}$ is a stationary weakly linear process, then the one-step ahead ($H = 1$) *least squares* (LS), minimum mean squared error, forecast is given by

$$Y_{t+1|t}^{\text{LS}} \equiv \mathbb{E}(Y_{t+1}|Y_s, -\infty < s \leq t) = Y_t + \sum_{i=1}^{\infty} \psi_i Y_{t-i}, \qquad (4.52)$$

4.5 A MSFE-BASED LINEARITY TEST

where ψ_i ($i = 1, 2, \ldots$) are to be determined. The process $\{e_{t+1|t}\}$, with $e_{t+1|t} = Y_{t+1} - Y_{t+1|t}^{\text{LS}} \equiv \varepsilon_{t+1}$, is the one-step ahead forecast error, or *innovation process*. It fulfils the conditions:

$$\mathbb{E}(e_t|\mathcal{F}_{t-1}) = 0, \quad \mathbb{E}(e_t^2|\mathcal{F}_{t-1}) = \sigma_\varepsilon^2, \qquad (4.53)$$

where $\mathcal{F}_t$ is the σ-algebra generated by $\{e_s, s \leq t\}$.

Many nonlinear predictors exist which do not require an explicit specification of the type of nonlinearity. Among these predictors, Masani and Wiener (1959) show that the best forecast which minimizes the one-step ahead mean squared forecast error (MSFE), i.e. MSFE$(H) = \mathbb{E}(e_{t+H|t}^2)$ with $H = 1$, is given by a polynomial of the observed time series and, under some suitable conditions, can be constructed by using only the values of the moments. The resulting one-step ahead *quadratic (Q) forecast* is given by

$$Y_{t+1|t}^{\text{Q}} = Y_t + \sum_{j=1}^{\infty} c_j Y_{t-j} + \sum_{j,v=0}^{\infty} c_{jv} Y_{t-j} Y_{t-v}, \qquad (4.54)$$

where the coefficients c_j and c_{jv} are chosen such that minimum of MSFE(1) is achieved. If $\{Y_t, t \in \mathbb{Z}\}$ is non-Gaussian, then the one-step ahead quadratic forecast has a smaller asymptotic MSFE than the one-step ahead linear forecast (cf. Exercise 4.2(b)).

Null- and alternative hypotheses

For simplicity of notation, we denote the process $\{e_{t+1|t}\}$ by $\{e_t\}$, and we assume that $\{e_t\}$ is a strictly stationary process with ACVF satisfying similar conditions as given by (4.49). In this case it is easy to see that $\{e_t\}$ is an uncorrelated process, and therefore it will not necessarily satisfy condition (1.3). Now suppose that the best one-step ahead LS forecast $Y_{t+1|t}^{\text{LS}}$ has already been constructed and the objective is to check the assumption $Y_{t+1|t}^{\text{LS}} = Y_{t+1|t}^{\text{Q}}$. Thus, in terms of the one-step ahead forecast errors, the null- and alternative hypotheses of interest are:

$$\mathbb{H}_0: \; \mathbb{E}[\{Y_{t+1} - Y_{t+1|t}^{\text{Q}}\} - \{Y_{t+1} - Y_{t+1|t}^{\text{LS}}\}]^2 = \mathbb{E}[Y_{t+1|t}^{\text{LS}} - Y_{t+1|t}^{\text{Q}}]^2 = 0, \qquad (4.55)$$

$$\mathbb{H}_1: \; \mathbb{E}[Y_{t+1|t}^{\text{LS}} - Y_{t+1|t}^{\text{Q}}]^2 > 0. \qquad (4.56)$$

Assume that the fourth-order moments of $\{Y_t, t \in \mathbb{Z}\}$ exists, and let $f_Y(\omega)$ satisfy the so-called *Szegö condition*, i.e., $\int_0^1 \log f_Y(\omega) d\omega > -\infty$, and assume all finite-dimensional distributions of $\{Y_t, t \in \mathbb{Z}\}$ have a positive spectrum. Then, in view of the symmetry relations (4.2), it can be shown (Terdik and Máth, 1993) that a necessary and sufficient condition for equivalence of $Y_{t+1|t}^{\text{LS}}$ and $Y_{t+1|t}^{\text{Q}}$ is that the bispectrum $f_e(\omega_1, \omega_2)$ of the innovation process has the additive form

$$f_e(\omega_1, \omega_2) = H(\omega_1) + H(\omega_2) + H^*(\omega_1 + \omega_2), \qquad (4.57)$$

where $H(\omega) = \sum_{j=0}^{\infty} \gamma_e(j,j) \exp(-2\pi i \omega_j)$. The functions $f_e(\cdot,\cdot)$ which satisfy (4.57) are exactly those for which the following relation holds. For any triplet (α, β, γ)

$$f_e(\alpha, \beta) + f_e(\gamma, 0) + f_e(-\alpha + \gamma, -\beta - \gamma) = f_e(\beta, \gamma) + f_e(0, -\alpha - \beta) \\ + f_e(-\alpha + \gamma, -\gamma). \quad (4.58)$$

This relationship forms the basis of the proposed linearity test statistic.

Test statistic

Consider the third-order periodogram of $\{e_t\}_{t=1}^T$

$$F_e(\omega_1, \omega_2) = e(\omega_1)e(\omega_2)e^*(\omega_1 + \omega_2)/T,$$

where $e(\omega_j) = \sum_{t=0}^{T-1} e_t \exp\{-2\pi i \omega_j\}$ ($j = 1, 2$). Then, analogous to (4.17), an asymptotically unbiased and consistent estimator of $f_e(\omega_1, \omega_2)$ can be obtained by smoothing with a two-dimensional window $\lambda(\cdot, \cdot)$, satisfying the symmetry relations (4.42) while at all frequencies (ω_1, ω_2) its values are again in the principal domain $\mathcal{D}$, the triangle with vertices $(0, 0)$, $(0, 1/2)$, $(1/3, 1/3)$ (see Figure 4.4). Terdik and Máth (1998) choose $\lambda(\omega_1, \omega_2)$ to be zero for $|\omega_j| > 1/2$ ($j = 1, 2$). The smoothed version of $f_e(\omega_1, \omega_2)$ is defined by

$$\widehat{f}_e(\omega_1, \omega_2) = \frac{1}{(Tb_T)^2} \sum_{u,v=1}^{T-1} W_1(u, v) F_e(u/T, v/T), \quad (4.59)$$

where b_T denotes a scale parameter such that $b_T > 0$, $b_T \to 0$, $Tb_T^2 \to \infty$ as $T \to \infty$, and where $W_1(u, v) = \lambda\big(b_T^{-1}(\omega_1 - u/T), b_T^{-1}(\omega_2 - v/T)\big)$. Observe that Tb_T plays the same role as M in the previous sections.

The bispectral estimators $\widehat{f}_e(\omega_1, \omega_2)$ are asymptotically independent inside $\mathcal{D}$. On the boundary of $\mathcal{D}$ they are correlated (see, e.g., Brillinger, 1975). If $\omega_1 \neq \omega_2$, $\omega_1 \omega_2 \neq 0$, and $\omega_1 \neq -2\omega_2$, the variance of $\widehat{f}_e(\cdot, \cdot)$ is

$$\lim_{T \to \infty} Tb_T^2 \text{Var}\{\widehat{f}_e(\omega_1, \omega_2)\} = (\sigma_\varepsilon^2)^3 W_2, \quad (4.60)$$

which implies

$$\lim_{T \to \infty} Tb_T^2 \text{Var}\{\Re\big(\widehat{f}_e(\omega_1, \omega_2)\big)\} = \frac{\sigma_\varepsilon^6 W_2}{2} \text{ and } \lim_{T \to \infty} Tb_T^2 \text{Var}\{\Im\big(\widehat{f}_e(\omega_1, \omega_2)\big)\} = \frac{\sigma_\varepsilon^6 W_2}{2},$$

where W_2 is given by (4.44). If $0 < \omega_1 < 1/2$, then

$$\lim_{T \to \infty} Tb_T^2 \text{Var}\{\widehat{f}_e(\omega_1, 0)\} = \sigma_\varepsilon^6 (W_2 + W_{01}), \quad (4.61)$$

where $W_{01} = \int_{-\infty}^{\infty} \lambda(0, \omega) d\omega$.

4.5 A MSFE-BASED LINEARITY TEST

To obtain a practical test, all frequencies (ω_1, ω_2) must be mapped into $\mathcal{D}$. In view of the symmetry conditions, and without changing the value of the bispectrum except for complex conjugation, this can be done using the following transformations:

$$T_1(\omega_1, \omega_2) = (\omega_2, \omega_1), \quad T_2(\omega_1, \omega_2) = (\omega_1, -\omega_2 - \omega_1),$$
$$T_3(\omega_1, \omega_2) = (-\omega_1 - \omega_2, \omega_2), \quad T_4(\omega_1, \omega_2) = (-\omega_1, -\omega_2).$$

Now, let (α, β, γ) denote a fixed triplet such that the map of $T_i(\cdot, \cdot)$ ($i = 1, \ldots, 4$) of the six points

$$(\alpha, \beta), \ (\gamma, 0), \ (-\alpha + \gamma, -\beta - \gamma), \ (\beta, \gamma), \ (0, -\alpha - \beta), \ (-\alpha + \gamma, -\gamma)$$

is different in $\mathcal{D}$. Then, the following statistic can be defined

$$Q_T(\alpha, \beta, \gamma) = \widehat{f}_e(\alpha, \beta) + \widehat{f}_e(\gamma, 0) + \widehat{f}_e(-\alpha + \gamma, -\beta - \gamma) - \widehat{f}_e(\beta, \gamma)$$
$$- \widehat{f}_e(0, -\alpha - \beta) - \widehat{f}_e(-\alpha + \gamma, -\gamma), \qquad (4.62)$$

with its asymptotic expectation

$$Q(\alpha, \beta, \gamma) = f_e(\alpha, \beta) + f_e(\gamma, 0) + f_e(-\alpha + \gamma, -\beta - \gamma) - \big(f_e(\beta, \gamma) + f_e(0, -\alpha - \beta)$$
$$+ f_e(-\alpha + \gamma, -\gamma)\big).$$

Under $\mathbb{H}_0$, we have $Q(\alpha, \beta, \gamma) = 0$. Moreover, under $\mathbb{H}_0$ and as $T \to \infty$, (4.62) is asymptotically complex normal distributed with mean zero and variance $\text{Var}\{Q_T(\alpha, \beta, \gamma)\} \approx 6\sigma_\varepsilon^6 W_2 / T b_T^2$.

Now, rather than using $Q_T(\alpha, \beta, \gamma)$ as a test statistic for linearity, Terdik and Máth (1998) use a standardized form of $Q_T(\alpha, \beta, \gamma)$. To this end they first define

$$R_{1,T}(\alpha, \beta, \gamma) = \Re\{Q_T(\alpha, \beta, \gamma)\} \left(\frac{1}{2}\text{Var}\{Q_T(\alpha, \beta, \gamma)\}\right)^{-1/2}$$
$$R_{2,T}(\alpha, \beta, \gamma) = \Im\{Q_T(\alpha, \beta, \gamma)\} \left(\frac{1}{2}\text{Var}\{Q_T(\alpha, \beta, \gamma)\}\right)^{-1/2}.$$

Next, the entire set of observations is divided into K separate stretches of length T. Let $R_{j,T}^{(i)}(\alpha, \beta, \gamma)$ ($i = 1, \ldots, K; j = 1, 2$) denote the (i, j)th statistic resulting from this approach. These $2K$ statistics are asymptotically independent with the same distribution as $R_{j,T}(\alpha, \beta, \gamma)$. From this, the standardized real and complex parts of $Q_T(\alpha, \beta, \gamma)$ are given by

$$M_{j,T}^{(K)}(\alpha, \beta, \gamma) = K^{-1/2} \sum_{i=1}^{K} R_{j,K}^{(i)}(\alpha, \beta, \gamma) \quad (j = 1, 2). \qquad (4.63)$$

Under $\mathbb{H}_0$, the expectation and variance of $M_{j,T}^{(K)}(\alpha, \beta, \gamma)$ ($j = 1, 2$) are respectively approximately equal to zero and unity. The resulting test statistic is given by

$$\mathcal{G}_T^{(K)} = \{M_{1,T}^{(K)}(\alpha,\beta,\gamma)\}^2 + \{M_{2,T}^{(K)}(\alpha,\beta,\gamma)\}^2. \tag{4.64}$$

Under $\mathbb{H}_0$, and as $T \to \infty$, $\mathcal{G}_T^{(K)}$ has a χ_2^2 distribution.

Computation
Clearly, (4.64) is computed for only one set of triplets in $\mathcal{D}$. Generalizing to n sets of triplets, each consisting of K stretches, is direct. The various stages in the computation of the resulting test statistic can be summarized as follows.

Algorithm 4.5: The MSFE-based linearity test statistic

(i) According to some order selection criterion determine p, and fit an AR(p) to the observed time series $\{Y_t\}_{t=1}^T$. Obtain the residuals $\{\widehat{\varepsilon}_t\}_{t=1}^T$.

(ii) Segment the series $\{\widehat{\varepsilon}_t\}_{t=1}^T$ into K stretches of length $N = 2^x$ ($x \geq 6, x \in \mathbb{Z}$), so $K = \lfloor T/N \rfloor$. Select a window-width Nb_N. A recommended choice for b_N is $N^{-0.49}$, so $Nb_N = N^{0.51}$ which parallels the choice of M in the bispectral estimator (4.27). Then compute the bispectral estimates $\widehat{f}_{\widehat{\varepsilon}}(\omega_j, \omega_k)$ ($j, k = 1, \ldots, N$).

(iii) Compute the bispectral estimates $\widehat{f}_{\widehat{\varepsilon}}(\omega_j, \omega_k)$ ($j, k = 1, \ldots, N$). A recommended choice for the weight function $\lambda(\cdot, \cdot)$ is

$$\lambda(\omega_1, \omega_2) = \begin{cases} \frac{4\sqrt{3}}{\pi}\{1 - 4(\omega_1^2 + \omega_2^2 + \omega_1\omega_2)\}, & (\omega_1^2 + \omega_2^2 + \omega_1\omega_2) < 1/4, \\ 0, & \text{otherwise.} \end{cases} \tag{4.65}$$

The above window is optimal in the sense that it minimizes the MSE of the bispectral estimate. For this window, evaluation of (4.44) gives $W_2 = 1.4628$. Figure 4.6(b) shows a plot of the profile of (4.65).

(iv) Using $n = 7$ triplets $(\alpha_i, \beta_i, \gamma_i)$, construct the two 3×2 matrices with indices

$$\frac{N}{64}\begin{pmatrix} \alpha_i & \beta_i \\ \gamma_i & 0 \\ -\alpha_i + \gamma_i & -\beta_i - \gamma_i \end{pmatrix}, \quad \frac{N}{64}\begin{pmatrix} \beta_i & \gamma_i \\ 0 & -\alpha_i - \beta_i \\ -\alpha_i + \gamma_i & -\gamma_i \end{pmatrix},$$

$$(i = 1, \ldots, n).$$

If an index is negative, then add N to its value. Let $(u, v)_i$ and $(u^*, v)_i$ ($u, u^* = 1, 2, 3; v = 1, 2$) denote the resulting index for the ith triplet, corresponding to either the first or the second matrix. For instance, for $N = 2^6 = 64$, it is recommended to use the set of $n = 7$ triplets given by

$$\{(\alpha_i, \beta_i, \gamma_i)\}_{i=1}^7 = \{(17, 27, 30), (17, 21, 10), (17, 24, 27), (18, 27, 14),$$
$$(18, 21, 24), (19, 30, 1), (21, 27, 9)\}. \tag{4.66}$$

4.5 A MSFE-BASED LINEARITY TEST   145

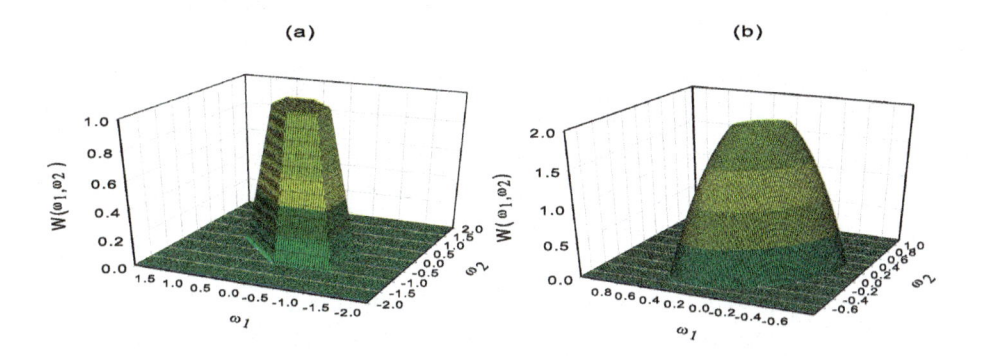

Figure 4.6: *(a) Profile of the flat-top two-dimensional window function (4.51) used with the bootstrap-based test statistics in Algorithm 4.4; (b) Profile of the two-dimensional lag window (4.65) used in (4.59).*

Algorithm 4.5: The MSFE-based linearity test statistic (Cont'd)

(v) Compute the complex-valued statistic

$$Q_i = \sum_{u=1}^{3} \widehat{f_\varepsilon}(\omega_{(u,1)_i+1}, \omega_{(u,2)_i+1}) - \sum_{u^*=1}^{3} \widehat{f_\varepsilon}(\omega_{(u^*,1)_i+1}, \omega_{(u^*,2)_i+1}),$$

$(i = 1, \ldots, n).$

(vi) Form the vector $\mathbf{Q} = (Q_1, \ldots, Q_n)'$, and compute the test statistic

$$\mathcal{G}_{n,T}^{(K)} = K \times \left(\frac{Nb_N^2}{3W_2}\right) \|\mathbf{Q}\|^2, \qquad (4.67)$$

where $\|\cdot\|$ denotes the Euclidean norm. Under $\mathbb{H}_0$, and as $T \to \infty$, the statistic (4.67) has an asymptotic central χ_ν^2 distribution with $\nu = 2n$ degrees of freedom.

Note that for the construction of the test it is assumed that the coefficients ψ_i in (4.52) and the coefficients c_j, c_{ju} in (4.54) are known. In practice these coefficients need to be estimated. However, under not too restrictive conditions on $\{e_t\}$, it can be shown (Matsuda and Huzii, 1997) that the quadratic predictor $Y_{t+1|t}^Q$ has a smaller asymptotic MSE than the LS predictor $Y_{t+1|t}^{LS}$, if $p \geq p^*$, where p and p^* are limits imposed on the infinite summations on the right-hand side of (4.52) and (4.54) respectively. Thus, $\mathbb{H}_0$ can still be tested using the statistic (4.67) if the unknown parameters are replaced by least squares estimates.

Discussion
One disadvantage of the above method of smoothing the bispectrum into K equal nonoverlapping records of size N is that information will be lost at lower frequencies,

the maximum cycle that we can now observe is for frequency N instead of frequency T. Also, since $K = \lfloor T/N \rfloor$ will not be an integer in general, some observations at the end of the series may be left out of the computation of the test statistic. Clearly, the alternative hypothesis $\mathbb{H}_1$ presents limitations in that it only examines second-order features in departures from the null hypothesis. Terdik and Máth (1998) compare the power of the test statistic (4.31) with Hinich's linearity test statistic for a number of (non)linear models, but $\mathcal{G}_{n,T}^{(K)}$ only shows an improvement for linear Hermite polynomial data. Applications of the Terdik–Máth test statistic are reported by, for instance, Terdik (1999), Terdik and Máth (1993), and Terdik et al. (2002).

4.6 Which Test to Use?

As stated earlier there are various strengths and weaknesses of frequency-domain test statistics. This section presents some additional information. Usually the overall performance of a test is obtained from a size and power study. A number of these studies have been carried out for the tests discussed above; see Table 4.1 for a summary. Some general observations are in order.

- The empirical rejection levels (sizes) for linear DGPs with Gaussian distributed errors from many simulation studies are not always at the nominal rejection level, which in most studies is preset at 5%. Hence, it is somewhat unfair to compare the powers of test statistics that have different sizes.

- The bootstrap test statistics give generally better power results than Hinich's Gaussianity and linearity tests. The classical Hinich linearity test, $Z_{\text{IQR}}^{\text{L}}$, gives poor answers for very short series as it often has too few independent values to form an IQR.

- Of the three maximal linearity test statistics the maximal IDR test statistic, $Z_{\text{IDR}}^{\text{L}}$, has the largest power improvement over the Hinich linearity test, which reinforces the conjecture that by carefully tweaking the user-specified parameters some improvement of the Hinich linearity test can be obtained. However, the overall performance of the IDR test statistic is quite limited for data generated from a two-state Markov(2, 1) model, an EAR(2, 1) model, and a rational nonlinear AR model.

- The power of the $\text{AD}_{\text{m}}^{\text{G}}$ and $\text{CvM}_{\text{m}}^{\text{G}}$ test statistics is comparable with that achieved by the Hinich test statistic T^{G}, but often higher, especially in the case of data generated from a SETAR(2; 1, 1) model.

Although there is no frequency-domain test statistic which uniformly outperforms all other tests for all DGPs and sample sizes considered in the literature, we recommend the use of the model-based bootstrap method jointly with the direct estimation method of the bispectrum. The method is more powerful than the Hinich

4.6 WHICH TEST TO USE?

Table 4.1: *Summary of size and power MC simulation studies for some frequency-domain Gaussianity* (G) *and linearity* (L) *test statistics.*

DGPs	T	M	Tests	Reference
BL(0,0,2,1), NLMA, extended NLMA, NLAR, SETAR(2;1,1), NL-TAR, ExpAR(2)	$\begin{cases}256\\512\\1{,}024\end{cases}$	12 16 23	Z^L_{IQR}, $Z^L_{80\%}$, T^G	Ashley et al. (1986)
AR(2), MA(2), ExpAR(1), BL(1,0,1,1), SETAR(2;1,1), 2 NLMAs, BL(2;1,1,1)	104	11	Z^L_{IQR}	W.S. Chan and Tong (1986)[1]
AR(2), Hermite polynomial of order 2, BL(2,0,1,1), BL(0,0,2,1), homogeneous BL with Hermite degree 2, homogeneous BL with polynomials	512	12	Z^L_{IQR}, $\mathcal{G}^{(4)}_{7,128}$	Terdik and Máth (1998)
i.i.d. $\mathcal{N}(0,1)$, AR(2), MA(2), NLMA, BL(2,1,1,1), SETAR(2;1,1), ESTAR(1), ExpAR(1), NLAR	$\begin{cases}100\\500\end{cases}$	10 22	T^G, AD^G_m, CvM^G_m	Jahan and Harvill (2008)[2]
NLMA, BL(0,0,2,1), ARCH(4), GARCH(1,1), SETAR(2;1,1), two state Markov(2,1), EAR(2,1), rational NLAR, exp. damped AR(2), logistic(4) map	350	34 [8 – 45]	Z^L_{IQR}, Z^L_{IDR}, $Z^L_{80\%}$, $\begin{cases}MD^L_{IQR},\\MD^L_{IDR},\\MD^L_{80\%}\end{cases}$	Rusticelli et al. (2009)
i.i.d. $\mathcal{N}(0,1)$, i.i.d. χ^2_1, AR(1), ARMA(2,2), BL(1,0,1,1), ARCH(1), GARCH(1,3), SETAR(4;1,2,1,1)$_2$	$\begin{cases}250\\500\\1{,}000\end{cases}$	4 [4] 6 8	$\begin{cases}T^G_{IQR},\\Z^{L+nG}_{IQR},\\Z^{L+S}_{IQR}\end{cases}$	Berg et al. (2010)[3]

[1] The paper includes a comparison with four time-domain nonlinearity tests.
[2] The paper includes a comparison with five time-domain nonlinearity tests.
[3] The study makes a distinction between the spectral bandwidth (M_s), and the bispectral bandwidth ($M_b \equiv M$). Asymptotically $M_s > M_b$.
[4] Other user-defined parameters are $K = 21$, $M_s = 8$, $p = 15$ for $T = 250$; $K = 36$, $M_s = 12$, $p = 20$ for $T = 500$; and $K = 55$, $M_s = 15$, $p = 30$ for $T = 1{,}000$.

Table 4.2: *Indicator pattern of p-values of the Gaussianity (G) and linearity (L) test statistics;* ** *marks a p-value* < 0.01, * *marks a p-value in the range* 1% − 5%, *and* † *a p-value* > 0.05.

	Gaussianity (G)			Linearity (L)					MSFE[3]
	GOF Tests[1]		Btstrp[2]	GOF Tests[1]		Btstrp[2]			
Series	AD_m^G	CvM_m^G	T^G	AD_m^L	CvM_m^L	Z_{IQR}^L	Z_{IDR}^L	$Z_{80\%}^L$	$\mathcal{G}_{7,T}^{(K)}$
Unemployment rate[4]	**	*	*	†	†	**	**	*	†
EEG recordings	**	**	**	†	†	**	**	**	**
Magnetic field data	**	*	†	**	†	†	†	†	**
ENSO phenomenon	**	†	†	†	†	†	†	†	**
Climate change: $\delta^{13}C$	†	**	†	†	†	†	†	†	**
$\delta^{18}O$	**	*	†	†	†	**	**	**	†

[1] $M = 18$ for all series.
[2] Based on 1,000 bootstrap replicates, and $M = \lfloor T \rfloor^{0.6}$ for all series.
[3] Based on stretch lengths $N = 2^7$ (Unemployment, $\delta^{13}C$, and $\delta^{18}O$), $N = 2^8$ (ENSO) $N = 2^9$ (EEG), $N = 2^{10}$ (Magnetic field data); window-width $Nb_N = 8$, $p_{max} = 24$.
[4] First differences of original series.

test statistics based on the asymptotic properties of the bispectrum. An obvious extension of the bootstrap method is to allow for an automatic grid search over the admissible M values, as for instance discussed in Section 4.4.2, to reduce the sensitivity of the tests to the choice of this parameter. Another extension of this method is to use fourth, or higher-order, polyspectra as a test statistic, using the same test framework.

4.7 Application: A Comparison of Linearity Tests

We now apply some of the above test statistics to the time series introduced earlier in Chapter 1. Table 4.2 shows the test results. We see that the GOF test statistics reject Gaussianity in almost all cases. On the other hand, the bootstrap version of the Hinich test statistic only rejects Gaussianity for the first differences of the U.S. unemployment series, and the EEG recordings. Recall from Table 1.2 (Example 1.7), that the parametric normality test statistic $\hat{\pi}_{34,Y}$ flat-out rejected Gaussianity for the EEG recordings and the magnetic field data. So, in summary, there seems to be some inconsistencies between the results of these test statistics.

When testing for linearity, we see that all GOF test statistics do not indicate that the series are nonlinear, except for the magnetic field data. However, the three bootstrap-based test statistics Z_{IQR}^L, Z_{IDR}^L, and $Z_{80\%}^L$ identify the first differences of the U.S. unemployment rate, the EEG recordings, and the $\delta^{18}O$ series to be nonlinear. So also in this case the test results vastly differ among the test statistics. To some extent these differences may be attributable to the choice of user-defined parameters as, e.g., deciding on an appropriate value of M. This comment also applies to the MSFE-based test statistic $\mathcal{G}_{7,T}^{(K)}$ which in addition to the choice of the window bandwidth, also depends on the stretch length N, and the order of the fitted autoregression.

4.8 Summary, Terms and Concepts

Summary

In this chapter we introduced the bispectrum and third-order moment as useful tools for detecting non-symmetry (in terms of the marginal distribution), nonlinearity, and possibly time-reversibility. We discussed two main estimates of the bispectrum, namely the direct and indirect method. We reviewed two "traditional" bispectrum-based test statistics for Gaussianity and nonlinearity, i.e., the Subba Rao–Gabr tests and the Hinich tests. Further, we indicated some strengths and weaknesses of these test statistics.

Various modifications and improvements of the Hinich test statistics have been considered, including two bootstrap-based versions. Also, we provided a brief literature review of MC simulation studies, comparing the size and power of the Gaussianity and linearity test statistics. Finally, we used several test statistics to investigate the nonlinear properties of the time series previously introduced in Chapter 1.

An important advantage of bispectral analysis is that tests discussed in this chapter can be applied either to the raw (original) series or to the residuals of a fitted model; see, e.g., Ashley et al. (1986). Hence, there is no need to prefilter the data first, using a fixed causal linear filter, in order to remove possible autocorrelations. This reduces the possibility of a misspecified nonlinear model and distorted statistical inference.

Terms and Concepts

aliasing, 150
bispectrum, 121
bootstrapping, 136
designated frequencies, 126
(in)direct method, 123
Fourier transform (FT), 120
frequency bicoherence, 123
goodness-of-fit (GOF) tests, 133
Hinich's tests, 130
interdecile range (IDR), 136
interquartile range (IQR), 132
leakage, 140
linear (L) forecast, 140

maximal tests, 136
mean squared forecast error (MSFE), 140
normalized bispectrum, 122
polyspectrum, 121
principal domain, 121
quadratic (Q) forecast, 141
spectrum, 120
Subba Rao–Gabr tests, 126
third-order cumulant, 124
third-order periodogram, 123
transfer function, 123
truncation point, 124

4.9 Additional Bibliographical Notes

Section 4.1: A rigorous treatment of the bispectrum is given by Brillinger and Rosenblatt (1967). Van Ness (1966) proves, under general conditions, that the bispectrum is asymptotically complex normal. There are several definitions of power spectra in the case of nonstationary processes; see Priestley (1988) for a review and Priestley and Gabr (1993) for a time-dependent definition. Subba Rao and Gabr (1984) update their original frequency domain tests to include frequencies along the manifold $\omega_j = 0$. Zoubir and Iskander (1999) propose a bootstrap-based approach for testing departures from Gaussianity. Their simulation results confirm that the Subba Rao–Gabr test statistic is a test of symmetry and not pure Gaussianity. Nichols et al. (2009) provide an analytical expression for the bispectrum and bicoherence functions for quadratically nonlinear DGPs subject to stationary, jointly non-Gaussian distributed error processes possessing an arbitrary ACF.

Lii and Masry (1995) and Lii (1996) consider estimation of the bispectral density function of continuous stationary DGPs when the data are obtained on unequally spaced time intervals. Subba Rao (1997) gives an illustration of the usefulness of bispectra to analyze nonlinear, unequally spaced, astronomical time series. Related to the analysis of continuous time series, the problem of *aliasing* may arise when a real frequency in the series is not matched by a Fourier frequency in the observed data. Testing for aliasing can be performed by an amended version of the Hinich bispectrum test statistic for Gaussianity; see Hinich and Wolinsky (1988).

Harvill et al. (2013) propose a bispectral-based procedure to distinguish among various nonlinear time series processes and between nonlinear and linear time series processes through application of a hierarchical clustering algorithm.

Barnett and Wolff (2005) advocate the time-domain third-order moment $\gamma_Y(\ell_1, \ell_2)$ for testing nonlinearity over using the bispectrum. For a linear stationary time series the estimated values of the third-order moment are correlated. This complicates the construction of a parametric test. They overcome this problem by using the so-called *phase scrambled* bootstrap procedure (Theiler et al., 1992), a frequency domain procedure. The method is computationally less intensive and more powerful than the Hinich test statistic. Three MATLAB files are available at http://www.mathworks.nl/matlabcentral/fileexchange/16062-test-of-non-linearity. These files are: third.m (calculates the 3rd-order moment for a time series), aaft.m (calculates the Amplitude Adjusted FT), and boot.m (calculates a bootstrap test for nonlinearity).

Section 4.2: Based on the evolutionary second-order spectrum and bispectrum (see, e.g., Priestley and Gabr (1993)), Tsolaki (2008) proposes test statistics for Gaussianity and linearity of nonstationary slowly varying time series processes. These test statistics are generalizations of the Subba Rao–Gabr tests for stationary processes.

Section 4.3: The use of a square shaped uniform smoothing window in the direct estimator of the bispectrum in Hinich's linearity and Gaussianity test statistics may introduce severely biased estimates in relatively small areas of the bispectrum, and hence may lead to a false acceptance of the null hypothesis with large probability. To ameliorate this problem, Birkelund and Hanssen (2009) obtain an improved version of Hinich's tests by proposing a hexagonal shaped smoothing window. Yuan (2000a) investigates the effect of estimating the noncentrality parameter λ_0 on the asymptotic level of Hinich's linearity test, and he introduces a modification. The modified test also uses the IQR, but it tests the equality of location parameters and its critical value does not depend on any unknown parameters. In another paper, Yuan (2000b) extends Hinich's Gaussianity and linearity test statistics to stationary random fields on $\mathbb{Z}^m$ ($m = 1, 2, \ldots$).

Section 4.7: Ashley and Patterson (1989), and Hinich and Patterson (1985) apply the Subba Rao–Gabr test statistics and the Hinich test statistics to various real economic time series. Brockett et al. (1988) and Patterson and Ashley (2000) present applications of these tests with series taken from other areas, including examples from, finance, engineering, and geophysics. Teles and Wei (2000) investigate the performance of various linearity test statistics, including Hinich's linearity test, on time series aggregates. Temporal aggregation greatly hampers the detection of nonlinear DGPs.

Drunat et al. (1998) compare the Hinich and the Subba Rao–Gabr linearity tests on a set of exchange rates. A modified version of the original Hinich linearity test statistic forms a part of a single-blind controlled competition among five linearity tests, and results are reported by Barnett et al. (1997). Hinich et al. (2005) examine the performance of Hinich's Gaussianity and linearity tests and the Hinich–Rothman test statistic for time-reversibility (Chapter 8), using bootstrap and surrogate data simulation methods. Using knowledge of the asymptotic distribution of the bispectral density function under the null hypothesis of Gaussianity, Epps (1987) proposes a large-sample GOF-type test statistic based on the difference between the sample mean estimate and the ensemble averaged value of the characteristic function of the time series, measured at some specific points. The AR-sieve bootstrap, discussed briefly in Section 4.4.3, is reviewed in detail in Kreiss and Lahiri (2011).

4.10 Software References

Section 4.2: A FORTRAN77 program for computing the Subba Rao–Gabr linearity test is listed as Program 4 on pp. 263 – 269 of Subba Rao and Gabr (1984). An extended version of this program can be downloaded from the website of this book.

Section 4.3: A public domain FORTRAN77 code for computing the Hinich test statistics can be downloaded from `http://www.la.utexas.edu/hinich/`. A user-friendly executable version of this code is contained in the nonlinear toolkit for detecting and identifying nonlinear time series, and detailed in Patterson and Ashley (2000); see `http://ashleymac.econ.vt.edu`. The toolkit was used to calculate the bootstrap results for the test statistics T^G and Z^L in Table 4.2. The MATLAB toolbox HOSA contains the file GLSTAT that can be used to calculate Hinich's Gaussianity and linearity test statistics with the approximation of the noncentral $\chi_2^2(\cdot)$ distribution as discussed in Section 4.4.1.

Section 4.4: The empirical results of the AD- and CvM-type Gaussianity and linearity test statistics (Table 4.2) can be reproduced with the goodnessfit.m MATLAB function available at the website of this book. Also available is R code for computing the bootstrapped form of Hinich's Gaussianity and linearity test statistics of Section 4.4.3; see Exercise 4.4. Furthermore, Gyorgy Terdik made available TerM.m, a MATLAB module for calculating the Terdik–Máth test statistic.

Exercises

Theory Questions

4.1 Prove that the triangular principal domain (4.7) of the bispectral density function $f_Y(\omega_1, \omega_2)$ is bounded by the manifolds $\omega_1 = \omega_2$, $\omega_1 = 0$, and $\omega_1 = (1 - \omega_2)/2$.

4.2 Consider the subdiagonal BL process $Y_t = \beta Y_{t-2} \varepsilon_{t-1} + \varepsilon_t$, where $\{\varepsilon_t\} \stackrel{\text{i.i.d.}}{\sim} \mathcal{N}(0, \sigma_\varepsilon^2)$ with $\beta^2 \sigma_\varepsilon^2 < 1$.

(a) Prove that

$$\gamma_Y(k) = \begin{cases} \sigma_\varepsilon^2/(1-\beta^2\sigma_\varepsilon^2), & k=0, \\ 0, & \text{otherwise,} \end{cases}$$

$$\mathbb{E}(Y_t Y_{t-k} Y_{t-\ell}) = \begin{cases} \beta\sigma_\varepsilon^4/(1-\beta^2\sigma_\varepsilon^2), & k=1,\ \ell=2, \\ 0, & \text{otherwise,} \end{cases}$$

and

$$\mathbb{E}(Y_t^2 Y_{t-1}^2) = \frac{\sigma_\varepsilon^4(1+2\beta^2\sigma_\varepsilon^2)}{(1-\beta^2\sigma_\varepsilon^2)^2}.$$

(b) The best one-step ahead quadratic predictor for $\{Y_t, t \in \mathbb{Z}\}$ is given by

$$Y_{t+1|t}^Q = c_{1,2} Y_t Y_{t-1}.$$

Using the moment results in part (a), prove that the coefficient $c_{1,2}$ is given by

$$c_{1,2} = \beta \frac{1-\beta^2\sigma_\varepsilon^2}{1+2\beta^2\sigma_\varepsilon^2}.$$

(c) Show that the maximum reduction of the one-step ahead MSFE of $Y_{t+1|t}^Q$, relative to $\mathbb{E}(Y_t^2) = \sigma_Y^2$, is reached at $\beta^2\sigma_\varepsilon^2 = (\sqrt{3}-1)/2$.

4.3 By assuming that the bispectrum is non-zero over the entire region $\mathcal{D}$, and that $f_Y(\omega_1, \omega_2)$ is partially differentiable once with respect to ω_1, Sakaguchi (1991) shows that for any triplet (α, β, γ) the bispectrum $f_Y(\omega_1, \omega_2)$ satisfies the relation

$$f_Y(\alpha,\beta) f_Y(\gamma,0) f_Y(-\alpha+\gamma,-\beta-\gamma) = f_Y(\beta,\alpha) f_Y(0,-\alpha-\beta) f_Y(-\alpha+\gamma,-\gamma). \quad (*)$$

This relation may be viewed as an alternative to (4.58).

(a) Consider the stationary nonlinear process defined by

$$Y_t = \varepsilon_t(1+\varepsilon_{t-1}) + (\eta_t^2 - 1),$$

where $\{\varepsilon_t\}$ and $\{\eta_t\}$ are independent and Gaussian i.i.d. processes with zero mean and unit variance. Show that the bispectrum is given by

$$f_Y(\omega_1, \omega_2) = 2[\exp\{-2\pi i(\omega_1+\omega_2)\} + \exp(2\pi i \omega_1) + \exp(2\pi i \omega_2)] + 8,$$
$$(\omega_1, \omega_2) \in [0,1]^2.$$

(b) Let $\alpha = \beta = 1/4$ and $\gamma = 0$. Show that for the above nonlinear process the left-hand side of $(*)$ is equal to 728 while the right-hand side is equal to 600, indicating that the series is nonlinear.

Empirical and Simulation Questions

4.4 Consider the first differences (USunemplmnt_first_dif.dat) of the quarterly U.S. unemployment rate, earlier introduced in Example 1.1.

(a) Using the R functions in the file Exercise44.r, write an MC simulation program to compare Hinich's Gaussianity test and Hinich's linearity test with bootstrapped forms of these tests. To evaluate the test statistics consider 1,000 BS replicates, and take 20 MC simulations across all tests.

Compare the percentage of rejections of the test statistics at the 5% nominal significance level. Are the results sensitive to the user-specified parameters (inputs) in the simulations?

[*Inputs:* The number of gridpoints K, a discrete uniform random variable taking values in the set $\{3, 4, 5\}$. The spectral bandwidth $M_s = cM_b$ where $c \sim U[1.5, 3]$ and the bispectral bandwidth $M_b = 4$. The bootstrap AR order parameter p, a discrete uniform random variable taking values in the set $\{4, 5, \ldots, 15\}$.]

(b) Compare part (a) with the corresponding test results reported in Table 4.2.

4.5 Consider the set of R functions in the file Exercise45.r.

(a) Generate 100 series of length $T = 250$ for the linear Gaussian processes $\{Y_t\} \stackrel{i.i.d.}{\sim} \mathcal{N}(0, 1)$, and for the linear, but non-Gaussian, process $\{Y_t\} \stackrel{i.i.d.}{\sim} \chi_1^2$. Compute and compare the percentages of rejections of Hinich's Gaussianity test and Hinich's linearity test with bootstrapped forms of these tests similar as in Exercise 4.4. Take $B = 200$, $M_b = 4$, $M_s = 8$, $p = 15$, and set the nominal significance level at 5%.

[*Note:* The computations can be time demanding.]

(b) Generate 100 series of length $T = 250$ for the diagonal BL process (4.19) with $\beta = 0.4$ and $\{\varepsilon_t\} \stackrel{i.i.d.}{\sim} \mathcal{N}(0, 1)$. Compute the percentages of rejections of the test statistics similar as in part (a). Comment on the obtained results.

Chapter 5

TIME-DOMAIN LINEARITY TESTS

Time-domain linearity test statistics are parametric; that is, they test the null hypothesis that a time series is generated by a linear process against a pre-chosen particular nonlinear alternative. Using the classical theory of statistical hypothesis testing, time-domain test nonlinearity tests can be based on three principles – the likelihood ratio (LR), Lagrange multiplier (LM), and Wald (W) principles. LR-based test statistics require estimation of the model parameters under both the null and the alternative hypothesis, whereas tests statistics based on the LM principle require estimation only under the null hypothesis. Application of W-based test statistics implies that the model parameters under the alternative hypothesis need to be estimated. Hence, in the case of complicated nonlinear alternatives, containing many more parameters than the model under the null hypothesis, test statistics constructed from the LM principle are often preferred over test statistics based on the other two testing principles.

In the first three sections that follow, we introduce these three principles briefly and show how they yield the most commonly known test statistics for nonlinearity. In Section 5.4, we discuss three test statistics based on a second-order Volterra expansion. These tests rely on an *added variable approach*, i.e., nonlinearity can be seen by examining the strength of the relationship of the residuals of a fitted linear model with nonlinear terms from a Volterra expansion via an F ratio of sums of squares of residuals. Evidently, this approach is linked to some of the LM test statistics proposed in Section 5.1. In Section 5.5, we first introduce the *arranged autoregression principle*. Based on this principle, we discuss two test statistics for SETARs. Then we discuss an F test statistic that combines the added variable approach with the arranged autoregression principles. Section 5.6 introduces a simple test procedure for discriminating among different nonlinear time series models.

Two appendices are added to the chapter. Appendix 5.A presents percentiles of the LR-SETAR test statistic. Appendix 5.B provides a summary of size and power studies. It includes some remarks about the strengths and weaknesses of the test statistics.

5.1 Lagrange Multiplier Tests

General testing framework

Before we derive LM-based nonlinearity test statistics, it is good to discuss the general testing framework briefly. Let $\{Y_t\}_{t=1}^T$ be a realization of a strictly stationary and ergodic nonlinear process defined by

$$Y_t = g(Y_{t-1}, \ldots, Y_{t-p}, \varepsilon_{t-1}, \ldots, \varepsilon_{t-q}; \boldsymbol{\theta}) + \varepsilon_t, \tag{5.1}$$

where $g(\cdot)$ is a sufficiently well-behaved function on $\mathbb{R}$, and $\boldsymbol{\theta}$ is a vector of unknown parameters. We treat the initial values $\{Y_{-(p\wedge q)+1}, \ldots, Y_0\}$ as fixed constants. This will not affect the distribution of the test statistics in large samples. Furthermore, we assume that the form of (5.1) nests a linear time series model. This implies that $\boldsymbol{\theta}$ can be partitioned as $\boldsymbol{\theta} = (\boldsymbol{\theta}_1', \boldsymbol{\theta}_2')'$, where $\boldsymbol{\theta}_i$ denotes an $\nu_i \times 1$ parameter vector of the linear components ($i = 1, 2$) with $\nu = \nu_1 + \nu_2$.

The null hypothesis we wish to test is $\boldsymbol{\theta}_2 = \boldsymbol{0}$. The LM test statistic is based on parameter estimates of the restricted model. In particular, the Lagrange method states that the (nonlinear) LS estimates under the null hypothesis, denoted by $\widehat{\boldsymbol{\theta}} = (\widehat{\boldsymbol{\theta}}_1', \boldsymbol{0}')'$, are obtained by minimization of the (unrestricted) Lagrange function

$$\mathcal{L}(\boldsymbol{\theta}, \boldsymbol{\lambda}) = L_T(\boldsymbol{\theta}) + 2\boldsymbol{\lambda}'\boldsymbol{\theta}_2, \tag{5.2}$$

where

$$L_T(\boldsymbol{\theta}) = \sum_{t=1}^T \varepsilon_t^2(\boldsymbol{\theta}) \tag{5.3}$$

is the (conditional) sum of squares function and $\boldsymbol{\lambda}$ is an $\nu_2 \times 1$ vector of constants, called *Lagrange multipliers*. Then, one form of the LM (or score) test statistic for $\boldsymbol{\lambda} = \boldsymbol{0}$ is given by

$$\text{LM}_T = \Big(\frac{\partial L_T(\boldsymbol{\theta})}{\partial \boldsymbol{\theta}_2}\Big|_{\mathbb{H}_0}\Big)' (\boldsymbol{\Sigma}_{22} - \boldsymbol{\Sigma}_{21}\boldsymbol{\Sigma}_{11}^{-1}\boldsymbol{\Sigma}_{12})^{-1}\Big|_{\mathbb{H}_0} \Big(\frac{\partial L_T(\boldsymbol{\theta})}{\partial \boldsymbol{\theta}_2}\Big|_{\mathbb{H}_0}\Big), \tag{5.4}$$

where $\boldsymbol{\Sigma}_{11}$, $\boldsymbol{\Sigma}_{12}$, $\boldsymbol{\Sigma}_{21}$ and $\boldsymbol{\Sigma}_{22}$ are $p \times p$ matrices, representing the respective partitions of the Fisher information matrix.

The LM_T test statistic (5.4) is not very illuminating as it stands. It can, however, be rewritten in a much more illuminating way. Define $\mathbf{z}_t(\boldsymbol{\theta}) = \partial \varepsilon_t(\boldsymbol{\theta})/\partial \boldsymbol{\theta}$ and denote $\widehat{\mathbf{z}}_t = \mathbf{z}_t(\widehat{\boldsymbol{\theta}})$ and $\widehat{\varepsilon}_t = \varepsilon_t(\widehat{\boldsymbol{\theta}})$. Partitioning $\widehat{\mathbf{z}}_t$ conformably to the vector $\boldsymbol{\theta}$ yields $\widehat{\mathbf{z}}_t = (\widehat{\mathbf{z}}_{1,t}', \widehat{\mathbf{z}}_{2,t}')'$. Now, for T large, we can rewrite (5.4) as

$$\text{LM}_T = \widehat{\sigma}_\varepsilon^{-2} \Big(\sum_{t=1}^T \widehat{\mathbf{z}}_{2,t}\widehat{\varepsilon}_t\Big)' \Big(\widehat{\boldsymbol{\Sigma}}_{22} - \widehat{\boldsymbol{\Sigma}}_{21}\widehat{\boldsymbol{\Sigma}}_{11}^{-1}\widehat{\boldsymbol{\Sigma}}_{12}\Big)^{-1} \Big(\sum_{t=1}^T \widehat{\mathbf{z}}_{2,t}\widehat{\varepsilon}_t\Big), \tag{5.5}$$

where

$$\widehat{\boldsymbol{\Sigma}}_{21} = \widehat{\boldsymbol{\Sigma}}_{12}' = \sum_{t=1}^T \widehat{\mathbf{z}}_{2,t}\widehat{\mathbf{z}}_{1,t}', \quad \text{and} \quad \widehat{\boldsymbol{\Sigma}}_{ii} = \sum_{t=1}^T \widehat{\mathbf{z}}_{i,t}\widehat{\mathbf{z}}_{i,t}', \quad (i = 1, 2),$$

5.1 LAGRANGE MULTIPLIER TESTS

and $\widehat{\sigma}_\varepsilon^2 = T^{-1} L_T(\widehat{\boldsymbol{\theta}})$. If the linearity hypothesis holds and $\{Y_t, t \in \mathbb{Z}\}$ satisfies appropriate regularity conditions, (5.5) has an asymptotic chi-square distribution. In particular, under $\mathbb{H}_0$ and as $T \to \infty$, we have

$$\text{LM}_T \xrightarrow{D} \chi^2_{\nu_2}. \tag{5.6}$$

Computation of (5.5) can also be based on the *auxiliary regression*

$$\widehat{\varepsilon}_t = \widehat{\mathbf{z}}'_{1,t} \boldsymbol{\beta}_1 + \widehat{\mathbf{z}}'_{2,t} \boldsymbol{\beta}_2 + \eta_t, \tag{5.7}$$

where $\boldsymbol{\beta}_1$ and $\boldsymbol{\beta}_2$ are artificial parameter vectors of dimension ν_1 and ν_2 respectively, and $\{\eta_t, t \in \mathbb{Z}\}$ is an artificial error process. Let SSE be the residual sum of squares in the linear regression (5.7), and SSE$_0$ for the residual sum of squares under the null hypothesis $\boldsymbol{\beta}_2 = \mathbf{0}$. Then, applying standard least squares regression theory, (5.5) can be written as

$$\text{LM}_T = T\left(\frac{\text{SSE}_0 - \text{SSE}}{\text{SSE}_0}\right). \tag{5.8}$$

We use the above formulation as a first step to derive various variants of LM test statistics below. These variants depend on the form of the vector $\widehat{\mathbf{z}}_{2,t}$, which is determined by the type of nonlinearity investigated.

Bilinear case
Consider the BL(p, q, P, Q) model (2.12). This model reduces to a linear ARMA(p, q) model if the last term on the right-hand side of (2.12) is zero, i.e., if $\psi_{uv} = 0 \ \forall u, v$. Thus, the null hypothesis we wish to test is

$$\mathbb{H}_0^{(1)}: \psi_{uv} = 0, \quad (u = 1, \ldots, P; v = 1, \ldots, Q). \tag{5.9}$$

Consequently, the vectors $\widehat{\mathbf{z}}_{1,t}$ and $\widehat{\mathbf{z}}_{2,t}$ are given by

$$\widehat{\mathbf{z}}_{1,t} = \left(\frac{\partial \varepsilon_t(\widehat{\boldsymbol{\theta}})}{\partial \phi_0}, \frac{\partial \varepsilon_t(\widehat{\boldsymbol{\theta}})}{\partial \phi_1}, \ldots, \frac{\partial \varepsilon_t(\widehat{\boldsymbol{\theta}})}{\partial \phi_p}, \frac{\partial \varepsilon_t(\widehat{\boldsymbol{\theta}})}{\partial \theta_1}, \ldots, \frac{\partial \varepsilon_t(\widehat{\boldsymbol{\theta}})}{\partial \theta_q}\right)' \tag{5.10}$$

and

$$\widehat{\mathbf{z}}_{2,t} = \left(\frac{\partial \varepsilon_t(\widehat{\boldsymbol{\theta}})}{\partial \psi_{11}}, \ldots, \frac{\partial \varepsilon_t(\widehat{\boldsymbol{\theta}})}{\partial \psi_{PQ}}\right)', \tag{5.11}$$

where the partial derivatives can be obtained from the recursions

$$\frac{\partial \varepsilon_t(\widehat{\boldsymbol{\theta}})}{\partial \phi_0} = -\Big(1 + \sum_{\ell=1}^{q} \widehat{\theta}_\ell \frac{\partial \varepsilon_{t-\ell}(\widehat{\boldsymbol{\theta}})}{\partial \phi_0}\Big),$$

$$\frac{\partial \varepsilon_t(\widehat{\boldsymbol{\theta}})}{\partial \phi_i} = -\Big(Y_{t-i} + \sum_{\ell=1}^{q} \widehat{\theta}_\ell \frac{\partial \varepsilon_{t-\ell}(\widehat{\boldsymbol{\theta}})}{\partial \phi_i}\Big), \quad (i = 1, \ldots, p),$$

$$\frac{\partial \varepsilon_t(\widehat{\boldsymbol{\theta}})}{\partial \theta_j} = -\Big(\widehat{\varepsilon}_{t-j} + \sum_{\ell=1}^{q} \widehat{\theta}_\ell \frac{\partial \varepsilon_{t-\ell}(\widehat{\boldsymbol{\theta}})}{\partial \theta_j}\Big), \quad (j = 1, \ldots, q),$$

$$\frac{\partial \varepsilon_t(\widehat{\boldsymbol{\theta}})}{\partial \psi_{uv}} = -\Big(Y_{t-v}\widehat{\varepsilon}_{t-u} + \sum_{\ell=1}^{q} \widehat{\theta}_\ell \frac{\partial \varepsilon_{t-\ell}(\widehat{\boldsymbol{\theta}})}{\partial \psi_{uv}}\Big), \quad (u = 1, \ldots, P; v = 1, \ldots, Q),$$

and where the necessary initial values are set to zero.

The above quantities can only be used if the inverses in (5.5) exist, at least for T sufficiently large. If this is not the case an identification problem occurs, i.e. there is a perfect linear dependence among the components of $\widehat{\mathbf{z}}_{2,t}$. A natural solution is to reduce the number of ψ_{ij} coefficients in the model, i.e. restrict some of them to zero. This means that the dimension of the vector $\widehat{\mathbf{z}}_{2,t}$ is reduced by deleting some of its components when necessary. To solve the identification problem it suffices to impose the following restrictions (Saikkonen and Luukkonen, 1988) on the BL model.

(i) If $Q - p \leq P - q$ then $\phi_p \neq 0$ and either $Q \leq p+1$ or the vector $\widehat{\mathbf{z}}_{2,t}$ does not contain partial derivatives $\partial \varepsilon_t(\widehat{\boldsymbol{\theta}})/\partial \psi_{ij}$ with i and j satisfying $1 \leq i < Q - p$, $p + i < j \leq Q$.

(ii) If $P - q \leq Q - p$ then $\theta_q \neq 0$ and either $P \leq q+1$ or the vector $\widehat{\mathbf{z}}_{2,t}$ does not contain partial derivatives $\partial \varepsilon_t(\widehat{\boldsymbol{\theta}})/\partial \psi_{ij}$ with i and j satisfying $1 \leq j < P - q$, $q + j < i \leq P$.

Now, the asymptotic distribution of the LM test statistic for BL(p, q, P, Q) models can formulated as follows. Let $\{Y_t, t \in \mathbb{Z}\}$ be generated by (5.1) with $\mathbb{E}(\varepsilon_t^4) < \infty$. Assuming conditions (i) and (ii) are fulfilled, define the LM-type test statistic, denoted by $\text{LM}_T^{(1)}$, by substituting (5.10) – (5.11), for the corresponding quantities in (5.5).[1] Assume that the hypothesis of interest is $\mathbb{H}_0^{(1)}$. Then, as $T \to \infty$,

$$\text{LM}_T^{(1)} \xrightarrow{D} \chi^2_{PQ - r(r+1)/2}, \qquad (5.12)$$

where $r = \max\{0, \min(P - q, Q - p) - 1\}$.

Note that for the special case of a BL$(p, 0, P, Q)$ model $\widehat{\mathbf{z}}_{2,t}$ is given by $\widehat{\mathbf{z}}_{2,t} = -(Y_{t-1}\widehat{\varepsilon}_t, Y_{t-2}\widehat{\varepsilon}_{t-1}, \ldots, Y_{t-Q}\widehat{\varepsilon}_{t-P})'$, and the sufficient condition is given by $\phi_p \neq 0$,

---

[1] Throughout Sections 5.1 – 5.3, we use the numbered superscript notation $(\cdot)$ to indicate the link between a particular linearity test statistic and its corresponding null hypothesis.

5.1 LAGRANGE MULTIPLIER TESTS

$Q \leq p+1$. Under $\mathbb{H}_0^{(1)}$, the corresponding LM-type test statistic is asymptotically distributed as χ^2_{PQ}. The additional assumption $\mathbb{E}(\varepsilon_t^4) < \infty$ is not necessary if it is assumed that $\{\varepsilon_t\}$ is Gaussian WN.

Exponential AR case

Consider the ExpARMA model in (2.20) with $q = 0$. There are two possibilities to reduce the resulting ExpAR(p) model to a linear AR(p). One can either set the scaling factor $\gamma = 0$ or take $\xi_i = 0$ ($i = 1, \ldots, p$). Since it appears that the first possibility is easier to work with, we introduce the null hypothesis

$$\mathbb{H}_0^{(2)} : \gamma = 0. \tag{5.13}$$

Unfortunately, from (2.20) one can immediately see that the ExpAR(p) model is not identified when $\mathbb{H}_0^{(2)}$ holds, i.e. the parameters $\xi_1, \ldots, \xi_p$ can take any values without changing the residual sum of squares. As a consequence the relevant inverses in (5.5) do not exist. To overcome this problem, the idea is to replace $\exp(\cdot)$ by a suitable linear approximation. The resulting test statistic is an LM-type test statistic which is identical to the LM test statistic for the hypothesis $\xi_1 = \cdots = \xi_p = 0$ in the auxiliary regression model (5.7). In this case the vectors $\widehat{\mathbf{z}}_{1,t}$ and $\widehat{\mathbf{z}}_{2,t}$ are defined as respectively

$$\widehat{\mathbf{z}}_{1,t} = -(1, Y_{t-1}, \ldots, Y_{t-p})' \text{ and } \widehat{\mathbf{z}}_{2,t} = -(Y_{t-1}Y_{t-d}^2, Y_{t-2}Y_{t-d}^2, \ldots, Y_{t-p}Y_{t-d}^2)'. \tag{5.14}$$

Let $\text{LM}_T^{(2)}$ denote the resulting linearity test statistic. Under $\mathbb{H}_0^{(2)}$, and provided $\mathbb{E}(\varepsilon_t^6) < \infty$,

$$\text{LM}_T^{(2)} \xrightarrow{D} \chi^2_p, \text{ as } T \to \infty. \tag{5.15}$$

STAR model

Consider the STAR(2; p, p) model (2.42) with the transition function $G(Y_{t-d}; \gamma, c) = \Phi(\gamma\{Y_{t-d} - c\})$, i.e.

$$Y_t = \phi_0 + \sum_{i=1}^p \phi_i Y_{t-i} + \left\{\xi_0 + \sum_{i=1}^p \xi_i Y_{t-i}\right\} G(Y_{t-d}; \gamma, c) + \varepsilon_t. \tag{5.16}$$

The null hypothesis we wish to test is given by

$$\mathbb{H}_0^{(3)} : \xi_0 = \xi_1 = \cdots = \xi_p = 0. \tag{5.17}$$

Note that the parameters γ, d ($1 \leq d \leq p$), and c are generally unknown. Hence, under $\mathbb{H}_0^{(3)}$, the STAR(2; p, p) model is not identified. Analogous to the LM-type test statistic for the ExpAR(p) model one can solve this problem by replacing $G(\cdot)$ by a suitable linear approximation. In fact, it turns out that LM-type test statistics can be obtained for a wide class of smooth transition functions $G(\cdot)$ provided the following conditions are satisfied (Luukkonen et al., 1988a).

(a) The functions $G(\cdot)$ are odd, monotonically increasing, and possess a nonzero derivative of order $(2s+1)$ in an open interval $(-a, a)$, for $a > 0$, $s \geq 0$.

(b) The functions $G(\cdot)$ are such that $G(0) = 0$ and $(d^k G(z)/dz^k)|_{z=0} \neq 0$ for k odd and $1 \leq k \leq 2s+1$.

Condition (b) is not restrictive. Its purpose is to provide a convenient parameterization for deriving the test statistic. In the case $G(0) \neq 0$ one can always redefine $G(\cdot)$ and use $\widetilde{G}(\cdot) = G(\cdot) - G(0)$ instead so that (b) is again satisfied. The condition is not required for parameter estimation.

STAR model: First-order test procedure

Assume that conditions (a) and (b) hold for $s = 0$. Let $g_1 = (dG(z)/dz)|_{z=0}$. The idea is to linearize the STAR$(2; p, p)$ model by using the first-order Taylor series approximation

$$T_1(z) \approx g_1 z. \tag{5.18}$$

Substituting (5.18) for $G(z_t) \equiv G(Y_{t-d}; \gamma, c)$ into (5.16) yields the auxiliary linear regression model

$$Y_t = a_0 + \sum_{i=1}^{p} a_i Y_{t-i} + c_0 (Y_{t-d} - c) + \sum_{i=1}^{p} c_i u_{i,t} + \eta_t, \tag{5.19}$$

where $c_j = \gamma g_1 \xi_j$ ($j = 0, 1, \ldots, p$), and $u_{i,t} = Y_{t-i}(Y_{t-d} - c)$ ($i = 1, \ldots, p$). Under the null hypothesis, $c_j = 0$ ($j = 0, 1, \ldots, p$) in (5.19) and $\eta_t = \varepsilon_t$. Note, however, that model (5.19) is not identified, i.e. Y_{t-1} appears twice on the right-hand side. One way to overcome this problem is to reorder the components of (5.19) first; this yields

$$Y_t = \alpha_0 + \sum_{i=1}^{p} \alpha_i Y_{t-i} + \sum_{i=1}^{p} \sum_{j=i}^{p} \beta_{ij} Y_{t-i} Y_{t-j} + \eta_t. \tag{5.20}$$

Thus, the null hypothesis of interest is

$$\mathbb{H}_0^{(3*)}: \beta_{ij} = 0, \quad (i = 1, \ldots, p; j = i, \ldots, p). \tag{5.21}$$

The steps for computing the corresponding LM-type test statistic are as follows.

5.1 LAGRANGE MULTIPLIER TESTS

> **Algorithm 5.1: $\mathrm{LM}_T^{(3^*)}$ test statistic**
>
> (i) Regress Y_t on $\{1, Y_{t-1}, \ldots, Y_{t-p}\}$ using LS; compute the residuals $\{\widehat{\varepsilon}_t\}_{t=1}^T$, and the residual sum of squares $\mathrm{SSE}_0 = \sum_t \widehat{\varepsilon}_t^2$.
>
> (ii) Regress $\widehat{\varepsilon}_t$ on $\{1, Y_{t-i}, Y_{t-i}Y_{t-j}; i=1, \ldots, p; j=i, \ldots, p\}$; compute the residuals $\{\widehat{\eta}_t\}_{t=1}^T$, and the residual sum of squares $\mathrm{SSE}_1 = \sum_t \widehat{\eta}_t^2$.
>
> (iii) Compute the LM-type test statistic
>
> $$\mathrm{LM}_T^{(3^*)} = T(\mathrm{SSE}_0 - \mathrm{SSE}_1)/\mathrm{SSE}_0. \tag{5.22}$$
>
> Under $\mathbb{H}_0^{(3^*)}$,
>
> $$\mathrm{LM}_T^{(3^*)} \xrightarrow{D} \chi^2_{\frac{1}{2}p(p+1)}, \quad \text{as } T \to \infty. \tag{5.23}$$

STAR model: Third-order test procedure

Clearly, the test statistic (5.22) does not depend on the form of the function $G(\cdot)$ but only on the variables Y_{t-i} ($i = 1, \ldots, p$) and Y_{t-d}. Thus, the same test is obtained for a wide range of nonlinear models so that its power against some particular alternative may be questioned. One way to improve the performance of the test statistic is to replace the function $G(\cdot)$ by appropriate higher order approximations. A second-order Taylor expansion is not useful because $G(\cdot)$ is odd and thus its second derivative evaluated under the null hypothesis is zero. However, the use of a third-order approximation is possible, if conditions (a) and (b) are assumed to hold with $s = 1$. Then the third-order Taylor series approximation of $G(\cdot)$ evaluated at $z = 0$ is given by

$$T_3(z) \approx g_1 z + g_3 z^3, \quad g_3 = (3!)^{-1} [\mathrm{d}^3 G(z)/\mathrm{d}z^3]\big|_{z=0}.$$

Now, replacing $G(\cdot)$ in (5.16) by $T_3(\gamma\{Y_{t-d} - c\})$ gives the auxiliary model

$$Y_t = a_0 + \sum_{i=1}^p a_i Y_{t-i} + c_0(Y_{t-d} - c) + \sum_{i=1}^p c_i u_{i,t} + d_0(Y_{t-d} - c)^3 + \sum_{i=1}^p d_i w_{i,t} + \eta_t,$$

where $c_j = \gamma g_1 \xi_j$, $d_j = \gamma^3 g_3 \xi_j$ ($j = 0, 1, \ldots, p$), $u_{i,t} = Y_{t-i}(Y_{t-d} - c)$, and $w_{i,t} = Y_{t-i}(Y_{t-d} - c)^3$ ($i = 1, \ldots, p$). Similar as in the case of the first-order test procedure the above model is not identified. Again, we can circumvent this problem by expanding $(Y_{t-d} - c)^3$ and reordering terms. The result is the auxiliary regression model

$$Y_t = \alpha_0 + \sum_{i=1}^p \alpha_i Y_{t-i} + \sum_{i=1}^p \sum_{j=i}^p \beta_{ij} Y_{t-i} Y_{t-j} + \sum_{i,j=1}^p \psi_{ij} Y_{t-i} Y_{t-j}^2 + \sum_{i,j=1}^p \kappa_{ij} Y_{t-i} Y_{t-j}^3 + \eta_t. \tag{5.24}$$

Thus, the null hypothesis to be tested can be rewritten as

$$\mathbb{H}_0^{(3^{**})}: \beta_{ij}=0,\ (i=1,\ldots,p; j=i,\ldots,p),\ \psi_{ij}=\kappa_{ij}=0,\ (i,j=1,\ldots,p). \tag{5.25}$$

The test procedure consists of the following steps.

Algorithm 5.2: $\text{LM}_T^{(3^{**})}$ test statistic

(i) Repeat step (i) of the first-order test procedure (Algorithm 5.1).

(ii) Regress $\widehat{\varepsilon}_t$ on $\{1, Y_{t-i}, Y_{t-i}Y_{t-j}; i = 1,\ldots,p; j = i,\ldots,p; Y_{t-i}Y_{t-j}^k, i,j = 1,\ldots,p; k = 2, 3\}$; compute the residuals $\{\widehat{\eta}_t\}_{t=1}^T$ and the residual sum of squares $\text{SSE}_2 = \sum_t \widehat{\eta}_t^2$.

(iii) Compute the LM-type test statistic

$$\text{LM}_T^{(3^{**})} = T(\text{SSE}_0 - \text{SSE}_2)/\text{SSE}_0. \tag{5.26}$$

Under $\mathbb{H}_0^{(3^{**})}$, and as $T \to \infty$,

$$\text{LM}_T^{(3^{**})} \xrightarrow{D} \chi^2_{\frac{1}{2}p(p+1)+2p^2}. \tag{5.27}$$

STAR Model: Augmented first-order test procedure

A problem with the $\text{LM}_T^{(3^{**})}$ test is that in small samples it uses $2p^2$ more degrees of freedom than the $\text{LM}_T^{(3^*)}$ test statistic. On the other hand, it may be noted that β_{dd} and ψ_{dd} are the only parameters in (5.24) which are functions of ξ_0. This suggests that one might in essence retain the first-order approximation of $G(\cdot)$ and augment by p third-order terms only when absolutely necessary. This means that instead of the auxiliary regression model (5.24) we have

$$Y_t = \alpha_0 + \sum_{i=1}^p \alpha_i Y_{t-i} + \sum_{i=1}^p \sum_{j=i}^p \phi_{ij} Y_{t-i} Y_{t-j} + \sum_{i=1}^p \psi_i Y_{t-i}^3 + \eta_t^*.$$

The null hypothesis of interest is

$$\mathbb{H}_0^{(4)}: \phi_{ij} = 0,\ (i=1,\ldots,p; j=i,\ldots,p),\ \psi_i = 0,\ (i=1,\ldots,p). \tag{5.28}$$

The corresponding LM-type test statistic is given by

$$\text{LM}_T^{(4)} = T(\text{SSE}_0 - \text{SSE}_3)/\text{SSE}_0, \tag{5.29}$$

where SSE_0 is as before and SSE_3 is the residual sum of squares from the least squares regression of $\widehat{\varepsilon}_t$ on $\{1, Y_{t-i}, Y_{t-i}Y_{t-j}; i = 1,\ldots,p, j = i\ldots,p; Y_{t-i}^3; i = 1,\ldots,p\}$. Under $\mathbb{H}_0^{(4)}$, and as $T \to \infty$,

$$\text{LM}_T^{(4)} \xrightarrow{D} \chi^2_{\frac{1}{2}p(p+1)+p}. \tag{5.30}$$

5.1 LAGRANGE MULTIPLIER TESTS

Note that the above three LM-type test statistics do not assume that the delay parameter d is known. If, however, if d is known, then it can be shown that the number of degrees of freedom of $\text{LM}_T^{(3*)}$, $\text{LM}_T^{(3**)}$, and $\text{LM}_T^{(4)}$ are p, $3p$, and $p+1$, respectively. In that case the resulting test statistics will be different from the ones given above since the residual sum of squares SSE_i ($i=1,2,3$) will be based on far fewer independent variables. Hence, prior knowledge about d can be quite valuable in testing linearity against $\text{STAR}(2;p,p)$ models.

AsMA and SETMA models

Recall the asARMA(p,q) model (2.37) with $p=0$, denoted by asMA(q), and compactly written in the form

$$Y_t = \mu + \varepsilon_t + \sum_{j=1}^{q} \theta_j^+ \varepsilon_{t-j} + \sum_{j=1}^{q} \delta_j I(\varepsilon_{t-j} \leq 0)\varepsilon_{t-j}, \quad (5.31)$$

where $\delta_j = \theta_j^- - \theta_j^+$. In addition, consider as a special case of the SETARMA model (2.29), the SETMA$(2;q,q)$ model given by

$$Y_t = \mu + \varepsilon_t + \sum_{j=1}^{q} \theta_j \varepsilon_{t-j} + \sum_{j=1}^{q} \delta_j I(Y_{t-d} \leq r)\varepsilon_{t-j}. \quad (5.32)$$

A notable difference between (5.31) and (5.32) is that with (5.31) the regime switching is in $\{\varepsilon_t\}$ whereas the threshold variable in the SETMA model is $\{Y_{t-d}\}$ ($d \in \mathbb{Z}^+$) itself. However, within the LM testing framework, this difference between both models does not play a role in the development of a linearity test. Hence, below we consider testing a linear MA model against an asMA(q) model. The procedure for testing SETMA$(2;q,q)$ types of nonlinearity is completely identical.

Define the parameter vectors $\boldsymbol{\theta} = (\theta_1, \ldots, \theta_q)'$, $\boldsymbol{\delta} = (\delta_1, \ldots, \delta_q)'$, and $\boldsymbol{\psi} = (\mu, \boldsymbol{\theta}', \boldsymbol{\delta}', \sigma_\varepsilon^2)'$, where $\theta_j \equiv \theta_j^+$. Furthermore, assume that there are q starting values $Y_{-q+1}, \ldots, Y_0$, and let $\{\varepsilon_t\} \stackrel{\text{i.i.d.}}{\sim} \mathcal{N}(0, \sigma_\varepsilon^2)$ which is needed to specify the log-likelihood function. For the asymptotic distribution of the LM-type test statistic this latter assumption can be relaxed by requiring the existence of certain moments higher than order two of the process $\{\varepsilon_t, t \in \mathbb{Z}\}$. Given these specifications, it is apparent from (5.31) that the null hypothesis of linearity is given by

$$\mathbb{H}_0^{(5)}: \boldsymbol{\delta} = \mathbf{0}. \quad (5.33)$$

Assume that under $\mathbb{H}_0^{(5)}$ the roots of $\theta(z) = 1 + \sum_k \theta_k z^k$ lie outside the unit circle to guarantee (global) invertibility. To derive an LM-type test statistic of $\mathbb{H}_0^{(5)}$ we need the components of the gradient, or score, vector $\partial L_T(\boldsymbol{\psi})/\partial \boldsymbol{\psi}$. They are

$$\frac{\partial L_T(\boldsymbol{\psi})}{\partial \theta_j} = -\frac{1}{\sigma_\varepsilon^2} \sum_{t=1}^{T} \varepsilon_t \left[\varepsilon_{t-j} + \sum_k \left(\theta_k + \delta_k I(\varepsilon_{t-k} \leq 0) \right) \frac{\partial \varepsilon_{t-k}}{\partial \theta_j} \right], \quad (j=1,\ldots,q),$$

$$(5.34)$$

$$\frac{\partial L_T(\psi)}{\partial \delta_j} = -\frac{1}{\sigma_\varepsilon^2}\sum_{t=1}^{T}\varepsilon_t\Big[I(\varepsilon_{t-j}\leq 0)\varepsilon_{t-j} + \sum_k\big(\theta_k + \delta_k I(\varepsilon_{t-k}\leq 0)\big)\frac{\partial \varepsilon_{t-k}}{\partial \delta_j}\Big], \quad (5.35)$$

$$\frac{\partial L_T(\psi)}{\partial \mu} = -\frac{1}{\sigma_\varepsilon^2}\sum_{t=1}^{T}\varepsilon_t\Big[1 + \sum_k\big(\theta_k + \delta_k I(\varepsilon_{t-k}\leq 0)\big)\frac{\partial \varepsilon_{t-k}}{\partial \mu}\Big], \quad (5.36)$$

and

$$\frac{\partial L_T(\psi)}{\partial \sigma_\varepsilon^2} = -\frac{T}{2\sigma_\varepsilon^2} + \frac{1}{2\sigma_\varepsilon^4}\sum_{t=1}^{T}\varepsilon_t^2. \quad (5.37)$$

Under $\mathbb{H}_0^{(5)}$, (5.34) has the form

$$\frac{\partial L_T(\psi)}{\partial \theta_j} = -\frac{1}{\sigma_\varepsilon^2}\sum_{t=1}^{T}\varepsilon_t\Big[\varepsilon_{t-j} + \sum_k\theta_k\frac{\partial \varepsilon_{t-k}}{\partial \theta_j}\Big], \quad (j=1,\ldots,q). \quad (5.38)$$

From (5.38) it follows that $(1 + \sum_k \theta_k B^k)(\partial \varepsilon_t/\partial \theta_j) = -\varepsilon_{t-j}$ $(j=1,\ldots,q)$, so that $\partial \varepsilon_t/\partial \theta_j = -\theta^{-1}(B)\varepsilon_{t-j}$ where B is the backward shift operator. Moreover, $\partial \varepsilon_t/\partial \delta_j = -\theta^{-1}(B)I(\varepsilon_{t-j}\leq 0)\varepsilon_{t-j}$ $(j=1,\ldots,q)$ and $\partial \varepsilon_t/\partial \mu = -\theta^{-1}(1) = $ constant, under $\mathbb{H}_0^{(5)}$. The actual testing can be performed by the following steps.

Algorithm 5.3: $F_T^{(5)}$ **test statistic**

(i) Estimate the parameters of the asMA(q) model (5.31) with $\delta_j = 0$ ($j = 1,\ldots,q$) consistently; compute the residuals $\{\widehat{\varepsilon}_t\}_{t=1}^T$. The Hannan and Rissanen (1982) procedure, based on first estimating a long AR, is recommended for computing the MA parameters.

(ii) Regress $\widehat{\varepsilon}_t$ on 1 and $\xi(B)\widehat{\varepsilon}_{t-j}$ ($j = 1,\ldots,q$), where $\xi(B) = \sum_{k=0}^{K}\xi_k B^k$ ($\xi_0 = 1$) is the Kth order approximation of $\widehat{\theta}^{-1}(B)$; compute the residuals $\{\widehat{v}_t\}_{t=1}^T$, and $\text{SSE}_0 = \sum_t \widehat{v}_t^2$.

(iii) Regress $\widehat{v}_t$ on 1, $\xi(B)\widehat{\varepsilon}_{t-j}$ and $\xi(B)I(\widehat{\varepsilon}_{t-j}\leq 0)\widehat{\varepsilon}_{t-j}$ ($j = 1,\ldots,q$); compute the residual sum of squares SSE.

(iv) Compute the test statistic

$$F_T^{(5)} = \frac{(\text{SSE}_0 - \text{SSE})/q}{\text{SSE}/(T - K - 2q - 1)}. \quad (5.39)$$

Under $\mathbb{H}_0^{(5)}$, and as $T \to \infty$,

$$F_T^{(5)} \xrightarrow{D} F_{\nu_1,\nu_2} \quad (5.40)$$

with $\nu_1 = q$ and $\nu_2 = T - K - 2q - 1$.

5.1 LAGRANGE MULTIPLIER TESTS

An F test is recommended because in small samples its empirical size usually is close to the nominal significance level while the power is good. The empirical size of the corresponding χ_q^2 distributed test statistic, based directly on asymptotic theory, may be too large if q happens to be large and T is small.

Note that (5.39) is computed by conditioning on the K first residuals $\widehat{\varepsilon}_1, \ldots, \widehat{\varepsilon}_K$. Another way to proceed is to obtain the estimates of the partial derivatives in (5.38) from the recursion

$$\frac{\partial \varepsilon_t}{\partial \theta_j} = -\left(\varepsilon_{t-j} + \sum_k \theta_k \frac{\partial \varepsilon_{t-k}}{\partial \theta_j}\right), \quad (j = 1, \ldots, q).$$

Analogously,

$$\frac{\partial \varepsilon_t}{\partial \delta_j} = -\left(I(\varepsilon_{t-j} \leq 0)\varepsilon_{t-j} + \sum_k \theta_k \frac{\partial \varepsilon_{t-k}}{\partial \delta_j}\right), \quad (j = 1, \ldots, q),$$

$$\frac{\partial \varepsilon_t}{\partial \mu} = -\left(1 + \sum_k \theta_k \frac{\partial \varepsilon_{t-k}}{\partial \mu}\right),$$

where the required initial values are set to zero. The second and third steps of the testing procedure can be modified as follows.

(ii*) Regress $\widehat{\varepsilon}_t$ on $\partial \widehat{\varepsilon}_t/\partial \widehat{\mu}$ and $\partial \widehat{\varepsilon}_t/\partial \widehat{\theta}_j$ $(j = 1, \ldots, q)$ to obtain $\{\widehat{v}_t\}$ and SSE_0.

(iii*) Regress $\widehat{v}_t$ on $\partial \widehat{\varepsilon}_t/\partial \widehat{\mu}$, $\partial \widehat{\varepsilon}_t/\partial \widehat{\theta}_j$ and $\partial \widehat{\varepsilon}_t/\partial \widehat{\delta}_j$ $(j = 1, \ldots, q)$ to get SSE.

In this case the F test statistic has q and $T - 1 - 2q - 1$ degrees of freedom.

ASTMA model

Consider the ASTMA model (2.45) which, for ease of exposition, we reproduce as

$$Y_t = \varepsilon_t + \sum_{j=1}^{q} \left(\theta_j + \delta_j G_j(\gamma \varepsilon_{t-j})\right) \varepsilon_{t-j}. \tag{5.41}$$

If we want to test a linear MA(q) against an ASTMA(q) model it is not necessary to parameterize the transition functions $G_j(\cdot)$ $(j = 1, \ldots, q)$ in detail. Following Luukkonen et al. (1988a), it suffices to assume that conditions (a) and (b) for the STAR model hold. Note that an ASTMA model is not identified under the null hypothesis of linearity

$$\mathbb{H}_0^{(6)}: \gamma = 0. \tag{5.42}$$

If $\mathbb{H}_0^{(6)}$ holds so that $G_j(0) \equiv 0$, the δ_j's in (5.41) are not estimable. We can, however, adopt a similar approach as introduced for the STAR model and approximate $G_j(\gamma \varepsilon_{t-j})$ by its first-order Taylor expansion at the origin. With $z = \gamma \varepsilon_{t-j}$ this expansion yields $T_j(z) = G'_j(0)z$. Substitute T_j for $I(\varepsilon_{t-j} \leq 0)$ in relations (5.34) – (5.38). Keep the unidentified $\delta_1, \ldots, \delta_q$ fixed and replace (5.35) by

$$\frac{\partial L_T(\psi)}{\partial \gamma} = -\frac{1}{\sigma_\varepsilon^2} \sum_{t=1}^{T} \varepsilon_t \sum_k \left[\theta_k \frac{\partial \varepsilon_{t-k}}{\partial \gamma} + \delta_k G_k'(0) \varepsilon_{t-k}^2 + \gamma G_k'(0) \delta_k \frac{\partial \varepsilon_{t-k}^2}{\partial \gamma} \right].$$

Thus, under $\mathbb{H}_0^{(6)}$,

$$\frac{\partial L_T(\psi)}{\partial \gamma} = -\frac{1}{\sigma_\varepsilon^2} \sum_{t=1}^{T} \varepsilon_t \sum_k \left[\theta_k \frac{\partial \varepsilon_{t-k}}{\partial \gamma} + \delta_k G_k'(0) \varepsilon_{t-k}^2 \right]$$

and

$$\frac{\partial \varepsilon_t}{\partial \gamma} = -\sum_{k=1}^{q} \delta_k G_k'(0) \theta^{-1}(B) \varepsilon_{t-k}^2 \approx -\sum_{k=1}^{q} \delta_k G_k'(0) \xi(B) \varepsilon_{t-k}^2. \quad (5.43)$$

Substituting (5.43), evaluated under $\mathbb{H}_0^{(6)}$, for $\partial \varepsilon_t / \partial \delta_j$ ($j = 1, \ldots, q$) at step (iii*) of the asMA testing procedure leads to the following modification.

(iii′) Regress $\widehat{v}_t$ on 1, $\partial \widehat{\varepsilon}_t / \partial \widehat{\theta}_j$ ($j = 1, \ldots, q$) and $\partial \widehat{\varepsilon}_t / \partial \widehat{\gamma}$; compute the residual sum of squares SSE$_\delta$.

This does not yield a practicable test because the resulting test statistic, say F_δ, depends on the unknown nuisance parameters δ_j ($j = 1, \ldots, q$). We may, however, replace SSE$_\delta$ by $\inf_\delta$ SSE$_\delta$ so that the test statistic becomes $\sup_\delta F_\delta$. The asymptotic null distribution of $\sup_\delta F_\delta$ is χ_q^2. This is done by treating the q elements in the last sum in (5.43) as separate variables and performing the following step.

(iii″) Regress $\widehat{v}_t$ on 1, $\xi(B)\widehat{\varepsilon}_{t-j}$ and $\xi(B)\widehat{\varepsilon}_{t-j}^2$ ($j = 1, \ldots, q$); compute the residual sum of squares SSE*. Replace SSE by SSE* in step (iv) of Algorithm 5.3.

The resulting test statistic is given by

$$F_T^{(6)} = \frac{(\text{SSE}_0 - \text{SSE}^*)/q}{\text{SSE}^*/(T - K - 2q - 1)}. \quad (5.44)$$

Under $\mathbb{H}_0^{(6)}$, and as $T \to \infty$, (5.44) has the same asymptotic distribution as $F_T^{(5)}$.

NCTAR and AR-NN models

Consider the NCTAR$(k;p)_q$ ($1 \leq q \leq p$) model (2.65) with the logistic activation-level function $G(\cdot)$ redefined as

$$G(\widetilde{\mathbf{X}}_{t-1}; \gamma_j, \widetilde{\boldsymbol{\omega}}_j, c_j) = \frac{1}{1 + \exp(-\gamma_j [\widetilde{\boldsymbol{\omega}}_j' \widetilde{\mathbf{X}}_{t-1} - c_j])} - \frac{1}{2}, \quad (j = 1, \ldots, k), \quad (5.45)$$

where

$$\widetilde{\mathbf{X}}_{t-1} = (Y_{t-1}, \ldots, Y_{t-q})', \text{ and } \widetilde{\boldsymbol{\omega}}_j = (\widetilde{\omega}_{1j}, \ldots, \widetilde{\omega}_{qj})'.$$

5.1 LAGRANGE MULTIPLIER TESTS

A possible null hypothesis for linearity is $\mathbb{H}_0\colon \gamma_j = 0$, $(j = 1, \ldots, k)$.

In principle, we can proceed in the same spirit as in the case of the STAR model, by introducing first- and third-order Taylor approximations of (5.45) under $\mathbb{H}_0$ and redefining the null hypothesis. However, similar to the $\text{LM}_T^{(3*)}$-type test statistic, all the information about nonlinearity will be lost if a first-order Taylor expansion is used. Instead, a third-order Taylor expansion of $G(\cdot)$ is recommended. To this end, consider for simplicity the case $k = 1$ (i.e. one node). Then, taking a third-order Taylor expansion of (5.45) about $\gamma_1 = 0$ and substitution in (2.65) gives, after rearranging and merging terms, the auxiliary regression model

$$Y_t = \alpha_0 + \sum_{i=1}^{p} \alpha_i Y_{t-i} + \sum_{i=1}^{q}\sum_{j=i}^{q} \beta_{ij} Y_{t-i} Y_{t-j} + \sum_{i=1}^{p-q}\sum_{j=1}^{q} \psi_{ij} Y^*_{t-i} Y_{t-j}$$

$$+ \sum_{i=1}^{q}\sum_{j=i}^{q}\sum_{u=j}^{q} \beta_{iju} Y_{t-i} Y_{t-j} Y_{t-u} + \sum_{i=1}^{p-q}\sum_{j=1}^{q}\sum_{u=j}^{q} \psi_{iju} Y^*_{t-i} Y_{t-j} Y_{t-u}$$

$$+ \sum_{i=1}^{q}\sum_{j=i}^{q}\sum_{u=j}^{q}\sum_{v=u}^{q} \beta_{ijuv} Y_{t-i} Y_{t-j} Y_{t-u} Y_{t-v}$$

$$+ \sum_{i=1}^{p-q}\sum_{j=1}^{q}\sum_{u=j}^{q}\sum_{v=u}^{q} \psi_{ijuv} Y^*_{t-i} Y_{t-j} Y_{t-u} Y_{t-v} + \eta_t, \qquad (5.46)$$

where the vector $\mathbf{Y}^*_t \in \mathbb{R}^{p-q}$ is formed by the elements of $\mathbf{X}_{t-1} = (Y_{t-1}, \ldots, Y_{t-p})'$ that are not contained in $\widetilde{\mathbf{X}}_{t-1}$. The corresponding null hypothesis of linearity is defined by

$$\mathbb{H}_0^{(7)}\colon \beta_{ij} = 0,\ \psi_{ij} = 0,\ \beta_{iju} = 0,\ \psi_{iju} = 0,\ \beta_{ijuv} = 0,\ \psi_{ijuv} = 0. \qquad (5.47)$$

Recall that an NCAR$(k;p)_q$ model with $p = q$ and $\xi_{0j} = 0$ $(j = 1, \ldots, k)$, is equivalent to an AR–NN$(k;p)$ model (see, e.g., Figure 2.16). Then the auxiliary regression (5.46) reduces to

$$Y_t = \alpha_0 + \sum_{i=1}^{q} \alpha_i Y_{t-i} + \sum_{i=1}^{q}\sum_{j=i}^{q} \beta_{ij} Y_{t-i} Y_{t-j} + + \sum_{i=1}^{p}\sum_{j=i}^{p}\sum_{u=j}^{p} \beta_{iju} Y_{t-i} Y_{t-j} Y_{t-u} +$$

$$+ \sum_{i=1}^{p}\sum_{j=i}^{p}\sum_{u=j}^{p}\sum_{v=u}^{p} \beta_{ijuv} Y_{t-i} Y_{t-j} Y_{t-u} Y_{t-v} + \eta_t, \qquad (5.48)$$

with similar modifications in the specification of the null hypothesis $\mathbb{H}_0^{(7)}$, and the degrees of freedom of the resulting tests statistics. Given (5.46) and (5.47), a third-order LM-type test statistic can be computed by the following steps.

> **Algorithm 5.4: $\text{LM}_T^{(7)}$ test statistic**
>
> (i) Regress Y_t on $\{1, Y_{t-1}, \ldots, Y_{t-p}\}$ using LS; compute the residuals $\{\widehat{\varepsilon}_t\}_{t=1}^T$, and the residual sum of squares $\text{SSE}_0 = \sum_t \widehat{\varepsilon}_t^2$.
>
> (ii) Regress $\widehat{\varepsilon}_t$ on $\{1, Y_{t-1}, \ldots, Y_{t-p}\}$ and on each of the nonlinear regressors of (5.46); compute the residuals $\{\widehat{\eta}_t\}_{t=1}^T$, and $\text{SSE}_2 = \sum_t \widehat{\eta}_t^2$.
>
> (iii) Compute the LM-type test statistic
> $$\text{LM}_T^{(7)} = T(\text{SSE}_0 - \text{SSE}_2)/\text{SSE}_0. \tag{5.49}$$
>
> Under $\mathbb{H}_0^{(7)}$ and standard regularity conditions, complemented with the assumption $\mathbb{E}(Y_{t-i}^\delta) < \infty$ ($i = 1, \ldots, p$) for some $\delta > 8$, the limiting distribution of (5.49) is given by
> $$\text{LM}_T^{(7)} \xrightarrow{D} \chi_\nu^2, \tag{5.50}$$
> where
> $$\nu = \frac{q}{2!}(q+1) + \frac{q}{3!}(q+1)(q+2) + \frac{q}{4!}(q+1)(q+2)(q+3)$$
> $$+ (p-q)\big(q + \frac{q}{2!}(q+1) + \frac{q}{3!}(q+1)(q+2)\big).$$
>
> (iv) Alternatively, compute the associated test statistic:
> $$F_T^{(7)} = \frac{(\text{SSE}_0 - \text{SSE})/\nu}{\text{SSE}/(T - p - 1 - \nu)}, \tag{5.51}$$
> which, as $T \to \infty$, has an approximate $F_{\nu, T-p-1-\nu}$ distribution under $\mathbb{H}_0^{(7)}$.

The asymptotic properties of the above two test statistics do not crucially depend on the assumption that the activation-level $G(\cdot)$ function is logistic, provided conditions (a) and (b) given with the STAR model are satisfied. In practice, the test statistic (5.51) is preferred over (5.49) since the asymptotic χ_ν^2 distribution is likely to be a poor approximation to the finite sample distribution of the LM-type test statistic if the degrees of freedom ν is large.

5.2 Likelihood Ratio Tests

SETAR

Let $\{Y_t, t \in \mathbb{Z}\}$ be a strictly stationary and ergodic time series. Assume for simplicity, but without generality, that $\{Y_t, t \in \mathbb{Z}\}$ is generated by the SETAR$(2; p, p)$ model

5.2 LIKELIHOOD RATIO TESTS

with delay d, i.e.

$$Y_t = \phi_0^{(1)} + \sum_{i=1}^{p} \phi_i^{(1)} Y_{t-i} + \left\{ \phi_0^{(2)} + \sum_{i=1}^{p} \phi_i^{(2)} Y_{t-i} \right\} I(Y_{t-d} \leq r) + \varepsilon_t. \quad (5.52)$$

Suppose, for the moment, that p and d are known ($1 \leq d \leq p$). Further, we assume that the unknown threshold parameter r takes a value inside a known bounded closed subset of $\mathbb{R}$, say $\widetilde{\mathbb{R}} = [\underline{r}, \overline{r}]$, with $\underline{r}$ and $\overline{r}$ finite constants.

Let $\boldsymbol{\phi}_i = (\phi_0^{(i)}, \ldots, \phi_p^{(i)})'$ ($i = 1, 2$), and $\boldsymbol{\theta} = (\boldsymbol{\phi}_1', \boldsymbol{\phi}_2')'$. We denote the parameter space by $\boldsymbol{\Theta} = \boldsymbol{\Theta}_{\phi_1} \times \boldsymbol{\Theta}_{\phi_2}$, where $\boldsymbol{\Theta}_{\phi_1}$ and $\boldsymbol{\Theta}_{\phi_2}$ are compact subsets of $\mathbb{R}^{p+1}$. Suppose the true parameter vector $\boldsymbol{\theta}_0 = (\boldsymbol{\phi}_{10}', \boldsymbol{\phi}_{20}')'$, is an interior point of $\boldsymbol{\Theta}$. The hypotheses of interest are

$$\mathbb{H}_0^{(8)}: \boldsymbol{\phi}_{20} = \mathbf{0}, \quad \mathbb{H}_1^{(8)}: \boldsymbol{\phi}_{20} \neq \mathbf{0} \text{ for some } r \in \widetilde{\mathbb{R}}. \quad (5.53)$$

By temporarily setting $\{\varepsilon_t\} \overset{\text{i.i.d.}}{\sim} \mathcal{N}(0, \sigma_\varepsilon^2)$, the conditional log-likelihood functions under $\mathbb{H}_0^{(8)}$ and $\mathbb{H}_1^{(8)}$ are, respectively,

$$L_{0T}(\boldsymbol{\phi}_1) = \sum_{t=1}^{T} \widehat{\varepsilon}_t^2(\boldsymbol{\phi}_1), \text{ and } L_{1T}(\boldsymbol{\phi}_2, r) = \sum_{t=1}^{T} \widehat{\varepsilon}_t^2(\boldsymbol{\phi}_2, r), \quad (5.54)$$

where $\widehat{\varepsilon}_t(\boldsymbol{\phi}_1) = \widehat{\varepsilon}_t(\boldsymbol{\theta}, -\infty)$, and $\widehat{\varepsilon}_t(\boldsymbol{\phi}_2, r)$ is defined based on the iterative equation (5.52). For a given r, let

$$\widehat{\boldsymbol{\phi}}_{1T} = \arg\min_{\boldsymbol{\phi}_1 \in \boldsymbol{\Theta}_{\phi_1}} L_{0T}(\boldsymbol{\phi}_1) \text{ and } \widehat{\boldsymbol{\phi}}_{2T} = \arg\min_{\boldsymbol{\phi}_2 \in \boldsymbol{\Theta}} L_{1T}(\boldsymbol{\phi}_2, r).$$

The quasi-LR statistic for testing $\mathbb{H}_0^{(8)}$ against $\mathbb{H}_1^{(8)}$ is then defined as

$$\text{LR}_T(r) = \text{LR}_{0T}(\widehat{\boldsymbol{\phi}}_{1T}) - \text{LR}_{1T}(\widehat{\boldsymbol{\phi}}_{2T}(r), r).$$

Since r is unknown, a natural choice for a test statistic is $\sup_{r \in \mathbb{R}} \text{LR}_T(r)$. This choice, however, is undesirable since the test diverges to infinity in probability as $T \to \infty$. An appropriate alternative test statistic is

$$\text{LR}_T^{(8)} = \left(\sup_{r \in \widetilde{\mathbb{R}}} \{ \text{LR}_{0T}(\boldsymbol{\phi}_{1T}) - \text{LR}_{1T}(\widehat{\boldsymbol{\phi}}_{2T}(r), r) \} \right) / \text{LR}_{0T}(\widehat{\boldsymbol{\phi}}_{1T}). \quad (5.55)$$

To describe the asymptotic null distribution of (5.55), we introduce the matrices

$$\boldsymbol{\Omega}(r) = \begin{pmatrix} \boldsymbol{\Sigma} & \boldsymbol{\Sigma}_{12}(r) \\ \boldsymbol{\Sigma}_{21}(r) & \boldsymbol{\Sigma}_{22}(r) \end{pmatrix} = \mathbb{E}\left\{ \frac{\partial \widehat{\varepsilon}_t(\boldsymbol{\theta}_0, r)}{\partial \boldsymbol{\theta}} \frac{\partial \widehat{\varepsilon}_t(\boldsymbol{\theta}_0, r)}{\partial \boldsymbol{\theta}'} \right\}, \quad (5.56)$$

and

$$\boldsymbol{\Omega}_1(r) = \left(\boldsymbol{\Sigma}_{21}(r) - \boldsymbol{\Sigma}_{21}(r) \boldsymbol{\Sigma}_{22}^{-1}(r) \boldsymbol{\Sigma}_{12}(r) \right)^{-1},$$

where $\boldsymbol{\Sigma}(\cdot)$, $\boldsymbol{\Sigma}_{21}(\cdot) = \boldsymbol{\Sigma}'_{12}(\cdot)$, and $\boldsymbol{\Sigma}_{22}(\cdot)$ are $(p+1)\times(p+1)$ matrices. Let $\{\mathcal{G}_{2(p+1)}(r)\}$ denote a $2(p+1)$-dimensional vector Gaussian process with zero mean and covariance kernel $\boldsymbol{\Sigma}_{(r\wedge s)} - \boldsymbol{\Sigma}_{21}(r)\boldsymbol{\Sigma}^{-1}\boldsymbol{\Sigma}_{12}(r)$; almost all its paths are continuous. Then, under $\mathbb{H}_0^{(8)}$, standard regularity conditions, it can be shown (Chan, 1991) that

$$\mathrm{LR}_T^{(8)} \xrightarrow{D} \frac{1}{\sigma_\varepsilon^2} \sup_{r\in\widetilde{\mathbb{R}}} \{\mathcal{G}'_{2(p+1)}(r)\boldsymbol{\Omega}_1(r)\mathcal{G}_{2(p+1)}(r)\}, \quad \text{as } T \to \infty. \tag{5.57}$$

Using the Poisson clumping heuristic (Aldous, 1989), it follows that the limiting null distribution for the test statistic (5.57) is given by

$$\mathbb{P}\Big(\sup_{r\in\widetilde{\mathbb{R}}} \{\mathcal{G}'_{2(p+1)}(r)\boldsymbol{\Omega}_1(r)\mathcal{G}_{2(p+1)}(r)\} \leq \alpha\Big) \sim \exp\Big\{-2\chi^2_{p+1}(\alpha)\Big(\frac{\alpha}{p+1} - 1\Big)$$

$$\times \sum_{i=1}^{p+1} \int_{\widetilde{\mathbb{R}}} \frac{\mathrm{d}t_i}{\mathrm{d}r}\mathrm{d}r\Big\}, \tag{5.58}$$

where $t_i = \frac{1}{2}\log\{\mathcal{L}_i/(1-\mathcal{L}_i)\}$, $\forall i$, $\mathcal{L}_i \equiv \mathcal{L}_i(r) = \mathbb{E}[I(Y_t \leq r)]$, $1 \leq i \leq (p-1)$, $\mathcal{L}_p$ and $\mathcal{L}_{p+1}$ are the roots of $x^2 - ux + v = 0$ with $u = \mathbb{E}[(1 + Y_t^2/\sigma_Y^2)I(Y_t \leq r)]$ and $v = \mathbb{E}[I(Y_t \leq r)]\mathbb{E}[Y_t^2 I(Y_t \leq r)/\sigma_Y^2] - \mathbb{E}^2[Y_t^2 I(Y_t \leq r)/\sigma_Y]$. Here, $\mathcal{L}_p$ and $\mathcal{L}_{p+1}$ are chosen such that they are continuous functions of r.

Note from (5.58) that for $p \geq 1$, and assuming $d \leq p$, the asymptotic null distribution of $\mathrm{LR}_T^{(8)}$ is independent of d. For the special case $p = 0$, Chan and Tong (1990) show that the asymptotic distribution of (5.58) reduces to the distribution of

$$\sup_{a\leq s\leq b} W^\circ(s)/(s - s^2), \quad (0 < a < b < 1), \tag{5.59}$$

where $\{W^\circ(s), 0 \leq s \leq 1\}$ is a one-dimensional Brownian bridge (a Gaussian random function) on $(0, 1)$. By introducing the well-known characterization $W^\circ(s) = W(s) - sW(1)$, where $\{W(s), s \geq 0\}$ is the Wiener–Lévy process, and using Doob's transformation $U_t = e^{-t}W(e^{2t})$ the distribution function of (5.59) is available in closed form; see Appendix 5.A. This appendix also contains asymptotic critical values of the LR test statistic (5.57) for $p \geq 1$.

The assumption $\{\varepsilon_t\} \overset{\text{i.i.d.}}{\sim} \mathcal{N}(0, \sigma_\varepsilon^2)$ is not necessary for the derivation of the asymptotic distribution of $\mathrm{LR}_T^{(8)}$. In fact, its asymptotics also holds when $\{\varepsilon_t\} \sim \mathrm{WN}(0, \sigma_\varepsilon^2)$; see, e.g., Chan (1990). Indeed, if this is the case we can treat (5.52) as a regression model with the $p+1$ vector of added variables $\mathbf{X}'_t I(Y_{t-d} \leq r)$, with $\mathbf{X}_t = (1, Y_{t-1}, \ldots, Y_{t-p})'$, and replace (5.55) by

$$F_T^{(8)} = T\Big(\frac{\sup_{r\in\widetilde{\mathbb{R}}}\{\mathrm{SSE}_0 - \mathrm{SSE}_1(\widehat{\boldsymbol{\phi}}_{2T}(r), r)\}}{\inf_{r\in\widetilde{\mathbb{R}}} \mathrm{SSE}_1(\widehat{\boldsymbol{\phi}}_{2T}(r), r)}\Big), \tag{5.60}$$

where SSE_0 and $\mathrm{SSE}_1(\cdot)$ are the sum of squares of residuals under $\mathbb{H}_0^{(8)}$ and $\mathbb{H}_1^{(8)}$, respectively.

5.2 LIKELIHOOD RATIO TESTS

Nested SETARs

It is straightforward to generalize the F test statistic (5.60) to a SETAR$(k; p, \ldots, p)$ model ($k \geq 2$). Let $\mathbf{X}_t = (1, Y_{t-1}, \ldots, Y_{t-p})'$ be a $(p+1) \times 1$ vector. Using the notation introduced in Section 2.6, a convenient way of writing the k-regime SETAR model is

$$Y_t = \phi_1' \mathbf{X}_t I_t^{(1)}(\mathbf{r}, d) + \cdots + \phi_k' \mathbf{X}_t I_t^{(k)}(\mathbf{r}, d) + \varepsilon_t, \quad \{\varepsilon_t\} \sim \text{WN}(0, \sigma_\varepsilon^2), \qquad (5.61)$$

where $\mathbf{r} = (r_1, \ldots, r_{k-1})'$, $r_0 = -\infty$, $r_k = \infty$, and $I_t^{(i)}(\mathbf{r}, d) = I(r_{i-1} < Y_{t-d} \leq r_i)$ ($i = 1, \ldots, k$).

When $k = 1$, (5.61) reduces to a linear AR(p), or a SETAR$(1; p)$, model with zero thresholds, being the most restrictive within the class of k-regime SETAR models. The models within this class are strictly *nested*. This simply means that the i-regime SETAR model being tested, the null hypothesis, is a special case of the alternative SETAR$(j; p, \ldots, p)$ model ($i < j; i = 1, \ldots, k$) against which it is being tested. Here, we implicitly assume that there are no additional different constraints on the parameters ϕ_i, and the delay d is the same for both models.

Suppose the parameters of (5.61) are collected in the vector $\boldsymbol{\theta} = (\phi_1', \ldots, \phi_k', \mathbf{r}', d)'$ belonging to the parameter space Θ. The LS estimator, say $\widehat{\boldsymbol{\theta}}$, of $\boldsymbol{\theta}$ solves the minimization problem

$$\widehat{\boldsymbol{\theta}} = \arg\min_{\boldsymbol{\theta} \in \Theta} \sum_{t=1}^T \Big\{ Y_t - \sum_{j=1}^k \phi_j' \mathbf{X}_t I_t^{(j)}(\mathbf{r}', d) \Big\}^2. \qquad (5.62)$$

Let SSE$_i$ be the residual sum of squares corresponding to an i-regime SETAR model. Then the natural analogue of (5.60) for testing an i-regime SETAR against a j-regime SETAR model is defined by

$$F_T^{(i,j)} = T\Big(\frac{\text{SSE}_i - \text{SSE}_j}{\text{SSE}_j}\Big), \quad (i < j; i = 1, \ldots, k). \qquad (5.63)$$

This is equivalent to the conventional LM-type test statistic (5.8).

We can solve the minimization problem (5.62) sequentially through concentration. For instance, for the case $k = 2$, minimization over $\phi = (\phi_1', \phi_2')'$ is an LS regression of Y_t on $\big(\mathbf{X}_t' I_t^{(1)}(r, d), \mathbf{X}_t' I_t^{(2)}(r, d)\big)$ with $r \in \widetilde{\mathbb{R}}$. Let SSE$_2(r, d)$ be the corresponding residual sum of squares for a given (r, d). Then

$$(\widehat{r}, \widehat{d}) = \arg\min_{\substack{r \in \widetilde{\mathbb{R}} \\ 1 \leq d \leq p}} \text{SSE}_2(r, d). \qquad (5.64)$$

Next, we can find the LS estimates of ϕ as $\widehat{\phi} = \widehat{\phi}(\widehat{r}, \widehat{d})$, and obtain SSE$_2 \equiv$ SSE$_2(\widehat{r}, \widehat{d})$. A natural by-product is the test statistic $F_T^{(1,2)} = T((\text{SEE}_1 - \text{SSE}_2)/\text{SSE}_2)$ with SSE$_1$ the residual sum of squares of the SETAR$(1; p)$ model.

Hansen (1996) derives the asymptotic null distribution of $F_T^{(1,2)}$, say $\mathcal{T}$, which is a vector mean-zero Gaussian process. To obtain a practical procedure for calculating

p-values, he replaces all population moments of the asymptotic distribution of $\mathcal{T}$ by their sample counterparts. Let $\mathbf{u}$ denote a random $\mathcal{N}(\mathbf{0}, \mathbf{I}_T)$ vector. Then the random variable of interest is defined as

$$\mathcal{T}_T = \max_{\substack{r \in \widetilde{\mathbb{R}}, \\ 1 \leq d \leq p}} \widehat{\mathbf{u}}'(r,d) \mathbf{X}_1(r,d) \mathbf{M}_T^{-1}(r,d) \mathbf{X}_1'(r,d) \widehat{\mathbf{u}}(r,d), \tag{5.65}$$

where

$$\widehat{\mathbf{u}}(r,d) = \mathbf{u} - \mathbf{X}(\mathbf{X}'\mathbf{X})^{-1}\mathbf{X}'\mathbf{u},$$
$$\mathbf{M}_T(r,d) = \mathbf{X}_1'(r,d)\mathbf{X}_1(r,d) - \big(\mathbf{X}_1'(r,d)\mathbf{X}_1'(r,d)\big)(\mathbf{X}'\mathbf{X})^{-1}\big(\mathbf{X}_1'(r,d)\mathbf{X}_1(r,d)\big),$$

with $\mathbf{X}_1(r,d) \equiv \mathbf{X}_t' I_t^{(1)}(r,d)$ and $\mathbf{X}$ is the $T \times (p+1)$ matrix whose ith row is $\mathbf{X}_t$.

The asymptotic null distribution of $\mathcal{T}_T$ follows from a large number of independent draws from (5.65).[2] It can be used to calculate critical values from the quantiles of these draws. We can also calculate an approximation to the asymptotic p-value of the test statistic by counting the percentage of draws which exceed the observed $F_T^{(1,2)}$. For $k > 2$ the procedure is similar, with the additional requirement that each regime contains at least a sufficient number of observations, say T_i ($i = 1, \ldots, k$).

Alternatively, the steps to bootstrap p-values of the test statistic are as follows.

Algorithm 5.5: Bootstrapping p-values of $F_T^{(1,i)}$ test statistic

(i) Select a subset $\widetilde{\mathbb{R}}$ of values of Y_t falling between the $\underline{r} \times 100$ lower and $\overline{r} \times 100$ upper percentiles of the EDF of $\{Y_t\}_{t=1}^T$.

(ii) Fit a SETAR$(1;p)$ model and a SETAR$(i;p,\ldots,p)$ ($i = 2, 3$) model to the data. Let $\widehat{\boldsymbol{\theta}}_i$ be the vector of parameter estimates as in (5.62) and SSE$_i$ the corresponding residual sum of squares. Compute the test statistic $F_T^{(1,i)}$.

(iii) Generate $\{\varepsilon_t^*\}_{t=1}^T$ random draws (with replacement) from the LS residuals of the fitted SETAR$(1;p)$ model.

(iv) With fixed initial values $\{Y_0, Y_{-1}, \ldots, Y_{-p+1}\}$, recursively generate $\{Y_t^*\}_{t=1}^T$ using the SETAR$(1;p)$ model with $\widehat{\boldsymbol{\theta}}_1$. Select a new set $\widetilde{\mathbb{R}}^*$ falling between the $\underline{r} \times 100$ lower and $\overline{r} \times 100$ upper percentiles of the EDF of $\{Y_t^*\}_{t=1}^T$.

(v) Given $\{Y_t^*\}$, calculate the test statistic $F_T^{(b)}$ using the same method as to calculate $F_T^{(1,i)}$.

(vi) Repeat steps (iii) – (v) B times to obtain $\{F_T^{(b)}\}_{b=1}^B$. The bootstrap p-value is the percentage of simulated $F_T^{(b)}$ values which exceeds the observed $F_T^{(1,i)}$.

---

[2] Hansen (1999) shows how to calculate the asymptotic distribution of $F_T^{(1,i)}$ for the case of a stationary process with possibly heteroskedastic error terms. Several minor modifications in the formula for the asymptotic approximation (5.65) are needed. Also, for this case, he proposes an adjusted version of the bootstrap procedure.

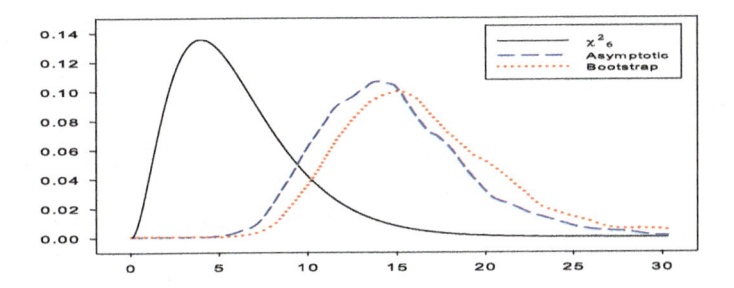

Figure 5.1: ENSO phenomenon. *Asymptotic and bootstrap distribution of the $F_T^{(1,2)}$ test statistic.*

The above procedures, i.e. via the asymptotic null distribution and bootstrapping, can be extended to the case of testing a two-regime SETAR model against a three-regime SETAR model. Some caution is needed, however. The problem is that under the null hypothesis, the parameter $\widehat{r}_1$ has a non-standard asymptotic distribution (Chan, 1993).

Example 5.1: ENSO Phenomenon (Cont'd)

We illustrate the use of the test statistic (5.63) with an application to the monthly ENSO series ($T = 748$) introduced in Example 1.4. After some initial exploration, we set $p = 5$. The estimated AR(5) model is given by

$$Y_t = -0.00_{(0.01)} + 1.41_{(0.04)}Y_{t-1} - 0.55_{(0.07)}Y_{t-2} + 0.15_{(0.07)}Y_{t-3}$$
$$+ 0.02_{(0.06)}Y_{t-4} - 0.11_{(0.04)}Y_{t-5} + \varepsilon_t, \tag{5.66}$$

where the sample variance of the residuals is given by $\widehat{\sigma}_\varepsilon^2 = 4.89 \times 10^{-2}$, and asymptotic standard errors are given in parentheses. Using (5.64), we find $\widehat{d} = 2$ and $\widehat{r}_1 = 0.21$. The associated SETAR(2; 5, 5) model is given by

$$Y_t = \begin{cases} -0.02_{(0.02)} + 1.34_{(0.04)}Y_{t-1} - 0.54_{(0.08)}Y_{t-2} + 0.14_{(0.09)}Y_{t-3} \\ +0.05_{(0.08)}Y_{t-4} - 0.09_{(0.05)}Y_{t-5} + \varepsilon_t^{(1)} & \text{if } Y_{t-2} \leq 0.21, \\ 0.06_{(0.02)} + 1.46_{(0.07)}Y_{t-1} - 0.60_{(0.12)}Y_{t-2} + 0.16_{(0.12)}Y_{t-3} \\ -0.02_{(0.10)}Y_{t-4} - 0.15_{(0.06)}Y_{t-5} + \varepsilon_t^{(2)} & \text{if } Y_{t-2} > 0.21, \end{cases} \tag{5.67}$$

where the sample variances of $\{\varepsilon_t^{(i)}\}$ ($i = 1, 2$) are 4.72×10^{-2} ($T_1 = 455$) and 4.70×10^{-2} ($T_2 = 288$) respectively. The $F_T^{(1,2)}$ statistic for the test of (5.66) against (5.67) equals 27.99. The asymptotic distribution, based on 1,000 independent draws, gives a p-value of 0.009. The bootstrapped p-value ($B = 1,000$) equals 0.014. So, there is sufficient evidence to reject the AR(5) model.

Next, we fit a SETAR(3; 5, 5, 5) model to the data, i.e.

$$Y_t = \begin{cases} -0.19_{(0.07)} + 1.25_{(0.07)}Y_{t-1} - 0.60_{(0.15)}Y_{t-2} + 0.17_{(0.16)}Y_{t-3} \\ +0.00_{(0.13)}Y_{t-4} - 0.06_{(0.07)}Y_{t-5} + \varepsilon_t^{(1)} \quad \text{if } Y_{t-2} \leq -0.78, \\ -0.02_{(0.02)} + 1.40_{(0.05)}Y_{t-1} - 0.64_{(0.10)}Y_{t-2} + 0.20_{(0.10)}Y_{t-3} \\ -0.00_{(0.10)}Y_{t-4} - 0.08_{(0.06)}Y_{t-5} + \varepsilon_t^{(2)} \quad \text{if } -0.78 < Y_{t-2} \leq 0.27, \\ 0.08_{(0.02)} + 1.44_{(0.07)}Y_{t-1} - 0.54_{(0.12)}Y_{t-2} + 0.04_{(0.12)}Y_{t-3} \\ +0.06_{(0.10)}Y_{t-4} - 0.17_{(0.06)}Y_{t-5} + \varepsilon_t^{(3)} \quad \text{if } Y_{t-2} > 0.27, \end{cases}$$
(5.68)

where the sample variances of $\{\varepsilon_t^{(i)}\}$ ($i = 1, 2, 3$) are 5.69×10^{-2} ($T_1 = 140$), 4.17×10^{-2} ($T_2 = 334$), and 4.71×10^{-2} ($T_3 = 269$) respectively. The $F_T^{(1,3)}$ test statistic equals 38.21. Both the asymptotic and bootstrapped p-values are 0.09. So, there is insufficient evidence to reject the AR(5) model in favor of the three-regime SETAR model. The $F_T^{(2,3)}$ test statistic equals 9.85, with a large bootstrapped p-value. Thus, in summary, it appears that an appropriate model for the ENSO data is the SETAR(2; 5, 5) model.

Figure 5.1 shows the asymptotic and bootstrap distributions of $F_T^{(1,2)}$. For fixed (r, d), the test statistic $F_T^{(i,j)}$ has an asymptotic χ_{p+1}^2 distribution. Its density function is plotted for reference. Clearly, the χ_6^2 distribution is highly misleading relative to the other two distributions. The bootstrap procedure properly approximates the asymptotic distribution in this case.

SETARMA model
Recall the SETARMA(2; p, p, q, q) model with delay d:

$$Y_t = \phi_0^{(1)} + \sum_{i=1}^{p} \phi_i^{(1)} Y_{t-i} + \sum_{j=1}^{q} \phi_j^{(2)} \varepsilon_{t-j}$$

$$+ \left\{ \psi_0^{(1)} + \sum_{i=1}^{p} \psi_i^{(1)} Y_{t-i} + \sum_{j=1}^{q} \psi_j^{(2)} \varepsilon_{t-j} \right\} I(Y_{t-d} \leq r) + \varepsilon_t, \qquad (5.69)$$

where, following Li and Li (2011), we assume that $\varepsilon_t = \eta_t \sigma_t$, where $\{\eta_t\} \stackrel{\text{i.i.d.}}{\sim} (0, \sigma_\varepsilon^2)$. and $\sigma_t > 0$ is a measurable function with respect to the information set $\mathcal{F}_t = \sigma(\eta_t, \eta_{t-1}, \ldots)$. So, $\{\varepsilon_t\}$ is an *uncorrelated* error sequence rather than an i.i.d. sequence. Along the same lines as above, quasi-LR test statistics for SETMA(2; q, q) (Ling and Tong, 2005) and SETMA–TGARCH models (Li and Li, 2008) can be defined. Not surprisingly, explicit expressions for the asymptotic null distribution of these LR-based test statistics take a very complicated form even for some simple cases. Only in the special case when $q < d$, the limiting distribution of the quasi-LR test statistic for SETMA(2; q, q) models is that of (5.59) with $W^\circ(s)$ replaced by $W_q^\circ(s)$, a q-dimensional Gaussian process with mean zero and covariance kernel

$(r \wedge s - rs)\mathbf{I}_q$. For more general SETARMA models bootstrap-based approximations are recommended to calculate p-values.

To avoid a time-consuming optimization in searching for the quasi-LR estimate for each bootstrapped sample, we discuss a so-called *stochastic permutation-based bootstrap procedure* only. First, however, we introduce the following notations. Let $\boldsymbol{\phi} = (\phi_0^{(1)}, \phi_1^{(1)}, \ldots, \phi_p^{(1)}, \phi_1^{(2)}, \ldots, \phi_q^{(2)})'$, $\boldsymbol{\psi} = (\psi_0^{(1)}, \psi_1^{(1)}, \ldots, \psi_p^{(1)}, \psi_1^{(2)}, \ldots, \psi_q^{(2)})'$, and $\boldsymbol{\theta} = (\boldsymbol{\phi}', \boldsymbol{\psi}')'$. Denote the parameter space by $\boldsymbol{\Theta} = \boldsymbol{\Theta}_\phi \times \boldsymbol{\Theta}_\psi$, where $\boldsymbol{\Theta}_\phi$ and $\boldsymbol{\Theta}_\psi$ are compact subsets of $\mathbb{R}^{p+q+1}$. Suppose the true parameter vector $\boldsymbol{\theta}_0 = (\boldsymbol{\phi}_0', \boldsymbol{\psi}_0')'$ is an interior point of the parameter space $\boldsymbol{\Theta}$. The hypotheses of interest are

$$\mathbb{H}_0^{(9)}: \boldsymbol{\psi}_0 = \mathbf{0}, \quad \mathbb{H}_1^{(9)}: \boldsymbol{\psi}_0 \neq \mathbf{0} \text{ for some } r \in \widetilde{\mathbb{R}}. \tag{5.70}$$

Similar to (5.55), by temporarily assuming normality for $\{\varepsilon_t\}$, the quasi-LR test statistic for testing $\mathbb{H}_0^{(9)}$ against $\mathbb{H}_1^{(9)}$ is defined as

$$\text{LR}_T^{(9)} = \frac{1}{\widehat{\sigma}_\varepsilon^2}\Big(\sup_{r \in \widetilde{\mathbb{R}}}\{\text{LR}_{0T}(\widehat{\boldsymbol{\phi}}_T) - \text{LR}_{1T}(\widehat{\boldsymbol{\theta}}(r), r)\}\Big), \tag{5.71}$$

where

$$\widehat{\sigma}_\varepsilon^2 = \text{LR}_{0T}(\widehat{\boldsymbol{\phi}}_T)/T$$

with $\widehat{\boldsymbol{\phi}}_T = \arg\min_{\boldsymbol{\phi} \in \boldsymbol{\Theta}_\phi} L_{0T}(\boldsymbol{\phi})$, and $\widehat{\boldsymbol{\theta}}_T(r) = \arg\min_{\boldsymbol{\theta} \in \boldsymbol{\Theta}} L_{1T}(\boldsymbol{\theta}, r)$. Denote $\boldsymbol{\Omega}(r)$ as in (5.56) with

$$\boldsymbol{\Omega}_1(r) = \boldsymbol{\Omega}^{-1}(r) - \text{diag}(\boldsymbol{\Sigma}^{-1}, \mathbf{0}),$$

where $\boldsymbol{\Sigma}(\cdot)$, $\boldsymbol{\Sigma}_{21}(\cdot) = \boldsymbol{\Sigma}_{12}'(\cdot)$, $\boldsymbol{\Sigma}_{22}(\cdot)$, and $\mathbf{0}$ are $(p+q+1) \times (p+q+1)$ matrices, and where $\widehat{\varepsilon}_t(\boldsymbol{\theta}_0, r)$ is defined based on the iterative equation (5.69).

Let $\{\mathcal{G}_{2(p+q+1)}(r), r \in \mathbb{R}\}$ denote a $2(p+q+1)$-dimensional vector Gaussian process with zero mean and covariance kernel $\mathbb{E}\{\varepsilon_t^2 \frac{\partial \varepsilon_t(\boldsymbol{\theta}_0, r)}{\partial \boldsymbol{\theta}} \frac{\partial \varepsilon_t(\boldsymbol{\theta}_0, s)}{\partial \boldsymbol{\theta}'}\}$, and almost all its paths are continuous. Assume that all roots of the polynomials $1 - \sum_{i=1}^p \phi_i^{(1)} z^i$ and $1 + \sum_{j=1}^q \phi_j^{(2)} z^j$ are outside the unit circle, and these polynomials are coprime. In addition, assume that the polynomials $1 - \sum_{i=1}^p \psi_i^{(1)} z^i$ and $1 + \sum_{j=1}^q \psi_j^{(2)} z^j$ are also coprime. The coprime nature of the polynomials is necessary to uniquely identify the parameters of the SETARMA model, i.e., the assumption makes the matrix $\boldsymbol{\Omega}(r)$ positive definite. Then, under $\mathbb{H}_0^{(9)}$, some standard regularity conditions, complemented with conditions on the moments of the random variable ε_t, it can be shown (Li and Li, 2011) that, as $T \to \infty$,

$$\text{LR}_T^{(9)} \xrightarrow{D} \frac{1}{\sigma_\varepsilon^2} \sup_{r \in \widetilde{\mathbb{R}}}\{\mathcal{G}'_{2(p+q+1)}(r)\boldsymbol{\Omega}_1(r)\mathcal{G}_{2(p+q+1)}(r)\}. \tag{5.72}$$

Because distribution theory is not available for the $\text{LR}_T^{(9)}$ test statistic for general SETARMA models, classical bootstrap methods can in principle be used to obtain p-values. However, computing time will be huge if, for each bootstrap replicate, (5.71) needs to be computed. Li and Li (2011) offer a bootstrap procedure that leads to substantial computational savings since optimization of the SETARMA model is required only once. Fundamental to the proposed procedure is the established results that, under $\mathbb{H}_0^{(9)}$,

$$\sup_{r \in \mathbb{\tilde{R}}} |\{\text{LR}_{0T}(\widehat{\boldsymbol{\phi}}_T) - \text{LR}_{1T}(\widehat{\boldsymbol{\theta}}_T(r), r)\} - \boldsymbol{\xi}'_T(r)\boldsymbol{\Omega}_1(r)\boldsymbol{\xi}_T(r)| = o_p(1), \qquad (5.73)$$

where $\boldsymbol{\xi}_T(r) = \frac{1}{\sqrt{T}} \sum_{t=1}^T \varepsilon_t \frac{\partial \varepsilon_t(\boldsymbol{\theta}_0, r)}{\partial \boldsymbol{\theta}}$. Clearly, the quantity $\boldsymbol{\xi}'_T(r)\boldsymbol{\Omega}_1(r)\boldsymbol{\xi}_T(r)$ is a quadratic form. Provided any possible dependence on the threshold structure in a observed time series is removed first, we can obtain a bootstrap approximation of $\text{LR}_T^{(9)}$ by randomly permuting the summand in $\boldsymbol{\xi}_T(r)$. In particular, the bootstrapping takes place as follows.

Algorithm 5.6: Bootstrapping p-values of $\text{LR}_T^{(9)}$ statistic

(i) Generate $\{\varepsilon_t\}_{t=1}^{T+n} \overset{\text{i.i.d.}}{\sim} \mathcal{N}(0, 1)$ random draws, with n the number of initial observations. Generate $\{Y_t\}_{t=1}^{T+n}$ from a SETARMA$(2; p, p, q, q)$ model, with or without possible dependence structure in the errors, using $\{\varepsilon_t\}$.

(ii) Select a subset $\widetilde{\mathbb{R}}$ of values Y_t falling between the $\underline{r} \times 100$ lower and $\overline{r} \times 100$ upper percentiles of the empirical distribution of $\{Y_t\}_{t=1}^T$.

(iii) Fit an ARMA(p, q) model to $\{Y_t\}_{t=1}^T$. Denote the resulting estimate of $\boldsymbol{\phi}$ by $\widehat{\boldsymbol{\phi}}_T$. Compute $\text{LR}_{0T}(\widehat{\boldsymbol{\phi}}_T) = \sum_{t=1}^T \widehat{\varepsilon}_t^2(\widehat{\boldsymbol{\phi}}_T)$.

(iv) For each value $Y_t \in \widetilde{\mathbb{R}}$ set $r = Y_t$, and fit a SETARMA$(2; p, p, q, q)$ model to $\{Y_t\}_{t=1}^T$. Let $\widehat{\boldsymbol{\theta}}_T(r)$ be the resulting estimate of $\boldsymbol{\theta}$. Also, for each r, compute $\text{LR}_{1T}(\widehat{\boldsymbol{\theta}}_T(r), r) = \sum_{t=1}^T \{\widehat{\varepsilon}_t(\widehat{\boldsymbol{\theta}}_T(r), r)\}^2$. Set $\text{LR}_{1T}(\widehat{\boldsymbol{\theta}}_T(\widehat{r}), \widehat{r}) = \min_{r \in \widetilde{\mathbb{R}}} \text{LR}_{1T}(\widehat{\boldsymbol{\theta}}_T(r), r)$.

(v) Compute the test statistic

$$\text{LR}_T^{(9)}(\widehat{r}) = T\Big(\text{LR}_{0T}(\widehat{\boldsymbol{\phi}}_T) - \text{LR}_{1T}(\widehat{\boldsymbol{\theta}}_T(\widehat{r}), \widehat{r})\Big) / \text{LR}_{0T}(\widehat{\boldsymbol{\phi}}_T). \qquad (5.74)$$

(vi) Generate a sequence $\{\varepsilon_t^*\}$ of i.i.d. random variables with mean zero, variance unity, and finite fourth moment. Suggested distribution functions are $\mathcal{N}(0, 1)$ and the Rademacher distribution, which takes values ± 1 with probability 0.5.

(vii) Let $\widetilde{\varepsilon}_t = \widetilde{\varepsilon}_t(\widehat{\boldsymbol{\theta}}_T(\widehat{r}), \widehat{r})$. Remove any possible threshold structure in a time series by generating $\widetilde{Y}_t = \widehat{\boldsymbol{\theta}}' \widetilde{\mathbf{Z}}_t + \widetilde{\varepsilon}_t$, where $\widetilde{\mathbf{Z}}_t = (1, \widetilde{Y}_{t-1}, \ldots, \widetilde{Y}_{t-p}, \widetilde{\varepsilon}_{t-1}, \ldots, \widetilde{\varepsilon}_{t-q})'$ with $\widetilde{\varepsilon}_t = 0$ for $t \leq 0$.

(viii) Select a new set $\widetilde{\mathbb{R}}^*$ falling between the $\underline{r} \times 100$ lower and $\overline{r} \times 100$ upper percentiles of the distribution of $\{\widetilde{Y}_t\}$. Let r be the new threshold parameter.

5.2 LIKELIHOOD RATIO TESTS

> **Algorithm 5.6: Bootstrapping p-values of $\text{LR}_T^{(9)}$ statistic (Cont'd)**
>
> (ix) Set $r = \widetilde{Y}_t \in \widetilde{\mathbb{R}}^*$, and compute the vector functions
>
> $$\frac{\partial \widetilde{\varepsilon}_t(r)}{\partial \boldsymbol{\phi}} = -\widetilde{\mathbf{Z}}_t - \sum_{j=1}^{q} \widehat{\phi}_j^{(2)} \frac{\partial \widetilde{\varepsilon}_{t-j}}{\partial \boldsymbol{\phi}},$$
>
> $$\frac{\partial \widetilde{\varepsilon}_t(r)}{\partial \boldsymbol{\psi}} = -\widetilde{\mathbf{Z}}_t I(\widetilde{Y}_{t-d} \leq r) - \sum_{j=1}^{q} \widehat{\phi}_j^{(2)} \frac{\partial \widetilde{\varepsilon}_{t-j}(r)}{\partial \boldsymbol{\psi}}, \quad \frac{\partial \widetilde{\varepsilon}_t(r)}{\partial \boldsymbol{\theta}} = \Big(\frac{\partial \widetilde{\varepsilon}_t(r)}{\partial \boldsymbol{\phi}'}, \frac{\partial \widetilde{\varepsilon}_t(r)}{\partial \boldsymbol{\psi}'} \Big)',$$
>
> where the necessary initial values in the recursions are set to zero. Moreover, as an estimator of $\boldsymbol{\Omega}(r)$, compute the outer product of the vector functions, i.e. $\widetilde{\boldsymbol{\Omega}}(r) = \frac{1}{T} \sum_{t=1}^{T} \big(\frac{\partial \widetilde{\varepsilon}_t(r)}{\partial \boldsymbol{\theta}} \frac{\partial \widetilde{\varepsilon}_t(r)}{\partial \boldsymbol{\theta}'} \big)$.
>
> (x) Compute the vector function $\boldsymbol{\xi}_T(\varepsilon^*, r) = \frac{1}{\sqrt{T}} \sum_{t=1}^{T} \varepsilon_t^* \widetilde{\varepsilon}_t \frac{\partial \widetilde{\varepsilon}_t(r)}{\partial \boldsymbol{\theta}}$, and the statistic
>
> $$\text{LR}_T^{(b)}(\varepsilon^*, r) = \frac{\boldsymbol{\xi}_T'(\varepsilon^*, r) \big(\widetilde{\boldsymbol{\Omega}}^{-1}(r) - \text{diag}(\widetilde{\boldsymbol{\Sigma}}^{-1}, \mathbf{0}) \big) \boldsymbol{\xi}_T(\varepsilon^*, r)}{\widehat{\sigma}_\varepsilon^2 \widehat{\sigma}_{\varepsilon^*}^2},$$
>
> where $\widehat{\sigma}_\varepsilon^2 = T^{-1} \text{LR}_{1T}(\widehat{\boldsymbol{\theta}}_T(\widehat{r}), \widehat{r})$ and $\widehat{\sigma}_{\varepsilon^*}^2 = T^{-1} \sum_{t=1}^{T} \{\varepsilon_t^*\}^2$.
>
> (xi) Repeat step (x) B times, to obtain $\{\text{LR}_T^{(b)}(\varepsilon^*, r)\}_{b=1}^{B}$.
>
> (xii) Repeat steps (ix) – (xi) for different values of r. Compute $\text{LR}_T^{(b)}(\varepsilon^*) = \max_{r \in \widetilde{\mathbb{R}}^*} \{\text{LR}_T^{(b)}(\varepsilon^*, r)\}$ $(b = 1, \ldots, B)$.
>
> (xiii) Transform the values $\{\text{LR}_T^{(b)}(\varepsilon^*)\}_{b=1}^{B}$ into p-values by computing the bootstrap statistic
>
> $$\frac{1}{B} \sum_{b=1}^{B} I\big(\text{LR}_T^{(9)}(\widehat{r}) < \text{LR}_T^{(b)}(\varepsilon^*)\big).$$

Example 5.2: U.S. Unemployment Rate (Cont'd)

Recall, in Example 1.1 we introduced the quarterly U.S. unemployment rate. Using the first differences of the original series, say $\{Y_t\}_{t=1}^{251}$, we fit the following ARMA(1, 1) model to the data

$$Y_t = 0.53_{(0.07)} Y_{t-1} + \varepsilon_t + 0.22_{(0.08)} \varepsilon_{t-1}, \tag{5.75}$$

where the sample variance of the residuals is given by $\widehat{\sigma}_\varepsilon^2 = 8.91 \times 10^{-2}$, and asymptotic standard errors are given in parentheses. The p-value of the Ljung–Box (LB) test statistic is 0.15, based on 40 lags. Although this specification can be improved (see Chapter 6), it can well serve as a benchmark for testing the ARMA(1,1) model against a SETARMA(2; 1, 1, 1, 1) model with delay $d \in [1, \ldots, 6]$. Setting $B = 10{,}000$, $\underline{r} = 0.1$, $\overline{r} = 0.9$, and generating $\{\varepsilon_t^*\}$ (step (vi)) from an $\mathcal{N}(0,1)$ distribution, we fitted various two-regime SETARMA models to the data. For $d = 2$ the p-value (0.049) of the $\text{LR}_T^{(9)}$ test statistic is smaller than the 5% nominal significance level. The associated model is given by

$$Y_t = 0.44_{(0.08)} Y_{t-1} + 0.48_{(0.07)} \varepsilon_{t-1}$$
$$+ (0.24_{(0.10)} Y_{t-1} - 0.71_{(0.12)} \varepsilon_{t-1}) I(Y_{t-2} \leq 1.01 \times 10^{-2}) + \varepsilon_t, \quad (5.76)$$

where the sample variance of the residuals is given by $\widehat{\sigma}_\varepsilon^2 = 8.34 \times 10^{-2}$. So, in terms of residual variances, (5.76) provides a better fit than the linear model (5.75).

5.3 Wald Test

ARasMA model

In Section 5.1, we introduced an LM-type test statistic for testing symmetry against an asMA(q) model. For the more general autoregressive-asymmetric moving average model (ARasMA) of order (p, q) with a linear AR(p) polynomial, an asymmetric MA polynomial of order q, and a constant term ϕ_0 (Brännäs and De Gooijer, 1994), the null hypothesis of symmetry is equivalent to testing the restriction $\boldsymbol{\theta}^+ = \boldsymbol{\theta}^-$, where $\boldsymbol{\theta}^+ = (\theta_1^+, \ldots, \theta_q^+)'$, and $\boldsymbol{\theta}^- = (\theta_1^-, \ldots, \theta_q^-)'$. Let $\boldsymbol{\theta} = (\phi_0, \boldsymbol{\phi}', (\boldsymbol{\theta}^+)', (\boldsymbol{\theta}^-)')'$ denote the $(1 + p + 2q) \times 1$ vector of parameters, with $\boldsymbol{\phi} = (\phi_1, \ldots, \phi_p)'$. Further, let $\mathbf{R}$ denote a restriction matrix of dimension $q \times (1 + p + 2q)$ such that $\mathbf{R}\boldsymbol{\theta} = \mathbf{r}$, and $\mathbf{r}$ is a $(1 + p + 2q)$-vector. Next, from the partition $\mathbf{R} = (\mathbf{R}_1 : \mathbf{R}_2)$, where $\mathbf{R}_1 = \mathbf{0}$ and $\mathbf{R}_2$ is a $q \times 2q$ matrix, the problem becomes one of testing the null hypothesis

$$\mathbb{H}_0^{(10)}: \mathbf{R}_2 \boldsymbol{\theta} = \mathbf{0} \quad \text{against} \quad \mathbb{H}_1^{(10)}: \mathbf{R}_2 \boldsymbol{\theta} \neq \mathbf{0}. \quad (5.77)$$

The third classical test, the Wald (W) test, is based exclusively on the unrestricted estimates $\widehat{\boldsymbol{\theta}}$ of $\boldsymbol{\theta}$. Assume that the ARasMA model is invertible, and let $\{\varepsilon_t\} \overset{\text{i.i.d.}}{\sim} \mathcal{N}(0, \sigma_\varepsilon^2)$. Then, for the unrestricted model, the log-likelihood function at time t (apart from an additive constant term), is given by

$$\ell_t(\boldsymbol{\theta}) = -\frac{1}{2\sigma_\varepsilon^2} \sum_t \varepsilon_t^2(\boldsymbol{\theta}) - \frac{1}{2} \log \sigma_\varepsilon^2, \quad (5.78)$$

where summation is over the range $(\max(p, q) + 1, T)$, and

$$\varepsilon_t(\boldsymbol{\theta}) = Y_t - \phi_0 - \sum_{i=1}^p \phi_i Y_{t-i} - \sum_{j=1}^q \theta_{t-j}^+ \varepsilon_{t-j}^+ - \sum_{j=1}^q \theta_{t-j}^- \varepsilon_{t-j}^-.$$

Let $\varepsilon_t \equiv \varepsilon_t(\boldsymbol{\theta})$. Then the score vector at time t is given by $\mathbf{G}_t(\boldsymbol{\theta}) = \partial \ell_t(\boldsymbol{\theta})/\partial \boldsymbol{\theta} = -\sigma_\varepsilon^2 \varepsilon_t \partial \varepsilon_t/\partial \boldsymbol{\theta}$, where

$$\frac{\partial \varepsilon_t}{\partial \boldsymbol{\theta}'} = -\Big(1 + v_{t,1}^{\phi_0} \vdots Y_{t-1} + v_{t,1}^{\phi} \cdots Y_{t-p} + v_{t,p}^{\phi} \vdots \varepsilon_{t-1}^{+} + v_{t,1}^{+} \cdots \varepsilon_{t-q}^{+} + v_{t,q}^{+} \vdots$$
$$\varepsilon_{t-1}^{-} + v_{t,1}^{-} \cdots \varepsilon_{t-q}^{-} + v_{t,q}^{-}\Big),$$

with

$$v_{t,j}^\theta = \sum_{k=1}^{q} \Big(\theta_k^{+} I(\varepsilon_{t-k} > 0) + \theta_k^{-} I(\varepsilon_{t-k} \leq 0)\Big) \partial \varepsilon_{t-k}/\partial \theta_j.$$

Here, the superscript on v_t together with the second subscript indicate the appropriate element within the $\boldsymbol{\theta}$ vector. The empirical Hessian $\widehat{\mathcal{H}}_T$ associated with the log-likelihood function can be approximated by the summed outer product of $\mathbf{G}_t$, i.e. $\widehat{\mathcal{H}}_T = \sum_{t=1}^{T} \mathbf{G}_t \mathbf{G}_t'$. Let $\widehat{\boldsymbol{\theta}}$ be the vector of parameter estimates of $\boldsymbol{\theta}$, and $\widehat{\mathcal{H}}_T^{-1}(\widehat{\boldsymbol{\theta}})$ the estimate of the corresponding covariance matrix. Then the W test statistic can be expressed as

$$W_T^{(10)} = \Big(\mathbf{R}_2 \widehat{\boldsymbol{\theta}}\Big)' \Big[\mathbf{R} \widehat{\mathcal{H}}_T^{-1}(\widehat{\boldsymbol{\theta}}) \mathbf{R}'\Big]^{-1} \mathbf{R}_2 \widehat{\boldsymbol{\theta}}. \tag{5.79}$$

Under $\mathbb{H}_0^{(10)}$, and as $T \to \infty$, (5.79) has an asymptotic χ_q^2 distribution.

5.4 Tests Based on a Second-order Volterra Expansion

In this section we discuss time-domain diagnostic tests statistics. For ease of representation we assume that $\{Y_t, t \in \mathbb{Z}\}$ is generated by a stationary linear AR(p) process ($\mathbb{H}_0$). The alternative hypothesis ($\mathbb{H}_1$) states that the process can be adequately approximated by a second-order Volterra expansion of the form

$$Y_t = \mu + \varepsilon_t + \sum_{u=-\infty}^{\infty} \psi_u \varepsilon_{t-u} + \sum_{u,v=-\infty}^{\infty} \psi_{uv} \varepsilon_{t-u} \varepsilon_{t-v}, \quad \{\varepsilon_t\} \stackrel{\text{i.i.d.}}{\sim} (0, \sigma_\varepsilon^2). \tag{5.80}$$

Thus $\mathbb{H}_1$ is quite general. Therefore the resulting test statistics are often termed *portmanteau-type* tests.

Obviously, if $\{Y_t, t \in \mathbb{Z}\}$ is linear, i.e., if $\psi_{uv} = 0 \; \forall u, v$, then ε_t will be independent of $\varepsilon_{t-u} \varepsilon_{t-v}$. If, however, $\{Y_t, t \in \mathbb{Z}\}$ is nonlinear, i.e., if any of the second-order coefficients ψ_{uv} are non-zero, this is not so. Then this nonlinearity will be reflected in the relationship of the residuals of a fitted linear model with, for instance, $Y_{t-1} Y_{t-2}$, a quadratic nonlinear term. This is called the *added variable* approach. Below, we discuss three variants.

The Tukey nonadditivity-type test
This test was developed by Keenan (1985) and is an analogue of Tukey's (T) (1949)

one degree of freedom test for nonadditivity in analysis of variance. The mechanisms for computing the test statistic are as follows.

> **Algorithm 5.7: Tukey's nonadditivity-type test statistic**
>
> (i) Choose an appropriate value $p \in [4, 8]$. Regress Y_t on $\{1, Y_{t-1}, \ldots, Y_{t-p}\}$; compute the fitted values $\{\widehat{Y}_t\}$, the residuals $\{\widehat{\varepsilon}_t\}_{t=p+1}^{T}$, and SSE=$\sum_t \widehat{\varepsilon}_t^2$.
>
> (ii) Regress $\{\widehat{Y}_t^2\}$ on $\{1, Y_{t-1}, \ldots, Y_{t-p}\}$; compute the residuals $\{\widehat{\xi}_t\}_{t=p+1}^{T}$.
>
> (iii) Regress $\widehat{\varepsilon}_t$ on $\widehat{\xi}_t$.
>
> (iv) From the regression in (iii) calculate the test statistic
>
> $$F_T^{(\text{T})} = \frac{\widehat{\eta}^2}{(\text{SSE} - \widehat{\eta}^2)/(T - 2p - 2)}, \qquad (5.81)$$
>
> where $\widehat{\eta} = \widehat{\eta}_0 \left(\sum_t \widehat{\xi}_t^2 \right)^{1/2}$ with $\widehat{\eta}_0$ the regression coefficient in step (ii).
>
> Under $\mathbb{H}_0$, and as $T \to \infty$, $F_T^{(\text{T})} \xrightarrow{D} F_{\nu_1,\nu_2}$ with $\nu_1 = 1$ and $\nu_2 = (T - p) - (p+1) - 1$. The estimated size of (5.81) can be improved by using $T - p$ instead of $T - 2p - 2$ in the denominator of $F_T^{(\text{T})}$ (Luukkonen et al., 1988b). This improvement also applies to the next two F test statistics.

Keenan (1985) shows that $F_T^{(\text{T})}$ is approximately distributed as χ_1^2 but the F-version may be preferred in practice because it is computationally convenient and reasonably powerful in finite samples. An advantage of (5.81) is that it is easy and quick to implement involving little subjective choice of parameters. On the other hand, the $F_T^{(\text{T})}$ test statistic is only valid for the Volterra expansion, but not all nonlinear processes possess this expansion.

Original F test

This F test statistic is a direct modification of the *original* (O) Tukey nonadditivity-type test statistic (5.81), and hence its name; see Tsay (1986).[3] The test considers the residuals of regressions that include the individual nonlinear terms and quadratic terms up to third order $\{Y_{t-1}^2, Y_{t-1}Y_{t-2}, \ldots, Y_{t-1}Y_{t-p}, Y_{t-2}^2, Y_{t-2}Y_{t-3}, \ldots, Y_{t-p}^3\}$ while $F_T^{(\text{T})}$ considers the residuals of regressions on only the squared terms.

Let $\mathbf{X}_t = (Y_{t-1}, \ldots, Y_{t-p})'$, and define the $P = \frac{1}{2}p(p+1)$-dimensional vector $\mathbf{Z}_t = \text{vech}(\mathbf{X}_t \mathbf{X}_t')$. Further, assume that $\{\varepsilon_t\} \sim \text{WN}(0, \sigma_\varepsilon^2)$ with $\mathbb{E}(\varepsilon_t^4) < \infty$. The procedure for performing the original F test statistic is outlined in the following steps.

---

[3]The name given to this test statistic is taken from Tsay (1991). This reference serves also as the source for the names given to the *original*, the *augmented*, and the *new* F test statistic (Section 5.5) which are discussed below.

5.4 TESTS BASED ON A SECOND-ORDER VOLTERRA EXPANSION

> **Algorithm 5.8:** $F_T^{(O)}$ test statistic
>
> (i) Choose an appropriate even value of p, e.g. $p = 4$ or $p = 8$. Regress Y_t on $\{1, Y_{t-1}, \ldots, Y_{t-p}\}$; compute the residuals $\{\widehat{\varepsilon}_t\}_{t=p+1}^{T}$.
>
> (ii) Regress the *first* $p+1$ elements of $\mathbf{Z}_t$ on $\{1, Y_{t-1}, \ldots, Y_{t-p}\}$ and obtain the residuals $\{\widehat{\xi}_{1,t}\}_{t=p+1}^{T}$.
>
> (iii) Then regress the *next* $p+1$ elements of $\mathbf{Z}_t$ on $\{1, Y_{t-1}, \ldots, Y_{t-p}\}$ and obtain the residuals $\{\widehat{\xi}_{2,t}\}_{t=p+1}^{T}$.
>
> (iv) Continue with steps (ii) – (iii) until the residuals from all $p/2$ regressions have been obtained. From these residuals, form the $(p/2) \times 1$ vector $\{\widehat{\boldsymbol{\xi}}_t\}_{t=p+1}^{T}$.
>
> (v) Regress $\widehat{\varepsilon}_t$ on $\widehat{\boldsymbol{\xi}}_t$; compute the residual sum of squares $\sum_t \widehat{\omega}_t^2$.
>
> (vi) From the regression in (v) calculate the test statistic $F_T^{(O)}$ as the F ratio of the mean square of regression to the mean square error, i.e.
>
> $$F_T^{(O)} = \frac{(\sum_t \widehat{\boldsymbol{\xi}}_t \widehat{\varepsilon}_t)'(\sum_t \widehat{\boldsymbol{\xi}}_t \widehat{\boldsymbol{\xi}}_t')^{-1}(\sum_t \widehat{\boldsymbol{\xi}}_t \widehat{\varepsilon}_t)/P}{\sum_t \widehat{\omega}_t^2/(T-p-P-1)}. \quad (5.82)$$
>
> Under $\mathbb{H}_0$, and as $T \to \infty$, $F_T^{(O)} \xrightarrow{D} F_{\nu_1,\nu_2}$ with degrees of freedom $\nu_1 = p(p+1)/2$ and $\nu_2 = T - \frac{1}{2}p(p+3) - 1$; Tsay (1986).

Note that the test statistic $PF_T^{(O)}$ is asymptotically distributed as χ_P^2. Using the LM testing procedure of Section 5.1, it can be easily shown (Luukkonen et al., 1988a) that both tests (5.81) and (5.82) are LM-type test statistics. Simulation results show that the $F_T^{(O)}$ is more powerful than the $F_T^{(T)}$ test statistic in identifying BL-type nonlinearity.

Augmented F test

The *augmented* (A) F test (Luukkonen et al., 1988a) extends the $F_T^{(O)}$ test statistic by including the regression of the cubic terms $\{Y_t^3\}$ on $(1, Y_{t-1}, \ldots, Y_{t-p})$ in the set of regressions in steps (ii) – (iv) of Algorithm 5.7. The $((p/2)+1)$th set of residuals $\{\widehat{\xi}_{(p/2)+1,t}\}_{t=p+1}^{T}$ are included in $\widehat{\boldsymbol{\xi}}_t$. Call the resulting vector $\widehat{\boldsymbol{\xi}}_t^{(A)}$. Perform a linear regression of $\widehat{\varepsilon}_t$ on $\widehat{\boldsymbol{\xi}}_t^{(A)}$, and obtain the residual sum of squares $\sum_t \{\widehat{\omega}_t^{(A)}\}^2$. Then the associated F test statistic is given by

$$F_T^{(A)} = \frac{\left(\sum_t \widehat{\boldsymbol{\xi}}_t^{(A)} \widehat{\varepsilon}_t\right)'\left(\sum_t \widehat{\boldsymbol{\xi}}_t^{(A)}(\widehat{\boldsymbol{\xi}}_t^{(A)})'\right)^{-1}\left(\sum_t \widehat{\boldsymbol{\xi}}_t^{(A)} \widehat{\varepsilon}_t\right)/P}{\sum_t \{\widehat{\omega}_t^{(A)}\}^2/(T-p-P-1)}. \quad (5.83)$$

Under $\mathbb{H}_0$ of linearity, and as $T \to \infty$, $F_T^{(A)} \xrightarrow{D} F_{\nu_1,\nu_2}$, where $\nu_1 = \frac{1}{2}p(p+1) + p$ and $\nu_2 = T - p(p+3)/2 - 2p$. Clearly, if $p = 1$, the asymptotic distribution of

(5.83) is identical to the asymptotic distribution of the Tukey nonadditivity-type test statistic (5.82).

5.5 Tests Based on Arranged Autoregressions

An *arranged autoregression* is an autoregression where the observed values of the "dependent variable" and the associated design matrix are sorted, or rearranged, according to the values of a particular regressor. For SETARMA processes, the regressor on which to sort is the threshold variable. For example, consider a SETAR(2; p,p) model with delay parameter d, and nontrivial threshold r;

$$Y_t = \begin{cases} \phi_0^{(1)} + \sum_{u=1}^p \phi_u^{(1)} Y_{t-u} + \varepsilon_t & \text{if } Y_{t-d} \leq r, \\ \phi_0^{(2)} + \sum_{u=1}^p \phi_u^{(2)} Y_{t-u} + \varepsilon_t & \text{if } Y_{t-d} > r. \end{cases} \quad (5.84)$$

Given the set of observations $\{Y_t\}_{t=1}^T$, the threshold variable Y_{t-d} can assume the values $\{Y_i\}_{i=h}^{T-d}$, where $h = \max\{1, p+1-d\}$. Let τ_j be the time index of the jth smallest observation among $\{Y_i\}_{i=h}^{T-d}$. Assume that the recursive autoregressions begin with a minimum number of start-up values, say $n_{\min} > p+1$. Denote the resulting ordered time series by $\{Y_{\tau_j}\}_{j=n_{\min}+1}^{T-d-h+1}$. Then we can write (5.84) as

$$Y_{\tau_j+d} = \begin{cases} \phi_0^{(1)} + \sum_{i=1}^p \phi_i^{(1)} Y_{\tau_j+d-i} + \varepsilon_{\tau_j+d}, & (j = n_{\min}+1, \ldots, s), \\ \phi_0^{(2)} + \sum_{i=1}^p \phi_i^{(2)} Y_{\tau_j+d-i} + \varepsilon_{\tau_j+d}, & (j = s+1, \ldots, T-d-h+1), \end{cases} \quad (5.85)$$

where s satisfies $Y_{\tau_s} < r \leq Y_{\tau_{s+1}}$.

This is an arranged autoregression with the first s observations in the first regime and the remaining observations in the second regime. This effectively separates the two regimes and also provides a means by which the data points fall into two groups where all of the observations in each group are generated from the same linear AR(p) model. If the value of the threshold parameter r is known, consistent estimates of the parameters can easily be obtained; see Chapter 6. Since, however, in most cases the value of r is not known, estimation of (5.85) is performed sequentially through recursive LS.

Let the $(p+1) \times 1$ vector $\widehat{\boldsymbol{\phi}}_m$ represent estimates of the parameters in (5.85) based on the first m cases. Also, denote the corresponding $(\mathbf{X}'\mathbf{X})^{-1}$ matrix by $\mathbf{P}_m$. Let $\mathbf{x}_{m+1}$ be the vector of regressors of the next observation to enter the arranged autoregression, namely $Y_{\tau_{m+1}+d}$. Then recursive LS estimates can be computed by (Ertel and Fowlkes, 1976; Tsay, 1989):

$$\widehat{\boldsymbol{\phi}}_{m+1} = \widehat{\boldsymbol{\phi}}_m + \mathbf{P}_{m+1}\mathbf{x}_{m+1}\left[1.0 + \mathbf{x}'_{m+1}\mathbf{P}_m\mathbf{x}_{m+1}\right]^{-1}\left[Y_{\tau_{m+1}+d} - \mathbf{x}'_{m+1}\widehat{\boldsymbol{\phi}}_m\right], \quad (5.86)$$

$$\mathbf{P}_{m+1} = \mathbf{P}_m - \mathbf{P}_m\mathbf{x}_{m+1}\left[1.0 + \mathbf{x}'_{m+1}\mathbf{P}_m\mathbf{x}_{m+1}\right]^{-1}\mathbf{x}'_{m+1}\mathbf{P}_m. \quad (5.87)$$

5.5 TESTS BASED ON ARRANGED AUTOREGRESSIONS

The predictive residuals $\widehat{\varepsilon}_{\tau_{m+1}+d}$ and standardized predictive residuals $\widehat{e}_{\tau_{m+1}+d}$ are given by

$$\widehat{\varepsilon}_{\tau_{m+1}+d} = Y_{\tau_{m+1}+d} - \mathbf{x}'_{m+1}\widehat{\boldsymbol{\phi}}_m, \tag{5.88}$$

$$\widehat{e}_{\tau_{m+1}+d} = \widehat{\varepsilon}_{\tau_{m+1}+d}\left[1 + \mathbf{x}'_{m+1}\mathbf{P}_m\mathbf{x}_{m+1}\right]^{-1/2}. \tag{5.89}$$

The LS estimates for the coefficients $\phi_u^{(1)}$ ($u = 1, \ldots, p$) are consistent if there are a large number of observations in the first regime. Moreover, the predictive residuals are asymptotically WN and independent of the regressors. When, however, j arrives at and exceeds s, the predictive residuals for the observation with index $\tau_{s+1} + d$ will become biased as a result of the model change at time τ_{s+1+d}, and the predictive residuals now become a function of the regressors $\{Y_{\tau_j+d-i}; i = 1, \ldots, p\}$. That is to say, the independence between the predictive residuals and the regressors is destroyed once the arranged autoregression includes observations whose threshold value exceeds r. In other words, there is a change at an unknown time-point in the cumulative sums of the standardized predictive residuals. This calls for a test statistic having its roots in the analysis of change-points. Typically, the first test statistic discussed below uses the change-point framework. The mechanics of the next two test statistics are based on the properties of the one-step ahead predictive residuals.

CUSUM test for SETAR nonlinearity

Petruccelli and Davies (1986) propose a cumulated sums (CUSUM) test statistic for SETAR models, using the above recursive LS estimation procedure. The test statistic can be computed as follows.

Algorithm 5.9: CUSUM test statistic

(i) Choose the AR order p, the lag d, and a minimum number $n_{\min} > p+1$ of start-up values. In practice $n_{\min} = [T/10] + p$ is recommended to have a sufficiently large number of observations in the first regime.

(ii) Then, for $n_{\min} \leq r \leq T - p$, find the recursive LS estimates; compute the standardized predictive residuals e_{τ_j+d} ($j = n_{\min} + 1, \ldots, T - d - h + 1; h = \max\{1, p+1-d\}$).

(iii) Compute the cumulative sums $Z_j = \sum_{i=n_{\min}+1}^{j} \widehat{e}_i$, ($j = n_{\min}+1, \ldots, T - d - h + 1$), and the associated CUSUM test statistic

$$Q_T = \max_{n_{\min}+1 \leq j \leq T-d-h+1} |Z_j|/\sqrt{T^*}, \tag{5.90}$$

> **Algorithm 5.9: CUSUM test statistic (Cont'd)**
>
> (iii) (Cont'd)
> where $T^* = T - d - h + 1 - n_{\min}$. Clearly, this is a Kolmogorov–Smirnov type statistic. Under mild conditions on the noise process $\{\varepsilon_t\}$, it follows (MacNeill, 1971) that the limiting distribution of Q_T is given by
>
> $$\mathbb{P}\big((Q_T/\sqrt{T^*}) \leqslant \alpha\big) = \Delta_\alpha$$
> $$\equiv \sum_{j=-\infty}^{\infty} (-1)^j \big[\Phi\big(\alpha(2j+1)\big) - \Phi\big(\alpha(2j-1)\big)\big], \quad (5.91)$$
>
> where $\Phi(\cdot)$ is the normal distribution function, and α the nominal significance level.
>
> (iv) Some upper quantiles are 0.2309 (90%), 0.3011 (92.5%), 0.3245 (95%), 0.3478 (97.5%), and 0.3616 (99%); see Grenander and Rosenblatt (1984, Chapter 6, Table 1) for a partial tabulation. If $Q_T > \Delta_\alpha$, then we reject the null hypothesis of linearity.

It is fairly obvious that the CUSUM test statistic is very simple to implement since it does not require the estimation of the SETAR model under the alternative hypothesis. The test statistic can be used to determine both the number and location of the thresholds. To avoid underfitting, it is recommended to iterate the recursive LS estimation procedure for different pairs (d, p).

TAR F test for SETAR models

The TAR F test statistic for threshold nonlinearity was developed in Tsay (1989). The alternative hypothesis is that the series is generated by a two-regime SETAR model as given in (5.84). The testing procedure consists of the following steps.

> **Algorithm 5.10: TAR F test statistic**
>
> (i) Perform the arranged autoregression, and calculate $\widehat{e}_{\tau_{j+1}+d}$.
>
> (ii) Compute a second regression with the predictive residuals on Y_{τ_j+d}; i.e.
>
> $$\widehat{e}_{\tau_j+d} = \beta_0 + \sum_{i=1}^{p} \beta_i Y_{\tau_j+d-i} + \omega_{\tau_j+d}, \quad (j = n_{\min}+1, \ldots, T-d-h+1).$$
>
> (iii) Next, compute the associated test statistic
>
> $$F_T^* = \frac{[\sum_t \widehat{e}_t^2 - \sum_t \widehat{\omega}_t^2]/(p+1)}{\sum_t \widehat{\omega}_t^2/(T-d-n_{\min}-p-h)}, \quad (5.92)$$

5.5 TESTS BASED ON ARRANGED AUTOREGRESSIONS

> **Algorithm 5.10: TAR F test statistic (Cont'd)**
>
> (iii) (Cont'd)
> where $\widehat{\omega}_t$ is the LS residual of the regression in step (ii). Then it can be shown (Tsay, 1989) that under the null hypothesis of linearity, and as $T \to \infty$,
>
> $$F_T^* \xrightarrow{D} F_{\nu_1, \nu_2},$$
>
> with degrees of freedom $\nu_1 = p+1$ and $\nu_2 = T-d-n_{\min}-p-h$. Furthermore, $(p+1)F_T^*$ is asymptotically a χ^2_ν random variable with $\nu = p+1$ degrees of freedom.

Simulation studies show that the TAR F test statistic has consistently higher empirical power than the portmanteau CUSUM test statistic.

New F test for BL, STAR, and ExpAR models

The *new* F test statistic combines the idea of an arranged autoregression along with an added variable approach resulting in a test procedure for detecting three types of nonlinear behavior. The $\mathbb{H}_0$ states that the time series is generated by a stationary linear AR(p) process. The resulting F test statistic can be computed as follows.

> **Algorithm 5.11: New F test statistic**
>
> (i) For a given delay d, fit recursively an arranged autoregression of order p to $\{Y_t\}_{t=1}^T$ and calculate the standardized predictive residuals $\{\widehat{e}_t\}_{t=n_{\min}+1}^T$.
>
> (ii) Calculate $\text{SSE}_0 = \sum_t \widehat{e}_t^2$.
>
> (iii) Regress $\widehat{\varepsilon}_t$ on $\{1, Y_{t-1}, \ldots, Y_{t-p}\}$, $\{Y_{t-i}\widehat{\varepsilon}_{t-i}, \widehat{\varepsilon}_{t-i}\widehat{\varepsilon}_{t-i-1}\}$ ($i = 1, \ldots, p$), and $\{Y_{t-1}\exp(-\gamma Y_{t-1}), \Phi(z_{t-d}), Y_{t-1}\Phi(Y_{t-d})\}$, where $z_t = (Y_{t-d} - \bar{Y}_d)/s_d$ with $\bar{Y}_d, s_d$ are the sample mean and standard deviation of the Y_{t-d}, respectively. Calculate the residual sum of squares from this regression, $\text{SSE}_1 = \sum_t \widehat{\omega}_t^2$.
>
> (iv) The associated test statistic is given by
>
> $$F_T^{(N)} = \frac{(\text{SSE}_1 - \text{SSE}_0)/[3(p+1)]}{\text{SSE}_0/[T - n_{\min} - 3(p+1)]}. \tag{5.93}$$
>
> It can be shown (Tsay, 1991) that under $\mathbb{H}_0$, and as $T \to \infty$,
>
> $$F_T^{(N)} \xrightarrow{D} F_{\nu_1, \nu_2},$$
>
> with $\nu_1 = 3(p+1)$ and $\nu_2 = T - n_{\min} - 3(p+1)$ degrees of freedom.

5.6 Nonlinearity vs. Specific Nonlinear Alternatives

Li (1993) proposes an LM-type test statistic for discriminating between different non-nested nonlinear models. Let $\{\varepsilon_{i,t}\} \stackrel{\text{i.i.d.}}{\sim} \mathcal{N}(0, \sigma_{i,\varepsilon}^2)$ ($i = 1, 2$) with $\varepsilon_{1,t}$ independent of $\varepsilon_{2,t}$. Let $\mathbf{Y}_{i,t}$ be a p_i-dimensional state vector ($i = 1, 2$). For simplicity, we consider the following two hypotheses:

$$\mathbb{H}_0\colon Y_t = f(\mathbf{Y}_{1,t}; \boldsymbol{\theta}_1) + \varepsilon_{1,t}, \quad \mathbb{H}_a\colon Y_t = g(\mathbf{Y}_{2,t}; \boldsymbol{\theta}_2) + \varepsilon_{2,t},$$

where $f(\cdot)$ and $g(\cdot)$ are two known nonlinear, real-valued functions, having continuous second-order derivatives with respect to the $p_i \times 1$ unknown parameter vector $\boldsymbol{\theta}_i$. To avoid identification problems, we assume that both families of nonlinear models are non-overlapping.

Let $\widehat{\boldsymbol{\theta}}_i$ be a consistent estimator of $\boldsymbol{\theta}_i$. Denote the corresponding residuals by $\widehat{\varepsilon}_{i,t}$ ($i = 1, 2$), and let $\widehat{Y}_t = g(\mathbf{Y}_{2,t}; \widehat{\boldsymbol{\theta}}_2)$ be the fitted values under $\mathbb{H}_a$. Then a test of $\mathbb{H}_0$ against $\mathbb{H}_a$ can be based on considering the null hypothesis $\mathbb{H}_0^*\colon \lambda = 0$, where λ is a parameter (the Lagrange multiplier) in the model

$$Y_t = f(\mathbf{Y}_{1,t}; \boldsymbol{\theta}_1) + \lambda g(\mathbf{Y}_{2,t}; \boldsymbol{\theta}_2) + \varepsilon_t,$$

where $\{\varepsilon_t\} \stackrel{\text{i.i.d.}}{\sim} \mathcal{N}(0, \sigma_\varepsilon^2)$. Thus, the adequacy of the model under $\mathbb{H}_0$ is tested versus a possible deviation in the direction of $\mathbb{H}_a$. Using the LM testing principle, it follows that the corresponding score form of the LM-type test statistic is given by

$$\text{LM}_T^* = T\widehat{\boldsymbol{\varepsilon}}' \mathbf{X}'(\mathbf{X}\mathbf{X}')^{-1}\mathbf{X}\widehat{\boldsymbol{\varepsilon}} \Big/ \sum_{t=1}^T \widehat{\varepsilon}_t^2, \tag{5.94}$$

where $\mathbf{X}'$ is a $T \times (p_1 + 1)$ matrix of regressors formed by stacking $(\partial \varepsilon_t(\boldsymbol{\theta})/\partial \boldsymbol{\theta}_1', \widehat{Y}_t)$, with $\partial \varepsilon_t(\boldsymbol{\theta})/\partial \boldsymbol{\theta}_1'$ evaluated under $\mathbb{H}_0$, and $\widehat{\boldsymbol{\varepsilon}} = (\widehat{\varepsilon}_{1,1}, \ldots, \widehat{\varepsilon}_{1,T})'$. Under $\mathbb{H}_0$ the test statistic LM_T^* has a χ_1^2 distribution, as $T \to \infty$. As before the above test statistic can also be written as TR^2, where R^2 is the coefficient of determination from the auxiliary regression of $\widehat{\varepsilon}_{1,t}$ on $\partial \varepsilon_t(\boldsymbol{\theta})/\partial \boldsymbol{\theta}_1'|_{\mathbb{H}_0}$ and $\widehat{Y}_t$. Thus, (5.94) is relatively straightforward to apply, provided $\partial \varepsilon_t(\boldsymbol{\theta})/\partial \boldsymbol{\theta}_1'$ can be obtained in a simple (recursive) way.

In practice, it will often be desirable to interchange the role of $\mathbb{H}_0$ and $\mathbb{H}_a$. It may, however, result in a situation where both or neither of the hypotheses will be rejected, giving interpretation problems. On the other hand, this information may well be used to look for alternative model specifications.

Example 5.3: Interpretation of the LM_T^*-type test statistic (Li, 1993)

One attraction of the LM_T^*-type test statistic in this context is its ease of interpretation following from a direct relation with the method of residual sum of squares. Consider the two auxiliary linear regressions

$$\widehat{\varepsilon}_{1,t} = \alpha \widehat{Y}_t + \omega_t, \quad \widehat{Y}_t = \frac{\partial f(\mathbf{Y}_{t-1}; \boldsymbol{\theta}_1)}{\partial \boldsymbol{\theta}_1'} \boldsymbol{\beta} + \eta_t, \tag{5.95}$$

where ω_t, η_t are independent zero mean normal random variables; α and β are the respective artificial parameters. For simplicity, let $\sigma_{1,\varepsilon}^2 = 1$ and $f_t \equiv f(\mathbf{Y}_{t-1}; \boldsymbol{\theta}_1)$.

In this case, the score vector under $\mathbb{H}_0$ is given by $-(\mathbf{0}', \sum_{t=1}^{T} \widehat{\varepsilon}_{1,t} \widehat{Y}_t)'$. Now, with the respective partitions of the observed information matrix, the LM-type test statistic under the null hypothesis will take the following form

$$\mathrm{LM}_T^* = \Big(\sum_t \widehat{\varepsilon}_{1,t} \widehat{Y}_t\Big)^2 \Big[\sum_t \widehat{Y}_t^2 - \Big(\sum_t \widehat{Y}_t \frac{\partial f_t}{\partial \boldsymbol{\theta}_1'}\Big)\Big(\sum_t \frac{\partial f_t}{\partial \boldsymbol{\theta}_1}\frac{\partial f_t}{\partial \boldsymbol{\theta}_1'}\Big)^{-1}\Big(\sum_t \widehat{Y}_t \frac{\partial f_t}{\partial \boldsymbol{\theta}_1}\Big)\Big]^{-1}\Big|_{\mathbb{H}_0}$$

$$= \Big(\frac{\sum_t \widehat{\varepsilon}_{1,t} \widehat{Y}_t}{\sum_t \widehat{Y}_t^2}\Big)^2 \frac{\sum_t \widehat{Y}_t^2}{1 - R^2},$$

where R^2 is the coefficient of determination for the second auxiliary regression in (5.95). Note that $\sum_t \widehat{\varepsilon}_{1,t} \widehat{Y}_t / \sum_t \widehat{Y}_t^2 = \widehat{\alpha}$, the LS estimate of α in the first auxiliary regression. Suppose the residual sum of squares from the first regression is denoted by $\sum_t \widehat{\omega}_t^2$. Then from standard linear regression theory it follows that

$$\mathrm{LM}_T^* = \frac{\sum_t \widehat{\alpha}^2 \widehat{Y}_t^2}{1 - R^2} = \frac{\sum_t \widehat{\varepsilon}_{1,t}^2 - \sum_t \widehat{\omega}_t^2}{1 - R^2}.$$

Hence, if $\mathbb{H}_0$ is true, the difference between the two residual sums of squares should be small if T is sufficiently large, and $\sum_t \widehat{\varepsilon}_{1,t}^2$ should be small. On the other hand, if $\mathbb{H}_a$ is true $\sum_t \widehat{\varepsilon}_{1,t}^2$ should be large while $\sum_t \widehat{\omega}_t^2$ should be small.

5.7 Summary, Terms and Concepts

Summary

In this chapter we have seen a large number of time-domain statistics for testing nonlinearity. A practitioner may be somewhat bewildered by the wide range of possibilities. To be of some help, Appendix 5.B reports some strengths and weaknesses of the available test statistics through reported simulation studies of their size and power. On the whole a test statistic is effective at identifying the type of nonlinearity it is designed to detect. This is a pleasing result. In addition, the form of the nonlinear functional relationship in the state-dependent model seems to be less important with test statistics based on the classical hypothesis testing principles, LR, LM, and W. Finding the correct dimension (order) of the state vector is more likely to be the key factor (see, e.g., Pitarakis, 2006). Nevertheless, one should always consider a linear model first. Occam's razor tells us that we should not introduce complexities unless absolutely necessary. Indeed, all the hypothesis tests discussed in this chapter are concerned with a *simple* null hypothesis which asserts that the given data set is a random realization of a specified unique linear DGP.

We have not discussed a testing framework where the null hypothesis is *composite*. The composite null hypothesis specifies a family of processes, and asserts that the actual DGP is a member of that family, but does not specify which one. This latter situation occurs when artificial, or *surrogate*,[4] data are created with MC simulation methods. Surrogate data sets are often used in studies of nonlinear dynamical systems; see, e.g., Theiler et al. (1992), and Theiler and Prichard (1996) for further insights into this topic.

Terms and Concepts

added variable, 179
arranged autoregression, 182
auxiliary regression, 157
simple (composite) hypothesis, 187
Lagrange multiplier, 156

nested, 171
Occam's razor, 187
portmanteau-type test, 179
stochastic permutation, 175
surrogate data, 188

5.8 Additional Bibliographical Notes

Section 5.1: The LM-type test statistics for BL, ExpAR, and STAR are due to Saikkonen and Luukkonen (1988), and Luukkonen et al. (1988a,b); see also Weiss (1986) for an early contribution. Brännäs et al. (1998) propose the LM-type test statistics for asMA and TMA nonlinearities. Wong and Li (1997, 2000a) study LM-type test statistics of so-called double-threshold ARCH models, which may be applied to situations where both the conditional mean and the conditional variance of the time series process are assumed to be piecewise linear, given time-delayed observations. Guégan and Wandji (1996) study the local (theoretical) power of the LM-type test statistic for a simple subdiagonal BL model.

The $LM_T^{(7)}$-type test statistic for NCTAR is due to Medeiros and Veiga (2005). Medeiros et al. (2006) apply sequentially LM-type test statistics within the context of AR–NN modeling. Lee et al. (1993) present an LM-type test statistic for AR–NN models. The test is a special case of the LM-type test statistic for NCTAR models. MC simulation results show a good performance in power compared to other competitors. However, the presence of an intercept in the nonlinear, hidden layer, causes a loss of power compared with other LM-type test statistics; see, e.g., Lee et al. (1993) and Teräsvirta et al. (1993). Also, various versions of the White (1989, 1992) dynamic information matrix test, a test statistic for neglected nonlinearity, are commonly used within the NN context.

Kiliç (2016) investigates the Taylor series approximations of STAR models around the null hypothesis of linearity. The approximations may not accurately describe the specific nonlinearity of the DGP and, as a result, the LM-type test statistics may fail to detect the correct form of nonlinearity.

Tong and Yeung (1991a) discuss the identification and estimation of continuous-time two-regime SETAR models. Tai and Chan (2000) consider a more general class of nonlinear continuous-time AR (NLCAR) models. In addition, they develop an LM-type test statistic for this class of models with the linear CAR model under the null hypothesis; see Tai and Chan (2002) for an extension.

---

[4]Surrogate data have no dynamical nonlinearities. By construction a surrogate is equivalent to passing i.i.d. Gaussian WN through a linear filter that reproduces the linear properties of one realization of the strictly stationary process $\{Y_t, t \in \mathbb{Z}\}$.

5.8 ADDITIONAL BIBLIOGRAPHICAL NOTES

Section 5.2: Asymptotic critical values of the LR test statistic for SETMA$(2; q, q)$ models with $d > q$ are the same as that of test statistics for change-points in Andrews (1993). Empirical implementations of the LR testing approach are reported by K.S. Chan and Tong (1986). Ling and Tong (2005) suggest a computationally intensive bootstrap method to calculate p-values of a quasi-LR test for SETMA$(2; q, q)$ models with $d < q$. Li and Li (2008) generalize the test in Ling and Tong (2005) to a quasi-LR test statistic for TMA models with GARCH errors.

Hansen (2000) recommends inverting the LR test statistic to construct confidence intervals for the threshold parameter of a SETAR process. If the error process in (5.61) is conditionally heteroskedastic, it is necessary to replace the $F_T^{(1,i)}$ test statistic with a heteroskedasticity-consistent Wald or LM-type test statistic; Hansen (1997).

Chen et al. (2012b) propose a LR test statistic to determine the number of regimes in SETAR models with two regimes.

Section 5.3: The Wald test statistic for symmetry of ARasMA models is due to Brännäs and De Gooijer (1994). For asMA(1) models, the size properties are best for the LM-type test statistic followed by, in order, the Wald and LR test statistics. The latter two tests are more powerful than the LM-type test statistic; see also Brännäs et al. (1998).

Testing for a linear (near) unit root against (stationary) TAR models is the topic of a large number of papers in the econometrics literature. For instance, Caner and Hansen (2001) propose a Wald statistic for testing a two-regime SETAR with stationary but unknown threshold parameter, Enders and Granger (1998) focus on an F test statistic for an M–TAR model with known threshold parameter, Lanne and Saikkonen (2002) introduce a stability test statistic for a TAR model with threshold effects only in the intercept term, Kapetanios and Shin (2006) consider a Wald statistic for testing a three-regime SETAR model with a random walk in the middle regime. Pitarakis (2008) comments on the limiting distribution of the Wald test statistic in Caner and Hansen (2001). Bec et al. (2008) propose a SupWald test statistic for SETARs with an adaptive set of thresholds, and Seo (2008) considers a residual-based block bootstrap algorithm for testing the null hypothesis of a unit root in SETARs.

Charemza et al. (2005) introduce a Student t-type test statistic for detecting unit root bilinearity in a simple BL$(1, 0, 1, 1)$ process. The linearity coefficient in this model may be estimated by the Kalman filter algorithm, following an approach suggested by Hristova (2005).

Section 5.4: The RESET test statistic of Ramsey (1969) may be viewed as an earlier, and more general, version of the Tukey nonadditivity-type test statistic.

Section 5.5: It is easy to verify that (5.91) is identical to the approximate large sample distribution given by Petruccelli and Davies (1986). Petruccelli (1990) introduces another CUSUM test statistic for linearity using the reversed predictive residuals, denoted by Q_T^{rev} in Table 5.2. Similarly, Sorour and Tong (1993) examine the performance of the LR test statistic for SETAR and the CUSUM test statistics in building a TARSO model.

Tong and Yeung (1990, 1991b) apply the CUSUM tests (original and reversed) and the TAR F test statistic to investigate nonlinearities in partially observed time series; see also Tsai and Chan (2000, 2002).

Following the basic structure of Algorithm 5.10, Liang et al. (2015) propose an F-type test statistic for testing linear MA models versus (rearranged) SETMA models. The procedure requires the subjective use of scatter plots to identify the number and locations of potential threshold values. The MA order follows from inspection of the sample ACF.

Section 5.6: Many studies have been performed investigating power properties of the test statistics considered in this Chapter. Important contributions published prior to the year 1992 are summarized in the review paper by De Gooijer and Kumar (1992, Exhibit 1). Teräsvirta et al. (1993) study and compare the power of LM-type and ANN test statistics (see also Lee et al., 1993). de Lima (1997) investigates the robustness of several portmanteau-type nonlinearity test statistics (e.g. Hinich's bispectrum test) to moment condition failure. More recently, Vavra (2013, Chapter 2) examines the robustness of eight nonlinearity test statistics against non-Gaussian innovations by MC simulation. Overall, there is no clear link between the performance of the test statistics and their moments requirements. However, some of the test statistics are not very trustworthy for DGPs with heavy-tailed innovations.

5.9 Software References

Section 5.1: The website https://www.estima.com/procs_perl/mainproclistwrapper.shtml contains freely available RATS[5] code (star and regstrtest) for LM-type testing of STAR models. Also, the website has RATS code for the arranged AR test statistic (tsaytest), the $F_T^{(O)}$ test statistic (tsaynltest), the F test statistic of Hansen (threshtest), and the Hinich (frequency-domain) linearity and Gaussianity test statistics (hinichtest). GAUSS code for computing the LM-type test statistic $F_T^{(6)}$ is available at the website of this book.

Section 5.2: A FORTRAN77 program (written by K.S. Chan) for computing the percentiles of the LR-SETAR test statistic $\text{LR}_T^{(8)}$ is available at the website of this book. The R-TSA package contains the $F_T^{(T)}$ test (Keenan.test), the $F_T^{(O)}$ test (Tsay.test), the $F_T^{(8)}$ test (tlrt). Bruce Hansen's web page at http://www.ssc.wisc.edu/~bhansen/ offers MATLAB, GAUSS and R code (and data) to replicate some of the empirical work reported in his papers on SETAR model selection and estimation. Based on papers written by Hansen and his co-authors, the R-tsDyn package has a host of test statistics for various forms of SETAR nonlinearity, including the bootstrapped version of the $F_T^{(i,j)}$ test statistic. A special file at the website of this book contains MATLAB programs to replicate the results of Example 5.1.

Two FORTRAN90 programs (written by Guodong Li) for replicating the results in Li and Li (2011) and using the $\text{LR}_T^{(9)}$-SETARMA test statistic summarized in Algorithm 5.6, are available at the website of this book.

Section 5.4: The function lin.test in the R-nlts package computes the $F_T^{(O)}$ test statistic of Algorithm 5.8 for AR(p) processes up to order $p = 5$. The nlts.f FORTRAN77 library (largely written by Jane L. Harvill), available at the website of this book, contains an extensive set of subroutines for nonlinear time series analysis, including Hinich's test for linearity, the CUSUM, TAR-F, New-F, and the Original- and Augmented F test statistics.

---

[5]RATS, also called WinRATS, is a registered trademark of Estima, Inc.

Appendix

5.A Percentiles of LR–SETAR Test Statistic

Critical values c_α, at the nominal significance level α, depend on p and on $\underline{r}$ and $\overline{r}$ only. In practice, $\widetilde{\mathbb{R}}$ can be taken as a closed interval with $\underline{r} \times 100$ and $\overline{r} \times 100$ percentiles of the empirical distribution of $\{Y_t\}_{t=1}^T$ as end points. Table 5.1 provides values of c_α for $\alpha = 0.01$, 0.05, and 0.10, $p = 1, \ldots, 10$, and $\widetilde{\mathbb{R}} = [r_0, 1 - r_0]$ for an array of r_0 values between 0.05 and 0.40. In addition, Table 5.1 covers a much wider range of intervals $\widetilde{\mathbb{R}}$ than just the symmetric interval $[r_0, 1 - r_0]$ through the parameter $\lambda = \overline{r}(1 - \underline{r})/(\underline{r}(1 - \overline{r}))$. Given a value of $p \geq 1$, this allows one to obtain critical values for some other interval $[\underline{r}, \overline{r}]$ either directly or by interpolation.

For the special case $p = 0$, we noted in Section 5.2 that an explicit expression for the asymptotic distribution of the $\mathrm{LR}_T^{(8)}$ test statistic is available. In particular, Chan and Tong (1990) show that, for $z \to \infty$,

$$\mathbb{P}\Big(\sup_{0 \leq t \leq t^*} |U_t| > z \Big) \sim (2/\pi)^{1/2} \exp(-z^2/2)\Big(t^* z - \frac{t^*}{z} + \frac{1}{z} \Big), \qquad (5.96)$$

where

$$t^* = \frac{1}{2} \log\Big\{ \frac{b(1-a)}{a(1-b)} \Big\}, \quad (0 < a < b < 1),$$

and $\{U_t\}$ is a so-called stationary Ornstein–Uhlenbeck process with $\mathbb{E}(U_t) = 0$ and $\mathbb{E}(U_s U_t) = \exp(-|t - s|)$.

Tables 1 and 2 in Chan (1991) contain upper 10%, 5%, 2.5%, 1% and 0.1% percentage points for the null distribution of the $\mathrm{LR}_T^{(8)}$ test statistic for $0 \leq p \leq 18$ and $(a, b) = (0.25, 0.75)$ and $(0.1, 0.9)$. For $p = 0$, it can be seen that the percentage points are close to that of a χ_3^2 distribution, which also follows from comparing (5.96) with the asymptotic distribution function $\mathbb{P}(\chi_3^2 > z^2) \sim (2/\pi)^{1/2} \exp(-z^2/2)(z + \frac{1}{z})$.

5.B Summary of Size and Power Studies

Usually the overall performance of a test statistic is obtained from an MC simulation study of its size and power. A number of these studies have been carried out for the tests discussed in this Chapter. Table 5.2 summarizes the main findings in this area. In general one can say that when a test statistic is used against the alternative hypothesis, which it is designed to reveal, it is more powerful than when it is used against other alternative hypotheses $\mathbb{H}_a$. Clearly, there is no test which can be used as an overall tool against any type of nonlinearity. Nevertheless, all LM-type test statistics seem to have reasonable size and power properties. These tests do not require estimation of the model under $\mathbb{H}_a$ nor do they depend on the particular form of $\mathbb{H}_a$. Thus, one might expect that for finite sample sizes tests which explicitly make use of the form of $\mathbb{H}_a$, like for example the LR test statistic, are more powerful. This seems to be the case for SETAR models, but evidence for other types of nonlinear models is lacking. In addition, it is important to realize that the presence and size of an intercept in a nonlinear model seems to have a considerable influence on the size and power of the test statistics when T is not large. Centering data, i.e. analyzing deviations

Table 5.1: *Asymptotic critical values of the* $\mathrm{LR}_T^{(8)}$ *test statistic for* $\mathrm{SETAR}(2;p,p)$ *models;* $\lambda = (1-r_0)^2/r_0^2$.

		$p=1$			$p=2$			$p=3$			$p=4$			$p=5$		
r_0	λ	10%	5%	1%	10%	5%	1%	10%	5%	1%	10%	5%	1%	10%	5%	1%
0.40	2.25	6.20	8.52	12.79	7.97	10.51	15.09	9.65	12.37	17.20	11.25	14.13	19.19	12.81	15.83	21.10
0.35	3.45	7.63	9.69	13.81	9.55	11.78	16.17	11.34	13.72	18.34	13.04	15.56	20.38	14.69	17.31	22.32
0.30	5.44	8.56	10.52	14.55	10.56	12.67	16.96	12.42	14.66	19.16	14.19	16.54	21.24	15.88	18.34	23.21
0.25	9.00	9.27	11.18	15.16	11.34	13.38	17.60	13.25	15.42	19.83	15.07	17.33	21.93	16.80	19.16	23.93
0.20	16.00	9.89	11.75	15.69	12.01	14.00	18.16	13.96	16.07	20.42	15.81	18.02	22.55	17.59	19.88	24.57
0.15	32.11	10.46	12.29	16.20	12.63	14.58	18.70	14.62	16.68	20.98	16.51	18.66	23.13	18.31	20.54	25.17
0.10	81.00	11.05	12.85	16.72	13.26	15.18	19.25	15.30	17.31	21.57	17.21	19.32	23.73	19.05	21.23	25.79
0.05	361.00	11.74	13.51	17.35	14.01	15.89	19.92	16.10	18.07	22.27	18.06	20.11	24.47	19.93	22.06	26.56

		$p=6$			$p=7$			$p=8$			$p=9$			$p=10$		
r_0	λ	10%	5%	1%	10%	5%	1%	10%	5%	1%	10%	5%	1%	10%	5%	1%
0.40	2.25	14.32	17.47	22.93	15.80	19.07	24.70	17.25	20.63	26.43	18.68	22.16	28.13	20.09	23.67	29.78
0.35	3.45	16.28	19.01	24.19	17.84	20.66	26.01	19.36	22.28	27.77	20.85	23.86	29.50	22.32	25.41	31.19
0.30	5.44	17.53	20.08	25.11	19.13	21.77	26.95	20.69	23.42	28.74	22.23	25.03	30.49	23.74	26.61	32.20
0.25	9.00	18.48	20.93	25.85	20.12	22.65	27.71	21.72	24.32	29.52	23.28	25.96	31.29	24.82	27.57	33.02
0.20	16.00	19.30	21.67	26.51	20.96	23.41	28.39	22.58	25.11	30.21	24.17	26.76	32.00	25.74	28.39	33.74
0.15	32.11	20.05	22.36	27.13	21.74	24.12	29.02	23.39	25.84	30.87	25.00	27.52	32.67	26.59	29.16	34.43
0.10	81.00	20.82	23.07	27.77	22.53	24.86	29.69	24.21	26.60	31.55	25.84	28.30	33.36	27.45	29.96	35.14
0.05	361.00	21.73	23.93	28.56	23.48	25.74	30.49	25.18	27.51	32.37	26.84	29.23	34.21	28.48	30.92	36.00

from the sample mean, is not recommended since then the asymptotic null distributions are no longer valid.

Some additional remarks are in order:

(i) With the test statistics Q_T, Q_T^{rev} and LR_T one must fix p and d. The selection of the order p can be done via, e.g., AIC. Also, the number of thresholds need to be pre-specified.

(ii) The selection of the added variables with many of the LM-type and F-type test statistics is somewhat arbitrary. For example, one uses p added variables specifically for the ExpAR(p) model and $p+1$ for the STAR($2;p,p$) model.

(iii) Test statistics based on the recursive LS method require a minimum number of observations $n_{\min}$ used to start the method. However, $n_{\min}$ depends on the order p and the sample size T.

(iv) The recursive estimation can be done via various algorithms such as the one given by (5.86) – (5.87), or by the Kalman filter. The latter method appears to be preferable when there are missing observations in the data.

(v) The empirical power studies in Table 5.2 have been carried out under a wide variety of alternatives (see the footnotes at the bottom of the table). No fixed set of DGPs has been used across all studies with the same sample size. So, comparison of the reported results is difficult. Moreover, power studies are criticized for the fact that test results are determined by the sample size, i.e. as T increases the empirical power goes to one under the alternative hypothesis. In contrast, local alternatives make its difference

APPENDIX 5.B

Table 5.2: *Summary of size and power studies for some time-domain linearity test statistics; equation numbers in parentheses refer to the particular test statistic in the main text.*

DGP		Test statistic	T	Power	Reference
BL[(1)]:	(i)	Q_T (5.90)	50, 100	• marginally outperforms $F_T^{(T)}$	Petruccelli and Davies (1986)
		$F_T^{(T)}$ (5.81)	<200	• reasonable only for extreme BL-DGPs	Davies and Petruccelli (1986)
			>200	good for wide range of BL-DGPs	
	(ii)	$F_T^{(T)}$ (5.81)	50, 100, 200	• good	Saikkonen and Luukkonen (1988)
		$\text{LM}_T^{(1)}$ (5.12)		• outperforms $F_T^{(T)}$	
	(iii)	$\text{LM}_T^{(1)}$ (5.12)	50, 75, 100, 150	• good for BL-DGPs	Saikkonen and Luukkonen (1991)
	(iv)	$F_T^{(O)}$ (5.82)	70, 140, 204	• outperforms $F_T^{(T)}$	Tsay (1986)
	(v)	$F_T^{(O)}$ (5.82), $F_T^{(N)}$ (5.93), $F_T^{(A)}$ (5.83)	100	• all tests have good power	Tsay (1991)
ExpAR[(2)]:		Q_T (5.90), $F_T^{(N)}$ (5.93), F_T^* (5.92)	100	• good • not powerful	Tsay (1991)
		$\text{LM}_T^{(2)}$ (5.15)	50, 100, 200	• outperforms $F_T^{(T)}$ and $\text{LM}_T^{(1)}$	Saikkonen and Luukkonen (1988)
SETAR[(3)]:	(i)	Q_T (5.90)	50, 100	• less powerful than $F_T^{(T)}$	Petruccelli and Davies (1986)
	(ii)	Q_T (5.90),	50, 100, 150, 200, 250	• less powerful than Q_T^{rev}	Moeanaddin and Tong (1988)
		$F_T^{(8)}$ (5.60)	50, 100	• outperforms Q_T and Q_T^{rev}	
	(iii)	Q_T^{rev} and $\text{LM}_T^{(3^{**})}$ (5.26)	100	• outperforms $F_T^{(8)}$ and F_T^*	Petruccelli (1990)
	(iv)	$F_T^{(T)}$ (5.81)	<100	• reasonable only for nearly nonstationary DGPs	Davies and Petruccelli (1986)
			>100	more satisfactory	
	(v)	F_T^* (5.92)	50, 100	• outperforms Q_T	Tsay (1989)
	(vi)	$\text{LM}_T^{(3^*)}$ (5.22), $\text{LM}_T^{(3^{**})}$ (5.26), $\text{LM}_T^{(4)}$ (5.29)	50, 100	• $\text{LM}_T^{(4)}$ is more powerful; $\text{LM}_T^{(3^*)}$ and Q_T are poor	Luukkonen et al. (1988b)
LSTAR[(4)]:	(i)	$F_T^{(O)}$ (5.82), $F_T^{(A)}$ (5.83), Q_T (5.90), $F_T^{(N)}$ (5.93),	100	• all tests have low power	Tsay (1991)
	(ii)	$\text{LM}_T^{(3^*)}$ (5.22), $\text{LM}^{(3^{**})}$ (5.26), $\text{LM}_T^{(4)}$ (5.29)	50, 100	• $\text{LM}_T^{(3)}$ is inferior to $\text{LM}_T^{(3^*)}$ and $\text{LM}_T^{(4)}$; $\text{LM}_T^{(3)}$, Q_T low power	Luukkonen et al. (1988a)

[(1)] (i) $Y_t = (\phi + \psi\varepsilon_t)Y_{t-1} + \varepsilon_t$; (ii) (2.13); (iii) $Y_t = \mu + \psi\varepsilon_{t-1}Y_{t-i} + \varepsilon_t$ ($i = 1,2$); (iv) $Y_t = \varepsilon_t - 0.4\varepsilon_{t-1} + 0.3\varepsilon_{t-2} + 0.5\varepsilon_t\varepsilon_{t-2}$; (v) $Y_t = 0.5Y_{t-1} + \psi Y_{t-1}\varepsilon_{t-1} + \varepsilon_t$ and $Y_t = \varepsilon_t + 0.5\varepsilon_{t-1} + \psi\varepsilon_{t-1}^2$.

[(2)] $Y_t = \{\phi + \xi\exp(-Y_{t-1}^2)\}Y_{t-1} + \varepsilon_t$.

[(3)] (i) SETAR(2;1,1) (no intercept); (ii) SETAR(2;1,1) (no intercept); (iii) SETAR(2;1,1), SETAR(2;3,2) and SETAR(3;1,1,1) (all with intercept); (iv) SETAR(2;1,1) (no intercept); (v) SETAR(2;1,1) (no intercept); (vi) SETAR(2;1,1) (with intercept).

[(4)] (i) $Y_t = 1 - \frac{1}{2}Y_{t-1} + (\phi + \xi Y_{t-1})G(\gamma Y_{t-1}) + \varepsilon_t$ with $G(z) = 1/(1 + \exp(-z))$; (ii) $Y_t = -\frac{1}{2}Y_{t-2} - \phi Y_{t-2}G(\frac{1}{2}Y_{t-1}) + \varepsilon_t$ with $G(z) = 1/(1 + \exp(-z))$.

with the null hypothesis shrink as T increases. Only a few papers investigate the local power of linearity tests; see, e.g., Guégan and Pham (1992) for the LM-type test statistic against a general diagonal BL model.

Exercises

Theory Questions

5.1 Let $\gamma_Y^{(1,2)}(\ell) = \mathrm{Cov}(Y_t, Y_{t-\ell}^2)$ denote the bicovariance at lag ℓ of a time series $\{Y_t, t \in \mathbb{Z}\}$ generated by an MA(ℓ) model with mean $\mathbb{E}(Y_t) = 0$, and with $\{\varepsilon_t\} \stackrel{\mathrm{i.i.d.}}{\sim} \mathcal{N}(0, \sigma_\varepsilon^2)$. Given an observed time series $\{Y_t\}_{t=1}^T$, the moment estimator of $\gamma_Y^{(1,2)}(\ell)$ equals $\widehat{\gamma}_Y^{(1,2)}(\ell) = (T - \ell)^{-1} \sum_{t=\ell+1}^T Y_t Y_{t-\ell}^2$. Under the null hypothesis $\mathbb{H}_0: \gamma_Y^{(1,2)}(\ell) = 0$ ($\ell = 1, 2, \ldots$), Welsh and Jernigan (1983) show that, as $T \to \infty$, the large sample distribution of the standardized bicovariance is given by

$$\mathrm{WJ} = \sum_{t=\ell+1}^T Y_t Y_{t-\ell}^2 / \sqrt{3(T-\ell)} \xrightarrow{D} \mathcal{N}(0, 1).$$

Show that the WJ test statistic is a special case of the LM-type test statistic of testing an MA(k) model against an ASTMA(k) model.

5.2 Suppose that the $T \times 1$ vector of observations $\mathbf{y} = (Y_1, \ldots, Y_T)'$ satisfies the asAR(p) model

$$Y_t = \sum_{i=1}^p (\phi_i + \alpha_i I(\varepsilon_{t-i} \geq 0)) Y_{t-i} + \varepsilon_t, \quad \{\varepsilon_t\} \stackrel{\mathrm{i.i.d.}}{\sim} \mathcal{N}(0, \sigma_\varepsilon^2).$$

Let $\boldsymbol{\theta} = (\boldsymbol{\phi}', \boldsymbol{\alpha}')'$ with $\boldsymbol{\phi} = (\phi_1, \ldots, \phi_p)'$, $\boldsymbol{\alpha} = (\alpha_1, \ldots, \alpha_p)'$, $\boldsymbol{\varepsilon} = (\varepsilon_1, \ldots, \varepsilon_T)'$, $\mathbf{I}_{\varepsilon,T} = \mathrm{diag}(I(\varepsilon_1 > 0), \ldots, I(\varepsilon_T > 0))$ and $\boldsymbol{\varepsilon}^+ = \mathbf{I}_{\varepsilon,T}\boldsymbol{\varepsilon}$. Construct an LM-type test statistic for the null hypothesis $\mathbb{H}_0: \boldsymbol{\alpha} = \mathbf{0}$.

5.3 Consider the nonlinear time series model

$$Y_t = \sum_{i=1}^p \big(a_i + \phi_i f_i(\boldsymbol{\alpha}_i' \mathbf{Y}_t)\big) Y_{t-i} + \sum_{j=1}^q \big(b_j + \theta_j g_j(\boldsymbol{\beta}_j' \mathbf{Y}_t)\big) W_{j,t} + \varepsilon_t, \quad \{\varepsilon_t\} \stackrel{\mathrm{i.i.d.}}{\sim} \mathcal{N}(0, \sigma_\varepsilon^2),$$

where $W_{j,t}$ is an observable regressor, and $\mathbf{Y}_t$ is a state vector. Assume that $\mathbf{W}_t = (W_{1,t}, \ldots, W_{q,t})'$ as well as $\mathbf{Y}_t$ are independent of ε_{t+s} ($s \geq 0$). Furthermore, assume that the functions $f_i(\cdot)$ and $g_j(\cdot)$ are real-valued possessing continuous derivatives of at least the first order in some neighborhood of the origin.

(a) The null hypothesis under study is

$$\mathbb{H}_0: \boldsymbol{\alpha}_i = \mathbf{0} \ (i = 1, \ldots, p), \text{ and } \boldsymbol{\beta}_j = \mathbf{0} \ (j = 1, \ldots, q).$$

How would you carry out an LM-type test?

(b) Suppose the parameter restrictions $\alpha_1 = \cdots = \alpha_p \equiv \alpha$ and $\beta_1 = \cdots = \beta_q \equiv \beta$ are already imposed on the above nonlinear model. The null hypothesis in part (a) is obviously replaced by

$$\mathbb{H}_0^*: \alpha = 0 \quad \text{and} \quad \beta = 0.$$

How would you carry out an LM-type test in this case?

Simulation Question

5.4 In this exercise we evaluate by simulation the power of the $F_T^{(1,2)}$ test statistic, defined by (5.63), under model-selection uncertainty. The SETAR(2; 2, 2) model for the observed time series is formulated as

$$Y_t = \begin{cases} \phi_0^{(1)} + \phi_1^{(1)} Y_{t-1} + \phi_2^{(1)} Y_{t-2} + \varepsilon_t & \text{if } Y_{t-2} \leq 0, \\ \phi_0^{(2)} + \phi_1^{(2)} Y_{t-1} + \phi_2^{(2)} Y_{t-2} + \varepsilon_t & \text{if } Y_{t-2} > 0, \end{cases}$$

where $\{\varepsilon_t\} \overset{\text{i.i.d.}}{\sim} \mathcal{N}(0,1)$. Consider the following two DGPs:

(i) $\phi_0^{(1)} = 0.5$, $\phi_1^{(1)} = -\phi_1^{(2)} = 0.2$, $\phi_0^{(2)} = 0.3$, $\phi_2^{(1)} = -\phi_2^{(2)} = -0.1$; and

(ii) $\phi_0^{(1)} = 0.5$, $\phi_1^{(1)} = -\phi_1^{(2)} = -0.1$, $\phi_2^{(1)} = -\phi_2^{(2)} = 0.1$.

(a) For $T = 200$ and 500, generate 2,000 MC replications of the DGPs (i) and (ii). Next, compute the empirical power of the $F_T^{(1,2)}$ test statistic, at the 5% nominal significance level, using (i) a correctly specified SETAR model (setting the true lag length at two), and (ii) the AIC and BIC order selection criteria (setting the maximum allowed lag order $p_{\max} = 6$). You should find the results given in Table 5.3 (approximately).

Table 5.3: *Empirical power (in %) of the $F_T^{(1,2)}$ test statistic, at the 5% nominal significance level, for two SETAR(2; 2, 2) models; 2,000 MC replications.*

DGP	T = 200			T = 500		
	True	AIC	BIC	True	AIC	BIC
(i)	55.15	33.40	18.35	97.20	83.35	51.00
(ii)	13.45	8.20	5.65	33.75	21.95	12.95

Compare and interpret the results in Table 5.3.

[*Hint:* Use Bruce Hansen's GAUSS, R, or MATLAB codes to compute the $F_T^{(1,2)}$ test statistic.]

(b) Gonzalo and Pitarakis (2002) introduce the following penalty-based model selection approach for deciding between an AR(p) and a SETAR(2; p, p) model:

- Select the best AR model that minimizes AIC, and the best SETAR model that minimizes the order selection criterion $\text{SC}(p, d; r) = T \log \widehat{\sigma}_\varepsilon^2 + C(T)(2p+2)$ with $C(T) = 2$ and $\widehat{\sigma}_\varepsilon^2$ the residual variance of the SETAR model.

Table 5.4: *Model-selection based correct decision frequencies (in %) under two SETAR models; 1,000 MC replications.*

DGP	$T = 200$		$T = 500$	
	AIC	BIC	AIC	BIC
(i)	99.8	48.0	100.0	91.2
(ii)	98.9	13.9	99.4	16.4

- Then select the AR(p) model if $\min_p \text{AIC}(p) < \min_{p,r,d} \text{SC}(p,d;r)$ ($1 \leq p \leq p_{\max}, r \in \widetilde{\mathbb{R}}, d \leq p$).

A similar approach can be based on BIC with $C(T) = \log T$.

For $T = 200$ and 500, generate 1,000 replications of the DGPs (i) and (ii). Next, apply the above two model-selection approaches (AIC and BIC) and record the number of correct decision frequencies. Table 5.4 provides a summary of the results you will find.

Compare and contrast the results in Tables 5.4 and 5.3.

Chapter 6

MODEL ESTIMATION, SELECTION, AND CHECKING

Model estimation, selection, and diagnostic checking are three interwoven components of time series analysis. If, within a specified class of nonlinear models, a particular linearity test statistics indicates that the DGP underlying an observed time series is indeed a nonlinear process, one would ideally like to be able to select the correct lag structure and estimate the parameters of the model. In addition, one would like to know the asymptotic properties of the estimators in order to make statistical inference. Moreover, it is evident that a good, perhaps automatic, order selection procedure (or criterion) helps to identify the most appropriate model for the purpose at hand. Finally, it is common practice to test the series of standardized residuals for white noise via a residual-based diagnostic test statistic.

In this chapter, we focus on these three themes within the context of parametric nonlinear modeling. Specifically, we consider the class of identifiable parametric stochastic models

$$Y_t = g(Y_{t-1}, \ldots, Y_{t-p}, \varepsilon_{t-1}, \ldots, \varepsilon_{t-q}; \boldsymbol{\theta}_g) + \eta_t \tag{6.1}$$

where

$$\eta_t = h(Y_{t-1}, \ldots, Y_{t-u}, \varepsilon_{t-1}, \ldots, \varepsilon_{t-v}; \boldsymbol{\theta}_h)^{1/2} \varepsilon_t.$$

Here $\{Y_t, t \in \mathbb{Z}\}$ is a strictly stationary and ergodic univariate stochastic process; $g(\cdot; \boldsymbol{\theta}_g)$ and $h(\cdot; \boldsymbol{\theta}_h)$ are two real-valued measurable (known) functions on $\mathbb{R}^{p+q}$ and $\mathbb{R}^{u+v}$ ($u \leq p$), respectively; and $\boldsymbol{\theta} = (\boldsymbol{\theta}_g', \boldsymbol{\theta}_h')'$ is a vector of unknown parameters that we wish to estimate, and we have available a set of observations $\{Y_t\}_{t=1}^{T}$ with which to do so. Further, we assume that $h(\cdot; \boldsymbol{\theta})$ is a non-negative function of past Y_t's and ε_t's.

The class of models (6.1) covers a wide range of nonlinear models, including many models introduced earlier in this book. Numerous methods have been proposed for estimating models contained within this class. Here, we do not provide a full technical treatment of the subject. Rather we elaborate on some commonly used estimation methods and, in some cases, their practical implementation. Throughout the discussion, we assume that (6.1) is completely known. In practice, however, this is seldom the case and the model structure needs to be specified first. This is a model selection problem, and there are several ways to approach it. One is to develop model selection criteria on the basis of the asymptotic properties of the estimated parameters, and we will therefore spend some time discussing these criteria here. Alternatively, model selection criteria have been suggested on the basis of sample reuse such as *cross-validation* (CV). Since several of the latter criteria are (asymptotically) linked to criteria in the first group, we include them as well in this chapter. Similarly, the effect of parameter estimation errors becomes relevant when checking for model adequacy.

Given the above themes, the chapter consists of three interrelated parts. First, in Section 6.1.1, we discuss the method of quasi maximum likelihood (QML) estimation and, in particular, nonlinear least squares (NLS) estimation within the general framework of model (6.1). In Section 6.1.2, we consider the method of conditional least squares (CLS) estimation tailor-made for SETARMA, subset SETARMA, STAR, and BL models. In Section 6.1.3, we present an iteratively weighted least squares algorithm for QML estimation of double threshold ARCH models.

In the second part, we concentrate on model selection rules that are associated with the QML and NLS estimation methods. Both estimation methods are likely the most commonly used in practice. Consequently, the associated order selection rules are of quite general interest. In the third part, we discuss a general class of standardized-residuals-based correlation test statistics. The proposed tests avoid potential "size distortion" problems due to estimation uncertainty. Finally, in Section 6.4, we bring together elements of (subset) TARSO model estimation, TARSO model selection and checking, to analyze an important nonlinear time series problem from the area of hydrology.

6.1 Model Estimation

6.1.1 Quasi maximum likelihood estimator

Consider model (6.1). Let $p^* = p \vee u$, $q^* = q \vee v$, $\mathbf{Y}_0 = (Y_0, \ldots, Y_{1-p^*})'$ be the initial starting values of the process $\{Y_t, t \in \mathbb{Z}\}$, and $\boldsymbol{\varepsilon}_0 = (\varepsilon_0, \ldots, \varepsilon_{1-q^*})'$ be the starting innovations. In addition, let $\boldsymbol{\theta}_0 = (\boldsymbol{\theta}_{0,g}', \boldsymbol{\theta}_{0,h}')'$ denote the true value of the parameter vector $\boldsymbol{\theta}$, and $\mathbf{Y}_t = (Y_1, \ldots, Y_t)'$. We assume that $\boldsymbol{\theta}_0$ belongs to $\boldsymbol{\Theta} = \boldsymbol{\Theta}_{\theta_g} \times \boldsymbol{\Theta}_{\theta_h} \subset \mathbb{R}^{p+q} \times \mathbb{R}^{u+v}$.

Under the above assumptions, it is easily seen that the conditional mean and variance of $\{Y_t, t \in \mathbb{Z}\}$ given $\mathbf{Y}_{t-1}$ and $\boldsymbol{\Theta}$ are

$$\mathbb{E}(Y_t|\mathbf{Y}_{t-1}, \boldsymbol{\Theta}) = g(Y_{t-1}, \ldots, Y_{t-p}, \varepsilon_{t-1}, \ldots, \varepsilon_{t-q}; \boldsymbol{\theta}_{0,g}) \equiv \mu_t(\boldsymbol{\theta}_{0,g})$$
$$\text{Var}(Y_t|\mathbf{Y}_{t-1}, \boldsymbol{\Theta}) = h(Y_{t-1}, \ldots, Y_{t-u}, \varepsilon_{t-1}, \ldots, \varepsilon_{t-v}; \boldsymbol{\theta}_{0,h})\varepsilon_t \equiv \sigma_t^2(\boldsymbol{\theta}_{0,h}).$$

6.1 MODEL ESTIMATION

Assume that $\{\varepsilon_t\}$ has density function $f_\varepsilon(\cdot)$. Given $\mathbf{Y}_0$, the (conditional) likelihood function evaluated at $\boldsymbol{\theta} \in \boldsymbol{\Theta}$, is equal to

$$L_T(\boldsymbol{\theta}) = \prod_{t=1}^{T} \frac{1}{\sigma_t(\boldsymbol{\theta}_h)} f_\varepsilon\left(\frac{Y_t - \mu_t(\boldsymbol{\theta}_g)}{\sigma_t(\boldsymbol{\theta}_h)}\right),$$

assuming $\sigma_t(\boldsymbol{\theta}_h) \neq 0$.

The above objective function is not operational because $f_\varepsilon(\cdot)$ and $\mathbf{Y}_0$ are generally unknown. The initial values can be replaced by some fixed constants, e.g., zeros. More generally, one can treat $\mathbf{Y}_0$ and $\boldsymbol{\varepsilon}_0$ as unknown, additional, parameter vectors and estimate them jointly with other parameters. This approach requires more intensive computation. In finite samples, it may result in different parameter estimates, but it will not affect the asymptotic properties of the estimator of $\boldsymbol{\theta}_0$.

Replacing $f_\varepsilon(\cdot)$ by the $\mathcal{N}(0,1)$ density function, and approximating $\mu_t(\boldsymbol{\theta}_g)$ by $\widetilde{\mu}_t(\boldsymbol{\theta}_g) = g(Y_{t-1}, \ldots, Y_1, 0, \ldots; \boldsymbol{\theta}_g)$ and $\sigma_t(\boldsymbol{\theta}_h)$ by $\widetilde{\sigma}_t(\boldsymbol{\theta}_h) = h^2(Y_{t-1}, \ldots, Y_1, 0, \ldots; \boldsymbol{\theta}_h)$, the minimizer $\widehat{\boldsymbol{\theta}}_T$ of $L_T(\boldsymbol{\theta})$ is called the *quasi* ML (QML) estimator of $\boldsymbol{\theta}_0$. That is,

$$\widehat{\boldsymbol{\theta}}_T = \arg\min_{\boldsymbol{\theta} \in \boldsymbol{\Theta}} \widetilde{Q}_T(\boldsymbol{\theta}), \tag{6.2}$$

where

$$\widetilde{Q}_T(\boldsymbol{\theta}) = \frac{1}{T}\sum_{t=1}^{T} \widetilde{\ell}_t \quad \text{and} \quad \widetilde{\ell}_t \equiv \widetilde{\ell}_t(\boldsymbol{\theta}) = \left(\frac{Y_t - \widetilde{\mu}_t(\boldsymbol{\theta}_g)}{\widetilde{\sigma}_t(\boldsymbol{\theta}_h)}\right)^2 + \log\widetilde{\sigma}_t^2(\boldsymbol{\theta}_h),$$

with $\widetilde{\ell}_t$ the *log-likelihood* function at time t. Furthermore, if $\widetilde{\sigma}_t^2(\boldsymbol{\theta}_h) \equiv \sigma_0^2 > 0$, i.e. a constant, the QML estimator coincides with the classical NLS estimator.

It is known that a solution to (6.2) exists when the parameter space $\boldsymbol{\Theta}$ is compact, and the functions $\boldsymbol{\theta}_g \to \widetilde{\mu}_t(\boldsymbol{\theta}_g)$ and $\boldsymbol{\theta}_h \to \widetilde{\sigma}_t(\boldsymbol{\theta}_h)$ are continuous. Moreover, under some regularity conditions, it follows that the QML estimator is strongly consistent, and asymptotic normally distributed; see, e.g., Tjøstheim (1986b). More precisely, with $\ell_t(\boldsymbol{\theta}) = (Y_t - \mu_t(\boldsymbol{\theta}_g))^2 \sigma_t^{-2}(\boldsymbol{\theta}_h) + \log\sigma_t^2(\boldsymbol{\theta}_h)$, and as $T \to \infty$,

$$\sqrt{T}(\widehat{\boldsymbol{\theta}}_T - \boldsymbol{\theta}_0) \xrightarrow{D} \mathcal{N}\big(\mathbf{0}, \mathcal{H}^{-1}(\boldsymbol{\theta}_0)\mathcal{I}(\boldsymbol{\theta}_0)\mathcal{H}^{-1}(\boldsymbol{\theta}_0)\big), \tag{6.3}$$

where

$$\mathcal{H}(\boldsymbol{\theta}_0) = \mathbb{E}\left(\frac{\partial^2 \ell_t(\boldsymbol{\theta}_0)}{\partial\boldsymbol{\theta}\partial\boldsymbol{\theta}'}\right), \quad \text{and} \quad \mathcal{I}(\boldsymbol{\theta}_0) = \mathbb{E}\left(\frac{\partial \ell_t(\boldsymbol{\theta}_0)}{\partial\boldsymbol{\theta}} \frac{\partial \ell_t(\boldsymbol{\theta}_0)}{\partial\boldsymbol{\theta}'}\right).$$

Here $\mathcal{H}(\cdot)$ denotes the *expected Hessian matrix*, and $\mathcal{I}(\cdot)$ is the *expected information matrix* with $\ell_t(\cdot)$ evaluated at $\boldsymbol{\theta}_0$.

Consistent estimates of the standard errors of the QML estimator $\widehat{\boldsymbol{\theta}}_T$ are obtained as the square root of the diagonal elements of the estimated covariance matrix of $\widehat{\boldsymbol{\theta}}_T$, that is

$$\widehat{\mathrm{Var}}(\widehat{\boldsymbol{\theta}}_T) = \frac{1}{T}\big(\widehat{\mathcal{H}}_T \widehat{\mathcal{I}}_T^{-1} \widehat{\mathcal{H}}_T\big)^{-1},$$

where the *empirical Hessian* and *average information matrix* for a sample of size T are defined as, respectively,

$$\widehat{\mathcal{H}}_T = \frac{1}{T}\sum_{t=1}^{T} \frac{\partial^2 \widetilde{\ell}_t(\widehat{\boldsymbol{\theta}}_T)}{\partial \boldsymbol{\theta} \partial \boldsymbol{\theta}'}, \quad \widehat{\mathcal{I}}_T = \frac{1}{T}\sum_{t=1}^{T} \frac{\partial \widetilde{\ell}_t(\widehat{\boldsymbol{\theta}}_T)}{\partial \boldsymbol{\theta}} \frac{\partial \widetilde{\ell}_t(\widehat{\boldsymbol{\theta}}_T)}{\partial \boldsymbol{\theta}'}. \tag{6.4}$$

Optimal values of $\boldsymbol{\theta}_0$ are characterized by the *likelihood equation*, which is just the first-order conditions: $G(\boldsymbol{\theta}_0) \equiv \mathbf{0}$, where the *gradient vector*, or *score vector*, $G \in \mathbb{R}^{p+q+u+v}$ is defined by

$$G(\boldsymbol{\theta}) = \sum_{t=1}^{T} \frac{\partial \widetilde{\ell}_t(\boldsymbol{\theta})}{\partial \boldsymbol{\theta}}.$$

In practice, it is usually not possible to obtain an analytic solution for $\widehat{\boldsymbol{\theta}}_T$, especially when the objective function involves many parameters. In such a situation, estimates of $\boldsymbol{\theta}_0$ must be sought numerically using nonlinear optimization algorithms. The basic idea of nonlinear optimization is to quickly find optimal parameters that maximize the log-likelihood. This is done by searching much smaller sub-sets of the multi-dimensional parameter space rather than exhaustively searching the whole parameter space, which becomes intractable as the number of parameters increases. Numerical optimization algorithms often involve the following steps.

Algorithm 6.1: Nonlinear iterative optimization

(i) Provide an initial estimate of $\boldsymbol{\theta}_0$, say $\widehat{\boldsymbol{\theta}}_{T,0}$. For instance, these estimates can be chosen at random or by guessing.

(ii) By an "intelligent" search over the parameter space $\boldsymbol{\Theta}$, determine an improved estimate of $\widehat{\boldsymbol{\theta}}_{T,0}$, say $\widehat{\boldsymbol{\theta}}_{T,1}$.

(iii) Taking into account the results from step (ii), obtain a new set of estimates $\widehat{\boldsymbol{\theta}}_{T,i}$ ($i = 2, 3, \ldots$) by adding small changes to the previous estimates in such a way that the new parameter estimates are likely to lead to improved performance.

(iv) Stop the iterative process in step (iii) if parameters estimates are judged to have converged, using an appropriately predefined criterion. For instance, if the relative improvement $\{\widetilde{Q}(\widehat{\boldsymbol{\theta}}_{T,i+1}) - \widetilde{Q}(\widehat{\boldsymbol{\theta}}_{T,i})\}/\widetilde{Q}(\widehat{\boldsymbol{\theta}}_{T,i})$ is a small prefixed number.

It is worth noting that the optimization algorithm does not necessarily guarantee that the final estimate $\widehat{\boldsymbol{\theta}}_T$ uniquely maximizes the log-likelihood. Even if $G(\widehat{\boldsymbol{\theta}}_T) \approx \mathbf{0}$, the algorithm can prematurely stop and return a sub-optimal set of parameter values. This is called the *local maxima problem*. Unfortunately, there exists no general solution to the local maximum problem. Instead, a variety of remedies have been

6.1 MODEL ESTIMATION

developed in an attempt to avoid the problem (see, e.g., Teräsvirta et al., 2010, Chapter 12), though there is no guarantee of their effectiveness. For example, one may choose different starting values over multiple runs of the iteration procedure and then examine the results to see whether the same solution is obtained repeatedly. When that happens, one can conclude with some confidence that $\widehat{\boldsymbol{\theta}}_T$ is close to a global optimum. If, however, the changes in the parameter estimates remain large in multiple iterations the parameters of the model may not be *identified*.

To assess the performance of the QML estimator of $\boldsymbol{\theta}_0$ in finite samples, the next example shows a simulation experiment.

Example 6.1: NLS Estimation

Consider, as a special case of the general ExpARMA model (2.20), an ExpAR(1) model with $p = d = 1$, i.e.,

$$Y_t = \{\phi + \xi \exp(-\gamma Y_{t-1}^2)\}Y_{t-1} + \varepsilon_t, \quad \{\varepsilon_t\} \overset{\text{i.i.d.}}{\sim} (0, \sigma_\varepsilon^2), \qquad (6.5)$$

where $|\phi| < 1$ and $\gamma > 0$. Thus, we have $\widetilde{\mu}_1(\boldsymbol{\theta}_g) = 0$, $\widetilde{\mu}_t(\boldsymbol{\theta}_g) = (\phi + \xi e^{-\gamma Y_{t-1}^2})Y_{t-1}$ $\forall t > 1$, and $\widetilde{\sigma}_t^2(\boldsymbol{\theta}_h) = \sigma_\varepsilon^2$ $\forall t \geq 1$. The gradient vector is $G(\boldsymbol{\theta}_g) = \sum_{t=2}^T (Y_t - \widetilde{\mu}_t(\boldsymbol{\theta}_g))\sigma_\varepsilon^{-2}(Y_{t-1}, Y_{t-1}e^{-\gamma Y_{t-1}^2}, -\xi Y_{t-1}^3 e^{-\gamma Y_{t-1}^2})'$.

The DGP is characterized by the parameter vector $\boldsymbol{\theta}_0 = (\phi_0, \xi_0, \gamma_0)'$ (introducing the subscript 0), and the so-called *nuisance parameter* $\sigma_{\varepsilon,0}^2$, that we are not interested in estimating. We assume that $\boldsymbol{\theta}_0$ belongs to the interior $\overset{\circ}{\boldsymbol{\Theta}}$ of the parameter space $\boldsymbol{\Theta} = [-\overline{\phi}, \overline{\phi}] \times [-\overline{\xi}, \overline{\xi}] \times [\underline{\gamma}, \overline{\gamma}]$ with $|\phi_0| < \overline{\phi} \leq 1$, $|\xi_0| < \overline{\xi}$ and $0 < \underline{\gamma} < \gamma_0 < \overline{\gamma}$. Note that the parameter γ_0 is not identified if $\xi_0 = 0$. That is, there exist parameter vectors $\boldsymbol{\theta}_{1,g} \neq \boldsymbol{\theta}_{2,g}$ with $\widetilde{\mu}_t(\boldsymbol{\theta}_{1,g}) = \widetilde{\mu}_t(\boldsymbol{\theta}_{2,g})$ $\forall Y_t$, then $\widetilde{Q}_T(\boldsymbol{\theta}_{1,g}) = \widetilde{Q}_T(\boldsymbol{\theta}_{2,g})$ in (6.2), in which case minima need not be unique.

Nonlinear estimation of (6.5) is easier if good initial parameter values are available. To this end it is convenient to express the model in matrix form. Let $\mathbf{Y} = (Y_2, \ldots, Y_T)'$, $\boldsymbol{\beta} = (\phi, \xi)'$, $\boldsymbol{\varepsilon} = (\varepsilon_2, \ldots, \varepsilon_T)'$, and

$$\mathbf{X} = \begin{pmatrix} Y_1 & Y_1 e^{-\gamma Y_1^2} \\ \vdots & \vdots \\ Y_{T-1} & Y_{T-1} e^{-\gamma Y_{T-1}^2} \end{pmatrix}.$$

Then we can write (6.5) as

$$\mathbf{Y} = \mathbf{X}\boldsymbol{\beta} + \boldsymbol{\varepsilon},$$

which, *conditional* on γ, is a simple linear regression model. The CLS estimate $\widehat{\boldsymbol{\beta}} = (\widehat{\beta}_1, \widehat{\beta}_2)'$ of $\boldsymbol{\beta}$ can be obtained in the usual manner as $\widehat{\boldsymbol{\beta}} = (\mathbf{X}'\mathbf{X})^{-1}\mathbf{X}'\mathbf{Y}$. Its associated covariance matrix is given by $\text{Var}(\widehat{\boldsymbol{\beta}}) = \sigma_\varepsilon^2(\mathbf{X}'\mathbf{X})^{-1}$. It is easily checked that $\widehat{\boldsymbol{\beta}}$ is $\sqrt{T}$-consistent. Thus, the above approach yields an efficient initial estimate $\widehat{\boldsymbol{\theta}}_{T,0}$.

In preparation for the MC simulation experiment, it is useful to consider the deterministic skeleton of (6.5), i.e. the difference equation

$$Y_t = \{\phi + \xi \exp(-\gamma Y_{t-1}^2)\} Y_{t-1}.$$

From (2.22) it follows that, if $|\phi + \xi| < 1$, $\{Y_t, t \in \mathbb{Z}\}$ will converge to a stable limit point at zero as $t \to \infty$. Otherwise, we may distinguish two cases in the dynamic behavior of Y_t:

- For $\xi > 1 - \phi > 0$, $\{Y_t, t \in \mathbb{Z}\}$ has twin limiting points at

$$Y = \pm \{\gamma^{-1} \log (\xi/(1-\phi))\}^{1/2}, \qquad (6.6)$$

which for $\xi < (1-\phi) \exp\{1/(1-\phi)\}$ will be stable;

- For $\xi < -(1+\phi) < 0$, $\{Y_t, t \in \mathbb{Z}\}$ has a limit cycle between the points

$$Y = \pm \{\gamma^{-1} \log (-\xi/(1+\phi))\}^{1/2}, \qquad (6.7)$$

which for $-\xi < (1+\phi) \exp\{1/(1+\phi)\}$ will be stable.

Consider model (6.5) with $\phi = -0.8$, $\xi = 2$, $\gamma = 2$, and $\{\varepsilon_t\} \stackrel{\text{i.i.d.}}{\sim} \mathcal{N}(0,1)$. So, by (6.6), the skeleton of $\{Y_t, t \in \mathbb{Z}\}$ has alternative limiting points at ± 0.2295 which are stable ($\xi < 3.1372$).

In step (i) of the numerical optimization procedure, we use 101 equidistant grid points of γ in the interval $[1.75, 2.25]$ to obtain CLS estimates of $\boldsymbol{\beta}$. Conditional on a value of γ, we select the 'best' estimate of $\boldsymbol{\beta}$, say $\widehat{\boldsymbol{\beta}}^* = (\widehat{\beta}_1^*, \widehat{\beta}_2^*)'$, for which the residual sum of squares attains a minimum, resulting in an initial estimate $\widehat{\boldsymbol{\theta}}_{T,0}$. Next, in step (ii), we set $[-\overline{\phi}, \overline{\phi}] = [\phi - 2\{\widehat{\text{Var}}(\widehat{\beta}_1^*)\}^{1/2}, \phi + 2\{\widehat{\text{Var}}(\widehat{\beta}_1^*)\}^{1/2}]$ and $[-\overline{\xi}, \overline{\xi}] = [\xi - 2\{\widehat{\text{Var}}(\widehat{\beta}_2^*)\}^{1/2}, \xi + 2\{\widehat{\text{Var}}(\widehat{\beta}_2^*)\}^{1/2}]$. Thus, with $[\underline{\gamma}, \overline{\gamma}] = [1.75, 2.25]$, $\boldsymbol{\Theta} \in \overset{\circ}{\boldsymbol{\Theta}}$, which is essential to obtain the asymptotic normality of the QML estimates.

Figure 6.1 shows boxplots of the NLS values of $(\widehat{\phi} - \phi)$, $(\widehat{\xi} - \xi)$, and $(\widehat{\gamma} - \gamma)$, using the gradient vector $G(\boldsymbol{\theta}_g)$. The plots indicate the consistency of the estimators and evidence of symmetry. Note the differences between the scales on the vertical axis for both sample sizes.

6.1.2 Conditional least squares estimator

SETARMA models
Chapter 2 introduced the k-regime SETARMA model (2.29). To economize on notation, we focus on a special case, i.e., the SETARMA$(2; p_1, q_1, p_2, q_2)$ model with

6.1 MODEL ESTIMATION

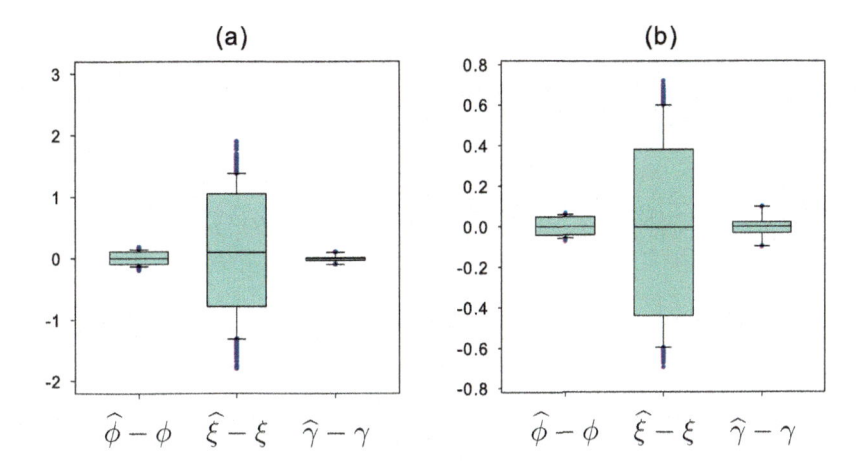

Figure 6.1: Boxplots of $(\widehat{\phi} - \phi)$, $(\widehat{\xi} - \xi)$, and $(\widehat{\gamma} - \gamma)$; (a) $T = 100$, and (b) $T = 500$; 1,000 MC replications.

all white noise variances being equal. The latter model is defined as

$$Y_t = \begin{cases} \phi_0^{(1)} + \sum_{i=1}^{p_1} \phi_i^{(1)} Y_{t-i} + \varepsilon_t + \sum_{j=1}^{q_1} \psi_j^{(1)} \varepsilon_{t-j} & \text{if } Y_{t-d} \leq r, \\ \phi_0^{(2)} + \sum_{i=1}^{p_2} \phi_i^{(2)} Y_{t-i} + \varepsilon_t + \sum_{j=1}^{q_2} \psi_j^{(2)} \varepsilon_{t-j} & \text{if } Y_{t-d} > r, \end{cases} \quad (6.8)$$

where $\{\varepsilon_t\} \overset{\text{i.i.d.}}{\sim} (0, \sigma_\varepsilon^2)$, $r \in \mathbb{R}$, p_i and q_i ($i = 1, 2$) are known nonnegative integers, and $d \in \mathbb{Z}^+$. Although (6.8) serves as a benchmark to study CLS estimation, the asymptotic results presented below can be easily extended to $k > 2$ thresholds.

Without loss of generality, we assume that the unknown threshold parameter $r \in [\underline{r}, \overline{r}] \subset \mathbb{R}$ with $\underline{r}$ and $\overline{r}$ finite constants. In addition, the delay variable d is an unknown parameter to be estimated, and its true value is d_0 with $1 \leq d_0 \leq D_0$, where D_0 is known. Let $\boldsymbol{\phi}_i = (\phi_0^{(i)}, \ldots, \phi_{p_i}^{(i)})'$ and $\boldsymbol{\psi}_i = (\psi_1^{(i)}, \ldots, \psi_{q_i}^{(i)})'$ ($i = 1, 2$) and $\boldsymbol{\tau} = (\boldsymbol{\phi}_1', \boldsymbol{\psi}_1', \boldsymbol{\phi}_2', \boldsymbol{\psi}_2')'$. Then, $\boldsymbol{\theta}_0 = (\boldsymbol{\tau}_0', r_0, d_0)' \equiv (\boldsymbol{\phi}_{1,0}', \boldsymbol{\psi}_{1,0}', \boldsymbol{\phi}_{2,0}', \boldsymbol{\psi}_{2,0}', r_0, d_0)'$ is the true value of the parameter vector $\boldsymbol{\theta} = (\boldsymbol{\tau}', r, d)'$. Denote the parameter space by $\boldsymbol{\Theta} = \boldsymbol{\Theta}_\tau \times [\underline{r}, \overline{r}] \times \{1, \ldots, D_0\}$, where $\boldsymbol{\Theta}_\tau$ is a compact subset of $\mathbb{R}^{p_1+p_2+q_1+q_2+2}$.

Suppose that a sample $\{Y_t\}_{t=1}^T$ is available from (6.8) with the true value $\boldsymbol{\theta}_0$. Let $p = p_1 \vee p_2$ and $q = q_1 \vee q_2$. Then, given the vector with initial values $\mathbf{Y}_0 = (Y_0, \ldots, Y_{1-(p \vee D_0)})'$, the (conditional) sum of squared errors function $L_T(\boldsymbol{\theta})$ is defined as

$$L_T(\boldsymbol{\theta}) = \sum_{t=1}^T \varepsilon_t^2(\boldsymbol{\theta}), \quad (6.9)$$

where

$$\varepsilon_t(\boldsymbol{\theta}) = Y_t - \Big(\phi_0^{(1)} + \sum_{i=1}^{p_1} \phi_i^{(1)} Y_{t-i} + \sum_{j=1}^{q_1} \psi_j^{(1)} \varepsilon_{t-j}(\boldsymbol{\theta})\Big) I(Y_{t-d} \le r)$$

$$- \Big(\phi_0^{(2)} + \sum_{i=1}^{p_2} \phi_i^{(2)} Y_{t-i} + \sum_{j=1}^{q_2} \psi_j^{(2)} \varepsilon_{t-j}(\boldsymbol{\theta})\Big) I(Y_{t-d} > r).$$

The CLS estimator $\widehat{\boldsymbol{\theta}}_T = (\widehat{\boldsymbol{\tau}}'_T, \widehat{r}_T, \widehat{d})'$ of $\boldsymbol{\theta}_0$ are the values which globally minimize (6.9), that is,

$$\widehat{\boldsymbol{\theta}}_T = \arg\min_{\boldsymbol{\theta} \in \Theta} L_T(\boldsymbol{\theta}). \qquad (6.10)$$

In practice, the vector of initial values $\mathbf{Y}_0$ is not available and can be replaced by constants. This will not affect the asymptotic properties of $\widehat{\boldsymbol{\theta}}_T$. For simplicity, we assume hereafter that $\mathbf{Y}_0$ is from model (6.8). Since $L_T(\boldsymbol{\theta})$ is discontinuous in r and d, the minimization in (6.10) can be done as follows.

Algorithm 6.2: A multi-parameter grid search

(i) Fix $r \in \mathbb{R}$ and $d \in \{1, \ldots, D_0\}$. Then minimize $L_T(\boldsymbol{\theta})$, and get its minimizer $\widehat{\boldsymbol{\tau}}_T(r, d)$ and the minimum value $L_T^*(r, d) \equiv L_T(\boldsymbol{\theta})|_{\boldsymbol{\tau} = \widehat{\boldsymbol{\tau}}_T(r,d)}$.

(ii) Since $L_T^*(r, d)$ takes finite possible values only, perform a grid search over the set of order statistics $\{Y_{(1)}, \ldots, Y_{(T)}\}$ of $\{Y_1, \ldots, Y_T\}$ and $\{1, \ldots, D_0\}$ to get the minimizer $(\widehat{r}_T, \widehat{d}_T)'$ of $L_T^*(r, d)$.

(iii) Use a plug-in method to obtain $\widehat{\boldsymbol{\tau}}_T(\widehat{r}_T, \widehat{d}_T)$ and $\widehat{\boldsymbol{\theta}}_T$.

Generally, there are infinitely many values r at which $L_T(\cdot)$ attains its global minimum, the one with the smallest r can be chosen as the estimator of r_0. It is easy to see that $\widehat{\boldsymbol{\theta}}_T$ is the CLS estimator of $\boldsymbol{\theta}_0$. For instance, with a SETAR(2; p, p) model, simple computation shows that for a given value of r the CLS estimator of $\boldsymbol{\theta}_0$ is given by

$$\widehat{\boldsymbol{\theta}}_T(r) = \Big(\sum_{t=1}^T \mathbf{X}_t(r) \mathbf{X}_t'(r)\Big)^{-1} \Big(\sum_{t=1}^T \mathbf{X}_t(r) Y_t\Big), \qquad (6.11)$$

where $\mathbf{X}_t(r) = (\mathbf{X}_t' I(Y_{t-d} \le r), \mathbf{X}_t' I(Y_{t-d} > r))'$ with $\mathbf{X}_t = (1, Y_{t-1}, \ldots, Y_{t-p})'$. With residuals $\widehat{\varepsilon}_t(r) = Y_t - \mathbf{X}_t'(r) \widehat{\boldsymbol{\theta}}_T(r)$, the corresponding (conditional) residual variance is given by $\widehat{\sigma}_T^2(r) = T^{-1} \sum_{t=1}^T \widehat{\varepsilon}_t^2(r)$.

SETARMA models: Asymptotic properties

Li et al. (2011), discuss (a) the consistency of the CLS estimator $\widehat{\boldsymbol{\theta}}_T$; (b) the limiting

6.1 MODEL ESTIMATION

distributions of $\hat{r}_T$ (a super-consistent estimator) and $\hat{\boldsymbol{\theta}}_T$; and (c) the convergence rate of $T(\hat{r}_T - r_0)$. A rigorous treatment of the conditions under which these authors prove the above issues is beyond the scope of this book. However, in case of (c), we introduce some notation to discuss the numerical method for tabulating the limiting distribution of $\hat{r}_T$.

Consider the profile sum of squares errors function

$$\tilde{L}_T(z) = L_T\left(\hat{\boldsymbol{\tau}}_T\left(r_0 + \frac{z}{T}\right), r_0 + \frac{z}{T}\right) - L_T(\hat{\boldsymbol{\tau}}_T(r_0), r_0), \quad z \in \mathbb{R}.$$

Let $\mathbf{e} = (1, 0, \ldots, 0)'$ be a $q \times 1$ vector, and

$$\mathbf{H}_{t,j}(\boldsymbol{\theta}) = \prod_{i=1}^{j}[\boldsymbol{\psi}_2 + (\boldsymbol{\psi}_1 - \boldsymbol{\psi}_2)I(Y_{t-d-i+1} \leq r)], \quad (j \geq 0),$$

with the convention $\prod_{i=1}^{0} = \mathbf{I}_q$, and

$$\boldsymbol{\psi}_i = \begin{pmatrix} -\psi_1^{(i)} & \cdots & -\psi_q^{(i)} \\ \mathbf{I}_{q-1} & \mathbf{0}_{(q-1) \times 1} \end{pmatrix}, \quad (i = 1, 2).$$

Using the asymptotic result in (b) and Taylor expansion, $\tilde{L}_T(z)$ can be approximated (Li et al., 2011) by

$$\wp_T(z) = I(z<0) \sum_{t=1}^{T} \zeta_t^{(1)} I\left(r_0 + \frac{z}{T} < Y_{t-d} \leq r_0\right) + I(z \geq 0) \sum_{t=1}^{T} \zeta_t^{(2)} I\left(r_0 < Y_{t-d} \leq r_0 + \frac{z}{T}\right),$$

where

$$\zeta_t^{(i)} = \left\{\sum_{j=0}^{\infty}[\mathbf{e}'\mathbf{H}_{t+j,j}(\boldsymbol{\theta}_0)\mathbf{e}]^2\right\}\delta_t^2 + 2(-1)^{i+1}\left\{\sum_{j=0}^{\infty}\varepsilon_{t+j}[\mathbf{e}'\mathbf{H}_{t+j,j}(\boldsymbol{\theta}_0)\mathbf{e}]\right\}\delta_t, \quad (i = 1, 2),$$

(6.12)

and

$$\delta_t = (\phi_{0,0}^{(1)} - \phi_{0,0}^{(2)}) + \sum_{i=1}^{p}(\phi_{i,0}^{(1)} - \phi_{i,0}^{(2)})Y_{t-i} + \sum_{i=1}^{q}(\psi_{i,0}^{(1)} - \psi_{i,0}^{(2)})\varepsilon_{t-i}.$$

Let $F_k(\cdot|r_0)$ be the conditional distribution of $\zeta_{d+1}^{(k)}$ ($k = 1, 2$) given $Y_1 = r_0$. To describe the limiting distribution of $\hat{r}_T$, consider two independent *compound Poisson processes* (CPPs) $\{\wp^{(1)}(z), z \geq 0\}$ and $\{\wp^{(2)}(z), z \geq 0\}$ with $\wp^{(1)}(0) = \wp^{(2)}(0) = 0$ a.s., and with the same jump rate $\pi(r_0) > 0$, where $\pi(\cdot)$ is the pdf of Y_1, and with the jump distributions $F_1(\cdot|r_0)$ and $F_2(\cdot|r_0)$, respectively. Define a two-sided CPP $\{\wp(z), z \in \mathbb{R}\}$ as follows

$$\wp(z) = I(z < 0)\wp^{(1)}(-z) + I(z \geq 0)\wp^{(2)}(z). \tag{6.13}$$

Observe that $\wp(z)$ goes to ∞ a.s. when $|z| \to \infty$ since $\int x dF_k(x|r_0) > 0$. Therefore, there exists a unique random interval $[M_-, M_+)$ on which the process (6.13) attains its global minimum and nowhere else. Then, under some mild conditions, it can be proved (Li et al., 2011) that: (i) $T(\hat{r}_T - r_0) \xrightarrow{D} M_-$, as $T \to \infty$; and (ii) $T(\hat{r}_T - r_0)$ is asymptotically independent of $\sqrt{T}(\hat{\tau}_T - \tau_0)$ and their asymptotic distributions are the same, regardless whether r_0 is known or not. In particular,

$$\sqrt{T}(\hat{\tau}_T - \tau_0) = \sqrt{T}(\hat{\tau}_T(r_0) - \tau_0) + o_p(1) \xrightarrow{D} \mathcal{N}(\mathbf{0}_{(p \vee d)+q}, \sigma_\varepsilon^2 \mathbf{\Sigma}^{-1}) \text{ as } T \to \infty,$$

where $\mathbf{\Sigma} = \mathbb{E}[(\partial \varepsilon_t(\boldsymbol{\theta}_0)/\partial \boldsymbol{\tau})(\partial \varepsilon_t(\boldsymbol{\theta}_0)/\partial \boldsymbol{\tau}')]$.

SETARMA models: Numerical implementation of M_-
The pdf of M_- (left jump) can be obtained as follows.

Algorithm 6.3: The density function of M_-

(i) Generate two independent Poisson random variables N_1 and N_2 with the same intensity parameter $\pi(r_0)N$, and $N > 0$ is a prefixed integer.

(ii) Generate two independent jump time sequences $\{U_1, \ldots, U_{N_1}\}$ and $\{V_1, \ldots, V_{N_2}\}$, where $\{U_i\} \overset{i.i.d.}{\sim} U[-N, 0]$ and $\{V_i\} \overset{i.i.d.}{\sim} U[0, N]$.

(iii) Generate two independent jump-size sequences: $\{Y_1, \ldots, Y_{N_1}\}$ and $\{Z_1, \ldots, Z_{N_2}\}$ from $F_1(\cdot|r_0)$ and $F_2(\cdot|r_0)$, respectively.

(iv) Create a set of equidistant points over the interval $[-N, N]$. For $z \in [-N, N]$, compute the trajectory of (6.13), i.e., $\wp(z) = I(z < 0) \sum_{i=1}^{N_1} I(U_i > z)Y_i + I(z \geq 0) \sum_{j=1}^{N_2} I(V_j < z)Z_j$. Find the smallest minimizer of $\wp(z)$ on $[-N, N]$ and call it $M_-^{(b)}$.

(v) Repeat step (iv) B times, to obtain $\{M_-^{(b)}\}_{b=1}^B$.

(vi) Use a nonparametric kernel-based estimation method, to obtain the density function of M_- numerically.

Algorithm 6.3 depends crucially on step (iii). When $\boldsymbol{\theta}_0$, $\pi(r_0)$, the distribution $F_\varepsilon(\cdot)$ of $\{\varepsilon_t\}$, and the distribution $G_{\mathbf{Z}_0}(\cdot)$ of $\mathbf{Z}_0 = (Y_0, \ldots, Y_{1-(p \vee d)}, \varepsilon_0, \ldots, \varepsilon_{1-q})'$ are known, the appropriate way to proceed is to first sample $\{\varepsilon_t\}_{t=2}^{d+1+L}$ independently from $F_\varepsilon(\cdot)$ where L is some large integer. Next, draw a sample $(\mathbf{z}_1, \ldots, \mathbf{z}_K)$ from $G_{\mathbf{Z}_0}(\cdot)$ where K is another large integer, and $\mathbf{z}_i = (Y_i, \ldots, Y_{i-(p \vee d)+1}, \varepsilon_0, \ldots, \varepsilon_{1-q})' \in \mathbb{R}^{(p \vee d)+q}$ ($i = 1, \ldots, K$). Then, generate $\{Y_t\}_{t=2}^{d+1+L}$ by iterating model (6.8) with the initial values $Y_1 = r_0$, $\mathbf{Z}_0 = \mathbf{z}_i$, and $\varepsilon_1 = r_0 - g(\mathbf{z}_i, \boldsymbol{\theta}_0)$ ($i = 1, \ldots, K$).

Obtain an approximation, say $\zeta_{d+1,k}^{(1)}$, of $\zeta_{d+1}^{(1)}$ ($k = 1, \ldots, K$) by truncating the infinite sums in (6.12) after L terms. Since $\|\mathbf{e}'\mathbf{H}_{d+1+j,j}(\boldsymbol{\theta}_0)\mathbf{e}\|_2 = \mathcal{O}(\rho^j)$ a.s., the remaining term is negligible when L is large enough. Calculate the conditional

6.1 MODEL ESTIMATION

density function of Y_1 given $\mathbf{Z}_0 = \mathbf{z}_k$, i.e. $\pi(r_0|\mathbf{z}_k) = f_\varepsilon(r_0 - g(\mathbf{z}_k, \boldsymbol{\theta}_0))$. Draw a U from a random sample, with replacement, from the integers 1 to $T - p + 1$, using a vector of positive weights $\pi(r_0|\mathbf{z}_k)/\sum_{k=1}^{K} \pi(r_0|\mathbf{z}_k)$ ($k = 1, \ldots, K$). Finally, obtain $Y_1 = \zeta_{d+1,U}^{(1)}$. This last step is asymptotically equivalent to obtaining one observation from $F_1(\cdot|r_0)$; Li et al. (2011). In an obvious manner the above procedure can be modified to obtain one observation from $F_2(\cdot|r_0)$.

It remains to discuss estimation of the pdf of M_- given $\{Y_t\}_{t=1}^{T}$. We can use the estimators $\widehat{\boldsymbol{\theta}}_T$, and $\widehat{\pi}(\widehat{r}_T)$ in place of the true values since they are consistent. Here, $\widehat{\pi}(\cdot)$ is the kernel density estimator of Y_t at r_0. Next, calculate the mean-deleted residuals $\{\widehat{\varepsilon}_t^*\}_{t=k_0+1}^{T}$ where $k_0 = \max(p, d, q)$. Then, compute $\widehat{F}_\varepsilon(x) = (T - k_0)^{-1} \sum_{t=k_0+1}^{T} I(\widehat{\varepsilon}_t^* \leq x)$ as the estimator of $F_\varepsilon(\cdot)$, and $\widehat{f}_\varepsilon(\cdot)$ as the kernel density estimator of $f_\varepsilon(\cdot)$. Now step (iii) of Algorithm 6.3 can be modified as follows.

Algorithm 6.4: Sampling Y_1 from an estimate of $F_1(\cdot|r_0)$

(i) Set $\widehat{\mathbf{z}}_i = (Y_i, \ldots, Y_{i-(p\vee d)+1}, \widehat{\varepsilon}_i, \ldots, \widehat{\varepsilon}_{i-q+1})'$ ($i = k_0 + 1, \ldots, T$).

(ii) Sample $\{\widetilde{\varepsilon}_t\}_{t=2}^{d+1+L}$ independently from $\widehat{F}_\varepsilon(\cdot)$ given $\{Y_t\}_{t=1}^{T}$.

(iii) Generate $\{\widetilde{Y}_t\}_{t=2}^{d+1+L}$ by iterating model (6.8) with the initial values $Y_1 = \widehat{r}_T$, $\mathbf{Z}_0 = \widehat{\mathbf{z}}_i$, and $\varepsilon_1 = \widehat{r}_T - g(\widehat{\mathbf{z}}_i; \widehat{\boldsymbol{\theta}}_T)$. Compute $\widehat{H}_{d+1+j,j}(\widehat{\boldsymbol{\theta}}_T) = \prod_{i=1}^{j}[\widehat{\psi}_2 + (\widehat{\psi}_1 - \widehat{\psi}_2)I(\widetilde{Y}_{i+1} \leq \widehat{r}_T)]$ as an estimate of $H_{d+1+j,j}(\cdot)$.

(iv) Calculate $\widetilde{\zeta}_{d+1,k}^{(1)}$ ($k = 1, \ldots, K$), as an estimate of $\zeta_{d+1}^{(1)}$, where

$$\widetilde{\zeta}_{d+1,k}^{(1)} = \Big\{ \sum_{j=0}^{L} [\mathbf{e}'\widehat{\mathbf{H}}_{d+1+j,j}(\widehat{\boldsymbol{\theta}}_T)\mathbf{e}]^2 \Big\}(\delta_{d+1}^*)^2$$
$$+ 2\Big\{ \sum_{j=0}^{L} \widetilde{\varepsilon}_{d+1+j}[\mathbf{e}'\widehat{\mathbf{H}}_{d+1+j,j}(\widehat{\boldsymbol{\theta}}_T)\mathbf{e}] \Big\}\delta_{d+1}^*,$$

with

$$\delta_{d+1}^* = (\widehat{\phi}_0^{(1)} - \widehat{\phi}_0^{(2)}) + \sum_{s=1}^{p}(\widehat{\phi}_s^{(1)} - \widehat{\phi}_s^{(2)})Y_{d+1-s}^* + \sum_{s=1}^{q}(\widehat{\psi}_s^{(1)} - \widehat{\psi}_s^{(2)})\varepsilon_{d+1-s}^*,$$

and

$$Y_j^* = \begin{cases} \widetilde{Y}_j & j \geq 2, \\ \widehat{r}_T & j = 1, \\ Y_{i+j} & j \leq 0, \end{cases} \qquad \varepsilon_j^* = \begin{cases} \widetilde{\varepsilon}_j & j \geq 2, \\ \widehat{r}_T - g(\widehat{\mathbf{z}}_i; \widehat{\boldsymbol{\theta}}_T) & j = 1, \\ \widehat{\varepsilon}_{i+j} & j \leq 0. \end{cases}$$

(v) Draw a U from a random sample, with replacement, from the integers 1 to $T - p + 1$, using a vector of positive weights $\widehat{\pi}(\widehat{r}_T|\widehat{\mathbf{z}}_i)/\sum_{i=k_0+1}^{K} \widehat{\pi}(\widehat{r}_T|\widehat{\mathbf{z}}_i)$ ($i = k_0 + 1, \ldots, K$).

(vi) Obtain $\widetilde{Y}_1 = \widetilde{\zeta}_{d+1,U}^{(1)}$.

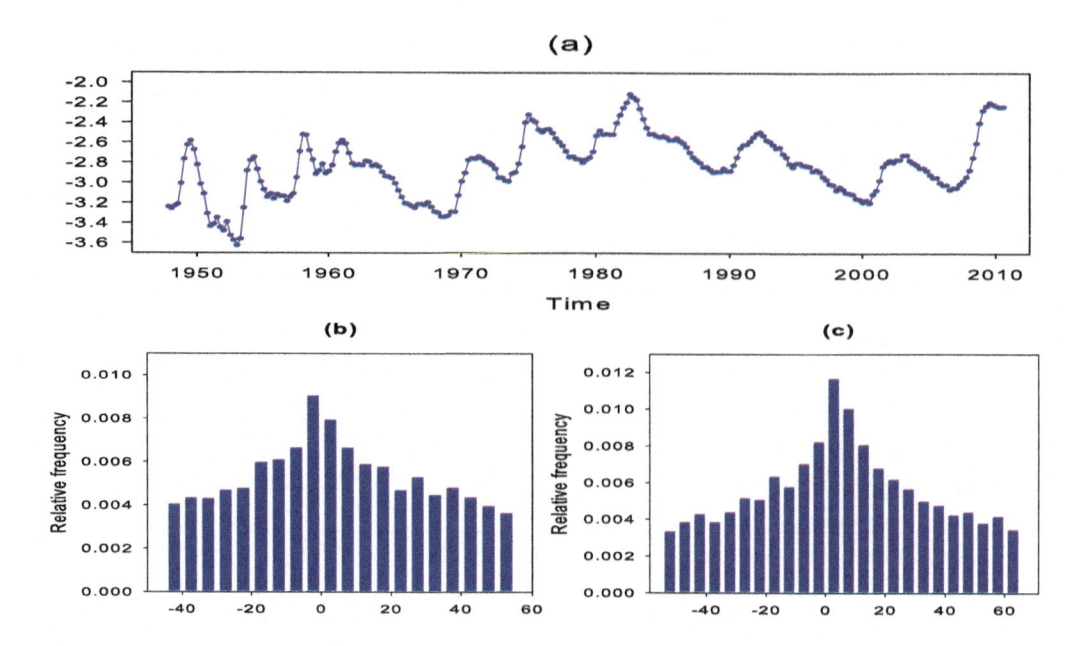

Figure 6.2: *(a) Plot of the logistic transformed* U.S. unemployment rate $\{Y_t\}_{t=1}^{252}$; *(b) and (c) relative frequency histograms of* $T(\widehat{r}_i - r_{i,0})$ $(i = 1, 2)$ *with* $r_{i,0}$ *the true threshold value.*

A probability density estimate, say $\widehat{M}_-$, of the density function of M_- follows from repeating Algorithm 6.3, with the modification in Algorithm 6.4, a large number of times. It can be shown that, as $K \to \infty$ (first) and $L \to \infty$ (second), $\widehat{M}_-$ weakly converges to M_-; Li et al. (2011) and Li and Ling (2012).

Example 6.2: U.S. Unemployment Rate (Cont'd)

Consider the quarterly U.S. unemployment rate in Example 1.1. In Exercise 2.10, we analyzed the logistic transformation of the original data, and denoted the resulting series by $\{Y_t\}_{t=1}^{252}$. Figure 6.2(a) shows a plot of the transformed series.

Some researchers suggested that a two-regime SETAR model is appropriate for characterizing the asymmetric behavior in the U.S. unemployment data. Others (e.g., Koop and Potter, 1999) consider a three-regime SETAR. With this specification, the model allows for the dynamics of the unemployment rate to differ in "good" times (expansion), "bad" times (recession), or change little in "normal" (stable) times. Following this suggestion, we fit a SETAR$(3; p_1, p_2, p_3)$ model, with threshold values r_1 and r_2, to the time series $\{Y_t\}$.

Setting $m_0 = \max\{p_1, p_2, p_3\} \leq 8$ and $1 \leq d \leq \max\{1, m_0\}$, we use the AIC below to determine the order in each regime,

$$\text{AIC}(p_1, p_2, p_3) = \sum_{i=1}^{m_0} \left\{ T_i \log \widehat{\sigma}_{T_i}^2 + 2(p_i + 1) \right\}, \qquad (6.14)$$

6.1 MODEL ESTIMATION

where T_i denotes the number of observations that belong to the ith regime, and $\widehat{\sigma}_{T_i}^2$ is the corresponding residual variance. The final SETAR model specification is given by

$$Y_t = \begin{cases} -0.55_{(0.17)} + 1.69_{(0.12)} Y_{t-1} - 0.81_{(0.14)} Y_{t-2} + \varepsilon_t^{(1)} & \text{if } Y_{t-5} \leq -3.14, \\ 1.47_{(0.50)} + 2.16_{(0.17)} Y_{t-1} - 1.11_{(0.30)} Y_{t-2} - 0.38_{(0.27)} Y_{t-3} \\ \quad + 0.57_{(0.29)} Y_{t-4} + 0.25_{(0.27)} Y_{t-5} + \varepsilon_t^{(2)} & \text{if } -3.14 < Y_{t-5} \leq -2.97, \\ -0.05_{(0.05)} + 1.47_{(0.07)} Y_{t-1} - 0.45_{(0.14)} Y_{t-2} + 0.07_{(0.14)} Y_{t-3} \\ \quad -0.28_{(0.13)} Y_{t-4} + 0.18_{(0.07)} Y_{t-5} + \varepsilon_t^{(3)} & \text{if } Y_{t-5} > -2.97, \end{cases} \quad (6.15)$$

where the sample variances of $\{\varepsilon_t^{(i)}\}$ ($i = 1, 2, 3$) are 0.63×10^{-2} ($T_1 = 44$), 0.19×10^{-2} ($T_2 = 34$), and 0.17×10^{-2} ($T_3 = 172$), and where the asymptotic standard errors of the parameter estimates are in parentheses. The coefficient estimates of $\phi_3^{(2)}$, $\phi_5^{(2)}$, $\phi_0^{(3)}$, and $\phi_3^{(3)}$ are not statistically different from zero at the 5% nominal significance level. The p-values of the LB test statistic at lags 6, 12, and 18 are, respectively, 0.54, 0.17 and 0.08, which suggests that the fitted SETAR$(2; 5, 5)$ model is adequate.

To run the simulation approach, we need some additional specifications. In step (i) of Algorithm 6.3, we set $N = 100$ and estimate $\pi(r_{i,0})$ ($i = 1, 2$) by $\widehat{\pi}(\widehat{r}_{i,0}) = T^{-1} \sum_{t=1}^T \mathcal{K}_h(\widehat{r}_{i,0}; Y_t)$, where $\mathcal{K}_h(\widehat{r}_{i,0}; Y_t) = (\sqrt{2\pi}h)^{-2} \exp\{-(\widehat{r}_{i,0} - Y_t)^2/2h^2\}$ with $h \equiv h_T > 0$ the bandwidth from a Gaussian kernel density estimate of $f_Y(\cdot)$.[1] In step (iv), we create $K = 1{,}000$ equidistant points, and in step (v) we use $B = 10{,}000$ replicates. In step (ii) of Algorithm 6.4, we construct the kernel density estimator $\widehat{f}_\varepsilon(\cdot)$ of $f_\varepsilon(\cdot)$ as follows

$$\widehat{f}_\varepsilon(x) = \frac{1}{T - k_0} \sum_{t=k_0+1}^T \mathcal{K}_{\widehat{h}_{\text{opt}}^*}(x; \widehat{\varepsilon}_t^*).$$

Here, we use a Gaussian kernel with an improved bandwidth (see, e.g., Fan and Yao, 2003, p. 201)

$$\widehat{h}_{\text{opt}}^* = \widehat{h}_{\text{opt}, T} \left(1 + \frac{35}{48} \widehat{\kappa} + \frac{35}{32} \widehat{\tau} + \frac{385}{1024} \widehat{\kappa}^2 \right)^{-1/5},$$

where $\widehat{h}_{\text{opt}, T} = 1.06 \widehat{\sigma} (T - k_0)^{-1/5}$ is the normal reference bandwidth, and $\widehat{\sigma}$, $\widehat{\tau}$, $\widehat{\kappa}$ are respectively the sample standard deviation, skewness, and kurtosis of the residuals $\{\widehat{\varepsilon}_t\}_{t=k_0+1}^T$.

Based on the simulation approach, the 95% confidence intervals of $r_{1,0}$ and $r_{2,0}$ are $(-3.54, -2.75)$ and $(-3.36, -2.58)$, respectively. The (normalized) relative frequency histograms of the estimated thresholds are given in Figures 6.2(b) and (c). We see that $T(\widehat{r}_i - r_{i,0})$ is very small, indicating the superconsistency of the CLS estimators of $r_{i,0}$ ($i = 1, 2$).

---
[1] See Appendix 7.A, for details on kernel estimation.

Subset SETARMA models

Finding a well-specified, while parsimonious, threshold model for a time series is practically difficult, if not infeasible, due to the variety of model options, the complexity in partitioning the parameter space by appropriate single or multivariate threshold values, as well as the conventional problems in model structure selection. Consider, for instance, a SETARMA$(2; 6, 6, 6, 6)$ model with maximum delay $d_{\max} = 6$, the total number of potentially useful models is $d_{\max} \times 2^{p_1+p_2+q_1+q_2+2} = 402{,}653{,}184$. This huge number increases even further if seasonal SETARMA models are considered. To overcome this problem, several local search techniques have been proposed to efficiently examine the parameter space and find the best subset of parameters that corresponds to the optimal solution for a given model selection criterion (objective function). One approach is to use *Markov chain Monte Carlo* (MCMC) methods for Bayesian subset model selection; see, e.g., Chen et al. (2011a).

Another approach can be based on *genetic algorithms* (GAs). GAs are randomized global search techniques that emulate natural genetic operators, such as reproduction, crossover, and mutation. At each iteration, a GA explores different areas of the parameter space and then directs the search to a region where there is a high probability of finding improved performance as measured by a positive real-valued objective function, called a *fitness function*, $g(\cdot)$. Following Baragona et al. (2004a), we briefly outline the working principles of the GA procedure only for subset SETARMA models. With a few simple modifications the GA-based SETARMA procedure can be applied to PLTAR models (Baragona et al., 2004b), DT(G)ARCH, and multivariate SETAR models.

A k-regime subset SETARMA model takes the form of (2.29) with some of the intermediate AR and MA parameters set to zero. To formalize, assume that

$$\phi^{(i)}_{j^{(i)}_1}, \ldots, \phi^{(i)}_{j^{(i)}_{p_i}}, \psi^{(i)}_{h^{(i)}_1}, \ldots, \psi^{(i)}_{h^{(i)}_{q_i}} \quad (i = 1, \ldots, k)$$

are non-zero parameters and that $\{j_1^{(i)}, \ldots, j_{p_i}^{(i)}\}$ ($p_i \leq p$) and $\{h_1^{(i)}, \ldots, h_{q_i}^{(i)}\}$ ($q_i \leq q$) are two subsets of the integers $1, \ldots, p_i$ and $1, \ldots, q_i$ respectively, with $p = \max_i p_i$ and $q = \max_i q_i$. Then we write a k-regime subset SETARMA model as

$$Y_t = \sum_{i=1}^{k} \Big(\phi_0^{(i)} + \sum_{u=1}^{p_i} \phi^{(i)}_{j^{(i)}_u} Y_{t-j^{(i)}_u} + \sum_{v=1}^{q_i} \psi^{(i)}_{h^{(i)}_v} \varepsilon^{(i)}_{t-h^{(i)}_v} \Big) I(Y_{t-d} \in \mathbb{R}^{(i)}), \qquad (6.16)$$

where $\varepsilon_t^{(i)} = \sigma_i^2 \varepsilon_t$ ($i = 1, \ldots, k$), $\{\varepsilon_t\} \stackrel{\text{i.i.d.}}{\sim} (0, 1)$, and $\mathbb{R}^{(i)} = (r_{i-1}, r_i]$ with $r_0 = -\infty$ and $r_k = \infty$. The delay d, the thresholds r_i, and the AR and MA lags in each regime are called *structural parameters*. They are collected together into the long vector

$$\mathbf{x}^* = \big(d, r_1, \ldots, r_{k-1}; \{p_i; j_1^{(i)}, \ldots, j_{p_i}^{(i)} | q_i; h_1^{(i)}, \ldots, h_{q_i}^{(i)}, i = 1, \ldots, k\}\big)'. \qquad (6.17)$$

Estimating (6.16) by CLS is computationally demanding since for each subset a nonquadratic optimization has to be done. Partly for this reason, it is recommended

6.1 MODEL ESTIMATION

to use an ARMA–LS estimation method due to Hannan and Rissanen (1982); see, e.g., step (i) in Algorithm 6.3. Given a set of observations $\{Y_t\}_{t=1}^{T}$, and assuming $\mathbf{x}^*$ is known, the CLS estimation procedure is as follows.

Algorithm 6.5: k-regime subset SETARMA–CLS estimation

(i) For each regime i, fit a high-order AR(n) ($1 \leq n \leq n_{\max}$) model to the series using the Yule–Walker equations. Select n by AIC, and set $n_{\max} = (\log T)^a$ ($0 < a < \infty$). Calculate $\{\widehat{\varepsilon}_t^{(i)}\}_{t=n+1}^{T}$ ($i = 1, \ldots, k$).

(ii) Set the maximum orders P and Q of respectively the AR and MA lags sufficiently large such that $p_i \leq p \leq P \leq n$ and $q_i \leq q \leq Q$.

(iii) Calculate the LS estimates of the ARMA parameters in (6.16) replacing $\{\varepsilon_{t-h_1^{(i)}}^{(i)}, \ldots, \varepsilon_{t-h_{q_i}^{(i)}}^{(i)}\}$ by $\{\widehat{\varepsilon}_{t-h_1^{(i)}}^{(i)}, \ldots, \widehat{\varepsilon}_{t-h_{q_i}^{(i)}}^{(i)}\}$, and using observations $\{Y_t\}_{t=n_0}^{T}$ where $n_0 = n + \max(P, Q)$, and subject to a minimum number of observations $T_{\min}$ per regime.

(iv) Find the optimal structural parameter vector by minimizing the *normalized* AIC (NAIC) values, that is

$$\text{NAIC}(\mathbf{x}^*) = \sum_{i=1}^{k} \{T_i \log \widehat{\sigma}_{T_i}^2 + 2(p_i + q_i + 1)\}/(\text{effective sample size}),$$

where T_i is the number of observations that belong to the ith regime, and $\widehat{\sigma}_{T_i}^2$ denotes the corresponding residual variance.

(v) Repeat steps (i) – (iv) for each $d \in [1, d_{\max}]$, with $d_{\max}$ a pre-specified integer.

Any vector $\mathbf{x}^*$, as defined by (6.17), represents a tentative solution to the problem of specifying the structural parameters of a k-regime subset SETARMA model leading to the best choice. The GA has the task of simultaneously finding the optimal model coefficients, as well as partitioning the parameter space by finding the number of regimes, and the threshold parameters $r_1, \ldots, r_{k-1}$. A solution is represented by a binary coding string, i.e. a transformation of $\mathbf{x}^*$ to the vector $\mathbf{x} = (x_1, \ldots, x_T)'$ where $x_j = 1$ if $Y_{(\tau_j)}$ is a threshold parameter, while $x_j = 0$ otherwise, and $Y_{(\tau_j)}$ is the value at time τ_j of the ordered time series $\{Y_{(\tau_j)}\}_{j=1}^{T}$. The number of regimes is given by $k = 1 + \sum_{j=2}^{T-1} x_j$; a string is not admissible if $k > k_{\max}$, where $k_{\max}$ is the maximum number of regimes, a pre-specified integer. Below are some guidelines for developing a simple GA.

Algorithm 6.6: A simple genetic algorithm

(i) Randomly generate an initial population of admissible binary strings $\{\mathbf{x}^{(1)}, \mathbf{x}^{(2)}, \ldots, \mathbf{x}^{(s)}\}$.

Algorithm 6.6: A simple genetic algorithm (Cont'd)

(ii) Calculate the fitness function $g(\cdot)$ for each string in the population. For instance, in view of step (iv) in Algorithm 6.5, one may choose $g(\mathbf{x}) = \exp(-\text{NAIC}(\mathbf{x})/C)$, where $C > 0$ is used to scale $g(\cdot)$.

(iii) Keep the best string intact for the next generation and create offspring strings by three evolutionary operators:

- *Selection*: Select s times a string from the population with probability $g(\mathbf{x}^{(i)})/\sum_{i=1}^{s} g(\mathbf{x}^{(i)})$. Replace the population by the selected strings. This part may include an elitist step by substituting the best string from the past population for the string having the smallest value of $g(\cdot)$ in the new population.

- *Crossover*: Adopt a simple crossover operator to change candidate solutions into new candidate solutions. In particular, with the single point crossover, $[s/2]$ string pairs are selected at random, and the crossover operator is applied to each of them with a pre-specified, usually large (0.8 or 0.9), probability p_c. If no crossover takes place, two offspring strings are formed that are exact copies of their "parents chromosomes".

- *Mutation*: Allow any bit x_j ($j = 2, \ldots, T-1$) of any string to flip with probability p_m, usually small $(0.001, \ldots, 0.01)$.

(iv) Form the new population using the results of step (iii). If the search aim is achieved, stop; else go to step (ii).

Example 6.3: U.S. Real GNP

We illustrate the GA procedure by analyzing the first differences of the logarithm of quarterly U.S. real GNP, say $\{X_t\}$ (seasonally unadjusted data). The data covers the time period 1947(i) – 2009(iv). Thus, we consider $\{Y_t = \log X_t - \log X_{t-1}\}_{t=2}^{252}$; see Figure 6.3 for a time plot. The series is viewed as a "test-case" for many nonlinear models and methods. Indeed, quite some attention has focused on fitting pure SETAR models to the data, albeit covering shorter time periods.

As for the specification of the GA parameters, we set the size of the population at $s = 50$, the crossover probability $p_c = 0.9$, the mutation probability $p_m = 0.01$, the adjusting constant is set $C = 1$ in the NAIC-based fitness function, and the maximum allowed number of iterations is equal to 300. Further, we set $d_{\max} = 5$, $k_{\max} = 3$, $n_{\max} = 20$, $T_{\min} = 30$, and the maximum allowed order of P and Q is set at 10. The number of bits ν for the binary representation of p_i and q_i ($i = 1, \ldots, k$) varies between 0 and $2^\nu - 1$. We set $\nu = 3$ so that the maximum allowed number of parameters p and q is 8. The number of bits μ

6.1 MODEL ESTIMATION

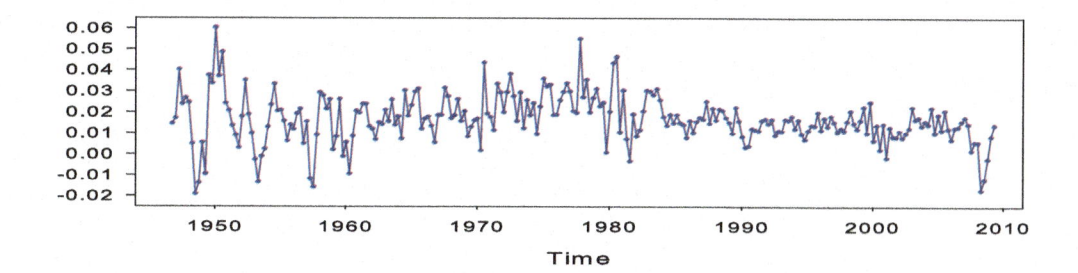

Figure 6.3: *Growth rates of quarterly real U.S. GNP; $T = 252$.*

($\mu \geq \nu$) for the lag values binary representation is constrained to the interval $[1, 2^\mu - 1]$. With $\mu = 3$, the maximum allowed lag is 7. The length of the chromosome can be computed as $(T - 2T_{\min}) + 2k\nu + \mu\{\sum_{i=1}^{k}(p_i + q_i)\}$.

The best subset SETARMA model with $k = 3$ regimes and delay $d = 2$ is given by

$$Y_t = \begin{cases} 0.45_{(0.28)} + 0.36_{(0.10)}Y_{t-10} + \varepsilon_t^{(1)} & \text{if } Y_{t-2} \leq 0.82_{(0.29)}, \\ 0.78_{(0.12)} + 0.46_{(0.12)}Y_{t-1} - 0.21_{(0.13)}Y_{t-3} + 0.16_{(0.39)}Y_{t-9} \\ \quad + \varepsilon_t^{(2)} - 0.19_{(0.09)}\varepsilon_{t-4}^{(2)} & \text{if } 0.82_{(0.29)} < Y_{t-2} \leq 1.64_{(0.12)}, \\ 1.12_{(0.10)} + 0.27_{(0.09)}Y_{t-1} + 0.11_{(0.08)}Y_{t-7} \\ \quad + 0.10_{(0.00)}Y_{t-9} + \varepsilon_t^{(3)} & \text{if } Y_{t-2} \geq 1.64_{(0.12)}, \end{cases}$$

(6.18)

where the sample variances of $\{\varepsilon_t^{(i)}\}$ ($i = 1, 2, 3$), are 1.34 ($T_1 = 34$), 0.31 ($T_2 = 85$), and 0.82 ($T_3 = 102$), respectively, bootstrap-calculated (1,000 replicates) standard errors of the parameter estimates are given in parentheses as subscripts, and NAIC = -0.3955.

For comparison, we repeated the GA-based subset SETARMA procedure with $k = 2$ regimes. The resulting model, in obvious short-hand notation, has the form SETARMA$(2; (9), (1, 6, 10); (1, 4, 6, 10), (0))$ with NAIC = -0.3069. On the other hand, if we perform a grid search among pure SETAR$(3; p_1, p_2, p_3)$ models with $\max\{p_1, p_2, p_3\} \leq 12$ and $d_{\max} \leq 12$, the best fitted model is a three-regime SETAR model with order $(6, 7, 10)$, delay $d = 6$, and AIC = -0.2998. These results illustrate that the selected subset SETARMA models are adequate and more parsimonious compared to the selected pure SETAR model.

STAR models

Efficient estimation of STAR-type nonlinear models can be carried out by NLS or, assuming the errors are normally distributed, by QML. Under certain regularity conditions both methods will result in estimates that are consistent and asymptotically normally distributed. Below we outline nonlinear CLS estimation of LSTAR models, but the issues that are addressed also apply to ESTAR, time-varying STAR, and multiple-regime STAR models.

Recall from Section 2.7 that for a stationary and ergodic time series process $\{Y_t, t \in \mathbb{Z}\}$ the LSTAR$(2;p,p)$ model is defined by

$$Y_t = \phi_0 + \sum_{i=1}^{p} \phi_i Y_{t-i} + \Big\{\xi_0 + \sum_{i=1}^{p} \xi_i Y_{t-i}\Big\} G(Y_{t-d}; \gamma, c) + \varepsilon_t,$$
$$= \phi' \mathbf{X}_t + \xi' \mathbf{X}_t G(Y_{t-d}; \gamma, c) + \varepsilon_t, \qquad (6.19)$$

where

$$\phi = (\phi_0, \ldots, \phi_p)', \quad \xi_j = (\xi_0, \ldots, \xi_p)', \quad \mathbf{X}_t = (1, Y_{t-1}, \ldots, Y_{t-p})',$$

with $\{\varepsilon_t\} \overset{\text{i.i.d.}}{\sim} (0,1)$, and $G(\cdot)$ is a logistic function defined by (2.43). Then, subject to some initial values, the problem is to minimize the ordinary least squares function

$$L_T(\boldsymbol{\theta}) = \sum_{t=1}^{T} \big\{Y_t - \phi' \mathbf{X}_t - \xi' \mathbf{X}_t G(Y_{t-d}; \gamma, c)\big\}^2 \qquad (6.20)$$

with respect to $\boldsymbol{\theta} = (\phi', \xi', \gamma, c)'$. However, joint estimation of $\boldsymbol{\theta}$ is not an easy task in general and can result in large γ values. One reason is that γ is not scale invariant, making it difficult to find a good starting value. To overcome this problem, and to improve the stability and speed of the numerical optimization procedure, it is usually preferred to estimate LSTAR models using the following transition function

$$G(Y_{t-d}; \gamma, c) = \big\{1 + \exp(-\gamma[Y_{t-d} - c]^2/\widehat{\sigma}_Y^2)\big\}^{-1}, \quad \gamma > 0, \qquad (6.21)$$

where $\widehat{\sigma}_Y^2$ is the sample variance of $\{Y_{t-d}\}$. Thus, the original slope parameter γ is transformed into a scale-free parameter.

Note that when the parameters γ and c are known and fixed, the LSTAR model is linear in the AR parameters ϕ and ξ. Hence, assuming d and p are known, the parameter vector $\boldsymbol{\tau} = (\phi', \xi')'$ can be estimated by CLS as

$$\widehat{\boldsymbol{\tau}}(\gamma, c) = \Big(\sum_{t=1}^{T} \widetilde{\mathbf{X}}_t(\gamma, c) \widetilde{\mathbf{X}}_t'(\gamma, c)\Big)^{-1} \Big(\sum_{t=1}^{T} \widetilde{\mathbf{X}}_t(\gamma, c) Y_t\Big), \qquad (6.22)$$

where $\widetilde{\mathbf{X}}_t(\gamma, c) = \big(\mathbf{X}_t', \mathbf{X}_t' G(Y_{t-d}; \gamma, c)\big)'$. Consequently, minimizing (6.20) can be simplified by concentrating the sum of squares function with respect to $\boldsymbol{\tau}$ as

$$L_T(\gamma, c) = \sum_{t=1}^{T} \big\{Y_t - \boldsymbol{\tau}'(\gamma, c) \widetilde{\mathbf{X}}_t(\gamma, c)\big\}^2. \qquad (6.23)$$

So, minimization of (6.20) is only performed over γ and c, which helps to reduce the computational burden considerably.

Using (6.23) some cautionary remarks are in order. It is apparent from Figure 2.9 that when the true slope parameter γ is relatively large, the slope of $G(\cdot)$ at c is steep. In that case a meaningful set of grid values for the location parameter c is needed (e.g., the sample percentiles of the transition variable Y_{t-d}) so that the value of the transition function $G(\cdot)$ varies sufficiently across the whole sample, and the optimization algorithm converges. Otherwise, the moment matrix of the regression (6.22) is ill-conditioned and the estimation fails. It is also recommended to have a large number of observations in the neighborhood of c to estimate γ accurately. If there are not many data values near c, γ will be poorly estimated, and so convergence may be slow. This situation may well result in a parameter estimate of γ which is not statistically different from zero as judged by, for instance, a large standard error and a small Student t-statistic. The calculated t-statistic, however, will not have an exact Student t distribution under the null hypothesis $\gamma = 0$, since then the LSTAR model is no longer identified; see Section 2.7. One implication is that in practice one should focus upon the end use of the LSTAR model when attempting to evaluate it and not necessarily on the parameter estimates.

Example 6.4: ENSO Phenomenon (Cont'd)

Recall Example 1.4 where the monthly ENSO series refers to the abnormal warming (cooling) of the ocean-atmosphere system in the eastern Pacific. Figure 1.4(b) shows that ENSO dynamics follow a nonlinear process that is mean-reverting, with the speed of adjustment toward equilibrium varying directly with the extent of the SST anomaly from its long-run mean. Changes between El Niño and La Niña events, however, occur gradually rather than abruptly. Within the bands $(-0.5°C, 0.5°C)$, when no ENSO events are identified, small deviations will not be corrected through the DGP. Ubilava and Helmers (2013) capture this type of behavior by a reparameterized form of the LSTAR process, called *logistic smooth transition error correction* (LSTEC),

$$\Delta Y_t = \alpha_0 + \beta_0 Y_{t-1} + \sum_{i=1}^{p-1} \psi_{0i} \Delta Y_{t-i} + \boldsymbol{\delta}' \mathbf{D}_t$$
$$+ \left\{ \alpha_1 + \beta_1 Y_{t-1} + \sum_{i=1}^{p-1} \psi_{1i} \Delta Y_{t-i} + \boldsymbol{\delta}' \mathbf{D}_t \right\} G(Y_{t-d}; \gamma, c) + \varepsilon_t, \quad (6.24)$$

where $\Delta Y_t \equiv Y_t - Y_{t-1}$ denotes the first-difference of the time series $\{Y_t\}$, $\mathbf{D}_t$ is a vector of monthly dummy variables, and $\boldsymbol{\delta}$ the corresponding parameter vector.

When $Y_{t-d} = c$, the adjustment process is given by the first term on the right-hand side of (6.24), and as $Y_{t-d} \to \pm\infty$, the adjustment process is given by (6.24) with $G(\cdot) = 1$. Here, the crucial parameters are β_0 and β_1. Since large deviations are mean-reverting, it implies that $\beta_1 < 0$ and $\beta_0 + \beta_1 < 0$, while $\beta_0 \geq 0$ is possible. A linear version of the regression in (6.24), called *error correction model* (ECM), is given by

$$\Delta Y_t = \alpha_0 + \beta_0 Y_{t-1} + \sum_{i=1}^{p-1} \psi_i \Delta Y_{t-i} + \boldsymbol{\delta}' \mathbf{D}_t + \varepsilon_t. \qquad (6.25)$$

Below we show estimation results for the series covering the time period January 1952 – December 1990 ($T = 468$). Later, in Chapter 10, we employ the remaining part of the series for a rolling out-of-sample forecasting experiment. Using a battery of time-domain nonlinearity tests, we obtain the following best-fitting (in terms of minimum AIC) model for the series

$$\begin{aligned}\Delta Y_t = &-0.19_{(0.21)} - 0.13_{(0.11)}Y_{t-1} + 0.21_{(0.18)}\Delta Y_{t-1} - 0.07_{(0.17)}\Delta Y_{t-2} \\ &+ 0.11_{(0.16)}\Delta Y_{t-3} + 0.11_{(0.16)}\Delta Y_{t-4} + 0.06_{(0.13)}\Delta Y_{t-5} \\ &+ 0.22_{(0.14)}D_{1t} + 0.52_{(0.26)}D_{2t} + 0.29_{(0.17)}D_{3t} + 0.19_{(0.14)}D_{4t} \\ &+ 0.11_{(0.12)}D_{5t} + 0.15_{(0.11)}D_{6t} + 0.10_{(0.12)}D_{7t} - 0.19_{(0.14)}D_{8t} \\ &- 0.26_{(0.17)}D_{9t} - 0.65_{(0.39)}D_{10,t} - 0.23_{(0.15)}D_{11,t} \\ &+ \{0.25_{(0.24)} - 0.02_{(0.09)}Y_{t-1} + 0.28_{(0.20)}\Delta Y_{t-1} - 0.02_{(0.19)}\Delta Y_{t-2} \\ &+ 0.11_{(0.19)}\Delta Y_{t-3} + 0.06_{(0.18)}\Delta Y_{t-4} + 0.10_{(0.16)}\Delta Y_{t-5} \\ &- 0.22_{(0.17)}D_{1t} - 0.71_{(0.29)}D_{2t} - 0.42_{(0.19)}D_{3t} - 0.32_{(0.16)}D_{4t} \\ &- 0.11_{(0.14)}D_{5t} - 0.13_{(0.13)}D_{6t} - 0.10_{(0.15)}D_{7t} + 0.24_{(0.18)}D_{8t} \\ &+ 0.29_{(0.20)}D_{9t} + 0.87_{(0.43)}D_{10,t} + 0.28_{(0.18)}D_{11,t}\}G(Y_{t-1};\gamma,c) \\ &+ \varepsilon_t,\end{aligned}$$

where

$$G(Y_{t-1};\gamma,c) = \left\{1 + \exp\left[(-1.95_{(0.83)}/0.82)(Y_{t-1} - (-0.77)_{(0.33)})\right]\right\}^{-1}, \quad (6.26)$$

with asymptotic standard errors in parentheses. The residual variance $\widehat{\sigma}_\varepsilon^2$ is 88.8% of that of a corresponding AR(8) model. The JB test statistic (1.6) does not reject normality of the residuals at the 5% nominal significance level (p-value = 0.612).

Figure 6.4(a) displays the transition function (6.26) as a function of the transition variable Y_{t-1}. The red medium dashed line denotes the estimate of the threshold value c, which is centered around $-0.77°$C of the SST anomaly. We observe that the majority of observations belongs to the upper regime (El Niño phase). From (6.26) it is apparent that the low value of γ results in a relatively slow speed of transition. Figure 6.4(b) shows the SST anomaly and the transition function as a function of time. Clearly, the ENSO dynamics are captured well by the transition function.

Bilinear models
There are many methods for estimating coefficients of BL models. Among them is the LS method, which is one of the most frequently applied. However, apart from some simple BL models, the asymptotic properties of the LS estimates are unknown.

6.1 MODEL ESTIMATION

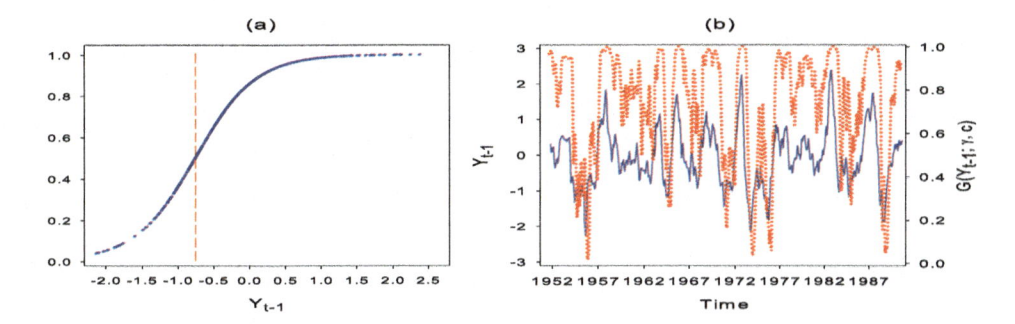

Figure 6.4: *(a) Transition function (6.26) as a function of Y_{t-1} (blue dots), and an estimate of the threshold value (red medium dashed line); (b) SST anomaly (blue solid line) and transition function (6.26) (red dotted line) as a function of time.*

In this section, we discuss a CLS approach with known asymptotic properties and proposed by Grahn (1995) for a special case of (2.12). In particular, we want to estimate the BL model:

$$Y_t = \phi_0 + \sum_{i=1}^{p} \phi_i Y_{t-i} + \varepsilon_t + \sum_{j=1}^{q} \psi_j \varepsilon_{t-j} + \sum_{i=1}^{k} \sum_{j=w}^{r} \tau_{ij} \varepsilon_{t-i} Y_{t-j}, \qquad (6.27)$$

where $w = (q \vee k) + 1$, and $\{\varepsilon_t\} \stackrel{\text{i.i.d.}}{\sim} (0, \sigma_\varepsilon^2)$. Below we assume, without loss of generality, that the process $\{Y_t, t \in \mathbb{Z}\}$ is standardized such that $\mathbb{E}(Y_t) = 0$.

The first step of the CLS procedure consists of estimating the parameter vector $\boldsymbol{\phi} = (\phi_1, \ldots, \phi_p)'$ by the Yule–Walker equations, given a set of observations $\{Y_t\}_{t=1}^T$. It can be shown that these equations hold for lags $s > w^*$ with $w^* = (q+1) \vee k$. In the second step, estimates of the other coefficients of (6.27) are obtained using conditional covariances of the AR-residual process, say $\{v_t, t \in \mathbb{Z}\}$. Assuming $\{Y_t, t \in \mathbb{Z}\}$ is a stationary, causal and invertible process with $\mathbb{E}(Y_t^4) < \infty$, Grahn (1995) deduces the following equation

$$\text{Cov}(v_t, v_{t-s} | \varepsilon_{t-w}, \varepsilon_{t-w-1}, \ldots) = \mathbb{E}(v_t, v_{t-s} | \varepsilon_{t-w}, \varepsilon_{t-w-1}, \ldots)$$

$$= \gamma_Y(s) + \sum_{j=w}^{r+s} d_j(s) Y_{t-j} + \sum_{j=w}^{r} \sum_{n=w}^{r} h_{j,n}(s) Y_{t-j} Y_{t-s-n}, \qquad (6.28)$$

where $\gamma_Y(s)$ is the ACVF of an MA(q) process with parameters ψ_j ($j = 1, \ldots, q$) and σ_ε^2, and where

$$d_j(s) \equiv \tau_{sj} \sigma_\varepsilon^2 + \sum_{i=s+1}^{w-1+s} (\psi_i \tau_{i-s,j-s} + \psi_{i-s} \tau_{ij}) \sigma_\varepsilon^2 \quad \text{and} \quad h_{j,n}(s) \equiv \sum_{i=s+1}^{k} \tau_{ij} \tau_{i-s,n} \sigma_\varepsilon^2,$$

$$(j = w, \ldots, r+s; n = w, \ldots, r),$$

and $\psi_i \equiv 0$ for $i > q$ and $\tau_{ij} \equiv 0$ $\forall i, j$ taking values outside the summation domain. Thus, $\text{Cov}(v_t, v_{t-s} | \varepsilon_{t-w}, \varepsilon_{t-w-1}, \ldots)$ depends on the parameters and a finite set of observations $\{Y_t\}_{t=1}^T$ only. As we will see in Algorithm 6.7, this property will be the basis for the proposed CLS estimation procedure.

Let $\beta_0(s)$ be the true value of the parameter vector $\beta(s)$ at lag s, i.e.

$$\beta(s) = \big(\gamma_Y(s), d_w(s), \ldots, d_{r+s}(s), h_{ww}(s), \ldots, h_{wr}(s), \ldots, h_{rw}(s), \ldots, h_{rr}(s)\big)'. \tag{6.29}$$

Hence, in the second step, the aim is to find an estimator $\widehat{\beta}(s)$ of $\beta_0(s)$. Now, summarizing the above results, the computation of CLS estimates goes as follows.

Algorithm 6.7: CLS estimation of the BL model (6.27)

(i) Calculate $\widehat{\phi}$ as an estimate of ϕ by solving the Yule–Walker equations

$$\widehat{\mathbf{C}}_p \widehat{\phi} = \widehat{\mathbf{c}},$$

where $\widehat{\mathbf{C}}_p$ is a $p \times p$ matrix with elements $\{\widehat{c}_Y(w^* - 1 + i - j)\}_{1 \leq i,j, \leq p}$, $\widehat{\mathbf{c}} = (\widehat{c}_Y(w^*), \ldots, \widehat{c}_Y(w^* + p))'$, and $\widehat{c}_Y(\cdot)$ is the sample ACVF of $\{Y_t\}_{t=1}^T$. Obtain the AR residuals by $\widehat{v}_t = Y_t - \sum_{i=1}^p \widehat{\phi}_i Y_{t-i}$.

(ii) Minimize the conditional sum of squares

$$\sum_{t=(r+s)\vee(p+1)}^T \big\{\widehat{v}_t \widehat{v}_{t-s} - \mathbb{E}(v_t v_{t-s} | \varepsilon_{t-w}, \varepsilon_{t-w-1}, \ldots)\big\}^2 \tag{6.30}$$

with respect to $\beta(s)$ ($s = 0, 1, \ldots, w-1$), giving rise to $\widehat{\beta}(s)$. It can be shown (Grahn, 1995) that $\widehat{\beta}(s) \to \beta_0(s)$ a.s., as $T \to \infty$.

The remaining task is to identify the parameters τ_{ij} ($i = 1, \ldots, k; j = w, \ldots, r$), ψ_j ($j = 1, \ldots, q$), and σ_ε^2 from $\beta_0(s)$ ($s = 0, 1, \ldots, w-1$). Regarding the identification of the MA parameters, consider the MA(q) process $Z_t = \sum_{j=0}^q \psi_j \varepsilon_{t-j}$, ($\psi_0 = 1$) where $\{\varepsilon_t\} \overset{\text{i.i.d.}}{\sim} (0, \sigma_\varepsilon^2)$, and assuming the process $\{Z_t, t \in \mathbb{Z}\}$ is invertible. The function $\gamma_Y(s)$ can be interpreted as the ACVF of this process. Therefore, $\gamma_Y(s) = \sigma_\varepsilon^2 \sum_{j=0}^{q-s} \psi_j \psi_{j+s}$. The equations which must be solved to obtain the MA parameters can be written, in two alternative ways, as

$$\begin{pmatrix} \gamma_Y(0) \\ \gamma_Y(1) \\ \vdots \\ \gamma_Y(q) \end{pmatrix} = \sigma_\varepsilon^2 \begin{pmatrix} \psi_0 & \psi_1 & \cdots & \psi_{q-1} & \psi_q \\ \psi_1 & \psi_0 & \cdots & \psi_q & 0 \\ \vdots & \vdots & \ddots & \vdots & \vdots \\ \psi_{q-1} & \psi_q & \cdots & 0 & 0 \\ \psi_q & 0 & \cdots & 0 & 0 \end{pmatrix} \begin{pmatrix} \psi_0 \\ \psi_1 \\ \vdots \\ \psi_{q-1} \\ \psi_q \end{pmatrix}$$

$$= \sigma_\varepsilon^2 \begin{pmatrix} \psi_0 & \psi_1 & \cdots & \psi_{q-1} & \psi_q \\ 0 & \psi_0 & \cdots & \psi_{q-2} & \psi_{q-1} \\ \vdots & \vdots & \ddots & \vdots & \vdots \\ 0 & 0 & \cdots & \psi_0 & \psi_1 \\ 0 & 0 & \cdots & 0 & \psi_0 \end{pmatrix} \begin{pmatrix} \psi_0 \\ \psi_1 \\ \vdots \\ \psi_{q-1} \\ \psi_q \end{pmatrix}.$$

6.1 MODEL ESTIMATION

These equations may be written in summary notation as

$$\boldsymbol{\gamma}_Y = \sigma_\varepsilon^2 \mathbf{A}^\# \boldsymbol{\psi} = \sigma_\varepsilon^2 \mathbf{A}' \boldsymbol{\psi}, \tag{6.31}$$

where $\mathbf{A}^\#$ is a $(q+1) \times (q+1)$ matrix with constant skew-diagonals, called *Hankel matrix*, $\boldsymbol{\gamma}_Y = (\gamma_Y(0), \gamma_Y(1), \ldots, \gamma_Y(q))'$, and $\boldsymbol{\psi} = (\psi_0, \psi_1, \ldots, \psi_q)'$.

Now, the objective is to solve

$$f(\boldsymbol{\psi}) = \boldsymbol{\gamma}_Y - \sigma_\varepsilon^2 \mathbf{A}^\# \boldsymbol{\psi} = \mathbf{0}. \tag{6.32}$$

Since (6.32) is nonlinear in $\boldsymbol{\psi}$, its solution must be found via an iterative procedure. For instance, we can use the Newton–Raphson algorithm (see, e.g., Wilson, 1969). In this case the $(u+1)$th approximation, say $\boldsymbol{\psi}^{(u+1)}$, to the final solution obtained from the uth approximation $\boldsymbol{\psi}^{(u)}$ ($u \geq 0$) is given by

$$\boldsymbol{\psi}^{(u+1)} = \boldsymbol{\psi}^{(u)} - \{\partial f(\boldsymbol{\psi}^{(u)})/\partial \boldsymbol{\psi}\}^{-1} f(\boldsymbol{\psi}^{(u)}),$$

which is equivalent to

$$\boldsymbol{\psi}^{(u+1)} = \boldsymbol{\psi}^{(u)} + \{\sigma_\varepsilon^2 (\mathbf{A}^\# + \mathbf{A}')\}_u^{-1} (\boldsymbol{\gamma}_Y - \sigma_\varepsilon^2 \mathbf{A}^\# \boldsymbol{\psi})_u,$$

where the subscript u indicates that the elements are to be evaluated at $\boldsymbol{\psi} = \boldsymbol{\psi}^{(u)}$. The equation for $\gamma_Y(s)$ can be normalized either by setting $\sigma_\varepsilon^2 = 1$ or by setting $\psi_0 = 1$. In the first case, it is reasonable to choose $\psi_0 = \gamma_Y(0)$ and $\psi_1 = \cdots = \psi_q = 0$ as starting values of the iterative procedure. Once it has converged, the equation for $\gamma_Y(s)$ can be re-normalized so that $\psi_0 = 1$.

Below we present a procedure for identifying the BL parameters τ_{ij} from $d_j(s)$ ($j = w, \ldots, r + s; s = 0, 1, \ldots, w - 1$). For simplicity, we assume that the equation for $d_j(s)$ is normalized either by setting $\sigma_\varepsilon^2 = 1$ or by considering $d_j(s)/\sigma_\varepsilon^2$. Define the following two $\frac{1}{2}w(2r - w + 1) \times 1$ vectors

$$\widetilde{\boldsymbol{\tau}} = (\tau_{0,w}, \tau_{0,w+1}, \ldots, \tau_{0,r}, \tau_{1,w}, \tau_{1,w+1}, \ldots, \tau_{1,r+1}, \ldots, \tau_{w-1,w},$$
$$\psi_{w-1,w+1}, \ldots, \psi_{w-1,r+w-1})',$$
$$\mathbf{d} = (d_w(0), d_{w+1}(0), \ldots, d_r(0), d_w(1), d_{w+1}(1), \ldots, d_{r+1}(1), \ldots, d_w(w-1),$$
$$d_{w+1}(w-1), \ldots, d_{r+w-1}(w-1))'.$$

Then the equation for $d_j(s)$ can be written as

$$\mathbf{T}\widetilde{\boldsymbol{\tau}} = \mathbf{d}, \tag{6.33}$$

where

$$\mathbf{T} = \begin{pmatrix} \mathbf{D}_0 & \mathbf{U}_{0,1} & \cdots & \mathbf{U}_{0,w-2} & \mathbf{U}_{0,w-1} \\ \mathbf{L}_{1,0} & \mathbf{D}_1 & \cdots & \mathbf{U}_{1,w-2} & \mathbf{U}_{1,w-1} \\ \vdots & \vdots & & \vdots & \vdots \\ \mathbf{L}_{w-1,0} & \mathbf{L}_{w-1,1} & \cdots & \mathbf{L}_{w-1,w-2} & \mathbf{D}_{w-1} \end{pmatrix},$$

with

$$\underbrace{\mathbf{D}_i}_{\substack{(h+i)\times(h+i)\\ 0\leq i\leq w-1}} = \begin{pmatrix} 1 & & & & & \\ 0 & \ddots & & & & \\ \vdots & & \ddots & & & \\ 0 & & & \ddots & & \\ \psi_{2i} & & & & \ddots & \\ & \ddots & & & & \ddots \\ & & \psi_{2i} & 0 & \cdots & 0 & 1 \\ & & \uparrow & & & & \\ & & i+1 & & & & \end{pmatrix}, \quad \underbrace{\mathbf{U}_{0,j}}_{\substack{(h+i)\times(h+j)\\ 0\leq j\leq w-1}} = \begin{pmatrix} 2\psi_j & 0 & \cdots & & 0 \\ & \ddots & \ddots & & \vdots \\ & & 2\psi_j & 0 & \cdots & 0 \end{pmatrix},$$

$$\underbrace{\mathbf{U}_{i,j}}_{\substack{(h+i)\times(h+j)\\ 0\leq i<j\leq w-1}} = \begin{pmatrix} \psi_{j-i} & & & & & \\ 0 & \ddots & & & & \\ \vdots & & \ddots & & & \\ 0 & & & \ddots & & \\ \psi_{j+i} & & & & \ddots & \\ & \ddots & & & & \ddots \\ & & \psi_{j+i} & 0 & \cdots & 0 & \psi_{j-i} \\ & & \uparrow & & & & \\ & & i+1 & & & & \end{pmatrix}, \quad \underbrace{\mathbf{L}_{i,j}}_{\substack{(h+j)\times(h+i)\\ 0\leq j<i\leq w-1}} = \begin{pmatrix} 0 & & & & \\ \vdots & \ddots & & & \\ 0 & & \ddots & & \\ \psi_{i+j} & & & \ddots & \\ & \ddots & & & \ddots \\ & & \psi_{i+j} & 0 & \cdots & 0 \\ & \uparrow & & & & \\ & i+1 & & & & \end{pmatrix},$$

and with $\mathbf{L}_{i,0} = \mathbf{0}$, and $h = r - (w-1) = r - (q \vee k)$.

The solution to the system of equations (6.33) can, for instance, be obtained by the method of Gaussian elimination which reduces $\mathbf{T}$ to an upper-triangular matrix $\mathbf{U}$ whilst $\mathbf{d}$ is transformed into some vector $\mathbf{x}$. Once $\mathbf{x}$ is available, the transformed system $\mathbf{U}\tilde{\boldsymbol{\tau}} = \mathbf{x}$ can be solved for $\tilde{\boldsymbol{\tau}}$ by a process of back-substitution. Following this approach, it is easy to prove that the coefficients τ_{ij} are uniquely determined by the system (6.33). Thus, asymptotically, we can define an estimator $\hat{\boldsymbol{\tau}}$ of $\tilde{\boldsymbol{\tau}}$ by solving the system

$$\widehat{\mathbf{T}}\hat{\boldsymbol{\tau}} = \widehat{\mathbf{d}}, \qquad (6.34)$$

where $\widehat{\mathbf{T}}$ and $\widehat{\mathbf{d}}$ are the estimators of $\mathbf{T}$ and $\mathbf{d}$, respectively.

Let $\boldsymbol{\theta} = (\boldsymbol{\phi}', \boldsymbol{\psi}', \boldsymbol{\tau}')'$ denote the parameter vector defined by the BL model (6.27) with $\boldsymbol{\tau} = (\tau_{ij}, 1 \leq i \leq k, w+1 \leq j \leq r)'$. The DGP is characterized by the true parameter vector $\boldsymbol{\theta}_0 = (\boldsymbol{\phi}_0', \boldsymbol{\psi}_0', \boldsymbol{\tau}_0')'$, ignoring the nuisance parameter $\sigma_{\varepsilon,0}^2$. We assume that $\boldsymbol{\theta} \in \Theta$ where Θ is an open subset of $\mathbb{R}^{p+q+k(r-w)}$. If $\widehat{\boldsymbol{\theta}}$ denotes the estimator of $\boldsymbol{\theta}_0$, where $\widehat{\boldsymbol{\theta}}$ is defined by the estimation procedure described by Algorithm 6.7 and equations (6.30) – (6.33). Then, under some mild regularity conditions and assuming $\{\varepsilon_t\}$ is an $8k$-th order symmetric innovation sequence, it can be proved (Grahn, 1995, Thm. 3.3) that

(i) $\widehat{\boldsymbol{\theta}} \to \boldsymbol{\theta}_0$ a.s.

(ii) $\sqrt{T}(\widehat{\boldsymbol{\theta}} - \boldsymbol{\theta}_0)$ is asymptotically normally distributed with mean zero. Moreover, the law of iterated logarithm holds, i.e. $(\widehat{\boldsymbol{\theta}} - \boldsymbol{\theta}_0) = \mathcal{O}(S_T)$ a.s., with $S_T = \{T/\log\log T\}^{-1/2}$.

6.1 MODEL ESTIMATION

In principle it is possible to derive an analytical expression for the asymptotic covariance matrix of $\widehat{\boldsymbol{\theta}}$ for BL models. However, as the order of the model increases, the algebra becomes rather involved. Hence, bootstrapping is recommended in practice. Below we present a simple example of CLS-based BL model estimation.

Example 6.5: CLS-based Estimation of a BL Model

Consider (6.27) with $p = q = 0$, $k = 2$, $r = 1$, and Gaussian innovations. That is

$$Y_t = \tau Y_{t-2}\varepsilon_{t-1} + \varepsilon_t, \quad \{\varepsilon_t\} \overset{\text{i.i.d.}}{\sim} \mathcal{N}(0, \sigma_\varepsilon^2), \qquad (6.35)$$

where $\tau \equiv \tau_{12}$. It is easy to see that $\{Y_t, t \in \mathbb{Z}\}$ is a stationary, ergodic and causal process if $\sigma_Y^2 = \sigma_\varepsilon^2/(1 - \tau^2\sigma_\varepsilon^2)$ exists, i.e., if $\tau^2\sigma_\varepsilon^2 < 1$. In that case it can be shown that $\{Y_t, t \in \mathbb{Z}\}$ has the unique representation

$$Y_t = \varepsilon_t + \sum_{k=1}^{\infty} \tau^k \varepsilon_{t-2k} \prod_{j=0}^{k-1} \varepsilon_{t-2j-1}, \qquad (6.36)$$

in L_2 sense. Moreover, from Chapter 2 it is easily seen that $\{Y_t, t \in \mathbb{Z}\}$ is invertible if $\tau^2\sigma_\varepsilon^2 < 1/2$. From (6.36) it follows that the necessary and sufficient condition of existence of the $2n$th moment of $\{Y_t, t \in \mathbb{Z}\}$ is $(2n-1)!!\tau^{2n}\sigma_\varepsilon^{2n} < 1$. If $n = 2$ then the condition for strong consistency, i.e. $\mathbb{E}(Y_t^4) < \infty$, becomes $\tau^4\sigma_\varepsilon^4 < 1/3$.

From Algorithm 6.7, step (ii), the CLS estimator of τ follows from minimizing (6.30) with respect to $\boldsymbol{\beta}(s)$ where, with $\widehat{v}_t = Y_t$, we have

$$\mathbb{E}(Y_tY_{t-s}|\varepsilon_{t-2}, \varepsilon_{t-3}, \ldots) = \begin{cases} \sigma_\varepsilon^2 + \tau^2\sigma_\varepsilon^2 Y_{t-2}^2 & \text{if } s = 0, \\ \tau\sigma_\varepsilon^2 Y_{t-2} & \text{if } s = 1, \\ 0 & \text{if } s \geq 2. \end{cases}$$

Thus, in accordance with (6.30), $\boldsymbol{\beta}(0) \equiv (\beta_1(0), \beta_2(0))' = (d_2(0), h_{22}(0))'$, and $\boldsymbol{\beta}(1) \equiv \beta_2(1) = d_2(1)$. This means that for $s = 0$, step (ii) in Algorithm 6.7 becomes

$$\widehat{\boldsymbol{\beta}}(0) = \arg\min_{\boldsymbol{\beta}(0)} \sum_{t=3}^{T} \{Y_t^2 - (\beta_1(0) + \beta_2(0)Y_{t-2}^2)\}^2. \qquad (6.37)$$

Similarly, for $s = 1$, step (ii) consists in estimating

$$\widehat{\beta}_2(1) = \arg\min_{\beta_2(1)} \sum_{t=3}^{T} \{Y_tY_{t-1} - \beta_2(1)Y_{t-2}\}^2. \qquad (6.38)$$

Hence, $\widehat{\boldsymbol{\beta}}(0) = (\widehat{\beta}_1(0), \widehat{\beta}_2(0))'$ estimates $(\sigma_\varepsilon^2, \tau^2\sigma_\varepsilon^2)'$ while $\widehat{\beta}_2(1)$ is an estimator of $\tau\sigma_\varepsilon^2$. Combining these results, the CLS estimator of τ is given by

$$\widehat{\tau} = \frac{\widehat{\beta}_2(1)}{\widehat{\beta}_1(0)} = \frac{\sum_{t=3}^{T} Y_tY_{t-1}Y_{t-2}}{\widehat{\sigma}_\varepsilon^2 \sum_{t=3}^{T} Y_{t-2}^2}. \qquad (6.39)$$

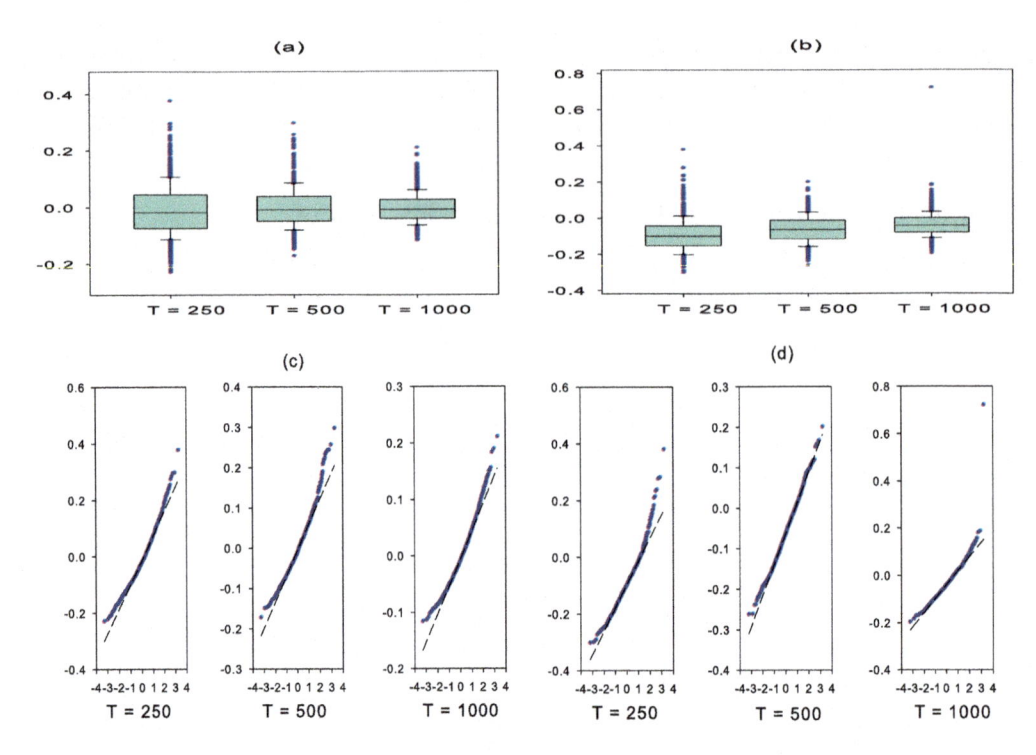

Figure 6.5: *Boxplots and Q-Q plots of $\sqrt{T}(\hat{\tau} - \tau)$ for $\tau = 0.3$ (panels (a) and (c)), and $\tau = 0.5$ (panels (b) and (d)); 1,000 MC replications.*

Clearly, we use three estimators $(\hat{\beta}_1(0), \hat{\beta}_2(0),$ and $\hat{\beta}_2(1))$ to estimate two unknown parameters (τ and σ_ε^2). Moreover, we neglect information contained in the product $\tau^2 \sigma_\varepsilon^2$. Instead of coding this term as $\beta_{11}\beta_2^2$, it is only included as the additional parameter $\beta_2(0)$ in (6.37). These somewhat unfavorable features of Algorithm 6.7 can be amended by trying to minimize the conditional sum of squares

$$\left\{ \sum_{t=3}^{T} \{Y_t^2 - (\theta_2 + \theta_1^2 \theta_2 Y_{t-2}^2)\}^2 + \sum_{t=3}^{T} \{Y_t Y_{t-1} - \theta_1 \theta_2 Y_{t-2}\}^2 \right\}$$

with respect to $\boldsymbol{\theta} = (\theta_1, \theta_2)'$. Obviously, such a refinement overcomes the disadvantages mentioned above – but the price we have to pay is solving a nonlinear minimization problem which needs more effort. Hence, in practical situations, Algorithm 6.7 may be adopted to obtain an estimate of $\boldsymbol{\theta}$, which may serve as a starting guess for a nonlinear optimization algorithm.

To assess the performance of the CLS estimator, we perform a small simulation experiment with the BL model (6.35). The DGP has parameters $\tau = 0.3, 0.5$, and $\sigma_\varepsilon^2 = 1$. Figure 6.5 shows boxplots and Q-Q plots of $\sqrt{T}(\hat{\tau} - \tau)$ for sample sizes $T = 250, 500,$ and $1,000$. Figure 6.6 shows boxplots of $\sqrt{T}(\hat{\sigma}_\varepsilon^2 - \sigma_\varepsilon^2)$ for 1,000 MC replications.

6.1 MODEL ESTIMATION

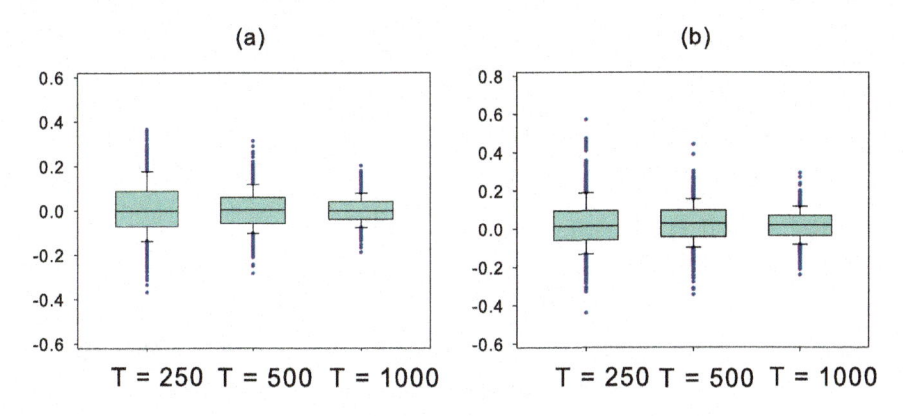

Figure 6.6: *Boxplots of $\sqrt{T}(\widehat{\sigma}_\varepsilon^2 - \sigma_\varepsilon^2)$ for (a) $\tau = 0.3$, and (b) $\tau = 0.5$; 1,000 MC replications.*

Clearly, for increasing values of $|\tau|$ the nonlinearity of the generated time series becomes more prominent, and as a consequence CLS estimation becomes more difficult. Still, for all values of T, the boxplots in Figure 6.5 look almost symmetric and most of them can be interpreted as being sampled from a Gaussian distribution. The Q-Q plots confirm this observation. However, all distributions tend to have negative medians as well as negative means. This tendency reduces with increasing values of T and is due to the interaction between values of $\widehat{\tau}$ and values of $\widehat{\sigma}_\varepsilon^2$. From Figure 6.6 we see that $\widehat{\sigma}_\varepsilon^2$ overestimates the parameter σ_ε^2, and this phenomenon is more present as τ increases from 0.3 to 0.5. According to its definition $\widehat{\beta}_1(0)$ is a positive quantity, but $\widehat{\beta}_2(1)$ can be either positive or negative. If $\widehat{\beta}_2(1) > 0$, overestimating σ_ε^2 will imply that $\widehat{\tau} < \tau$. On the other hand, if $\widehat{\beta}_2(1) \leq 0$, $\widehat{\tau} \leq 0$. Hence, in both cases, overestimating σ_ε^2 results in underestimation of the parameter τ.

6.1.3 Iteratively weighted least squares

Mak (1993) considers an efficient and easy-to-use procedure for *iteratively weighted least squares* (IWLS) estimation of general nonlinear models. Below we first summarize the theory. Next, following Mak et al. (1997), we consider an IWLS algorithm for QML estimation of DTARCH models.

General formulation
Let $\boldsymbol{\theta}$ be an m-dimensional parameter vector of interest. Assume that the actual value $\boldsymbol{\theta}_0$ generating $\mathbf{y}$, an $T \times 1$ random vector of observations with corresponding density function $f(\mathbf{y}; \boldsymbol{\theta})$, belongs to an open parameter space $\Theta \subseteq \mathbb{R}^m$. The ML estimate $\widehat{\boldsymbol{\theta}}$ of $\boldsymbol{\theta}_0$ follows from solving

$$G(\mathbf{y}, \boldsymbol{\theta}) \equiv \partial \log f(\mathbf{y}; \boldsymbol{\theta}) / \partial \boldsymbol{\theta} = 0.$$

For any $\boldsymbol{\theta}, \widetilde{\boldsymbol{\theta}} \in \Theta$, let $g(\widetilde{\boldsymbol{\theta}}, \boldsymbol{\theta}) = \mathbb{E}\{f(\mathbf{y}; \boldsymbol{\theta}) | \widetilde{\boldsymbol{\theta}}\}$. Then:

(i) Fisher's information matrix is given by $\partial g(\widetilde{\boldsymbol{\theta}}, \boldsymbol{\theta})/\partial \widetilde{\boldsymbol{\theta}}|_{\widetilde{\boldsymbol{\theta}}=\boldsymbol{\theta}}$.

(ii) If $\boldsymbol{\theta}^{(0)}$ is a given starting value, and define in the $(u+1)$th iteration $\boldsymbol{\theta}^{(u+1)}$ $(u \geq 0)$ as a root of the equation, $g(\widetilde{\boldsymbol{\theta}}, \boldsymbol{\theta}^{(u)}) = G(\mathbf{y}, \boldsymbol{\theta}^{(u)})$, then $\boldsymbol{\theta}^{(u)} \to \widehat{\boldsymbol{\theta}}$ as $u \to \infty$. Furthermore, it can be shown that $|\boldsymbol{\theta}^{(u)} - \widehat{\boldsymbol{\theta}}| = \mathcal{O}_p(T^{-u/2})$.

Thus, (ii) implies that if the equation

$$g(\widetilde{\boldsymbol{\theta}}, \boldsymbol{\theta}) = G(\mathbf{y}, \boldsymbol{\theta}) \tag{6.40}$$

can be solved explicitly for $\widetilde{\boldsymbol{\theta}}$, the algorithm in (ii) provides sufficient numerical accuracy in a few iterations. When (6.40) does not have an explicit solution, it is recommended to use the following linearization

$$G(\mathbf{y}, \boldsymbol{\theta}) \simeq g(\boldsymbol{\theta}, \boldsymbol{\theta}) + \Big(\frac{\partial g(\widetilde{\boldsymbol{\theta}}, \boldsymbol{\theta})}{\partial \widetilde{\boldsymbol{\theta}}}\Big|_{\widetilde{\boldsymbol{\theta}}=\boldsymbol{\theta}}\Big)'(\widetilde{\boldsymbol{\theta}} - \boldsymbol{\theta}) = \Big(\frac{\partial g(\widetilde{\boldsymbol{\theta}}, \boldsymbol{\theta})}{\partial \widetilde{\boldsymbol{\theta}}}\Big|_{\widetilde{\boldsymbol{\theta}}=\boldsymbol{\theta}}\Big)'(\widetilde{\boldsymbol{\theta}} - \boldsymbol{\theta}).$$

Hence,

$$\widetilde{\boldsymbol{\theta}} \approx \boldsymbol{\theta} + \Big\{\Big(\frac{\partial g(\widetilde{\boldsymbol{\theta}}, \boldsymbol{\theta})}{\partial \widetilde{\boldsymbol{\theta}}}\Big|_{\widetilde{\boldsymbol{\theta}}=\boldsymbol{\theta}}\Big)'\Big\}^{-1} G(\mathbf{y}, \boldsymbol{\theta}), \tag{6.41}$$

and at the $(u+1)$th step

$$\widehat{\boldsymbol{\theta}}^{(u+1)} = \widehat{\boldsymbol{\theta}}^{(u)} + \Big\{\Big(\frac{\partial g(\widetilde{\boldsymbol{\theta}}, \boldsymbol{\theta})}{\partial \widetilde{\boldsymbol{\theta}}}\Big|_{\widetilde{\boldsymbol{\theta}}=\widehat{\boldsymbol{\theta}}^{(u)}}\Big)'\Big\}^{-1} G(\mathbf{y}, \widehat{\boldsymbol{\theta}}^{(u)}).$$

In other words, the ML estimate of $\boldsymbol{\theta}_0$ is constructed via an IWLS algorithm.

IWLS for QML of DTARCH models

In Appendix 2.B, we briefly characterized the general class of (k_1, k_2)-regime double self-exciting threshold ARMA conditional heteroskedastic (DTARMACH) model. The specification consists of a k_1-regime SETARMA conditional mean process combined with a k_2-regime TGARCH conditional variance. Here, we consider IWLS estimation of a special case, i.e. the two-regime DTARCH model also called SETAR$(2; p_1, p_2)$–ARCH$(2; q_1, q_2)$ model, which is given by

$$Y_t = \begin{cases} \phi_0^{(1)} + \sum_{i=1}^{p_1} \phi_i^{(1)} Y_{t-i} + \varepsilon_t & \text{if } Y_{t-d} \leq r, \\ \phi_0^{(2)} + \sum_{i=1}^{p_2} \phi_i^{(2)} Y_{t-i} + \varepsilon_t & \text{if } Y_{t-d} > r, \end{cases} \tag{6.42}$$

$$\sigma_t^2 = \begin{cases} \alpha_0^{(1)} + \sum_{i=1}^{q_1} \alpha_i^{(1)} \varepsilon_{t-i}^2 & \text{if } Y_{t-d} \leq r, \\ \alpha_0^{(2)} + \sum_{i=1}^{q_2} \alpha_i^{(2)} \varepsilon_{t-i}^2 & \text{if } Y_{t-d} > r, \end{cases} \tag{6.43}$$

6.1 MODEL ESTIMATION

where $\{\varepsilon_t|\mathcal{F}_{t-1}\} \stackrel{\text{i.i.d.}}{\sim} \mathcal{N}(0, \sigma_t^2)$ with $\mathcal{F}_{t-1} = \{Y_{t-1}, Y_{t-2}, \ldots\}$ the available information set at time $t-1$. The conditional mean and conditional variance of $\{Y_t, t \in \mathbb{Z}\}$ are given by

$$\mu_t = \sum_{i=1}^{2} \left(\phi_0^{(i)} + \sum_{j=1}^{p_i} \phi_j^{(i)} Y_{t-j} \right) I_t^{(i)}, \quad \sigma_t^2 = \sum_{i=1}^{2} \left(\alpha_0^{(i)} + \sum_{j=1}^{q_i} \alpha_j^{(i)} \varepsilon_{t-j}^2 \right) I_t^{(i)},$$

where $I_t^{(1)} = I(Y_{t-d} \leq r)$ and $I_t^{(2)} = I(Y_{t-d} > r)$, and $\boldsymbol{\theta} = (\boldsymbol{\phi}_1', \boldsymbol{\alpha}_1', \boldsymbol{\phi}_2', \boldsymbol{\alpha}_2', r)'$ with $\boldsymbol{\phi}_i = (\phi_0^{(i)}, \ldots, \phi_{p_i}^{(i)})'$, and $\boldsymbol{\alpha}_i = (\alpha_0^{(i)}, \ldots, \alpha_{q_i}^{(i)})'$ $(i = 1, 2)$.

Assume d is known. Let $p = \max(p_1, p_2, q_1, q_2)$. Then, given the initial values $\mathbf{Y}_0 = (Y_0, \ldots, Y_{1-p})'$ and the set of observations $\{Y_t\}_{t=1}^T$, the conditional log-QML function (omitting a constant), under conditional normality is

$$Q_T(\boldsymbol{\theta}) = -\frac{1}{2} \sum_{t=1}^{T} \sum_{i=1}^{2} \left(\log \sigma_t^2 + \frac{\varepsilon_t^2}{\sigma_t^2} \right) I_t^{(i)},$$

where $\varepsilon_t = Y_t - \mu_t(\boldsymbol{\theta})$. For fixed r, differentiating $Q_T(\boldsymbol{\theta})$ with respect to $\boldsymbol{\theta}$ gives (cf. Exercise 6.3) expressions for $G(\mathbf{y}, \boldsymbol{\theta})$ and $\partial g(\boldsymbol{\theta}, \boldsymbol{\theta})/\partial \boldsymbol{\theta}|_{\widetilde{\boldsymbol{\theta}}=\boldsymbol{\theta}}$. Substituting these expressions in (6.41), it can be shown (Li and Li, 1996) that

$$\sum_{t=1}^{T} \mathbf{Z}_t' \mathbf{W}_t \mathbf{X}_t \widetilde{\boldsymbol{\theta}}(r) = \sum_{t=1}^{T} \mathbf{Z}_t' \mathbf{W}_t \mathbf{Z}_t \boldsymbol{\theta}(r) + \sum_{t=1}^{T} \mathbf{Z}_t \mathbf{W}_t \mathbf{X}_t, \quad (6.44)$$

where

$$\mathbf{Z}_t = \begin{pmatrix} \partial \sigma_t^2/\partial \boldsymbol{\theta} \\ \partial \mu_t/\partial \boldsymbol{\theta} \end{pmatrix}, \quad \mathbf{W}_t = \begin{pmatrix} 1/2\sigma_t^4 & 0 \\ 0 & 1/\sigma_t^2 \end{pmatrix}, \quad \mathbf{X}_t = \begin{pmatrix} (Y_t - \mu_t)^2 - \sigma_t^2 \\ Y_t - \mu_t \end{pmatrix}.$$

Next, stacking up by t and denoting the corresponding matrices by $\mathbf{Z}$, $\mathbf{W}$, and $\mathbf{X}$ respectively, the (conditional) IWLS equation is given by

$$\widetilde{\boldsymbol{\theta}}(r) = \boldsymbol{\theta}(r) + (\mathbf{Z}'\mathbf{W}\mathbf{Z})^{-1}(\mathbf{Z}'\mathbf{W}\mathbf{X}), \quad (6.45)$$

where an explicit expression for $\mathbf{Z}$ follows from direct differentiation.

Example 6.6: Daily Hong Kong Hang Seng Index

The well-known (G)ARCH model has the ability to capture stylized facts of financial and economic time series, such as excess kurtosis and volatility clustering where large positive and negative returns follow each other. SETARMA models, on the other hand, can accommodate structural changes or regime shifts, but they cannot generate volatility pooling or leverage effects. A combination of both models, as in the sub-class of DT(G)ARCH models, can incorporate the important facets of both.

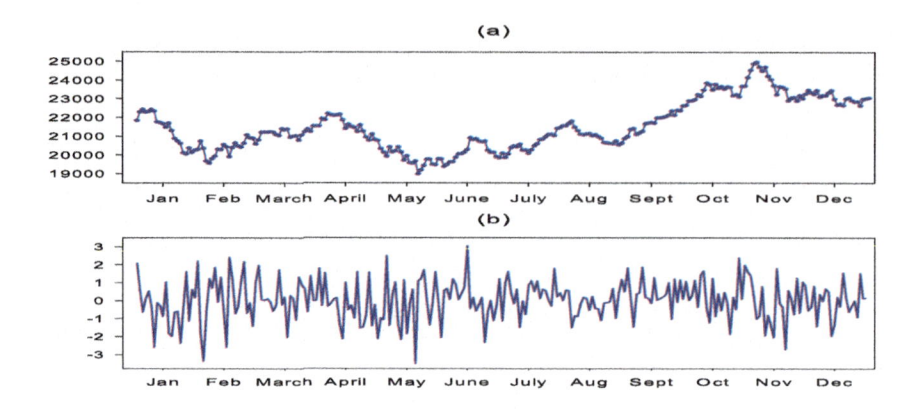

Figure 6.7: *Time plots of (a) the daily closing prices, and (b) the log-returns for the* Hong Kong Hang Seng Index (HSI) *for the year* 2010.

To illustrate the application of DTARCH models in financial time series analysis, we consider the Hong Kong Hang Seng Index (HSI) for the year 2010. Let $\{P_t\}_{t=1}^{253}$ be the daily closing prices at time t. The log-return Y_t is defined as $\{Y_t = 100(\log P_t - \log P_{t-1})\}_{t=1}^{252}$. Figure 6.7 shows time plots of $\{P_t\}$ and $\{Y_t\}$, respectively. The LR–SETAR test statistic suggests that $\{Y_t\}$ contains SETAR nonlinearity, and the McLeod–Li test statistic indicates that there are ARCH effects in the residuals.

We use the IWLS algorithm, combined with the GA-subset threshold model selection procedure to fit DTARCH models to the data. For the GA parameters and the model parameters, we use the same specification as reported in Example 6.3. Based on minimizing the NAIC, we obtain the following SETAR(3; 1, 5, 6)–TARCH(3; 1, 1, 3) model

$$Y_t = \begin{cases} 0.13 + 0.07 Y_{t-1} + \varepsilon_t^{(1)} & \text{if } Y_{t-1} \leq 0.16, \\ -0.47 + 0.69 Y_{t-1} + 0.02 Y_{t-2} + 0.19 Y_{t-3} \\ \quad -0.37 Y_{t-4} + 0.41 Y_{t-5} + \varepsilon_t^{(2)} & \text{if } 0.16 < Y_{t-1} \leq 1.03, \quad (6.46) \\ 0.61 - 0.39 Y_{t-1} + 0.10 Y_{t-2} + 0.08 Y_{t-3} \\ \quad +0.09 Y_{t-4} - 0.16 Y_{t-5} + 0.23 Y_{t-6} + \varepsilon_t^{(3)} & \text{if } Y_{t-1} > 1.03, \end{cases}$$

with

$$\sigma_t^2 = \begin{cases} 1.29 + 0.02 \varepsilon_{t-1}^2 & \text{if } Y_{t-1} \leq 0.16, \\ 0.91 + 0.73 \varepsilon_{t-1}^2 & \text{if } 0.16 < Y_{t-1} \leq 1.03, \quad (6.47) \\ 0.24 + 0.02 \varepsilon_{t-1}^2 + 0.07 \varepsilon_{t-2}^2 + 0.13 \varepsilon_{t-3}^2 & \text{if } Y_{t-1} > 1.03, \end{cases}$$

where $\varepsilon_t^{(i)} = \sigma_t^2 \varepsilon_t$ $(i = 1, 2, 3)$ and $\{\varepsilon_t\} \stackrel{\text{i.i.d.}}{\sim} \mathcal{N}(0, 1)$. The sample variances of $\{\varepsilon_t^{(i)}\}$ are 1.31 ($T = 138$), 1.18 ($T = 58$), and 57 ($T = 49$), respectively. The sample variances of the volatility equation are 3.41, 1.87, and 76.3, respectively.

The most important feature is clearly the difference in the behavior of the series in each regime. When Y_{t-1} is between 0.16 and 1.03 the behavior is slower in adjusting to shocks than in the third regime. In the first regime the series $\{P_t\}$ closely approximates a random walk process with a drift term. The behavior of the conditional variance also varies considerably between regimes; shocks to the conditional variance are more persistent in the second and third regime, and weakly persistent in the first regime. Observe, all estimated coefficients in σ_t^2 are nonnegative. Negative coefficients are counter-intuitive in (6.43) which implies that the IWLS algorithm needs to be constrained.

6.2 Model Selection Tools

6.2.1 Kullback–Leibler information

Let $f(\mathbf{y}; \boldsymbol{\theta}_{0,m})$ denote the true pdf of the observed observations $\{Y_t\}_{t=1}^T$, where $\theta_{0,m} \in \Theta \subset \mathbb{R}^m$ is an m-dimensional parameter vector, Θ denotes the parameter space, and with $\mathbf{y} = (Y_1, \ldots, Y_T)'$. Furthermore, assume that some generic (or candidate) model $\mathcal{M}_m$ gives a density function $f_m(\cdot; \boldsymbol{\theta}_m)$ to the observations, where $\boldsymbol{\theta}_m$ is a p_m-dimensional parameter. Recall from Section 1.3.3 that the "discrepancy" between $f(\cdot; \boldsymbol{\theta}_{0,m})$ and $f_m(\cdot; \boldsymbol{\theta}_m)$ can be measured by the *Kullback–Leibler* (KL) *divergence*, defined by

$$\begin{aligned} I^{\text{KL}}(\boldsymbol{\theta}_{0,m}, \boldsymbol{\theta}_m) &= \mathbb{E}_0\{\log f(\mathbf{y}; \boldsymbol{\theta}_{0,m})\} - \mathbb{E}_0\{\log f_m(\mathbf{y}; \boldsymbol{\theta}_m)\} \\ &= \mathbb{E}_0\{\log f(\mathbf{y}; \boldsymbol{\theta}_{0,m})\} + \frac{1}{2}\{-2\mathbb{E}_0\{\log f_m(\mathbf{y}; \boldsymbol{\theta}_m)\}\}, \end{aligned} \quad (6.48)$$

where $\mathbb{E}_0(\cdot)$ denotes the expectation with respect to $\mathbf{y}$ evaluated by the true density. Hereby it is assumed that $\mathbb{E}_0\{\log f_m(\cdot; \boldsymbol{\theta}_m)\}$ exists $\forall \boldsymbol{\theta}_m \in \Theta$.

The main property of the KL divergence is that $I^{\text{KL}}(\cdot) \geq 0$ with equality when $f(\cdot; \boldsymbol{\theta}_{0,m}) = f_m(\cdot; \boldsymbol{\theta}_m)$ a.e. As we have seen in Exercise 1.4, this property can be obtained from Jensen's inequality: if x is a non-degenerate random variable and $h(x)$ is a strictly convex function, then $\mathbb{E}\{h(x)\} > h\{\mathbb{E}(x)\}$, while an equality holds when x is degenerate at $\mathbb{E}(x)$. As $-\log(x)$ is a strictly convex function of x, we find

$$\mathbb{E}_0\Big\{-\log\Big(\frac{f_m(\mathbf{y}; \boldsymbol{\theta}_m)}{f(\mathbf{y}; \boldsymbol{\theta}_{0,m})}\Big)\Big\} \geq -\log \mathbb{E}_0\Big\{\Big(\frac{f_m(\mathbf{y}; \boldsymbol{\theta}_m)}{f(\mathbf{y}; \boldsymbol{\theta}_{0,m})}\Big)\Big\}. \quad (6.49)$$

The expectation on the right-hand side employs the density function $f(\cdot; \boldsymbol{\theta}_{0,m})$, so that the right-hand side of (6.49) equals $-\log 1 = 0$, and (6.48) is equivalent to

$$I^{\text{KL}}(\boldsymbol{\theta}_{0,m}, \boldsymbol{\theta}_m) \geq 0, \quad \forall \boldsymbol{\theta}_m \in \Theta. \quad (6.50)$$

The equality in (6.49) and (6.50) arises if and only if $f_m(\cdot; \boldsymbol{\theta}_m)/f(\cdot; \boldsymbol{\theta}_{0,m})$ is degenerate at $\mathbb{E}_0\{f_m(\cdot; \boldsymbol{\theta}_m)/f(\cdot; \boldsymbol{\theta}_{0,m})\}$ $(= 1)$, in other words if and only if $f_m(\cdot; \boldsymbol{\theta}_m) = f(\cdot; \boldsymbol{\theta}_{0,m})$ a.e. In particular, the equality in (6.49) and (6.50) holds when $\boldsymbol{\theta}_m = \boldsymbol{\theta}_0$.

The application of Jensen's inequality clarifies that $I^{\text{KL}}(\cdot)$ is determined by the dispersion of $f_m(\cdot; \boldsymbol{\theta}_m)/f(\cdot; \boldsymbol{\theta}_{0,m})$, and this explains why $I^{\text{KL}}(\cdot)$ can serve as a measure of the divergence between the density function $f_m(\cdot; \boldsymbol{\theta}_m)$ and the true density function $f(\cdot;\boldsymbol{\theta}_{0,m})$. Sometimes (6.49) is referred to as a measure of the distance between $f(\cdot;\boldsymbol{\theta}_{0,m})$ and $f_m(\cdot;\boldsymbol{\theta}_m)$, but we remark that $I^{\text{KL}}(\cdot)$ is not a metric on the space of probability densities, because $I^{\text{KL}}(\boldsymbol{\theta}_{0,m},\boldsymbol{\theta}_m) \neq I^{\text{KL}}(\boldsymbol{\theta}_m,\boldsymbol{\theta}_{0,m})$ and $I^{\text{KL}}(\cdot)$ does not satisfy the triangle inequality. Nevertheless, the choice of $I^{\text{KL}}(\cdot)$ as the loss function is firmly supported by a most relevant information-theoretic interpretation, namely $I^{\text{KL}}(\cdot)$ can be interpreted as the surprise experienced on average when we believe that $f_m(\cdot;\boldsymbol{\theta}_m)$ describes a given phenomenon and we are then informed that in fact the phenomenon is described by $f(\cdot;\boldsymbol{\theta}_{0,m})$ (Rényi, 1961).

6.2.2 The AIC, AIC$_c$, and AIC$_u$ rules

AIC rule

Given (6.48) as the loss function, the objective is narrowed down to minimizing $I^{\text{KL}}(\cdot)$ or, equivalently, minimizing $-2\mathbb{E}_0\{\log f_m(\mathbf{y};\boldsymbol{\theta}_m)\}$ subject to $\boldsymbol{\theta}_m \in \Theta$. When the density functions $f_m(\cdot;\boldsymbol{\theta}_m)$ and $f(\cdot;\boldsymbol{\theta}_{0,m})$ are equal (for almost all $\mathbf{y}$) only for a unique vector in Θ (necessarily $\boldsymbol{\theta}_m = \boldsymbol{\theta}_{0,m}$). Then, under perfect knowledge, such optimization would yield $\boldsymbol{\theta}_{0,m}$. In practice, however, either objective function is unknown, because $\mathbb{E}_0(\cdot)$ is evaluated by the unknown density function $f(\cdot;\boldsymbol{\theta}_{0,m})$. To overcome this hurdle, we introduce a *fictitious* vector of observations $\mathbf{x} = (X_1,\ldots,X_T)'$ with the same pdf as $\mathbf{y}$, but which is independent of $\mathbf{y}$. Let $\widehat{\boldsymbol{\theta}}_{T,m}$ denote a QML estimator of $\boldsymbol{\theta}_{0,m}$ based on $\mathbf{y}$. So, instead of $-2\mathbb{E}_0\{\log f_m(\mathbf{y};\boldsymbol{\theta}_m)\}$ itself, we want to minimize the function

$$I(m) = -2\mathbb{E}_y \mathbb{E}_x\{\log f_m(\mathbf{x};\widehat{\boldsymbol{\theta}}_{T,m})\}, \tag{6.51}$$

where $\mathbb{E}_y$ refers to the dependence of $\widehat{\boldsymbol{\theta}}_{T,m}$ on the data vector $\mathbf{y}$. Note that (6.51) has an interesting cross-validatory interpretation: the sample $\mathbf{y}$ is used for estimation and the independent sample $\mathbf{x}$ for validation of the so-obtained model's pdf.

Now, to derive a model selection criterion we decompose $I(m)$ as follows

$$I(m) = -2\mathbb{E}_y\{\log f_m(\mathbf{y};\widehat{\boldsymbol{\theta}}_{T,m})\} \underbrace{-2\mathbb{E}_y\{\log f_m(\mathbf{y};\boldsymbol{\theta}_{0,m})\} + 2\mathbb{E}_y\{\log f_m(\mathbf{y};\widehat{\boldsymbol{\theta}}_{T,m})\}}_{\text{A1}}$$

$$\underbrace{-2\mathbb{E}_y\mathbb{E}_x\{\log f_m(\mathbf{x};\widehat{\boldsymbol{\theta}}_{T,m})\} + 2\mathbb{E}_y\{\log f_m(\mathbf{y};\boldsymbol{\theta}_{0,m})\}}_{\text{A2}}. \tag{6.52}$$

The term A1 on the right-hand side of (6.52) measures the *average overfitting* of the QML estimator, since $\log f_m(\mathbf{y};\widehat{\boldsymbol{\theta}}_{T,m}) \geq \log f_m(\mathbf{y};\boldsymbol{\theta}_{0,m})$. The term A2 can be interpreted as an *average cost* for using $\widehat{\boldsymbol{\theta}}_{T,m}$ in lieu of the true parameter vector $\boldsymbol{\theta}_{0,m}$, when the model is fitted to an independent replication of the DGP.

Consider the term A1 in (6.52). Under assumptions similar to those made in Section 6.1.1, and in particular the uniqueness of the parameter $\boldsymbol{\theta}_{0,m}$, we can expand $2\mathbb{E}_y\{\log f_m(\mathbf{y};\widehat{\boldsymbol{\theta}}_{T,m})\}$ in a second-order Taylor expansion around $\boldsymbol{\theta}_{0,m}$. The estimator $\widehat{\boldsymbol{\theta}}_{T,m}$ converges to $\boldsymbol{\theta}_{0,m}$ a.s. Moreover, analogous to (6.3), we have

$$\sqrt{T}(\widehat{\boldsymbol{\theta}}_{T,m} - \boldsymbol{\theta}_{0,m}) \xrightarrow{D} \mathcal{N}\left(\mathbf{0}, \mathcal{H}_m^{-1}(\mathbf{y})\mathcal{I}_m(\mathbf{y})\mathcal{H}_m^{-1}(\mathbf{y})\right), \tag{6.53}$$

6.2 MODEL SELECTION TOOLS

where

$$\mathcal{H}_m(\mathbf{y}) \stackrel{a.s.}{=} \lim_{T\to\infty} \frac{1}{T} \frac{\partial^2 \log f_m(\mathbf{y};\boldsymbol{\theta}_{0,m})}{\partial \boldsymbol{\theta} \partial \boldsymbol{\theta}'}, \quad \mathcal{I}_m(\mathbf{y}) = \lim_{T\to\infty} \frac{1}{T} \text{Var}\left(\frac{\partial \log f_m(\mathbf{y};\boldsymbol{\theta}_{0,m})}{\partial \boldsymbol{\theta}}\right).$$

Hence, the third term on the right-hand side of (6.52) becomes

$$\mathbb{E}_y\left\{\sqrt{T}(\widehat{\boldsymbol{\theta}}_{T,m} - \boldsymbol{\theta}_{0,m})' \frac{1}{T}\left(\frac{\partial^2 \log f_m(\mathbf{y};\boldsymbol{\theta})}{\partial \boldsymbol{\theta} \partial \boldsymbol{\theta}'}\right)\Big|_{\boldsymbol{\theta}=\boldsymbol{\theta}_{0,m}} \sqrt{T}(\widehat{\boldsymbol{\theta}}_{T,m} - \boldsymbol{\theta}_{0,m})\right\} =$$

$$\text{tr}\left(\mathcal{H}_m(\mathbf{y})\mathbb{E}_y\{T(\widehat{\boldsymbol{\theta}}_{T,m} - \boldsymbol{\theta}_{0,m})(\widehat{\boldsymbol{\theta}}_{T,m} - \boldsymbol{\theta}_{0,m})'\}\right) = \text{tr}\left(\mathcal{I}_m(\mathbf{y})\mathcal{H}_m^{-1}(\mathbf{y})\right) + o_p(1). \quad (6.54)$$

Substituting (6.54) into (6.52), we get

$$2\mathbb{E}_y\{\log f_m(\mathbf{y};\widehat{\boldsymbol{\theta}}_{T,m})\} = 2\mathbb{E}_y\{\log f_m(\mathbf{y};\boldsymbol{\theta}_{0,m})\} + \text{tr}\left(\mathcal{I}_m(\mathbf{y})\mathcal{H}_m^{-1}(\mathbf{y})\right) + o_p(1). \quad (6.55)$$

Recall that $\mathbf{y}$ and $\mathbf{x}$ have the same pdf (which implies that $\mathcal{H}_m(\mathbf{y}) = \mathcal{H}_m(\mathbf{x})$) and that they are independent of each other. Consider the term $2\mathbb{E}_y\mathbb{E}_x\{\log f_m(\mathbf{x};\widehat{\boldsymbol{\theta}}_{T,m})\}$ in (6.52). Assuming that $\mathbb{E}_x(\cdot)$ is sufficiently smooth, and its derivatives under the expectation sign exist, a second-order Taylor expansion of $2\mathbb{E}_x\{\log f_m(\mathbf{x};\widehat{\boldsymbol{\theta}}_{T,m})\}$ around $\boldsymbol{\theta}_{0,m}$ yields

$$2\mathbb{E}_x\{\log f_m(\mathbf{x};\widehat{\boldsymbol{\theta}}_{T,m})\} = 2\mathbb{E}_x\{\log f_m(\mathbf{x};\boldsymbol{\theta}_{0,m})\}$$

$$+ 2(\widehat{\boldsymbol{\theta}}_{T,m} - \boldsymbol{\theta}_{0,m})'\left(\frac{\partial \mathbb{E}_x\{\log f_m(\mathbf{x};\boldsymbol{\theta})\}}{\partial \boldsymbol{\theta}}\right)\Big|_{\boldsymbol{\theta}=\boldsymbol{\theta}_{0,m}}$$

$$+ \sqrt{T}(\widehat{\boldsymbol{\theta}}_{T,m} - \boldsymbol{\theta}_{0,m})' \frac{1}{T}\left(\frac{\partial^2 \mathbb{E}_x\{\log f_g(\mathbf{x};\boldsymbol{\theta})\}}{\partial \boldsymbol{\theta} \partial \boldsymbol{\theta}'}\right)\Big|_{\boldsymbol{\theta}=\boldsymbol{\theta}_{0,m}} \sqrt{T}(\widehat{\boldsymbol{\theta}}_{T,m} - \boldsymbol{\theta}_{0,m}) + o_p(1)$$

$$= 2\mathbb{E}_x\{\log f_m(\mathbf{x};\boldsymbol{\theta}_{0,m})\} + T(\widehat{\boldsymbol{\theta}}_{T,m} - \boldsymbol{\theta}_{0,m})\mathcal{H}_m(\mathbf{y})(\widehat{\boldsymbol{\theta}}_{T,m} - \boldsymbol{\theta}_{0,m})' + o_p(1). \quad (6.56)$$

We deduce from (6.56) that

$$2\mathbb{E}_y\{\mathbb{E}_x\{\log f_m(\mathbf{x};\boldsymbol{\theta}_{T,m})\}\} = 2\mathbb{E}_x\{\log f_m(\mathbf{x};\boldsymbol{\theta}_{0,m})\} + \text{tr}\left(\mathcal{I}_m(\mathbf{y})\mathcal{H}_m^{-1}(\mathbf{y})\right). \quad (6.57)$$

Inserting (6.55) and (6.57) in (6.52), yields

$$I(m) = -2\mathbb{E}_y\{\log f_m(\mathbf{y};\widehat{\boldsymbol{\theta}}_{T,m})\} + 2\text{tr}\left(\mathcal{I}_m(\mathbf{y})\mathcal{H}_m^{-1}(\mathbf{y})\right) + o_p(1), \quad (6.58)$$

which completes the asymptotic approximation of (6.52).

It can be shown (Findley, 1993) that, under some regularity conditions, the trace term in (6.58) can be approximated by p_m, i.e. the dimension of $\boldsymbol{\theta}_m$. Hence, minimizing (6.51) is equivalent to

$$\min_{\boldsymbol{\theta}_m \in \Theta} \{\text{AIC}(m) = -2\log f_m(\mathbf{y};\widehat{\boldsymbol{\theta}}_{T,m}) + 2p_m\}, \quad (6.59)$$

where the acronym AIC stands for *Akaike information criterion*. Clearly, this model selection criterion establishes a certain balance between the model-size p_m and the lack-of-fit measured by $-2\log f_m(\mathbf{y};\widehat{\boldsymbol{\theta}}_{T,m})$. In other words, it is beneficial to simplify the model, by leaving out the less important aspects, as long as the reduction in model-size outweighs the deterioration of the fit.

The performance of the AIC rule can be judged in different ways. One reasonable scenario is to assume that the approximating parametric family of models $\mathcal{M}_m$ includes the DGP. This is a strong assumption, but it is also used in the derivation of AIC. Then it can be shown (see, e.g., McQuarrie and Tsai, 1998) that, under quite general conditions, the AIC rule is inconsistent and the asymptotic probability of overfitting is not insignificant, as $T \to \infty$. A more practical scenario is to assume that the DGP is more complex than any of the candidate models. In such a case the selected model can be viewed as an *approximation* of the DGP, and we can consider, for instance, the model's average prediction error as a performance measure of the AIC rule.

AIC$_c$ rule

Hurwich and Tsai (1989) obtain an approximation of (6.58) for univariate linear regression and AR time series models that reduces the small sample bias of the AIC rule. This so-called *corrected* AIC (AIC$_c$) is given by

$$\mathrm{AIC}_c(m) = -2\log f_m(\mathbf{y}; \widehat{\boldsymbol{\theta}}_{T,m}) + \frac{2T p_m}{T - p_m - 1}. \qquad (6.60)$$

Due to the second term in (6.60), AIC$_c$ has a smaller risk of overfitting than AIC for finite values of T. With this fact in mind, and being pragmatic rather than theoretical, AIC$_c$ can be used as an order selection criterion for more general linear and nonlinear time series models.

AIC$_u$ rule

McQuarrie et al. (1997) introduce an alternative criterion for linear regression time series models which is an approximate *unbiased* (u) estimate of the KL information $I(m)$ defined in (6.51). This criterion, denoted by AIC$_u$, is given by

$$\mathrm{AIC}_u(m) = -2\log f_m(\mathbf{y}; \widehat{\boldsymbol{\theta}}_{T,m}) + \frac{2T p_m}{T - p_m - 1} + 2T\log\left\{\frac{T}{T - p_m}\right\}. \qquad (6.61)$$

However, AIC$_u$ is neither a consistent nor an asymptotically efficient criterion. The criterion has a good performance in finite samples, and hence can be adopted for more general models than just linear regressions.

6.2.3 Generalized information criterion: The GIC rule

Note that in (6.51) the validation sample $\mathbf{x}$ has the same length as the estimation sample $\mathbf{y}$. Intuitively, the risk of overfitting will decrease if the length T_x of $\mathbf{x}$ is much larger than T_y, the length of $\mathbf{y}$. Specifically, assume that $T_x = \nu T_y$ with $\nu \geq 1$. Since $\mathcal{H}_m(\mathbf{x}) = \nu \mathcal{H}_m(\mathbf{y})$, it is easily seen that an asymptotic approximation of (6.51) is given by

$$I(m) = -2\mathbb{E}_y\big\{\log f_m(\mathbf{y}; \widehat{\boldsymbol{\theta}}_{T,m}) + (\nu + 1)p_m\big\} + o_p(1). \qquad (6.62)$$

6.2 MODEL SELECTION TOOLS

In practice, the term on the right-hand side of (6.62) can be replaced by an unbiased estimator. The resulting criterion, called *generalized information criterion* (GIC), is given by

$$\text{GIC}(m) = -2\log f_m(\mathbf{y}; \widehat{\boldsymbol{\theta}}_{T,m}) + (\nu + 1)p_m. \tag{6.63}$$

Clearly, when $\nu = 1$, GIC reduces to AIC. Extensive simulation studies (see, e.g., Bhansali and Downham, 1977) have empirically shown that for $\nu \in [2, 5]$ the correct order is found more frequently than AIC. The Bayesian approach of the next section provides an explicit expression for the term $(\nu + 1)$.

6.2.4 Bayesian approach: The BIC rule

From a Bayesian point of view it is natural to choose among models by selecting the one that maximizes the posterior probability $f(\mathcal{M}_m|\mathbf{y})$. Assume that the parameter vector $\boldsymbol{\theta}_m$ is a random variable with a given *a priori* pdf denoted by $f(\boldsymbol{\theta}_m|\mathcal{M}_m)$ which does not depend on T. Now, modifying our previous notation, $f((\mathbf{y}; \boldsymbol{\theta}_m)|\mathcal{M}_m)$ denotes the joint pdf of the random variables $\mathbf{y}$ and $\boldsymbol{\theta}_m$. Furthermore, let $f(\mathbf{y}|\boldsymbol{\theta}_m, \mathcal{M}_m)$ denote the conditional distribution. Using this notation and Bayes' rule, we can write

$$f(\mathcal{M}_m|\mathbf{y}) \propto f(\mathbf{y}|\mathcal{M}_m)f(\mathcal{M}_m),$$

where

$$f(\mathbf{y}|\mathcal{M}_m) = \int f(\mathbf{y}|\boldsymbol{\theta}_m, \mathcal{M}_m)f(\boldsymbol{\theta}_m|\mathcal{M}_m)\mathrm{d}\boldsymbol{\theta}_m,$$

and where the symbol $\propto$ denotes proportionality. Assuming the same prior probability for all models, Schwarz (1978) derives the following large sample approximation

$$\log f(\mathbf{y}|\mathcal{M}_m) \approx \log f_m(\mathbf{y}; \widehat{\boldsymbol{\theta}}_{T,m}) - \frac{p_m}{2}\log T. \tag{6.64}$$

Hence, maximizing (6.64) is equivalent to minimizing the *Bayesian information criterion* (BIC):

$$\text{BIC}(m) = -2\log f_m(\mathbf{y}; \widehat{\boldsymbol{\theta}}_{T,m}) + p_m \log T, \tag{6.65}$$

independently of the chosen prior. It is an interesting fact that the BIC rule can also be derived within the KL framework. Moreover, it can be shown (see, e.g., McQuarrie and Tsai, 1998) that the BIC rule is consistent, that is the probability of correct detection approaches one as $T \to \infty$.

All five order selection criteria AIC, AIC$_c$, AIC$_u$, BIC and GIC have a common form, that is they are members of the family of criteria

$$\min_{\boldsymbol{\theta}_m \in \Theta} \left\{ -2\log f_m(\mathbf{y}; \widehat{\boldsymbol{\theta}}_{T,m}) + p_m C(T, p_m) \right\}, \tag{6.66}$$

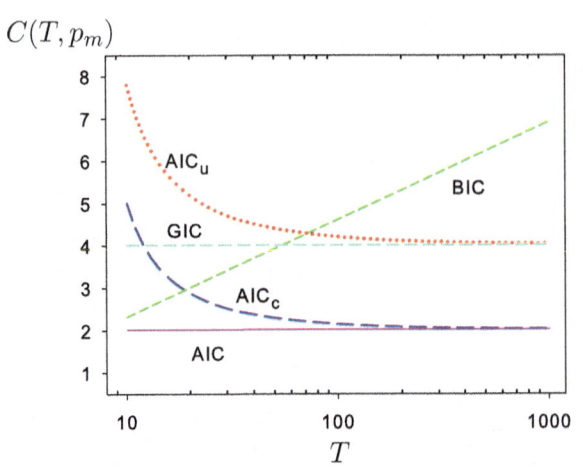

Figure 6.8: *Penalty functions $C(T, p_m)$ of AIC (pink solid line), AIC_c with $p_m = 5$ (blue long dashed line), AIC_u with $p_m = 5$ (red dotted line), BIC (green short dashed line), and GIC with $\nu = 3$ (cyan medium dashed line).*

but with a different *penalty function* $C(T, p_m)$. Figure 6.8 shows the behavior of $C(T, p_m)$ as a function of T for each selection rule.

Given the above model selection criteria, an obvious question is: Which criterion to use in practice? Unfortunately, within the context of nonlinear time series this question has been the subject of only a few papers (cf. Section 6.2.6). Overall, AIC_c outperforms AIC and BIC in small samples. BIC penalizes models which are over-parameterized and so gives some value to parsimony. For this reason one may prefer BIC over other criteria. On the other hand, if parsimony is not considered to be really important, one may use a criterion which picks up any subtle nuance in the data and as a result the fitted nonlinear model will be inclined to overfit in sample. In fact, we recommend that any model should be evaluated in terms of its out-of-sample forecasting ability, and compared with forecasts from linear and other nonlinear time series models.

6.2.5 Minimum descriptive length principle

The *minimum descriptive length* (MDL) principle (Rissanen, 1986) allows comparisons between nested, non-nested and misspecified models without requiring restrictive assumptions. The MDL criterion chooses $\boldsymbol{\theta}_m$ so as to minimize

$$\text{MDL}(m) = -\log f_m(\mathbf{y}; \widehat{\boldsymbol{\theta}}_{T,m}) + \frac{p_m}{2} \log \frac{T}{2\pi} + \log \int \sqrt{|\widehat{\mathcal{I}}(\boldsymbol{\theta}_m)|}\, d\boldsymbol{\theta}_m, \qquad (6.67)$$

where $\widehat{\mathcal{I}}(\cdot)$ denotes an estimate of the expected Fisher information matrix. The second- and third term in (6.67) are often referred to as a *complexity penalty*. When the density function $f(\cdot)$ is known, both the MDL and BIC criteria have reasonable explanations, though the results may not be the same. When, however, $f(\cdot)$ depends

6.2 MODEL SELECTION TOOLS

on a functional form, e.g. a conditional mean function $g(\cdot;\boldsymbol{\theta}_g)$, BIC does not take this extra complexity into account, while in MDL, this extra bit of uncertainty is reflected in $\widehat{\mathcal{I}}(\cdot)$. For parametric models an estimator of $\mathcal{I}(\cdot)$ is given by (6.4). The integration in the last term of (6.67) can be well approximated by MC simulation methods (see, e.g., Robert and Casella, 2004).

6.2.6 Model selection in threshold models

As k-regime SETAR models are piecewise linear, it seems natural to extend the various order selection criteria for linear AR models to this class of models, using knowledge of the asymptotic properties of CLS estimator given in Section 6.1.2. Indeed, within this context a number of relevant rules arise which can help to decide how large the number of AR lags should be. First, we consider four members of the family of *order selection* criteria (SC) defined by

$$\text{SC}(p_1,\ldots,p_k) = \min_{p_1,\ldots,p_k}\left\{\sum_{i=1}^{k}\left\{T_i\log\widehat{\sigma}^2_{T_i}+(p_i+1)C(T_i,p_i+1)\right\}\right\}, \qquad (6.68)$$

where T_i $(i=1,\ldots,k)$ denotes the number of observations in each regime, $\widehat{\sigma}^2_{T_i}$ the corresponding (conditional) residual variance, and with penalty function

$$C(T_i, p_i+1) = \begin{cases} 2 & \text{for AIC,} \\ \frac{1}{p_i+1}\frac{T_i(T_i+p_i+1)}{T_i-(p_i+1)-2} & \text{for AIC}_\text{c}, \\ \frac{1}{p_i+1}\left[\frac{T_i(T_i+p_i+1)}{T_i-(p_i+1)-2}+T_i\log\left\{\frac{T_i}{T_i-(p_i+1)-1}\right\}\right] & \text{for AIC}_\text{u}, \\ \log T_i & \text{for BIC.} \end{cases}$$

The generalization of (6.68) to SETARMA models is obvious.

For simplicity of presentation, we consider a SETAR$(2;p_1,p_2)$ model with unknown threshold r and delay parameter d. In that case the order selection procedure can be entertained within the following framework.

> **Algorithm 6.8: Minimum order selection**
>
> (i) Fix the maximum orders (p_1^*, p_2^*), and the maximum delay $d_{\max}$.
>
> (ii) Assume $r \in [\underline{r}, \overline{r}] \subset \mathbb{R}$ with $\underline{r}$ the $0.25\times 100\%$ percentile and $\overline{r}$ the $0.75\times 100\%$ percentile of $\{Y_t\}_{t=1}^T$.
>
> (iii) Let $\{Y_{(j)}(d)\}_{j=1}^T$ denote the order statistics of $\{Y_t\}_{t=1}^T$ for a fixed $d \in [1, d_{\max}]$. Let $I_r = \{[0.25T], [0.25T]+1,\ldots,[0.75T]\}$. Set $r = Y_{(j)}(d)$.

> **Algorithm 6.8: Minimum order selection (Cont'd)**
>
> (iv) Calculate $\min_{1\leq k_1 \leq p_1^*, 1\leq k_2 \leq p_2^*} \{SC(k_1, k_2)\}$. Let $SC(Y_{(j)}(d))$ be the minimum. Denote the corresponding model orders giving this minimum as $k_i^*(Y_{(j)}(d))$ $(i = 1, 2)$. Note, in the calculation the first $\max(d, p_1^*, p_2^*)$ observations should be discarded to make the comparison meaningful.
>
> (v) Calculate $\min_{j \in I_r} SC(Y_{(j)}(d))$, and denote the value of $Y_{(j)}(d)$ giving this minimum as $Y_{(j)}^*(d)$.
>
> (vi) Calculate $\min_{1 \leq d \leq d_{\max}} SC(Y_{(j)}^*(d))$, and denote the value of d giving this minimum as $\widehat{d}$.
>
> (vii) The selected delay parameter is $\widehat{d}$, the estimate of the threshold parameter is $\widehat{r} = Y_{(j)}^*(\widehat{d})$, the selected orders are $k_i(Y_{(j)}^*(\widehat{d}))$ $(i = 1, 2)$.

The second set of order selection criteria is based on the concept of CV. This comes down to dividing the available data set into two subsets: a *calibration* set for estimating a model, and a *validation* set for evaluating its performance, as we briefly explained in Section 6.2.3. In principle these subsets may contain different number of observations. Within the context of SETAR$(2; p_1, p_2)$ model selection, however, we focus on the so-called *leave-one-out* CV-criterion. In that case the order selection procedure goes as follows.

> **Algorithm 6.9: Leave-one-out CV order selection**
>
> (i) Follow steps (i) – (iii) of Algorithm 6.8.
>
> (ii) Omit one observation from the available data set $\{Y_t\}_{t=1}^T$, and with the remaining data set obtain the CLS estimates of a SETAR model, using Algorithm 6.2. Let $\widehat{r}^{(t)}$ be the corresponding estimate of r, and $\widehat{\phi}_{T-1,i}^{(t)}$ an estimate of $\phi = (\phi_0^{(i)}, \ldots, \phi_{p_i}^{(i)})'$ $(i = 1, 2)$.
>
> (iii) Predict the omitted observation and obtain the predictive residual $\widehat{\varepsilon}_t(\widehat{\phi}_{T-1,i}^{(t)}, \widehat{r}^{(t)})$.
>
> (iv) Repeat steps (ii) – (iii) for all remaining observations.
>
> (v) The final model is the one which minimizes the MSFE over all SETAR models:
>
> $$\min_{p_1, p_2} \Big\{ C(p_1, p_2) = \sum_{t=s}^{T} \sum_{i=1}^{2} \widehat{\varepsilon}_t^2(\widehat{\phi}_{T-1,i}^{(t)}, \widehat{r}^{(t)}) \Big\}, \qquad (6.69)$$
>
> where $s = \max(d, p_1^*, p_2^*) + 1$.

Under fairly weak conditions it can be proved (Stoica et al., 1986) that for

6.2 MODEL SELECTION TOOLS

linear time series regressions $T\log\{T^{-1}\mathrm{C}(\cdot)\} = \mathrm{AIC}(\cdot) + \mathcal{O}(T^{-1/2})$. Using this relationship, De Gooijer (2001) proposes the following CV model selection criteria for SETAR$(k; p, \ldots, p)$ models

$$\mathrm{C_c} = T\log\Big(\sum_{t=s}^{T}\sum_{i=1}^{k}\widehat{\varepsilon}_t^2(\widehat{\phi}_{T-1,i}^{(t)}, \widehat{r}^{(t)})\Big) + \sum_{i=1}^{k}\frac{T_i(T_i + p_i + 1)}{T_i - (p_i + 1) - 2}, \qquad (6.70)$$

$$\mathrm{C_u} = T\log\Big(\sum_{t=s}^{T}\sum_{i=1}^{k}\widehat{\varepsilon}_t^2(\widehat{\phi}_{T-1,i}^{(t)}, \widehat{r}^{(t)})\Big) + \sum_{i=1}^{k}\frac{T_i(T_i + p_i + 1)}{T_i - (p_i + 1) - 2}$$
$$+ T_i\log\Big\{\frac{T_i}{T_i - (p_i + 1) - 1}\Big\}. \qquad (6.71)$$

De Gooijer (2001) and Galeano and Peña (2007) compare by simulation the performance of various CV- and AIC-type (including BIC) criteria for two-regime SETAR model selection in case both d and r are unknown. Their results indicate that AIC$_\mathrm{u}$ and C$_\mathrm{u}$ have larger frequencies in detecting the true AR orders and delay parameters than AIC, AIC$_\mathrm{c}$, and BIC, when the sample size is small to moderate ($T \in [30, 75]$). Since AIC$_\mathrm{u}$ and C$_\mathrm{u}$ will tend to select a more parsimonious two-regime SETAR model than AIC, we recommend to use both criteria rather than AIC for relatively small samples. The extra computing time C$_\mathrm{u}$ needs, as opposed to the time it takes to estimate a "conventional" criterion like AIC, is negligible for $T \leq 75$. Otherwise, i.e., in situations with $T \geq 100$, the improvement of the modified criteria over AIC diminishes.

Example 6.7: U.S. Unemployment Rate (Cont'd)

It is interesting to compare the performance of the above model selection criteria using the transformed quarterly U.S. unemployment rate series $\{Y_t\}_{t=1}^{252}$ plotted in Figure 6.2(a). For two-regime SETAR models, we set the maximum allowable orders $p_{1,\max} = p_{2,\max} = 10$. For three-regime SETAR models, we take $p_{1,\max} = p_{2,\max} = p_{3,\max} = 6$. In both cases, we prefix the maximum value of the delay at $d_{\max} = 10$. Parameter estimates are based on CLS. Candidate threshold values are searched between the 25th and 75th percentiles of the empirical distribution of $\{Y_t\}$.

Table 6.1 contains the orders of the selected SETAR models, jointly with selected values of d and estimates of the threshold parameters. We see that AIC prefers a model with relatively high AR orders in each regime while almost all other criteria tend to select a more parsimonious model. Of course, the preference for a less parsimonious or a parsimonious criterion largely depends on how one weighs these overfitting or underfitting tendencies in a given empirical situation. Note, that AIC$_\mathrm{u}$ and BIC favor a SETAR$(2; 2, 2)$ model with delay $d = 5$ while CV$_\mathrm{c}$ and CV$_\mathrm{u}$ choose the same model with $d = 10$. Also, in the case of selecting a three-regime SETAR model, there is hardly any difference between the orders selected by AIC$_\mathrm{c}$, AIC$_\mathrm{u}$, BIC, CV, and CV$_\mathrm{c}$. One interesting situation occurs with CV$_\mathrm{u}$ with all orders equal one and $d = 1$. Clearly,

Table 6.1: SETAR *orders selected for the transformed quarterly* U.S. unemployment rate.

| | Two-regime SETAR |||| | Three-regime SETAR |||||
Criterion	p_1	p_2	d	$\widehat{r}$	p_1	p_2	p_3	d	$\widehat{r}_1$	$\widehat{r}_2$
AIC	3	5	5	-2.98	2	6	5	10	-3.64	-2.72
AIC$_c$	2	5	5	-2.99	2	3	2	10	-3.64	-2.72
AIC$_u$	2	2	5	-2.99	2	3	2	10	-3.64	-2.72
BIC	2	2	5	-2.88	2	1	2	10	-3.64	-2.72
CV	3	10	5	-2.88	2	1	2	10	-3.64	-2.96
CV$_c$	2	2	10	-3.02	2	1	2	10	-3.64	-2.96
CV$_u$	2	2	10	-3.02	1	1	1	1	-3.64	-3.58

the estimated threshold parameter values are quite near to each other, suggesting that a two-regime rather than a three-regime SETAR model is more appropriate in this case.

6.3 Diagnostic Checking

6.3.1 Pearson residuals

It is well known that the LB test statistic, can serve as a diagnostic check to see if the residuals from an estimated ARMA model behave as a (weak) WN process. Given an estimator $\widehat{\boldsymbol{\theta}}_T$ of the true parameter value $\boldsymbol{\theta}_0$, the test is based on the sample ACF of the standardized residuals, also called *Pearson residuals*, defined by

$$\widehat{\varepsilon}_t \equiv \widehat{\varepsilon}_t(\widehat{\boldsymbol{\theta}}_T) = \big(Y_t - \mathbb{E}(Y_t|\mathcal{F}_{t-1}, \widehat{\boldsymbol{\theta}}_T)\big)/\sqrt{\text{Var}(Y_t|\mathcal{F}_{t-1}, \widehat{\boldsymbol{\theta}}_T)}. \qquad (6.72)$$

Unfortunately, the LB test statistic has certain features one may consider undesirable in a nonlinear time series context. One problem is that the test has a high tendency to let through models with interesting dependencies (e.g., GARCH) in the residuals. Interests in a diagnostic tool based on the sample ACF of residuals from nonlinear relationships started off with the McLeod–Li test statistic which is based on the sample ACF of the squared standardized residuals of a linear time series model. The McLeod–Li test statistic has high power against departures from linearity that have apparent ARCH structures. The test statistic has little power in detecting other types of (non)linear dependencies in the residuals; see, e.g., Li and Mak (1994), and Tse and Zuo (1998). Li (1992) derives the asymptotic distribution of residual autocorrelations for a general stationary NLAR process with strict WN errors; cf. Exercise 6.4.

Chen (2008) presents a general framework for testing Pearson residuals from the pth-order NLAR model with conditional heteroskedasticity. This model, as a special case of (6.1), is given by

$$Y_t = g(\mathbf{Y}_{t-1}; \boldsymbol{\theta}) + \eta_t, \quad \eta_t = h(\mathbf{Y}_{t-1}; \boldsymbol{\theta})^{1/2}\varepsilon_t, \qquad (6.73)$$

6.3 DIAGNOSTIC CHECKING

where $\mathbf{Y}_{t-1} = (Y_{t-1}, Y_{t-2}, \ldots, Y_{t-p})'$, and $\boldsymbol{\theta} \in \boldsymbol{\Theta}$ denotes a parameter vector in a compact parameter space $\boldsymbol{\Theta}$. Here, $g(\cdot; \boldsymbol{\theta})$ and $h(\cdot; \boldsymbol{\theta})$ are twice continuously differentiable functions, and $\{\varepsilon_t\}$ is an i.i.d. WN process with moments $\mu_{1,\varepsilon} = 0$, $\mu_{2,\varepsilon} = 1$, and $\mu_{4,\varepsilon} < \infty$, where $\mu_{r,\varepsilon} = \mathbb{E}(\varepsilon_t^r)$.

Using residual autocorrelations, the objective is to test the null hypothesis

$$\mathbb{H}_0: \{\varepsilon_t\} \text{ is an i.i.d. sequence for some } \boldsymbol{\theta}_0 \in \boldsymbol{\Theta}. \tag{6.74}$$

The resulting test statistic may be based on transformed (e.g. squared) or untransformed standardized (Pearson) residuals. Since we wish to remain agnostic about the precise form of transformation for the moment, we introduce the following notation. Let $u_i(\cdot)$ and $v_j(\cdot)$ be two continuously differentiable functions of $\{\varepsilon_t\}$ with the finite moments $\mu_{u_i} = \mathbb{E}[u_i(\varepsilon_t)]$, $\mu_{v_j} = \mathbb{E}[v_j(\varepsilon_t)]$, $\sigma_{u_i}^2 = \text{Var}[u_i(\varepsilon_t)]$, and $\sigma_{v_j}^2 = \text{Var}[v_j(\varepsilon_t)]$ ($i = 1, \ldots, P; j = 1, \ldots, Q$). Moreover, we introduce the standardized random variables $u_i^*(\varepsilon_t) = (u_i(\varepsilon_t) - \mu_{u_i})/\sigma_{u_i}$, $v_j^*(\varepsilon_t) = (v_j(\varepsilon_t) - \mu_{v_j})/\sigma_{v_j}$. Then, under $\mathbb{H}_0$, the lag ℓ ($\ell \in \mathbb{Z}$) cross-correlation, defined as

$$\rho_\varepsilon^{(i,j)}(\ell) = \mathbb{E}[u_i^*(\varepsilon_t) v_j^*(\varepsilon_{t-\ell})], \quad (i = 1, \ldots, P; j = 1, \ldots, Q), \tag{6.75}$$

is zero $\forall i, j, \ell$. Similarly, under $\mathbb{H}_0$, the $PQ \times 1$ vector $\boldsymbol{\rho}(\ell) = \mathbb{E}[\mathbf{U}(\varepsilon_t) \otimes \mathbf{V}(\varepsilon_{t-\ell})] = \left(\rho_\varepsilon^{(1,1)}(\ell), \ldots, \rho_\varepsilon^{(1,Q)}(\ell), \ldots, \rho_\varepsilon^{(P,1)}(\ell), \ldots, \rho_\varepsilon^{(P,Q)}(\ell)\right)'$ is zero $\forall \ell$, where $\mathbf{U}(\varepsilon_t) = \left(u_1^*(\varepsilon_t), \ldots, u_P^*(\varepsilon_t)\right)'$ and $\mathbf{V}(\varepsilon_t) = \left(v_1^*(\varepsilon_t), \ldots, v_Q^*(\varepsilon_t)\right)'$.

Naturally, given $\{Y_t\}_{t=1}^T$, we replace the above quantities by their corresponding sample statistics with $\widehat{\boldsymbol{\theta}}_T$ the QML or CLS estimator of $\boldsymbol{\theta}$. Denote the estimated Pearson residuals by $\widehat{\varepsilon}_t \equiv \widehat{\varepsilon}_t(\widehat{\boldsymbol{\theta}}_T) = (Y_t - \widehat{g}_t)/\widehat{h}_t^{1/2}$ in which $\widehat{g}_t \equiv g(\mathbf{Y}_{t-1}; \widehat{\boldsymbol{\theta}}_T)$ and $\widehat{h}_t \equiv h(\mathbf{Y}_{t-1}; \widehat{\boldsymbol{\theta}}_T)$. Let $\widehat{\mu}_{u_i}$ and $\widehat{\mu}_{v_j}$ ($\widehat{\sigma}_{u_i}^2$ and $\widehat{\sigma}_{v_j}^2$) be, respectively, the sample means (variances) of $u_i(\cdot)$ and $v_j(\cdot)$. Moreover, let $\nabla_{\boldsymbol{\theta}} g_t$ and $\nabla_{\boldsymbol{\theta}} h_t$ be, respectively, the column vectors of partial derivatives of g_t and h_t with respect to $\boldsymbol{\theta}$. Denote $\mathbf{w}_t = (\nabla_{\boldsymbol{\theta}} g_t) h_t^{-1/2}$, $\mathbf{z}_t = (\nabla_{\boldsymbol{\theta}} h_t) h_t^{-1}$, $\widehat{\mathbf{w}}_t = \mathbf{w}_t|_{\boldsymbol{\theta}=\widehat{\boldsymbol{\theta}}_T}$, $\widehat{\mathbf{z}}_t = \mathbf{z}_t|_{\boldsymbol{\theta}=\widehat{\boldsymbol{\theta}}_T}$, $u_i^*(\widehat{\varepsilon}_t) = (u_i(\widehat{\varepsilon}_t) - \widehat{\mu}_{u_i})/\widehat{\sigma}_{u_i}$, and $v_j^*(\widehat{\varepsilon}_t) = (v_j(\widehat{\varepsilon}_t) - \widehat{\mu}_{v_j})/\widehat{\sigma}_{v_j}$. The lag ℓ sample cross-correlation of $u_i(\widehat{\varepsilon}_t)$ and $v_j(\widehat{\varepsilon}_{t-\ell})$ is given by $\widehat{\rho}_{\widehat{\varepsilon}}^{(i,j)}(\ell) = (T - \ell)^{-1} \sum_{t=\ell+1}^T \widehat{u}_i^*(\widehat{\varepsilon}_t) \widehat{v}_j^*(\widehat{\varepsilon}_{t-\ell})$ and the sample analogue of $\boldsymbol{\rho}(\ell)$ is $\widehat{\boldsymbol{\rho}}(\ell) = \left(\widehat{\rho}_{\widehat{\varepsilon}}^{(1,1)}(\ell), \ldots, \widehat{\rho}_{\widehat{\varepsilon}}^{(1,Q)}(\ell), \ldots, \widehat{\rho}_{\widehat{\varepsilon}}^{(P,1)}(\ell), \ldots, \widehat{\rho}_{\widehat{\varepsilon}}^{(P,Q)}(\ell)\right)'$. Finally, to describe the asymptotic behavior of a finite set of $\widehat{\boldsymbol{\rho}}(\ell)$ vectors, we define a $PQM \times 1$ ($M \ll T$) vector $\widehat{\boldsymbol{\Pi}}(M) = \left(\widehat{\boldsymbol{\rho}}(1), \ldots, \widehat{\boldsymbol{\rho}}(M)\right)'$

Under $\mathbb{H}_0$, and certain regularity conditions, it can be shown (Chen, 2008) that

$$\sqrt{T-k}\, \widehat{\boldsymbol{\rho}}(\ell) = \frac{1}{\sqrt{T-\ell}} \sum_{t=k+1}^T \boldsymbol{\Psi}(\varepsilon_t, \varepsilon_{t-\ell}) + o_p(1),$$

where

$$\boldsymbol{\Psi}(\varepsilon_t, \varepsilon_{t-\ell}) = \mathbf{U}(\varepsilon_t) \otimes \mathbf{V}(\varepsilon_{t-\ell}) - \boldsymbol{\Lambda}(\ell) \boldsymbol{\Upsilon}^{-1} [\mathbf{w}_t \varepsilon_t + \frac{1}{2} \mathbf{z}_t(\varepsilon_t^2 - 1)],$$

$$\boldsymbol{\Upsilon} = \mathbb{E}[\mathbf{w}_t \mathbf{w}_t'] + \frac{1}{2} \mathbb{E}[\mathbf{z}_t \mathbf{z}_t'],$$

and

$$\boldsymbol{\Lambda}(\ell) = \mathbb{E}[\nabla \mathbf{U}(\varepsilon_t)] \otimes \mathbb{E}[\mathbf{V}(\varepsilon_{t-\ell})\mathbf{w}_t'] + \frac{1}{2}\mathbb{E}[\nabla \mathbf{U}(\varepsilon_t)] \otimes \mathbb{E}[\mathbf{V}(\varepsilon_{t-\ell})\mathbf{z}_t'],$$

where $\nabla \mathbf{U}(\cdot)$ denotes the $PQ \times 1$ vector of first derivatives of $\mathbf{U}(\cdot)$ with respect to $\boldsymbol{\theta}$. So, under $\mathbb{H}_0$, $\sqrt{T-\ell}\,\widehat{\boldsymbol{\rho}}(\ell)$ is *not* asymptotically equivalent to its standardized-errors-based counterpart $\sum_{t=\ell+1}^{T} \mathbf{U}(\varepsilon_t) \otimes \mathbf{V}(\varepsilon_{t-\ell})/\sqrt{T-\ell}$ unless $\boldsymbol{\Lambda}(\ell) = \mathbf{0}$, due to the effect of estimation uncertainty. Furthermore, it can be shown that

$$\text{Cov}\Big[\sum_{t=\ell+1}^{T} \boldsymbol{\Psi}(\varepsilon_t, \varepsilon_{t-\ell}), \sum_{t=\ell'+1}^{T} \boldsymbol{\Psi}(\varepsilon_t, \varepsilon_{t-\ell'})\Big] = (T-\ell')[\delta_{\ell\ell'}\mathbf{I}_{PQ} + \mathbf{A}(\ell, \ell')], \quad (6.76)$$

where

$$\mathbf{A}(\ell, \ell') = \boldsymbol{\Lambda}(\ell)\boldsymbol{\Upsilon}^{-1}\boldsymbol{\Omega}\boldsymbol{\Upsilon}^{-1}\boldsymbol{\Lambda}'(\ell') - \boldsymbol{\Delta}(\ell)\boldsymbol{\Upsilon}^{-1}\boldsymbol{\Lambda}'(\ell') - \boldsymbol{\Lambda}(\ell)\boldsymbol{\Upsilon}^{-1}\boldsymbol{\Delta}'(\ell'),$$

$$\boldsymbol{\Omega} = \mathbb{E}[\mathbf{w}_t\mathbf{w}_t'] + \frac{1}{2}\mu_{3,\varepsilon}\mathbb{E}[\mathbf{w}_t\mathbf{z}_t'] + \mathbb{E}[\mathbf{z}_t\mathbf{w}_t'] + \frac{1}{4}(\mu_{4,\varepsilon} - 1)\mathbb{E}[\mathbf{z}_t\mathbf{z}_t'],$$

and

$$\boldsymbol{\Delta}(\ell) = \mathbb{E}[\mathbf{U}(\varepsilon_t)\varepsilon_t] \otimes \mathbb{E}[\mathbf{V}(\varepsilon_{t-\ell})\mathbf{w}_t'] + \frac{1}{2}\mathbb{E}[\mathbf{U}(\varepsilon_t)\varepsilon_t^2] \otimes \mathbb{E}[\mathbf{V}(\varepsilon_{t-\ell})\mathbf{z}_t'].$$

From the proof of this last result it can be deduced that $\{\boldsymbol{\Psi}(\varepsilon_t, \varepsilon_{t-\ell})\}$ is a sequence of uncorrelated elements. Then it follows that the asymptotic null distribution is given by

$$\sqrt{T-\ell}\,\widehat{\boldsymbol{\rho}}(\ell) \xrightarrow{D} \mathcal{N}_{PQ}\big(\mathbf{0}, \boldsymbol{\Sigma}(l)\big), \quad \boldsymbol{\Sigma}(\ell) = \mathbf{I}_{PQ} + \mathbf{A}(\ell, \ell), \quad (6.77)$$

for any fixed ℓ. In addition, as $T \to \infty$, it follows that under $\mathbb{H}_0$:

$$\sqrt{T}\,\widehat{\boldsymbol{\Pi}}(M) \xrightarrow{D} \mathcal{N}_{PQM}\big(\mathbf{0}, \boldsymbol{\Xi}(M)\big), \quad \boldsymbol{\Xi}(\ell) = \mathbf{I}_{PQM} + \mathbf{B}(M), \quad (6.78)$$

for any fixed $M \in \mathbb{Z}^+$, where $\mathbf{B}(M)$ is a $PQM \times PQM$ matrix with elements $\{\mathbf{A}(i, j)\}$ $(i, j = 1, \ldots, M)$.

Given (6.77) and (6.78), the proposed test statistics are

$$C_T(\ell) = (T - \ell)\,\widehat{\boldsymbol{\Gamma}}'(\ell)\widehat{\boldsymbol{\Sigma}}_T^{-1}(\ell)\widehat{\boldsymbol{\Gamma}}(\ell), \quad (6.79)$$

$$Q_T(M) = T\,\widehat{\boldsymbol{\Pi}}'(M)\widehat{\boldsymbol{\Xi}}_T^{-1}(M)\widehat{\boldsymbol{\Pi}}(M), \quad (6.80)$$

where $\widehat{\boldsymbol{\Sigma}}_T(\ell)$ and $\widehat{\boldsymbol{\Xi}}_T(M)$ are consistent estimates of $\boldsymbol{\Sigma}(\ell)$ and $\boldsymbol{\Xi}(M)$, respectively. Under $\mathbb{H}_0$, and as $T \to \infty$, it follows that for any fixed ℓ, $C_T(\ell) \xrightarrow{D} \chi^2_{PQ}$, and for any fixed M, $Q_T(M) \xrightarrow{D} \chi^2_{PQM}$.

6.3 DIAGNOSTIC CHECKING

Table 6.2: *Standardized-residuals-based test statistics for diagnostic checking of three SETAR-type models fitted to the log-returns of the daily* Hong Kong Hang Seng Index. *The blue-typed number indicates rejection of* $\mathbb{H}_0$ *at the 5% nominal significance level.*[1]

Model	(i,j)	$C_T^{(i,j)}(\ell)$ $\ell=1$	$\ell=3$	$\ell=5$	$Q_T^{(i,j)}(M)$ $M=5$
SETAR(2; 1, 1)	(1,1)	0.56	0.31	0.14	2.26
	(1,2)	0.17	0.58	0.00	3.68
	(2,1)	2.00	3.21	0.60	6.51
	(2,2)	0.52	0.59	2.16	4.05
SETAR(2; 1, 1)–GARCH(1, 1)	(1,1)	0.07	0.52	0.26	1.78
	(1,2)	0.00	0.14	0.06	1.89
	(2,1)	2.07	2.32	0.41	6.40
	(2,2)	4.68	0.03	0.76	7.59
SETAR(2; 1, 1)–EGARCH(1, 1)	(1,1)	0.14	0.63	0.30	2.03
	(1,2)	0.02	0.60	0.19	2.62
	(2,1)	0.83	1.03	0.07	4.10
	(2,2)	3.67	0.03	0.61	7.36

[1] The 95% critical values of the χ_1^2, χ_3^2, χ_5^2, χ_{10}^2, and χ_{20}^2 distribution are approximately 3.84, 7.81, 11.07, 18.31, and 31.41.

We note that under $\mathbb{H}_0$, the asymptotic variance of $\sqrt{T-\ell}\,\widehat{\rho}(\ell)$ is exactly the same as the variance of $\boldsymbol{\Psi}(\varepsilon_t, \varepsilon_{t-\ell})$, so that we have a simple estimate of $\boldsymbol{\Sigma}(\ell)$, i.e.

$$\widehat{\boldsymbol{\Sigma}}_T(\ell) = \frac{1}{T-\ell} \sum_{t=\ell+1}^{T} \widehat{\boldsymbol{\Psi}}_t(\ell) \widehat{\boldsymbol{\Psi}}'_t(\ell), \qquad (6.81)$$

where $\widehat{\boldsymbol{\Psi}}_t(\ell)$ denotes the sample analogue of $\boldsymbol{\Psi}(\varepsilon_t, \varepsilon_{t-\ell})$ evaluated at $\boldsymbol{\theta} = \boldsymbol{\theta}_T$. In addition, $\sqrt{T}\,\widehat{\boldsymbol{\Pi}}(M)$ is exactly the same as the variance-covariance matrix of $\big(\boldsymbol{\Psi}'(\varepsilon_t, \varepsilon_{t-1}), \ldots, \boldsymbol{\Psi}'(\varepsilon_t, \varepsilon_{t-M})\big)'$. So, it can be consistently estimated by

$$\widehat{\boldsymbol{\Xi}}_T(M) = \frac{1}{T-M} \sum_{t=M+1}^{T} \big(\widehat{\boldsymbol{\Psi}}'_t(1), \ldots, \widehat{\boldsymbol{\Psi}}'_t(M)\big)' \big(\widehat{\boldsymbol{\Psi}}'_t(1), \ldots, \widehat{\boldsymbol{\Psi}}'_t(M)\big). \qquad (6.82)$$

Example 6.8: Daily Hong Kong Hang Seng Index (Cont'd)

To illustrate the performance of the diagnostic test statistics (6.79) and (6.80), we reconsider the log-returns of the daily Hong Kong Hang Seng Index introduced in Example 6.6, and denoted by $\{Y_t\}_{t=1}^{253}$. Assuming $\{\varepsilon_t\} \stackrel{\text{i.i.d.}}{\sim} \mathcal{N}(0, \sigma_\varepsilon^2)$, we fitted three SETAR-type models to the data.[2] In order to compute the

---
[2] As an approximation of $I(Y_{t-1} \leq r)$, we use the continuously differentiable logistic transition function (2.43) with $c = r$ and $\gamma = 1{,}000$.

test, we consider the class of power-transformed-based correlations $\rho_\varepsilon^{(i,j)}(\ell)$'s with

$$\left(u_i(\varepsilon_t), v_j(\varepsilon_{t-\ell})\right) = (\varepsilon_t^i, \varepsilon_{t-\ell}^j), \quad (i,j = 1, 2). \tag{6.83}$$

Replacing $\rho_\varepsilon^{(i,j)}(\ell)$ by $\widehat{\rho}_{\widehat\varepsilon}^{(i,j)}(\ell)$, Table 6.2 shows values of the test statistics $C_T^{(i,j)}(\ell)$ for $\ell = 1$, 3, and 5 and $Q_T^{(i,j)}(5)$ $(i, j = 1, 2)$. Except for $C_T^{(2,2)}(1)$ in the case of a SETAR(2; 1, 1)–GARCH(1, 1) model, none of the reported values are significant at the 5% nominal level; hence, we conclude that the standardized residuals are serially uncorrelated. This suggests that a simple SETAR model is capable of describing the DGP. The fit of a more complicated model, as in Example 6.6, does not seem to be needed.

6.3.2 Quantile residuals

When the conditional distribution of the residual process is asymmetric or multimodal, $\mathbb{E}(Y_t|\mathcal{F}_{t-1}, \widehat{\boldsymbol{\theta}}_T)$ in (6.72) may not be the best forecast of the process $\{Y_t, t \in \mathbb{Z}\}$. Moreover, some nonlinear models may involve unobservable random variables.[3] In that case, Pearson residuals will not be the empirical counterparts of the process $\{\varepsilon_t, t \in \mathbb{Z}\}$. In fact, assuming the model is correctly specified, the residual process $\{\widehat{\varepsilon}_t, t \in \mathbb{Z}\}$ is a martingale difference sequence with zero mean and unit variance, and its asymptotic distribution differs from that of the noise process $\{\varepsilon_t, t \in \mathbb{Z}\}$. As an alternative, various diagnostic test statistics for parametric nonlinear time series models can be based on *quantile residuals*. These quantities are defined as follows.

Following the notation introduced in Section 6.2.1, let $f(\mathbf{y}; \boldsymbol{\theta}_{0,m})$ be the true pdf of the observations $\{Y_t\}_{t=1}^T$, $\boldsymbol{\theta}_{0,m} \in \boldsymbol{\Theta} \subset \mathbb{R}^m$, and $\mathbf{y} = (Y_1, \ldots, Y_T)'$. For each $f: \boldsymbol{\Theta} \times \mathbb{R}^T \to \mathbb{R}^+$, we can write

$$f(\mathbf{y}; \boldsymbol{\theta}_m) = \prod_{t=1}^T f_{t-1}(Y_t; \boldsymbol{\theta}_m), \tag{6.84}$$

where $f_{t-1}(Y_t; \boldsymbol{\theta}_m) \equiv f(Y_t; \boldsymbol{\theta}_m|\mathcal{F}_{t-1})$ is the conditional density function of $\{Y_t, t \in \mathbb{Z}\}$ given $\mathcal{F}_{t-1} = \sigma(\mathbf{Y}_0, Y_1, \ldots, Y_{t-1})$, the σ-algebra generated by the random variables $\{\mathbf{Y}_0, Y_1, \ldots, Y_{t-1}\}$, $\boldsymbol{\theta}_m \subset \mathbb{R}^m$ an m-dimensional parameter vector, and where $\mathbf{Y}_0$ represents the initial model values. Then, according to Dunn and Smyth (1996), the *theoretical quantile residual* is defined by

$$R_{t,\boldsymbol{\theta}_m} = \Phi^{-1}\big(F_{t-1}(Y_t; \boldsymbol{\theta}_m)\big), \tag{6.85}$$

where $\Phi^{-1}(\cdot)$ is the inverse CDF of the $\mathcal{N}(0,1)$ distribution, and $F_{t-1}(Y_t; \boldsymbol{\theta}_m) = \int_{-\infty}^{Y_t} f_{t-1}(u; \boldsymbol{\theta}_m)du$ is the conditional CDF of $\{Y_t, t \in \mathbb{Z}\}$, also called the *probability*

---

[3]This is, for instance, the case with the mixture AR (MAR) model (see, Exercise 7.7), and the MAR–GARCH model (Wong and Li, 2000b, 2001).

6.3 DIAGNOSTIC CHECKING

integral transform (PIT). The corresponding *sample quantile residual* is

$$r_{t,\widehat{\boldsymbol{\theta}}_T} = \Phi^{-1}\big(F_{t-1}(Y_t;\widehat{\boldsymbol{\theta}}_T)\big), \tag{6.86}$$

where $\widehat{\boldsymbol{\theta}}_T$ (dropping the subscript m) is a QML estimate of $\boldsymbol{\theta}_{0,m}$. Observe that quantile residuals of linear and nonlinear AR models with normal errors are identical to Pearson residuals.

General testing framework

Kalliovirta (2012) develops a general testing framework for detecting different potential departures from the characteristic properties of quantile residuals ($\mathbb{H}_0$). The framework is based on transformations of $R_{t,\boldsymbol{\theta}_0}$ by a continuously differentiable function $g: \mathbb{R}^d \to \mathbb{R}^n$ such that $\mathbb{E}\big(g(\mathbf{R}_{t,\boldsymbol{\theta}_0})\big) = \mathbf{0}$, where $\mathbf{R}_{t,\boldsymbol{\theta}_0} = (R_{t,\boldsymbol{\theta}_0},\ldots,R_{t-d+1,\boldsymbol{\theta}_0})'$, and d and n are the dimensions of the domain and range of g. Different choices of g lead to different test statistics.

Conditional on a vector with initial values $\mathbf{Y}_0$, and assuming that the conditional density functions $f_{t-1}(Y_t;\boldsymbol{\theta}_m)$ exist, the log-likelihood function $\ell_T(\mathbf{y},\boldsymbol{\theta}) = \sum_{t=1}^{T} \ell_t(Y_t,\boldsymbol{\theta}) = \sum_{t=1}^{T} \log f_{t-1}(Y_t;\boldsymbol{\theta})$ of the sample follows directly. Then, under some fairly standard regularity conditions, Kalliovirta (2012) proves the following CLT

$$\frac{1}{\sqrt{T}}\sum_{t=1}^{T} g(\mathbf{R}_{t,\widehat{\boldsymbol{\theta}}_T}) \xrightarrow{D} \mathcal{N}_d(\mathbf{0},\boldsymbol{\Omega}), \tag{6.87}$$

where

$$\boldsymbol{\Omega} = \mathbf{G}\mathcal{I}(\boldsymbol{\theta}_0)^{-1}\mathbf{G}' + \boldsymbol{\Psi}\mathcal{I}(\boldsymbol{\theta}_0)^{-1}\mathbf{G}' + \mathbf{G}\mathcal{I}(\boldsymbol{\theta}_0)^{-1}\boldsymbol{\Psi}' + \mathbf{H}, \tag{6.88}$$

with $\mathbf{G} = \mathbb{E}\big(\partial g(\mathbf{R}_{t,\boldsymbol{\theta}_0})/\partial \boldsymbol{\theta}'\big)$, $\mathbf{H} = \mathbb{E}\big(g(\mathbf{R}_{t,\boldsymbol{\theta}_0})g(\mathbf{R}_{t,\boldsymbol{\theta}_0})'\big)$, and where $\mathcal{I}(\boldsymbol{\theta}_0)$ denotes the expected information matrix evaluated at $\boldsymbol{\theta}_0$, and $\boldsymbol{\Psi}$ is a constant matrix. The first three terms in the asymptotic covariance matrix $\boldsymbol{\Omega}$ represent model uncertainty due to the effect of parameter estimation. If $\mathbf{G} = \mathbf{0}$, there is (asymptotically) no need to take this uncertainty into account in the resulting test statistic. In general, however, $\mathbf{G} \neq \mathbf{0}$ which resembles the case $\Lambda(\ell) \neq \mathbf{0}$ in Section 6.3.1.

Assume that the nonlinear model under study is correctly specified, so that $\{R_{t,\boldsymbol{\theta}_0}\} \overset{\text{i.i.d.}}{\sim} \mathcal{N}(0,1)$ holds. Let $\widehat{\mathcal{I}}_T$ be a consistent estimator of $\mathcal{I}(\boldsymbol{\theta}_0)$. Then a consistent estimator for $\boldsymbol{\Omega}$ is

$$\widehat{\boldsymbol{\Omega}}_T = \widehat{\mathbf{G}}_T \widehat{\mathcal{I}}_T^{-1} \widehat{\mathbf{G}}_T' + \widehat{\boldsymbol{\Psi}}_T \widehat{\mathcal{I}}_T^{-1} \widehat{\mathbf{G}}_T' + \widehat{\mathbf{G}}_T \widehat{\mathcal{I}}_T^{-1} \widehat{\boldsymbol{\Psi}}_T' + \widehat{\mathbf{H}}_T, \tag{6.89}$$

where $\widehat{\mathbf{G}}_T = T^{-1}\sum_{t=1}^{T} \partial g(\mathbf{r}_{t,\widehat{\boldsymbol{\theta}}_T})/\partial \boldsymbol{\theta}'$, $\widehat{\boldsymbol{\Psi}}_T = T^{-1}\sum_{t=1}^{T} g(\mathbf{r}_{t,\widehat{\boldsymbol{\theta}}_T})\partial \ell_t(Y_t,\widehat{\boldsymbol{\theta}}_T)/\partial \boldsymbol{\theta}'$, and $\widehat{\mathbf{H}}_T = T^{-1}\sum_{t=1}^{T} g(\mathbf{r}_{t,\widehat{\boldsymbol{\theta}}_T})g(\mathbf{r}_{t,\widehat{\boldsymbol{\theta}}_T})'$. Based on (6.87), a *general test statistic* is defined as

$$S_{T,d} = \frac{1}{T-d+1}\sum_{t=1}^{T-d+1} g(\mathbf{r}_{t,\widehat{\boldsymbol{\theta}}_T})' \widehat{\boldsymbol{\Omega}}_T^{-1} \sum_{t=1}^{T-d+1} g(\mathbf{r}_{t,\widehat{\boldsymbol{\theta}}_T}), \tag{6.90}$$

Table 6.3: *Three diagnostic test statistics based on univariate quantile residuals, as special cases of the general test statistic $S_{T,d}$.*

Null hypothesis $\mathbb{H}_0$	Transformation function g	Test statistic
$\rho_{R_{t,\theta_0}}(\ell) = 0, \forall t,$ $(\ell = 1, \ldots, K_1; K_1 \ll T)$ (Autocorrelation)	$g: \mathbb{R}^{K_1+1} \to \mathbb{R}^{K_1}$ $g(\mathbf{r}_{t,\theta}) =$ $(r_{t,\theta} r_{t+1,\theta}, \ldots, r_{t,\theta} r_{t+K_1,\theta})'$	$A_{T,K_1} = S_{T,d}$ with $d = K_1 + 1$
$\rho_{R^2_{t,\theta_0}}(\ell) = 0, \forall t,$ $(\ell = 1, \ldots, K_2; K_2 \ll T)$ (Heteroskedasticity)	$g: \mathbb{R}^{K_2+1} \to \mathbb{R}^{K_2}$ $g(\mathbf{r}_{t,\theta}) =$ $((r^2_{t,\theta} - 1) r^2_{t+1,\theta}, \ldots, (r^2_{t,\theta} - 1) r^2_{t+K_2,\theta})'$	$H_{T,K_2} = S_{T,d}$ with $d = K_2 + 1$
$\mathbb{E}(R^2_{t,\theta_0} - 1, R^3_{t,\theta_0}, R^4_{t,\theta_0} - 3)' = \mathbf{0}, \forall t$ (Normality)	$g: \mathbb{R} \to \mathbb{R}^3$ $g(r_{t,\theta}) = (r^2_{t,\theta} - 1, r^3_{t,\theta}, r^4_{t,\theta} - 3)'$	$N_T = S_{T,d}$ with $d = 1$

where $\mathbf{r}_{t,\widehat{\boldsymbol{\theta}}_T} = (r_{t,\widehat{\boldsymbol{\theta}}_T}, \ldots, r_{t-d+1,\widehat{\boldsymbol{\theta}}_T})'$.[4] Under $\mathbb{H}_0$, and as $T \to \infty$, (6.90) has an asymptotic χ^2_n distribution; Kalliovirta (2012).

Table 6.3 shows three diagnostic test statistics, as special cases of (6.90). Note, that the test statistic for residual autocorrelation is based on uncentered sample autocovariances $(T-\ell)^{-1} \sum_{t=1}^{T-\ell} r_{t,\widehat{\boldsymbol{\theta}}_T} r_{t+\ell,\widehat{\boldsymbol{\theta}}_T}$. The test statistic for conditional heteroskedasticity is based on the sample autocovariances $(T-\ell)^{-1} \sum_{t=1}^{T-\ell} (r^2_{t,\widehat{\boldsymbol{\theta}}_T} - 1) r^2_{t+\ell,\widehat{\boldsymbol{\theta}}_T}$, while the normality test statistic builds on ideas suggested by Lomnicki (1961); see, e.g., Section 1.3.1. Under $\mathbb{H}_0$ these test statistics are asymptotically distributed as respectively $\chi^2_{K_1}$, $\chi^2_{K_2}$, and χ^2_3.

6.4 Application: TARSO Model of a Water Table

In lowland areas such as the Netherlands or Belgium, structural changes in the water table fluctuation will often have impact on agricultural land use and ecology. To support decision making in these areas, water managers need reliable predictions of the effects of interventions in the hydrological regime on the water table fluctuations. Preferably, these effects are expressed in terms of risks or probabilities, which implies the use of stochastic models and methods. Water table depths $\{Y_t\}$ (output) can be related to precipitation surplus $\{X_t\}$ (input). Both linear and nonlinear time series models can be used for this purpose. One form of nonlinearity is caused by the presence of thresholds which divide the relationship between precipitation surplus and water table depth into several regimes. These thresholds are, for instance, soil physical boundaries or drainage levels; see Figure 6.9 for a schematic view.

SSTARSO model

Knotters and De Gooijer (1999) show that subset TARSO (SSTARSO) models for

---

[4] It is known that under $\mathbb{H}_0$, $\mathbb{E}((R_{t,\theta_0})^n) = \prod_{i=1}^{n/2}(2i-1)$ $(n = 2, 4, 6, \ldots)$, and 0 elsewhere. Using this result, it is straightforward to obtain explicit expressions for the matrix $\mathbf{H}$ for each of the three hypotheses in Table 6.3.

6.4 APPLICATION: TARSO MODEL OF A WATER TABLE

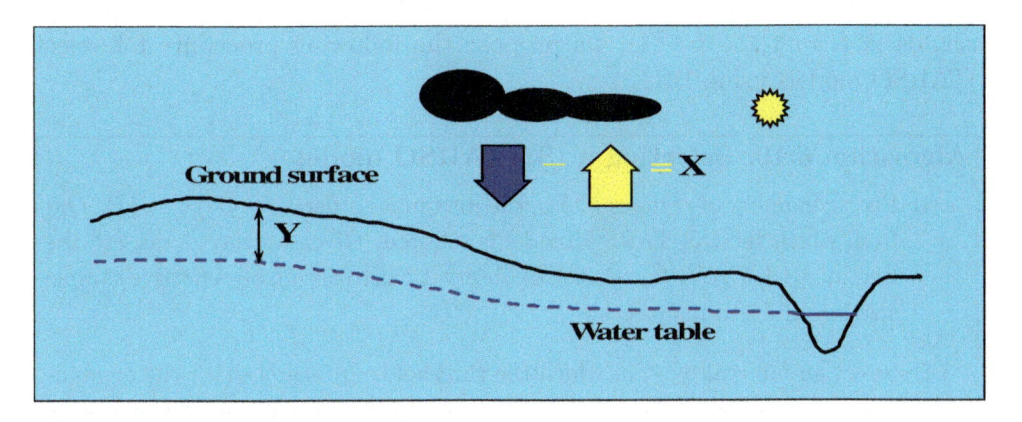

Figure 6.9: *Schematic view of a water table relative to the ground surface elevation, called "water table depth" (denoted by Y_t), with as input variable "precipitation excess" (denoted by X_t), i.e. the difference between precipitation and evapotranspiration.*

the process $\{(Y_t, X_t), t \in \mathbb{Z}\}$, with the regime switching depending on Y_t rather than X_t, can capture the nonlinear relationships of the hydrologic system successfully. Adopting a similar notation as for the subset SETARMA model in (6.16), a k-regime SSTARSO model is defined as

$$Y_t = \sum_{i=1}^{k} \left(\phi_0^{(i)} + \sum_{u=1}^{p_i} \phi_{j_u^{(i)}}^{(i)} Y_{t-j_u} + \sum_{v=0}^{q_i} \psi_{h_v^{(i)}}^{(i)} X_{t-h_i} + \varepsilon_t^{(i)} \right) I(Y_{t-d} \in \mathbb{R}^{(i)}), \quad (6.91)$$

where $\varepsilon_t^{(i)} = \sigma_i^2 \varepsilon_t$ $(i = 1, \ldots, k)$, $\{\varepsilon_t\} \stackrel{\text{i.i.d.}}{\sim} (0, 1)$, and $\mathbb{R}^{(i)} = (r_{i-1}, r_i]$ with $r_0 = -\infty$ and $r_k = \infty$. Below we focus on a time series of a semi-monthly observed water table depth covering the time period 1982 – 1992.

The $\{Y_t\}$ series is measured relative to the ground surface elevation nearby the observation well. The well is situated in a drained loamy, fine sandy soil. Drains are present at about -80 centimeter (cm), relative to the ground surface at the well location. Moreover, at a distance of 50 cm to the well a trench with a bottom at about -50 cm is present. Therefore, we assume $k = 3$.

Model selection

We divide the series into a validation and a calibration set,[5] each set consists of $T = 120$ observations. As a model selection criterion we adopt BIC, which for the SSTARSO model (6.91) is defined as

$$\text{BIC} = \min_{\substack{p_1, \ldots, p_k \\ q_1, \ldots, q_k}} \left\{ \sum_{i=1}^{k} \{T_i \log \widehat{\sigma}_{T_i}^2 + (p_i + q_i + 1) \log T_i\} \right\}, \quad (6.92)$$

where T_i is the number of observations that belong to the ith regime, and $\widehat{\sigma}_{T_i}^2$ the corresponding residual variance. If no prior information is used on the values of the

---

[5] *Calibration* refers to the statistical consistency between the distributional forecasts and the observations, and is a joint property of the forecasts and the observed values.

thresholds r_i $(i = 1, \ldots, k - 1)$, we propose the following procedure for selecting (SS)TARSO models using BIC.

Algorithm 6.10: Selecting a (SS)TARSO model

(i) Fix the number of regimes k. Fix the maximum orders $(P_1, Q_1), \ldots, (P_k, Q_k)$ from which the (SS)TARSO model is selected. Given a delay d, discard the first $\max_i\{d, P_i, Q_i\}$ $(i = 1, \ldots, k)$ observations to obtain one effective sample size for all fitted models.

(ii) Select an interval $[\underline{r}, \overline{r}]$ in which the thresholds are searched, or the combination of threshold values if there are more than two regimes. For instance, take the 10th percentile and the 90th percentile of the empirical distribution of $\{Y_t\}_{t=1}^T$ respectively.

(iii) To guarantee that there are enough observations in each regime, search r's at a fixed interval (here 1 cm) between $\underline{r}$ and $\overline{r}$ such that within each ith regime $T_i \geq 20$. This results in a set of, say R (combinations of) candidate threshold values $r_1, \ldots, r_{k-1}$.

(iv) Select candidate subsets for the non-zero coefficients $\phi_u^{(i)}$ and $\psi_v^{(i)}$, say subsets $\{s_j\}$, where $j = 1, \ldots, K$ denotes the jth of K subsets. Assign to these subsets the lags $j_1^{(i)}, \ldots, j_{P_i}^{(i)}, h_0^{(i)}, h_1^{(i)}, \ldots, h_{q_i}^{(i)}$ of the AR terms in the output and input series in the ith regime. Given k regimes, fixed threshold values, and a fixed delay, there are $S = K^k$ candidate SSTARSO models to represent the process $\{Y_t, X_t\}$. Below we set $P_i = 3$, $Q_i = 2$ $(i = 1, 2, 3)$, and $K = 25$.

(v) Calculate (6.92) over all $R \times S$ candidate models using CLS.

Model selection results

The final model fitted to the data in the calibration set is given by

$$Y_t = \begin{cases} -16.10_{(4.17)} + 0.58_{(0.06)} Y_{t-1} + 0.24_{(0.05)} Y_{t-3} + 6.81_{(0.43)} X_t \\ \quad + 1.86_{(0.53)} X_{t-2} + \varepsilon_t^{(1)} \qquad \text{if } Y_{t-1} \leq -57_{(-87, -56)}, \\ -64.07_{(2.00)} + 7.69_{(1.09)} X_t + \varepsilon_t^{(2)} \text{ if } -57_{(-87, -56)} < Y_{t-1} \leq -47_{(-70, -44)}, \\ -19.10_{(9.06)} + 0.29_{(0.28)} Y_{t-1} + 0.39_{(0.12)} Y_{t-3} \\ \quad + 3.01_{(0.91)} X_t + \varepsilon_t^{(3)} \qquad \text{if } Y_{t-1} > -47_{(-70, -44)}. \end{cases} \quad (6.93)$$

The sample standard deviations of the residuals are 7.15, 8.65, and 6.13, respectively. Thresholds are estimated at -57 cm and -47 cm. The 95% asymptotic confidence intervals of $\widehat{r}_i$ $(i = 1, 2, 3)$ are estimated from 10,000 BS replicates. The skewness of the intervals is a result of the short distance of the threshold at -47 cm to the upper limit of the range in which thresholds are searched; only 21 observations are present in regime 3. Similarly, thresholds are selected more often below than above -57 cm.

6.4 APPLICATION: TARSO MODEL OF A WATER TABLE

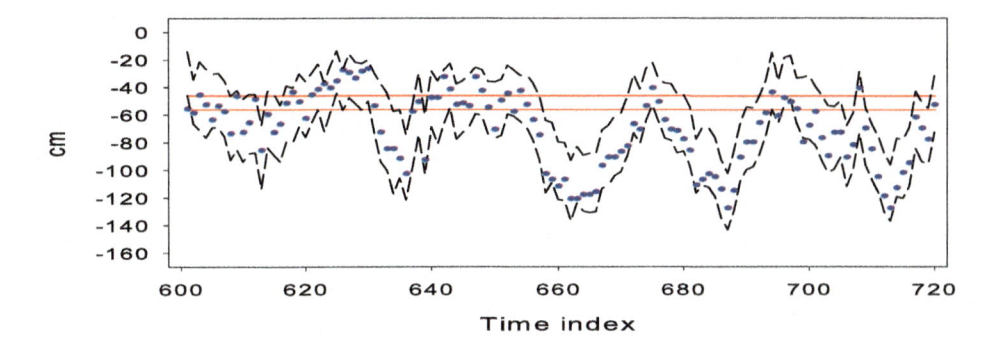

Figure 6.10: *Results of SSTARSO model selection in the calibration period. Observed water table depth (blue dots), intervals in which 95% of the simulated water table depths fall (black dashed lines), and selected thresholds (red solid lines). From* Knotters and De Gooijer (1999).

It is interesting to note that the estimated threshold values are possibly related to the drainage level of the trench at about -40 cm. The estimated AR–coefficient for $\{X_t\}$ in regime 3 is small as compared with those in the other two regimes (3.01 versus 6.81, 7.69). In physical terms the value 3.01 means that, starting from equilibrium conditions, a unit change of the precipitation excess at time t causes a change of 3.01 units in the water table depth $\{Y_t\}$. Further, note that $\{X_t\}$ is the average daily precipitation excess between $t-1$ and t. A physical explanation of the relatively small AR–coefficient for $\{X_t\}$ in regime 3 may be that the fluctuation of the water table in regime 3 is damped by the drainage to the trench. This effect can be seen in Figure 6.10, which shows a plot of the observed water table depth in the calibration period and the interval in which 95% of the simulated water table depths fall, using a set of 720 BS replicates of $\{Y_t\}$. Note that the graph shows a clear seasonal behavior, with a seasonality of 24 semi-monthly time steps.

Model-validation
To compare the performance of the SSTARSO model, we employ a transfer function model with added noise (TFN). Within the present context, it consists of a functional relationship between Y_t^F and a noise process N_t^F. Here Y_t^F denotes that part of the water table depth Y_t which is explained by the precipitation surplus X_t, and N_t^F is modeled in its own right by an ARMA process. More specifically, the TFN model fitted to the data in the calibration period (minimizing BIC) is given by

$$Y_t = Y_t^F + N_t^F, \tag{6.94}$$

where

$$Y_t^F = 0.84_{(0.03)} Y_{t-1} + 6.48_{(0.44)} X_t - 1.78_{(0.56)} X_{t-1},$$
$$(N_t^F - 91.20_{(1.93)}) = 0.56_{(0.08)}(N_{t-1}^F - 91.20_{(1.93)}) + \varepsilon_t,$$

with residual sample standard deviation $\widehat{\sigma}_\varepsilon = 8.57$, and asymptotic standard errors are given in parentheses.

Based on (6.91) and (6.94), we generate 1,000 series of length $T = 120$ and compute the mean error (ME), the root mean squared error (RMSE) and the mean absolute error (MAE) using data on $\{Y_t\}$ from the validation period.[6] The values of these measures for the SSTARSO model, and in parentheses the fitted TFN model, are: ME = -0.3 (1.7), RMSE = 15.3 (16.3), and MAE = 12.3 (13.2). Clearly, the fitted SSTARSO model performs better than the fitted linear TFN model. The percentages of observations outside the interval in which 95% of the simulated water table depths fall are 8 (SSTARSO) and 13 (TFN), respectively. Thus, the fitted SSTARSO model provides an adequate representation. Moreover, the model can be interpreted with respect to the hydrological conditions at the well location.

6.5 Summary, Terms and Concepts

Summary

In the first part of this chapter, we focused on QML, NLS, and CLS estimation methods within the framework of model (6.1), with emphasis on the CLS estimator. Subsequently, we specialized some of these methods to a number of classic nonlinear time series models. We have not attempted to give a full treatment to the fairly large literature on the computation of nonlinear estimation methods. Rather, in Section 6.6, we offer some references to methods not covered by this chapter.

Our treatment of the CLS estimation method was perhaps somewhat detailed. However, anyone who intends to use this method in empirical work should be aware of the underlying assumptions. For example, the finite-sample properties of the CLS method of the threshold parameter in SETAR models depend crucially on the assumption of symmetry of the error process, and the magnitude and signs of SETAR coefficients; see, e.g., Kapetanios (2000) and Norman (2008). Another point worth mentioning is that the CLS estimator is not asymptotically efficient in general. Chandra and Taniguchi (2001) explore this point via MC simulation. Nevertheless, there is still a need for simulation studies which are designed to shed light on the finite-sample properties of CLS and other estimation methods, and their impact on nonlinear model selection, diagnostic checking, and forecasting.

As we have seen in the second part of this chapter, all estimation methods are directly tied to a host of model selection criteria. With nonlinear models, the curse of model complexity and model over-parameterization seems much more prominent when using AIC than in the linear case. If parsimony is considered to be really important, then perhaps a "super-parsimonious" order selection criterion may be helpful; see Granger (1993) for a suggestion.

Finally, within the unifying theme of model estimation, we have discussed residuals-based diagnostic test statistics for remaining serial correlation. The pro-

---

[6]See Knotters and De Gooijer (1999) for details about the design of the MC simulation experiment.

posed test statistics make an explicit correction for effects of estimation uncertainty. Modified versions of these test statistics may also be used to check the null hypothesis of serial independence in the original series because the estimation error's effect is irrelevant in this case. In the next chapter, we will take up the topic of testing for serial independence in time series again, this time in a nonparametric setting.

Terms and Concepts

Akaike information criterion (AIC), 229
average information matrix, 200
Bayesian information criterion (BIC), 231
calibration, 234
compound Poisson process (CPP), 205
conditional least squares (CLS), 202
crossover, 212
cross-validation (CV), 198
empirical Hessian, 199
expected Hessian matrix, 200
expected information matrix, 199
fitness function, 210
genetic algorithm (GA), 210
generalized information criterion (GIC), 231
gradient vector, 200
Hankel matrix, 219
Hellinger distance, 248
iteratively weighted LS (IWLS), 223
Jensen's inequality, 227

Kullback-Leibler (KL) divergence, 227
leave-one-out CV, 234
likelihood equation, 200
local maxima problem, 200
log-likelihood, 199
Markov chain Monte Carlo (MCMC), 210
minimum descriptive length (MDL), 232
mutation, 212
nonlinear least squares (NLS), 200
normalized AIC (NAIC), 211
nuisance parameter, 201
Pearson residuals, 236
penalty function, 232
probability integral transform (PIT), 241
quantile residuals, 240
quasi maximum likelihood (QML), 198
score vector, 200
selection, 212
structural parameter, 210

6.6 Additional Bibliographical Notes

Sections 6.1.1 and 6.1.2: Petruccelli (1986) proves strong consistency of the CLS estimator in the case of a SETAR(2; 1, 1) model. Pham et al. (1991) establish strong consistency of the CLS estimator for a simple non-ergodic SETAR model, so relaxing the stationarity and ergodicity condition. Chan (1993) develops strong consistency and asymptotic normality of the CLS estimator in the general SETAR(2; p, p) model, and Qian (1998) obtains strong consistency of the QML estimate for this model. Asymptotic properties of NLS estimates, under a set of explicit and easy to check conditions, are discussed in Mira and Escribano (2006), Suárez–Fariñas et al. (2004), and Medeiros and Veiga (2005) for a general class of nonlinear dynamic regression models, including STAR–GARCH models.

Liu et al. (2011) study the limiting distribution of the CLS estimators in the case of a SETAR(2; 1, 1) model (no intercept) with a unit root in one regime, and in the case of an explosive SETAR(2; 1, 1) model (no intercept). In both cases, the limiting behavior of the estimators is quite different from the CLS estimators based on the linear counterpart of these models.

De Gooijer (1998) considers ML estimation of TMA models. Under some moderate conditions, Li et al. (2013) show that the estimator of the threshold parameter in a TMA model, is n-consistent and its limiting distribution is related to a two-sided CPP, while the estimators of the other coefficients are strongly consistent and asymptotically normal.

Using the rearranged autoregressions, Coakley et al. (2003) introduce an efficient SETAR model estimation approach which relies on the computational advantages of QR factorization of matrices. Aase (1983) considers recursive estimation of nonlinear AR models. Zhang et al. (2011) discuss QML estimation of a two-regime SETAR–ARCH model with the conditional variance process depending on past time series observations. Koul and Schick (1997) propose adaptive estimators for the SETAR(2; 1, 1) and the ExpAR(1) model with known parameter γ, without sample splitting. These estimators have better performance (i.e. smaller MSEs) than estimators based on the sampling splitting technique.

Hili (1993, 2001, 2003, 2008a,b) considers the minimum Hellinger distance (MHD) (see Chapter 7) for estimating the parameters of the ExpARMA model (2.20), the simultaneous switching AR model, the general BL model (2.12), the SETAR($k; p, \ldots, p$) model, and nonlinear dynamical systems, respectively. Under some mild conditions he establishes consistency and asymptotic normality of the resulting parameter estimates. It is interesting to note that the practical feasibility of employing the MHD method covers many areas, including nonparametric ML estimation, and model selection criteria.

The theory of asymptotically optimal estimating function for stochastic models proposed by Godambe (1960, 1985) has been used as a framework for finite-sample nonlinear time series estimation. Thavaneswaran and Abraham (1988) construct G estimators (named after Godambe) for RCAR, doubly stochastic time series, and SETAR models; see also Chandra and Taniguchi (2001). These latter authors show that G estimators are better than CLS estimation by simulation. Amano (2009) obtains similar results for NLAR, RCAR, and GARCH models. Here, it is also appropriate to mention the generalized method of moments (GMM) developed by Hansen (1982) which is a widely used estimation method in econometrics. In fact, GMM estimation and Godambe's estimation function method are essentially the same. Caner (2002) obtains the asymptotic distribution for the least absolute deviation estimator of the threshold parameter in a threshold regression model.

For the CLS-based estimator of the BL model in (6.35), an expression for the asymptotic variance is given by Giordano (2000) and Giordano and Vitale (2003), assuming $\mathbb{E}(Y_t^8) < \infty$. This condition restricts the permissible parameter space considerably. Kim and Billard (1990) derive the asymptotic properties of the moment estimators of the parameters in a first-order diagonal BL model extended with a linear AR(1) term. This model is also the focus of a study by Ling et al. (2015). These authors propose a GARCH-type ML estimator for parameter estimation which is consistent and asymptotically normal under only finite fourth moment of the errors.

Outliers pose serious problems in time series model identification and estimation procedures. Gabr (1998) investigates the effect of additive outliers (AO) on the CLS estimation of BL models. For SETAR models, Chan and Cheung (1994) modify the class of generalized M-estimates. Their approach, however, can lead to inconsistent and very inefficient estimates of the threshold parameter even when the model is correctly specified and the errors

6.6 ADDITIONAL BIBLIOGRAPHICAL NOTES

are normally distributed (Giordani, 2006). Battaglia and Orfei (2005) propose a model-based method for detecting AO and innovational outliers (IO) in general NLAR time series processes.

Traditional likelihood analysis of threshold models is complicated because the threshold parameters can give rise to unknown shifts at arbitrary time points. On the other hand, the problem of estimating these parameters may be formulated into a Bayesian framework, and apply the Gibbs sampler (Geman and Geman, 1984), an MC simulation method, to obtain posterior distributions from conditional distributions. Amendola and Francq (2009, Section 7) briefly review MCMC methods, in particular the Metropolis–Hastings algorithm (Metropolis et al. (1953) and Hastings (1970)) and the Gibbs sampler for fitting STAR models. These authors also provide tools and approaches for nonlinear time series modeling in econometrics; see the website of this book. The function metrop in the R-mcmc package, and the function MCMCmetrop1R in the R-MCMCpack package can be used to perform a Bayesian analysis. Gibbs sampling, being a special case of the Metropolis–Hastings algorithm, is included in the R-gibbs.met package; see Robert and Casella (2004) for more information on MCMC methods.

Section 6.2: Sub-section 6.2.2 is partly based on Van Casteren and De Gooijer (1997). Using knowledge of the asymptotic properties of the CLS estimator for the SETAR model, Wong and Li (1998) show that AIC_c is an asymptotically unbiased estimator for the KL information. Kapetanios (2001) compares the small-sample performance of KL information-based model selection criteria for Markov switching, EDTAR, and two-regime SETAR models. A similar, but more extensive study, is undertaken by Psaradakis et al. (2009). Hamaker (2009) investigates six information criteria for determining the number of regimes in two-regime SETAR models. For small samples AIC_u should be preferred. Rinke and Sibbertsen (2016) compare regime weighted and equally weighted information criteria for simultaneous lag order and model class selection of SETAR and STAR models. Overall, in large samples, equally weighted criteria perform well.

Simonoff and Tsai (1999) derive and illustrate the AIC_c criterion for general regression models, including semiparametric and additive models. The MDL principle has been successfully applied to a wide variety of model selection problems in the fields of computer science, electrical engineering, and database mining; see, e.g., Grünwald et al. (2005). Good tutorial introductions are provided by Bryant and Cordero-Braña (2000), Hansen and Yu (2001), and Lanterman (2001). Qi and Zhang (2001) investigate the performance of AIC and BIC in selecting ANNs.

Öhrvik and Schoier (2005) propose three bootstrap criteria for two-regime SETAR model selection. Chen (1995) considers threshold variable selection in TARSO models. Chen et al. (1997) propose a unified, but computationally intensive, approach for model estimation via Gibbs sampling and to select an appropriate (non-nested) nonlinear model; see also Chen et al. (2011a). However, the correct specification of potentially non-nested nonlinear models and/or priors is not an easy task (Koop and Potter, 2001).

Based on the superconsistency of the SETAR–CLS threshold estimate established by Chan (1993), Strikholm and Teräsvirta (2006) provide a simple sequential method for determining the number of thresholds using general linearity tests. In addition, they compare their method with the approaches suggested by Gonzalo and Pitarakis (2002) (cf. Exercise 5.4(b)) and Hansen (1999).

Olteanu (2006) uses Kohonen maps and hierarchical clustering of arranged autoregressions to determine the number of regimes in switching AR (TAR and Markov switching) models.

Bermejo et al. (2011) propose an automatic procedure to identify SETAR models and to specify the values of thresholds. The method is based on recursive estimation of time-varying parameters in an arranged autoregression.

Dey et al. (1994) and Holst et al. (1994) consider ML estimation via recursive EM algorithms of switching AR(MAX) processes with a Markov regime. Krishnamurthy and Yin (2002) study the convergence and rate of convergence issues of these algorithms; see also Douc et al. (2014, Chapter 13 and Appendix D) on stochastic approximation EM algorithms.

Section 6.3: Li (2004, Sections 6.3 and 6.4) provides a comprehensive review on various diagnostic test statistics for ARCH and multivariate ARCH models. Li (1992) derives the asymptotic distribution of residual autocorrelations for a general NLAR model with strict WN errors. Hwang et al. (1994) extend this result to NLAR with random coefficients. Baek et al. (2012) derive the joint limit distribution of the sample residual ACF for NLAR time series models with unspecified heteroskedasticity. Based on this result they propose a test statistic which is an analogue of the test statistic $C_T^{(1,1)}(\ell)$.

An and Cheng (1991) introduce a KS-type test statistic based on the predicted residuals obtained by the best linear predictor for a NLAR process where the noise process follows a stationary martingale difference. The limiting distribution of the test statistic depends on the estimates of the unknown parameters of the AR(p) model considered under the null hypothesis. As an alternative, Kim and Lee (2002) propose a new KS test statistic and an associated BS procedure, which outperforms the original one. Hjellvik and Tjøstheim (1995, 1996) develop a nonparametric test statistic based on the distance between the best linear predictor and a nonlinear predictor obtained by kernel estimates of the conditional mean and conditional variance. However, to avoid the "curse-of-dimensionality", the conditional mean and variance functions only depend on $\{Y_{t-i}\}$ ($i = 1, \ldots, p$) rather than on $\{Y_{t-1}, \ldots, Y_{t-p}\}$. The difficulty which then emerges is that consistency of the resulting test statistic no longer holds. Also, Hjellvik et al. (1998) consider local polynomial estimation as a useful alternative to kernel estimation. Deriving asymptotic properties of the resulting linearity test statistic is, however, complicated.

An and Cheng (1991) and An et al. (2000) construct a CvM type test statistic which is simple to compute and partly avoids the curse of dimensionality problem when p is large. For time series generated by (6.73), Ling and Tong (2011) develop GOF test statistics that are based on empirical processes marked by certain scores. The tests are easy to implement, and are more powerful than other, residuals-based, test statistics.

6.7 Data and Software References

Data
Example 6.6: The daily HSI closing prices, adjusted for dividends and splits, for the year 2010 can be downloaded from the website of this book. For the estimation of the DTARCH model by GAs we used Double Threshold, a C++ executable program made available by Roberto Baragona and Domenico Cucina.

Software References
Sections 6.1.1: Tong (1983, Appendices A7 – A21) offers FORTRAN77 functions for testing, estimation, and evaluation of SETAR models. Some of these functions are rather dated. They are included in the interactive STAR package, to accompany the book by Tong

(1990). Unfortunately, the STAR package is no longer available for sale. However, with the consent of Howell Tong, the DOS-STAR3.2 program as an executable file (32-bit) is made available at the website of this book. Alternatively, the R-TSA package, supporting results in the textbook by Cryer and Chan (2008, Chapter 15), may be adopted for analyzing SETAR models; see also the R-tsDyn package mentioned earlier in Section 2.14.

RSTAR is a package for smooth transition AR modeling and forecasting; see https://www.researchgate.net/publication/293486017_RSTAR_A_Package_for_Smooth_Transition_Autoregressive_STAR_Modeling_Using_R. Alternatively, smooth transition regression (STR) models can be specified, estimated and checked in the freely available, and menu-driven, computer package JMulTi; see also Section 9.5. An EViews[7] add-in for STR analysis is available at http://forums.eviews.com/viewtopic.php?f=23&t=11597&sid=e01abc77f3732bfcdebcf2bce8dd1888. Another option is the Ox-STR2 package[8] (see http://www.doornik.com/download.html) based on Timo Teräsvirta's GAUSS code; see, also, http://people.few.eur.nl/djvandijk/nltsmef/nltsmef.htm.

Section 6.2.6: MATLAB code for comparing the performance of the various order selection criteria discussed in this section is available at the website of this book.

Section 6.3.1: The test results in Table 6.2 are computed using a GAUSS code provided by Yi-Ting Chen. The code is also available at the *Journal of Applied Econometrics* Data Archive.

Section 6.3.2: MATLAB codes for computing the test statistics A_{T,K_1} and H_{T,K_2} are available at the website of this book (file: Exercise 77b.zip).

Section 6.4: The paper by Knotters and De Gooijer (1999) contains (SS)TARSO models for time series of semi-monthly observed water table depths from six observation wells. The application only shows (SS)TARSO results for the first well. As a companion to the above paper, the website of this book offers FORTRAN77 codes for (SS)TARSO model identification and estimation.

Exercises

Theory Questions

6.1 Consider the simple BL model (6.35). Given the series of observation $\{Y_t\}_{t=1}^T$, the CLS estimator $\widehat{\tau}$ of the model parameter τ is defined by (6.39). Giordano (2000) proposes another estimator of τ, defined as

$$\widetilde{\tau} = \widehat{\gamma}_Y(1,2)/\sigma_\varepsilon^2 \text{Var}(Y_t),$$

where $\widehat{\gamma}_Y(i,j) = T^{-1}\sum_{t=1}^T Y_t Y_{t-i} Y_{t-j}$ ($Y_t = 0, t < 0$) is an estimator of the third-order cumulant $\mathbb{E}(Y_t Y_{t-i} Y_{t-j})$ ($i = 1, 2$), and $\text{Var}(Y_t) = \sigma_\varepsilon^2/(1 - \tau^2 \sigma_\varepsilon^2)$. Assume σ_ε^2 and σ_Y^2 are known, and let $\tau^4 \sigma_\varepsilon^4 < 1/3$. Then show that

$$|\widetilde{\tau} - \widehat{\tau}| \to 0 \quad \text{a.s., and} \quad |\widetilde{\tau} - \widehat{\tau}| = \mathcal{O}(S_T),$$

---

[7]EViews® (Econometric Views) is a software package for Windows, used mainly for econometric time series analysis. It was developed by Quantitative Micro Software, now a part of IHS.

[8]OxMetrics® is a commercial package using an object-oriented matrix programming language with a mathematical and statistical function library; published and distributed by http://www.timberlake.co.uk/software/oxmetrics.html. The downloadable Ox Console may be freely used for academic research and teaching purposes.

where $S_T = \{T/\log\log T\}^{-1/2}$.

6.2 Consider the diagonal BL(0,0,1,1) model $Y_t = \tau Y_{t-1}\varepsilon_{t-1} + \varepsilon_t$ with $\{\varepsilon_t\} \stackrel{\text{i.i.d.}}{\sim} \mathcal{N}(0, \sigma_\varepsilon^2)$. Let $\lambda = \tau\sigma_\varepsilon$. Assume that the stationarity condition holds, i.e., $|\lambda| < 1$. Then, by repeated substitution, the process $\{Y_t, t \in \mathbb{Z}\}$ can be written as

$$Y_t = U_{t,m} + W_{t,m},$$

where

$$U_{t,m} = \varepsilon_t + \sum_{j=1}^{m}\left(\prod_{\ell=1}^{j}\tau\varepsilon_{t-\ell}\right)\varepsilon_{t-j}, \quad W_{t,m} = \sum_{j=m+1}^{\infty}\left(\prod_{\ell=1}^{j}\tau\varepsilon_{t-\ell}\right)\varepsilon_{t-j}, \quad (m=1,2,\ldots).$$

(a) Show that $\mathbb{E}(Y_t) = \tau\sigma_\varepsilon^2$ and

$$\gamma_Y(\ell) = \begin{cases} \sigma_\varepsilon^2(1+\lambda^2+\lambda^4)/(1-\lambda^2), & \ell = 0, \\ \sigma_\varepsilon^2\lambda^2, & |\ell| = 1, \\ 0, & |\ell| \geq 2. \end{cases}$$

(b) Compare the ACF of the BL(0,0,1,1) process with the ACF of an invertible MA(1) process having the same innovation process as above. What do you conclude?

(c) Show that the BL process is invertible if the condition $|\lambda| < 0.605$ holds.

(d) Given the observations $\{Y_t\}_{t=1}^{T}$. Let $\overline{U}_T = T^{-1}\sum_{t=1}^{T} U_{t,m}$. Prove that, as $T \to \infty$,

$$\sqrt{T}(\overline{U}_T - \mu_U) \xrightarrow{D} \mathcal{N}\left(0, \sigma_\varepsilon^2\left(1+\lambda^2+3\sum_{j=1}^{m}\lambda^{2j}\right)\right),$$

where $\mathbb{E}(U_{t,m}) = \mu_U$.

(e) Assume σ_ε^2 is known. Kim et al. (1990) estimate the parameter τ by the method of moments. Their moment estimator $\widehat{\tau}$ is given by

$$\widehat{\tau} = \overline{Y}_T/\sigma_\varepsilon^2,$$

where $\overline{Y}_T = T^{-1}\sum_{t=1}^{T} Y_t$. Using the results in steps (a) and (c), prove that as $T \to \infty$,

$$\sqrt{T}(\widehat{\tau} - \tau) \xrightarrow{D} \mathcal{N}\left(0, \frac{1+3\tau^2-\tau^4}{1-\tau^2}\right).$$

[Hint: Define $Q_{m,T} = T^{-1/2}\sum_{t=1}^{T}(U_{t,m} - \mu_Y)$ and $R_{m,T} = T^{-1/2}\sum_{t=1}^{T} W_{t,m}$, with $\mu_Y = \mathbb{E}(Y_t) = \tau\sigma_\varepsilon^2$. Then consider the asymptotic distribution of $\sqrt{T}(\overline{Y}_T - \mu_Y)$.]

6.3 (a) Verify (6.44).

(b) Derive an explicit expression for the matrix $\mathbf{Z}$ in (6.45).

EXERCISES

6.4 Consider, as a special case of (6.73), the NLAR(p) model

$$Y_t = g(\mathbf{Y}_{t-1}; \boldsymbol{\theta}) + \varepsilon_t, \quad \{\varepsilon_t\} \stackrel{\text{i.i.d.}}{\sim} (0, \sigma_\varepsilon^2), \qquad (6.95)$$

where $\mathbf{Y}_{t-1} = (Y_{t-1}, Y_{t-2}, \ldots, Y_{t-p})'$, and $\boldsymbol{\theta} \in \boldsymbol{\Theta}$ is a parameter vector in a compact parameter space $\boldsymbol{\Theta}$. Take $P = Q = 1$ in (6.75), and set $(u_1(\varepsilon_t), v_1(\varepsilon_{t-\ell})) = (\varepsilon_t, \varepsilon_{t-\ell})$.

(a) Show the (i,j)th element of the asymptotic variance-covariance matrix $\boldsymbol{\Sigma}(\ell) = \mathbf{I}_{PQ} + \mathbf{A}(\ell, \ell)$ in (6.77) becomes

$$\boldsymbol{\Sigma}_{i,j}(\ell) = \delta_{ij} - \sigma_\varepsilon^{-2} \mathbf{m}_i' \mathbf{V}^{-1} \mathbf{m}_j,$$

with the $p \times 1$ vector $\mathbf{m}_i = \mathbb{E}[\varepsilon_t \nabla g(\mathbf{Y}_{t+i-1}; \boldsymbol{\theta})]$, $(i = 1, \ldots, \ell)$, and where $\mathbf{V}$ is a $p \times p$ matrix defined by $\mathbf{V} = \mathbb{E}[(\nabla g_t)(\nabla g_t)']$.

(b) Using part (a), suggest a general residuals-based diagnostic test statistic for non-linearity.

Empirical and Simulation Questions

6.5 Consider the BL model in (6.35). Let $\lambda = \tau \sigma_\varepsilon$, and in view of the moment condition when $\{\varepsilon_t\} \stackrel{\text{i.i.d.}}{\sim} \mathcal{N}(0, \sigma_\varepsilon^2)$ assume $\lambda^8 < 1/105$. Using the results in Exercise 6.1 it can be shown (Giordano and Vitale, 2003) that $\widehat{\tau}$, defined by (6.39), and $\widetilde{\tau}$ are asymptotically normally distributed with mean τ and variances respectively given by

$$\text{Var}(\widehat{\tau}) \approx \frac{1}{T} \frac{1}{\sigma_\varepsilon^2} \frac{1}{1 - 15\lambda^6} \Big(\frac{1 - \lambda^2}{1 - 3\lambda^4} (183\lambda^6 + 42\lambda^4 + 14\lambda^2 + 1) \Big),$$

$$\text{Var}(\widetilde{\tau}) \approx \frac{1}{T}(1 - \lambda^2)\Big(1 + 22\lambda^2 + 9\tau^2 \sigma_\varepsilon^2 - 6\frac{\lambda^2}{1 - \lambda^2}\Big).$$

Assume $\sigma_\varepsilon^2 = 1$. Based on 1,000 MC replications, compute 95% coverage probabilities of both estimators $\widehat{\tau}$ and $\widetilde{\tau}$ for $T = 1{,}000$, using $\tau = \pm 0.1$, ± 0.4 and ± 0.6. In addition, with the above specifications, compute the average length of the 95% confidence interval for both estimators. Compare and contrast the two estimators on the basis of the simulation results.

6.6 Consider the BL model of Exercise 6.2. If σ_ε^2 is known, it follows from $\mathbb{E}(Y_t) = \tau \sigma_\varepsilon^2$ that the moment estimator of τ is given by $\overline{Y}_T / \sigma_\varepsilon^2$. The solution of Exercise 6.2(c), contains an expression for σ_ε^2 in terms of $\gamma_Y(0)$ and $\gamma_Y(1)$. Using this expression, and assuming σ_ε^2 is *unknown*, Kim et al. (1990) propose the following method of moment estimator $\widehat{\tau}^*$ of τ

$$\widehat{\tau}^* = \frac{2\overline{Y}_T}{\{\widehat{\gamma}_Y(0) - \widehat{\gamma}_Y(1)\} + \{\widehat{\gamma}_Y^2(0) - 6\widehat{\gamma}_Y(0)\widehat{\gamma}_Y(1) - 3\widehat{\gamma}_Y^2(1)\}^{1/2}},$$

where $\widehat{\gamma}_Y(\ell) = T^{-1} \sum_{t=1}^{T-\ell}(Y_t - \overline{Y}_T)(Y_{t+\ell} - \overline{Y}_T)$ is the lag ℓ sample ACVF, with normalizing constant T^{-1} instead of $(T-\ell)^{-1}$. They show that $T^{1/2}(\widehat{\tau}^* - \tau)$ is asymptotically normally distributed with mean zero and with a lengthy expression for the variance.

(a) Based on 1,000 MC replications, compute the mean of the moment estimator $\widehat{\tau}^*$ for $T = 500$ and $1{,}000$, using $\tau = \pm 0.2$ and ± 0.4 as the parameters of the DGP. Also, compute the mean of the CLS estimator $\widehat{\tau}$ of τ.

(b) For comparison purposes, compute the bootstrap mean and standard deviation of $\widehat{\tau}^*$ and $\widehat{\tau}$, using 1,000 BS replicates and with the same data sets and specifications as in part (a). Comment on the obtained simulation results.

6.7 Consider the BL model (6.35) with $\tau = 0.6$, and $\sigma_\varepsilon^2 = 1$.

(a) Let $\widehat{\tau}$ be the estimator of τ as defined by (6.39). Based on 1,000 MC simulations obtain the distribution of $\sqrt{T}(\widehat{\tau}-\tau)$ and $\sqrt{T}(\widehat{\sigma}_\varepsilon^2-\sigma_\varepsilon^2)$ for $T = 250$ and $T = 1,000$. Investigate whether $\widehat{\tau}$ is an unbiased and/or consistent estimator of τ.

(b) Also, argue whether or not $\widehat{\sigma}_\varepsilon^2$ will be an unbiased and/or consistent estimator of σ_ε^2.

6.8 Consider the following LSTAR(2; 1, 1) model

$$Y_t = 1 + 0.9Y_{t-1} + (3 - 1.7Y_{t-1})/(1 + \exp(-10(Y_{t-1} - 5))) + \varepsilon_t, \quad \{\varepsilon_t\} \stackrel{\text{i.i.d.}}{\sim} \mathcal{N}(0,1).$$

(a) Using the R-tsDyn package, generate 100 times series of length $T = 200$ of this model, with starting condition $Y_0 = 0$. Check the local stationarity of the LSTAR model.

(b) Compute the sample distribution of the six parameter estimates. Comment on the outcomes.

(c) Optional: If the S-Plus FinMetrics commercial software package is available, repeat part (a). Compare the outcomes with those obtained in part (b).

6.9 As a part of the diagnostic checking stage, it is common to check the normality assumption. The data file Example62_res.dat contains the SETAR residuals of model (6.15).

(a) Using the Lin–Mudholkar test statistic (1.7), test the SETAR residuals for normality.

(b) Doornik and Hansen (2008) propose an omnibus test statistic for testing univariate or multivariate normality; see, e.g., the function normality.test1 in the R-normwhn.test package. Using this test statistic, investigate the normality assumption of the SETAR residuals.

Also, perform the Doornik–Hansen test using the function normality.test2. The associated test statistic allows for time series variables which are weakly dependent rather than i.i.d. Explain the differences with the results from part (a) if there are any?

(c) Relatively little is known about the finite-sample performance of diagnostic test statistics applied to residuals of fitted nonlinear time series models. This question explores this issue through a small MC simulation experiment. In particular, consider the SETAR(2; 1, 1) model

$$Y_t = \begin{cases} 0.3 - 0.5Y_{t-1} + \sigma_1\varepsilon_t & \text{if } Y_{t-1} \leq 0, \\ -0.1 + 0.5Y_{t-1} + \sigma_2\varepsilon_t & \text{if } Y_{t-1} > 0, \end{cases}$$

where (i) $\sigma_1 = \sigma_2 = 1$ (homoskedastic case), and (ii) $\sigma_1 = \sqrt{2}$, $\sigma_2 = 1$ (heteroskedastic case), and $\{\varepsilon_t\} \stackrel{\text{i.i.d.}}{\sim} \mathcal{N}(0,1)$.

EXERCISES

Using bootstrapped CLS–SETAR residuals, compare the empirical size of the Lin–Mudholkar normality test statistic and the Doornik–Hansen omnibus normality test statistics for $T = 100$ and $T = 300$, and at nominal significance levels $\alpha = 0.01$, 0.025, and 0.05. Set the number of BS replicates at $B = 10{,}000$, and assume that the threshold parameter $r = 0$ and the delay $d = 1$ are known. Also, as a benchmark, compute the empirical size of both test statistics for pure i.i.d. $\mathcal{N}(0, 1)$ errors.

Chapter 7

TESTS FOR SERIAL INDEPENDENCE

Testing for randomness of a given finite time series is one of the basic problems of statistical analysis. For instance, in many time series models the noise process is assumed to consist of i.i.d. random variables, and this hypothesis should be testable. Also, it is the first issue that gets raised when checking the adequacy of a fitted time series model through observed "residuals", i.e. are they approximately i.i.d. or are there significant deviations from that assumption. In fact, many inference procedures apply only to i.i.d. processes.

In Section 1.3.2, we noted that the traditional sample ACF and sample PACF are rather limited in measuring nonlinear dependencies in strictly stationary time series processes. As a result a wide variety of alternative dependence measures have been proposed, often resulting in test statistics which have appealing statistical properties. Broadly, these test statistics can be divided into two categories: those designed with a specific nonlinear alternative in mind – such as the time-domain test statistics discussed in Chapter 5 – and serial independence tests. When the parameters of the fitted model are known, these latter tests are useful to detect neglected structure in residuals. In reality, however, the model parameters are unknown. This has motivated the development of nonparametric test statistics for serial independence. In fact, over the past few years, enormous progress has been made in this area.

In this chapter, we consider both historic and more recent work in the area of nonparametric serial independence tests for conditional mean models. In the next section, we start off by expressing the null hypothesis of interest in various forms. In Section 7.2, we introduce a number of distance measures and dependence functionals. Jointly with a particular form of the null hypothesis, these measures and functionals are the "backbone" for constructing the test statistics in Sections 7.3 and 7.4. Here, we distinguish between procedures for testing first-order, or single-lag, serial dependence (two dimensions), and high-dimensional tests. Throughout the chapter, a number of examples illustrate the performance of the proposed test statistics on empirical data. In Section 7.5, this is complemented with an application of high-dimensional serial independence test statistics to a famous data set.

To facilitate reading, technical details will be kept to a minimum. They are only provided to understand the main premises underlying the construction of the test statistics. In particular, three technical appendices are added to the chapter. In Appendix 7.A, we briefly discuss kernel-based density and regression estimation in the simple setting of i.i.d. DGPs. Many of the nonparametric methods discussed in this chapter are direct generalizations of this case. In Appendix 7.B, we present a general overview of copula theory. Finally, in Appendix 7.C, we provide some information about the theory of U- and V-statistics. These notions are often mentioned in this chapter as useful ways to derive asymptotic theory of certain test statistics.

7.1 Null Hypothesis

Let $\{Y_t, t \in \mathbb{Z}\}$ be a strictly stationary time series process with values in $\mathbb{R}$. The null hypothesis of interest is

$$\mathbb{H}_0: \{Y_t\} \stackrel{\text{i.i.d.}}{\sim} \mu, \tag{7.1}$$

where μ is some probability measure on the real line associated with $\{Y_t, t \in \mathbb{Z}\}$. In practice, it will not be easy to uniquely determine dependencies in a set of observed time series data given the above setup. Rather than focusing on a single time series in $\mathbb{R}$, it is practical to consider a time series process in $\mathbb{R}^m$, which at lag ℓ, is given by

$$\mathbf{Y}_t^{(\ell)} = (Y_{1,t}, \ldots, Y_{m,t})' = (Y_t, Y_{t-\ell}, \ldots, Y_{t-(m-1)\ell})', \quad (m \in \mathbb{Z}^+, \ell \in \mathbb{Z}),$$

with probability measure, say $\mu_m^{(1)}$. Then the null hypothesis of serial independence can be rephrased as

$$\mathbb{H}_0: \mu_m^{(1)} = \mu_m^{(2)} \quad (m \in \mathbb{N}^+), \tag{7.2}$$

where for any Borel-measurable set $A \in \mathbb{R}^m$

$$\mu_m^{(2)}(A) = \int_A d\mu(y_1) \times \cdots \times d\mu(y_m)$$

which is invariant under permutations of the m coordinates.[1]

Alternatively, a more direct formulation of the null hypothesis of serial independence, follows from assuming that $\{\mathbf{Y}_t^{(\ell)}, t \in \mathbb{Z}\}$ admits a common continuous joint density function $f_m(A)$. Denote the marginal density function by $f(y)$. Then, if $\{Y_t, t \in \mathbb{Z}\}$ is i.i.d., the joint density function will be equal to the product of the individual marginals, and the hypothesis of interest is

$$\mathbb{H}_0: f_m(\mathbf{y}) = f(y_1) \times \cdots \times f(y_m), \quad \forall \mathbf{y} \in \mathbb{R}^m. \tag{7.3}$$

---

[1] For continuous distributions, the measure $\mu(\mathbf{y})$ is zero at a single point $\mathbf{y} = (y_1, \ldots, y_m)'$, so we should consider $\mu(\cdot)$ on measurable (compact) subsets A of $\mathbb{R}^m$.

7.1 NULL HYPOTHESIS

Moreover, if $\{\mathbf{Y}_t^{(\ell)}, t \in \mathbb{Z}\}$ admits a continuous distribution function $F_m(\mathbf{y})$, the above hypothesis can also be formulated in terms of joint and marginal distribution functions, i.e.,

$$\mathbb{H}_0: F_m(\mathbf{y}) = F(y_1) \times \cdots \times F(y_m), \quad \forall \mathbf{y} \in \mathbb{R}^m, \tag{7.4}$$

where $F(y_i)$ is the marginal distribution of $\{Y_{t-(i-1)\ell}\}$ $(i = 1, \ldots, m)$.

In view of the one-to-one correspondence between distribution functions and characteristic functions, it is natural to construct serial independence test statistics on the basis of the difference between the joint characteristic function of $\{\mathbf{Y}_t^{(\ell)}, t \in \mathbb{Z}\}$ and the product of its marginal characteristic functions. Specifically, let $\phi_\ell(\mathbf{u}) = \mathbb{E}\{\exp\left(i(\sum_{k=1}^m u_k Y_{t-(k-1)|\ell|})\right)\}$ be the joint characteristic function where $\mathbf{u} = (u_1, \ldots, u_m)' \in \mathbb{R}^m$. Then the difference between $\phi_\ell(\cdot)$ and the product of the marginal characteristic functions $\phi(u_k) = \mathbb{E}\{\exp(iu_k Y_t)\}$ $(k = 1, \ldots, m)$ can be expressed as

$$D_\ell(\mathbf{u}) = \phi_\ell(\mathbf{u}) - \prod_{k=1}^m \phi(u_k), \quad \ell = 0, \pm 1, \ldots. \tag{7.5}$$

This expression is zero $\forall \mathbf{u} \in \mathbb{R}^m$, if and only if there is no serial dependence of order $m - 1$ or, equivalently,

$$\mathbb{H}_0: D_\ell(\mathbf{u}) = 0, \quad \forall \mathbf{u} \in \mathbb{R}^m. \tag{7.6}$$

Finally, an equivalent formulation of the null hypothesis of serial independence can be based on copula functions. To be more specific, consider an m-dimensional joint CDF $F_m(\mathbf{y}): \mathbb{R}^m \to [0, 1]$, with marginal distributions $F(y_i)$ which are assumed to be absolutely continuous. According to Sklar's theorem (see Appendix 7.B), there exists an m-copula function $\mathbb{C}(\cdot)$ of $\{\mathbf{Y}_t^{(\ell)}, t \in \mathbb{Z}\}$, such that $\forall \mathbf{y} \in \mathbb{R}^m$, $F_m(\mathbf{y}) = \mathbb{C}(F(y_1), \ldots, F(y_m))$. The corresponding joint pdf is

$$f_m(\mathbf{y}) = c(F(y_1), \ldots, F(y_m)) \prod_{i=1}^m f(y_i), \tag{7.7}$$

where $c(\mathbf{u})$, the density of the copula $\mathbb{C}(\mathbf{u})$, is given by

$$c(\mathbf{u}) = \frac{\partial^m \mathbb{C}(\mathbf{u})}{\partial u_1 \times \cdots \times \partial u_m} = \frac{f_m(\mathbf{u})}{\prod_{i=1}^m f(u_i)}, \quad \mathbf{u} \in [0, 1]^m. \tag{7.8}$$

Hence, in terms of copulas, (7.3) corresponds to testing the null hypothesis

$$\mathbb{H}_0: c(\mathbf{u}) = 1. \tag{7.9}$$

For each of the null hypotheses specified above any deviation from the corresponding equality is evidence of serial dependence.

7.2 Distance Measures and Dependence Functionals

7.2.1 Correlation integral

In view of the null hypothesis (7.2), Grassberger and Procaccia (1983) propose the so-called *correlation integral* as a measure of spatial correlation in $\{\mathbf{Y}_t^{(\ell)}, t \in \mathbb{Z}\}$ with $\ell = 1$, which we denote by $\{\mathbf{Y}_t, t \in \mathbb{Z}\}$. This measure of distance is characterized by

$$C_{m,Y}(h) = \int_{\mathbb{R}^m} \int_{\mathbb{R}^m} I(\|\mathbf{y} - \mathbf{x}\| \leq h) \mathrm{d}\mu_m(\mathbf{y}) \mathrm{d}\mu_m(\mathbf{x}), \tag{7.10}$$

where h is a bandwidth, depending on T, and $\|\cdot\|$ a norm (e.g., Euclidean norm).[2] If the m-dimensional time series process $\{\mathbf{Y}_t, t \in \mathbb{Z}\}$ clusters in any dimension, then $C_{m,Y}(h)$ will take on relatively large values. If, however, the time series process is i.i.d. the correlation integral factorizes, i.e.

$$C_{m,Y}(h) = \{C_{1,Y}(h)\}^m, \tag{7.11}$$

and this equality can be used as a basis for a test of serial independence. Note that for (7.11) no moments of $\{\mathbf{Y}_t, t \in \mathbb{Z}\}$ are required.

7.2.2 Quadratic distance

Model fit assessment for i.i.d. (time-independent) data is usually based, explicitly, or implicitly, on measures of distance $\Delta(\mu_F, \mu_G)$ between probability measures μ_F and μ_G. One particular class of measures is the kernel-based quadratic distance defined as

$$\Delta_{\mathcal{K}}(\mu_F, \mu_G) = \iint \mathcal{K}(s,t) \mathrm{d}(\mu_F - \mu_G)(s) \mathrm{d}(\mu_F - \mu_G)(t), \tag{7.12}$$

where $\mathcal{K}(s,t)$ (possibly depending on G) is a bounded, symmetric kernel function on the two-dimensional sample space. This form is asymmetric in μ_F and μ_G, but it is symmetric with respect to interchanging μ_F and μ_G. For computational purposes (7.12) can be written in the form

$$\Delta_{\mathcal{K}}(\mu_F, \mu_G) = \mathcal{K}(\mu_F, \mu_F) - \mathcal{K}(\mu_F, \mu_G) - \mathcal{K}(\mu_G, \mu_F) + \mathcal{K}(\mu_G, \mu_G),$$

where $\mathcal{K}(A, B) = \iint \mathcal{K}(s,t) \mathrm{d}A(s) \mathrm{d}B(t)$.

Clearly, the building block of (7.12) is the kernel function $\mathcal{K}(\cdot, \cdot)$. This function is assumed to be bounded, absolutely integrable, and consequently it has an FT which does not vanish on any interval. Then, in analogy with matrix theory, its associated *quadratic form* $\iint \mathcal{K}(s,t) \mathrm{d}\sigma(s) \mathrm{d}\sigma(t)$ is called *nonnegative definite*, for all bounded signed measures σ.

---

[2] Within the information theoretic literature the symbol ϵ is often used for the bandwidth, also called *tolerance distance* or *cut-off threshold*.

7.2 DISTANCE MEASURES AND DEPENDENCE FUNCTIONALS

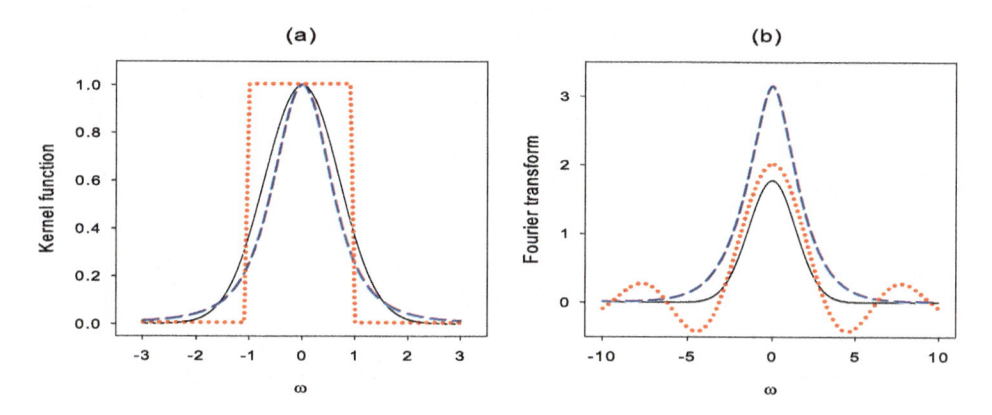

Figure 7.1: *Three kernel functions (left panel) and their associated FTs (right panel): Gaussian (black solid line), squared Cauchy (blue medium dashed line), and uniform (red dotted line).*

Example 7.1: Some Kernel Functions and their FTs

Figure 7.1 shows plots of three kernel functions and their associated FTs. In particular, we have (i) the Gaussian kernel $\mathcal{K}(x) = e^{-x^2}$ and its FT $\widetilde{\mathcal{K}}(\omega) = \sqrt{\pi} e^{-\omega^2/4}$; (ii) the squared Cauchy kernel $\mathcal{K}(x) = 1/(1+x^2)^2$ and its FT $\widetilde{\mathcal{K}}(\omega) = \pi(|\omega|+1)e^{-|\omega|}$; and (iii) the uniform kernel $\mathcal{K}(x) = I(|x| \leq 1)$ and its FT $\widetilde{\mathcal{K}}(\omega) = (2/\omega)\sin(\omega)$. Note, that the Gaussian kernel has a Gaussian density as its FT, which is everywhere positive. Hence, the Gaussian product kernel is positive definite and defines a quadratic form suitable for detecting any differences between a pair of distributions. Similarly, (ii) corresponds, after normalizing, to a density function. On the other hand, (iii) is not a positive definite kernel, as its FT takes negative values for certain frequencies.

A number of classically distances such as Pearson's chi-square or Cramér–von Mises (CvM), are quadratic distances; see Lindsay et al. (2008). For instance, within the context of serial correlation tests, the L_2-norm can be used. Specifically, given the m-dimensional process $\{\mathbf{Y}_t, t \in \mathbb{Z}\}$, a quadratic (Q) form measuring the serial dependence in this process is given by

$$\Delta^Q(m) = \|\mu_m^{(1)} - \mu_m^{(2)}\|^2 = (\mu_m^{(1)}, \mu_m^{(1)}) - 2(\mu_m^{(1)}, \mu_m^{(2)}) + (\mu_m^{(2)}, \mu_m^{(2)}), \qquad (7.13)$$

where

$$(\mu_m^{(i)}, \mu_m^{(j)}) = \int_{\mathbb{R}^m} \int_{\mathbb{R}^m} \mathcal{K}_h(\mathbf{y} - \mathbf{x}) d\mu_m^{(i)}(\mathbf{y}) d\mu_m^{(j)}(\mathbf{x}), \quad (i, j = 1, 2),$$

with $\mathcal{K}_h(\cdot)$ a nonnegative definite, spherically symmetric m-variate kernel function, and $h > 0$ a bandwidth parameter. To make the distance calculation explicit and fast, we recommend kernels that factorize as $\mathcal{K}_h(\mathbf{z}) = \prod_{i=1}^m K(z_i)/h$. Here, $K(\cdot)$ is a one-dimensional kernel function, which is symmetric around zero. It is easily seen that the functional $(\mu^{(1)}, \mu^{(1)}) - (\mu^{(2)}, \mu^{(2)})$ with the 'naive' or identity kernel function $K_h(z) = I(|z| < h)$ corresponds to (7.11).

Because FTs leave the L_2-norm invariant by *Parseval's identity* (loosely speaking the sum or integral of the square of a function is equal to the sum or integral of the square of its FT), we can express (7.13) as

$$\Delta^Q(m) = \int_{\mathbb{R}^m} \int_{\mathbb{R}^m} \mathcal{K}_h(\mathbf{y}-\mathbf{x}) \mathrm{d}(\mu_m^{(1)} - \mu_m^{(2)})(\mathbf{y}) \mathrm{d}(\mu_m^{(1)} - \mu_m^{(2)})(\mathbf{x})$$
$$= \int_{\mathbb{R}^m} \widetilde{\mathcal{K}}_h(\boldsymbol{\xi}) |\phi(\mu_m^{(1)}(\boldsymbol{\xi})) - \phi(\mu_m^{(2)}(\boldsymbol{\xi}))|^2 \mathrm{d}\boldsymbol{\xi}, \qquad (7.14)$$

where $\widetilde{\mathcal{K}}_h(\cdot)$ is the FT of $\mathcal{K}_h(\cdot)$, $\phi(\mu_m^{(i)}(\cdot))$ the characteristic function of $\mu_m^{(i)}(\cdot)$, and $|\cdot|$ the modulus.

Example 7.2: An Explicit Expression for $\Delta^Q(\cdot)$ (Diks, 2009)

Let $\{Y_t, t \in \mathbb{Z}\}$ be a strictly stationary time series process with a standard normal marginal distribution. The joint density function of $\{\mathbf{Y}_t, t \in \mathbb{Z}\}$ is of the form $f_m(\mathbf{y}) = (2\pi)^{-m/2} |\mathbf{R}|^{-1/2} \exp(-\frac{1}{2}\mathbf{y}'\mathbf{R}^{-1}\mathbf{y})$ where $\mathbf{y} = (y_1, \ldots, y_m)'$ and $\mathbf{R}$ is the $m \times m$ correlation matrix of $\{\mathbf{Y}_t, t \in \mathbb{Z}\}$, which is assumed to be positive definite. The Gaussian density product kernel is given by $\mathcal{K}_h(\mathbf{y}-\mathbf{x}) = (2\sqrt{\pi}h)^{-m} \prod_{i=1}^m \exp\big(-(y_i - x_i)^2/(4h^2)\big)$, where the factor 4 is chosen for convenience as it simplifies some of the results given below.

Evaluating the multivariate normal integral in (7.14) can be simplified by making the transformation $\mathbf{z} = \mathbf{V}\mathbf{y}$, where $\mathbf{V}$ is an orthogonal matrix and where, by the spectral decomposition of a positive definite symmetric matrix, $\mathbf{R} = \mathbf{V}\mathbf{D}\mathbf{V}'$, with $\mathbf{D} = \mathrm{diag}(\lambda_1^2, \ldots, \lambda_m^2)$ giving the joint pdf $f_m^*(\mathbf{z}) = (2\pi)^{-m/2} \prod_{i=1}^m \lambda_i^{-1} \exp\big(-z_i^2/(2\lambda_i^2)\big)$, with the Jacobian of the transformation equal to unity. Denote the product of the marginal pdfs of the transformed process by $f^0(\cdot)$. Then, replacing $\mathrm{d}\mu_m(\mathbf{y})$ by $\mathrm{d}\mathbf{y} f_m(\mathbf{y})$, it is easy to see that

$$(\mu_m^{(1)}, \mu_m^{(1)}) = \int_{\mathbb{R}^m} \int_{\mathbb{R}^m} \mathcal{K}_h(\mathbf{r}-\mathbf{s}) \mathrm{d}\mathbf{r} f_m^*(\mathbf{r}) \mathrm{d}\mathbf{s} f_m^*(\mathbf{s}) = \frac{1}{(2\sqrt{\pi})^m} \prod_{i=1}^m \frac{1}{\sqrt{h^2 + \lambda_i^2}},$$

$$(\mu_m^{(1)}, \mu_m^{(2)}) = \int_{\mathbb{R}^m} \int_{\mathbb{R}^m} \mathcal{K}_h(\mathbf{r}-\mathbf{s}) \mathrm{d}\mathbf{r} f_m^*(\mathbf{r}) \mathrm{d}\mathbf{s} f_m^0(\mathbf{s})$$
$$= \frac{1}{(2\sqrt{\pi})^m} \prod_{i=1}^m \frac{1}{\sqrt{h^2 + (\lambda_i^2 + 1)/2}},$$

$$(\mu_m^{(2)}, \mu_m^{(2)}) = \int_{\mathbb{R}^m} \int_{\mathbb{R}^m} \mathcal{K}_h(\mathbf{r}-\mathbf{s}) \mathrm{d}\mathbf{r} f_m^0(\mathbf{r}) \mathrm{d}\mathbf{s} f_m^0(\mathbf{s}) = \frac{1}{(2\sqrt{\pi})^m} \prod_{i=1}^m \frac{1}{\sqrt{h^2 + 1}}.$$

Combining terms gives an explicit, no-integration needed, formula for $\Delta^Q(m)$. If, for example $m = 2$, $\lambda_{1,2}^2 = 1 \pm \rho$, where ρ is the correlation coefficient

7.2 DISTANCE MEASURES AND DEPENDENCE FUNCTIONALS

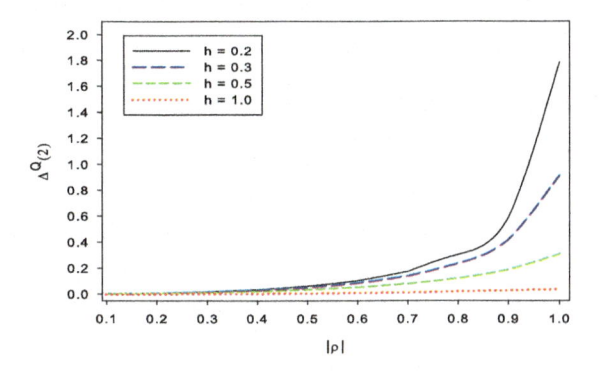

Figure 7.2: *Distance $\Delta^Q(2)$ between a bivariate standard normal distribution and a correlated bivariate normal distribution with correlation coefficient ρ, for different values of h.*

between Y_t and Y_{t-1}. Consequently,

$$\Delta^Q(2) = \frac{1}{4\pi}\left(\frac{1}{\sqrt{(h^2+1)^2 - \rho^2}} - \frac{2}{\sqrt{(h^2+1)^2 - \rho^2/4}} + \frac{1}{h^2+1}\right). \quad (7.15)$$

Figure 7.2 shows $\Delta^Q(2)$ for bandwidths $h = 0.2, 0.3, 0.5$, and 1.0 as a function of $|\rho|$. Note from (7.15) that, as $h \to 0$, the limiting squared distance function is well-defined which need not be the case for other combinations of kernel functions and pdfs.

7.2.3 Density-based measures

Several density-based measures can be used for testing (7.3). Here, we consider the case of pairwise ($m = 2$) serial dependence, and suppress the dependence on m for notational clarity. That is, for a strictly stationary time series process $\{Y_t, t \in \mathbb{Z}\}$ with marginal density function $f(\cdot)$ and joint pdf $f_\ell(\cdot, \cdot)$ of $(Y_t, Y_{t-\ell})'$ ($\ell \in \mathbb{Z}$), we measure the degree of dependence by $\Delta(\ell) \equiv \Delta\big(f_\ell(x,y), f(x)f(y)\big)$. It is natural to require that $\Delta(\cdot)$ has the following basic properties: (i) nonnegativity, (ii) maximal information, and (iii) invariance under continuous monotonic increasing transformations. For divergence measures not satisfying (iii), one can obtain scale and location invariance by simply standardizing $\{Y_t, t \in \mathbb{Z}\}$, assuming that the second moments exist. Or retain invariance under monotonic transformations by transforming the data to any given marginal density function (e.g. take ranks or transform to a standard normal marginal). The second moment then doesn't even need to exists.

The functionals considered below are all of the type

$$\Delta(\ell) = \int_{S^2} B\{f_\ell(x,y), f(x), f(y)\} f_\ell(x,y) \mathrm{d}x\mathrm{d}y, \quad (7.16)$$

where $B(\cdot,\cdot,\cdot)$ is a real-valued function, and the integrals are taken over the support, say S^2, of $(Y_t, Y_{t-\ell})'$.

Several functionals have been proposed in the information theory literature. Roughly, the resulting measures can be classified in four major categories:

- *Generalized Kolmogorov* (K) *divergence measure*

$$\Delta_q^{\mathrm{K}}(\ell) = \Big\{\int_{S^2} \big|f_\ell(x,y) - f(x)f(y)\big|^q \mathrm{d}x\mathrm{d}y\Big\}^{1/q}, \quad (q>0),$$

which for $q=1$ is the L_1-norm. $\Delta_q^{\mathrm{K}}(\cdot)$ satisfies properties (i) – (ii), but not (iii).

- *Csiszár* (C) (1967) *divergence measure*

$$\Delta^{\mathrm{C}}(\ell) = \int_{S^2} \phi\Big\{\frac{f_\ell(x,y)}{f(x)f(y)}\Big\} f_\ell(x,y) \mathrm{d}x\mathrm{d}y,$$

where $\phi(\cdot)$ is some strictly convex function on $[0,\infty)$. Thus, $B\{z_1, z_2, z_2\} \equiv \phi(z_1/z_2 z_3)$.

- *Rényi* (R) (1961) *divergence measure*

$$\Delta_q^{\mathrm{R}}(\ell) = \frac{1}{q-1} \log \int_{S^2} \big\{f_\ell(x,y)\big\}^{q-1} \big\{f(x)f(y)\big\}^q \mathrm{d}x\mathrm{d}y, \quad (0<q<1).$$

- *Tsallis* (T) (1998) *divergence measure*

$$\Delta_q^{\mathrm{T}}(\ell) = \begin{cases} \dfrac{1}{1-q} \int_{S^2} \Big\{1 - \Big(\dfrac{f(x)f(y)}{f_\ell(x,y)}\Big)^{1-q}\Big\} f_\ell(x,y) \mathrm{d}x\mathrm{d}y & (q\neq 1), \\ \displaystyle\int_{S^2} \log\Big(\dfrac{f_\ell(x,y)}{f(x)f(y)}\Big) f_\ell(x,y) \mathrm{d}x\mathrm{d}y & (q=1). \end{cases}$$

For testing purposes, both Rényi's measure and Tsallis' measure satisfy properties (i) – (iii).

The above list is far from exhaustive. Other possible candidates for measuring statistical (serial) dependence include the difference functional (Skaug and Tjøstheim, 1993a) which, if we set $B\{z_1, z_2, z_3\} = z_1 - z_2 z_3$ in (7.16), is given by

$$\Delta^*(\ell) = \int_{S^2} \{f_\ell(x,y) - f(x)f(y)\} f_\ell(x,y) \mathrm{d}x\mathrm{d}y, \qquad (7.17)$$

and the *Hellinger* (H) (1909) *distance* which, with $B\{z_1, z_2, z_3\} = \big(1-(z_1/z_2 z_3)^{-1/2}\big)^2$, is defined as

$$\Delta^{\mathrm{H}}(\ell) = \int_{S^2} \Big\{f_\ell^{1/2}(x,y) - \big(f(x)f(y)\big)^{1/2}\Big\}^2 \mathrm{d}x\mathrm{d}y$$

$$= 2 - 2 \int_{S^2} \Big(\frac{f(x)f(y)}{f_\ell(x,y)}\Big)^{1/2} f_\ell(x,y) \mathrm{d}x\mathrm{d}y.$$

7.2 DISTANCE MEASURES AND DEPENDENCE FUNCTIONALS

It is easy to see that the Hellinger distance is symmetric, and hence it can serve as a distance measure contrary to other divergences.[3]

In addition, various relations exist between the divergence measures. For instance, Rényi's information divergence follows from Csiszár's measure by taking $\phi(u) = \text{sign}(u-1)u^q$ ($u \geq 0; q \neq 1$) which yields $\Delta_q^{\text{R}}(\cdot) = (q-1)^{-1}\log|\Delta_q^{\text{C}}(\cdot)|$. The connection between Rényi's measure and Tsallis' measure is given by $\Delta_q^{\text{R}}(\cdot) = (q-1)^{-1}\log[1+(1+q)\log\Delta_q^{\text{T}}(\cdot)]$. Clearly, when $\phi(\cdot)$ is taken as the logarithmic function, Csiszár's measure is equivalent to the KL information measure $I^{\text{KL}}(\cdot)$. defined in (1.18). Moreover, $I^{\text{KL}}(\cdot) \equiv \Delta_1^{\text{T}}(\cdot)$ and $\Delta_{1/2}^{\text{T}}(\cdot) \equiv \Delta^{\text{H}}(\cdot)$.

7.2.4 Distribution-based measures

In view of (7.4), test statistics for pairwise serial independence also have been proposed on appropriate functionals measuring the distance between the joint distribution function $F_\ell(x,y)$, suppressing the dependence on m, and the product of the marginal distributions $F(x)F(y)$. Two useful types of functionals for this purpose are

$$C_q(\ell) = \int_{S^2} \Delta_q^{\text{CR}}(\ell) \mathrm{d}w_\ell(x,y), \text{ and } C_q^{\max}(\ell) = \sup_{S^2}[\Delta_q^{\text{CR}}(\ell)w_\ell(x,y)], \quad (7.18)$$

where $w_\ell(\cdot,\cdot)$ is a positive weight function and $\Delta_q^{\text{CR}}(\cdot)$ is the so-called Cressie–Read (CR) (1984) divergence measure which, in a time series setting, is defined by

$$\Delta_q^{\text{CR}}(\ell) = \frac{2}{q+1}\Big\{F(x)F(y)\Big(\frac{F(x)F(y)}{F_\ell(x,y)}\Big)^q + \Big(1-F(x)F(y)\Big)\Big(\frac{1-F(x)F(y)}{1-F_\ell(x,y)}\Big)^q - 1\Big\}.$$

The Cressie–Read measure and Rényi's divergence measure are related:

$$\Delta_q^{\text{CR}}(\ell) = \frac{2}{q+1}\Big\{\exp\Big[q\Big(\Delta_{q+1}^{\text{R}}\Big(\frac{F(x)F(y)}{F_\ell(x,y)}\Big) + \Delta_{q+1}^{\text{R}}\Big(\frac{1-F(x)F(y)}{1-F_\ell(x,y)}\Big)\Big)\Big] - 1\Big\}.$$

By choosing different weight functions in (7.18), a number of "classical" functionals follow. For instance, using $q=1$ and $w_\ell(x,y) = F_\ell(x,y)(1-F_\ell(x,y))\mathrm{d}F_\ell(x,y)$ in $C_q(\cdot)$ gives the CvM functional

$$\Delta^{\text{CvM}}(\ell) = \int_{S^2}\Big\{F_\ell(x,y) - F(x)F(y)\Big\}^2 \mathrm{d}F_\ell(x,y).$$

This measure satisfies the properties of nonnegativity and maximal information, but is not invariant under continuous monotonic increasing transformations. By evaluating the integral and replacing the distribution functions by their empirical

---

[3]The Hellinger (H) distance satisfies the inequality $0 \leq \Delta^{\text{H}}(\ell) \leq 2$. Some authors prefer to have an upper bound of 1; they include an extra factor of $1/2$ in the definition of $\Delta^{\text{H}}(\ell)$.

counterparts, the CvM–GOF test statistic (4.38) can be obtained. Another well-known functional follows from setting $q = 1$ and $w_\ell(x,y) = F_\ell(x,y)(1 - F_\ell(x,y))$ in $C_q^{\max}(\cdot)$, i.e.,

$$\left(\Delta^{\mathrm{KS}}(\ell)\right)^2 = \left(\sup_{S^2} |F_\ell(x,y) - F(x)F(y)|\right)^2,$$

where $\Delta^{\mathrm{KS}}(\cdot)$ is the *Kolmogorov–Smirnov* (KS) *divergence measure*. This measure satisfies the basic properties (i) – (iii). Setting $q = 1$ and $w_\ell(x,y) = \mathrm{d}F_\ell(x,y)$ in $C_q(\cdot)$ generates the *Anderson–Darling* (AD) functional

$$\Delta^{\mathrm{AD}}(\ell) = \int_{S^2} \left(F(x)F(y) - F_\ell(x,y)\right)^2 F_\ell^{-1}(x,y)\left(1 - F_\ell(x,y)\right)^{-1} \mathrm{d}F_\ell(x,y),$$

which, after evaluating the integral and some algebra, leads to (4.39).

All the above measures consider the distance between two-dimensional densities or two-dimensional distribution functions at a *single-lag* ℓ. However, for testing $\mathbb{H}_0^{(\ell)}: f(Y_t, Y_{t-\ell}) = f(Y_t)f(Y_{t-\ell})$, it is possible that two different lags ℓ may give conflicting conclusions. It is thus desirable to have a *multiple-lag* testing procedure. One simple procedure is to form M linear combinations of single-lag two-dimensional test functionals $\Delta(\ell)$, i.e.

$$\mathcal{Q}(M) = \frac{1}{\sqrt{M}} \sum_{\ell=1}^{M} \Delta(\ell), \quad (M \in \mathbb{N}^+), \tag{7.19}$$

with corresponding null hypothesis

$$\mathbb{H}_0^{\mathrm{P}}: \cap_{\ell=1}^{M} \mathbb{H}_0^{(i_\ell)}, \quad (i_1 < \cdots < i_M). \tag{7.20}$$

Test statistics derived from (7.19) are *portmanteau*-type tests. Alternatively, one may use the Bonferroni correction procedure, based on the p-values of the individual single-lag serial correlation test statistics. Notice, however, that pairwise (serial) independence for all combinations of paired random variables does not imply joint (serial) independence in general. Hence, methods for the detection of serial dependence in $m > 2$ dimensions are needed; see Section 7.4.

7.2.5 Copula-based measures

From (7.7), we see that factorization of the joint pdf in the product of marginals is a property of the copula. In this sense the copula contains all relevant information regarding the dependence structure of $\{\mathbf{Y}_t^{(\ell)}, t \in \mathbb{Z}\}$. Thus, similar as the two-dimensional density-based measures, it is natural to define m-dimensional copula-based measures for serial dependence. Moreover, if the invariance property (iii) of Section 7.2.3 holds, the dependence structure of $\{\mathbf{Y}_t^{(\ell)}, t \in \mathbb{Z}\}$ is completely captured by the copula.

Recall that Tsallis' divergence satisfies (i) – (iii). In line with its definition in Section 7.2.3, it is easy to see that an m-dimensional copula-based (denoted by the superscript c) version of $\Delta_q^T(\cdot)$ is defined as

$$\Delta_{m,q}^{T,c}(\ell) = \begin{cases} \dfrac{1}{1-q} \displaystyle\int_{[0,1]^m} \left\{1 - \left(\dfrac{1}{c(\mathbf{u})}\right)^{1-q}\right\} c(\mathbf{u}) d\mathbf{u} & (q \neq 1), \\ \displaystyle\int_{[0,1]^m} c(\mathbf{u}) \log[c(\mathbf{u})] d\mathbf{u} & (q = 1), \end{cases} \quad (7.21)$$

where $c(\mathbf{u})$ is the *copula density* of $\{\mathbf{Y}_t^{(\ell)}, t \in \mathbb{Z}\}$. It can be shown that $\Delta_{m,q}^{T,c}(\ell) \geq 0$ and $\Delta_{m,q}^{T,c}(\ell) = 0$ if and only if the process $\{\mathbf{Y}_t^{(\ell)}, t \in \mathbb{Z}\}$ is serially independent. Equivalently, $\Delta_{m,q}^{T,c}(\mathbb{C}) = 0$ if and only if $\mathbb{C}(\mathbf{u}) = \Pi(\mathbf{u})$, where $\Pi(\mathbf{u}) \equiv \prod_{i=1}^m u_i$ being the *independence copula* ($m \geq 2$).

Other m-variate copula-based measures can be obtained in a similar manner as we previously applied to introduce the four major density-based measures as special cases of the general functional (7.16). In particular, in terms of the m-dimensional copula density, we have

$$\Delta_m^c(\ell) = \int_{[0,1]^m} B\{c(\mathbf{u}, 1, \ldots, 1)\} d\mathbf{u} = \int_{[0,1]^m} B^c\{c(\mathbf{u})\} c(\mathbf{u}) d\mathbf{u} \quad (7.22)$$

as the copula-based version of (7.16).

7.3 Kernel-Based Tests

The distance measures and dependence functionals introduced in Sections 7.2.3 – 7.2.5 are central to many serial independence test statistics. However, the devil is in the details; i.e., in the way these measures and functionals are made "operational". Clearly, the foundation stone is the dependence functional in (7.16). Depending on the assumptions made on the joint and the univariate marginal distributions, three general methods for estimating this functional are: parametric, semiparametric (cf. Exercise 7.3), and nonparametric. In this section, we solely consider nonparametric testing methods for which $f(\cdot)$ and $f_\ell(\cdot, \cdot)$ are assumed to be unknown under the null hypothesis of serial independence. Within this framework we need to ask, among other things:

- What is the most appropriate technique to estimate the densities?

- Which divergence measure should we adopt?

- Should we compute the functional estimates directly, or can we approximate the integration by a summation?

- Is there a need to include a trimming (weighting) function in the test functional, that is, screening off outliers by bounding the set of observations to some compact set?

- What is the most appropriate method of computing p-values: a bootstrap approach or an MC permutation (random shuffle) approach of the data at hand?

Searching for answers to these questions, the work of Bagnato et al. (2014) provides useful guidelines. These authors present an exhaustive MC simulation comparison of the performance of ten nonparametric serial independence tests, both single-lag and multiple-lag test procedures, using a wide class of linear and non-linear models. They conclude that the integrated estimator of the KL functional (recall $I^{\mathrm{KL}} \equiv \Delta_1^{\mathrm{T}}$) combined with Gaussian kernel density estimation, provides the best performance in terms of empirical size and power. Also, a permutation-based approach is to be preferred over BS, and trimming functions are not needed. Below, we discuss each of these observations and elaborate briefly on possible alternatives.

7.3.1 Density estimators

The Gaussian kernel-based estimator is commonly adopted in the context of nonparametric serial independence testing. For the univariate density function $f(\cdot)$ it is defined as

$$\widehat{f}(y) = \frac{1}{T} \sum_{t=1}^{T} \mathcal{K}_h(y; Y_t), \qquad (7.23)$$

where $\mathcal{K}_h(y; Y_t) = (\sqrt{2\pi}h)^{-2} \exp\{-(y - Y_t)^2/2h^2\}$ with $h > 0$ the bandwidth. Similarly, the Gaussian product kernel density is often used for estimating the bivariate density function $f_\ell(\cdot, \cdot)$, i.e.,

$$\widehat{f}_\ell(x, y) = \frac{1}{T - \ell} \sum_{t=1}^{T-\ell} \mathcal{K}_h(x; Y_t) \mathcal{K}_h(y; Y_{t+\ell}). \qquad (7.24)$$

Common assumptions on the bandwidth are $h \equiv h_T \to 0$, and $Th_T \to \infty$ as $T \to \infty$. Using the same bandwidth for (7.23) and (7.24) is not necessary, but often simplifies asymptotic analysis.

One approach to find the optimal bandwidth h is via likelihood cross-validation (CV) (Silverman, 1986, p. 52). For a marginal density estimator, this approach comes down to maximizing the loss-function

$$CV(h) = \frac{1}{T} \sum_{t=1}^{T} \log \left\{ \frac{1}{T - 1} \sum_{s=1}^{T} \mathcal{K}_h(Y_t; Y_s) I(s \neq t) \right\}, \qquad (7.25)$$

where the term in curly brackets represents the kernel-based "leave-one-out" density estimator.[4] This produces a density estimate which is "close" to the true density in terms of the KL information divergence.

---

[4] As an aside, note that the local marginal density is usually not the main object of interest in a testing context.

7.3 KERNEL-BASED TESTS

The boundedness of the support set S of $(Y_t, Y_{t-\ell})$ in the nonparametric entropy-based divergence measures $\Delta_q^{\text{CR}}(\cdot)$ and $\Delta_1^{\text{T}}(\cdot)$ is a key assumption to establish the asymptotic distribution theory of the resulting test statistics. Gaussian kernel estimation suffers from so-called *boundary effects* with parts of the window devoid of data. Such an effect can be diminished by, for instance, modifying the divergence measures with a trimming function $w(x,y) = I\{(x,y) \in C\}$ which selects only a compact set $C \subseteq S = S^X \times S^Y$. Two simple trimming functions, adopted by Fernandes and Néri (2010) and Bagnato et al. (2014), are based respectively on the compact sets

$$C_1^u = \{u : |u - \bar{u}| \leq 2\hat{\sigma}_u\} \quad \text{and} \quad C_2^u = \{u : \hat{\xi}_{0.1}(u) \leq u \leq \hat{\xi}_{0.9}(u)\},$$

where $\bar{u}$ and $\hat{\sigma}_u$ denote the sample mean and sample standard deviation, while $\hat{\xi}_q(\cdot)$ ($q \in (0,1)$) denotes the q-quantile of the empirical distribution. In addition, the boundary effect can be corrected by using special boundary kernel density estimators. Another widely-known way of nonparametric density estimation is to use histogram methods. In the next section we discuss the histogram estimator within the framework of high-dimensional copula estimation.

7.3.2 Copula estimators

Nonparametric estimates of the m-copula function $\mathbb{C}(\mathbf{u})$ can obtained in three steps. First, every univariate marginal distribution function $F(y_i)$ of $\{Y_{i,t}\}_{t=1}^T$ ($i = 1, \ldots, m$) is estimated by its rescaled empirical counterpart, i.e.,

$$\widehat{F}_{i,T}(y) = \frac{1}{T+1} \sum_{t=1}^T I(Y_{i,t} \leq y), \quad \forall y \in \mathbb{R}. \tag{7.26}$$

Next, the estimated marginal distribution functions are used to obtain the so-called *pseudo-observations*, or PITs, $\widehat{\mathbf{U}}_t = (\widehat{U}_{1,t}, \ldots, \widehat{U}_{m,t})'$ with $\widehat{U}_{i,t} = \widehat{F}_{i,T}(Y_{i,t})$. Note, residuals are just a special case of pseudo-observations. Finally an estimator of $\mathbb{C}(\mathbf{u})$, called the *empirical copula*, is defined as

$$\widehat{\mathbb{C}}_T(\mathbf{u}) = \frac{1}{T} \sum_{t=1}^T I(\widehat{\mathbf{U}}_t \leq \mathbf{u}), \quad \mathbf{u} \in [0,1]^m. \tag{7.27}$$

The factor $T+1$ in the denominator of (7.26) guarantees that the pseudo-observations are strictly located in the interior of $[0,1]^m$. Observe that $\widehat{\mathbb{C}}_T(\mathbf{u})$ is actually a function of the rank $R_{i,t}$ of $Y_{i,t}$ in the vector $(Y_{i,1}, \ldots, Y_{i,T})'$, since

$$(T+1)\widehat{F}_{i,T}(Y_{i,t}) \equiv R_{i,t} = \sum_{j=1}^T I(Y_{j,t} \leq Y_{i,t}), \quad (1 \leq i \leq m; 1 \leq t \leq T).$$

Hence, any rank test of serial independence is a function of $\widehat{\mathbb{C}}_T(\mathbf{u})$. Due to the invariance property of the ranks, the empirical copula is invariant under strictly monotonic increasing transformations of the margins.

In the one-dimensional case, classical histogram methods may be used to construct root-n consistent density estimators with compact support. For $m \geq 2$, a conceptually easy way to obtain a copula-based histogram estimator is to divide the sample space into hyper-rectangular regions (bins or cells) of equal size. To this end, let $(q_1, \ldots, q_m)'$ be an m-dimensional vector of integers, let $(v_1, \ldots, v_m)'$ denote any fixed m-vector, and let

$$B_q = \{\mathbf{u} : |u_i - (v_i + q_i h_b)| \leq \frac{1}{2} h_b, \ 1 \leq i \leq m\}$$

represent the histogram bin-centered at $v_i + q_i h_b$. Here, h_b is the *binwidth*, a number which decreases to zero as $T \to \infty$. Write N_q for the number of sample points $\widehat{\mathbf{U}}_t$ which fall into bin B_q. Of course, $\sum_{q=1}^Q N_q = T$ with Q the total number of bins. Then, for $\mathbf{u} \in B_q$, the equidistant histogram estimate of the copula density $c(\mathbf{u})$ is given by

$$\widehat{c}_{h_b}(\mathbf{u}) = \frac{N_q}{T h_b^m}, \tag{7.28}$$

and

$$\widehat{\Delta}_1^{\text{T},c}(\ell) = \frac{1}{T} \sum_{q=1}^Q N_q \log\left(\frac{N_q}{T h_b^m}\right) = \frac{1}{T} \sum_{q=1}^Q N_q \log N_q - \log(T h_b^m) \tag{7.29}$$

is a copula-based estimator of $\Delta_1^{\text{T}}(\cdot)$. The optimal value of h_b, minimizing the mean squared error, is of order $\mathcal{O}(T^{-1/(2+m)})$; cf. Silverman (1986).

7.3.3 Single-lag test statistics

Table 7.1 offers a list of eight pairwise (single-lag) serial independence test statistics along with their corresponding divergence measures. For completeness, we add the following details.

- The test statistic $\widehat{\Delta}_{T,1}^{\text{CT}}(\cdot)$ employs histogram-based density estimators with equidistant cells while all other tests use kernel-based density estimators.

- The test statistics $\widehat{\Delta}_{T,\gamma}^{\text{R}}(\cdot)$, $\widehat{\Delta}_T^{\text{ST1}}(\cdot)$, $\widehat{\Delta}_T^{\text{ST2}}(\cdot)$, $\widehat{\Delta}_T^{\text{GL}}(\cdot)$, and $\widehat{\Delta}_{T,q}^{\text{FN}}(\cdot)$ all make use of the Gaussian kernel density estimator. In contrast, $\widehat{\Delta}_T^{\text{HW}}(\cdot)$ uses "leave-one-out" marginal and bivariate kernel density estimators, with a special provision in the kernel function to avoid boundary effects; see Hong and White (2005).

- Apart from $\widehat{\Delta}_{T,1}^{\text{CT}}(\cdot)$, the tests have an asymptotic normal distribution under the null hypothesis of pairwise serial independence. Under weak regularity conditions, it can be shown (see the cited references) that all tests are consistent against lag one dependent alternatives. No limiting distribution theory is available for $\widehat{\Delta}_{T,1}^{\text{CT}}(\cdot)$ which has hindered its application in practice.

7.3 KERNEL-BASED TESTS

Table 7.1: Single-lag ($m = 2$) serial independence tests.

Reference	Divergence measure	Test statistic[1][2][3]
	Density functions	
Chan and Tran (1992)	Δ_1^K	$\widehat{\Delta}_{T,1}^{CT}(\ell) = \sum_{t \in S_T(\ell)} |\widehat{f}(Y_t, Y_{t-\ell}) - \widehat{f}(Y_t)\widehat{f}(Y_{t-\ell})|$
Robinson (1991) [4]	$I^{KL} \equiv \Delta_1^T$	$\widehat{\Delta}_{T,\gamma}^{R}(\ell) = \dfrac{1}{T-\ell} \sum_{t \in S_T(\ell)} C_t(\gamma) \log\left(\dfrac{\widehat{f}(Y_t, Y_{t-\ell})}{\widehat{f}(Y_t)\widehat{f}(Y_{t-\ell})}\right)$
Skaug and Tjøstheim (1993a)	Δ^H	$\widehat{\Delta}_T^{ST1}(\ell) = \dfrac{1}{T-\ell} \sum_{t \in S_T(\ell)} 2\left\{1 - \sqrt{\dfrac{\widehat{f}(Y_t, Y_{t-\ell})}{\widehat{f}(Y_t)\widehat{f}(Y_{t-\ell})}}\right\} w_t(\ell)$
Skaug and Tjøstheim (1996)	Δ^*	$\widehat{\Delta}_T^{ST2}(\ell) = \dfrac{1}{T-\ell} \sum_{t \in S_T(\ell)} \{\widehat{f}(Y_t, Y_{t-\ell})$ $- \widehat{f}(Y_t)\widehat{f}(Y_{t-\ell})\} w_t(\ell)$
Granger and Lin (1994)	$1 - e^{-2I^{KL}}$	$\widehat{\Delta}_T^{GL}(\ell) = 1 - \exp\left[\dfrac{-2}{T-\ell} \sum_{t \in S_T(\ell)} \log\left(\dfrac{\widehat{f}(Y_t, Y_{t-\ell})}{\widehat{f}(Y_t)\widehat{f}(Y_{t-\ell})}\right)\right]$
Hong and White (2005)	$I^{KL} \equiv \Delta_1^T$	$\widehat{\Delta}_T^{HW}(\ell) = \dfrac{1}{T-\ell} \sum_{t \in S_T(\ell)} \log\left(\dfrac{\widehat{f}(Y_t, Y_{t-\ell})}{\widehat{f}(Y_t)\widehat{f}(Y_{t-\ell})}\right)$
Fernandes and Néri (2010)	$\Delta_{q \in \{\frac{1}{2}, 1, 2, 4\}}^T$	$\widehat{\Delta}_{T,q}^{FN}(\ell) = \dfrac{1}{(1-q)(T-\ell)} \times$ $\sum_{t \in S_T(\ell)} \left\{1 - \left(\dfrac{\widehat{f}(Y_t)\widehat{f}(Y_{t-\ell})}{\widehat{f}(Y_t, Y_{t-\ell})}\right)^{1-q}\right\} w_t(\ell)$
	Distribution functions	
Skaug and Tjøstheim (1993b)	Δ^{CvM}	$\widehat{\Delta}_T^{ST3}(\ell) = \dfrac{1}{T-\ell} \sum_{t=1}^{T-\ell} \{\widehat{F}(Y_t, Y_{t+\ell})$ $- \widehat{F}(\infty, Y_t)\widehat{F}(Y_{t+\ell}, \infty)\}^2$

[1] $S_T(\ell) \equiv \{t \in \mathbb{N} : \ell < t \leq T, \widehat{f}_2(Y_t, Y_{t-\ell}) > 0, \widehat{f}(Y_t) > 0, \widehat{f}(Y_{t-\ell}) > 0\}$.
[2] $C_t(\gamma) = 1 - \gamma$ if t is odd, and $C_t(\gamma) = 1 + \ell\gamma$ if $t = 1, \bmod(\ell+1)$ and $C_t(\gamma) = 1 - \gamma$ otherwise, with $\gamma \in (0,1)$.
[3] $w_t(\ell) = I\{(Y_t, Y_{t-\ell}) \in S^2\}$ is a trimming (weight) function.
[4] When $\ell = 1$, Robinson's (1991) test has the form $\mathcal{R}(\ell) \equiv (1/[2(\ell+1)(T-1)\gamma^2 \widehat{v}_\delta])^{1/2} \widehat{\Delta}_{T,\gamma}^R(\ell)$ where $\widehat{v}_\delta \equiv T^{-1} \sum_{t \in S_T} C_t(\delta)(\log \widehat{f}(Y_t))^2 - [T^{-1} \sum_{t \in S_T} C_t(\delta) \log \widehat{f}(Y_t)]^2$, $S_T \equiv \{t \in \mathbb{N} : 1 \leq t \leq T, \widehat{f}(Y_t) > 0\}$ with $\delta \in [0, 1)$.

- The trimming function $w_t(\ell)$ is generally not needed for $\widehat{\Delta}_T^{ST1}(\cdot)$ and $\widehat{\Delta}_T^{ST2}(\cdot)$. For i.i.d. data from the uniform distribution, $w_t(\ell)$ is needed to prevent degeneracy, because otherwise the asymptotic variance of the test statistics would vanish to 0.

- The test statistic $\widehat{\Delta}_T^{ST3}(\cdot)$ utilizes the following unbiased estimators of the one- and two-dimensional EDF of $\{Y_t\}_{t=1}^T$, respectively,

$$\widehat{F}_T(y) = \frac{1}{T} \sum_{t=1}^T I(Y_t \leq y), \quad \widehat{F}_{\ell,T}(x,y) = \frac{1}{T-\ell} \sum_{t=1}^{T-\ell} I(Y_t \leq x) I(Y_{t+\ell} \leq y).$$

Observe, all test statistics in Table 7.1 have an equivalent integral representation. Also, using the copula-based measure (7.21) in conjunction with the copula estimators of Section 7.3.2, the construction of copula-based serial independence test statistics is entirely obvious.

The results in Table 7.1 prompt the question: is there a test statistic preferable over others? Partly, the answer comes from the MC simulation study of Bagnato et al. (2014) to which we already alluded earlier. These authors recommend using the KL functional $\Delta_1^{\mathrm{KL}}(\cdot)$ combined with Gaussian kernel density estimation, and with a slight preference for the integral representation of the resulting test statistic over its summed counterpart. Simulation results reported by Hong and White (2005) show that $\widehat{\Delta}_T^{\mathrm{HW}}(\cdot)$ has much lower power than $\widehat{\Delta}_T^{\mathrm{ST2}}(\cdot)$, but it is always better than or equal to the power of $\widehat{\Delta}_{T,\gamma}^{\mathrm{R}}(\cdot)$ for all DGPs and sample sizes under consideration.

7.3.4 Multiple-lag test statistics

The test statistics in Section 7.3.3 are informative in revealing serial dependence at individual lags. On the other hand, as already mentioned in Section 7.2.4, the pairwise approach depends on the choice of the lag order. To mitigate this problem, we introduce the two-dimensional test functional $\mathcal{Q}(M)$ jointly with the null hypothesis (7.20). A portmanteau-type estimator of $\mathcal{Q}(M)$ can be defined as

$$\widehat{\mathcal{Q}}(M) = \frac{1}{\sqrt{M}} \sum_{\ell=1}^{M} \widehat{\Delta}(\ell), \quad (M \in \mathbb{N}^+), \tag{7.30}$$

where, except for the test statistic proposed by Chan and Tran (1992), $\widehat{\Delta}(\cdot)$ can be one of the single-lag test statistics listed in Table 7.1. Hong and White (2005) consider (7.30) with $\widehat{\Delta}(\cdot)$ replaced by $\widehat{\Delta}_T^{\mathrm{HW}}(\cdot)$, $\mathcal{R}(\cdot)$ (see Table 7.1, footnote (4)), and $\widehat{\Delta}_T^{\mathrm{ST2}}(\cdot)$. In each case the resulting portmanteau-type test statistic has an asymptotic normal null distribution. Bagnato et al. (2014) only focus on the integrated Gaussian kernel estimator of $\Delta_1^{\mathrm{T}}(\cdot)$. These authors conclude that, as opposed to a simultaneous test based on the Bonferroni procedure, the portmanteau-type test statistic is the best choice since it preserves size across lags.

Using the CvM functional, Hong (1998) considers a modified version of the portmanteau-type pairwise serial independence test statistic of Skaug and Tjøstheim (1993b). That is,

$$\widehat{\mathcal{Q}}^{\mathrm{H1}}(M) = \sum_{\ell=1}^{M}(T-\ell)\widehat{\Delta}_T^{\mathrm{ST3}}(\ell). \tag{7.31}$$

Thus, similar as the well-known LB portmanteau-type test statistic for joint significance of the first M serial autocorrelation coefficients, the test statistics $\widehat{\Delta}_T^{\mathrm{ST3}}(\ell)$ ($\ell = 1, \ldots, M$) are weighted. A sensible generalization of (7.31) is to include a symmetric continuous window kernel $\lambda(\cdot)$ with $\lambda(0) = 1$. This ensures that the asymptotic bias of the test statistic vanishes.

7.3 KERNEL-BASED TESTS

Under the null hypothesis of serial independence $\{(T-\ell)\widehat{\Delta}_T^{\mathrm{ST3}}(\ell); \ell = 1, \ldots, T-1\}$ can be viewed as an asymptotically i.i.d. sequence with mean $1/6^2$ and variance $2/90^2$. These results suggest the test statistic

$$\widehat{\mathcal{Q}}^{\mathrm{H2}}(M) = \frac{\sum_{\ell=1}^{T-1} \lambda^2(\ell/M)\{(T-\ell)\widehat{\Delta}_T^{\mathrm{ST3}}(\ell) - 1/6^2\}}{\sqrt{2\sum_{\ell=1}^{T-2} \lambda^4(\ell/M)/90^2}}, \qquad (7.32)$$

with the *Daniell lag window* $\lambda(u) = \sin(\pi u)/\pi u$, which is optimal over a class of window kernels that includes the Parzen window; see (4.18). Based on the theory of degenerate V-statistics, it can be shown that (7.32) has a limiting $\mathcal{N}(0,1)$ distribution, under the null hypothesis of serial independence. A simple way to obtain p-values is via the smoothed BS or permutation method; see Section 7.3.6 for details.

Example 7.3: Magnetic Field Data (Cont'd)

In Example 1.3, we saw that the magnetic field data is highly nonlinear. Terdik (1999, p. 207) fits the following diagonal BL model to the series $\{Y_t\}_{t=1}^{1,962}$

$$Y_t = 0.5421 Y_{t-1} + 0.0541 Y_{t-1}\varepsilon_{t-1} + \varepsilon_t,$$

with residual variance $\widehat{\sigma}_\varepsilon^2 = 0.2765$. The sample residual ACF shows significant (5% level) values at lags $\ell = 3, 4, 6, 7, 9$, and 10. Clearly, it is likely that the fitted model is not appropriate. To investigate this in more detail, we consider $\widehat{\Delta}_T^{\mathrm{ST2}}(\ell)$ ($\ell \ll T$) and a standardized version of this test statistic, namely

$$\mathcal{J}_T(\ell) = \widehat{S}^{-1}(T-\ell)^{1/2}\widehat{\Delta}_T^{\mathrm{ST2}}(\ell)$$
$$= \widehat{S}^{-1}(T-\ell)^{-1/2} \sum_{t=\ell+1}^{T} \{\widehat{f}(Y_t, Y_{t-\ell}) - \widehat{f}(Y_t)\widehat{f}(Y_{t-\ell})\} w_t(\ell),$$

where $\widehat{S}^2$ is a consistent asymptotic variance estimator. Under $\mathbb{H}_0$, $\mathcal{J}_T(\ell) \xrightarrow{D} \mathcal{N}(0,1)$, as $T \to \infty$. For the Gaussian kernel density estimators, we obtain the bandwidth h through a data-driven bandwidth method; see, e.g., Hong and White (2005, p. 859) and Bagnato et al. (2014).

Based on 1,000 bootstrap replicates, both test statistics $\widehat{\Delta}_T^{\mathrm{ST2}}(\ell)$ and $\mathcal{J}_T(\ell)$ have nearly zero p-values for all lags ℓ from 1 to 10. Moreover, the multiple-lag portmanteau-type test statistics have p-values less than 0.05 for $M = 2$, 4, 6, and 8. All these test results indicate that the residuals are not serially independent, suggesting that the fitted BL model is far from adequate.

7.3.5 Generalized spectral tests

Recall from Chapter 4 that the dependence of a strictly stationary time series $\{Y_t, t \in \mathbb{Z}\}$ can be characterized by the spectral density function $f_Y(\omega)$ defined by (4.3), or alternatively by its spectral distribution function $F_Y(\omega)$ defined by

$$F_Y(\omega) = 2\int_0^{\omega\pi} f_Y(\omega)\mathrm{d}\omega = \omega + 2\sum_{\ell=1}^{\infty} \gamma_Y(\ell)\frac{\sin(\pi\omega\ell)}{\ell\pi}, \quad \omega \in [0,1]. \qquad (7.33)$$

Thus, under the null hypothesis of serial independence $F_Y(\omega) = \omega$, which is analogous to a flat spectrum. Flat spectra, however, can result from nonlinear processes which would be accepted as WN by a test statistic based on (7.33) with a high probability. For example, the BL process $Y_t = \beta\varepsilon_{t-1}\varepsilon_{t-2} + \varepsilon_t$, where $\{\varepsilon_t\} \sim \text{WN}(0,\sigma_\varepsilon^2)$, has $\gamma_Y(\ell) = 0$ for $\ell > 0$, hence estimates of the spectrum will be constant over all frequencies ω.

As an alternative, Hong (2000) introduces two test statistics (denoted by the superscripts H_1 and H_2) for pairwise serial independence using a *generalized spectrum*. The key idea of the generalized spectrum is to transform $\{Y_t, t \in \mathbb{Z}\}$ via a complex-valued exponential function

$$Y_t \longrightarrow \exp(iuY_t), \quad u \in \mathbb{R},$$

and then consider the spectrum of the transformed process. Specifically, let $\phi(u_1) = \mathbb{E}\{\exp(iu_1 Y_t)\}$ be the marginal characteristic function of the process $\{Y_t, t \in \mathbb{Z}\}$, and let $\phi_\ell(u_1, u_2) = \mathbb{E}\{\exp(i(u_1 Y_t + u_2 Y_{t-|\ell|}))\}$ ($\ell = 0, \pm 1, \ldots$) be the pairwise joint characteristic function of $\{(Y_t, Y_{t-|\ell|})\}$. Then the lag ℓ ACVF of the transformed processes is given by

$$\gamma_{u_1, u_2}(\ell) \equiv \text{Cov}\big(e^{iu_1 Y_t}, e^{iu_2 Y_{t-|\ell|}}\big) = \phi_\ell(u_1, u_2) - \prod_{k=1}^{2}\phi(u_k) \equiv D_\ell(u_1, u_2), \qquad (7.34)$$

where $D_\ell(\cdot,\cdot)$ is defined by (7.5). If $\gamma_{u_1,u_2}(\ell) = 0\ \forall (u_1, u_2) \in \mathbb{R}^2$, then there is no serial dependence between Y_t and $Y_{t-|\ell|}$, otherwise there is. In other words, the null hypothesis of interest is given by (7.6) with $m = 2$.

Now, suppose that $\sup_{(u_1, u_2) \in \mathbb{R}^2} \sum_{\ell=-\infty}^{\infty} |\gamma_{u_1, u_2}(\ell)| < \infty$, which holds under a proper mixing condition. Then the FT of $\gamma_{u_1, u_2}(\ell)$

$$f_Y(\omega, u_1, u_2) = \sum_{\ell=-\infty}^{\infty} \gamma_{u_1, u_2}(\ell)\exp(-2\pi i\omega\ell), \quad \omega \in [0,1], \qquad (7.35)$$

exists. Because $-\partial^2 f_Y(\omega, u_1, u_2)/\partial u_1 \partial u_2|_{(0,0)} = f_Y(\omega)$, (7.35) is called a *generalized spectral density* of $\{Y_t, t \in \mathbb{Z}\}$, although it does not have the mathematical properties of a pdf. Similarly, a generalization of (7.33), is given by

$$F_Y(\omega, u_1, u_2) = \gamma_{u_1, u_2}(0)\omega + 2\sum_{\ell=1}^{\infty}\gamma_{u_1, u_2}(\ell)\frac{\sin(\pi\omega\ell)}{\ell\pi}, \quad \omega \in [0,1], \qquad (7.36)$$

which is called a *generalized spectral distribution function*. However, unlike higher-order spectra, (7.35) and (7.36) do not require any moment conditions on $\{Y_t, t \in \mathbb{Z}\}$.

7.3 KERNEL-BASED TESTS

A plausible estimator for $F_Y(\cdot)$ is

$$\widehat{F}_T(\omega, x, y) = \widehat{\gamma}_{x,y}(0)\omega + 2\sum_{\ell=1}^{T-1}\left(1 - \frac{\ell}{T}\right)^{1/2}\widehat{\gamma}_{x,y}(\ell)\frac{\sin(\pi\omega\ell)}{\ell\pi}, \qquad (7.37)$$

where

$$\widehat{\gamma}_{x,y}(\ell) = \widehat{F}_{\ell,T}(x,y) - \widehat{F}_T(x,\infty)\widehat{F}_T(\infty,y), \quad (\ell = 1, \ldots, T-1),$$

with

$$\widehat{F}_{\ell,T}(x,y) = \frac{1}{T-\ell}\sum_{t=1}^{T-\ell} I(Y_t \leq x)I(Y_{t+\ell} \leq y).$$

The factor $(1 - \ell/T)^{1/2}$ in (7.37) is a small sample correction for weighting down higher order lags ℓ.

Utilizing the CvM functional, the "summed version" of a test statistic for pairwise serial independence is given by

$$\widehat{\Delta}^{\mathbb{H}_1}_{F_Y} = \sum_{\ell=1}^{T-1}\frac{T-\ell}{(\ell\pi)^2}\left(\frac{1}{T^2}\sum_{t=1}^{T}\sum_{s=1}^{T}\widehat{\gamma}^2_{Y_t,Y_s}(\ell)\right). \qquad (7.38)$$

A second test statistic, based on the KS functional, is given by

$$\widehat{\Delta}^{\mathbb{H}_2}_{F_Y} = \max_{1\leq t,s\leq T}\sup_{\omega\in[0,1]}\left|\sum_{\ell=1}^{T-1}(T-\ell)^{1/2}\widehat{\gamma}_{Y_t,Y_s}(\ell)\frac{\sqrt{2}\sin(\pi\omega\ell)}{\ell\pi}\right|. \qquad (7.39)$$

Note that both test statistics do not assume that the lag order M is known *a priori*. This may be appealing, since for certain DGPs it is not obvious how to choose the optimal lag order leading to the highest power of a particular serial independence test statistic.

Under $\mathbb{H}_0$, and assuming that the stationary process $\{Y_t, t \in \mathbb{Z}\}$ has a continuous marginal distribution function $F_Y(\cdot)$, it can be shown (Hong, 2000) that the test statistics (7.38) and (7.39) are asymptotically distributed as, respectively,

$$\widehat{\Delta}^{\mathbb{H}_1}_{F_Y} \xrightarrow{D} \sum_{i,j,l=1}^{\infty}\frac{1}{(i\pi)^2}\frac{1}{(j\pi)^2}\frac{1}{(l\pi)^2}Z^2_{ijl} \qquad (7.40)$$

and

$$\widehat{\Delta}^{\mathbb{H}_2}_{F_Y} \xrightarrow{D} \sup_{(\omega_1,\omega_2,\omega_3)\in[0,1]^3}\left|\sum_{i,j,l=1}^{\infty}\frac{\sqrt{2}\sin(i\pi\omega_1)}{(i\pi)^2}\frac{\sqrt{2}\sin(j\pi\omega_2)}{(j\pi)^2}\frac{\sqrt{2}\sin(l\pi\omega_3)}{(l\pi)^2}Z_{ijl}\right|, \qquad (7.41)$$

where $\{Z_{ijl}; i,j,l \geq 1\}$ are i.i.d. $\mathcal{N}(0,1)$ random variables. Both test statistics enjoy the *nuisance-parameter-free property*, which ensures that their critical values and/or p-values can be obtained by directly simulating $\widehat{\Delta}^{\mathbb{H}_1}_{F_Y}$ and $\widehat{\Delta}^{\mathbb{H}_2}_{F_Y}$.

Example 7.4: U.S. Unemployment Rate (Cont'd)

In this example we explore residual serial dependence using the test statistics (7.38) and (7.39). To this end, we continue our analysis of the quarterly U.S. unemployment rate (original data), but now for the subperiod 1948 – 1993. Montgomery et al. (1998) fit the following SETAR(2; 2, 2) model to the first differences $\{\Delta Y_t = Y_t - Y_{t-1}\}_{t=2}^{184}$ (asymptotic standard errors are in parentheses):

$$\Delta Y_t = \begin{cases} 0.01_{(0.03)} + 0.73_{(0.10)}\Delta Y_{t-1} + 0.10_{(0.12)}\Delta Y_{t-2} + \varepsilon_t^{(1)} & \text{if } \Delta Y_{t-2} \leq 0.1, \\ 0.18_{(0.09)} + 0.80_{(0.12)}\Delta Y_{t-1} - 0.56_{(0.16)}\Delta Y_{t-2} + \varepsilon_t^{(2)} & \text{if } \Delta Y_{t-2} > 0.1. \end{cases}$$

The residual variances are respectively 0.076 and 0.165. Note that, apart from the constant and the AR(2) term in the lower regime, all coefficients are significantly different from zero at the 5% nominal level.

Significant (5% nominal level) residual autocorrelations were noticed at lags $\ell = 4$ and 5, suggesting that the above model specification is not adequate. To follow along, we selected 100 grid points for computing the frequencies ω and 1,000 BS samples. Using the naive bootstrap, and with 181 observations, the p-values of $\widehat{\Delta}_{F_Y}^{\text{HW}1}$ and $\widehat{\Delta}_{F_Y}^{\text{HW}2}$ are respectively 0.09 and 0.03. Thus, only the second test statistic reveals that the residuals are not serially independent.

7.3.6 Computing p-values

It has been extensively documented that the normal approximation based on the asymptotic distribution of many kernel-based test statistics does not perform well in finite samples. As a possible alternative, one can simulate a large number of time series satisfying the null hypothesis, and calculate empirical quantiles and/or p-values from the null distribution of the sampled test statistic. This approach is suitable only if the marginal distribution under the null hypothesis is known, or if the distribution of the test statistic is (asymptotically) independent of the (unknown) marginal distribution. Since these options are generally not available in practice, it is better to reflect the nonparametric nature of the null hypothesis through the use of either random permutation or BS approaches.

Bootstrapping

Unfortunately, the naive nonparametric bootstrap cannot be used with many entropy-based serial independence test statistics (e.g., Δ_T^{GL}, Δ_T^{HW}, and $\Delta_{T,q}^{\text{FN}}$) since their leading term is a degenerate U-statistic under $\mathbb{H}_0$. Consequently, the bootstrap fails to mimic the limiting distribution of the test statistic. Instead, the following practical procedure is recommended.

> **Algorithm 7.1: Bootstrapped p-values for single-lag tests**
>
> (i) Compute $\widehat{\Delta}^{(0)}(\ell)$ ($\ell = 1, \ldots, T-1$) using the original data $\{Y_t\}_{t=1}^T$, and a kernel density estimator with a fixed bandwidth h. Here $\widehat{\Delta}^{(0)}(\ell)$ is any of the test statistics defined above.

7.3 KERNEL-BASED TESTS

> **Algorithm 7.1: Bootstrapped p-values for single-lag tests (Cont'd)**
>
> (ii) Draw a bootstrap sample $\{Y_t^*\}_{t=1}^T$ from the smoothed kernel density (7.23) where $\mathcal{K}_h(\cdot)$ and h are the same as used for the computation of $\widehat{\Delta}^{(0)}(\ell)$. Then, compute a bootstrap statistic $\widehat{\Delta}^{*,(0)}(\ell)$, in the same way as $\widehat{\Delta}^{(0)}(\ell)$, using $\{Y_t^*\}_{t=1}^T$.
>
> (iii) Repeat step (ii) B times, to obtain $\{\widehat{\Delta}^{*,(b)}(\ell)\}_{b=1}^B$.
>
> (iv) Compute the one-sided bootstrap p-value as
>
> $$\widehat{p}(\ell) = \frac{1 + \sum_{b=1}^B I\big(\widehat{\Delta}^{*,(b)}(\ell) \geq \widehat{\Delta}^{(0)}(\ell)\big)}{1 + B}.$$

This procedure maintains the asymptotically pivotal character of the entropy-based test statistics. That is, the distribution of the tests does not depend on any unknown parameters under the null hypothesis of pairwise serial independence.

Permutation

When testing a composite hypothesis, an exact level MC test statistic can be obtained by conditioning on an observed value of a minimal sufficient statistic under the null hypothesis (Engen and Lillegård, 1997). By definition, the resulting distribution does not depend on unknown parameters so that it can be used to simulate data that have the same (exact) conditional distribution as the DGP under the null hypothesis, given the sufficient statistic. Under the null hypothesis of pairwise serial independence, the order statistics provide a minimal and sufficient statistic. To be specific, let $\widehat{\Delta}^{(0)}(\cdot)$ denote the value of the dependence functional conditioned on the original data, and let $\{\widehat{\Delta}^{(i)}(\cdot)\}_{i=1}^B$ be the set of "bootstrapped" test statistics obtained from a random permutation of the original data. Then calculate the one-sided p-value as

$$\widehat{p}(\cdot) = \frac{1 + \sum_{i=1}^B I\big(\widehat{\Delta}^{(i)}(\cdot) \geq \widehat{\Delta}^{(0)}(\cdot)\big)}{1 + B}. \tag{7.42}$$

Thus, reject the null hypothesis of pairwise serial independence if $\widehat{p}(\cdot) < \alpha$, where α is some pre-specified nominal significance level.

For multiple-lag tests, Diks and Panchenko (2007) advocate the following algorithm.

> **Algorithm 7.2: Permutation-based p-values for multiple-lag tests**
>
> (i) Compute $\widehat{\Delta}^{(0)}(\ell)$ ($\ell = 1, \ldots, M$) using $\{Y_t\}_{t=1}^T$ and a kernel-based density estimator with a fixed bandwidth h. Next, construct the $1 \times M$ vector $\widehat{\boldsymbol{\Delta}}^{(0)} = (\widehat{\Delta}^{(0)}(1), \ldots, \widehat{\Delta}^{(0)}(M))$.

> **Algorithm 7.2: Permutation-based p-values (Cont'd)**
>
> (ii) Randomly permute B times the data, and build the $B \times M$ matrix $\widetilde{\mathbf{B}}$ whose $b\ell$th element is $\widetilde{\Delta}^{(b)}(\ell)$ ($b = 1,\ldots,B; \ell = 1,\ldots,M$). Then assemble $\widehat{\mathbf{\Delta}}^{(0)}$ and $\widetilde{\mathbf{B}}$ into the $(B+1) \times M$ matrix
>
> $$\mathbf{B} = \begin{pmatrix} \widehat{\mathbf{\Delta}}^{(0)} \\ \cdots \\ \widetilde{\mathbf{B}} \end{pmatrix}.$$
>
> (iii) Transform $\mathbf{B}$ into the $(B+1) \times M$ matrix $\mathbf{P}$ of p-values with elements
>
> $$\widehat{p}_i(\ell) = \frac{1 + \sum_{k=0}^{B} I(\widehat{\Delta}^{(k)} > \widehat{\Delta}^{(i)}(\ell))}{1+B}, \quad (i=0,\ldots,B; \ell=1,\ldots,M).$$
>
> (iv) For each row of $\mathbf{P}$ select the smallest $\widehat{p}_i(\ell)$ and call it $\widehat{T}_i$, i.e.
>
> $$\widehat{T}_i = \inf_{\ell \in (1,\ldots,M)} \widehat{p}_i(\ell), \quad (i = 0, \ldots, B).$$
>
> (v) Adopt, $\widehat{T}$ say, as a test statistic. Interpret $\widehat{T}_0$ as its observed value and the set $\{\widehat{T}_1,\ldots,\widehat{T}_B\}$ as the values associated with each permutation. Then calculate an "overall" p-value of $\widehat{T}$, i.e.
>
> $$\widehat{p} = \frac{1 + \sum_{i=0}^{B} I(\widehat{T}_i > \widehat{T}_0)}{1+B}.$$

For multiple bandwidth selection, the multiple-lag testing procedure can be easily modified. In particular, in step (i) calculate the vector of values $\widehat{\mathbf{\Delta}}_h^{(0)} = (\widehat{\Delta}_h^{(0)}(1),\ldots,\widehat{\Delta}_h^{(0)}(M))'$ for a range of bandwidths $h \in \{h_1,\ldots,h_n\}$ with n the number of elements. With appropriate changes in steps (ii) – (iii), step (iv) becomes "... select the smallest p-values among all bandwidths and all lags ...", while step (v) remains the same. As in the single bandwidth case, the multiple bandwidth procedure yields an exact α-level ($0 < \alpha < 1$) test statistic if the null hypothesis (7.20) is rejected, whenever $\widehat{p} \leq \alpha$.

7.4 High-Dimensional Tests

7.4.1 BDS test statistic

Assume that the m-dimensional process $\{\mathbf{Y}_t, t \in \mathbb{Z}\}$ admits a common continuous joint pdf $f_m(\mathbf{y})$ for $\mathbf{y} = (y_1,\ldots,y_m)'$. Hence, $C_{m,Y}(h)$ in (7.10) can be rewritten as $\mathbb{E}[I(\|\mathbf{Y}_i - \mathbf{Y}_j\| \leq h)]$. An estimator of $C_{m,Y}(h)$ is $\widehat{C}_{m,Y}(h)$, which is a U-statistic

7.4 HIGH-DIMENSIONAL TESTS

of the following form:

$$\widehat{C}_{m,Y}(h) = \binom{N}{2}^{-1} \sum_{1 \leq i < j \leq N} I(\|\mathbf{Y}_i - \mathbf{Y}_j\| < h), \qquad (7.43)$$

where $N = T - m + 1$ is the number of vectors obtained from a time series $\{Y_t\}_{t=1}^T$. Now, given the divergence measure $C_{m,Y}(h) - \{C_{1,Y}(h)\}^m$, a test statistic for serial independence in $\{Y_t\}_{t=1}^T$ is defined as

$$S_{m,Y}(h) = \sqrt{N} \, \frac{\widehat{C}_{m,Y}(h) - \{\widehat{C}_{1,Y}(h)\}^m}{\widehat{\sigma}_{m,Y}(h)}, \qquad (7.44)$$

where $\widehat{\sigma}_{m,Y}^2(h)$ is a consistent estimator of the variance of $\sqrt{N}\big(C_{m,Y}(h) - \{C_{1,Y}(h)\}^m\big)$. The specific estimator proposed by Brock et al. (1996) is

$$\frac{1}{4}\widehat{\sigma}_{m,Y}^2(h) = m(m-2)\widehat{C}_{m,Y}^{2m-2}(K_{m,Y} - \widehat{C}_{m,Y}^2) + K_{m,Y}^m - \widehat{C}_{m,Y}^{2m}$$

$$+ 2\sum_{j=1}^{m-1} [\widehat{C}_{m,Y}^{2j}(K_{m,Y}^{m-j} - \widehat{C}_{m,Y}^{2m-2j}) - m\widehat{C}_{m,Y}^{2m-2}(K_{m,Y} - \widehat{C}_{m,Y}^2)], \qquad (7.45)$$

where

$$K_{m,Y} = \frac{2}{N(N-1)(N-2)} \sum_{i=1}^{N-2} \sum_{s=i+1}^{N-1} \sum_{t=s+1}^{N} I(|Y_i - Y_s| < h) I(|Y_s - Y_t| < h),$$

and where the dependence of the terms in (7.45) on T and h has been suppressed for notational clarity. Under the null hypothesis of serial independence, and by exploiting the asymptotic theory for U-statistics, it can be shown that, as $T \to \infty$,

$$S_{m,Y}(h) \xrightarrow{D} \mathcal{N}(0,1), \quad \forall h \in (0,\infty). \qquad (7.46)$$

The test statistic (7.44) is stated in terms of the data series $\{Y_t\}_{t=1}^T$. Brock et al. (1996) show that the limiting behavior of $S_{m,Y}(h)$, under $\mathbb{H}_0$ of no serial dependence, remains the same whether the model parameters are known or estimated in a root-n consistent fashion. Thus, (7.44) can be adapted to test situations involving "residuals" $\{e_t\}_{t=1}^T$. The resulting diagnostic test, called BDS test statistic after its three originators Brock, Dechert, and Scheinkman, is defined as

$$S_{m,e}(h) = \sqrt{T} \, \frac{\widehat{C}_{m,e}(h) - \{\widehat{C}_{1,e}(h)\}^m}{\widehat{\sigma}_{m,e}(h)}, \qquad (7.47)$$

where in this case the sample correlation integral is given by

$$\widehat{C}_{m,e}(h) = \binom{T-m+1}{2}^{-1} \sum_{t=m+1}^{T} \sum_{s=m}^{t-1} \prod_{j=0}^{m-1} I(|e_{t-j} - e_{s-j}| < h),$$

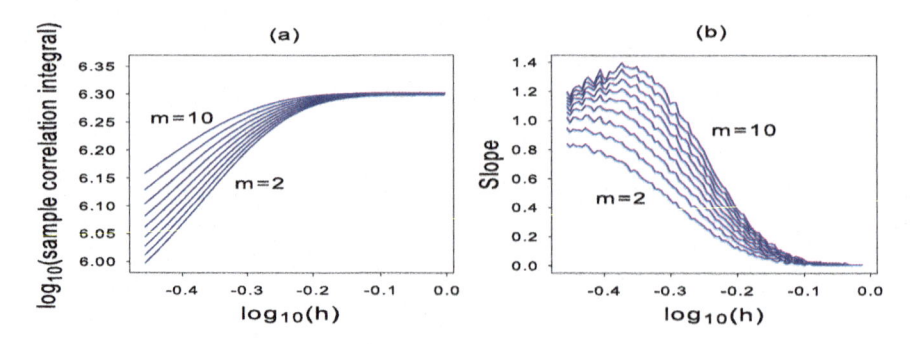

Figure 7.3: *(a) Estimated correlation integral* $\log_{10} \widehat{C}_{m,Y}(h)$; *(b) Slope estimates* $\widehat{\beta}_m$ *for a simulated* ExpAR(1) *process;* $T = 2{,}000$.

and where $\widehat{\sigma}^2_{m,e}(h)$ follows from (7.45). Under $\mathbb{H}_0$, the test statistic (7.47) is again asymptotically standard normal distributed.

The *correlation dimension* of $\{e_t\}_t^T$ is defined as

$$D_m = \lim_{h \to 0} \lim_{T \to \infty} \frac{\log \widehat{C}_{m,e}(h)}{\log h}, \tag{7.48}$$

indicating that $\widehat{C}_{m,e}(h) \propto h^{D_m}$. Notice, the dimensionality of the distribution of $\{Y_t, t \in \mathbb{Z}\}$ need not be an integer number, which in chaos theory is an indication of a fractal structure. For a given value m, the relationship between $\log \widehat{C}_{m,e}(h)$ and $\log h$ can be illustrated as the slope of $\log \widehat{C}_{m,e}(h) = D_m \times \log h$. The slope will converge to a stationary value for increasing lengths m of the delay vector $\mathbf{Y}_t$, when the dynamic system is deterministic; when the limit in (7.48) is finite. When the dynamical system is stochastic, the slope continually increases as m increases; the limit in (7.48) is infinite.

Rather than using an estimator of the slope for a single value h, Kočenda and Briatka (2005) propose to use an estimator of the average slope across a range of values h, which means calculating $\widehat{\beta}_m$ as a consistent estimate of the slope coefficient β_m from the LS regression

$$\log \widehat{C}_{m,e}(h_i) = \alpha_m + \beta_m \log h_i + u_i, \quad (i = 1, \ldots, n), \tag{7.49}$$

where α_m is an intercept, u_i an error term, and n the number of h_i's taken into consideration. However, these authors ignore the fact that $\widehat{C}_{m,e}(\cdot)$ is an empirical CDF (of distances between pairs of points). A regression ignoring this will be inefficient, as it leads to correlated residuals.

Example 7.5: Dimension of an ExpAR(1) process

Similar as in Example 2.4, we consider the ExpAR(1) process

$$Y_t = \{-0.9 - 0.95 \exp(-Y_{t-1}^2)\} Y_{t-1} + \varepsilon_t, \quad \{\varepsilon_t\} \stackrel{\text{i.i.d.}}{\sim} \mathcal{N}(0, 0.36). \tag{7.50}$$

We showed that the skeleton (deterministic part) of this particular ExpAR process has a limit cycle $(-1.50043, 1.50043)$ which suggests that the dimensionality of the distribution of $\{Y_t, t \in \mathbb{Z}\}$ equals two. To investigate this issue, we generate $T = 2,000$ observations from the above process. Next, we compute $\widehat{C}_{m,Y}(h)$ ($m = 2, \ldots, 10$) for 100 consecutive h-values in the range $[0.349, 0.990]$.

Figure 7.3(a) shows a plot of $\log_{10} \widehat{C}_{m,Y}(h)$ versus $\log_{10} h$ for $m = 2, \ldots, 10$. We see that for approximately values of $\log_{10} h < -0.17$ there is a clear linear relationship, indicating that $\{Y_t, t \in \mathbb{Z}\}$ is concentrated in a low-dimensional space. Figure 7.3(b) shows $\widehat{\beta}_m$ as estimates of β_m. These estimates are calculated by taking the LS values of the lines through three subsequent points, corresponding to $\log_{10} h_i$, $\log_{10} h_{i+1}$, and $\log_{10} h_{i+2}$ ($i = 1, \ldots, 98$). For i.i.d. time series processes β_m is equal to m, for small values of h. This is not the case here, with slope estimates $\widehat{\beta}_m < m$. In fact, it can be shown that $\mathbb{E}(\widehat{\beta}_m) \leq m$; cf. Exercise 7.1(d).

At this point it is appropriate to mention that in finite samples the asymptotic normality of the BDS test statistic may not be accurate. A naturally alternative is to use BS methods to approximate the distribution of the test statistic. One fast way of computing p-values of (7.44) is by randomizing (permuting) the order of the observed time series values. Because $\widehat{\sigma}_{m,T}(e; h)$ is a positive constant under randomization, simulation can be restricted to the non-normalized statistic $\widehat{C}_{m,e}(h) - \{\widehat{C}_{1,e}(h)\}^m$. For the observed p-values, which are invariant under a scale transformation, this does not make a difference. Similarly, $\{\widehat{C}_{1,e}(h)\}^m$ is a constant under permutations. Thus, one may determine p-values by computing the statistic $\widehat{C}_{m,e}(h)$ only. The resulting procedure is as follows.

Algorithm 7.3: Bootstrapping p-values of the BDS test statistic

(i) Compute $\widehat{C}_{m,e}(h)$ for the standardized residuals $\{e_t\}_{t=1}^T$, and permute $\{e_t\}$, to obtain the series $\{\widetilde{e}_t\}_{t=1}^T$.

(ii) Compute $\widehat{C}_{m,\widetilde{e}}(h)$.

(iii) Repeat steps (i) – (ii) B times, to obtain $\{\widehat{C}_{m,\widetilde{e}}^{(b)}(h)\}_{b=1}^B$.

(iv) Compute the one-sided p-value as

$$\widehat{p}^{\text{BDS}} = \frac{1 + \sum_{b=1}^B I\bigl(\widehat{C}_{m,\widetilde{e}}^{(b)}(h) \geq \widehat{C}_{m,e}(h)\bigr)}{1 + B}.$$

The nuisance-parameter-free property that any root-n consistent estimator of the model parameters has no impact on the null limit distribution of the BDS test statistic, under a class of linear and nonlinear conditional mean models, makes the test statistic a useful diagnostic tool in the context of nonlinear time series analysis. On the other hand, the BDS test statistic suffers from some problems (Brock et al., 1991).

- There is arbitrariness in the choice of h, which may affect both the power and size of the test. In fact, some choices of h may render the BDS test statistic inconsistent against certain alternatives. Thus, the probability of rejecting $\mathbb{H}_0$ does not always approach 1, as $T \to \infty$. In practice, h is usually taken as a fraction of the standard deviation of the time series under study.

- Another problem is that the BDS test statistic, though asymptotically normal under the null hypothesis, has high rates of Type I error, especially for non-Gaussian data.[5]

In the next section various extensions of the BDS test statistic are considered that are freed from some or all of these drawbacks.

7.4.2 Rank-based BDS test statistics

In an attempt to mitigate the problems with the BDS test statistic Genest et al. (2007) propose a number of rank-based extensions. The first test statistic is a circular version of the BDS test statistic $S_{m,e}(h)$ defined in (7.47). In particular, let $e_{t+T} = e_t \; \forall t \in \mathbb{N}^+$. Write $\mathbf{W}_t = (W_{1,t}, \ldots, W_{m,t})' = (e_t, \ldots, e_{t-m+1})'$ ($m \in \mathbb{Z}^+$). Then a circular version of the BDS test statistic (7.43) is given by

$$S_{m,W}(h) = \sqrt{T} \frac{\widehat{C}_{m,W}(h) - \{\widehat{C}_{1,W}(h)\}^m}{\widehat{\sigma}_{m,W}(h)}, \qquad (7.51)$$

where $\widehat{C}_{m,W}(h)$ and $\widehat{\sigma}_{m,W}(h)$ are defined in a similar way as respectively (7.43) and (7.44). In analogy with $S_{m,e}(h)$ it can be shown that the large-sample distribution of $S_{m,W}(h)$ is standard normal under the null hypothesis of no serial dependence.

In a similar fashion, Genest et al. (2007) propose a rank-based analogue of the BDS test statistic. Let $\widetilde{e}_t = \text{rank}(e_t)/(T+1)$ denote the normalized ranks of the time series $\{e_t\}_{t=1}^T$. Write $\widetilde{\mathbf{W}}_t = (\widetilde{W}_{1,t}, \ldots, \widetilde{W}_{m,t})' = (\widetilde{e}_t, \ldots, \widetilde{e}_{t-m+1})'$. Then a rank-based version of $S_{m,W}(h)$ may be defined as

$$S_{m,\widetilde{W}}(h) = \sqrt{T} \frac{\widehat{C}_{m,\widetilde{W}}(h) - \{\widehat{C}_{1,\widetilde{W}}(h)\}^m}{\widehat{\sigma}_{m,\widetilde{W}}(h)}. \qquad (7.52)$$

Again, under the $\mathbb{H}_0$ of no serial dependence, it follows that $S_{m,\widetilde{W}}(h) \xrightarrow{D} \mathcal{N}(0,1)$, $\forall h \in (0, \infty)$, as $T \to \infty$.

---

[5]This problem does not occur with the permutation-based BDS test statistic (Algorithm 7.2), as it has exact size.

Table 7.2: *Rank-based* BDS *test statistics of serial independence using three functionals (direct integration* (D)*, Kolmogorov–Smirnov* (KS)*, and Cramér–von Mises* (CvM)*), and two empirical processes.*

Functional	Empirical processes [1][2]	
	$\widetilde{\mathbb{D}}_T(\mathbf{u}) = \sqrt{T}\{\widetilde{B}_T(\mathbf{u}) - \prod_{k=1}^m \widetilde{G}_T(u_k)\}$	$\widetilde{\mathbb{B}}_T^*(\mathbf{u}) = 2\sqrt{T}\{\widetilde{B}_T^*(\mathbf{u}) - \widetilde{B}_T(\mathbf{u})\}$
D	$\widetilde{I}_{m,\widetilde{W}} = \int_0^1 \widetilde{\mathbb{D}}_T(h,\ldots,h)\mathrm{d}\widetilde{G}(h)$	$\widetilde{I}^*_{m,\widetilde{W}} = \int_0^1 \widetilde{\mathbb{B}}_T^*(h,\ldots,h)\mathrm{d}\widetilde{G}(h)$
KS	$\widetilde{M}_{m,\widetilde{W}} = \max_{i\in\{1,\ldots,T\}}\left|\widetilde{\mathbb{D}}_T\left(\frac{i}{T+1},\ldots,\frac{i}{T+1}\right)\right|$	$\widetilde{M}^*_{m,\widetilde{W}} = \max_{i\in\{1,\ldots,T\}}\left|\widetilde{\mathbb{B}}_T^*\left(\frac{i}{T+1},\ldots,\frac{i}{T+1}\right)\right|$
CvM	$\widetilde{T}_{m,\widetilde{W}} = \int_{[0,1]^m}|\widetilde{\mathbb{D}}_T(\mathbf{u})|^2\mathrm{d}\widetilde{B}(\mathbf{u})$	$\widetilde{T}^*_{m,\widetilde{W}} = \int_{[0,1]^m}|\widetilde{\mathbb{B}}_T^*(\mathbf{u})|^2\mathrm{d}\widetilde{B}(\mathbf{u})$

[1] $\widetilde{B}_T(\mathbf{u}) = \binom{T}{2}^{-1}\sum_{1\leq i\leq j\leq T}\prod_{k=1}^m I(|\widetilde{W}_{k,j} - \widetilde{W}_{k,i}| \leq u_k)$ with $\mathbf{u} = (u_1,\ldots,u_m)' \in [0,1]^m$; $\widetilde{G}_T(h) = \widetilde{B}_T(h,1,\ldots,1)$ with $h \in (0,1]$.

[2] $\widetilde{B}_T^*(\mathbf{u}) = T^{-1}\sum_{i=1}^T \prod_{k=1}^m\{\widetilde{F}(\widetilde{w}_{k,i} + u_k) - \widetilde{F}(\widetilde{w}_{k,i} - u_k)\}$, where $\widetilde{F}(\cdot)$ is the distribution of a $U(0,1)$ random variable; $\widetilde{B}_T^*(\mathbf{u}) = \prod_{k=1}^m \widetilde{G}(u_k)$ with $\widetilde{G}(\cdot)$ a Beta(1,2) distribution.

Clearly, the finite-sample performances of the test statistics (7.51) and (7.52) depend on the choice of h. A common way to get around this problem is to integrate out h with regard to some empirical process using various continuous functionals. Adopting direct integration (D), the KS and CvM functionals, and two empirical processes, Genest et al. (2007) propose six rank-based BDS test statistics; see Table 7.2. Moreover, they show that under $\mathbb{H}_0$, all six test statistics converge in distribution to centered Gaussian variables.

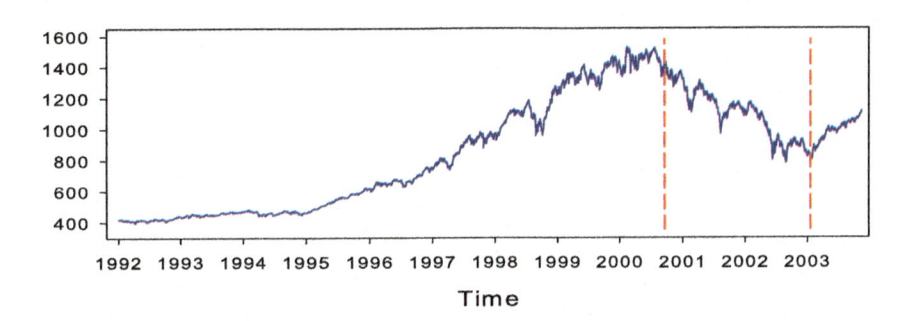

Figure 7.4: S&P 500 *daily stock price index for the time period* 1992 – 2003 (3,102 *observations) with two subperiods, denoted by vertical red medium dashed lines, from* November 2000 – February 2003 ($T = 608$) *and* March 2003 – December 2003 ($T = 218$).

Example 7.6: S&P 500 daily stock price index

Figure 7.3 shows the daily S&P 500 stock price (closing) index from 1992 – 2003. It has long been hypothesized that stock prices, say $\{P_t\}$, follow a

Table 7.3: *Bootstrap p-values of seven test statistics for serial independence applied to daily* S&P 500 *stock returns. Time period November 2000 – February 2003 ($T = 608$), and March 2003 – December 2003 ($T = 218$); $B = 1,000$. Blue-typed numbers indicate rejection of $\mathbb{H}_0$ at the 5% nominal significance level.*

		BDS	\multicolumn{6}{c}{Rank-based BDS test statistics}
Period	m	$S_{m,T}$	$\widetilde{I}^*_{m,\widetilde{R}}$	$\widetilde{M}^*_{m,\widetilde{R}}$	$\widetilde{T}^*_{m,\widetilde{R}}$	$\widetilde{I}_{m,\widetilde{R}}$	$\widetilde{M}_{m,\widetilde{R}}$	$\widetilde{T}_{m,\widetilde{R}}$
11/2000 –	2	0.21	0.07	0.14	0.08	0.57	0.53	0.91
02/2003	4	0.29	0.00	0.02	0.00	0.30	0.59	0.09
	6	0.36	0.00	0.02	0.00	0.30	0.58	0.01
	8	0.43	0.00	0.02	0.00	0.29	0.76	0.00
03/2003 –	2	0.21	0.91	0.31	0.89	0.33	0.22	0.00
12/2003	4	0.30	0.91	0.49	0.80	0.10	0.85	0.00
	6	0.36	0.41	0.34	0.48	0.12	0.88	0.00
	8	0.46	0.13	0.15	0.15	0.31	0.75	0.00

(geometric) random walk possibly with drift. We consider two sample sub-periods. The first one (11/2000 – 02/2003; $T = 608$), corresponds to the worst decline in the S&P 500 index since 1931, with the end of the "dot-com bubble" around November 2000. The second time period (03/2003 – 12/2003; $T = 218$) corresponds to an upward trend with moderate volatility, indicating the start of a new bull market in the first quarter of 2003. Using the circular version of the BDS test statistic, we test for serial independence in the series of daily stock returns, $R_t = \log(P_t/P_{t-1})$, with $h = \widehat{\sigma}_R$, i.e. the standard deviation of $\{R_t\}_{t=1}^T$. In addition, using the six ranked-based test statistics, we investigate $\widetilde{R}_t = \text{rank}(R_t)/(T+1)$.

Table 7.3 reports bootstrapped p-values, based on $B = 1,000$ bootstrap replicates, for each of the seven test statistics. Note that for the first, downward, period the results of almost all test statistics suggest that the underlying DGP is not i.i.d. On the other hand, the p-values of the circular BDS test statistic $S_{m,R}$, and the rank-based test statistics $\widetilde{I}_{m,\widetilde{R}}$ and $\widetilde{M}_{m,\widetilde{R}}$ are insignificant at the 5% nominal level for all values of m. The second, upward, period shows a very different picture. There, except for the test statistics $\widetilde{T}_{m,\widetilde{R}}$, almost all test results suggest that the process $\{R_t, t \in \mathbb{Z}\}$ is i.i.d., i.e., the S&P 500 daily stock price index follows a random walk.

7.4.3 Distribution-based test statistics

The pairwise test statistic $\widehat{\Delta}_T^{\text{ST3}}$ is a special case of a test statistic of multivariate independence proposed by Blum et al. (1961). These authors consider the difference between the nonparametric estimator of the joint EDF and the product of

7.4 HIGH-DIMENSIONAL TESTS

the nonparametric marginals. In a time series context, with a set of observations $\{Y_t\}_{t=1}^T$ drawn from a strictly stationary m-dimensional process $\{\mathbf{Y}_t, t \in \mathbb{Z}\}$, the corresponding empirical process is

$$\mathbb{H}_{m,T}(\mathbf{y}) = \sqrt{T}\{\widehat{F}_{m,T}(\mathbf{y}) - \prod_{i=1}^m \widehat{F}(y_i)\}, \quad \mathbf{y} \in \mathbb{R}^m, \tag{7.53}$$

where

$$\widehat{F}_{m,T}(\mathbf{y}) = \frac{1}{T}\sum_{t=1}^{T-m+1}\prod_{i=1}^m I(Y_{t+i-1} \leq y_i), \text{ and } \widehat{F}(y_i) = \frac{1}{T}\sum_{t=1}^{T-m+1} I(Y_{t+i-1} \leq y_i),$$
$$(i = 1, \ldots, m).$$

Various functionals of (7.53) can be used for testing the null hypothesis (7.4). Delgado (1996) proposes the CvM functional. When $m = 2$, the resulting test statistic $\widehat{\Delta}_{m,T}^{\text{D}}$ (see Table 7.4) has the same asymptotic null distribution as the test statistic of Blum et al. (1961) in the bivariate case. However, for $m > 2$, the asymptotic covariance function of $\widehat{\Delta}_{m,T}^{\text{D}}$ is not convenient for the tabulation of critical values, due to the complex nature of the limiting distribution of $\mathbb{H}_{m,T}(\cdot)$.

High-dimensional test statistics leading to considerably simpler asymptotic covariances under the null hypothesis than $B_{m,T}^{\text{CvM}}$ can be based on the Möbius *transformation* (Rota, 1964), or decomposition, of the process $\mathbb{H}_{m,T}(\cdot)$. Consider an index set $\mathcal{S}_m = \{A \subseteq \{1, \ldots, m\}; |A| > 1\}$, where $|A|$ is the cardinality of the index set A. Since $|A| = m$, $\mathcal{S}_m$ contains $2^m - m - 1$ elements. Now, the Möbius transformation $\mathcal{M}$ decomposes $\mathbb{H}_{m,T}(\cdot)$ into $2^m - m - 1$ sub-processes $\mathbb{G}_{A,T} = \mathcal{M}_A(\mathbb{H}_{m,T})$, namely

$$\mathbb{G}_{A,T}(\mathbf{y}) = \sum_{B \subseteq A}(-1)^{|A\setminus B|}\mathbb{H}_{m,T}(\mathbf{y})\prod_{i\in A\setminus B}\widehat{F}(y_i)$$
$$= \frac{1}{\sqrt{T}}\sum_{t=1}^{T-m+1}\prod_{i\in A}\{I(Y_{t+i-1} \leq y_i) - \widehat{F}(y_i)\}, \quad \mathbf{y} \in \mathbb{R}^m, \tag{7.54}$$

where $\prod_{i\in\emptyset} = 1$ by convention. In this case, the characterization of serial independence of $(Y_{1,t}, \ldots, Y_{m,t})'$ is equivalent to having $\mathcal{M}_A(\cdot) \equiv 0$, for all $A \subseteq \{1, \ldots, m\}$.

It follows from standard theory (see, e.g., Shorack and Wellner, 1984) that under the null hypothesis of (serial) independence, $\mathbb{G}_{A,T}(\cdot)$ converges weakly to a continuous centered Gaussian process with covariance function

$$\text{Cov}_A(\mathbf{x}, \mathbf{y}) = \prod_{i\in A}\Big\{\min\{F(x_i), F(y_i)\} - F(x_i)F(y_i)\Big\}, \quad \mathbf{x}, \mathbf{y} \in \mathbb{R}^m,$$

whose eigenvalues, given by

$$\lambda_{(i_1,\ldots,i_{|A|})} = \frac{1}{\pi^{2|A|}(i_1\cdots i_{|A|})^2}, \quad (i_1,\ldots,i_{|A|}) \in \mathbb{N},$$

may be deduced from the Karhunen–Loève decomposition of the Brownian bridge. Moreover, $\mathbb{G}_{A,T}(\cdot)$ and $\mathbb{G}_{A',T}(\cdot)$ are mutually independent asymptotically whenever $A \neq A'$.

Using the CvM functional, Ghoudi et al. (2001) propose $2^m - m - 1$ test statistics of the form

$$M_{A,T}^{\text{CvM}} = \int \{\mathbb{G}_{A,T}(\mathbf{y})\}^2 \mathrm{d} F_{m,T}(\mathbf{y}). \tag{7.55}$$

When $m = 2$, (7.55) simplifies to the single test statistic $M_{\{1,2\},T}^{\text{CvM}}$ which, interestingly, coincides with the test statistic $\widehat{\Delta}_T^{\text{ST3}}(\ell)$ at lag $\ell = 1$. Thus, a Möbius transformation is not needed in this particular case. Under the null hypothesis of (serial) independence, the limiting distribution of $M_{A,T}^{\text{CvM}}$ is given by

$$\sum_{(i_1,\ldots,i_{|A|}) \in \mathbb{N}} \lambda_{(i_1,\ldots,i_{|A|})} Z_{(i_1,\ldots,i_{|A|})}^2,$$

where the $Z_{(i_1,\ldots,i_{|A|})}$'s are independent $\mathcal{N}(0,1)$ random variables; Deheuvels (1981).

Observe that the sets A contribute differently to each of the test statistics $M_{A,T}^{\text{CvM}}$, with the biggest contribution coming from small-sized sets. To avoid this problem, it is convenient to standardize $M_{A,T}^{\text{CvM}}$ by the asymptotic mean and variance of $\xi_{|A|}$ which are, respectively, given by $\mathbb{E}(\xi_{|A|}) = 1/6^{|A|}$ and $\text{Var}(\xi_k) = 2/90^{|A|}$. The lower part of Table 7.4 displays the two resulting test statistics, denoted by the short-hand notation GKR$_1$ and GKR$_2$.

An obvious limitation of tests based on the above approach is the dependence of the asymptotic null distribution of the $\mathbb{G}_{A,T}(\cdot)$'s on the marginals of $\mathbb{H}_{m,T}(\cdot)$. To alleviate this problem, the original observations are replaced by their associated ranks in Section 7.4.4.

7.4.4 Copula-based test statistics

Univariate

Similar as in Section 7.4.3, empirical stochastic processes can be based on the pseudo-observations $\{\widehat{\mathbf{U}}_t = (\widehat{U}_{1,t}, \ldots, \widehat{U}_{m,t})'\}_{t=1}^T$ (see Section 7.3.2). To be specific, the natural analogue of (7.53) is defined as

$$\mathbb{C}_T(\mathbf{u}) = \frac{1}{\sqrt{T}} \sum_{t=1}^{T-m+1} \left\{ \prod_{i=1}^m I\{R_{t+i-1} \leq (T+1)u_i\} - \prod_{i=1}^m u_i \right\}, \quad \mathbf{u} \in [0,1]^m, \tag{7.56}$$

where $\{R_t\}_{t=1}^T$ are the ranks of $\{Y_t\}_{t=1}^T$. Using the Möbius transformation, Genest and Rémillard (2004) define the $2^m - m - 1$ stochastic processes

$$\mathbb{G}_{A,T}^c(\mathbf{u}) = \frac{1}{\sqrt{T}} \sum_{t=1}^{T-m+1} \prod_{i \in A} \left\{ I\{R_{t+i-1} \leq (T+1)u_i\} - U_T(u_i) \right\}, \tag{7.57}$$

7.4 HIGH-DIMENSIONAL TESTS

Table 7.4: *High-dimensional ($m \geq 2$) serial independence test statistics.*

Reference	Test statistic
Delgado (1996)	$\widehat{\Delta}_{m,T}^{\mathrm{D}} = \sum_{t=1}^{T}\{\mathbb{H}_{m,T}(\mathbf{Y}_t)\}^2,$ where $\mathbb{H}_{m,T}(\mathbf{y}) = \frac{1}{T}\sum_{t=1}^{T-m+1}\prod_{i=1}^{m}I(Y_{t+i-1}\leq y_i) - \prod_{i=1}^{m}\{\frac{1}{T}\sum_{t=1}^{T-m+1}I(Y_{t+i-1}\leq y_i)\}$
Ghoudi et al. (2001)	$\widehat{\Delta}_{m,T}^{\mathrm{GKR_1}} = \sum_{A}(M_{A,T}^{\mathrm{CvM}} - (1/6^{|A|}))/\sqrt{2/90^{|A|}},$ $\widehat{\Delta}_{m,T}^{\mathrm{GKR_2}} = \max_{A}\left|(M_{A,T}^{\mathrm{CvM}} - (1/6^{|A|}))/\sqrt{2/90^{|A|}}\right|,$ where $M_{A,T}^{\mathrm{CvM}} = \int\{\mathbb{G}_{A,T}(\mathbf{y})\}^2 \mathrm{d}F_{m,T}(\mathbf{y})$ with $\mathbb{G}_{A,T}(\mathbf{y}) = \frac{1}{\sqrt{T}}\sum_{t=1}^{T-m+1}\prod_{i\in A}\{I(Y_{t+i-1}\leq y_i) - \widehat{F}(y_i)\}$

where $U_T(\cdot)$ is the distribution of a discrete random variable U uniformly distributed on the set $\{1/(T+1), 2/(T+1), \ldots, T/(T+1)\}$, that is $U_T(t) = \min\{\lfloor(T+1)t\rfloor/T, 1\}$. Most conveniently, using the CvM functional, the copula-based version of $M_{A,T}^{\mathrm{CvM}}$ is

$$M_{A,T}^{\mathrm{CvM},c} = \int_{[0,1]^m}\{\mathbb{G}_{A,T}^c(\mathbf{u})\}^2 \mathrm{d}\mathbf{u}. \tag{7.58}$$

Some algebra shows that (7.58) can be computed directly from the ranks as

$$M_{A,T}^{\mathrm{CvM},c} = \frac{1}{T}\sum_{t=1}^{T-m+1}\sum_{s=1}^{T-m+1}\prod_{i\in A}\left\{\frac{2T+1}{6T} + \frac{R_{t+i-1}(R_{t+i-1}-1)}{2T(T+1)}\right.$$
$$\left. + \frac{R_{s+i-1}(R_{s+i-1}-1)}{2T(T+1)} - \frac{(R_{t+i-1}\vee R_{s+i-1})}{T+1}\right\}. \tag{7.59}$$

Since the subset A and its δ-translate, say $A+\delta$, generate basically the same process, computation of the test statistic (7.59) can be restricted to subsets $A \in \mathcal{A}_m = \{A \subset \mathcal{I}_m; 1 \in A, |A| > 1\}$ with cardinality $2^{m-1} - 1$. The limiting distribution of $M_{A,T}^{\mathrm{CvM},c}$ is the same as that of $M_{A,T}^{\mathrm{CvM}}$.

Multivariate

Kojadinovic and Yan (2011) address the generalization of the univariate serial copula correlation test to the case of continuous multivariate time series. Consider a strictly stationary ergodic sequence of q-dimensional random vectors $\mathbf{Y}_1, \mathbf{Y}_2, \ldots$, where the common distribution function of each $\mathbf{Y}_t = (Y_{1,t}, \ldots, Y_{q,t})'$ is denoted by $F(\cdot)$ and the associated copula by $\mathbb{C}(\cdot)$. Furthermore, let $m > 1$ be an integer, let $T' = T + m - 1$, and, for any $i \in \mathbb{R}^q$, let $R_{i,1}, \ldots, R_{i,T'}$ be the ranks associated with the univariate sequence $\{Y_{i,t}\}_{t=1}^{T'}$. The ranks are related to the univariate empirical marginal distribution function $\widehat{F}_{i,T}(Y_{i,t})$ through the equalities $R_{i,t} = T'\widehat{F}_{i,T}(Y_{i,t})$ $(t = 1, \ldots, T'; i = 1, \ldots, q)$.

To build an empirical copula in the multivariate case, we need to introduce some notation. First, given the index set $B \subseteq \{1, \ldots, m\}$, we define the vector $\mathbf{u}_B \in [0, 1]^{mq}$ by

$$u_B^{(j)} = \begin{cases} u^{(j)} & \text{if } j \in \bigcup_{i \in B}\{(i-1)q + 1, \ldots, iq\}, \\ 1 & \text{otherwise.} \end{cases}$$

Next, given $\mathbf{u} \in [0, 1]^{mq}$ and $i \in \{1, \ldots, m\}$, define the sub-vector $\mathbf{u}_{\langle i \rangle} \in [0, 1]^q$ of $\mathbf{u}$ by

$$u_{\langle i \rangle}^{(j)} = u^{(j+(i-1)q)}, \quad (i = 1, \ldots, m; j = 1, \ldots, q).$$

Finally, we form the mq-dimensional random vector $\widetilde{\mathbf{Y}}_t = (\mathbf{Y}_t', \ldots, \mathbf{Y}_{t+m-1}')'$ ($t = 1, \ldots, T$). Then, given $\{\widetilde{\mathbf{Y}}_t\}_{t=1}^T$, and in analogy with (7.26), the serial (s) empirical (multivariate) copula is defined as

$$\widehat{\mathbb{C}}_T^s(\mathbf{u}) = \frac{1}{T}\sum_{t=1}^T \prod_{i=1}^m \prod_{j=1}^q I\left(\widehat{F}_{j,T}(Y_{j,t+i-1}) \leq u_{\langle i \rangle}^{(j)}\right) = \frac{1}{T}\sum_{t=1}^T \prod_{i=1}^m \prod_{j=1}^q I(R_{j,t+i-1} \leq T' u_{\langle i \rangle}^{(j)}).$$

A multivariate extension of the empirical process (7.56) is then

$$\mathbb{C}_T^s(\mathbf{u}) = \sqrt{T}\Big\{\widehat{\mathbb{C}}_T^s(\mathbf{u}) - \prod_{i=1}^m \widehat{\mathbb{C}}_T^s(\mathbf{u}_{\langle i \rangle})\Big\}, \quad \mathbf{u} \in [0, 1]^{mq}. \tag{7.60}$$

As noticed by Ghoudi et al. (2001) in the univariate case, it follows from the Möbius decomposition (transformation) of $\mathbb{C}_T^s(\cdot)$, that the limiting distribution of the processes $\sqrt{T}\mathcal{M}_A(\mathbb{C}_T^s)$ and $\sqrt{T}\mathcal{M}_{A+\delta}(\mathbb{C}_T^s)$ are roughly the same. Hence, attention can be restricted to the $2^{m-1} - 1$ processes $\sqrt{T}\mathcal{M}_A(\mathbb{C}_T^s)$ for $A \in \mathcal{A}_m$. Then, after some tedious algebra, the resulting CvM test statistics are given by

$$M_{A,q,T}^{\text{CvM},c} = \frac{1}{T}\sum_{t=1}^T \sum_{s=1}^T \prod_{i \in A} \Big\{\prod_{j=1}^q \Big[1 - \frac{(R_{j,t+i-1} \vee R_{j,s+i-1})}{T'}\Big]$$

$$- \frac{1}{T}\sum_{l=1}^T \prod_{j=1}^m \Big[1 - \frac{(R_{j,t+i-1} \vee R_{j,l+i-1})}{T'}\Big] - \frac{1}{T}\sum_{l=1}^T \prod_{j=1}^m \Big[1 - \frac{(R_{j,s+i-1} \vee R_{j,l+i-1})}{T'}\Big]$$

$$+ \frac{1}{T^2}\sum_{r=1}^T \sum_{s=1}^T \prod_{k=1}^m \Big[1 - \frac{(R_{k,r+i-1} \vee R_{k,s+i-1})}{T'}\Big]\Big\}. \tag{7.61}$$

Unfortunately, adopting the KS functional, an explicit expression for multivariate serial independence tests statistics is far more difficult to derive. Hence, we focus on $M_{A,q,T}^{\text{CvM},c}$.

7.4 HIGH-DIMENSIONAL TESTS

For $q = 1$, and using the approximation $T \approx T'$, (7.61) coincides with (7.59). In contrast with $M_{A,T}^{\text{CvM},c}$, however, the asymptotic null distribution of $M_{A,q,T}^{\text{CvM},c}$ ($q > 1$) is no longer distribution free. To overcome this problem, a bootstrap procedure is recommended. Below we distinguish between computing p-values for each $A \in \mathcal{A}_m$, and combined p-values across all index sets. In the latter case, and following Kojadinovic and Yan (2011), two p-value combination methods are considered, one due to Fisher (F) and one to Tippett (T). For ease of reading, we remove the superscripts CvM and c from $M_{A,m,T}^{\text{CvM},c}$.

Algorithm 7.4: Bootstrap-based p-values for multivariate serial independence tests

(i) Compute the test statistic $M_{A,q,T}^{(0)}$ for $|A| \leq h$ with h fixed in $\{2, \ldots, m-1\}$, using the *original* time series data $\{Y_t\}_{t=1}^T$, and $A \in \mathcal{A}_m$.

(ii) Generate B pseudo-random samples of size T' from a $U[0,1]$ distribution, and let $M_{A,q,T}^{(b)}$ ($b = 1, \ldots, B; A \in \mathcal{A}_m$) denote the value of the test statistics $M_{A,q,T}$, where B is some large integer.

(iii)
- **p-values for each $A \in \mathcal{A}_m$:**
  Compute an approximate p-value for the test statistic $M_{A,q,T}^{(i)}$ ($A \in \mathcal{A}_m$) as follows

  $$\widehat{p}(M_{A,q,T}^{(i)}) = \frac{\frac{1}{2} + \sum_{b=1}^{B} I(M_{A,q,T}^{(b)} \geq M_{A,q,T}^{(i)})}{1 + B}, \quad i \in \{0, 1, \ldots, B\}.$$

  The factor $1/2$ ensures that the p-values are in the open interval $(0, 1)$ so that transformations by inverse CDFs of continuous distributions are always well-defined.

- **Combined p-values:**
  For all $i \in \{0, 1, \ldots, B\}$, compute

  $$\text{F}_T^{(i)} = -2 \sum_{A \in \mathcal{A}_m} \log\{\widehat{p}(M_{A,q,T}^{(i)})\},$$

  and

  $$\text{T}_T^{(i)} = \min_{A \in \mathcal{A}_m} \log\{\widehat{p}(M_{A,q,T}^{(i)})\}.$$

  Approximate "global" p-values are then given by

  $$\widehat{p}_{\text{F}} = \frac{1}{B}\sum_{b=1}^{B} I(\text{F}_T^{(b)} \geq \text{F}_T^{(0)}), \text{ and } \widehat{p}_{\text{T}} = \frac{1}{B}\sum_{b=1}^{B} I(\text{T}_T^{(b)} \geq \text{T}_T^{(0)}).$$

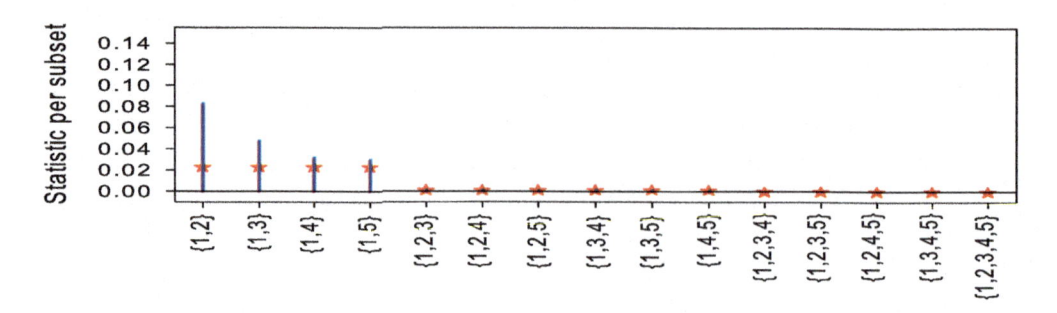

Figure 7.5: *Dependogram summarizing the results of the multivariate test of serial independence for the* climate change *data set; $q = 2$, $m = 5$. A red star denotes the approximate critical value.*

Example 7.7: Climate Change (Cont'd)

We illustrate the use of the preceding test statistics by revisiting the climate change data of Example 1.5. It can be verified that the δ^{13}C and δ^{18}O time series take only 149 and 133 unique values out of $T = 216$ observations, which means that there is a non-negligible number of ties in the data. Hence, some artificial smoothing of the series is needed in order to meet the assumption of continuous marginal distributions of the proposed test statistics. For instance, the method of *jittering* (adding random uniform noise to the series) can deal with this problem. For simplicity, we ignore the ties and focus on the original data.

To visualize the results of the serial independence tests it is convenient to use a graphical display, called *dependogram*. For each subset A, a vertical bar is drawn of height corresponding to the value of the subset test statistic $M_{A,q,T}^{\text{CvM},c}$. A star denotes the approximate, bootstrapped, critical values of $M_{A,q,T}^{\text{CvM},c}$. Subsets for which the bar exceeds the critical value are considered to be composed of serially dependent variables.

Figures 7.5 displays a serial dependogram with $q = 2$ and $m = 5$ for $M_{A,q,T}^{\text{CvM},c}$ applied to the time series δ^{13}C and δ^{18}O jointly. The global test statistic takes the value 0.878×10^{-3} with p-value 0.500×10^{-3}. The combined tests à la Fisher (F_T) and à la Tippett (T_T) both have a p-value of 0.500×10^{-3}. Thus, there is evidence of serial dependence. In fact, the rejection of the null hypothesis of serial independence appears to be essentially due to subsets $\{1, 2\}, \ldots, \{1, 5\}$, while the test statistics are not significant for other subsets.

7.4.5 A test statistic based on quadratic forms

In view of the quadratic form $\Delta^Q(\cdot)$ given by (7.13), a natural way of forming a high-dimensional test statistic for serial independence is to replace the integrals by empirical averages of $(\mu_m^{(i)}, \mu_m^{(j)}) = \int_{\mathbb{R}^m} \int_{\mathbb{R}^m} \mathcal{K}_h(\mathbf{y} - \mathbf{x}) d\mu_m^{(i)}(\mathbf{y}) d\mu_m^{(j)}(\mathbf{x})$ $(i, j = 1, 2)$. For two independent m-dimensional processes $\{\mathbf{Y}_t, t \in \mathbb{Z}\} \sim \mu_m^{(1)}$ and $\{\mathbf{Y}_{t'}, t' \in \mathbb{Z}\} \sim \mu_m^{(2)}$ ($t \neq t'$) the first term $(\mu_m^{(1)}, \mu_m^{(1)})$ can be consistently estimated by the U-statistic estimator

$$(\widehat{\mu}_m^{(1)}, \widehat{\mu}_m^{(1)}) = \binom{T-m+1}{2}^{-1} \sum_{i=2}^{T-m+1} \sum_{s=1}^{i-1} \prod_{j=0}^{m-1} \mathcal{K}_h(Y_{i+j}, Y_{s+j}),$$

using a product kernel. Similarly, the terms $(\mu_m^{(1)}, \mu_m^{(2)})$ and $(\mu_m^{(2)}, \mu_m^{(2)})$ can be consistently estimated by

$$(\widehat{\mu}_m^{(1)}, \widehat{\mu}_m^{(2)}) = \frac{1}{T-m+1} \sum_{t=1}^{T-m+1} \prod_{j=0}^{m-1} \widehat{C}_h(Y_{t+j}),$$

$$(\widehat{\mu}_m^{(2)}, \widehat{\mu}_m^{(2)}) = \frac{1}{(T-m+1)^m} \prod_{j=0}^{m-1} \Big(\sum_{t=1}^{T-m+1} \widehat{C}_{h,T}(Y_{t+j}) \Big),$$

where

$$\widehat{C}_{h,T}(\mathbf{y}) = \frac{1}{T-m+1} \sum_{i=1}^{T-m+1} \mathcal{K}_h(\mathbf{y}, Y_i)$$

is a kernel-based estimate of the one-dimensional correlation integral associated with the marginal distribution function. Collecting the above expressions together, Diks and Panchenko (2007) propose the test statistic

$$\widehat{\Delta}_{m,T}^{\text{DP}} = (\widehat{\mu}_m^{(1)}, \widehat{\mu}_m^{(1)}) - 2(\widehat{\mu}_m^{(1)}, \widehat{\mu}_m^{(2)}) + (\widehat{\mu}_m^{(2)}, \widehat{\mu}_m^{(2)}). \tag{7.62}$$

Note that, for $\mathcal{K}_h(y) = I(|y| < h)$, the estimator $(\widehat{\mu}_m^{(1)}, \widehat{\mu}_m^{(1)})$ coincides with $C_{m,T}(Y;h)$ given by (7.32) as an estimator of the correlation integral. So, using the uniform kernel with the functional $(\mu_m^{(1)}, \mu_m^{(1)}) - (\mu_m^{(2)}, \mu_m^{(2)})$ will lead to the BDS test statistic (7.43), after standardizing. The theory of U-statistics can be used to prove the asymptotic normality of $\widehat{\Delta}_{m,T}^{\text{DP}}$ under the null hypothesis of serial independence. An alternative way of obtaining critical values and p-values involves using the bootstrap or the permutation methodology as outlined in Section 7.3.6.

7.5 Application: Canadian Lynx Data

The Canadian annual lynx trappings records (1821 – 1934; $T = 114$) in the MacKenzie River district of North–West Canada (i.e. the number of furs harvested by the Hudson Bay Company), plotted in the upper panel of Figure 7.6, provide an interesting basis for many nonlinear time series techniques. The data exhibits irregular

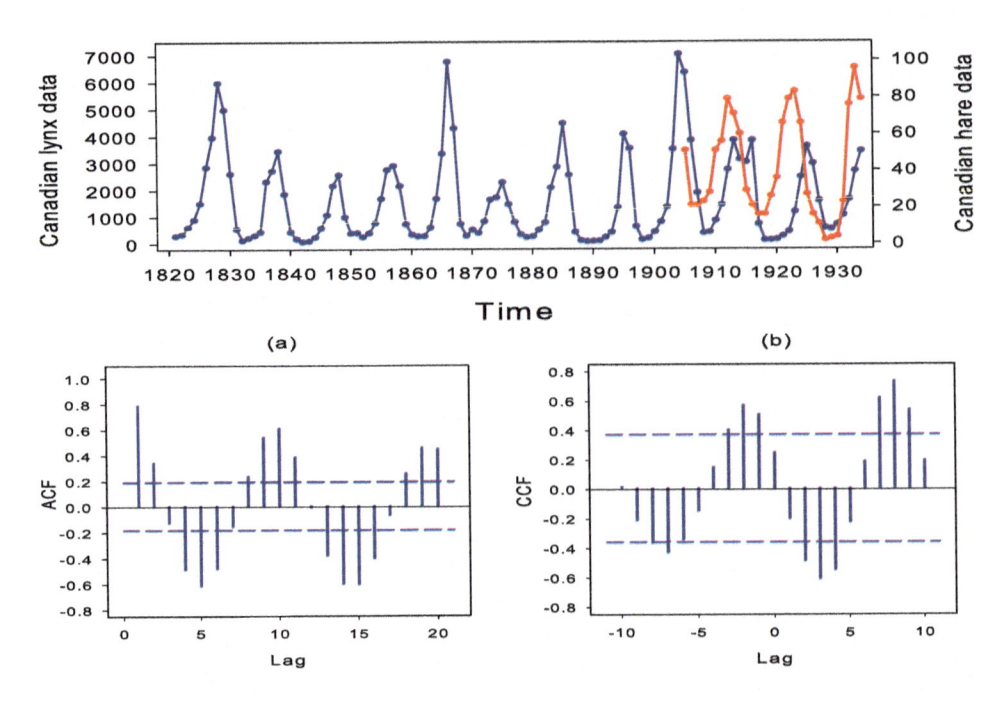

Figure 7.6: *Upper panel: yearly* Canadian lynx *data for the time period 1821 – 1934 (blue solid line), and yearly* Canadian snowshoe hare *data (in thousands) for the time period 1905 – 1934 (red solid line). Lower panel: (a) the sample ACF for the complete lynx series, and (b) the sample cross-correlation function (CCF) between the lynx series and the snowshoe hare series for the time period 1905 – 1934. Both plots contain 95% asymptotic confidence limits (blue medium dashed lines).*

periodic fluctuations with sharp and large peaks and relatively small troughs. As shown in Figure 7.6(a), the pattern of the sample ACF of the data indicates a cyclical behavior of about ten years (a 9.61- year periodicity). The data set is assumed to represent the relative magnitude of the lynx population and, hence, is of great interests to ecological researchers. To understand the cyclical behavior in the Canadian lynx series, the upper panel of Figure 7.6 also shows 30 yearly observations of the Canadian snowshoe hare series for the time period 1905 – 1934. Snowshoe hares (prey) constitute a major part of the lynx's (predator) diet. Note that the hare series lags behind the lynx series. Indeed, as can be seen from the sample CCF in Figure 7.6(b) there is a significant relationship between both series, but the lynx–hare interaction is not instantaneous, rather there is a time delay of about 2 years. According to McCarthy (2005), a possible cause of the cyclical fluctuations is that hare populations increase and eat vegetation. In response, the vegetation produces secondary defence compounds which are less palatable and nutritious. This triggers a crash of the hare population – hares die in great numbers. However, the lynx continue to feed on hares, but run out of prey eventually. This is followed by a decline in the lynx population. Next, the vegetation slowly recovers and this rejuvenates

7.5 APPLICATION: CANADIAN LYNX DATA

Table 7.5: *Five models fitted to the Canadian lynx data set; $T = 114$.*

Reference	Model	(Pooled) $\widehat{\sigma}_\varepsilon^2$
Moran (1953)	$Y_t = 1.0549 + 1.4101 Y_{t-1} - 0.7734 Y_{t-2} + \varepsilon_t$	0.0459
Tong (1990, p. 387)	$Y_t = \begin{cases} 0.546 + 1.032_{t-1} - 0.173 Y_{t-2} + 0.171 Y_{t-3} \\ \quad - 0.431 Y_{t-4} + 0.332 Y_{t-5} - 0.284 Y_{t-6} \\ \quad + 0.210 Y_{t-7} + \varepsilon_t^{(1)}, \qquad\qquad\quad Y_{t-2} \leq 3.116 \\ 2.632 + 1.492 Y_{t-1} - 1.324 Y_{t-2} + \varepsilon_t^{(2)}, \quad Y_{t-2} > 3.116 \end{cases}$	$0.0358^{(1)}$
Tsay (1989)	$Y_t = \begin{cases} 0.083 + 1.096 Y_{t-1} + \varepsilon_t^{(1)}, \qquad\qquad Y_{t-2} \leq 2.373 \\ 0.63 + 0.96 Y_{t-1} - 0.11 Y_{t-2} \\ \quad + 0.23 Y_{t-3} - 0.61 Y_{t-4} + 0.48 Y_{t-5} \\ \quad - 0.39 Y_{t-6} + 0.28 Y_{t-7} + \varepsilon_t^{(2)}, \quad 2.373 < Y_{t-2} \leq 3.154 \\ 2.323 + 1.530 Y_{t-1} - 1.266 Y_{t-2} + \varepsilon_t^{(3)}, \quad 3.154 < Y_{t-2} \end{cases}$	$0.0348^{(2)}$
Ozaki (1982) [3]	$Y_t = [1.167 + (0.316 + 0.982 Y_{t-1}) \exp(-3.89 Y_{t-1}^2)] Y_{t-1}$ $\quad - [0.437 + (0.659 + 1.26 Y_{t-1}) \exp(-3.89 Y_{t-1}^2)] Y_{t-2} + \varepsilon_t$	0.0433
Teräsvirta (1994)	$Y_t = 1.17 Y_{t-1} + (-0.92 Y_{t-2} + 1.00 Y_{t-3} - 0.41 Y_{t-4} + 0.27 Y_{t-9}$ $\quad - 0.21 Y_{t-11}) \times [1 + \exp\{-1.73 \times 1.8 (Y_{t-3} - 2.73)\}]^{-1} + \varepsilon_t$	0.0350

[1] $\mathrm{Var}(h_t^{(1)}) = 0.0259$ and $\mathrm{Var}(h_t^{(2)}) = 0.0505$.
[2] $\mathrm{Var}(h_t^{(1)}) = 0.015$, $\mathrm{Var}(h_t^{(2)}) = 0.025$, and $\mathrm{Var}(h_t^{(3)}) = 0.053$.
[3] As suggested by Tong (1990), the parameter 1.167 in the ExpAR(2) model replaces the original parameter 0.138 given by Ozaki.

the hare population, and so the cycle continues.

It is generally believed that the lynx series is nonlinear, but there is no agreement on which nonlinear model is most appropriate for the data. Lim (1987) summarizes the work done in analyzing this time series. Five estimated time series models, for the log-transformed data (base 10), are reproduced in Table 7.5. The SETAR(2; 7, 2) model admits nice biological interpretation; see, e.g., Stenseth et al. (1997). Below the threshold value the lynx population roughly increases. But above the threshold value, the population decreases due to the complex interplay between the available food, the mortality due to overall predation, and the indirect effects of predation by a suite of predators.

Table 7.6 shows p-values, based on 1,000 BS replicates, of eight high-dimensional tests for serial independence applied to the residuals of the fitted models. We see that $S_{m,T}$, $\widetilde{M}_{m,T}^*$, $\widetilde{I}_{m,T}$, and $\widetilde{M}_{m,T}$ fail to reject $\mathbb{H}_0$ at the 5% nominal significance level for all models, and all values of m. A similar conclusion emerges from the p-values of $\widehat{\Delta}_{m,T}^{\mathrm{DP}}$, except for the ExpAR(2) model with $m = 2$. Interestingly, all p-values suggest that the SETAR(2; 7, 2) and SETAR(3; 1, 7, 2) models adequately capture the nonlinear phenomena in the data. This result confirms earlier observations made in the literature; see, e.g., Tong (1990). For the ExpAR(2) model, we observe that $\mathbb{H}_0$ is rejected at the 5% nominal significance level on the basis of the reported p-values of the test statistics $\widetilde{T}_{m,T}^*$, and $\widetilde{T}_{m,T}$. For the LSTAR(11) model, evidence of residual dependence can be noted from the p-values of $\widetilde{I}_{m,T}^*$, and $\widetilde{T}_{m,T}^*$.

Table 7.6: *Bootstrap p-values of eight test statistics for high-dimensional serial independence applied to the residuals of five time series models fitted to the log of the Canadian lynx time series (see* Table 7.5*); $T = 114$, $B = 1,000$. Blue-typed numbers indicate rejection of $\mathbb{H}_0$ at the 5% nominal significance level.*

		BDS	\multicolumn{6}{c}{Rank-based BDS test statistics}	
Model	m	$S_{m,T}$	$\widetilde{I}^*_{m,T}$	$\widetilde{M}^*_{m,T}$	$\widetilde{T}^*_{m,T}$	$\widetilde{I}_{m,T}$	$\widetilde{M}_{m,T}$	$\widetilde{T}_{m,T}$	$\widehat{\Delta}^{\text{DP}}_{m,T}$
AR(2)	2	0.25	0.07	0.55	0.04	0.67	0.54	0.01	0.23
	4	0.31	0.01	0.38	0.01	0.40	0.12	0.01	0.29
	6	0.43	0.01	0.62	0.01	0.56	0.04	0.02	0.50
SETAR(2; 7, 2)	2	0.26	0.33	0.64	0.34	0.59	0.81	0.21	0.25
	4	0.34	0.15	0.67	0.13	0.94	0.28	0.09	0.44
	6	0.44	0.25	0.58	0.21	0.63	0.15	0.08	0.60
SETAR(3; 1, 7, 2)	2	0.25	0.66	0.38	0.63	0.98	0.56	0.13	0.50
	4	0.32	0.40	0.27	0.32	0.92	0.27	0.15	0.52
	6	0.41	0.44	0.17	0.41	0.62	0.15	0.14	0.38
ExpAR(2)	2	0.25	0.12	0.32	0.01	0.12	0.15	0.01	0.04
	4	0.33	0.14	0.39	0.01	0.14	0.68	0.02	0.15
	6	0.43	0.38	0.55	0.00	0.38	0.32	0.04	0.33
LSTAR(11)	2	0.25	0.02	0.41	0.03	0.23	0.91	0.26	0.37
	4	0.32	0.01	0.20	0.01	0.19	0.99	0.24	0.09
	6	0.42	0.04	0.18	0.04	0.08	0.95	0.30	0.15

Not surprisingly, the lack of fit of Moran's AR(2) model, and Ozaki's ExpAR(2) model has been noted by other researchers. However, the fact that the residuals of the LSTAR(11) model do not pass all test statistics is new. It suggests that the model may be further improved. Finally, note that for the AR(2) model no evidence of residual dependence is detected by $\widetilde{I}^*_{m,T}$ when $m = 2$, while for $m = 4$ and $m = 6$ the p-value of this test statistic is smaller than the 5% nominal significance level. Thus, it is recommended not to rely completely on low-dimensional test results.

7.6 Summary, Terms and Concepts

Summary
Serial independence is central to time series analysis, especially within the context of checking the adequacy of fitted nonlinear time series models. In this chapter, we highlighted influential research on nonparametric test statistics for serial dependence in conditional mean. We have not said anything about other types of serial dependence, for instance, through the conditional variance or through conditional higher order moments. Readers interested in this topic should consult Su and White (2008), Huang et al. (2015) and the references therein.

An obvious question is, which serial independence test should one adopt in practice? Within the context of single-lag and multiple-lag test procedures, we have already dwelt upon conclusions emerging from the extensive MC simulation study by Bagnato et al. (2014). Generally speaking, the tests considered by these authors have reasonable size and power properties compared with many nonlinear alternatives. We should emphasize, however, that adopting the limiting null distribution of a test statistic can be hazardous, except for very large sample sizes T. When using random permutation or bootstrapping approaches the size of a test statistic is often much closer to its nominal significance level for $T < 500$.

On the other hand, it is now generally believed that many empirical time series, while nonlinear, are generated by high-dimensional processes. Hence, it is natural to consider test statistics designed for this purpose. In this case, several of the rank-based extensions of the BDS test statistic discussed in Section 7.4.2, and the copula-based test statistics of Section 7.4.4 are useful. In particular, these test statistics are more powerful than their single-lag and multiple-lag counterparts, with $\widetilde{T}_{m,T}$ as the best performing rank-based BDS test.

Terms and Concepts

binwidth, 270
boundary effects, 269
copula density, 267
correlation dimension, 280
correlation integral, 260
Cressie–Read (CR) divergence, 265
Csiszár (C) divergence, 264
Daniell window, 273
dependogram, 290
empirical copula, 269
Gaussian copula, 307
generalized spectral density, 273
Hellinger (H) distance, 264
high-dimensional tests, 278
independence copula, 267
jittering, 290

Kolmogorov (K) divergence, 264
mixing proportions, 313
Möbius transformation, 285
multiple-lag tests, 272
nuisance-parameter-free property, 275
Parseval's identity, 262
permutation, 277
portmanteau-type test, 266
pseudo-observations, 269
quadratic (Q) distance, 261
Rényi (R) divergence, 264
single-lag tests, 270
Student t copula, 307
Tsallis (T) divergence, 264

7.7 Additional Bibliographical Notes

Sections 7.1 – 7.3: Tjøstheim (1994, 1996) reviews the early literature on (non)parametric tests of serial independence. An extensive bibliography of permutation, sign, and rank-based test statistics for serial independence is provided by Dufour et al. (1982). Hallin and Puri (1992) cover the literature of rank tests. In the context of econometric applications, Ullah (1996) provides a unified treatment of various entropy, divergence and distance measures. Giannerini et al. (2015) propose test statistics for pairwise nonlinear dependence under the null hypothesis of general linear dependence rather than serial independence. The R-package that implements these latter test statistics is available at CRAN (tseriesEntropy) and at http://www2.stat.unibo.it/giannerini/software.html.

The asymptotic properties of nonparametric estimators of copulas for time series processes are considered by Fermanian and Scaillet (2003), and Ibragimov (2009), among others.

Section 7.4: Matilla–Garcia and Ruiz–Marin (2008) propose a test statistic for high-dimensional serial independence using symbolic dynamics and permutation entropy. The test requires unrealistic large sample sizes for dimensions $m \geq 6$. De Gooijer and Yuan (2016) explore a link between the correlation integral and the Shannon entropy, or second order Rényi entropy, to derive two nonparametric portmanteau-type test statistics for serial independence. In commonly used samples, both tests performed similarly as the best performing rank-based BDS test statistics of Section 7.4.2.

Baek and Brock (1992a) extend the BDS test statistic to vector time series. Wolff and Robinson (1994) observe that the estimator of the unnormalized correlation integral has a limiting Poisson distribution under some moderate assumptions regarding the marginal distribution. This motivated a nonparametric test procedure with slightly reduced size distortion compared with the BDS test statistic. de Lima (1996) formulates five conditions under which the BDS test statistic is asymptotically nuisance-parameter-free.

Within the context of independent component analysis, a concept that is important in signal processing and neural networks, a subsampling pairwise test statistic for serial independence has been suggested by Karvanen (2005), based on the test of total independence by Kankainen and Ushakov (1998). Related to this, is the paper by Wu et al. (2009). They propose a smoothed bootstrap-based test statistic for high-dimensional serial independence in multivariate time series data by combining pairwise independence tests for all pairs. Other recently proposed test statistics suitable for both time-independent and time-dependent component analysis have been derived by, among others, Achard (2008), Baringhaus and Franz (2004), Fernández et al. (2008), Székely et al. (2007) (see the R-energy package), Gretton et al. (2005), and Zhou (2012).

Evidently, many density-based serial correlation tests require the data come from a continuous population. Although they will no longer be distribution free, some of the discussed test statistics can also be used in the discrete case. For instance, the Skaug–Tjøstheim (1993b) test statistic Δ_T^{ST1} can be applied to continuous as well as to discrete (or discretized) data, after some slight adjustment of the form of the test. For a stationary sequence of a categorical variable, high-dimensional serial independence can be checked via a test statistic developed by Bilodeau and Lafaye de Micheaux (2009).

The so-called *k-nearest neighbor* density estimator avoids the problem of a pre-defined grid required to compute the multi-dimensional copula-based histogram estimator discussed in Section 7.3.2; see Blumentritt and Schmid (2012). Alternatively, for estimating the copula density, a nonparametric method proposed by Kallenberg (2009) may be adopted.

Exercise 7.7: Various MAR models are available in the literature. Le et al. (1996), and Wong and Li (2000b, 2001) assume that the mixing proportions are time invariant. More general (Gaussian) MAR and MAR–GARCH models follow by assuming that the mixing proportions are functions of observed variables; see, e.g., Lanne and Saikkonen (2003), and Kalliovirta et al. (2015) and the references therein. Sufficient conditions for strict and second order stationarity are given by, among others, Zeevi et al. (2000), Wong and Li (2000b), and Saikkonen (2008).

7.8 Data and Software References

Data

Section 7.5: The Canadian snowshoe hare data derive from the main drainage of the Hudson Bay, based on trappers' questionnaires. The hare data used in this section are taken from the R-TSA package, and first published by D.A. MacLulich (1937) in the paper "Fluctuations in the Number of the Varying Hare (Lepus americanus)" (Univ. of Toronto Press, Ontario, Stud. Biol. Ser. No. 43, 136 pp.) which is not widely available. The paper by E.L. Leigh (1968) published in M. Gerstenhaber (Ed.) *Some Mathematical Problems in Biology* (American Mathematical Society, Providence, pp. 1 – 61) contains yearly hare data for the time period 1847 – 1903. There are slight differences between this data set and the data contained in the TSA package. The main source for the Canadian lynx data is Table 4 in the paper by C. Elton and M. Nicholson (*J. Anim. Ecol.*, 1942, 11, pp. 215 – 244). The data set is on DataMarket (http://data.is/TSDLdemo) at http://data.is/Ky69xY and can be read directly into R using the rdatamarket package.

Software references

Section 7.3: The entire R code for replicating the simulation study of Bagnato et al. (2014) is available at the website of this book.

Section 7.4: A windows executable file for computing the values of the slope coefficient in (7.49) can be downloaded from http://kocenda.fsv.cuni.cz/software.htm. The copula-based univariate and multivariate serial independence test statistics are implemented as separate functions in the R-copula package; see, e.g., Exercise 7.5. These functions are briefly described by Kojadinovic and Yan (2010). Partly overlapping the content of the R-copula package are the functions for nonparametric testing of mutual serial independence contained in the R-IndependenceTests package. When applying BS methods to functionals based on the empirical copula, standard ranking procedures are computationally expensive. Blumentritt and Grothe (2013) present a pseudocode algorithm that reduces the running time of these procedures considerably.

A fast MATLAB code for computing the traditional BDS test statistic was developed by Ludwig Kanzler; see http://papers.ssrn.com/paper.taf?abstract_id=151669. The code is available at http://econpapers.repec.org/software/bocbocode/t891501.htm. Also BDS C++, and BDS MATLAB source codes are available at the address http://people.brandeis.edu/~blebaron/.

C++ code for computing the rank-based BDS test statistics (made available by Kilani Ghoudi), Gauss code for computing the Hong–White, the Skaug–Tjøstheim, and Hong's generalized spectral test statistics (made available by Yongmia Hong) can be downloaded from the website of this book. Based on a generalized spectral approach (Section 7.3.5) of nonlinear model residuals, Hong and Lee (2003) propose some new diagnostic test statistics for serial independence. Their GAUSS code is available at the website of this book. Also available is a set of C++ computer routines written by Hans J. Skaug which are based on the various test statistics introduced in the papers by Skaug and Tjøstheim (1993a,b), and Skaug and Tjøstheim (1996).

Section 7.4.5: The C++ code of the $\widehat{\Delta}_{m,T}^{\mathrm{DP}}$ test statistic (7.62) can be downloaded from Cees Diks' web page located at http://cendef.uva.nl/people.

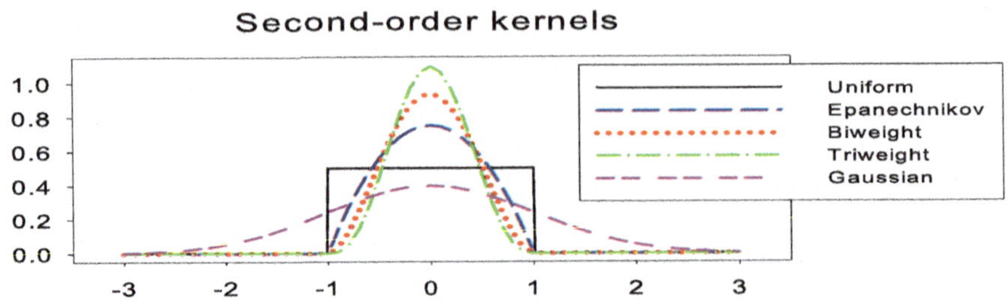

Figure 7.7: *Selected second-order kernel functions.*

Appendix

7.A Kernel-based Density and Regression Estimation

In this Appendix, we review some major concepts of kernel density and regression estimation in the i.i.d. case. Out of necessity, the discussion is cursory. The interested reader can, for instance, consult Härdle (1990), Wand and Jones (1995), or Li and Racine (2007) for accounts with greater detail.

Univariate density estimation

Let $X \in \mathbb{R}$ be a random variable with continuous distribution function $F(\cdot)$ and a proper density $f(\cdot)$. The goal of kernel density estimation is to approximate $f(\cdot)$ from a random sample $\{X_i\}_{i=1}^n$. Given this set of realizations, a natural estimator of $F(\cdot)$ is given by $\widehat{F}_n(x) = n^{-1} \sum_{i=1}^n I(X_i \leq x) \; \forall x \in \mathbb{R}$. However, differentiating $\widehat{F}_n(\cdot)$ with respect to x would not lead to a useful estimator of a smooth density function $f(\cdot)$. Instead, for small values of $h_n > 0$, a two-sided finite difference approximation to $f(\cdot)$ follows from

$$\widehat{f}_{h_n}(x) = \frac{\widehat{F}_n(x+h_n) - \widehat{F}_n(x-h_n)}{2h_n}$$
$$= \frac{1}{nh_n} \sum_{i=1}^n I(x - h_n \leq X_i \leq x + h_n) = \frac{1}{2nh_n} \sum_{i=1}^n I\left(\frac{|X_i - x|}{h_n} \leq 1\right). \quad (\text{A.1})$$

Clearly, $\widehat{f}_{h_n}(\cdot)$ counts the proportion of observations falling in the neighborhood of x. The parameter h_n, (*bandwidth*), controls the degree of smoothing: the greater h_n, the greater the smoothing.

Equation (A.1) is a special case of what is called *kernel density estimator* with a weight function, or kernel, $K(\cdot) = \frac{1}{2} I(|\cdot| \leq 1)$. The general, basic, kernel estimator may be written compactly as

$$\widehat{f}_{h_n}(x) = \frac{1}{nh_n} \sum_{i=1}^n K\left(\frac{x - X_i}{h_n}\right) = \frac{1}{n} \sum_{i=1}^n \mathcal{K}_{h_n}(x - X_i), \quad (\text{A.2})$$

where $\mathcal{K}_{h_n}(\cdot) = K(\cdot/h_n)/h_n$. Here, $K(\cdot)$ is a so-called *kernel function*.

Kernel functions

A kernel function $K: \mathbb{R} \to \mathbb{R}$ is any function for which $\int_{\mathbb{R}} K(u) \mathrm{d}u = 1$. A *non-negative*

APPENDIX 7.A

Table 7.7: *Some second-order ($\nu = 2$) kernel functions.* [1]

Kernel	Equation	$R(K)$	$\mu_2(K)$	eff(K)	$C_2(K)$
Uniform	$K_{[2],0}(u) = \frac{1}{2}I(|u| \leq 1)$	1/2	1/3	1.0758	1.84
Epanechnikov	$K_{[2],1}(u) = \frac{3}{4}(1-u^2)I(|u| \leq 1)$	3/5	1/5	1.0000	2.34
Biweight	$K_{[2],2}(u) = \frac{15}{16}(1-u^2)^2 I(|u| \leq 1)$	5/7	1/7	1.0061	2.78
Triweight	$K_{[2],3}(u) = \frac{35}{32}(1-u^2)^3 I(|u| \leq 1)$	350/429	1/9	1.0135	3.15
Gaussian	$K_{[2],\infty}(u) = \frac{1}{\sqrt{2\pi}}\exp(-\frac{1}{2}u^2)$	$1/2\sqrt{\pi}$	1	1.0513	1.06

[1] All kernels are supported on the interval $[-1, 1]$ except for the Gaussian kernel which has infinite support.

kernel satisfies $K(u) \geq 0\ \forall u$ which ensures that $K(\cdot)$ is a pdf. A *symmetric* kernel satisfies $K(u) = K(-u)\ \forall u$. In this case all odd moments of a kernel are zero, where the moments of $K(\cdot)$ are defined by

$$\mu_j(K) = \int_{\mathbb{R}} u^j K(u) \mathrm{d}u.$$

The use of symmetric and unimodal kernels is standard in nonparametric estimation, and will henceforth be adopted. The order of a kernel, say ν, is defined as the first non-zero moment, i.e. if $\mu_0(K) = 1$ and $\mu_j(K) = 0$ $(j = 1, \ldots, \nu - 1)$, but $\mu_\nu(K) \neq 0$. Some common second-order kernel functions are listed in Table 7.7 and exhibited in Figure 7.7. The first four second-order kernels are special cases of the polynomial family

$$K_{[2],p}(u) = \frac{(2p+1)!!}{2^{p+1}p!}(1-u^2)^p I(|u| \leq 1), \quad (p = 0, 1, 2, 3).$$

The Gaussian kernel follows by taking the limit $p \to \infty$ after re-scaling. Higher-order kernels are smoother, reducing the order of the bias of the curve estimator provided large sample sizes ($n \gg 1,000$) are available. The basic shape of the kernels are similar. Since, however, higher-order kernel functions take on negative values, the resultant estimate of $f(\cdot)$ also can have negative values.

Distance measures and relative efficiency

A common and convenient measure of evaluating the estimation precision of $\widehat{f}_{h_n}(\cdot)$ is the MSE, which at a single point x, is given by

$$\mathrm{MSE}(\widehat{f}_{h_n}(x)) = \mathbb{E}\left[\left(\widehat{f}_{h_n}(x) - f(x)\right)^2\right] = \mathrm{Bias}(\widehat{f}_{h_n}(x))^2 + \mathrm{Var}(\widehat{f}_{h_n}(x)). \tag{A.3}$$

If we want to minimize (A.3) with respect to h_n, we are confronted with a bias-variance trade-off as mentioned earlier. Rather than measuring the distance of the kernel density estimator in terms of the pointwise MSE, a "global" measure is often preferred in practice. Two most popular measures are the *integrated squared error* (ISE) and the *mean integrated squared error* (MISE), where

$$\text{ISE}(\widehat{f}_{h_n}(x)) = \int_{\mathbb{R}} \left(\widehat{f}_{h_n}(x) - f(x)\right)^2 \mathrm{d}x,$$

$$\text{MISE}(\widehat{f}_{h_n}(x)) = \mathbb{E}[\text{ISE}(\widehat{f}_{h_n}(x))] = \mathbb{E}\Big[\int_{\mathbb{R}} \left(\widehat{f}_{h_n}(x) - f(x)\right)^2 \mathrm{d}x\Big].$$

Since we can reverse the order of integration (over the support of X and over the probability space of X), we have $\text{MISE}(\widehat{f}_{h_n}(x)) = \int_{\mathbb{R}} \text{MSE}(\widehat{f}_{h_n}(x))\mathrm{d}x$ so that MISE equals to the integrated MSE, a measure which does not depend upon the data.

Ideally, we want to pick a bandwidth value h_n such that it minimizes the MISE. However, the optimal bandwidth that minimizes the MISE depends on the unknown pdf $f(\cdot)$. In order to make progress under this distance measure, it is usual to employ asymptotic approximations to bias and variance of the kernel density estimator. The result is called asymptotic MISE (AMISE), i.e., $\text{AMISE}(\widehat{f}_{h_n}(x)) = \int_{\mathbb{R}} \text{AMSE}(\widehat{f}_{h_n}(x))\mathrm{d}x$ with AMSE the asymptotic MSE of $\widehat{f}_{h_n}(\cdot)$. The optimal bandwidth, say h_{opt}, is the one that minimizes the $\text{AMISE}(\cdot)$, giving rise to $\text{AMISE}_{\text{opt}}(\cdot)$.

Now, given that we have selected the kernel order ν, which kernel should we use? It is straightforward to verify (cf. Exercise 7.7) that the kernel's contribution to the optimal AMISE is the following dimensionless factor:

$$\text{AMISE}_{\text{opt}}(K) \propto \left(\mu_\nu^2(K) R(K)^{2\nu}\right)^{1/(2\nu+1)}, \tag{A.4}$$

where $R(g) = \int_{\mathbb{R}} g^2(z)\mathrm{d}z$ is the *roughness* penalty of the function $g(\cdot)$ (column three of Table 7.7). Then, to compare kernels, the *efficiency* (eff) of kernel $K(\cdot)$ relative to kernel $K^*(\cdot)$ is defined as

$$\text{eff}(K) = \left(\frac{\text{AMISE}_{\text{opt}}(K)}{\text{AMISE}_{\text{opt}}(K^*)}\right)^{(2\nu+1)/2\nu} = \left(\frac{\mu_\nu^2(K)}{\mu_\nu^2(K^*)}\right)^{1/2\nu} \frac{R(K)}{R(K^*)}. \tag{A.5}$$

Usually, the Epanechnikov kernel is taken as a reference kernel since it is optimal in a minimal variance sense.

The fifth column of Table 7.7 shows the asymptotic relative efficiency of estimating $f(\cdot)$ with kernel $K(\cdot)$ as compared to estimating it with $K_{[\nu],1}(\cdot)$. We see, for instance, that relative to $K_{[\nu],1}(\cdot)$ the uniform kernel has an asymptotic efficiency loss of about 7% when $\nu = 2$. Similar observations follow for the other kernels. In general, there is no single kernel that can be recommended for all purposes. One serious candidate is the Gaussian kernel; however, it is relatively inefficient and has infinite support. Even the Epanechnikov kernel is not so attractive because it has a discontinuous first derivative, and hence it is inappropriate for density derivative estimation.

Bandwidth selection

For practical problems the choice of the kernel is not so critical, as compared to the choice of the bandwidth. The bandwidth depends on the sample size n and has to fulfill $h_n \to 0$ and $nh_n \to \infty$ when $n \to \infty$ as a necessary condition for consistency of the density estimator. Clearly, this result is not very helpful for finite-sample application. Rather, we may use the AMISE-optimal bandwidth with $R(f^{(\nu)}(\cdot))$ replaced by $R(g_{\widehat{\sigma}_X}^{(\nu)}(\cdot))$ where $g_{\sigma_X}(\cdot)$ is a plausible reference density, $\widehat{\sigma}_X$ is the sample standard deviation, and $f^{(\nu)}(\cdot)$ is the νth derivative of

APPENDIX 7.A

$f(\cdot)$, assuming it exists. Assume $g_{\sigma_X}(\cdot) = \varphi_{\widehat{\sigma}_X}$, the $\mathcal{N}(0, \widehat{\sigma}_X^2)$ density. It can be shown (cf. Exercise 7.7) that

$$R(\varphi_{\widehat{\sigma}_X}^{(\nu)})^{-1/(2\nu+1)} = 2\widehat{\sigma}_X \left(\frac{\sqrt{\pi}\nu!}{(2\nu)!}\right)^{1/(2\nu+1)}. \tag{A.6}$$

Then a *rule-of-thumb* (rot) bandwidth is given by

$$h_{\mathrm{rot}} = \widehat{\sigma}_X C_\nu(K) n^{-1/(2\nu+1)}, \tag{A.7}$$

where

$$C_\nu(K) = 2\left(\frac{\sqrt{\pi}(\nu!)^3 R(K)}{2\nu(2\nu)!\mu_\nu^2(K)}\right)^{1/(2\nu+1)}.$$

The last column of Table 7.7 shows values of $C_\nu(\cdot)$ when $\nu = 2$. If a Gaussian second-order kernel is used, (A.7) is often simplified to $h_{\mathrm{rot}} = \widehat{\sigma}_X n^{-1/5}$. Rule-of-thumb bandwidths are sensitive to outliers. A robust version of the rule-of-thumb bandwidth rule is $h_{\mathrm{rot}} = \min\{\widehat{\sigma}_X, (\mathrm{IQR}_X/1.34)\}n^{-1/5}$ where IQR_X is the interquartile range computed from the sample distribution of X.

Rule-of-thumb bandwidths are "pilot" bandwidths, i.e. they are a useful starting point. A more flexible way for obtaining bandwidths is to use a so-called *plug-in* bandwidth procedure. This method is based on considering some type of quadratic error between the true function and its estimator. Minimizing an asymptotic approximation of the resulting error and replacing the unknown parameters by estimates gives the optimal (plug-in) bandwidth. Plug-in methods have been extensively studied for nonparametric univariate density estimation, but for multivariate data the choice of a method is less clear. A flexible and generally applicable alternative, is CV.

Multivariate density estimation

Multivariate kernel density estimation is a straightforward extension of plain univariate estimation. Now, suppose that $\mathbf{X}_i$ is a p-variate i.i.d. random variable and we want to estimate its density $f(\mathbf{x}) = f(x_1, \ldots, x_p)$ ($\mathbf{x} \in \mathbb{R}^p$), given a set of observations $\{X_i\}_{i=1}^n$ from $f(\cdot)$. Analogue to (A.2), the multivariate kernel density estimator takes the form

$$\widehat{f}_{\mathbf{H}}(\mathbf{x}) = \frac{1}{n|\mathbf{H}|}\sum_{i=1}^n K\big(\mathbf{H}^{-1}(\mathbf{x} - \mathbf{X}_i)\big) = \frac{1}{n}\sum_{i=1}^n \mathcal{K}_{\mathbf{H}}(\mathbf{x} - \mathbf{X}_i), \tag{A.8}$$

where $\mathbf{H}$ is a $p \times p$ symmetric positive definite matrix of bandwidths, and

$$\mathcal{K}_{\mathbf{H}}(\mathbf{x}) = |\mathbf{H}|^{-1/2} K(\mathbf{H}^{-1/2}\mathbf{x}).$$

Here, $K(\cdot)$ is a p-dimensional kernel function satisfying $\int K(\mathbf{x})d\mathbf{x} = 1$. In practice, a product of p univariate kernels $K_{\mathrm{univ}}(u_j)$, such as a univariate standard Gaussian density function, is commonly used for $K(\cdot)$, i.e., $K(\mathbf{u}) = \prod_{j=1}^p K_{\mathrm{univ}}(u_j)$. The matrix $\mathbf{H}$ is often taken to be a diagonal matrix with common diagonal elements h_n. As in the univariate case, one additionally desires that $K(\cdot) \geq 0$ so that $K(\cdot)$ is a proper pdf.

Suppose $\mathbf{H} = \mathrm{diag}(h_n, \ldots, h_n)$. Then, with some algebra, it can be shown that the optimal (in the sense of minimizing the AMISE) bandwidth is given by

$$h_{\mathrm{opt}} = R(\nabla^\nu f)^{-1/(2\nu+p)}\left(\frac{(\nu!)^2 p R(K)^p}{2\nu\mu_\nu^2(K)}\right)^{1/(2\nu+p)} n^{-1/(2\nu+p)}, \tag{A.9}$$

where

$$\nabla^\nu f(\mathbf{x}) = \sum_{j=1}^{p} \frac{\partial^\nu}{\partial x_j^\nu} f(\mathbf{x}).$$

When the observed data set is from a multivariate normal density φ, an explicit expression for $R(\nabla^\nu f)$ can be calculated straightforwardly. By replacing $R(\nabla^\nu f)$ by $R(\nabla^\nu \varphi)$ in (A.9), we obtain the rot-bandwidth

$$h_{\text{rot}} = \sigma_j C_{\nu,p}(K) n^{-1/(2\nu+p)} \quad (j = 1, 2, \ldots, p), \tag{A.10}$$

where

$$C_{\nu,p}(K) = \left(\frac{\pi^{p/2} 2^{p+\nu-1} (\nu!)^2 R(K)^p}{\nu \big((2\nu-1)!! + (p-1)((\nu-1)!!)^2 \big) \mu_\nu^2(K)} \right)^{1/(2\nu+p)},$$

and with σ_j the standard deviation of the jth variable, which can be replaced by its sample estimator in practical applications. The constant $C_{\nu,p}(\cdot)$ is exactly 1 in the bivariate case ($p = 2$), with a second-order Gaussian kernel. Numerical values of $C_{\nu,p}(\cdot)$ for other combinations of kernel functions, p, and ν can be obtained directly using the results for $R(\cdot)$ and $\mu_\nu(\cdot)$ given in Table 7.7.

Note from (A.8) that, unless $\mathbf{X}_i$ is distributed more or less uniformly in the p-dimensional space, there is the risk that for a given bandwidth, no data lies in the neighborhood specified by $\mathbf{H}$. This problem becomes worse as p increases, and is known as the "curse of dimensionality". Hence, in practice, multivariate kernel density estimation is often restricted to dimension $p = 2$.

Nadaraya–Watson estimator

Let $\{(\mathbf{X}_i, Y_i)\}_{i=1}^n$ represent n independent observations of the random pair $(\mathbf{X}, Y)$, where $\mathbf{X} = (X_{1,i}, \ldots, X_{p,i})'$ is a p-variate random variable. To keep things simple, we assume that such data is generated by the process

$$Y_i = \mu(\mathbf{X}_i) + \varepsilon_i, \tag{A.11}$$

where $\{\varepsilon_i\}$ is a sequence of i.i.d. zero mean and finite variance random variables such that ε_i is independent of $\mathbf{X}_i$, and $\mu \colon \mathbb{R}^p \to \mathbb{R}$ is an "arbitrary" function called the *nonparametric regression function* and it satisfies $\mu(\mathbf{x}) = \mathbb{E}(Y|\mathbf{X} = \mathbf{x})$ $(\mathbf{x} \in \mathbb{R}^p)$.

We wish to estimate $\mu(\cdot)$. If $\mu(\cdot)$ is a smooth function at point $\mathbf{x} = (x_1, \ldots, x_p)'$, responses corresponding to $\mathbf{X}_i$'s near $\mathbf{x}$ should contain some information about the value of $\mu(\cdot)$. Therefore, local averaging of the responses about $\mathbf{X} = \mathbf{x}$ may yield a meaningful estimate of $\mu(\cdot)$. One particular formulation, called *Nadaraya–Watson* (NW) kernel estimator and attributed to Nadaraya (1964) and Watson (1964), uses a kernel function to vary the weights given to the responses. In particular, a kernel estimate of $\mu(\cdot)$ is a weighted average of observations in the neighborhood of $\mathbf{x}$, and is defined as

$$\widehat{\mu}_\mathbf{H}^{\text{NW}}(\mathbf{x}) = \frac{\sum_{i=1}^n \mathcal{K}_\mathbf{H}(\mathbf{x} - \mathbf{X}_i) Y_i}{\sum_{i=1}^n \mathcal{K}_\mathbf{H}(\mathbf{x} - \mathbf{X}_i)} = \sum_{i=1}^n W_i(\mathbf{x}) Y_i, \tag{A.12}$$

with the weights $W_i(\mathbf{x}) = \mathcal{K}_\mathbf{H}(\mathbf{x} - \mathbf{X}_i) / \sum_{i=1}^n \mathcal{K}_\mathbf{H}(\mathbf{x} - \mathbf{X}_i)$ summing up to one, and where $\mathbf{H}$ is a $p \times p$ symmetric positive definite matrix of bandwidths.

APPENDIX 7.A

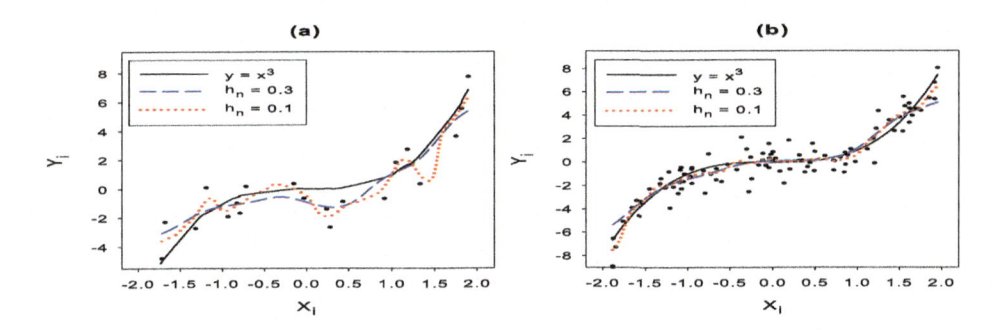

Figure 7.8: *Local averages: (a) based on $n = 20$ observations from the DGP $Y_i = X_i^3 + \varepsilon_i$ with $\{\varepsilon_i\} \stackrel{i.i.d.}{\sim} \mathcal{N}(0,1)$, and $\{X_i\} \stackrel{i.i.d.}{\sim} U[-2,2]$; (b) based on $n = 100$ observations from the same DGP as in part (a).*

The kernel regression estimate can be more formally derived from the regression of $\mathbf{X}$ to Y, i.e., $\mu(\mathbf{x}) = \int_{\mathbb{R}} y f(y|\mathbf{x}) dy = \int_{\mathbb{R}} y f(\mathbf{x},y) dy / g(\mathbf{x})$ where the density $g(\cdot)$ is assumed positive at $\mathbf{x}$. Indeed, estimating these densities using univariate and multivariate kernel density estimates (all with the same kernel) results in a kernel regression estimate which matches (A.12). Alternatively, the kernel regression estimator (A.12) can be viewed as a local constant fit about $\mathbf{x}$ which minimizes the weighted sum of squares of the residuals (weighted by the product kernel $\mathcal{K}_{h_n}(\mathbf{v}) = h_n^{-p} \prod_{i=1}^{p} K(v_i/h_n)$).

Example A.1: NW Kernel Regression Estimation

Figure 7.8(a) shows two NW kernel smoothed averages based on the series $\{(X_i, Y_i)\}_{i=1}^{20}$ generated from the model $Y_i = X_i^3 + \varepsilon_i$ with $\{\varepsilon_i\} \stackrel{i.i.d.}{\sim} \mathcal{N}(0,1)$, and $\{X_i\} \stackrel{i.i.d.}{\sim} U[-2,2]$. The true regression function $y = x^3$ is shown by the black solid line. Using a Gaussian kernel with $h_n = 0.3$, the local averages are shown as a blue medium dashed line, and the local average corresponding to $h_n = 0.1$ by the red dotted line.

The kernel discriminates each Y_i according to the distance of its corresponding X_i from x and has its greatest value at the origin. Generally, it is positive and symmetric, and decreases from the origin. In this way, the kernel has the effect of reducing bias without increasing variance. The bandwidth h_n controls the 'width' of the kernel and is used to 'tune' the degree of smoothing: the greater h_n, the greater the smoothing. Clearly, the blue medium dashed line is less 'wiggly', and hugs closer to the true regression curve than the red dotted line. Overall, the NW estimator with $h_n = 0.3$ is to be preferred because, intrinsically, its variance and squared bias are better balanced.

As n increases, variance will decrease as more averaging is performed. Then h_n should be decreased to reduce the amount of local smoothing – thus reducing bias – but not so much as to effect a comparable increase to the variance, i.e. $h_n \to 0$ as $n \to \infty$. As n becomes large, we may expect the estimate to converge to the true curve at every point $\mathbf{x}$. Figure 7.8(b) illustrates convergence effects and shows local averages computed for $n = 100$.

Optimum convergence of the kernel estimate can be achieved by selecting the bandwidth h_n using CV. It uses the aptly named *leave-one-out* estimator $\widehat{\mu}_{h_n}^{-i}(\cdot)$ of $\mu(\cdot)$. At $\mathbf{X}_i = \mathbf{x}$, this estimator is defined as

$$\widehat{\mu}_{h_n}^{-i}(\mathbf{X}_i) = \sum_{\substack{j=1 \\ j \neq i}}^{n} W_j^{-i}(\mathbf{X}_i) Y_j, \qquad (A.13)$$

with weights $W_j^{-i}(\cdot)$ as defined in (A.12); superscript $-i$ indicates the absence of Y_i in the averaging, and h_n the explicit dependence on the bandwidth. The CV function is then defined as the sample-average MSE that results from adopting the leave-one-out estimator, i.e.,

$$\text{CV}(h_n) = \frac{1}{n} \sum_{i=1}^{n} \{Y_i - \widehat{\mu}_{h_n}^{-i}(\mathbf{X}_i)\}^2. \qquad (A.14)$$

The (global) bandwidth $\widehat{h}_{\text{CV}}$ that minimizes (A.14) across a pre-specified range of values h_n is then used to compute the kernel estimate $\widehat{\mu}_{h_n}(\cdot)$. Typically, $\text{CV}(\cdot)$ has one unique minimum with no other local minima. In the i.i.d. case, the CV routine produces asymptotically optimal kernel estimates. For dependent data, convergence results of the CV bandwidth selection method have been obtained for certain types of mixing processes and univariate regression functions.

Note that the computation of one value of $\text{CV}(\cdot)$ requires n^2 kernel evaluations, which may be unacceptable when n is large. A variety of refinements of the CV bandwidth selection method are available to address this problem. For instance, minimizing a generalized CV function, or minimizing the final prediction error. Another way for obtaining global bandwidths is to use a plug-in bandwidth procedure.

Local polynomial regression

The *locally constant*, or NW kernel smoothing method can be extended to allow local polynomial estimation of $\mu(\cdot)$ and its partial derivatives. The resulting estimator is obtained by fitting locally to the data a polynomial of degree d, using multivariate weighted least squares. Assume that $\mu(\cdot)$ has derivatives of total order $p+1$ at point $\mathbf{x}$. Then, from a standard Taylor argument, it follows that for (A.11) the local polynomial estimator of $\mu(\cdot)$ is defined as $\widehat{\beta}_0$, where $(\widehat{\beta}_0, \widehat{\beta}_{m_1}, \ldots, \widehat{\beta}_{m_p})'$ minimizes

$$\sum_{i=1}^{n} \Big(Y_t - \big(\beta_0 + \sum_{1 \leq m_1 + \cdots + m_p \leq d} \beta_{m_1, \ldots, m_p} \prod_{j=1}^{p} (X_{j,i} - x_j)^{m_j}\big)\Big)^2 \mathcal{K}_{h_n}(\mathbf{x} - \mathbf{X}_i), \qquad (A.15)$$

with $\mathcal{K}_{h_n}(\mathbf{v}) = h_n^{-p} \prod_{i=1}^{p} K(v_i/h_n)$. The above minimization problem can be rephrased in matrix notation to allow for direct computation using weighted least squares. For instance, with $d = 1$, the so-called *local linear* (LL) estimator is given by

$$\widehat{\mu}_{h_n}^{\text{LL}}(\mathbf{x}) = \mathbf{e}'(\mathbf{X}_\mathbf{x}' \mathbf{W}_\mathbf{x} \mathbf{X}_\mathbf{x})^{-1} \mathbf{X}_\mathbf{x}' \mathbf{W}_\mathbf{x} \mathbf{y}, \qquad (A.16)$$

where $\mathbf{e}$ is a $(d+1) \times 1$ vector having 1 in the first entry and zeros elsewhere, $\mathbf{y} = (Y_1, \ldots, Y_n)'$ is the vector of responses,

$$\mathbf{X}_\mathbf{x} = \begin{pmatrix} 1 & (\mathbf{x} - \mathbf{X}_1)' \\ \vdots & \vdots \\ 1 & (\mathbf{x} - \mathbf{X}_n)' \end{pmatrix}$$

the $n \times (d+1)$ design matrix, and

$$\mathbf{W_x} = \text{diag}\big(\mathcal{K}_{h_n}(\mathbf{x} - \mathbf{X}_1), \ldots, \mathcal{K}_{h_n}(\mathbf{x} - \mathbf{X}_n)\big),$$

is an $n \times n$ matrix of weights.

In general, the local polynomial estimator is more attractive than the NW estimator because of its better asymptotic bias performance. Moreover, the estimator does not suffer from boundary effects, and hence does not require modifications in regions near the end points of the support set. Another useful feature is that the method immediately estimates the rth derivative, $\mu^{(r)}(\cdot)$ $(r = 1, \ldots, d)$, via the relationship $\widehat{\mu}_{h_n}^{(r)}(\cdot) = r! \widehat{\beta}_{m_r}(\cdot)$.

Some selective background information
The class of kernel estimators was originally defined by Rosenblatt (1956) and generalized by Parzen (1962) for pdf estimation. Marron (1994) provides a visual understanding of higher-order kernels. For standard second-order normal kernels, the bandwidth (A.7) is often termed Silverman's (1986, p. 48) rule-of-thumb. Härdle and Marron (1995) show that the CV routine yields bandwidths which produce asymptotically optimal kernel estimates. Hansen (2005) derives the exact MISE of several higher-order kernel density estimators. For multivariate kernel density estimation Zhang et al. (2006) provide a posterior estimate of the full bandwidth matrix via the use of the MCMC technique. Their technique is applicable to data of any dimension.

7.B Copula Theory

Let $\mathbf{X} = (X_1, \ldots, X_m)'$ be an m-dimensional random vector with joint CDF $F(x_1, \ldots, x_m) = \mathbb{P}(X_1 \leq x_1, \ldots, X_m \leq x_m)$ with univariate marginal CDFs $F_i(x_i)$ $(i = 1, \ldots, m)$. Since it is usually easier to handle marginal distributions separately, our interests is in a function that can reconstruct the joint distribution function from its marginals. Such a function is called *copula* (Sklar, 1959), i.e. it "couples' (or links) univariate marginal distributions to a multivariate joint distribution. Excellent introductions to copulae and related concepts are given in Nelsen (2006) and Joe (1997), where most of the material below can be found. We start with the definition of copulas.

Definition B.1 (Copula) *Let* $\mathbb{C} \colon [0, 1]^m \to [0, 1]$ *be an m-dimensional distribution function on $[0, 1]^m$. Then $\mathbb{C}$ is a copula if it has uniformly distributed univariate marginal CDFs on the interval $[0, 1]$.*

Another interpretation of a copula function follows from the *probability integral transform* (PIT), $U_i \equiv F_i(X_i)$. If the marginal distribution functions $F_1, \ldots, F_m$ of F are continuous, the random variable U_i will have the $U(0, 1)$ distribution regardless of the original distribution F_i, i.e.

$$U_i \equiv F_i(X_i) \sim U(0, 1), \quad (i = 1, \ldots, m).$$

Thus, the copula $\mathbb{C}$ of $\mathbf{X}$ represents the joint CDF of the vector of PITs of the random vector $\mathbf{U} = (U_1, \ldots, U_m)'$ and thus is a joint CDF with $U(0, 1)$ marginals.

The next theorem is cardinal to the theory of copulas.

Theorem B.1 (Sklar's (1959) theorem) *Let F be an m-dimensional joint CDF on $\mathbb{R}^m$ with univariate marginal distribution functions $F_1,\ldots,F_m$. Then there exists an m-dimensional copula $\mathbb{C}$ such that for all $\mathbf{x} = (x_1,\ldots,x_m)' \in \mathbb{R}^m$,*

$$F(x_1,\ldots,x_m) = \mathbb{C}\big(F_1(x_1),\ldots,F_m(x_m)\big). \tag{B.1}$$

Moreover, if $F_1,\ldots,F_m$ are continuous, then $\mathbb{C}$ is unique; otherwise $\mathbb{C}$ is uniquely determined on $\operatorname{Ran} F_1 \times \cdots \times \operatorname{Ran} F_m$.

As a direct consequence of Theorem B.1, one can derive a method to specify a parametric copula, known as the *inversion method*.

Corollary B.1 (Inversion method) *Let F be an m-dimensional distribution function with univariate marginal distribution functions $F_1,\ldots,F_m$ and corresponding copula $\mathbb{C}$ satisfying (B.1). Assume that $F_1,\ldots,F_m$ are continuous. Then an explicit representation of $\mathbb{C}$ is given by*

$$\mathbb{C}(\mathbf{u}) = F\big(F_1^{-1}(u_1),\ldots,F_m^{-1}(u_m)\big), \quad \mathbf{u} = (u_1,\ldots,u_m)' \in [0,1]^m, \tag{B.2}$$

where $F_i^{-1}(u_i) = \inf\{x | F_i(x) \geq u_i\}$ $(i = 1,\ldots,m)$.

The behavior of the copulas with respect to strictly monotonic transformations is established in the next theorem; see Embrechts et al. (2003, Thm. 2.6). It forms the basis for the role of copulas in the study of (multivariate) measures of association (dependence).

Theorem B.2 (Invariance) *Let $\mathbf{X} = (X_1,\ldots,X_m)'$ be an m-dimensional continuous random variable with copula $\mathbb{C}$ and let $T_1,\ldots,T_m$ be strictly increasing functions on $\operatorname{Ran} X_1,\ldots,\operatorname{Ran} X_m$, respectively. Then the transformed random variable $T(\mathbf{X}) = \big(T_1(X_1),\ldots,T_m(X_m)\big)'$ has exactly the same copula $\mathbb{C}$ as $\mathbf{X}$.*

According to Nelsen (2006, Thm. 2.2.7), the partial derivatives $\partial \mathbb{C}(\mathbf{u})/\partial u_i$ of $\mathbb{C}$ exist for almost all u_i $(i = 1,\ldots,m)$. Then we may define a copula density as follows.

Definition B.2 (Copula density) *Suppose $\mathbb{C}(\mathbf{u})$ is a copula function of a continuous m-dimensional random variable, then the copula density $c(\mathbf{u})$ is defined as $c(\mathbf{u}) \equiv \partial^m C(\mathbf{u})/(\partial u_1 \cdots \partial u_m)$.*

Differentiating (B.1) with respect to x_i $(i = 1,\ldots,m)$, yields the joint pdf:

$$f(\mathbf{x}) = c\big(F_1(x_1),\ldots,F_m(x_m)\big) \prod_{i=1}^m f_i(x_i), \tag{B.3}$$

where $f_i(x_i)$ is the density associated with the marginal CDF $F_i(x_i)$. This representation is particularly useful for copula ML parameter estimation because it provides an explicit expression for the likelihood function in terms of the copula density and the product marginal densities.

Every m-dimensional copula $\mathcal{C}$ $(m \geq 2)$ is bounded in the following sense:

$$W(\mathbf{u}) \equiv \max\{u_1 + \cdots + u_m - (m-1), 0\} \leq \mathcal{C}(\mathbf{u}) \leq \min\{u_1,\ldots,u_m\} \equiv M(\mathbf{u}),$$
$$\forall \mathbf{u} \in [0,1]^m, \tag{B.4}$$

APPENDIX 7.B

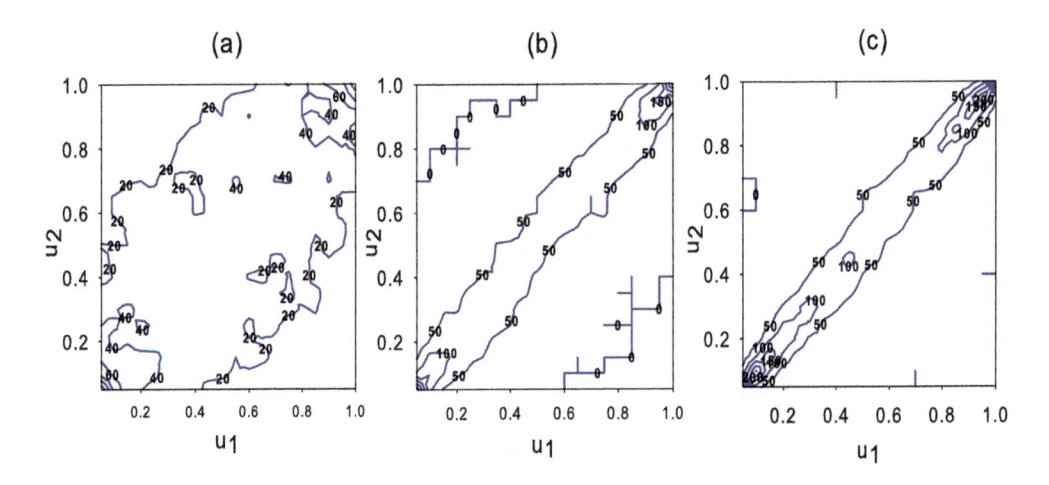

Figure 7.9: *Contour plots of three bivariate copula densities: (a) Gaussian copula with $\rho = 0.5$, (b) Student t_ν copula with $\rho = 0.9$ and $\nu = 15$ degrees of freedom, and (c) Student t_ν copula with $\rho = 0.9$ and $\nu = 1$ degree of freedom.*

where $M(\cdot)$ and $W(\cdot)$ are the *Fréchet–Hoeffding bounds*. The upper bound $M(\cdot)$ is also known as the *comonotonic* copula. It represents the copula of **X**, if each of the random variables $X_1, \ldots, X_m$ can (a.s.) be represented as a strictly functional relationship between X_i and X_j ($i \neq j$). This copula is also said to describe perfect positive dependence. The lower bound $W(\cdot)$ is a copula only for dimension $m = 2$.

Example B.1: Gaussian and Student t copulas

A wide range of copulas exists. The most commonly used copulae are the *Gumbel copula* for extreme distributions, the *Gaussian copula* for linear correlation, and the *Archimedean copula* and the *Student t copula* for dependence in the tail. A multivariate Gaussian distribution $\Phi(\cdot)$ with $m \times m$ correlation matrix **R** yields the Gaussian copula

$$\mathbb{C}^G(\mathbf{u}) = \Phi\big(\Phi^{-1}(u_1), \ldots, \Phi^{-1}(u_m)\big)$$
$$= \int_{-\infty}^{\Phi^{-1}(u_1)} \cdots \int_{-\infty}^{\Phi^{-1}(u_m)} \frac{1}{(2\pi)^{m/2}|\mathbf{R}|^{1/2}} \exp\big(-\tfrac{1}{2}\mathbf{y}'\mathbf{R}^{-1}\mathbf{y}\big) \mathrm{d}\mathbf{y},$$

where $\Phi^{-1}(\cdot)$ is the quantile function of an $\mathcal{N}(0,1)$ distribution.

The t copula provides a more sophisticated model to analyze the association between a multivariate distribution and its univariate marginal distribution functions. In the same way as $\mathbb{C}^G(\mathbf{u})$, the t copula is derived from the multivariate t distribution with correlation matrix **R** and degrees of freedom ν, i.e.

$$\mathbb{C}^t(\mathbf{u}) = t_\nu\big(t_\nu^{-1}(u_1), \ldots, t_\nu^{-1}(u_m)\big)$$
$$= \int_{-\infty}^{t_\nu^{-1}(u_1)} \cdots \int_{-\infty}^{t_\nu^{-1}(u_m)} \frac{\Gamma(\tfrac{\nu+m}{2})|\mathbf{R}|^{-1/2}}{\Gamma(\tfrac{\nu}{2})(\nu\pi)^{m/2}} \Big(1 + \tfrac{1}{\nu}\mathbf{y}'\mathbf{R}^{-1}\mathbf{y}\Big)^{-\tfrac{\nu+m}{2}} \mathrm{d}\mathbf{y},$$

where $t_\nu^{-1}(\cdot)$ denotes the quantile function of a standard univariate Student t_ν distribution. The multivariate Gaussian copula may be thought of as a limiting case of the multivariate t copula as $\nu \to \infty$ $\forall \mathbf{u} \in [0,1]^m$.

Based on three MC simulation samples of $T = 10{,}000$ observations, Figure 7.9 shows contour plots of (a) a bivariate Gaussian copula density with correlation coefficient $\rho = 0.5$, (b) a bivariate t copula density with $\rho = 0.9$ and $\nu = 15$, and (c) a bivariate Student t_ν copula density with $\rho = 0.9$ and $\nu = 1$. We see that the copulas have symmetric tail dependencies. The lower- and upper tail dependencies are better captured with the $t_{\nu=1}$ copula than the one with $\nu = 15$ degrees of freedom.

7.C U- and V-statistics

In this appendix, we briefly introduce the notions of U- and V-statistics which are mentioned throughout the book as a mean to derive consistent estimators of certain parameters of interest. For a more thorough discussion on these notions, we refer the reader to the originating papers cited below, and to the books by Serfling (1980, Chapters 5 and 6) and Lee (1990).

Definitions

Let $X_1, X_2, \ldots$ be i.i.d. random variables with distribution function F taking values in an m-dimensional Euclidean space $\mathbb{R}^m$. Consider a measurable kernel function $h \colon \mathbb{R}^r \to \mathbb{R}$ ($r \in \mathbb{N}$), that is symmetric in its arguments. Suppose we wish to derive a minimum-variance unbiased estimator of an estimable parameter (alternatively, statistical functional), say $\theta = \theta(F)$. That is,

$$\theta(F) \equiv \mathbb{E}[h(X_1, \ldots, X_r)] = \int_{\mathbb{R}^r} h(x_1, \ldots, x_r) \mathrm{d}F(x_1) \cdots \mathrm{d}F(x_r).$$

Then, given a (possibly multivariate) sequence $\{X_i\}_{i=1}^n$ ($n \geq r$), the U-statistic of order r (the letter U stands for unbiased) is given by

$$\mathrm{U}_n = \binom{n}{r}^{-1} \sum_{1 \leq i_1 < i_2 < \cdots < i_r \leq n} h(X_{i_1}, \ldots, X_{i_r}).$$

The basic theory of U-statistics is due to Hoeffding (1948) as a generalization of the notion of forming an average. One well-known example is the sample variance with $h(x_1, x_2) = (x_1 - x_2)^2/2$. Another example is Kendall's τ statistic (1.13) with $h\big((x_1, y_1), (x_2, y_2)\big) = 2I(x_1 < x_2, y_1 < y_2) + 2I(x_2 < x_1, y_2 < y_1) - 1$. Also, it is easy to see that the correlation integral (7.10) is a U-statistic with $h(\mathbf{x}, \mathbf{y}) = I(\|\mathbf{x} - \mathbf{y}\| < h)$.

Closely related to the U-statistic is the V-statistic for estimating $\theta(F)$, defined by

$$\mathrm{V}_n = n^{-r} \sum_{i_1, \ldots, i_r = 1}^n h(X_{i_1}, \ldots, X_{i_r}).$$

Observe that

$$\mathrm{V}_n = \theta(F_n) = \int_{\mathbb{R}^r} h(x_1, \ldots, x_r) \mathrm{d}F_n(x_1) \cdots \mathrm{d}F_n(x_r),$$

where $F_n(x) = n^{-1}\sum_{i=1}^n I(X_i \leq x)$. This is an example of a differentiable statistical functional, a class of statistics introduced by von Mises (1947) (hence the letter V). Clearly, V_n is a biased statistic for $r > 1$, because the sum in the defining equation contains some terms in which $i_1,\ldots,i_r$ are not all distinct. However, the bias of V_n is asymptotically negligible ($\mathcal{O}(n^{-1})$). Also, for a fixed sample size n, the variance of V_n satisfies $V_n = U_n + \mathcal{O}(n^{-2})$. So, in terms of MSE, V_n may be preferred over U_n.

A U-statistic (or V-statistic) of order r and variances $\sigma_1^2 \leq \sigma_2^2 \leq \cdots \leq \sigma_r^2$ has a *degeneracy* of order k if $\sigma_1^2 = \cdots = \sigma_k^2 = 0$ and $\sigma_{k+1}^2 > 0$ ($k < r$). Many examples exist of exact or approximate (as $n \to \infty$) degenerate U- or V-statistics. For instance, it is easy to prove that CvM–GOF type test statistics (see, e.g., Section 4.4.1) are degenerate V-statistics, i.e. $\int_{-\infty}^{\infty} h(x,y)dF(y) = 0 \;\forall x$, where $h(x,y) = \int_{-\infty}^{\infty}\bigl(I(x \leq z) - F(z)\bigr)\bigl(I(y \leq z) - F(z)\bigr)\bigl(w(F(z))dF(z)$ with $w(\cdot)$ a non-negative weight function on $(0,1)$.

Asymptotic distribution theory

As a prelude to discussing the asymptotic distribution theory of the U- and V-statistics, we introduce some notation. For a given estimable parameter, $\theta = \theta(F)$, and corresponding symmetric kernel, $h(x_1,\ldots,x_r)$ satisfying $\mathrm{Var}\bigl(h(X_1,\ldots,X_r)\bigr) < \infty$, we define a sequence of functions $h_c(\cdot)$ ($c = 0, 1, \ldots, r$) related to $h(\cdot)$ as follows

$$h_c(x_1,\ldots,x_c) = \mathbb{E}[h(x_1,\ldots,x_c,X_{c+1},\ldots,X_r)],$$

where $X_{c+1},\ldots,X_r$ are i.i.d. random variables from the distribution F. In fact, $h_c(\cdot)$ is (a version of) the conditional (hence the subscript letter c) expectation of $h(X_1,\ldots,X_r)$ given $X_1,\ldots,X_c$.

Since $h_0 = \theta$ and $h_r(x_1,\ldots,x_r) = h(x_1,\ldots,x_r)$, the functions $h_c(\cdot)$ all have expectation θ. Further, note that the variance of the U-statistic U_n depends on the variances of the $h_c(\cdot)$. Without loss of generality we may take $\sigma_0^2 = 0$. Moreover, for $c = 1,\ldots,r$, we define

$$\sigma_c^2 = \mathrm{Var}\bigl(h_c(X_1,\ldots,X_c)\bigr),$$

so that $\sigma_r^2 = \mathrm{Var}\bigl(h(X_1,\ldots,X_r)\bigr)$. Using these preliminaries, it can be shown (Hoeffding, 1948) that the variance of U_n is given by

$$\mathrm{Var}(U_n) = \binom{n}{r}^{-1}\sum_{c=1}^{r}\binom{r}{c}\binom{n-r}{r-c}\sigma_c^2.$$

If $\sigma_r^2 < \infty$, then $\mathrm{Var}(U_n) \sim r^2\sigma_1^2/n + \mathcal{O}(n^{-2})$ as $n \to \infty$.

Asymptotic theory for U-statistics is based on the so-called "projection" of U_n, say $\widehat{U}_n$, which is in terms of $h_1(\cdot)$ is defined as

$$\widehat{U}_n = \theta + \frac{r}{2}\sum_{i=1}^{n}\bigl(h_1(X_i) - \theta\bigr).$$

With the projection $\widehat{U}_n$, one can decompose U_n as

$$U_n = \widehat{U}_n + R_n,$$

where the remainder $R_n \to 0$, as $n \to \infty$. Thus, U_n can be approximated by a sum of i.i.d. random variables, so that the asymptotic distribution of U_n follows from classical limit theory for sums.

Yoshihara (1976, Thm. 1) and Denker and Keller (1983, Thm. 1(c)) relax the assumption of i.i.d. random variables X_i to accommodate strictly stationary weakly dependent processes. Specifically, for a non-degenerate symmetric kernel $h\colon \mathbb{R}^r \to \mathbb{R}$, and assuming that $\{X_i\}$ is β-mixing, these authors showed that

$$\sqrt{n}\,(U_n - \theta) \xrightarrow{D} \mathcal{N}(0, r^2\sigma_1^2), \quad \text{as } n \to \infty.$$

This result can easily be applied to the correlation integral (7.10). As before, consider the m-dimensional time series $\{\mathbf{Y}_t, t \in \mathbb{Z}\}$ for which each random variable is assumed to be generated from the distribution $F_m(\cdot)$. Likewise, let the kernel be the indicator function, and note then that

$$h_1(\mathbf{Y}_t) = \mathbb{E}[h(\mathbf{Y}_t, \mathbf{X}_s | \mathbf{X}_s = \mathbf{x})] = \int_{\mathbb{R}^m} I(\|\mathbf{y} - \mathbf{x}\| \leq h)\mathrm{d}F_m(\mathbf{x}).$$

Let $h_1(\mathbf{y}; h) \equiv h_1(\mathbf{y})$, so that the dependence on the bandwidth h of $h_1(\cdot)$ is made explicit. Then the asymptotic distribution of the estimator $\widehat{C}_{m,T}(Y; h)$, defined by (7.43), can be expressed as

$$\sqrt{n}\widehat{C}_{m,T}(Y; h) \sim \mathcal{N}\big(C_{m,Y}(h), 4\sigma^2_{m,T}(Y; h)\big),$$

where

$$\sigma^2_{m,T}(Y; h) = \mathbb{E}\Big[\big(h_1(\mathbf{Y}_1; h) - C_m(Y, h)\big)^2$$
$$+ 2\sum_{t=1}^{T}\big(h_1(\mathbf{Y}_1; h) - C_{m,Y}(h)\big)\big(h_1(\mathbf{Y}_t; h) - C_{m,Y}(h)\big)\Big].$$

In the case of a degenerate symmetric kernel $h(\cdot)$ of order c ($c = 1, \ldots, r-1$), the asymptotic distribution of U_n is given by

$$n(U_n - \theta) \xrightarrow{D} \binom{r}{c}\sum_{j=1}^{\infty}\lambda_j(Z_j^2 - 1), \quad \text{as } n \to \infty,$$

where Z_j are independent $\mathcal{N}(0,1)$ random variables, and λ_j are the eigenvalues for the kernel $h_2(x_1, x_2) - \theta$. This result also applies to the V-statistic, since $\sqrt{n}(U_n - V_n) \xrightarrow{P} 0$, under the additional assumption that $\sum_{j=1}^{\infty}\lambda_j < \infty$. A more general version of this asymptotic result is given by Beutner and Zähle (2014) using a new representation for U- and V-statistics. In fact, their continuous mapping approach not only encompasses most of the results on the asymptotic distribution known in literature, but also allows for the first time a unifying treatment of non-degenerate and degenerate U- and V-statistics.

Exercises

Theory Questions

7.1 Let $\{Y_t\}$ be an i.i.d. process with distribution function $F(y)$. An equivalent form of the one-dimensional correlation integral is given by $C_{1,Y}(h) = \mathbb{P}(|Y_t - Y_s| < h)$ ($t \neq s$).

EXERCISES

(a) Show that
$$C_{1,Y}(h) = C \equiv \int_{-\infty}^{\infty} [F(y+h) - F(y-h)] dF(y).$$

(b) Show that
$$\mathbb{P}(|Y_t - Y_s| < h, |Y_{t+1} - Y_{s+1}| < h) = \begin{cases} N & \text{if } |t-s| = 1, \\ C^2 & \text{if } |t-s| > 1, \end{cases}$$

where $N \equiv \int_{-\infty}^{\infty} [F(y+h) - F(y-h)]^2 dF(y)$.

(c) Show that $\lim_{T \to \infty} \mathbb{E}[\widehat{C}_{2,Y}(h)] = \{C_{1,Y}(h)\}^2$, where

$$\widehat{C}_{2,Y}(h) = \frac{2}{(T-1)(T-2)} \sum_{i=2}^{T-1} \sum_{j=1}^{i-1} I(|Y_i - Y_j| < h) I(|Y_{i+1} - Y_{j+1}| < h).$$

7.2 Suppose $\{Y_t, t \in \mathbb{Z}\}$ is a strictly stationary process generated by the following two models:

$$\text{ARCH}(1): \quad Y_t = \sigma_t \varepsilon_t, \quad \sigma_t^2 = 1 + \theta Y_{t-1}^2,$$
$$\text{sign AR}(1): \quad Y_t = \theta \, \text{sign}(Y_{t-1}) + \sqrt{1-\theta} \, \varepsilon_t,$$

where $0 < \theta < 1$, and $\{\varepsilon_t\} \stackrel{\text{i.i.d.}}{\sim} \mathcal{N}(0,1)$. Given a set of observations $\{Y_t\}_{t=1}^T$, the parameter θ can be estimated semiparametrically by maximizing the pseudo log-likelihood for the copula density $c(\widehat{F}(Y_t; \theta), \widehat{F}(Y_{t-1}; \theta); \theta)$ where $\widehat{F}(Y_t; \theta)$ is the EDF. For testing the null hypothesis of serial independence the associated semiparametric (denoted by the superscript SP) score-type test statistic, apart from a normalizing-factor, is defined as

$$Q^{\text{SP}} = \sum_{t=2}^{T} \frac{\partial \log c(\widehat{u}_t, \widehat{u}_{t-1}; \theta)}{\partial \theta} \bigg|_{\theta=0},$$

where $\widehat{u}_t$ are the realizations of $\widehat{U}_t \equiv \widehat{F}(Y_t; \theta)$.

(a) Show for the ARCH(1) model, that the SP score-type test statistic is given by

$$Q_{\text{ARCH}}^{\text{SP}} = \sum_{t=2}^{T} \left(\Phi^{-1}(\widehat{u}_t)\right)^2 \left(\Phi^{-1}(\widehat{u}_{t-1})\right)^2,$$

where $\Phi^{-1}(\cdot)$ is the quantile function of a standard normal distribution.

(b) Similar as in part (a), show that for the sign AR(1) (sAR) model

$$Q_{\text{sAR}}^{\text{SP}} = \sum_{t=2}^{T} \text{sign}\left(\Phi^{-1}(\widehat{u}_{t-1})\right) \Phi^{-1}(\widehat{u}_t).$$

7.3 $\widehat{\Delta}_T^{\text{ST}^2}(\ell)$ is the weighted functional $\Delta^*(\ell) = \int_{S^2} \{f_\ell(x,y) - f(x)f(y)\} f_\ell(x,y) dx dy$ given in Section 7.2.3. Let $\{Y_t, t \in \mathbb{Z}\}$ be a Gaussian zero-mean stationary process. Show that $\Delta^*(\cdot)$ satisfies the nonnegativity property $\Delta^*(\cdot) \geq 0$, where the equality holds if and only if Y_t and $Y_{t-\ell}$ are independent.

(Skaug and Tjøstheim, 1993a)

7.4 Let $\{e_t\}_{t=1}^T$ be the residuals from a fitted time series model. Consider the least squares regression (7.49). The slope coefficient β_m can be estimated as

$$\widehat{\beta}_m = \frac{\sum_h \left(\log h - \overline{\log h}\right)\left(\log C_{m,T}(e;h) - \overline{\log C_{m,T}(e;h)}\right)}{\sum_h \left(\log h - \overline{\log h}\right)^2},$$

where $\log h$ is the logarithm of the tolerance distance, $\log C_{m,T}(e;h)$ is the logarithm of the sample correlation integral, m is the embedding dimension, and where the bars denote the means of their counterparts without bars. Show that

$$\mathbb{E}[\widehat{\beta}_m] \leq m.$$

(This was first proved by Cutler (1991), and later by Kočenda (2001)).

Empirical and Simulation Questions

7.5 In Section 2.11 we fitted a RBF–AR(8) model to the EEG recordings (epilepsy data). The data file epilepsyMR.dat contains the residual series $\{e_t\}_{t=1}^{623}$.

(a) Make a time series plot of the residuals. Also make a plot of the sample ACF of the residuals (30 lags), and a histogram. What conclusions do you draw from these graphs?

(b) The R-copula package contains the copula-based CvM test statistic $M_{A,T}^{\text{CvM},c}$ for testing univariate serial independence $M_{A,T}^{\text{CvM},c}$ introduced in Section 7.4.4; see Ghoudi et al. (2001) and Genest and Rémillard (2004). In this part, we investigate the null hypothesis of serial independence of the residuals in a more formal way.

- First, simulate the distribution of the CvM test statistic, the distribution of the combined test statistic à la Fisher, and the distribution of the combined test statistic à la Tippett. Use the function serialIndepTestSim with lag.max=5, and fix the number of bootstrap replicates at 1,000 (default value). [*Note:* The computations can be time demanding.]
- Next, using the function serialIndepTest, compute approximate p-values of the test statistics with respect to the EDFs obtained in the previous step.
- Finally, display the dependogram.

Use the above results, to investigate the type of departure from residual serial independence, if any.

7.6 Tong (1990, p. 178) fits the following SETAR$(2;2,2)$ model to the ($\log_{10}$) Canadian lynx data of Section 7.5:

$$Y_t = \begin{cases} 0.62 + 1.25 Y_{t-1} - 0.43 Y_{t-2} + \varepsilon_t^{(1)} & \text{if } Y_{t-2} \leq 3.25, \\ 2.25 + 1.52 Y_{t-1} - 1.24 Y_{t-2} + \varepsilon_t^{(2)} & \text{if } Y_{t-2} > 3.25, \end{cases}$$

where $\{\varepsilon_t^{(1)}\}$ and $\{\varepsilon_t^{(2)}\}$ are independent sequences of i.i.d. random variables with $\{\varepsilon_t^{(1)}\} \stackrel{\text{i.i.d.}}{\sim} \mathcal{N}(0, 0.0381)$ and $\{\varepsilon_t^{(2)}\} \stackrel{\text{i.i.d.}}{\sim} \mathcal{N}(0, 0.0621)$.

(a) Obtain the residual series $\{\widehat{\varepsilon}_t\}_{t=1}^{T=112}$ for this model. Next, compute p-values, based on 100 BS replicates, using the rank-based BDS test statistics defined in Section 7.4.2 with $m = 2, 4$, and 6.

(b) What conclusions do you draw from the obtained p-values for each computed test statistic?

7.7 Wong and Li (2000b) fit a so-called *Gaussian mixture* AR (MAR) model to the log-transformed Canadian lynx series $\{Y_t\}_{t=1}^{114}$. For a time series process $\{Y_t, t \in \mathbb{Z}\}$, the K-component MAR model of order $(p_1, \ldots, p_K)$, denoted by $\text{MAR}(K; p_1, \ldots, p_K)$, is defined by

$$F(Y_t|\mathcal{F}_{t-1}) = \sum_{i=1}^{K} \pi_i \Phi\left(\frac{Y_t - \phi_{i,0} - \phi_{i,1} Y_{t-1} - \cdots - \phi_{i,p_i} Y_{t-p_i}}{\sigma_i}\right),$$

where $\mathcal{F}_t$ is the σ-algebra generated by $\{Y_t, s \leq t\}$, $\Phi(\cdot)$ is the CDF of the $\mathcal{N}(0,1)$ distribution, $\phi_{i,0}, \phi_{i,1}, \ldots, \phi_{i,p_i}$ and σ_i are the AR parameters of the ith component of the mixtures, and $\{\pi_i\}_{i=1}^{K}$ is a set of so-called *mixing proportions* which satisfy $\pi_i > 0$ and $\sum_{i=1}^{K} \pi_i = 1$. A characteristic feature of the MAR model is that both its conditional and unconditional marginal distributions are nonnormal and they can be multimodal.

The BIC model selection criterion is given by BIC $= -2\ell_T(\mathbf{y}; \widehat{\boldsymbol{\theta}}_T) + m\log(T - n)$, where $\ell_T(\mathbf{y}; \widehat{\boldsymbol{\theta}}_T)$ is the value of the maximized log-likelihood function of the sample, m is the dimension of the parameter vector $\boldsymbol{\theta}$, and n is the number of initial values. Using this criterion, the best fitted MAR model is

$$F(Y_t|\mathcal{F}_{t-1}, \widehat{\boldsymbol{\theta}}_T) = 0.3163_{(0.0810)} \Phi\left(\frac{Y_t - 0.7107_{(0.1798)} - 1.1022_{(0.0621)} Y_{t-1} + 0.2835_{(0.0826)} Y_{t-2}}{0.0887_{(0.0202)}}\right)$$
$$+ 0.6837_{(0.0810)} \Phi\left(\frac{Y_t - 0.9784_{(0.1564)} - 1.5279_{(0.0884)} Y_{t-1} + 0.8817_{(0.0869)} Y_{t-2}}{0.0887_{(0.0202)}}\right),$$

where asymptotic standard errors of the parameter estimates are given in parentheses, and the value of BIC is -198.82.

(a) Check the adequacy of the fitted MAR model by computing the first 20 sample autocorrelations of the Pearson residuals defined by (6.72). Repeat this step for the squared Pearson residuals.

(b) Check the adequacy of the fitted MAR model by computing the first two diagnostic test statistics in Table 6.3 (A_{T,K_1} and H_{T,K_2}) using quantile residuals, and with $K_1 = K_2 = \{5, 10, 15, 20, 25, 30\}$. Compare and contrast the results with those obtained in part (a).

[*Hint:* Replace the covariance estimator $\widehat{\boldsymbol{\Omega}}_T$ in (6.89) by an estimator $\widetilde{\boldsymbol{\Omega}}_{\widetilde{T}}$ using numerical derivatives for both the log-likelihood function and quantile residuals given a set of $\widetilde{T} = 20{,}000$ simulated observations (Kalliovirta, 2012, p. 365)].

Theoretical Question for Appendix 7.A

7.8 Assume that: (i) the density $f(\cdot)$ has $(\nu + 1)$ continuous derivatives, which are square integrable and monotone; (ii) the bandwidth $h \equiv h_n$ is a non-random sequence of positive numbers such that $\lim_{n\to\infty} h = 0$, and $\lim_{n\to\infty} nh^\nu = \infty$; (iii) the kernel $K(\cdot)$ is a bounded pdf having finite jth $(j < \nu)$ order moment and symmetric about the origin.

(a) Show that the bias and variance of $\widehat{f}_h(x)$, defined in (A.2), satisfy

$$\text{Bias}\big(\widehat{f}_h(x)\big) = \mathbb{E}\big(\widehat{f}_h(x)\big) - f(x) = \frac{1}{\nu!} f^{(\nu)}(x) h^\nu \mu_\nu(K) + o(h^\nu),$$

$$\text{Var}\big(\widehat{f}_h(x)\big) = \frac{1}{nh} f(x) R(K) + o\Big(\frac{1}{nh}\Big),$$

where $f^{(\nu)}(\cdot)$ denotes the νth derivative of $f(\cdot)$, assuming it exists. Comment on the difference in bias between second- and higher-order kernels.

(b) Combine the results in part (a), to obtain the asymptotic MSE (AMSE) of $\widehat{f}_h(\cdot)$. Comment on the bias-variance trade-off.

(c) Derive an expression for the AMISE of $\widehat{f}_h(\cdot)$.

(d) Show that by differentiating $\text{AMISE}\big(\widehat{f}_h(x)\big)$ with respect to h, and setting the derivative equal to zero, the optimal bandwidth is given by

$$h_{\text{opt}} = R\big(f^{(\nu)}\big)^{-1/(2\nu+1)} \Big(\frac{(\nu!)^2 R(K)}{2\nu \mu_\nu^2(K)}\Big)^{1/(2\nu+1)} n^{-1/(2\nu+1)}.$$

Comment on the difference between the optimal bandwidth for second-order kernels and for higher-order kernels.

(e) Verify (A.5).

(f) Verify (A.6).

Chapter 8

TIME-REVERSIBILITY

Time-reversibility (TR) amounts to temporal symmetry in the probabilistic structure of a strictly stationary time series process. In other words, a stochastic process is said to be TR if its probabilistic structure is unaffected by reversing ("mirroring") the direction of time. Otherwise, the process is said to be *time-irreversible*, or *non-reversible*. Confirmation of time-irreversibility is important because, according to Cox (1981), it is a symptom of nonlinearity and/or non-Gaussianity. In the analysis of business cycles, for instance, the peaks and troughs of a business time series differ in magnitude, not just in sign, as the dynamics of contractions in an economy are more violent but also more short-lived than the expansions, indicating asymmetric cycles. Time irreversible behavior may also naturally arise in stochastic processes considered in, for instance, quantum mechanics, biomedicine, queuing theory, system engineering, and financial economics. Time-irreversibility automatically excludes Gaussian linear processes, or static nonlinear transformations of such processes, as possible DGPs.

In Example 1.2, we discussed a graphical technique to detect departures from TR, at least in extreme cases. In this chapter we follow a more formal approach, that is, the focus is on test statistics for assessing TR. First, in Section 8.1, we review various general definitions of TR for stationary DGPs. In Section 8.2, we introduce time-domain TR tests which satisfy certain symmetry conditions of the probability distribution of the stochastic process under study. In Section 8.3, we consider two frequency-domain TR tests. These tests are motivated by the property that the imaginary part of all polyspectra is zero for TR processes; see Chapter 4. In Section 8.4, we discuss three nonparametric tests statistics. First, in Section 8.4.1, we present a copula-based TR test statistic applicable to stationary Markov chains. Next, in Section 8.4.2 and Section 8.4.3 respectively, we discuss a kernel-based and a sign TR test statistic for high-dimensional stationary DGPs. We illustrate the use of various TR test statistics in Section 8.5, with an application to the set of time series introduced in Chapter 1. We conclude with a short summary, and offer some concluding remarks.

8.1 Preliminaries

A strictly stationary stochastic process $\{Y_t, t \in \mathbb{Z}\}$ is defined to be TR if, for any integer m and for all integers $t_1, \ldots, t_n$ ($-\infty < t_1 < \cdots < t_n < \infty$), the vectors $(Y_{-t_1}, Y_{-t_2}, \ldots Y_{-t_n})'$ and $(Y_{-t_1+m}, Y_{-t_2+m}, \ldots Y_{-t_n+m})'$ have the same joint probability distribution. Letting $m = t_1 + t_n$, we see that for a strictly stationary process $\{Y_t, t \in \mathbb{Z}\}$ time reversibility implies that

$$(Y_{t_1}, Y_{t_2}, \ldots, Y_{t_n})' \stackrel{D}{\sim} (Y_{t_n}, Y_{t_n+(t_1-t_2)}, \ldots, Y_{t_1})', \tag{8.1}$$

where $\stackrel{D}{\sim}$ denotes equal in distribution. For causal linear ARMA processes, it is well known that TR is essentially restricted to processes having Gaussian innovations. For stationary univariate and multivariate non-Gaussian linear processes, TR requires some regularity conditions on the coefficients of the model representing the DGP.

Test statistics for TR are often devised for bivariate or trivariate random variables because of the complexities associated with multi-dimensional distributions. Indeed, several proposed tests statistics are based on the following, less exhaustive, definition of TR. That is, $\{Y_t, t \in \mathbb{Z}\}$ is said to be a TR process if $(Y_t, Y_{t-\ell})' \stackrel{D}{\sim} (Y_{t-\ell}, Y_t)'$ ($\ell \in \mathbb{N}$). In consequence, for any $(a, b) \in \mathbb{R}^2$, and each $\ell \in \mathbb{N}$ we have $F_{Y_t, Y_{t-\ell}}(a, b) = F_{Y_t, Y_{t-\ell}}(b, a)$. Let $A(x) = \{(a, b): b - a \leq x\}$, and $B(x) = \{(a, b): b - a \geq -x\}$, where x is a real number. Then, for every x, we can write the distribution of the stochastic process $\{X_t(\ell) \equiv Y_t - Y_{t-\ell}, t \in \mathbb{Z}\}$ as

$$F_{X_t(\ell)}(x) = \int_{A(x)} \mathrm{d}F_{Y_t, Y_{t-\ell}}(a, b) = \int_{B(x)} \mathrm{d}F_{Y_t, Y_{t-\ell}}(a, b)$$

$$= 1 - \int_{A(-x)} \mathrm{d}F_{Y_t, Y_{t-\ell}}(a, b) = 1 - F_{X_t(\ell)}(-x). \tag{8.2}$$

Thus, the one-dimensional marginal distribution of $\{X_t(\ell), t \in \mathbb{Z}\}$ is symmetric about zero, i.e., $X_0(\ell) \stackrel{D}{=} -X_0(\ell)$. This implication of TR is the basis of the two test statistics introduced in Section 8.2.

It is well known that many nonlinear DGPs are stationary Markov chains or can be rephrased as a Markov chain. The dynamic properties of Markov chains may be conveniently modeled via *copula functions*. Let $\{Y_t, t \in \mathbb{Z}\}$ be a stationary real-valued Markov chain with invariant CDF $F_Y: \mathbb{R} \to [0, 1]$ which is assumed to be continuous. Sklar's theorem (Appendix 7.B) ensures the existence of a unique bivariate copula function $\mathbb{C}: [0, 1]^2 \to [0, 1]$ characterizing the relationship between Y_t and Y_{t+1} for any $t \in \mathbb{Z}$. Let $H: \mathbb{R}^2 \to [0, 1]$ denote the joint CDF of $\mathbf{Y}_t = (Y_t, Y_{t+1})'$. Then we have $H(y_1, y_2) = \mathbb{C}(F_Y(y_1), F_Y(y_2))$, $\forall (y_1, y_2) \in \mathbb{R}^2$ and all $t \in \mathbb{Z}$. Therefore, the following two statements provide equivalent formulations of TR for stationary first-order Markov chains:

(i) $H(y_1, y_2) = H(y_2, y_1)$, $\forall (y_1, y_2) \in \mathbb{R}^2$,
(ii) $\mathbb{C}(u, v) = \mathbb{C}(v, u)$, $\forall (u, v) \in [0, 1]^2$.

8.2 TIME-DOMAIN TESTS

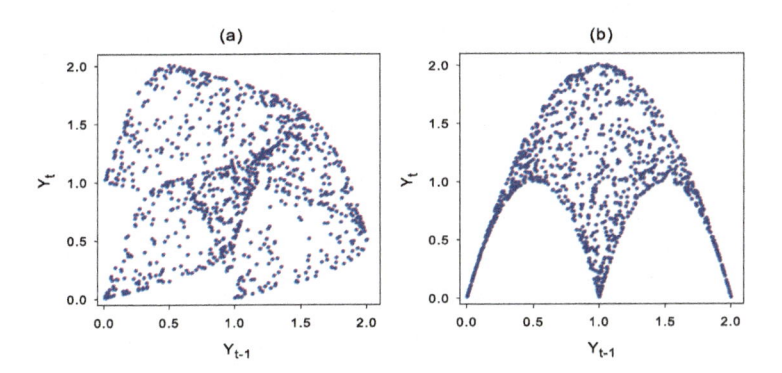

Figure 8.1: *(a) Scatter plot at lag 1 of the time series* $\{X_t = Y_{1,t} + Y_{2,1001-t}\}_{t=1}^{1,000}$, *where* $\{Y_{i,t}, t \in \mathbb{Z}\}$ $(i = 1, 2)$ *are two independent realizations of the logistic map (1.22) with* $a = 4$; *(b) Scatter plot at lag 1 of the time series* $\{X_t^* = Y_{1,t} + Y_{2,t}\}_{t=1}^{1,000}$.

Property (i) is sometimes referred to as *detailed balance equations*. A copula satisfying (ii) is said to be *exchangeable, commutative* or *symmetric*.

Example 8.1: Exploring a Logistic Map for TR

Figure 8.1(a) shows a scatter plot at lag 1 of the time series $\{X_t = Y_{1,t} + Y_{2,1001-t}\}_{t=1}^{1,000}$, where $\{Y_{i,t}, t \in \mathbb{Z}\}$ $(i = 1, 2)$ are two independent realizations of the logistic map (1.22) with $a = 4$. Note that the scatter plot is symmetric along the main diagonal, suggesting that the DGP is symmetric. For the same logistic map, Figure 8.1(b) shows a scatter plot at lag 1 of a time series $\{X_t^* = Y_{1,t} + Y_{2,t}\}_{t=1}^{1,000}$. We see that the distribution of $\{X_t^*\}$ is asymmetric. Hence, the series $\{X_t^*\}$ is not a realization of a static transformation of a linear Gaussian DGP.

8.2 Time-Domain Tests

8.2.1 A bicovariance-based test

Since the condition of TR implies the equivalence of various distributions, it also implies the equality of various subsets of moments from the joint distribution of $(Y_{t_1}, \ldots, Y_{t_n})'$, where they exist. Autocovariances, however, are by definition symmetric. Also the spectral density function and its time-reversed version are identical. So, we need higher-order moments to detect irreversibility. Assume, for ease of notation, that $\{Y_t, t \in \mathbb{Z}\}$ has mean zero. Then a sufficient, but not necessary, condition for TR is the equality

$$\mathbb{E}(Y_t^i Y_{t-\ell}^j) = \mathbb{E}(Y_t^j Y_{t-\ell}^i), \quad \forall (i,j) \in \mathbb{N} \text{ and } \forall \ell \in \mathbb{Z}. \tag{8.3}$$

Pomeau (1982) and Steinberg (1986) use (8.3) with $i = 1$ and $j = 3$ to examine TR. Later, Ramsey and Rothman (1996) consider the case $i = 1$, $j = 2$. In particular these authors investigate the difference between two bicovariances, termed the *symmetric-bicovariance function*, and defined as follows

$$\psi_Y(\ell) = \gamma_Y^{(2,1)}(\ell) - \gamma_Y^{(1,2)}(\ell), \tag{8.4}$$

where $\gamma_Y^{(i,j)}(\ell) = \mathbb{E}(Y_t^i Y_{t-\ell}^j)$. If a strictly stationary process $\{Y_t, t \in \mathbb{Z}\}$ is TR, then $\psi_Y(\ell) = 0 \ \forall \ell \in \mathbb{Z}$.

Ramsey and Rothman (1996) note that, within the context of stationary DGPs, TR can stem from two sources. First, the model representing the DGP may be nonlinear even though the innovations $\{\varepsilon_t\}$ follow a symmetric (perhaps Gaussian) probability distribution. They refer to this case as *"Type I"* time-irreversibility. Second, $\{\varepsilon_t\}$ is a sequence of i.i.d. non-Gaussian random variables while the model is linear. This latter case is called *"Type II"* time-irreversibility. Note, however, that nonlinearity does not imply Type I time-irreversibility; there exist stationary reversible nonlinear time series models; see, e.g., McKenzie (1985), Lewis et al. (1989), and Exercise 8.4. So, a test for Type I time-irreversibility is not fully equivalent to a test for nonlinearity.

Using moment estimates of the bicovariances, the TR test statistic is based on the estimator

$$\widehat{\psi}_Y(\ell) = \widehat{\gamma}_Y^{(2,1)}(\ell) - \widehat{\gamma}_Y^{(1,2)}(\ell), \quad (\ell \in \mathbb{Z}),$$

where $\widehat{\gamma}_Y^{(i,j)}(\ell) = (T-\ell)^{-1} \sum_{t=\ell+1}^T Y_t^i Y_{t-\ell}^j$ with $(i,j) = (1,2)$.[1] One can easily show that $\widehat{\gamma}_Y^{(i,j)}(\ell)$ is an unbiased and consistent estimator of $\gamma_Y^{(i,j)}(\ell)$. Moreover, if $\{Y_t, t \in \mathbb{Z}\}$ is a zero-mean i.i.d. process with $\mathbb{E}(Y_t^4) < \infty$, it is easy to verify (Exercise 8.2(a)) that an exact expression of the variance of $\widehat{\psi}_Y(\ell)$ is given by

$$\mathrm{Var}\{\widehat{\psi}_Y(\ell)\} = \frac{2(\mu_{4,Y}\mu_{2,Y} - \mu_{3,Y}^2)}{(T-\ell)} - \frac{2\mu_{2,Y}^3(T-2\ell)}{(T-\ell)^2}. \tag{8.5}$$

Replacing $\mu_{3,Y}$ and $\mu_{4,Y}$ by their sample counterparts leads to $\widehat{\mathrm{Var}}\{\widehat{\psi}_Y(\ell)\}$, i.e., the sample analogue of (8.5). Then the TR test statistic is defined by

$$\mathrm{TR}(\ell) = \widehat{\psi}_Y(\ell) \big/ \sqrt{\widehat{\mathrm{Var}}\{\widehat{\psi}_Y(\ell)\}}. \tag{8.6}$$

Under $\mathbb{H}_0: \psi_Y(\ell) = 0$, it can be shown that $\mathrm{TR}(\ell) \xrightarrow{D} \mathcal{N}(0,1)$ as $T \to \infty$. The pre-requisite of the test statistic is that $\{Y_t, t \in \mathbb{Z}\}$ must possess at least a finite six-order moment. Note that this condition may often be viewed as too restrictive for DGPs without higher-order moments, which typically is the case with financial data.

Ramsey and Rothman (1996) recommend the following two-stage procedure for testing Type I and II time-irreversibility.

---

[1] The idea of using the difference $\widehat{\gamma}_Y^{(2,1)}(\ell) - \widehat{\gamma}_Y^{(1,2)}(\ell)$ as a measure for TR is comparable to using the difference between lag ℓ sample cross-correlations of standardized residuals, e.g. $\widehat{\rho}_{\widehat{\varepsilon}}^{(2,1)}(\ell) - \widehat{\rho}_{\widehat{\varepsilon}}^{(1,2)}(\ell)$ (see Example 6.8) as an alternative (omnibus-type) test statistic for diagnostic checking.

8.2 TIME-DOMAIN TESTS

> **Algorithm 8.1: The Ramsey–Rothman TR test**
>
> *Stage 1:* **Type I and II time-irreversibility**
>
> (i) Standardize the time series under study, and compute $\widehat{\psi}_Y(\ell)$ for $\ell = 1, 2, \ldots$.
>
> (ii) Fit a causal ARMA(p,q) model to the standardized series $\{Y_t\}_{t=1}^T$, using an order selection criterion to find the optimal values of p and q. Obtain the residuals and compute (8.5), replacing $\mu_{r,Y}$ by $\widehat{\mu}_{r,Y} = T^{-1} \sum_{t=1}^T Y_t^r$ ($r = 2, 3, 4$).
>
> (iii) Generate a new time series $\{Y_t^*\}_{t=1}^T$ using the fitted model in step (ii), and with $\{\varepsilon_t\}_{t=1}^T$ generated as a sequence of i.i.d. $\mathcal{N}(0,1)$ random variables. Obtain the corresponding value of $\widehat{\psi}_{Y^*}(\ell)$. Repeat this step a large number of times.
>
> (iv) Compute the sample standard deviation of $\widehat{\psi}_{Y^*}(\ell)$ via its simulated distribution. Using the result in step (i), compute TR(ℓ) for $\ell = 1, 2, \ldots$.
>
> (v) To avoid possible interdependence among the computed test statistics at different lags, estimate the p-value of $\max_\ell |\text{TR}(\ell)|$ running a second MC simulation. Rejection of $\mathbb{H}_0$ is consistent with both Type I and II time-irreversibility.
>
> *Stage 2:* **Distinguishing Type I and Type II time-irreversibility**
>
> (vi) Given a rejection in Stage 1, repeat steps (i) and (ii) above. Next, compute TR(ℓ) ($\ell = 1, 2, \ldots$). Finally, estimate the p-value of $\max_\ell |\text{TR}(\ell)|$ running a single MC simulation. If the DGP is Type II, i.e., the model is ARMA with non-Gaussian innovations, the residuals will be approximately TR. Thus, $\mathbb{H}_0$ will not be rejected.

Two comments are in order. First, with some fitted linear ARMA models, direct computation of the variance formula (8.5) may result in negative estimates. Step (iii) overcomes this potential problem by simulating the distribution function of $\widehat{\psi}_Y(\ell)$. A second, and more serious problem, is that the ARMA prewhitening in step (ii) may destroy TR since it induces a phase shift in the series; see Hinich et al. (2006). As a consequence, the TR test statistic (8.6) could lead to false rejections of the null hypothesis.

8.2.2 A test based on the characteristic function

A distribution of a continuous random variable X is symmetric if and only if the imaginary part of its characteristic function, $\Im\{\phi_X(\omega)\}$ say, is zero for all real numbers ω. In view of (8.2), and using the fact that there is a one-to-one correspondence between distribution functions and characteristic functions, it seems natural to construct a TR test statistic for the null hypothesis

$$\mathbb{H}_0\colon \Im\{\phi_{X,\ell}(\omega)\} = \mathbb{E}\{\sin\left(\omega(X_t(\ell))\right)\} = 0, \quad \forall \omega \in \mathbb{R}^+. \tag{8.7}$$

This result forms the basis of a TR test statistic proposed by Chen et al. (2000).

Let $g(\cdot)$ be a weighting function such that $\int_0^\infty g(\omega)d\omega < \infty$. More specifically, $g(\cdot)$ should be chosen such that $\phi_{X,\ell}(\cdot)$ will not be integrated to zero when the distribution of $\{X_t(\ell), t \in \mathbb{Z}\}$ is asymmetric. A necessary condition is

$$\int_0^\infty \phi_{X,\ell}(\omega)g(\omega)d\omega = \int_{-\infty}^\infty \Big(\int_0^\infty \sin(\omega X_t(\ell))g(\omega)d\omega\Big)dF_{X_{t,\ell}} = 0, \quad \forall \ell \in \mathbb{Z}. \tag{8.8}$$

By changing the order of integration, (8.8) is equivalent to

$$\mu_g(\ell) \equiv \mathbb{E}[\psi_g(X_t(\ell))] = \int_{-\infty}^\infty \psi_g(x)dF_{X_{t,\ell}}(x) = 0, \tag{8.9}$$

where $\psi_g(x) = \int_0^\infty \sin(\omega x)g(\omega)d\omega$. Given an observable segment $\{Y_t\}_{t=1}^T$ of $\{Y_t, t \in \mathbb{Z}\}$, and by abuse of notation, a natural point estimator of (8.9) is given by

$$\overline{\psi}_g(\ell) = \frac{1}{T-\ell}\sum_{t=\ell+1}^T \psi_g(Y_t(\ell)). \tag{8.10}$$

Because $\psi_g(\cdot)$ is a static transformation, $\{X_t(\ell)\}$ and $\{\psi_g(X_t(\ell))\}$ are also strictly stationary processes for each fixed $\ell \in \mathbb{Z}$. Then, under a minimal mixing condition (see, e.g., White, 1984, Thm. 5.15), it is easy to show that, as $T \to \infty$,

$$\sqrt{T-\ell}\left(\overline{\psi}_g(\ell) - \mu_g(\ell)\right) \xrightarrow{D} \mathcal{N}(0, \sigma^2_{\psi_g}(\ell)), \tag{8.11}$$

where

$$\sigma^2_{\psi_g}(\ell) = \lim_{T\to\infty} \mathrm{Var}\Big(\frac{1}{\sqrt{T-\ell}}\sum_{t=\ell+1}^T \psi_g(X_t(\ell))\Big)$$
$$= \mathrm{Var}\{\psi_g(X_t(\ell))\}$$
$$+ 2\lim_{T\to\infty}\Big(\sum_{i=1}^{T-\ell-1}\Big(1-\frac{i}{T-\ell}\Big)\mathrm{Cov}\{\psi_g(X_t(\ell)), \psi_g(X_{t-i}(\ell))\}\Big).$$

This leads to the following test statistic for $\mathbb{H}_0$:

$$\mathcal{C}_g(\ell) = \sqrt{T-\ell}\Big(\frac{\overline{\psi}_g(\ell)}{\widehat{\sigma}_{\psi_g}(\ell)}\Big), \tag{8.12}$$

where $\widehat{\sigma}^2_{\psi_g}(\ell)$ is a consistent estimator for $\sigma^2_{\psi_g}(\ell)$. Its form is given by

$$\widehat{\sigma}^2_{\psi_g}(\ell) = \widehat{\gamma}_{\psi_g}(0) + 2\sum_{j=1}^{T-\ell-1} W_{T,\ell}(j)\widehat{\gamma}_{\psi_g}(j),$$

where $\widehat{\gamma}_{\psi_g}(j)$ is the lag-j sample autocovariance of $\{\psi_g(X_t(l)); \ell+1 \leq t \leq T\}$ and

$$W_{T,\ell}(j) = \left(1 - \frac{j}{T-\ell}\right)\left\{1 - \frac{1}{2(T-\ell)^{1/3}}\right\}^j$$
$$+ \frac{j}{T-\ell}\left\{1 - \frac{1}{2(T-\ell)^{1/3}}\right\}^{T-\ell-j}, \quad (j \in \mathbb{N}). \qquad (8.13)$$

The weight function (8.13) ensures that $\widehat{\sigma}^2_{\psi_g}(\ell)$ is always non-negative. Its form is motivated by the lag window used in the stationary bootstrap method of Politis and Romano (1994) and adopted by Chen et al. (2000) and Chen (2003). These latter authors further suggest to take $g(\omega) = (1/\beta)\exp(-\omega/\beta)$ ($\omega > 0$), for some $\beta \in (0, \infty)$, so that $\psi_g(x) = \beta x/(1 + \beta^2 x^2)$. By adjusting the parameter β, the resulting test statistic is flexible to capture various types of asymmetry. The test statistic (8.12) seems to have high empirical power with $\beta = 1$ and $\beta = 2$.

Observe that (8.12) essentially is a general test statistic for detecting symmetry of the marginal distribution of the observed time series $\{Y_t\}_{t=1}^T$. It is a TR test statistic when applied to $\{X_t(\ell)\}_{t=\ell+1}^T$. A useful feature of $\mathcal{C}_g(\ell)$ is that the test statistic can be used without any moment assumptions.[2] Indeed, simulations provided by Chen et al. (2000) confirm that this test statistic is quite robust to the moment property of the DGP being tested.

Unfortunately, the test statistic (8.12) is a check for *un*conditional symmetry using the observed time series $\{Y_t\}_{t=1}^T$. From an application perspective, however, conditional symmetry is often of more interest. This implies that we need to replace $\{Y_t, t \in \mathbb{Z}\}$ by some residual series $\{\widehat{\varepsilon}_t\}$. In that case, Chen and Kuan (2002) suggest to modify the computation of $\widehat{\sigma}^2_{\psi_g}(\ell)$ by bootstrapping from the standardized residuals of a time series model, using a model-free bootstrap approach. Provided the first four moments of the error process $\{\varepsilon_t\}$ exist, the resulting TR test statistic is still asymptotically normally distributed under the null hypothesis that $\mathbb{E}(\psi_g(\varepsilon_t)) = 0$.

Example 8.2: Exploring a Simulated SETAR Process for TR

A simple way to explore an observed time series $\{Y_t\}_{t=1}^T$ for TR is to detect asymmetries in plots of the sample distributions of $X_t(\ell) = Y_t - Y_{t-\ell}$ ($\ell = 1, 2, \ldots$). As an illustration, consider the stationary SETAR(2; 1, 1) process

$$Y_t = \begin{cases} 0.5 Y_{t-1} + \varepsilon_t & \text{if } Y_{t-1} \leq 0, \\ -0.4 Y_{t-1} + \varepsilon_t & \text{if } Y_{t-1} > 0, \end{cases} \qquad (8.14)$$

where $\{\varepsilon_t\} \overset{\text{i.i.d.}}{\sim} \mathcal{N}(0, 1)$. Figure 8.2(a) shows a plot of a typical subset of length $T = 100$ of a simulated time series of 10,000 observations. Figure 8.2(b) displays the kernel smoothed densities of $\{W_t(\ell) = Y_t - Y_{t-\ell}\}_{t=1}^{10,000}$ ($\ell = 1, \ldots, 5$), using a normal kernel. It is visually clear that the distributions are not symmetric about the origin, indicating the SETAR process is time-irreversible.

---

[2]This feature trivially holds for the kernel-based TR test statistic $S_{h,T}(m)$ of Diks et al. (1995), to be discussed in Section 8.4.2, since the adopted Gaussian kernel is bounded.

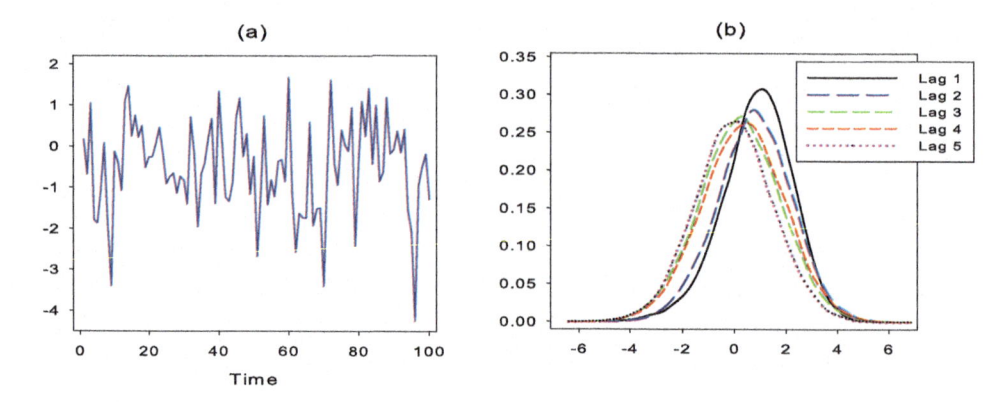

Figure 8.2: *(a) A typical subset $\{Y_t\}_{t=1}^{100}$ of the simulated* SETAR$(2;1,1)$ *process (8.14); (b) Simulated marginal distributions of $\{W_t(\ell) = Y_t - Y_{t-\ell}\}_{t=1}^{10,000}$ for $\ell = 1,\ldots,5$.*

8.3 Frequency-Domain Tests

8.3.1 A bispectrum-based test

In Section 4.1, we showed that, under the null hypothesis of TR, $\Im\{f_Y(\omega_1,\omega_2)\} = 0$ $\forall (\omega_1,\omega_2) \in \mathcal{D}$ where $\mathcal{D}$ is the principal domain (4.7). Hinich and Rothman (1998) use this result to define a frequency-domain TR test statistic based on the imaginary part of the normalized estimated bispectrum $\widehat{B}_Y(\omega_1,\omega_2)$, say $\Im\{\widehat{B}_Y(\omega_1,\omega_2)\}$. The computation of the corresponding test statistic involves the following steps.

Algorithm 8.2: The bispectrum-based TR test

(i) Divide the series $\{Y_t\}_{t=1}^T$ into K nonoverlapping stretches, or frames, of length N so that $K = \lfloor T/N \rfloor$. Define the discrete Fourier frequencies $\omega_j = j/N$ $(j = 1,\ldots,N)$.

(ii) Calculate the discrete FT $Y_k(\omega_j) = \sum_{t=1}^{N} Y_{t+(k-1)N} \exp\{-2\pi i \omega_j (t+(k-1)N)\}$, and the periodogram of the kth frame $N^{-1}|Y_k(\omega_j)|^2 = N^{-1}Y_k(\omega_j)Y_k(\omega_{-j})$, $(k = 1,\ldots,K)$.

(iii) Compute the averaged estimate of the spectrum at frequency ω_j, i.e., $\widehat{f}_Y(\omega_j) = T^{-1}\sum_{k=1}^{K}|Y_k(\omega_j)|^2$, since $T \approx KN$. In addition, calculate $\widehat{f}_Y(\omega_{j_1},\omega_{j_2}) = N^{-1}\sum_{k=1}^{K} Y_k(\omega_{j_1})Y_k(\omega_{j_2})Y_k(-\omega_{j_1}-\omega_{j_2})$. Then the normalized estimated bispectrum is

$$\widehat{B}_Y(\omega_{j_1},\omega_{j_2}) = \frac{\widehat{f}_Y(\omega_{j_1},\omega_{j_2})}{\sqrt{\widehat{f}_Y(\omega_{j_1})\widehat{f}_Y(\omega_{j_2})\widehat{f}_Y(\omega_{j_1}+\omega_{j_2})}}.$$

> **Algorithm 8.2: The bispectrum-based TR test (Cont'd)**
>
> (iv) Compute the test statistic
>
> $$S_{\text{TR}} = 2T^{2c-1} \sum_{(\omega_{j_1},\omega_{j_2}) \in \mathcal{D}} |\Im\{\widehat{B}_Y(\omega_{j_1},\omega_{j_2})\}|^2. \quad (8.15)$$
>
> Under $\mathbb{H}_0$: $\Im\{B_Y(\omega_{j_1},\omega_{j_2})\} = 0$, and as $T \to \infty$,
>
> $$S_{\text{TR}} \xrightarrow{D} \chi^2_M, \quad (8.16)$$
>
> with degrees of freedom $M = [N^2/16]$. Hinich and Rothman (1998) prove consistency of S_{TR}.

8.3.2 A trispectrum-based test

Similar to the bispectrum (4.4), we can define the *trispectrum* as the triple FT of the fourth-order cumulant function of a stationary time series process $\{Y_t, t \in \mathbb{Z}\}$, i.e.,

$$f_Y(\omega_1, \omega_2, \omega_3) = \sum_{\ell_1,\ell_2,\ell_3=-\infty}^{\infty} \gamma_Y(\ell_1, \ell_2, \ell_3) \exp\{-2\pi i(\omega_1\ell_1 + \omega_2\ell_2 + \omega_3\ell_3)\}, \quad (8.17)$$

where $(\omega_1, \omega_2, \omega_3) \in [0,1]^3$ are normalized frequencies, and the third-order cumulant function is defined as $\gamma_Y(\ell_1, \ell_2, \ell_3) = \mathbb{E}(Y_t Y_{t+\ell_1} Y_{t+\ell_2} Y_{t+\ell_3})$. Owing to symmetry relations, the trispectrum need to be calculated only in a subset of the complete $(\omega_1, \omega_2, \omega_3)$-space; see, e.g., Dalle Molle and Hinich (1995) for a description of nonredundant regions of (8.17), including its principal domain.

The normalized magnitude of the trispectrum, known as the *squared tricoherence*, can be expressed as

$$|T_Y(\omega_1,\omega_2,\omega_3)|^2 = \frac{|f_Y(\omega_1,\omega_2,\omega_3)|^2}{f_Y(\omega_1,-\omega_1)f_Y(\omega_2,-\omega_2)f_Y(\omega_3,-\omega_3)f_Y(\omega_1+\omega_2+\omega_3,-\omega_1-\omega_2-\omega_3)}. \quad (8.18)$$

If a stationary DGP can be represented as a linear convolution of a sequence of i.i.d. random variables, then (8.18) is a constant for all points in the stationary set. If, moreover, the process is Gaussian, then this constant is equal to zero for all points belonging to the principal domain, say Ω. Thus, as in Chapter 4, global test statistics for Gaussianity and linearity can be defined at a particular frequency triple $(\omega_1, \omega_2, \omega_3) \in \Omega$.

Dalle Molle and Hinich (1995) consider a frame-averaging procedure for estimating (8.17), similar as the one given in Section 8.3.1 for the bispectrum-based TR test statistic. In particular, start with steps (i) and (ii) of Algorithm 8.2. Also, compute $\widehat{f}_Y(\omega_j) = T^{-1} \sum_{k=1}^{K} |Y_k(\omega_j)|^2$ with $T \approx KN$. Next, replace steps (iii) and (iv) in Algorithm 8.2 by the following steps.

Algorithm 8.3: The trispectrum-based TR test

(iii*) Compute, as a consistent estimator of (8.17),

$$\widehat{f}_Y(\omega_{j_1}, \omega_{j_2}, \omega_{j_3}) = \frac{1}{T} \sum_{k=1}^{K} Y_k(\omega_{j_1}) Y_k(\omega_{j_2}) Y_k(\omega_{j_3}) Y_k(-\omega_{j_1} - \omega_{j_2} - \omega_{j_3}).$$

Then the normalized estimated trispectrum is

$$\widehat{T}_Y(\omega_{j_1}, \omega_{j_2}, \omega_{j_3}) = \frac{\widehat{f}_Y(\omega_{j_1}, \omega_{j_2}, \omega_{j_3})}{\sqrt{\widehat{f}_Y(\omega_{j_1})\widehat{f}_Y(\omega_{j_2})\widehat{f}_Y(\omega_{j_3})\widehat{f}_Y(\omega_{j_1} + \omega_{j_2} + \omega_{j_3})}}.$$

This normalization standardizes the variance of the trispectrum estimate using the estimated asymptotic variance in place of the true variance.

(iv*) Compute the TR test statistic

$$S_{\text{TR}}^* = 2T^{2c-1} \sum_{\omega_{j_1}, \omega_{j_2}, \omega_{j_3} \in \Omega} |\Im\{\widehat{T}_Y(\omega_{j_1}, \omega_{j_2}, \omega_{j_3})\}|^2, \quad (\tfrac{1}{2} < c < 1). \quad (8.19)$$

Under $\mathbb{H}_0$: $\Im\{T_Y(\omega_{j_1}, \omega_{j_2}, \omega_{j_3})\} = 0$, and as $T \to \infty$,

$$S_{\text{TR}}^* \xrightarrow{D} \chi^2_{M^*} \quad (8.20)$$

with M^* the number of frequency triples in Ω. This number is automatically computed in the available software code; see Section 8.7.

The test statistic S_{TR}^* is applicable if the one-dimensional marginal distribution of $\{Y_t, t \in \mathbb{Z}\}$ has a finite eighth moment. Like the bispectrum-based TR test statistic S_{TR}, this moment requirement rules out many economic and financial time series encountered in practice.

8.4 Other Nonparametric Tests

The frequency-domain TR test statistics discussed in Section 8.3 are nonparametric in nature. They may be computationally demanding, and require special care when the boundary (nonredundant) bispectral lags are included. Here, we discuss three nonparametric TR test statistics which are computationally more attractive.

8.4.1 A copula-based test for Markov chains

In Section 8.1, we briefly introduced the notion of exchangeability. A measure for the "amount" or "degree" of nonexchangeability of each pair (X, Y) of identically distributed random variables (see, e.g., Klement and Mesiar, 2006; Nelsen, 2007) is given by

$$\delta_\mathbb{C} = 3 \sup_{(u,v)\in[0,1]^2} |\mathbb{C}(u,v) - \mathbb{C}(v,u)|. \tag{8.21}$$

This measure takes values in $[0, 1]$ for any copula with the lower and upper bounds attainable. Based on (8.21), Beare and Seo (2014) propose a TR test statistic for the null hypothesis $\mathbb{H}_0 : \delta_\mathbb{C} = 0$. Using the notation in Section 8.1, let $\theta \in [0, 1/3]$ be given by

$$\theta = \sup_{(y_1,y_2)\in\mathbb{R}^2} |H(y_1,y_2) - H(y_2,y_1)|,$$

which, in view of (8.21), implies that $\theta = \frac{1}{3}\delta_\mathbb{C}$. Given a set of observations $\{Y_t\}_{t=1}^T$, a natural empirical analogue of θ is

$$\theta_T = \sup_{(y_1,y_2)\in\mathbb{R}^2} |H_T(y_1,y_2) - H_T(y_2,y_1)|, \tag{8.22}$$

where $H_T(\cdot,\cdot)$ is the joint EDF

$$H_T(y_1,y_2) = \frac{1}{T-1} \sum_{t=1}^{T-1} I(Y_t \leq y_1, Y_{t+1} \leq y_2).$$

Under $\mathbb{H}_0$ and fairly weak regularity conditions, it can be shown (Beare and Seo, 2014) that θ_T is asymptotically distributed as

$$\sqrt{T}\theta_T \xrightarrow{D} \sup_{(y_1,y_2)\in\mathbb{R}^2} |\mathbb{B}(y_1,y_2) - \mathbb{B}(y_2,y_1)|, \text{ as } T \to \infty, \tag{8.23}$$

where $\mathbb{B}(\cdot,\cdot)$ is a continuous centered Gaussian process with covariance kernel

$$\text{Cov}\{\mathbb{B}(y_1,y_2), \mathbb{B}(y_1',y_2')\} = \sum_{t\in\mathbb{Z}} \text{Cov}\{I(Y_0 \leq y_1, Y_1 \leq y_2), I(Y_t \leq y_1', Y_{t+1} \leq y_2')\}.$$

In addition, $\forall c \in \mathbb{R}$, $T^{-1/2}\theta_T > c$ with probability approaching one, as $T \to \infty$. Thus, for a fixed value c, $\sqrt{T}\theta_T$ is consistent against any violation of TR. One can easily generalize (8.23) so that it applies to stationary pth-order ($p \geq 2$) Markov chains. But the factor of 3 in (8.21) does not hold for higher-dimensional copulas, and a different constant is needed.

For practical implementation critical values of the limiting distribution of $\sqrt{T}\theta_T$ are required. These values can be obtained via the *local bootstrap* for strictly stationary pth-order Markov processes of Paparoditis and Politis (2002). In particular, conditional on the observed data $\{Y_t\}_{t=1}^T$, the objective is to generate bootstrap pseudo-replicates $Y_1^*, \ldots, Y_T^*$ from which the statistic of interest, in the present case (8.22), can be calculated.

For a first-order Markov chain the local resampling algorithm generating the bootstrap replicates may be applied in the following way.

Algorithm 8.4: Resampling scheme

(i) (*Initialization step*)
Select an initial state Y_1^*, and the so-called *resampling width* $b \equiv b_T > 0$ of the neighborhood of a given state.

(ii) Let us suppose that for some $t \in \{1, \ldots, T-1\}$ that $Y_1^*, \ldots, Y_t^*$ is already sampled. Now, for the $(t+1)$th bootstrap observation set $Y_{t+1}^* = Y_{J+1}$, where J is a discrete random variable with probability mass function (pmf)

$$\mathbb{P}(J = j) = \mathcal{K}_h(Y_t^* - Y_j) / \sum_{i=1}^{T-1} \mathcal{K}_h(Y_t^* - Y_i), \quad (j = 1, \ldots, T-1).$$

Here, $\mathcal{K}_h(\cdot) = K(\cdot/h)/h$ with $K(\cdot)$ a one-dimensional, nonnegative and symmetric kernel function with mean zero.

Recursive application of step (ii) yields the pseudo-time series $\{Y_t^*\}_{t=1}^T$. Notice that the above procedure resamples the observed time series in a way according to which the probability of Y_j being selected is higher the closer is its preceding value Y_{j-1} to the last generated bootstrap replicate Y_{t-1}^*.

One practical aspect is the choice of the initial bootstrap observation Y_1^*. A simple approach is to draw at random from the entire set of observations $\{Y_t\}_{t=1}^T$ with equal probability. Another issue concerns the selection of h. One simple rule-of-thumb approach is to use the 'optimal' resampling width, in the sense of minimizing the AMSE of the bootstrap one-step transition distribution function; see Paparoditis and Politis (2002). Assume that $\{Y_t\}_{t=1}^T$ is generated by an AR(1) process $Y_t = \phi_0 + \phi_1 Y_{t-1} + \varepsilon_t$ with $\{\varepsilon_t\}$ an i.i.d. sequence of random variables. Then, under the simplifying assumption that $\{\varepsilon_t\} \stackrel{\text{i.i.d.}}{\sim} \mathcal{N}(0, \sigma_\varepsilon^2)$, it can be proved that the optimal resampling width $h \equiv h(\mathbf{y})$ is given by

$$h(\mathbf{y}) = \left[\frac{\sigma_\varepsilon^4 W_1}{T f_Y(\mathbf{y}) \{2\sigma_\varepsilon^2 C_1^2(\mathbf{y}) + 0.25 C_2^2\}} \right]^{1/5}, \quad (\mathbf{y} \in \mathbb{R}), \qquad (8.24)$$

where, with a Gaussian kernel, $K_1 = 1/(2\sqrt{\pi})$, $C_1(\mathbf{y}) = \phi_1 \sigma_Y^{-2}(\mathbf{y} - \mu_Y)$ and $C_2 = \phi_1^2$. A sample version of $h(\mathbf{y})$ can be easily obtained by fitting an AR(1) model to the data, and replacing the unknown quantities in (8.24) by their sample estimates.

8.4.2 A kernel-based test

The above TR test statistics are all devised in a two-dimensional state space by considering only distributions, or higher-order moments, of pairs $(Y_t, Y_{t-\ell})$. Using the delay vector $\mathbf{Y}_t^{(\ell)} = (Y_t, Y_{t-\ell}, \ldots, Y_{t-(m-1)\ell})'$ ($m \in \mathbb{Z}^+, \ell \in \mathbb{Z}$), TR can also be formulated in a state space framework via the joint density function $f_m(\mathbf{y})$ of $\{\mathbf{Y}_t^{(\ell)}, t \in \mathbb{Z}\}$, i.e., the process is invariant under time reversal for all m and ℓ if and only if,

$$f_m(\mathbf{P}\mathbf{y}) = f_m(\mathbf{y}), \quad \forall \mathbf{y} \in \mathbb{R}^m, \tag{8.25}$$

where $\mathbf{P}$ denotes an $m \times m$ matrix operator with elements $P_{ij} = \delta_{i,m+1-j}$, and $\delta_{i,j}$ is Kronecker's delta. Note that this characterization of TR is related to the classical two-sample problem of testing the equivalence of two multi-dimensional distributions for independent samples. This equivalence suggests a test statistic based on the distance between $f_m(\mathbf{y})$ and $f_m(\mathbf{P}\mathbf{y})$. Diks et al. (1995) develop such a test using a quadratic measure of dependence.

Assume that the delay vectors $\{\mathbf{Y}_t^{(\ell)}\}_{t=1}^N$, with finite variance, are sampled independently according to $f_m(\mathbf{y})$, with $N = T - (m-1)\ell$. Let $f_m^*(\mathbf{y})$ be a smoothed pdf defined as the convolution of $f_m(\mathbf{y})$ with a multivariate Gaussian kernel $\mathcal{K}_h(\cdot)$, i.e.,

$$f_m^*(\mathbf{y}) = \int_{\mathbb{R}^m} \mathcal{K}_h(\mathbf{y} - \boldsymbol{\xi}) f_m(\boldsymbol{\xi}) d\boldsymbol{\xi}, \tag{8.26}$$

where

$$\mathcal{K}_h(\mathbf{x}) = (\sqrt{2\pi}h)^{-m} \exp\{-\|\mathbf{x}\|^2/2h^2\},$$

with $h > 0$ the bandwidth, and $\|\cdot\|$ the Euclidean norm. The convolution process has the symmetry-preserved property that $f_m^*(\mathbf{y}) = f_m^*(\mathbf{P}\mathbf{y}) \ \forall \mathbf{y} \in \mathbb{R}^m$ under the null hypothesis $\mathbb{H}_0$: $f_m(\mathbf{y}) = f_m(\mathbf{P}\mathbf{y})$. Then a quadratic measure to evaluate the difference between the smoothed densities is defined as

$$\begin{aligned} Q_h(m) &= \frac{1}{2}(2h\sqrt{\pi})^m \int_{\mathbb{R}^m} \left(f_m^*(\mathbf{y}) - f_m^*(\mathbf{P}\mathbf{y})\right)^2 d\mathbf{y} \\ &= (2h\sqrt{\pi})^m \int_{\mathbb{R}^m} \left(f_m^*(\mathbf{y})f_m^*(\mathbf{y}) - f_m^*(\mathbf{y})f_m^*(\mathbf{P}\mathbf{y})\right)d\mathbf{y}, \end{aligned} \tag{8.27}$$

which is always positive-semidefinite and equals zero if and only if $f_m^*(\mathbf{y}) = f_m^*(\mathbf{P}\mathbf{y})$.

Substituting (8.26) in (8.27), using integration by parts and a change of variables, gives the expression

$$Q_h(m) = \int_{\mathbb{R}^m} f_m(\mathbf{r}) \int_{\mathbb{R}^m} \Big(\exp\{-\|\mathbf{r} - \mathbf{s}\|^2/(4h^2)\} \\ - \exp\{-\|\mathbf{r} - \mathbf{P}\mathbf{s}\|^2/(4h^2)\} \Big) f_m(\mathbf{s}) d\mathbf{s} d\mathbf{r}. \tag{8.28}$$

Replacing the integrals by an average of contributions from different pairs of m-dimensional delay vectors $\{\mathbf{Y}_i\}$ and $\{\mathbf{Y}_j\}$ ($i \neq j$) results in the following, unbiased, estimator $\widehat{Q}$ (a U-statistic)[3] of Q:

$$\widehat{Q}_{h,T}(m) = \binom{N}{2}^{-1} \sum_{i<j} w_{ij}, \qquad (8.29)$$

where

$$w_{ij} = \exp\{-\|\mathbf{y}_i - \mathbf{y}_j\|^2/(4h^2)\} - \exp\{-\|\mathbf{y}_i - \mathbf{P}\mathbf{y}_j\|^2/(4h^2)\}. \qquad (8.30)$$

Under $\mathbb{H}_0$, the expected value of $\widehat{Q}_{h,T}(m)$ is zero and its variance is given by

$$\mathrm{Var}(\widehat{Q}_{h,T}(m)) = \binom{N}{2}^{-2} \sum_{i<j} w_{ij}^2.$$

Therefore, the test statistic is defined as follows

$$S_{h,T}(m) = \widehat{Q}_{h,T}(m) \big/ \sqrt{\mathrm{Var}(\widehat{Q}_{h,T}(m))}, \qquad (8.31)$$

which, approximately, has a mean zero and a standard deviation one, if the m-dimensional processes $\{\mathbf{Y}_i, i \in \mathbb{Z}\}$ and $\{\mathbf{Y}_j, j \in \mathbb{Z}\}$ are independent.

In applications of the test statistic $S_{h,T}(m)$, an important question is how to select the bandwidth h. In kernel-based estimation it is well known that selecting h too small leads to a higher variance of the kernel estimator, called *undersmoothing*. On the other hand, choosing a bandwidth that is too large increases the bias (*oversmoothing*) of the estimator. In practice, both factors are often balanced via CV.

Another issue concerns the dependence among delay vectors. Diks et al. (1995) suppress this effect by dividing the (i,j) plane of indices into squares of size $\tau \times \tau$, with τ some fixed number larger than the typical time scale, and next replacing w_{ij} by $w'_{i',j'} = \tau^{-2} \sum_{p=1}^{\tau} \sum_{q=1}^{\tau} w_{i'\tau+p, j'\tau+q}$. This method is supposed to provide more reliable estimates of the standard deviation of $S_{h,T}(m)$. Clearly, the influence of the parameter τ on the performance of this test statistic is comparable to the bandwidth influence. Moreover, since the parameters τ and h are bound together, the selection of their optimal values should be carried out simultaneously, for instance by using CV.

8.4.3 A sign test

The projection of the m-dimensional delay vectors on each bi-dimensional plane $(Y_t, Y_{t-\ell})$ ($\ell = 1, \ldots, m-1$) can be readily evaluated by exploiting the fact that for

---

[3] Strictly speaking this U-statistic is unbiased for a finite sample size only if the $\{\mathbf{Y}_i, i \in \mathbb{Z}\}$ are independent.

8.4 OTHER NONPARAMETRIC TESTS

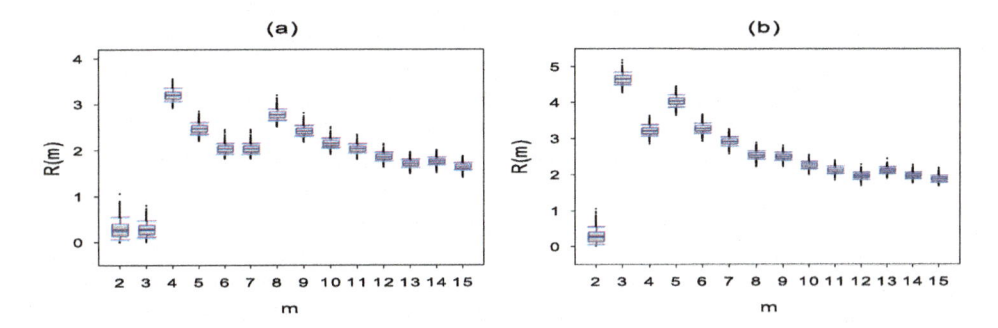

Figure 8.3: *Boxplots of* $\mathcal{R}(m)$ *based on* $1{,}000$ *MC replications of series of length* $T = 5{,}000$ *generated from the time-delayed* Hénon *map with dynamic noise process* (8.35), *and with* (a) $\ell = 1$ *and* (b) $\ell = 2$.

a strictly stationary and TR stochastic process $\{X_t(\ell) = Y_t - Y_{t-\ell}, t \in \mathbb{Z}\}$, we have

$$\mathbb{P}(X_0(\ell) > 0) = \mathbb{P}(X_0(\ell) < 0) = \frac{1}{2}, \quad (\ell = 1, \ldots, m-1).$$

The object of interest is thus the probability $\pi(\ell) \equiv \mathbb{P}(X_0(\ell) > 0)$, which may be thought of as a simple measure of deviation from zero of the one-dimensional distribution of $\{X_t(\ell), t \in \mathbb{Z}\}$. A natural point estimator of $\pi(\ell)$ is

$$\widehat{\pi}(\ell) = \frac{1}{T - \ell} \sum_{t=\ell+1}^{T} I(X_t(\ell) > 0), \quad (\ell = 1, \ldots, m-1). \tag{8.32}$$

Psaradakis (2008) proves that, for each fixed $\ell \in \mathbb{N}$, as $T \to \infty$,

$$\sqrt{T - \ell}\,(\widehat{\pi}(\ell) - \pi(\ell)) \xrightarrow{D} \mathcal{N}(0, \sigma_X^2(\ell)), \tag{8.33}$$

where

$$\sigma_X^2(\ell) = \pi(\ell)(1 - \pi(\ell)) + 2\pi(\ell) \sum_{t=1}^{\infty} \{\mathbb{P}(X_t(\ell) > 0) | X_0(\ell) > 0) - \pi(\ell)\}. \tag{8.34}$$

The circular block bootstrap procedure of Politis and Romano (1992) for stationary processes may be used to obtain an estimate of (8.34). A practical difficulty with this approach is the choice of the block length. Another possibility is to approximate the sampling distribution of (8.32) by subsampling, which requires the selection of a subsample size. Below we present an example of the TR test statistic $\widehat{\pi}(\ell)$ applied to data generated by a nonlinear high-dimensional stochastic process.

Example 8.3: Exploring a Time-delayed Hénon Map for TR

Consider the stochastic process

$$Y_t = 1 - 1.4 Y_{t-\ell}^2 + 0.3 Y_{t-2\ell-1} + \varepsilon_t, \quad \{\varepsilon_t\} \stackrel{\text{i.i.d.}}{\sim} U(-0.01, 0.01). \tag{8.35}$$

Table 8.1: *P-values of six TR test statistics. Blue-typed numbers indicate rejection of the null hypothesis of TR at the 5% nominal significance level.*

	Time domain		Frequency		Nonparametric [4]			
	$\max_{\ell=1,\ldots,10} |TR(\ell)|$		domain [3]			$S_{h,T}(m)$		
Series	Type I & II [1]	Type II [2]	S_{TR}	S^*_{TR}	θ_T	$m=2$ $m=3$ $m=4$		$m=5$
Unemployment rate[5]	0.000	0.010	0.000 0.000		0.338	0.164 0.123 0.239		0.190
EEG recordings	0.004	0.000	0.133 0.000		1.000	0.639 0.085 0.022		0.008
Magnetic field data	0.000	0.004	0.000 0.000		0.010	0.445 0.203 0.176		0.120
ENSO phenomenon	0.026	0.010	0.000 0.000		0.713	0.217 0.193 0.401		0.639
Climate change: δ^{13}C	0.516	0.815	0.739 0.000		0.780	0.806 0.977 0.999		0.999
δ^{18}O	0.002	0.016	0.095 0.000		0.828	0.086 0.130 0.405		0.483

[1] Based on 1,000 MC estimated standard errors, and 1,000 MC simulations to estimate the *p*-value.
[2] Test results are based on i.i.d. standard errors using (8.5), and 1,000 MC simulations to estimate the *p*-value.
[3] $M = 25$ (see Chapter 4) for all series and both test statistics; no prewhitening.
[4] *p*-values of θ_T are based on 400 bootstrap replicates, using the resampling scheme of Section 8.4.1. *p*-values of $S_{h,T}(m)$ are based on 1,000 MC simulations with $h = 0.5$, and $\tau = 20$.
[5] First differences of the original data.

This is a "clothed", or randomized, version of the time-delayed deterministic (*its skeleton*) Hénon map. Time series generated by the Hénon map are known to be irreversible. We generated 1,000 replications of (8.35) for series of length $T = 5,000$. Subsequently, with $m = 2, \ldots, 15$, we computed the measure

$$\mathcal{R}(m) = \frac{1}{m-1} \sum_{\ell=1}^{m-1} |0.5 - \widehat{\pi}(\ell)| \times 100, \quad (8.36)$$

where $\widehat{\pi}(\ell)$ is given by (8.32).

Figures 8.3(a) and (b) show boxplots at lags $\ell = 1$ and 2, respectively, of 1,000 $\mathcal{R}(m)$ values. In the case $\ell = 1$, the median values of $\mathcal{R}(2)$ and $\mathcal{R}(3)$ are approximately equal to zero, and hence irreversibility is not detected. In contrast, all median values of $\mathcal{R}(m)$ ($m > 3$) depart from zero significantly, indicating that the DGP (8.35) is actually time-irreversible. A similar picture emerges from Figure 8.3(b). Thus, TR cannot be consistently tested by considering only distributions of pairs $(Y_t, Y_{t-\ell})$.

8.5 Application: A Comparison of TR Tests

Table 8.1 presents *p*-values of six TR test statistics. Columns 2 – 3 provide evidence of time-irreversibility, using the Ramsey–Rothman statistic $\max_{\ell=1,\ldots,10} |TR(\ell)|$. The AR order selection was done using BIC with $p_{\max} = 10$. The only series that fails to display evidence of both Type I and Type II time-irreversibility is the climate

8.5 APPLICATION: A COMPARISON OF TR TESTS

Table 8.2: *Results of* TR *test statistic* $\mathcal{C}_g(\ell)$, *as defined by* (8.12), *for lags* $\ell = 1, \ldots, 10$.[1] *Blue-typed numbers indicate rejection of the null hypothesis of* TR *at the 5% nominal significance level.*

	\multicolumn{10}{c}{Time lag ℓ}
Series	1	2	3	4	5	6	7	8	9	10
Unemployment rate[2]	1.512	2.122	2.183	1.684	1.605	0.809	0.407	0.226	0.622	0.489
EEG recordings	-0.257	-0.241	-0.285	-0.224	-0.222	-0.173	-0.104	-0.019	0.081	0.116
Magnetic field data	-0.610	-0.479	-0.541	0.040	-0.286	-0.397	-0.334	-0.081	0.757	0.549
ENSO phenomenon	1.258	1.182	1.282	1.195	1.141	1.209	1.321	1.378	1.362	1.287
Climate change: δ^{13}C	0.571	-0.299	-0.122	0.370	-0.016	-0.469	-0.384	-0.730	-0.622	-0.342
δ^{18}O	-0.548	-1.288	-1.660	-1.620	-1.320	-1.156	-1.104	-0.971	-0.574	-0.593

[1] Based on the exponential density function $g(\omega) = (1/\beta)\exp(-\omega/\beta)$ ($\omega > 0$) with β set at the reciprocal of the sample standard deviation of each series.
[2] First differences of original series.

change δ^{13}C time series. For the remaining five series, TR is rejected at the 5% nominal significance level. The p-values of the frequency-domain test statistic S_{TR} (column 4) differ considerably from those of S^*_{TR} (column 5). For all time series TR is strongly rejected on the basis of S^*_{TR}, while with S_{TR}, evidence of time-irreversibility is restricted to three series. Thus, the p-values of S^*_{TR} rule out linear models with Gaussian distributions for all series. Note, however, that these test results can be sensitive to the choice of M; see also the discussion in Section 4.4.4.

Except for the magnetic field data, the copula-based test statistic θ_T (column 6) does not reveal evidence of time-irreversibility, at the 5% nominal significance level. This may be due to the first-order Markov chain assumption used in the construction of the test statistic; that is higher-order Markov chains may well provide a better representation of the DGP underlying the time series, and consequently may change the outcome of the test statistic.

The p-values of $S_{h,T}(m)$ differ considerably across the values of m. For $m = 2$ and 3 all p-values do not reject TR at the 5% nominal significance level. For $m = 4$ and 5, we see that there is evidence of time-irreversibility in the EEG recordings. Thus, it seems worthwhile not to rely completely on low-dimensional test results.

Table 8.2 presents test results of $\mathcal{C}_g(\ell)$ for $\ell = 1, \ldots, 10$. Only in one case the test statistic rejects the TR null hypothesis, i.e. the U.S. unemployment series at lags $\ell = 2$ and 3. In all other cases, the null hypothesis is not rejected at the 5% nominal significance level. Characterization of the U.S. unemployment series as time-irreversible through the various TR test statistics suggest asymmetric behavior consistent with the steepness asymmetry business cycle hypothesis, elaborated upon in the introductory paragraph of this chapter. Also time-irreversibility of the EEG recordings, as we observed in Table 8.1, is an indicator of nonlinear dynamics.

8.6 Summary, Terms and Concepts

Summary

Gaussianity and TR suggest a linear model for the data under study. These are two fundamental properties of DGPs which must be checked before adopting a nonlinear model. A large number of potential approaches to testing for TR have been proposed in the literature. In this chapter, we provided a brief overview of some of the major developments in this area. Broadly, the TR test statistics were divided into three categories. The first of these is those based on higher-order cumulants and characteristic functions in the time domain, having close relationships with general, non-temporal, tests of symmetry. In the second category we included test statistics based on the symmetry property of cumulants in the frequency domain. These latter tests are computationally more demanding than time-domain TR tests, and are applicable only if high-order moments exist. In addition, we focused on nonparametric TR test statistics which have been designed to avoid specific assumptions about the underlying marginal distribution of the DGP under the null hypothesis of TR. Finally, we provided empirical evidence comparing the performance of various TR test statistics.

In closing this chapter, we should mention that practically all existing test statistics are only able to detect specific forms of TR. Moreover, many test procedures regard time-irreversibility as a "complementary test hypothesis". Few papers, consider the notion of TR in its own right, and try to characterize the nature of TR when it is present. One notable exception is McCausland (2007) who proposes an index for certain types of TR, applicable to finite regular stationary Markov chains. Another exception is Beare and Seo (2014) who use a so-called *circulation density function* to measure the degree of temporal irreversibility in a stationary Markov chain.

Terms and Concepts

Anosov diffeomorphism, 335
BGAR(1) process, 335
Beta-Gamma transformation, 335
commutative, 317
copula functions, 316
directionality, 333
detailed balance equations, 317
exchangeability, 317
local bootstrap, 325

oversmoothing, 328
resampling width, 326
squared tricoherence, 323
symmetric-bicovariance function, 318
time-irreversible, 315
trispectrum, 323
Type I and II time-irreversibility, 318
undersmoothing, 328

8.7 Additional Bibliographical Notes

The literature of TR is quite large and dates back to the mid–1930s, starting with Hostinsky and Potocek (1935) and Kolmogorov (1936). As Dobrushin et al. (1988) note, the founder of the theory of temporal reversibility for Markov processes is considered to be Kolmogorov. Reversibility, or *directionality*, appears to be mentioned first by Daniels (1946) in the context of analyzing time series processes. Lawrance (1991) reviews the state of the theoretical research up to 1990s. Breidt and Davis (1992) and Cheng (1992, 1999) study TR and related problems in the context of general linear processes. Tong and Zhang (2005) and Chan et al. (2006) derive conditions of TR of multivariate non-Gaussian linear processes.

Hoover (1999) describes TR from the perspective of computer simulation with many examples and concepts taken from dynamical-systems theory. Also, time-irreversibility has gained a lot of attention in the analysis of human heart rate variability (beat-to-beat time series); see, e.g., Casali et al. (2008) and Hou et al. (2011).

Rothman (1992) compares the power of the Ramsey–Rothman TR test statistic with the power of the BDS and Hinich's bispectrum test against some simple SETAR alternatives. In a similar vein, the study by Belaire-Franch and Contreras (2003) compares the Ramsey–Rothman TR test statistic and the Chen et al. (2000) TR test statistics for time series generated by BL, SETAR, and GARCH models. Fong (2003) applies the Chen et al. (2000) TR test statistic to daily stock closing prices and trading volume of the 30 component series representing the Dow Jones Industrial Index. Giannakis and Tsatsanis (1994) propose a time-domain analogue of the trispectrum-based TR test statistic of Section 8.3. Their simulation study includes comparisons with the TR test statistic of Algorithm 8.3, and application to real seismic data.

In addition to the test statistics reviewed in this chapter, several alternative test statistics of TR have been put forward in the literature. Both Robinson (1991) and Racine and Maasoumi (2007) introduce entropy-based test statistics which can be used for testing TR; see, e.g., Exercise 8.6. The asymptotic distribution associated with these test statistics, however, imposes strong regularity conditions on the DGP. Darolles et al. (2004) propose a test statistic based on nonlinear canonical correlation analysis. Their approach comes down to testing whether a given pair of canonical directions are equal to one another. Sharifdoost et al. (2009) design a test statistic of TR applicable to finite state Markov chains. Kessler and Sørensen (2005) study the case when martingale estimating functions and other unbiased estimating functions have the same structure as the score function for a TR Markov process.

Symbolization converts continuous-valued time series observations into a stream of discrete symbols. Using this concept, Daw et al. (2000) propose a specific method for TR without the need for generating surrogate data. Steuber et al. (2012) introduce two Markov chain-based time reversibility tests. The test statistics are based on observed deviations of transition sample counts between each pair of states in a sequence sampled from a stationary time-homogeneous Markov chain.

8.8 Software References

Section 8.2: Philip Rothman contributed FORTRAN77 code to calculate the first and second stage of the Ramsey–Rothman TR test statistic, which can be found at the website of this book; see Rothman (1996) for documentation. A GAUSS program for running the Chen–Chou–Kuan TR test statistic $C_g(\ell)$ was kindly made available by Yi-Ting Chen.

Section 8.3: The Hinich–Rothman bispectrum-based test and the trispectrum-based test can be computed using the BISPEC and TRISPEC programs, respectively, both coded in FORTRAN77 by the late Melvin J. Hinich; see http://www.la.utexas.edu/hinich/.

Section 8.4: Brendan Beare and Juwon Seo have made available MATLAB code for computing the copula-based TR test statistic for Markov chains. The C++ source code and a Linux/Windows executable of the kernel-based TR test statistic $S_{h,T}(m)$ (Section 8.4.2) can be downloaded from Cees Diks' web page, located at `http://cendef.uva.nl/people`.

Exercises

Theory Questions

8.1 Let $\{Y_t, t \in \mathbb{Z}\}$ be a strictly stationary i.i.d. process with mean zero, $\mu_{3,Y} = \mathbb{E}(Y_t^3) = 0$, and finite moments $\mu_{2,Y} = \mathbb{E}(Y_t^2)$ and $\mu_{4,Y} = \mathbb{E}(Y_t^4)$. Verify (8.5).

8.2 Suppose that $\{f(t), t \in \mathbb{Z}\}$ is a strictly stationary time series process with mean zero, defined on the interval $[T_1, T_2]$. The bicovariance function of $f(t)$ can be approximated by

$$\gamma^{(i,j)}(\ell) = \frac{1}{(T_2 - \ell) - T_1} \int_{T_1}^{T_2 - \ell} f^i(t) f^j(t+\ell) \mathrm{d}t, \quad (i \neq j; \ell \in \mathbb{Z}).$$

Show that the bicovariance function $\gamma_{\mathrm{TR}}^{(i,j)}(\ell)$ of the time-reversed stochastic function is not necessarily equal to $\gamma^{(i,j)}(\ell)$, except when $f(t)$ obeys time reversal, i.e. $f_{\mathrm{TR}}(t) = f(-t) = f(t+\xi)$, where ξ is an adjustable parameter that fixes the origin of the time axis.

8.3 Consider the strictly stationary, zero-mean, stochastic process $\{X_t(\ell) \equiv Y_t - Y_{t-\ell}, t \in \mathbb{Z}, \ell \in \mathbb{N}\}$. Let $\rho_Y^{(2,1)}(\ell) = \mathbb{E}(Y_t^2 Y_{t-\ell})/\mathbb{E}(Y_t^2)^{3/2}$, and $\rho_Y^{(1,1)}(\ell) = \mathbb{E}(Y_t Y_{t-\ell})/\mathbb{E}(Y_t^2)$.

(a) Show the standardized third-order cumulant of $\{X_t, t \in \mathbb{Z}\}$ can be expressed as

$$\frac{\mathbb{E}(X_t^3)}{\mathbb{E}(X_t^2)^{3/2}} = \frac{3}{2\sqrt{2}} \frac{\rho_Y^{(2,1)}(\ell) - \rho_Y^{(2,1)}(-\ell)}{\{1 - \rho_Y^{(1,1)}(\ell)\}^{3/2}}.$$

(b) Assume that the functions $\rho_Y^{(2,1)}(\ell)$ and $\rho_Y^{(1,1)}(\ell)$ are differentiable on $[0, \infty)$. Show the above expression is approximately given by

$$\frac{\mathbb{E}(X_t^3)}{\mathbb{E}(X_t^2)^{3/2}} \approx -\frac{3}{\sqrt{2}} \frac{\rho'_{21}(0)}{\{-\rho'_{11}(0)\}^{3/2} \ell^{1/2}},$$

where $\rho'_{21}(0)$ and $\rho'_{11}(0)$ denote the first non-zero derivatives of $\rho_Y^{(2,1)}(\ell)$ and $\rho_Y^{(1,1)}(\ell)$ at the origin, respectively.

(c) Using part (b), argue that as $\ell \downarrow 0$ time-irreversibility is most apparent for small values of ℓ.

(Cox, 1991)

8.4 The Gamma distribution is often used to model a wide variety of positive valued time series variables. Applications include fields such as hydrology (river flows), meteorology (rainfall, wind velocities), and finance (intraday durations between trades).

Within this context, Lewis et al. (1989) introduce the simple first-order *Beta-Gamma autoregressive* (BGAR(1)) process

$$Y_t = B_t Y_{t-1} + G_t, \quad (t \in \mathbb{Z}),$$

where $\{B_t\}$ and $\{G_t\}$ are mutually independent sequences of i.i.d. random variables with Beta$(k\rho, k(1-\rho))$ and Gamma$(k(1-\rho), \beta)$ distributions, respectively, with shape parameter $k > 0$, rate parameter $\beta > 0$, and ρ ($0 \leq \rho < 1$) describes the dependency structure of the process. It is easily established, using moments of Beta variables, that $\rho(\ell) = \rho^{|\ell|}$ ($\ell \in \mathbb{Z}$).

(a) Let Y and B be independent Gamma(k, β) and Beta$(k\rho, k(1-\rho))$ random variables respectively. Then it can be shown that BY and $(1-B)Y$ are independent Gamma$(k\rho, \beta)$ and Gamma$(k(1-\rho), \beta)$ variables. Using this result, prove that the Laplace–Stieltjes transform of the random variable $(v + Bu)X$ ($v \geq 0, u \geq 0$) is given by

$$\mathbb{E}\left(e^{-(v+Bu)X}\right) = \left(\frac{\beta}{\beta+v}\right)^{k(1-\rho)} \left(\frac{\beta}{\beta+v+u}\right)^{k\rho}.$$

When $v = 0$, this result is known as the *Beta–Gamma transformation*.

(b) For the stationary BGAR(1) process $\{Y_t, t \in \mathbb{Z}\}$, let $\mathcal{L}_{Y_t, Y_{t-1}}(u, v)$ denote the joint Laplace–Stieltjes transform of (Y_t, Y_{t-1}). Then, using part (a), show that

$$\mathcal{L}_{Y_t, Y_{t-1}}(u, v) = \left(\frac{\beta}{\beta+u} \times \frac{\beta}{\beta+v}\right)^{k(1-\rho)} \left(\frac{\beta}{\beta+v+u}\right)^{k\rho}.$$

(c) Given the result in part (b), state your conclusion about the TR of the BGAR(1) process.

8.5 Consider the stationary stochastic process $\{Y_t, t \in \mathbb{Z}\}$

$$Y_t = (Y_{t-1} + Y_{t-2} + \varepsilon_t) \pmod{1},$$

where $\{\varepsilon_t\}$ is a sequence of i.i.d. random variables with a continuous marginal distribution. The process $\{Y_t, t \in \mathbb{Z}\}$ may be viewed as a stochastic version of the so-called *Anosov diffeomorphism* on a two-dimensional torus, i.e.

$$\begin{pmatrix} y_{i+1} \\ x_{i+1} \end{pmatrix} = \begin{pmatrix} 1 & 1 \\ 1 & 0 \end{pmatrix} \begin{pmatrix} y_i \\ x_i \end{pmatrix} \pmod{1},$$

which is a chaotic nonlinear deterministic system.

Let $f_m(y_1, \ldots, y_m)$ be the joint pdf of $\mathbf{Y}_t = (Y_t, Y_{t-1}, \ldots, Y_{t-m+1})'$ ($m \in \mathbb{N}^+$). The following statements are claimed.

(a) $\{Y_t, t \in \mathbb{Z}\}$ has a unique invariant joint probability measure.

(b) The process is time-irreversible, as the joint distribution of the process $\{\mathbf{Y}_t, t \in \mathbb{Z}\}$ for dimension $m > 2$ is not symmetric with respect to reversing the time order of the variables. So, (8.25) does not hold for $m \geq 3$.

(c) The joint distribution of each of the pairs $(Y_{t-\ell}, Y_t)$ ($\ell \geq 1$) *is* symmetric with respect to the matrix operator $\mathbf{P}$, defined as $\mathbf{P}(y_1, y_2) = (y_2, y_1)$.

Sketch a proof of each of the above statements.

(Based on private communication with C. Diks)

Empirical and Simulation Question

8.6. Let $\{Y_t, t \in \mathbb{Z}\}$ be a strictly stationary time series process with marginal density function $f(y)$ and joint pdf $f_\ell(x, y)$ of $(Y_t, Y_{t-\ell})'$ ($\ell \in \mathbb{Z}$). Granger et al. (2004) consider a normalization of the Hellinger distance of dependence (Section 7.2.3) given by $S(\ell) = (1/2) \int_{-\infty}^{\infty} \int_{-\infty}^{\infty} \{f_\ell^{1/2}(x, y) - (f(x) f(y))^{1/2}\}^2 \mathrm{d}x \mathrm{d}y$.[4] Replacing the unknown densities in $S(\ell)$ with kernel-based estimators yields the test statistic $\widehat{S}(\ell)$; see the function npunitest in the R-np package.

(a) Investigate the six time series in Table 8.1 for the presence of TR using $\widehat{S}(\ell)$, i.e. test the null hypothesis $\mathbb{H}_0^{(0)} : f(y) = f(-y) \; \forall y$. To reduce the computational burden, set the number of BS replicates at 99.

(b) Repeat part (a), but now test the null hypothesis $\mathbb{H}_0^{(1)} : f(Y_t, Y_{t-1}) = f(Y_{t-1}, Y_t)$. Are there any marked difference between the test results in parts (a) and (b)?

---

[4] Also known as the Bhattacharyya–Matusita–Hellinger measure of dependence; see Bhattacharyya (1943), and Matusita (1955).

Chapter 9

SEMI- AND NONPARAMETRIC FORECASTING

The time series methods we have discussed so far can be loosely classified as parametric (see, e.g., Chapter 5), and semi- and nonparametric (see, e.g., Chapter 7). For the parametric methods, usually a quite flexible but well-structured family of finite-dimensional models are considered (Chapter 2), and the modeling process typically consists of three iterative steps: identification, estimation, and diagnostic checking. Often these steps are complemented with an additional task: out-of-sample forecasting. Within this setting, specification of the functional form of a parametric time series model generally arrives from theory or from previous analysis of the underlying DGP; in both cases a great deal of knowledge must be incorporated in the modeling process. Semi- and nonparametric methods, on the other hand, are infinite-dimensional. These methods assume very little a priori information and instead base statistical inference mainly on data. Moreover, they require "weak" (qualitative) assumptions, such as smoothness of the functional form, rather than quantitative assumptions on the global form of the model.

For all these reasons, a practitioner is often steered into the realm of semi- and nonparametric function estimation or "smoothing". However, the price to be paid is that parametric estimates typically converge at a root-n rate, while nonparametric estimates usually converge at a slower rate. Also, semi- and nonparametric methods acknowledge that fitted models are inherently misspecified, which implies specification bias. Increasing the complexity of a fitted model typically decreases the absolute value of this bias, but increases the estimation variance: a feature known as the *bias-variance trade-off*. The bandwidth or tuning parameter controls this trade-off, i.e. its choice is often critical to implementation and practical consideration.

In this chapter, we deal with various aspects of semi- and nonparametric models/methods with a strong focus on forecasting. The desire for forecasting future time series values, along with frequent misuse of methods based on linear or Gaussian assumptions, motivates this area of interest. Based on results in Appendix 7.A, the first half of this chapter is concerned with kernel-based methods for estimating the conditional mean, median, mode, variance, and the complete conditional density of a time series process. We examine and compare the use of single-stage versus multi-stage quantile prediction. Further, we describe kernel-based methods for jointly estimating the conditional mean and the conditional variance. This part also includes methods for estimating multi-step density forecasts using bootstrapping, and methods for nonparametric lag selection.

The second half of the chapter deals with semiparametric models/methods. It is well known that conventional nonparametric estimators can suffer poor accuracy for data of dimension two and higher. In fact, the number of observations needed to attain a fixed level of estimate confidence grows exponentially with the number of dimensions. This problem is called the *curse of dimensionality* and presents a dilemma for the effective and practical use of nonparametric forecast methods. One way to circumvent this "curse" is to use additive models. These models make the assumption that the underlying regression function may have a simpler, additive structure, comprising of several lower-dimensional functions. As such, they fall in the class of semiparametric models/methods, combining parametric and nonparametric features. In Section 9.2, we discuss several additive (semiparametric) models for time series prediction with emphasis on conditional mean and conditional quantile forecasts. Then, in Sections 9.2.5 and 9.2.6, we introduce two restricted, and closely related, forms of a semiparametric AR model.

9.1 Kernel-based Nonparametric Methods

9.1.1 Conditional mean, median, and mode

Preliminaries

In what follows, we are going to discuss kernel-based predictors for a strictly stationary time series process $\{Y_t, t \in \mathbb{Z}\}$ which is assumed to be a Markovian process of order p.[1] Let $\{Y_t\}_{t=1}^T$ be a sequence of observations on the process $\{Y_t, t \in \mathbb{Z}\}$. Our objective is to predict the unobserved real random variable Y_{T+H} where H ($1 \leq H \leq T - p$) denotes the forecast horizon. For this purpose, we construct the associated process $\{(\mathbf{X}_{p,t}, Z_{p,t}), t \in \mathbb{Z}\}$ denoted as $\{(\mathbf{X}_t, Z_t)\} \in \mathbb{R}^p \times \mathbb{R}$ where

$$\mathbf{X}_t = (Y_t, Y_{t+1}, \ldots, Y_{t+p-1})', \ Z_t = Y_{t+H+p-1}, \ (t = 1, \ldots, n; n = T - H - p + 1). \tag{9.1}$$

Let $\{(\mathbf{X}_t, Z_t), t \in \mathbb{Z}\}$ be a sequence of random variable with common probability density function with respect to the Lebesgue measure on $\mathbb{R}^{p+1}$. Now the problem of predicting Y_{T+H}, or equivalently Z_{T-p+1}, consists of finding the closest (with respect to a certain norm) random variable knowing all the past. Suppose that there exists a function $\mu(\cdot)$ modeling the relationship between the response Z_t and

---

[1] Bosq (1998, Section 3.4.2) notes that kernel-based prediction methods can still be used if there is a simple form of nonstationarity in the data. For instance in case the data exhibit a slowly varying trend and/or there is a periodic function with a known period (seasonal component).

9.1 KERNEL-BASED NONPARAMETRIC METHODS

the covariate $\mathbf{X}_t$, and that $\mu(\cdot)$ is defined through the conditional distribution. Given a loss function $L(\cdot)$ with a unique minimum, define $\mu(\cdot)$ such that it minimizes the conditional mean $\mathbb{E}(L(Z_t - a)|\mathbf{X}_t = \mathbf{x})$ with respect to a, i.e.

$$\mu(a) = \arg\min_{a \in \mathbb{R}} \mathbb{E}(L(Z_t - a)|\mathbf{X}_t = \mathbf{x}). \tag{9.2}$$

Then estimating nonparametrically $\mu(\cdot)$ by $\widehat{\mu}(\cdot)$ and calculating $\widehat{\mu}(\mathbf{X}_{T-p+1})$ gives $\widehat{Z}_{T-p+1}$. In this way, we obtain the H-step ahead forecast value $\widehat{Y}_{T+H|T}$ as an estimator of $Y_{T+H|T} = \mathbb{E}(Y_{T+H}|\mathbf{X}_T)$.

Using the above principle, we define three predictors, i.e. the conditional mean, the conditional median, and the conditional mode, each depending on a particular form of the function $L(\cdot)$. These predictors will be expressed as a sum of products between functions of $\{Y_t\}$ and weights $W_t(\mathbf{x})$, depending on the values of $\mathbf{X}_t$, i.e. the weights are defined as

$$W_t(\mathbf{x}) = K\Big(\frac{\mathbf{x} - \mathbf{X}_t}{h_n}\Big) \Big/ \sum_{t=1}^{n} K\Big(\frac{\mathbf{x} - \mathbf{X}_t}{h_n}\Big), \quad (n = T - H - p + 1). \tag{9.3}$$

In practice, $K(\cdot)$ is often assumed to be a product kernel. For ease of readability, we denote the bandwidth by h without explicitly indicating its dependence on n.

It is well known that $L(u) = u^2$ leads to the conditional mean function $\mu(\mathbf{x}) = \mathbb{E}(Z_t|\mathbf{X_t} = \mathbf{x})$. Using the NW kernel density approach (see, e.g., Chapter 7, expression (A.12)), an estimator of $\mu(\mathbf{x})$ can be constructed as

$$\widehat{\mu}^{\text{NW}}(\mathbf{x}) = \sum_{t=1}^{n} Z_t W_t(\mathbf{x}). \tag{9.4}$$

Hence, given $\{Y_t, t \leq T\}$, the H-step ahead nonparametric estimator of the conditional mean is defined as

$$\widehat{Y}_{T+H|T}^{\text{Mean}} = \sum_{t=1}^{n} Z_t W_t(\mathbf{X}_{T-p+1}). \tag{9.5}$$

Under certain mixing conditions of the process $\{(\mathbf{X}_t, Z_t), t \in \mathbb{Z}\}$, Collomb (1984) shows uniform convergence of $\widehat{Y}_{T+H|T}^{\text{Mean}}$.

Conditional median

When the conditional distribution of Z_t given $\mathbf{X}_t$ is heavy-tailed or asymmetric, it may be sensible to use the conditional median rather than the conditional mean to generate future values, as the median is highly resistant against outliers. In this case the loss function is given by $L(u) = |u|$, and the solution of (9.2) leads to the conditional median function $\xi(\mathbf{x}) = \inf\{z \colon F(z|\mathbf{x}) \geq 1/2\}$. Here, $F(\cdot|\cdot)$ is the CDF of Z_t given $\mathbf{X}_t = \mathbf{x}$. Estimating $\xi(\cdot)$ nonparametrically gives

$$\widehat{\xi}(\mathbf{x}) = \inf\Big\{z \colon \sum_{t=1}^{n} W_t(\mathbf{x}) I(Z_t \leq z) \geq 1/2 \Big\}. \tag{9.6}$$

Hence, given $\{Y_t, t \leq T\}$, the H-step ahead nonparametric estimator of the conditional median, denoted by $\widehat{Y}_{T+H|T}^{\text{Mdn}}$, is defined as

$$\widehat{Y}_{T+H|T}^{\text{Mdn}} = \inf\left\{z: \sum_{t=1}^{n} W_t(\mathbf{X}_{T-p+1}) I(Z_t \leq z) \geq 1/2\right\}. \tag{9.7}$$

Under certain mixing conditions, uniform convergence of $\widehat{Y}_{T+H|T}^{\text{Mdn}}$ can be proved; see, e.g., Gannoun (1990), and Boente and Fraiman (1995).

Conditional mode

Collomb et al. (1987) propose a method to produce nonparametric predictions based on the conditional mode function. In this case, we have a non-convex loss function with a unique minimum $L(u) = 0$ when $u = 0$, and $L(u) = 1$ otherwise. The solution of (9.2) leads to the conditional mode function $\tau(\mathbf{x}) = \arg\max_{z \in \mathbb{R}} f(z|\mathbf{x})$, where $f(\cdot|\mathbf{x})$ denotes the conditional density function of Z_t given $\mathbf{X}_t = \mathbf{x}$. Estimating $\tau(\cdot)$ nonparametrically gives

$$\widehat{\tau}(\mathbf{x}) = \arg\max_{z \in \mathbb{R}} \sum_{t=1}^{n} K\left(\frac{z - Z_t}{h}\right) W_t(\mathbf{x}). \tag{9.8}$$

Consequently, given $\{Y_t, t \leq T\}$, the H-step ahead nonparametric estimator of the conditional mode is given by

$$\widehat{Y}_{T+H|T}^{\text{Mode}} = \arg\max_{z \in \mathbb{R}} \sum_{t=1}^{n} K\left(\frac{z - Z_t}{h}\right) W_t(\mathbf{X}_{T-p+1}). \tag{9.9}$$

Under some mixing conditions on $\{(\mathbf{X}_t, Z_t), t \in \mathbb{Z}\}$, Collomb et al. (1987) show the uniform convergence of $\widehat{Y}_{T+H|T}^{\text{Mode}}$.

The predictors defined above are *direct* estimators since they use direct smoothing techniques. Clearly, these predictors are point estimates of a particular loss function $L(\cdot)$ at some $\mathbf{x}$. However, they do not estimate the *whole* loss function. In fact, the H-step ahead conditional mean, median, and mode all ignore information contained in the intermediate variables $\mathbf{X}_{t+1}, \ldots, \mathbf{X}_{t+(H-1)}$. In Section 9.1.2, we introduce a nonparametric kernel smoother which uses such information.

Choice of the bandwidth

As we saw in Appendix 7.A, the main problem in the implementation of nonparametric kernel-based smoothing methods is the selection of the bandwidth in finite samples. Let us suppose that the kernel function $K(\cdot)$ is symmetric, second-order, *Lipschitz continuous* and has absolutely integrable FT.[2] Under the assumption that

---

[2] A function $f: \mathbb{R}^p \to \mathbb{R}$ is said Lipschitz continuous on $\mathcal{D} \subset \mathbb{R}^p$ if there exists a finite constant C, such that $|f(x_1) - f(x_2)| \leq C|x_1 - x_2| \; \forall x_1, x_2 \in \mathcal{D}$. The Lipschitz requirement is necessary for proving uniform convergence results.

the DGP is Markovian, and imposing proper (regularity) conditions, the leave-one-out CV method can be extended to time series processes.

Table 9.1 gives leave-one-out estimators of the conditional mean, median, and mode with corresponding CV measures. The optimal bandwidth follows from $h_{\text{opt}} = \arg\min_h\{CV^{(\cdot)}(h)\}$, where the superscript $(\cdot)$ denotes one of the three predictors. Then, given h_{opt}, the H-step ahead nonparametric predictor follows directly. When a time series is strongly correlated, it is reasonable to leave out more than just one observation. For nonparametric density estimation of i.i.d. observations, the plug-in bandwidth $h_d = \widehat{\sigma}_Y T^{-1/(p+4)}$ can be used with $\widehat{\sigma}_Y$ the standard deviation of $\{Y_t\}_{t=1}^T$. This choice is a simplified version of expression (A.10) in Chapter 7, with $\nu = 2$. It guarantees an optimal rate of convergence with respect to the MISE. However, h_d is not optimal in all cases since it does not take into account the mixing condition of the stochastic process. Nevertheless, it may serve as an initial pilot for CV methods.

Choice of the Markov coefficient

The performance of a kernel-based forecasting method depends on the Markov coefficient p. Intuitively, we would like to have p as large as possible in order not to lose too much information about the past. However, as p increases, the data available for forecasting decreases. Matzner–Løber et al. (1998) propose the following empirical procedure. For $p \in \{1, \ldots, p_{\max}\}$ compute the functions

$$f_1(p) = \sum_{t=T-k}^{T} \left|Y_t - \widehat{Y}_{t+1|t}^{(\cdot)}(p, h)\right|, \quad f_2(p) = \sum_{t=T-k}^{T} \left(Y_t - \widehat{Y}_{t+1|t}^{(\cdot)}(p, h)\right)^2,$$

and

$$f_3(p) = \sup_t \left|Y_t - \widehat{Y}_{t+1|t}^{(\cdot)}(p, h)\right|, \tag{9.10}$$

where $\widehat{Y}_{t+1|t}^{(\cdot)}(p, h)$ denotes the one-step ahead kernel-based predictor (i.e. conditional mean, median, or mode) depending on the Markov coefficient p and the bandwidth h. The value of p is chosen as follows. For a fixed h, obtain $p_j = \arg\min_p f_j(p)$ for each j, and subsequently $p = \max_j p_j$ ($j = 1, 2, 3$). For series with $T \geq 100$ observations, it is recommended to take $k = [T/5]$, and $k = [T/4]$ otherwise. This procedure is simple and quick. Nevertheless, there is a need for its theoretical underpinning. Section 9.1.6 discusses alternative methods of lag selection.

9.1.2 Single- and multi-stage quantile prediction

In addition to the three conditional predictors introduced in Section 9.1.1, conditional quantiles are of interest in various time series applications. Suppose that the conditional distribution function of Z_t given $\mathbf{X}_t = \mathbf{x}$, $F(\cdot|\mathbf{x})$, has a unique quantile of order $q \in (0, 1)$ at a point $\xi_q(\mathbf{x})$. Then the conditional qth quantile is defined by

$$\xi_q(\mathbf{x}) = \inf\{z : F(z|\mathbf{x}) \geq q\}. \tag{9.11}$$

Table 9.1: *Leave-one-out estimators of the conditional mean, the conditional median, and the conditional mode with corresponding CV measures.*

Predictor	Leave-one-out estimator [1]	Cross-validation
Mean	$\widehat{\mu}^{-i}(\mathbf{X}_t) = \sum_{\substack{j=1 \\ j \neq i}}^{n} Z_j W_j^{-i}(\mathbf{X}_t)$	$\mathrm{CV}^{\mathrm{Mean}}(h) = \frac{1}{n}\sum_{t=1}^{n}\{Z_t - \widehat{\mu}^{-t}(\mathbf{X}_t)\}^2$
Median (Mdn)	$\widehat{\xi}^{-i}(\mathbf{X}_t) = \inf\{z \mid \widehat{F}^{-i}(z\mid\mathbf{X}_t) \geq 1/2\}$ with $\widehat{F}^{-i}(z\mid\mathbf{X}_t) = \sum_{\substack{j=1 \\ j \neq i}}^{n} I\{Z_j \leq z\} W_j^{-i}(\mathbf{X}_t)$	$\mathrm{CV}^{\mathrm{Mdn}}(h) = \frac{1}{n}\sum_{t=1}^{n}\{Z_t - \widehat{\xi}^{-t}(\mathbf{X}_t)\}^2$
Mode	$\widehat{\tau}^{-i}(\mathbf{X}_t) = \arg\max_{z \in \mathbb{R}} \widehat{f}^{-i}(z\mid\mathbf{X}_t)$ with $\widehat{f}^{-i}(z\mid\mathbf{X}_t) = \frac{1}{h}\sum_{\substack{j=1 \\ j \neq i}}^{n} K\left(\frac{z - Z_j}{h}\right) W_j^{-i}(\mathbf{X}_t)$	$\mathrm{CV}^{\mathrm{Mode}}(h) = \frac{1}{n}\sum_{t=1}^{n}\{Z_t - \widehat{\tau}^{-t}(\mathbf{X}_t)\}^2$

[1] $W_j^{-t}(\mathbf{X}_t) = K\left(\frac{\mathbf{X}_t - \mathbf{X}_j}{h}\right) \big/ \sum_{j=1; j \neq t}^{n} K\left(\frac{\mathbf{X}_t - \mathbf{X}_j}{h}\right)$; $n = T - H - p + 1$.

Equivalently, $\xi_q(\mathbf{x})$ can also viewed as any solution to the following problem

$$\xi_q(\mathbf{x}) = \arg\min_{a \in \mathbb{R}} \mathbb{E}\{\rho_q(Z_t - a) \mid \mathbf{X}_t = \mathbf{x}\},$$

where $\rho_q(u) = |u| + (2q - 1)u$ is the so-called *check function*. Note that $\xi_{1/2}(\mathbf{x}) \equiv \xi(\mathbf{x})$, i.e. the conditional median.

Now, given the observations $\{(\mathbf{X}_t, Z_t)\}_{t=1}^{n}$, an estimator $\widehat{\xi}_q(\mathbf{x})$ of $\xi_q(\mathbf{x})$ can be defined as the root of the equation $\widehat{F}(z\mid\mathbf{x}) = q$ where $\widehat{F}(\cdot\mid\mathbf{x})$ is an estimator of $F(\cdot\mid\mathbf{x})$. Thus, a predictor of the qth conditional quantile of Y_{T+H} is given by $\widehat{\xi}_q(\mathbf{X}_{T-H-p+1})$. Of course, in practice a nonparametric estimate of the conditional distribution function is needed. One possible estimator is the NW smoother which in a time series setting is given by

$$\widetilde{F}(z\mid\mathbf{x}) = \frac{\sum_{t=1}^{n} K\{(\mathbf{x} - \mathbf{X}_t)/h\} I(Z_t \leq z)}{\sum_{t=1}^{n} K\{(\mathbf{x} - \mathbf{X}_t)/h\}}, \quad (n = T - H - p + 1). \tag{9.12}$$

We shall refer to the solution of the equation

$$\widetilde{F}(z\mid\mathbf{x}) = q \tag{9.13}$$

as the *single-stage* conditional quantile predictor and denote this by $\widetilde{\xi}_q^{\mathrm{NW}}(\mathbf{x})$. Alternatively, we may use the local linear (LL) conditional quantile estimator; see Section 9.1.3 for its definition.

Note that the conditional quantile predictor in (9.13) uses only the information in the pairs $\{(\mathbf{X}_t, Z_t)\}_{t=1}^{n}$ and ignores the information contained in

$$\mathbf{W}_t^{(1)} = \mathbf{X}_{t+1}, \quad \mathbf{W}_t^{(2)} = \mathbf{X}_{t+2}, \quad \ldots, \quad \mathbf{W}_t^{(H-1)} = \mathbf{X}_{t+(H-1)}. \tag{9.14}$$

9.1 KERNEL-BASED NONPARAMETRIC METHODS

Below we illustrate the impact of the data contained in (9.14) on multi-step ahead prediction accuracy.

Let $\mathcal{G}_1(\mathbf{w}) = \mathbb{E}(I(Z_t \leq z)|\mathbf{W}_t^{(H-1)} = \mathbf{w})$. For $j = 2, \ldots, H-1$, also define $\mathcal{G}_j(\mathbf{w}) = \mathbb{E}(\mathcal{G}_{j-1}(\mathbf{W}_t^{(H-(j-1))})|\mathbf{W}_t^{(H-j)} = \mathbf{w})$. Hence,

$$\mathrm{Var}[\mathcal{G}_j(\mathbf{W}_t^{(H-j)})] = \mathrm{Var}[\mathbb{E}(\mathcal{G}_j(\mathbf{W}_t^{(H-j)})|\mathbf{W}_t^{(H-j-1)})]$$
$$+ \mathbb{E}[\mathrm{Var}(\mathcal{G}_j(\mathbf{W}_t^{(H-j)})|\mathbf{W}_t^{(H-j-1)})].$$

For $j = 1, \ldots, H-2$, we have $\mathcal{G}_{j+1}(\mathbf{W}_t^{(H-j-1)}) = \mathbb{E}(\mathcal{G}_j(\mathbf{W}_t^{(H-j)})|\mathbf{W}_t^{(H-j-1)})$. Thus,

$$\mathrm{Var}[\mathcal{G}_{j+1}(\mathbf{W}_t^{(H-j-1)})] \leq \mathrm{Var}[\mathcal{G}_j(\mathbf{W}_t^{(H-j)})]. \tag{9.15}$$

Likewise, it is easy to see that

$$\mathrm{Var}[\mathcal{G}_1(\mathbf{W}_t^{(H-1)})|\mathbf{X}_t = \mathbf{x}] \leq \mathrm{Var}[I(Z_t \leq z)|\mathbf{X}_t = \mathbf{x}]. \tag{9.16}$$

Exploiting the Markovian property of $\{Y_t, t \in \mathbb{Z}\}$, we can rewrite $\mathbb{E}(I(Z_t \leq z)|\mathbf{X}_t = \mathbf{x})$ in such a way that the information in (9.14) is incorporated, i.e.

$$\begin{aligned}\mathbb{E}(I(Y_t^* \leq y)|\mathbf{X}_t = \mathbf{x}) &= \mathbb{E}(\mathcal{G}_1(\mathbf{W}_t^{(H-1)})|\mathbf{X}_t = \mathbf{x}), \\ &= \mathbb{E}(\mathcal{G}_2(\mathbf{W}_t^{(H-2)})|\mathbf{X}_t = \mathbf{x}), \\ &\quad \vdots \\ &= \mathbb{E}(\mathcal{G}_{H-1}(\mathbf{W}_t^{(1)})|\mathbf{X}_t = \mathbf{x}).\end{aligned} \tag{9.17}$$

Observe that as we go down line by line in (9.17) more and more information is utilized. Recalling the two previous inequalities, (9.15) and (9.16), we can see that as more information is used, the prediction variance gets smaller and hence prediction accuracy in terms of MSFE improves. Thus, at least in theory, it pays off to use all the ignored information.

Based on the above recursive setup, we now introduce a kernel-based estimator of $F(z|\mathbf{x})$. First the estimators of $\mathcal{G}_1(\mathbf{w})$ and $\mathcal{G}_j(\mathbf{w})$, $(j = 2, \ldots, H-2)$ are defined, respectively, as follows.

Stage 1: $\quad \widehat{\mathcal{G}}_1(\mathbf{w}) = \dfrac{\sum_{t=1}^n K\{(\mathbf{w} - \mathbf{W}_t^{(H-1)})/h_1\} I(Z_t \leq z)}{\sum_{t=1}^n K\{(\mathbf{w} - \mathbf{W}_t^{(H-1)})/h_1\}}$,

Stage j: $\quad \widehat{\mathcal{G}}_j(\mathbf{w}) = \dfrac{\sum_{s=1}^n K\{(\mathbf{w} - \mathbf{W}_s^{(H-j)})/h_j\} \widehat{\mathcal{G}}_{j-1}(\mathbf{W}_s^{(H-(j-1))})}{\sum_{s=1}^n K\{(\mathbf{w} - \mathbf{W}_s^{(H-j)})/h_j\}}$.

Then, using $\widehat{\mathcal{G}}_{H-1}(\mathbf{w})$, compute $\widehat{F}(z|\mathbf{x})$ by

Stage H: $\quad \widehat{F}(z|\mathbf{x}) = \dfrac{\sum_{k=1}^n K\{(\mathbf{x} - \mathbf{X}_k)/h_H\} \widehat{\mathcal{G}}_{H-1}(\mathbf{W}_k^{(1)})}{\sum_{k=1}^n K\{(\mathbf{x} - \mathbf{X}_k)/h_H\}}. \tag{9.18}$

We shall refer to the root of the equation $\widehat{F}(z|\mathbf{x}) = q$ as the *multi-stage qth conditional quantile predictor* $\widehat{\xi}_q^{\text{NW}}(\mathbf{x})$.

To compare the AMSE of $\widehat{\xi}_q^{\text{NW}}(\mathbf{x})$ (multi-stage) with the AMSE of $\widetilde{\xi}_q^{\text{NW}}(\mathbf{x})$ (single-stage), we assume for simplicity of notation that $H = 2$, and $p = 1$. From $\{Y_t, t \in \mathbb{Z}\}$, let us construct the associated process $\mathbf{U}_t = (X_t, W_t, Z_t)'$ defined by

$$X_t = Y_t, \quad W_t = W_t^{(1)} = Y_{t+1}, \quad Z_t = Y_{t+2}.$$

We suppose that the random variables $\{(X_t, W_t)\}$, respectively $\{(W_t, Z_t)\}$, have joint densities $f_{X,W}(\cdot,\cdot)$, respectively $f_{W,Z}(\cdot,\cdot)$. Let $g(x)$, $g(z)$, and $g(w)$ be the marginal densities of $\{X_t\}$, $\{Z_t\}$, and $\{W_t\}$, and $f(\cdot|x) = f_{X,Z}(x,\cdot)/g(x)$ be the conditional density function. Furthermore, we assume that some regularity conditions on the process $\{\mathbf{U}_t, t \in \mathbb{Z}\}$ are satisfied, and that $nh \to \infty$ as $n \to \infty$, $nh_1 \to \infty$ as $n \to \infty$ and $h_1 = o(h_2)$.

For $y \in \mathbb{R}$, define $\sigma^2(y,x) = \text{Var}(Y_t \leq y | X_t = x)$, $v_1(y,x) = \text{Var}(\mathcal{G}_1(W_t)|X_t = x)$ and $v_2(y,x) = \mathbb{E}[\text{Var}(I(Y_t \leq y)|W_t)|X_t = x]$. Then it can be shown (De Gooijer et al., 2001) that for all $x \in \mathbb{R}$ the best possible asymptotic MSE of $\widetilde{\xi}_q^{\text{NW}}(x)$ and $\widehat{\xi}_q^{\text{NW}}(x)$ are respectively given by

$$\text{AMSE}\{\widetilde{\xi}_q^{\text{NW}}(x)\} \simeq \frac{5n^{-4/5}}{4f^2(\xi_q(x)|x)} D_2^{4/5}(\xi_q(x),x) D_1^{1/5}(\xi_q(x),x), \qquad (9.19)$$

$$\text{AMSE}\{\widehat{\xi}_q^{\text{NW}}(x)\} \simeq \frac{5n^{-4/5}}{4f^2(\xi_q(x)|x)} D_3^{4/5}(\xi_q(x),x) D_1^{1/5}(\xi_q(x),x), \qquad (9.20)$$

where

$$D_1(y,x) = \mu_2^2(K)\left\{F^{(2,0)}(y|x) + \frac{2F^{(1,0)}(y|x)g^{(1)}(x)}{g(x)}\right\}^2,$$

$$D_2(y,x) = \frac{R(K)\sigma^2(y,x)}{g(x)}, \quad D_3(y,x) = R(K)\frac{v_1(y,x)}{g(x)},$$

with

$$F^{(i,j)}(t|s) = \frac{\partial^{i+j} F(t|s)}{\partial s^i \partial t^j}, \quad \text{and} \quad g^{(1)}(x) = \frac{dg(x)}{dx},$$

and where $R(K)$ is the roughness function, as defined in Appendix 7.A. Consequently, the ratio of the best possible AMSEs of the single-stage estimator $\widetilde{\xi}_q^{\text{NW}}(x)$ and the two-stage estimator $\widehat{\xi}_q^{\text{NW}}(x)$ is given by

$$r(\xi_q(x),x) = \left\{1 + \frac{v_2(\xi_q(x),x)}{v_1(\xi_q(x),x)}\right\}^{4/5}, \qquad (9.21)$$

which takes values ≥ 1.

9.1 KERNEL-BASED NONPARAMETRIC METHODS

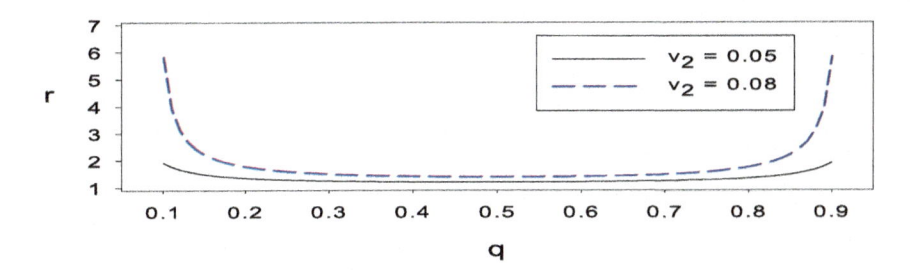

Figure 9.1: *Ratio of asymptotic best possible AMSEs (r) versus the quantile level q. From De Gooijer et al. (2001).*

It is easy to verify that $\text{Var}(\xi_q(x), x) = q(1-q)$. Further, note that $\text{Var}(\xi_q(x), x) = v_1(\xi_q(x), x) + v_2(\xi_q(x), x)$ with $v_2 \leq q(1-q)$. Thus, we may re-express (9.21) as follows: $r(\xi_q(x), x) = \{q(1-q)/(q(1-q) - v_2(\xi_q(x), x))\}^{4/5}$. Figure 9.1 shows a plot of r versus q ($0.1 \leq q \leq 0.9$) for $v_2 = 0.05$ and 0.08. Clearly, r increases sharply as we go to the edge of the conditional distribution. This illustrates theoretically that the improvement achieved by $\widehat{\xi}_q(x)$ is more pronounced for quantiles in the tails of $F(\cdot|x)$.

From asymptotic theory it follows that the optimal bandwidth for both predictors depends on q. Thus, the amount of smoothing required to estimate different parts of $F(\cdot|\mathbf{x})$ may differ from what is optimal to estimate the whole conditional distribution function. This is particularly the case for the tails of $F(\cdot|\mathbf{x})$. We can, however, turn to the following rule-of-thumb calculations based on assuming a normal (conditional) distribution as an appropriate approach:

(a) Select a primary bandwidth, say h_{mean}, suitable for conditional mean estimation. For instance, one may use h_{rot} as given by (A.7) in Appendix 7.A with a Gaussian second-order kernel. Alternatively, various ready-made bandwidth selection methods for kernel-type estimators of $\mu(\cdot)$ are available in the literature.

(b) Adjust h_{mean} according to the following rule-of-thumb

$$h_q = h_{\text{mean}}[\{q(1-q)\}/\{\varphi(\Phi^{-1}(q))^2\}]^{1/(p+4)}, \tag{9.22}$$

where $\varphi(\cdot)$ and $\Phi(\cdot)$ are the standard normal density and distribution functions, respectively, and p refers to the order of the Markovian process. In particular, when $q = 1/2$, $h_{1/2} = h_{\text{mean}}(2/\pi)^{1/(p+4)}$ using $\varphi(\Phi^{-1}(1/2))^2 = (2\pi)^{-1}$.

Example 9.1: A Comparison Between Conditional Quantiles

Consider the simple, Markovian-type, NLAR(1) process

$$Y_t = 0.23 Y_{t-1}(16 - Y_{t-1}) + 0.4\varepsilon_t, \tag{9.23}$$

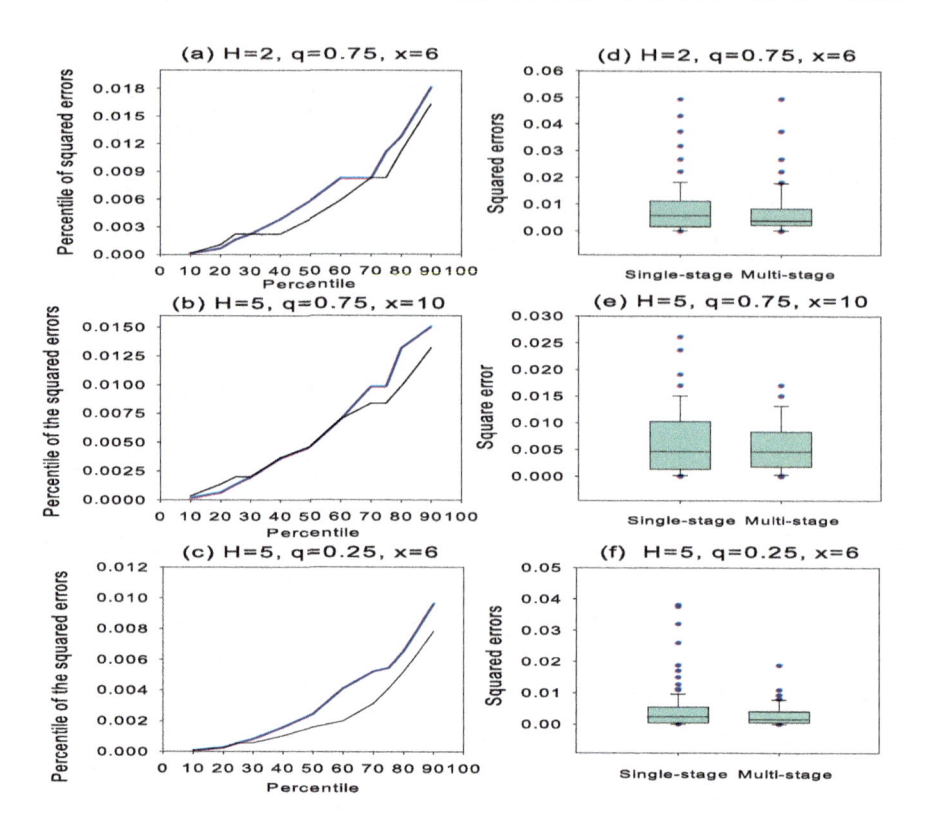

Figure 9.2: *(a) – (c) Percentile plots of the empirical distribution of the squared errors for model (9.23) for the single-stage predictor $\widetilde{\xi}_q^{NW}(\cdot)$ (blue solid line), and the multi-stage (here two) predictor $\widehat{\xi}_q^{NW}(\cdot)$ (black solid line); (d) – (f) Boxplots corresponding to the percentile plots (a) – (c), respectively; $T = 150$, and 150 MC replications. From De Gooijer et al. (2001).*

where $\{\varepsilon_t\} \stackrel{i.i.d.}{\sim} \mathcal{N}(0,1)$ random variables with the standard normal distribution truncated in the interval $[-12, 12]$. The objective is to estimate two and five steps ahead q-conditional quantiles using both $\widetilde{\xi}_q^{NW}(x)$ and $\widehat{\xi}_q^{NW}(x)$ ($q = 0.25$ and 0.75; $x = 6$ and 10), and compare their prediction accuracy.

Clearly, a proper evaluation of the accuracy of both predictors requires knowledge about the "true" conditional quantile $\xi_q(x)$. This information is obtained by generating 10,000 independent realizations of $(Y_{t+H}|Y_t = x)$ ($H = 2$ and 5) iterating the DGP (9.23) and computing the appropriate quantiles from the empirical conditional distribution function of the generated observations.

From (9.23), we generate 150 samples of size $T = 150$. Based on these estimates, we compute for each replication j ($j = 1, \ldots, 150$) the following error measures:

$$e_{\widetilde{\xi}_q(x)}^{(j)} = \frac{\{\widetilde{\xi}_q^{(j)}(x) - \xi_q(x)\}^2}{\xi_q(x)^2} \quad \text{and} \quad e_{\widehat{\xi}_q(x)}^{(j)} = \frac{\{\widehat{\xi}_q^{(j)}(x) - \xi_q(x)\}^2}{\xi_q(x)^2},$$

where $\widetilde{\xi}_q^{(j)}(x)$ and $\widehat{\xi}_q^{(j)}(x)$ denote the jth estimators $\widetilde{\xi}_q^{\text{NW}}(x)$ and $\widehat{\xi}_q^{\text{NW}}(x)$, respectively. Next, we compute percentile values from the empirical distributions of these two error measures. Figures 9.2(a) – (c) show that the percentiles of the squared errors from the 2-stage predictions (black solid line) lie overall below the corresponding percentiles of the squared errors from the single-stage predictions (blue solid line). This implies that the conditional quantile predictions made by $\widehat{\xi}_q^{\text{NW}}(x)$ are more accurate than those made by $\widetilde{\xi}_q^{\text{NW}}(x)$. Boxplots corresponding to the percentile plots (a) – (c) are given in Figures 9.2(d) – (f). It is clear from these plots that the multi-stage quantile predictor has a much smaller variability while its bias is nearly the same as that of the single-stage quantile estimator, supporting asymptotic results.

9.1.3 Conditional densities

Let $\{(\mathbf{X}_t, Y_t), t \in \mathbb{Z}\}$ be a $\mathbb{R}^p \times \mathbb{R}$ valued strictly stationary process with a common pdf $f(\cdot)$ as $(\mathbf{X}, Y)$. In a univariate time series context, $\mathbf{X}_t$ typically denotes lagged values of $\{Y_t\}$. Also assume that $\mathbf{X}_t$ admits a marginal density $g(\cdot)$. Suppose we are given $\{(\mathbf{X}_t, Y_t)\}_{t=1}^n$ observations of $\{(\mathbf{X}, Y), t \in \mathbb{Z}\}$ with $n = T - p$. We wish to estimate the conditional density function of Y_t given $\mathbf{X}_t = \mathbf{x}$, i.e. $f(y|\mathbf{x}) = f(\mathbf{x}, y)/g(\mathbf{x})$, where $g(\cdot)$ is assumed positive at $\mathbf{x}$. The conditional density function can be a useful statistical tool in several ways. The most obvious need for estimating conditional densities arises when exploring relationships between a response and potential covariates.

Example 9.2: Old Faithful Geyser

To motivate ideas and as an illustration we consider, as a classical example for the analysis of bimodal time series data, the waiting time between the starts of successive eruptions and the duration of the subsequent eruption for the Old Faithful geyser in Yellowstone National Park, Wyoming, USA. The average interval between eruptions is about 72.3 minutes (median = 76 minutes) with a standard deviation of about 13.9 minutes. Figure 9.3(a) shows a scatter plot of the duration time and the waiting time. Both variables are transformed to have mean zero and variance one. From the plot it is clear than when there has been a relatively short waiting time between eruptions, the duration of the next eruption is relatively long. When, however, the waiting time between eruptions is longer than about -0.17 (or 70 minutes in the scale of the untransformed data), the duration of the next eruption is more or less a mixture of short and long durations. This interesting observation can be nicely summarized by the conditional density function.

Figure 9.3(b) gives the estimated conditional density. Notice that when the waiting time to eruption is more than -0.17, the conditional density function of eruption duration conditional on waiting time to eruption is bimodal. On the other hand, for waiting times below -0.17, the conditional density function is unimodal.

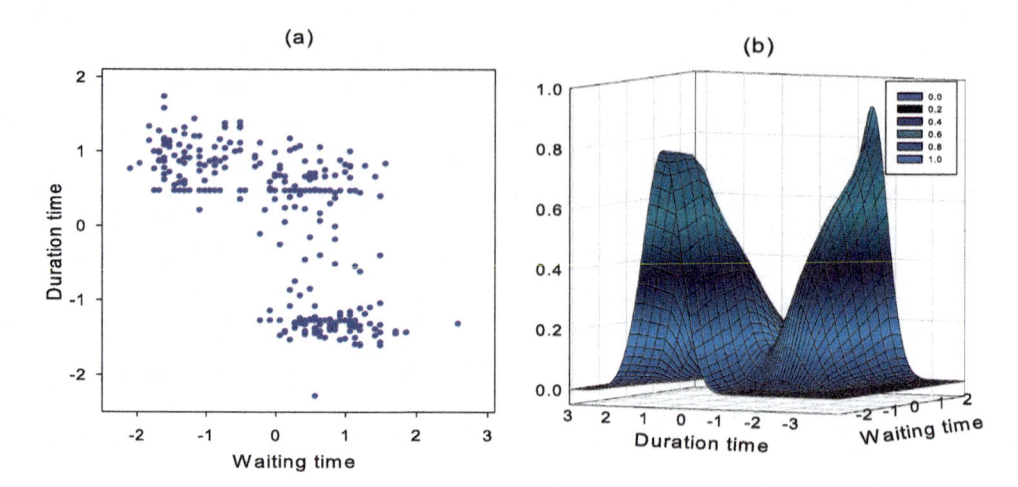

Figure 9.3: Old Faithful geyser *data set: (a) Duration of eruption plotted against waiting time to eruption, and (b) conditional density estimates of eruption duration conditional on the waiting time to eruption. Time period: August 1, 1985 – August 15, 1985 ($T = 299$). From De Gooijer and Zerom (2003).*

In the sequel, we first discuss two existing kernel-based smoothers of the conditional density: the NW estimator and the LL estimator. Next, following De Gooijer and Zerom (2003), we introduce a simple kernel smoother which combines the better sides of both estimators. For simplicity, we shall consider the case $p = 1$, i.e. $\{\mathbf{X}_t, t \in \mathbb{Z}\}$ is a univariate process.

Nadaraya–Watson (NW) and local linear (LL) estimators

Let the kernel $K(\cdot)$ be a symmetric density function on $\mathbb{R}$. Let h_1 and h_2 denote two bandwidths. As $h_1 \to 0$ when $n \to \infty$, it is easy to see from a standard Taylor argument that

$$\mathbb{E}\{\mathcal{K}_{h_1}(y - Y)|X = x\} \simeq f(y|x),$$

where $\mathcal{K}_h(\cdot) = K(\cdot/h)/h$. This suggests that the estimation of $f(y|x)$ can be viewed as a nonparametric regression of $\mathcal{K}_h(y - Y_t)$ on $\{X_t\}$. In fact, it is based on this particular idea that the NW kernel smoother of $f(y|x)$ was first proposed. Within the current setting, the natural NW estimator of $f(y|x)$ is given by

$$\widehat{f}^{\text{NW}}(y|x) = \sum_{t=1}^{n} \mathcal{K}_{h_1}(y - Y_t) W_t^{\text{NW}}(x), \quad (n = T - p), \tag{9.24}$$

where

$$W_t^{\text{NW}}(x) = \frac{\mathcal{K}_{h_2}(x - X_t)}{\sum_{t=1}^{n} \mathcal{K}_{h_2}(x - X_t)}.$$

Now, suppose that the second derivative of $f(y|x)$ exists. Also, introduce the short-hand notation $f^{(i,j)}(y|x) = \partial^{i+j} f(y|x)/\partial x^i \partial y^j$. In a small neighborhood of a

9.1 KERNEL-BASED NONPARAMETRIC METHODS

point x, we can approximate $f(y|z)$ locally by a linear term

$$f(y|z) \simeq f(y|x) + f^{(1,0)}(y|x)(z-x)$$
$$\equiv a + b(z-x).$$

In this sense, one can also regard the estimation of $f(y|x)$ as a nonparametric weighted regression of $\mathcal{K}_{h_1}(y - Y_t)$ against $(1, (x - X_t))$ using weights $\mathcal{K}_{h_2}(x - X_t)$. Considerations of this nature suggest the following LS problem. Let $(\hat{\beta}_0, \hat{\beta}_1)$ minimize

$$\sum_{t=1}^{n} \left(\mathcal{K}_{h_1}(y - Y_t) - \beta_0 - \beta_1(x - X_t)\right)^2 \mathcal{K}_{h_2}(x - X_t).$$

The LL estimator of $f(y|x)$, here denoted by $\hat{f}^{\text{LL}}(y|x)$, is defined as $\hat{\beta}_0$. Simple algebra (Fan and Gijbels, 1996) shows that $\hat{f}^{\text{LL}}(y|x)$ can be expressed as

$$\hat{f}^{\text{LL}}(y|x) = \sum_{t=1}^{n} \mathcal{K}_{h_1}(y - Y_t) W_t^{\text{LL}}(x), \quad (n = T - p), \tag{9.25}$$

where

$$W_t^{\text{LL}}(x) = \frac{\mathcal{K}_{h_2}(x - X_t)\{T_{n,2} - (x - X_t)T_{n,1}\}}{(T_{n,0}T_{n,2} - T_{n,1}^2)},$$

with $T_{n,j} = \sum_{t=1}^{n} \mathcal{K}_{h_2}(x - X_t)(x - X_t)^j$ $(j = 0, 1, 2)$.

From the definition of the two estimators, we can see that $\hat{f}^{\text{NW}}(y|x)$ approximates $f(y|x)$ locally by a constant while $\hat{f}^{\text{LL}}(y|x)$ approximates $f(y|x)$ locally by a linear model. To appreciate why the extension of the local constant fitting to the local linear alternative is interesting, we now compare the two estimators via their respective moments. To keep the presentation simple, we assume without loss of generality, that $h_1 = h_2 = h$. When the process $\{(X_t, Y_t), t \in \mathbb{Z}\}$ is α-mixing it can be shown (Chen et al., 2001) that the approximate asymptotic bias and variance of $\hat{f}^{\text{NW}}(y|x)$ is given by

$$\text{Bias}(\hat{f}^{\text{NW}}(y|x)) = \frac{1}{2}\mu_2(K)h^2\left[f^{(2,0)}(y|x) + f^{(0,2)}(y|x) + 2\frac{g^{(1,0)}(x)}{g(x)}f^{(1,0)}(y|x)\right] \tag{9.26}$$

and

$$\text{Var}(\hat{f}^{\text{NW}}(y|x)) = R^2(K)\frac{1}{nh^2}\frac{f(y|x)}{g(x)}, \tag{9.27}$$

where $\mu_2(K) = \int_{\mathbb{R}} u^2 K(u) du$ and $R(K) = \int_{\mathbb{R}} K^2(z) dz$ are defined earlier in Appendix 7.A. Similarly, it can be shown (Fan and Gijbels, 1996, Thm. 6.2) that the asymptotic bias and variance of $\widehat{f}^{\text{LL}}(y|x)$ are given by

$$\text{Bias}\big(\widehat{f}^{\text{LL}}(y|x)\big) = \frac{1}{2}\mu_2(K)h^2\Big[f^{(2,0)}(y|x) + f^{(0,2)}(y|x)\Big], \qquad (9.28)$$

$$\text{Var}\big(\widehat{f}^{\text{LL}}(y|x)\big) = R^2(K)\frac{1}{nh^2}\frac{f(y|x)}{g(x)}. \qquad (9.29)$$

Note that the two variances are identical and the differences in the AMSEs between the two estimators depend only on their respective biases. We see that the bias of $\widehat{f}^{\text{NW}}(y|x)$ has an extra term $\big(g^{(1,0)}(x)/g(x)\big)f^{(1,0)}(y|x)$. The bias of $\widehat{f}^{\text{NW}}(y|x)$ is large if either $|g^{(1,0)}(x)/g(x)|$ or $|f^{(1,0)}(y|x)|$ is large, but neither term appears in (9.28). For example, when the marginal density function of X (design density) is highly clustered, the term $|g^{(1,0)}(x)/g(x)|$ becomes large. Of course, when $g(x)$ is uniform, the biases of the two estimators are the same. Thus, the fact that $\widehat{f}^{\text{LL}}(y|x)$ does not depend on the density of X makes it *design adaptive* (see, e.g., Fan, 1992). Now, let's consider $|f^{(1,0)}(y|x)|$. For simplicity, suppose that the conditional density of Y depends on x only through a location parameter, say the conditional mean $\mu(\cdot)$ and hence $f(y|x) = f(y - \mu(x))$. Then $f^{(1,0)}(y|x) = \mu^{(1)}(x)f^{(1,0)}(y - \mu(x)|x)$ where $\mu^{(1)}(\cdot)$ denotes the first derivative of $\mu(\cdot)$. In this setup when, for example, $\mu(x) = a + bx$ with large coefficient b, the bias of $\widehat{f}^{\text{NW}}(\cdot|x)$ gets large. When, however, $\mu(x)$ is flat or has maximum or minimum, or inflection point at x, the biases of the two estimators become the same.

The above theoretical comparisons suggest that the LL estimator is more attractive than the NW alternative because of its better bias performance and design adaptation. It is also possible to show that both in the interior and near the boundary of the support of $g(\cdot)$, the asymptotic bias and the variance of $\widehat{f}^{\text{LL}}(\cdot|x)$ are of the same order of magnitude. On the other hand, $\widehat{f}^{\text{NW}}(\cdot|x)$ has a bias of order h for x in the boundary. So, at least in theory, the LL smoother does not suffer from boundary effects and hence does not require modifications at the boundaries.

Re-weighted Nadaraya–Watson (RNW) estimator

From LS theory, we see that the LL weights satisfy: $\sum_{t=1}^{n}(x - X_t)W_t^{\text{LL}}(x) = 0$. On the other hand, this moment condition is not fulfilled for the NW weights. One way to overcome this difficulty is to force the weights $W_t^{\text{NW}}(\cdot)$ to resemble $W_t^{\text{LL}}(\cdot)$. To this end, let $\tau_i(x)$ denote the "probability-like" weights with properties that $\tau_t(x) \geq 0$, $\sum_{t=1}^{n}\tau_t(x) = 1$, and

$$\sum_{t=1}^{n}\tau_t(x)(x - X_t)\mathcal{K}_h(x - X_t) = 0. \qquad (9.30)$$

Next, we define the RNW conditional density estimator as

$$\widehat{f}^{\text{RNW}}(y|x) = \sum_{t=1}^{n}\mathcal{K}_h(y - Y_t)W_t^{\text{RNW}}(x), \qquad (9.31)$$

9.1 KERNEL-BASED NONPARAMETRIC METHODS

where

$$W_t^{\text{RNW}}(x) = \frac{\tau_t(x)\mathcal{K}_h(x - X_t)}{\sum_{t=1}^n \tau_t(x)\mathcal{K}_h(x - X_t)}.$$

From a computational perspective the RNW smoother is easy to implement. In particular, we choose to look for the unique solution of $\tau_t(x)$ by maximizing its empirical likelihood $\sum_{t=1}^n \log \tau_t(x)$, subject to the constraints on $\tau_t(x)$, via Lagrange multipliers. That is,

$$\mathcal{L}_n(\kappa, \lambda) = \sum_{t=1}^n \log \tau_t(x) + \kappa\left(1 - \sum_{t=1}^n \tau_t(x)\right) - n\lambda \sum_{t=1}^n \tau_t(x)(x - X_t)\mathcal{K}_h(x - X_t).$$

Setting $\partial \mathcal{L}_n(\cdot, \cdot)/\partial \tau_t(x) = 0$, we obtain $\tau_t(x) = 1/\{\kappa + n\lambda(x - X_t)\mathcal{K}_h(x - X_t)\}$. In addition, summing $\partial \mathcal{L}_n(\cdot, \cdot)/\partial \tau_t(x)$ and employing (9.30), we can see that $\kappa = n$. Hence,

$$\tau_t(x) = n^{-1}\{1 + \lambda(x - X_t)\mathcal{K}_h(x - X_t)\}^{-1}. \tag{9.32}$$

Substituting (9.32) into (9.30), we obtain

$$0 = \sum_{t=1}^n \frac{(x - X_t)\mathcal{K}_h(x - X_t)}{1 + \lambda(x - X_t)\mathcal{K}_h(x - X_t)} \equiv G(\lambda).$$

Now, notice that $-G(\cdot)$ is just the gradient with respect to λ of

$$L_n(\lambda) = -\sum_{t=1}^n \log\{1 + \lambda(x - X_t)\mathcal{K}_h(x - X_t)\}.$$

So, a zero of $G(\cdot)$ is a stationary point of $L_n(\cdot)$. The implication is that, in practice, one can compute λ as the unique minimizer of $L_n(\cdot)$. De Gooijer and Zerom (2003) suggest that a line search algorithm is a suitable choice to compute λ. The conditional density function displayed in Figure 9.3(b) is computed via the RNW smoother.

It is straightforward to show (De Gooijer and Zerom, 2003) that $|\lambda| \leq \mathcal{O}_p(h)$. Moreover, the bias and variance of $\widehat{f}^{\text{RNW}}(\cdot)$ are identical to the bias and variance of the LL smoother respectively given by (9.28) and (9.29). Thus, the RNW smoother shares the better bias behavior of the LL smoother. If one chooses the optimal bandwidth, say h^*, such that it minimizes the AMSE of $\widehat{f}^{\text{RNW}}(\cdot)$, it is easy to see that

$$h^* = Bn^{-1/6},$$

where B is a functional of some unknowns such as $f(\cdot|x)$. In practice, B may be replaced by consistent estimates. Unlike the $n^{-1/5}$ rate from the univariate density estimation, notice that $h^* \sim n^{-1/6}$ as one needs to smooth in both x and

y directions. Recall that in defining the RNW smoother we used one bandwidth $h = h_1 = h_2$. However, in practice there may indeed arise a need to have different levels of smoothing for each direction. For example, in the Old Faithful geyser illustration, it is not advisable to have the same h for both variables because they have different levels of variability. In fact, that was the reason for standardizing the variables before using a single bandwidth for both. If the approach of pre-standardizing the data is found inadequate, the RNW smoother can be easily re-defined to involve two bandwidths.

9.1.4 Locally weighted regression

The classic kernel-based, methods depend on a real-valued non-random bandwidth sequence $\{h_n\}$. For locally weighted nonparametric estimation, however, the smoothing parameter depends on the number of neighbors around a point of interest using only data (training set) that are "local" to that point. There are several ways of performing nearest-neighbor estimation. Below we present two main approaches. As in the previous sections, we assume that $\{Y_t, t \in \mathbb{Z}\}$ is a strictly stationary process. Moreover, $\{Y_t, t \in \mathbb{Z}\}$ is allowed to follow a Markovian process of order p, and, given the observed time series $\{Y_t\}_{t=1}^T$, $\{\mathbf{X}_t, t \in \mathbb{Z}\}$ is obtained by the construct $\mathbf{X}_t = (Y_t, Y_{t+1}, \ldots, Y_{t+p-1})' \in \mathbb{R}^p$ ($t = 1, \ldots, n; n = T - p$). That is $H = 1$ in (9.1).

K-nearest neighbors

In an i.i.d. setting the method of k-nearest neighbors (k-NN) is a simple, yet powerful and versatile, nonparametric pattern recognition procedure. Within a time series context the intuition underlying the k-NN approach is that the DGP causes patterns of behavior to be repeated in $\{Y_t\}_{t=1}^n$ with $n = T - p$. If a previous pattern can be identified as most similar to the current behavior of Y_t, then the previous subsequent behavior of the series can be used to predict behavior in the immediate future. Here, the objective is to produce a nonparametric estimator of the conditional mean $\mu(\mathbf{x}) = \mathbb{E}(Y_{t+1}|\mathbf{X}_t = \mathbf{x})$ using the $k_n < n$ vectors closest to $\mathbf{X}_n$ in the training, or fitting set $\mathcal{F}_t = \{\mathbf{X}_t | t = 1, \ldots, n\}$. To this end, we define a neighborhood around $\mathbf{x} \in \mathbb{R}^m$ such that $N(\mathbf{x}) = \{i | i = 1, \ldots, k_n$ whose $\mathbf{X}_{(i)}$ represents the ith-nearest neighbor of $\mathbf{x}$ in the sense of a given semi-metric, say $D(\mathbf{x}, \mathbf{X}_{(i)})\}$.

Let $K(\cdot)$ denote a kernel function on $\mathbb{R}^m$. Then the k-NN estimator of $\mu(\mathbf{x})$ is defined as

$$\widehat{\mu}^{k\text{-NN}}(\mathbf{x}) = \sum_{\substack{\mathbf{X}_{(i)} \in \mathcal{F}_t \\ i \in N(\mathbf{x})}} Y_{(i)+1} W_{(i)}(\mathbf{x}), \tag{9.33}$$

where

$$W_{(i)}(\mathbf{x}) = \frac{K(H_{k_n}^{-1} D(\mathbf{x}, \mathbf{X}_{(i)}))}{\sum_{i=1}^n K(H_{k_n}^{-1} D(\mathbf{x}, \mathbf{X}_{(i)}))}, \text{ if } \sum_{i=1}^n K(H_{k_n}^{-1} D(\mathbf{x}, \mathbf{X}_{(i)})) \neq 0,$$

9.1 KERNEL-BASED NONPARAMETRIC METHODS

and where H_{k_n} is the bandwidth, defined as the distance to the furthest neighbor, i.e. $H_{k_n} \equiv D(\mathbf{x}, \mathbf{X}_{(k_n)})$. Two-step ahead forecasts can be obtained along the same lines as above using the data set $\{Y_1, \ldots, Y_n, \widehat{\mu}^{k\text{-NN}}(\mathbf{x})\}$.

Clearly, a weighting scheme is necessary to combine the forecasts implied by each neighbor. When $K(\mathbf{u}) = I(\|\mathbf{u}\|_p \leq 1)$, the kernel weights are just the uniform weights, i.e. $W_{(i)}(\mathbf{x}) = 1/k_n \; \forall i$. Using these weights, and some weak mixing conditions, Yakowitz (1987) shows that $\text{AMSE}\{\widehat{\mu}^{k\text{-NN}}(\mathbf{x})\} = \mathcal{O}(n^{-4/(p+4)})$. He also establishes asymptotic normality of $\widehat{\mu}^{k\text{-NN}}(\mathbf{x})$.

Note that the k-NN method can be thought of as a kernel regression in which the size of the local neighborhood around $\mathbf{x}$ is allowed to vary, thus providing a large window around $\mathbf{x}$ when the data are sparse. The k-NN kernel estimate is also automatically able to take into account the local structure of the data. This advantage, however, may turn into a disadvantage. If there is an outlier in the data, the local prediction may be bad; see, however, below for a robustification of the k-NN method. Typically, k_n is chosen on the order of magnitude $n^{1/2}$, but can be selected using a procedure such as (G)CV. Traditionally, the Euclidean semi-metric is chosen as a distance measure.

Loess/Lowess

The acronyms "loess" and "lowess" both refer to a nonparametric method to calculate an estimate of $\mu(\mathbf{x}) = \mathbb{E}(Y_{t+1}|\mathbf{X}_t = \mathbf{x})$ using *locally weighted regression* (LWR) to smooth data. LWR was first introduced by Cleveland (1979) and further developed by Cleveland and Devlin (1988). The basic underlying model supposes that

$$Y_t = \mu(\mathbf{X}_t) + \varepsilon_t, \tag{9.34}$$

where $\mu(\cdot)$ is a smooth function mapping $\mathbb{R}^p \to \mathbb{R}$, and $\{\varepsilon_t\} \overset{\text{i.i.d.}}{\sim} (0,1)$. LWR is a numerical approach that describes how $\widehat{\mu}(\mathbf{x}^*)$, the estimate of the unknown function $\mu(\cdot)$ at the specific value $\mathbf{x}^*$, is estimated using a local Taylor series approximation of order d. Let f be a "smoothing" parameter such that $0 < f \leq 1$, and let $q_f = [f \times n]$. Then the LWR uses the "window" of q_f observations nearest to $\mathbf{x}^*$, where proximity is defined by the distance $D(\cdot, \cdot)$, commonly taken as the Euclidean norm. In summary, the basic steps to calculate an estimate of $\mu(\mathbf{x}) = \mathbb{E}(Y_{t+1}|\mathbf{X}_t = \mathbf{x})$ are as follows.

Algorithm 9.1: Loess/Lowess

(i) Define a local weight function. For instance, use the tricube weighting function $W(u) = (1 - |u|^3)^3$ if $|u| < 1$, and 0 elsewhere.

(ii) For each $\{\mathbf{X}_t\}_{t=1}^n$ compute the ordered values of the distances $D(\mathbf{x}, \mathbf{X}_{(i)})$ with $\mathbf{X}_{(i)}$ the ith-nearest neighbor of $\mathbf{x}$ as in (9.33).

> **Algorithm 9.1: Loess/Lowess (Cont'd)**
>
> (iii) For any value of **x** compute the local weights
> $$w_{(i)}(\mathbf{x}) = W\big(h_{q_f}^{-1} D(\mathbf{x}, \mathbf{X}_{(i)})\big),$$
> where f is selected by the user.
>
> (iv) Perform a LWR over the span of values. For lowess, set the order of the polynomial at $d = 1$, i.e. the regressions are based on LL–fits. For loess, set $d = 2$ (local polynomial or quadratic fits). The estimate of $\mu(\cdot)$ is simply the estimate of the parameter β_0 from the corresponding LS regression.

Note, the parameter f indicates the fraction of data used in the LWR procedure, analogous to the bandwidth in kernel smoothing. As f increases much more smoothing is done. Since the LWR estimate of $\mu(\cdot)$ is linear in Y_t, the asymptotic properties (e.g. consistency) of the estimator can be derived (Stone, 1977) using standard techniques provided that as $n \to \infty$, $q_f \to \infty$, but $q_f/n \to 0$. If the data set contains outliers, it is generally recommended to use a robust variant of Algorithm 9.1. Basically, the robust LWR procedure involves the following steps.

> **Algorithm 9.2: Robust Loess/Lowess**
>
> (i) Compute the residuals $\{\widehat{\varepsilon}_t\}_{t=1}^n$ from a k-NN pilot estimate of $\mu(\cdot)$, and $s = \text{Mdn}\{|\widehat{\varepsilon}_t|\}$.
>
> (ii) Calculate the robustness weights δ_t which are defined as $\delta_t = K(\widehat{\varepsilon}_t/(6s))$, where $K(\cdot)$ denotes the biweight second-order kernel function given in Table 7.7 of Appendix 7.A.
>
> (iii) Set $d = 1$ or $d = 2$. Then, for each **x**, perform a weighted LS regression as in Algorithm 9.1, but with weights $\{\delta_i W_{(i)}(\mathbf{x}, \mathbf{X}_{q_f})\}$.
>
> (iv) Given the smoothed values from step (ii), compute the next set of residuals and a new set of robustness weights.
>
> (v) Repeat the previous two steps a few times (by default three times in the R and S-Plus implementations of loess/lowess). This produces the final estimate of $\mu(\cdot)$.

Example 9.3: Hourly River Flow Data

Figures 9.4(a) and (b) show the lowess and robust lowess curves fitted to an hourly river flow series $\{Y_t\}_{t=1}^{401}$ from a typical catchment in Wales, UK. The modeling of such processes is a major task of hydrologists who require models for applications such as runoff and flood forecasting. The data are known to exhibit short-term nonlinearity caused by 'soil moisture' effects. In that case

9.1 KERNEL-BASED NONPARAMETRIC METHODS

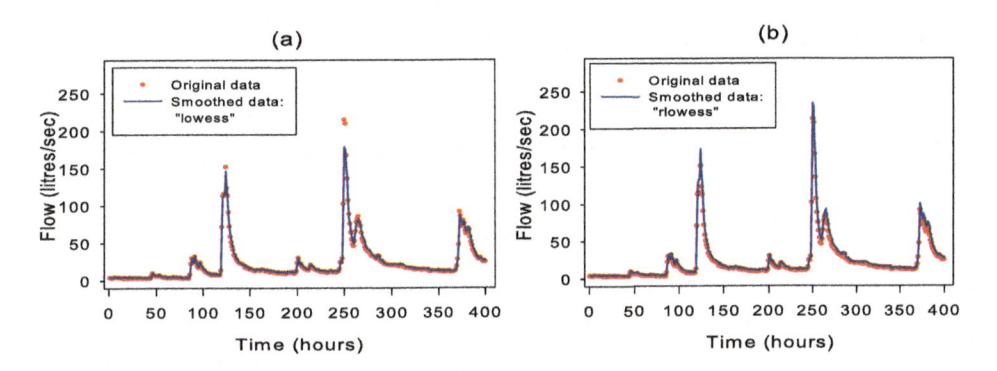

Figure 9.4: *(a) Lowess curve fitted to the hourly river flow data set; (b) Robust lowess curve fitted to the hourly river flow data set; $m = 1$ and $f = 0.1$.*

the soil is infiltrated to its full capacity due to prior rainfall or melting of snow and, as a consequence, river flow will be significantly higher than if the soil has dried out through lack of external sources. There is no discernible long term nonlinearity caused by evapotranspiration. Here, we ignore the information that the major effect on the river flow behavior comes from the amount of rainfall with a few hours delay.

Plot (a) suggests that the lowess method gives a very good identification of the base flow effects, but extreme peaks, or "outliers", are less well explained. Plot (b) shows that the robust lowess method reflects the outlier influences slightly better ($R^2 = 0.999$) than the non-robust lowess method ($R^2 = 0.996$) with smoothed values quite close to the observed data (red dots).

9.1.5 Conditional mean and variance

Let $\{Y_t, t \in \mathbb{Z}\}$ be a strictly stationary process. In this subsection it is convenient to start from the following functional relationship

$$Y_t = \mu(\mathbf{X}_t) + \sigma(\mathbf{X}_t)\varepsilon_t, \quad t \geq 1, \tag{9.35}$$

where $\mathbf{X}_t = (Y_{t-1}, \ldots, Y_{t-p})'$, $\sigma(\mathbf{x}) > 0 \; \forall \mathbf{x} \in \mathbb{R}^p$, $Y_0, \ldots, Y_p$ are initial conditions, $\{\varepsilon_t\} \stackrel{\text{i.i.d.}}{\sim} (0,1)$ random variables with $\{\varepsilon_t\}$ independent of past Y_t, $\mu(\cdot)$ and $\sigma(\cdot)$ are unknown functions on $\mathbb{R}$. The first objective is to estimate $\mu(\cdot)$ and $\sigma(\cdot)$ jointly from T available observations using methods analogous to those for estimating conditional means. In the second part, we focus on the complete conditional density.

Nadaraya–Watson (NW) estimation

Auestad and Tjøstheim (1990) and Tjøstheim and Auestad (1994a,b) propose the NW estimator with product kernels. In particular, as in (9.4), the NW estimator of

$\mu(\cdot)$ and $\sigma^2(\cdot)$ at point $\mathbf{x}$ are given by

$$\widehat{\mu}^{\text{NW}}(\mathbf{x}) = \frac{\sum_{t=p+1}^{T} \mathcal{K}_{\mathbf{H}}(\mathbf{x} - \mathbf{X}_t) Y_t}{\sum_{t=p+1}^{T} \mathcal{K}_{\mathbf{H}}(\mathbf{x} - \mathbf{X}_t)}, \quad \widehat{\sigma}^2(\mathbf{x}) = \frac{\sum_{t=p+1}^{T} \mathcal{K}_{\mathbf{H}}(\mathbf{x} - \mathbf{X}_t) Y_t^2}{\sum_{t=p+1}^{T} \mathcal{K}_{\mathbf{H}}(\mathbf{x} - \mathbf{X}_t)} - \{\widehat{\mu}^{\text{NW}}(\mathbf{x})\}^2.$$
(9.36)

Masry and Tjøstheim (1995) establish strong consistency and asymptotic normality of these estimators for α-mixing processes.

In an analogous fashion, we can adopt LL estimators and other nonparametric regression methods to estimate $\mu(\cdot)$ and $\sigma(\cdot)$ jointly. However, there is no a priori reason to assume that the only features of the conditional distribution that depend on $\mathbf{X}_t$ are the mean and the variance. Hence, it seems reasonable to obtain a complete conditional density estimate of Y_t given $\mathbf{X}_t = \mathbf{x}$. The basic setup is as in Section 9.1.3. Then, assuming a single bandwidth h, a kernel estimate of the conditional (one-step ahead) density $f(\cdot|\mathbf{x})$ associated with (9.35) is given by

$$\widehat{f}^{\text{NW}}(y|\mathbf{x}) = \frac{(Th^{p+1})^{-1} \sum_{t=p+1}^{T} K_{p+1}[\{(y, \mathbf{x}) - (Y_t, \mathbf{X}_t)\}/h]}{(Tp)^{-1} \sum_{t=p+1}^{T} K_p\{(\mathbf{x} - \mathbf{X}_t)/h\}},$$
(9.37)

where $K_{p+1}(\cdot)$ denotes a $p+1$ dimensional kernel function, commonly of the product form. Robinson (1983) establishes a CLT for this estimator. For $H \geq 2$, the forecast transition density can be obtained by applying an iterative scheme; see, e.g., Algorithm 9.3.

Singh and Ullah (1985) extend the above results to the estimation of the conditional density of a (jointly) strictly stationary real-valued bivariate process $\{(\mathbf{X}_t, \mathbf{Z}_t), t \in \mathbb{Z}\}$ with $\mathbf{Z}_t = (Z_t, \ldots, Z_{t-q})'$ $(q \geq 0)$. Moreover, they establish a CLT under far weaker mixing conditions than those used in Robinson (1983).

Bootstrapping conditional densities

Paparoditis and Politis (2001, 2002) combine the flexibility of nonparametric, kernel-based, estimators with bootstrap techniques for pth-order Markovian processes. We already explored this method, called *local resampling*, when discussing a nonparametric test statistic for TR; see Algorithm 8.4. Manzan and Zerom (2008) extend the local resampling (bootstrap) approach to the context of density forecasting. Using the previous framework, the objective is to estimate the out-of-sample H-step forecast density $f_{T+H}(\cdot|\mathbf{X}_T)$ where $\mathbf{X}_T = (Y_T, Y_{T-1}, \ldots, Y_{T-p+1})'$. Since the proposed estimation procedure is recursive in nature it is convenient to introduce the vectors $\mathbf{X}_t = (Y_t, Y_{t-1}, \ldots, Y_{t-p+1})'$ where $t \in S_{p,T}$ and $S_{p,T} = \{p, p+1, \ldots, T-1\}$. The strategy is to assign probability weights $W_t(\cdot) \in \mathbb{R}^p$ to each vector $\mathbf{X}_p, \ldots, \mathbf{X}_{T-1}$, and use these weights to resample from the successors of $\mathbf{X}_t$. The resulting algorithm for Markov forecast densities (MFDs) is as follows.

9.1 KERNEL-BASED NONPARAMETRIC METHODS

Algorithm 9.3: Resampling scheme for MFDs

$H = 1$ **(One-step ahead):**

1.1 Set $n = T$. For $t = p, p+1, \ldots, T-1$ compute the weights at $\mathbf{X}_n = \mathbf{x}$,

$$W_t(\mathbf{x}) = \mathcal{K}_{h_1}(\mathbf{x} - \mathbf{X}_t) \Big/ \sum_{t=p+1}^{T-1} \mathcal{K}_{h_1}(\mathbf{x} - \mathbf{X}_t), \qquad (9.38)$$

where $h_1 > 0$ is a bandwidth and $\mathcal{K}_{h_1}(\cdot) = K_1(\cdot/h_1)/h_1$ with $K_1(\cdot)$ a symmetric kernel function (e.g., the Gaussian product kernel).

1.2 Using (9.38), resample with replacement from the successors of $\mathbf{X}_t$, i.e., $Y^*_{T+1} = Y_{J+1}$ where J is a discrete random variable taking its value in the set $S_{p,T}$.

1.3 Repeat steps 1.1 – 1.2 B times, to obtain the bootstrap replicates $\{Y^{*,(b)}_{T+1}\}_{b=1}^{B}$.

$H \geq 2$ **(Multi-step ahead):**

2.1 Move n one period forward, i.e., $n = T+1$, and update $\mathbf{X}_n$ accordingly, i.e., $\mathbf{X}^*_n = (Y^*_n, Y_{n-1}, \ldots, Y_{n-p+1})'$. Compute new weights using an updated version of (9.38). Resample with replacement from the successors of $\mathbf{X}^*_t$, i.e., $Y^*_{T+2} = Y_{J+1}$.

2.2 Keep moving n forward one step. Repeat step 2.1 until $n = T + H - 1$ by updating $\mathbf{X}_t$.

2.3 Repeat steps 2.1 – 2.2 B times, to obtain $\{Y^{*,(b)}_{T+H}\}_{b=1}^{B}$.

Using another bandwidth $h_2 > 0$ (i.e., $h_2 \sim B^{-1/5}$) and kernel $K_2(\cdot)$, compute the H-step ahead MFD kernel estimator, say $\widehat{f}^{\mathrm{MFD}}_{T+H}(\cdot|\mathbf{X}_T)$, from the B-bootstrap replicates in steps 1.3 and 2.3.

By Algorithm 9.3 the values of the probability weights depend on how "close" the vectors $\mathbf{X}_t$ are to the conditioning vector $\mathbf{X}_n$. That is, the closer $\mathbf{X}_t$ is to $\mathbf{X}_n$ the larger weight it receives as compared to state vectors that are further away. In so doing the method actually defines for each time point $t \in S_{p,T}$ a local neighborhood from which the value Y^*_{T+H} is obtained, and hence its name *local bootstrap*. Under certain mixing conditions on the associated process $\{(\mathbf{X}_t, Z_t)\} \in \mathbb{R}^p \times \mathbb{R}$ where $Z_t = Y_{t+H}$ and some technical assumptions Manzan and Zerom (2008) demonstrate the asymptotic validity of MFD when $H \geq 2$.

To accurately capture the dependence structure of the data, the following approach for the selection of h_1 is recommended:

(i) Compute a pilot density estimate $\widehat{f}_{h_{\mathrm{rot}}}(\mathbf{X}_t) = (T-p)^{-1} \sum_{t \in S_{p,T}} \mathcal{K}_{h_{\mathrm{rot}}}(\mathbf{X}_T - \mathbf{X}_t)$, using $h_{\mathrm{rot}} = \widehat{\sigma}_Y N^{-1/5}$, where $\widehat{\sigma}_Y$ is the standard deviation of $\{Y_t\}_{t=1}^{N}$.

(ii) Compute the local bandwidth factor $\lambda_t = \{\widehat{f}_{h_{\mathrm{rot}}}(\mathbf{X}_t)/g\}^{-\gamma}$ where g is the geometric mean of $\widehat{f}_{h_{\text{opt}}}(\mathbf{X}_t)$, i.e., $\log g = (1/T) \sum_{t=1}^{T} \log \widehat{f}_{h_{\text{opt}}}(\mathbf{X}_t)$, and γ ($0 \leq \gamma \leq 1$) is a *sensitivity parameter* that regulates the amount of weight that is attributed to the observations in the low density regions. In terms of lowest MSE, a good choice is $\gamma = 1/2$; see Silverman (1986).

(iii) Compute the adaptive (A) bandwidth $h_{t,A} = \lambda_t h_{\text{rot}}$. The idea here is to adjust the pilot density estimate in such a way that areas of high (low) density use a smaller (larger) bandwidth.

9.1.6 Model assessment and lag selection

Assessment of the independence properties of residuals from nonparametric models can be carried out as in the linear case but using methods appropriate for assessing possible nonlinear dependence. For instance, residuals can be checked for independence using the mutual information mentioned in Section 1.3.3, or a test of nonlinearity can be applied to see if any nonlinear structure remains. In general, any of the test statistics of Chapter 7 that are not tied to a particular nonlinear model can be used to assess the GOF for nonparametric modeling procedures.

Related to these tests are methods of lag selection. They are often based on modifications of time series model selection criteria. For example, methods for variable selection based on minimization of a criterion such as AIC or *final prediction error* (FPE) have been investigated for kernel-based (i.e., NW and LL estimates) autoregression. To highlight the statistical ideas, we use the framework of (9.35). The goal of lag selection is to determine a proper subset $(Y_{t-i_1}, \ldots, Y_{t-i_p})'$ from $\mathbf{X}_t$ with p as small as possible such that $\mathbb{E}(Y_t|Y_{t-i_1}, \ldots, Y_{t-i_p}) \stackrel{\text{a.s.}}{=} \mathbb{E}(Y_t|\mathbf{X}_t)$. Thus, we assume that all lags are needed for specifying $\mu(\cdot)$, but not necessarily for $\sigma(\cdot)$. Moreover, we let $\{\varepsilon_t, t \geq i_p + 1\} \stackrel{\text{i.i.d.}}{\sim} (0, 1)$ with finite fourth moment.

Below we focus on the FPE criterion of a nonparametric estimate $\widehat{\mu}(\cdot)$ of $\mu(\cdot)$. Let $\{\widetilde{Y}_t, t \in \mathbb{Z}\}$ be a process independent of $\{Y_t\}$ but having identical properties. Then, using the notation $\widetilde{\mathbf{X}}_t = (\widetilde{Y}_{t-1}, \ldots, \widetilde{Y}_{t-p})'$, the FPE is defined as

$$\text{FPE}(\widehat{\mu}) = \mathbb{E}[\{\widetilde{Y}_t - \widehat{\mu}(\widetilde{\mathbf{X}}_t)\}^2 W(\widetilde{\mathbf{X}}_{M,t})], \tag{9.39}$$

where $\widetilde{\mathbf{X}}_{M,t} = (Y_{t-1}, \ldots, Y_{t-M})'$ ($M \geq i_p$) is the full lag vector process, and $W: \mathbb{R}^M \to \mathbb{R}$ is a suitably chosen weight function (usually a 0 – 1 function with compact support). Similar as AIC and its variants, the idea is to choose the lag combination which leads to the smallest FPE($\cdot$).

Tjøstheim and Auestad (1994a) derive a stepwise FPE criterion with a penalty term that is a complicated function of the chosen bandwidth and the selected kernel. For a DGP with correct lag vector $(i_1, \ldots, i_p)$ and bandwidth h, as $T \to \infty$, Tschernig and Yang (2000) obtain an expression for the *asymptotic* FPEs (AFPEs). Then, under some mild assumptions, and for both NW and LL estimators of $\mu(\cdot)$, they propose the estimated FPEs

$$\widehat{\text{FPE}}(h, i_1, \ldots, i_p) = \widehat{\text{AFPE}}(h, i_1, \ldots, i_p) + o\big(h^4 + (T - i_p)^{-1} h^{-p}\big), \tag{9.40}$$

9.1 KERNEL-BASED NONPARAMETRIC METHODS

in which the $\widehat{\text{AFPE}}$s are given by

$$\widehat{\text{AFPE}}(h, i_1, \ldots, i_p) = \widehat{A}_{h_{\text{opt}}} + \frac{2\{K(0)\}^p}{(T-i_p)h_{\text{opt}}} \widehat{B}_h, \qquad (9.41)$$

where, at $\mathbf{X}_{M,t} = \mathbf{x}_M$,

$$\widehat{A}_{h_{\text{opt}}} = \frac{1}{T-i_p}\sum_{t=i_p+1}^{T}\{Y_t - \widehat{\mu}(Y_t)\}^2 W_t(\mathbf{x}_M), \quad \widehat{B}_h = \frac{1}{T-i_p}\sum_{t=i_p+1}^{T}\{Y_t - \widehat{\mu}(Y_t)\}^2 \frac{W_t(\mathbf{x}_M)}{\widehat{f}_{h_{\text{opt}}}(Y_t)}, \qquad (9.42)$$

and where $\widehat{A}_{h_{\text{opt}}}$ and $\widehat{f}_{h_{\text{opt}}}$ (a kernel-based estimator of the density function $f(y)$) are evaluated at the optimal bandwidth h_{opt}, while $\widehat{B}_h$ uses any bandwidth of order $(T-i_p)^{-1/(p+4)}$. For a second-order Gaussian kernel h_{opt} is given as the rule-of-thumb (rot) bandwidth $h_{\text{rot}} = \widehat{\sigma}_Y\{4/(p+2)\}^{1/(p+4)}T^{-1/(p+4)}$, and $K(0) = (2\pi)^{-1/2}$.

Tschernig and Yang (2000) show that conducting lag selection on the basis of (9.40) is consistent if the underlying DGP is nonlinear. Nevertheless, they find that the $\widehat{\text{AFPE}}$ criterion tends to select too many lags in general, and suggest a correction to reduce the chance of overfitting. The resulting estimate of the *corrected* AFPE (CAFPE) is given by

$$\widehat{\text{CAFPE}} = \widehat{\text{AFPE}}\{1 + p(T-i_p)^{-4/(p+4)}\}. \qquad (9.43)$$

Fukuchi (1999) introduces a consistent CV-type method for checking the adequacy of a chosen lag vector, albeit in a linear parametric model setting. The set of candidate models can be correctly or incorrectly specified, nested or nonnested. The method also provides a valid approach for selecting the correct lag vector in (9.35). It uses a measure of forecast risk for each set of one-step ahead forecasts, with the forecast risk estimated from a growing subsample of the original series $\{Y_t\}_{t=1}^T$. Specifically, in the first step the data set is split into a sample for estimation that contains the first R values ($R \leq T-1$). The remaining $T-R$ observations are used to forecast Y_{R+1}, say $\widehat{Y}_{R+1}$. Next, the one-step ahead forecast $\widehat{Y}_{R+2}$ of Y_{R+2} is based on the sample $\{Y_t\}_{t=1}^{R+1}$. This procedure is repeated until the one-step ahead forecast of Y_T is based on $T-1$ observations. The so-called *rolling-over*, one-step ahead MSFE is

$$\text{MSFE} = \frac{1}{T-R}\sum_{t=1}^{T-R}\{Y_{t+R} - \widehat{Y}_{t+R}\}^2. \qquad (9.44)$$

The selected subset lag vector is the one giving the smallest MSFE. Clearly, if a lag selection is carried out for each forecast using e.g. $\widehat{\text{CAFPE}}$, the above method can be computationally demanding.

Example 9.4: Canadian Lynx Data (Cont'd)

Consider the $\log_{10}$-transformed Canadian lynx data introduced in Section 7.5 ($T = 114$). Based on the LL nonparametric estimation method with a Gaussian kernel, we conduct a full search over a wide set of lag combinations with

$M = 15$. The maximum number of lags entertained in the state vector is set at 4. Both methods, $\widehat{\text{AFPE}}$ and $\widehat{\text{CAFPE}}$ select the lag vector (1, 2, 10, 11) as the optimal one. Comparing this result with the specified lags of the five fitted models in Table 7.5, it is clear that a subset NLAR with only these four lags might be sufficient in describing the data. In fact, the residual variance in both cases is 0.0271 which is considerably lower than the corresponding values reported in the last column of Table 7.5.

Using the rolling-over, one-step ahead forecasts of the last 12 observations, we obtain a MSFE of 0.0165, with the pre-set lag vector (1, 2, 10, 11). This MSFE value remains the same if we apply the $\widehat{\text{CAFPE}}$-based criterion using the initial estimation sample up to and including time $t = 102$, and then maintain the selected lag vector for all remaining periods. If, however, we apply the $\widehat{\text{CAFPE}}$ lag selection criterion for each forecast separately, the overall MSFE is 0.0087. In this case, the forecasts are based on the selected lag vector (1, 2, 10, 11) for subsamples of observations up to and including time $t = 102, 110, 111$, and 112, and on the lag vector (1, 2, 3, 4) for subsamples of observations up to and including $t = 103, \ldots, 109$.

9.2 Semiparametric Methods

9.2.1 ACE and AVAS

As noted in Appendix 7.A, allowing $\mu(\cdot)$ to take any possible form using kernel estimation suffers from the curse of dimensionality. If $\mu(\cdot)$ is constrained in such a way that it still provides a flexible representation of the unknown underlying function yet does not suffer from excessive data requirements, a more stable estimate may be obtained. Several different methods have been used to construct such $\mu(\cdot)$. We describe two of them below.

ACE

Consider the multiple regression model in (9.35) with $\sigma(\cdot)$ constant. The *alternating conditioning expectations* (ACE), and *additive and variance stabilizing* (AVAS) transformations algorithms are methods designed to find nonlinear transformations of both the response variable, Y_t, and the predictor variables, $\mathbf{X}_t = (Y_{t-1}, \ldots, Y_{t-p})'$ with the number of lagged Y_t's limited by some fixed p. Specifically, the "workhorse" for these two methods is

$$\theta(Y_t) = \phi(\mathbf{X}_t) + \varepsilon_t$$
$$= \sum_{i=1}^{p} \phi_i(Y_{t-i}) + \varepsilon_t, \qquad (9.45)$$

where $\theta(\cdot)$ and $\phi_i(\cdot)$ are smooth real-valued, but unknown, functions. For identifiability reasons, we usually require that $\mathbb{E}[\phi_i(Y_t)] = 0$.

9.2 SEMIPARAMETRIC METHODS

The objective is to find the optimal transformations $\theta(\cdot)$ and $\phi(\cdot)$ of Y_t and $\mathbf{X}_t$, respectively, such that the squared-loss regression function

$$\frac{\mathbb{E}[\theta(Y_t) - \phi(\mathbf{X}_t)]^2}{\mathbb{E}[\theta^2(Y_t)]}$$

is minimized over all smooth real-valued functions $\theta(\cdot)$ and $\phi(\cdot)$. Clearly, if we fix $\phi(\mathbf{X}_t)$, the solution of $\theta(Y_t)$ is the conditional expectation $\theta_\phi(Y_t) = \mathbb{E}[\phi(\mathbf{x})|Y_t]/\parallel \mathbb{E}[\phi^2(\mathbf{X}_t)]\parallel$. If we fix $\theta(Y_t)$, then the solution of $\phi(\mathbf{X}_t)$ is $\phi_\theta(\mathbf{X}_t) = \mathbb{E}[\theta(Y_t)|\mathbf{X}_t]$. Assume that the joint distribution of the stochastic processes $\{Y_t\}$ and $\{\varepsilon_t\}$ is known. Then, combining the above steps, leads to an iterative procedure for finding the optimal transformation in the sense of minimizing the LS errors, that is

$$\arg\min_{\theta,\phi} \left\{ \mathbb{E}[\theta(Y_t) - \sum_{i=1}^p \phi_i(Y_{t-i})]^2 \right\}, \tag{9.46}$$

where, to avoid the trivial solution $\theta(\cdot) \equiv \phi_i(\cdot) \equiv 0$, we set $\mathbb{E}[\theta^2(Y_t)] = 1$.

In applications, the conditional expectations in (9.46) are replaced by suitable estimates obtained from the data. More specifically, within a time series context, the ACE algorithm works as follows.

Algorithm 9.4: ACE

(i) *Initialize*: Set $\widehat{\theta}(Y_t) = (Y_t - \overline{Y})/\widehat{\sigma}_Y$, where $\overline{Y} = T^{-1}\sum_{t=1}^T Y_t$ and $\widehat{\sigma}_Y^2 = T^{-1}\sum_{t=1}^T (Y_t - \overline{Y})^2$; compute $\widehat{\phi}_i(Y_{t-i})$ as the regression of Y_t on Y_{t-i} ($i = 1,\ldots,p$).

(ii) *New transformation of $\mathbf{X}_t$ (backfit)*: Using kernel estimation or a variant thereof, estimate each $\phi_i(\cdot)$ as a regression of $\theta(Y_t) - \sum_{j=1, j\neq i}^p \widehat{\phi}_j(Y_{t-j})$ on Y_{t-i} ($i = 1,\ldots,p$).

(iii) *New transformation of Y_t*: Compute $\widehat{\theta}(\cdot)$ as a regression of Y_t on $\sum_{i=1}^p \widehat{\phi}_i(Y_{t-i})$, and standardize $\widehat{\theta}(\cdot)$.

(iv) *Alternate*: Do steps (ii) and (iii) until a convergence criterion is reached. The resulting functions $\theta^*(\cdot), \phi_1^*(\cdot),\ldots,\phi_p^*(\cdot)$ are then taken as estimates of the corresponding optimal transformations.

For time series data, convergence may be slow due to the correlated nature of the observations. Also, if $\{Y_t, t \in \mathbb{Z}\}$ is close to unit root nonstationarity in the sense that the lag one serial correlation is close to unity, then the ACE algorithm tends to suggest linear transformations for Y_{t-1}. Nevertheless, the ACE procedure will converge to the optimal solution asymptotically, provided the serial dependence decays sufficiently fast. Besides, the ACE algorithm can handle variables other than continuous predictors such as categorical (ordered or unordered), integer, and indicator variables.

AVAS

AVAS differs from ACE in that $\theta(\cdot)$ is selected so that $\text{Var}\{\theta(Y_t)|\sum_i \widehat{\phi}_i(Y_{t-i})\}$ is constant. This modification removes the problem with heteroskedasticity which lies at the root of the ACE difficulties in multiple regression. It is known that if a random variable Z has mean μ and variance $V(\mu)$, then the asymptotic variance stabilizing transformation for Z is $h(t) = \int_0^1 V(s)^{-1/2} ds$. The resulting AVAS algorithm is like Algorithm 9.4 except that in step (iii) it applies the estimated variance stabilizing transformation to $\widehat{\theta}(\cdot)$ before standardization.

AVAS can be viewed as a generalization of the Box–Cox ML procedure for choosing power transformations of the response, Y_t. It also generalizes the Box–Tidwell procedure for choosing transformations of the predictor variables, $Y_{t-1}, Y_{t-2}, \ldots, Y_{t-p}$. Both ACE and AVAS are useful primarily as exploratory tools for determining which of the response Y_t and the predictors $Y_{t-1}, \ldots, Y_{t-p}$ are in need of nonlinear transformations and what type of transformation is needed.

Since both the ACE and AVAS algorithms are based on smoothing methods, prediction of $\theta(Y_t)$ based on the conditional mean function may be carried out in a manner similar to the simple kernel regression case. For example, to predict $\theta(Y_{T+1})$ as a function of p lagged values of the series, the functions $\phi_i(Y_{T+1-i})$ $(i = 1, \ldots, p)$ are estimated separately as

$$\widehat{\phi}_i(x) = \frac{\sum_{t=1}^n K\{(x - Y_t)/h\} Y_t^*}{\sum_{t=1}^n K\{(x - Y_t)/h\}}, \quad (n = T - i), \tag{9.47}$$

where $x = Y_{T+1-i}$ and $Y_t^* = Y_{t+i}$. Then the one-step ahead forecast of $\theta(Y_{T+1})$ is

$$\widehat{\theta}(Y_{T+1}) = \sum_{i=1}^p \widehat{\phi}_i(Y_{T+1-i}). \tag{9.48}$$

If the transformation of the response is constructed to be monotone, both ACE and AVAS enable prediction of $\{Y_t, t \in \mathbb{Z}\}$ itself by inverting $\theta(\cdot)$.

Example 9.5: Sea Surface Temperatures

Oceanographers are interested in modeling sea surface temperatures (SSTs) to understand what drives changes in temperatures and to obtain accurate predictions of SSTs. Short-term predictions (approximately 1 to 20 days) are used in large-scale weather models, whereas long-term predictions (2 to 3 years or more) are used to explore issues such as global warming and El Niño effects; see, e.g., Example 1.4.

Figure 9.5 shows 30 years of SSTs (in °C) measured at approximately 0800 hours each morning at a point on the California coast about thirty miles south of Monterey Bay, called Granite Canyon. The nonlinear behavior of SSTs has been studied extensively by, among others, Lewis and Ray (1993, 1997). The series, denoted by $\{Y_t\}_{t=1}^{7,361}$, has a sample mean (median) of 11.89 (11.80) and its values range between $[8.00, 18.70]$. The Jarque–Bera (JB) test statistic (1.6) rejects normality (p-value $= 0.00$).

9.2 SEMIPARAMETRIC METHODS

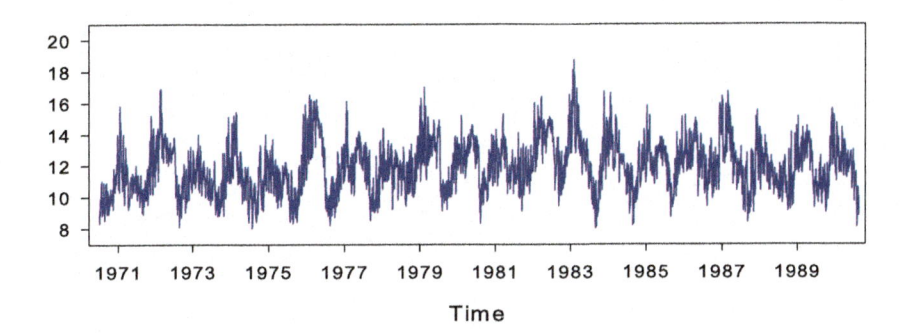

Figure 9.5: *Thirty years of daily* sea surface temperatures *(SSTs) in °C at Granite Canyon California measured from* March 1, 1971 – April 30, 1991; $T = 7{,}361$.

We illustrate the use of the ACE algorithm for approximating a functional relationship between SSTs and lagged SSTs. The ACE algorithm is applied to the raw SST data to approximate a nonlinear AR(7) model, i.e. lagged values of SSTs up to one week previous are used as predictor variables.

Figure 9.6 shows the estimated $\theta(\cdot)$ and $\phi_i(\cdot)$ ($i = 1, \ldots, 7$) obtained using the ACE algorithm with a symmetric k-NN linear least squares procedure for estimating $\theta(\cdot)$ and $\phi_i(\cdot)$, having bandwidth chosen using local CV. The estimated functions for Y_t, Y_{t-1} and Y_{t-2} are fairly linear, suggesting a positive linear relationship between SSTs on day t and SSTs on the previous day and a negative linear relationship for SSTs two days back. There is some suggestion of nonlinear relationships for SSTs at longer lags. The multiple R^2 value for the fitted data is 89.29%.

9.2.2 Projection pursuit regression

Whereas ACE and AVAS estimate the relation between Y_t and $\mathbf{X}_t$ using linear combinations of one-dimensional nonparametric functions operating on individual coordinates of the predictor space, the *projection pursuit regression* (PPR) method estimates the relation using a sum of M one-dimensional nonparametric functions of linear combinations of the predictors. PPR thus allows for the possibility of interactions between predictor variables. Within a time series context, the primary concept underlying PPR is as follows. Given the response and predictor variables Y_t and $\mathbf{X}_t = (Y_{t-1}, \ldots, Y_{t-p})'$, respectively, the PPR function locates the p-dimensional "directional" vector $\boldsymbol{\alpha}_i = (\alpha_{i,1}, \ldots, \alpha_{i,p})'$ and a univariate "activation-level" function $\phi_i(\cdot)$ ($i = 1, \ldots, M$) of the projection $\boldsymbol{\alpha}_i'\mathbf{X}_t$, such that the model

$$Y_t = \beta_0 + \sum_{i=1}^{M} \beta_i \phi_i(\boldsymbol{\alpha}_i'\mathbf{X}_t) + \varepsilon_t, \tag{9.49}$$

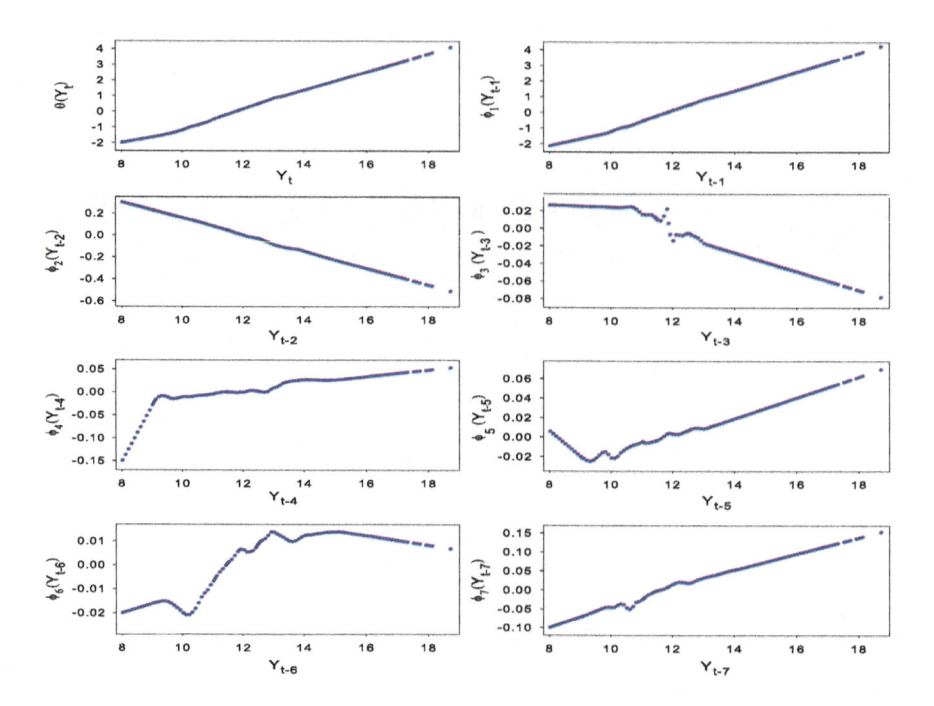

Figure 9.6: *Estimated additive functional relationships between SSTs and transformed lagged SSTs obtained using ACE.*

has the best predictive power, in terms of lowest MSFE. Each $\phi_i(\cdot)$ is estimated nonparametrically using a kernel-based smoothing method such that $\mathbb{E}[\phi_i(\boldsymbol{\alpha}_i'\mathbf{X}_t)] = 0$ and $\text{Var}[\phi_i(\boldsymbol{\alpha}_i'\mathbf{X}_t)] = 1$.

Model (9.49), with $p > 1$, is a generalization of the original PPR model introduced by Friedman and Stuetzle (1981), i.e. the series $\{Y_t\}_{t=1}^T$ is modeled as a (smooth, but otherwise unrestricted) function of a (usually) different linear combination of $\mathbf{X}_t$. For the case $p = 1$ both models have the same form, but the estimation algorithm differs in the sense that the original PPR algorithm chooses $\boldsymbol{\alpha}_i'$ ($i = 1, \ldots, M$) in a forward stepwise manner. This can result in considerably different model specifications. PPR, as specified by (9.49), is implemented in both R and S-Plus with the constraint $\sum_{j=1}^p \alpha_{i,j}^2 = 1$.

Example 9.6: Sea Surface Temperatures (Cont'd)

As mentioned in Section 9.2.1, the ACE algorithm constrains the functional relationship to operate on individual coordinates of the predictor space, which is quite restrictive. It is reasonable to believe that the behavior of SSTs depends on complex interactions between climate signals as captured in previous SST values.

Figure 9.7 shows the estimated functional relationship between Y_t and $\boldsymbol{\alpha}_i'\mathbf{X}_t$ obtained using PPR with $M = 2$ and $\mathbf{X}_t = (Y_{t-1}, \ldots, Y_{t-7})'$. Table 9.2 gives

9.2 SEMIPARAMETRIC METHODS

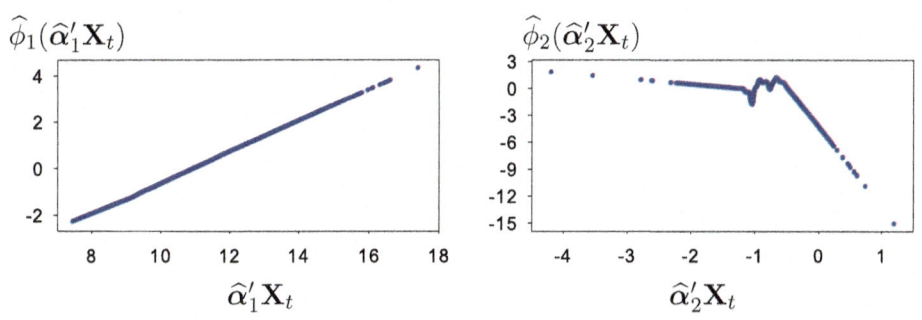

Figure 9.7: *Estimated functional relationships between SSTs and lagged SSTs obtained using PPR.*

the values of β_i and $\boldsymbol{\alpha}_i$ for the fitted PPR model.

Table 9.2: *The estimated coefficients in the PPR model for the SSTs.*

i	β_i	$\alpha_{i,1}$	$\alpha_{i,2}$	$\alpha_{i,3}$	$\alpha_{i,4}$	$\alpha_{i,5}$	$\alpha_{i,6}$	$\alpha_{i,7}$
1	1.51	0.99	-0.12	-0.02	0.02	0.00	0.03	0.03
2	0.03	0.40	-0.33	-0.24	0.30	-0.42	0.55	-0.32

Most of the weight in the first projection vector falls on Y_{t-1} and the estimated relationship is approximately linear. The second projection vector has weights on all lagged values of Y_t and the graph suggests that this projection is related to Y_t in a nonlinear fashion. The multiple R^2 value is 89.05%, similar to that for the ACE fitted model. A third-order projection makes little additional contribution to the prediction of SSTs.

9.2.3 Multivariate adaptive regression splines (MARS)

MARS (Friedman, 1991) is a global adaptive method for fitting nonlinear multivariate regression models using splines. In a time series context, MARS can be used to model nonlinear univariate series, with or without exogenous predictors, and is referred to as TSMARS.

Estimation
Although nonparametric methods do not require an explicit model, the TSMARS methodology is probably best understood through introducing the following setup. Let $\{Y_t, t \in \mathbb{Z}\}$ be a univariate stationary time series process that depends on p_1 ($p_1 \geq 0$) past values of Y_t and on q p_i-dimensional vectors of exogenous time series variables $\mathbf{X}_{i,t} = (X_{i,t-1}, \ldots, X_{i,t-p_i})'$, ($p_i \geq 0; i = 1, \ldots, q$). Assume that there are T observations on $\{Y_t\}$ and $\{\mathbf{X}_{i,t}\}$, and that the data is presumed to be described

by the semi-multivariate time series model

$$Y_t = \mu(1, \mathbf{Y}_{t-1}, \mathbf{X}_{1,t}, \ldots, \mathbf{X}_{q,t}) + \varepsilon_t \qquad (9.50)$$

over some domain $\mathcal{D} \subset \mathbb{R}^n$, $(n = 2 + \sum_{i=1}^{q} p_i)$, which contains the data. Here, 1 denotes a model constant, $\mathbf{Y}_{t-1} = (Y_{t-1}, \ldots, Y_{t-p_l})'$, $\mu(\cdot)$ is a measurable function from $\mathbb{R}^n$ to $\mathbb{R}$ which reflects the true, but unknown, relationship between Y_t and $\mathbf{Y}_{t-1}, \mathbf{X}_{1,t}, \ldots, \mathbf{X}_{q,t}$, and $\{\varepsilon_t\} \overset{\text{i.i.d.}}{\sim} (0, \sigma_\varepsilon^2)$ with ε_t independent of $\mathbf{X}_{i,t}$ ($i = 1, \ldots, q$). The goal is to construct a function $\widehat{\mu}(\cdot)$ that can serve as a reasonable approximation of $\mu(\cdot)$ over the domain $\mathcal{D}$.

We introduce the (TS)MARS methodology by first discussing the method of *recursive partitioning*. Let $\{\mathbb{R}^{(s)}\}_{s=1}^{S}$ be a set of S disjoint subregions representing a partitioning of $\mathcal{D}$. Given these subregions, recursive partitioning approximates the unknown function $\mu(\cdot)$ at $\mathbf{W}_t = (1, \mathbf{Y}_{t-1}', \mathbf{X}_{1,t}', \ldots, \mathbf{X}_{q,t}')'$ in terms of *basis functions* $B_s(\cdot)$ so that

$$\widehat{\mu}(\mathbf{W}_t) = \beta_0 + \sum_{s=1}^{S} \beta_s B_s(\mathbf{W}_t), \qquad (9.51)$$

where $B_s(\mathbf{W}_t) = I(\mathbf{W}_t \in \mathbb{R}^{(s)})$ ($s = 1, \ldots, S$). Each indicator function is a product of Heaviside or step functions: $H(z) = 1$, if $z \geq 0$; $H(z) = 0$, if $z < 0$, describing each subregion $\mathbb{R}^{(s)}$. The aim is to use the data to simultaneously estimate a good set of subregions, without enforcing continuity at the boundaries, and the parameters associated with the separate basis functions in each subregion.

The recursive partitioning follows a two-step procedure.

- *Forward step*: Start from the entire domain $\mathbb{R}^{(1)} = \mathcal{D}$. Split all existing subregions (parent) into two sibling subregions. Optimize the split jointly over all variables and all observed values using a GOF criterion on the resulting approximation $\widehat{\mu}(\cdot)$ to $\mu(\cdot)$. Continue this step until a large number of disjoint subregions $\{\mathbb{R}^{(s)}\}_{s=2}^{M}$, for some pre-specified $M \geq S$, are generated.

- *Backward step*: Recombine the subregions in a reverse manner until a good set of non-overlapping subregions is obtained, using a criterion that penalizes both for lack-of-fit and increasing number of regions.

The basis function of the (TS)MARS algorithm are usually described by linear splines of the form $(x - \tau)_+$ and $(\tau - x)_+$, where

$$(x - \tau)_+ = \begin{cases} x - \tau & \text{if } x \geq \tau, \\ 0 & \text{otherwise,} \end{cases} \quad \text{and} \quad (\tau - x)_+ = \begin{cases} \tau - x & \text{if } x \leq \tau, \\ 0 & \text{otherwise,} \end{cases}$$

which is a non-zero function; see the "hockeystick" graphs in Figure 9.8. For multivariate problems, products of the univariate basis functions are used. As a result the TSMARS estimate of the function $\mu(\cdot)$ takes the form

$$\widehat{\mu}(\mathbf{W}_t) = \beta_0 + \sum_{s=1}^{S} \beta_s \prod_{k=1}^{K_s} [u_{ks}(W^*_{v(ks),t} - t^*_{ks})]_+. \qquad (9.52)$$

9.2 SEMIPARAMETRIC METHODS

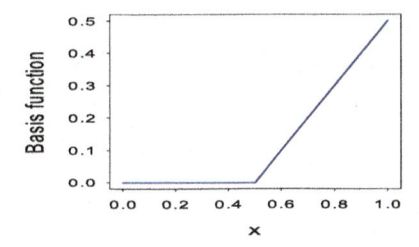

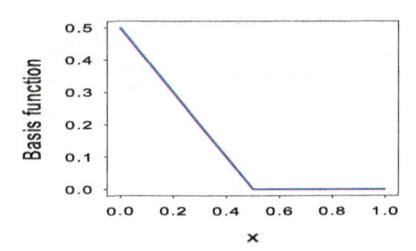

Figure 9.8: *Pair of one-dimensional basis functions used by the* MARS *method;* $(x-0.5)_+$ *(left panel) and* $(0.5-x)_+$ *(right panel).*

Here, β_0 is the coefficient of the constant basis function $B_0(\mathbf{W}_t) = 1$, and the sum is over all remaining basis functions produced by the forward step that survive the backwards deletion step, $u_{ks} = \pm 1$ and indicates the (left/right) sense of the associated step function. The quantity K_s is the number of factors or splits that give rise to the sth basis function $B_s(\cdot)$. The subscript $v(ks), t$ ($v = 1, \ldots, n$) labels the predictor variables at time t ($t = 1, \ldots, T$), and the $t^*_{k,s}$ represent values on the corresponding variables.

Model selection

To evaluate the GOF and compare partition points, (TS)MARS uses residual squared errors in the forward step. In the backward step, it uses a modified generalized CV (GCV) criterion that requires only one evaluation of the model and hence reduces some of the computational burden of (TS)MARS. That is

$$\text{GCV}(M) = \widehat{\sigma}_\varepsilon^2 \Big/ \Big[1 - \frac{C(M)}{T}\Big]^2, \qquad (9.53)$$

where $\widehat{\sigma}_\varepsilon^2 = T^{-1} \sum_{t=1}^T \{Y_t - \widehat{\mu}_M(\mathbf{W}_t)\}^2$ is an estimate of σ_ε^2, measuring the lack-of-fit to the training data. The term in the denominator of (9.53) penalizes over-parameterization, with

$$\begin{aligned} C(M) &= \text{(number of parameters, } c_j, \text{ being fit)} + \\ &\quad + \text{(number of non-constant basis functions)} \\ &= (M+1) + dM. \end{aligned}$$

The quantity d ($2 \leq d \leq 5$) represents an additional contribution by each basis function to the overall model complexity resulting from the (nonlinear) fitting of the basis function parameters to the data at each iterative step. It can be regarded as a smoothing parameter of the (TS)MARS procedure, and d is generally chosen to be 3. Larger values of d result in fewer partition points being placed and thereby smoother function estimates. Observe that TSMARS is more general than the SETAR-type models in the sense that in the SETAR approach interactions among lagged predictor variables (if present) are not allowed, whereas this is not the case with TSMARS.

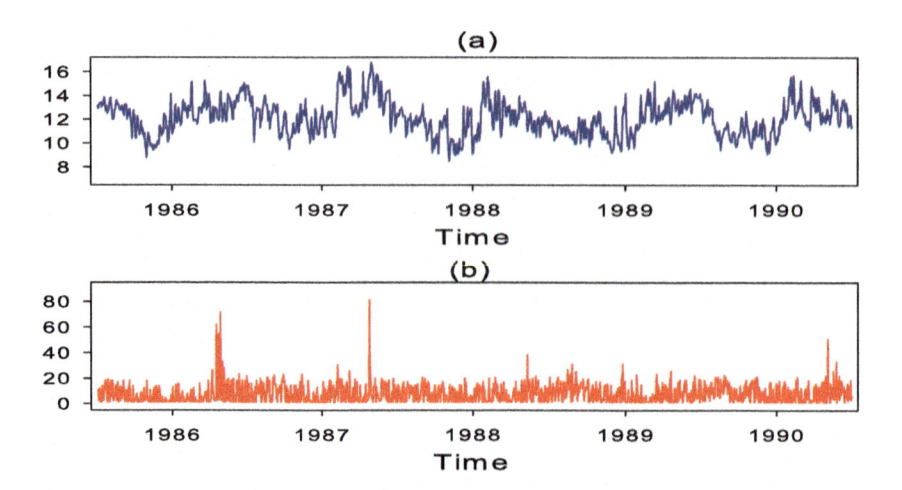

Figure 9.9: *(a) Five years of daily* SSTs *(°C); (b)* Wind speed *(in knots) at Granite Canyon;* $T = 1{,}825$.

On the other hand, the SETAR model allows for different error variances in different regimes, whereas homogeneity of error variances is assumed in TSMARS.

Forecasting

Forecasts for TSMARS models that involve no stochastic exogenous covariates may be obtained in two ways – iteratively or directly. Given Y_{t+1-j}, $(j = 1, \ldots, p)$, an iterated forecast of Y_{t+H} ($H \geq 1$) is computed as

$$\widehat{Y}_{t+H|t} = \widehat{\mu}(\widehat{Y}_{t+H-1|t}, \ldots, \widehat{Y}_{t+H-p|t}), \tag{9.54}$$

where $\widehat{Y}_{t+H-j|t} = Y_{t+H-j}$ when $H - j < 0$, beginning with $\widehat{Y}_{t+1|t}$. This is analogous to the iterative prediction of a parametric AR model, as $\widehat{\mu}(\cdot)$ can be considered as a parametric spline function. Direct forecasts of Y_{t+H} can be obtained by fitting a TSMARS model using only values of the series at lags greater than or equal to H as predictors, e.g., $\widehat{Y}_{t+H|t} = \widehat{\mu}(Y_{t+H-1}, \ldots, Y_{t+H-p})$. This is analogous to the methods of forecasting for kernel-based regression models. Using the direct method, a different model should be estimated for each value of H to be forecast, as the TSMARS model is selected to minimize a function of the forecast errors.

Example 9.7: Sea Surface Temperatures (Cont'd)

Figure 9.9(a) shows a subset of the daily SSTs at Granite Canyon introduced in Example 9.5, now covering the time period January 1986 – December 1990 ($T = 1{,}825$). The corresponding daily wind speeds are plotted in Figure 9.9(b). Lewis and Ray (1997) adopt the TSMARS methodology to approximate a nonlinear functional relationship between logged SSTs, 50 lags of logged SST, 10 lags of the logarithm of (1 + wind speed), say WS_{t-j}, and 10 lags of wind directions (WD_{t-j}). They use logs of the SSTs to remove the variance

9.2 SEMIPARAMETRIC METHODS

inhomogeneity in the series. Also, they remove a one-year cycle from the data before model fitting, i.e. $Y_t = \log \text{SST}_t - \{\widehat{b}_0 + \widehat{b}_1 \sin(2\pi t/365) + \widehat{b}_2 \cos(2\pi t/365)\}$ with LS estimates $\widehat{b}_0 = 2.4826$, $\widehat{b}_1 = -0.0907$, and $\widehat{b}_2 = 0.0460$. Further, they recode the WD_t series as a categorical variable representing the following four wind directions: 1 = East; 2 = North; 3 = West; and 4 = South. Days with no wind or only light airs receive a code of 5. The resulting fitted model is:

$$\widehat{Y}_t = \begin{cases} 2.19_{(0.00)} + 0.88_{(0.01)}(Y_{t-1} - 2.13)_+ + 1.62_{(0.28)}(2.22 - Y_{t-34})_+ \\ \\ +0.01_{(0.00)}(\text{WS}_{t-1} - 1.10)_+ I(\text{WD}_{t-1} \in \{1, 2\}) \\ -0.04_{(0.00)}(\text{WS}_{t-1} - 1.10)_+ I(\text{WD}_{t-1} \in \{2, 3\}) \\ \\ -0.50_{(0.01)}(Y_{t-1} - 2.13)_+(2.75 - Y_{t-7})_+(2.68 - Y_{t-17})_+ \\ \\ -0.58_{(0.10)}(2.27 - Y_{t-34})_+(\text{WS}_{t-1} - 1.10)_+ I(\text{WD}_{t-1} \in \{2, 3\}) \\ -0.52_{(0.12)}(Y_{t-49} - 2.51)_+(\text{WS}_{t-1} - 3.00)_+ I(\text{WD}_{t-1} \in \{1, 4, 5\}) \\ +4.67_{(1.03)}(2.51 - Y_{t-49})_+(2.26 - Y_{t-24})_+ I(\text{WD}_{t-1} \in \{2, 3\}), \end{cases} \quad (9.55)$$

where values in parentheses indicate standard errors of the coefficients obtained from regression theory, assuming that the model terms and threshold values are predetermined.

The model may be interpreted explicitly to obtain a better understanding of the nonlinear relationship between Y_t, WD_t, and WS_t. Consider, for instance, the second and third terms in (9.55). The second term, $0.88(Y_{t-1} - 2.13)_+$, indicates that when the value of Y_t one day ago is larger than 2.13, the next value of the series will be pulled up by a factor 0.88 multiplied by the amount that Y_{t-1} is larger than 2.13. Furthermore, the third term has a non-zero (positive) contribution to the value of $\widehat{Y}_t$ when $Y_{t-34} \leq 2.22$, which rarely happens since the minimum value of Y_t is 2.13. Another example, is the last term in (9.55) which shows that when the previous wind direction was toward the Northwest (categories 2 and 3), the next day's SST is decreased in all cases, except when $Y_{t-24} \leq 2.26$ and $Y_{t-49} \leq 2.51$. The relationship between SSTs and WS is more explicit in lines 2 and 3. In particular, for WD_{t-1} in categories 1, 2, or 3, the effect on $\widehat{Y}_t$ is to add either 0.01, (0.01 − 0.04), or −0.04 times the excess of WS_{t-1} over 1.10. In addition, the wind speed thresholds, which are selected automatically by the TSMARS algorithm, have a meteorological interpretation. For instance, a transformed wind speed threshold of 1.10 knots translates into 1.031 m/s, below which it is well known that wind speeds have little effect on SSTs.

9.2.4 Boosting

Boosting is a semiparametric forward stagewise algorithm that, in a time series context, iteratively estimates a multivariate nonlinear additive AR model, with or without exogenous variables. Let $\{Y_t, t \in \mathbb{Z}\}$ be a univariate stationary time series process which depends on the $(q+1)p$-dimensional vector $\mathbf{W}_t = (W_{1,t}, \ldots, W_{(q+1)p,t})'$

$= (\mathbf{Y}'_t, \mathbf{X}'_{1,t}, \ldots, \mathbf{X}'_{q,t})' \in \mathbb{R}^{(q+1)p}$, where $\mathbf{Y}_{t-1} = (Y_{t-1}, \ldots, Y_{t-p})'$ is the p-dimensional vector of lagged values, and $\mathbf{X}_{i,t} = (X_{i,t-1}, \ldots, X_{i,t-p})'$ ($i = 1, \ldots, q$) the q p-dimensional vectors of explanatory variables. Similar as with (TS)MARS, the goal is to obtain an estimate, or approximation, $\widehat{\mu}(\cdot)$ of the regression function $\mu(\cdot) \equiv \mathbb{E}(Y_t|\mathbf{W}_t = \mathbf{w})$. For a sample of T observations, this approximation comes down to minimizing the expected value of a loss function, say $L(\cdot)$, over all values $\{Y_t, \mathbf{W}_t\}_{t=1}^T$.

A common procedure that solves the above problem, and facilitates interpretation, is to restrict $\mu(\cdot)$ to be a member of a parameterized class of functions $\mu(\cdot; \boldsymbol{\beta})$. To be specific, we reformulate the original *function* optimization problem as a *parameter* optimization problem, i.e.

$$\widehat{\mu}(\mathbf{W}_t) \equiv \mu(\mathbf{W}_t; \widehat{\boldsymbol{\beta}}), \tag{9.56}$$

where

$$\widehat{\boldsymbol{\beta}} = \arg\min_{\boldsymbol{\beta}} \sum_{t=1}^T L(Y_t, \mu(\mathbf{W}_t; \boldsymbol{\beta})), \tag{9.57}$$

with $L(\cdot)$ a loss function which is assumed to be differentiable and convex with respect to the second argument. Two frequently used loss functions are the L_2 loss, and the absolute error or L_1 loss. The final solution is given by

$$\mu(\mathbf{W}_t; \widehat{\boldsymbol{\beta}}^{[M]}) = \sum_{m=0}^M \nu h(\mathbf{W}_t; \widehat{\boldsymbol{\gamma}}^{[m]}). \tag{9.58}$$

Here $h(\cdot)$, termed a *weak learner* or *base learner*, is characterized by the mth estimate $\widehat{\boldsymbol{\gamma}}^{[m]}$ of an M-dimensional parameter vector $\boldsymbol{\gamma}$; $\nu \in (0, 1)$ is a *shrinkage* parameter; and $\widehat{\boldsymbol{\gamma}}^{[0]}$ is an initial guess of $\boldsymbol{\gamma}$. Thus, the underlying structure in the parameters is assumed to be of an "additive" form

$$\widehat{\boldsymbol{\beta}}^{[M]} = \sum_{m=0}^M \nu \widehat{\boldsymbol{\gamma}}^{[m]}.$$

The shrinkage parameter ν can be regarded as controlling the learning rate of the boosting procedure. It provides the base learner to be "weak" enough, i.e. the base learner has large bias, but low variance.

Now, solving (9.58) directly is infeasible. One practical way to proceed is to use *greedy* (stepwise) optimization to estimate the additive terms one at a time. Jointly with a steepest-descent step, the resulting (generic) algorithm, called *gradient descent boosting*, can be summarized as follows.

9.2 SEMIPARAMETRIC METHODS

> **Algorithm 9.5: Gradient descent boost**
>
> (i) Set $m = 0$. Initialize $\mu(\mathbf{W}_t; \widehat{\boldsymbol{\beta}}^{[m]}) = \overline{Y} = T^{-1} \sum_{t=1}^T Y_t$ for each t.
>
> (ii) Set $m = m + 1$. Compute the negative gradient:
> $$-g^{[m]}(\mathbf{W}_t) = \left[\frac{\partial L(Y_t, \mu(\mathbf{W}_t))}{\partial \mu(\mathbf{W}_t)}\right]\bigg|_{\mu(\cdot) = \mu(\cdot; \widehat{\boldsymbol{\beta}}^{[m-1]})}, \quad (t = 1, \ldots, T).$$
>
> (iii) Perform a simple regression of the weak learner on the negative gradient vector, i.e. $\widehat{\boldsymbol{\gamma}}^{[m]} = \arg\min_{\boldsymbol{\gamma}} \sum_{t=1}^T \left(g^{[m]}(\mathbf{W}_t) - h(\mathbf{W}_t; \boldsymbol{\gamma})\right)^2$.
>
> (iv) Update $\mu(\mathbf{W}_t; \widehat{\boldsymbol{\beta}}^{[m]}) = \mu(\mathbf{W}_t; \widehat{\boldsymbol{\beta}}^{[m-1]}) + \nu \cdot h(\mathbf{W}_t; \widehat{\boldsymbol{\gamma}}^{[m]})$.
>
> (v) Iterate steps (ii) – (iv) until $m = M$, where M may be chosen by GCV, as in (TS)MARS, or an AIC-type stopping criterion, e.g., AIC$_C$.

The parameter ν is often taken to be small ($\nu \in [0.01, 0.3]$); Bühlmann and Yu (2003). A small value of ν typically implies a larger number of boosting iterations. Hence, in step (iv), the estimate $\widehat{\mu}(\cdot)$ is continuously improved by the little *boosts* $\nu \cdot h(\mathbf{W}_t; \widehat{\boldsymbol{\gamma}}^{[m]})$. Observe that for the L_2 loss, gradient boosting is equivalent to repeated LS fitting of residuals $\{Y_t - \mu(\mathbf{W}_t; \widehat{\boldsymbol{\beta}}^{[m-1]})\}_{t=1}^T$. Bühlmann and Yu (2003) also show that the addition of new terms in the model does not linearly increase its "complexity", but rather by an exponentially diminishing amount as m gets larger. This result partly explains the "overfitting resistance" of boosting.

High-dimensional models

For regression problems with a large number of predictor variables, Bühlmann (2006) proposes component boosting, where the base learner is applied to one variable at a time. The simplest weak learner is linear. For this learner $\widehat{\boldsymbol{\gamma}}^{[m]} = (0, \ldots, 0, \widehat{\gamma}_{\widehat{s}_m}, 0, \ldots, 0)' \in \mathbb{R}^{(q+1)p}$ where $\widehat{s}_m \in \{1, 2, \ldots, (q+1)p\}$ denotes the respective component at the mth boosting iteration. Then, for *componentwise L_2 boosting*, the modification of $h(\cdot)$ in Algorithm 9.5 is as follows:

$$h(\mathbf{W}_t; \widehat{\boldsymbol{\gamma}}^{[m]}) = \mathbf{W}_t' \widehat{\boldsymbol{\gamma}}^{[m]},$$
$$\widehat{\gamma}_j = \text{OLS}\{\gamma_j\}, \quad \forall j \in \mathcal{J} \equiv \{1, 2, \ldots, (q+1)p\}, \tag{9.59}$$
$$\widehat{s}_m = \arg\min_{j \in \mathcal{J}} \sum_{t=1}^T \left(g^{[m]}(\mathbf{W}_t) - h(\mathbf{W}_t, \widehat{\boldsymbol{\gamma}}^{[j]})\right)^2, \tag{9.60}$$

where OLS$\{\gamma_j\}$ denotes the ordinary LS estimator of γ_j with the negative gradient of the loss function as a T-dimensional pseudo-response vector. The resulting algorithm is called *generalized linear model boosting* (glmboost).

Table 9.3: *Comparison of MSFEs for $H = 1, 4, 8$, and 12-steps ahead predictions made with* glmboost, gamboost, BRUTO, *and* MARS *for the quarterly U.S. unemployment rate. For each H, blue-typed numbers indicate the lowest MSFE.*

H	glmboost	gamboost	BRUTO	MARS
1	7.85×10^{-4}	8.23×10^{-4}	5.99×10^{-4}	8.66×10^{-4}
4	1.48×10^{-2}	1.68×10^{-2}	1.26×10^{-2}	1.53×10^{-2}
8	3.72×10^{-2}	4.23×10^{-2}	4.62×10^{-2}	4.60×10^{-2}
12	7.79×10^{-2}	8.08×10^{-2}	7.72×10^{-2}	6.65×10^{-2}

Example 9.8: Quarterly U.S. Unemployment Rate (Cont'd)

We consider further the quarterly U.S. unemployment rate $\{U_t\}_{t=1}^{252}$ introduced in Example 1.1. In Example 6.2, we fitted a SETAR model to the series $Y_t = \log\{U_t/(1-U_t)\}$. Here, we apply glmboost, gamboost, as well as BRUTO and TSMARS to $\{Y_t\}$ and obtain forecasts for $H = 1, 4, 8$, and 12-steps ahead. Gamboost is a boosting procedure which employs penalized B-splines, called P-splines, with evenly spaced knots as weak learners (Eilers and Marx, 1996). This implies that the weak learner representation is a *generalized additive model* (Hastie and Tibshirani, 1990) with P-splines. BRUTO (Hastie, 1989) is a variation on ACE that uses a step-wise procedure for selecting predictors.

For both boosting algorithms we choose the L_2 loss function, and we set $\nu = 0.1$. The stopping criterion M is determined by AIC_c, with its upper bound fixed at 500. Additionally, for gamboost the degrees of freedom in the smoothing spline base learner was set at 3.5. The R implementation of BRUTO and MARS both have a tuning parameter (denoted by a) for the cost per degree of freedom change. Following Huang and Yang (2004) we set $a = \log T$ (a BIC type of penalty). These authors noted that the default $a = 2$ (an AIC type of penalty) always yielded substantial overfitting.

The initial information set covers the time period 1948(i) – 2001(iv). The maximal number of lags p is set at 12. Next, we generate twelve forecasts from the four prediction methods with a recursive approach. That is, at the first stage, twelve forecasts are calculated for the time period 2002(i) – 2004(iii). At the next stage, the information set is enlarged with one observation and the corresponding horizon is re-estimated. We continue with this approach until 2007(i), and then we compute the final twelve forecasts. Thus, the recursive scheme consists of 21 stages in total. From Figure 1.1, we see that the total forecasting period includes two subperiods of economic contraction (or recession) with rapidly rising unemployment, and one subperiod with economic expansion. In general, interest in forecasting unemployment will be greater during contractionary periods.

Table 9.3 summarizes the forecast results in terms of MSFEs. BRUTO has the

9.2 SEMIPARAMETRIC METHODS

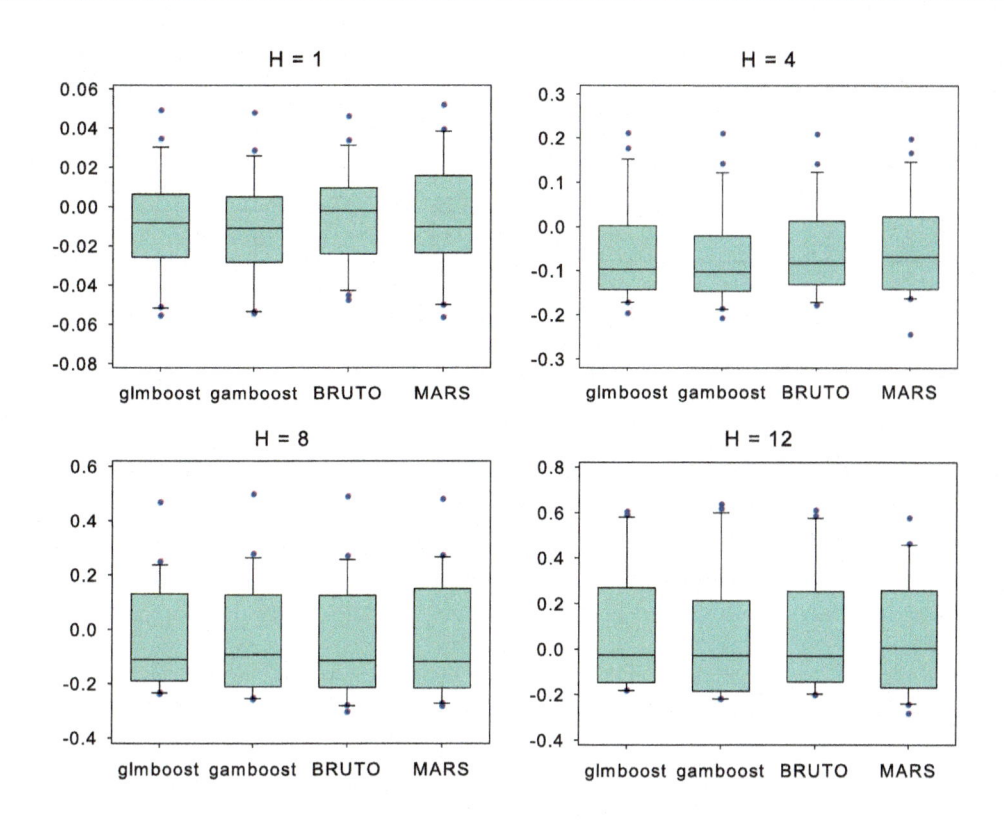

Figure 9.10: *Boxplots of the averaged squared forecast errors, based on 21 forecasts, for $H = 1, 4, 8,$ and 12-steps ahead forecasts of the quarterly U.S. unemployment rate.*

lowest MSFEs for $H = 1$ and 4. Relative to glmboost, the reduction in MSFE of BRUTO is about 24% ($H = 1$) and 15% ($H = 4$). For eight-quarter ahead forecasts, glmboost outperforms all other methods. MARS seems to be more efficient for long-term ($H = 12$) forecasting. Thus, apart from gamboost, each semiparametric method has some forecasting merits over the other methods during certain forecasting horizons. Of course, a real benchmark comparison is needed to support these empirical findings.

Figure 9.10 shows the differences between the four semiparametric methods as boxplots of the average squared forecast errors, based on 21 forecasts. Surprisingly, there is no clear "winner" among the methods, each approach has comparable forecast results. The selected model lags, however, differ at the 21 forecasting stages. Table 9.4 shows the selected lags for the first quarter of 2007 when the available forecast information set reaches its maximum, and hence is the most representative. Clearly, with only three lags the glmboost model is easier to interpret than the more complicated gamboost, BRUTO, and MARS models with the latter two methods selecting lag variables via GCV. Interestingly, the gamboost model uses many lag variables in spite of its relatively poor forecasting performance.

Table 9.4: *Selected lags for the first quarter of 2007 when the available information set reaches its maximum; Quarterly* U.S. unemployment rate.

	glmboost	gamboost	BRUTO	MARS
Selected lags	1, 5, 6	1, 3, 4, 5, 6, 7, 9, 10, 12	1, 2, 6, 8, 9, 10	1, 2, 5, 6, 8, 10

9.2.5 Functional-coefficient AR models

In Section 9.1.5, we introduced the functional relationship (9.35). One restricted functional form that allows for practical implementation is the so-called *functional-coefficient* AR (FCAR) model of Chen and Tsay (1993b) and its adaptive version (Cai et al., 2000b; Fan et al., 2003, among others). Here, and in the next section, we discuss two of these approaches briefly. We refer to Fan and Yao (2003, Chapter 8) who provide an excellent and detailed overview of the many developments in this area.

A strictly stationary time series process $\{Y_t, t \in \mathbb{Z}\}$ is said to follow a FCAR model of order p if it satisfies

$$Y_t = \phi_1(Y_{t-d})Y_{t-1} + \cdots + \phi_p(Y_{t-d})Y_{t-p} + \varepsilon_t, \quad (d \leq p), \tag{9.61}$$

where $\{\varepsilon_t\} \overset{\text{i.i.d.}}{\sim} (0, \sigma_\varepsilon^2)$ with ε_t independent of Y_{t-i} $\forall i > 0$. Model (9.61) is a special case of the state-dependent model (2.10), hence has all the nice properties of a SDM. The model encompasses the SETAR and STAR models. A direct generalization follows from introducing functional-coefficient MA terms (Wang, 2008). If $d > p$, a coefficient term $\phi_0(Y_{t-d})$ may be included in the model. For $d = p$, such a term creates ambiguity and is generally omitted. Clearly, the FCAR(p) model forces each function of Y_{t-i} ($i = 1, \ldots, p$) to be of the form $\phi_i(Y_{t-d})Y_{t-i}$, whereas the more general NLAR model allows $\phi_i(\cdot)$ to vary freely.

The functional form of the coefficients can be simply estimated at time t using an arranged local regression with a fixed-length moving window, and a minimum data size. The resulting estimates $\widehat{\phi}_i(\cdot)$ of $\phi_i(\cdot)$ are consistent under geometric ergodicity conditions (Chen and Tsay, 1993b). By plotting $\widehat{\phi}_i(\cdot)$ versus the threshold variables Y_{t-i} ($i = 1, \ldots, p$) one may infer good candidates for the functional form.

Generalized FCAR

Cai et al. (2000b) propose a generalized FCAR(p) model, given by

$$Y_t = \phi_1(\mathbf{X})Z_{1,t} + \cdots + \phi_p(\mathbf{X})Z_{p,t} + \varepsilon_t, \tag{9.62}$$

where $\mathbf{X} \in \mathbb{R}^q$ can consist of possibly more than one lagged value of the time series process $\{Y_t, t \in \mathbb{Z}\}$ or some other exogenous variable. In addition, the $Z_{i,t}$ ($i = 1, \ldots, p$) can be lagged values of $\{Y_t, t \in \mathbb{Z}\}$ or can be a different exogenous variable,

9.2 SEMIPARAMETRIC METHODS

although commonly $Z_{i,t} = Y_{t-i}$ is used. The $\phi_i(\cdot)$ are assumed to have a continuous second derivative. The functional form can be estimated nonparametrically using kernel-based methods. In this sense, analysis of (9.62) may be thought of as a hybrid of parametric and nonparametric methods. In the following, we discuss the LL smoother for the case $q = 1$.

Let $\{Y_t, X_t, \mathbf{Z}_t = (Z_{1,t}, \ldots, Z_{p,t})'\}_{t=1}^T$ denote process observations. We approximate $\phi_i(\cdot)$ locally at a point $x_0 \in \mathbb{R}$ as $\phi_i(x) \approx a_i + b_i(x - x_0)$. Then (a_i, b_i) are estimated to minimize the weighted sum of squares

$$\sum_{t=1}^T W_t \Big(Y_t - \sum_{i=1}^p \{a_i + b_i(x_0 - X_t)\} Z_{i,t}\Big)^2, \qquad (9.63)$$

where $W_t = \mathcal{K}_{h_T}(x_0 - X_t)$ and h_T is a bandwidth. The LL estimator of $\phi_i(\cdot)$ is then defined as $\widehat{\phi}_i(x_0) = \widehat{a}_i$. For q-dimensional $\mathbf{X}_t$ ($q > 1$), a q-dimensional kernel and a $q \times q$ bandwidth matrix may be used.

The one-step ahead forecast of $\{Y_t, t \in \mathbb{Z}\}$ given (X_t, Z_t) is given by $\widehat{Y}_{t+1|t} = \sum_{i=1}^p \widehat{\phi}_i(X_t) Z_{i,t}$. The bandwidth, h_T, may be selected to minimize a measure of out-of-sample one-step ahead forecast errors for the fitted model. Specifically, let

$$\text{MSFE}_s(h_T) = \frac{1}{n} \sum_{t=T-sn+1}^{T-sn+n} \Big\{Y_t - \sum_{i=1}^p \widehat{\phi}_{i,s}(X_t) Z_{i,t}\Big\}^2, \qquad (9.64)$$

where n denotes the length of the sth subseries of $\{Y_t\}$ ($s = 1, \ldots, S$), and the $\phi_{i,s}(\cdot)$ are computed from the series up to observation $T - sn$ using bandwidth $h_T = [T/(T-sn)]^{1/5}$. The optimal bandwidth is defined to minimize

$$\text{MSFE}(h_T) = \sum_{s=1}^S \text{MSFE}_s(h_T). \qquad (9.65)$$

This measure can be regarded as a modified form of multifold CV, appropriate for stationary time series processes. In practical applications, it is recommended to set $n = [0.1T]$ and $S = 4$. The same criterion can be used to select among different X and different model orders, p.

Model assessment

Cai et al. (2000b) propose a bootstrap LR-type test for FCAR models to determine whether the coefficient functions are constant or take a particular parametric form. Suppose, for some parameter vector $\boldsymbol{\theta} \in \boldsymbol{\Theta}$, where $\boldsymbol{\Theta}$ denotes the space of allowed values of $\boldsymbol{\theta}$, we have the null hypothesis

$$\mathbb{H}_0: \phi_i(x) = \phi_i(x; \boldsymbol{\theta}), \quad (i = 1, \ldots, p),$$

where $\phi_i(\cdot; \boldsymbol{\theta})$ is a specified family of functions parameterized by $\boldsymbol{\theta}$. The bootstrap procedure consists of the following steps.

> **Algorithm 9.6: Bootstrap-based LR-type test**
>
> (i) Estimate $\boldsymbol{\theta}$ for the specified parametric model and construct the residuals, $\widehat{\varepsilon}_t = Y_t - \sum_{i=1}^{p} \widehat{\phi}_i(x; \widehat{\boldsymbol{\theta}}) Z_{i,t}$ and the residual sum-of-squares, $\text{RSS}_0 = \sum_{t=1}^{T} \widehat{\varepsilon}_t^2$.
>
> (ii) Estimate the FCAR model nonparametrically and construct the residuals, $\widetilde{\varepsilon}_t = Y_t - \sum_{i=1}^{p} \widehat{\phi}_i(x) Z_{i,t}$ and the residual sum-of-squares, $\text{RSS}_1 = \sum_{t=1}^{T} \widetilde{\varepsilon}_t^2$.
>
> (iii) Compute the test statistic
>
> $$\text{LR}_T = (\text{RSS}_0 - \text{RSS}_1)/\text{RSS}_1. \qquad (9.66)$$
>
> Large values of LR_T indicate that $\mathbb{H}_0$ should be rejected.
>
> (iv) Generate the bootstrap residuals $\{\varepsilon_t^*\}$ from the EDF of the centered residuals $\{\widetilde{\varepsilon}_t - \bar{\widetilde{\varepsilon}}\}$ from the nonparametric FCAR-fit, and construct bootstrap process values as $Y_t^* = \sum_{i=1}^{p} \widehat{\phi}_j(x; \widehat{\boldsymbol{\theta}}) Z_{i,t} + \varepsilon_t^*$. (Note that if z or $Z_{i,t}$ are functions of the original $\{Y_t, t \in \mathbb{Z}\}$ process, the original values are used, not values obtained from the bootstrapped process. This corresponds to a fixed-design nonparametric regression method.)
>
> (v) Compute the test statistic $\text{LR}_T^{(0)}$ based on the bootstrapped sample in the same way as (9.66).
>
> (vi) Repeat step (v) B times, to obtain $\{\text{LR}_T^{*,(b)}\}_{b=1}^{B}$.
>
> (vii) Compute the one-sided bootstrap-based p-value as
>
> $$\widehat{p} = \frac{1 + \sum_{b=1}^{B} I(\text{LR}_T^{*,(b)} \geq \text{LR}_T^{(0)})}{1 + B}.$$

Note that the above test statistic can be used to test for constant coefficients by letting $\widehat{\phi}_i(x; \widehat{\boldsymbol{\theta}}) = \widehat{\phi}_i$. The residuals are bootstrapped from the nonparametric fit to ensure that the estimated residuals are consistent, no matter whether the null hypothesis or the alternative hypothesis is correct.

Example 9.9: Quarterly U.S. Unemployment Rate (Cont'd)

We reconsider the transformed U.S. unemployment rate $\{Y_t\}_{t=1}^{252}$ of Examples 6.2 and 9.8. To find the optimum FCAR model among the class of FCAR models defined in (9.61), we set $p_{\max} = 11$ (the largest model considered). In the MSFE criterion (9.64) we let $S = 4$ (the number of multi-folds), $n = [0.1T] = 25$ (the length of the sth subseries ($s = 1, \ldots, S$)). Figure 9.11(a) plots MSFE values against a range of bandwidth values. The optimal

9.2 SEMIPARAMETRIC METHODS

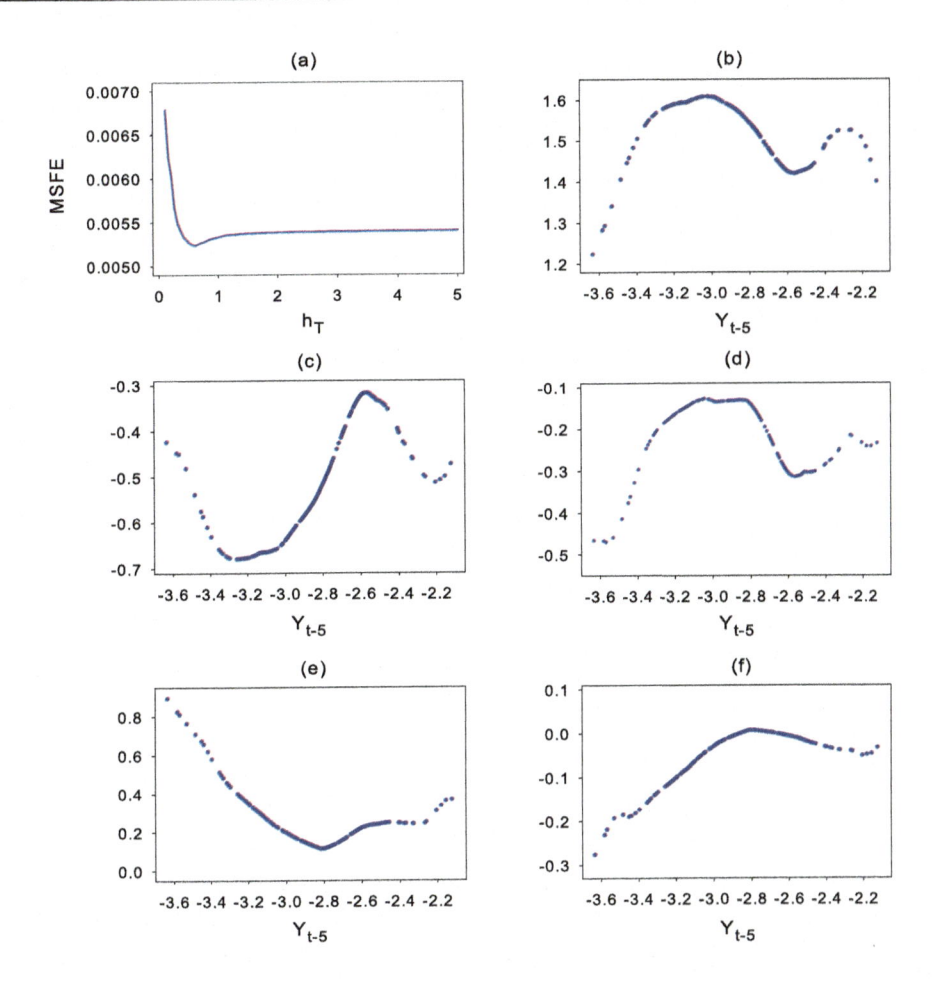

Figure 9.11: *(a) Plot of the MSFE versus h_T for estimation of model (9.67); (b)-(f) Estimated functional-coefficients $\widehat{\phi}_i(\cdot)$ in model (9.67) for the quarterly U.S. unemployment rate.*

bandwidth, which minimizes the AMSE, is $h_T = 0.60$. Moreover, MSFE identifies a FCAR with $d = 5$ and $p = 10$ as the best model. Recall that we set $d = 5$ in the final three-regime SETAR model (6.15). Combining this with the specified lag structure in (6.15), we fit the data with the FCAR model

$$Y_t = \phi_1(Y_{t-5})Y_{t-1} + \phi_2(Y_{t-5})Y_{t-2} + \phi_4(Y_{t-5})Y_{t-4} + \phi_5(Y_{t-5})Y_{t-5}$$
$$+ \phi_{10}(Y_{t-5})Y_{t-10} + \varepsilon_t. \tag{9.67}$$

Figure 9.11(b) – (f) shows the estimated functional-coefficient functions. We see that these functions behave differently for Y_{t-5} around approximately -3.10, which is close to the threshold values at -3.14 identified in (6.15). There also seems to be a changing point around -2.60 which, however, corresponds less well with the obtained threshold value -2.97. Clearly, these figures indicate that most functions $\phi_i(\cdot)$ are either quadratic or sine functions. Finally, we apply the bootstrap LR-type test statistic in Algorithm 9.6 (500 replicates) to test (6.15) against the FCAR model in (9.67). The p-value is 0.00, which reinforces that the three-regime SETAR model is adequate with a residual variance $\widehat{\sigma}_\varepsilon^2 = 0.26 \times 10^{-2}$ versus $\widehat{\sigma}_\varepsilon^2 = 0.43 \times 10^{-2}$ for the fitted FCAR model.

9.2.6 Single-index coefficient model

A model related to the FCAR model is the *single-index coefficient* model, discussed by Ichimura (1993) in a regression setting and extended to the dependent time series setting by Xia and Li (1999). For a strictly stationary time series process $\{Y_t, t \in \mathbb{Z}\}$, the model is formulated as

$$Y_t = \phi_0\big(g(\mathbf{X};\boldsymbol{\theta})\big) + \sum_{i=1}^{p} \phi_i\big((g(\mathbf{X};\boldsymbol{\theta}))Y_{t-i} + \varepsilon_t, \quad (9.68)$$

where $\{\varepsilon_t\} \stackrel{\text{i.i.d.}}{\sim} (0, \sigma_\varepsilon^2)$ with ε_t independent of X and Y_{t-i} $\forall i > 0$. Here, $\phi_i(\cdot)$ ($i = 0, 1, \ldots$) are unknown (arbitrary) coefficient functions, $\mathbf{X}$ is a random q-covariate, and $g(\mathbf{X};\boldsymbol{\theta}): \mathbb{R}^{k+q} \to \mathbb{R}$ is known up to a parameter vector $\boldsymbol{\theta} \in \boldsymbol{\Theta}$, where $\boldsymbol{\Theta} \subset \mathbb{R}^k$ is usually a convex subset.

Model (9.68) is quite general and encompasses various existing nonlinear time series models. The idea is that the nonlinear functions $\phi_j\big(g(\mathbf{X};\boldsymbol{\theta})\big)$ "single index" the threshold variable $\mathbf{X}$, hence its name. When $g(\mathbf{X};\boldsymbol{\theta}) = \boldsymbol{\theta}'\mathbf{X}$ with $\|\boldsymbol{\theta}\| = 1$, it is considered a *linear single-index* model and is related to the projection pursuit AR model (9.49) when $\mathbf{X} = (Y_{t-1}, \ldots, Y_{t-p})'$. As the coefficients $\phi_i(\cdot)$ are functions of a random variable $\mathbf{X}$, it is a type of random coefficient model. When $\mathbf{X} = Y_{t-d}$, $g(X;\theta) = \exp(-\theta X^2)$ and $\phi_i\big(g(X;\theta)\big) = \alpha_i + \beta_i g(X;\theta)$, the model has the form of an ExpAR model. When $\theta = 1$, $g(X;\theta) = Y_{t-d}$ ($d \leq p$), it is a FCAR(p) model. The model having only two terms with $\phi_1(\cdot)$ restricted to be linear and $g(\mathbf{X};\boldsymbol{\theta}) = \boldsymbol{\theta}'\mathbf{X}$ is the *extended partially linear single-index* model of Xia et al. (1999).

One advantage of the single-index model over the FCAR model is that the coefficient functions $\phi_i(\cdot)$ are one-dimensional. This avoids the curse of dimensionality in estimating $\phi_i(\cdot)$ nonparametrically. On the other hand, some nonlinear models cannot be expressed in the form of a single-index model. Xia et al. (1999) give the example of a Hénon map with dynamic noise. Additionally, there does not appear to be any general guidance as to the appropriate choice of $g(\cdot)$ in the single-index model for describing different types of nonlinearity.

Once $\boldsymbol{\theta}$ and a bandwidth h_T are specified, the coefficient functions can be estimated using LL regression in the neighborhood of $g(\mathbf{X}_t;\boldsymbol{\theta})$ ($t = 1, \ldots, T$) as discussed in the previous section, provided the inverse of $\mathbf{W}_\mathbf{x}$, the weight (or design) matrix in the LL regression at the point $\mathbf{x}$, exists and is not large. If this is not the case, then only $\mathbf{X}_t$ values in a subset $\mathcal{A}$ of $\mathbb{R}^q$ so that $\mathbf{W}_\mathbf{x}$ tends to a positive definite

9.2 SEMIPARAMETRIC METHODS

matrix, are used for estimation. Xia et al. (1999) suggest selecting $\boldsymbol{\theta}$ and h_T using a leave-one-out CV method, as follows.

Algorithm 9.7: Estimating $\boldsymbol{\theta}$ and h_T for the single-index model

(i) For a range of $\boldsymbol{\theta}$ and h_T values, compute

$$\widehat{S}(\boldsymbol{\theta}, h_T) = \sum_{\mathbf{X}_t \in \mathcal{A}} \left(Y_t - \widehat{\boldsymbol{\Phi}}'_{\boldsymbol{\theta},t}(g(\mathbf{X}_t; \boldsymbol{\theta}))Y_t \right)^2, \quad (9.69)$$

where $\widehat{\boldsymbol{\Phi}}_{\boldsymbol{\theta},t}(\cdot)$ denotes the LL regression estimate of $\boldsymbol{\Phi}_{\boldsymbol{\theta}}(\mathbf{x}) = (\phi_{0,\boldsymbol{\theta}}(\mathbf{x}), \ldots, \phi_{p,\boldsymbol{\theta}}(\mathbf{x}))'$ obtained using kernel regression when the point (Y_t, X_t) is omitted from the data.

(ii) Choose h_T and $\boldsymbol{\theta}$ to minimize (9.69), and estimate σ_ε^2 by

$$\widehat{\sigma}_\varepsilon^2 = \frac{1}{\sum_{t=1}^T I\{\mathbf{X}_t \in \mathcal{A}\}} \widehat{S}(\widehat{\boldsymbol{\theta}}, \widehat{h}_T),$$

where $(\widehat{\boldsymbol{\theta}}, \widehat{h}_T)$ is a pair of solutions.

Xia et al. (1999) prove the asymptotic normality of the estimator of $\widehat{\boldsymbol{\theta}}$ and the consistency of the estimators for $\phi_i(\cdot)$ under some regularity conditions. They also show that the estimated bandwidth, $\widehat{h}_T$, is asymptotically efficient and is proportional to $T^{-1/5}$.

Example 9.10: A Monte Carlo Simulation Experiment

Consider the following partial linear single-index coefficient regression model

$$Y_t = 0.45X_t - 0.6X_{t-1} + \exp\{-2(0.8X_t + 0.6X_{t-1})^2\} + 0.1\varepsilon_t, \quad (9.70)$$

where $\{\varepsilon_t\}, \{X_t\} \stackrel{\text{i.i.d.}}{\sim} \mathcal{N}(0, 1)$, and $\{\varepsilon_t\}$ and $\{X_t\}$ are mutually independent processes. Alternatively, (9.70) corresponds to the model

$$Y_t = \boldsymbol{\beta}'\mathbf{X}_t + \phi_1(g(\mathbf{X}_t; \boldsymbol{\theta})) + 0.1\varepsilon_t, \quad (9.71)$$

where

$$g(\mathbf{X}_t; \boldsymbol{\theta}) = \cos(\alpha)X_t + \sin(\alpha)X_{t-1}$$

with $\boldsymbol{\beta} = (\lambda\cos(\alpha), \lambda\sin(\alpha))'$, $\boldsymbol{\theta} = (\cos(\alpha), \sin(\alpha))'$, $\mathbf{X}_t = (X_t, X_{t-1})'$, $\boldsymbol{\beta} \perp \boldsymbol{\theta}$ (to ensure estimability), $\|\boldsymbol{\theta}\| = 1$, $\alpha = 0.9273$, and $\lambda = 0.75$.

For sample sizes $T = 50$, 100 and 200, we simulate 1,000 independent samples. We take $\mathcal{A}$ such that it includes all observations, and use a Gaussian kernel. We minimize $\widehat{S}(\boldsymbol{\theta}, h_T)$ within $\boldsymbol{\theta} \in [0.2, 1.3]$, and $h_T \in [0.01, 0.2]$. Table 9.5

Table 9.5: *Sample mean and standard deviation (in parentheses) of estimated $\boldsymbol{\theta}$, $\boldsymbol{\beta}$ and σ_ε^2 for different sample sizes T; based on 1,000 MC replications.*

T	$\widehat{\boldsymbol{\theta}}$	$\widehat{\boldsymbol{\beta}}$	$\widehat{\sigma}_\varepsilon^2$
50	0.7978 (0.0331) 0.5994 (0.0569)	0.4427 (0.0447) -0.5875 (0.0593)	0.0261 (0.0189)
100	0.7987 (0.0170) 0.6010 (0.0226)	0.4484 (0.0212) -0.5959 (0.0219)	0.0194 (0.0158)
200	0.7996 (0.0091) 0.6003 (0.0122)	0.4492 (0.0112) -0.5983 (0.0112)	0.0169 (0.0072)

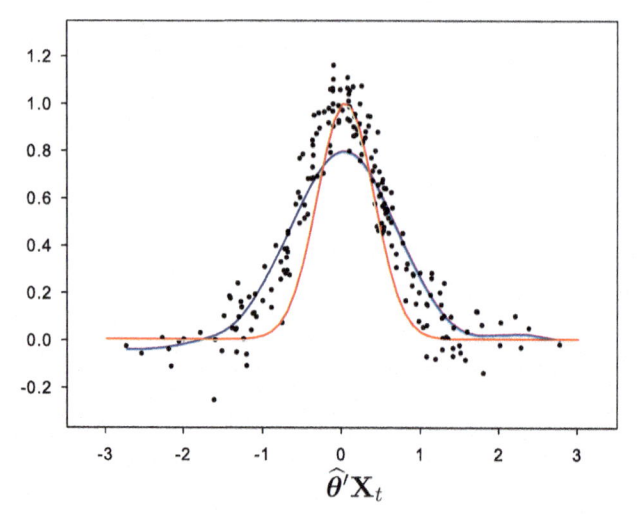

Figure 9.12: *Simulation result from a typical data set of size $T = 200$. The blue sold line denotes the estimated nonlinear relation between Y_t and $\boldsymbol{\theta}'\mathbf{X}_t$. Black dots denote $Y_t - \boldsymbol{\beta}'\mathbf{X}_t$ against $\boldsymbol{\theta}'\mathbf{X}_t$. The red solid line denotes the real nonlinear part of relation (9.70).*

confirms the theoretical results; stable estimates of $\boldsymbol{\theta}$, $\boldsymbol{\beta}$, and σ_ε^2 are obtained even for $T = 50$. Figure 9.12 shows the estimated nonlinear relation between Y_t and $\boldsymbol{\theta}'\mathbf{X}_t$ from a typical simulated data set of size $T = 200$. We see that the estimated function (blue solid line) is relatively close to the real one (red solid line).

9.3 Summary, Terms and Concepts

Summary
This chapter has focused on some of the many methods available for semi- and nonparametric time series forecasting. Because there is a rich literature in this area, we have restricted attention to the principal methods which have demonstrated good prediction performance in practice and comparative MC simulation studies. As such the chapter is somewhat "selective", although it does not imply that a particular

9.3 SUMMARY, TERMS AND CONCEPTS

Table 9.6: *Some applications of semi- and nonparametric methods to univariate time series.*

Section	Method	Reference	Applications
9.1.1	Mean, Mdn, Mode	De Gooijer and Zerom (2000)	U.S. weekly T-bill rate
9.1.4	k-NN	Lall and Sharma (1996)	Monthly streamflow data
		Rajagopalan and Lall (1999)	Daily weather data
	Loess	Barkoulas et al. (1997)	U.S. quarterly T-bill rate
9.2.1	ACE, BRUTO	Chen and Tsay (1993a)	Daily river flow data
	BRUTO	Shafik and Tutz (2009)	Monthly unemployment index
9.2.2	PPR	Xia and An (1999)	Australian blowfly data
		Lin and Pourahmadi (1998)	Canadian lynx data
9.2.3	TSMARS	Lewis and Stevens (1991)	Annual sunspot numbers
		Lewis and Ray (1997)	Daily sea surface temperatures
		Chen et al. (1997)	Eight environmental time series
		De Gooijer et al. (1998)	Weekly exchange rates
9.2.4	Glmboost/Gamboost	Robinzonov et al. (2012)	German monthly industrial production
	Glmboost	Buchen and Wohlrabe (2011)	U.S. monthly industrial production

method is unimportant if it is not included. Much of the material we have discussed is quite new. To facilitate further reading, we have summarized some applications in Table 9.6.

Adapting semi- and nonparametric methods for forecasting is more convenient than using parametric models (Chapter 10) because the functional form of the underlying DGP is unknown or indeterminable in practice. Additionally, semi- and nonparametric approaches offer much greater flexibility to capture variations in the conditional second- and higher-order moments of the noise process than linear and other specific parametric nonlinear models. Additive semiparametric methods have a host of applications, especially in engineering where online analysis of possibly (locally) nonstationary data is often required. A typical example is the magnetic field data of Example 1.3. Hence, we foresee further investigations of semiparametric forecasting methods in real-world applications.

Terms and Concepts

backward step, 366
base (weak) learner, 370
basis function, 366
boosting, 369
check function, 342
curse of dimensionality, 338
design adaptive, 350
forward step, 366
gradient descent boosting, 370

leave-one-out CV, 341
Lipschitz continuous, 340
locally weighted regression (LWR), 353
multi-stage, 344
plug-in bandwidth, 341
projection pursuit regression (PPR), 363
rolling-over MSFE, 359

9.4 Additional Bibliographical Notes

Section 9.1: The use of kernel regression for time series data has been extensively discussed in the literature, going back to Rosenblatt (1969). A useful but slightly outdated source of information on this topic is the review article by Härdle et al. (1997); see also Heiler (2001) and Fan and Yao (2003). Recursive schemes (not a part of this Chapter) for kernel-based regression estimation have been proposed by many authors; see, e.g., Härdle (1990) for some of these. For mixing and ergodic stationary processes, a good starting point for recursive kernel density estimators is Györfi et al. (1989). Franke et al. (2002) show that bootstrap procedures can be used for estimating the distribution of kernel smoothers in NLAR–ARCH processes.

Section 9.1.1: Using strong mixing conditions (α-mixing), Berlinet et al. (2001) prove that the conditional median is asymptotically normally distributed. Similarly, for α-mixing stationary processes, Berlinet et al. (1998) prove that the conditional mode is asymptotically normally distributed.

Härdle and Vieu (1992) extend the leave-one-out CV bandwidth selector to time series processes. Deheuvels (1977) proposes the plug-in bandwidth h_d for density estimation. Matzner–Løber et al. (1998) apply a modified version of h_d, in conjunction with a local and global CV procedure, within the context of an empirical nonparametric forecast setting. These authors also compare nonparametric forecasts based on kernel estimation of the conditional mean, median, and mode.

Section 9.1.2: Direct, or single-stage, kernel-based multi-step predictors for the mean are given by, among others, Auestad and Tjøstheim (1990), Härdle (1990), and Härdle and Vieu (1992). Chen (1996) and Chen et al. (2004) consider the problem of multi-stage kernel prediction for the conditional mean. As special cases of (9.19) and (9.20), De Gooijer et al. (2002) derive the AMSE properties of the kernel-based multi-stage median predictor for α-mixing time series of Markovian structure. Using the LL regression method, Zhou and Wu (2009) estimate quantile curves of a special class of nonstationary processes, called locally stationary processes.

Section 9.1.3: Hyndman and Yao (2002) also introduce two alternative kernel smoothers of the conditional density, both aimed at producing non-negative estimators. In practice, however, the RNW approach is computationally more feasible than the smoothers proposed by these authors.

Section 9.1.4: Fan and Gijbels (1996) provide a detailed study of the asymptotic properties of the local polynomial estimator. Masry (1996a,b) presents similar theory for the LL estimator under dependence. Vilar–Fernandez and Cao (2007) compare nonparametric forecasts of the conditional mean using the NW estimator, and the LL estimator with forecasts obtained from parametric ARIMA specifications.

The method of k-NN for time series prediction was introduced by Yakowitz (1985, 1987) in the context of predicting river runoff for flood warnings. Lall and Sharma (1996) provide a nearest neighbor bootstrap algorithm for resampling hydrologic time series. Application of the k-NN method to predicting GDP and stock returns have been considered by respectively Guégan and Rakotomarolahy (2010) and Kim et al. (2002).

Section 9.1.5: Yang et al. (1999) consider nonparametric local polynomial estimation of (9.35), where they assume that the mean function is additive and the volatility function is multiplicative. Fan and Yim (2004) propose a CV method for estimating a conditional density. The bandwidth selection rule optimizes the estimated conditional density by minimizing the ISE. Fan et al. (1996) provide a similar, but ad – hoc method. McKeague and Zhang (1994) study cumulative versions of one-step lagged conditional mean and variance functions.

Section 9.1.6: Early studies on CV nonparametric lag selection consider functional relationships with conditional homoskedasticity; see, e.g., Cheng and Tong (1992), Yao and Tong (1994), and Vieu (1994, 1995). Guo and Shintani (2011) investigate the properties of the FPE lag selection procedure for nonlinear additive AR models. Also, there is an extensive literature on CV methods for the simultaneous selection of the parametric and nonparametric components in a partially linear model; see, e.g., Gao and Tong (2004), and Avramidis (2005) and the references therein. Chen et al. (1995) propose three procedures for testing additivity in nonlinear ARs of the form (9.45).

Section 9.2.1: The ACE and AVAS algorithms were originally introduced for regression modeling by Breiman and Friedman (1985); see also Hastie and Tibshirani (1990) and Tibshirani (1988).

Section 9.2.2: Following Hall (1989), a kernel-based PPR estimation method for time series has been proposed by Xia and An (1999), and applied to real data. Granger and Teräsvirta (1992b) report results of a small experiment in which linear models, PPR models, and models containing both linear and PPR terms are fitted to nonlinear time series under a variety of signal to noise cases. They conclude that when nonlinearity is strong, PPR models fit and forecast quite well, but tend to overfit the data when nonlinearity is weak.

Section 9.2.3: Lewis and Ray (2002) use TSMARS to model nonlinear threshold-type AR behavior in periodically correlated time series. A Bayesian nonparametric implementation of nonlinear AR model fitting using splines has been discussed by Wong and Kohn (1996). A Bayesian implementation of MARS, with application to time series prediction, has been given by Denison et al. (1998). In both cases, Bayesian estimation is carried out by MCMC methods. These methods generate enormous combinations of basis functions from which it is difficult to extract information on the regression structure. Sakamoto (2007) solves this problem by proposing an empirical Bayes method to select basis functions and the position of the knots. Porcher and Thomas (2003) propose a penalized least squares approach to order determination in TSMARS.

Section 9.2.4: Robinzonov et al. (2012) perform a nonlinear time series Monte Carlo comparison of glmboost, gamboost, TSMARS, BRUTO, and an algorithm due to Huang and Yang (2004). These latter authors use a stepwise procedure for the identification of nonlinear additive AR models based on spline estimation and BIC. Robinzonov et al. (2012) conclude that boosting is superior to its rivals in discovering the true nonlinear DGP. From a computational point of view, Schmid and Hothorn (2008) advocate the use of component P-splines based learners with the shrinkage parameter vector estimated via penalized least squares; see also Shafik and Tutz (2009) for the corresponding boosting algorithm. Some ideas to address the multivariate generalization of boosting are provided by Lutz et al. (2008). Assaad et al. (2008) adopt the boosting algorithm for predicting future time series values using recurrent NNs as base learners. For an overview on boosting in general, we refer to Bühlmann and Hothorn (2007).

Section 9.2.5: Chen and Liu (2001) place the estimation of (9.61) in the smoothing context, proposing an LL regression estimate of $\phi_i(\cdot)$ $(i = 1, \ldots, p)$. In addition, these authors give two test statistics. One for assessing whether all the coefficient functions are constant. The second one tests if all the coefficient functions are continuous. A small MC simulation study complements the paper. Chen and Wang (2011) investigate some probabilistic properties (stationarity and invertibility) of combined AR–FCMA models. Chen and Huo (2009) provide an approach that generalizes smoothing splines to high dimensions (> 3 covariates) and is relatively free from formulational assumptions such as the restricted number of covariates in the FCAR models; MATLAB and R codes are available at http://www.tandfonline.com/doi/suppl/10.1198/jcgs.2009.08040?scroll=top.

Matsuda (1998) proposes an alternative GOF test statistic to determine whether the coefficient functions are constant or take a particular parametric form. Although the test statistic has asymptotically a χ^2 distribution under certain regularity conditions, he finds that a bootstrap method provides better significance levels in practice. Cai et al. (2000a) provide details of estimating varying-coefficient models in a regression setting. Cai et al. (2009) consider the estimation of a generalized functional coefficient regression model with nonstationary covariates.

Section 9.2.6: Wu et al. (2011) recommend to estimate the univariate varying-coefficient functions in the single-index model by P-splines. This approach provides an explicit fit which allows the authors to conduct multi-step ahead out-of-sample forecasting. The paper includes implementation details of the proposed estimation algorithm. Wu et al. (2010) introduce LL estimation for quantile regression via single-index models as well as some computational algorithms.

9.5 Data and Software References

Data

Example 9.2: The data on the Old Faithful geyser in Yellowstone National Park, Wyoming, USA, are taken from Azzalini and Bowman (1990, Table 1). The data set, containing 299 observations on the duration of eruptions and the waiting time between the starts of the successive eruptions, can be downloaded from the website of this book. The duration measurements with codes L (long), M (medium), and S (short) are recoded as 4, 3, and 2 minutes, respectively. This data set is more complete than the one in the R-datasets package, and the numbers are slightly different. The stacked conditional density plot can be obtained using the R-hdrcde package.

Example 9.3: The river flow data were made available by Peter C. Young of Lancaster University. Previous analysis of this series can be found in Young (1993) and Young and Beven (1994) and references therein; see also De Gooijer and Gannoun (2000), and Polinik and Yao (2000). The data set, including hourly observations on rainfall, can be downloaded from the website of this book.

Example 9.5: The SST data set can be downloaded from the website of this book. Previous studies of daily SSTs at Granite Canyon include Breaker and Lewis (1988), Lewis and Ray (1993, 1997), and Breaker (2006).

Example 9.7: The subset SST data set includes time series on (interpolated) water salinity, and sine and cosine terms. These series can be used in the TSMARS model as potential predictors to investigate whether the observed cyclic effects (see Figure 9.9(a)) are wind driven. Missing values in the wind direction series are filled in using the wind direction series from the same date of a different year.

Exercise 9.2: The monthly GSL data were made available by David Tarboton (Utah State University). The measured dates are reported by the U.S. Geological Survey (USGS).

9.5 DATA AND SOFTWARE REFERENCES

Software References

Section 9.1: A kernel smoothing MATLAB toolbox is available at http://nl.mathworks.com/matlabcentral/linkexchange/links/3551-kernel-smoothing-toolbox as a part of the book by Horová et al. (2012). The toolbox contains menu-driven functions for the estimation of: univariate densities, distribution functions, quality indices, hazard functions, regression functions, and multivariate densities. Various alternative software codes can be downloaded from MATLAB Central. For instance, ksr (Gaussian kernel smoothing regression), ksrlin (local linear Gaussian kernel regression), and smoothing (Nadaraya–Watson smoothing with GCV). Also, several R packages for kernel smoothing are available. For instance, ksmooth {stats} (NW estimator (local constant fit), univariate x only, no automatic bandwidth selection), and sm (nonparametric smoothing methods described in Bowman and Azzalini (1997)).

KDE is a general MATLAB class for k-dimensional kernel density estimation (written in a mix of "m" files and MEX/C++ code); see http://www.ics.uci.edu/~ihler/code/kde.html. There are various R-packages available. For instance, sskernel (kernel density estimation with an automatic bandwidth selection), gkde (Gaussian kernel density estimation with bounded support), kerdiest (kernel estimators of the distribution function and related functionals, with several CV bandwidth methods), KernSmooth (local linear or quadratic kernel smoothing; up to bivariate density estimation with restricted bandwidths; see Wand and Jones (1995)), and ks (kernel smoothing; kernel density estimation; kernel discriminant analysis; two- to six-dimensional data; general bandwidths).

An extensive set of semi- and nonparametric methods comes with the interactive commercial statistical computing environment XploRe. Using this software, it is easy to reproduce many of the examples in the book by Härdle (1990). XploRe is not sold anymore. However, the last version, 4.8, can be freely downloaded from the website http://sfb649.wiwi.hu-berlin.de/fedc_homepage/xplore.php.

MATLAB code (mean_median.m) for obtaining the conditional mean and the conditional median forecasts, using single- and multi-stage methods, can be downloaded from the website of this book. The solutions manual (Exercise 9.4) contains MATLAB code for computing the conditional mean, median, and mode.

Section 9.1.4: The R-packages knn, class, and FNN (fast nearest neighbor) contain k-NN implementations. A related package is knnflex; see http://cran.r-project.org/src/contrib/Archive/knnflex/. The R-kknn package performs weighted k-NN. The R-KODAMA (KnOwledge Discovery by Accuracy MAximization) package contains the function KNN.CV which performs a 10-fold CV bandwidth selection on a given data set using k-NN.

The MATLAB function knn.m is available at MATLAB Central. Related MATLAB functions are kNearestNeighbors, knnsearch, and knnclassify. Alternatively, a MATLAB package for obtaining one-step ahead k-NN forecasts is available at https://sites.google.com/site/marceloperlin/.

The working paper "Computing nonparametric functional estimates in semiparametric problems" by Miguel A. Delgado (http://orff.uc3m.es/bitstream/handle/10016/5821/we9217.PDF) offers a set of FORTRAN77 routines including k-NN, kernel regression with symmetric and possibly non-symmetric kernels, and nonparametric k-NN regression.

The Loess/Lowess methodology of Cleveland (1979) is implemented in the R (S-Plus) functions lowess and loess including their iterative robust versions. The loess function (local linear or quadratic fits, multivariate x's, no automatic bandwidth selection) is more flexible and powerful. For large sample sizes, however, the computations can be time-consuming. Cleveland et al. (1990) develop a seasonal adjustment algorithm based on robust loess. It is implemented in the R (S-Plus) function stl. The curve fitting toolbox in MATLAB contains the function smooth with the loess/lowess methods and their robust variants.

Section 9.1.5: Ox code (Test-Algorithm-93.ox) for obtaining the Markov forecast densities, as summarized in Algorithm 9.3, is available at the website of this book.

Section 9.1.6: The lag selection methods $\widehat{\text{AFPE}}$, $\widehat{\text{CAFPE}}$, and MSFE are options within the freely available computer package JMulTi, a JAVA application designed for the specific needs of time series econometrics. The package can be downloaded from http://www.jmulti.de/download.html.

Section 9.2.1: ACE and AVAS are implemented in the R-acepack package. S-Plus has an implementation of both algorithms too, called ace and avas, respectively. The FORTRAN77 source codes of Friedman's ACE algorithm, and PPR are available from http://www-stat.stanford.edu/~jhf/ftp/progs/. A FORTRAN90 version of the ACE algorithm (mace.f90) can be downloaded from Alan Miller's FORTRAN software webpage at http://jblevins.org/mirror/amiller/. The MATLAB–ACE algorithm, using adaptive partitioning to calculate the conditional expectations, and the supersmoother algorithm are available from the MATLAB archive. The function areg in the R-Hmisc package offers the option to control the smoothness of the transformation in ACE.

Section 9.2.2: PPR is implemented in the R-stats package as the function ppr, and within S-Plus it is called ppreg. Both functions are based on the so-called *smooth multiple additive regression technique* (SMART) of Friedman (1984). As explained in Section 9.2.2, SMART modeling is a generalization of PPR (Friedman and Stuetzle, 1981).

Section 9.2.3:
MARS and BRUTO are provided in the R-mda package. A new, slightly more flexible alternative implementation of MARS (fast MARS) is in the R-earth package. A commercial version of MARS is available from http://www.salford-systems.com/products/mars. ARESLab is an Adaptive Regression Splines toolbox for MATLAB/Octave, which can be downloaded from Gints Jekabsons' webpage at http://www.cs.rtu.lv/jekabsons/regression.html.

Section 9.2.4: There are several implementations of boosting techniques, available as add-ons for R. Both procedures glmboost and gamboost are contained in the packages mboost and GAMBoost. The first package provides an implementation for fitting GLMs, as well as additive gradient-based boosting. GAMBoost contains an implementation of likelihood boosting as proposed by Tutz and Binder (2006).

Section 9.2.5: The results in Example 9.9 were obtained with the S-Plus code to accompany the book by Fan and Yao (2003); see http://orfe.princeton.edu/~jqfan/fan/nls.html.

Section 9.2.6: The simulation results in Example 9.10 were obtained using SAS code,[3] provided by Yingcun Xia. The epls.sas code is available at the website of this book.

---

[3]SAS is a registered trademark of SAS Institute, Inc.

Exercises

Empirical and Simulation Questions

9.1 Consider the NLAR(1) process

$$Y_t = \sin(Y_{t-1}) + \varepsilon_t, \quad \{\varepsilon_t\} \overset{\text{i.i.d.}}{\sim} \mathcal{N}(0,1).$$

(a) The file Yt-n500-sinus.dat contains $T = 500$ simulated data points from the above process. Compute the NW local constant smoother $\widehat{\mu}_h^{\text{NW}}(\mathbf{x})$ with $\mathbf{x} = Y_{t-1}$ equally spaced in the range $[-2, 2]$, $h \equiv h_T = 0.02$, and with the Epanechnikov kernel (see Table 7.7). If $h \to 0$, what happens with $\widehat{\mu}_h^{\text{NW}}(\cdot)$?

(b) Repeat part (a) using the local linear smoother $\widehat{\mu}_h^{\text{LL}}(\cdot)$, with a Gaussian kernel.

(c) Plot both kernel regression estimates jointly with the true regression function, and the generated data. Comment on the results.

(d) Repeat part (a) using the plug-in bandwidth $h_{\text{rot}} = \widehat{\sigma}_Y T^{-1/5}$. Compare all kernel regression estimates. Is there any observable difference? Why?

[*Hint*: Use the MATLAB-ksrgress function or a similar interactive package.]

9.2 A simple algorithm (Jaditz and Sayers, 1998) for NN estimation of $\mu(\mathbf{x}) = \mathbb{E}(Y_{t+1}|\mathbf{X}_t = \mathbf{x})$ goes as follows. For a given lag length p, let $\{(Y_t, \mathbf{X}_t)\}_{t=1}^T$ be a set of available observations where $\mathbf{X}_t = (Y_t, Y_{t+1}, \ldots, Y_{t+p-1})'$. Divide the data in a *prediction* set $\mathcal{P} = \{(Y_t, \mathbf{X}_t) : N_f < t \leq T\}$ and, for some $N_f < T$ a *fitting* (training) set $\mathcal{F}_t$. For each $Y_t \in \mathcal{P}$ calculate the distance between $\mathbf{X}_t = \mathbf{x}$ and $\mathbf{X}_i \; \forall i \in \mathcal{F}_t$ using the supremum norm. Sort the data according to the distance. Then, for a given number of NNs, select the k_n ($n = T - p$) nearest pairs to estimate the parameters α_{0,k_n} and $\boldsymbol{\alpha}_{p,k_n} = (\alpha_{1,k_n}, \ldots, \alpha_{p,k_n})'$ in the local linear regression model, $Y_{(i)} = \alpha_{0,k_n} + \mathbf{X}_{(i)}'\boldsymbol{\alpha}_{p,k_n} + \varepsilon_{(i),k_n}$ with $\{\varepsilon_{(i),k_n}\}$ a zero-mean WN process. Next, use the estimated parameters $\widehat{\alpha}_{0,k_n}$ and $\widehat{\boldsymbol{\alpha}}_{p,k_n}$ to calculate the one-step ahead forecast $\widehat{Y}_{(i)+1|(i)} = \widehat{\alpha}_{0,k_n} + \mathbf{X}_{(i)}'\widehat{\boldsymbol{\alpha}}_{p,k_n}$, and the associated one-step ahead forecast error $e_{(i)+1|(i)} = Y_{(i)+1} - \widehat{Y}_{(i)+1|(i)}$. Pick the value of k_n that minimizes the MSFE. Finally, given the specified number of NNs, say k_n^*, rebuild the data set to replicate the regression. Then, in the present setting, the k-NN estimator for $\mu(\mathbf{x})$ is defined as $\widehat{\mu}^{k\text{-NN}}(\mathbf{x}) = (1/k_n^*) \sum_{\mathbf{x}_{(i)} \in \mathcal{F}_t, i \in N(\mathbf{x})} Y_{(i)+1}$; see (9.33) for a more general case.

(a) Using your favorite programming language, write a computer code to obtain H one-step ahead forecasts for the above k-NN regression algorithm. Include a "robust" matrix inversion routine as a provision for near-singular matrices $\mathbf{X}_{(i)}'\mathbf{X}_{(i)}$.

The Great Salt Lake (GSL) of Utah is the fourth largest, perennial, closed basin, saline lake in the world. Monthly measurements of the volume (in m^3) in the north arm of the lake from October 1949 to December 2012 (756 observations) are given in the file gsl.dat. These measurements have been investigated in an effort to understand the dynamics of the precipitous rise of the lake during the years 1983 – 1987 and its consequent rapid retreat; see, e.g., Lall et al. (1996) and Moon et al. (2008) for background information on recent analyzes. Such behavior is typical of nonlinear systems driven by large scale, persistent, climatic fluctuations.

(b) Assume the GSL time series is generated by a NLAR(2) process. Based on the first $T = 507$ observations (training set) of the standardized GSL data, obtain twelve one-step ahead forecasts. Re-estimate the model before each forecast is computed (expanding the training set) and use the following estimation methods.
- k-NN regression with the computer code from part (a). Given a fixed sample size n, comment on the choice of k_n in the limiting case $k_n = n$ and $k_n = 1$.
- locally constant kernel regression with a Gaussian product kernel and a single bandwidth obtained by CV.
[*Hint:* Use the functions npregbw and npksum in the R-np package.]
- AVAS estimation with bandwidth obtained by CV and no weights. Comment on the selected transformation of the GSL time series.
[*Hint:* Use the AVAS function in the R-acepack package.]

Comment on which method produces forecasts with smallest MSFEs over the course of the year.

9.3 The data set ExpAR2.dat contains 200 simulated data points from an ExpAR(2) model of the form

$$Y_t = \{0.9 + 0.1 \exp(-Y_{t-1}^2)\} Y_{t-1} - \{0.2 + 0.1 \exp(-Y_{t-1}^2)\} Y_{t-2} + \varepsilon_t, \ \{\varepsilon_t\} \stackrel{\text{i.i.d.}}{\sim} N(0,1).$$

(a) Check the strict stationarity of the ExpAR(2) process.
(b) Use PPR to fit a model containing $M = 2$ terms, with $p = 2$ lagged predictor variables, to the first 189 observations.
[*Hint*: Use the R-fRegression package for answering questions (b) – (d).]
(c) Fit an $m - k - 1 = 2 - 2 - 1$ ANN model to the first $T = 189$ observations using LS.
(d) Compute the one-step ahead forecasts at times $t = 190, \ldots, 200$ using a fixed, but rolling (cf. Section 10.4.1) sample size of 188 observations for the PPR and ANN models. Compare the in-sample residual variances obtained in parts (a) and (b) with the one-step ahead MSFE for the two models.

9.4 Consider the Old Faithful Geyser data introduced in Example 9.2. Here, we explore some aspects of the data that were not investigated previously. In particular, we focus on forecasting the last ten ($H_{\max} = 10$) observations of the waiting time $\{Y_t\}_{t=1}^{299}$ where t denotes the eruption number (geyser_waiting.dat). If the time to next eruption can be predicted accurately, visitors to the Yellowstone National Park could use this information to organize their visit.

(a) Recall the empirical method for selecting the Markov coefficient p in (9.10). Set $p_{\max} = 10$, $k = 60$, and take $h = \widehat{\sigma}_Y T^{-1/(p+4)}$ ($p = 1, \ldots, p_{\max}$). Verify that for the conditional mean the most appropriate order of the NLAR process equals $p = 1$, using the function $f_2(p)$ with $\{Y_t\}_{t=230}^{289}$.
(b) Using the specification in part (a), compute the conditional mean, median, and mode for $h = 1, \ldots, H_{\max}$ given the observations up to and including the waiting time at $t = 289$ ($Y_{289} = 47$). Summarize the forecast performance in terms of the MSFE and RMAFE and comment on the results.
(c) Suggest an empirical method to construct forecast intervals on the basis of the nonparametric estimates.

(d) Until now we have not used information on the eruption duration time. Based on descriptive statistics and boxplots of the waiting and duration times, the following simple (naive) forecasting rule has been suggested.[4] An eruption with a duration < 3 minutes will be followed by a waiting time of about 55 minutes, while an eruption with a duration > 3 minutes will be followed by a waiting time of about 80 minutes. For the last ten observations, compare and contrast the forecasting performance of this rule with the results obtained in part (b).

9.5 Consider the river flow data set, consisting of the hourly river flow time series $\{Y_t\}_{t=1}^{401}$ introduced in Example 9.3 (file name: flow.dat) and the hourly rainfall time series $\{X_t\}_{t=1}^{401}$ (file name: rain.dat). Following the forecasting procedure described in Example 9.8, obtain forecasts $\widehat{Y}_{t+H|t}$ from past values of $\{(Y_t, X_t)\}$ for $H = 1, 10$, and 20, with the initial information set defined from $t = 1$ until $t = 366$.

(a) Use the following methods to produce the 15 out-of-sample forecasts: glmboost, gamboost, MARS, and VAR (unrestricted). Summarize the forecasts in terms of MSFEs and discuss the results.

[*Hint*: Modify the Forecasting-USunemplmnt.r function (file: example_9-8.zip), available at the website of this book. Note, the computations can be time demanding.]

(b) Using the four forecasting methods mentioned above, obtain the MSFEs of $\{Y_t\}$ in a univariate setting. Compare your results with those obtained in part (a).

---

[4]See Chatterjee, Handcock, and Simonoff (1995, pp. 224 – 226), *A Casebook for a First Course in Statistics and Data Analysis*, Wiley.

Chapter 10

FORECASTING

As we saw in Chapter 9, it is fairly straightforward to forecast future values of a time series process using semi- and nonparametric methods, given data up to a certain time t. In contrast, the situation becomes more complicated when real out-of-sample forecast are computed from parametric nonlinear time series models; in particular, as we explain below, this is a difficult issue for $H \geq 2$ steps ahead.

To be more specific, recall that for a strictly stationary stochastic process $\{Y_t, t \in \mathbb{Z}\}$ the least squares (LS), or minimum mean squared error (MMSE), forecast of Y_{t+H} ($H = 1, 2, \ldots$), given a finite or semi-finite past history $Y_t, Y_{t-1}, \ldots$ is given by $\mathbb{E}(Y_{t+H}|Y_s, -\infty < s \leq t)$ when this exists. When we restrict attention to a pth order Markov process the MMSE forecast of Y_{t+H} equals the conditional mean, i.e. $Y_{t+H|t}^{\text{LS}} = \mathbb{E}(Y_{t+H}|\mathbf{X}_t)$, where $\mathbf{X}_t = (Y_t, Y_{t-1}, \ldots, Y_{t-p+1})'$. Calculation of $Y_{t+H|t}^{\text{LS}}$ requires knowledge of the conditional pdf of $\{Y_t, t \in \mathbb{Z}\}$, which is a substantial task in general. The task becomes easier for a NLAR(p) model

$$Y_t = \mu(\mathbf{X}_{t-1}; \boldsymbol{\theta}) + \varepsilon_t, \tag{10.1}$$

where $\{\varepsilon_t\} \overset{\text{i.i.d.}}{\sim} (0, \sigma_\varepsilon^2)$ such that ε_t is independent of $\mathbf{X}_{t-1}$, $\boldsymbol{\theta}$ is a finite-dimensional vector of unknown parameters, and $\mu: \mathbb{R}^p \to \mathbb{R}$. Given (10.1), the one-step ahead LS forecast at time t equals

$$Y_{t+1|t}^{\text{LS}} = \mathbb{E}(Y_{t+1}|\mathbf{X}_t) = \mathbb{E}\{\mu(\mathbf{X}_t; \boldsymbol{\theta}) + \varepsilon_{t+1}|\mathbf{X}_t\} = \mu(\mathbf{X}_t; \boldsymbol{\theta}). \tag{10.2}$$

So, for $H = 1$, the conditional mean is independent of the distribution of ε_{t+1} which is an important property for both linear and NLAR models. When $H \geq 2$, however, this is true only for linear models.

For example, the two-step ahead LS forecast for model (10.1) is given by

$$\begin{aligned} Y_{t+2|t}^{\text{LS}} &= \mathbb{E}(Y_{t+2}|\mathbf{X}_t) = \mathbb{E}\{\mu(\mathbf{X}_{t+1}; \boldsymbol{\theta}) + \varepsilon_{t+2}|\mathbf{X}_t\} \\ &= \mathbb{E}\{\mu(\mu(\mathbf{X}_t; \boldsymbol{\theta}) + \varepsilon_{t+1})|\mathbf{X}_t\} = \int_{-\infty}^{\infty} \mu(\mu(\mathbf{X}_t; \boldsymbol{\theta}) + \varepsilon)\mathrm{d}F(\varepsilon), \end{aligned} \tag{10.3}$$

where $F(\cdot)$ is the distribution function of $\{\varepsilon_t\}$. Thus, the second term on the right-hand side of (10.3) depends on $F(\cdot)$, and cannot further be reduced as in (10.2). The reason is that, in general, the conditional expectation of a nonlinear function is not equal to the function evaluated at the expected value of its argument.

From the above results one may erroneously conclude that it is not possible to obtain closed-form analytical expressions for $H \geq 2$ forecasts. However, by using the so-called *Chapman–Kolmogorov* recurrence relationship, "exact" LS multi-step ahead forecasts for general NLAR models can, in principle, be obtained through complex numerical integration as we will see in Section 10.1.[1] The section also describes two "exact" forecast strategies for SETARMA models.

An alternative way to obtain more than one-step ahead forecasts, and possibly the nearest one can get to an explicit analytical form, is a numerical approximation (Monte Carlo simulation, bootstrap and related methods), a series expansion, or by assuming that the innovation distribution is known. Applying these and some other approaches, we discuss seven approximate methods for making point forecasts in Section 10.2.

With point forecasts, the accuracy is often measured by the forecast error variance or by a forecast interval. In Section 10.3, we address the problem of constructing (bootstrap) forecast intervals and regions for nonlinear and nonparametric ARs. We make a distinction between percentile- and density-based forecast intervals. The latter intervals are often more informative than the former when, for instance, the forecast distribution is asymmetric or multimodal. In Section 10.4, we provide a limited review of measures evaluating the accuracy of competing point forecasts. In the same vein, this section gives a description of methods for interval and density evaluation. Finally, in Section 10.5, we briefly discuss methods for optimal forecast combination. By combining forecasts of different models/methods instead of relying on individual forecasts, forecast accuracy can often be improved.

10.1 Exact Least Squares Forecasting Methods

10.1.1 Nonlinear AR model

Consider the NLAR(p) model as given by (10.1) and assume that the process $\{Y_t, t \in \mathbb{Z}\}$ is strictly stationary. Let $g(\cdot)$ be the pdf of $\{\varepsilon_t\}$. By using the Chapman–Kolmogorov relation, the conditional pdf of Y_{t+H} given $\mathbf{X}_t = \mathbf{x}_t$ can be written as

$$f(y_{t+H}|\mathbf{x}_t) = \int_{-\infty}^{\infty} f(y_{t+H}|\mathbf{x}_{t+1}) f(y_{t+1}|\mathbf{x}_t) \mathrm{d}x_{t+1}, \qquad (10.4)$$

---

[1] As noted above, the solution of the Chapman–Kolmogorov recurrence relationship requires numerical integration techniques. The quotation marks around "exact" are put there to emphasize that the numerical accuracy of $H \geq 2$ forecasts depends on certain tuning parameters. For instance, a change of variable of integration to get a finite range, and the judicious choice of weights and abscissae of a numerical integration method.

10.1 EXACT LEAST SQUARES FORECASTING METHODS

where

$$f(y_{t+1}|\mathbf{x}_t) = g(y_{t+1} - \mu(\mathbf{x}_t;\boldsymbol{\theta})).$$

Alternatively, this equation can be obtained by considering the joint pdf of Y_{t+H}, $Y_{t+H-1}, \ldots, Y_{t+1}$ conditional on $\mathbf{X}_t = \mathbf{x}$ and integrating out the unwanted variables.[2] Introducing the short-hand notation $f_H(\cdot) = f_{Y_{t+H}|Y_t}(\cdot|\mathbf{x})$, equation (10.4) immediately gives

$$f_H(x) = \int_{-\infty}^{\infty} f_{H-1}(x) g(z - \mu(\mathbf{x};\boldsymbol{\theta})) dz. \tag{10.5}$$

Thus, starting from $f_1(x) = g(x - \mu(\mathbf{X}_t;\boldsymbol{\theta}))$, equation (10.5) is a recursive formula for evaluating the conditional density. Given $f_H(\cdot)$ at step $H = 1$, the conditional mean for $H \geq 2$ can be calculated using

$$Y_{t+H|t} = \int_{-\infty}^{\infty} f_{H-1}(Y_{t+1}) g(Y_{t+1} - \mu(\mathbf{X}_t;\boldsymbol{\theta})) dY_{t+1}. \tag{10.6}$$

Similarly, a recurrence relation for the jth ($j = 1, 2, \ldots$) conditional moment is given by

$$\mathbb{E}(Y_{t+H}^j | \mathbf{X}_t = \mathbf{x}) = \int_{-\infty}^{\infty} f_{H-1}(Y_{t+1}^j) g(Y_{t+1} - \mu(\mathbf{X}_t;\boldsymbol{\theta})) dY_{t+1}. \tag{10.7}$$

Except for some special cases of $\mu(\cdot;\boldsymbol{\theta})$, the integral equations (10.5) and (10.6) do not readily admit explicit analytic solutions. To evaluate (10.6) numerically, each forecasting step requires $p + 1$ numerical integrations. Standard numerical integration methods can be used for this purpose, but care must be taken to handle accumulation of rounding errors; see, e.g., Pemberton (1987), Al-Qassem and Lane (1989), and Cai (2005).

Example 10.1: Forecast Density

Consider the SETAR$(2;0,0)$ model

$$Y_t = \begin{cases} \alpha + \varepsilon_t & \text{if } Y_{t-1} \leq 0, \\ -\alpha + \varepsilon_t & \text{if } Y_{t-1} > 0, \end{cases} \tag{10.8}$$

where $\{\varepsilon_i\} \overset{\text{i.i.d.}}{\sim} \mathcal{N}(0,1)$. In the sequel, $\varphi(\cdot)$ denotes the pdf and $\Phi(\cdot)$ the CDF of $\mathcal{N}(0,1)$. Then the stationary marginal pdf of $\{Y_t, t \in \mathbb{Z}\}$ is given by

$$f(y_t) = \{\varphi(y_t + \alpha) + \varphi(y_t - \alpha)\}/2. \tag{10.9}$$

---

[2] For economy of notation, we suppress the dimension of the information set on which the conditional density forecast is conditioned.

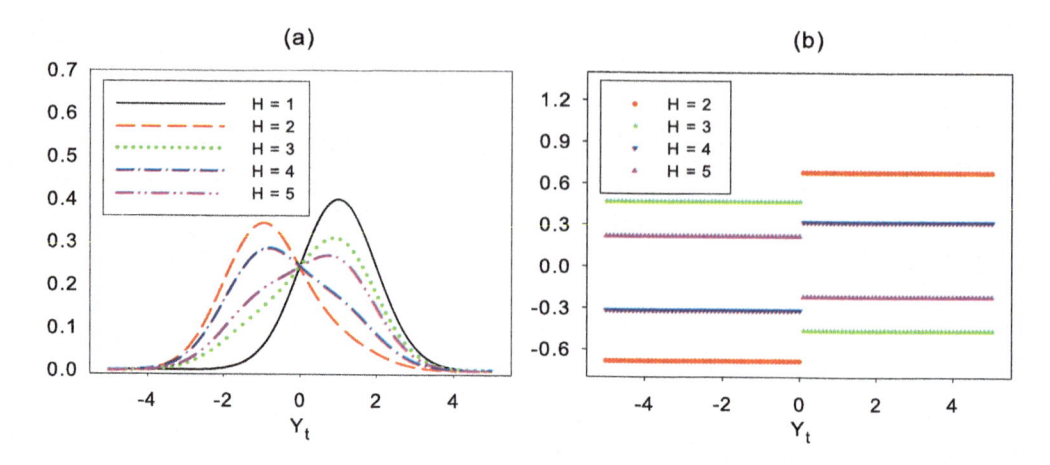

Figure 10.1: *(a) Forecast density $f(y_{t+H}|\mathbf{x}_t)$ ($H = 1, \ldots, 5$) for the SETAR(2; 0, 0) model (10.8); (b) Conditional mean $\mathbb{E}(Y_{t+H}|\mathbf{X}_t)$ ($H = 2, \ldots, 5; \alpha = 1$).*

The exact (LS) conditional pdf of Y_{t+H} ($H = 1, 2, \ldots$) given $\mathbf{X}_t = \mathbf{x}$ has the form

$$f(y_{t+H}|\mathbf{x}) = w_1^{(H)}(\beta)\varphi\big(Y_{t+H} - I(Y_t \leq 0)\alpha\big) + w_2^{(H)}(\beta)\varphi\big(Y_{t+H} + I(Y_t > 0)\alpha\big), \tag{10.10}$$

where $w_1^{(H)}(\beta) = (1 - \beta^{H-1})/2$, $w_2^{(H)}(\beta) = 1 - w_1^{(H)}(\beta)$, and $\beta = 1 - 2\Phi(\alpha)$; cf. Exercise 10.3. From (10.10), the conditional mean and the conditional variance are given by respectively

$$\mathbb{E}(Y_{t+H}|\mathbf{X}_t = \mathbf{x}) = \alpha\beta^{H-1}I(Y_t \leq 0) - \alpha\beta^{H-1}I(Y_t > 0),$$
$$\text{Var}(Y_{t+H}|\mathbf{X}_t = \mathbf{x}) = 1 + \alpha - \mathbb{E}^2(Y_{t+H}|\mathbf{X}_t = \mathbf{x}).$$

Note that the skewness of $f(y_{t+H}|\mathbf{x})$ is affected by both H and β which determine the weights $w_i^{(H)}(\beta)$ ($i = 1, 2$) of the linear combination of $\varphi(Y_{t+H} + \alpha)$ and $\varphi(Y_{t+H} - \alpha)$; see Figure 10.1(a). Figure 10.1(b) shows plots of the H-step ahead conditional mean.

10.1.2 Self-exciting threshold ARMA model

It will often be the case that $\mu(\cdot; \boldsymbol{\theta})$ in (10.1) has a much more complicated functional form than, for instance, the SETAR model considered in Example 10.1. So the analytic solution to (10.6) is not available. Still, after some algebra, the stationary k-regime SETARMA model introduced in (2.29) allows for explicit expressions of the multi-step forecast and the variance of the forecast error, assuming the model is invertible. To reduce the burden of notation, we focus on the SETARMA$(2; p_1, q_1, p_2, q_2)$ model (6.8) with the same error distribution in both regimes.

10.1 EXACT LEAST SQUARES FORECASTING METHODS

From (6.8), we observe that the two-regime SETARMA model can be written as

$$Y_t = \{\phi_0^{(1)} + \phi_{p_1}^{(1)}(B)Y_t + \psi_{q_1}^{(1)}(B)\varepsilon_t\}I(Y_{t-d} \leq r) \\ + \{\phi_0^{(2)} + \phi_{p_2}^{(2)}(B)Y_t + \psi_{q_2}^{(2)}(B)\varepsilon_t\}(1 - I(Y_{t-d} \leq r)), \quad (10.11)$$

where $\phi_{p_i}^{(i)}(B) = \sum_{j=1}^{p_i} \phi_j^{(i)} B^j$ and $\psi_{q_i}^{(i)}(B) = 1 + \sum_{j=1}^{q_i} \psi_j^{(i)} B^j$ ($i = 1, 2$). Denote the indicator process by $I_{t-d} \equiv I(Y_{t-d} \leq r)$, and the ARMA process in the ith regime by $Y_t^{(i)} \sim \text{ARMA}(p_i, q_i)$. Then (10.11) can be written more compactly as

$$Y_t = Y_t^{(1)} I_{t-d} + Y_t^{(2)} (1 - I_{t-d}). \quad (10.12)$$

Now assume that the joint process $\{(Y_t^{(1)}, Y_t^{(2)}, I_{t-d}), t \in \mathbb{Z}\}$ is strictly stationary, invertible, and ergodic. The exact H-step ahead ($H \geq 2$) LS forecast of (10.11) is given by

$$Y_{t+H|t}^{\text{LS}} = Y_{t+H|t}^{(1)} \mathbb{E}(I_{t+H-d}|\mathcal{F}_t) + Y_{t+H|t}^{(2)} (1 - \mathbb{E}(I_{t+H-d}|\mathcal{F}_t)), \quad (10.13)$$

where $Y_{t+H|t}^{(i)}$ is the ARMA forecast in regime i, and $\mathcal{F}_t = \{Y_t, Y_{t-1}, \ldots\}$ denotes the information set up to time t. Depending on the case $H \leq d$ or $H > d$, there are various approaches to calculate the forecast and the forecast error variance.

Case $H \leq d$: It is easy to see that $Y_{t+H|t}^{\text{LS}}$ is an unbiased estimator of Y_{t+H}. Moreover, the variance of the LS forecast error $e_{t+H|t}^{\text{LS}} = Y_{t+H} - Y_{t+H|t}^{\text{LS}}$ is given by

$$\text{Var}(e_{t+H|t}^{\text{LS}}) = \sigma_\varepsilon^2 \sum_{j=1}^{H-1} \left[(\omega_j^{(1)})^2 I_{t+H-d} + (\omega_j^{(2)})^2 (1 - I_{t+H-d}) \right], \quad (10.14)$$

where $\omega_j^{(i)} = \sum_{s=0}^{j-1} \phi_s^{(i)} \omega_{j-s}^{(i)} - \psi_j^{(i)}$ ($i = 1, 2; j \geq 1$) with $\omega_0^{(i)} = 1$, and $\psi_j^{(i)} = 0$ for $j > q_i$.

Case $H > d$: Observe that $Y_{t+H-d} \notin \mathcal{F}_t$. So the value of the threshold variable is unknown. This makes the computation of the LS forecast more complicated. For this case Amendola et al. (2006b) suggest the following forecast strategies.

- *Least squares* (LS) *forecast*: Clearly, under the stationarity assumption, I_{t+H-d} becomes a Bernoulli random variable i_{H-d} according to

$$i_{H-d} = \begin{cases} 1 & \text{with } \mathbb{P}(Y_{t+H-d} \leq r | \mathcal{F}_t) \equiv p_{(H-d)}, \\ 0 & \text{with } \mathbb{P}(Y_{t+H-d} > r | \mathcal{F}_t) \equiv 1 - p_{(H-d)}. \end{cases} \quad (10.15)$$

Thus, the indeterminacy regarding the future now hinges on $p_{(H-d)}$. In this case, the LS forecast in (10.13) reduces to

$$Y_{t+H|t}^{\text{LS}} = Y_{t+H|t}^{(2)} + p_{(H-d)} (Y_{t+H|t}^{(1)} - Y_{t+H|t}^{(2)}), \quad (10.16)$$

and the LS forecast error variance becomes

$$\text{Var}(e^{\text{LS}}_{t+H|t}) = \text{Var}(e^{(2)}_{t+H|t}) + p \cdot [\text{Var}(e^{(1)}_{t+H|t}) - \text{Var}(e^{(2)}_{t+H|t})]$$
$$+ (p + p^2_{(H-d)} - 2p \cdot p_{(H-d)})$$
$$\times [\text{Var}(Y^{(1)}_{t+H|t}) + \text{Var}(Y^{(2)}_{t+H|t}) - 2\sigma_\varepsilon^2 \sum_{j=h}^{\infty} \omega_j^{(1)} \omega_j^{(2)}], \quad (10.17)$$

where $e^{(i)}_{t+H|t}$ is the forecast error in regime i, p the unconditional expected value of I_{t+H-d}, $\text{Var}(Y^{(i)}_{t+H|t}) = \sigma_\varepsilon^2 \sum_{j=h}^{\infty} (\omega_j^{(i)})^2$ the forecast variance in regime i ($i = 1, 2$), and the last term in squared brackets in (10.17) denotes the covariance between the forecasts generated from the two regimes.

- *Plug-in* (PI) *(or naive, or skeleton) forecast:* Assume that the last predicted values are the true values $Y^{\text{PI}}_{t+H|t} = \mathbb{E}(Y_{t+H}|\mathcal{F}_{t+H-d})$ where $\mathcal{F}_{t+H-d} = \{Y_1, \ldots, Y_t, Y_{t+1|t}, \ldots, Y_{t+H|t}\}$ is the augmented information set. Then the indicator function I_{t+H-d} becomes

$$i_{t+H-d} = [I_{t+H-d}|\mathcal{F}_{t+H-d}] = \begin{cases} 1 & \text{if } Y_{t+H-d} \leq r, \\ 0 & \text{if } Y_{t+H-d} > r. \end{cases} \quad (10.18)$$

So, on replacing $p_{(H-d)}$ in (10.16) by i_{t+H-d}, we obtain the PI forecast with corresponding forecast error variance $\text{Var}(e^{\text{PI}}_{t+H|t})$.

We note that the LS and PI forecasts strategies make use of the available information set differently. Nevertheless, it is easy to prove that both $Y^{\text{LS}}_{t+H|t}$ and $Y^{\text{PI}}_{t+H|t}$ are unbiased estimators of Y_{t+H}. However, in terms of minimum MSFE, the gain in using one method over the other comes from their forecast error variances. Since $p_{(H-d)} \to p$, as $T \to \infty$, it can be deduced that $\text{Var}(e^{\text{LS}}_{t+H|t}) \geq \text{Var}(e^{\text{PI}}_{t+H|t})$ if $Y_{t+H-d} \geq r$ and $\text{Var}(e^{\text{LS}}_{t+H|t}) \leq \text{Var}(e^{\text{PI}}_{t+H|t})$ if $Y_{t+H-d} < r$. As an immediate result Amendola et al. (2006b) propose the *combined* (C) forecast

$$Y^{\text{C}}_{t+H|t} = Y^{\text{PI}}_{t+H|t} i_{t+H-d} + Y^{\text{LS}}_{t+H|t}(1 - i_{t+H-d}), \quad (10.19)$$

with i_{t+H-d} the indicator function given by (10.18). Accordingly, the combined forecast is as good as the best of the two forecast methods LS and PI. Note that in practice a reasonable approximation of $p_{(H-d)}$ ($H > d$) is needed for all three forecast strategies, and hence the quotation marks around "exact'.

Example 10.2: Comparing LS and PI Forecast Strategies

To evaluate the performance of the LS and PI forecast strategies, we consider the SETARMA$(2; 1, 1, 1, 1)$ model with $d = 1$ and parameter vectors $\boldsymbol{\theta} = (\phi_0^{(1)}, \phi_1^{(1)}, \psi_1^{(1)}, \phi_0^{(2)}, \phi_1^{(2)}, \psi_1^{(2)}, r)' = (0, 0.6, -0.7, 0, 0.4, 0.5, 0)'$, and $\boldsymbol{\theta} = (0.6, 0.6, -0.7, -1, 0.4, 0.5, 0)'$. So the difference between these models is

10.1 EXACT LEAST SQUARES FORECASTING METHODS

Table 10.1: *Averaged MSFEs and MAFEs for the least squares (LS), plug-in (PI), and combined (C) forecast strategies for the SETARMA(2; 1, 1, 1, 1) models specified in Example 10.2; $T = 250$, and 1,000 MC replications.*

Strategy	\multicolumn{4}{c}{SETARMA without intercept}				\multicolumn{4}{c}{SETARMA}			
	$H=2$	$H=3$	$H=4$	$H=5$	$H=2$	$H=3$	$H=4$	$H=5$
	\multicolumn{8}{c}{MSFE}							
LS	1.382	1.248	1.191	1.155	2.258	1.914	1.732	1.633
PI	1.388	1.255	1.197	1.160	3.076	2.832	2.629	2.490
C	1.399	1.264	1.203	1.165	2.620	2.292	2.049	1.896
	\multicolumn{8}{c}{MAFE}							
LS	0.944	0.884	0.862	0.846	1.223	1.116	1.053	1.016
PI	0.948	0.887	0.865	0.848	1.482	1.403	1.343	1.297
C	0.953	0.891	0.867	0.850	1.348	1.235	1.161	1.104

that the second model has intercept terms while the first one has not. It is well known that non-zero intercepts can greatly extenuate or attenuate the relative forecast performance of the SETARMA model. The number of MC replications is set to 1,000 with $\{\varepsilon_i\} \overset{\text{i.i.d.}}{\sim} \mathcal{N}(0,1)$, and $T = 250$. The forecast horizon H ranges from 1 to 5. The probability $p_{(H-d)}$ $(H > d)$ is estimated as $\sum_{t=d+1}^{T} I(Y_{t-d} \leq r)/(T-d)$.

Figure 10.2 shows boxplots of the forecast errors $e_{t+H|t}$ of the LS and PI forecast strategies for the SETARMA models. Observe that the variability in $e_{t+H|t}$ differs for the SETARMA model with and without intercept. This phenomenon also appears in the sample means of the forecast errors, which for the LS strategy are ranging between $[-0.027, -0.080]$ and $[0.083, 0.414]$, respectively. For the PI strategy the range of the two sets of forecast errors are given by $[0.025, -0.083]$ and $[0.083, 0.429]$. Clearly, there is a difference between the forecasts from the two SETARMA models. This confirms results in other studies: the sign and magnitude of the intercept in the SETARMA model have a large effect on the forecast performance of a particular method.

Table 10.1 shows the averaged (over all replications) MSFEs and MAFEs for $H = 2, \ldots, 5$ of $Y_{t+H|t}^{\text{LS}}$, $Y_{t+H|t}^{\text{PI}}$, and $Y_{t+H|t}^{\text{C}}$ with starting-point $t = 250$. For the SETARMA model without intercept, there is not much to be gained in terms of out-of-sample forecasting by using the LS, PI, or C forecast strategy. We also see that for the SETARMA model with intercept term, the LS forecast strategy renders superior forecasts for all forecast horizons. The combined method performs second best, whilst the PI method is generally the worst over the horizons considered.

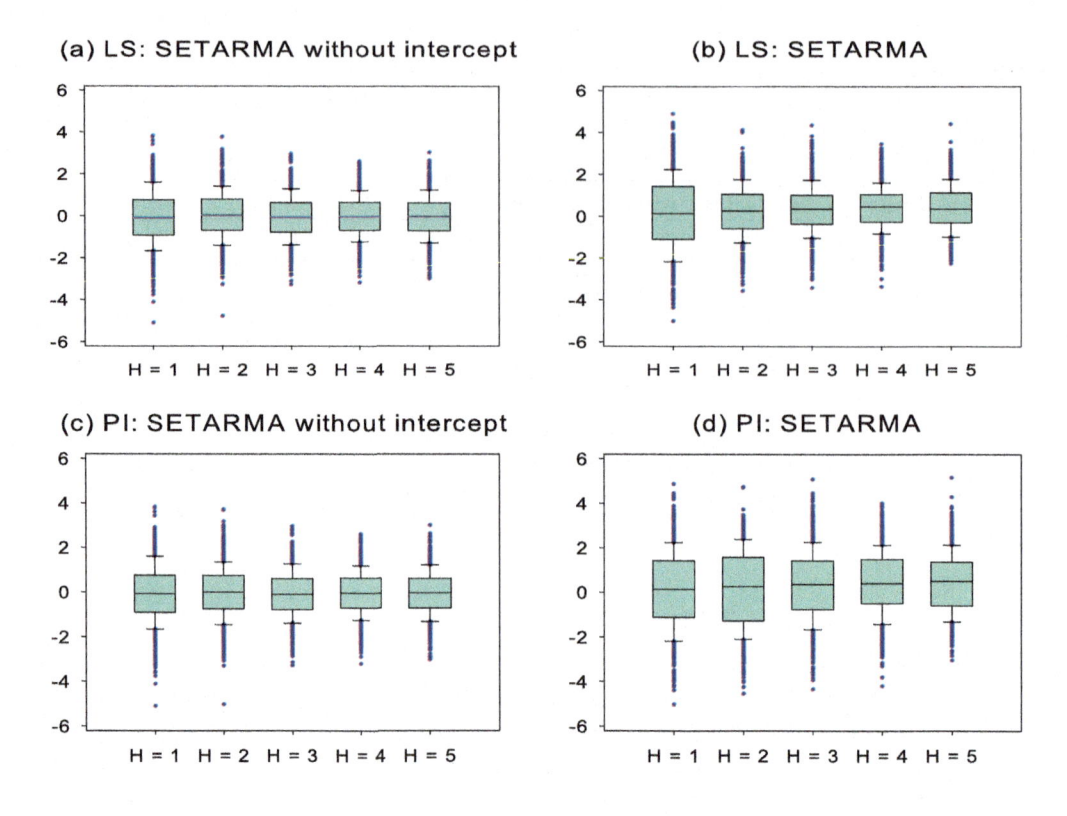

Figure 10.2: *Boxplots of the forecast errors of the* LS *and* PI *forecast strategies;* $T = 250$, *1,000 MC replications.*

10.2 Approximate Forecasting Methods

In this section, we briefly outline a number of approximate methods for obtaining multi-step ahead forecasts from a NLAR(1) model. The methods can all be generalized in a fairly straightforward manner to the NLAR(p) model (10.1).

10.2.1 Monte Carlo

Given a one-step ahead forecast at time t, the *Monte Carlo* (MC) method is a simple recursive simulation method to approximate the expectation of Y_{t+H} ($H \geq 2$) conditional upon $\mathcal{F}_t$. From (10.3) the two-step ahead MC forecast can be constructed as

$$Y_{t+2|t}^{\text{MC}} = \frac{1}{N} \sum_{i=1}^{N} Y_{t+2|t}^{\text{MC}_i}, \qquad (10.20)$$

where

$$Y_{t+2|t}^{\text{MC}_i} = \mu\big((Y_{t+1|t}; \boldsymbol{\theta}) + \varepsilon_{2,i}\big),$$

10.2 APPROXIMATE FORECASTING METHODS

with $\{\varepsilon_{2,i}\}_{i=1}^{N}$ a set of pseudo-random numbers drawn from the presumed distribution of $\{\varepsilon_{t+1}\}$, and with N some large number. In general, the H-step ahead forecast is given by

$$Y_{t+H|t}^{\text{MC}} = \frac{1}{N} \sum_{i=1}^{N} Y_{t+H|t}^{\text{MC}_i}, \qquad (10.21)$$

where

$$\begin{aligned} Y_{t+H|t}^{\text{MC}_i} &= \mu\big((Y_{t+H-1|t}^{\text{MC}_i}; \boldsymbol{\theta}) + \varepsilon_{H,i}\big) \\ &= \mu\big(\mu(\cdots(\mu(Y_{t+1|t}; \boldsymbol{\theta}) + \varepsilon_{2,i}) + \cdots) + \varepsilon_{H,i}\big), \end{aligned}$$

with $\varepsilon_{j,i}$ ($j = 2, \ldots, H; i = 1, \ldots, N$) independent pseudo-random numbers drawn from some pre-specified distribution of $\{\varepsilon_{t+H}\}$, usually the Gaussian distribution. In the case of a SETARMA model the pseudo-random drawings in period $t + H$ are often taken from a distribution with a variance appropriate for the regime the process $\{Y_t, t \in \mathbb{Z}\}$ is in, determined by the MC forecast value of the process at time $t + H - 1$.

10.2.2 Bootstrap

Forecasts obtained from the *bootstrap* (BS) method are similar to the MC simulation method except that the $e_{j,i}^*$ are drawn randomly (with replacement) from the within-sample residuals e_i ($i = 2, 3, \ldots, T$), assuming a set of T historical data is available to obtain some consistent estimate of $\boldsymbol{\theta}$. In this case the H-step ahead ($H \geq 2$) forecast is given by

$$Y_{t+H|t}^{\text{BS}} = \frac{1}{T-1} \sum_{i=2}^{T} Y_{t+H|t}^{\text{BS}_i}, \qquad (10.22)$$

where

$$\begin{aligned} Y_{t+H|t}^{\text{BS}_i} &= \mu\big((Y_{t+H-1|t}^{\text{BS}_i}; \boldsymbol{\theta}) + e_{H,i}^*\big) \\ &= \mu\big(\mu(\cdots(\mu(Y_{t+1|t}; \boldsymbol{\theta}) + e_{2,i}^*) + \cdots) + e_{H,i}^*\big). \end{aligned}$$

The advantage of this method over the MC method is that no assumptions are made about the distribution of the innovation process.

10.2.3 Deterministic, naive, or skeleton

The deterministic, or naive, or *skeleton* (SK) method amounts to approximating $\mathbb{E}\big(\mu(\cdot; \boldsymbol{\theta})\big)$ by $\mu\big(\mathbb{E}(\cdot; \boldsymbol{\theta})\big)$, and can be viewed as a special case of the MC method in which we 'switch off the white noise' in (10.1). Thus, the two-step ahead forecast is given by

$$Y^{\text{SK}}_{t+2|t} = \mu(Y_{t+1|t}; \boldsymbol{\theta}).$$

Note that this approach leads to biased predictions since $Y^{\text{SK}}_{t+2|t} \neq \mathbb{E}(Y_{t+2|t})$. By induction, the H-step ahead forecast can be computed as

$$Y^{\text{SK}}_{t+H|t} = \mu\big(\mu(\cdots \mu(Y^{\text{SK}}_{t+1|t}; \boldsymbol{\theta}))\big). \tag{10.23}$$

Clearly, the SK method is computationally inexpensive. However, unlike the other methods discussed in this section, the SK forecasts do not necessarily converge to the mean of the process. Moreover, as σ_ε^2 increases there is the possibility that the deterministic component of the model ceases to dictate the behavior of the process and the noise part starts to be dominant, causing for instance switches between different limit/oscillation points, etc.; see Tong (1990, Section 6.2.2) for an example.

10.2.4 Empirical least squares

Assume that the NLAR(1) model is known and correctly specified for the DGP, but the innovation distribution is unspecified. This is the setup introduced in Section 10.2.2. However, rather than bootstrapping the empirical distribution of the within-sample residuals e_i ($i = 2, \ldots, T$), the *empirical least squares* (ELS) forecast method of Guo et al. (1999) uses $\widehat{F}_T(x) = (T-1)^{-1} \sum_{i=2}^{T} I(e_i < x)$ as an estimate of the innovation distribution. Then, given (10.3), the two-step ahead ELS forecast can be defined as

$$Y^{\text{ELS}}_{t+2|t} = \frac{1}{T-1} \sum_{i=2}^{T} \mu\big(\mu(Y_{t+1|t}; \boldsymbol{\theta}) + e_i\big). \tag{10.24}$$

The ELS method can be readily extended to the case $H > 2$. For instance, the exact three-step ahead LS forecast is given by

$$Y^{\text{LS}}_{t+3|t} = \int_{-\infty}^{\infty} \mu\big(\mu(\mu(Y_{t+1|t}; \boldsymbol{\theta}) + \varepsilon) + \varepsilon'\big) \mathrm{d}F(\varepsilon) \mathrm{d}F(\varepsilon').$$

Thus, as a three-stage ELS forecast, we may take

$$Y^{\text{ELS}}_{t+3|t} = \frac{1}{(T-1)(T-2)} \sum_{2 \leq i \neq j \leq T} \mu\big(\mu(\mu(Y_{t+1|t}; \boldsymbol{\theta}) + e_i) + e_j\big).$$

In general, the exact H-step ahead LS forecast is given by

$$Y^{\text{LS}}_{t+H|t} = \int_{-\infty}^{\infty} \cdots \int_{-\infty}^{\infty} \mu\big(\mu(\cdots(\mu(Y_t; \boldsymbol{\theta}) + \varepsilon_1) + \cdots) + \varepsilon_{H-1}\big) \mathrm{d}F(\varepsilon_1) \cdots \mathrm{d}F(\varepsilon_{H-1}),$$

10.2 APPROXIMATE FORECASTING METHODS

and the proposed ELS forecast can be written as

$$Y_{t+H|t}^{\text{ELS}} = \frac{(T-H)!}{(T-1)!} \sum_{(H-1,T)} \mu\big(\mu(\cdots(\mu(Y_{t+1|t};\boldsymbol{\theta}) + e_{1,i}) + \cdots) + e_{H-1,i}\big), \quad (10.25)$$

where the summation $\sum_{(H-1,T)}$ runs over all possible $(H-1)$-tuples of distinct $(i_1, \ldots, i_{H-1})$. Guo et al. (1999) show that the above prediction scheme is asymptotically equivalent to the exact LS forecast.

The ELS method can be easily generalized to NLAR models with conditional heteroskedasticity. For instance, consider the model

$$Y_t = \mu(Y_{t-1}; \boldsymbol{\theta}_1) + \varepsilon_t \sigma(Y_{t-1}; \boldsymbol{\theta}_2),$$

where $\boldsymbol{\theta}_i$ ($i=1,2$) is a vector of unknown parameters, $\mu(\cdot; \boldsymbol{\theta}_1)$ and $\sigma(\cdot; \boldsymbol{\theta}_2)$ are two real-valued known functions on $\mathbb{R}$, and the ε_t's are assumed to satisfy $\mathbb{E}(\varepsilon_t) = 1$ for identification purposes. Given T observations, the series $\{e_i\}$ can be calculated exactly from the model based on particular estimates of $\boldsymbol{\theta}_i$. Next, we use these residuals as proxies for the disturbance term instead of random draws from some assumed parametric distribution as in Section 10.2.1. Then, using the same idea as above, the H-step ahead predictor follows directly. It is apparent that, in comparison with the MC predictor, the ELS predictor is less sensitive to distributional assumptions about the error process.

10.2.5 Normal forecasting error

An alternative to the H-step ahead ($H \geq 2$) exact LS predictor in (10.6) is to assume as an approximation that all $(H-1)$ forecast errors $e_{t+H-1|t}$ ($H \geq 2$) are normally distributed with mean zero and variance $\sigma_{e,H-1}^2 \equiv \text{Var}(e_{t+H-1|t})$. The resulting method is known as the *normal forecasting error* (NFE) method. As we will see, for both the ExpAR(1) model (Al-Qassem and Lane, 1989) and the SETAR(2; 1, 1) model (De Gooijer and De Bruin, 1998) the normality assumption avoids the use of numerical methods. However, as $\mu(\cdot; \boldsymbol{\theta})$ is a nonlinear function the multi-step ahead forecast errors $e_{t+H-1|t}$ will not equal the linear innovations, nor will they follow an i.i.d. Gaussian process.

ExpAR(1) model
To obtain the NFE forecast value for any step, we employ the following result. Let $r(Z)$ be a function of the random variable $Z \stackrel{\text{i.i.d.}}{\sim} \mathcal{N}(0, \sigma_Z^2)$, and M and c are constants. Then

$$\mathbb{E}\{r(Z)\exp\big(-c(Z+M)^2\big)\} = A^{-1/2}\exp(-c_1 M^2)\mathbb{E}\big(r(V)\big), \quad (10.26)$$

where $A = 1 + 2c\sigma_Z^2$, $c_1 = cA^{-1}$, and $V \stackrel{\text{i.i.d.}}{\sim} \mathcal{N}(-2c_1 \sigma_Z^2 M, \sigma_Z^2/A)$; cf. Exercise 10.6. Consider the ExpAR(1) model at time $t + H$, i.e.,

$$Y_{t+H} = \{\phi + \xi\exp\big(-\gamma(Y_{t+H-1|t} + e_{t+H-1|t})^2\big)\}(Y_{t+H-1|t} + e_{t+H-1|t}) + \varepsilon_{t+H},$$

substituting $Y_{t+H-1|t} + e_{t+H-1|t}$ for Y_{t+H-1}.

The one-step ahead forecast is the conditional expectation of the ExpAR(1) model given the available data at time t, $Y_{t+1|t} = \mu(Y_t; \boldsymbol{\theta})$. By applying (10.26) with $Z = e_{t+H-1|t}$, $M = Y_{t+H-1|t}$, and $c = 1$, the H-step ahead ($H \geq 2$) NFE forecast is given by

$$Y_{t+H|t}^{\text{NFE}} = \mathbb{E}(Y_{t+H}|\mathcal{F}_t) = \{\phi + \xi_{H-1}\exp(-\gamma_{H-1}Y_{t+H-1|t}^2)\}Y_{t+H-1|t}, \qquad (10.27)$$

where $A_{H-1} = 1 + 2\sigma_{e,H-1}^2$, $c_{H-1} = A_{H-1}^{-1}$, $\xi_{H-1} = \xi A_{H-1}^{-3/2}$. After substitution and some algebra, the forecast error is given by

$$e_{t+H|t} = \phi e_{t+H-1|t} + \xi\{Y_{t+H-1}\exp(-\gamma Y_{t+H-1}^2) - \mathbb{E}(Y_{t+H-1}\exp(-\gamma Y_{t+H-1}^2)|\mathcal{F}_t)\}$$
$$+ \varepsilon_{t+H},$$

so that $\mathbb{E}(e_{t+H|t}) = 0$. Since $e_{t+H-1|t}$ does not depend on future noise ε_{t+H}, the forecast error variance is given by

$$\sigma_{e,H}^2 = \phi^2 \sigma_{e,H-1}^2 + \xi^2 v_{H-1} + 2\phi\xi u_{H-1} + \sigma_\varepsilon^2, \qquad (10.28)$$

where $\sigma_{e,1}^2 = \sigma_\varepsilon^2$ and, using (10.26) with $c = 2$,

$$v_{H-1} \equiv \text{Var}\{(Y_{t+H-1|t} + e_{t+H-1|t})\exp(-\gamma(Y_{t+H-1|t} + e_{t+H-1|t})^2)\}$$
$$= B_{H-1}^{-3/2}\left(\sigma_{e,H-1}^2 + \frac{Y_{t+H-1|t}^2}{B_{H-1}}\right)\exp(-d_{H-1}Y_{t+H-1|t}^2)$$
$$- A_{H-1}^{-3}Y_{t+H-1|t}^2\exp(-2c_{H-1}Y_{t+H-1|t}^2)$$

with $B_{H-1} = 1 + 4\sigma_{e,H-1}^2$, and $d_{H-1} = 2B_{H-1}^{-1}$. Moreover, it can be deduced that

$$u_{H-1} \equiv \text{Cov}\{e_{t+H-1|t}, (Y_{t+H-1|t} + e_{t+H-1|t})\exp(-\gamma(Y_{t+H-1|t} + e_{t+H-1|t})^2)\}$$
$$= \mathbb{E}\{e_{t+H-1|t}(Y_{t+H-1|t} + e_{t+H-1|t})\exp(-\gamma(Y_{t+H-1|t} + e_{t+H-1|t})^2)\}$$
$$= \sigma_{e,H-1}^2 A_{H-1}^{-3/2}(1 - 2c_{H-1}Y_{t+H-1|t}^2)\exp(-c_{H-1}Y_{t+H-1|t}^2),$$

where the last equation follows from (10.26) by defining $U = V + M$ with $U \stackrel{\text{i.i.d.}}{\sim} \mathcal{N}(M/A, \sigma_Z^2/A)$.

To generalize the above results to an ExpAR(p) model, requires the assumption that the $p \times 1$ vector $(e_{t+H|t}, \ldots, e_{t+H-p+1|t})'$ is jointly multivariate normally distributed. Moreover, depending on the order p of the model, we also need various generalizations of (10.26). Altogether, however, the algebra involved is manageable.

SETAR$(2;1,1)$ model

Consider, as a special case of (10.11), the SETAR$(2;1,1)$ model

$$Y_t = \{\phi^{(1)}Y_{t-1} + \varepsilon_t\}I(Y_{t-1} \leq r) + \{\phi^{(2)}Y_{t-1} + \varepsilon_t\}(1 - I(Y_{t-1} \leq r)), \qquad (10.29)$$

10.2 APPROXIMATE FORECASTING METHODS

where $\{\varepsilon_t\} \overset{\text{i.i.d.}}{\sim} \mathcal{N}(0, \sigma_\varepsilon^2)$. Assume that the $(H-1)$-step $(H \geq 2)$ ahead forecast errors are $\mathcal{N}(0, \sigma_{e,H-1}^2)$ distributed. Then, as in (10.13), the H-step ahead NFE forecast is a weighted average of the forecasts from the first regime $Y_{t+H|t}^{(1)} = \phi^{(1)} Y_{t+H-1|t}$ and the second regime $Y_{t+H|t}^{(2)} = \phi^{(2)} Y_{t+H-1|t}$ with weights equal to the probability of being in a particular regime at time $t+H-1$ under normality of the forecast errors, *plus* an additional correction factor. In particular, the H-step ahead NFE forecast follows from the recursion

$$Y_{t+H|t}^{\text{NFE}} = p_{(H-1)} Y_{t+H|t}^{(1)} + (1 - p_{(H-1)}) Y_{t+H|t}^{(2)} + (\phi^{(2)} - \phi^{(1)}) \sigma_{e,H-1} \varphi(z_{t+h-1|t})$$
$$= \left(\phi^{(1)} + (\phi^{(1)} - \phi^{(2)}) \Phi(z_{t+H-1|t})\right) Y_{t+H-1|t} + (\phi^{(2)} - \phi^{(1)}) \sigma_{e,H-1} \varphi(z_{t+H-1|t}), \quad (10.30)$$

where $p_{(H-1)} = \Phi(z_{t+H-1|t})$ and $z_{t+H-1|t} = (r - Y_{t+H-1|t})/\sigma_{e,H-1}$. The corresponding forecast error variance is given by the recursive relation

$$\sigma_{e,H}^2 = 2\sigma_\varepsilon^2 \Phi(z_{t+H-1|t})$$
$$+ \{(\phi^{(1)})^2 + ((\phi^{(1)})^2 - (\phi^{(2)})^2) \Phi(z_{t+H-1|t})\} \{Y_{t+H-1|t}^2 + \sigma_{e,H-1}^2\}$$
$$+ \{(\phi^{(1)})^2 - (\phi^{(2)})^2\} \sigma_{e,H-1}(r + Y_{t+H-1}) \phi(z_{t+H-1|t}) - Y_{t+H|t}^2. \quad (10.31)$$

For $H = 2$, it can be shown that (10.30) is identical to the two-step ahead exact MMSE forecast; cf. Exercise 10.1. The above results can be directly extended to more general SETAR models, including models with multiple regimes, and to situations where the delay has a value greater than one. An additional advantage is that for both ExpAR(1) and SETAR(2; 1, 1) models the NFE method can be rapidly calculated using, for instance, a spreadsheet.

Example 10.3: Comparing NFE and MC Forecasts

To quantify the accuracy of (10.30) consider the SETAR(2; 1, 1) model (10.29) with $r = 0$, $Y_0 = 0$, and $\{\varepsilon_t\} \overset{\text{i.i.d.}}{\sim} \mathcal{N}(0, 1)$. Necessary and sufficient conditions for stationarity are $\phi^{(1)} < 1$, $\phi^{(2)} < 1$, and $\phi^{(1)} \phi^{(2)} < 1$; see Table 3.1. Subject to these conditions, we compute $Y_{t+H|t}^{\text{NFE}}$ for $H = 3, \ldots, 10$ with parameter values $\phi^{(1)} = -1.50, -1.25, \ldots, 0.50, 0.75$ and $\phi^{(2)} = -1.75, -1.50, \ldots, 0.50, 0.75$. Also, we obtain H-step ahead forecasts by the MC method, generating for each step H 100,000 realizations of Y_{t+H}. Next, for each parameter combination, we calculate the relative mean absolute forecast error (RMAFE):

$$\text{RMAFE}_t = \frac{1}{8} \sum_{H=3}^{10} |(Y_{t+H} - Y_{t+H|t}^{\text{MC}})/Y_{t+H|t}^{\text{MC}}|. \quad (10.32)$$

Figure 10.3 shows a contour plot of (10.32). The results indicate good agreement between the NFE and the MC method over a wide range of parameter values. More generally, MC simulations show that for values of $\sigma_\varepsilon^2 = 0.4$ and 1

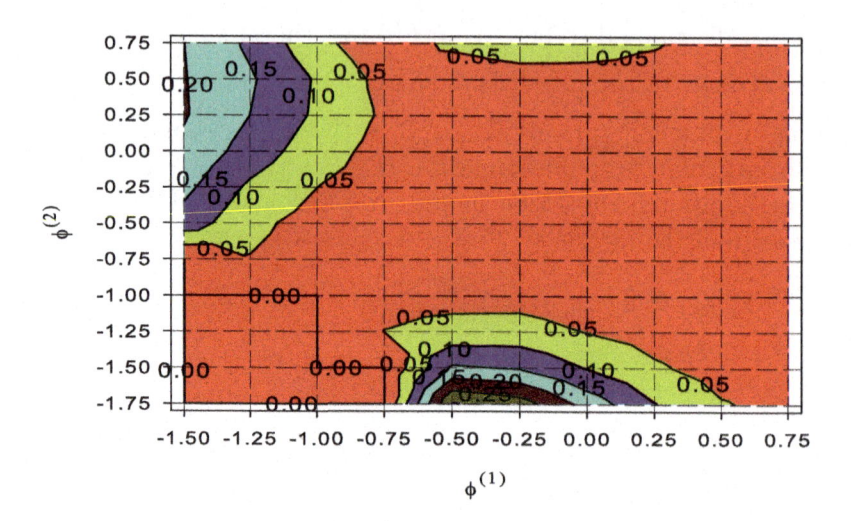

Figure 10.3: *Contour plot of* (10.32) *for the* SETAR(2;1,1) *model* (10.29) *with* $r = 0$, $Y_0 = 0$, $\{\varepsilon_t\} \stackrel{i.i.d.}{\sim} \mathcal{N}(0,1)$. From De Gooijer and De Bruin (1998).

the SETAR–NFE method performs well as opposed to the exact and the MC forecasting method. For $\sigma_\varepsilon^2 = 2$ NFE is quite reliable for forecasts up to, say, five- or six-steps ahead.

10.2.6 Linearization

Another approach to approximate the exact forecast $Y_{t+H|t}$ is to linearize the problem. In particular, Taylor's expansion up to order two of $\mu(\cdot; \boldsymbol{\theta})$ about the point $Y_{t+H-1|t}$ (ignoring the remainder term), is

$$\mu(Y_{t+H-1}; \boldsymbol{\theta}) \simeq \mu(Y_{t+H-1|t}; \boldsymbol{\theta}) + e_{t+H-1|t}\mu^{(1)}(Y_{t+H-1|t}; \boldsymbol{\theta}) + \frac{1}{2}e_{t+H-1|t}^2 \mu^{(2)}(Y_{t+H-1|t}; \boldsymbol{\theta}), \quad (10.33)$$

where $\mu^{(i)}(\cdot; \boldsymbol{\theta})$, $(i = 1, 2)$ denotes the ith derivative of $\mu(Y_{t+H-1|t}; \boldsymbol{\theta})$ with respect to $Y_{t+H-1|t}$, and $e_{t+H-1|t}$ is the $(H-1)$-step ahead forecast error ($H \geq 2$). We refer to this approach as the *linearization* (LN) method.

Assume, for simplicity, that the forecasting error process $\{e_{t+H-1|t}\} \stackrel{i.i.d.}{\sim} \mathcal{N}(0, \sigma_{e,H-1}^2)$ distributed. Then, substituting (10.33) in the NLAR(1) model and taking the conditional expectation of the resulting specification, gives the H-step ahead LN forecast, i.e.

$$Y_{t+H|t}^{\text{LN}} \simeq \mu(Y_{t+H-1|t}; \boldsymbol{\theta}) + \frac{1}{2}\sigma_{e,H-1}^2 \mu^{(2)}(Y_{t+H-1|t}; \boldsymbol{\theta}). \quad (10.34)$$

10.2 APPROXIMATE FORECASTING METHODS

Substituting (10.34) in the corresponding H-step ahead forecast error and simplifying gives

$$e_{t+H|t} = \varepsilon_{t+H} + e_{t+H-1|t}^2 \mu^{(1)}(Y_{t+H-1|t};\boldsymbol{\theta}) + \frac{1}{2}\{e_{t+H-1|t}^2 - \sigma_{e,H}^2\}\mu^{(2)}(Y_{t+H-1|t};\boldsymbol{\theta}).$$

The forecast error variance for this step is given by the recurrence relation

$$\sigma_{e,H}^2 = \sigma_\varepsilon^2 + \sigma_{e,H-1}^2 \big(\mu^{(1)}(Y_{t+H-1|t};\boldsymbol{\theta})\big)^2 + \frac{1}{2}\sigma_{e,H-1}^4 \big(\mu^{(2)}(Y_{t+H-1|t};\boldsymbol{\theta})\big)^4. \quad (10.35)$$

Forecasts obtained from this method can be quite different from the exact prediction method or from the NFE method for moderate or large σ_ε^2 (mainly $\geq 10^{-2}$). Al-Qassem and Lane (1989) provide a discussion on the limiting behavior of (10.33) in the case of the ExpAR(1) model. They emphasize the need for great caution in using linearized forecasts in nonlinear models.

Extension of the LN method to ExpAR(p) is straightforward with a Taylor expansion of $\mu(\cdot;\boldsymbol{\theta})$ around the point $\mathbf{Y}_{t+H-1|t} = (Y_{t+H-1|t}, Y_{t+H-2|t}, \ldots, Y_{t+H-p|t})'$ where $\mathbf{Y}_{t+j|t} = Y_{t+j}$ if $j \leq 0$. Similarly, an expression for the H-step ahead forecast error variance can be obtained by assuming that the forecast errors have a multivariate normal distribution.

Example 10.4: Forecasts from an ExpAR(1) Model

Consider the ExpAR(1) model with nonlinear function of the form

$$\mu(X;\boldsymbol{\theta}) = \{\phi + \xi \exp(-\gamma X^2)\}X,$$

where $\boldsymbol{\theta} = (\phi, \xi, \gamma)'$. The function $\mu(\cdot;\boldsymbol{\theta})$ has the following partial derivatives with respect to X

$$\mu^{(1)}(X;\boldsymbol{\theta}) = \phi + \xi(1 - 2\gamma X^2)\exp(-\gamma X^2),$$
$$\mu^{(2)}(X;\boldsymbol{\theta}) = 2\xi\gamma X(2\gamma X^2 - 3)\exp(-\gamma X^2).$$

Substituting $\mu^{(2)}(\cdot;\boldsymbol{\theta})$ into (10.34), we get

$$Y_{t+H|t}^{\text{LN}} = \phi + \xi f_{H-1}\exp\big(-\gamma(Y_{t+H-1|t})^2\big)Y_{t+H-1|t},$$

where

$$f_{H-1} = 1 + \gamma\sigma_{e,H-1}^2\big(2\gamma(Y_{t+H-1|t})^2 - 3\big).$$

Thus, f_{H-1} is increasing with $\sigma_{e,H-1}^2$. We also see that if $\sigma_{e,H-1}^2$ is large and $Y_{t+H-1|t}$ is near zero, f_{H-1} can be negative. It seems that this is the root cause of the instability of the LN method.

Figure 10.4(a) shows 50 forecasts obtained by the NFE, SK, and LN methods applied to a typical single simulation of an ExpAR(1) model with $\phi = 0.8$,

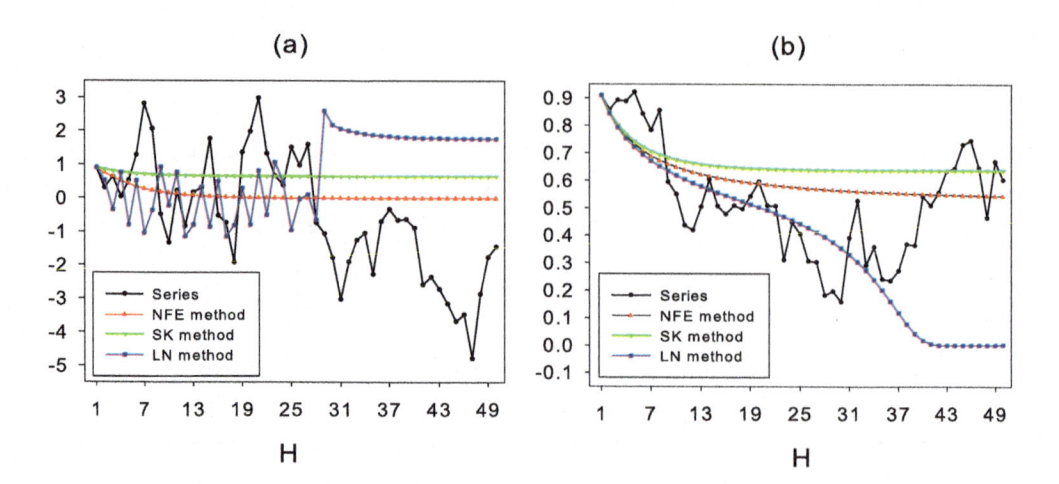

Figure 10.4: *Forecasts from the* ExpAR(1) *model in Example 10.4 with the NFE, SK, and LN methods; (a)* $\sigma_\varepsilon^2 = 1$, *and (b)* $\sigma_\varepsilon^2 = 0.01$.

$\xi = 0.3$, $\{\varepsilon_t\} \stackrel{\text{i.i.d.}}{\sim} \mathcal{N}(0,1)$, and starting value $Y_0 = 1$. By relation (6.6) the process has two limit points at ± 0.6368. It is clear that the NFE forecasts go to a limit point zero, SK forecasts go to the upper limit point 0.6368, while the series of LN forecast are unstable up to about $H = 30$, then stabilize to a point far off the upper limit point. Four more plots are given in Figure 10.4(b) for $\sigma_\varepsilon^2 = 0.01$.

For short-term forecasting ($H \leq 5$) there is hardly any noticeable difference between the three forecasting methods, provided σ_ε^2 is small. On the other hand, for long-term ($H \geq 30$) forecasting the LN method may go to the "wrong" limit point.

10.2.7 Dynamic estimation

In the spirit of *dynamic estimation* (DE) applied to linear models, the next method is based on the in-sample relationship between Y_t and Y_{t+H}, ignoring contributions of intermediate values, to produce H-step ahead forecasts. In other words, for H-step ahead forecasts we replace the NLAR(1) model by the following specification

$$Y_{t+H} = \mu(Y_t; \boldsymbol{\theta}_H^*) + \varepsilon_{t+H}^*, \tag{10.36}$$

where $\boldsymbol{\theta}_H^*$ is a vector of parameters depending upon the forecast horizon H. These parameters can, for instance, be estimated by minimizing the sum of squares of ε_{T+H}^* over $\boldsymbol{\theta}_H^*$ for the sample period $t = 1, \ldots, T$. So that, given the parameter estimates $\widehat{\boldsymbol{\theta}}_H^*$, the corresponding H-step ahead DE forecast can be written as

$$Y_{t+H|t}^{\text{DE}} = \mu(Y_t; \widehat{\boldsymbol{\theta}}_H^*). \tag{10.37}$$

In a linear setting, there are no gains in terms of increased forecast accuracy using DE over the traditional minimization of in-sample sum of squares of one-step ahead errors when the model is correctly specified. When a nonlinear model, however, is correctly specified, the DE method may result in better out-of-sample forecasts due to its simplicity. An obvious drawback of the method is that the nonlinear model needs to be estimated for each forecasting horizon.

Example 10.5:   Forecasts from a SETAR$(2;1,1)$ Model

Recall the SETAR$(2;1,1)$ model (10.29) with $\mu(Y_{t-1};\boldsymbol{\theta}) = \phi^{(1)}Y_{t-1}I(Y_{t-1} \leq r) + \phi^{(2)}Y_{t-1}(1 - I(Y_{t-1} \leq r))$ and $\boldsymbol{\theta} = (\phi^{(1)}, \phi^{(2)})'$. The two-step ahead version of the model can be written as

$$\begin{aligned}Y_{t+2} &= \phi^{(2)}\{\phi^{(2)}Y_t + (\phi^{(1)} - \phi^{(2)})Y_t I(Y_t \leq r) + \varepsilon_{t+1}\} \\ &\quad + (\phi^{(1)} - \phi^{(2)})\{\phi^{(2)}Y_t + (\phi^{(1)} - \phi^{(2)})Y_t I(Y_t \leq r) + \varepsilon_{t+1}\}I(Y_{t+1} \leq r) \\ &\quad + \varepsilon_{t+2} \\ &\approx \phi^{(2)}\{\phi^{(2)}Y_t + (\phi^{(1)} - \phi^{(2)})Y_t I(Y_t \leq r)\} + \varepsilon^*_{t+2} \\ &= \mu(Y_t;\boldsymbol{\theta}^*_2) + \varepsilon^*_{t+2},\end{aligned} \qquad (10.38)$$

where $\boldsymbol{\theta}^*_2 = (\theta^{(1)}_2, \theta^{(2)}_2)' = (\phi^{(1)}\phi^{(2)}, (\phi^{(2)})^2)'$. Observe that in the second equation terms multiplied by $I(Y_{t+1} \leq r)$ are missing. So, the DE method is just a projection of Y_{t+2} on the period t information, but using the form of nonlinearity in the "one-step ahead" model.

The parameter estimates $\widehat{\theta}^{(i)}_2$ ($i = 1, 2$) follow from minimizing the sum of squares of ε_{t+2} for the in-sample period, using the CLS estimation procedure outlined in Section 6.1.2. This requires a grid search over r; see Algorithm 6.2. Denote the resulting estimate by $\widehat{r}_2$. Then the two-step ahead DE forecast is given by

$$Y^{\text{DE}}_{t+2|t} = \mu(Y_t; \widehat{\boldsymbol{\theta}}^*_2) = \widehat{\theta}^{(1)}_2 Y_t I(Y_t \leq \widehat{r}_2) + \widehat{\theta}^{(2)}_2 Y_t (1 - I(Y_t \leq \widehat{r}_2)). \qquad (10.39)$$

The generalization to H-step ahead ($H > 2$) forecasts entails minimizing the sum of squares of ε^*_{t+H} over $\boldsymbol{\theta}^*_H = (\theta^{(1)}_H, \theta^{(2)}_H)' = (\phi^{(1)}(\phi^{(2)})^{H-1}, (\phi^{(2)})^H)'$, and r, where

$$Y_{t+H} = \mu(Y_t; \boldsymbol{\theta}^*_H) + \varepsilon^*_{t+H}. \qquad (10.40)$$

The corresponding H-step ahead DE forecast is given by

$$Y^{\text{DE}}_{t+H|t} = \mu(Y_t; \widehat{\boldsymbol{\theta}}^*_H) = \widehat{\theta}^{(1)}_H Y_t I(Y_t \leq \widehat{r}_H) + \widehat{\theta}^{(2)}_H Y_t (1 - I(Y_t \leq \widehat{r}_H)). \qquad (10.41)$$

Note that $\{\varepsilon^*_{t+H}\}$ is not a WN process, but has temporal relationships. So, in general, the forecasts are biased.

In an MC simulation experiment Clements and Smith (1997) conclude that the DE method is worse than the BS, MC and NFE forecasting methods for SETAR$(2;1,1)$ models with Gaussian disturbances and zero intercepts.

10.3 Forecast Intervals and Regions

10.3.1 Preliminaries

The forecast methods discussed in the previous two sections produce a single approximation for Y_{T+H}. Ideally, forecast intervals/regions are more informative than point predictions as they indicate the likely range of forecast outcomes. As such, a forecast interval/region is a measure of the inherent model accuracy. The conditional distribution of Y_{T+H} given $\mathcal{F}_t = \{Y_t, Y_{t-1}, \ldots\}$ forecast interval/region for Y_{T+H}. Given $\mathbf{X}_t = \mathbf{x}$, $Q_\alpha \equiv Q_\alpha(\mathbf{x}) \subset \mathbb{R}$ is such an interval with coverage probability $1 - \alpha$ ($\alpha \in [0, 1]$). That is $\mathbb{P}\{Y_{T+H} \in Q_\alpha(\mathbf{x}) | \mathbf{X}_{T-H-p+1} = \mathbf{x}\} = 1 - \alpha$, assuming the DGP is strictly stationary and Markovian of order p. The set Q_α will be called *forecast region* (FR). When Q_α is a connected set, we call it a *forecast interval* (FI). Obviously, such a region/interval can be constructed in an infinite number of ways. For instance, a natural FI for the conditional median of Y_{T+H} is the so-called *conditional percentile interval* (CPI) given by

$$\text{CPI}_{1-\alpha} = [\widehat{\xi}_{\alpha/2}(\mathbf{x}), \widehat{\xi}_{1-\alpha/2}(\mathbf{x})], \tag{10.42}$$

where $\widehat{\xi}_\alpha(\cdot)$ is the αth conditional percentile of $\xi_\alpha(\cdot)$ defined by (9.11) with $\alpha \equiv q$, changing the notation of the quantile level q to the symbol α.

In the context of linear ARMA models, we normally construct a FI for $H \geq 1$ steps ahead by using an estimate of the conditional mean, an estimate of the conditional variance, and, in addition, a certain critical value taken from either the normal or the Student t distribution. For some nonparametric methods, FIs can be constructed on the basis of available asymptotic theory of the forecast under study (Yao and Tong, 1995). In general, however, some form of resampling is necessary because of non-normality of the forecast errors and/or nonlinearity of the forecast. Below, we consider both approaches, making a distinction between FI/FRs based on percentiles and on conditional densities where in the latter case the shape of the densities may change over the domain of $\mathbf{X}_t$.

10.3.2 Conditional percentiles

As it is informative to provide general theory covering all (non)parametric nonlinear models/methods, we discuss FIs for two prominent cases: (i) the Nadaraya–Watson (NW) and local linear (LL) estimators of the conditional mean function, and (ii) the SETAR-based estimator of the conditional mean.

FIs for the NW and LL estimators of the conditional mean
Consider a strictly stationary and real-valued stochastic process $\{Y_t, t \in \mathbb{Z}\}$ that follows the functional relationship defined in (9.35) which, for ease of reference, we re-introduce as

$$Y_t = \mu(\mathbf{X}_t) + \sigma(\mathbf{X}_t)\varepsilon_t, \quad t \geq 1, \tag{10.43}$$

10.3 FORECAST INTERVALS AND REGIONS

where $\mathbf{X}_t = (Y_{t-1}, \ldots, Y_{t-p})'$, $\sigma(\mathbf{x}) > 0 \ \forall \mathbf{x} \in \mathbb{R}^p$, $Y_0, \ldots, Y_p$ are initial conditions, $\{\varepsilon_t\} \stackrel{\text{i.i.d.}}{\sim} (0, 1)$ random variables with $\{\varepsilon_t\}$ independent of past Y_t, $\mu(\cdot)$ and $\sigma(\cdot)$ are unknown functions on $\mathbb{R}$. Let $f(\mathbf{x})$ denote the density function of the lag vector at the point $\mathbf{X}_t = \mathbf{x}$. Recall $\widehat{\mu}^{\text{NW}}(\mathbf{x})$, the NW estimator of the conditional mean function $\mu(\mathbf{x})$, is given by (9.36). Under certain mixing conditions it can be shown (see, e.g., Fan and Gijbels, 1996, Thm. 6.1) that $\widehat{\mu}^{\text{NW}}(\mathbf{x})$ is asymptotically normally distributed with asymptotic bias and variance given by

$$\text{Bias}(\widehat{\mu}^{\text{NW}}(\mathbf{x})) = \frac{1}{2}\mu_2(K)h^2 \Big[\mu^{(2)}(\mathbf{x}) + 2\mu^{(1)}(\mathbf{x})\frac{f^{(1)}(\mathbf{x})}{f(\mathbf{x})}\Big], \qquad (10.44)$$

$$\text{Var}(\widehat{\mu}^{\text{NW}}(\mathbf{x})) = R(K)\frac{1}{nh}\frac{\sigma^2(\mathbf{x})}{f(\mathbf{x})}, \quad (n = T - H - p + 1), \qquad (10.45)$$

where $\mu_2(K) = \int_{\mathbb{R}} u^2 K(u) du$ and $R(K) = \int_{\mathbb{R}} K^2(z) dz$. Similarly, based on the LL regression approach, the estimator $\widehat{\mu}^{\text{LL}}(\mathbf{x})$ of $\mu(\mathbf{x})$ is asymptotically normally distributed with asymptotic mean and variance

$$\text{Bias}(\widehat{\mu}^{\text{LL}}(\mathbf{x})) = \frac{1}{2}\mu_2(K)h^2\mu^{(2)}(\mathbf{x}), \quad \text{Var}(\widehat{\mu}^{\text{LL}}(\mathbf{x})) = R(K)\frac{1}{nh}\frac{\sigma^2(\mathbf{x})}{f(\mathbf{x})}. \qquad (10.46)$$

We see that the bias of the NW estimator does not only depend on the first- and second derivatives of $\mu(\mathbf{x})$, but also on the score function $-f^{(1)}(\mathbf{x})/f(\mathbf{x})$. This is the reason why an unbalanced design may lead to an increased bias, especially when p is large and T is small. Clearly, consistent bias estimates of the NW and LL estimators of $\mu(\mathbf{x})$ require estimates of $\mu^{(2)}(\mathbf{x})$. Such estimates will possibly reduce the bias, and hence improve forecast accuracy in small samples. On the other hand, the variance may increase since more parameters have to estimated. Thus, it is reasonable to construct FIs for both nonparametric conditional mean estimators without a small-sample bias correction. Since the expression for the asymptotic variance is the same for $\widehat{\mu}^{\text{NW}}(\mathbf{x})$ and $\widehat{\mu}^{\text{LL}}(\mathbf{x})$, the resulting FI with coverage probability $(1 - \alpha)$ is defined as

$$\text{FI}_\alpha = \Big[\widehat{\mu}^{(\cdot)}(\mathbf{x}) - z_{\alpha/2}\sqrt{\sigma^2(\mathbf{x}) + \frac{\text{Var}(\widehat{\mu}^{(\cdot)}(\mathbf{x}))}{nh}}, \widehat{\mu}^{(\cdot)}(\mathbf{x}) + z_{\alpha/2}\sqrt{\sigma^2(\mathbf{x}) + \frac{\text{Var}(\widehat{\mu}^{(\cdot)}(\mathbf{x}))}{nh}}\Big]. \qquad (10.47)$$

Here, $z_{\alpha/2}$ denotes the $(1 - \alpha/2)$th percentile of the standard normal distribution, and the notation $\widehat{\mu}^{(\cdot)}(\mathbf{x})$ denotes the NW or the LL conditional mean forecast.

Bootstrap FIs for SETAR models

Consider the stationary SETAR$(2; p, p)$ model with $d \leq p$:

$$Y_t = \Big\{\phi_0^{(1)} + \sum_{i=1}^{p} \phi_i^{(1)} Y_{t-i}\Big\} I(Y_{t-d} \leq r) + \Big\{\phi_0^{(2)} + \sum_{i=1}^{p} \phi_i^{(2)} Y_{t-i}\Big\} I(Y_{t-d} > r) + \varepsilon_t, \qquad (10.48)$$

where $\{\varepsilon_t\} \overset{\text{i.i.d.}}{\sim} (0,1)$ random variables, and p is assumed to be known. Given the initial, pre-sample, values $(Y_{-p+1},\ldots,Y_0)$ and the set of observations $\{Y_t\}_{t=1}^T$, an estimate $\widehat{r}_T$ of r follows from using Algorithm 6.2. We have seen in Section 6.1.2 that this estimator is super-consistent with the rate of convergence of $\mathcal{O}_p(T^{-1})$.

Bootstrap FIs for linear ARs have received quite some attention; see, e.g., Pan and Politis (2016) for a recent review. Within this context, BS can be based on the backward and forward time representation of an AR(p) model. For SETAR models there is no immediate way of inverting the lag polynomial augmented with indicator variables. Hence, the so-called *backward* BS procedure does not apply in this case. In contrast, the *forward* BS generates bootstrap series conditionally on the first p observations of the observed series as the initial values of the bootstrap replicates. Both Li (2011) and Pan and Politis (2016) use forward BS in a SETAR forecasting context. One simple algorithm to construct the FI for Y_{t+H} is as follows.

Algorithm 10.1: Bootstrap FI

1.1 Using Algorithm 6.2, compute the CLS estimates $\widehat{\phi}_i^{(j)}$ ($i=0,\ldots,p;\ j=1,2$), conditional on $\widehat{r}_T$. Compute the EDF, say $\widehat{F}_{\widetilde{\varepsilon}}$, of the mean-deleted residuals $\{\widetilde{\varepsilon}_t = \widehat{\varepsilon}_t - \overline{\widehat{\varepsilon}}\}_{t=p+1}^T$, where $\overline{\widehat{\varepsilon}} = (T-p)^{-1}\sum_{t=p+1}^T \widehat{\varepsilon}_t$ and
$$\widehat{\varepsilon}_t = Y_t - \{\widehat{\phi}_0^{(1)} + \sum_{i=1}^p \widehat{\phi}_i^{(1)} Y_{t-i}\} I(Y_{t-d} \leq \widehat{r}) + \{\widehat{\phi}_0^{(2)} + \sum_{i=1}^p \widehat{\phi}_i^{(2)} Y_{t-i}\} I(Y_{t-d} > \widehat{r}).$$

1.2 Draw (with replacement) BS pseudo-residuals $\{\varepsilon_t^*\}$ from $\widehat{F}_{\widetilde{\varepsilon}}$, and generate the BS replicate of Y_t, denoted by Y_t^*, as $Y_t^* = Y_t$, ($t=1,\ldots,p$),
$$Y_t^* = \{\widehat{\phi}_0^{(1)} + \sum_{i=1}^p \widehat{\phi}_i^{(1)} Y_{t-i}^*\} I(Y_{t-d}^* \leq \widehat{r}_T) + \{\widehat{\phi}_0^{(2)} + \sum_{i=1}^p \widehat{\phi}_i^{(2)} Y_{t-i}^*\} I(Y_{t-d}^* > \widehat{r}_T)$$
$$+ \varepsilon_t^*, \quad (t=p+1,\ldots,T+H). \tag{10.49}$$

1.3 Based on the pseudo-data $\{Y_t^*\}_{t=1}^T$, and using $\widehat{r}_T$, re-estimate the coefficients $\phi_i^{(j)}$. Obtain a new set of BS coefficients $\widehat{\phi}_i^{*,(j)}$.

1.4 Compute the BS H-step ahead forecast, denoted by Y_{t+H}, as $Y_t^* = Y_t$, ($t=T, T-1,\ldots,T-p+1$),
$$Y_{t+H}^* = \{\widehat{\phi}_0^{*,(1)} + \sum_{i=1}^p \widehat{\phi}_i^{*,(1)} Y_{t+H-i}^*\} I(Y_{t+H-d}^* \leq \widehat{r}_T) +$$
$$\{\widehat{\phi}_0^{*,(2)} + \sum_{i=1}^p \widehat{\phi}_i^{*,(2)} Y_{t+H-i}^*\} I(Y_{t+H-d}^* > \widehat{r}_T) + \varepsilon_{t+H}^*,$$

where $\widehat{\varepsilon}_{t+H}^*$ is a random draw (with replacement) from $\widehat{F}_{\widetilde{\varepsilon}}$. So, the BS forecasts are all conditioned on the forecast origin data.

1.5 Repeat steps 1.1 – 1.4 B times, and obtain the BS forecasts $\{Y_{t+H}^{*,(b)}\}_{b=1}^B$.

10.3 FORECAST INTERVALS AND REGIONS

> **Algorithm 10.1: Bootstrap FI (Cont'd)**
>
> 1.5 (Cont'd)
>
> Then the bootstrap FI (BFI) with coverage probability $(1 - \alpha)$ is given by
>
> $$\text{BFI}_{H,\alpha} = [\widehat{Y}_{t+H}^{(\alpha/2)}, \widehat{Y}_{t+H}^{(1-\alpha/2)}], \qquad (10.50)$$
>
> where $\widehat{Y}_{t+H}^{(\alpha/2)}$ and $\widehat{Y}_{t+H}^{(1-\alpha/2)}$ are, respectively, the $(\alpha/2)$th and $(1 - \alpha/2)$th percentiles of the EDF of $\{Y_{t+H}^{*,(b)}\}_{b=1}^{B}$.

Note that Algorithm 10.1 ignores the sampling variability of $\widehat{r}_T$. To adjust for this, step 1.3 can be repeated many times with BS threshold values obtained from Algorithm 6.2; see Li (2011). Another modification follows from using bias-corrected estimators of the coefficients $\phi_i^{(j)}$; see, e.g., Kilian (1998). In the context of linear AR models, Kim (2003) provides a BS mean bias-corrected estimator which can simply be adopted to correct for biases of SETAR coefficient estimators. In particular, Algorithm 10.1 needs to be modified as follows.

> **Algorithm 10.2: Bootstrap bias-corrected FI**
>
> 2.1 Same as step 1.1.
>
> 2.2 Re-estimate (10.8) using $\{Y_t^*\}_{t=1}^T$ and $\widehat{r}_T$, and obtain the BS coefficients $\widehat{\phi}_i^{*,(j)}$ ($i = 0, \ldots, p; j = 1, 2$). Repeat this step C times to get a set of BS coefficients $\{\widehat{\phi}_i^{*,(c),(j)}\}_{c=1}^C$.
>
> 2.3 Compute the bias of $\widehat{\phi}_i^{(j)}$ as $\text{Bias}(\widehat{\phi}_i^{(j)}) = \overline{\widehat{\phi}}_i^{*,(j)} - \widehat{\phi}_i^{(j)}$ where $\overline{\widehat{\phi}}_i^{(j)}$ is the sample mean of $\{\widehat{\phi}_i^{*,(c),(j)}\}_{c=1}^C$. Next, compute the bias-corrected coefficients as $\widehat{\phi}_i^{c,(j)} = \widehat{\phi}_i^{(j)} - \text{Bias}(\widehat{\phi}_i^{(j)})$.
>
> 2.4 Then, analogously to (10.49), generate the bias-corrected BS replicates $\{Y_t^{c*}\}$ using $\widehat{\phi}_i^{c,(j)}$.
>
> 2.5 Re-estimate (10.8) using $\{Y_t^{c*}\}_{t=1}^T$ and $\widehat{r}_T$, and obtain the BS coefficients $\widehat{\phi}_i^{*,(c),(j)}$. Next, compute the bias-corrected BS forecasts as $Y_t^{c*} = Y_t$ ($t = T, T-1, \ldots, T-p+1$),
>
> $$Y_{t+H}^{c*} = \{\widehat{\phi}_0^{*,(c),(1)} + \sum_{i=1}^p \widehat{\phi}_i^{*,(c),(1)} Y_{t+H-i}^{c*}\} I(Y_{t+H-d}^{c*} \leq \widehat{r}_T) +$$
>
> $$\{\widehat{\phi}_0^{*,(c)(2)} + \sum_{i=1}^p \widehat{\phi}_i^{*,(c),(2)} Y_{t+H-i}^{c*}\} I(Y_{t+H-d}^{c*} > \widehat{r}_T) + \varepsilon_{t+H}^*.$$
>
> 2.6 Repeat steps 2.1 – 2.4 B times and obtain a set of bias-corrected forecasts $\{Y_{t+H}^{c*,(b)}\}_{b=1}^B$. The bias-corrected BFI (BFI^c) with coverage probability $(1-\alpha)$ is given by

> **Algorithm 10.2: Bootstrap bias-corrected FI (Cont'd)**
>
> 2.6 (Cont'd)
>
> $$\text{BFI}^c_{H,\alpha} = [\widehat{Y}^{(\alpha/2),c}_{t+H}, \widehat{Y}^{(1-\alpha/2),c}_{t+H}], \quad (10.51)$$
>
> where $\widehat{Y}^{(\alpha/2),c}_{t+H}$ and $\widehat{Y}^{(1-\alpha/2),c}_{t+H}$ are, respectively, the $(\alpha/2)$th and $(1-\alpha/2)$th percentiles of the EDF of $\{Y^{c*,(b)}_{t+H}\}^B_{b=1}$.

Note that the bias-correction in step 2.3 can push the coefficients into the non-stationary region of the parameter space; see, e.g. Clements (2005, Section 4.2.4), Kilian (1998), and Li (2011) for a stationarity correction procedure which can easily be implemented in Algorithm 10.2. Another modification is to replace the fitted residuals by *predictive residuals* (Politis, 2013, 2015). For a SETAR$(2; p, p)$ model these residuals can be computed as follows: Delete the row $(1, Y_{t-1}, \ldots, Y_{t-p})$ in the $T \times (p+1)$ design matrix $\mathbf{X}_t(r)$ (see (6.11)), and delete Y_t^{-t} from the series $\{Y_t\}_{t=1}^T$. Next, compute the leave-one-out CLS estimator of the model coefficients using (6.11), and obtain the leave-one-out fitted value $\widehat{Y}_t^{-t}$ Then the predictive residuals are given by $\widehat{\varepsilon}_t^{-t} = Y_t - \widehat{Y}_t^{-t}$. The key idea here is that the distribution of the one-step-ahead forecast errors can be approximated better by the EDF of $\{\widehat{\varepsilon}_t^{-t}\}_{t=p+1}^T$ than by the EDF of $\{\widehat{\varepsilon}_t\}_{t=p+1}^T$; cf. Exercise 10.7.

Example 10.6: FIs for a Simulated SETAR Process

Consider the stationary SETAR$(2; 1, 1)$ process of Example 8.2, i.e.

$$Y_t = 0.5 Y_{t-1} I(Y_{t-1} \leq 0) - 0.4 Y_{t-1} I(Y_{t-1} > 0) + \varepsilon_t, \quad (10.52)$$

where $Y_0 = 0$ and $\{\varepsilon_t\} \overset{\text{i.i.d.}}{\sim} \mathcal{N}(0, 1)$. We set $T = 100$, $B = 1{,}000$, $C = 200$, and $\alpha = 0.05$. To assess the performance of the BFIs, we use the empirical *coverage rate* (CVR) defined by

$$\text{CVR}_{H,\alpha} = \frac{1}{m} \sum_{i=1}^m I\big(Y_{i,T+H} \in \text{FI}^{(\cdot)}_\alpha\big), \quad (10.53)$$

where $Y_{i,T+H}$ denotes the H-step ahead forecast made at time $t = T$ from the ith data set, and $\text{FI}^{(\cdot)}_\alpha$ denotes either $\text{BFI}_{H,\alpha}$ or $\text{BFI}^c_{H,\alpha}$.

Figure 10.5 shows boxplots of the $\text{CVR}_{H,\alpha}$ for $H = 1, \ldots, 5$ and $m = 100$. There are no serious size distortions in coverage rates; both BFIs have an IQR of about 0.03, on average, across all values of H. This implies that the BFIs generally work well. The variability of the threshold variable estimator does not seem to cause higher CVRs in the case of $\text{BFI}^c_{H,\alpha}$. Moreover, the CVRs seem to remain fairly constant as H increases with average standard deviation

10.3 FORECAST INTERVALS AND REGIONS

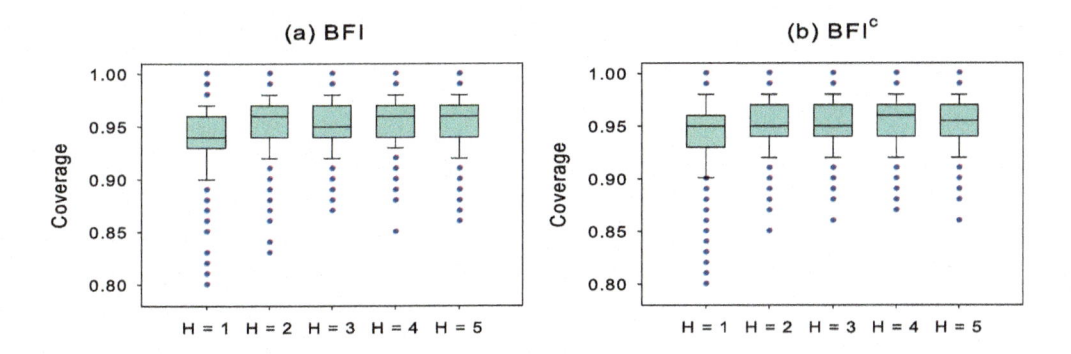

Figure 10.5: *Empirical CVRs for (a)* $\text{BFI}_{H,\alpha}$ *and (b)* $\text{BFI}^c_{H,\alpha}$ *for the* SETAR$(2;1,1)$ *model* (10.52); $T = 100$, $\alpha = 0.05$, $B = 1{,}000$, $m = 100$, *and 500 MC replications.*

of about 0.02 in both cases (a) and (b). Observe that (10.53) represents an *unconditional* coverage probability since Y_T is different for each simulated data set.

10.3.3 Conditional densities

For nonlinear DGPs, the width of the CPI in (10.42) is no longer a constant, as in the case of linear DGPs, but may vary with respect to the position in the state space from which forecasts are being made.[3] Unfortunately, CPI's are not always efficient (in the sense of having the smallest width) when the forecast distribution is asymmetric or multi-modal. To overcome this problem Yao and Tong (1995), De Gooijer and Gannoun (2000), and Polinik and Yao (2000) advocate the use of the following two alternative methods.

Shortest conditional modal interval (SCMI)
For any given $\alpha \in [0,1]$ and $\mathbf{x} \in \mathbb{R}^p$, we define the minimal conditional density region as

$$b_\alpha(\mathbf{x}, y) = \min\left\{ b > 0 \,\Big|\, \int_{y-b}^{y+b} f(u|\mathbf{x})du \leq 1 - \alpha \right\}, \qquad y \in \mathbb{R}, \qquad (10.54)$$

where $f(\cdot|\mathbf{x})$ denotes the conditional density function of Y_t given $\mathbf{X}_t = \mathbf{x}$. Let

$$b_\alpha(\mathbf{x}) = \min_{y \in \mathbb{R}} b_\alpha(\mathbf{x}, y), \qquad m_\alpha(\mathbf{x}) = \arg\min_{y \in \mathbb{R}} b_\alpha(\mathbf{x}, y). \qquad (10.55)$$

The so-called *shortest conditional modal interval* (SCMI) with coverage probability $1 - \alpha$ is defined as

$$\text{SCMI}_\alpha(\mathbf{x}) = [m_{\alpha/2}(\mathbf{x}) - b_{\alpha/2}(\mathbf{x}), b_{1-\alpha/2}(\mathbf{x}) + b_{1-\alpha/2}(\mathbf{x})], \qquad \alpha \in [0,1]. \qquad (10.56)$$

---

[3]The property of variable-size FIs is commonly named *sharpness* or *resolution*. Sometimes a subtle difference is made between both terms in the sense that sharpness relates to the average size of FIs and resolution to their associated variability; cf. Exercise 10.8.

It follows from (10.54) and (10.55) that the SCMI can also be defined as

$$[a, b] = \arg\min\{\text{Leb}\{[c, d]\} \,|\, F(d|\mathbf{x}) - F(c|\mathbf{x}) \leq 1 - \alpha\}, \quad \alpha \in [0, 1], \quad (10.57)$$

where $\text{Leb}(\mathcal{C})$ denotes the Lebesgue measure of the set $\mathcal{C}$, which is a measurable subset of $\mathbb{R}^p$, and $F(\cdot|\mathbf{x})$ the conditional distribution function of Y_t given $\mathbf{X}_t = \mathbf{x}$. Thus, the idea is to search for the set with the minimum length among all predictive sets; see Fan and Yao (2003, Section 10.4) for a more thorough discussion.

Of course, in practice, a natural estimator for the SCMI is obtained by replacing $F(\cdot|\mathbf{x})$ by a consistent estimate, e.g. the NW or the LL kernel-based estimator. For symmetric and unimodal conditional predictive distributions SCMI reduces to CPI.

Maximum or highest conditional density region (HDR)

The second method, initially called *maximum conditional density region* (MCDR), but better known as the *highest* (conditional) *density region* (HDR), is the smallest region (i.e., Lebesgue measure) of the sample space to a given coverage probability. More formally, for $\alpha \in [0, 1]$, define

$$l_\alpha(\mathbf{x}) \equiv l_\alpha(f(y|\mathbf{x})) = \inf\left\{l \in (0, \infty) \,\Big|\, \int_{-\infty}^{\infty} f(y|\mathbf{x}) I\big(f(y|\mathbf{x}) \geq l\big) dy \leq 1 - \alpha\right\}.$$

We call the subset R_α the $100(1 - \alpha)\%$ HDR of $f(\cdot|\mathbf{x})$ (cf. Hyndman, 1995, 1996) such that

$$R_\alpha = \{\mathbf{x} \in \mathbb{R}^p : f(y|\mathbf{x}) \geq l_\alpha(\mathbf{x})\}, \quad \alpha \in [0, 1]. \quad (10.58)$$

Thus, the HDR is naturally related to the conditional mode since they are both based on points of highest density. The HDR can be equivalently defined as

$$\bigcup_{i=1}^{\ell} [a_i, b_i] = \arg\min\left\{\text{Leb}\left(\bigcup_{i=1}^{\ell} [c_i, d_i]\right) \,\Big|\, c_1 < d_1 \leq c_2 < d_2 \leq \cdots \leq c_\ell < d_\ell,\right.$$

$$\left.\text{and } \sum_{i=1}^{\ell} \{F(d_i|\mathbf{x}) - F(c_i|\mathbf{x}) \leq 1 - \alpha\}\right\},$$

where $\ell \geq 1$ denotes the number of sub-intervals. Replacing $F(\cdot|\mathbf{x})$ by, for instance, the NW smoother gives an estimator of the HDR. By definition, HDR is of the smallest Lebesgue measure among all FRs with the same α. The HDR may consist of less than ℓ disconnected intervals even though $f(\cdot|\mathbf{x})$ has ℓ modes. Equivalently as the SCMI, the HDR reduces to the CPI when $f(\cdot|\mathbf{x})$ is unimodal and also symmetric with respect to its mode.

Example 10.7: Hourly River Flow Data (Cont'd)

We reconsider the hourly river flow series $\{Y_t\}_{t=1}^{401}$ introduced in Example 9.3. The series is stationary and positively autocorrelated. We predict the flow

10.4 FORECAST EVALUATION

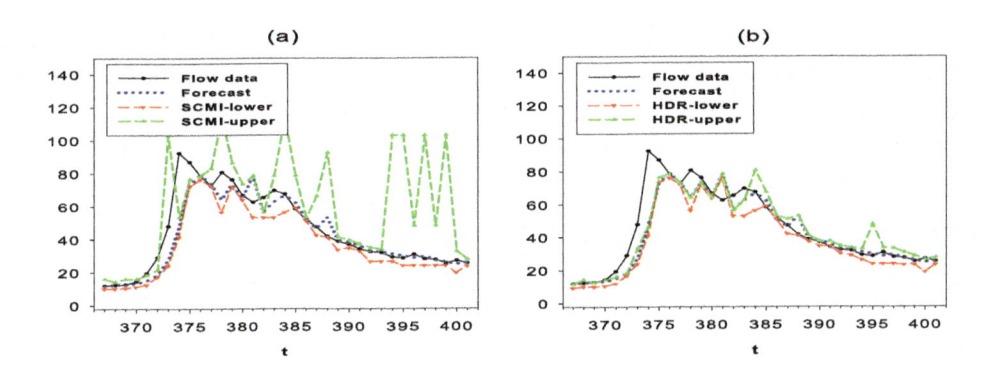

Figure 10.6: Hourly river flow *data set.* (a) *One-step ahead forecast* $\widehat{Y}_{t+1|t}^{\text{Mdn}}$ *and estimated* SCMI's *(with coverage probability 0.9);* (b) *One-step ahead forecast* $\widehat{Y}_{t+1|t}^{\text{Mdn}}$ *and estimated* HDRs *(with the highest coverage probability). From* De Gooijer and Gannoun (2000).

at the tth hour Y_t from the observed values of Y_{t-1} using the nonparametric predictor $\widehat{Y}_{t+H|t}^{\text{Mdn}}$, defined in (9.7). We use a Gaussian kernel, and set $p = 1$. As a starting-point we select $t = 366$ which is just located before the large peak in $\{Y_t\}$ at time $t = 374$. Next, we predict Y_{368} using the observed values up to an including the one at $t = 367$. This procedure is repeated till the end of the series. Hence, in total 35 one-step ahead predictions are available. Further, with coverage probability $(1 - \alpha) = 0.9$, we estimate the SCMI and the HDR in each step. The bandwidths follow from minimizing $\text{CV}^{\text{Mdn}}(H)$; see Table 9.1.

Figures 10.6(a) and 10.6(b) show plots of the last 35 observations of $\{Y_t\}$ with one-step ahead forecasts $\widehat{Y}_{t+1|t}^{\text{Mdn}}$ for the SCMI and the HDR. Clearly, the SCMI is very wide and asymmetric whereas the HDR is much tighter. Note, however, that at $t = 370 - 375, 378 - 380, 382 - 383, 391$, and 400 the realizations do not fall within the HDR. On the other hand, the SCMI does not cover the corresponding observed values at $t = 371 - 372, 374 - 375, 382$, and 385. Similar observations were noted for FRs based on $\widehat{Y}_{t+H|t}^{\text{Mean}}$, defined in (9.5). For the time period $t = 370 - 375$ this is due to a steep rise in river flow, due to heavy rainfall (3.2 mm/hour at $t = 374$). Thus, the width of both FRs can be quite sensitive to the position in the state space from which predictions are being made.

10.4 Forecast Evaluation

10.4.1 Point forecast

Classical, stand-alone, accuracy measures for comparing forecasts are the MSFE and the MAFE. The smaller the value of these measures, the better is a particular forecast. More generally, it frequently happens that two (or more) forecasts of the same quantity are available via rival forecast methodologies. Then the question naturally arises as how likely it is that differences between the two forecasts is due to chance or whether they are "significant". Below we review various tests for comparing the accuracy of competing point forecasts. First, we describe the basic forecast setup.

Setup

Let $\{Y_t\}_{t=1}^{T+H}$ be the sample of observation, where $H \equiv H_{\max} \geq 1$ denotes the longest forecast horizon of interest. We assume that the available data set is divided into in-sample and out-sample portions, with R (R as in Regress) the total number of in-sample observations and P the number of H-step ahead forecasts. Thus, $R + P + H - 1 \equiv T + H$ is the size of the available sample. Note that this setup implies that P out-of-sample forecasts depend on the same parameter vector estimated on the first R observations. So, the forecast scheme is based on a single, fixed, estimation sample.

Alternatively, a *rolling* or a *recursive* forecasting scheme can be employed. In the latter case, the first forecast is based on a model with parameter vector estimated using $\{Y_t\}_{t=1}^{R}$, the second on a parameter vector estimated using $\{Y_t\}_{t=1}^{R+1}, \ldots,$ the last on a parameter vector estimated using $\{Y_t\}_{t=1}^{R+P-1}$, where $T \equiv R + P - 1$. In the rolling scheme, the sequence of parameter estimates is always generated from a fixed, but rolling, sample of size R: The first forecast is based on parameter estimates obtained from the set of observations $\{Y_t\}_{t=1}^{R}$, the next on parameter estimates obtained from $\{Y_t\}_{t=2}^{R+1}$, and so on.

Diebold–Mariano (DM) test

Diebold and Mariano (1995) propose a test statistic based on the null hypothesis that two forecasts are the same in terms of forecasting accuracy, for some arbitrary loss function $L(e_{i,t+H|t})$ where $e_{i,t+H|t} = Y_{t+H} - Y_{i,t+H|t}$ is the H-step ahead forecast error with $Y_{i,t+H|t}$ the forecasts from model i ($i = 1, 2$). The so-called H-step ahead *loss differential* is defined as

$$d_t = L(e_{i,t+H|t}) - L(e_{j,t+H|t}), \quad (i, j = 1, 2; \; i \neq j).$$

So, the null hypothesis entails

$$\mathbb{E}[L(e_{i,t+H|t})] = \mathbb{E}[L(e_{j,t+H|t})], \quad (i, j = 1, 2; \; i \neq j), \tag{10.59}$$

or $\mu_d \equiv \mathbb{E}(d_t) = 0$. Typically, $L(\cdot)$ is the squared-error loss or the absolute error loss. Still one may consider other loss functions, including ones based on economic rather than statistical criteria.

Suppose that a sample realization $\{d_t\}_{t=R+H}^{R+P+H-1}$ of a covariance stationary process $\{d_t, t \in \mathbb{Z}\}$ is available. Then, as $R \to \infty$ at a faster rate than $P \to \infty$, as $T \to \infty$, it is easy to deduce that the asymptotic distribution of the sample mean loss differential, $\overline{d} = P^{-1} \sum_{t=R+H}^{R+P+H-1} d_t$, is given by

$$\sqrt{P}(\overline{d} - \mu_d) \xrightarrow{D} \mathcal{N}\big(0, \operatorname{Var}(\overline{d})\big), \tag{10.60}$$

10.4 FORECAST EVALUATION

where

$$\text{Var}(\overline{d}) \approx \frac{1}{P} \sum_{\ell=-(H-1)}^{H-1} \gamma_d(\ell), \qquad (10.61)$$

with $\gamma_d(\cdot)$ the ACVF of $\{d_t, t \in \mathbb{Z}\}$. The lag ℓ autocovariance can be estimated by

$$\widehat{\gamma}_d(\ell) = \frac{1}{P} \sum_{t=R+H+\ell}^{R+P+H-1} (d_t - \overline{d})(d_{t-\ell} - \overline{d}), \quad \ell \in \mathbb{Z}.$$

Then a consistent estimate $\widehat{\text{Var}}(\overline{d})$ of $\text{Var}(\overline{d})$ follows directly. The resulting asymptotic distribution of the DM test statistic is then

$$\text{DM} = \frac{\overline{d}}{\sqrt{\widehat{\text{Var}}(\overline{d})}} \xrightarrow{D} \mathcal{N}(0,1), \quad \text{as } P \to \infty. \qquad (10.62)$$

It is apparent that, for fixed H, relevant applications of the DM test statistic are those in which $H \ll R, P$. The DM test statistic is "model-free", i.e., the forecast models are assumed to be correctly specified, but unknown, and the associated loss function $L(\cdot)$ does not rests on additional, conditioning, information. In other words, only a set of forecasts and actual values of the predictand are considered. Furthermore, it is implicitly assumed that the competing forecasts $Y_{1,t+H|t}$ and $Y_{2,t+H|t}$ are obtained from non-nested models. With nested models the limiting distribution of the DM test statistic and other existing tests for comparing forecast accuracy are non-standard, can be difficult to compute or are context-specific (see, e.g., Clark and McCracken, 2001; Clark and West, 2007). Motivated by the above observations, Giacomini and White (2006) present a general framework for out-of-sample forecast evaluation. It applies to multi-horizon point, interval, probability, and density forecasts for general loss functions applicable to both nested and non-nested models. The resulting tests can be viewed as extensions to the DM test statistic. Moreover, the asymptotic standard normal distribution of the DM test statistic remains unchanged for nested models and finite in-sample sizes; see also Table 10.2.[4]

Modified DM test

When the forecast errors are Gaussian distributed or fat tailed, MC simulation results (Diebold and Mariano, 1995) indicate that the DM test statistic, under quadratic loss, is robust to contemporaneous and serial correlation in large samples, but the test is oversized in small samples. Indeed, for a small number of forecasts it is

---

[4]It is good to mention that the null hypothesis of the Giacomini–White approach is different from that of West and his co-authors in two respects: (i) the loss function $L(\cdot)$ depends on estimates rather than their probability limits; and (ii) the expectation in (10.59) is conditional on some information set.

recommended to use the modified DM (MDM) test statistic proposed by Harvey et al. (1997). The modification follows from replacing (10.61) by the exact variance

$$\text{Var}(\overline{d}) = \frac{1}{P}\left(\gamma_d(0) + 2P^{-1}\sum_{\ell=1}^{H-1}(P-\ell)\gamma_d(\ell)\right). \qquad (10.63)$$

Then $\widehat{\text{Var}}(\overline{d})$ can be written as

$$\widehat{\text{Var}}(\overline{d}) = \frac{1}{P}\left(\widehat{\gamma}_d^*(0) + 2P^{-1}\sum_{\ell=1}^{H-1}(P-\ell)\widehat{\gamma}_d^*(\ell)\right), \quad (P \geq 2), \qquad (10.64)$$

where

$$\widehat{\gamma}_d^*(\ell) = \frac{1}{P-\ell}\sum_{t=R+H+\ell}^{R+P+H-1}(d_t - \overline{d})(d_{t-\ell} - \overline{d}).$$

Assume the mean of $\{d_t, t \in \mathbb{Z}\}$ is known and, without loss of generality, can be taken to be zero. With a little algebra (cf. Exercise 10.5), it follows that for $\ell \ll P$

$$\mathbb{E}\big(\widehat{\gamma}_d^*(\ell)\big) = \gamma_d(\ell) - (P-\ell)^{-1}(P+\ell)\text{Var}(\overline{d}) + \mathcal{O}(P^{-2})$$
$$\approx \gamma_d(\ell) - \text{Var}(\overline{d}). \qquad (10.65)$$

Taking expectations in (10.64) and substituting (10.65), we have

$$\mathbb{E}\big(\widehat{\text{Var}}(\overline{d})\big) \approx \frac{P + 1 - 2H + P^{-1}H(H-1)}{P}\text{Var}(\overline{d}). \qquad (10.66)$$

The term $P^{-1}H(H-1)$ is included here, since (10.66) is exact in the special case where the process $\{d_t, t \in \mathbb{Z}\}$ is WN.

As an implication of (10.66), the DM test statistic can be modified (m) for its finite sample oversizing by using an approximately unbiased variance estimate, say $\widehat{\text{Var}}_m(\overline{d})$. The resulting MDM test statistic is therefore simply

$$\text{MDM} = \frac{\overline{d}}{\sqrt{\widehat{\text{Var}}_m(\overline{d})}} = \left\{\frac{P + 1 - 2H + P^{-1}H(H-1)}{P}\right\}^{1/2}\text{DM}, \qquad (10.67)$$

where

$$\widehat{\text{Var}}_m(\overline{d}) = [P + 1 - 2H + P^{-1}H(H-1)]^{-1}\big(\widehat{\gamma}_d(0) + 2\sum_{\ell=1}^{H-1}\widehat{\gamma}_d(\ell)\big).$$

Significance may be assessed using the Student t distribution with $P - 1$ degrees of freedom.

10.4.2 Interval evaluation

Using the forecast setup introduced in the previous subsection, let $L_{t+H|t}^{(1-\alpha)}$ and $U_{t+H|t}^{(1-\alpha)}$ denote the lower and upper limits of the $H(\equiv H_{\max})$-step ahead interval forecasts of Y_{t+H} made at time t, for a coverage probability $(1-\alpha)$, and given the sample of observations $\{Y_t\}_{t=1}^{T+H}$. We define the sequence of indicator functions $\{i_t^{(\alpha)}\}_{t=R+H}^{R+P+H-1}$ as

$$i_t^{(\alpha)} = \begin{cases} 1 & \text{if } Y_{t+H} \in [L_{t+H|t}^{(1-\alpha)}, U_{t+H|t}^{(1-\alpha)}], \\ 0 & \text{otherwise,} \end{cases} \quad (10.68)$$

where P denotes the number of H-step ahead forecasts, and R the total number of in-sample observations. Thus, the indicator (or "hit") function tells whether the actual value Y_{t+H} lies (a "hit") or does not lie (a "miss" or a "violation") in the FI for that lead time H. The sequence of interval forecasts is said to be "well-specified' with respect to the past information set $\Psi_t = \{i_t^{(\alpha)}, i_{t-1}^{(\alpha)}, \ldots\}$ if $\mathbb{E}(i_t^{(\alpha)}|\Psi_{t-1}) = 1 - \alpha \equiv p$. Within this framework, Christoffersen (1998) proposes the following, widely used, LR-based test statistics.

Unconditional (uc) coverage LR test statistic

The easiest way to evaluate FIs is to compare the coverage probability p with the sample proportion of times that the FI includes Y_{t+H}, ignoring the dependence in $\{i_t^{(\alpha)}\}$. Hence, the null hypothesis $\mathbb{H}_0^{(uc)}$ of interest is $\mathbb{E}(i_t^{(\alpha)}) = p$, while the alternative hypothesis is $\mathbb{E}(i_t^{(\alpha)}) \equiv \pi \neq p$. For a given H and α, denote

$$n_1 = \#\{i_t^{(\alpha)} = 1\} = \sum_{t=1}^{P} i_t^{(\alpha)} \text{ and } n_0 = \#\{i_t^{(\alpha)} = 0\} = P - n_1.$$

The likelihoods of the data under the null and alternative hypotheses are, respectively,

$$L_p \equiv L(p; i_1^{(\alpha)}, \ldots, i_P^{(\alpha)}) = (1-p)^{n_0} p^{n_0} \text{ and } L_{\widehat{\pi}} \equiv L(\widehat{\pi}; i_1^{(\alpha)}, \ldots, i_P^{(\alpha)}) = (1-\widehat{\pi})^{n_1} \widehat{\pi}^{n_1},$$

where the relative hit frequency $\widehat{\pi} = n_1/(n_0 + n_1)$ is the ML estimate of π. Then the LR-based test statistic is given by

$$\text{LR}_{uc} = -2\log(L_p/L_{\widehat{\pi}}). \quad (10.69)$$

Under $\mathbb{H}_0^{(uc)}$, and as $P \to \infty$, LR_{uc} has a χ_1^2 distribution.

Independence (ind) LR test statistic[5]

The test statistic (10.69) will have very low power when there are discernible, time-dependent, patterns in $\{i_t^{(\alpha)}\}$. To overcome this problem, Christoffersen (1998)

---

[5]The term "independence" is a misnomer, because only second-order properties will be considered.

suggests testing for independence by modeling the process $\{i_t^{(\alpha)}, t \in \mathbb{Z}\}$ as a two-state (i.e., $k = 2$ in the notation of Section 2.10) first-order Markov chain with transition probability matrix

$$\mathbf{P}_1 = \begin{pmatrix} 1 - p_{12} & p_{12} \\ 1 - p_{22} & p_{22} \end{pmatrix}, \tag{10.70}$$

where $p_{ij} = \mathbb{P}(i_t^{(\alpha)} = j | i_{t-1}^{(\alpha)} = i)$ and $\sum_{j=1}^{2} p_{ij} = 1$ $(i, j = 1, 2)$. Let n_{ij} denote the number of events that a state i is followed by a state j. Then the approximate likelihood function under the alternative hypothesis for the whole process is

$$L(\widehat{\mathbf{P}}_1) = (1 - \widehat{p}_{12})^{n_{11}} \widehat{p}_{12}^{n_{12}} (1 - \widehat{p}_{22})^{n_{21}} \widehat{p}_{22}^{n_{22}}, \tag{10.71}$$

with $\widehat{p}_{ij} = n_{ij}/(n_{i1} + n_{i2})$ $(i, j = 1, 2)$ the ML estimate of p_{ij}. Under the null hypothesis $\mathbb{H}_0^{(\text{ind})}$: $p_{12} = p_{22}$, the state of the process at time t conveys no information on the relative likelihood of it being in one state as opposed to another at time $t + 1$. Thus, when the outcome, say $i_t^{(\alpha)}$, of the chain lies in state j, the nearest outcome $i_{t-1}^{(\alpha)}$ has the same probability of lying in any state. We can write this as $p_{1j} = p_{2j} = \pi_j$, where $\pi_j = \mathbb{P}(i_t^{(\alpha)} = j)$ $(j = 1, 2)$. Let n_j denote the corresponding number of outcomes. Then the ML estimate of π_j is given by $\widehat{\pi}_j = n_j/N$ with $N = \sum_{i,j=1}^{2} n_{ij}$. Hence, the approximate likelihood function under $\mathbb{H}_0^{(\text{ind})}$ is $L_{\widehat{P}_0} \equiv L(\widehat{P}_0; i_1^{(\alpha)}, \ldots, i_P^{(\alpha)}) = \prod_{j=1}^{2} (n_j/N)^{n_j}$, and the unrestricted likelihood function is $L_{\widehat{P}_1} \equiv L(\widehat{P}_1; i_1^{(\alpha)}, \ldots, i_P^{(\alpha)}) = \prod_{i=1}^{2} \prod_{j=1}^{2} (n_{ij}/\sum_{j=1}^{2} n_{ij})^{n_{ij}}$. Then the LR-based test statistic for independence is given by

$$\text{LR}_{\text{ind}} = -2 \log(L_{\widehat{P}_1}/L_{\widehat{P}_0}). \tag{10.72}$$

Under $\mathbb{H}_0^{(\text{ind})}$, and as $P \to \infty$, LR_{ind} has a $\chi^2_{(2-1)^2}$ distribution. Similarly, it is straightforward to show that for a k-state $(k \geq 2)$ first-order Markov chain, the corresponding LR-based test statistic has (asymptotically) a $\chi^2_{(k-1)^2}$ distribution under the null hypothesis.

Conditional coverage (cc) LR test statistic

Note the LR_{uc} and LR_{ind} test statistics do not affect each other. To test whether the FI has the correct coverage in the form of the null hypothesis $\mathbb{H}_0^{(\text{cc})}$: $p_{12} = p_{22} = p$ with $p = \mathbb{E}(i_t^{(\alpha)} | \Psi_{t-1})$, it is sensible to combine both test statistics. In particular, a test statistic of correct conditional coverage is given by

$$\text{LR}_{\text{cc}} = -2 \log(L_p/L_{\widehat{P}_1}). \tag{10.73}$$

Under $\mathbb{H}_0^{(\text{cc})}$ it follows (Christoffersen, 1998) that, as $P \to \infty$, the test statistic LR_{cc} has a χ^2_2 distribution. For a $k \geq 2$ state first-order Markov chain, the corresponding

LR-based test statistic is asymptotically $\chi^2_{k(k-1)}$ distributed. Moreover, when ignoring the first observation $i_1^{(\alpha)}$, the three LR test statistics are numerically related by the identity $\mathrm{LR}_{cc} = \mathrm{LR}_{uc} + \mathrm{LR}_{ind}$ (an additivity property).

Note that the above LR test statistics do not take into account time-dependencies in the information set Ψ_{t-1} of order higher than one. So, in some cases, these tests may ignore patterns of clustering in Ψ_{t-1}. Furthermore, within the Markov chain framework, it is not possible to extend the information set with information contained in another exogenous variable. The list with additional bibliographical notes given at the end of this chapter contains references to papers which discuss test statistics aimed at avoiding these and other drawbacks; see also below.[6]

Detecting clustering effects

When dealing with linear and nonlinear ARCH-type DGPs it is likely that FIs are too small in turbulent periods compared to relatively tranquil times. This will result in clustering of misses (violations) at high volatility times. Araújo Santos and Fraga Alves (2012) propose a new class of test statistics for explicitly testing $\mathbb{H}_0^{(ind)}$ against an alternative hypothesis expressing a *tendency to clustering patterns*. They define this notion more formally as follows.

Let $\{D_j = t_j - t_{j-1}\}_{j=1}^N$ ($t_0 = 0$) be the sample of N *durations* between two consecutive violations in the sequence $\{i_t^{(\alpha)}\}_{i=1}^P$ where t_j denotes the time-index of violation j. If $\mathbb{H}_0^{(cc)}$ is valid, then the process $\{i_t^{(\alpha)}, t \in \mathbb{Z}\} \stackrel{i.i.d.}{\sim} \mathrm{Bernoulli}(p)$ ($0 < p < 1$). Consequently, the random variable D_j is geometrically distributed with pmf $f_D(d) = (1-p)^{d-1} p$ ($d \in \mathbb{N}$). Hence, $\mathbb{H}_0^{(ind)}$ can be written as $\{D_j, j \in \mathbb{Z}^+\} \stackrel{i.i.d.}{\sim} \mathrm{Geometric}(p)$. Furthermore, let $D_{1:N} \leq \cdots \leq D_{N:N}$ be the order statistics of $\{D_j\}_{j=1}^N$. Then a hit function is said to have a tendency to clustering of violations if $\mathrm{Mdn}(D_{N:N}/D_{[N/2]:N})$ is higher than the median of the process $\{D_j, j \in \mathbb{Z}^+\}$ under $\mathbb{H}_0^{(ind)}$.

Next, as a special case of the proposed class of independence tests, Araújo Santos and Fraga Alves (2012) define the test statistic

$$T_{N,[N/2]} = \log 2 \, \frac{D_{N:N} - 1}{D_{[N/2]:N}} - \log N. \tag{10.74}$$

The test statistic is pivotal in the sense that its distribution does not depend on an unknown parameter. However, (10.74) is a test statistic for $\mathbb{H}_0^{(ind)}$, not for testing $\mathbb{H}_0^{(cc)}$. The decision rule for rejecting $\mathbb{H}_0^{(ind)}$ can be based on critical values (using an exact distribution) provided by Araújo Santos and Fraga Alves (2012, Appendix) or by simulating p-values (cf. Exercise 10.11).

---

[6] Within the Value-at-Risk (VaR) evaluation literature of FIs these test statistics are often called *backtesting* procedures.

10.4.3 Density evaluation

As we mentioned earlier, in stationary time series the conditional density function provides the most informative characterization of the possible future values of a time series variable, conditional on the information available at the time the forecast is made. Interest in the topic has recently surged in the literature (see, e.g., Clements, 2005, Chapter 5) with, for instance, the MFD method in Algorithm 9.3 as a particular contribution. Here, we consider methods of evaluating the performance of density forecasts using the PIT of the actual realizations of the variable with respect to the forecast densities.

Suppose we have a set of P one-step ahead forecast densities for the future value of a process $\{Y_t, t \in \mathbb{Z}\}$, denoted by $\{\widehat{f}_t(Y_t|\mathcal{F}_{t-1})\}_{t=1}^P$, made at time t with $f_1(Y_1|\mathcal{F}_0) \equiv f(y_1)$. The PIT, denoted by U_t, is defined as

$$U_t \equiv \int_{-\infty}^{Y_t} \widehat{f}_t(u|\mathcal{F}_{t-1})du, \quad (t=1,\ldots,P). \tag{10.75}$$

Under the null hypothesis ($\mathbb{H}_0$) that the model forecasting density corresponds to the true conditional density, given by the DGP which is denoted by $f_t(\cdot|\mathcal{F}_{t-1})$, that is $\widehat{f}_t(\cdot|\mathcal{F}_{t-1}) = f_t(\cdot|\mathcal{F}_{t-1})$, the process $\{U_t, t \in \mathbb{Z}\}$ is i.i.d. $U(0,1)$ distributed (Rosenblatt, 1952).

A simple way of testing the uniformity part of the null hypothesis *conditional* on the i.i.d. assumption is by using a nonparametric GOF test like the KS, AD or CvM test statistics; see, e.g., Chapter 7. Alternatively, a plot of the CDF of the U_t may be used and visually compared with a line at an angle of 45° representing the cumulative uniform distribution. The independence part of the null hypothesis may be tested by using an LM-type test for serial correlation in the sequences $\{(U_t - \overline{U})^u\}_{t=1}^P$ ($u = 1, 2$), where $\overline{U}$ is the sample mean of the U_t. For the case $u = 2$, the sample ACF may indicate some form of nonlinear dependence such as heteroskedasticity. Similar evaluation techniques can be applied to the transformed sequence $\{\Phi^{-1}(U_t)\}_{t=1}^P$ which is i.i.d. $\mathcal{N}(0,1)$ distributed under the null hypothesis (Berkowitz, 2001). Other ways of testing forecast densities are given in the next chapter, albeit in a vector nonlinear time series framework.

Example 10.8: ENSO Phenomenon (Cont'd)

Recall the monthly ENSO time series discussed in Examples 1.4, 5.1, and 6.4. We proceed by evaluating the out-of-sample forecast performance of the nonlinear LSTEC model (6.24) as opposed to its linear (AR-type) counterpart (6.25) using a rolling forecasting approach. In (6.24) an LSTEC model was fitted to $\{\Delta Y_t\}_{t=1}^{468}$, covering the time period January 1952 – December 1990. This period will serve as the first in-sample set. The last estimation window ends with December 2008 ($T = 684$). Hence, in total, we estimate 216 linear and nonlinear models on a monthly basis while, following Ubilava and Helmers

10.4 FORECAST EVALUATION

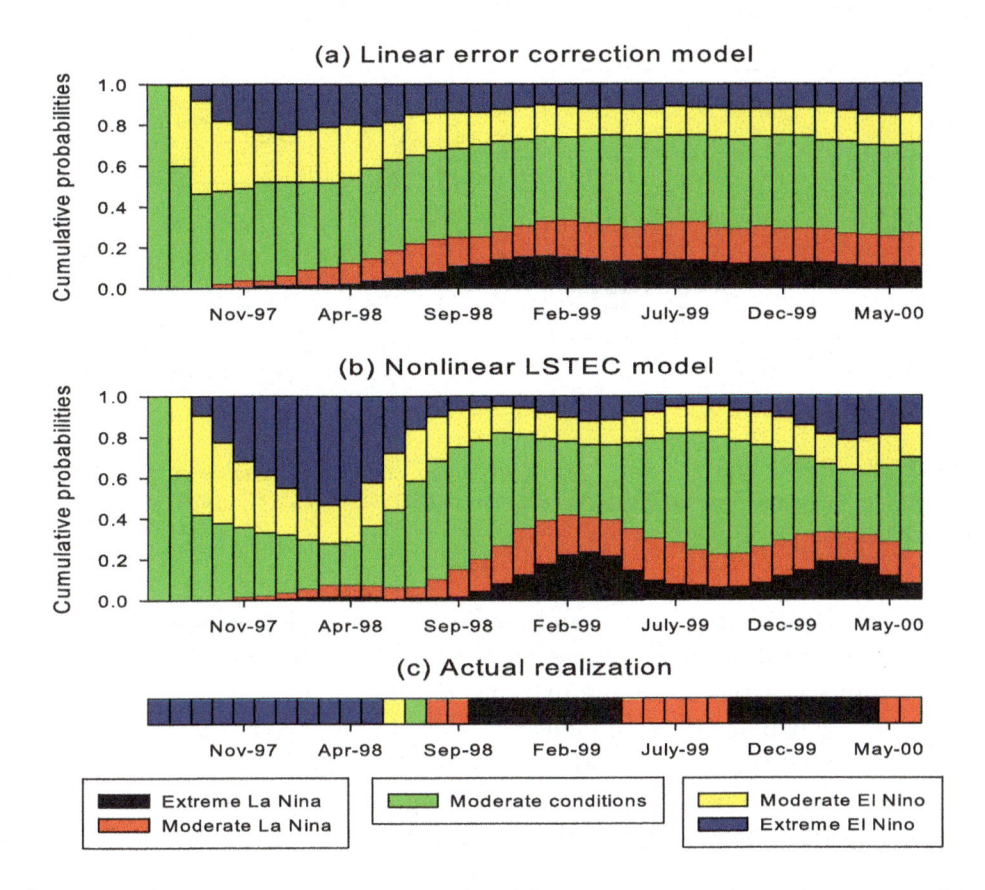

Figure 10.7: *Predictive probabilities of ENSO events, using information up to and including June 1997. (a) Linear ECM (6.25), (b) LSTEC model (6.24), and (c) actual realization.*

(2013), the AR order p and the delay lag d of the transition variable are re-examined on an annual basis with $d = 1, \ldots, 6$ and $p = 1, \ldots, 24$ as possible candidate values. We set $H_{max} = 36$ (months).

Genuine out-of-sample forecasts are obtained via a block bootstrap approach to mitigate for the effects of potential residual autocorrelation and heteroskedasticity, and we fix the number of BS replicates at 1,000.

To assess the accuracy of the fitted time series models in forecasting El Niño and La Niña events, we introduce five thresholds windows: SST $\leq -0.9°$C ("Extreme" La Niña), $-0.9°$C $<$ SST $\leq -0.5°$C ("Moderate" La Niña), $-0.5°$C $<$ SST $< 0.5°$C (Normal conditions), $0.5°$C $<$ SST $< 0.9°$C ("Moderate" El Niño), and SST $\geq 0.9°$C ("Extreme" El Niño). For each window, and each forecast horizon, we compile probability forecasts of ENSO events using empirical forecast densities.

Figure 10.7 shows probability forecasts using information up to and including June 1997 – when ENSO conditions are normal. For short-term, 3 months

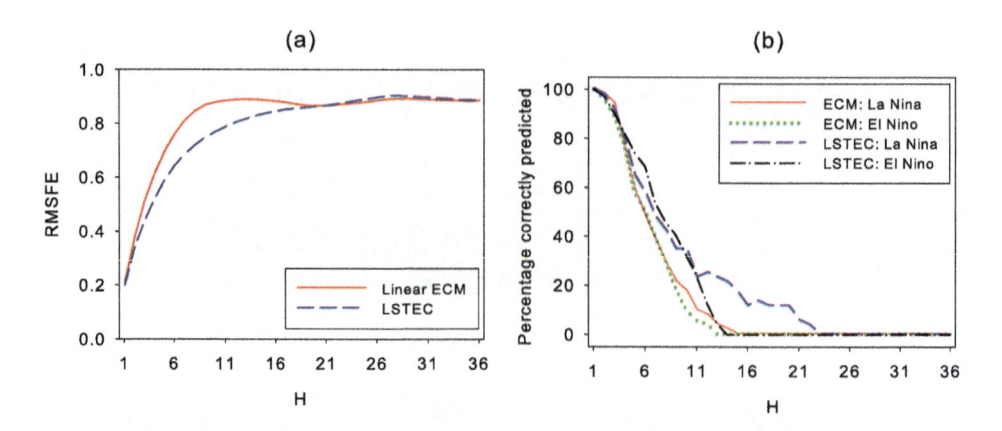

Figure 10.8: ENSO phenomenon. (a) Root mean squared forecast errors (RMSFEs); (b) Percentage correctly predicted La Niña events (SST < −0.5°C), and El Niño events (SST > 0.5°C).

ahead, forecasting both linear and nonlinear models yield comparable results. Note, the overall picture changes for 6 – 12 months ahead when the LSTEC model forecasts the upcoming extreme El Niño episode with about a twice as large probability than the linear model. In reality the 1997 – 1998 time period showed the strongest El Niño event since 1950. Note that this period was followed by a period of extreme La Niña, starting in the Fall of 1998 and continuing into 1999 and 2000. Again, the LSTEC model is able to forecast the beginning of this episode with a relatively high forecast accuracy (about 24% probability) as compared to the linear model, which forecasts this up-coming event with a modest 14% probability.

In addition, the DM test statistic rejects the null hypothesis of equality of MSFEs for $H = 1, \ldots, 10, 14, 20, \ldots, 27, 31, \ldots, 37$ with p-values < 0.03. For $H = 11, 12, 13, 28, 29$, and 30 the DM test statistic indicates that there is no statistically significant improvement in forecast accuracy of the nonlinear model over the linear model. Moreover, for $H = 15, \ldots, 19$ negative variance estimates of $\overline{d}$ were obtained. Diebold and Mariano (1995) suggest that the variance estimate should then be treated as zero and the null hypothesis of equal forecast accuracy be rejected. All these results indicate a preference for the LSTEC model in ENSO forecasting.

The above observation is further supported by Figure 10.8(a) displaying the RMSFEs from both models, and by Figure 10.8(b) showing the percentage correctly predicted ENSO events. As we see, up to $H = 20$ the LSTEC model shows the largest improvement in forecast accuracy as measured by the RMSFE. Figure 10.8(b) reveals that La Niña events are more accurately predicted by the LSTEC model than El Niño events. In addition, the LSTEC model is more effective in forecasting La Niña over a notably longer time period.

10.5 Forecast Combination

Point forecasts

Combining H-step ahead point forecasts $\{Y_{i,t+H|t}\}_{i=1}^{n}$ of n different time series models, representing different information sets, instead of relying on a forecast from an ex-ante best individual model is, on average, an effective way of improving the forecast accuracy of a certain target variable Y_{t+H}. The central question here is to determine the optimal weights for the calculation of combined forecasts. For instance, in the case of SETARMA models, we explored the performance of the combined forecast $Y_{t+H|t}^{C}$ in (10.19) with weights based on the same information set.

If the individual forecasts are unbiased then common practice is to obtain a weighted average of forecasts, with weights $w_i \geq 0$ and $\sum_{i=1}^{n} w_i = 1$.[7] The weights follow from minimizing some loss function, usually the MSFE.[8] However, in empirical applications equal-weighting (ew) often outperforms estimated optimal forecast combinations[9], i.e.

$$Y_{t+H|t}^{\text{ew}} = \frac{1}{n} \sum_{i=1}^{n} Y_{i,t+H|t}. \qquad (10.76)$$

Indeed, for short samples n, estimating forecast combination weights is unlikely to lead to any improvements in forecast accuracy.

Interval forecasts

FIs are frequently too narrow, i.e. too many observations are in the tails of the forecast distribution; Chatfield (1993) discusses seven reasons for this problem occurring. One most likely reason is that forecast errors are not normally distributed because the underlying DGP is nonlinear. Granger (1989) suggests a simple method to construct realistic, non-symmetrical FIs. The method combines the H-step ahead conditional quantile predictor $\{\widehat{\xi}_{i,q}(\mathbf{x})\}_{i=1}^{n}$ ($q \in (0, 1)$) obtained from n different time series models with weights $w_{i,q}(\mathbf{x})$ based upon within-sample estimation. That is,

$$\widehat{\xi}_{q}^{C}(\mathbf{x}) = \sum_{i=1}^{n} w_{i,q}(\mathbf{x})\widehat{\xi}_{i,q}(\mathbf{x}), \qquad (10.77)$$

where the weights are chosen to minimize the (local linear) "check" function; see Section 9.1.2. If the conditional quantile estimators are unbiased, then we might expect that $\sum_{i=1}^{n} w_{i,q}(\mathbf{x}) \approx 1$, and this constraint could be used for simplification, assuming the individual conditional quantile functions $\xi_{i,q}(\mathbf{x})$ are sufficiently smooth.

---

[7] The weights may change through time; see Deutsch et al. (1994) for an example. Note that the underlying DGP may or may not be second-order stationary.

[8] If the component forecasts are biased, it is recommended (Granger and Ramanathan, 1984) to add a constant to the combined forecasting model and not to constrain the weights to add to unity.

[9] This is known as the *forecast combination puzzle*; see, e.g, Huang and Lee (2010), Smith and Wallis (2009), Aiolfi et al. (2011), and Claeskens et al. (2016), for some answers to this puzzle.

A combined conditional percentile interval then follows from (10.42); see Granger et al. (1989b) for an application.

Density forecasts

Generalizing the notation in Section 10.4.3, we denote n sequences of P individual one-step ahead forecast densities of a process $\{Y_t, t \in \mathbb{Z}\}$ at some time t, as $\{\widehat{f}_{i,t}(Y_t|\mathcal{F}_{i,t-1})\}_{t=1}^{P}$, where $\mathcal{F}_{i,t-1}$ represents the ith information set ($i = 1,\ldots,n$). Then, assuming the density forecasts are continuous, the combined density forecast is defined as

$$\widehat{f}_t^C(Y_t) = \sum_{i=1}^{n} w_i \widehat{f}_{i,t}(Y_t|\mathcal{F}_{i,t-1}), \quad (t = 1,\ldots,P), \tag{10.78}$$

with $w_i \geq 0$ and $\sum_{i=1}^{n} w_i = 1$.[10] This combined density satisfies certain properties such as the "unanimity" property which amounts to saying that if all forecasters agree on the probability of a certain event then the combined probability agrees also. Further characteristics of $\widehat{f}_t^C(\cdot)$ can be drawn out by, for instance, defining the forecast mean $\mu_{i,t} = \int_{-\infty}^{\infty} y_t \widehat{f}_{i,t}(y_t|\mathcal{F}_{i,t-1}) dy_t$ and variance $\sigma_{i,t}^2 = \int_{-\infty}^{\infty} (y_t - \mu_{i,t})^2 \widehat{f}_{i,t}(y_t|\mathcal{F}_{i,t-1}) dy_t$ of the ith density sequence at time t. The combined one-step ahead density has mean and variance

$$\mathbb{E}(\widehat{f}_t^C(Y_t)) = \mu_t^C = \sum_{i=1}^{n} w_i \mu_{i,t}, \quad \text{Var}(\widehat{f}_t^C(Y_t)) = \sum_{i=1}^{n} w_i \sigma_i^2 + \sum_{i=1}^{n} w_i (\mu_{i,t} - \mu_t^C)^2. \tag{10.79}$$

The second equation of (10.79) indicates that the variance of the combined density equals the average individual uncertainty ("within" model variance) plus a measure of the dispersion of the individual forecast ("between" model variance). This result stands in contrast to the combined, optimal point forecast which has the smallest MSFE within the particular set of individual point forecasts (cf. Exercise 10.6).

Clearly, as before, the key issue is to find w_i. Most simply, various authors (see, e.g., Hendry and Clements, 2004) advocate the use of equal weights $w_i = 1/n$. A related topic is finding the set of weights in (10.78) that minimize the Kullback–Leibler divergence (see (6.48)) between the combined density forecast and the true, but unknown, conditional density $f_t(\cdot|\mathcal{F}_{t-1})$; see, among others, Bao et al. (2007) and Hall and Mitchell (2007).

10.6 Summary, Terms and Concepts

Summary

This chapter has covered quite a lot of important material related to the topic of obtaining forecasts from parametric nonlinear models. We started off by discussing

---

[10]The restriction that the weights are positive can be relaxed; see Genest and Zidek (1986).

10.6 SUMMARY, TERMS AND CONCEPTS

various exact and approximate methods for the generation of point forecasts. We then described general methods for constructing forecast intervals and regions. We also considered methods and test statistics for the evaluation of sequences of subsequent point, interval, and density forecasts. Finally, we discussed some weighting schemes for the optimal combination of model-based forecasts.

We would like to stress that this chapter introduced the major forecasting, evaluation and combination methods. As such, the chapter may well serve as a starting point for anyone who intends to do empirical work. Table 10.2 can be helpful in choosing an appropriate test statistic for forecast evaluation. Kock and Teräsvirta (2011) provide additional literature on nonlinear forecasts (conditional means) of economic time series obtained from parametric models, including NNs. Cheng et al. (2015) summarize the "state-of-the-art" of forecasting models for complex (nonlinear and nonstationary) biological, physical, and engineering dynamic systems.

Table 10.2: *Overview of some forecast evaluation tests: Forecast errors are denoted by $e_{i,t} \equiv e_{i,t+H|t}$ ($i = 1, 2$), PEE = parameter estimation error, and HLN = Harvey, Leybourne, and Newbold (1998). Based on Clark (2007).*

Forecast evaluation	No parameter estimation Nonnested models	Parameter estimation Nonnested models	Parameter estimation Nested models
		Point forecasts	
Equal MSE	DM test (10.62) with loss $d_t = e_{1,t}^2 - e_{2,t}^2$. For $H > 1$: • use MDM test (10.67) • use a resampled version of $e_{i,t}$ (White, 2000).	West (1996): Asymptotically, the effect of PEE on forecast uncertainty cancel out (recursive and rolling schemes)$^{(1)}$. Giacomini and White (2006): Despite PEE, DM $\xrightarrow{D} \mathcal{N}(0,1)$ (rolling scheme).	Giacomini and White (2006): DM $\xrightarrow{D} \mathcal{N}(0,1)$ (rolling scheme). Clark and McCracken (2005): DM has a non-standard distribution (recursive and rolling schemes).
Encompassing$^{(2)}$	Harvey et al. (1998): t test with loss $d_t = e_{1,t}(e_{1,t} - e_{2,t})$. HLN test = $\bar{d}/(\widehat{\text{Var}}(\bar{d}))^{1/2}$ $\xrightarrow{D} \mathcal{N}(0,1)$.	West (2001): Given a recursive or rolling scheme, use the HLN test with a specific estimate of the asymptotic variance of d_t. Giacomini and White (2006): Despite PEE, HLN $\xrightarrow{D} \mathcal{N}(0,1)$ (rolling scheme).	Clark and McCracken (2001; 2005): HLN has a non-standard distribution (recursive and rolling schemes). Giacomini and White (2006): HLN $\xrightarrow{D} \mathcal{N}(0,1)$ (rolling scheme).
		Density forecasts	
Accuracy	PITs $\{U_t\}$ as in (10.75): • $H = 1$: $\{U_t\} \stackrel{i.i.d.}{\sim} U(0,1)$, • $H > 1$: $\{U_t\} \sim U(0,1)$. "Tests": • Histogram of $\{U_t\}$, • EDF against 45° line.	PITs with some adjustments for PEE (only applicable for $H = 1$): • use the out-of-sample version of Bai's (2003) test; see Corradi and Swanson (2006a) • use max distance between EDF and 45° line and bootstrap the resulting distribution (Corradi and Swanson, 2006b).	From pairs of models: LR based test based on log predictive density score; see Amisano and Giacomini (2007). (model estimation: rolling scheme).

$^{(1)}$ Recursive scheme: sample expands. Rolling scheme: constant sample size, rolled forward.
$^{(2)}$ A forecast is said to encompass another when the optimal weight attached with one forecast is zero in a linear combination of two out-of-sample forecasts of the same variable.

Terms and Concepts

backward (forward) bootstrapping, 410
bootstrap (BS) forecasting, 399
bootstrap forecast interval (BFI), 409
Chapman–Kolmogorov, 392
conditional coverage (cc), 420
conditional percentile interval (CPI), 408
Diebold–Mariano (DM) test, 416
direct forecasting, 429
duration, 421
dynamic estimation (DE), 406
empirical least squares (ELS) forecasting, 400
(forecast) encompassing, 427
equal-weighting (ew), 425
forecast interval (FI), 408
forecast region (FR), 408
highest density region (HDR), 414
least squares (LS) forecasting, 392

linearization (LN) method, 404
modified DM (MDM) test, 417
Monte Carlo (MC) forecasting, 398
minimum MSE (MMSE), 391
normal forecasting error (NFE), 401
parameter estimation error (PEE), 427
plug-in (PI) forecasting, 396
predictive residuals, 412
recursive forecasting scheme, 416
relative mean absolute forecast error (RMAFE), 403
rolling forecasting scheme, 416
shortest conditional modal interval (SCMI), 413
skeleton (SK) forecasting, 399
unconditional coverage (uc), 419

10.7 Additional Bibliographical Notes

Section 10.1: Jones (1978) considers power-series expansions for the moments of the stationary distribution of NLAR(1) processes. One method also enables the corresponding expansions for conditional distributions to be found. Both Pemberton (1987) and Al-Qassem and Lane (1989) arrive at (10.6) independently. The approach followed by the first author is to look at H-steps ahead as one step followed by $(H-1)$ steps whereas the latter authors consider H-steps ahead prediction as $(H-1)$ steps followed by a single step. Tong and Moeanaddin (1988) observe that the forecast error function of the nonlinear LS predictor is not necessarily a monotonic non-decreasing function of the forecast horizon. Similar as in Example 10.2, one may use the Markovian structure of SETAR models jointly with the assumption that the errors are Gaussian distributed, to estimate the probability $p_{(H-d)}$; see De Gooijer and Kumar (1992, Section 6.2.1).

Cai (2003) presents a convergence theory for a particular numerical method to solve the Chapman–Kolmogorov relation. It is, however, unclear whether the accuracy of the predictive CDF, mean, and variance can be guaranteed by the proposed accuracy check on the calculation of the predictive pdf.

Section 10.2: In this chapter, and particularly this section, most approximate forecasting methods are for time series of a Markovian structure. Although the assumption of Markov dependence is satisfied by a large class of linear and nonlinear models that are of interest in time series analysis and forecasting, there exist non-Markovian processes, e.g. nonlinear MA models. Guégan (1993) gives analytic expressions for the LS forecasts from some simple non-Markovian processes. Fassò and Negri (2002) obtain multi-step ahead MC forecasts of hourly ozone concentration using a seasonal fractionally integrated SETARX–ARCH model.

10.7 ADDITIONAL BIBLIOGRAPHICAL NOTES

Section 10.2.4: Lai and Zhu (1991) consider adaptive multi-step ahead MMSE predictors for NLAR models when the parameters are unknown, and provide a numerical comparison between their forecast method and the exact LS forecast $Y_{t+H|t}^{\text{LS}}$.

Section 10.2.5: Clements and Smith (1997) compare a number of alternative methods of obtaining multi-step SETAR forecasts, including the NFE method. They conclude that the MC method performs reasonably well. The BS forecast method is preferred when the errors in the SETAR model come from a highly asymmetric distribution. Other comparisons include Amendola and Niglio (2004), Brown and Mariano (1984), Clements and Krolzig (1998), and Clements and Smith (1999, 2001). Niglio (2007) investigates forecasts from SETARMA models under asymmetric (linex) loss.

Section 10.2.6: Linearization is often used by control engineers in filtering and nonlinear system analysis. Apart from the Taylor series expansion there exists several other linearization methods of nonlinear state equations; see, e.g., Jordan (2006).

Section 10.2.7: The DE forecasting method was first introduced by Granger (1993, p. 132) and called *direct* forecasting.

Section 10.3: Similar to the construction of kernel-type nonparametric BS confidence intervals, nonparametric BFIs can also be based on *pivotal* statistics which are more conducive for theoretical analysis. De Brabanter et al. (2005) construct such an interval. Moreover, they provide an algorithm for the wild bootstrap.

The modal interval SCMI was originally proposed by Lientz (1970, 1972) for unconditional distribution functions. Hyndman (1995, 1996) was the first to construct HDRs for unconditional densities. Yao and Tong (1995) and De Gooijer and Gannoun (2000) provide applications of FRs and FIs with both real and simulated time series. Polinik and Yao (2000) establish various asymptotic properties of the conditional HDR, called *minimum volume predictive region*. The HDR estimation problem has been the focus of many papers; see, e.g., Samsworth and Wand (2010) who study the asymptotic and optimal bandwidth selection for nonparametric HDR estimation of a sequence of i.i.d. random variables.

Section 10.4.1: There is a myriad of theoretical papers dealing with extensions and modifications of the DM test statistic; see, e.g., Harvey et al. (1997), Corradi et al. (2001), Clements et al. (2003), Van Dijk and Franses (2003), and White (2000). West (2006) and Corradi and Swanson (2012) provide surveys of the "state-of-the-art". Two well-received empirical studies dealing with forecast evaluation are by Swanson and White (1997a,b). Recently, Diebold (2015) gives some personal reflections about the history of the DM test statistic. The test was originally developed to compare the accuracy of model-free forecasts. Mariano and Preve (2012) consider a multivariate version of the DM test statistic with multiple forecasts and forecast errors from more than two alternative models.

Note, the section does not include nonparametric techniques. For instance, assuming that the loss differentials are i.i.d., a standard sign test may be performed to test the null hypothesis that the median of the loss-differential distribution is equal to zero. Alternatively, Wilcoxon's signed rank sum test for matched pairs can be used for this purpose. Also, Pesaran and Timmermann (1992) propose a nonparametric test statistic for the null hypothesis that there are no predictable relationships between the actual and predicted sign changes of the predictand. Swanson and White (1997a,b), Chung and Zhou (1996) and Jaditz and Sayers (1998) each construct nonparametric test statistics for out-of-sample forecasting.

Gneiting (2011) demonstrates that averaging individual point forecasts, summarized in measures such as the MAFE and MSFE, can lead to grossly misguided inferences, unless there is a careful matching between the evaluation (loss) function and the forecasting task.

Section 10.4.2: There exists a large number of studies (see, e.g., Clements and Taylor, 2003; Engle and Manganelli, 2004; Berkowitz et al., 2011; Dumitrescu et al., 2013, and the references therein) offering alternative approaches to testing for independence; see also Campbell (2007) for a review.

Section 10.4.3: Diebold et al. (1998, 1999a,b) popularize the idea of using PITs in the context of macro-econometrics; see Tay and Wallis (2000) for a survey. Wallis (2003) suggests another way of evaluating density forecasts. Mainly it recasts the LR_{uc} and LR_{ind} test statistics into the framework of a Pearson χ^2 test. Unfortunately, this approach lacks the additivity property of the likelihoods. In fact, it is easy to see that the LR and Markov chain based FI evaluation approach can be directly extended to the case of evaluating density forecasts.

A number of empirical studies have shown that nonlinear models produce superior interval and density forecasts (see, e.g., Clements and Smith, 2000; Ma and Wohar, 2014). Rapach and Wohar (2006) compare out-of-sample point, interval and density forecasts generated by the Band–TAR, ESTAR, and linear AR models. The quality (i.e. the statistical performance) and the operational value of probabilistic forecasts is a primary requirement of many studies of atmospheric variables. Within this context nonparametric evaluation methods play an important role; see, e.g., Pinson et al. (2009) and the reference therein.

Section 10.5: Since the seminal work of Bates and Granger (1969) a voluminous literature has emerged on combining; see Timmermann (2006) for a recent review, and Granger (1989) and Wallis (2011) for some extensions. One recent paper is Adhikari (2015) who proposes a linear combination method for point forecasts that determines the combining weights through a novel NN structure.

Software References

Section 10.1.1: FORTRAN77 code, written by Yuzhi Cai, to find the "exact" conditional pdf of two-regime SETAR models and STAR models is available at the website of this book.

Section 10.2.1: The PI and LS SETARMA forecast results presented in Table 10.1 of Example 10.2 are obtained by the LS-PI-forecast.r function, available at the website of this book. The computer code was provided by Marcella Niglio, who also supplied the Linux-Procedure.r function related to the generation of forecasts using the linex asymmetric loss function.

Clements (2005, Chapter 8) contains sample GAUSS code for the estimation and forecasting (MC method) of $SETAR(2; 1, 1)$ models.

Section 10.3: The BS forecast intervals in Example 10.6 are computed using a RATS code provided by Jing Li. A MATLAB function for computing BFIs is available at the website of this book. The R-BootPR package provides a way to obtain BS bias-corrected coefficients for forecasting linear AR models. The code can easily be adapted to SETAR-type models.

EXERCISES

The R-hdrcde package contains computer code for the calculation and plotting of HDRs. GAUSS and MATLAB codes for computing the conditional mean, median, mode, SCMI and HDR are available at the website of this book.

R codes for the estimation, forecasting, and out-of-sample evaluation of the ENSO series are available in the file Example_6-4.zip.

Section 10.4.1: The MATLAB function dmtest retrieves the DM test statistic (under quadratic loss) using the Newey–West (1987) estimator for the covariance matrix of the loss differential. The R-forecast package contains the function dm.test. Some old R code for the DM test statistic is available at the R-help forum: http://r.789695.n4.nabble.com/R-help-f789696.html.

The URL http://qed.econ.queensu.ca/jae/datasets/alquist001/ has MATLAB code used in Alquist and Killian (2010) to calculate the DM test statistic under both quadratic and absolute loss, the Clark–West (2006) test statistic, and the Pesaran–Timmermann (1992) test statistic.

Section 10.4.2: MATLAB code for computing the three LR-based test statistics is available from http://www.runmycode.org/companion/view/93. GAUSS code for MC evaluation of interval lengths and coverages is given by Clements (2005, Chapter 8).

Exercises

Theory Questions

10.1 Consider the strictly stationary NLAR(1) process

$$Y_t = \omega Y_{t-1}^{1/2} + \varepsilon_t,$$

where $\omega > 0$, and $\{\varepsilon_t\} \overset{\text{i.i.d.}}{\sim} U(a,b)$ distributed with $0 \leq a < b < \infty$. Recall from Section 10.1.1 that the exact H-step ahead point forecast is given by $\mathbb{E}(Y_{t+H}|Y_t) = f_H(Y_t)$ ($H \geq 1$) using the short-hand notation $f_H(\cdot) = f_{Y_{t+H}|Y_t}(\cdot|\mathbf{x})$. Moreover, it is convenient to introduce the functions $g_0(x) = x$, $g_H(x) = \omega(g_{H-1}(x)) + \mu_\varepsilon$ for $H \geq 1$. Then $Y_{t+H|t}^{\text{Naive}} = g_H(Y_t)$ is the naive H-step forecast of Y_{t+H}, i.e. an SK (skeleton) forecast with additive WN.

(a) Show that the exact three-step ahead LS conditional pdf is given by

$$f_3(x) = \int_{-\infty}^{\infty} f_2(y)g(y-\mu(x))\mathrm{d}y = \frac{a+b}{2}$$

$$+ \frac{8}{105\omega(b-a)^2}\Big[Q(a,b,x)R(a,b,x) + Q(b,a,x)R(b,a,x)$$

$$- Q(a,a,x)R(a,a,x) - Q(b,b,x)R(b,b,x)\Big],$$

where

$$Q(u,v,x) = \sqrt{u + \omega\sqrt{v + \omega\sqrt{x}}},$$

$$R(u,v,x) = 2u^3 - u^2\omega\sqrt{v + \omega\sqrt{x}} - 8u\omega^2(v + \omega\sqrt{x}) - 5\omega^3(v + \omega\sqrt{x})^{3/2}.$$

[*Hint*: Use a software package for algebraic manipulation.]

(b) Let $z \geq 0$ be a given number. Then the equation $x = \omega x^{1/2} + z$ has a unique positive root x_z. Especially, if $z = 0$, $x_0 = \omega^2$. Furthermore, x_z is an increasing function of z. Define $\alpha = x_a$ and $\beta = x_b$. It is easy to verify that $Y_{t-1} \in [\alpha, \beta]$ implies $Y_t \in [\alpha, \beta]$ $\forall t > 0$. It can also be proved that for arbitrary $Y_0 \geq 0$ the process $\{Y_t, t \in \mathbb{Z}\}$, after a finite number of steps, falls with probability 1 into $[\alpha, \beta]$ and remains there.

Compute the functions $f_H(Y_t)$ and $g_H(Y_t)$ for $H = 2$ and 3, with $Y_t \in [\alpha, \beta]$ for the following two cases:

(i) $\omega = 1$, $a = 0$, and $b = 1$;

(ii) $\omega = 1$, $a = 0$, and $b = 100$.

Comment on the results.

(Anděl, 1997)

10.2 Consider the stationary SETAR(2; 1, 1) process

$$Y_t = \begin{cases} \phi_1 Y_{t-1} + \varepsilon_t & \text{if } Y_{t-1} \leq r, \\ \phi_2 Y_{t-1} + \varepsilon_t & \text{if } Y_{t-1} > r, \end{cases}$$

where $\{\varepsilon_t\} \stackrel{\text{i.i.d.}}{\sim} \mathcal{N}(0, \sigma_\varepsilon^2)$. The one-step ahead MMSE forecast is given by $Y_{t+1|t} = \mathbb{E}(Y_{t+1}|Y_t) = \phi_1 Y_t$, if $Y_t \leq r$, and by $Y_{t+1|t} = \phi_2 Y_t$, if $Y_t > r$. The one-step ahead forecast variance $\sigma_{e,1}^2 = \mathbb{E}(Y_{t+1}^2|Y_t) - Y_{t+1|t}^2 = \sigma_\varepsilon^2$. Let $z_{t+1|t} = (r - Y_{t+1|t})/\sigma_{e,t+1}$.

(a) Show that the exact two-step ahead MMSE forecast is given by

$$Y_{t+2|t} = \{\phi_1 \Phi(z_{t+1|t}) + \phi_2 \Phi(-z_{t+1|t})\} Y_{t+1|t} + (\phi_2 - \phi_1)\sigma_{e,t+1}\varphi(z_{t+1|t}),$$

where $\Phi(\cdot)$ and $\varphi(\cdot)$ are respectively the CDF and the pdf of the standard normal distribution.

(b) Show that the exact two-step ahead forecast variance is given by

$$\sigma_{e,2}^2 = 2\sigma_\varepsilon^2 \Phi(z_{t+t|t}) + \{\phi_1^2 \Phi(z_{t+1|t}) + \phi_2^2 \Phi(-z_{t+1|t})\}\{Y_{t+1|t}^2 + \sigma_{e,t+1}^2\}$$
$$+ (\phi_2^2 - \phi_1^2)(r + Y_{t+1|t})\sigma_{e,t+1}\varphi(z_{t+1|t}) - Y_{t+2|t}^2.$$

(c) Explore the limiting behavior of $\sigma_{e,2}^2$ as $Y_t \to \pm\infty$.

(De Gooijer and De Bruin, 1998)

10.3 Consider predicting from a stationary AR(1) process $Y_t = \phi Y_{t-1} + \varepsilon_t$ with $\{\varepsilon_t\} \stackrel{\text{i.i.d.}}{\sim} \mathcal{N}(0, 1)$ when the true process factually is the SETAR(2; 0, 0) process in Example 10.1.

(a) Verify (10.9) and (10.10).

(b) Using (10.9) show that $\mathbb{E}(Y_t) = 0$, $\text{Var}(Y_t) = 1 + \alpha^2$, and $\gamma_Y(1) = \mathbb{E}(Y_t Y_{t-1}) = -\alpha(\varphi(\alpha) - \alpha\beta)$ with $\beta = 1 - 2\Phi(\alpha)$.

(c) Show that the ratio of the MSFE of the H-step ahead forecast $Y_{t+H|t}^{\text{AR}}$ from the AR(1) process to the MSFE of the H-step ahead forecast $Y_{t+H|t}^{\text{SETAR}}$ from the SETAR(2; 0, 0) process in Example 10.1 can be expressed as

$$\text{Ratio-MSFE}(H) \equiv \frac{\text{MSFE}(Y_{t+H|t}^{\text{AR}})}{\text{MSFE}(Y_{t+H|t}^{\text{SETAR}})}$$
$$= 1 + \frac{\phi^H Y_t + \alpha\beta^{H-1}I(Y_t \leq 0) - \alpha\beta^{H-1}I(Y_t > 0)}{1 + \alpha^2(1 - \beta^{2H-2})}.$$

(d) Obtain a value for the AR(1) parameter ϕ by equating the lag 1 autocorrelations of the AR(1) process and the SETAR(2; 0, 0) process for $\alpha = 1.5$. Next, using part (c), plot Ratio-MSFE(H) versus $Y_t \in [-5, 5]$ for $H = 1, 2, 3$, and 5. Comment on the shape of the line plots.

(Guo and Tseng, 1997)

10.4 With reference to Section 2.8.1, we recall that the EAR(1) model is defined as

$$Y_t = \begin{cases} \alpha Y_{t-1} & \text{with prob. } \alpha, \\ \alpha Y_{t-1} + E_t & \text{with prob. } 1 - \alpha, \end{cases}$$

where $\{E_t\}$ are i.i.d. exponentially distributed random variables with mean μ. Gaver and Lewis (1980) show that Y_{t+j} can be expressed as

$$Y_{t+j} = \alpha^j Y_t + \alpha^{j-1} \varepsilon_{t+1} + \alpha^{j-2} \varepsilon_{t+2} + \cdots + \varepsilon_{t+j}, \quad (j = 0, 1, 2, \ldots), \tag{10.80}$$

where $\varepsilon_t = 0$ with probability α, and $\varepsilon_t = E_t$ with probability $1 - \alpha$.

(a) Using (10.80), show that the MSFE(H) of the least squares (LS) forecast is given by

$$\text{MSFE}(H) = \mathbb{E}(e_{t+H|t}^2) = \mathbb{E}\{(Y_t - Y_{t+H|t}^{\text{LS}})^2\} = \mu^2(1 - \alpha^{2H}), \quad (H = 1, 2, \ldots).$$

(b) Show that the MAFE of the one-step ahead LS forecast, denoted by MAFE(1), is given by

$$\text{MAFE}(1) = 2\mu(1 - \alpha)e^{-(1-\alpha)}.$$

10.5 With reference to the point forecast evaluation measures in Section 10.4.1:

(a) Verify (10.65).

(b) Verify the statement below (10.66) about the exactness of $\mathbb{E}(\text{Var}(\overline{d}))$ in the case the process $\{d_t, t \in \mathbb{Z}\}$ is WN.

10.6 Let $\{Y_t\}_{t=1}^T$ be an observed time series with T observations. Suppose that we have two unbiased one-step ahead forecasts $Y_{1,T+1|T}$ and $Y_{2,T+1|T}$, obtained from two different models for time $t = T + 1$. The corresponding forecast errors are $e_{i,T+1|T} = Y_{T+1} - Y_{i,T+1|T}$ ($i = 1, 2$). The one-step ahead forecast errors have variances $\sigma_{1,e}^2$ and $\sigma_{2,e}^2$ with $\sigma_{2,e}^2 \leq \sigma_{1,e}^2$. The covariance between $e_{1,T+1|T}$ and $e_{2,T+1|T}$ is equal σ_{12}.

Consider the following linear combination of the two forecasts

$$Y_{T+1|T}^C = w Y_{1,T+1|T} + (1 - w) Y_{2,T+1|T},$$

for some weight w. The corresponding forecast error is $e_{T+1|T}^C = Y_{T+1} - Y_{T+1|T}^C$.

(a) Show that $\text{Var}(e_{T+1|T}^C)$ is minimal for $w = w^*$, with

$$w^* = \frac{\sigma_{2,e}^2 - \sigma_{12}}{\sigma_{1,e}^2 + \sigma_{2,e}^2 - 2\sigma_{12}}.$$

(b) Let $\sigma_C^2(w^*)$ denote the variance of the combined forecast error evaluated at w^*. Show that $\sigma_C^2(w^*) \leq \sigma_{2,e}^2$, and thus $\sigma_C^2(w^*) \leq \sigma_{1,e}^2$.

(c) How does the optimal weight w^*, obtained via the combined forecast $Y_{T+1|T}^C$, behave as a function of the correlation $\rho_{12} = \sigma_{12}/\sigma_{1,e}\sigma_{2,e}$ using values $\rho_{12} = 0$ and $\rho_{12} = \pm 1$?

(d) In practice, the variances $\sigma_{1,e}^2$, $\sigma_{2,e}^2$ are unknown. Also, the covariance σ_{12} is unknown. How would you suggest to estimate the optimal weight w^*?

10.7 Consider the stationary SETAR(2; 1, 1) process

$$Y_t = \begin{cases} \phi_1 Y_{t-1} + \sigma_1 \varepsilon_t & \text{if } Y_{t-1} \leq 0, \\ \phi_2 Y_{t-1} + \sigma_2 \varepsilon_t & \text{if } Y_{t-1} > 0, \end{cases}$$

where $\{\varepsilon_t\} \stackrel{\text{i.i.d.}}{\sim} \mathcal{N}(0,1)$. Let $\{Y_t\}_{t=1}^T$ be a time series satisfying the above model. Suppose that the ith observation $(2 \leq i \leq T-1)$ is missing from the series, but that $\{Y_t\}_{t=1}^{i-1}$ and $\{Y_t\}_{t=i+1}^T$ are known. Let $\mathbf{Y}_t^{-i}$ denote the vector of known observations, and $\boldsymbol{\theta}$ the vector of unknown parameters. Then show that the best minimum MSE (MMSE) forecast for Y_i is given by

$$\widehat{Y}_i = \mathbb{E}(Y_i | \mathbf{Y}_t^{-i}; \boldsymbol{\theta})$$
$$= \{(c_1^{(1)} c_2^{(1)} \Phi(c_2^{(1)}) + c_3^{(1)} c_4^{(1)}) c_5^{(1)} + (c_1^{(2)} c_2^{(2)} \Phi(-c_2^{(2)}) - c_3^{(2)} c_4^{(2)}) c_5^{(2)}\} / f(Y_{i+1}|Y_{i-1}),$$

where, for $j = 1, 2$,

$$c_1^{(j)} = 1/\sqrt{2\pi(\sigma_j^2 + \phi_j \sigma_{(j)}^2)}$$
$$c_2^{(j)} = (\phi_{(j)} \sigma_j^2 Y_{i-1} + \phi_j Y_{i+1})/(\sigma_j^2 + \phi_j^2 \sigma_{(j)}^2)$$
$$c_3^{(j)} = \sigma_j \sigma_{(j)}/2\pi(\sigma_j^2 + \phi_j \sigma_{(j)}^2)$$
$$c_4^{(j)} = \exp\{-(\phi_{(j)} \sigma_j^2 Y_{i-1} + \phi_j \sigma_{(j)}^2 Y_{i+1})^2 / 2\sigma_j^2 \sigma_{(j)}^2 (\sigma_j^2 + \phi_j \sigma_{(j)}^2)\}$$
$$c_5^{(j)} = \exp\{-(Y_{i+1} - \phi_j \phi_{(j)} Y_{i-1})^2 / 2(\sigma_j^2 + \phi_j \sigma_{(j)}^2)\},$$

with

$$\phi_{(j)} = \phi_j + (-1)^{j+1}(\phi_2 - \phi_1) I(Y_{i-1} \geq 0)$$
$$\sigma_{(j)} = \sigma_j + (-1)^{j+1}(\sigma_2 - \sigma_1) I(Y_{i-1} \geq 0),$$

and where $\Phi(\cdot)$ is the CDF of the standard normal distribution.

Empirical and Simulation Questions

10.8 Reconsider the SETAR(2; 1, 1) process in Exercise 10.1. Figures 10.9(a) and 10.9(b) show the exact two-step ahead forecast function and the exact two-step ahead forecast variance functions for two SETAR(2; 1, 1) processes each having a threshold at $r = -2$.

(a) Construct a tree diagram of all possible paths from Y_t to Y_{t+2}. Explain qualitatively the maxima in the two variance functions.

EXERCISES

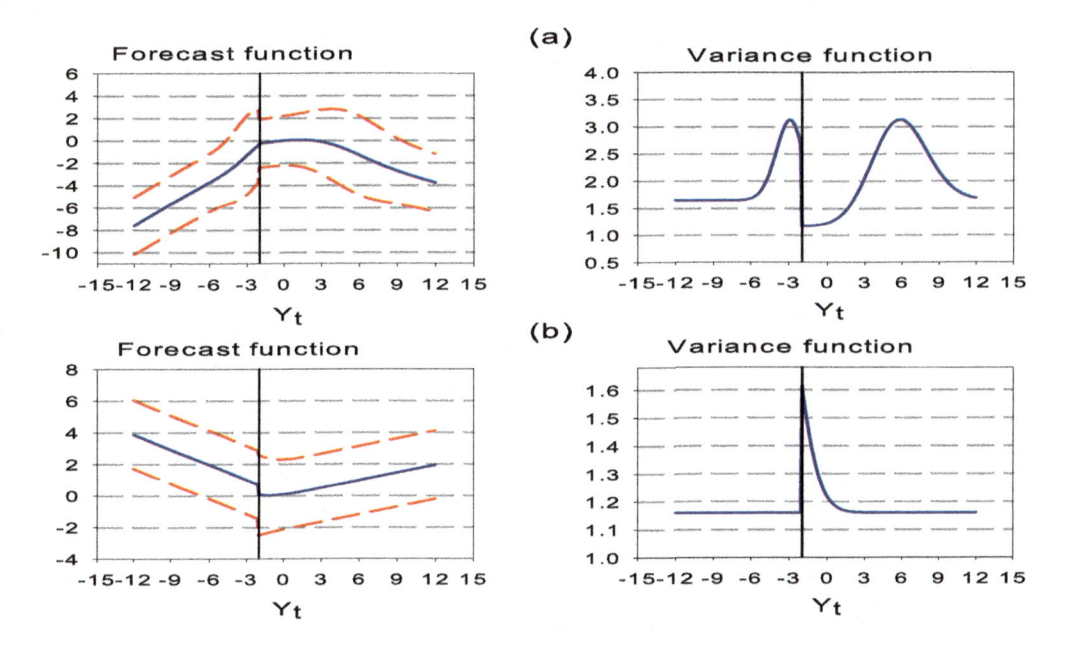

Figure 10.9: Two-step ahead forecast function (with $\pm 2\sigma_{e,2}$) and variance function (blue solid lines) for the SETAR(2; 1, 1) process in Exercise 10.1 with (a) $(\phi_1, \phi_2) = (0.8, -0.4)$, and (b) $(\phi_1, \phi_2) = (-0.8, 0.4)$; $r = -2$ (black solid vertical line) and $\sigma_\varepsilon^2 = 1$.

(b) Consider the SETAR process in Figure 10.9(a). Locate the two maxima of the two-step ahead forecast variance function by solving $d\sigma_{e,2}^2/du = 0$ numerically.

10.9 (a) Verify (10.26). In addition, using this result, prove that

$$\mathbb{E}\big(r(Z+M)\exp(-c(Z+M)^2)\big) = A^{-1/2}\exp(-c_1 M^2)\mathbb{E}\big(r(U)\big),$$

where $U \stackrel{\text{i.i.d.}}{\sim} \mathcal{N}(M/A, \sigma_Z^2/A)$ with the notation introduced in Section 10.2.5.

The next part consists of a small MC simulation experiment. Consider an EXPAR(1) model with parameters $\phi = -0.8$, $\xi = 2$, $\gamma = 2$, $\{\varepsilon_t\} \stackrel{\text{i.i.d.}}{\sim} \mathcal{N}(0, 1)$. Generate 50 samples of length $T = 130$ of the above process. Discard the first 99 values of each realization, and use Y_{100} as a starting value. Next, given the last 30 values, make forecast 30-steps ahead with the NFE and SK methods.

(b) Suppose that $e_{i,j}$ represents the forecasting error for the jth-step ahead in the ith replication ($i = 1, \ldots, 50; j = 1, \ldots, 30$). Analyze and compare the two forecasting methods in terms of short-term ($H = 5$), medium-term ($H = 15$), and long-term ($H = 30$) forecasting accuracy via the measures

$$\text{MSFE}(H) = \frac{1}{50}\sum_{i=1}^{50}\frac{1}{H}\sum_{j=1}^{H} e_{i,j}^2, \quad \text{and} \quad \text{MAFE}(H) = \frac{1}{50}\sum_{i=1}^{50}\frac{1}{H}\sum_{j=1}^{H} |e_{i,j}|.$$

10.10 Consider the SETAR(2; 1, 1) model in Example 10.6. In addition to the empirical coverage rate (CVR) given by (10.53), two other measures for evaluating the sharpness

Table 10.3: *Average CVR (coverage rate) and $\overline{\text{FI}}$ (asymptotic standard errors are in parentheses) for the SETAR(2; 1, 1) model in Example 10.6 with (a) $\{\varepsilon_t\} \stackrel{\text{i.i.d.}}{\sim} \mathcal{N}(0,1)$, and (b) $\{\varepsilon_t\} \stackrel{\text{i.i.d.}}{\sim} t_5$ distribution; $T = 100$, $B = 1{,}000$, $H = 1$, $\alpha = 0.05$, and $m = 500$.*

$\text{FI}_{0.95}$	$\{\varepsilon_t\} \stackrel{\text{i.i.d.}}{\sim} \mathcal{N}(0,1)$		$\{\varepsilon_t\} \stackrel{\text{i.i.d.}}{\sim} t_5$	
	CVR	$\overline{\text{FI}}$	CVR	$\overline{\text{FI}}$
BFI (fitted residuals)	0.937	$3.890_{(0.375)}$	0.978	$5.188_{(0.785)}$
BFI (predictive residuals)	0.947	$4.060_{(0.382)}$	0.983	$5.416_{(0.802)}$

and resolution are the size and standard error of the length of the FIs, i.e. $\widehat{Y}_{T+H,i}^{(1-\alpha/2)} - \widehat{Y}_{T+H,i}^{(\alpha/2)}$, where $\widehat{Y}_{T+H,i}^{(1-\alpha/2)}$ and $\widehat{Y}_{T+H,i}^{(\alpha/2)}$ are based on m MC replications.

(a) Consider Algorithm 10.1 with $H = 1$, $B = 1{,}000$, $\alpha = 0.05$, and $T = 100$. Moreover, set $m = 500$. Compute the CVR, the average length $\overline{\text{FI}}_H^{(1-\alpha)} = m^{-1}\sum_{i=1}^m (\widehat{Y}_{T+H,i}^{(1-\alpha/2)} - \widehat{Y}_{T+H,i}^{(\alpha/2)})$, and the associated standard error when the error terms of the SETAR process are simulated using (a) the standard normal distribution, and (b) the fat-tailed Student t_5 distribution re-scaled to unit variance. You will obtain (approximately) the results in Table 10.3.

(b) Compare and contrast the results in Table 10.3.

10.11 Consider the river flow data set; see Examples 9.3 and 10.7. The file SCMI-HDR.dat contains the last 35 observations of the river flow data set (column 1), the SCMI-lower and upper FI (columns 2 – 3), and the HDR-lower and upper FI (columns 4 – 5). Both FIs are shown in Figure 10.6, with coverage probability $1 - \alpha = 0.9$.

(a) Evaluate the two FIs using the test statistics LR_{uc}, LR_{ind}, and LR_{cc}. In the case of LR_{uc} and LR_{cc}, take $p = [0.5, 0.525, \ldots, 0.95]$ (19 values).

(b) Test for independence of the process $\{i_t^{(\alpha)}, t \in \mathbb{Z}\}$ using the test statistic (10.74). Calculate the rejection frequency under the null hypothesis over 25,000 replications. Compare the outcome of the test with the test result of LR_{ind} in part (a).

10.12 Consider a certain strictly stationary and invertible time series process $\{Y_t, t \in \mathbb{Z}\}$ whose ACF is identically zero. Therefore, it is reasonable to use $Y_{t+H|t}^{\text{LS}} = \mathbb{E}(Y_{t+H}|Y_s, -\infty < s \le t) = 0$ $(H \ge 1)$ as the best (in the MSE sense) least squares (LS) forecast of Y_{t+H}. Yet, assume that in reality $\{Y_t, t \in \mathbb{Z}\}$ is nonlinear. If this fact is known, the forecast accuracy may be improved using a nonlinear (NL) forecast based on a proper nonlinear model. This is a starting-point for the following forecast comparison.

Suppose that the time series $\{Y_t\}_{t=1}^T$ is generated by the subdiagonal BL model $Y_t = \psi Y_{t-2}\varepsilon_{t-1} + \varepsilon_t$, where $\{\varepsilon_t\} \stackrel{\text{i.i.d.}}{\sim} \mathcal{N}(0,1)$. The coefficient ψ is assumed to be known, and ψ satisfies the invertibility condition $|\psi| < 1/\sqrt{2}$. Of course, the assumption that the BL model is completely known is not very realistic in practice. However, under

EXERCISES

not too restrictive assumptions, Matsuda and Huzii (1997) show that the LS and NL predictors with LS estimated parameters converge to their asymptotic values.

(a) Using your favorite programming language, write a computer code to obtain estimates of the relative MSFE of the LS and NL forecasts for $H = 1$ and 2, and $\psi = -0.65, -0.55, \ldots, 0.65$. That is

$$\text{MSFE}(Y_{t+H|t}^{\text{LS}})/\text{MSFE}(Y_{t+H|t}^{\text{NL}}), \quad (H = 1, 2),$$

where the nonlinear two-step ahead forecasts are computed by the MC simulation method in Section 10.2.1. Set the number of replications $N = 100$. Moreover, set the number of MC replications at 2,000, and take $T = 50$.

In addition, consider the quadratic (Q) predictor introduced in (4.54). For the subdiagonal BL model the one-step ahead forecast is given by $Y_{t+1|t}^{\text{Q}} = \psi Y_{t-2} Y_{t-1}$. This approximation follows from replacing ε_{t-1} by its definition $\varepsilon_{t-1} = Y_{t-1} - \psi Y_{t-3} \varepsilon_{t-2}$ and ignoring the term containing ψ^2 in the subsequent expression. The two-step ahead quadratic predictor can be obtained by MC simulation.

(b) Write a computer code to obtain estimates of the one- and two-step ahead MSFE of the quadratic predictor using the same MC setup as in part (a). Compute $\text{MSFE}(Y_{t+H|t}^{\text{Q}})/\text{MSFE}(Y_{t+H|t}^{\text{NL}})$ $(H = 1, 2)$ for $\psi = -0.65, -0.55, \ldots, 0.65$. Compare the estimates of the relative MSFEs with those obtained under (a).

Chapter 11

VECTOR PARAMETRIC MODELS AND METHODS

In this chapter, we extend the univariate nonlinear parametric time series framework to encompass multiple, related time series exhibiting nonlinear behavior. Over the past few years, many multivariate (vector) nonlinear time series models have been proposed. Some of them are "ad - hoc", with a special application in mind. Others are direct multivariate extensions of their univariate counterparts. Within the latter class, a definition of a multivariate nonlinear time series model is often proposed with the following objectives in mind. First, the definition should contain the most general linear vector model as a special case when the nonlinear part is not present. This is analogous to univariate nonlinear time series models embedding linear ones. Second, the definition should contain the most general univariate nonlinear model within its class of models. Also, a potential candidate for a multivariate nonlinear time series model should possess some specified properties in order to permit estimation of the unknown model parameters and allow statistical inference. Moreover, because one of the main uses of time series analysis is forecasting, it is reasonable to restrict consideration to models which are capable of producing forecasts.

In Section 11.1, we give a general parametric multivariate nonlinear model in the context of a vector Volterra series expansion, extending the discussion in Section 2.1.1. However, with this specification an enormous range of possible models emerges. The obvious way to avoid this problem is to impose some sensible restrictions on the structure of the model. This has led to a wealth of "restricted" vector nonlinear models. Our treatment in Section 11.2 covers only a few of the most basic ones. Each subsection provides a definition of the model, and discusses conditions for stationarity and invertibility, if available. In contrast, we will not say much about estimating these vector nonlinear models. In most cases, QML and CLS estimation methods may be employed. In Sections 11.3 and 11.4, we then discuss a number of time-domain test statistics for nonlinearity. Most of these tests are generalizations of similar tests discussed in Chapter 5. In Section 11.5, we briefly address the problem of choosing the proper structure of a model using two model selection criteria.

To check the model adequacy, we discuss two portmanteau-type test statistics in Section 11.6. Section 11.7 deals with the calculation of forecasts, and we consider a method for forecast density evaluation using PITs. Finally, in Section 11.8, we apply some of the modeling and testing procedures to two Icelandic river flow series.

Two appendices are added to the chapter: Appendix 11.A contains selected percentiles for the LR test statistic introduced in Section 11.4. Appendix 11.B provides a step-by-step algorithm for the estimation of GIRFs in nonlinear VAR processes.

11.1 General Multivariate Nonlinear Model

Consider an m-dimensional stochastic process $\mathbf{Y}_t = (Y_{1,t}, \ldots, Y_{m,t})'$. Let $\mathbf{g}(\cdot) = (g_1(\cdot), \ldots, g_m(\cdot))'$ denote a sufficiently smooth vector function on $\mathbb{R}^m$, and $\boldsymbol{\theta}$ a vector of unknown parameters. Then, following the discussion of Section 2.1, a general nonlinear vector (multivariate) time series model can be written as

$$\mathbf{Y}_t = \mathbf{g}(\mathbf{Y}_{t-1}, \ldots, \mathbf{Y}_{t-p}, \boldsymbol{\varepsilon}_{t-1}, \ldots, \boldsymbol{\varepsilon}_{t-q}; \boldsymbol{\theta}) + \boldsymbol{\varepsilon}_t, \qquad (11.1)$$

where $\boldsymbol{\varepsilon}_t = (\varepsilon_{1,t}, \ldots, \varepsilon_{m,t})'$ is an m-variate i.i.d. random sequence with mean zero and positive definite covariance matrix $\boldsymbol{\Sigma}_\varepsilon$, independent of $\mathbf{Y}_t$.

As in (2.3), we can express $g_i(\cdot)$ ($i = 1, \ldots, m$) by a multivariate discrete-time Volterra series representation. The ith component of the resulting expression is given by

$$Y_{i,t} = \mu_i + \varepsilon_{i,t} + \sum_{u=1}^{m} \sum_{k=1}^{\infty} b_{i,u,k} \varepsilon_{u,t-k} + \sum_{u,v=1}^{m} \sum_{k,\ell=1}^{\infty} b_{i,u,v,k,\ell} \varepsilon_{v,t-k} \varepsilon_{u,t-\ell} + \cdots, \quad (11.2)$$

$$(i = 1, \ldots, m).$$

In practice, a truncated representation involving a finite number of parameters is used to approximate this structure. In particular, the ith component of a vector BL model results if all the coefficients of the second- and higher-order terms in (11.2) equal zero. Furthermore, we introduce the $m(p+q)$-dimensional state vector $\mathbf{S}_t$ defined by

$$\mathbf{S}_t = (\mathbf{Y}'_t, \ldots, \mathbf{Y}'_{t-p+1}, \boldsymbol{\varepsilon}'_t, \ldots \boldsymbol{\varepsilon}'_{t-q+1})'. \qquad (11.3)$$

Then we can define a multivariate SDM of order (p, q) which is locally linear, just as in (2.10). Its ith component is given by

$$Y_{i,t} = \mu_i(\mathbf{S}_{t-1}) + \sum_{j=1}^{p} \phi_{i,j}(\mathbf{S}_{t-1}) Y_{i,t-j} + \varepsilon_{i,t} + \sum_{\ell=1}^{q} \theta_{i,\ell}(\mathbf{S}_{t-1}) \varepsilon_{i,t-\ell}, \quad (i = 1, \ldots, m).$$

$$(11.4)$$

If all the parameters are constant, we have the ith component of the well-known vector autoregressive moving average (VARMA) model. Clearly, an obvious generalization of (11.1) is to allow for exogenous regressors in the function $\mathbf{g}(\cdot)$.

11.2 Vector Models

11.2.1 Bilinear models

An m-dimensional vector BL model follows as a special case of the Volterra representation in (11.2). Its ith component ($1 \leq i \leq m$) is given by

$$Y_{i,t} = \varepsilon_{i,t} + \sum_{u=1}^{m}\sum_{j=1}^{p} \phi_{i,u}^{j} Y_{u,t-j} + \sum_{u=1}^{m}\sum_{j=1}^{q} \theta_{i,u}^{j} \varepsilon_{u,t-j} + \sum_{k,\ell=1}^{m}\sum_{u=1}^{P}\sum_{v=1}^{Q} \psi_{i,k,\ell}^{uv} Y_{k,t-u}\varepsilon_{\ell,t-v}, \tag{11.5}$$

where $\{\phi_{i,u}^{j}\}$, $\{\theta_{i,u}^{j}\}$, and $\{\psi_{i,k,\ell}^{uv}\}$ are sequences of constants.

By introducing matrix notation and the Kronecker product, we can write the system of equations defined by (11.5) in vector form as

$$\mathbf{Y}_t = \sum_{j=1}^{p} \mathbf{\Phi}^j \mathbf{Y}_{t-j} + \boldsymbol{\varepsilon}_t + \sum_{j=1}^{q} \boldsymbol{\Theta}^j \boldsymbol{\varepsilon}_{t-j} + \sum_{u=1}^{P}\sum_{v=1}^{Q} \boldsymbol{\Psi}^{uv}\{\boldsymbol{\varepsilon}_{t-v} \otimes \mathbf{Y}_{t-u}\}. \tag{11.6}$$

Here, $\mathbf{\Phi}^j = \{\phi_{i,u}^j, 1 \leq j \leq p\}$ and $\boldsymbol{\Theta}^j = \{\theta_{i,u}^j, 1 \leq j \leq q\}$ are $m \times m$ matrices, and $\boldsymbol{\Psi}^{uv}$ ($1 \leq u \leq P; 1 \leq v \leq Q$) is an $m \times m^2$ matrix with the ith row obtained by vectorizing the $m \times m$ matrix $\boldsymbol{\psi}_i^{uv} = \{\psi_{i,k,\ell}^{uv}, 1 \leq k, \ell \leq m\}$, where k is the row index and ℓth column index, i.e.

$$\boldsymbol{\Psi}^{uv} = \big((\text{vec}(\boldsymbol{\psi}_1^{uv}))', \ldots, (\text{vec}(\boldsymbol{\psi}_m^{uv}))'\big)'.$$

Note that (11.5) involves $PQm^2 + m(p+q)$ parameters, making it too general to be of use in practice. As for the univariate BL model, special cases of (11.5) include the

- superdiagonal case: $\psi_{i,k,\ell}^{uv} = 0$, $\forall u > v$.
- subdiagonal case: $\psi_{i,k,\ell}^{uv} = 0$, $\forall u < v$.
- diagonal case: $\psi_{i,k,\ell}^{uv} = 0$, $\forall u \neq v$.

Stationarity

Stensholt and Tjøstheim (1987) give sufficient conditions for strict stationarity of vector subdiagonal BL models, and obtain expressions for the mean and higher-order autocovariance matrices.[1] For simplicity, we assume that $P = p$ and $Q = q$, and $q \leq p$. This is not an essential assumption, since it can be fulfilled by introducing

---

[1] Our use of the term "subdiagonal" is in line with the definition given by Granger and Andersen (1978a) and Stensholt and Tjøstheim (1987).

a suitable number of zero matrices. Now, we can rewrite (11.6) in a state space form. That is

$$\mathbf{S}_t = \mathbf{F}\boldsymbol{\varepsilon}_t + \mathbf{A}\mathbf{S}_{t-1} + \sum_{v=1}^{q} \mathbf{C}_v[\boldsymbol{\varepsilon}_{t-v} \otimes \mathbf{I}_{m(p+q)}]\mathbf{S}_{t-1}, \qquad (11.7)$$

where we define the $m(p+q) \times m$ matrix $\mathbf{F}$ and the $m(p+q) \times m(p+q)$ matrix $\mathbf{A}$ as follows

$$\mathbf{F} = \begin{pmatrix} \mathbf{I}_m \\ \mathbf{0}_{m(p-1) \times m} \\ \hdashline \mathbf{I}_m \\ \mathbf{0}_{m(q-1) \times m} \end{pmatrix}, \quad \mathbf{A} = \left(\begin{array}{ccc:ccc} \boldsymbol{\Phi}^1 & \cdots & \boldsymbol{\Phi}^p & \boldsymbol{\Theta}^1 & \cdots & \boldsymbol{\Theta}^q \\ \mathbf{I}_{m(p-1)} & \mathbf{0}_{m(p-1) \times m} & & \mathbf{0}_{m(p-1) \times m(q-1)} & & \\ \hdashline & & & \mathbf{0}_{m \times m} & \cdots & \mathbf{0}_{m \times m} \\ \mathbf{0}_{mq \times mp} & & & \mathbf{I}_{m(q-1)} & & \mathbf{0}_{m(q-1) \times m} \end{array} \right).$$

We also define the $m(p+q) \times m^2(p+q)$ matrices $\mathbf{C}_v$ ($v = 1, \ldots, q$) as

$$\mathbf{C}_v = \begin{pmatrix} \mathbf{C}_1^v & \cdots & \mathbf{C}_m^v \\ & \mathbf{0}_{m(p+q-1) \times m^2(p+q)} & \end{pmatrix}, \quad \text{where } \mathbf{C}_j^v = \left(\begin{array}{c:c} (\text{vec}(\mathbf{C}_{1,j}^v))' & \\ \vdots & \mathbf{0}_{m \times mq} \\ (\text{vec}(\mathbf{C}_{m,j}^v))' & \end{array} \right),$$

with the $m \times p$ matrices $\mathbf{C}_{i,j}^v$ ($1 \leq i \leq m$) defined by

$$\mathbf{C}_{i,j}^v = \{\psi_{i,u,v}^{k\ell}, 1 \leq u \leq m, 1 \leq k \leq p\},$$

and where for simplicity we assume that $\psi_{i,u,v}^{k\ell} = 0$ for $k < \ell$ in the sequel.

Following Stensholt and Tjøstheim (1987), we shall use the above matrices to formulate a strictly stationary solution of (11.7). Let $\mathbf{H} = \mathbb{E}[\{\boldsymbol{\varepsilon}_t \otimes \mathbf{I}_{m(p+q)}\} \otimes \{\boldsymbol{\varepsilon}_t \otimes \mathbf{I}_{m(p+q)}\}]$. We further introduce the $m^2(p+q)^2 \times m^2(p+q)^2$ matrices $\boldsymbol{\Gamma}_v$ ($1 \leq v \leq q$) defined by

$$\boldsymbol{\Gamma}_1 = \mathbf{A} \otimes \mathbf{A} + (\mathbf{C}_1 \otimes \mathbf{C}_1)\mathbf{H},$$

$$\boldsymbol{\Gamma}_v = \sum_{i=1}^{v-1} \{(\mathbf{A}^{v-i}\mathbf{C}_i) \otimes \mathbf{C}_v\}\mathbf{H}(\mathbf{A}^{i-1} \otimes \mathbf{A}^{v-1}) + (\mathbf{C}_v \otimes \mathbf{C}_v)\mathbf{H}(\mathbf{A}^{v-1} \otimes \mathbf{A}^{v-1})$$

$$+ \sum_{i=1}^{v-1} \{\mathbf{C}_v \otimes (\mathbf{A}^{v-i}\mathbf{C}_i)\}\mathbf{H}(\mathbf{A}^{v-1} \otimes \mathbf{A}^{i-1}), \quad (2 \leq v \leq q),$$

where $\mathbf{A}^0 = \mathbf{I}_{m(p+q)}$. Moreover, let $\mathbf{L}$ be the $qm^2(p+q)^2 \times qm^2(p+q)^2$ matrix defined by

$$\mathbf{L} = \begin{pmatrix} \boldsymbol{\Gamma}_1 & \boldsymbol{\Gamma}_2 & \cdots & \boldsymbol{\Gamma}_q \\ \mathbf{I}_{(q-1)m^2(p+q)^2} & & \mathbf{0}_{(q-1)m^2(p+q)^2 \times m^2(p+q)^2} \end{pmatrix}.$$

Then, if

$$\rho(\mathbf{L}) < 1, \qquad (11.8)$$

11.2 VECTOR MODELS

equation (11.7) has a unique strictly stationary and ergodic solution (Stensholt and Tjøstheim, 1987, Thm. 4.1) given by

$$\mathbf{S}_t = \mathbf{F}\boldsymbol{\varepsilon}_t + \sum_{j=1}^{\infty}\prod_{r=1}^{j}\left(\mathbf{A} + \sum_{v=1}^{q}\mathbf{C}_v[\boldsymbol{\varepsilon}_{t-v-r+1}\otimes \mathbf{I}_{m(p+q)}]\right)\mathbf{F}\boldsymbol{\varepsilon}_{t-j}, \qquad (11.9)$$

where the expression on the right-hand side of (11.9) converges absolutely almost surely as well as in the mean for every fixed t in $\mathbb{Z}$.

Liu (1989b) derives a sufficient condition for the existence of a strictly stationary solution of the general vector BL model (11.6). The condition has the same form as (11.8) except that the order and entries of the matrix $\boldsymbol{\Gamma}_j$ follow from another state space representation than (11.7) with fewer dimensions. By assuming $\{\boldsymbol{\varepsilon}_t\}$ is an i.i.d. sequence satisfying $\mathbb{E}(\varepsilon_{i,t})^{2Q} < \infty$ $(i = 1, \ldots, m)$ and $\mathbb{E}(\boldsymbol{\varepsilon}_t) = \mathbf{0}$, the condition for strict stationarity reduces to (11.8).

A potentially useful result (Stensholt and Tjøstheim, 1987) for the identification of vector superdiagonal BL models is that the autocovariance matrix of $\{\mathbf{S}_t, t \in \mathbb{Z}\}$ at lag ℓ $(\ell > q)$ is given by

$$\text{Cov}(\mathbf{S}_t, \mathbf{S}_{t-\ell}) = \sum_{i=1}^{p}\mathbf{A}^i \text{Cov}(\mathbf{S}_{t-i}, \mathbf{S}_{t-\ell}), \qquad (11.10)$$

assuming the existence of the fourth moments of $\{\boldsymbol{\varepsilon}_t\}$. Thus, the process (11.9) has the same autocovariance structure as for a VARMA(p,q) process. This result suggests that p and q selected by standard linear model selection techniques such as AIC or BIC, can also serve as upper bounds on the lag orders P and Q in the specification of BL models.

Invertibility

Here, we discuss invertibility of the process $\{\mathbf{Y}_t, t \in \mathbb{Z}\}$ given by (11.6) with $P = p$, $Q = q$ and $q \leq p$. Define the $mp \times 1$ vectors

$$\mathbf{S}_t = (\mathbf{Y}'_{t-1}, \ldots, \mathbf{Y}'_{t-p+1})', \quad \mathbf{U}_t = (\boldsymbol{\varepsilon}'_{t-1}, \ldots, \boldsymbol{\varepsilon}'_{t-q+1}, \mathbf{0}', \ldots, \mathbf{0}')'.$$

Then, in matrix notation, we can write $\{\mathbf{Y}_t, t \in \mathbb{Z}\}$ as follows

$$\mathbf{S}_t = \mathbf{U}_t + \boldsymbol{\Phi}\mathbf{S}_{t-1} + \boldsymbol{\Theta}\mathbf{U}_{t-1} + \boldsymbol{\Psi}[\mathbf{S}_{t-1}\otimes \mathbf{I}_{mp}]\mathbf{U}_{t-1}, \qquad (11.11)$$

where we define the $mp \times mp$ matrices $\boldsymbol{\Phi}$ and $\boldsymbol{\Theta}$ as follows

$$\boldsymbol{\Phi} = \begin{pmatrix} \boldsymbol{\Phi}^1 & \cdots & \boldsymbol{\Phi}^p \\ \mathbf{I}_{m(p-1)} & & \end{pmatrix}, \quad \boldsymbol{\Theta} = \begin{pmatrix} \boldsymbol{\Theta}^1 & \cdots & \boldsymbol{\Theta}^q & \mathbf{0} & \cdots & \mathbf{0} \\ & -\mathbf{I}_{m(q-1)} & & & \vdots & \vdots \\ \mathbf{0} & \cdots & & \mathbf{0} & \mathbf{0} & \cdots & \mathbf{0} \end{pmatrix}.$$

We also define the following matrices

$$\Psi_{i,u} = \begin{pmatrix} \psi^{u1}_{i,1,1} & \cdots & \psi^{u1}_{i,m,1} & \psi^{u2}_{i,1,1} & \cdots & \psi^{u2}_{i,m,1} & \cdots & \psi^{uq}_{i,1,1} & \cdots & \psi^{uq}_{i,m,1} & 0 & \cdots & 0 \\ \vdots & & & & & & & & & \vdots & \vdots & & \vdots \\ \psi^{u1}_{i,1,m} & \cdots & \psi^{u1}_{i,m,m} & \psi^{u2}_{i,1,m} & \cdots & \psi^{u2}_{i,m,m} & \cdots & \psi^{uq}_{i,1,m} & \cdots & \psi^{uq}_{i,m,m} & 0 & \cdots & 0 \end{pmatrix}_{m \times mp}$$

$$\Psi_u = [\Psi_{1,u}, \cdots, \Psi_{m,u}]_{m \times m^2 p}, \quad \Psi = \begin{pmatrix} \Psi_1 & \cdots & \Psi_p \\ \mathbf{0}_{m(p-1) \times m^2 p} & & \end{pmatrix}_{mp \times m^2 p^2}.$$

Now the process satisfying (11.6) is invertible (both by the classical concept of invertibility and by the Granger–Andersen invertibility concept), if

$$\exp \mathbb{E}\{\log \|\prod_{v=1}^{q} \Theta + \Psi[\mathbf{S}_{t-v} \otimes \mathbf{I}_{mp}]\|\} < 1. \tag{11.12}$$

Using Jensen's inequality, we obtain

$$\exp \mathbb{E}\{\log \|\prod_{j=1}^{q} \Theta + \Psi[\mathbf{S}_{t-j} \otimes \mathbf{I}_{mp}]\|\} \le \mathbb{E}\{\|\prod_{j=1}^{q} \Theta + \Psi[\mathbf{Y}_{t-j} \otimes \mathbf{I}_{mp}]\|\}.$$

Hence, a stronger condition for invertibility than (11.12) is given by

$$\mathbb{E}\{\|\prod_{j=1}^{q} \Theta + \Psi[\mathbf{Y}_{t-j} \otimes \mathbf{I}_{mp}]\|\} < 1. \tag{11.13}$$

It is clear that conditions (11.12) and (11.13) do not depend on the coefficients of the linear VAR(p) submodel. Nevertheless, these conditions are hard to verify in practice since they depend on the distribution of $\{\mathbf{Y}_t, t \in \mathbb{Z}\}$. However, we can replace (11.12) by a stronger condition which assumes only the existence of second moments of $\{\mathbf{Y}_t, t \in \mathbb{Z}\}$. As an example, we consider a multivariate BL model with a single lag in the noise term and $P = p$. First, we define the $m \times 1$ vectors

$$\mathbf{Y}_t = (Y_{1,t}, \ldots, Y_{m,t})', \text{ and } \boldsymbol{\varepsilon}_t = (\varepsilon_{1,t}, \ldots, \varepsilon_{m,t})'.$$

Then the representation of the multivariate BL model with just one lag in the noise term is given by

$$\mathbf{Y}_t = \sum_{i=1}^{p} \Phi^i \mathbf{Y}_{t-i} + \boldsymbol{\varepsilon}_t + \{\Theta^v + \sum_{u=1}^{p} \Psi^{uv}[\mathbf{Y}_{t-u} \otimes \mathbf{I}_m]\}\boldsymbol{\varepsilon}_{t-v}, \tag{11.14}$$

$$(v \in \{1, \ldots, q\}; q \le p),$$

where

$$\Psi^{uv} = \begin{pmatrix} \psi^{uv}_{1,1,1} & \cdots & \psi^{uv}_{m,1,1} & \cdots & \phi^{uv}_{m,1,1} & \cdots & \psi^{uv}_{m,1,m} \\ \vdots & & & & & & \vdots \\ \psi^{uv}_{1,m,1} & \cdots & \psi^{uv}_{m,m,1} & \cdots & \phi^{uv}_{m,m,1} & \cdots & \psi^{uv}_{m,m,m} \end{pmatrix}_{m \times m^2}. \tag{11.15}$$

11.2 VECTOR MODELS

Now, it can be shown (cf. Exercise 11.1) that $\{\mathbf{Y}_t, t \in \mathbb{Z}\}$ is invertible if

$$\|\boldsymbol{\Theta}^v\| + \sum_{u=1}^{p} \|\boldsymbol{\Psi}^{uv}\|\sqrt{\mathbb{E}\|\mathbf{Y}_t\|^2} < 1, \quad (v \in \{1,\ldots,q\}; q \le p). \tag{11.16}$$

This criterion is sufficient, but not necessary.

Example 11.1: Stationarity and Invertibility of a Bivariate BL Model

Consider a bivariate ($m = 2$) BL model with $p = P = Q = 1$ and a single lag in the noise term, say at lag $q = 1$. Then the state space representation (11.7) is given by

$$\mathbf{S}_t = \mathbf{F}\boldsymbol{\varepsilon}_t + \mathbf{A}\mathbf{S}_{t-1} + \mathbf{C}_1[\boldsymbol{\varepsilon}_{t-1} \otimes \mathbf{I}_{4)}]\mathbf{S}_{t-1},$$

where

$$\mathbf{F} = \begin{pmatrix} 0.5 & 0 \\ 0 & -0.7 \\ 0.5 & 0 \\ 0 & -0.7 \end{pmatrix}, \quad \mathbf{A} = \begin{pmatrix} 0.2 & 0.3 & 0 & 0 \\ 0.1 & -0.5 & 0 & 0 \\ 0 & 0 & 0 & 0 \\ 0 & 0 & 0 & 0 \end{pmatrix}, \quad \mathbf{C}_1 = \begin{pmatrix} 0.2 & -0.1 & 0 & 0 & 0.1 & 0.3 & 0 & 0 \\ 0.4 & -0.3 & 0 & 0 & -0.3 & 0.4 & 0 & 0 \\ 0 & 0 & 0 & 0 & 0 & 0 & 0 & 0 \\ 0 & 0 & 0 & 0 & 0 & 0 & 0 & 0 \end{pmatrix}.$$

The stationarity condition (11.8) becomes $\rho\{(\mathbf{A} \otimes \mathbf{A}) + (\mathbf{C}_1 \otimes \mathbf{C}_1)\mathbf{H}\} < 1$ with $\mathbf{H} = \mathbb{E}[\{\boldsymbol{\varepsilon}_t \otimes \mathbf{I}_{4)}\} \otimes \{\boldsymbol{\varepsilon}_t \otimes \mathbf{I}_{4)}\}]$, and by (11.16) the invertibility condition becomes $\|\boldsymbol{\Psi}^{11}\|\sqrt{\mathbb{E}\|\mathbf{Y}_t\|^2} < 1$, where

$$\boldsymbol{\Psi}^{11} = \begin{pmatrix} 0.2 & -0.1 & 0.1 & 0.3 \\ 0.4 & -0.3 & -0.3 & 0.4 \end{pmatrix}.$$

We can obtain the stationarity condition by simple calculation. Since $\sqrt{\mathbb{E}\|\mathbf{Y}_t\|^2}$ is unknown, we replace the expression for the invertibility condition by the approximation $\|\boldsymbol{\Psi}^{11}\|(1{,}000)^{-1}\sum_{t=1}^{1{,}000}\|\mathbf{Y}_t\|^2$.

When we fix the covariance matrix of the vector time series process $\{\boldsymbol{\varepsilon}_t\}$ at $\boldsymbol{\Sigma}_\varepsilon = \mathbf{I}_2$, the value of the stationarity condition equals 0.57, and the values of the approximate invertibility condition are in the range (0.71, 1.19) with an average of 0.89. When $\boldsymbol{\Sigma}_\varepsilon = \begin{pmatrix} 2 & 0.5 \\ 0.5 & 2 \end{pmatrix}$ the value of the stationarity condition is 0.70. On the other hand, the values of the approximate invertibility condition are in the range (1.26, 1.62), so indicating that the process is non-invertible. Figures 11.1(a) – (b) show the pattern of a typical realization of $\{\mathbf{Y}_t, t \in \mathbb{Z}\}$, for each covariance matrix $\boldsymbol{\Sigma}_\varepsilon$. Overall these time series are rather stable in both cases, with larger changes in the variance of $\{\mathbf{Y}_t, t \in \mathbb{Z}\}$ in Figure 11.1(b) than in Figure 11.1(a). In general, the stationarity condition (11.8) works well for a wide range of parameter matrices. However, one has to be careful in using condition (11.16) since it seems to be too strong, i.e. the invertibility domain is smaller than the exact invertibility domain. We discussed this point earlier in Section 3.5 for the univariate case.

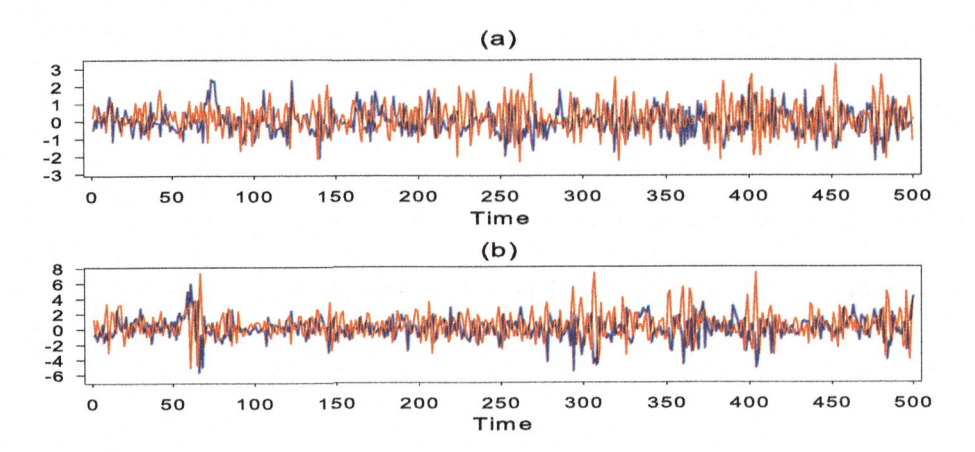

Figure 11.1: *A typical realization of a bivariate BL process ($T = 500$; blue solid line $Y_{1,t}$, red solid line $Y_{2,t}$); (a) $\boldsymbol{\Sigma}_\varepsilon = \mathbf{I}_2$, and (b) $\boldsymbol{\Sigma}_\varepsilon = \begin{pmatrix} 2 & 0.5 \\ 0.5 & 2 \end{pmatrix}$.*

11.2.2 General threshold ARMA (TARMA) model

The general form (11.1) is not really useful in practice. A model that includes a wide range of, but not all, multivariate possibilities, while still retaining practical significance, is a more worthwhile object. One way to accommodate this consideration, is to assume that the function $\mathbf{g}(\cdot)$ in (11.1) is additive while retaining a vector linear model as a special case. Under the additivity setup, we present some special cases of the resulting models in this and the next five subsections.

Let $\{\mathbf{X}_t, t \in \mathbb{Z}\}$ denote a weakly stationary m-variate continuous process in $\mathbb{R}^m$. Assume that $\mathbb{R}^m$ can be partitioned into $k > 1$ non-overlapping subspaces $\mathbb{R}^m_i$, i.e. $\mathbb{R}^m_i \cap \mathbb{R}^m_{i'} = \emptyset \ \forall i \neq i' \ (i, i' = 1, \ldots, k)$ determined by the values of $\{\mathbf{X}_{t-d}\}$, where $d > 0$ is the threshold lag or delay parameter. Then, for an m-dimensional strictly stationary time series process $\{\mathbf{Y}_t, t \in \mathbb{Z}\}$, a VTARMA model of order $(k; p, \ldots, p, q, \ldots, q)$ is defined as

$$\mathbf{Y}_t = \sum_{i=1}^{k} \left(\boldsymbol{\Phi}_0^{(i)} + \sum_{u=1}^{p} \boldsymbol{\Phi}_u^{(i)} \mathbf{Y}_{t-u} + \boldsymbol{\varepsilon}_t^{(i)} + \sum_{v=1}^{q} \boldsymbol{\Psi}_v^{(i)} \boldsymbol{\varepsilon}_{t-v}^{(i)} \right) I\left((\boldsymbol{\omega}^{(i)})' \mathbf{X}_{t-d} \in \mathbb{R}^m_i \right), \quad (11.17)$$

where $\boldsymbol{\Phi}_0^{(i)}$ are $m \times 1$ constant vectors, $\boldsymbol{\Phi}_u^{(i)}$ and $\boldsymbol{\Psi}_u^{(i)}$ are $m \times m$ matrix parameters, and $\boldsymbol{\omega}^{(i)} = (\omega_1^{(i)}, \ldots, \omega_m^{(i)})'$ is a pre-specified m-dimensional vector. When $\boldsymbol{\omega}^{(i)} = (1, 0, \ldots, 0)'$, the threshold variable is simply $X_{1,t-d}$. The error process in the ith regime satisfies $\boldsymbol{\varepsilon}_t^{(i)} = (\boldsymbol{\Sigma}_\varepsilon^{(i)})^{1/2} \boldsymbol{\varepsilon}_t$, where $(\boldsymbol{\Sigma}_\varepsilon^{(i)})^{1/2}$ are symmetric positive definite matrices and $\{\boldsymbol{\varepsilon}_t\}$ is an m-variate serially uncorrelated process with mean $\mathbf{0}$ and covariance matrix $\mathbf{I}_m$. The process $\{\mathbf{X}_t, t \in \mathbb{Z}\}$ can include lagged values of the time series process $\{\mathbf{Y}_t, t \in \mathbb{Z}\}$, or lagged values of an exogenous (independent or explanatory) variable. Additionally, the order $(p, \ldots, p, q, \ldots, q)$ can be different in each regime. Also, the threshold regimes may include lagged exogenous variables.

11.2 VECTOR MODELS

Note that (11.17) is very general, in the sense that the regimes are defined by arbitrary subspaces of $\mathbb{R}^m$. However, identification of such regimes can be difficult in practice. Tsay (1998) discusses a VTAR model in which values of a single exogenous variable $X_{1,t-d} \equiv X_{t-d}$ are used to determine the different regimes. That is, with $\mathbb{R}^{(i)} = (r_{i-1}, r_i]$, where $-\infty = r_0 < r_1 < \cdots < r_{k-1} < r_k = \infty$, (11.17) simplifies to

$$\mathbf{Y}_t = \sum_{i=1}^{k} \Big(\mathbf{\Phi}_0^{(i)} + \sum_{u=1}^{p} \mathbf{\Phi}_u^{(i)} \mathbf{Y}_{t-u} + \boldsymbol{\varepsilon}_t^{(i)} \Big) I(X_{t-d} \in \mathbb{R}^{(i)})$$

$$= \sum_{i=1}^{k} \Big(\mathbf{\Phi}_0^{(i)} + \sum_{u=1}^{p} \mathbf{\Phi}_u^{(i)} \mathbf{Y}_{t-u} + \boldsymbol{\varepsilon}_t^{(i)} \Big) (I_{t-d}^{(i-1)} - I_{t-d}^{(i)}), \qquad (11.18)$$

where

$$(I_{t-d}^{(i-1)} - I_{t-d}^{(i)}) \equiv I(X_{t-d} > r_{i-1}) - I(X_{t-d} \geq r_i), \quad (I_{t-d}^{(0)} = 1, \ I_{t-d}^{(k)} = 0).$$

Analogous to properties described for univariate TAR models, it can be shown (Lai and Wei, 1982) that under mild regularity conditions the CLS estimates of $(\mathbf{\Phi}_u^{(i)}, r_j, d)$ $(i = 1, \ldots, k; j = 1, \ldots, k-1)$ are strongly consistent and the LS estimates of $\mathbf{\Phi}_u^{(i)}$ are asymptotically normally distributed and independent of r_j and d. These results apply whenever the conditional expectation $\mathbb{E}(\mathbf{Y}_t | \mathcal{F}_{t-1})$ has a discontinuity at the threshold $X_{t-d} = r_j$ $(j = 1, \ldots, k-1)$ where $\mathcal{F}_{t-1}$ is the information set available at time $t-1$. When the expectation is continuous at the threshold values, the process will be the multivariate version of the CSETAR model described in Section 2.6.3. When $X_{t-d} = Y_{1,t-d}$, (11.18) reduces to a vector SETAR (VSETAR) model.

Stationarity

To present stationarity conditions for the VTARMA process, we first define the $m(p+q)$-dimensional vector $\mathbf{U}_t = (\boldsymbol{\varepsilon}_t', \mathbf{0}_{m(p-1) \times 1}', \boldsymbol{\varepsilon}_t', \mathbf{0}_{m(q-1) \times 1}')'$. We set $\boldsymbol{\omega} = (1, \ldots, 1)'$ and $\mathbf{\Phi}_0^{(i)} = \mathbf{0}, \forall i$, in (11.17). We also need the state space vector $\mathbf{S}_t$ defined in (11.3). Then we can re-write the VTARMA model compactly as a VTAR$(k; 1, \ldots, 1)$ process. That is,

$$\mathbf{S}_t = \mathbf{\Phi}^{(i)} \mathbf{S}_{t-1} + \mathbf{U}_t, \quad \text{if } X_{t-d} \in \mathbb{R}^{(i)}, \ (i = 1, \ldots, k), \qquad (11.19)$$

where $\mathbf{\Phi}^{(i)}$ is an $m(p+q) \times m(p+q)$ matrix, partitioned as follows

$$\mathbf{\Phi}^{(i)} = \begin{pmatrix} \mathbf{\Phi}_{11}^{(i)} & \vdots & \mathbf{\Phi}_{12}^{(i)} \\ \hdashline \mathbf{0}_{mq \times mp} & \vdots & \mathbf{\Phi}_{22}^{(i)} \end{pmatrix},$$

with

$$\mathbf{\Phi}_{11}^{(i)} = \begin{pmatrix} \mathbf{\Phi}_1^{(i)} & \cdots & \mathbf{\Phi}_p^{(i)} \\ \mathbf{I}_{m(p-1)} & \mathbf{0}_{m(p-1) \times m} \end{pmatrix}, \ \mathbf{\Phi}_{12}^{(i)} = \begin{pmatrix} \mathbf{\Psi}_1^{(i)} & \cdots & \mathbf{\Psi}_q^{(i)} \\ \mathbf{0}_{m(p-1) \times mq} & \end{pmatrix}, \ \mathbf{\Phi}_{22}^{(i)} = \begin{pmatrix} \mathbf{0}_{m \times m} & \cdots & \mathbf{0}_{m \times m} \\ \mathbf{I}_{m(q-1)} & \mathbf{0}_{m(q-1) \times m} \end{pmatrix}.$$

Observe that (11.19) is identical to the SRE in (3.1) with $\mathbf{A}_t \equiv \mathbf{\Phi}^{(i)}$ if $X_{t-d} \in \mathbb{R}^{(i)}$ ($i = 1, \ldots, k$). After s iterations, and similar to (3.4), (11.19) can be written as

$$\mathbf{S}_t = \Big(\prod_{i=0}^{s} \big(\mathbf{A}_{t-i}\big)\Big)\mathbf{S}_{t-s-1} + \sum_{i=0}^{s} \Big(\prod_{j=0}^{i-1} \mathbf{A}_{t-j}\Big)\mathbf{U}_{t-i}, \quad \forall s \in \mathbb{N}, \tag{11.20}$$

where $\prod_{j=0}^{-1} \mathbf{A}_{t-j} = \mathbf{I}_m$. Then, under some mild conditions, Niglio and Vitale (2014) show that the process $\{\mathbf{S}_t, t \in \mathbb{Z}\}$ is strictly stationary and ergodic if

$$\prod_{i=1}^{k} \rho(\mathbf{\Phi}^{(i)})^{p_i} < 1, \tag{11.21}$$

where $p_i = \mathbb{E}[I(X_{t-d} \in \mathbb{R}^{(i)})] < 1$ with $\sum_{i=1}^{k} p_i = 1$, and $\rho(\mathbf{\Phi}^{(i)})$ is the dominant eigenvalue (or spectral radius) of $\mathbf{\Phi}^{(i)}$ ($i = 1, \ldots, k$).

Invertibility

Consider, as a special case of (11.17), the VTMA($k; q, \ldots, q$) model

$$\mathbf{Y}_t = \sum_{i=1}^{k} \Big(\sum_{v=1}^{q} \mathbf{\Psi}_v^{(i)} \boldsymbol{\varepsilon}_{t-v}\Big) I\big(X_{t-d} \in \mathbb{R}^{(i)}\big) + \boldsymbol{\varepsilon}_t. \tag{11.22}$$

It is convenient to rewrite (11.22) as an mq-dimensional TVMA($k; 1, \ldots, 1$) process, using the state vector $\mathbf{S}_t$ and the vector of errors $\mathbf{U}_t$, respectively defined as

$$\mathbf{S}_t = (\mathbf{Y}_t', \boldsymbol{\varepsilon}_{t-1}', \ldots, \boldsymbol{\varepsilon}_{t-q+1}')', \quad \mathbf{U}_t = (\boldsymbol{\varepsilon}_t', \boldsymbol{\varepsilon}_{t-1}', \ldots, \boldsymbol{\varepsilon}_{t-q+1}')'.$$

Thus, the model is given by

$$\mathbf{S}_t = \mathbf{U}_t + \sum_{i=1}^{k} \mathbf{\Psi}^{(i)} \mathbf{U}_{t-1} I(X_{t-d} \in \mathbb{R}^{(i)}), \tag{11.23}$$

where $\mathbf{\Psi}^{(i)}$ is an $mq \times mq$ matrix defined by

$$\mathbf{\Psi}^{(i)} = \begin{pmatrix} \mathbf{\Psi}_1^{(i)} & \cdots & \mathbf{\Psi}_q^{(i)} \\ & \mathbf{0}_{(m-1)q \times mq} & \end{pmatrix}.$$

Then, under some mild conditions, Niglio and Vitale (2013) show that (11.23) is globally invertible if

$$\prod_{i=1}^{k} \rho(\mathbf{\Psi}^{(i)})^{p_i} < 1. \tag{11.24}$$

11.2 VECTOR MODELS

Further, they show that under condition (11.24), the VTMA($k; q, \ldots, q$) model can be written as a VTAR of infinite order with conditionally time dependent parameters, i.e.

$$\varepsilon_t = \mathbf{Y}_t + \sum_{j=1}^{\infty} \mathbf{\Pi}_{j,t} \mathbf{Y}_{t-j}, \qquad (11.25)$$

where

$$\mathbf{\Pi}_{j,t} = -\sum_{i=1}^{q} \mathbf{\Pi}_{j-i,t} \Big(\sum_{\ell=1}^{k} \mathbf{\Psi}_i^{(\ell)} I(X_{t-d-(j-i)} \in \mathbb{R}^{(\ell)}) \Big),$$

with $\mathbf{\Pi}_{j-i,t} = \mathbf{0}$ if $j < i$, and $\mathbf{\Pi}_{j-i,t} = \mathbf{I}_m$ if $i = j$. The above condition is sufficient and global. So, the VTMA model can still have locally non-invertible regimes while the full model satisfies condition (11.24).

11.2.3 VSETAR with multivariate thresholds

Model (11.18) describes a specification with a single threshold variable partition. A more general formulation follows when we allow up to, say, $k_{max} = \max\{k_1, k_2, \ldots, k_m\}$ partitions for each variable in the m-dimensional threshold sample space. This parallels a similar model specification introduced in Section 2.6.4.

More specifically, let $(k_1, k_2, \ldots, k_m) \in \mathbb{Z}^+$, and let $(\mathbb{R}_j^{(i)})_{(i=1,2,\ldots,k_j)}$ be a disjunctive decomposition of the real line $\mathbb{R}$, such that $\mathbb{R} = \cup_{i=1}^{k_j} \mathbb{R}_j^{(i)}$ with $\mathbb{R}_j^{(i)} = \emptyset$ ($i = k_j + 1, \ldots, k_{max}; j = 1, \ldots, m$). Then, for a strictly stationary m-dimensional process $\{\mathbf{Y}_t, t \in \mathbb{Z}\}$, a VSETAR process of order (k_{max}, p_J) and delay d, is defined as

$$\mathbf{Y}_t = \sum_{J \in \{1, \ldots, k_{max}\}^m} \Big(\mathbf{\Phi}_0^{(J)} + \sum_{u=1}^{p_J} \mathbf{\Phi}_u^{(J)} \mathbf{Y}_{t-u} + \varepsilon_t^{(J)} \Big) I(\mathbf{Y}_{t-d} \in R_J), \qquad (11.26)$$

where $R_J \equiv \mathbb{R}_1^{(J)} \times \cdots \times \mathbb{R}_m^{(J)}$, $\mathbf{\Phi}_0^{(J)}$ is an $m \times 1$ constant vector, $\mathbf{\Phi}_u^{(J)}$ are $m \times m$ parameter matrices with elements $\{(\phi_{u;r,s}^{(J)}); 1 \leq r, s \leq m, u = 1, \ldots, p_J\}$. The process $\{\varepsilon_t^{(J)}\}$ is an m-dimensional vector martingale difference sequence satisfying $\mathbb{E}(\varepsilon_t^{(J)}|\mathcal{F}_{t-1}) = \mathbf{0}$, $\text{Cov}(\varepsilon_t^{(J)}, \varepsilon_t^{(J)}|\mathcal{F}_{t-1}) = \mathbf{\Sigma}_\varepsilon$ and $\text{Cov}(\varepsilon_t^{(J)}, \varepsilon_s^{(J)}|\mathcal{F}_{t \wedge s}) = \mathbf{0}$ ($t \neq s$), where $\mathcal{F}_t$ is the information set generated by $\{\mathbf{Y}_s, s \leq t\}$.

Analogous to univariate SETAR models, parameter estimation of VSETARs, with and without multivariate thresholds, can be performed by CLS assuming the order of the model, the delay, and the number of threshold parameters are known. Alternatively, one may use an algorithm for recursive LS estimation by, for instance, adopting a multivariate version of the recursions (5.86) – (5.87); see, e.g., Arnold and Günther (2001). Finally, note that we may extend (11.18) and (11.26) to a V(SE)TARX model by introducing eXogenous variables.

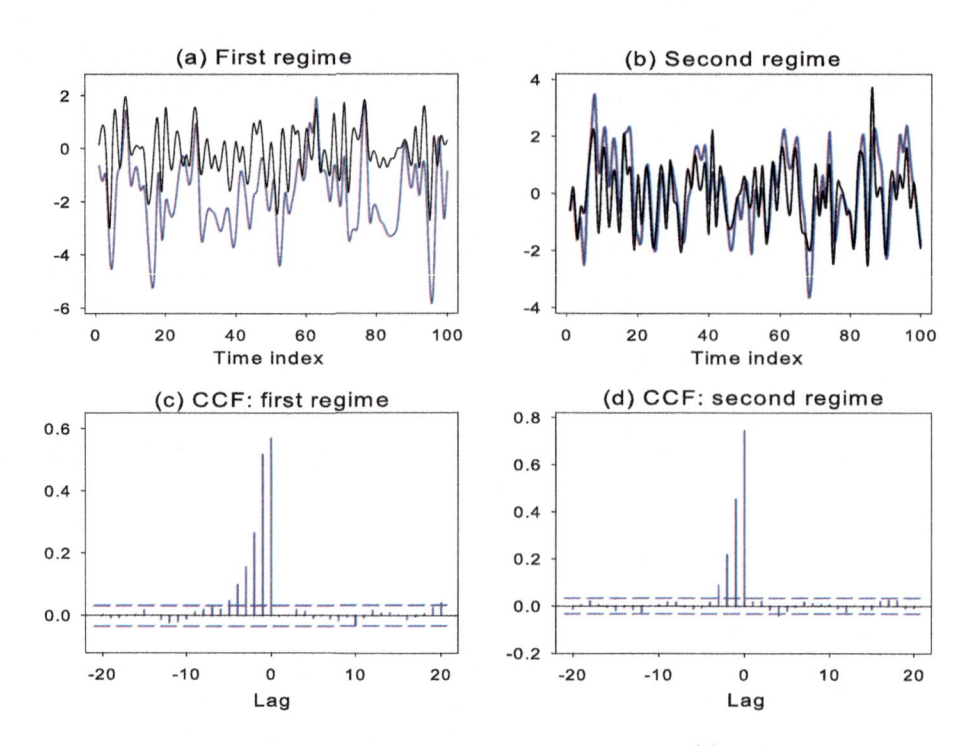

Figure 11.2: *(a) – (b) Regime-specific realizations of $\{Y_{j,t}^{(J)}\}_{t=1}^{100}$ ($j = 1,2; J = 1,2$) obtained from model (11.27) with $T = 7{,}000$; (c) – (d) Sample CCFs for $\{(Y_{1,t}^{(J)}, Y_{2,t}^{(J)})\}_{t=1}^{T_J}$ ($T_1 = 3{,}539$ and $T_2 = 3{,}460$) with 95% asymptotic confidence limits (blue medium dashed lines).*

Example 11.2: A Two-regime Bivariate VSETAR(2; 1, 1) Model

Consider the following two-regime bivariate VSETAR(2; 1, 1) model with $\mathbb{R}_1^{(J)} = \mathbb{R}$ ($J = 1, 2$), $\mathbb{R}_2^{(1)} = (-\infty, 0)$, and $\mathbb{R}_2^{(2)} = [0, \infty)$, i.e.

$$\mathbf{Y}_t = \begin{cases} \mathbf{\Phi}_1^{(1)} \mathbf{Y}_{t-1} + \boldsymbol{\varepsilon}_t & \text{if } \mathbf{Y}_{t-1} \in R_1 \equiv \mathbb{R} \times (-\infty, 0), \\ \mathbf{\Phi}_1^{(2)} \mathbf{Y}_{t-1} + \boldsymbol{\varepsilon}_t & \text{if } \mathbf{Y}_{t-1} \in R_2 \equiv \mathbb{R} \times [0, \infty), \end{cases} \quad (11.27)$$

where

$$\mathbf{\Phi}_1^{(1)} = \begin{pmatrix} 0.8 & 0.5 \\ 0 & 0 \end{pmatrix}, \quad \mathbf{\Phi}_1^{(2)} = \begin{pmatrix} 0.5 & 0.3 \\ 0 & 0 \end{pmatrix}, \quad \{\boldsymbol{\varepsilon}_t\} \stackrel{\text{i.i.d.}}{\sim} \mathcal{N}(\mathbf{0}, \mathbf{I}_2).$$

Figures 11.2(a) – (b) show plots of the series $\{Y_{j,t}^{(J)}\}_{t=1}^{100}$ ($J = 1, 2; j = 1, 2$) for each regime obtained as subseries from two typical regime-specific realizations of length $T_1 = 3{,}539$ and $T_2 = 3{,}460$ respectively. Both plots provide information of a possible feedback relationship from earlier values of $Y_{2,t}^{(J)}$ (black solid lines) to $Y_{1,t}^{(J)}$ (blue solid lines). The sample CCFs in Figures 11.2(c) – (d) support this observation with significant values at lag $\ell = -1$, as indicated by the Bartlett 95% asymptotic confidence limits $\pm 1.96/\sqrt{T_J}$.[2]

Unfortunately, Bartlett's confidence limits are no longer valid for nonlinear DGPs as we have remarked earlier. Thus, it may be safer to follow another route to detect whether one time series is leading another. In particular, Granger's causality concept, or rather its opposite, Granger non-causality, may be used for this purpose. The concept is well known in the context of VAR models; see Section 12.5 for a definition. Evidently, for (11.27) a Granger causality test based on the parameter restriction $\phi_{1;1,2}^{(J)} = 0 \; \forall J$ is no longer sufficient due to across-regime interactions.

An immediate but approximate solution is to compute a Granger causality measure for each regime J separately; Leistritz et al. (2006). Let $\mathbf{Y}_t^{-u} = (Y_{1,t}, \ldots, Y_{u-1,t}, Y_{u+1,t}, \ldots, Y_{m,t})'$, where the superscript $-u$ denotes omission of the uth variable in $\mathbb{R}^m$, with corresponding restricted information set $R_j^{-u} = \prod_{\mathbb{R}_1,\ldots,\mathbb{R}_{u-1},\mathbb{R}_{u+1},\ldots,\mathbb{R}_m} R_j$ $(j = 1, \ldots, m)$. Further, for each regime J $(J = 1, \ldots, k_{\max})$, let $e_{j,t+1}^{(J)}|\mathbf{Y}_{t-d}$ denote the one-step ahead forecast error for $Y_{j,t+1}^{(J)}$ $(j = 1, \ldots, m)$ conditional on $\mathbf{Y}_{t-d} \in R_j$ when the forecast is given by the conditional mean. In addition, let $\widetilde{e}_{j,t+1}^{(J)}|\mathbf{Y}_{t-d}^{-u}$ denote the one-step ahead forecast error for $Y_{j,t+1}^{(J)}$ $(i = 1, \ldots, m)$ conditional on $\mathbf{Y}_{t-d}^{-u} \in R_j^{-u}$ with similar properties. Then $Y_{u,t}^{(J)}$ does not Granger cause in variance $Y_{j,t}^{(J)}$ $(j \neq u)$, denoted by $Y_{u,t}^{(J)} \overset{V}{\nrightarrow} Y_{j,t}^{(J)}$, if and only if

$$\mathbb{E}\big((e_{j,t+1}^{(J)})^2|\mathbf{Y}_{t-d}\big) = \mathbb{E}\big((\widetilde{e}_{j,t+1}^{(J)})^2|\mathbf{Y}_{t-d}^{-u}\big) < \infty \quad \forall t. \qquad (11.28)$$

In view of (11.28), the Granger *causality index* (GCI) for regime J is defined as

$$\gamma_{u \to j}^{(J)} = \log\left(\frac{\mathbb{E}\big((\widetilde{e}_{j,t+1}^{(J)})^2|\mathbf{Y}_{t-d}^{-u}\big)}{\mathbb{E}\big((e_{j,t+1}^{(J)})^2|\mathbf{Y}_{t-d}\big)}\right). \qquad (11.29)$$

In practice, we replace $\gamma_{u \to j}^{(J)}$ by an estimate $\widehat{\gamma}_{u \to j}^{(J)}$ using consistent estimates of $\mathbb{E}\big((e_{j,t+1}^{(J)})^2|\mathbf{Y}_{t-d}\big)$ and $\mathbb{E}\big((\widetilde{e}_{j,t+1}^{(J)})^2|\mathbf{Y}_{t-d}^{-u}\big)$. Thus, if the series $Y_{u,t}^{(J)}$ does not improve the prediction of $Y_{j,t+1}^{(J)}$, $\widehat{\gamma}_{u \to j}^{(J)}$ will be close to zero. Any improvement in prediction of $Y_{j,t+1}^{(J)}$ by the inclusion of $Y_{u,t}^{(J)}$ in the information set leads to an increase in $\widehat{\gamma}_{u \to j}^{(J)}$.

In order to evaluate the performance of $\widehat{\gamma}_{u \to j}^{(J)}$ in the case of the two-regime VSETAR(2; 1, 1) model (11.27) we bootstrapped the EDF of $\widehat{\gamma}_{u \to j}^{(J)}$ (100 BS replicates) and computed 95% critical values for each regime. Next, based on 500

---

[2] If $\{X_t\}_{t=1}^T$ and $\{Y_t\}_{t=1}^T$ are two time series normalized to have zero-mean and unit-variance, their lag ℓ sample CCF is given by $c_{XY}(\ell) = (T - \ell)^{-1} \sum_{t=1}^{T-\ell} X_{t+\ell} Y_t$ $(\ell = 0, 1, 2, \ldots)$.

MC simulations, we explored the interrelationship between the components of $\{\mathbf{Y}_t, t \in \mathbb{Z}\}$. The results permit the following observations: $Y_{1,t}^{(J)} \overset{V}{\nrightarrow} Y_{2,t}^{(J)}$ in 95.4% ($J = 1$) and 94.6% ($J = 2$) of the cases. Moreover, $Y_{2,t}^{(J)} \overset{V}{\nrightarrow} Y_{1,t}^{(J)}$ in 0% ($J = 1, 2$) of the cases. Thus, there are unidirectional causal relationships from past values of $Y_{2,t}$ to $Y_{1,t}$ in both regimes. In contrast, there is no evidence of a time-lagged feedback from $Y_{1,t}$ to $Y_{2,t}$. These findings confirm earlier observations based on the sample CCF.

The GCI can be easily extended to cases where $m > 2$. Nevertheless, one serious limitation of the above analysis is that the GCI is defined for pairwise comparison of time series. However, a bivariate GCI for each pair of time series from a multivariate process of dimension $m > 2$ does not account for all the covariance structure information from the full data set. Also, the definition is for conditional second-order moments of the one-step ahead residuals rather than in terms of conditional pdfs. In Section 12.5, we will return to these issues when we discuss nonparametric Granger causality testing.

11.2.4 Threshold vector error correction

Preamble

Before introducing the threshold vector error correction model, we briefly discuss the notion of "long-term equilibrium" between the components of an m-dimensional nonstationary time series process $\{\mathbf{Y}_t, t \in \mathbb{Z}\}$ of order 1, or simply $I(1)$ (I as in Integrated). Assume there exists an $m \times 1$ vector of parameters $\boldsymbol{\beta}$, called *cointegration* vector. Then $\{\mathbf{Y}_t, t \in \mathbb{Z}\}$ is said to be an *equilibrium error* process if $X_t = \boldsymbol{\beta}'\mathbf{Y}_t$ is stationary in the mean, or $I(0)$. When long-run components of $\{\mathbf{Y}_t, t \in \mathbb{Z}\}$ obey equilibrium constraints, it is often sensible to isolate these components from those which are nonstationary. A model which can be used for this purpose is the linear *vector error correction* (VEC) model of order p. It can be compactly written as

$$\Delta \mathbf{Y}_t = \mathbf{a} + \boldsymbol{\alpha}\boldsymbol{\beta}'\mathbf{Y}_{t-1} + \sum_{i=1}^{p-1} \mathbf{A}_i \Delta \mathbf{Y}_{t-i} + \boldsymbol{\varepsilon}_t, \tag{11.30}$$

where $\Delta \mathbf{Y}_t \equiv \mathbf{Y}_t - \mathbf{Y}_{t-1}$ is $I(0)$, $\mathbf{a}$ and $\boldsymbol{\alpha}$ are both $m \times 1$ parameter vectors, $\mathbf{A}_i$ are $m \times m$ matrices of coefficients, and $\{\boldsymbol{\varepsilon}_t\}$ is a sequence of i.i.d. random variables with mean zero and positive definite covariance matrix $\boldsymbol{\Sigma}_\varepsilon$, independent of $\mathbf{Y}_t$. If the time series are not cointegrated, then a VAR in $\Delta \mathbf{Y}_t$ with $p - 1$ lags is appropriate. The partition of the matrix $\boldsymbol{\alpha}\boldsymbol{\beta}'$ in (11.30) is not unique, a convenient normalization condition is to set one element of $\boldsymbol{\beta}$ equal to unity.

Threshold

Assume $(\sum_{u=1}^{p} \boldsymbol{\Phi}_u^{(i)} - \mathbf{I}_m)$ has rank $m - 1$. Then, after rearranging some terms, we can write (11.17) as a k-regime *threshold vector error correction* (TVEC) model:[3]

---
[3]The acronym TVEC is commonly used in the literature. Adopting the short-hand notation VTEC would have been more in line with abbreviations introduced in Sections 11.2.2 and 11.2.3.

11.2 VECTOR MODELS

$$\Delta \mathbf{Y}_t = \sum_{i=1}^{k}\Big(\boldsymbol{\phi}_0^{(i)} + \boldsymbol{\alpha}^{(i)}((\boldsymbol{\beta}^{(i)})'\mathbf{Y}_{t-1}) + \sum_{u=1}^{p-1}\mathbf{A}_u^{(i)}\Delta \mathbf{Y}_{t-u} + \boldsymbol{\varepsilon}_t^{(i)}\Big)I(\boldsymbol{\beta}'\mathbf{Y}_{t-1}\in\mathbb{R}^{(i)}),$$

(11.31)

where $\sum_{u=1}^{p}\boldsymbol{\Phi}_u^{(i)} - \mathbf{I}_m = \boldsymbol{\alpha}^{(i)}(\boldsymbol{\beta}^{(i)})'$, with $\boldsymbol{\alpha}^{(i)}$ and $\boldsymbol{\beta}^{(i)}$ $m\times 1$ vectors, and $\mathbf{A}_u^{(i)} = -\sum_{j=u+1}^{p}\boldsymbol{\Phi}_j^{(i)}$. Note that the delay d is set at one with $\mathbf{X}_{t-1}\equiv \mathbf{Y}_{t-1}$. Making the delay a part of the set of unknown parameters is in principle possible, but would make the estimation and identification process much more involved.

Model (11.31) implies that there exists a regime-specific stationary equilibrium solution. To achieve identification of (11.31), some normalization must be imposed on $\boldsymbol{\beta}$ and $\boldsymbol{\beta}^{(i)}$ ($i=1,\ldots,k$). In the bivariate case, we recommend to do this by setting one element of these vectors equal to one. Note that if in regime i $(\boldsymbol{\beta}^{(i)})'\mathbf{Y}_{t-1}$ is $I(0)$, then the threshold variable $\boldsymbol{\beta}'\mathbf{Y}_{t-1}$ will not be stationary when $\boldsymbol{\beta}^{(i)}\neq \boldsymbol{\beta}$.

Estimation of TVEC models can be performed by recursive CLS, assuming that the order of the model and the value of the threshold cointegration parameters are known. Another way to proceed is by adopting a QML procedure. Third, two-stage LS can be used in a conditional way; see De Gooijer and Vidiella–i–Anguera (2005) for a finite-sample comparison of these estimation procedures. El-Shagi (2011) compares various genetic algorithms to optimize the likelihood function of TVEC models. It is beyond the scope of this book to discuss these and other estimation methods in detail.

11.2.5 Vector smooth transition AR

The VTARMA model has abrupt transitions from one regime to another. In contrast, an m-dimensional analogue of the STAR$(2;p,p)$ model discussed in Section 2.7, allows the conditional expectation of the model to change smoothly over time. Let $\mathbf{Z}_t = (1, \mathbf{Y}'_{t-1}, \ldots, \mathbf{Y}'_{t-p})'$ be an $(mp+1)\times 1$ vector. Then an m-dimensional k-regime vector smooth transition AR model of order $(k;p,\ldots,p)$, called VSTAR, is defined as

$$\mathbf{Y}_t = \sum_{i=1}^{k}\Big\{\big(\boldsymbol{\Phi}_0^{(i)} + \sum_{u=1}^{p}\boldsymbol{\Phi}_u^{(i)}\mathbf{Y}_{t-u}\big)\big(\mathbf{G}_t^{(i-1)} - \mathbf{G}_t^{(i)}\big)\Big\} + \boldsymbol{\varepsilon}_t$$

$$= \Big\{\sum_{i=1}^{k}\big(\mathbf{G}_t^{(i-1)} - \mathbf{G}_t^{(i)}\big)(\boldsymbol{\Phi}^{(i)})'\Big\}\mathbf{Z}_t + \boldsymbol{\varepsilon}_t, \qquad (11.32)$$

where $\boldsymbol{\Phi}^{(i)}$ is an $(mp+1)\times m$ matrix given by

$$\boldsymbol{\Phi}^{(i)} = ((\boldsymbol{\Phi}_0^{(i)})', (\boldsymbol{\Phi}_1^{(i)})', \ldots, (\boldsymbol{\Phi}_p^{(i)})')',$$

and where $\mathbf{G}_t^{(i)} \equiv \mathbf{G}(\mathbf{X}_t^{(i)}; \boldsymbol{\gamma}^{(i)}, \mathbf{c}^{(i)})$ is an $m\times m$ diagonal matrix of transition functions

$$\mathbf{G}_t^{(i)} = \mathrm{diag}\{G(X_{1,t}^{(i)}; \gamma_1^{(i)}, c_1^{(i)}), \ldots, G(X_{m,t}^{(i)}; \gamma_m^{(i)}, c_m^{(i)})\},\quad (i=1,\ldots,k-1),\quad (11.33)$$

with $\mathbf{G}_t^{(0)} = \mathbf{I}_m$, $\mathbf{G}_t^{(k)} = \mathbf{0}$, $\boldsymbol{\gamma}^{(i)} = (\gamma_1^{(i)}, \ldots, \gamma_m^{(i)})'$ (the slope parameters), $\mathbf{c}^{(i)} = (c_1^{(i)}, \ldots, c_m^{(i)})'$ (the location parameters), and $\gamma_j^{(i)} > 0$, $\forall i, j$. The sequence $\{\boldsymbol{\varepsilon}_t\}$ is an m-dimensional vector WN process with mean zero and $m \times m$ positive definite covariance matrix $\boldsymbol{\Sigma}_\varepsilon$, independent of $\mathbf{Y}_t$. The transition variable $\mathbf{X}_t^{(i)} = (X_{1,t}^{(i)}, \ldots, X_{m,t}^{(i)})'$ ($i = 1, \ldots, k-1$) can take many forms, for example a lagged variable of one of the components of $\{\mathbf{Y}_t, t \in \mathbb{Z}\}$, a linear combination of the m series, a weakly stationary exogenous variable, or a deterministic time trend.

When $k = 2$, (11.32) becomes

$$\mathbf{Y}_t = \{(\mathbf{I}_m - \mathbf{G}_t^{(1)})(\boldsymbol{\Phi}^{(1)})' + \mathbf{G}_t^{(1)}(\boldsymbol{\Phi}^{(2)})'\}\mathbf{Z}_t + \boldsymbol{\varepsilon}_t$$
$$= \boldsymbol{\Phi}_0 + \sum_{u=1}^p \boldsymbol{\Phi}_u \mathbf{Y}_{t-u} + \Big\{\widetilde{\boldsymbol{\Phi}}_0 + \sum_{u=1}^p \widetilde{\boldsymbol{\Phi}}_u \mathbf{Y}_{t-u}\Big\}\mathbf{G}(\mathbf{X}_t; \boldsymbol{\gamma}, \mathbf{c}) + \boldsymbol{\varepsilon}_t, \qquad (11.34)$$

where $\boldsymbol{\Phi}_u = \boldsymbol{\Phi}_u^{(1)}$, $\widetilde{\boldsymbol{\Phi}}_u = \boldsymbol{\Phi}_u^{(2)} - \boldsymbol{\Phi}_u^{(1)}$ ($u = 1, \ldots, p$), $\boldsymbol{\Phi}_0 = \boldsymbol{\Phi}_0^{(1)}$, $\widetilde{\boldsymbol{\Phi}}_0 = \boldsymbol{\Phi}_0^{(2)} - \boldsymbol{\Phi}_0^{(1)}$, and with $\mathbf{X}_t \equiv \mathbf{X}_t^{(1)}$, $\boldsymbol{\gamma} \equiv \boldsymbol{\gamma}^{(1)}$, and $\mathbf{c} \equiv \mathbf{c}^{(1)}$. From the first expression we see that each location parameter $c_j^{(1)}$ ($j = 1, \ldots, m$) represents the inflection point in which the transition function has value $1/2$, i.e. the process is halfway through the transition from $\mathbf{G}_t^{(1)}$ to $\mathbf{G}_t^{(2)}$.

When the diagonal elements of $\mathbf{G}_t^{(i)}$ are logistic functions, (11.32) becomes the so-called logistic vector STAR (LVSTAR) model. On the other hand, when $\gamma_j^{(i)} \to \infty$, $\forall j$, and when also $X_{1,t}^{(i)} = \cdots = X_{m,t}^{(i)}$, $c_1^{(i)} = \cdots = c_m^{(i)}$, the resulting model approaches an m-dimensional VTAR$(k; p, \ldots, p)$ model. If the form of (11.32) assumes that the transition functions are common to the m component series, we have $\mathbf{G}_t^{(i)} = G(X_t^{(i)}; \gamma^{(i)}, c^{(i)})\mathbf{I}_m$ with $\gamma_1^{(i)} = \cdots = \gamma_m^{(i)} = \gamma^{(i)}$, $c_1^{(i)} = \cdots = c_m^{(i)} = c^{(i)}$, and $X_{1,t}^{(i)} = \cdots = X_{m,t}^{(i)} = X_t^{(i)}$.

Once the transition variable $\mathbf{X}_t^{(i)}$ and the form of $\mathbf{G}(\cdot)$ have been specified, parameters in the VSTAR model can be estimated using NLS. The VSTAR model is identified if we restrict the location parameters $c_j^{(i)}$ in equation j such that they are in monotonically increasing order during the estimation.

Stationarity

The transition functions $\mathbf{G}_t^{(i)}$ are continuous and bounded between 0 and 1 for all values of $\mathbf{X}_t^{(i)}$ ($i = 1, \ldots, k-1$). This implies that the VSTAR model has the same stability condition as the linear VAR model. Unfortunately, explicit necessary and sufficient conditions for weak stationarity of LVSTAR models are not available yet. Nevertheless, a "rough-and-ready" check for stationarity of nonlinear models in general is to determine whether the skeleton is stable, using MC simulation. If the skeleton is such that the observed vector time series tends to explode for certain initial values, the process is likely to be nonstationary.

11.2.6 Vector smooth transition error correction

Following the discussion in Section 11.2.4, an m-dimensional (vector) two-regime smooth transition error correction (VSTEC) model is defined as

$$\Delta \mathbf{Y}_t = \boldsymbol{\phi}_0^{(1)} + \boldsymbol{\alpha}^{(1)}\big((\boldsymbol{\beta}^{(1)})'\mathbf{Y}_{t-1}\big) + \sum_{u=1}^{p-1} \mathbf{A}_u^{(1)} \Delta \mathbf{Y}_{t-u}$$

$$+ \Big(\boldsymbol{\phi}_0^{(2)} + \boldsymbol{\alpha}^{(2)}\big((\boldsymbol{\beta}^{(2)})'\mathbf{Y}_{t-1}\big) + \sum_{u=1}^{p-1} \mathbf{A}_u^{(2)} \Delta \mathbf{Y}_{t-u}\Big) \mathbf{G}(\mathbf{X}_t; \boldsymbol{\gamma}, \mathbf{c}) + \boldsymbol{\varepsilon}_t, \quad (11.35)$$

where $\Delta \mathbf{Y}_t$ is $I(0)$, the $m \times 1$ vectors $\boldsymbol{\alpha}^{(i)}$ and $\boldsymbol{\beta}^{(i)}$ ($i = 1, 2$) are as in (11.30), and $\mathbf{G}(\cdot)$ is an $m \times m$ diagonal matrix defined in (11.33). One way of keeping the computational aspects tractable, is to assume that the transition variables as well as the transition functions in (11.35) are the same for each model equation. In that case, $\mathbf{G}(\mathbf{X}_t; \boldsymbol{\gamma}, \mathbf{c}) = G(X_t; \gamma, \mathbf{c})\mathbf{I}_m$.

Stationarity
Saikkonen (2005, 2008) considers conditions for stationarity and ergodicity of a general three-regime nonlinear error correction model that encompasses the VSTEC model. The m-dimensional process $\{\mathbf{Y}_t, t \in \mathbb{Z}\}$ is transformed to a process $\{\mathbf{Z}_t, t \in \mathbb{Z}\}$ which can be viewed as a Markov chain. The Markov chain $\mathbf{Z}_t$ is geometrically ergodic when the joint spectral radius of a (finite) set $\mathcal{A} \subset \mathbb{R}^{mp \times mp}$ of square matrices is less than one. The set $\mathcal{A}$ consists of companion matrices defined through the transformed representation of $\mathbf{Y}_t$. If $\mathcal{A}$ only contains a single matrix then the *joint spectral radius* $\rho(\mathcal{A})$ (see (B.7) for its definition) coincides with the spectral radius of a square matrix. Clearly, the condition $\rho(\mathcal{A}) < 1$ is hard to verify analytically. An alternative method is to use one of the many algorithms for approximating the joint spectral radius; see Chang and Blondel (2013) for an overview and a comparison of these algorithms.

11.2.7 Other vector nonlinear models

It is easily seen how other parametric univariate nonlinear ARMA models in Chapter 2 can be extended to the vector case. For instance, Nicholls and Quinn (1981, 1982) investigate vector RCAR models. Another example is given in Exercise 11.1, where we introduce a vector asMA model as a generalization of the univariate asMA model of Section 2.6.5. Some of these models are restricted to low-dimensional ($m \leq 3$) time series processes due to the fast increase of parameters. Below, we discuss two options within the framework of a two-regime m-dimensional VSTAR model.

Smooth transition cointegration
In general, modeling and forecasting multivariate time series can be improved by imposing parameter restrictions that are driven by so-called *common features* in

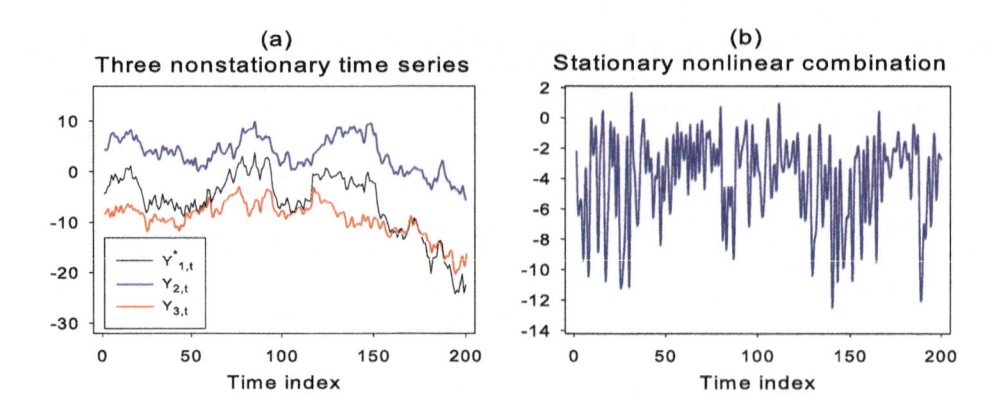

Figure 11.3: *(a) Three nonstationary, $I(1)$, time series $\{Y_{1,t}^*\}$, $\{Y_{2,t}\}$ and $\{Y_{3,t}\}$ of length $T = 200$; (b) A stationary nonlinear combination of the time series $\{Y_{2,t}\}$ and $\{Y_{3,t}\}$ in plot (a).*

the individual component series.[4] These features may, for instance, be a common stochastic trend as with linear cointegration; see the brief exposition in the preamble of Section 11.2.4. A less restrictive specification arises under the assumption that the cointegrating vector $\boldsymbol{\beta} = (\beta_1, \ldots, \beta_m)'$ is not a constant; e.g. $\boldsymbol{\beta}$ depends on time t, or $\boldsymbol{\beta}$ is assumed to be a vector of random variables. This prompted Li and He (2012a) to propose the following definition. The vector time series process $\{\mathbf{Y}_t, t \in \mathbb{Z}\}$ is said to contain *smooth transition cointegration* if there exists an $m \times 1$ time-varying vector $\boldsymbol{\beta}_t = (\beta_{1,t}, \ldots, \beta_{m,t})'$ such that the nonlinear combination of $\mathbf{Y}_t$ is $I(0)$, that is

$$\boldsymbol{\beta}_t' \mathbf{Y}_t \sim I(0), \tag{11.36}$$

where $\beta_{i,t} = \beta_i G(\mathbf{X}_t; \gamma, \mathbf{c})$, and $G(\cdot)$ is a logistic transition function given by

$$G(\mathbf{X}_t; \gamma, \mathbf{c}) = \frac{1}{1 + \exp\{-\gamma \prod_{j=1}^{q}(X_{j,t} - c_j)\}}, \tag{11.37}$$

with $\mathbf{X}_t = (X_{1,t}, \ldots, X_{q,t})'$ a $q \times 1$ vector of transition variables, $\gamma > 0$ a slope parameter, and $\mathbf{c} = (c_1, \ldots, c_q)'$ the vector of location parameters.

Example 11.3: An LVSTAR Model with Nonlinear Cointegration

Consider the following LVSTAR process $\{\mathbf{Y}_t = (Y_{1,t}, Y_{2,t}, Y_{3,t})', t \in \mathbb{Z}\}$ with

$$Y_{1,t} = \beta_{2,t} Y_{2,t} + \beta_{3,t} Y_{3,t} + \varepsilon_{1,t}, \quad Y_{2,t} = Y_{2,t-1} + \varepsilon_{2,t}, \quad Y_{3,t} = Y_{3,t-1} + \varepsilon_{3,t},$$

where $\beta_{2,t} = -0.8\big(1 + \exp\{-2(X_t - 0.3)\}\big)^{-1}$, $\beta_{3,t} = \big(1 + \exp\{-(X_t - 1)\}\big)^{-1}$, $\{\boldsymbol{\varepsilon}_t = (\varepsilon_{1,t}, \varepsilon_{2,t}, \varepsilon_{3,t})'\} \stackrel{\text{i.i.d.}}{\sim} \mathcal{N}(\mathbf{0}, \mathbf{I}_3)$, and $\{X_t\} \stackrel{\text{i.i.d.}}{\sim} \mathcal{N}(0,1)$. Both $\{Y_{2,t}, t \in \mathbb{Z}\}$

---
[4]A feature that is present in each group of individual time series is said to be *common* to those series if there exists a non-zero linear combination of the series that does not have the feature; Engle and Kozicki (1993).

11.2 VECTOR MODELS

and $\{Y_{3,t}, t \in \mathbb{Z}\}$ are random walks, or $I(1)$ processes. Their linear combination, $Y_{1,t}^* = Y_{2,t} + Y_{3,t}$ is also a nonstationary, or $I(1)$ process. Figure 11.3(a) shows plots of the three series $Y_{1,t}^*$, $Y_{2,t}$ and $Y_{3,t}$ over a sample period of length $T = 200$. Figure 11.3(b) shows a plot of the stationary nonlinear combination $\{Y_{1,t}\}_{t=1}^{200}$ with $\boldsymbol{\beta}_t = (\beta_{2,t}, \beta_{3,t})'$ the time-dependent cointegration vector.

Common nonlinear features (CNFs)

Another way to reduce model complexity is by investigating whether an m-dimensional stationary time series process $\{\mathbf{Y}_t, t \in \mathbb{Z}\}$ has CNFs. Anderson and Vahid (1998) introduced this concept within the context of LVSTAR and VSETAR modeling. Let $\mathbf{Z}_t = (1, \mathbf{Y}_{t-1}', \ldots, \mathbf{Y}_{t-p}')'$ be an $(mp + 1) \times 1$ vector. Consider the specification

$$\mathbf{Y}_t = \boldsymbol{\Phi}_0 + \sum_{u=1}^{p} \boldsymbol{\Phi}_u \mathbf{Y}_{t-u} + \mathbf{g}(\mathbf{Z}_t; \boldsymbol{\theta}) + \boldsymbol{\varepsilon}_t, \qquad (11.38)$$

where $\boldsymbol{\Phi}_0$ and $\boldsymbol{\theta}$ are vectors of parameters, $\boldsymbol{\Phi}_u$ is an $m \times m$ parameter matrix $(u = 1, \ldots, p)$, $\mathbf{g}(\cdot)$ is an $m \times 1$ vector of nonlinear functions, defined in a similar way as in (11.1), and $\{\boldsymbol{\varepsilon}_t\} \stackrel{\text{i.i.d.}}{\sim} (\mathbf{0}, \boldsymbol{\Sigma}_\varepsilon)$, independent of $\mathbf{Y}_t$. Suppose that there are r $(r < m)$ linearly independent linear combinations of the components of $\mathbf{Y}_t$ whose conditional expectation is linear in $\mathbf{Z}_t$. Consequently, there is an $m \times r$ matrix $\mathbf{A}$, of full column rank, such that

$$\mathbf{A}' \mathbf{g}(\mathbf{Z}_t; \boldsymbol{\theta}) = \mathbf{0}. \qquad (11.39)$$

The matrix $\mathbf{A}$ is not unique, a convenient normalization is to rearrange $\mathbf{A}$ such that its first $r \times r$ block is the identity matrix. Then we can partition $\mathbf{g}(\cdot)$ accordingly. That is, in partitioned form we have

$$\mathbf{A} = \begin{pmatrix} \mathbf{I}_r \\ \mathbf{A}^{**} \end{pmatrix}, \text{ and } \mathbf{g}(\mathbf{Z}_t; \boldsymbol{\theta}) = \begin{pmatrix} \mathbf{g}^*(\mathbf{Z}_t; \boldsymbol{\theta}) \\ \mathbf{g}^{**}(\mathbf{Z}_t; \boldsymbol{\theta}) \end{pmatrix}.$$

Clearly, (11.39) implies that $\mathbf{g}^*(\cdot) = -(\mathbf{A}^{**})' \mathbf{g}^{**}(\cdot)$, an $r \times 1$ vector. Moreover, it implies the following relation:

$$\mathbf{g}(\mathbf{Z}_t; \boldsymbol{\theta}) = \begin{pmatrix} -(\mathbf{A}^{**})' \\ \mathbf{I}_{m-r} \end{pmatrix} \mathbf{g}^{**}(\mathbf{Z}_t; \boldsymbol{\theta}).$$

Hence, we can write the conditional expectation of $\{\mathbf{Y}_t, t \in \mathbb{Z}\}$ in terms of $m - r$ common nonlinear components $\mathbf{g}^{**}(\cdot)$, i.e.

$$\mathbb{E}(\mathbf{Y}_t | \mathbf{Z}_t, \boldsymbol{\theta}) = \boldsymbol{\Phi}' \mathbf{Z}_t + \mathbf{A}_\perp \mathbf{g}^{**}(\mathbf{Z}_t; \boldsymbol{\theta}), \qquad (11.40)$$

where $\boldsymbol{\Phi} = (\boldsymbol{\Phi}_0', \boldsymbol{\Phi}_1', \ldots, \boldsymbol{\Phi}_p')'$ is an $(mp + 1) \times m$ parameter matrix, and $\mathbf{A}_\perp = \begin{pmatrix} -(\mathbf{A}^{**})' \\ \mathbf{I}_{m-r} \end{pmatrix}$ such that $\mathbf{A}' \mathbf{A}_\perp = \mathbf{0}$, an $r \times (m-r)$ matrix. Model (11.38) is said to have $m - r$ *common nonlinear features* when it is possible to rewrite the conditional

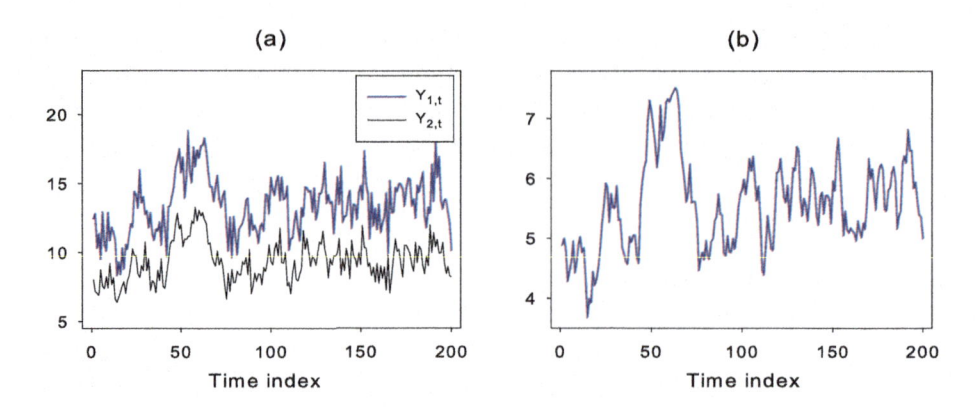

Figure 11.4: *(a) Two stationary nonlinear time series with a single CNF; (b) A stationary linear combination of the time series in plot (a).*

expectation of (11.38) in the form (11.40). Often, it is convenient to split the $m \times r$ matrix $\mathbf{A}$ into its columns, i.e. $\mathbf{A} = (\boldsymbol{\alpha}_1 \; \boldsymbol{\alpha}_2 \; \cdots \; \boldsymbol{\alpha}_r)$, where $\boldsymbol{\alpha}_i$ $(i = 1, \ldots, r)$ is an $m \times 1$ vector.

Example 11.4: An LVSTAR Model with a single CNF

Consider the following bivariate $(m = 2)$ LVSTAR(1) model with a single CNF

$$\begin{pmatrix} Y_{1,t} \\ Y_{2,t} \end{pmatrix} = \begin{pmatrix} 0.8 \\ -0.3 \end{pmatrix} + \begin{pmatrix} -0.3 & 0.5 \\ 0.2 & 0.1 \end{pmatrix} \begin{pmatrix} Y_{1,t-1} \\ Y_{2,t-1} \end{pmatrix}$$
$$+ \begin{pmatrix} 2 \\ 1 \end{pmatrix} (0.5 + 0.2 Y_{1,t-1} + 0.3 Y_{2,t-1}) G(Y_{2,t-1}; \gamma, c) + \boldsymbol{\varepsilon}_t, \quad (11.41)$$

where $G(Y_{2,t-1}; \gamma, c) = (1 + \exp\{-(Y_{2,t-1} - 1)\})^{-1}$, and $\{\boldsymbol{\varepsilon}_t\} \overset{\text{i.i.d.}}{\sim} \mathcal{N}(\mathbf{0}, \mathbf{I}_2)$. In this case the processes $\{Y_{1,t}, t \in \mathbb{Z}\}$ and $\{Y_{2,t}, t \in \mathbb{Z}\}$ share a single ($r = 1$) linear combination $\widetilde{Y}_t = \boldsymbol{\beta}' \mathbf{Z}_t = 0.5 + 0.2 Y_{1,t-1} + 0.3 Y_{2,t-1}$, where $\boldsymbol{\beta} = (0.5, 0.2, 0.3)'$ is a 3×1 vector. As a result, (11.41) has a common nonlinear component $(0.5 + 0.2 Y_{1,t-1} + 0.5 Y_{2,t-1}) G(Y_{2,t-1}; \gamma, c)$. Moreover, $\boldsymbol{\alpha}_\perp = \binom{2}{1}$, a non-zero 2×1 vector. Multiplying both sides of (11.41) by the 1×2 vector $\boldsymbol{\alpha}' = (-1, 2)$ leads to a linear VAR(1) process, since $\boldsymbol{\alpha}' \boldsymbol{\alpha}_\perp = 0$. Figure 11.4(a) shows two generated time series $\{Y_{i,t}\}_{t=1}^{200}$ ($i = 1, 2$) with a CNF. Figure 11.4(b) shows a plot of the stationary linear combination $\{\widetilde{Y}_t\}_{t=1}^{200}$.

11.3 Time-Domain Linearity Tests

Nonadditivity-type test statistics

Recall the nonadditivity-type test statistics $F_T^{(\text{T})}$ and $F_T^{(\text{O})}$ discussed in Section 5.4, with the superscripts (T) and (O) referring to Tukey and *original* respectively. It is straightforward to generalize these test statistics to the multivariate framework. For convenience, we assume that each component of $\mathbf{Y}_t = (Y_{1,t}, \ldots, Y_{m,t})'$ has mean zero.

11.3 TIME-DOMAIN LINEARITY TESTS

The null hypothesis states that $\{\mathbf{Y}_t, t \in \mathbb{Z}\}$ is generated by an m-dimensional stationary VAR(p) process. The alternative hypothesis states that the underlying process can be adequately approximated by a truncated multivariate second-order Volterra expansion with ℓth component given by (11.2). The tests determine whether at least one of the component series is nonlinear. Several computational procedures can be used for this purpose, each depending on different approximations and asymptotic expansions of the F distribution. The first test statistic, as proposed by Harvill and Ray (1999), uses an approximation due to Rao (R); see Rao (1973, Section 8c.5).

Algorithm 11.1: A nonadditivity-type test for nonlinearity

(i) Fit a VAR(p) model to $\{\mathbf{Y}_t\}_{t=1}^T$ by regressing $\mathbf{Y}_t$ on $\mathbf{Z}_t = (\mathbf{Y}'_{t-1}, \ldots, \mathbf{Y}'_{t-p})'$. Compute the $m \times 1$ vector of residuals $\{\widehat{\boldsymbol{\varepsilon}}_t\}_{t=p+1}^T$.

(ii) Let $\mathbf{U}_t = \text{vech}(\mathbf{Z}_t \otimes \mathbf{Z}'_t)$ be an $\nu_U \equiv mp(mp+1)/2$-dimensional vector which contains all second-order cross-product terms of lagged values of the process up to order p. So ν_U is the degrees of freedom for the hypothesis. Regress $\mathbf{U}_t$ on $\mathbf{Z}_t$. Obtain the residuals $\mathbf{W}_t = (W_{1,t}, \ldots, W_{\nu_U,t})'$.

(iii) Regress $\widehat{\boldsymbol{\varepsilon}}_t$ from step (i) on $\mathbf{W}_t$ from step (ii). Compute the corresponding $m \times m$ sum of squared regression matrix, **SSR**, and the sum of squared error matrix, **SSE**.

(iv) For $m > 1$, let
$$w = (\nu_E - \nu_U) - \frac{1}{2}(m - \nu_U + 1) \text{ and } v = \sqrt{\frac{m^2 \nu_U^2 - 4}{m^2 + \nu_U^2 - 5}},$$

where $\nu_E = T - p - mp$ is the degrees of freedom for error. Compute the F test statistic
$$F_{T,p}^{(\text{R})}(m) = \left(\frac{wv - \frac{1}{2}m\nu_U + 1}{m\nu_U}\right)\left(\frac{1 - (\Lambda^{(\text{W})})^{1/2}}{(\Lambda^{(\text{W})})^{1/2}}\right), \tag{11.42}$$

where
$$\Lambda^{(\text{W})} = |\mathbf{SSE}|/\{|\mathbf{SSR} + \mathbf{SSE}|\} \tag{11.43}$$

is Wilks' (W) lambda statistic. If $\{\mathbf{Y}_t, t \in \mathbb{Z}\}$ follows a strictly stationary zero-mean Gaussian VAR(p) process ($\mathbb{H}_0$), then from standard theory of multivariate linear regression models it follows that
$$F_{T,p}^{(\text{R})}(m) \xrightarrow{D} F_{\nu_1, \nu_2}, \text{ as } T \to \infty, \tag{11.44}$$

with $\nu_1 = m\nu_U$ and $\nu_2 = wv - m\nu_U/2 + 1$.

If $m = 1$ or $\nu_U = 1$, v is set equal to 1. Note that ν_2 need not be integral. The approximation is exact if $\min(m, \nu_U) \leq 2$. A (less accurate) approximate test statistic (Bartlett, 1954) is given by

$$\lambda_{T,p}(m) = -[(\nu_E - \frac{1}{2}(m - \nu_U + 1)]\log \Lambda^{(W)} \qquad (11.45)$$

which, under $\mathbb{H}_0$ and as $T \to \infty$, has an approximate $\chi^2_{m\nu_U}$ distribution.

Just as for the univariate test statistic $F_T^{(T)}$, Algorithm 11.1 reduces to a multivariate version of Tukey's nonadditivity-type test statistic if the $\mathbf{U}_t$ in step (ii) are aggregated using weights based on the LS coefficients in step (i); i.e. the fitted values $\widehat{\mathbf{Y}}_t$ from step (i) are used as the dependent variable in (ii). The resulting test statistic can be computed as follows.

Algorithm 11.2: Tukey's nonadditivity-type test for nonlinearity

(i) Fit a VAR(p) model to $\{\mathbf{Y}_t\}_{t=1}^T$ by regressing $\mathbf{Y}_t$ on $\mathbf{Z}_t = (\mathbf{Y}'_{t-1}, \ldots, \mathbf{Y}'_{t-p})'$. Compute the $m \times 1$ vector of fitted values $\{\widehat{\mathbf{Y}}_t\}_{t=p+1}^T$, the $m \times 1$ vector of residuals $\widehat{\varepsilon}_t$, and the corresponding $m \times m$ matrix $\mathbf{SSR}_1$ of sum of squared and cross-product terms.

(ii) Compute an $m \times 1$ vector of squares of fitted values, say $\mathbf{X}_t$, from the m-variate AR(p) regression in step (i). Remove the linear dependence of $\mathbf{X}_t$ on $\mathbf{Z}_t$ by a second m-variate AR(p) regression of $\mathbf{X}_t$ on $\mathbf{Z}_t$. Obtain the $m \times 1$ vector of fitted values $\widehat{\mathbf{X}}_t$, and the $m \times 1$ vector of residuals $\mathbf{U}_t = \mathbf{X}_t - \widehat{\mathbf{X}}_t$.

(iii) Regress $\widehat{\varepsilon}_t$ from step (i) on the vector of residuals $\mathbf{U}_t$ from step (ii). Compute the corresponding $m \times m$ sum of squared regressions matrix, $\mathbf{SSR}_2$, and the sum of squared errors matrix, $\mathbf{SSE}_2$. Let $\mathbf{SSR}_{2|1} = \mathbf{SSR}_2 - \mathbf{SSR}_1$, i.e. $\mathbf{SSR}_{2|1}$ is the extra sum of squares due to the addition of the second-order terms to the model.

(iv) Compute the F test statistic:

$$F_{T,p}^{(T)}(m) = \Big(\frac{T - p - m(p+1)}{m}\Big)\Big(\frac{1 - (\Lambda^{(W)})^{1/2}}{(\Lambda^{(W)})^{1/2}}\Big), \qquad (11.46)$$

where

$$\Lambda^{(W)} = |\mathbf{SSE}_2|/\{|\mathbf{SSR}_{2|1} + \mathbf{SSE}_2|\}. \qquad (11.47)$$

If $\{\mathbf{Y}_t, t \in \mathbb{Z}\}$ follows a strictly stationary zero-mean Gaussian VAR(p) process ($\mathbb{H}_0$),

$$F_{T,p}^{(T)}(m) \xrightarrow{D} F_{\nu_1, \nu_2}, \text{ as } T \to \infty, \qquad (11.48)$$

with $\nu_1 = m$ and $\nu_2 = T - p - mp - m$.

The proof of (11.48) follows from standard multivariate regression theory. It

11.3 TIME-DOMAIN LINEARITY TESTS

may be noted that for $m = 1$, the degrees of freedom of ν_1 and ν_2 are nearly the same as those reported in Algorithm 5.7; recall that $\mathbb{E}(\mathbf{Y}_t) = \mathbf{0}$ while in Algorithm 5.7 the univariate second-order Volterra expansion has a non-zero mean. Clearly, computation of (11.46) requires fewer degrees of freedom; i.e. the response variable is an m-variate vector as compared to an $\nu_U = mp(mp+1)/2$-variate vector in Algorithm 11.1. This may be preferable for short series.

Original F test

The multivariate generalization of the $F_T^{(O)}$ test statistic (Algorithm 5.8) employs disaggregated variables in step (ii) of Algorithm 11.2. The test statistic is based on the following model

$$\mathbf{Y}_t = \sum_{j=1}^{p} \mathbf{\Phi}_j \mathbf{Y}_{t-j} + \mathbf{\Psi}\, \text{vech}(\mathbf{Z}_t \otimes \mathbf{Z}_t') + \boldsymbol{\varepsilon}_t, \qquad (11.49)$$

where $\mathbf{Z}_t = (\mathbf{Y}_{t-1}', \ldots, \mathbf{Y}_{t-p}')'$ is an $mp \times 1$ vector, and $\mathbf{\Psi}$ is an $m \times mp(mp+1)/2$ parameter matrix. Thus, the null hypothesis of interest is given by $\mathbb{H}_0 : \mathbf{\Psi} = \mathbf{0}$. The computation of the corresponding test statistic goes as follows.

Algorithm 11.3: $F_T^{(O)}$ **test statistic for nonlinearity**

(i) Follow step (i) of Algorithm 11.2.

(ii) Compute $\mathbf{U}_t = \text{vech}(\mathbf{Z}_t \otimes \mathbf{Z}_t')$. Thus, the $\nu_U = mp(mp+1)/2$-dimensional vector $\mathbf{U}_t$ contains all second-order cross-product terms of lagged values of the process up to order p. Regress $\mathbf{U}_t$ on $\mathbf{Z}_t$. Obtain the residuals $\mathbf{W}_t = (W_{1,t}, \ldots, W_{\nu_U, t})'$.

(iii) Regress $\widehat{\boldsymbol{\varepsilon}}_t$ from step (i) on $\mathbf{W}_t$ from step (ii). Compute the $m \times m$ sum of squared regressions matrix, $\mathbf{SSR}_2$, and the sum of squared errors matrix, $\mathbf{SSE}_2$. Let $\mathbf{SSR}_{2|1} = \mathbf{SSR}_2 - \mathbf{SSR}_1$.

(iv) Compute the F test statistic:

$$F_{T,p}^{(O)}(m) = \Big(\frac{T - p - \tfrac{1}{2}(mp(mp+3))}{\nu_U}\Big)\Big(\frac{1 - (\Lambda^{(W)})^{1/2}}{(\Lambda^{(W)})^{1/2}}\Big), \qquad (11.50)$$

where

$$\Lambda^{(W)} = |\mathbf{SSE}_2|/\{|\mathbf{SSR}_{2|1} + \mathbf{SSE}_2|\}. \qquad (11.51)$$

Under $\mathbb{H}_0$,

$$F_{T,p}^{(O)}(m) \xrightarrow{D} F_{\nu_1, \nu_2}, \text{ as } T \to \infty, \qquad (11.52)$$

with $\nu_1 = \nu_U$ and $\nu_2 = T - p - \tfrac{1}{2}(mp(mp+3))$.

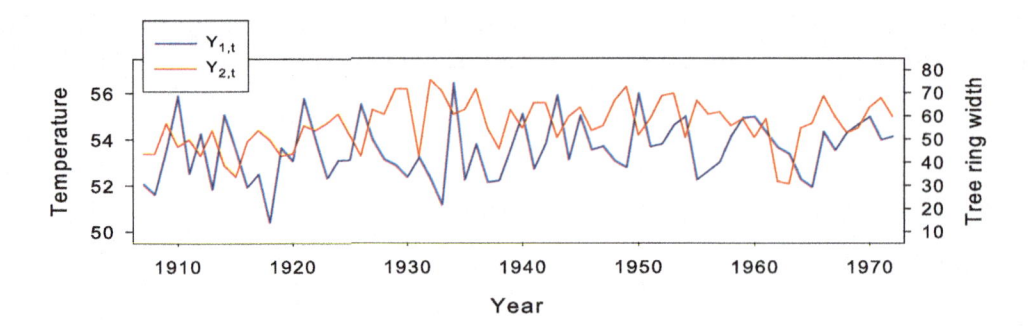

Figure 11.5: Annual temperatures ($Y_{1,t}$) and tree ring widths ($Y_{2,t}$) for the years 1907 – 1972 ($T = 66$) at Campito Mountain, California.

Harvill and Ray (1999) also consider a semi-multivariate version of the test statistics in Algorithm 11.3 in which each component of the vector series is regressed individually on $\mathbf{U}_t$ in step (ii). The individual test statistics for this semi-multivariate version have a simple F distribution under the null hypothesis of linearity with $\nu_1 = mp(mp+1)/2$ and $\nu_2 = T - mp(mp+3)/2$ degrees of freedom. In this case, however, possible cross-correlation in the error terms is not accounted for by the procedure. On the other hand, the semi-multivariate test may be more powerful when only one of the component series of $\{\mathbf{Y}_t, t \in \mathbb{Z}\}$ is nonlinear.

The Wilks' $\Lambda^{(W)}$ test statistics in Algorithms (11.1) – (11.3) are formulated as LR-type tests. Other test statistics can be defined directly in terms of the sum of squared errors matrix **SSE** and the sum of squared regression matrix **SSR**, or in terms of their non-zero eigenvalues; see, e.g., Johnson and Wichern (2002, Chapter 7). Two well known multivariate test statistics are the Hotelling–Lawley (HL) trace test statistic and Pillai's (P) trace test statistic, respectively defined by:

$$U^{(\mathrm{HL})} = \mathrm{tr}[\mathbf{SSE}^{-1}\,\mathbf{SSR}], \tag{11.53}$$

$$V^{(\mathrm{P})} = \mathrm{tr}[\mathbf{SSR}(\mathbf{SSR} + \mathbf{SSE})^{-1}]. \tag{11.54}$$

The test statistic (11.53) is valid when **SSE** is positive definite. The test statistic (11.54) requires a less restrictive assumption: **SSR+SSE** is positive definite. Wilks' lambda and the Hotelling–Lawley trace test statistics are nearly equivalent for large sample sizes.

The test statistic (11.50) can be extended to include cubic terms, as in the augmented $F_T^{(A)}$ test statistic of Section 5.4. However, the proliferation of additional terms in the multivariate case is expected to result in a loss of power due to fewer degrees of freedom for the F test statistic, unless m is small and T is large. Also, as in the univariate case, a VARMA(p,q) model can be fit to the data initially (using, e.g., QML estimation) to allow for linear MA structure. In that case the test statistic (11.50) is modified by letting $\mathbf{Z}_t$ in step (i) be $(\mathbf{Y}'_{t-1}, \ldots, \mathbf{Y}'_{t-p}, \widehat{\boldsymbol{\varepsilon}}'_{t-1}, \ldots, \widehat{\boldsymbol{\varepsilon}}'_{t-q})'$, where $\widehat{\boldsymbol{\varepsilon}}_t$ denotes the series of residuals from the VARMA fit.

11.3 TIME-DOMAIN LINEARITY TESTS

Table 11.1: *Values of the multivariate nonlinearity test statistics for the annual temperatures ($Y_{1,t}$) and tree ring widths ($Y_{2,t}$) time series; $T = 66$, $p = 4$, and $m = 2$.*

				Degrees of freedom				
Test	Wilks	HL-Trace	P-Trace	Num.	Den.	p-value	p-value	p-value
$F_{T,p}^{(O)}(m)$	0.042	8.888	1.539	36	18	0.028	0.023	0.047
$F_{T,p}^{(T)}(m)$	0.572	0.747	0.429	2	52	0.000	0.000	0.000
$F_T^{(O)}(Y_{1,t})$	3.191			10	52	0.003		
$F_T^{(O)}(Y_{2,t})$	3.146			10	52	0.003		
Semi ($Y_{1,t}$)	2.691			36	22	0.008		
Semi ($Y_{2,t}$)	1.358			36	22	0.227		

Example 11.5: Tree Ring Widths

The rings of trees in certain cites of western North America provide a unique source on past variations of climatic and other environmental factors which prevail over North America and the adjoining oceans. Figure 11.5 shows plots of annual temperatures (in °F) and annual tree ring widths (in 0.01 mm) measured at Campito Mountain in California for the years 1907 – 1972 ($T = 66$). Below, we use this data set as an illustration of the nonlinearity test statistics discussed above.

The sample ACF and PACF matrices both identify an association between tree ring widths in year t ($Y_{2,t}$) and tree ring widths one, three, and four years back, while changes in temperature ($Y_{1,t}$) are associated with the previous year's tree growth; cf. Exercise 12.2. So, as a first step, we fitted a VAR(4) model to the data. Next, we computed the test statistics in Algorithms 11.2 and 11.3 using appropriate versions of Wilks' lambda statistic, the HL test statistic, and the P test statistic. In addition, based on the Wilks' lambda statistic, we applied the semi-multivariate version of the $F_{T,p}^{(O)}(m)$ test statistic and its univariate analogue, $F_T^{(O)}$ (Algorithm 5.8).

Table 11.1 contains the values of the test statistics, p-values, and degrees of freedom. The p-values for the multivariate nonlinearity test statistics $F_{T,p}^{(O)}(m)$ and $F_{T,p}^{(T)}(m)$, for the Wilks' lambda statistic, the HL test statistic, and the P test statistic, all indicate that the null hypothesis of linearity should be rejected at the 5% nominal significance level. The same conclusion emerges for each series from the p-values of the $F_T^{(O)}$ test statistic based on the Wilks' lambda test statistic. On the other hand, the p-value of the semi-multivariate version of Tsay's original test statistic does not reject linearity for the tree ring widths $Y_{2,t}$. However, as stated above, the semi-multivariate test statistics do not account for significant, at the 5% nominal level, sample cross-correlations between the time series $\{Y_{1,t}\}$ and $\{Y_{2,t}\}$.

11.4 Testing Linearity vs. Specific Nonlinear Alternatives

A test for VSETAR nonlinearity

Tsay (1998) provides a generalization of the TAR F_T^* test statistic (Algorithm 5.10) to the VSETAR case. Given a strictly stationary m-variate time series process $\{\mathbf{Y}_t, t \in \mathbb{Z}\}$, assume that this process follows a VSETAR$(2; p, \ldots, p)$ model with regimes determined by the threshold variable X_{t-d}. Let $\mathbf{Z}_t = (1, \mathbf{Y}'_{t-1}, \ldots, \mathbf{Y}'_{t-p})'$ be an $(mp+1)$-dimensional regressor. Placing the model in a regression framework gives

$$\mathbf{Y}'_t = \mathbf{Z}'_t \mathbf{\Phi} + \boldsymbol{\varepsilon}'_t, \quad (t = h+1, \ldots, T), \tag{11.55}$$

where $h = p \vee d$, and $\mathbf{\Phi}$ denotes the parameter matrix. Ordering $\mathbf{Y}_t$ and $\mathbf{Z}_t$ according to increasing values of X_{t-d} gives

$$\mathbf{Y}'_{\tau_i+d} = \mathbf{Z}'_{\tau_i+d} \mathbf{\Phi} + \boldsymbol{\varepsilon}'_{\tau_i+d}, \quad (i = 1, \ldots, T-h), \tag{11.56}$$

where τ_i denotes the time index of $X_{(i)}$, the ith smallest value of $\{X_{t-d}\}_{t=h+1-d}^{T}$.

If $\{\mathbf{Y}_t, t \in \mathbb{Z}\}$ is linear, the predictive residuals of (11.56) are an m-variate WN process, whereas if $\{\mathbf{Y}_t, t \in \mathbb{Z}\}$ follows an m-dimensional VTAR$(2; p, p)$ model with threshold variable X_{t-d}, the predictive residuals are correlated with $\mathbf{Z}'_{\tau_i+d}$. Based on this idea, the computation of the test statistic goes as follows.

Algorithm 11.4: Multivariate test statistic for VSETAR

(i) Given d, fit an arranged VAR(p) to $\{\mathbf{Y}_t\}_{t=1}^T$ using data points associated with the s smallest values of X_{t-d}, obtaining $\{\widehat{\mathbf{\Phi}}_s\}_{s=n_{\min}+1}^{T-h}$, where $n_{\min}$ is a minimum number for starting the multivariate version of the recursive LS estimation procedure given by (5.86)–(5.87). For unit root time series, Tsay (1998) recommends taking $n_{\min} \approx 5\sqrt{T}$, and $n_{\min} \approx 3\sqrt{T}$ for the stationary case.

(ii) Compute the predictive residuals

$$\widehat{\boldsymbol{\varepsilon}}_{\tau_{s+1}+d} = \mathbf{Y}_{\tau_{s+1}+d} - \widehat{\mathbf{\Phi}}'_s \mathbf{Z}_{\tau_{s+1}+d}$$

and the standardized predictive residuals

$$\widehat{\mathbf{e}}_{\tau_{s+1}+d} = \widehat{\boldsymbol{\varepsilon}}_{\tau_{s+1}+d} / [1 + \mathbf{Z}'_{\tau_{s+1}+d} \mathbf{P}_s \mathbf{Z}_{\tau_{s+1}+d}]^{1/2},$$

where $\mathbf{P}_s = [\sum_{i=1}^s \mathbf{Z}_{\tau_{s+1}+d} \mathbf{Z}'_{\tau_{s+1}+d}]^{-1}$.

(iii) Regress $\widehat{\mathbf{e}}_{\tau_\ell+d}$ on $\mathbf{Z}_{\tau_\ell+d}$ ($\ell = n_{\min}+1, \ldots, T-h$).

(iv) Compute the test statistic

$$C_{T,p}(d,m) = [T - h - n_{\min} - (mp+1)]\{\log|\mathbf{SSE}_0| - \log|\mathbf{SSE}_1|\}, \tag{11.57}$$

> **Algorithm 11.4: Multivariate test statistic for VSETAR (Cont'd)**
>
> (iv) (Cont'd)
>   where d signifies that the test depends on the threshold variable X_{t-d}, and
>
> $$\mathbf{SSE}_0 = \frac{1}{T^*}\sum_{\ell=n_{\min}+1}^{T-h}\widehat{\mathbf{e}}_{\tau_\ell+d}\widehat{\mathbf{e}}'_{\tau_\ell+d}, \quad \mathbf{SSE}_1 = \frac{1}{T^*}\sum_{\ell=n_{\min}+1}^{T-h}\widehat{\boldsymbol{\omega}}_{\tau_\ell+d}\widehat{\boldsymbol{\omega}}'_{\tau_\ell+d}$$
>
> with $T^* = T - h - n_{\min}$ and $\widehat{\boldsymbol{\omega}}_{\tau_\ell+d}$ denotes the LS residual from step (ii).
>
> (v) Under the null hypothesis that $\{\mathbf{Y}_t, t \in \mathbb{Z}\}$ follows a strictly stationary VAR(p) process, and some regularity conditions, Tsay (1998) shows that
>
> $$C_{T,p}(d,m) \xrightarrow{D} \chi^2_{m(mp+1)}, \quad \text{as } T \to \infty. \tag{11.58}$$

The test statistic has good power when the delay d is correctly specified; Tsay (1998). The power deteriorates when the delay used in the test is different from the actual delay. Note, $\mathbb{H}_0$ includes a zero intercept for all predictive residuals. In theory, a non-zero intercept signifies a systematic bias in the estimation of (11.56), indicating possible change points. So, due to the possibility of finite-sample bias, one may wish to exclude the intercept term from the nonlinearity test statistic (11.57) which can be achieved by mean-correcting $\mathbf{SSE}_0$. In this case, the resulting test statistic has an asymptotical $\chi^2_{m^2p}$ distribution under the null hypothesis.

Likelihood ratio test statistic for VSETAR

Recall, in Section 5.2 we introduced a LR test statistic for SETAR models. Using similar arguments, Liu (2011) proposes a LR test statistic for an m-dimensional strictly stationary time series $\{\mathbf{Y}_t, t \in \mathbb{Z}\}$ generated by the VSETAR($2;p,p$) model with (exogenous) threshold variable $\{X_{t-d}\}$ ($d \leq p$):

$$\mathbf{Y}_t = \boldsymbol{\Phi}_0 + \sum_{i=1}^p \boldsymbol{\Phi}_i \mathbf{Y}_{t-i} + \Big(\boldsymbol{\Psi}_0 + \sum_{i=1}^p \boldsymbol{\Psi}_i \mathbf{Y}_{t-i}\Big)I(X_{t-d} \leq r) + \boldsymbol{\varepsilon}_t, \tag{11.59}$$

where $\boldsymbol{\Phi}_0$ and $\boldsymbol{\Psi}_0$ are $m \times 1$ parameter vectors, and $\boldsymbol{\Phi}_i$ and $\boldsymbol{\Psi}_i$ ($1 \leq i \leq p$) are $m \times m$ parameter matrices. The process $\{\boldsymbol{\varepsilon}_t\}$ is an m-dimensional vector martingale difference sequence satisfying

$$\mathbb{E}(\boldsymbol{\varepsilon}_t|\mathcal{F}_{t-1}) = \mathbf{0}, \quad \text{Cov}(\boldsymbol{\varepsilon}_t,\boldsymbol{\varepsilon}_t|\mathcal{F}_{t-1}) = \boldsymbol{\Sigma}_\varepsilon, \quad \text{and } \text{Cov}(\boldsymbol{\varepsilon}_t,\boldsymbol{\varepsilon}_s|\mathcal{F}_{t\wedge s}) = \mathbf{0}, \quad (t \neq s), \tag{11.60}$$

with $\mathcal{F}_t$ the information set, and $\boldsymbol{\Sigma}_\varepsilon$ is a positive definite matrix. It is also assumed that p and d are unknown, and that r belongs to a known bounded subset $\widetilde{\mathbb{R}} = [\underline{r},\overline{r}]$ of $\mathbb{R}$.

For simplicity it is convenient to rewrite (11.59) in a regression form first and vectorize the resulting equation next. To this end, we introduce the following notation. Let $\boldsymbol{\Phi}_{(U)} = (\boldsymbol{\Phi}_0, \boldsymbol{\Phi}_1, \ldots, \boldsymbol{\Phi}_p)'$, $\boldsymbol{\Psi}_{(U)} = (\boldsymbol{\Psi}_0, \boldsymbol{\Psi}_1, \ldots, \boldsymbol{\Psi}_p)'$ be two $m \times (mp+1)$ matrices, with the subscript (U) denoting an unrestricted parameter vector,

$$\mathbf{Y} = \begin{pmatrix} \mathbf{Y}'_{p+1} \\ \mathbf{Y}'_{p+2} \\ \vdots \\ \mathbf{Y}'_T \end{pmatrix}, \quad \mathbf{X} = \begin{pmatrix} 1 & \mathbf{Y}'_p & \cdots & \mathbf{Y}'_1 \\ 1 & \mathbf{Y}'_{p+1} & \cdots & \mathbf{Y}'_2 \\ \vdots & \vdots & & \vdots \\ 1 & \mathbf{Y}'_{T-1} & \cdots & \mathbf{Y}'_{T-p} \end{pmatrix}, \quad \boldsymbol{\varepsilon} = \begin{pmatrix} \varepsilon'_{p+1} \\ \varepsilon'_{p+2} \\ \vdots \\ \varepsilon'_T \end{pmatrix}, \quad \text{and}$$

$$\mathbf{Y}_r = \begin{pmatrix} I(X_{p+1-d} \leq r) & I(X_{p+1-d} \leq r)\mathbf{Y}'_p & \cdots & I(X_{p+1-d} \leq r)\mathbf{Y}'_1 \\ I(X_{p+2-d} \leq r) & I(X_{p+2-d} \leq r)\mathbf{Y}'_{p+1} & \cdots & I(X_{p+2-d} \leq r)\mathbf{Y}'_2 \\ \vdots & \vdots & & \vdots \\ I(X_{T-d} \leq r) & I(X_{T-d} \leq r)\mathbf{Y}'_{T-1} & \cdots & I(X_{T-d} \leq r)\mathbf{Y}'_{T-p} \end{pmatrix}.$$

Now, we can rewrite (11.59) in a regression framework as

$$\mathbf{Y} = \mathbf{X}\boldsymbol{\Phi}_{(U)} + \mathbf{Y}_r\boldsymbol{\Psi}_{(U)} + \boldsymbol{\varepsilon}. \tag{11.61}$$

Let $\mathbf{A}^{\text{v}} \equiv \text{vec}(\mathbf{A})$. Then a vectorization of (11.61) is given by

$$\mathbf{Y}^{\text{v}} = (\mathbf{I}_m \otimes \mathbf{X})\boldsymbol{\Phi}^{\text{v}}_{(U)} + (\mathbf{I}_m \otimes \mathbf{Y}_r)\boldsymbol{\Psi}^{\text{v}}_{(U)} + \boldsymbol{\varepsilon}^{\text{v}}. \tag{11.62}$$

The hypotheses of interest are

$$\mathbb{H}_0: \boldsymbol{\Psi}^{\text{v}}_{(U)} = \mathbf{0}, \quad \mathbb{H}_1: \boldsymbol{\Psi}^{\text{v}}_{(U)} \neq \mathbf{0}, \text{ for some } r \in \widetilde{\mathbb{R}}. \tag{11.63}$$

Note, under $\mathbb{H}_0$ equation (11.62) reduces to the linear regression

$$\mathbf{Y}^{\text{v}} = (\mathbf{I}_m \otimes \mathbf{X})\boldsymbol{\Phi}^{\text{v}}_{(R)} + \boldsymbol{\eta}^{\text{v}}, \tag{11.64}$$

where $\boldsymbol{\eta}^{\text{v}} \equiv \text{vec}(\boldsymbol{\eta})$ is defined in the same way as $\boldsymbol{\varepsilon}^{\text{v}}$ with $\boldsymbol{\eta} = (\boldsymbol{\eta}'_{p+1}, \ldots, \boldsymbol{\eta}'_T)'$. Here, $\{\boldsymbol{\eta}_t\}$ is an m-dimensional vector martingale difference sequence that is strictly stationary and ergodic with covariance matrix $\boldsymbol{\Sigma}_\eta$. Also, the subscript (R) in $\boldsymbol{\Phi}^{\text{v}}_{(R)}$ reflects the fact that the parameter vector of the original VSETAR model is "restricted".

Given (11.64), the CLS estimate of the restricted parameter vector $\boldsymbol{\Phi}^{\text{v}}_{(R)}$ and the corresponding estimate of $\boldsymbol{\Sigma}_\eta$ are given by

$$\widehat{\boldsymbol{\Phi}}^{\text{v}}_{(R)} = \{\mathbf{I}_m \otimes (\mathbf{X}'\mathbf{X})^{-1}\mathbf{X}'\}\mathbf{Y}^{\text{v}} \quad \text{and} \quad \widehat{\boldsymbol{\Sigma}}_\eta = \widehat{\boldsymbol{\eta}}'\widehat{\boldsymbol{\eta}}/(T-p),$$

with $\widehat{\boldsymbol{\eta}}^{\text{v}} = \mathbf{Y}^{\text{v}} - (\mathbf{I}_m \otimes \mathbf{X})\widehat{\boldsymbol{\Phi}}^{\text{v}}_{(R)}$ a vector of residuals. Similarly, given (11.61), the CLS estimates of the unrestricted parameter vectors $\boldsymbol{\Phi}^{\text{v}}_{(U)}$ and $\boldsymbol{\Psi}^{\text{v}}_{(U)}$, and the corresponding estimate of $\boldsymbol{\Sigma}_\varepsilon$ are respectively given by

$$\widehat{\boldsymbol{\Phi}}^{\text{v}}_{(U)} = \{\mathbf{I}_m \otimes (\mathbf{X}'\mathbf{X})^{-1}\mathbf{X}'[\mathbf{I}_{T-p} - \mathbf{Y}_r\mathbf{G}^{-1}\mathbf{Y}'_r(\mathbf{I}_{T-p} - \mathbf{P}_X)]\}\mathbf{Y}^{\text{v}},$$

$$\widehat{\boldsymbol{\Psi}}^{\text{v}}_{(U)} = \{\mathbf{I}_m \otimes \mathbf{G}^{-1}\mathbf{Y}'_r(\mathbf{I}_{T-p} - \mathbf{P}_X)\}\mathbf{Y}^{\text{v}} \quad \text{and} \quad \widehat{\boldsymbol{\Sigma}}_\varepsilon = \widehat{\boldsymbol{\varepsilon}}'\widehat{\boldsymbol{\varepsilon}}/(T-p),$$

11.4 TESTING LINEARITY VS. SPECIFIC NONLINEAR ALTERNATIVES

where $\mathbf{P}_X = \mathbf{X}(\mathbf{X}'\mathbf{X})^{-1}\mathbf{X}'$, $\mathbf{G} = \mathbf{Y}'_r(\mathbf{I}_{T-p} - \mathbf{P}_\mathbf{X})\mathbf{Y}_r$, and $\widehat{\boldsymbol{\varepsilon}}^{\text{v}} = \mathbf{Y}^{\text{v}} - (\mathbf{I}_m \otimes \mathbf{X})\widehat{\boldsymbol{\Phi}}^{\text{v}}_{(\text{U})} - (\mathbf{I}_m \otimes \mathbf{Y}_r)\widehat{\boldsymbol{\Psi}}^{\text{v}}_{(\text{U})}$.

Now, let $\widehat{\boldsymbol{\Lambda}}_\eta = (\widehat{\boldsymbol{\Sigma}}_\varepsilon^{-1/2} \otimes \mathbf{I}_{T-p})\widehat{\boldsymbol{\eta}}^{\text{v}}$ and $\widehat{\boldsymbol{\Lambda}}_\varepsilon = (\widehat{\boldsymbol{\Sigma}}_\varepsilon^{-1/2} \otimes \mathbf{I}_{T-p})\widehat{\boldsymbol{\varepsilon}}^{\text{v}}$ be the rescaled residual vectors. Then the LR statistic for testing $\mathbb{H}_0$ against $\mathbb{H}_1$ is defined in terms of the residual sum of squares matrices as

$$\text{LR}_{T,p}(m,r) = \sup_{r \in \widetilde{\mathbb{R}}} \{\widehat{\boldsymbol{\Lambda}}'_\eta \widehat{\boldsymbol{\Lambda}}_\eta - \widehat{\boldsymbol{\Lambda}}'_\varepsilon \widehat{\boldsymbol{\Lambda}}_\varepsilon\},$$

$$= \sup_{r \in \widetilde{\mathbb{R}}} \Big(\{[\widehat{\boldsymbol{\Sigma}}_\varepsilon^{-1/2} \otimes \mathbf{Y}'_r(\mathbf{I}_{T-p} - \mathbf{P}_\mathbf{X})]\mathbf{Y}^{\text{v}}\}'[\mathbf{I}_{T-p} \otimes \mathbf{G}^{-1}]$$

$$\{[\widehat{\boldsymbol{\Sigma}}_\varepsilon^{-1/2} \otimes \mathbf{Y}'_r(\mathbf{I}_{T-p} - \mathbf{P}_\mathbf{X})]\mathbf{Y}^{\text{v}}\} \Big), \quad (11.65)$$

where the second expression on the right-hand side follows from some simple algebra. Note that for a fixed r and $m = 1$, (11.65) reduces to the LR test statistic $\text{LR}_T^{(8)}$ defined by (5.55).

The asymptotic null distribution follows in a similar way as described in Section 5.2. Suppose the following assumption holds

$$\frac{1}{T}\begin{pmatrix} \mathbf{X}'\mathbf{X} & \mathbf{X}'\mathbf{Y}_r \\ \mathbf{Y}'_r\mathbf{X} & \mathbf{Y}'_r\mathbf{Y}_r \end{pmatrix} \xrightarrow[T \to \infty]{\text{a.s.}} \begin{pmatrix} \boldsymbol{\Sigma} & \boldsymbol{\Sigma}_{12}(r) \\ \boldsymbol{\Sigma}_{21}(r) & \boldsymbol{\Sigma}_{22}(r) \end{pmatrix},$$

where $\boldsymbol{\Sigma}(\cdot)$, $\boldsymbol{\Sigma}_{21}(\cdot) = \boldsymbol{\Sigma}'_{12}(\cdot)$, and $\boldsymbol{\Sigma}_{22}(\cdot)$ are $(mp+1) \times (mp+1)$ matrices. Under $\mathbb{H}_0$, standard regularity conditions, and as $T \to \infty$, it can be shown (Liu, 2011) that

$$\text{LR}_{T,p}(m,r) \xrightarrow{D} \sup_{r \in \widetilde{\mathbb{R}}}\{\mathcal{G}'_{2(mp+1)}(r)\boldsymbol{\Omega}(r)\mathcal{G}_{2(mp+1)}(r)\}, \quad (11.66)$$

where

$$\boldsymbol{\Omega}(r) = \mathbf{I}_m \otimes \big(\boldsymbol{\Sigma}_{21}(r) - \boldsymbol{\Sigma}_{21}(r)\boldsymbol{\Sigma}_{22}^{-1}(r)\boldsymbol{\Sigma}_{12}(r)\big)^{-1},$$

and $\{\mathcal{G}_{2(mp+1)}(r)\} \sim \mathcal{N}_{2(mp+1)}\big(\mathbf{0}, \mathbf{I}_m \otimes (\boldsymbol{\Sigma}_{(r \wedge s)} - \boldsymbol{\Sigma}_{21}(r)\boldsymbol{\Sigma}^{-1}\boldsymbol{\Sigma}_{12}(r))\big)$ distributed. Then, for large α, and using the Poisson clumping heuristic method, we have

$$\mathbb{P}(\sup_{r \in \widetilde{\mathbb{R}}} \mathcal{G}'_{2(mp+1)}(r)\boldsymbol{\Omega}_1(r)\mathcal{G}_{2(mp+1)}(r) \leq \alpha) \sim \exp\Big\{-2\chi^2_{m(mp+1)}(\alpha)\Big(\frac{\alpha}{mp+1} - 1\Big)$$

$$\times \sum_{i=1}^{mp+1}\big(t_i(\overline{r}) - t_i(\underline{r})\big)\Big\}, \quad (11.67)$$

where $\chi^2(\cdot)_{m(mp+1)}$ denotes the pdf of the χ^2 distribution with $m(mp+1)$ degrees of freedom, $t_i(r) = \frac{1}{2}\log\{\mathcal{L}_i(r)/(1 - \mathcal{L}_i(r))\}$ $\forall i$, and $\mathcal{L}_i(r)$ are eigenvalues of $\boldsymbol{\Sigma}_{21}^{-1/2}(r)\boldsymbol{\Sigma}_{22}(r)\boldsymbol{\Sigma}_{12}^{-1/2}(r)$. Appendix 11.A contains a table with selected percentiles of the LR-VATR test statistic when $m = 2$.

LM-type test statistics for VSTAR

Recall the two LM-type test statistics for STAR nonlinearity in Section 5.1. Their construction is based on respectively a first-order and a third-order Taylor expansion of the univariate transition function $G(\cdot)$ around the slope parameter γ. This approach is also applicable to VSTAR models with a single transition variable X_t.

As an example, consider the two-regime m-dimensional VSTAR(p) model (11.32) with the matrix of transition functions given by $\mathbf{G}_t \equiv G(X_t; \gamma, \mathbf{c})\mathbf{I}_m$:

$$\mathbf{Y}_t = \mathbf{B}_1'\mathbf{Z}_t + \mathbf{G}_t\mathbf{B}_2'\mathbf{Z}_t + \boldsymbol{\varepsilon}_t, \tag{11.68}$$

where $\mathbf{B}_i$ ($i = 1, 2$) are $(mp+1) \times m$ matrices given by

$$\mathbf{B}_1 = \boldsymbol{\Phi}^{(1)}, \quad \mathbf{B}_2 = \boldsymbol{\Phi}^{(2)} - \boldsymbol{\Phi}^{(1)},$$

$\mathbf{Z}_t = (1, \mathbf{Y}_{t-1}', \ldots, \mathbf{Y}_{t-p}')'$ is an $(mp+1) \times 1$ vector, and $\{\boldsymbol{\varepsilon}_t\} \overset{\text{i.i.d.}}{\sim} \mathcal{N}_m(\mathbf{0}, \boldsymbol{\Sigma}_\varepsilon)$. We wish to test the null hypothesis $\mathbb{H}_0: \gamma = 0$, versus the alternative hypothesis $\mathbb{H}_1: \gamma > 0$. However, as in the univariate case, model (11.68) contains nuisance parameters that are not identified under the null hypothesis. To circumvent this problem, it is common to replace $\mathbf{G}_t$ by a suitable linear approximation.

For instance, in case the alternative is an LVSTAR model with $\mathbf{G}_t$ a diagonal matrix of transition functions, a first-order Taylor expansion around $\gamma = 0$ yields the auxiliary regression model

$$\mathbf{Y}_t = \boldsymbol{\Theta}_0'\mathbf{Z}_t + \boldsymbol{\Theta}_1'\mathbf{Z}_t X_t + \boldsymbol{\eta}_t, \quad (t = 1, \ldots, T), \tag{11.69}$$

where $\boldsymbol{\Theta}_0 = \mathbf{B}_1 + \mathbf{B}_2\mathbf{B}$, $\boldsymbol{\Theta}_1 = \mathbf{B}_2\mathbf{A}$, $\boldsymbol{\eta}_t = \mathbf{R}_t\mathbf{B}_2'\mathbf{Z}_t + \boldsymbol{\varepsilon}_t$, with $\mathbf{A} = \text{diag}(a_1, \ldots, a_m)$ and $\mathbf{B} = \text{diag}(b_1, \ldots, b_m)$ having, respectively elements $a_j = (1/4)\gamma$ and $b_j = (1/2) - a_j c_j$ ($j = 1, \ldots, m$), and $\mathbf{R}_t$ denotes an $m \times m$ diagonal matrix containing the remainder terms. The null hypothesis implies $\mathbf{G}_t = (1/2)\mathbf{I}_m$. Clearly, model (11.69) is linear when $\boldsymbol{\Theta}_0 = \mathbf{B}_1 + (1/2)\mathbf{B}_2$ and $\boldsymbol{\Theta}_1 = \mathbf{0}$. Thus, the original null hypothesis of linearity is equivalent to testing $\mathbb{H}_0: \boldsymbol{\Theta}_1 = \mathbf{0}$ versus the alternative hypothesis $\mathbb{H}_1: \boldsymbol{\Theta}_1 \neq \mathbf{0}$.

We begin our discussion of the score form of the first-order LM-type test statistic by introducing the following notation:

$$\mathbf{Y} = \begin{pmatrix} \mathbf{Y}_1' \\ \vdots \\ \mathbf{Y}_T' \end{pmatrix}, \quad \mathbf{X} = \begin{pmatrix} \mathbf{Z}_1' \\ \vdots \\ \mathbf{Z}_T' \end{pmatrix}, \quad \mathbf{U} = \begin{pmatrix} \mathbf{Z}_1'X_1 \\ \vdots \\ \mathbf{Z}_T'X_T \end{pmatrix}.$$

Also, let $\boldsymbol{\theta}$ denote the vector of available parameters. As $\{\boldsymbol{\varepsilon}_t\} \overset{\text{i.i.d.}}{\sim} \mathcal{N}_m(\mathbf{0}, \boldsymbol{\Sigma}_\varepsilon)$, the conditional log-likelihood function of the data, evaluated at $\boldsymbol{\theta} \in \boldsymbol{\Theta}$ (a compact parameter space) and apart from some additional constants, is equal to

$$\log L_T(\boldsymbol{\theta}) = -(1/2)\sum_{t=1}^{T}(\mathbf{Y}_t - \boldsymbol{\Psi}_t'\mathbf{B}'\mathbf{Z}_t)'\boldsymbol{\Sigma}_\varepsilon^{-1}(\mathbf{Y}_t - \boldsymbol{\Psi}_t'\mathbf{B}'\mathbf{Z}_t),$$

11.4 TESTING LINEARITY VS. SPECIFIC NONLINEAR ALTERNATIVES

where $\boldsymbol{\Psi}_t = (\mathbf{I}_m, \mathbf{G}_t)'$ is a $2m \times m$ full rank matrix, and $\mathbf{B} = (\mathbf{B}_1, \mathbf{B}_2)$ is an $(mp+1) \times 2m$ matrix. Assume some standard regularity conditions are satisfied. Then the LM-type test statistic follows from the score matrix $\partial \log L_T(\widehat{\boldsymbol{\theta}})/\partial \boldsymbol{\Theta}_1$, where $\widehat{\boldsymbol{\theta}}$ is an estimate of $\boldsymbol{\theta}$ under the null hypothesis. In particular, the chi-square version of the LM-type test statistic is given by

$$\mathrm{LM}_{T,p}^{(1)}(m) = \mathrm{tr}\{\widehat{\boldsymbol{\Sigma}}_\varepsilon^{-1}(\mathbf{Y}-\mathbf{X}\widehat{\mathbf{B}}_1)'\mathbf{U}\left[\mathbf{U}'(\mathbf{I}_T - \mathbf{P}_X)\mathbf{U}\right]^{-1}\mathbf{U}'(\mathbf{Y}-\mathbf{X}\widehat{\mathbf{B}}_1)\}, \quad (11.70)$$

where $\mathbf{P}_X = \mathbf{X}(\mathbf{X}'\mathbf{X})^{-1}\mathbf{X}'$, $\widehat{\mathbf{B}}_1$ and $\widehat{\boldsymbol{\Sigma}}_\varepsilon$ are parameter estimates under the null hypothesis, i.e. the restricted model specification. Here, the superscript (1) indicates that the test is based on the first-order Taylor expansion of the logistic transition function. Similar as Algorithm 5.1, the test procedure consists of the following steps.

Algorithm 11.5: $\mathrm{LM}_{T,p}^{(1)}(m)$**-type test statistic for LVSTAR**

(i) Fit a VAR(p) model to $\{\mathbf{Y}_t\}_{t=1}^T$ using, e.g., CLS or NLS. Obtain the $T \times m$ matrix of residuals $\widehat{\mathbf{E}} = (\mathbf{I}_T - \mathbf{P}_X)\mathbf{Y}$, and compute the corresponding sum of squared errors matrix, $\mathbf{SSE}_0 = \widehat{\mathbf{E}}'\widehat{\mathbf{E}}$.

(ii) Regress $\widehat{\mathbf{E}} = (\widehat{\varepsilon}_1, \ldots, \widehat{\varepsilon}_T)'$ on $(\mathbf{X}, \mathbf{U})$, i.e. an auxiliary regression. Obtain the matrix of residuals $\widehat{\boldsymbol{\Xi}}$ and compute the corresponding sum of squared errors matrix, $\mathbf{SSE}_1 = \widehat{\boldsymbol{\Xi}}'\widehat{\boldsymbol{\Xi}}$.

(iii) Compute the LM-type test statistic

$$\mathrm{LM}_{T,p}^{(1)}(m) = T\,\mathrm{tr}[(\mathbf{SSE}_0 - \mathbf{SSE}_1)\mathbf{SSE}_0^{-1}]. \quad (11.71)$$

Under $\mathbb{H}_0$, it easy to show (Teräsvirta and Yang (2014a) and Exercise 11.3) that

$$\mathrm{LM}_{T,p}^{(1)}(m) \xrightarrow{D} \chi^2_{m(mp+1)}, \text{ as } T \to \infty. \quad (11.72)$$

The degrees of freedom correspond to the number of restrictions m multiplied by the column dimension $mp+1$ of $\mathbf{U}$.

The $\mathrm{LM}_{T,p}^{(1)}(m)$-type test statistic can also be used to help select an appropriate transition variable X_t by computing the statistic for various X_t's and selecting the one for which the p-value of the test statistic is smallest. MC simulation studies (e.g., Teräsvirta and Yang, 2014a) indicate that the power of the above test is good when the transition variable is correctly specified.

In small samples, it is recommended to compute an F-version of the test statistic (11.71) to improve its empirical size. Also, Bartlett and Bartlett-type corrections have been suggested. One is the so-called Laitinen–Meisner correction, which is a simple degrees of freedom rescaling of an LM-type test statistic. Within the setup

Table 11.2: *Values of the multivariate nonlinearity test statistics for the tree ring widths data set; $T = 66$, $p = 4$, and $m = 2$ (p-values are given in parentheses).*

d	$C_{T,p}(d,m)$	$\text{LM}^{(1)}_{T,p}(m)$	$F^{(1)}_{T,p}(m)$	$\lambda_{T,p}(m)$	$F^{(R)}_{T,p}(m)$
1	21.587 (0.251)	39.263 (0.006)	1.384 (0.153)	40.565 (0.004)	2.297 (0.005)
2	13.758 (0.745)	35.613 (0.017)	1.255 (0.232)	33.969 (0.026)	1.852 (0.028)
3	34.222 (0.012)	37.997 (0.009)	1.339 (0.178)	36.078 (0.015)	1.991 (0.016)
4	23.655 (0.167)	34.876 (0.021)	1.229 (0.252)	34.134 (0.025)	1.863 (0.026)
5	19.315 (0.373)	34.622 (0.022)	1.220 (0.258)	34.138 (0.025)	1.863 (0.026)

of Algorithm 11.5, the test statistic is given by

$$F^{(1)}_{T,p}(m) = \frac{mT - (m + (mp+1))}{T \times m(mp+1)} \text{LM}^{(1)}_{T,p}(m), \quad (11.73)$$

where $m(mp+1)$ denotes the number of regression parameters in the auxiliary regression model, and $m+(mp+1)$ represents the total number of restrictions. Under $\mathbb{H}_0$ the rescaled test statistic (11.73) will asymptotically follow an F_{ν_1,ν_2} distribution with $\nu_1 = m(mp+1)$ and $\nu_2 = mT - (m + (mp+1))$ degrees of freedom, as $T \to \infty$. Another test statistic follows by modifying Rao's (R) approximation of the F distribution in step (iv) of Algorithm 11.1 to the present situation with ν_U replaced by $mp + 1$, i.e. the dimension of $\mathbf{Z}_t$.

Moreover, it is easy to extend the test procedure in Algorithm 11.5 to incorporate equation-specific transition functions; see Teräsvirta and Yang (2014a). The limiting null distribution of the resulting LM-type test statistic is, however, unknown and has to be obtained by MC simulation. These authors also modify Algorithm 11.5 by augmenting the first-order test (11.71) with regressors $\mathbf{Z}'_t X_t^2$ and $\mathbf{Z}'_t X_t^3$ to accommodate a third-order, rather than a first-order, Taylor expansion of the logistic transition function around $\gamma = 0$.

Example 11.6: Tree Ring Widths (Cont'd)

We continue our analysis of the annual temperature $(Y_{1,t})$ and tree ring widths $(Y_{2,t})$ data introduced in Example 11.5. AIC indicates that a VAR(4) model best describes the interdependencies between the two time series. So, for all test statistics, we fix p at 4. The second column of Table 11.2 gives the results of the test statistic $C_{T,p}(d,m)$ for delay $d = 1, \ldots, 5$. The recursive estimation starts with $n_{\min} = 25$, which is about $3\sqrt{T}$ with $T = 66$. For $d = 3$ the test statistic suggests threshold nonlinearity, but in all other cases there is no evidence to reject $\mathbb{H}_0$.

Many studies indicate that ring width growth relates to climatic factors at different period during the growing season. In fact, when temperatures exceed a physiological threshold value the long-run effect is that tree ring widths decline. Hence, it is reasonable to take $Y_{1,t-d}$ as a transition variable in the four test procedures considered here. The test results are summarized in Table 11.2, columns 3 – 6. We see that the test statistics attain their largest value for delay $d = 1$. Except for $F_{T,p}^{(1)}(m)$, the p-values of the test statistics are all close to zero. Thus, the $\mathbb{H}_0$ of linearity is rejected against the alternative of LVSTAR nonlinearity.

11.5 Model Selection Tools

For parametric vector nonlinear models, standard information theoretic criteria, such as AIC and BIC can be used for variable selection, including identifying the appropriate lag length. For instance, consider a strictly stationary m-dimensional VTARMA$(k; p_1, \ldots, p_k, q_1, \ldots, q_k)$ process $\{\mathbf{Y}_t, t \in \mathbb{Z}\}$ with a single weakly stationary threshold variable X_{t-d}. Assume that $\{X_t\}$, and the maximum number of regimes k are known. It is obvious that for a fixed delay d, the number of data points in regime i equals $T_i = \sum_{t=h+1}^{T}(I_{t-d}^{(i-1)} - I_{t-d}^{(i)})$, where $h = \max(p_1, \ldots, p_k, d)$, and T denotes the total number of observations. Setting $p = (p_1, \ldots, p_k)$ and $q = (q_1, \ldots, q_k)$, the multivariate versions of AIC and BIC are defined as

$$\text{AIC}(p, q, d, k) = \sum_{i=1}^{k} \left(T_i \log |\widehat{\mathbf{\Sigma}}_\varepsilon^{(i)}| + 2m(mp_i + vq_i + 1)\right), \tag{11.74}$$

$$\text{BIC}(p, q, d, k) = \sum_{i=1}^{k} \left(T_i \log |\widehat{\mathbf{\Sigma}}_\varepsilon^{(i)}| + \log(T_i) m(mp_i + vq_i + 1)\right), \tag{11.75}$$

where $\widehat{\mathbf{\Sigma}}_\varepsilon^{(i)}$ is an estimate of the residual covariance matrix in each regime i ($i = 1, \ldots, k$).

Clearly, the explosion of parameters for VSETARMA models can be problematic in practice. Therefore, one often restricts the number of regimes k to a small number such as 2 or 3 to keep the analysis manageable. In addition, it is useful to divide the available multivariate data set into subsets according to the empirical percentiles of $\{X_t\}_{t=1}^T$, and adopt a vector time-domain nonlinearity test statistic to detect any model change within each subset. This approach may also provide some tentative information on the location of the threshold intervals $\mathbb{R}^{(i)}$ ($i = 1, \ldots, k$). Moreover, in the case of VSETAR model identification, a regression subset selection method based on GAs may be considered as an attractive, and easily implemented, alternative; see Baragona and Cucina (2013).

Evidently, for an m-dimensional VSTAR$(k; p, \ldots, p)$ model with a single transition variable X_{t-d}, we can use both AIC and BIC. Then the regime-specific number of observations is not necessarily an integer, i.e. $T_i = \sum_{t=h+1}^{T}(G_{t-d}^{(i-1)} - G_{t-d}^{(i)})$, where $G_{t-d}^{(i)} \equiv G(X_{t-d}; \gamma^{(i)}, c^{(i)})$ is the transition function corresponding to the ith regime, $G_{t-d}^{(0)} = 1$ and $G_{t-d}^{(k)} = 0$.

11.6 Diagnostic Checking

Preamble

After model selection and model estimation, diagnostic checking is the next important step before we can use the model for forecasting, control, and other purposes. In Section 7.4, we introduced a number of high-dimensional nonparametric test statistics for serial correlation. Valuable as these test statistics can sometimes be, they have implicitly or explicitly relied on the assumption that the error terms are independent, and some results have depended on the further assumption that they are normally distributed. In this section, we discuss two portmanteau-type test statistics proposed by Chabot–Hallé and Duchesne (2008), which allow us to handle the more realistic situation that the error terms follow a stationary martingale difference sequence.

Asymptotics

Let $\{\mathbf{Y}_t, t \in \mathbb{Z}\}$ be a stationary and ergodic m-dimensional stochastic process defined by the nonlinear model

$$\mathbf{Y}_t = \mathbf{g}(\mathcal{F}_{t-1}; \boldsymbol{\theta}_0) + \boldsymbol{\varepsilon}_t, \tag{11.76}$$

where $\mathcal{F}_{t-1}$ represents the information set generated by $\{\mathbf{Y}_s, s < t\}$, $\mathbf{g}(\cdot; \boldsymbol{\theta}_0)$ is a known real-valued measurable function on $\mathbb{R}^m$, and $\boldsymbol{\theta}_0$ denotes the true, but unknown, value of the $K \times 1$ parameter vector $\boldsymbol{\theta}$. The vector function $\mathbf{g}(\cdot; \cdot)$ is supposed to have continuous second-order derivatives with respect to $\boldsymbol{\theta}$ a.s. The process $\{\boldsymbol{\varepsilon}_t\}$ is an m-dimensional vector martingale difference sequence satisfying (11.60).

Let $\{\mathbf{Y}_t\}_{t=1}^T$ be a finite set of realizations of the process $\{\mathbf{Y}_t, t \in \mathbb{Z}\}$. Given a vector of initial values, the CLS estimator $\widehat{\boldsymbol{\theta}}_T$ of $\boldsymbol{\theta}_0$ is obtained by minimizing the sum of squared errors

$$L_T(\boldsymbol{\theta}) = \sum_{t=1}^T \big(\mathbf{Y}_t - \mathbf{g}(\mathcal{F}_{t-1}; \boldsymbol{\theta}_0)\big)' \boldsymbol{\Sigma}_\varepsilon^{-1} \big(\mathbf{Y}_t - \mathbf{g}(\mathcal{F}_{t-1}; \boldsymbol{\theta}_0)\big). \tag{11.77}$$

Under appropriate regularity conditions, it is straightforward to show (Tjøstheim, 1986b) that $\widehat{\boldsymbol{\theta}}_T$ is strongly consistent and asymptotical normally distributed. That is, in the notation of Chapter 6, and as $T \to \infty$,

$$\sqrt{T}(\widehat{\boldsymbol{\theta}}_T - \boldsymbol{\theta}_0) \xrightarrow{D} \mathcal{N}_k\big(\mathbf{0}, \mathcal{H}^{-1}(\boldsymbol{\theta}_0)\mathcal{I}(\boldsymbol{\theta}_0)\mathcal{H}^{-1}(\boldsymbol{\theta}_0)\big), \tag{11.78}$$

where

$$\mathcal{H}(\boldsymbol{\theta}_0) = \mathbb{E}\Big(\frac{\partial \mathbf{g}'_{t-1}}{\partial \boldsymbol{\theta}} \boldsymbol{\Sigma}_\varepsilon^{-1} \frac{\partial \mathbf{g}_{t-1}}{\partial \boldsymbol{\theta}'}\Big),$$

$$\mathcal{I}(\boldsymbol{\theta}_0) = \mathbb{E}\Big(\frac{\partial \mathbf{g}_{t-1}}{\partial \boldsymbol{\theta}} \boldsymbol{\Sigma}_\varepsilon^{-1} (\mathbf{Y}_t - \mathbf{g}_{t-1})(\mathbf{Y}_t - \mathbf{g}_{t-1})' \boldsymbol{\Sigma}_\varepsilon^{-1} \frac{\partial \mathbf{g}_{t-1}}{\partial \boldsymbol{\theta}'}\Big),$$

11.6 DIAGNOSTIC CHECKING

with $\mathbf{g}_{t-1} \equiv \mathbf{g}(\mathcal{F}_{t-1}; \boldsymbol{\theta}_0)$. Let $\boldsymbol{\Gamma}_\varepsilon(\ell) = \text{Cov}(\boldsymbol{\varepsilon}_t, \boldsymbol{\varepsilon}_{t-\ell})$ be the lag ℓ theoretical autocovariance matrix. Its sample analogue is defined as

$$\mathbf{C}_\varepsilon(\ell) = \begin{cases} T^{-1} \sum_{t=\ell+1}^{T} \boldsymbol{\varepsilon}_t \boldsymbol{\varepsilon}'_{t-\ell}, & (\ell = 0, 1, \ldots, T-1), \\ \mathbf{C}'_\varepsilon(-\ell), & (\ell = -1, \ldots, -T+1). \end{cases} \quad (11.79)$$

Let $\mathbf{c}_\varepsilon = \big(\mathbf{c}'_\varepsilon(1), \ldots, \mathbf{c}'_\varepsilon(M)\big)'$, with $\mathbf{c}_\varepsilon(j) = \text{vec}\big(\mathbf{C}_\varepsilon(j)\big)$ $(j = 1, \ldots, M)$ an $m^2 \times 1$ vector of sample autocovariances, and M denotes a fixed positive integer $(M \ll T)$ chosen large enough to cover all lags of interest. Then, under the assumptions for $\{\boldsymbol{\varepsilon}_t\}$, it can be shown (Chabot–Hallé and Duchesne, 2008) that the limiting distribution of $\mathbf{c}_\varepsilon$ is given by

$$\sqrt{T} \mathbf{c}_\varepsilon \xrightarrow{D} \mathcal{N}_{Mm^2}(\mathbf{0}, \boldsymbol{\Delta}_M),$$

where

$$\boldsymbol{\Delta}_M = \big(\boldsymbol{\Delta}_{ij}\big)_{i,j=1,\ldots,M} = \mathbb{E}(\boldsymbol{\varepsilon}_{t-i} \boldsymbol{\varepsilon}'_{t-j} \otimes \boldsymbol{\varepsilon}_t \boldsymbol{\varepsilon}'_t). \quad (11.80)$$

From standard matrix differentiation, and the martingale difference property of $\{\boldsymbol{\varepsilon}_t\}$, it follows that $\partial \mathbf{c}_\varepsilon(\ell)/\partial \boldsymbol{\theta}' \xrightarrow{P} -\mathbf{J}_\ell$, where $\mathbf{J}_\ell = \mathbb{E}(\boldsymbol{\varepsilon}_{t-\ell} \otimes \partial \mathbf{g}_{t-1}/\partial \boldsymbol{\theta}')$ $(\ell = 1, \ldots, M)$ is an $m^2 \times K$ matrix.

Now, consider the case that the parameters of the model are estimated by CLS. Let $\widehat{\mathbf{g}}_{t-1} \equiv \mathbf{g}(\mathcal{F}_{t-1}; \widehat{\boldsymbol{\theta}}_T)$. Denote the $m \times 1$ vector of estimated residuals by $\widehat{\boldsymbol{\varepsilon}}_t = \mathbf{Y}_t - \widehat{\mathbf{g}}_{t-1}$. Then, replacing $\boldsymbol{\varepsilon}_t$ by $\widehat{\boldsymbol{\varepsilon}}_t$ in (11.79), the $m^2 \times 1$ vector of residual autocovariances $\mathbf{c}_{\widehat{\varepsilon}}$ is defined naturally. By expanding $\mathbf{c}_{\widehat{\varepsilon}}$ in a Taylor series expansion, it is easy to see that $\mathbf{c}_{\widehat{\varepsilon}} = \mathbf{c}_\varepsilon - \mathbf{J}(\widehat{\boldsymbol{\theta}}_T - \boldsymbol{\theta}_0) + \mathbf{o}_p(T^{-1/2})$, where $\mathbf{J} = (\mathbf{J}'_1, \ldots, \mathbf{J}'_M)'$ is an $Mm^2 \times K$ matrix. Furthermore, it can be shown (Tjøstheim, 1986b, Thm. 2.2) that the asymptotic distribution of $T^{-1/2} \partial L_T(\boldsymbol{\theta}_0)/\partial \boldsymbol{\theta}$ is normal. Also, using the martingale difference property of $\{\boldsymbol{\varepsilon}_t\}$ it can be proved (Chabot–Hallé and Duchesne, 2008) that $T^{1/2}\big((\widehat{\boldsymbol{\theta}}_T - \boldsymbol{\theta}_0)', \mathbf{c}'_\varepsilon\big)'$ converges in distribution to a Gaussian random vector. Combining these results, it follows that, as $T \to \infty$,

$$\sqrt{T} \mathbf{c}_{\widehat{\varepsilon}} \xrightarrow{D} \mathcal{N}_{Mm^2}(\mathbf{0}, \boldsymbol{\Omega}), \quad (11.81)$$

where

$$\boldsymbol{\Omega} = \boldsymbol{\Delta}_M - \mathbf{J}^* \mathcal{H}^{-1}(\boldsymbol{\theta}_0) \mathbf{J}' - \mathbf{J} \mathcal{H}^{-1}(\boldsymbol{\theta}_0) \mathbf{J}^{*\prime} + \mathbf{J} \mathcal{H}^{-1}(\boldsymbol{\theta}_0) \mathcal{I}(\boldsymbol{\theta}_0) \mathcal{H}^{-1}(\boldsymbol{\theta}_0) \mathbf{J}',$$

and

$$\mathbf{J}^* = \big(\mathbf{J}_1^{*\prime}, \ldots, \mathbf{J}_M^{*\prime}\big)', \text{ with } \mathbf{J}_\ell^* = \mathbb{E}\big(\boldsymbol{\varepsilon}_{t-\ell} \otimes \boldsymbol{\varepsilon}_t \boldsymbol{\varepsilon}'_t \boldsymbol{\Sigma}_\varepsilon^{-1} \partial \mathbf{g}_{t-1}/\partial \boldsymbol{\theta}'\big), \quad (\ell = 1, \ldots, M).$$

If $\{\boldsymbol{\varepsilon}_t\}$ is a strict WN process, it follows that $\mathbf{J}^* = \mathbf{J}$ and $\mathcal{H}(\boldsymbol{\theta}_0) = \mathcal{I}(\boldsymbol{\theta}_0)$. This implies that $\sqrt{T} \mathbf{c}_{\widehat{\varepsilon}}$ converges to a Gaussian random vector with mean $\mathbf{0}$ and covariance matrix $\mathbf{I}_M \otimes \boldsymbol{\Sigma}_\varepsilon \otimes \boldsymbol{\Sigma}_\varepsilon - \mathbf{J} \mathcal{H}^{-1}(\boldsymbol{\theta}_0) \mathbf{J}'$; see Hosking (1980).

Portmanteau-type test statistics

The null hypothesis of model adequacy is

$$\mathbb{H}_0: \boldsymbol{\Gamma}_\varepsilon(\ell) = \mathbf{0}, \quad (\ell = 1, 2, \ldots). \tag{11.82}$$

Let $\widehat{\boldsymbol{\Omega}}$ be a consistent estimator of $\boldsymbol{\Omega}$. Then a multivariate portmanteau-type test statistic may be written as $T\mathbf{c}'_{\widehat{\varepsilon}}\widehat{\boldsymbol{\Omega}}^{-1}\mathbf{c}_{\widehat{\varepsilon}}$, which has a limiting $\chi^2_{Mm^2}$ distribution under $\mathbb{H}_0$. As in the univariate case, the "Ljung–Box variant" of this test statistic is preferable in practice. In that case, replace $\mathbf{c}_{\widehat{\varepsilon}}$ by $\mathbf{c}^*_{\widehat{\varepsilon}} = \big(\sqrt{T/(T-1)}\mathbf{c}'_{\widehat{\varepsilon}}(1), \ldots, \sqrt{T/(T-M)}\mathbf{c}'_{\widehat{\varepsilon}}(M)\big)'$ to obtain a level-adjusted test statistic. In other words, we can calculate

$$\mathcal{Q}(M) = T\mathbf{c}^{*'}_{\widehat{\varepsilon}}\widehat{\boldsymbol{\Omega}}^{-1}\mathbf{c}^*_{\widehat{\varepsilon}}, \quad (M \in \mathbb{Z}^+), \tag{11.83}$$

and its null distribution is also asymptotically $\chi^2_{Mm^2}$, as $T \to \infty$.

Clearly, (11.83) is a *multiple-lag* test statistic. The test does not provide insight in the possible residual dependence at each individual lag ℓ. It that case one may consider the following level-adjusted *single-lag* test statistic

$$\mathcal{Q}(\ell) = \frac{T^2}{T-\ell}\mathbf{c}'_{\widehat{\varepsilon}}(\ell)\widehat{\boldsymbol{\Omega}}^{-1}_\ell\mathbf{c}_{\widehat{\varepsilon}}(\ell), \quad (\ell = 1, \ldots, M), \tag{11.84}$$

where

$$\widehat{\boldsymbol{\Omega}}_\ell = \widehat{\boldsymbol{\Delta}}_{\ell\ell} - \widehat{\mathbf{J}}^*_\ell\widehat{\mathcal{H}}^{-1}_T\widehat{\mathbf{J}}'_\ell - \widehat{\mathbf{J}}_\ell\widehat{\mathcal{H}}^{-1}_T\widehat{\mathbf{J}}^{*'}_\ell + \widehat{\mathbf{J}}_\ell\widehat{\mathcal{H}}^{-1}_T\widehat{\mathcal{I}}_T\widehat{\mathcal{H}}^{-1}_T\widehat{\mathbf{J}}'_\ell$$

be a consistent estimator of the asymptotic covariance matrix, say $\boldsymbol{\Omega}_\ell$, of $\mathbf{c}_{\widehat{\varepsilon}}(\ell)$. Under $\mathbb{H}_0$, it follows that $\mathcal{Q}(\ell) \xrightarrow{D} \chi^2_{m^2}$, as $T \to \infty$. For testing several lags simultaneously, one may use Bonferroni-type adjustments; see, e.g., Section 7.2.4.

11.6.1 Quantile residuals

Recall the definition of univariate quantile residuals in Section 6.3.2. In a vector framework, we denote by $f_{t-1}(\mathbf{Y}_t; \boldsymbol{\theta})$ the conditional density function of an m-dimensional stochastic process $\{\mathbf{Y}_t, t \in \mathbb{Z}\}$ with $\boldsymbol{\theta}$ a vector of unknown parameters. Assume the components of $\mathbf{Y}_t = (Y_{1,t}, \ldots, Y_{m,t})'$ are independent. Then the conditional CDF of $\{\mathbf{Y}_t, t \in \mathbb{Z}\}$ has the product form $F_{t-1}(\mathbf{Y}_t; \boldsymbol{\theta}) = \prod_{j=1}^m F_{j,t-1}(Y_{j,t}; \boldsymbol{\theta})$, where $F_{j,t-1}(Y_{j,t}; \boldsymbol{\theta})$ is the marginal distribution of the jth component. Similarly, by conditioning with respect to any chosen order of the components of $\{\mathbf{Y}_t, t \in \mathbb{Z}\}$, we can write $f_{t-1}(\mathbf{Y}_t; \boldsymbol{\theta})$ in the product form, that is

$$f_{t-1}(\mathbf{Y}_t; \boldsymbol{\theta}) = \prod_{j=1}^m f_{i_j,j-1,t-1}(Y_{i_j,t}; \boldsymbol{\theta}|\mathcal{A}_{j-1}), \tag{11.85}$$

11.6 DIAGNOSTIC CHECKING

Table 11.3: *Three diagnostic test statistics based on multivariate quantile residuals.*

Null hypothesis $\mathbb{H}_0$	Transformation function g	Test statistic
$\mathbb{E}(\mathbf{R}_{t,\theta_0}\mathbf{R}'_{t-\ell,\theta_0}) = \mathbf{0}_{m\times m}, \ \forall t,$ ($\ell = 1,\ldots,K_1; K_1 \ll T$) (Autocorrelation)	$g\colon \mathbb{R}^{m(K_1+1)} \to \mathbb{R}^{m^2 K_1}$ $g(\mathbf{u}_{t,\theta}) =$ $\text{vec}(\mathbf{r}_{t,\theta}\mathbf{r}'_{t+1,\theta},\ldots,\mathbf{r}_{t,\theta}\mathbf{r}'_{t+K_1,\theta})$	$\widetilde{A}_{T,K_1} = \widetilde{S}_{T,d}$ with $d = K_1 + 1$
$\mathbb{E}(R^2_{i,t,\theta_0}, R^2_{j,t-\ell,\theta_0}) = 0, \ \forall t,$ and $\forall i,j \in \{1,\ldots,m\}$ ($\ell = 1,\ldots,K_2; K_2 \ll T$) (Heteroskedasticity)	$g\colon \mathbb{R}^{m(K_2+1)} \to \mathbb{R}^{m^2 K_2}$ $g(\mathbf{u}_{t,\theta}) =$ $\text{vec}(\mathbf{v}_{t,\theta}\mathbf{v}'_{t+1,\theta},\ldots,\mathbf{v}_{t,\theta}\mathbf{v}'_{t+K_2,\theta})$ with $\mathbf{v}_{t-\ell,\theta} = (r^2_{1,t,\theta}-1,\ldots,r^2_{m,t+K_2,\theta}-1)'$	$\widetilde{H}_{T,K_2} = \widetilde{S}_{T,d}$ with $d = K_2 + 1$
$\mathbb{E}(R^2_{j,t,\theta_0}-1, R^3_{j,t,\theta_0}, R^4_{j,t,\theta_0}-3)' = \mathbf{0},$ $\forall t,$ and $\forall j \in \{1,\ldots,m\}$ (Normality)	$g\colon \mathbb{R}^m \to \mathbb{R}^{3m}$ $g(r_{j,t,\theta}) = (r^2_{j,t,\theta}-1, r^3_{t,\theta}, r^4_{t,\theta}-3)'$	$\widetilde{N}_T = \widetilde{S}_{T,d}$ with $d = 1$

where $\mathcal{A}_{j-1} = \sigma(Y_{i_1,t},\ldots,Y_{i_{j-1},t})$ is the σ-algebra generated by the jth component variable. Interpret $f_{i_1,0,t-1}(Y_{i_1,t};\boldsymbol{\theta}) = f_{i_1,t-1}(Y_{i_1,t};\boldsymbol{\theta})$, and $F_{i_1,j-1,t-1}(Y_{i_1,t};\boldsymbol{\theta}) = \int_{-\infty}^{Y_{i_j,t}} f_{i_1,j-1,t-1}(u;\boldsymbol{\theta})du$. Thus, generalizing (6.85), the $m \times 1$ vector of theoretical quantile residuals at time point t is defined by

$$\widetilde{\mathbf{R}}_{t,\boldsymbol{\theta}} = \begin{pmatrix} R_{1,t,\boldsymbol{\theta}} \\ \vdots \\ R_{m,t,\boldsymbol{\theta}} \end{pmatrix} = \begin{pmatrix} \Phi^{-1}(F_{i_1,t-1}(Y_{i_1,t};\boldsymbol{\theta})) \\ \vdots \\ \Phi^{-1}(F_{i_m,t-1}(Y_{i_m,t};\boldsymbol{\theta})) \end{pmatrix}, \quad (11.86)$$

and the corresponding $m \times 1$ vector of sample quantile residuals is $\widetilde{\mathbf{r}}_{t,\widehat{\boldsymbol{\theta}}_T} = (r_{1,t,\widehat{\boldsymbol{\theta}}_T},\ldots,r_{m,t,\widehat{\boldsymbol{\theta}}_T})'$, where $\widehat{\boldsymbol{\theta}}_T$ is the QML of the true parameter vector $\boldsymbol{\theta}_0$.

Following a similar approach as in Section 6.3.2, Kalliovirta and Saikkonen (2010) introduce a general testing framework based on transformations of $\widetilde{\mathbf{R}}_{t,\boldsymbol{\theta}}$ by a continuously differentiable function $g\colon \mathbb{R}^{dm} \to \mathbb{R}^n$ such that $\mathbb{E}(g(\mathbf{U}_{t,\boldsymbol{\theta}_0})) = \mathbf{0}$, where $\mathbf{U}_{t,\boldsymbol{\theta}_0} = (\widetilde{\mathbf{R}}'_{t,\boldsymbol{\theta}_0},\ldots,\widetilde{\mathbf{R}}'_{t-d+1,\boldsymbol{\theta}_0})' \in \mathbb{R}^{dm}$ and with d given in Table 11.3. Conditional on a vector with initial values, and assuming the conditional density function $f_{t-1}(\mathbf{Y}_t;\boldsymbol{\theta})$ exists, the log-likelihood function $\ell_T(\mathbf{y},\boldsymbol{\theta}) = \sum_{t=1}^T \ell_t(\mathbf{Y}_t,\boldsymbol{\theta}) = \sum_{t=1}^T \log f_{t-1}(\mathbf{Y}_t;\boldsymbol{\theta})$ of the set of observations $\{\mathbf{Y}_t\}_{t=1}^T$ follows directly. Then, under some mild conditions, Kalliovirta and Saikkonen (2010) prove a CLT for transformed vector quantile residuals. Next, they define the general test statistic

$$\widetilde{S}_{T,d} = \frac{1}{T-d+1} \sum_{t=1}^{T-d+1} g(\mathbf{u}_{t,\widehat{\boldsymbol{\theta}}_T})' \widehat{\boldsymbol{\Omega}}_T^{-1} \sum_{t=1}^{T-d+1} g(\mathbf{u}_{t,\widehat{\boldsymbol{\theta}}_T}), \quad (11.87)$$

where $\mathbf{u}_{t,\widehat{\boldsymbol{\theta}}_T} = (\widetilde{\mathbf{r}}'_{t,\widehat{\boldsymbol{\theta}}_T},\ldots,\widetilde{\mathbf{r}}'_{t-d+1,\widehat{\boldsymbol{\theta}}_T})'$, and $\widehat{\boldsymbol{\Omega}}_T$ is a consistent estimator of the asymptotic covariance matrix $\boldsymbol{\Omega}$. Specifically,

$$\widehat{\boldsymbol{\Omega}}_T = \widehat{\mathbf{G}}_T \widehat{\mathcal{I}}_T^{-1} \widehat{\mathbf{G}}'_T + \widehat{\boldsymbol{\Psi}}_T \widehat{\mathcal{I}}_T^{-1} \widehat{\mathbf{G}}'_T + \widehat{\mathbf{G}}_T \widehat{\mathcal{I}}_T^{-1} \widehat{\boldsymbol{\Psi}}'_T + \widehat{\mathbf{H}}_T, \quad (11.88)$$

where $\widehat{\mathbf{G}}_T = T^{-1}\sum_{t=1}^{T}\partial g(\mathbf{u}_{t,\widehat{\boldsymbol{\theta}}_T})/\partial\boldsymbol{\theta}'$, $\widehat{\boldsymbol{\Psi}}_T = T^{-1}\sum_{t=1}^{T} g(\mathbf{u}_{t,\widehat{\boldsymbol{\theta}}_T})\partial\ell_t(\mathbf{Y}_t,\widehat{\boldsymbol{\theta}}_T)/\partial\boldsymbol{\theta}'$, $\widehat{\mathbf{H}}_T = T^{-1}\sum_{t=1}^{T} g(\mathbf{u}_{t,\widehat{\boldsymbol{\theta}}_T})g(\mathbf{u}_{t,\widehat{\boldsymbol{\theta}}_T})'$, and $\widehat{\mathcal{I}}_T$ is a consistent estimator of $\mathcal{I}(\boldsymbol{\theta}_0)$, the expected information matrix evaluated at $\boldsymbol{\theta}_0$. In practice, one can compute these matrices by simulation. Moreover, given the above null hypotheses, explicit expressions for $\mathbf{H} = \mathbb{E}\big(g(\mathbf{U}_{t,\boldsymbol{\theta}})g(\mathbf{U}_{t,\boldsymbol{\theta}})'\big)$ follow in a straightforward way.

Assume that the vector nonlinear model under study is correctly specified. Then (11.87) has an asymptotic χ_n^2 distribution; Kalliovirta and Saikkonen (2010). This result does not depend on the chosen order of conditioning of $\widetilde{\mathbf{R}}_{t,\boldsymbol{\theta}}$. Table 11.3 shows three diagnostic test statistics, as special cases of (11.87). Under $\mathbb{H}_0$, these test statistics are asymptotically distributed as respectively $\chi_{m^2 K_1}^2$, $\chi_{m^2 K_2}^2$, and χ_{3m}^2.

11.7 Forecasting

11.7.1 Point forecasts

Calculating a point forecast (conditional mean) from multivariate nonlinear time series with correlated errors is a far more substantial task than in the univariate case (Chapter 10). Generally, explicit forecast expressions for the forecast density do not exist for any horizon H, even for one-step ahead forecasts. To see this, consider the general multivariate nonlinear model in (11.1) with $q = 0$, i.e. a vector NLAR(p) model. Then, the one-step ($H = 1$) ahead LS forecast of the m-dimensional time series process $\{\mathbf{Y}_t, t \in \mathbb{Z}\}$ at time t is given by

$$\mathbf{Y}_{t+1|t}^{\mathrm{LS}} = \mathbb{E}(\mathbf{Y}_{t+1}|\mathcal{F}_t) = \mathbb{E}\{g(\mathbf{Y}_t;\boldsymbol{\theta}) + \boldsymbol{\varepsilon}_{t+1}|\mathcal{F}_t\} \neq g(\mathbf{Y}_t;\boldsymbol{\theta}), \qquad (11.89)$$

since $\mathbb{E}(\boldsymbol{\varepsilon}_{t+1}|\mathcal{F}_t) \neq \mathbf{0}$. When $H = 2$, the two-step ahead LS forecast is given by

$$\mathbf{Y}_{t+2|t}^{\mathrm{LS}} = \mathbb{E}(\mathbf{Y}_{t+2}|\mathcal{F}_t) = \mathbb{E}\{g(\mathbf{Y}_{t+1};\boldsymbol{\theta}) + \boldsymbol{\varepsilon}_{t+2}|\mathcal{F}_t\}$$
$$= \int_{-\infty}^{\infty}\cdots\int_{-\infty}^{\infty} g\big(g^*(\mathbf{Y}_t;\boldsymbol{\theta}) + \boldsymbol{\eta}_{t+1}\big) + \boldsymbol{\varepsilon}_{t+2}|\mathcal{F}_t\big)\mathrm{d}F(\boldsymbol{\eta},\boldsymbol{\varepsilon}), \qquad (11.90)$$

where $\boldsymbol{\eta}_t$ and $g^*(\cdot)$ are defined in a similar way as $\boldsymbol{\varepsilon}_t$ and $g(\cdot)$ respectively, and $F(\cdot)$ is the joint distribution function of the dependent processes $\{\boldsymbol{\eta}_t\}$ and $\{\boldsymbol{\varepsilon}_t\}$. Thus, just as in the univariate case, one can only obtain forecasts by numerical methods.

Two common approaches to computing multi-step ahead forecasts is to use MC simulation and BS. Often, however, a BS procedure is preferred in practice since no assumptions need to be made about the distribution of $\{\boldsymbol{\varepsilon}_t\}$. One option is to use some form of block bootstrapping by resampling from non-overlapping blocks of consecutive centered residuals, say $\{\widehat{\boldsymbol{\varepsilon}}_t\}$. Another option is to use a model-based bootstrap. By this it is meant that a finite-order VAR model is first fitted to $\{\widehat{\boldsymbol{\varepsilon}}_t\}$, assuming that the vector error process is i.i.d. and its components are mutually uncorrelated. Then, assuming that the VAR residuals are i.i.d., and using the recursive structure of the VAR model, it is straightforward to obtain the H-step ahead forecast $\mathbb{E}(\mathbf{Y}_{t+H}|\mathcal{F}_t)$ via block bootstrapping.

11.7 FORECASTING

Example 11.7: Forecasting an LVSTAR(1) Model with CNFs

Consider a two-dimensional LVSTAR(1) strictly stationary process $\{\mathbf{Y}_t, t \in \mathbb{Z}\}$ with CNFs. This implies that there exists a non-zero 2×1 vector $\boldsymbol{\alpha}$ such that the LSTAR(1) nonlinearity vanishes in the linear combination $\boldsymbol{\alpha}'\mathbf{Y}_t$. More formally, using the notation of (11.34), we have

$$\mathbf{Y}_t = \boldsymbol{\Phi}_0 + \boldsymbol{\Phi}_1 \mathbf{Y}_{t-1} + (\widetilde{\boldsymbol{\Phi}}_0 + \widetilde{\boldsymbol{\Phi}}_1 \mathbf{Y}_{t-1}) G(\mathbf{X}_t; \gamma, \mathbf{c}) + \boldsymbol{\varepsilon}_t, \qquad (11.91)$$

where $\boldsymbol{\alpha}_\perp \phi_0^* = \widetilde{\boldsymbol{\Phi}}_0$, $\boldsymbol{\alpha}_\perp \boldsymbol{\beta}' = \widetilde{\boldsymbol{\Phi}}_1$ with ϕ_0^* a scalar parameter, $\boldsymbol{\beta}$ is a 2×1 parameter vector, $\boldsymbol{\alpha}'\boldsymbol{\alpha}_\perp = 0$, $G(\cdot)$ is a logistic transition function given by (11.37), and $\{\boldsymbol{\varepsilon}_t\} \stackrel{\text{i.i.d.}}{\sim} (\mathbf{0}, \boldsymbol{\Sigma}_\varepsilon)$ independent of $\mathbf{Y}_t$, and $\mathbf{X}_t \equiv \mathbf{Y}_{t-d}$ $(d > 0)$.

The one-step ahead LS forecast for model (11.91) is given by

$$\mathbf{Y}^{\text{LS}}_{t+1|t} = \mathbb{E}(\mathbf{Y}_{t+1}|\mathcal{F}_t) = \boldsymbol{\Phi}_0 + \boldsymbol{\Phi}_1 \mathbf{Y}_t + \boldsymbol{\alpha}_\perp(\phi_0^* + \boldsymbol{\beta}'\mathbf{Y}_t) G(\mathbf{Y}_{t+1-d}; \gamma, \mathbf{c}). \qquad (11.92)$$

Using (11.92), the two-step ahead LS forecast is given by

$$\begin{aligned}\mathbf{Y}^{\text{LS}}_{t+2|t} = \mathbb{E}(\mathbf{Y}_{t+2}|\mathcal{F}_t) &= \boldsymbol{\Phi}_0 + \boldsymbol{\Phi}_1 \boldsymbol{\Phi}_0 + \boldsymbol{\Phi}_1 \boldsymbol{\alpha}_\perp(\phi_0^* + \boldsymbol{\beta}'\mathbf{Y}_t) G(\mathbf{Y}_{t+1-d}; \gamma, \mathbf{c}) \\ &\quad + \phi_0^* \boldsymbol{\alpha}_\perp \mathbb{E}[G(\mathbf{Y}_{t+2-d}; \gamma, \mathbf{c})] + \boldsymbol{\alpha}_\perp \boldsymbol{\beta}' \boldsymbol{\Phi}_0 \mathbb{E}[G(\mathbf{Y}_{t+2-d}; \gamma, \mathbf{c})] \\ &\quad + \boldsymbol{\alpha}_\perp \boldsymbol{\beta}' \boldsymbol{\Phi}_1 \mathbf{Y}_t \mathbb{E}[G(\mathbf{Y}_{t+2-d}; \gamma, \mathbf{c})] + \boldsymbol{\alpha}_\perp \boldsymbol{\beta}' \mathbb{E}[\boldsymbol{\varepsilon}_{t+1} G(\mathbf{Y}_{t+2-d}; \gamma, \mathbf{c})] \\ &\quad + \boldsymbol{\alpha}_\perp \boldsymbol{\beta}' \boldsymbol{\alpha}_\perp (\phi_0^* + \boldsymbol{\beta}'\mathbf{Y}_t) G(\mathbf{Y}_{t+1-d}; \gamma, \mathbf{c}) \mathbb{E}[G(\mathbf{Y}_{t+2-d}; \gamma, \mathbf{c})]. \end{aligned} \qquad (11.93)$$

For further evaluation of (11.93) we need to distinguish between the two cases $d < 2$ and $d \geq 2$. When $d = 1$, explicit expressions for $\mathbb{E}[G(\mathbf{Y}_{t+2-d}; \gamma, \mathbf{c})]$ and $\mathbb{E}[\boldsymbol{\varepsilon}_{t+1} G(\mathbf{Y}_{t+2-d}; \gamma, \mathbf{c})]$ are not directly available; then we need to replace them by estimates obtained via MC simulation or BS. However, when $d \geq 2$, we see that $G(\mathbf{Y}_{t+2-d}; \gamma, \mathbf{c})$ is available at time t. In this case (11.93) reduces to

$$\begin{aligned}\mathbf{Y}^{\text{LS}}_{t+2|t} = \mathbb{E}(\mathbf{Y}_{t+2}|\mathcal{F}_t) &= \boldsymbol{\Phi}_0 + \boldsymbol{\Phi}_1 \boldsymbol{\Phi}_0 + \boldsymbol{\Phi}_1 \boldsymbol{\alpha}_\perp(\phi_0^* + \boldsymbol{\beta}'\mathbf{Y}_t) G(\mathbf{Y}_{t+1-d}; \gamma, \mathbf{c}) \\ &\quad + \big(\phi_0^* \boldsymbol{\alpha}_\perp + \boldsymbol{\alpha}_\perp \boldsymbol{\beta}' \boldsymbol{\Phi}_0 + \boldsymbol{\alpha}_\perp \boldsymbol{\beta}' \boldsymbol{\Phi}_1 \mathbf{Y}_t\big) G(\mathbf{Y}_{t+2-d}; \gamma, \mathbf{c}) \\ &\quad + \boldsymbol{\alpha}_\perp \boldsymbol{\beta}' \boldsymbol{\alpha}_\perp (\phi_0^* + \boldsymbol{\beta}'\mathbf{Y}_t) G(\mathbf{X}_{t+1}; \gamma, \mathbf{c})[G(\mathbf{Y}_{t+2-d}; \gamma, \mathbf{c})]. \end{aligned} \qquad (11.94)$$

In general, when $H \leq d$, exact analytic expressions for $\mathbf{Y}^{\text{LS}}_{t+H|t}$ can be obtained. However, when $H > d$, one has to resort to MC or BS methods. For instance, in the case of block bootstrapping with a block size of one, $\mathbb{E}[\boldsymbol{\varepsilon}_{t+1} G(\mathbf{Y}_{t+2-d}; \gamma, \mathbf{c})]$ can be estimated by

$$\Big(\frac{1}{B}\sum_{b=1}^B \widehat{\varepsilon}^{(b)}_{1,t+1} G(\mathbf{Y}^{(b)}_{t+2-d}; \gamma, \mathbf{c}), \frac{1}{B}\sum_{b=1}^B \widehat{\varepsilon}^{(b)}_{2,t+1} G(\mathbf{Y}^{(b)}_{t+2-d}; \gamma, \mathbf{c})\Big)',$$

and $\mathbb{E}[G(\mathbf{Y}_{t+2-d}; \gamma, \mathbf{c})]$ by $B^{-1}\sum_{b=1}^B G(\mathbf{Y}^{(b)}_{t+2-d}, \gamma, \mathbf{c})$ with B the number of BS replicates. The steps to obtain the 2×1 vector $\widehat{\boldsymbol{\varepsilon}}^{(b)}_{t+1} = (\widehat{\varepsilon}^{(b)}_{1,t+1}, \widehat{\varepsilon}^{(b)}_{2,t+1})'$ are as follows.

(i) Compute the bias-corrected residuals $\widetilde{\boldsymbol{\varepsilon}}_t = \widehat{\boldsymbol{\varepsilon}}_t - \overline{\widehat{\boldsymbol{\varepsilon}}}_t$, where $\overline{\widehat{\boldsymbol{\varepsilon}}}_t$ is the sample mean of the "raw" residuals $\{\widehat{\boldsymbol{\varepsilon}}_t\}$.

(ii) Obtain the bootstrap residuals $\widetilde{\boldsymbol{\varepsilon}}_t^{(b)}$ as random draws with replacement from $\widetilde{\boldsymbol{\varepsilon}}_t$, taking account of serial correlation in $\{\boldsymbol{\varepsilon}_t\}$ via the Cholesky form of the sample estimate of $\boldsymbol{\Sigma}_\varepsilon$. Next, compute $\widehat{\boldsymbol{\varepsilon}}_{t+1}^{(b)}$ as $\widetilde{\boldsymbol{\varepsilon}}_{t+1}^{(b)} + \overline{\widehat{\boldsymbol{\varepsilon}}}_t$.

The value of $\mathbf{Y}_{t+1}^{(b)} = (Y_{1,t+1}^{(b)}, Y_{2,t+1}^{(b)})'$ follows from

$$\mathbf{Y}_{t+1}^{(b)} = \boldsymbol{\Phi}_0 + \boldsymbol{\Phi}_1 \mathbf{Y}_t + \boldsymbol{\alpha}_\perp (\phi_0^* + \boldsymbol{\beta}' \mathbf{Y}_t) G(\mathbf{Y}_{t+1-d}^{(b)}; \gamma, \mathbf{c}) + \widehat{\boldsymbol{\varepsilon}}_{t+1}^{(b)}.$$

Alternatively, one can use a fixed block size which depends on the forecast horizon H, or a random block size when the errors are serial correlated.

11.7.2 Forecast evaluation

RMSFE

Various measures to compare the forecasting accuracy of two or more alternative (nonlinear) multivariate models follow from direct generalizations of well known univariate measures. One ubiquitous measure is the multivariate version of the RMSFE which we define as follows. Let $\mathbf{e}_{t+h} = \mathbf{Y}_{t+h} - \mathbb{E}(\mathbf{Y}_{t+h}|\mathcal{F}_t)$ denote the forecast error from a certain model for forecast period h ($h = 1, \ldots, H$) associated with an m-dimensional time series process $\{\mathbf{Y}_t, t \in \mathbb{Z}\}$. Then, corresponding to the RMSFE in the univariate case, the RMSFE for the multivariate system is defined as the square root of the trace of the covariance matrix of out-of-sample forecast errors, i.e., by $\{\text{trace } \mathbb{E}(\mathbf{e}_{t+h}\mathbf{e}_{t+h}')\}^{1/2}$. Below, we make this concept operational within a rolling forecasting framework.

Let T be the total number of observations. Also, let n be the last in-sample observation, i.e. n is the first forecast origin. Then, for this particular origin, $T - n$ observations are retained as a hold-out or subsample for evaluating the forecast performance of a particular model. As explained in Chapter 10, by rolling it is meant t extends as far as $T - H$, where $H \leq T - 1$ is the maximum forecast horizon under consideration. At each time point t, the parameters of the forecast model are re-estimated as new observations become available in the subsample. Using this approach, evaluation is based on the dynamic out-of-sample forecasts. That is, the rolling method gives rise to $T - n$ one-step ahead forecasts and associated forecast errors, $T - n - 1$ two-step ahead forecasts and associated forecasts errors, ..., $T - H - n + 1$ H-step ahead forecast and associated forecast errors. Below we set $R \equiv T - H - n + 1$ for each forecast period h. So, the rolling forecasting method has a fixed-length R. The corresponding vector of forecast errors are $\{\mathbf{e}_{n+j+h|n+j}\}_{j=0}^{R-1}$. Then the RMSFE measure can be estimated by

$$\text{RMSFE}_R(h) = \left\{ \text{trace}\left[\frac{1}{R}\sum_{j=0}^{R-1} \mathbf{e}_{n+j+h|n+j}\mathbf{e}_{n+j+h|n+j}'\right] \right\}^{1/2}, \quad (h = 1, \ldots, H). \quad (11.95)$$

Generalized MSFE

One problem with using (11.95) is that $\mathbb{E}(\mathbf{e}_{t+h}\mathbf{e}'_{t+h})$ is not invariant to non-singular, scale preserving transformations. Hence, different models may yield the most accurate forecasts for different transformations. To avoid this problem, Clements and Hendry (1993) propose the so-called *generalized forecast error second moment* (GFESM). Let $\tilde{\mathbf{e}}_{n+h} = (\mathbf{e}'_{n+h|n}, \mathbf{e}'_{n+h+1|n+1}, \ldots, \mathbf{e}'_{n+h+(T-H-n)|n+(T-H-n)})'$ be the vector of h-step ahead forecast errors. Then the GFESM is defined as the determinant of the matrix $\mathbb{E}(\mathbf{E}_h \mathbf{E}'_h)$ where $\mathbf{E}_h = (\tilde{\mathbf{e}}'_{n+1}, \tilde{\mathbf{e}}'_{n+2}, \ldots, \tilde{\mathbf{e}}'_{n+h})'$. An estimate of this criterion is given by

$$\text{GFESM}_R(h) = \frac{1}{hR}|\widehat{\mathbf{E}}_h \widehat{\mathbf{E}}'_h|, \quad (h = 1, \ldots, H), \tag{11.96}$$

where $\widehat{\mathbf{E}}_h$ is defined in a similar way as $\mathbf{E}_h$ with $\tilde{\mathbf{e}}_{n+h}$ replaced by $\widehat{\mathbf{e}}_{t+h} = (\widehat{\mathbf{e}}'_{n+h|n}, \widehat{\mathbf{e}}'_{n+h+1|n+1}, \ldots, \widehat{\mathbf{e}}'_{n+h+(T-H-n)|n+(T-H-n)})'$ and where $\widehat{\mathbf{e}}_{(n+j)+h|n+j}$ is an estimate of $\mathbf{e}_{(n+j)+h|n+j}$ ($j = 0, \ldots, R-1; h = 1, \ldots, H$). One important difference between (11.95) and (11.96) is that the GFESM$_R(h)$ statistic reflects the interrelationships between the different forecast values whereas MSFE$_R(h)$ does not.

Forecast densities

Multivariate forecast densities can be evaluated in the same fashion as discussed in Section 10.4.3. For instance, suppose we have a series of $T - n$ one-step ahead forecasts of a bivariate time series $\mathbf{Y}_t = (Y_{1,t}, Y_{2,t})'$ obtained via the rolling forecasting scheme as we just described. Let $\widehat{f}_t(Y_{1,t}, Y_{2,t}|\mathcal{F}_{t-1})$ ($t = 1, \ldots, T - n$) denote the joint forecast density with $\widehat{f}_t(Y_{1,1}, Y_{2,1}|\mathcal{F}_0) \equiv f(y_1, y_2)$. Further, suppose this density function can be factorized into the product of the conditional (c) density and the marginal (m) density as, e.g., $\widehat{f}_t(Y_{1,t}, Y_{2,t}|\mathcal{F}_{t-1}) = \widehat{f}^{(c)}_t(Y_{1,t}|Y_{2,t}, \mathcal{F}_{t-1}) \times \widehat{f}^{(m)}_t(Y_{2,t}|\mathcal{F}_{t-1})$. We can transform each element $(Y_{1,t}, Y_{2,t})'$ by its corresponding PIT to give

$$U^{(c)}_{1|2,t} = \int_{-\infty}^{Y^{(c)}_{1|2,t+1}} \widehat{f}^{(c)}_t(u|Y_{2,t}, \mathcal{F}_{t-1})du, \quad U^{(m)}_{2,t} = \int_{-\infty}^{Y^{(m)}_{2,t+1}} \widehat{f}^{(m)}_t(u|\mathcal{F}_{t-1})du,$$
$$(t = 1, \ldots, T - n), \tag{11.97}$$

where $Y^{(c)}_{1|2,t+1}$ and $Y^{(m)}_{2,t+1}$ are respectively the conditional and marginal one-step ahead forecasts. The null hypothesis of interest is that the model forecasting density corresponds to the true conditional density. That is,

$$\mathbb{H}_0: \quad f_t(Y_{1,t}, Y_{2,t}|\mathcal{F}_{t-1}) = \widehat{f}_t(Y_{1,t}, Y_{2,t}|\mathcal{F}_{t-1}),$$

where $f_t(Y_{1,t}, Y_{2,t}|\mathcal{F}_{t-1})$ is the true joint forecast density. Then the two sequences $\{U^{(c)}_{1|2,t}\}_{t=1}^{T-n}$ and $\{U^{(m)}_{2,t}\}_{t=1}^{T-n}$ will each be i.i.d. $U(0,1)$; Rosenblatt (1952). Moreover, the two sequences of PITs will themselves be independent.

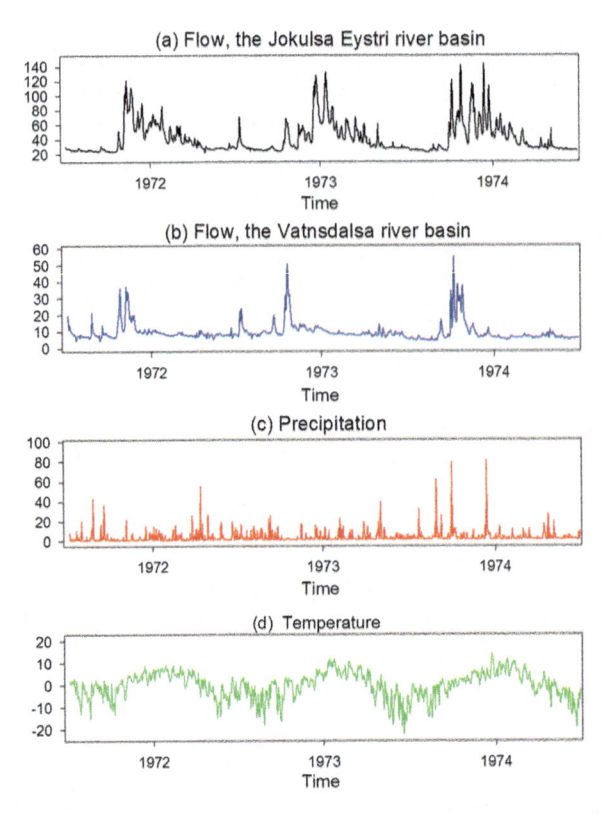

Figure 11.6: *Time plots of flow* (m³/s) *of (a)* Jökulsá Eystri *river and (b)* Vatnsdalsá *river, Iceland, (c)* precipitation *(mm), and (d)* temperature *(°C). Daily data covering the time period* January 1972 – December 1974; $T = 1{,}095$.

Various approaches can be used to assess whether a particular sequence of PITs is i.i.d. $U(0,1)$. Within this context, Clements and Smith (2002) show that the KS test statistic of uniformity has the highest empirical power for both the product (p) and ratio (r) of PITs, with typical elements $\{U_t^{(p)} = U_{1|2,t}^{(c)} \times U_{2,t}^{(m)}\}$ and $\{U_t^{(r)} = U_{1|2,t}^{(c)}/U_{2,t}^{(m)}\}$ respectively. Nevertheless, these results depend on the sign of the correlation coefficient ρ between $Y_{1,t}$ and $Y_{2,t}$. The power of the KS($U^{(p)}$) test statistic is markedly better for $\rho < 0$, and the test statistic using $U_t^{(r)}$ has power only when $\rho > 0$. The asymmetry in power comes from the functional form of the two test statistics. As an alternative to $\{U_t^{(p)}\}$, Ko and Park (2013) propose a location-adjusted transformation of $\{U_{1|2,t}^{(c)}\}$ and $\{U_{2,t}^{(m)}\}$. Under the null hypothesis, these two sequences are each i.i.d. $U(0,1)$. Thus, the sequence of modified PITs is given by $\{\widetilde{U}_t^{(p)} = (U_{1|2,t}^{(c)} - 1/2) \times (U_{2,t}^{(m)} - 1/2)\}$. Simulation results indicate that the resulting KS($\widetilde{U}^{(p)}$) test statistic delivers much more powerful test results than the KS($U^{(p)}$) test statistic, irrespective of the value of ρ.

So far in this subsection, we have focussed on one-step ahead forecasts. However, when interest is in $H > 1$-step ahead forecasts, the following simple provision should be applied for the usual $(H-1)$ dependence of the forecasts. That is, divide the forecasts into sets of independent series, taking the first, the $H+1$, the $2H+1$ etc. for set 1, and the second, the $H+2$, the $2H+2$ etc. for the second set, and so on. Thus, each of the sub-series of PITs $\{U_1, U_{1+H}, U_{1+2H}, \ldots\}$, $\{U_2, U_{2+H}, U_{2+2H}, \ldots\}$, and $\{U_H, U_{2H}, U_{3H}, \ldots\}$ should be i.i.d. $U(0, 1)$ under $\mathbb{H}_0$.

11.8 Application: Analysis of Icelandic River Flow Data

In this section, we reconsider the Jökulsá Eystri daily river flow data $(Q_{1,t})$, earlier introduced in Exercise 2.11, and measured in m^3/s for the years 1972 – 1974. The exogenous variables are precipitation (P_t), measured in mm, and temperature (T_t) in °C. As a second variable of interest, we use daily streamflow data for the Vatnsdalsá river $(Q_{2,t})$, also located in north-west Iceland. Jökulsá Eystri is the bigger river of the two, with a large drainage basin (1,200 km^2) that includes a glacier (155 km^2); as a result, the effect of temperature goes beyond producing spring snowmelt. Vatnsdalsá has a much smaller drainage area (450 km^2), and some of the flow is due to groundwater. Full description of this streamflow system is available in Tong et al. (1985) and the references cited there.

Figure 11.6 shows time plots of the four variables. We see sharp rises and slow declines with a more pronounced spring peak in the Vatnsdalsá flow than in the Jökulsá Eystri flow data due to the presence of the glacier in its drainage area. Since the recorded values of P_t represent the accumulated rain or snow at 9 a.m. from the time of the day before, we adjust the series P_t by a forward translation of one day. In total there are 1,095 observations for analysis.

VTARX model
Following Tsay (1998), we use T_t as a threshold variable for both flows. Furthermore, we focus on a two-regime model. Initially, the maximum AR-order of $Q_{i,t}$ ($i=1,2$) and the maximum order of the exogenous variables P_t and T_t were set at 15 and 3, respectively. After some fine tuning, using the multivariate F test statistic of Section 11.4 and AIC, Table 11.4 reports the final equations for the bivariate two-regime VTARX model with AIC $= 16,981.7$ and BIC $= 17,355.0$.[5] The corresponding threshold parameter estimate is given by $\hat{r} = -0.409$°C. The number of data points in each regime are 479 and 601, respectively.

Some observations are in order. First, the estimate of the threshold parameter for T_t is slightly below freezing, which effectively separates the histories of $Q_{1,t}$ and $Q_{2,t}$ into two regimes. However, only for $T_t > -0.409$°C (regime 2) the series $Q_{1,t}$ strongly depends on current and one day ago temperature. This phenomenon may be explained by the presence of the glacier in the basin. There is no effect of temperature on $Q_{i,t}$ ($i=1,2$) in the other three regimes. Second, lagged precipitation has effect on current flow for both series. The lags and amount, however, depend on T_t with

---
[5]The parameter estimates are not completely identical to those reported by Tsay (1998). This may be due to small differences in computer code.

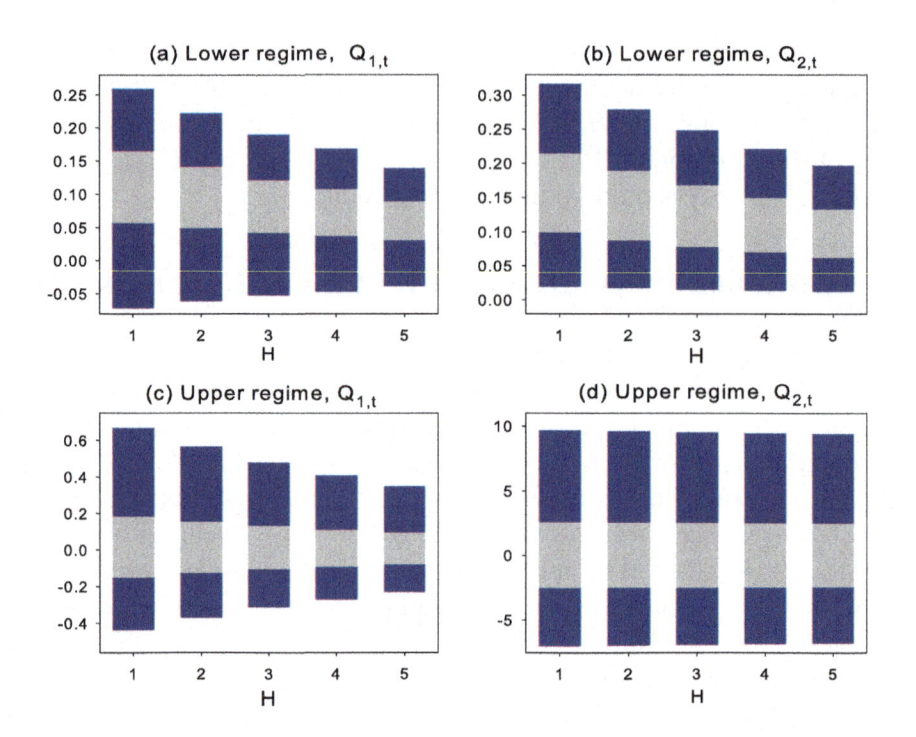

Figure 11.7: HDR's *based on* 50% *(grey) and* 90% *(blue) coverage probabilities for the* GIRF *of the* VTARX *model for a one-unit, system-wide shock; (a)* Jökulsá Eystri *river,* $T_t \leq -0.409°$C, *(b)* Jökulsá Eystri *river,* $T_t > -0.409°$C, *(c)* Vatnsdalsá *river,* $T_t \leq -0.409°$C, *and (d)* Vatnsdalsá *river,* $T_t > -0.409°$C.

a pronounced effect of P_t on $Q_{1,t}$ (as indicated by larger Student t values, not shown here) in the second regime. Third, the fitted model suggests a causal, but asymmetric, relationship between $Q_{1,t}$ and $Q_{2,t}$ in both regimes. According to Tsay (1998), this may be an indication of missing useful variables such as evaporation and ground moisture content.

Table 11.5 shows the sample residual cross-correlation matrices summarized by the symbols $+$, $-$, and $\bullet$ in the (i,j)th position, where $+$ denotes a value greater than 2 estimated standard errors, $-$ denotes a value less than -2 estimated standard errors, and $\bullet$ denotes a value within 2 estimated standard errors. The pattern indicates that the fitted model is adequate with no strong serial correlation in the residuals. We also see some significant CCF values at clusters of lags (3, 4, 5), (8, 10, 11), and (19, 20, 21). This suggests some minor periodic behavior in the series, likely to be caused by seasonality. Thus, it seems reasonable to complement the fitted VTARX model by a seasonal component.

Impulse response analysis

In order to illustrate the dynamic behavior of the fitted VTARX model, we estimate the GIRF defined in Appendix 2.A for single equation nonlinear time series models. For an m-dimensional strictly stationary vector nonlinear time series process $\{\mathbf{Y}_t, t \in$

11.8 APPLICATION: ANALYSIS OF ICELANDIC RIVER FLOW DATA

Table 11.4: CLS *estimates of a bivariate VTARX model for the* Iceland river flow *data set;* $T = 1{,}095$. *Blue-typed numbers denote significant parameter values at the 5% nominal significance level.*

	Lower regime $Q_{1,t}$	Lower regime $Q_{2,t}$	Upper regime $Q_{1,t}$	Upper regime $Q_{2,t}$
$\phi_0^{(i)}$	7.75	1.42	0.69	1.31
$Q_{1,t-1}$	0.52	-0.06	1.12	0.02
$Q_{1,t-2}$	-0.02	0.03	-0.42	-0.04
$Q_{1,t-3}$	0.06	-0.01	0.29	
$Q_{1,t-4}$	0.05	0.01	-0.27	
$Q_{1,t-5}$	-0.07	-0.02	0.17	
$Q_{1,t-6}$	0.12	0.03	-0.12	
$Q_{1,t-7}$	-0.05	-0.01	0.05	
$Q_{1,t-8}$	0.00	-0.01	0.04	
$Q_{1,t-9}$	0.01	0.02	-0.02	
$Q_{1,t-10}$	-0.03			
$Q_{1,t-11}$	0.05			
$Q_{1,t-12}$	0.01			
$Q_{1,t-13}$	0.04			
$Q_{1,t-14}$	-0.07			
$Q_{1,t-15}$	0.05			
$Q_{2,t-1}$	0.11	0.80	0.84	1.25
$Q_{2,t-2}$		-0.18	-1.05	-0.67
$Q_{2,t-3}$		0.09	0.19	0.24
$Q_{2,t-4}$		0.03	0.54	0.16
$Q_{2,t-5}$		-0.02	-0.21	-0.01
$Q_{2,t-6}$		0.02	0.14	-0.03
$Q_{2,t-7}$		-0.00	0.01	0.16
$Q_{2,t-8}$		0.02	-0.55	-0.30
$Q_{2,t-9}$		-0.02	0.47	0.17
$Q_{2,t-10}$		-0.04		
$Q_{2,t-11}$		-0.05		
$Q_{2,t-12}$		0.01		
$Q_{2,t-13}$		-0.08		
$Q_{2,t-14}$		0.09		
P_{t-1}	0.07	0.01	0.44	0.09
P_{t-2}	-0.03	-0.00	-0.25	-0.06
P_{t-3}	0.04	-0.01		0.05
T_t	0.03	0.00	1.33	
T_{t-1}	-0.02	-0.02	-0.54	

$$\widehat{\boldsymbol{\Sigma}}_\varepsilon^{(1)} = \begin{pmatrix} 1.72 & 0.13 \\ 0.13 & 0.46 \end{pmatrix} \qquad \widehat{\boldsymbol{\Sigma}}_\varepsilon^{(2)} = \begin{pmatrix} 48.71 & 2.44 \\ 2.44 & 5.96 \end{pmatrix}$$

Table 11.5: Icelandic river flow *data set. Indicator pattern of the statistically significant values of the residual sample cross-correlation matrices for the* $\{Q_{1,t}\}$ *and* $\{Q_{2,t}\}$ *time series.*

					Lag				
1	2	3	4	5	6	7	8	9	10
$\begin{pmatrix} \cdot & \cdot \\ + & + \end{pmatrix}$	$\begin{pmatrix} \cdot & \cdot \\ \cdot & \cdot \end{pmatrix}$	$\begin{pmatrix} + & \cdot \\ \cdot & \cdot \end{pmatrix}$	$\begin{pmatrix} + & \cdot \\ \cdot & \cdot \end{pmatrix}$	$\begin{pmatrix} + & \cdot \\ \cdot & \cdot \end{pmatrix}$	$\begin{pmatrix} \cdot & \cdot \\ \cdot & \cdot \end{pmatrix}$	$\begin{pmatrix} \cdot & \cdot \\ \cdot & \cdot \end{pmatrix}$	$\begin{pmatrix} \cdot & \cdot \\ + & \cdot \end{pmatrix}$	$\begin{pmatrix} \cdot & \cdot \\ \cdot & \cdot \end{pmatrix}$	$\begin{pmatrix} + & \cdot \\ \cdot & \cdot \end{pmatrix}$

$\mathbb{Z}$} the GIRF is defined as follows:

$$\text{GIRF}_Y(H, \boldsymbol{\varepsilon}_t^{(\delta)}, \boldsymbol{\Omega}_{t-1}) = \mathbb{E}[\mathbf{Y}_{t+H}|\boldsymbol{\varepsilon}_{\delta,t}, \boldsymbol{\Omega}_{t-1}] - \mathbb{E}[\mathbf{Y}_{t+H}|\boldsymbol{\Omega}_{t-1}], \quad (H \geq 1), \quad (11.98)$$

where $\boldsymbol{\varepsilon}_t^{(\delta)} = (\varepsilon_{1,t}^{(\delta)}, \ldots, \varepsilon_{m,t}^{(\delta)})'$ is an m-dimensional vector of shocks at time t, and $\boldsymbol{\Omega}_{t-1} = \{\boldsymbol{\omega}_{t-j}; j \geq 1\}$ is a set (or an appropriate subset) of possible histories. The conditioning variables $\boldsymbol{\varepsilon}_t^{(\delta)}$ and $\boldsymbol{\Omega}_{t-1}$ are assumed to be random, and hence $\text{GIRF}_Y(\cdot)$ is a random variable itself. As noted in Chapter 2, the GIRF can be estimated by either MC simulation, when the distribution of the shocks is known, or by bootstrapping the residuals when the distribution is unknown.[6]

Within the present setting the maximum horizon, H, is set to 5, and we average over 1,000 BS replicates. We define two separate sets of histories: one when the temperature $T_t \leq -0.409°\text{C}$ at the moment of a shock, and the other when $T_t > -0.409°\text{C}$. Since the maximum lag order of the VTARX model is 15, we examine only the effect of a positive, one-unit, "system-wide" shock from time $t = 16$ through $t = 20$. Figure 11.7 shows HDR's (50% and 90% coverage probabilities) of the GIRF. For the Jökulsá Eystri river, the effect of a positive shock is not very persistent and dies out gradually for both regimes. We see a similar dynamic effect for the Vatnsdalsá river, when $T_t \leq -0.409°\text{C}$; Figure 11.7(c). In contrast, when $T_t > -0.409°\text{C}$, shocks persist longer for the Vatnsdalsá river than for the Jökulsá Eystri river; Figure 11.7(d). Also, there is no indication of bimodality in the HDRs of the impulse responses for all values of H. The modes of the HDRs converge more quickly to zero in the summer than in the winter period. Note, however, that for the summer period the range of values of the HDR of the Vatnsdalsá river is much wider than that for the Jökulsá Eystri river. Indicating once more the completely different hydrological and meteorological conditions of the two rivers. We leave it to the reader to investigate the effect of a negative shock on the system.

11.9 Summary, Terms and Concepts

Summary
Vector nonlinear time series analysis will become more and more prominent in future applications. This chapter has covered quite a lot of aspects of the subject,

---

[6]The algorithm for estimating the multivariate GIRF is given in Appendix 11.B.

much of it taken from relatively recent reports and papers. Certainly, and despite various advantages of vector nonlinear methods over corresponding linear methods, we should mention that these methods are not free of caveats. For instance, if the multivariate nonlinear DGP is a "long way" from linearity (null hypothesis) due to outliers in the series, it is likely that asymptotic test theory will not work well. In that case, one would expect to reject the null hypothesis emphatically – with a large number of candidate models under the alternative hypothesis. Moreover, outliers can have a more serious effect on multivariate nonlinear conditional mean forecasts than on univariate forecasts due to complex interactions among simultaneously acquired time series. To some extent, these and other difficulties may be overcome by adopting the vector semi- and nonparametric methods/models discussed in Chapter 12. In any case, we have seen that vector parametric nonlinear time series analysis can be useful in giving insight into the interdependence between many time series met in practice. With an interplay between theory and practice, further research will no doubt result in a "nonlinearity toolkit" for vector time series.

Terms and Concepts

Cholesky decomposition, 490
cointegration, 452
common nonlinear features (CNF), 457
cross-correlation function (CCF), 450
equilibrium error process, 452
Granger causality index (GCI), 451
generalized forecast error
    second moment (GFESM), 479
joint spectral radius, 455

multivariate density forecast, 479
multiple-lag diagnostic test statistic, 474
root mean squared forecast error
    (RMSFE), 478
smooth transition (ST) cointegration, 456
threshold vector error (TVEC), 452
vector error correction (VEC), 452

11.10 Additional Bibliographical Notes

Section 11.1: Thavaneswaran and Abraham (1991) present methods for estimating general nonlinear multivariate time series models using optimal estimating functions, but do not provide any practical application of their method for specific nonlinear models. Nicholls and Quinn (1981, 1982) investigate vector RCAR models. Li and Racine (2007) introduce vector nonlinear AR models for panels of nonlinear time series, using reduced-rank regression.

Section 11.2.1: Terdik (1990) gives a sufficient condition and asymptotic results concerning the stationarity and second-order properties of superdiagonal vector BL models. Subba Rao and Terdik (2003) review recent developments both for univariate and multivariate versions of the BL model. For the analysis of spatial-temporal processes, Dai and Billard (1998, 2003) propose a space-time subdiagonal BL model, which is a direct generalization of the vector subdiagonal BL model.

In principle, parameter estimation of vector BL can be obtained in an analogous way as in the univariate case. For instance, in the time-domain one may use the ML method via the Newton–Raphson method by providing recursive equations for the gradient vector and the Hessian matrix. Alternatively, one may apply the Kalman filter to evaluate the likelihood function. Also, the repeated residual method of Subba Rao and Gabr (1984) may be adopted for the estimation of vector BL models. Within the frequency-domain, Subba Rao and Wong (1999) propose an extension of the method described by Sesay and Subba Rao (1992). Kumar (1988) investigates some moment properties of bivariate BL models.

Section 11.2.2: Nieto (2005) proposes a methodology for analyzing bivariate time series with missing data using a VSETAR model transformed into a state space form with regime switching. The identification and estimation of the model is based on a combination of MCMC and Bayesian approaches.

There is a wealth of literature applying VSETARs to empirical (financial) economic data. Three interesting publications outside the area of economics are: Bacigál (2004) (bivariate GPS data), Chan et al. (2004) (trivariate actuarial data), and Solari and Van Gelder (2011) (five-variate sea wave and wind data).

Section 11.2.3: Yi and Deng (1994) present sufficient conditions for geometric ergodicity of a first-order bivariate VSETAR model with two partitions in each regime. They assume that the structural parameters of a bivariate VSETAR model with multivariate regimes are unknown and jointly estimated with the other parameters of the model.

Section 11.2.4: Yang et al. (2007) suggest a hybrid algorithm for the estimation of TVEC models which combines aspects of GAs and elements of simulated annealing (SA). Simulation results show that the algorithm does a better job than either SA or GA alone.

Hansen and Seo (2002) propose a SupLM-type test statistic for testing a linear VEC model against a two-regime TVEC model; see the function **TVECM.HStest** in the R-tsDyn package. However, this test can suffer from substantial power loss (see, e.g., Pippenger and Goering, 2000 and Seo, 2006) when the alternative hypothesis is threshold cointegration. As an alternative, Seo (2006) adopts a SupWald-type test statistic, and derives its asymptotic null distribution. The power of the proposed test dominates the power of conventional cointegration tests.

Section 11.2.5: Many extensions of the VSTAR models have been proposed in the literature; see Hubrich and Teräsvirta (2013) for a survey. For instance, Dueker et al. (2011) propose a so-called vector contemporaneous-threshold STAR model. A key characteristic of the model is that regime weights depend on the ex-ante probabilities that latent regime-specific variables exceed certain threshold values. Several methods are available to find good starting-values for the estimation of VSTAR models. In an MC simulation study, Schleer–van Gellecom (2015) compares grid search algorithms and three heuristic procedures: differential evolution (DE), threshold accepting (TA), and simulated annealing (SA). It appears that SA and DE improve LVSTAR model estimation.

Section 11.3: Harvill and Ray (1998) compare the various nonlinearity test statistics in an MC simulation study. Their results indicate that the power of the test statistics is affected by cross-correlation between process errors terms. In general, the multivariate test statistics tend to perform better than their univariate counterparts when the cross-correlation is moderate or weak. For small sample sizes, the multivariate version of the Tukey nonadditivity-type test statistic is preferable, as the test requires fewer degrees of freedom.

Section 11.4: Li and He (2012a) develop an F-type test statistic to examine linear versus nonlinear cointegration in a bivariate LVSTAR model. In case the null hypothesis is rejected, they recommend to examine the time series for CNFs using an LM-type test statistic as

11.10 ADDITIONAL BIBLIOGRAPHICAL NOTES

proposed by Li and He (2012b). Within this context, Li and He (2013) propose a residual-based Wald-type test statistic for CNFs in LVSTAR models.

As noted earlier, tests for nonlinearity can be quite sensitive to extreme outliers. This is, for instance, the case with the multivariate test statistic in Algorithm 11.4. Chan et al. (2015) propose a new and robust VSETAR-nonlinearity test statistic, and derive its asymptotic null distribution.

There are many ways in which an estimated nonlinear vector model can be misspecified. Yang (2012) and Teräsvirta and Yang (2014b) consider three LM-type misspecification test statistics for possible VSTAR model extensions: a test of no serial correlation, a test of no additive nonlinearity, and a test for parameter constancy.

Section 11.5: Billings et al. (1989) propose a method for variable selection in general (including exogenous variables) nonlinear models based on a truncated multivariate, discrete-time, Volterra series representation; see also Billings (2013). The method uses a recursive orthogonal LS algorithm which efficiently combines model identification and parameter estimation. It can be tied to the subset model selection method for univariate nonlinear time series models of Rech et al. (2001); see Section 12.7 for details about the method in the multivariate case. Camacho (2004) presents a strategy for building (specification, estimation, and evaluation) bivariate STAR models; see Yang (2012) for the multivariate case.

Section 11.6: Ling and Li (1997) and Duchesne (2004), among others, present diagnostic test statistics for checking multivariate (G)ARCH errors.

Section 11.7: Using BS and MC simulation procedures, De Gooijer and Vidiella-i-Anguera (2003b) explore the long-term forecast ability of two threshold vector cointegrated systems via a rolling forecasting approach. For model comparison they apply several forecast accuracy measures, including forecast densities. Polanski and Stoja (2012) propose a test statistic for evaluating multi-dimensional time-varying density forecasts.

The KS test statistic of uniformity, and related GOF tests, are sometimes referred to as omnibus tests, i.e. they are sensitive to almost all alternatives to the null hypothesis. For evaluating forecast densities, this property implies that when an omnibus test fails to reject $\mathbb{H}_0$, we can conclude that there is not enough evidence that the time series is not generated from the joint forecasting density. On the other hand, a rejection would not provide any information about the form of the density. Test statistics that can be decomposed into interpretable components may be a solution. Such a test is Neyman's smooth test for testing uniformity. De Gooijer (2007) explores the properties of this test statistic in a bivariate VAR framework. Moreover, he applies the test to multivariate forecast densities obtained from the VSETAR model in Exercise 11.5 fitted to the S&P 500 stock index data.

Section 11.8: Teräsvirta and Yang (2014b) present another study of the Icelandic river flow data, using a VLSTAR model with a yearly sine and cosine term as input variable.

Table 11.6: *Asymptotic critical values of the* $\mathrm{LR}_{T,p}(m, r_0)$ *test statistic* (11.65) *for various bivariate* VTAR *models of order p;* $\lambda = (1-r_0)^2/r_0^2$.

		$p=1$			$p=2$			$p=3$			$p=4$			$p=5$		
r_0	λ	10%	5%	1%	10%	5%	1%	10%	5%	1%	10%	5%	1%	10%	5%	1%
0.40	2.25	13.31	16.20	21.40	19.10	22.47	28.37	24.67	28.40	34.86	29.92	33.98	40.95	35.07	39.42	46.84
0.35	3.45	15.10	17.67	22.63	21.19	24.14	29.74	26.99	30.25	36.34	32.40	35.94	42.51	37.79	41.56	48.53
0.30	5.44	16.26	18.68	23.51	22.54	25.30	30.72	28.49	31.53	37.42	34.03	37.33	43.65	39.51	43.01	49.73
0.25	9.00	17.16	19.49	24.22	23.57	26.21	31.51	29.63	32.54	38.29	35.28	38.42	44.59	40.84	44.17	50.71
0.20	16.00	17.93	20.19	24.85	24.44	27.00	32.21	30.62	33.42	39.06	36.34	39.37	45.41	41.97	45.18	51.58
0.15	32.11	18.64	20.85	25.45	25.25	27.75	32.88	31.50	34.23	39.76	37.32	40.25	46.18	43.01	46.13	52.40
0.10	81.00	19.36	21.53	26.07	26.09	28.52	33.57	32.43	35.09	35.63	38.30	41.15	46.98	44.08	47.10	53.25

		$p=6$			$p=7$			$p=8$			$p=9$			$p=10$		
r_0	λ	10%	5%	1%	10%	5%	1%	10%	5%	1%	10%	5%	1%	10%	5%	1%
0.40	2.25	40.10	44.73	52.56	45.17	50.02	58.21	50.08	55.17	63.71	55.03	60.31	69.17	59.86	65.33	74.52
0.35	3.45	43.01	47.00	54.35	48.16	52.35	60.04	53.28	57.65	65.65	58.29	62.84	71.15	63.28	67.99	76.58
0.30	5.44	44.86	48.56	55.63	50.11	53.99	61.38	55.27	59.32	67.02	60.41	64.62	72.59	65.46	69.82	78.06
0.25	9.00	46.26	49.78	56.65	51.59	55.28	62.46	56.83	60.68	68.15	62.00	65.99	73.74	67.12	71.26	79.26
0.20	16.00	47.45	50.84	57.56	52.83	56.38	63.39	58.15	61.85	69.14	63.37	67.21	74.77	68.54	72.51	80.31
0.15	32.11	48.56	51.84	58.42	54.00	57.43	64.30	59.35	62.92	70.06	64.62	68.33	75.73	69.85	73.69	81.32
0.10	81.00	49.70	52.88	59.33	55.20	58.52	65.26	60.61	64.07	71.06	65.94	69.53	76.77	71.21	74.92	82.39

11.11 Data and Software References

Data

Example 11.5: The tree ring widths has been used by Fritts et al. (1971) "Multivariate techniques for specifying tree-growth and climatic relationships and for reconstructing anomalies in Paleoclimate", *Journal of Applied Meteorology*, 10(5), pp. 845 – 864. The data were produced and assembled at the Tree Ring Laboratory at the University of Arizona, Tuscon. Both annual (monthly averaged) tree ring widths and temperature are included in the folder **LAMARCHE** in the mhsets.zip collection of data sets, available at http://www.stats.uwo.ca/faculty/mcleod/epubs/mhsets/readme-mhsets.html. Alternatively, one may visit the website of this book.

Exercise 11.5: Forbes et al. (1999) and Tsay (1998, 2010) provide detailed information about the intraday transaction data of the S&P 500 index. Similar to Tsay (1998, Section 5) we replaced 10 extreme values (5 on each side) in the series $Y_{1,t}$ and $Y_{2,t}$ by the simple average of their two nearest neighbors. The original data set (with outliers) can be downloaded from Ruey Tsay's teaching website http://faculty.chicagobooth.edu/ruey.tsay/teaching/fts2/, file: sp5may.dat. The data set (intraday.dat), corrected for outliers, is available at the website of this book.

Application: The complete data set of Icelandic river flow system (1,096 observations) is included in the file tsayjasa1998.zip, available at the Estima website (https://estima.com). This website provides links to a long list of RATS time series procedures. The zip file also contains RATS code to replicate the threshold parameter estimation results of Tsay (1998, Section 6). On the other hand, the simplest way is to download the file ice.dat from the website of this book.

Software References

Section 11.2.2: MATHEMATICA source code for testing and estimating bivariate TAR models can be downloaded from Tomáš Bacigál's web page at https://www.math.sk/bacigal/homepage/. The R-tsDyn-package contains various functions for bivariate TVAR estimation, simulation and linearity testing.

Section 11.2.3: The website http://repec.wirtschaft.uni-giessen.de/~repec/RePEc/jns/Datenarchiv/v233y2013i1/y233y2013i1p3_21/ provides access to C++ source code and executable files for multivariate threshold bivariate VSETAR analysis using GAs.

Section 11.3: The test results in Table 11.1 are computed with applytot.f, a FORTRAN77 program written by Jane L. Harvill and Bonnie K. Ray, and available at the website of this book.

Section 11.4: Yang (2012, Appendix) provides a collection of R functions for the specification and evaluation of VSTAR models; see http://pure.au.dk/portal/files/45638557/Yukai_Yang_PhD_Thesis.pdf.

Application: Several FORTRAN77 programs for threshold estimation and parameter estimation of VTARX models (three regimes at most), created by Ruey S. Tsay, are available at the website of this book.

Appendix

11.A Percentiles of the LR–VTAR Test Statistic

Using formula (11.67), we can tabulate the asymptotic critical values for the null distribution of the $\text{LR}_{T,p}(m,r)$ test statistic. The distribution of $\text{LR}_{T,p}(m,r)$ is parameter-free, only depending on the dimension m of $\mathbf{Y}_t$, the threshold value r, and the order p of the fitted VTAR(2; p,p) model. Ordinarily, $r \in \widetilde{\mathbb{R}} = [\underline{r}, \overline{r}]$ with $\underline{r} = 0.1 \times T$ and $\overline{r} = 0.9 \times T$. Table 11.6 lists the upper 10%, 5%, and 1% points for the asymptotic null distribution of the $\text{LR}_{T,p}(m,r_0)$ test statistic for $p = 1, \ldots, 10$, $\widetilde{\mathbb{R}} = [r_0, 1-r_0]$ with $r_0 = 0.05, 0.10, \ldots, 0.40$, and $m = 2$. Percentiles for another (non-symmetric) interval $[\underline{r}, \overline{r}]$ can be obtained through the parameter λ or by interpolation.

11.B Computing GIRFs

In this appendix, we describe the steps involved in computing the GIRF for a strictly stationary m-dimensional nonlinear VAR(p) process along the lines of Koop et al. (1996). Assume that the functional form of the fitted model is completely known. Given the set of m-dimensional vector residuals $\{\widehat{\varepsilon}_t\}_{t=p+1}^T$, Algorithm 11.6 summarizes the relevant steps.

Algorithm 11.6: Bootstrapping the GIRF

(i) Draw at random a history from the available set $\mathbf{\Omega}_{t-1} = \{\boldsymbol{\omega}_{t-j}; j \geq 1\}$. This set is used to initiate the simulation of the process in the subsequent steps.

Algorithm 11.6: Bootstrapping the GIRF (Cont'd)

(ii) Obtain a *Cholesky decomposition* of the residual covariance matrix: $\widehat{\boldsymbol{\Sigma}}_\varepsilon = \widehat{\mathbf{P}}'\widehat{\mathbf{P}}$, where $\widehat{\mathbf{P}}$ is an $m \times m$ non-singular upper triangular matrix. Then compute the set of orthogonal (transformed) vectors $\{\mathbf{e}_t = \widehat{\mathbf{P}}^{-1}\widehat{\boldsymbol{\varepsilon}}_t\}_{t=1}^T$.

(iii) Draw randomly (with replacement) a sequence of vector residuals from this set, i.e. $\{\mathbf{e}_t^*, \ldots, \mathbf{e}_{t+H}^*\}$, where H ($H \geq 1$) is the forecast horizon.

(iv) Suppose that the effect of a shock on the ith variable $Y_{i,t}$ ($i = 1, \ldots, m$) is of interest given the initial history $\boldsymbol{\omega}_{i,t-1}$ of this variable. Then replace the ith element of $\mathbf{e}_t^*$ by a shock of size $e_{i,t}^{(\delta)} = \delta$ drawn from a set of shocks. Alternatively, δ may be a pre-fixed number. Denote the resulting sequence of residuals by $\{\mathbf{e}_{i,t}^{(\delta)}, \mathbf{e}_{t+1}^*, \ldots, \mathbf{e}_{t+H}^*\}$, where $\mathbf{e}_{i,t}^{(\delta)} = (e_{1,t}, \ldots, e_{i-1,t}, \delta, e_{i+1,t}, \ldots, e_{m,t})'$.

(v) Recover the "original residuals" by the transformation $\widehat{\boldsymbol{\varepsilon}}_{t+j}^* = \widehat{\mathbf{P}}\mathbf{e}_{t+j}^*$ ($j = 1, \ldots, H$) and $\widehat{\boldsymbol{\varepsilon}}_{i,t}^{(\delta)} = \widehat{\mathbf{P}}\mathbf{e}_{i,t}^{(\delta)}$ ($i = 1, \ldots, m$).

(vi) For each $j = 1, \ldots, H$, and a history $\boldsymbol{\omega}_{i,t-1}$, generate two values of $Y_{i,t+j}$, one using $\widehat{\boldsymbol{\varepsilon}}_{t+j}^*$ and one using $\widehat{\boldsymbol{\varepsilon}}_{i,t}^{(\delta)}$ ($i = 1, \ldots, m$). Compute the differences, say $\mathrm{GIRF}_Y^{(b)}(H, \widehat{\boldsymbol{\varepsilon}}_{i,t}^{(\delta)}, \boldsymbol{\omega}_{i,t-1})$, between both values.

(vii) Repeat steps (iii) – (vi) B times, to obtain $\{\mathrm{GIRF}_Y^{(b)}(H, \widehat{\boldsymbol{\varepsilon}}_{i,t}^{(\delta)}, \boldsymbol{\omega}_{i,t-1})\}_{b=1}^B$ ($i = 1, \ldots, m$). Finally compute, as an estimate of the GIRF (11.98), the sample average $\overline{\mathrm{GIRF}}_{Y,i,H} = B^{-1} \sum_{b=1}^B \mathrm{GIRF}_Y^{(b)}(H, \widehat{\boldsymbol{\varepsilon}}_{i,t}^{(\delta)}, \boldsymbol{\omega}_{i,t-1})$ for each variable i and each horizon H.

Repeating steps (i) – (vi) a sufficiently large number of times (say R), an estimate of the unconditional pdf of the random GIRF, given $\boldsymbol{\omega}_{i,t-1}$ follows directly. So, each time a new set of histories is drawn from a given initial set of histories. If the size of the shock $e_{i,t}^{(\delta)}$ and/or subset of histories is restricted, a conditional estimate of the pdf can be obtained. In the application of Section 11.8, we set $H = 10$, $B = 1,000$ and $R = 1,000$. Finally, it is good to mention that for unrestricted linear VAR and cointegrated VAR models the computation of GIRFs do not require orthogonalization of shocks, as in step (ii) above, and they are invariant to the ordering of the variables in the VAR; see Pesaran and Shin (1998).

Exercises

Theory Questions

11.1 Given the BL model (11.14), verify condition (11.16).

[*Hint:* First show that $\exp \mathbb{E}\{ \log \|\boldsymbol{\Theta}^v + \sum_{u=1}^p \boldsymbol{\Psi}^{uv}[\mathbf{Y}_{t-u} \otimes \mathbf{I}_m]\|\} < 1$ ($v \in \{1, \ldots, q\}$; $q \leq p$). Next, prove (11.16), using Jensen's inequality, the Cauchy–Schwarz inequality, the strict stationarity of the process (ergodic theorem), and using the properties of vectors and matrices given in Appendix 7.A.]

EXERCISES

11.2 Let $\{\mathbf{X}_t, t \in \mathbb{Z}\} \overset{\text{i.i.d.}}{\sim} \mathcal{N}_m(\mathbf{0}, \boldsymbol{\Sigma}_X)$ with $\mathbf{X}_t = (X_{1,t}, \ldots, X_{m,t})' \in \mathbb{R}^m$. In addition, assume that $\mathbb{R}^m$ can be partitioned into two non-overlapping subspaces, i.e.

$$M_i = \{\mathbf{x} \in \mathbb{R}^m | \mathbf{1}'\mathbf{x} \in \mathbb{R}^{(i)}\}, \quad (i = 1, 2).$$

Here, $\mathbf{1} = (1, \ldots, 1)'$, and $\mathbb{R}^{(i)}$ denotes the support of the associated density function, assuming it exists. Then a multivariate analogue of the univariate asMA(1) model is defined as

$$\mathbf{Y}_t = \mathbf{X}_t + \sum_{i=1}^{2} \mathbf{B}_i I(\mathbf{X}_{t-1} \in M_i)\mathbf{X}_{t-1}$$
$$= \mathbf{X}_t + \mathbf{B}_1 \mathbf{X}_{t-1} + \mathbf{B} I(\mathbf{X}_{t-1} \in M_1)\mathbf{X}_{t-1},$$

where $\mathbf{B}_i$ ($i = 1, 2$) are $m \times m$ matrices with constants, and $\mathbf{B} = \mathbf{B}_2 - \mathbf{B}_1$.

(a) Now, let $\boldsymbol{\Sigma}_X = \{(\sigma_{ij})\} = \begin{pmatrix} \sigma_{11} & \boldsymbol{\Sigma}_{12} \\ \boldsymbol{\Sigma}'_{12} & \boldsymbol{\Sigma}_{22} \end{pmatrix}$, $(i, j = 1, \ldots, m)$, where $\boldsymbol{\Sigma}_{12}$ is an $1 \times (m-1)$ vector, $\boldsymbol{\Sigma}_{22}$ an $(m-1) \times (m-1)$ matrix, and $|\boldsymbol{\Sigma}_{22}| > 0$. Further, let $f_m(\mathbf{x})$ denote the density function of $\{\mathbf{X}_t, t \in \mathbb{Z}\}$. Show that

   (i) $\int_{A_i} x_j f_m(\mathbf{x}) d\mathbf{x} = \sigma_{j1}(\mu_i/\sigma_{11})$, $(i = 1, 2; j = 1, \ldots, m)$.
   (ii) $\int_{A_i} x_k x_j f_m(\mathbf{x}) d\mathbf{x} = (\sigma_{k1}\sigma_{j1}/\sigma_{11})((\sigma_i/\sigma_{11}) - \alpha_i) + \sigma_{kj}\alpha_i$, $(i = 1, 2; j, k = 1, \ldots, m)$.

   where $A_i = \{(z_1, \ldots, z_m) \in \mathbb{R}^m; z_1 \in \mathbb{R}^{(i)}, (z_2, \ldots, z_m) \in \mathbb{R}^{m-1}\}$, and where $\mu_i = \int_{\mathbb{R}^{(i)}} u f_1(u) du$ and $\sigma_i = \int_{\mathbb{R}^{(i)}} u^2 f_1(u) du$.

(b) Let $\mathbf{r}$ and $\mathbf{s}$ be two m-dimensional non-random vectors in $\mathbb{R}^m$. Using the results in part (a), show that

   (i) $\int_{A_i} \mathbf{r}'\mathbf{x} f_m(\mathbf{x}) d\mathbf{x} = (\mu_i/\sigma_{11})\mathbf{r}'\boldsymbol{\Sigma}^*_{12}$, $(i = 1, 2)$.
   (ii) $\int_{A_i} \mathbf{r}'\mathbf{x}\mathbf{x}'\mathbf{s} f_m(\mathbf{x}) d\mathbf{x} = \gamma_i \mathbf{r}'\boldsymbol{\Sigma}^*_{12}\boldsymbol{\Sigma}^{*'}_{12}\mathbf{s} + \alpha_i \mathbf{r}'\boldsymbol{\Sigma}_X \mathbf{s}$, $(i = 1, 2)$, where $\alpha_i = \int_{\mathbb{R}^{(i)}} f_1(u) du$, and

$$\gamma_i = \frac{\sigma_i}{\sigma_{11}^2} - \frac{\alpha_i}{\sigma_{11}}, \quad \boldsymbol{\Sigma}^*_{12} = (\sigma_{11}, \boldsymbol{\Sigma}_{12})'.$$

(c) Using the results in part (b), and assuming the process $\{\mathbf{Y}_t, t \in \mathbb{Z}\}$ is weakly stationary, show that

$$\mathbb{E}(\mathbf{Y}_t) = (2\pi \mathbf{1}'\boldsymbol{\Sigma}_X \mathbf{1})^{-1/2}\mathbf{B}\boldsymbol{\Sigma}_X \mathbf{1},$$
$$\text{Var}(\mathbf{Y}_t) = \boldsymbol{\Sigma}_X + \frac{1}{2}(\mathbf{B}_1 \boldsymbol{\Sigma}_X \mathbf{B}'_1 + \mathbf{B}_2 \boldsymbol{\Sigma}_X \mathbf{B}'_2) - \mathbb{E}(\mathbf{Y}_t)(\mathbb{E}(\mathbf{Y}_t))',$$
$$\text{Cov}(\mathbf{Y}_t, \mathbf{Y}_{t-1}) = \frac{1}{2}(\mathbf{B}_1 + \mathbf{B}_2)\boldsymbol{\Sigma}_X.$$

11.3 Consider the LM-type test statistic for testing linearity versus the LVSTAR model in (11.69).

(a) Verify (11.70).

(b) Show that under the null hypothesis $\mathbb{H}_0: \boldsymbol{\Theta}_1 = \mathbf{0}$, and as $T \to \infty$, the asymptotic distribution of the test statistic $\text{LM}^{(1)}_{T,p}(m)$ converges in probability to a χ^2 distribution with $m(mp + 1)$ degrees of freedom.

11.4 Let U_1 and U_2 be two independent random variables each $U(0,1)$ distributed.

(a) Show that the random variable $U^{(\text{p})} = U_1 \times U_2$ has a distribution function given by $F_{U^{(\text{p})}}(x) = x - x\log(x)$ if $0 < x < 1$.

(b) Show that the distribution function of $U^{(\text{r})} = U_1/U_2$ is given by $F_{U^{(\text{r})}}(x) = x/2$ if $0 < x < 1$, and $F_{U^{(\text{r})}}(x) = 1 - (1/2x)$ if $1 < x < \infty$.

(c) Show that the distribution function of $\widetilde{U}^{(\text{p})} = (U_1 - \tfrac{1}{2})(U_2 - \tfrac{1}{2})$ is given by

$$F_{\widetilde{U}^{(\text{p})}}(x) = \begin{cases} -2x\log 2 + 2x - 2x\log(2x) + \tfrac{1}{2}, & x > 0, \\ -2x\log 2 + 2x - 2x\log(-2x) + \tfrac{1}{2}, & x < 0. \end{cases}$$

(Clements and Smith, 2002; Ko and Park, 2013)

Empirical Questions

11.5 The V(SE)TAR model is a useful tool to study index futures arbitrage in finance. Tsay (1998) studies the intraday (1–minute) transactions for the S&P 500 stock index in May 1993 and its June futures contract traded at the Chicago Mercantile Exchange. Specifically, let $\{\mathbf{Y}_t = (Y_{1,t}, Y_{2,t}, X_t)'\}_{t=1}^{7,060}$ denote the data set under study (file: intraday.dat) with $Y_{1,t} = f_{t,\ell} - f_{t-1,\ell}$ and $Y_{2,t} = s_t - s_{t-1}$, where $f_{t,\ell}$ is the log price of the index futures at maturity ℓ, and s_t is the log of the security index cash prices.

(a) Check the threshold nonlinearity of the series $\{\mathbf{Y}_t\}$ using the test statistics $F_{T,p}^{(\text{T})}(m)$ (Algorithm 11.2), $F_{T,p}^{(\text{O})}(m)$ (Algorithm 11.3), and $C_{T,p}(d,m)$ (Algorithm 11.4). In all cases, assume that a VAR(8) model best describes the interdependencies between the two series.

(b) Using LS, estimate the parameters of the following bivariate VSTARX(2; 8, 8) model

$$\mathbf{Y}_t = \begin{cases} \boldsymbol{\phi}_0^{(1)} + \sum_{u=1}^{8} \boldsymbol{\Phi}_u^{(1)} \mathbf{Y}_{t-u} + \boldsymbol{\beta}_1 X_{t-1} + \boldsymbol{\varepsilon}_t^{(1)} & \text{if } X_{t-1} \leq r, \\ \boldsymbol{\phi}_0^{(2)} + \sum_{u=1}^{8} \boldsymbol{\Phi}_u^{(2)} \mathbf{Y}_{t-u} + \boldsymbol{\beta}_2 X_{t-1} + \boldsymbol{\varepsilon}_t^{(2)} & \text{if } X_{t-1} > r, \end{cases}$$

where X_t is an exogenous variable (column three of the available data set) controlling the switching dynamics, r is a real number, $\boldsymbol{\Phi}_u^{(i)}$ ($i = 1,2; u = 1,\ldots,p$) are 2×2 matrices of coefficients, $\boldsymbol{\phi}_0^{(i)}$ and $\boldsymbol{\beta}_i$ are 2×1 vectors of unknown parameters. The error process $\{\boldsymbol{\varepsilon}_t^{(i)}\}$ satisfies $\boldsymbol{\varepsilon}_t^{(i)} = (\boldsymbol{\Sigma}_\varepsilon^{(i)})^{1/2}\boldsymbol{\varepsilon}_t$, where $\boldsymbol{\Sigma}_\varepsilon^{(i)}$ ($i = 1, 2$) are 2×2 symmetric positive definite matrices, and $\{\boldsymbol{\varepsilon}_t\} \overset{\text{i.i.d.}}{\sim} \mathcal{N}(\mathbf{0}, \mathbf{I}_2)$. Provide an (economic) interpretation for the estimation results.

(c) Apply the LVSTAR nonlinearity test statistic $\text{LM}_{T,p}^{(1)}(m)$ (Algorithm 11.5), and the rescaled $F_{T,p}^{(1)}(m)$ test statistic (Expression (11.73)) to the intraday transaction series, letting $p = 8$. Compare the test results with those of part (a).

11.6 Consider the monthly percentage growth of personal consumption expenditures, and the percentage growth of personal disposable income in the U.S. for the time period January 1985 – December 2011 ($T = 324$). Both series are measured in millions of dollars, and months are seasonally adjusted at annual rates. Let $\{Y_{i,t}\}$ ($i = 1,2$) denote the logs of the first differences of the two series. Li and He (2013) use the first

EXERCISES

263 observations of the differenced series to fit an LVSTAR(3) model with a common CNF and transition variable $Y_{1,t-7}$ to the data. Using the notation introduced earlier in this chapter, the model is given by

$$\mathbf{Y}_t = \mathbf{\Phi}_0 + \sum_{u=1}^{3} \mathbf{\Phi}_u \mathbf{Y}_{t-u} + \boldsymbol{\alpha}_\perp \big(\phi_0^* + \sum_{u=1}^{3} \boldsymbol{\beta}_u' \mathbf{Y}_{t-u}\big) G(Y_{1,t-7}; \gamma, c). \qquad (11.99)$$

The last 60 observations are set aside for out-of-sample forecasting in a rolling forecasting framework. Thus, the first forecast origin is 264. Then h-step ahead forecasts ($h = 1, \ldots, H$) are obtained with maximum forecast horizon $H = 1, 3$, and 6. Next, at time $t = 264$, the parameters of the model are re-estimated as new observations become available, but the model structure remains unchanged. This process is repeated until t extends as far as 323.

The aim of this exercise is to compare the out-of-sample forecasting performance of (11.99) with forecasts obtained from a VAR(3) model fitted to the series $\{Y_{i,t}\}$ ($i = 1, 2$).

(a) The file con_inc.dat contains the original, untransformed data. Obtain H-step forecasts (with $H = 1, 3$ and 6) from a VAR(3) model in a similar manner to the rolling forecast experiment described above.

Collect the corresponding three series of forecast errors in appropriately named data files. The data files eNL1.dat ($T = 60$), eNL3.dat ($T = 176$), and eNL6.dat ($T = 335$) contain the H-step ahead forecast errors ($H = 1, 3$, and 6) from the LVSTAR(3)–CNF model.

(b) Evaluate the forecast performance of both models in terms of RMSFEs.

(c) Use the DM and MDM test statistics (see Chapter 10) to test for equal forecast accuracy. Take as benchmarks the following three series: (i) the forecast errors of $\{Y_{1,t}\}$ and $\{Y_{2,t}\}$ from the VAR model, (ii) the forecast errors of $\{Y_{1,t}\}$ from the VAR model, and (iii) the forecast errors of $\{Y_{2,t}\}$ from the VAR model.

Chapter 12

VECTOR SEMI- AND NONPARAMETRIC METHODS

Quite often it is not possible to postulate an appropriate parametric form for the DGP under study. In such cases, semi- and nonparametric methods are called for. Certain of these methods introduced in Chapter 9 can be easily extended to the multivariate (vector) framework. Specifically, let $\mathbf{Y}_t = (Y_{1,t}, \ldots, Y_{m,t})'$ denote an m-dimensional process. We consider again the general nonlinear VAR(p) model

$$Y_{\ell,t} = f_\ell(\mathbf{Y}_{t-1}, \ldots, \mathbf{Y}_{t-p}) + \varepsilon_{\ell,t}, \quad (\ell = 1, \ldots, m), \tag{12.1}$$

where $\boldsymbol{\varepsilon}_t = (\varepsilon_{1,t}, \ldots, \varepsilon_{m,t})'$ is an m-dimensional i.i.d. variable with mean vector $\mathbf{0}$ and $m \times m$ covariance matrix $\boldsymbol{\Sigma}_\varepsilon$, independent of $\mathbf{Y}_t$. In this chapter, we discuss various aspects related to data-driven estimation and forecasting methods, as well as to the detection of dependence structures and interrelationships in multivariate time series.

In Section 12.1, we start off by extending the theory of univariate kernel-based conditional quantile estimation to higher dimensions. In addition, we present a kernel-based forecasting method. Valuable as these methods can sometimes be, the increase in the dimensionality of the predictor space makes straightforward application of kernel-based methods impractical in practice unless both m and p are small and T is large. As an alternative, constraining the functions $f_\ell(\cdot)$ in (12.1) in such a way that they still provide flexible representations of the unknown underlying functions yet do not suffer from excessive data requirements results is often a more useful approach. Of the semiparametric methods discussed in Chapter 9, (TS)MARS, k-NN, PPR and FCAR are most easily extended to the multivariate framework; see Section 12.2. In Section 12.3, we discuss vector frequency-domain Gaussianity and linearity test statistics.

In Section 12.4, we turn our attention to an exploratory nonparametric test statistic for lag identification in vector nonlinear time series which is a multivariate analogue to the mutual information coefficient $R(\cdot)$ given by (1.20). Finding appropriate lags for inclusion in a vector nonlinear time series model can be based on this test statistic and hence it can serve as an initial way to infer causal relationships. In Section 12.5, we then introduce three formal nonlinear causality test statistics. These tests are closely related to test statistics for high-dimensional serial independence, which we discussed earlier in Chapter 7.

Two appendices are added to the chapter. Appendix 12.A provides information about the numerical computation of multivariate conditional quantiles. Appendix 12.B discusses how to compute percentiles of the vector based analogue of the univariate test statistic $\widehat{R}_Y(\cdot)$ introduced in Section 1.3.3.

12.1 Nonparametric Methods

12.1.1 Conditional quantiles

Suppose that data are available in the form of a strictly stationary stochastic process $\{(\mathbf{X}_t, \mathbf{Y}_t), t \in \mathbb{Z}\}$ with the same distribution as $(\mathbf{X}, \mathbf{Y})$ taking values in $\mathbb{R}^{mp}$ ($p \geq 1, m \geq 2$). Our aim is to generalize the univariate conditional quantile definition of Section 9.1.2 into a multivariate setting, i.e., $m \geq 2$. First, we introduce some notation.

Let $\|\cdot\|_{s,q} \colon \mathbb{R}^m \to \mathbb{R}$, be the application defined by

$$\|\mathbf{z}\|_{s,q} = \|(z_1, \ldots, z_m)\|_{s,q} = \left\| \frac{|z_1| + (2q-1)z_1}{2}, \ldots, \frac{|z_m| + (2q-1)z_m}{2} \right\|_s.$$

Although $\|\cdot\|_{s,q}$ is not a norm on $\mathbb{R}^m$, it has properties similar to those of a norm; see Abdous and Theodorescu (1992). Below, we consider the Euclidean norm. Furthermore, for notational simplicity, we write $\|\cdot\|_q$ for $\|\cdot\|_{2,q}$, and $\|\cdot\|$ for $\|\cdot\|_2$.

For a fixed $\mathbf{x} \in \mathbb{R}^p$, we define a vector function of $\boldsymbol{\theta}$ ($\boldsymbol{\theta} \in \mathbb{R}^m$) by

$$\varphi(\boldsymbol{\theta}, \mathbf{x}) = \mathbb{E}(\|\mathbf{Y} - \boldsymbol{\theta}\|_q - \|\mathbf{Y}\|_q | \mathbf{X} = \mathbf{x})$$
$$= \int_{\mathbb{R}^m} (\|\mathbf{y} - \boldsymbol{\theta}\|_q - \|\mathbf{y}\|_q) Q(\mathrm{d}\mathbf{y}|\mathbf{x}), \qquad (12.2)$$

where $Q(\cdot|\mathbf{x})$ is the conditional probability measure of $\mathbf{Y}_t$ given $\mathbf{X}_t = \mathbf{x}$. Because $\|\boldsymbol{\theta}\|_q < \|\boldsymbol{\theta}\|$, we have $|\varphi(\boldsymbol{\theta}, \mathbf{x})| \leq \|\boldsymbol{\theta}\| \ \forall \boldsymbol{\theta} \in \mathbb{R}^m$. Thus, $\varphi(\cdot, \mathbf{x})$ is well-defined. We shall call a q-conditional *multivariate quantile*, any point $\boldsymbol{\theta}_q(\mathbf{x})$ which assumes the infimum

$$\varphi(\boldsymbol{\theta}_q(\mathbf{x}), \mathbf{x}) = \inf_{\boldsymbol{\theta} \in \mathbb{R}^m} \varphi(\boldsymbol{\theta}, \mathbf{x}). \qquad (12.3)$$

Unless $Q(\cdot|\mathbf{x})$ is included into a straight line in $\mathbb{R}^m$, it can be shown (Kemperman, 1987, Thm. 2.17) that $\varphi(\boldsymbol{\theta}, \mathbf{x})$ must be a strictly convex function of $\boldsymbol{\theta}$, assuming $\|\cdot\|_q$ is a strictly convex norm (Appendix 3.A). This guarantees the existence and uniqueness of $\boldsymbol{\theta}_q(\mathbf{x})$. If the norm is not strictly convex, uniqueness of $\varphi(\cdot, \mathbf{x})$ is not guaranteed; see, e.g., Oja (1983). Also, when $\varphi(\cdot, \mathbf{x})$ is defined on an infinite-dimensional space, it may have no minimum (León and Massé, 1992).

12.1 NONPARAMETRIC METHODS

Now, we introduce a consistent nonparametric estimator of $\boldsymbol{\theta}_q(\mathbf{x})$. In particular, given observations $\{(\mathbf{X}_t, \mathbf{Y}_t)\}_{t=1}^T$, we define $\widehat{F}(\cdot|\mathbf{x})$ ($\mathbf{x} \in \mathbb{R}^p$), a nonparametric estimate of $F(\cdot|\mathbf{x})$ the conditional distribution function of $\mathbf{Y}$ given $\mathbf{X} = \mathbf{x}$, by

$$\widehat{F}(\mathbf{y}|\mathbf{x}) = \frac{\sum_{t=1}^T \mathcal{K}_h(\mathbf{x} - \mathbf{X}_t) I(\mathbf{Y}_t \leqslant \mathbf{y})}{\sum_{t=1}^T \mathcal{K}_h(\mathbf{x} - \mathbf{X}_t)}, \quad \mathbf{y} \in \mathbb{R}^m.$$

Here, h is the bandwidth, and $\mathcal{K}_h(\mathbf{v}) = h^{-p} \prod_{i=1}^p K(v_i/h)$ where $K(\cdot)$ is a kernel function. Further

$$I(\mathbf{Y}_t \leqslant \mathbf{y}) = I(Y_{1,t} \leqslant y_1) \times \cdots \times I(Y_{m,t} \leqslant y_m),$$

if $\mathbf{y} = (y_1, \ldots, y_m)' \in \mathbb{R}^m$ and $\mathbf{Y}_t = (Y_{1,t}, \ldots, Y_{m,t})'$ for $t \geq 1$.

For any Borel-measurable set $V \subset \mathbb{R}^m$, let $Q_T(\cdot|\mathbf{x}) = \int_V F_T(d\mathbf{y}|\mathbf{x})$ be the estimate of $Q(\cdot|\mathbf{x})$. Then, for $\boldsymbol{\theta} \in \mathbb{R}^m$, the natural estimate of $\varphi(\boldsymbol{\theta}, \mathbf{x})$ denoted by $\varphi_T(\boldsymbol{\theta}, \mathbf{x})$ can be defined by

$$\varphi_T(\boldsymbol{\theta}, \mathbf{x}) = \int_{\mathbb{R}^m} \left(\|\mathbf{y} - \boldsymbol{\theta}\|_q - \|\mathbf{y}\|_q\right) Q_T(d\mathbf{y}|\mathbf{x})$$

$$= \sum_{t=1}^T \left(\|\mathbf{Y}_t - \boldsymbol{\theta}\|_q - \|\mathbf{Y}_t\|_q\right) \frac{\mathcal{K}_h(\mathbf{x} - \mathbf{X}_t)}{\sum_{t=1}^T \mathcal{K}_h(\mathbf{x} - \mathbf{X}_t)}.$$

Finally, if we minimize $\varphi_T(\boldsymbol{\theta}, \mathbf{x})$ instead of $\varphi(\boldsymbol{\theta}, \mathbf{x})$, the minimizer is an estimator of $\boldsymbol{\theta}_q(\mathbf{x})$. Denoted by $\boldsymbol{\theta}_{q,T}(\mathbf{x})$, such an estimator is given by

$$\boldsymbol{\theta}_{q,T}(\mathbf{x}) = \arg\min_{\boldsymbol{\theta} \in \mathbb{R}^m} \sum_{t=1}^T \left(\|\mathbf{Y}_t - \boldsymbol{\theta}\|_q - \|\mathbf{Y}_t\|_q\right) \mathcal{K}_h(\mathbf{x} - \mathbf{X}_t), \qquad (12.4)$$

and the estimator is consistent (De Gooijer et al., 2006). In Appendix 12.A, we discuss the computation of (12.4).

Example 12.1: A Monte Carlo Experiment

Consider a vector time series process $\{\mathbf{W}_t = (W_{1,t}, W_{2,t})', t \in \mathbb{Z}\}$ which is strictly stationary and described by a NLAR(1) process of the form

$$\mathbf{W}_{t+1} = \boldsymbol{\theta} \mathbf{W}_t + \boldsymbol{\varepsilon}_{t+1}, \qquad (12.5)$$

where $\boldsymbol{\theta}(\cdot) \colon \mathbb{R}^2 \to \mathbb{R}^2$ is defined as

$$\begin{pmatrix} u \\ v \end{pmatrix} \to \begin{pmatrix} \theta_{(1)} \\ \theta_{(2)} \end{pmatrix} = \begin{pmatrix} -0.1 & 0.5 \\ -0.3 & 0.2 \end{pmatrix} \begin{pmatrix} u \\ v \end{pmatrix} + \begin{pmatrix} -2.5 & 0 \\ 0 & 2 \end{pmatrix} \begin{pmatrix} \exp(-3.89u^2)u \\ \exp(-3.89v^2)v \end{pmatrix}.$$

The innovations satisfy $\boldsymbol{\varepsilon}_t = \boldsymbol{\Sigma}_\varepsilon^{1/2} \boldsymbol{\eta}_t$ where $\boldsymbol{\Sigma}_\varepsilon^{1/2} = \text{diag}(0.2, 0.2)$ is a symmetric positive definite matrix, $\{\boldsymbol{\eta}_t\}$ is a sequence of serially uncorrelated bivariate

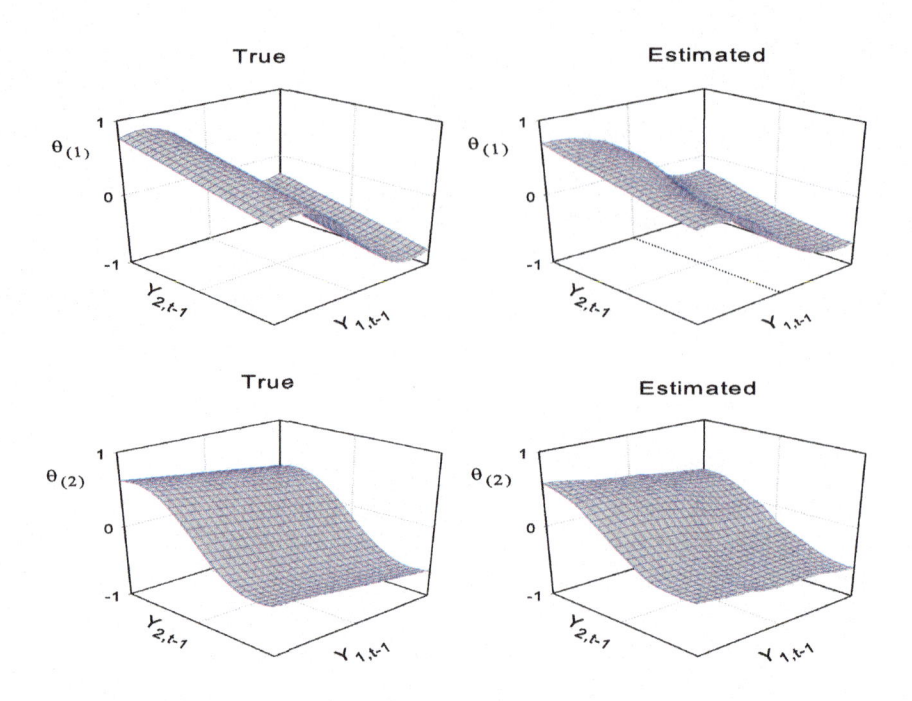

Figure 12.1: *True and estimated bivariate conditional quantile functions at $q = 0.5$ for a typical MC simulation of the NLAR(1) process (12.5).*

normally distributed random vectors with mean $\mathbf{0}$ and covariance matrix $\mathbf{I}_2$, and $\{\mathbf{W}_t\}$ is independent of $\{\boldsymbol{\eta}_t\}$.

Assume that the objective is to estimate the vector function $\boldsymbol{\theta}$ given the data points $\{\mathbf{W}_t\}_{t=1}^T$. Let $\mathbf{X}_t = (X_{1,t}, X_{2,t})' = \mathbf{W}_t$ and $\mathbf{Y}_t = (Y_{1,t}, Y_{2,t})' = \mathbf{W}_{t+1}$ ($t \in \{1, \ldots, T-1\}$). Then, using $(\mathbf{X}_t, \mathbf{Y}_t)$, we can directly apply the multivariate conditional quantile estimator (12.4) to approximate $\boldsymbol{\theta}$.

To gain some insight in the shape of the estimated conditional quantile function for model (12.5), we generate 101 random samples of size $T = 600$. With a Gaussian kernel function $K(\cdot)$, and choosing $h = 1.06\widehat{\sigma}_{i,W} T^{-1/5}$ ($i = 1, 2$), with $\widehat{\sigma}_{i,W}$ the estimated standard deviation of $\{W_{i,t}\}$, we compute $\boldsymbol{\theta}_{q,T}(\cdot)$ for each replication. Figure 12.1 shows the estimated conditional quantile functions at $q = 0.5$ along with the "true" functions $\theta_{(i)}(\cdot)$ as functions of $\mathbf{X}_t$ for a typical replication. Note that even without using any data-driven bandwidth choice criterion the shape as well as the values of each estimated conditional quantile estimator are fairly close to the corresponding true one.

12.1.2 Kernel-based forecasting

The multivariate conditional quantile estimator can be adapted to out-of-sample prediction problems from Markovian time series processes in a similar manner as we discussed in Section 9.1.2. Let $\{\mathbf{W}_t; t \in \mathbb{Z}\}$ be a strictly stationary process

12.1 NONPARAMETRIC METHODS

taking values in $\mathbb{R}^m$, with $m \geq 2$. Suppose that $\{\mathbf{W}_t, t \in \mathbb{Z}\}$ is α-mixing and p-Markovian. Consider the problem of predicting the qth quantile of the random vector $\mathbf{W}_{T+H}$ ($H \geq 1$) given the set of observations $\{\mathbf{W}_t\}_{t=1}^T$. This comes down to estimating the conditional quantile of $\mathbf{W}_{T+H}$ given $(\mathbf{W}_T', \ldots, \mathbf{W}_{T-p+1}')'$. Thus, using the associated process $\{(\mathbf{X}_t, \mathbf{Y}_t)\} \in \mathbb{R}^{mp} \times \mathbb{R}^m$ with

$$\mathbf{X}_t = (\mathbf{W}_t', \ldots, \mathbf{W}_{t+p-1}')' \text{ and } \mathbf{Y}_t = \mathbf{W}_{t+H+p-1}$$
$$(t = 1, \ldots, n; n = T - H - p + 1),$$

the problem of predicting the q-quantile of $\mathbf{W}_{T+H}$ is equivalent to estimating the q-quantile of $\mathbf{Y}_t$ conditional on $\mathbf{X}_t = \mathbf{X}_{T-p+1}$.

Example 12.2: Daily Returns of Exchange Rates

As an illustration of the multivariate forecasting approach, we consider two series of daily returns (differences of log spot rates): the Deutsche Mark/US Dollar (DEM/USD), and the Deutsche Mark/British Pound (DEM/GBP). The time period of interest is January 3, 1990 to December 28, 1994 ($T = 1{,}300$); see Figure 12.2.[1] The two series, denoted by $\{W_{i,t}\}_{t=1}^{1,300}$ ($i = 1, 2$), are correlated, the sample correlation equals 0.16, and $\{W_{1,t}^2\}$ and $\{W_{2,t}^2\}$ have a sample correlation of 0.11. Both correlations are statistically significant. The series are rescaled so that their range always has length 1. Also, we set the Markov order of the general nonlinear VAR(p) model in (12.1) at $p = 1$. The aim is to compute $H = 1, 2$, and 3-step ahead conditional quantiles for each return series using $\boldsymbol{\theta}_{q,T}$.

To see the relative performance of the multivariate conditional quantile predictor, we compare it against the univariate conditional quantile predictor,

$$\widetilde{\theta}_{q,T} = \arg\min_{\theta \in \mathbb{R}} \sum_{t=1}^n \rho_q(Y_t - \theta) K_h(\mathbf{X}_{T-p+1} - \mathbf{X}_t), \qquad (12.6)$$

where

$$\mathbf{X}_t = (W_{i,t}, \ldots, W_{i,t+p-1})' \text{ and } Y_t = W_{i,t+H+p-1} \quad (i = 1, 2), \qquad (12.7)$$

and $\rho_q(u) = 0.5(|u| + (2q-1)u)$, i.e. the check function. Note that in the univariate case, the series $\{W_{1,t}\}$ and $\{W_{2,t}\}$ are considered separately.

We need some measure to evaluate how well the quantile forecasts from the two methods are doing. To this end, we use a rolling forecast framework of 800 observations which gives a total of 498 conditional quantiles for each forecast step. Then, for each q, we calculate the following accuracy measure

---

[1] Härdle et al. (1998) discuss an LL kernel-based method for the estimation of (12.1) in the multivariate case, allowing for conditional heteroskedasticity of the error process. They use a longer version of the above bivariate data set.

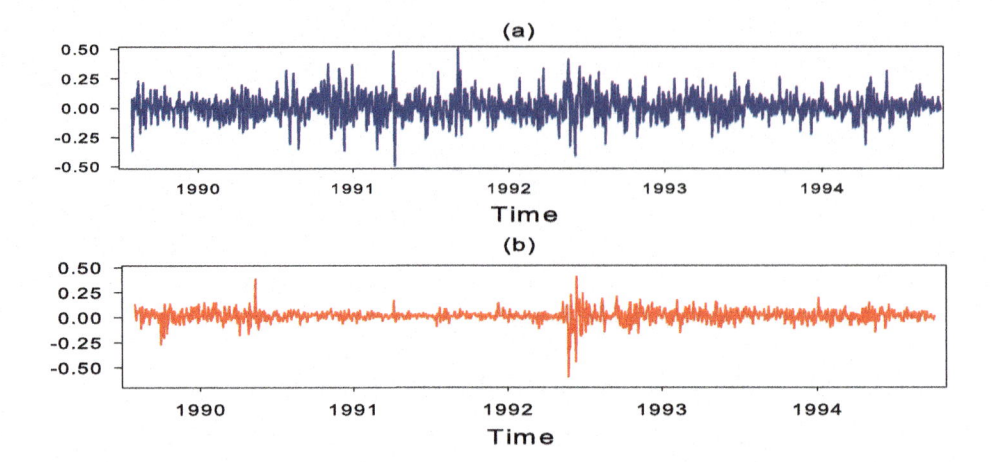

Figure 12.2: *Daily returns (rescaled) of the exchange rates data set; (a) $\{W_{1,t} = \text{DEM/USD}\}$ and (b) $\{W_{2,t} = \text{DEM/GBP}\}$ for the time period January 3, 1990 – December 28, 1994; $T = 1{,}300$.*

Table 12.1: *Exchange rates data set. Values of the accuracy measure $q_{i,H}$ based on 498 out-of-sample forecasts; $\boldsymbol{\theta}_{q,T}$ is the multivariate conditional quantile estimator, and $\widetilde{\boldsymbol{\theta}}_{q,T}$ is the univariate conditional quantile estimator. Blue-typed numbers indicate values which are statistically significantly different from q. From* De Gooijer et al. (2006).

q		$W_{1,t}$ (DEM/USD)			$W_{2,t}$ (DEM/GBP)		
		\multicolumn{6}{c}{H}
		1	2	3	1	2	3
0.01	$\boldsymbol{\theta}_{q,T}$	0.010	0.010	0.008	0.010	0.016	0.008
	$\widetilde{\theta}_{q,T}$	0.010	0.004	0.010	0.038	0.016	0.018
0.025	$\boldsymbol{\theta}_{q,T}$	0.016	0.020	0.020	0.024	0.030	0.026
	$\widetilde{\theta}_{q,T}$	0.032	0.032	0.030	0.074	0.060	0.054
0.05	$\boldsymbol{\theta}_{\alpha,n}$	0.042	0.042	0.044	0.058	0.060	0.050
	$\widetilde{\theta}_{q,T}$	0.060	0.064	0.052	0.108	0.104	0.100
0.95	$\boldsymbol{\theta}_{q,T}$	0.956	0.954	0.956	0.956	0.959	0.949
	$\widetilde{\theta}_{q,T}$	0.943	0.946	0.944	0.869	0.874	0.876
0.975	$\boldsymbol{\theta}_{q,T}$	0.984	0.976	0.972	0.976	0.978	0.979
	$\widetilde{\theta}_{q,T}$	0.974	0.969	0.976	0.932	0.942	0.939
0.99	$\boldsymbol{\theta}_{q,T}$	0.986	0.986	0.988	0.986	0.986	0.992
	$\widetilde{\theta}_{q,T}$	0.990	0.998	0.994	0.968	0.974	0.974

12.1 NONPARAMETRIC METHODS

$$q_{i,H} = \frac{1}{498}\sum_{j=1}^{498} I(W_{i,T+H+j-1} \leqslant \widehat{\theta}_{q,T}^{(H)}), \quad (i=1,2; H=1,2,3; T=800),$$

where $\widehat{\theta}_{q,T}^{(H)}$ is either $\boldsymbol{\theta}_{q,T}^{(H)}$ (multivariate) or $\widetilde{\theta}_{q,T}^{(H)}$ (univariate) with the superscript (H) denoting the H-step ahead prediction. If the conditional quantiles are accurate, we expect the value of $q_{i,H}$ to closely approximate q. Table 12.1 shows the results for $q_{i,H}$. The results of the significance test are obtained using the Gaussian assumption and using the well-known fact that the standard deviation for a set of $n = 498$ proportions equals $(q(1-q)/n)^{1/2}$.

Given their role in Value at Risk calculations, a type of risk in a financial market (see, e.g., Tsay, 2010), we only discuss the conditional quantile results for the lower tail quantile levels, $q = 0.01, 0.025$, and 0.05. The $q_{i,H}$ values from the calibration of the conditional quantiles of the $\{W_{2,t} = \text{DEM/GBP}\}$ series shows that $\widetilde{\theta}_{q,T}$ consistently underpredicts tail quantile values, with larger biases at $q = 0.025$ and $q = 0.05$. In contrast, for the $\{W_{1,t} = \text{DEM/USD}\}$ series, $\widetilde{\theta}_{q,T}$ performs as well as $\boldsymbol{\theta}_{q,T}$, in terms of its empirical q or $q_{i,H}$. The distribution of the DEM/GBP returns has a rather heavy tail with a standardized kurtosis of 18.2. Thus, it may not be a surprise when $\widetilde{\theta}_{q,T}$ underpredicts the tails. However, when the returns are jointly considered in a multivariate fashion, the tails of the DEM/GBP distribution are accurately tracked by $\boldsymbol{\theta}_{q,T}$ with no statistically significant bias.[2] In all cases, the bandwidths $h_{i,T}$ ($i = 1, 2$) are chosen according to the rule-of-thumb (9.22).

12.1.3 K-nearest neighbors

The univariate k-nearest neighbor method discussed in Section 9.1.4 extend most naturally to the vector framework. For ease of exposition, let $\{(Y_{1,t}, Y_{2,t}, Y_{3,t})'\}_{t=1}^T$ be a set of three observed time series on the strictly stationary time series process $\{(Y_{1,t}, Y_{2,t}, Y_{3,t}), t \in \mathbb{Z}\}$. Moreover, assume that each series can be transformed into an m-dimensional vector by the construct $\mathbf{X}_{i,t} = (Y_{i,t}, Y_{i,t+1}, \ldots, Y_{i,t+m-1})' \in \mathbb{R}^m$ ($i = 1, 2, 3$). As a first step, we are interested in producing a nonparametric estimator of the conditional mean $Y_{i,t+1|t} = \mathbb{E}(Y_{i,t+1}|\mathbf{X}_t = \mathbf{x})$, where $\mathbf{X}_t = (\mathbf{X}'_{1,t}, \mathbf{X}'_{2,t}, \mathbf{X}'_{3,t})' \in \mathbb{R}^{3m}$. To this end, we start by fixing an integer $1 \leq k_T < T$. Then, at time point $t = T$, we look for the k_T closest vectors $\mathbf{X}_{i,j}$ ($i = 1, 2, 3; j = j_1, \ldots, j_{k_T}$) to $\mathbf{X}_T = \mathbf{x}$ in the vector space $\mathbb{R}^{3m}$, in the sense that they minimize the function $\sum_{i=1}^3 \|\mathbf{X}_{i,j} - \mathbf{X}_T\|$

---

[2]In finance, it is common to assume normality of returns although it is well known that one of the stylized facts of many financial time series is their being heavy tailed and most often asymmetric. The most usual way of estimating quantile predictions is by first computing conditional variance (volatility) predictions and then make a normality assumption. Obviously, this parametric approach leads to a sizeable underprediction of tail events because in practice returns are not normally distributed. In contrast, the multivariate conditional quantile approach can be computed directly and no distributional assumptions about the process under study are needed.

($j = j_1, \ldots, j_{k_T}$), where $\|\cdot\|$ denotes the usual Euclidean norm.[3] In this way, we obtain a set of k_T simultaneous m-histories in the three series under study, i.e. $\{(Y_{1,j_1}, Y_{2,j_1}, Y_{3,j_1}), \ldots, (Y_{1,j_{k_T}}, Y_{2,j_{k_T}}, Y_{3,j_{k_T}})\}$. Then compute the one-step ahead forecasts $Y_{i,T+1|T}$ using linear regressions of Y_{i,j_r+1} on $(Y_{i,j_r}, Y_{i,j_r-1}, \ldots, Y_{i,j_r-m+1})'$ ($i = 1, 2, 3; r = 1, \ldots, k_T$). Alternatively, a VAR model may be used to obtain the joint vector of one-step ahead forecasts $\widehat{\mu}^{\text{k-NN}}(\mathbf{x})$. Next, the two-step ahead vector forecasts follow from the new information set $\{\mathbf{X}_1, \ldots, \mathbf{X}_T, \widehat{\mu}^{\text{k-NN}}(\mathbf{x})\}$.

As expected, the value of k_T controls the degree of smoothing. Again, there is an optimum choice for k_T that is neither too large nor too small. Given a value of the embedding dimension m, the number of neighbors k_T can be obtained from minimizing the RMSE. Note that the model produced by the nearest neighbors is not a true density model because the integrals over all vector spaces diverge.

12.2 Semiparametric methods

12.2.1 PolyMARS

PolyMARS, or for short PMARS, is an extension of the MARS procedure (see Section 9.2.3) that allows for multiple polychotomous regression; Kooperberg et al. (1997). The method was introduced primarily to extend the advantages of the (TS)MARS algorithm over simple recursive partitioning to the multiple classification problem, in which multinomial response data is considered as a set of 0 – 1 multiple responses. With PMARS, by letting the predictor variables be lagged values of multivariate time series, one obtains a new method for modeling vector threshold nonlinear time series with or without additional (lagged) exogenous predictors. The resulting specification, called *vector adaptive spline threshold AR (eXogenous)* (VASTAR(X)) model can be considered as a type of generalized VTAR model.

Description of PMARS

Let $\mathbf{Y}_t = (Y_{1,t}, \ldots, Y_{m,t})' \in \mathbb{R}^m$ be an m-dimensional time series which depends on q p_j-dimensional vectors of time series variables $\mathbf{X}_{j,t} = (X_{j,t-1}, \ldots, X_{j,t-p_j})'$ ($p_j \geq 0; j = 1, \ldots, q$). Assume that there are T observations on $\{\mathbf{Y}_t\}$ and $\{\mathbf{X}_{j,t}\}$ and that the data are presumed to be described by the time series regression model

$$Y_{\ell,t} = \mu^{(\ell)}(\mathbf{X}_{1,t}, \ldots, \mathbf{X}_{q,t}) + \varepsilon_{\ell,t}, \quad (\ell = 1, \ldots, m), \tag{12.8}$$

over some domain $\mathcal{D} \in \mathbb{R}^n$ ($n = \sum_{j=1}^q p_j$), which contains the data. Here, the superscript (ℓ) denotes that this is the ℓth component of m possible regressions, the $\mu^{(\ell)}(\cdot)$ are measurable functions from $\mathbb{R}^n$ to $\mathbb{R}$ which reflect the true, but unknown, relationship between $\mathbf{Y}_t$, and $\mathbf{X}_{1,t}, \ldots, \mathbf{X}_{q,t}$, and $\varepsilon_{\ell,t}$ ($\ell = 1, \ldots, m$) are mean zero

---

[3]Alternatively, one can minimize other functions like $\sum_{i=1}^{3}\{1 - \text{Corr}(\mathbf{X}_{i,j}, \mathbf{X}_T)\}$ ($j = j_1, \ldots, j_{k_T}$). Of course, other methods of determining the nearest neighbors in the multivariate framework exist. For instance, different (kernel) weights could be assigned to different components.

12.2 SEMIPARAMETRIC METHODS

random variables which are correlated with those from the other regressions, as specified in (12.10) below.

The goal of semiparametric multivariate regression modeling is to construct a data-driven procedure for simultaneous estimation of the unknown functions $\mu^{(\ell)}(\mathbf{X}_t)$ where $\mathbf{X}_t = (\mathbf{X}_{1,t}, \ldots, \mathbf{X}_{q,t})'$. Specifically, each regression function is modeled as a linear combination of $S > 0$ basis functions $B_s(\mathbf{X}_t)$, so that for a function $\mu^{(\ell)}(\cdot)$,

$$\widehat{\mu}^{(\ell)}(\mathbf{X}_t) = \sum_{s=1}^{S} \beta_s^{(\ell)} B_s(\mathbf{X}_t), \quad (\ell = 1, \ldots, m). \tag{12.9}$$

Here, S denotes the number of knots or thresholds τ_s, representing a partitioning of $\mathcal{D}$, and the $\beta_s^{(\ell)}$'s are regression parameters. To keep the PMARS methodology fast, and to allow for a better interpretable final model, the candidate basis functions $B_s(\mathbf{X}_t)$ ($s = 1, \ldots, S$) are limited to the following set:

- x_i;

- $(x_i - \tau_{is})_+$ if x_i is already a basis function in the model;

- $x_i(x_j - \tau_{js})_+$ if $x_i x_j$ and $(x_j - \tau_{js})_+$ are in the model;

- $(x_i - \tau_{is})_+(x_j - \tau_{js})_+$ if $x_i(x_j - \tau_{js})_+$ and $x_j(x_i - \tau_{is})_+$ are in the model.

This procedure is a little different from that of (TS)MARS, which constrains the set of candidate basis functions at each step in a slightly different way. PMARS thus creates a preference for linear models over nonlinear ones, while interactions are only considered if they are between predictors that are already in the model. Further note that PMARS, in contrast to (TS)MARS, does not allow basis functions of the form $(\tau_s - x)_+$.

Let $\mathbf{X}_{\ell,t} = (b_1(\mathbf{X}_t), \ldots, b_S(\mathbf{X}_t))$, and $\boldsymbol{\beta}_\ell = (\beta_1^{(\ell)}, \ldots, \beta_S^{(\ell)})'$ ($\ell = 1, \ldots, m$). Then, given a choice of a particular basis for the approximation at (12.9), (12.8) can be placed into vector notation as follows:

$$\mathbf{Y}_t = \mathbf{X}_t \boldsymbol{\beta} + \boldsymbol{\varepsilon}_t. \tag{12.10}$$

Here, $\mathbf{X}_t = \text{diag}(\mathbf{X}_{1,t}, \ldots, \mathbf{X}_{m,t})$, $\boldsymbol{\beta} = (\boldsymbol{\beta}_1', \ldots, \boldsymbol{\beta}_m')'$, and $\boldsymbol{\varepsilon}_t = (\varepsilon_{1,t}, \ldots, \varepsilon_{m,t})'$ is an m-dimensional vector of i.i.d. random variables with mean zero and $m \times m$ covariance matrix $\boldsymbol{\Sigma}_\varepsilon$, independent of $\mathbf{Y}_t$. In PMARS, estimates of $\boldsymbol{\beta}$ are obtained by the method of CLS. As in multivariate regression, simultaneous estimation of the $\boldsymbol{\beta}_\ell$ takes advantage of correlation among the $\varepsilon_{\ell,t}$ ($\ell = 1, \ldots, m$) for efficient estimation. Note that the fitted model has the same basis functions for each response; different structure in different component series is captured through the different coefficients.

Model selection

Analogous to the univariate (TS)MARS methodology, we can use a GCV criterion for model selection. Given a maximum number M of basis functions ($M \geq S$), the criterion is given by

$$\text{GCV}(M) = \frac{T^{-1} \sum_{\ell=1}^{m} \sum_{t=1}^{T} \{Y_{\ell,t} - \widehat{\mu}_M^{(\ell)}(\mathbf{X}_t)\}^2}{\{1 - (d \times M)/T\}^2}, \quad (12.11)$$

where d is a user-specified constant that penalizes for larger models. A value of d such that $2 \leq d \leq 5$ is recommended in practice. The value of M is commonly set equal to $\min([6T^{1/3}], [T/4], 100)$.

Alternatively, a test data set can be used for model selection by specifying the test response data, and the test predictor values. Then compute for each fitted model the residual sum of squared errors (RSS) of the test set. Next, select at each stage the VASTAR model with the smallest RSS. Fitting a VASTAR model to all data except a test set of length h and evaluating the model over the test set corresponds to a leave-out h CV method evaluated only for a single block of series.

Forecasting
Multi-step ahead forecasts for PMARS models can be made using a naive, or plug-in, iterative approach as a simple extension of (9.54). Specifically, correlations between the forecast errors of the component variables should be considered in a vector framework. In that case, the method of model-based block bootstrapping may be used as an alternative to the "plug-in" method.

12.2.2 Projection pursuit regression

Recall, in Section 9.2.2 we introduced the PPR method to estimate the relation between a univariate time series process $\{Y_t, t \in \mathbb{Z}\}$ and a specified p-dimensional vector of predictors, $\mathbf{X}_t$, using a linear combination of M one-dimensional non-parametric functions. In a vector framework, the PPR representation of the ℓth component of an m-dimensional time series process $\{\mathbf{Y}_t, t \in \mathbb{Z}\}$ is given by

$$Y_{\ell,t} = \beta_{\ell,0} + \sum_{i=1}^{M} \beta_{\ell,i} \phi_i(\boldsymbol{\alpha}_i' \mathbf{X}_t) + \varepsilon_{\ell,t}, \quad (\ell = 1, \ldots, m), \quad (12.12)$$

where each $\boldsymbol{\alpha}_i$ is a p-dimensional vector and $\boldsymbol{\alpha}'$ and $\beta_{\ell,i}$ are chosen using an LS criterion. Each $\phi_i(\cdot)$ is a univariate function of the projection $\boldsymbol{\alpha}' \mathbf{X}_t$ estimated non-parametrically using a kernel-based smoothing method such that $\mathbb{E}(\phi_i(\cdot)) = 0$ and $\text{Var}(\phi_i(\cdot)) = 1$. PPR thus searches for low-dimensional linear projections of a high-dimensional data cloud that can be transformed using nonlinear functions and added together to approximate the structure of $\{\mathbf{Y}_t, t \in \mathbb{Z}\}$.

Example 12.3: Sea Surface Temperatures (Cont'd)

Recall, in Example 9.7 we showed a TSMARS model fitted to a subset of the transformed daily SSTs at Granite Canyon, i.e. the series $\{Y_t\}_{t=1}^{1,825}$ with lagged values of Y_t, lagged values of wind speed data $\{\text{WS}_t\}$, and lagged values of wind

12.2 SEMIPARAMETRIC METHODS

Table 12.2: *Estimated β and α_i' values for the PPR model fitted to the SST and wind speed (WS) data set.*

i	\multicolumn{5}{c	}{Predictor weights ($\widehat{\alpha}_i'$)}	\multicolumn{2}{c}{Coefficients}
	Y_{t-1}	WS_{t-1}	WD_{t-1}	WS_{t-4}	WS_{t-9}	$\widehat{\beta}_{1,i}$	$\widehat{\beta}_{2,i}$
1	0.067	0.701	0.598	0.257	0.284	0.006	0.360
2	1.000	-0.003	-0.001	-0.004	0.001	0.077	-0.090
3	0.996	0.029	-0.076	0.012	-0.022	0.007	0.107

directions WD_t as predictors. First, we fit a PMARS model to the bivariate series $(Y_t, WS_t)'$ with (Y_{t-j}, WS_{t-j}) $(j = 1, \ldots, 10)$ and WD_{t-j} $(j = 1, \ldots, 5)$ as predictor variables, using default values to specify model selection (GCV with $M = 73$) and space between knots. Including only terms with absolute coefficient value more than twice their estimated standard error, we obtain the model

$$\widehat{Y}_t = 0.0030 WS_{t-1} + 0.8971 Y_{t-1} - 0.0050 I(WD_{t-1} = 2) \quad (12.13)$$

$$\widehat{WS}_t = 0.9079 + 0.1597 WS_{t-1} + 0.09690 WS_{t-9} + 0.0832 WS_{t-4}$$
$$+ 0.2232 I(WD_{t-1} = 2) + 0.5189 (WS_{t-1} - 2.445)_+. \quad (12.14)$$

The fitted PMARS model suggests that lagged values of WS_t have only a minimal effect on transformed SSTs. There is indication that winds blowing from the North (coded as 2) act to lower SSTs on the following day. Transformed wind speeds are modeled primarily as a function of lagged transformed wind speeds. Wind speeds greater than 2.445 act to increase the wind speed on the following day, as do winds blowing from the North. Taking the inverse transform, the threshold value translates into 10.53 knots, or about 12 mph (5.5 m/sec). The PMARS model explains about 80.5% of the observed variation in SSTs, while explaining only 11.4% of observed variability in wind speeds.

Based on the PMARS model fitting results, we apply PPR with $M = 3$ using Y_{t-1}, WS_{t-1}, WD_{t-1}, WS_{t-1}, and WS_{t-9} as predictor variables, giving $p = 5$. Figure 12.3 shows $\widehat{\phi}_i(\cdot)$ as a function of $\widehat{\alpha}_i' \mathbf{X}_t$. Table 12.2 gives the estimated values of α_i' and $\beta_{\ell,i}$.

The $\widehat{\alpha}_1'$ vector suggests that a combination of lagged wind speeds and lagged wind directions affect the responses. The $\widehat{\alpha}_i$ ($i = 2, 3$) vectors have most weight given to Y_{t-1}. The $\widehat{\phi}_1(\cdot)$ function is fairly linear, with a slope near 1. The coefficient of $\widehat{\phi}_2(\cdot)$ is 0.077 for the SST response, thus this term corresponds roughly to the term $0.8971 Y_{t-1}$ in (12.13). The nonlinear nature of $\widehat{\phi}_3(\cdot)$ suggests a nonlinear relation between SSTs and wind speeds and the SST of the previous day. The fitted PPR model explains about 75.2% of the

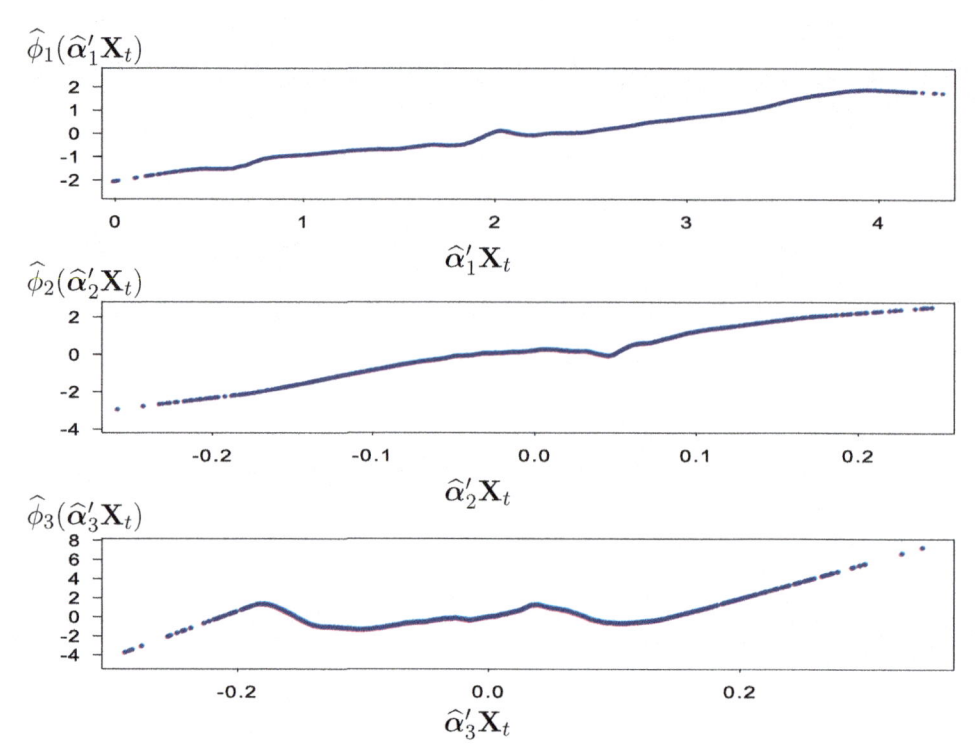

Figure 12.3: *Estimated functional relationships $\phi_i(\cdot)$ ($i = 1, 2, 3$) for the PPR model fitted to the SST and wind speed (WS) time series.*

variance in the SST series, while only 12.5% of the variability in wind speeds is explained, comparable to the PMARS model results. The wind direction predictor variable does not play a significant role in the fitted PPR model.

12.2.3 Vector functional-coefficient AR model

Harvill and Ray (2005, 2006) extend the FCAR idea of Section 9.2.5 to the vector AR framework. Consider the case where all functions $f_\ell(\cdot)$ in (12.1) are additive; that is,

$$f_\ell = \sum_{j=1}^{p} \phi_\ell^{(j)}(\mathbf{X}_t)\mathbf{Y}_{t-j}, \quad (\ell = 1, \ldots, m), \tag{12.15}$$

where (j) is a superscript, and $\mathbf{X}_t$ is a q-dimensional exogenous random variable, or lagged values of the series $\{\mathbf{Y}_t\}_{t=1}^{T}$. There is little or no information about the specific forms of the $\phi_\ell^{(j)}(\cdot)$. Specification of (12.15) with $\mathbf{X}_t = Y_{t-d}$ ($d \leq p$) gives a multivariate version of the FCAR model (9.62). More formally, combining (12.1) and (12.15), we define the vector FCAR model of order p, VFCAR(p), as

$$\mathbf{Y}_t = \mathbf{\Phi}_0(\mathbf{X}_t) + \sum_{j=1}^{p} \mathbf{\Phi}_j(\mathbf{X}_t)\mathbf{Y}_{t-j} + \boldsymbol{\varepsilon}_t, \quad (t = p+1, \ldots, T), \tag{12.16}$$

12.2 SEMIPARAMETRIC METHODS

where $\{\varepsilon_t\}$ is independent of $\mathbf{Y}_s$ and $\mathbf{X}_t$ $\forall s < t$. The $\mathbf{\Phi}_j(\cdot)$ ($j = 1, \ldots, p$) are $m \times m$ matrices with elements $\{\phi_{\ell,k}^{(j)}(\cdot)\}$ that are real-valued measurable functions that change as a function of a designated variable $\mathbf{X}_t$ and which have continuous second derivatives. If the variable $\mathbf{X}_t$ consists of lagged values of $\mathbf{Y}_{t-d}$, the intercept term, or the lag d term in the sum of (12.16) should be omitted to avoid a non-identifiable model, giving unstable estimates of the functional coefficients.

Estimation

The elements of the matrices $\mathbf{\Phi}_j(\cdot)$ can be estimated from the observations $\{(\mathbf{X}_t, \mathbf{Y}_t)\}_{t=1}^T$ using local constant or LL multivariate regression in a neighborhood of $\mathbf{X}_t$ with a specified kernel and bandwidth matrix. At time t, denote the AR fit order by p^*, and the mp^*-dimensional vector of predictors by $\mathbf{Z}_t$; that is, let

$$\mathbf{Z}_t = (\mathbf{1}', \mathbf{Y}_{t-1}', \ldots, \mathbf{Y}_{t-p^*}')',$$

where $\mathbf{Y}_{t-j} = (Y_{1,t-j}, \ldots, Y_{m,t-j})'$ ($j = 1, \ldots, p^*$). Define $\mathbf{\Phi}(\cdot)$ by

$$\mathbf{\Phi}(\mathbf{X}_t) = \big(\mathbf{\Phi}_0(\mathbf{X}_t), \mathbf{\Phi}_1(\mathbf{X}_t), \ldots, \mathbf{\Phi}_{p^*}(\mathbf{X}_t)\big)'.$$

Then model (12.16) can be written as

$$\mathbf{Y}_t = \mathbf{\Phi}(\mathbf{X}_t)\mathbf{Z}_t + \varepsilon_t, \quad (t = p^* + 1, \ldots, T).$$

For the sake of discussion, we temporarily restrict the dimension of the functional variable $\mathbf{X}_t$ to $q = 1$. Since all elements of $\mathbf{\Phi}(\cdot)$ have continuous second-order derivatives, we may approximate each $\phi_{\ell,k}^{(j)}(\cdot)$ locally at a point $x_0 \in \mathbb{R}$ by a linear function $\phi_{\ell,k}^{(j)}(x) = a_{\ell,k}^{(j)} + b_{\ell,k}^{(j)}(x - x_0)$. Partitioning the coefficient matrices in the form $(\mathbf{a} \,|\, \mathbf{b})$, the LL kernel-based estimator of $\mathbf{\Phi}(\cdot)$ is defined as $\widehat{\mathbf{\Phi}}(x_0) = \widehat{\mathbf{a}}$, where $(\widehat{\mathbf{a}} \,|\, \widehat{\mathbf{b}})$ is the solution to $(\mathbf{a} \,|\, \mathbf{b})$ that minimizes the weighted sum of squares

$$\sum_{t=p^*+1}^T \left[\mathbf{Y}_t - (\mathbf{a}\,|\,\mathbf{b}) \begin{pmatrix} \mathbf{Z}_t \\ \mathbf{U}_t \end{pmatrix}\right]\left[\mathbf{Y}_t - (\mathbf{a}\,|\,\mathbf{b}) \begin{pmatrix} \mathbf{Z}_t \\ \mathbf{U}_t \end{pmatrix}\right]' \mathcal{K}_h(x_0 - X_t). \quad (12.17)$$

Here, $\mathbf{U}_t$ is a partitioned matrix with the first partition being $(\mathbf{Z}_{p^*+1}, \ldots, \mathbf{Z}_T)'$, and the second partition is the result of the element-by-element product of $\mathbf{Z}_t$ and $(x_0 - X_t)$, $\mathcal{K}_h(\cdot) = K(\cdot/h)/h$ with $K(\cdot)$ a specified kernel function, and h is the bandwidth. From least squares theory, the solution of (12.17) is given by

$$\begin{pmatrix} \widehat{\mathbf{a}} \\ \widehat{\mathbf{b}} \end{pmatrix} = (\mathbf{U}'\mathbf{W}\mathbf{U})^{-1}\mathbf{U}'\mathbf{W}\mathbf{Y},$$

where

$$\mathbf{U} = \begin{pmatrix} \mathbf{Z}_{p^*+1} & \mathbf{Z}_{p^*+1}(x_0 - X_{p^*+1}) \\ \vdots & \vdots \\ \mathbf{Z}_T & \mathbf{Z}_T(x_0 - X_T) \end{pmatrix},$$

$\mathbf{U}'\mathbf{W}\mathbf{U}$ is non-singular, and

$$\mathbf{W} = \text{diag}\big(\mathcal{K}_h(x_0 - X_{p^*+1}), \ldots, \mathcal{K}_h(x_0 - X_T)\big).$$

If $q > 1$, the first mp^* rows of $\mathbf{U}_t$ are the element-by-element product of $\mathbf{Z}_t$ and $(x_{1,0} - X_{1,t})$, the second mp^* rows are that of $\mathbf{Z}_t$ and $(x_{2,0} - X_{2,t})$, etc. In this case $K(\cdot)$ is a specified q-variate kernel function. If the intent is to use the VFCAR model for testing, a boundary kernel is recommended to avoid trimming the functional coefficient estimates. In general, results given in Section 9.2.5 for the bandwidth selection carry over to the present vector framework.

Forecasting

Forecasting with VFCAR models can be based on, for instance, the naive, or plug-in, method, on MC simulation, and BS. For ease of discussion, consider the univariate FCAR model (9.61) with $\mathbf{Y}_t = (Y_{t-1}, \ldots, Y_{t-p+1})'$. The goal is to find the H-step ahead ($H \geq 1$) MMSE forecast of Y_{t+H}, i.e.

$$\mathbb{E}(Y_{t+H}|\mathbf{Y}_t) = \sum_{i=1}^{p} \phi_i(Y_{t+H-d})\mathbb{E}(Y_{t+H-d}|\mathbf{Y}_t), \qquad (12.18)$$

assuming $\phi_i(\cdot)$ is known. The BS forecast method is, by far, most commonly used for this purpose. That is, the H-step ahead ($H \geq 2$) forecast is given by

$$\widehat{Y}^{\text{BS}}_{t+H|t} = \frac{1}{B}\sum_{b=1}^{B} \widehat{Y}^{(b)}_{t+H|t}, \qquad (12.19)$$

where

$$\widehat{Y}^{(b)}_{t+H|t} = \sum_{i=1}^{p^*} \widehat{\phi}_i(\widehat{Y}_{t+H-d|t})\widehat{Y}_{t+H-i|t} + e^{(b)}, \qquad (12.20)$$

with $e^{(b)}$ ($b = 1, \ldots, B$) a bootstrapped value of the within-sample residuals from the fitted FCAR model. Extension of this approach to the vector framework is straightforward. One advantage of the bootstrapping forecast method is that the series $\{\widehat{Y}^{(b)}_{t+H|t}\}_{b=1}^{B}$ can be used to construct interval forecasts and density forecasts.

Model assessment

Specific choices for the elements of the matrices $\mathbf{\Phi}_j(\cdot)$ in (12.16) can result in parametric vector time series models. This feature is particularly useful, and can be assessed by testing the null hypothesis

$$\mathbb{H}_0: \quad \mathbf{\Phi}_j(\mathbf{X}) = \mathbf{G}_j(\mathbf{X};\boldsymbol{\theta}) \quad \text{versus} \quad \mathbb{H}_1: \quad \mathbf{\Phi}_j(\mathbf{X}) \neq \mathbf{G}_j(\mathbf{X};\boldsymbol{\theta}),$$

12.2 SEMIPARAMETRIC METHODS

where $\mathbf{G}_j(\cdot;\boldsymbol{\theta})$ $(j = 1,\ldots,p^*)$ is a given family of matrix functions indexed by an unknown parameter vector $\boldsymbol{\theta}$, and of the same dimension as $\boldsymbol{\Phi}_j(\cdot)$. The corresponding LR-type test statistic is given by

$$\mathrm{LR}_T = \frac{1-\Lambda}{\Lambda}, \quad \text{where } \Lambda = \left(\frac{\mathrm{tr}(\mathbf{RSS}_0)}{\mathrm{tr}(\mathbf{RSS}_1)}\right)^{1/2}, \qquad (12.21)$$

with $\mathbf{RSS}_i$ $(i = 0, 1)$ the matrix residual sum of squares obtained under $\mathbb{H}_i$, given an estimator $\widehat{\boldsymbol{\theta}}$ of $\boldsymbol{\theta}$ in the specified parametric model $\mathbf{G}_j(\cdot;\boldsymbol{\theta})$. Large values of LR_T indicate that $\mathbb{H}_0$ should be rejected.

Finding the distribution of the test statistic (12.21) in finite samples is a difficult problem. However, along the same lines as Algorithm 9.6, Harvill and Ray (2006) propose the following bootstrap procedure.

Algorithm 12.1: Bootstrap-based p-values for LR_T

(i) Sample bootstrap residuals $\{\boldsymbol{\varepsilon}_t^*\}_{t=1}^T$ from the EDF of the centered residuals $\{\widetilde{\boldsymbol{\varepsilon}}_t - \overline{\widetilde{\boldsymbol{\varepsilon}}}\}_{t=1}^T$, where $\overline{\widetilde{\boldsymbol{\varepsilon}}}$ is the mean of the m-dimensional residual vector $\widehat{\boldsymbol{\varepsilon}}_t = \mathbf{Y}_t - \widehat{\boldsymbol{\Phi}}(\mathbf{X}_t)\mathbf{Z}_t$ $(t = p^*+1,\ldots,T)$.

(ii) Construct the vector of pseudo-observations $\mathbf{Y}_t^* = \mathbf{G}(\mathbf{X}_t;\widehat{\boldsymbol{\theta}})\mathbf{Z}_t + \boldsymbol{\varepsilon}_t^*$. Next, compute a bootstrap statistic $\mathrm{LR}_T^{(0)}$ in the same way as LR_T using $\{Y_t^*\}_{t=1}^T$.

(iii) Repeat step (ii) B times, to obtain $\{\mathrm{LR}_T^{*,(b)}\}_{b=1}^B$.

(iv) Compute the one-sided bootstrap p-value as

$$\widehat{p} = \frac{1 + \sum_{b=1}^B I(\mathrm{LR}_T^{*,(b)} \geq \mathrm{LR}_T^{(0)})}{1+B}.$$

Example 12.4: Sea Surface Temperatures (Cont'd)

For illustration, we fit a VFCAR(1) model to the transformed daily SSTs at Granite Canyon and transformed WS data, i.e. $\{\mathbf{Y}_t = (Y_t, \mathrm{WS}_t)'\}_{t=1}^{1,825}$, letting $\mathbf{X}_t = \mathrm{WS}_{t-1}$. Figures 12.4(a) – (d) show the elements of the estimated $\boldsymbol{\Phi}_1(\cdot)$ matrix as a function of WS_{t-1}, using an Epanechnikov kernel with a single bandwidth across components, i.e. $h = 0.8T^{-1/5}$. The top left plot corresponds to the estimated FCAR coefficient of Y_{t-1} for the SST response, whereas the top right plot corresponds to the estimated FCAR coefficient of WS_{t-1}. The bottom plots are similar, but for the wind speed response.

For the SST response (Figure 12.4(a)), the coefficient of Y_{t-1} varies in the range [0.85 – 0.95], except when lagged values of WS are large. This corresponds roughly to the coefficient of 0.8971 for Y_{t-1} in the PMARS model (12.13). The estimated coefficient of Y_{t-1} for the lagged wind speed response (Figure 12.4(c)) is fairly constant around zero except in the boundary regions, possibly

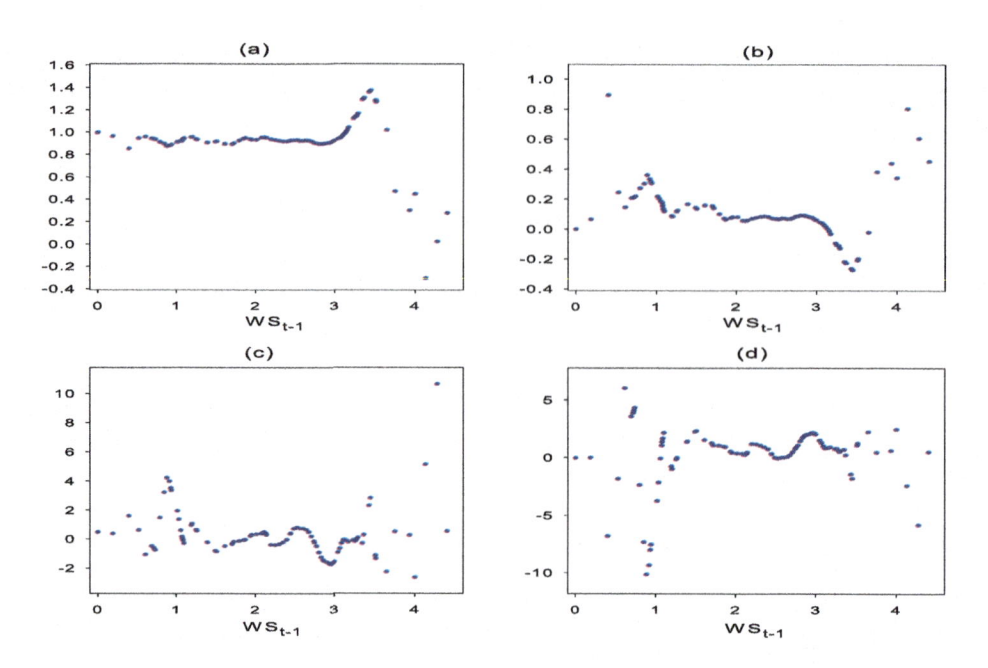

Figure 12.4: *Estimated* AR(1) *coefficients for the* VFCAR *model of* SST *and* wind speed (WS) *as a function of lag one wind speeds* (WS$_{t-1}$).

an artifact of boundary effects in the LL kernel-based smoothing method. The estimated coefficient of WS$_{t-1}$ for the wind speed response is close to the estimated coefficient of (0.1597+0.5189) from the PMARS model for WS$_{t-1}$ > 2.445, but does not correspond to the PMARS model coefficient of 0.1597 when WS$_{t-1}$ < 2.445. Of course, the fitted PMARS model (12.14) includes additional lagged wind speed terms, which are unaccounted for in the fitted VFCAR model.

12.3 Frequency-Domain Tests

Analogous to the univariate case (Section 1.1), we say that an m-dimensional stationary (up to the rth-order) time series process $\{\mathbf{Y}_t, t \in \mathbb{Z}\}$ is linear if it can be represented as

$$\mathbf{Y}_t = \sum_{j=-\infty}^{\infty} \mathbf{\Psi}_j \boldsymbol{\varepsilon}_{t-j}, \quad \sum_{j=-\infty}^{\infty} \|\mathbf{\Psi}_j\|^2 < \infty, \tag{12.22}$$

where $\{\mathbf{\Psi}_j\}$ is a sequence of $m \times m$ coefficient matrices and $\{\boldsymbol{\varepsilon}_t\}$ is a sequence of i.i.d. random vectors such that

$$\text{Cum}(\boldsymbol{\varepsilon}_t) = \mathbb{E}(\boldsymbol{\varepsilon}_t) = \mathbf{0},$$

$$\text{Cum}(\boldsymbol{\varepsilon}_{t_1}, \ldots, \boldsymbol{\varepsilon}_{t_r}) = \begin{cases} \mathbf{C}_{r,\varepsilon} & \text{if } t_1 = \cdots = t_r, \\ \mathbf{0} & \text{otherwise.} \end{cases}$$

12.3 FREQUENCY-DOMAIN TESTS

Here, $\mathbf{C}_{r,\varepsilon}$ is an $m^r \times 1$ column vector.

In view of (12.22), the second-order $m \times m$ spectral matrix $\mathbf{g}_Y(\omega)$ is defined as

$$\mathbf{g}_Y(\omega) = \sum_{\ell=-\infty}^{\infty} \mathbf{\Sigma}_Y(\ell) \exp(-2\pi i \omega \ell), \quad \omega \in [0, 1], \quad (12.23)$$

where

$$\mathbf{\Sigma}_Y(\ell) \equiv \mathrm{Cov}(\mathbf{Y}_t, \mathbf{Y}_{t+\ell}) = \sum_{j=-\infty}^{\infty} \mathbf{\Psi}_{j+\ell} \mathbf{\Sigma}_\varepsilon \mathbf{\Psi}'_j,$$

with $\mathbf{\Sigma}_\varepsilon = \mathbb{E}(\varepsilon_t \varepsilon'_t)$, and $\mathbf{C}_{2,\varepsilon} = \mathrm{vec}(\mathbf{\Sigma}_\varepsilon)$. Then the $(m^2 \times 1)$ second-order spectral vector, denoted by $\mathbf{f}_Y(\omega)$, is related to $\mathbf{g}_Y(\omega)$ by the expression

$$\begin{aligned}\mathbf{f}_Y(\omega) &= \mathrm{vec}\big(\mathbf{g}_Y(\omega)\big) \\ &= \big(\mathbf{H}(-\omega) \otimes \mathbf{H}(\omega)\big) \mathrm{vec}(\mathbf{\Sigma}_\varepsilon),\end{aligned} \quad (12.24)$$

where $\mathbf{H}(\omega) = \sum_{j=0}^{\infty} \mathbf{\Psi}_j \exp(-2\pi i \omega j)$ is the transfer function matrix, and $\mathbf{H}(-\omega) \equiv \mathbf{H}^*(\omega)$ the complex conjugate and transpose of $\mathbf{H}(\omega)$; cf. the univariate case in Section 4.1. In a similar manner, the rth-order $(r > 2)$ spectral density vector $(m^r \times 1)$ is given by (Wong, 1997; Subba Rao and Wong, 1999)

$$\mathbf{f}_Y(\omega_1, \ldots, \omega_{r-1}) = \{\mathbf{H}(\omega_1) \otimes \cdots \otimes \mathbf{H}(\omega_r)\} \mathbf{C}_{r,\varepsilon}, \quad (\omega_1, \ldots, \omega_r) \in [0, 1]^r, \quad (12.25)$$

where $\omega_r = -\sum_{j=1}^{r-1} \omega_j$.

If $\{\mathbf{Y}_t, t \in \mathbb{Z}\}$ is Gaussian distributed, $\mathbf{C}_{r,\varepsilon} = \mathbf{0}$ for $r > 2$, and all higher-order spectra are zero. On the other hand, if $\{\mathbf{Y}_t, t \in \mathbb{Z}\}$ has a linear (and non-Gaussian) representation of the form (12.22), Wong (1997) shows that

$$\begin{aligned}&\mathbf{f}_Y^*(\omega_1, \ldots, \omega_{r-1}) \big(\mathbf{g}_Y(\omega_1) \otimes \cdots \otimes \mathbf{g}_Y(\omega_r)\big)^{-1} \mathbf{f}_Y(\omega_1, \ldots, \omega_{r-1}) \\ &= \mathbf{C}'_{r,\varepsilon} \big(\mathbf{\Sigma}_\varepsilon \otimes \cdots \otimes \mathbf{\Sigma}_\varepsilon\big)^{-1} \mathbf{C}_{r,\varepsilon}.\end{aligned} \quad (12.26)$$

Note, the right-hand side of expression (12.26) is a constant, i.e. independent of $(\omega_1, \ldots, \omega_{r-1})$. Similar as in Section 4.1, this property forms the basis for testing linearity in the frequency domain as we explain below.

Let $X_t = \boldsymbol{\alpha}'\mathbf{Y}_t$ be a scalar time series process, where $\boldsymbol{\alpha}$ is an $m \times 1$ vector of constants and $\{\mathbf{Y}_t, t \in \mathbb{Z}\}$ is given by (12.22). Then the second-order spectral density function and the rth-order cumulant spectral density function of $\{X_t\}$ are given by

$$g_X(\omega) = \boldsymbol{\alpha}' \mathbf{g}_Y(\omega) \boldsymbol{\alpha} = (\boldsymbol{\alpha}^{[2]})' \mathbf{f}_Y(\omega) \quad (12.27)$$

$$f_X(\omega_1, \ldots, \omega_{r-1}) = (\boldsymbol{\alpha}^{[r]})' \mathbf{f}_Y(\omega_1, \ldots, \omega_{r-1}), \quad (12.28)$$

where $\mathbf{f}_Y(\omega_1, \ldots, \omega_{r-1})$ is given by (12.25), and $\boldsymbol{\alpha}^{[r]} = \boldsymbol{\alpha} \otimes \cdots \otimes \boldsymbol{\alpha}$ (r times). Using (12.27) and (12.28), it follows directly that the rth-order normalized spectral density function, defined by

$$\frac{|f_X(\omega_1, \ldots, \omega_{r-1})|^2}{g_X(\omega_1) \cdots g_X(\omega_{r-1}) g_X(\omega_1 + \cdots + \omega_{r-1})}, \qquad (12.29)$$

is not a constant, showing that a linear combination of $\{\mathbf{Y}_t, t \in \mathbb{Z}\}$ satisfying (12.22) is not linear (in contrast to Gaussianity). Note that for $m = 1$ and $r = 3$, (12.29) becomes the square modulus of the normalized bispectrum.

Clearly, linear combinations cannot be used for testing vector linearity. So, one has to test for vector linearity using (12.26). In fact, as a direct generalization of Hinich's test statistic for linearity in the univariate case (Section 4.2.2), Wong (1997) proposes the test statistic

$$\widehat{S}_Y = \sum_{(j,k) \in \mathcal{L}} \widehat{R}_{j,k}(\omega_j, \omega_k), \qquad (12.30)$$

where

$$\widehat{R}_{j,k}(\omega_j, \omega_k) = \widehat{f}_Y^*(\omega_j, \omega_k) \big(\widehat{g}_Y(\omega_j) \otimes \widehat{g}_Y(\omega_k) \otimes \widehat{g}_Y(-\omega_j - \omega_k)\big)^{-1} \widehat{f}_Y(\omega_j, \omega_k), \qquad (12.31)$$

with $\widehat{f}_Y(\omega_j, \omega_k)$ the bispectral vector estimator, and $\mathcal{L}$ a lattice in the principal domain $\mathcal{D}$ defined by (4.7). Then $\widehat{S}_Y$ is asymptotically distributed as $\chi^2_{2m^3 P}$ under the null hypothesis of Gaussianity, with P the number of $\widehat{R}_{j,k}$'s in $\mathcal{D}$.

Under the null hypothesis of linearity, and as $T \to \infty$, the test statistic $\widehat{S}_Y$ is asymptotically distributed as $\chi^2_{2m^3 P}(\widehat{\lambda}_0)$ where $\widehat{\lambda}_0 = P^{-1} \sum_{(j,k) \in \mathcal{L}} \widehat{R}_{j,k} - 2m^3$. Under the alternative hypothesis, the non-centrality parameter of the distribution is not constant. Thus, as in the univariate frequency-domain case, it is recommended to use the IQR of the EDF of $\{\widehat{R}_{j,k}\}$ to compare the dispersion of the $\widehat{R}_{j,k}$'s to that of $\chi^2_{2m^3 P}(\widehat{\lambda}_0)$.[4]

12.4 Lag Selection

Sample ACF, PACF, and CCF matrices are useful in specifying the lags to be used in linear VARMA models. In practice, these test statistics may not be helpful in weeding out nonsignificant variables with data generated by nonlinear processes. Recall, in Section 1.3 we introduced several test statistics for lag identification of univariate nonlinear time series models. It is straightforward to extend Kendall's $\widehat{\tau}(\ell)$ test statistic and Kendall's partial $\widehat{\tau}_p(\ell)$ test statistic to the multivariate case. In

---

[4] Apart from a very small MC simulation study by Wong (1997), the finite-sample behavior of both test statistics has not been investigated in detail. Since, however, (12.30) is a generalization of Hinich's linearity test statistic in the univariate case, Wong's multivariate test statistic may have the same general weaknesses; see Section 4.3.3.

12.4 LAG SELECTION

Table 12.3: *Climate change data set. Indicator pattern of the statistically significant values of the sample ACF, sample PACF, $\widehat{R}(\ell)$, Kendall's $\widehat{\tau}(\ell)$ and Kendall's partial $\widehat{\tau}_p(\ell)$ test statistics for the $\delta^{13}C$ and $\delta^{18}O$ time series; $T = 216$.*

Lag	ACF [1]	PACF [1]	$\widehat{R}(\ell)$ [2]	$\widehat{\tau}(\ell)$ [3]	$\widehat{\tau}_p(\ell)$ [3]
1	$\begin{pmatrix} + & - \\ - & + \end{pmatrix}$	$\begin{pmatrix} + & - \\ - & + \end{pmatrix}$	$\begin{pmatrix} \bullet & \bullet \\ \bullet & \bullet \end{pmatrix}$	$\begin{pmatrix} +^{**} & -^{**} \\ -^{**} & +^{**} \end{pmatrix}$	$\begin{pmatrix} +^{**} & -^{**} \\ -^{**} & +^{**} \end{pmatrix}$
2	$\begin{pmatrix} + & - \\ - & + \end{pmatrix}$	$\begin{pmatrix} + & \star \\ \star & \star \end{pmatrix}$	$\begin{pmatrix} \bullet & \bullet \\ \bullet & \bullet \end{pmatrix}$	$\begin{pmatrix} +^{**} & -^{**} \\ -^{**} & +^{**} \end{pmatrix}$	$\begin{pmatrix} +^{**} & -^{\dagger} \\ -^{**} & +^{**} \end{pmatrix}$
3	$\begin{pmatrix} + & - \\ - & + \end{pmatrix}$	$\begin{pmatrix} \star & + \\ \star & \star \end{pmatrix}$	$\begin{pmatrix} \bullet & \bullet \\ \bullet & \bullet \end{pmatrix}$	$\begin{pmatrix} +^{**} & -^{**} \\ -^{**} & +^{**} \end{pmatrix}$	$\begin{pmatrix} +^{**} & -^{\dagger} \\ +^{\dagger} & +^{\dagger} \end{pmatrix}$
4	$\begin{pmatrix} + & \star \\ \star & + \end{pmatrix}$	$\begin{pmatrix} \star & + \\ \star & \star \end{pmatrix}$	$\begin{pmatrix} \bullet & \bullet \\ \bullet & \bullet \end{pmatrix}$	$\begin{pmatrix} +^{**} & -^{*} \\ -^{\dagger} & +^{**} \end{pmatrix}$	$\begin{pmatrix} +^{*} & +^{\dagger} \\ +^{\dagger} & -^{\dagger} \end{pmatrix}$
5	$\begin{pmatrix} + & \star \\ \star & \star \end{pmatrix}$	$\begin{pmatrix} \star & \star \\ \star & \star \end{pmatrix}$	$\begin{pmatrix} \bullet & \bullet \\ \bullet & \bullet \end{pmatrix}$	$\begin{pmatrix} +^{**} & -^{\dagger} \\ -^{\dagger} & +^{\dagger} \end{pmatrix}$	$\begin{pmatrix} +^{\dagger} & +^{\dagger} \\ +^{\dagger} & +^{\dagger} \end{pmatrix}$

[1] $+$ indicates a value $> 1.96T^{-1/2}$, $-$ indicates a value $< -1.96T^{-1/2}$, and $\star$ indicates a value between $-1.96T^{-1/2}$ and $1.96T^{1/2}$.

[2] $\bullet$ indicates a value significantly different from zero at the 5% nominal level, and $\circ$ indicates a value *not* significantly different from zero at the 5% nominal level.

[3] ** marks a *p*-value smaller than 1%, * marks a *p*-value in the range 1% – 5%, and † marks a *p*-value larger than 5%.

a similar vein, Harvill and Ray (2000) define the multivariate version of the mutual information coefficient (1.20) at lag ℓ by

$$R(Y_{i,t}, Y_{j,t-\ell}) \equiv R_{i,j}(\ell), \quad (i, j = 1, \ldots, m; \ell \geq 1). \quad (12.32)$$

Simulation results indicate that the corresponding sample estimate of $R_{i,j}(\ell)$, say $\widehat{R}_{i,j}(\ell)$, identifies appropriate lagged nonlinear bivariate MA terms. Kendall's $\widehat{\tau}(\ell)$ and partial $\widehat{\tau}_p(\ell)$ test statistics have some power in identifying appropriate lagged nonlinear MA and AR terms, respectively, when the relationship between the lagged variables is monotonic. These test statistics fail when the nonlinear dependence is nonmonotonic, as with bivariate NLMA models.

Example 12.5: Climate Change (Cont'd)

As an example, we apply the lag identification techniques to the $\delta^{13}C$ and $\delta^{18}O$ ($T = 216$) time series introduced earlier in Example 1.5. Table 12.3 summarizes the significance of the sample ACF and PACF values at the 5% nominal level, in terms of three "indicator symbols"; see footnote (1) below the table. Similarly, we mark *p*-values of the test statistics $\widehat{R}(\ell)$, $\widehat{\tau}(\ell)$ and $\widehat{\tau}_p(\ell)$ through the symbols listed in footnote (2). To facilitate examination of $\widehat{R}(\ell)$, we obtain empirical significance levels by MC simulation using 1,000 replications of a bivariate Gaussian WN series of length $T = 216$; see Appendix 12.B for details.

The pattern of the sample ACF identifies a bi-directional association between $\delta^{13}C$ in year t and $\delta^{18}O$ one to three years back. The sample PACF takes almost all nonsignificant values after lag one, suggesting a VAR(1) model for *linearly* modeling the data. However, the pattern of indicator symbols for the $\widehat{R}(\ell)$ test statistic suggests a bi-directional nonlinear relationship between $\delta^{13}C$ and $\delta^{18}O$ up to lag three, and a uni-directional relationship from $\delta^{13}C$ to $\delta^{18}O$ at lags four and five involving no feedback. For Kendall's $\widehat{\tau}(\ell)$ test statistic, we see a significant bi-directional relationships between $\delta^{13}C$ and $\delta^{18}O$ up to and including lag three. Additionally, values of Kendall's partial $\widehat{\tau}_{\mathrm{p}}(\ell)$ test statistic are nonsignificant after lag one. In summary, these last three statistics suggest that a first-order NLAR model might be appropriate to model the interdependence between the two climate variables.

We have seen that the nonparametric test statistic $\widehat{R}(\ell)$ can serve as an initial way to infer causal nonlinear relationships. Some subjective interpretation problems, however, exist with this approach. We therefore need some more formal method to investigate causality, and we shall see in the next section how to achieve this.

12.5 Nonparametric Causality Testing

12.5.1 Preamble

Identifying causal relationships among a set of multivariate time series is important in fields ranging from physics to biology to economics. Indeed, using Granger's (1969) parametric causality test statistic there exists a large body of literature examining the presence of causal *linear* linkages between *bivariate* time series. On the other hand, there is substantially less literature on uncovering *nonlinear* causal relationships among strictly stationary *multivariate* time series variables. In this section, we discuss the concept of Granger causality in a more flexible nonparametric setting for both bivariate and multivariate time series processes. However, before doing so, we first introduce the general setting for testing causality.

Assume $\{(X_t, Y_t); t \in \mathbb{Z}\}$ is a strictly stationary bivariate time series process. We say that $\{X_t, t \in \mathbb{Z}\}$ is a strictly *Granger cause* of $\{Y_t, t \in \mathbb{Z}\}$ if past and current values of X_t contain additional information on future values of $\{Y_t\}$ that is not contained in the past and current Y_t-values alone. More formally, let $\mathcal{F}_{X,t}$ and $\mathcal{F}_{Y,t}$ denote the information sets consisting of past observations of X_t and Y_t up to and including time t. Then the process $\{X_t, t \in \mathbb{Z}\}$ is a Granger cause of $\{Y_t, t \in \mathbb{Z}\}$ if, for some $H \geq 1$,

$$(Y_{t+1}, \ldots, Y_{t+H})' | (\mathcal{F}_{X,t}, \mathcal{F}_{Y,t}) \stackrel{D}{\nsim} (Y_{t+1}, \ldots, Y_{t+H})' | \mathcal{F}_{Y,t}. \tag{12.33}$$

This definition is general and does not involve model assumptions. In practice one often assumes $H = 1$, i.e. testing for Granger non-causality (bivariate) comes down to comparing the one-step ahead conditional distribution of $\{Y_t, t \in \mathbb{Z}\}$, with and

12.5 NONPARAMETRIC CAUSALITY TESTING

without past and current observed values of $\{X_t, t \in \mathbb{Z}\}$. Note, the testing framework introduced above concerns conditional distributions given an infinite number of past observations. In practice, however, tests are usually confined to finite orders in $\{X_t, t \in \mathbb{Z}\}$ and $\{Y_t, t \in \mathbb{Z}\}$. To this end, we define the delay vectors

$$\mathbf{X}_t = (X_t, \ldots, X_{t-\ell_X+1})' \text{ and } \mathbf{Y}_t = (Y_t, \ldots, Y_{t-\ell_Y+1})', \quad (\ell_X, \ell_Y \geq 1).$$

If past observations of $\{\mathbf{X}_t, t \in \mathbb{Z}\}$ contain no information about future values, it follows from (12.33) that the null hypothesis of interest is given by

$$\mathbb{H}_0: \quad Y_{t+1}|(\mathbf{X}_t, \mathbf{Y}_t) \sim Y_{t+1}|\mathbf{Y}_t. \tag{12.34}$$

For a strictly stationary bivariate time series, (12.34) comes down to a statement about the invariant distribution of the $d_W = (\ell_X + \ell_Y + 1)$-dimensional vector $\mathbf{W}_t = (\mathbf{X}_t', \mathbf{Y}_t', Z_t)'$ where $Z_t = Y_{t+1}$. To simplify notation, we drop the time index t, and just write $\mathbf{W} = (\mathbf{X}', \mathbf{Y}', Z)'$.

Under $\mathbb{H}_0$, the conditional distribution of Z given $(\mathbf{X}', \mathbf{Y}')' = (\mathbf{x}', \mathbf{y}')'$ is the same as that of Z given $\mathbf{Y} = \mathbf{y}$. Then (12.34) can be restated in terms of ratios of joint distributions. Specifically, the joint pdf $f_{X,Y,Z}(\mathbf{x}, \mathbf{y}, z)$ and its marginals must satisfy the relationship

$$\frac{f_{X,Y,Z}(\mathbf{x}, \mathbf{y}, z)}{f_{X,Y}(\mathbf{x}, \mathbf{y})} = \frac{f_{Y,Z}(\mathbf{y}, z)}{f_Y(\mathbf{y})},$$

or equivalently

$$\frac{f_{X,Y,Z}(\mathbf{x}, \mathbf{y}, z)}{f_Y(\mathbf{y})} = \frac{f_{X,Y}(\mathbf{x}, \mathbf{y})}{f_Y(\mathbf{y})} \frac{f_{Y,Z}(\mathbf{y}, z)}{f_Y(\mathbf{y})}, \tag{12.35}$$

for each vector $(\mathbf{x}', \mathbf{y}', z)'$ in the support of $\mathbf{W}$.

12.5.2 A bivariate nonlinear causality test statistic

Along the lines of Baek and Brock (1992a,b) for testing conditional independence, Hiemstra and Jones (1994) devise a nonparametric Granger causality test statistic for bivariate relationships, sometimes called the HJ *test statistic*. The test employs ratios of correlation integrals to measure the discrepancy between the left- and right-hand sides of (12.35). Specifically, dropping the subscript m in the definition of the correlation integral (7.10), the test statistic is based on the equation

$$\frac{C_{X,Y,Z}(h)}{C_Y(h)} = \frac{C_{X,Y}(h)}{C_Y(h)} \frac{C_{Y,Z}(h)}{C_Y(h)}, \quad (h > 0). \tag{12.36}$$

Replacing the correlation integral $C_W(h)$ by its corresponding sample counterpart $\widehat{C}_W(h)$ defined in (7.43), the proposed test statistic is given by

$$Q_{T,W}(h) = \frac{\widehat{C}_{X,Y,Z}(h)}{\widehat{C}_Y(h)} - \frac{\widehat{C}_{X,Y}(h)}{\widehat{C}_Y(h)} \frac{\widehat{C}_{Y,Z}(h)}{\widehat{C}_Y(h)}, \tag{12.37}$$

where
$$\widehat{C}_W(h) = \binom{T}{2}^{-1} \sum_{1 \leq i \leq j \leq T} I(\|\mathbf{W}_i - \mathbf{W}_j\| < h).$$

Since the correlation integral is a U-statistic (Appendix 7.C), it can be shown (Hiemstra and Jones, 1994, Appendix) that, under $\mathbb{H}_0$,

$$\sqrt{T}\, Q_{T,W}(h) \xrightarrow{\mathcal{D}} \mathcal{N}(0, \sigma_W^2(h)), \quad \text{as } T \to \infty, \tag{12.38}$$

where $\sigma_W^2(h)$ is a lengthy expression, not given here. An autocorrelation consistent estimator of $\sigma_W^2(h)$ follows from using the theory of Newey and West (1987). In practice, it is recommend to use one-sided critical values of $Q_{T,W}(h)$. Bai et al. (2010) extend the HJ test statistic to the multivariate case.

12.5.3 A modified bivariate causality test statistic

Diks and Panchenko (2005, 2006) observe that, for a given nominal size, the actual rejection rate of $Q_{T,W}$ may tend to one as T increases, i.e. the test statistic over-rejects the null hypothesis. The reason is that equation (12.36) follows from (12.35) only in specific cases. For instance, when $\mathbf{X}$ and Z are independent conditionally on $\mathbf{Y} = \mathbf{y}$, for each fixed value of $\mathbf{y}$. To overcome this, and following Diks and Panchenko (2006), we rewrite the null hypothesis as

$$\mathbb{H}_0: \ \mathbb{E}\left[\left(\frac{f_{X,Y,Z}(\mathbf{X},\mathbf{Y},Z)}{f_Y(\mathbf{Y})} - \frac{f_{X,Y}(\mathbf{X},\mathbf{Y})}{f_Y(\mathbf{Y})}\frac{f_{Y,Z}(\mathbf{Y},Z)}{f_Y(\mathbf{Y})}\right) g(\mathbf{X},\mathbf{Y},Z)\right] = 0. \tag{12.39}$$

Here $g(\mathbf{x}, \mathbf{y}, z)$ is a positive weight function which for convenience is set at $g(\mathbf{x}, \mathbf{y}, z) = f_Y^2(\mathbf{y})$, giving more stable results than alternative weight functions. Thus, the corresponding functional is simply given by

$$\Delta \equiv \mathbb{E}[f_{X,Y,Z}(\mathbf{X},\mathbf{Y},Z) f_Y(\mathbf{Y}) - f_{X,Y}(\mathbf{X},\mathbf{Y}) f_{Y,Z}(\mathbf{Y},Z)] = 0. \tag{12.40}$$

Under $\mathbb{H}_0$ the term within square brackets vanishes, so that the expectation is zero. Clearly, (12.40) is a density-based distance measure similar in structure as the measures introduced in Section 7.2.3. In fact, Δ is closely related to the difference functional $\Delta^*(\cdot)$ given by (7.17).

Let $\widehat{f}_W(W_i)$ denote a local density estimator of a d_W-variate random vector $\mathbf{W}$ at W_i defined by

$$\widehat{f}_W(W_i) = \frac{(2h)^{-d_W}}{T-1} \sum_{j, j \neq i} I_{ij}^{(W)},$$

where $I_{ij}^{(W)} = I(\|\mathbf{W}_i - \mathbf{W}_j\| < h)$. Given this estimator, the proposed nonparametric Granger causality (bivariate) test statistic is given by

$$Q^*_{T,W}(h) = \frac{T-1}{T(T-2)} \sum_i \widehat{f}_Y^2(Y_i)\{\widehat{f}_{X,Z|Y}(X_i, Z_i|Y_i) - \widehat{f}_{X|Y}(X_i|Y_i)\widehat{f}_{Z|Y}(Z_i|Y_i)\}.$$

$$\tag{12.41}$$

12.5 NONPARAMETRIC CAUSALITY TESTING

For an appropriate sequence of bandwidths, the estimator $\widehat{f}_W(\cdot)$ of the pdf $f_W(\cdot)$ is consistent. So, $Q^*_{T,W}(h)$ consists of a weighted average of local contributions given by the expression in curly brackets, which tends to zero in probability under $\mathbb{H}_0$.

The test statistic (12.41) can be rearranged in terms of a U-statistic as follows,

$$Q^*_{T,W}(h) = \frac{1}{T(T-1)(T-2)} \sum_{i \neq j \neq k \neq i} K(W_i, W_j, W_k), \qquad (12.42)$$

where

$$K(W_j, W_j, W_k) = \frac{(2h)^{-d_X - 2d_Y - d_Z}}{3!} \Big(\\
(I^{(XYZ)}_{ik} I^{(Y)}_{ij} - I^{(XY)}_{ik} I^{(YZ)}_{ij}) + (I^{(XYZ)}_{ij} I^{(Y)}_{ik} - I^{(XY)}_{ij} I^{(YZ)}_{ik}) \\
+ (I^{(XYZ)}_{jk} I^{(Y)}_{ji} - I^{(XY)}_{jk} I^{(YZ)}_{ji}) + (I^{(XYZ)}_{ji} I^{(Y)}_{jk} - I^{(XY)}_{ji} I^{(YZ)}_{jk}) \\
+ (I^{(XYZ)}_{ki} I^{(Y)}_{kj} - I^{(XY)}_{ki} I^{(YZ)}_{kj}) + (I^{(XYZ)}_{kj} I^{(Y)}_{ki} - I^{(XY)}_{kj} I^{(YZ)}_{ki})\Big).$$

By exploiting the asymptotic theory for U-statistics, assuming that $h = cT^{-\beta}$ ($c > 0$, $\beta > 0$), and setting $d_X = d_Y = d_Z = 1$, it can be shown (Diks and Panchenko, 2006, Appendix A.1) that, as $T \to \infty$, (12.42) satisfies

$$\sqrt{T} \frac{Q^*_{T,W}(h) - \Delta}{\sigma_W(h)} \xrightarrow{\mathcal{D}} \mathcal{N}(0,1), \text{ iff } \frac{1}{2\nu} < \beta < \frac{1}{d_X + d_Y + d_Z}, \qquad (12.43)$$

where ν is the order of the density estimation kernel (Appendix 7.A), as opposed to the U-statistics kernel, and where

$$\sigma^2_W(h) = 9 \operatorname{Var}(r_0(W_i)), \quad \text{with} \quad r_0(w) = \lim_{h \to 0} \mathbb{E}(K(w_1, W_2, W_3)),$$

and W_i ($i = 1, 2, 3$) are i.i.d. random variables according to $\mathbf{W}$.

A consistent estimate of $r_0(W_i)$ is given by

$$\widehat{r}_0(W_i) = \frac{(2h)^{-d_X - 2d_Y - d_Z}}{(T-1)(T-2)} \sum_{j, j \neq i} \sum_{k, k \neq i} K(W_i, W_j, W_k).$$

An autocorrelation consistent estimator for $\sigma^2_W(h)$ (Newey and West, 1987) is given by

$$S^2_{T,W}(h) = \sum_{\ell=1}^{[T^{1/4}]} \widehat{\gamma}_W(\ell) \omega_T(\ell),$$

where $\widehat{\gamma}_W(\ell)$ is the lag ℓ sample ACVF, i.e.

$$\widehat{\gamma}_W(\ell) = \frac{1}{T-\ell} \sum_{i=1}^{T-\ell} (\widehat{r}_0(W_i) - Q_T)(\widehat{r}_0(W_{i+\ell}) - Q_T),$$

and $\omega_T(\ell)$ is a weight function given by $\omega_T(\ell) = 1$, if $\ell = 1$, and $\omega_T(\ell) = 2(1 - (\ell-1)/[T^{1/4}])$, otherwise, which declines as ℓ increases. Then, under suitable mixing conditions (Denker and Keller, 1983), it follows that the test statistic $Q^*_{T,W}(h)$ satisfies

$$\sqrt{T}\frac{Q^*_{T,W}(h) - \Delta}{S_{T,W}(h)} \xrightarrow{\mathcal{D}} \mathcal{N}(0,1), \text{ as } T \to \infty. \quad (12.44)$$

It is recommended to use a one-sided version of $Q^*_{T,W}(h)$, rejecting the null hypothesis when the left-hand side of (12.42) is too large, because in practice it is often found to have larger power than a two-sided test.

Example 12.6: Climate Change (Cont'd)

In Examples 1.5, 7.7, and 12.5 we analyzed the δ^{13}C ($Y_{1,t}$) and δ^{18}O ($Y_{2,t}$) climate change time series. Here, we consider an extended version of the ODP data set with insolation ($Y_{3,t}$) as an additional variable. Insolation is a measure of solar radiation energy received at a given latitude on Earth. Its value largely depends on astronomical, often called Milankovitch, parameters. The Milankovitch theory proposes that variation in the Earth's orbital elements and therefore changes in insolation are a driving force of climate change, a hypothesis that has been supported by various empirical studies. All series are rescaled to zero-mean and unit-variance.

Figure 12.5 shows path diagrams for the nonparametric causality test statistics $Q_{T,W}(h)$ (top row) and $Q^*_{T,W}(h)$ (bottom row), at lags $\ell_{Y_1} = \ell_{Y_2} = 1, \ldots, 5$, and bandwidth $h = 1.5$.[5] The absence of an arrow from a node i to a node j ($i \neq j$) means that $Y_{i,t}$ is a non-Granger-cause of $Y_{j,t}$, i.e. the null hypothesis (12.34) is not rejected. Both test statistics indicate a very strong nonlinear causal (often bi-directional) relationship from δ^{18}O ($Y_{2,t}$) to δ^{13}C ($Y_{1,t}$) at all lags. This confirms earlier results presented in Table 12.3. Furthermore, at lags 1 – 3, the modified test statistic $Q^*_{T,W}(h)$ suggests that insolation ($Y_{3,t}$) is an important driving force for global warming either directly, or mediated by δ^{18}O ($Y_{2,t}$) indirectly. The causality graph for the HJ test statistic $Q_{T,W}(h)$ only suggests this indirect relationship at lag two. Interestingly, for all other lags, there is a complete absence of significant nonlinear causal relationships between insolation on the one hand, and δ^{13}C ($Y_{1,t}$) and δ^{18}O ($Y_{2,t}$) on the other.

12.5.4 A multivariate causality test statistic

The above bivariate nonparametric test statistics allow for pairwise causality testing, as in Example 12.6. However, the outcome of the test statistics may be blurred by the

---

[5] Diks and Panchenko (2006) show that the estimator $Q^*_{T,W}(h)$ has the smallest MSE with the rate $\beta = 2/7$. This implies a bandwidth of approximately 1.5, with $C = 7$ and $T = 216$. The bias of the HJ test statistic $Q_{T,W}(h)$ cannot be removed by choosing a bandwidth smaller than 1.5.

12.5 NONPARAMETRIC CAUSALITY TESTING

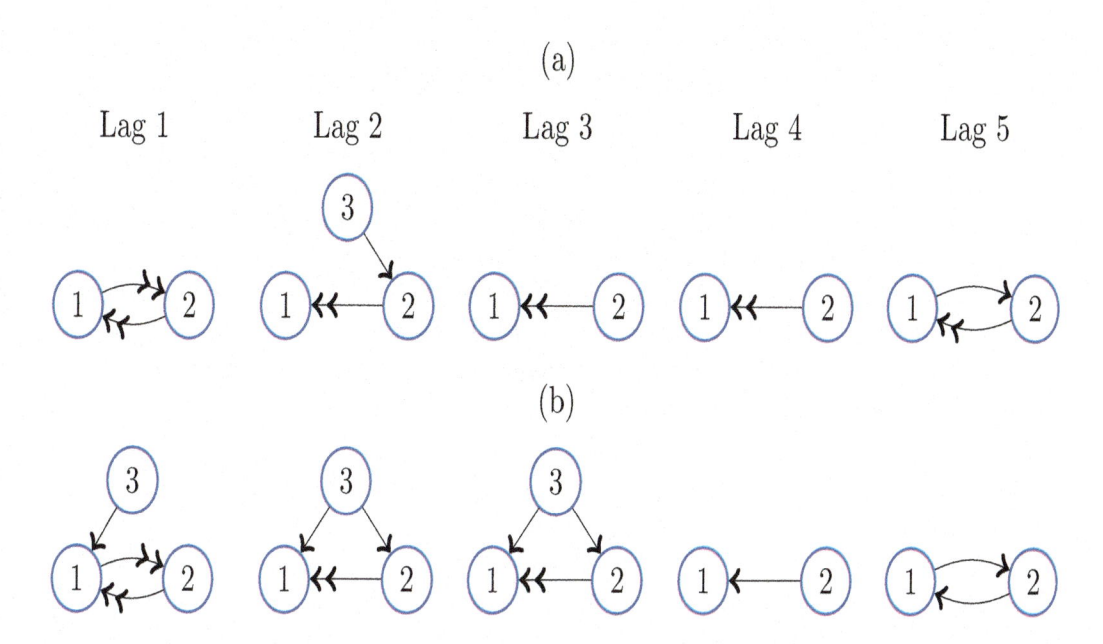

(a)

Lag 1　　　　Lag 2　　　　Lag 3　　　　Lag 4　　　　Lag 5

(b)

Figure 12.5: *Extended Climate change data set. Nonparametric causality testing at lags $\ell_{Y_1} = \ell_{Y_2} = 1, \ldots, 5$; with $h = 1.5$; (a) $Q_{T,W}(h)$ test statistic and (b) $Q^*_{T,W}(h)$ test statistic. The single arrow symbol marks a p-value in the range $1\% - 5\%$, and the double arrow symbol marks a p-value smaller than 1%; $T = 216$.*

confounding effect of other variables. One simple way to control these additional variables is by pre-filtering the multivariate data by a parametric model (e.g. a linear VAR model), and next performing a bivariate causality test of the residuals pairwise. As an alternative, Diks and Wolski (2016) generalize the bivariate test statistic $Q^*_{T,W}(h)$ to a multivariate setting. Following these authors, we first state a generalization of (12.33).

Consider the strictly stationary multivariate time series process $\{(X_t, Y_t, Q_t), t \in \mathbb{Z}\}$, where $\{X_t, t \in \mathbb{Z}\}$ and $\{Y_t, t \in \mathbb{Z}\}$ are univariate time series processes, and $\{Q_t, t \in \mathbb{Z}\}$ is a univariate or multivariate time series process. Then the process $\{X_t, t \in \mathbb{Z}\}$ is a Granger cause of $\{Y_t, t \in \mathbb{Z}\}$ if, for some $H \geq 1$,

$$(Y_{t+1}, \ldots, Y_{t+H})' | (\mathcal{F}_{X,t}, \mathcal{F}_{Y,t}, \mathcal{F}_{Q,t}) \overset{D}{\not\sim} (Y_{t+1}, \ldots, Y_{t+H})' | \mathcal{F}_{Y,t} \mathcal{F}_{Q,t}, \quad (12.45)$$

where $\mathcal{F}_{X,t}, \mathcal{F}_{Y,t}$, and $\mathcal{F}_{Q,t}$ are the corresponding information sets. Note, the assumption that both $\{X_t, t \in \mathbb{Z}\}$ and $\{Y_t, t \in \mathbb{Z}\}$ are scalar-valued time series processes makes it possible to determine whether the causal relationship between these two processes is direct or mediated by other variables.

Now, consider the same setup as in Section 12.5.1 with the delay vectors $\mathbf{X}_t, \mathbf{Y}_t$, and $\mathbf{Q}_t = (Q_t, \ldots, Q_{t-\ell_Q+1})'$. So, the multivariate analogue of the null hypothesis (12.34) is given by

$$\mathbb{H}_0: \quad Y_{t+1} | (\mathbf{X}_t, \mathbf{Y}_t, \mathbf{Q}_t) \sim Y_{t+1} | (\mathbf{Y}_t, \mathbf{Q}_t). \quad (12.46)$$

For simplicity, assume that the embedding dimensions are all equal to unity, i.e. $\ell_X = \ell_Y = \ell_Q = 1$. Thus, the dimensionality of the vector $\mathbf{W}_t = (\mathbf{X}'_t, \mathbf{Y}'_t, \mathbf{Q}'_t, Z_t)'$, where $Z_t = Y_{t+1}$, is a number $d_W \geq 4$. In this case, and following the same reasoning as in Section 12.5.3, the asymptotic normality condition becomes $1/(2\nu) < \beta < 1/d_W$. So, for a standard second-order kernel ($\nu = 2$) and $d_W \geq 4$, there is no feasible β-region which would endow the test statistic $Q^*_{T,W}(h)$ with asymptotic normality. The associated problem is the well-known curse of dimensionality.

One solution, followed by Diks and Wolski (2016), is to improve the precision of the density estimator by reducing the kernel estimator bias using *data-sharpening* (Hall and Minotte, 2002) as a bias reduction method. The sharpened (s) form of the plug-in density estimator is given by

$$\widehat{f}^s_W(W_i) = \frac{h^{-d_W}}{T-1} \sum_{j, j \neq i} K\left(\frac{W_i - \psi_p(W_j)}{h}\right),$$

where $\psi_p(\cdot)$ is a so-called sharpening function, with p the order of bias reduction. On replacing the data by their sharpened form in the definition of the kernel density estimator $\widehat{f}_W(\cdot)$ one obtains an estimator of $f_W(\cdot)$ of which the bias equals $\mathcal{O}(h^4)$, with $p \equiv d_W = 4$, rather than $\mathcal{O}(h^2)$ (Hall and Minotte, 2002) for $\widehat{f}_W(\cdot)$. In this case the sharpening function is of the form

$$\psi_4(W) = I + h^2 \frac{\mu_2(K)}{2} \frac{\widehat{f}'(W)}{\widehat{f}(W)},$$

where I denotes the identity function, $\mu_2(K) = \int_{\mathbb{R}} u^2 K(u) du$, and $\widehat{f}'$ is the estimator of the gradient of f. In practice, the NW kernel estimator may be used as an approximation for the ratio $\widehat{f}'(W)/\widehat{f}(W)$. Clearly, the lower order of the bias makes it possible to find a range of feasible β-values again, in this case $\beta \in (1/(2p), 1/d_w) = (1/8, 1/4)$.

The sharpened form of the test statistic is given by

$$Q^s_{T,W}(h) = \frac{T-1}{T(T-2)} \sum_i \left(\widehat{f}^s_{X,YZ}(X_i, Y_i, Z_i) \widehat{f}^s_Y(Y_i) - \widehat{f}^s_{X,Y}(X_i, Y_i) \widehat{f}^s_{Y,Z}(Y_i, Z_i)\right).$$

(12.47)

Under certain mixing conditions Diks and Wolski (2016, Appendix B) show that, as $T \to \infty$,

$$\sqrt{T} \frac{Q^s_{T,W}(h) - \Delta}{S_T} \xrightarrow{\mathcal{D}} \mathcal{N}(0,1), \text{ iff } \frac{1}{2p} < \beta < \frac{1}{d_W},$$

(12.48)

where S_T^2 is a consistent estimator of the asymptotic variance of $\sqrt{T}\left(Q^s_{T,W}(h) - \Delta\right)$.

12.6 Summary, Terms and Concepts

Summary

In the first part of this chapter, we focused on a multivariate conditional quantile estimator using a kernel-based method, and we explored its use in forecasting multivariate nonlinear time series. In addition, we discussed three semiparametric multivariate regression methods. Depending on the modeling goal, each of these methods can be used as an ends in itself, or as a technique for exploring the structure in the data to aid in proposing a particular parametric vector time series model. Nevertheless, issues such as stationarity, ergodicity, and variable selection of the fitted semiparametric models are still largely open for research.

In the second part, we discussed two nonlinear and nonparametric test statistics for investigating Granger noncausality in a bivariate setting: the HJ test statistic, and a test statistic proposed by Diks and Panchenko (2006). The second test statistic avoids the over-rejection problem of the first one. However, it lacks consistency in a multivariate setting. The problem is the result of the kernel density estimator bias, which does not converge to zero at a sufficiently fast rate when the number of conditioning variables is larger than one. One solution is to use a data-sharpening method which reduces the bias of the original estimator without affecting the order of its variance. Readers are invited to compare this approach with other methods to reduce the dimensionality problem; Scott (1992).

Terms and Concepts

data sharpening, 520
Granger cause, 514
Hiemstra–Jones (HJ) test, 515
multivariate conditional quantiles, 496
polyMARS (PMARS), 502

projection pursuit regression (PPR), 504
second order spectral vector, 511
spectral matrix, 511

12.7 Additional Bibliographical Notes

Section 12.1.1: There are many ways to define multivariate quantiles; see, e.g., Serfling (2002, 2004). Two different approaches based on norm minimization are by Abdous and Theodorescu (1992) and Chaudhuri (1996). Throughout this section, conditional quantiles are based on the definition of Chaudhuri (1996) for unconditional quantiles. In general, there has been a proliferation of research aimed at extending quantiles for multivariate data. Few studies, however, deal with the case where covariates are allowed to explain the distribution of the multivariate data. One notable exception is Chakraborty (2003) who proposes a technique for estimating "linear" conditional quantiles with multivariate responses. In contrast, the nonparametric method proposed in this chapter estimates conditional quantiles from multiple responses when no restriction (i.e. not necessarily linear) is imposed on the form of the conditional quantile function.

Section 12.1.2: The section is based on De Gooijer et al. (2006). Cheng and De Gooijer (2007) focus on an alternative formulation of multivariate conditional quantiles generalizing a notion of geometric or spatial quantile studied by Chaudhuri (1992, 1996).

Section 12.1.3: Yang and Shahabi (2007) present a similarity measure to efficiently perform k-NN searches for vector time series. Fernández–Rodríguez et al. (1997, 1999) apply the k-NN multivariate method to nine currencies participating in the European Monetary System.

Section 12.2.1: De Gooijer and Ray (2003) provide an extensive discussion of the various "tuning" parameters in the S-Plus and R implementations of PMARS. These authors also illustrate the use of PMARS by fitting various VASTAR(X) models to two series of half-hourly average electricity load data. The data (electricity.dat) can be downloaded from the website of this book.

Section 12.2.3: Harvill and Ray (2005) compare the forecasting performance of VFCAR models using a simple "plug-in" approach, a bootstrapped-based approach, and a multi-stage smoothing approach, where the functional coefficients are updated in a rolling framework. The BS approach outperforms the other two methods. Baniscescu et al. (2005, 2011) present an approach to the parallelization of VFCAR–MC simulations to reduce computational time when bandwidth selection and bootstrapped-based model assessment are parts of the analysis.

Section 12.3: Subba Rao and Wong (1998) propose frequency-domain test statistics for Gaussianity and linearity of multivariate stationary time series based on classical multivariate measures of skewness and kurtosis. Rao et al. (2006) present a unified and comprehensive approach for deriving expressions for higher-order cumulants of random vectors. It is used to study the asymptotic theory of test statistics for multivariate stationary nonlinear time series processes.

Quite some scientific work has been published on nonparametric test statistics for stationarity in the framework of so-called *locally stationary* univariate time series processes; see, e.g., Puchstein and Preuß (2016) and the references therein. Also, these authors present a nonparametric procedure for validating local-stationarity in the multivariate time series case.

Section 12.4: In principle, the FPE criterion of Tschernig and Yang (2000) (see Section 9.1.6) may be used as an alternative model lag selection method in the multivariate case. Unfortunately, the explosion in the number of possible lagged predictors results in the curse of dimensionality for kernel-based regression methods used in estimating the nonlinear ARs. So, model selection based on the nonparametric FPE criterion is not feasible. The regression subset method, a *parametric approach*, of Rech et al. (2001) provides an attractive and easily implemented alternative. The method goes as follows in a multivariate setting.

(i) For a given sample size T, select the polynomial order ℓ in the truncated Volterra representation for $\{\mathbf{Y}_t\}_{t=1}^{T}$. A larger ℓ is necessary for larger T.

(ii) Regress $\{\mathbf{Y}_t\}$ on all variables (lagged values of $\{\mathbf{Y}_t\}$, any exogenous variables, and products up to order ℓ of all lagged values and exogenous variables) and compute the value of an appropriate model selection criterion, such as AIC or BIC.

(iii) Omit one regressor from the original model, regress the time series $\{\mathbf{Y}_t\}$ on all remaining variables in the ℓth order Taylor series expansion and compute the value of the selection criterion.

(iv) Repeat, omitting one regressor each time. Continue, omitting two regressors at a time, etc. until the regression consists of only a constant term (all regressors removed, corresponding to $\{\mathbf{Y}_t, t \in \mathbb{Z}\}$ being WN).

(v) The combination of regressors resulting in the optimal model selection criterion value is selected.

Section 12.5: By exploiting the geometry of reproducing kernel Hilbert spaces, Marinazzo et al. (2008) develop a nonlinear Granger causality test statistic for bivariate time series. Gao and Tian (2009) consider the construction of Granger causality graphs for multivariate nonlinear time series. Péguin–Feissolle et al. (2013) propose two test statistics for bivariate Granger non-causality in a stationary nonlinear model of unknown functional form. The idea is to globally approximate the potential causal relationship between the variables by a Taylor series expansion. A few applications of the test statistics in Section 12.5.3 have been reported. For instance, Bekiros and Diks (2008) investigate linear and nonlinear causal linkages among six currencies. De Gooijer and Sivarajasingham (2008) apply both parametric and nonparametric Granger causality tests to determine linkages between international stock markets. Francis et al. (2010) use both linear and nonlinear causality tests to examine the relationship between the returns on large and small firms.

12.8 Data and Software References

Data
Example 16.2: The bivariate series of daily returns of exchange rates (ExchangeRates.dat) can be downloaded from the website of this book.

Software References
Section 12.2.1: PolyMARS (or PMARS) is available in the R-polspline package. The R-fRegression package has an option for computing a PMARS model as a part of the function regFit; see also the references to software packages in Section 9.5.

Section 12.2.2: The function ppr in the R-stat package, and the function ppreg in S-Plus both allow for PPR model fitting with multivariate responses.

Section 12.5: R codes for performing the HJ (hj.r) and the Diks–Panchenko (dp.r) nonparametric test statistics are available at the website of this book. The C source code, and an executable file, for computing both test statistics can be downloaded from http://www1.fee.uva.nl/cendef/upload/6/hjt2.zip. Alternatively, a windows version and C source code are available at http://research.economics.unsw.edu.au/vpanchenko/#software. C source code for the multivariate nonlinear nonparametric Granger causality test is available at http://qed.econ.queensu.ca/jae/datasets/diks001/.

Appendix

12.A Computing Multivariate Conditional Quantiles

To solve a highly discontinuous problem such as (12.4) numerically, the most obvious choice is the simplex algorithm. However, a simplex search becomes less efficient when for dimension $m > 2$. In fact, convergence becomes extremely slow. Thus, we suggest here a simple iteratively re-weighted least squares algorithm. The idea of the algorithm is to transform an L_1-like minimization problem into an L_2-minimization problem such that weighted least squares can be applied. First, we rewrite (12.4) as follows

$$\boldsymbol{\theta}_{q,T}(\mathbf{x}) = \arg\min_{\boldsymbol{\theta} \in \mathbb{R}^m} \sum_{t=1}^{T} \|\mathbf{Y}_t - \boldsymbol{\theta}\|_q \mathcal{K}_h(\mathbf{x} - \mathbf{X}_t)$$

$$= \arg\min_{\boldsymbol{\theta} \in \mathbb{R}^m} \sum_{t=1}^{T} (\|\mathbf{Y}_t - \boldsymbol{\theta}\|_q)^2 G_q(\mathbf{x}, \mathbf{X}_t, \mathbf{Y}_t; \boldsymbol{\theta}, h), \qquad (A.1)$$

where

$$G_q(\mathbf{x}, \mathbf{X}_t, \mathbf{Y}_t; \boldsymbol{\theta}, h) = \frac{\mathcal{K}_h(\mathbf{x} - \mathbf{X}_t)}{\|\mathbf{Y}_t - \boldsymbol{\theta}\|_q}.$$

Note that, for $\mathbf{Y}_t = (Y_{1,t}, \ldots, Y_{m,t})'$ and $\boldsymbol{\theta} = (\theta_1, \ldots, \theta_m)'$,

$$\|\mathbf{Y}_t - \boldsymbol{\theta}\|_q = \|0.5\,[\text{sign}(Y_{1,t} - \theta_1) + (2q-1)](Y_{1,t} - \theta_1), \ldots,$$
$$0.5\,[\text{sign}(Y_{m,t} - \theta_m) + (2q-1)](Y_{m,t} - \theta_m)\|.$$

We now follow an iterative approach to solve (A.1). Let $\boldsymbol{\theta}_{q,T}^{(1)}(\mathbf{x}), \ldots, \boldsymbol{\theta}_{q,T}^{(r)}(\mathbf{x})$ be successive approximations of $\boldsymbol{\theta}_{q,T}(\mathbf{x})$ obtained in consecutive iterations. Let $\mathbf{1} = (1, \ldots, 1)'$ denote a unity row vector with dimension m. First, we define the $T \times m$ matrix $\mathbf{W}_q(\cdot)$ as a direct (or Hadamard) product ($\odot$) of two $T \times m$ matrices, i.e.

$$\mathbf{W}_q(\mathbf{Y}, \mathbf{x}, \mathbf{X}; \boldsymbol{\theta}, h) = \mathbf{M}_q(\mathbf{Y}; \boldsymbol{\theta}) \odot \{\mathbf{G}_q(\mathbf{x}, \mathbf{X}, \mathbf{Y}; \boldsymbol{\theta}, h) \times \mathbf{1}\},$$

where the $T \times m$ matrix $\mathbf{M}_q(\mathbf{Y}; \boldsymbol{\theta})$ is given by

$$\mathbf{M}_q(\mathbf{Y}; \boldsymbol{\theta}) =$$
$$(0.5)^2 \begin{pmatrix} \{\text{sign}(Y_{1,1} - \theta_1) + (2q-1)\}^2, \ldots, \{\text{sign}(Y_{m,1} - \theta_m) + (2q-1)\}^2 \\ \{\text{sign}(Y_{1,2} - \theta_1) + (2q-1)\}^2, \ldots, \{\text{sign}(Y_{m,2} - \theta_m) + (2q-1)\}^2 \\ \cdots \\ \{\text{sign}(Y_{1,T} - \theta_1) + (2q-1)\}^2, \ldots, \{\text{sign}(Y_{m,T} - \theta_m) + (2q-1)\}^2 \end{pmatrix}$$

and the $T \times 1$ vector $\mathbf{G_q}(\cdot)$ is

$$\mathbf{G}_q(\mathbf{x}, \mathbf{X}, \mathbf{Y}; \boldsymbol{\theta}, h) = \big(G_q(\mathbf{x}, \mathbf{X}_1, \mathbf{Y}_1; \boldsymbol{\theta}, h), \ldots, G_q(\mathbf{x}, \mathbf{X}_T, \mathbf{Y}_n; \boldsymbol{\theta}, h)\big)'.$$

The vector $\mathbf{1}$ is used to resize the vector $\mathbf{G}_q(\cdot)$ into a $T \times m$ matrix. Then, at iteration step $(r+1)$, $\boldsymbol{\theta}_{q,T}^{(r+1)}(\mathbf{x})$ is simply computed by,

$$\boldsymbol{\theta}_{q,T}^{(r+1)}(\mathbf{x}) = \frac{\sum\{\mathbf{Y} \odot \mathbf{W}_q(\mathbf{Y}, \mathbf{x}, \mathbf{X}; \boldsymbol{\theta}_{q,T}^{(r)}, h)\}}{\sum\{\mathbf{W}_q(\mathbf{Y}, \mathbf{x}, \mathbf{X}; \boldsymbol{\theta}_{q,T}^{(r)}, h)\}}. \qquad (A.2)$$

The sum $\sum$ in the above formula refers to the sum for each column and the division is a direct division. Equation (A.2) shows that once $\boldsymbol{\theta}_{q,T}^{(r)}$ is given, the solution to (A.1) at iteration step $r+1$ simply follows from applying weighted least squares.

The iteration is continued until two successive approximations of $\boldsymbol{\theta}_{q,T}(\mathbf{x})$ are sufficiently close. For the numerical illustration in this chapter, convergence is assumed if $\|\boldsymbol{\theta}_{q,T}^{(r+1)}(\mathbf{x}) -$

$\boldsymbol{\theta}_{q,T}^{(r)}(\mathbf{x})\|_2 \leqslant 10^{-3}\|(\mathbf{Y} - \mathbf{1}' \times \boldsymbol{\theta}_{q,T}^{(1)}(\mathbf{x}))\|_2$. The above algorithm is fully vectorized so that it can be easily implemented in matrix oriented software packages like GAUSS or MATLAB (see, e.g., the file illustrate.m).

It is worth noting that the algorithm requires a good initial approximation of $\boldsymbol{\theta}_{q,T}(\mathbf{x})$ to start the iteration. We suggest the following approach. When $q = 0.5$, the conditional mean can be taken as the starting value. For $q > 0.5$ or $q < 0.5$, one may start from the optimal value for $q = 0.5$ and move upward or downward. For example, to estimate the conditional quantile at $q = 0.9$, one may first estimate this quantity for $q = 0.6$ starting from $q = 0.5$. Then estimate the conditional quantile for $q = 0.7$ starting from $q = 0.6$ and so on until the end. In doing so, convergence to local optimum is facilitated.

Finally, it is interesting to mention that the proposed estimator is more efficient in the sense that it requires less computing time than the corresponding univariate estimator. This is the case even for dimension m as high as 7 or 8. This empirical evidence may suggest that the fast converging property of the unconditional multivariate quantiles (see, e.g., Chaudhuri, 1996) may also be shared by the conditional estimator defined above.

12.B   Percentiles of the $\widehat{R}(\ell)$ Test Statistic

Following Harvill and Ray (2000, Section 2.2), we estimate the marginal densities by smoothing the standardized data with a (scaled) second-order Student t_ν kernel-based density, as given by

$$K(u) = \frac{\Gamma\big((\nu+1)/2\big)/\big(\sqrt{\pi\nu}\,\Gamma(\nu/2)\big)}{h\big(1 + u^2/(\nu h^2)\big)^{(\nu+1)/2}}, \qquad (B.1)$$

with $\nu = 4$ degrees of freedom, and adopting a bandwidth $h = 0.85T^{-1/5}$. We estimate the bivariate density of the pair of random variables (X, Y) by a product kernel of Student's t_4 distributions with bandwidth

$$h = 0.85(1 - \rho_{XY}^2)^{5/12}(1 + \rho_{X,Y}^2/2)^{-1/6}T^{-1/6},$$

where $\rho_{X,Y}$ is the correlation coefficient. Apart from the factor 0.85, this particular bandwidth follows from minimizing the AMISE using a bivariate Gaussian kernel; see Scott (1992, Section 6.3.1). The choice for the Student t_4 kernel is motivated by the work of Hall and Morton (1993). No boundary correction is needed in both kernel-based density computations since the Student t distribution has infinite support. In addition, we estimate the integrals in (1.18) numerically using a 30-point Gaussian quadrature. The limits of the integration are chosen conservatively, as the minimum and maximum of the observed data.

Table 12.3 shows the empirical mean, standard deviation, and 90%, 95%, and 99% percentile points of the $\widehat{R}(\ell)$ test statistic for various sample sizes T, and lags ℓ using 1,000 MC replications. The results for $T = 300$ are in agreement with percentiles reported by Harvill and Ray (2000, Table I). It is clear that $\widehat{R}(\ell)$ is biased in finite samples. As expected, the bias decreases as T increases. Joe (1989) and Hall and Morton (1993) show that a summation-based estimator of the Shannon entropy $H(\mathbf{X}) = -\int \log\{f_X(\mathbf{x})\} f_X(\mathbf{x})d\mathbf{x}$ of an m-dimensional random variable $\mathbf{X}$, and thus of $R(\ell)$, is root-n consistent in $m = 1, 2$ and 3 dimensions. This result requires certain properties of the tails of the underlying distribution.

Table 12.4: *Empirical mean, standard deviation, and percentile points of the $\widehat{R}(\ell)$ test statistic for dimension $m = 2$, various sample sizes T, and lags ℓ; 1,000 MC replications.*

Lag	90%	95%	99%	Mean	Std.dev	90%	95%	99%	Mean	Std.dev
			$T = 100$					$T = 200$		
1	0.2536	0.2604	0.2755	0.2282	0.0200	0.2214	0.2266	0.2370	0.2033	0.0139
2	0.2535	0.2620	0.2768	0.2287	0.0197	0.2225	0.2278	0.2361	0.2037	0.0142
3	0.2542	0.2636	0.2757	0.2290	0.0199	0.2228	0.2286	0.2388	0.2042	0.0144
4	0.2554	0.2629	0.2810	0.2298	0.0204	0.2241	0.2296	0.2385	0.2049	0.0144
5	0.2554	0.2616	0.2757	0.2294	0.0202	0.2233	0.2303	0.2421	0.2046	0.0145
			$T = 300$					$T = 400$		
1	0.2059	0.2101	0.2170	0.1894	0.0122	0.1925	0.1984	0.2074	0.1791	0.0109
2	0.2032	0.2082	0.2178	0.1885	0.0119	0.1928	0.1971	0.2042	0.1792	0.0107
3	0.2046	0.2088	0.2165	0.1891	0.0119	0.1936	0.1970	0.2067	0.1794	0.0108
4	0.2059	0.2102	0.2211	0.1901	0.0120	0.1934	0.1974	0.2046	0.1797	0.0108
5	0.2049	0.2101	0.2173	0.1900	0.0118	0.1935	0.1980	0.2042	0.1801	0.0103
			$T = 500$					$T = 1,000$		
1	0.1850	0.1888	0.1948	0.1717	0.0098	0.1888	0.1976	0.2059	0.1652	0.0196
2	0.1850	0.1881	0.1966	0.1712	0.0101	0.1900	0.1955	0.2033	0.1659	0.0201
3	0.1841	0.1881	0.1950	0.1717	0.0099	0.1873	0.1962	0.2071	0.1657	0.0188
4	0.1841	0.1880	0.1948	0.1719	0.0097	0.1880	0.1972	0.2066	0.1648	0.0193
5	0.1845	0.1883	0.1964	0.1717	0.0096	0.1851	0.1977	0.2069	0.1653	0.0193

Exercises

Theory Question

12.1 Consider the well-known property of the Kronecker product $(\mathbf{A} \otimes \mathbf{B})(\mathbf{C} \otimes \mathbf{D}) = \mathbf{AC} \otimes \mathbf{BD}$, if $\mathbf{AC}$ and $\mathbf{BD}$ exist. Using this property, verify (12.26).

Empirical and Simulation Questions

12.2 The file treering.dat contains the annual temperatures and tree ring widths series, denoted by $\{(Y_{1,t}, Y_{2,t})\}_{t=1}^{66}$; see, e.g., Examples 11.5 and 11.6.

(a) Compute the sample ACF and PACF matrices for lags $\ell = 1, \ldots, 5$. Discuss the overall pattern of these statistics. Verify your observations with those made in Example 11.5.

(b) Using the MATLAB code Rtest.m, compute the values of the $\widehat{R}(\ell)$ test statistic for $\ell = 1, \ldots, 5$. Determine the appropriate lags for inclusion in a vector NLAR model.

[*Note:* For $T = 66$, the 5% critical values of the $\widehat{R}(\ell)$ test statistic are given by 0.317 ($\ell = 1$), 0.315 ($\ell = 2$), 0.325 ($\ell = 3$), 0.315 ($\ell = 4$), and 0.326 ($\ell = 5$).]

12.3 The files earthP1.dat – earthP4.dat accompany the climate change data set of Example 1.5, but now covering each of the four climatic periods P1 – P4. Each file consists of four time series variables: δ^{13}C, δ^{18}O, dust flux, and insolation.

(a) Test for the presence of a nonlinear causal pairwise relationship between the four series (all re-scaled) in time periods P4, P3, and P2, using the modified bivariate nonparametric test statistic $Q^*_{T,W}(h)$ with bandwidth $h = 1.5$ (denoted by the variable "epsilon" in the C and R codes). Use nominal significance levels of 1% and 5% in all pairwise tests.

(b) Compare and contrast the test results in part (a) with those reported in Example 12.5 for time period P1.

12.4 Consider the Icelandic river flow data set introduced in Section 11.8. The dependent variables are the daily river flow measured in m^3/s, of the Jökulsá Eystri river ($Q_{1,t}$), and Vatnsdalsá river ($Q_{2,t}$), i.e. 1,095 observations for analysis. The exogenous variables used in the model specification are lagged values of streamflow ($Q_{1,t-\ell}, Q_{2,t-\ell}$) ($\ell = 1, \ldots, 20$), lagged values of precipitation ($P_{t-1}, P_{t-2}, P_{t-3}$), and contemporaneous and lagged values of temperature (T_t, T_{t-1}).

(a) Fit two PMARS models to the data: an unrestricted VARX model, and a restricted (additive) VARX model. Use the GCV criterion for model selection with default value $d = 4$. Find the unrestricted model with the lowest value of $|\widehat{\boldsymbol{\Sigma}}_\varepsilon|$, i.e. the determinant of the residual covariance matrix.

[*Hint:* Use the function polymars in the R-polspline package.]

(b) In part (a) you will notice that the "best" fitted unrestricted PMARS–VARX model is attained at lag $\ell = 15$. Compare the determinant of the residual covariance matrix of this particular model with the determinant of the pooled residual covariance matrix computed from $\widehat{\boldsymbol{\Sigma}}_\varepsilon^{(1)}$ and $\widehat{\boldsymbol{\Sigma}}_\varepsilon^{(2)}$ given in Table 11.4 for the VTARX model.

(c) Given the unrestricted PMARS–VARX model in part (b), consider only terms with absolute coefficient value more than twice the estimated standard error. Compare the resulting model with the nonlinear time series models presented in Exercise 2.11 and Table 11.4.

(d) Test for the presence of a nonlinear causal relationship between the series $\{Q_{1,t}\}$ and $\{Q_{2,t}\}$, using the modified bivariate nonparametric test statistic $Q^*_{T,W}(h)$ with $h = 1.5$ and embedding dimension $\ell_{Q_1} = \ell_{Q_2} = 1, \ldots, 8$.

References*

Pages on which each reference is cited are given in square brackets.

Aase, K.K. (1983). Recursive estimation in non-linear time series models of autoregressive type. *Journal of the Royal Statistical Society*, B 45(2), 228–237. [248]

Abdous, B. and Theodorescu, R. (1992). Note on the spatial quantile of a random vector. *Statistics & Probability Letters*, 13(4), 333–336.
DOI: 10.1016/0167-7152(92)90043-5. [496, 521]

Abraham, B. and Balakrishna, N. (2012). Product autoregressive models for non-negative variables. *Statistics & Probability Letters*, 82(8), 1530–1537.
DOI: 10.1016/j.spl.2012.04.022. [74]

Achard, S. (2008). Asymptotic properties of a dimension-robust quadratic dependence measure. *Comptes Rendus de l'Académie des Sciences, Paris* Series I 346, 213–216.
DOI: 10.1016/j.crma.2007.10.043. [296]

Adhikari, R. (2015). A neural network based linear ensemble framework for time series forecasting. *Neurocomputing*, 157(1), 231–242. DOI: 10.1016/j.neucom.2015.01.012. [430]

Aiolfi, M., Capistrán, C., and Timmermann, A. (2011). Forecast combinations. In M.P. Clements and D.F. Hendry (Eds.), *The Oxford Handbook of Economic Forecasting*, Oxford University Press, Oxford, UK, pp. 355–388.
DOI: 10.1093/oxfordhb/9780195398649.013.0013. [425]

Akamanam, S.I., Bhaskara Rao, M., and Subramanyam, K. (1986). On the ergodicity of bilinear time series models. *Journal of Time Series Analysis*, 7(3), 157–163.
DOI: 10.1111/j.1467-9892.1986.tb00499.x. [110]

Aldous, D. (1989). *Probability Approximation via the Poisson Clumping Heuristic*. Applied Mathematical Sciences 77, Springer-Verlag, New York. (Freely available at: http://en.booksee.org/book/1304840). [170]

---

*A DOI (Digital Object Identifier) number can be converted to a web address with the URL prefix http://dx.doi.org/. The URL will lead to the abstract of a paper or a book.

Al-Qassem, M.S. and Lane, J.A. (1989). Forecasting exponential autoregressive models of order 1. *Journal of Time Series Analysis*, 10(2), 95–113. DOI: 10.1111/j.1467-9892.1989.tb00018.x. [393, 401, 405, 428]

Alquist, R. and Killian, L. (2010). What do we learn from the price of crude oil futures? *Journal of Applied Econometrics*, 25(4), 539–573. DOI: 10.1002/jae.1159. [431]

Amano, T. (2009). Asymptotic efficiency of estimating function estimators for nonlinear time series models. *Journal of the Japan Statistical Society*, 39(2), 209–231. DOI: 10.14490/jjss.39.209. [73, 248]

Amendola, A. and Francq, C. (2009). Concepts of and tools for nonlinear time series modelling. In E. Kontoghiorghes and D. Belsley (Eds.) *Handbook of Computational Econometrics*, Wiley, New York, pp. 377-427. DOI: 10.1002/9780470748916. See also the MPRA working paper at http://mpra.ub.uni-muenchen.de/15140. [249]

Amendola, A. and Niglio, M. (2004). Predictor distribution and forecast accuracy of threshold models. *Statistical Methods & Applications*, 13(1), 3–14. DOI: 10.1007/s10260-003-0072-0. [429]

Amendola, A., Niglio, M., and Vitale, C. (2006a). The moments of SETARMA models. *Statistics & Probability Letters*, 76(6), 625–633. DOI: 10.1016/j.spl.2005.09.016. [111]

Amendola, A., Niglio, M., and Vitale, C. (2006b). Multi-step SETARMA predictors in the analysis of hydrological time series. *Physics and Chemistry of the Earth*, 31(18), 1118–1126. DOI: 10.1016/j.pce.2006.04.040. [395, 396]

Amendola, A., Niglio, M., and Vitale, C. (2007). The autocorrelation functions in SETARMA models. In E. Kontoghiorghes and C. Gatu (Eds.) *Optimisation, Econometric and Financial Analysis*. Springer-Verlag, New York, pp. 127–142. DOI: 10.1007/3-540-36626-1_7. [111]

Amendola, A, Niglio, M., and Vitale, C. (2009a). Statistical properties of threshold models. *Communications in Statistics: Theory and Methods*, 38(15), 2479–2497. DOI: 10.1080/03610920802571146. [100]

Amendola, A, Niglio, M., and Vitale, C. (2009b). Threshold moving average models invertibility. Available at: http://new.sis-statistica.org/wp-content/uploads/2013/09/RS10-Threshold-Moving-Average-Models-Invertibility.pdf. [109]

Amisano, G. and Giacomini, R. (2007). Comparing density forecasts via weighted likelihood ratio tests. *Journal of Business & Economic Statistics*, 25(2), 177–190. DOI: 10.1198/073500106000000332. [427]

An, H.Z. and Chen S.G. (1997). A note on the ergodicity of non-linear autoregressive model. *Statistics & Probability Letters*, 34(4), 365–372. DOI: 10.1016/s0167-7152(96)00204-0. [110]

An, H.Z. and Cheng, B. (1991). A Kolmogorov-Smirnov type statistic with application to test for nonlinearity in time series. *International Statistical Review*, 59(3), 287–307. DOI: 10.2307/1403689. [250]

References

An, H.Z., Zhu, L.X., and Li, R.Z. (2000). A mixed-type test for linearity in time series. *Journal of Statistical Planning and Inference*, 88(2), 339–353. DOI: 10.1016/S0378-3758(00)00087-2. [250]

Anděl, J. (1976). Autoregressive series with random parameters. *Mathematische Operationsforschung und Statistik*, Statistics 7(5), 735–741. DOI: 10.1080/02331887608801334. [39]

Anděl, J. (1984). On autoregressive models with random parameters. In P. Mandle and M. Hušová (Eds.) *Proceedings of the Third Prague Symposium on Asymptotic Statistics*. Elsevier, Amsterdam, pp. 17–30. [39]

Anděl, J. (1997). On extrapolation in some non-linear AR(1) processes. *Communications in Statistics: Theory and Methods*, 26(3), 581–587. DOI: 10.1080/03610929708831935. [432]

Anderson, H.M. and Vahid, F. (1998). Testing multiple equation systems for common nonlinear components. *Journal of Econometrics*, 84(1), 1–36. DOI: 10.1016/S0304-4076(97)00076-6. [457]

Anderson, H.M., Nam, K., and Vahid, F. (1999). Asymmetric nonlinear smooth transition Garch models. In P. Rothman (Ed.) *Non Linear Time Series Analysis of Economic and Financial Data*. Kluwer, Amsterdam, pp. 191–207. DOI: 10.1007/978-1-4615-5129-4_10. [80]

Andrews, D.W.K. (1993). Tests for parameter instability and structural change with unknown change point. *Econometrica*, 61(4), 821–856. DOI: 10.2307/2951764. [189]

Araújo Santos, P. and Fraga Alves, M.I. (2012). A new class of independence tests for interval forecasts evaluation. *Computational Statistics & Data Analysis*, 56(11), 3366–3380. DOI: 10.1016/j.csda.2010.10.002. [421]

Arnold, M. and Günther, R. (2001). Adaptive parameter estimation in multivariate self-exciting threshold autoregressive models. *Communications in Statistics: Simulation and Computation*, 30(2), 257–275. DOI: 10.1081/sac-100002366. [79, 449]

Ashley, R.A., Patterson, D.M., and Hinich, M.J. (1986). A diagnostic test for nonlinear serial dependence in time series fitting errors. *Journal of Time Series Analysis*, 7(3), 165–178. DOI: 10.1111/j.1467-9892.1986.tb00500.x. [133, 147, 149]

Ashley, R.A. and Patterson, D.M. (1989). Linear versus nonlinear macroeconomies: A statistical test. *International Economic Review*, 30(3), 685–704. DOI: 10.2307/2526783. [150]

Ashley R.A. and Patterson, D.M. (2002). Identification of coefficients in a quadratic moving average process using the generalized method of moments. Available at: http://ashleymac.econ.vt.edu/working_papers/E2003_5.pdf. [73]

Assaad, M., Boné, R., and Cardot, H. (2008). A new boosting algorithm for improved time-series forecasting with recurrent neural networks. *Information Fusion*, 9(1), 41–55. DOI: 10.1016/j.inffus.2006.10.009. [383]

Astatkie, T. (2006). Absolute and relative measures for evaluating the forecasting performance of time series models for daily streamflows. *Nordic Hydrology*, 37(3), 205–215. DOI: 10.2166/nh.2006.008. [74]

Astatkie, T., Watt, W.E., and Watts, D.G. (1996). Nested threshold autoregressive (NeTAR) models for studying sources of nonlinearity in streamflows. *Nordic Hydrology*, 27(5), 323–336. [74, 75]

Astatkie, T., Watts, D.G., and Watt, W.E. (1997). Nested threshold autoregressive (NeTAR) models. *International Journal of Forecasting*, 13(1), 105–116. DOI: 10.1016/s0169-2070(96)00716-9. [49, 84]

Aue, A., Horváth, L., and Steinebach, J. (2006). Estimation in random coefficient autoregressive models. *Journal of Time Series Analysis*, 27(1), 61–76. DOI: 10.1111/j.1467-9892.2005.00453.x. [73]

Auestad, B. and Tjøstheim, D. (1990). Identification of nonlinear time series: First order characterization and order estimation. *Biometrika*, 77(4), 669–687. DOI: 10.1093/biomet/77.4.669. [355, 382]

Avramidis, P. (2005). Two-step cross-validation selection method for partially linear models. *Statistica Sinica*, 15(4), 1033–1048. [383]

Aznarte, J.L. and Benítez, J.M. (2010). Equivalences between neural-autoregressive time series models and fuzzy systems. *IEEE Transactions on Neural Networks*, 21(9), 1434–1444. DOI: 10.1109/tnn.2010.2060209. [75]

Aznarte, J.L., Benítez, J.M., and Castro, J.L. (2007). Smooth transition autoregressive models and fuzzy rule-based systems: Functional equivalence and consequences. *Fuzzy Sets and Systems*, 158(4), 2734–2745. DOI: 10.1016/j.fss.2007.03.021. [74]

Azzalini, A. and Bowman, A.W. (1990). A look at some data on the Old Faithful geyser. *Applied Statistics*, 39(3), 357–365. DOI: 10.2307/2347385. [384]

Bacigál, T. (2004). Multivariate threshold autoregressive models in geodesy. *Journal of Electrical Engineering*, 55(2), 91–94. [486]

Bacon, D.W. and Watts, D.G. (1971). Estimating the transition between two intersecting straight lines. *Biometrika*, 58(3), 525–534. DOI: 10.1093/biomet/58.3.525. [74]

Baek, E.G. and Brock, W.A. (1992a). A nonparametric test for independence of a multivariate time series. *Statistica Sinica*, 2(1), 137–156. [296, 515]

Baek, E.G. and Brock, W.A. (1992b). A general test for nonlinear Granger causality: Bivariate model. Technical report, Department of Economics, University of Wisconsin. Available at: http://www.ssc.wisc.edu/~wbrock/. [515]

Baek, J.S., Park, J.A., and Hwang, S.Y. (2012). Preliminary test of fit in a general class of conditionally heteroscedastic nonlinear time series. *Journal of Statistical Computation and Simulation*, 82(5), 763–781. DOI: 10.1080/00949655.2011.558087. [250]

Bagnato, L., De Capitani, L., and Punzo, A. (2014). Testing serial independence via density-based measures of divergence. *Methodology and Computing in Applied Probability*, 16(3), 627–641. DOI: 10.1007/s11009-013-9320-4. [268, 269, 272, 273, 294, 297]

Bai, J. (2003). Testing parametric conditional distributions of dynamic models. *The Review of Economics and Statistics*, 85(3), 531–549. DOI: 10.1162/003465303322369704. [427]

References

Bai, J. and Ng, S. (2005). Tests for skewness, kurtosis, and normality for time series data. *Journal of Business & Economic Statistics*, 23(1), 49–60. DOI: 10.1198/073500104000000271. [13]

Bai, Z., Wong, W.K., and Zhang, B. (2010). Multivariate linear and nonlinear causality tests. *Mathematics and Computers in Simulation*, 81(1), 5–17. DOI: 10.1016/j.matcom.2010.06.008. [516]

Balke, N.S. and Fomby, T.B. (1997). Threshold cointegration. *International Economic Review*, 38(3), 627–645. DOI: 10.2307/2527284. [79, 80]

Banicescu, I., Carino, R.L., Harvill, J.L., and Lestrade, J.P. (2005). Simulation of vector nonlinear time series models on clusters. In *Proceedings of the 19th IEEE International Parallel and Distributed Processing Symposium (IPDPS'05)*, pp. 4–8. DOI: 10.1109/ipdps.2005.402. [522]

Banicescu, I., Carino, R.L., Harvill, J.L., and Lestrade, J.P. (2011). Investigating asymptotic properties of vector nonlinear time series models. *Journal of Computational and Applied Mathematics*, 236(3), 411–421. DOI: 10.1016/j.cam.2011.07.018. [522]

Bao, Y., Lee, T.-H., and Saltoğlu, B. (2007). Comparing density forecast models. *Journal of Forecasting*, 26(3), 203–225. DOI: 10.1002/for.1023. [426]

Baragona, R., Battaglia, F., and Cucina, D. (2004a). Fitting piecewise linear threshold autoregressive models by means of genetic algorithms. *Computational Statistics & Data Analysis*, 47(2), 277–295. DOI: 10.1016/j.csda.2003.11.003. [79, 210]

Baragona, R., Battaglia, F., and Cucina, D. (2004b). Estimating threshold subset autoregressive moving-average models by genetic algorithms. *Metron*, LXII, n. 1, 39–61. [80, 210]

Baragona, R. and Cucina, D. (2013). Multivariate self-exciting threshold autoregressive modeling by genetic algorithms. *Journal of Economics and Statistics* (Jahrbücher für Nationalökonomie und Statistik), 233(1), 3–21. [471]

Baringhaus, L. and Franz, C. (2004). On a new multivariate two-sample test. *Journal of Multivariate Analysis*, 88(1), 190–206. DOI: 10.1016/s0047-259x(03)00079-4. [296]

Barkoulas, J.T., Baum, C.F., and Onochie, J. (1997). A nonparametric investigation of the 90-day T-bill rate. *Review of Financial Economics*, 6(2), 187–198. DOI: 10.1016/s1058-3300(97)90005-7. [381]

Barnett, A.G. and Wolff, R.C. (2005). A time-domain test for some types of nonlinearity. *IEEE Transactions on Signal Processing*, 53(1), 26–33. DOI: 10.1109/tsp.2004.838942. [150]

Barnett, W.A., Gallant, A.R., Hinich, M.J., Jungeilges, J.A., Kaplan, D.T., and Jensen, M.J. (1997). A single-blind controlled competition among tests for nonlinearity and chaos. *Journal of Econometrics*, 82(1), 157–192. DOI: 10.1016/s0304-4076(97)00081-x. [151]

Barnett, W.A., Hendry, D.F., Hylleberg, S., Teräsvirta, T., Tjøstheim, D.J., and Würtz, A. (Eds.) (2006). *Nonlinear Econometric Modeling in Time Series*. Cambridge University Press, Cambridge, UK. [597]

Bartlett, M.S. (1954). A note on multiplying factors for various χ^2 approximations. *Journal of the Royal Statistical Society*, B 16(2), 296–298. [460]

Basrak, B., Davis, R.A., and Mikosch, T. (2002). Regular variation of GARCH processes. *Stochastic Processes and their Applications*, 99(1), 95–115. DOI: 10.1016/s0304-4149(01)00156-9. [97]

Bates, J.M. and Granger, C.W.J. (1969). The combination of forecasts. *Operational Research Quarterly*, 20(4), 451–468. DOI: 10.2307/3008764. [430]

Battaglia,, F. and Orfei, L. (2005). Outlier detection and estimation in nonlinear time series. *Journal of Time Series Analysis*, 26(1), 107–121. DOI: 10.1111/j.1467-9892.2005.00392.x. [249]

Bazzi, M., Blasques, F., Koopman, S.J., and Lucas, A. (2014). Time varying transition probabilities for Markov regime switching models. TI Discussion Paper, no. 14-072/III, Amsterdam. Available at: http://papers.tinbergen.nl/14072.pdf. DOI: 10.2139/ssrn.2456632. [75]

Beare, B.K. and Seo, J. (2014). Time-reversible copula-based Markov models. *Econometric Theory*, 30(5), 923–960. DOI: 10.1017/s0266466614000115 [325, 332]

Bec, F., Guay, A., and Guerre, E. (2008). Adaptive consistent unit root tests based on autoregressive threshold model. *Journal of Econometrics*, 142(1), 94–133. DOI: 10.1016/j.jeconom.2007.05.011. [189]

Becker, R.A., Clark, L.A., and Lambert, D. (1994). Cave plots: A graphical technique for comparing time series. *Journal of Computational and Graphical Statistics*, 3(3), 277–283. DOI: 10.2307/1390912. [24]

Bekiros, S.D. and Diks, C. (2008). The nonlinear dynamic relationship of exchange rates: Parametric and nonparametric causality testing. *Journal of Macroeconomics*, 30(4), 1641–1650. DOI: 10.1016/j.jmacro.2008.04.001. [523]

Belaire–Franch, J. and Contreras, D. (2003). Tests for time reversibility: A complementary analysis. *Economics Letters*, 81(2), 187–195. DOI: 10.1016/S0165-1765(03)00169-1. [333]

Berg, A., Paparoditis, E., and Politis, D.N. (2010). A bootstrap test for time series linearity. *Journal of Statistical Planning and Inference*, 140(12), 3841–3857. DOI: 10.1016/j.jspi.2010.04.047. [136, 139, 140, 147]

Berkowitz, J. (2001). Testing density forecasts, with applications to risk management. *Journal of Business & Economic Statistics*, 19(4), 465–474. DOI: 10.1198/073500101525967-18. [422]

Berkowitz, J., Christoffersen, P., and Pelletier, D. (2011). Evaluating value-at-risk models with desk-level data. *Management Science*, 57(12), 2213–2227. DOI: 10.1287/mnsc.1080.0964. [430]

Berlinet, A. and Francq, C. (1997). On Bartlett's formula for non-linear processes. *Journal of Time Series Analysis*, 18(6), 535–552. DOI: 10.1111/1467-9892.00067. [15]

Berlinet, A., Gannoun, A., and Matzner–Løber, E. (1998). Normalité asymptotique d'estimateurs convergents du mode conditionnel. *The Canadian Journal of Statistics*, 26(2), 365–380. DOI: 10.2307/3315517. [382]

Berlinet, A., Gannoun, A., and Matzner–Løber, E. (2001). Asymptotic normality of convergent estimates of conditional quantiles. *Statistics*, 35(2), 139–169. DOI: 10.1080/02331880108802728. [382]

Bermejo, M.A., Peña, D., and Sánchez, I. (2011). Identification of TAR models using recursive estimation. *Journal of Forecasting*, 30(1), 31–50. DOI: 10.1002/for.1188. [250]

Beutner, E. and Zähle, H. (2014). Continuous mapping approach to the asymptotics of U- and V-statistics. *Bernoulli*, 20(2), 846–877. DOI: 10.3150/13-bej508. [310]

Bhansali, R.J. and Downham, D.Y. (1977). Some properties of the order of an autoregressive model selected by a generalization of Akaike's EPF criterion. *Biometrika*, 64(3), 547–551. DOI: 10.1093/biomet/64.3.547. [231]

Bhattacharyya, A. (1943). On a measure of divergence between two statistical populations defined by their probability distribution. *Bulletin of the Calcutta Mathematical Society*, 35(1), 99–110. [336]

Bhattacharaya, R. and Lee, C. (1995). On geometric ergodicity of nonlinear autoregressive models. *Statistics & Probability Letters*, 22(4), 311–315. DOI: 10.1016/0167-7152(94)00082-j. Erratum: *Statist. Prob. Lett.*, 2009, 41(4), 439–440. [110]

Billings, S.A. (2013). *Nonlinear System Identification: NARMAX Methods in the Time, Frequency, and Spatio-Temporal Domains*. Wiley, New York. DOI: 10.1002/9781118535561. [487]

Billings, S.A., Chen, S., and Korenberg, M.J. (1989). Identification of MIMO non-linear systems using a forward regression orthogonal estimator. *International Journal of Control*, 49(6), 2157–2189. DOI: 10.1080/00207178908559767. [487]

Billingsley, P. (1995). *Probability and Measure* (3rd edn.). Wiley, New York. (Freely available at: http://www.math.uoc.gr/~nikosf/Probability2013/3.pdf). [98]

Bilodeau, M. and Lafaye de Micheaux, P. (2009). A dependence statistic for mutual and serial independence of categorical variables. *Journal of Statistical Planning and Inference*, 139(7), 2407–2419. DOI: 10.1016/j.jspi.2008.11.006. [296]

Birkelund, Y. and Hanssen, A. (2009). Improved bispectrum based tests for Gaussianity and linearity. *Signal Processing*, 89(12), 2537–2546. DOI: 10.1016/j.sigpro.2009.04.013. [150]

Blum, J.R., Kiefer, J., and Rosenblatt, M. (1961). Distribution free tests of independence based on the sample distribution function. *Annals of Mathematical Statistics*, 32(2), 485–498. DOI: 10.1214/aoms/1177705055. [284, 285]

Blumentritt, T. and Grothe, O. (2013). Ranking ranks: A ranking algorithm for bootstrapping from the empirical copula. *Computational Statistics*, 28(2), 455–462. DOI: 10.1007/s00180-012-0310-8. [297]

Blumentritt, T. and Schmid, F. (2012). Mutual information as a measure of multivariate association: Analytical properties and statistical estimation. *Journal of Statistical Computation and Simulation*, 82(9), 1257–1274. DOI: 10.1080/00949655.2011.575782. [296]

Boente, G. and Fraiman, R. (1995). Asymptotic distribution of data-driven smoothers in density and regression estimation under dependence. *The Canadian Journal of Statistics*, 23(4), 383–397. DOI: 10.2307/3315382. [340]

Bosq, D. (1998). *Nonparametric Statistics for Stochastic Processes* (2nd edn.). Springer-Verlag, New York. DOI: 10.1007/978-1-4684-0489-0. [338]

Bougerol, P. and Picard, D. (1992). Strict stationarity of generalized autoregressive processes. *The Annals of Probability*, 20(4), 1714–1729. DOI: 10.1214/aop/1176989526. [89]

Boutahar, M. (2010). Behaviour of skewness, kurtosis and normality tests in long memory data. *Statistical Methods & Applications*, 19(2), 193–215. DOI: 10.1007/s10260-009-0124-1. [23]

Bowman, A.W. and Azzalini, A. (1997). *Applied Smoothing Techniques for Data Analysis: The Kernel Approach with S-Plus Illustrations*. Oxford University Press, Oxford. [385]

Bowman, K.O. and Shenton, L.R. (1975). Omnibus contours for departures from normality based on $\sqrt{b_1}$ and b_2. *Biometrika*, 62(2), 243–250. DOI: 10.1093/biomet/62.2.243. [22]

Box, G.E.P., Jenkins, G.M., and Reinsel, G.C. (2008). *Time Series Analysis, Forecasting, and Control* (4th edn.). Wiley, New York. [1]

Brandt, A. (1986). The stochastic equation $Y_{n+1} = A_n Y_n + B_n$ with stationary coefficients. *Advances in Applied Probability*, 18(1), 211–220. DOI: 10.2307/1427243. [89]

Brännäs, K. and De Gooijer, J.G. (1994). Autoregressive-asymmetric moving average models for business cycle data. *Journal of Forecasting*, 13(6), 529–544. DOI: 10.1002/for.3980130605. [48, 74, 78, 178, 189]

Brännäs, K. and De Gooijer, J.G. (2004). Asymmetries in conditional mean and variance: Modelling stock returns by asMA-asQGARCH. *Journal of Forecasting*, 23(3), 155–171. DOI: 10.1002/for.910. [74, 80]

Brännäs, K., De Gooijer, J.G., and Teräsvirta, T. (1998). Testing linearity against nonlinear moving average models. *Communications in Statistics: Theory and Methods*, 27(8), 2025–2035. DOI: 10.1080/03610929808832207. [74, 188, 189]

Brännäs, K., De Gooijer, J.G., Lönnbark, C., and Soultanaeva, A. (2011). Simultaneity and asymmetry of returns and volatilities in the emerging Baltic state stock exchanges. *Studies in Nonlinear Dynamics & Econometrics*, 16:1. DOI: 10.1515/1558-3708.1855. [74]

Breaker, L.C. (2006). Nonlinear aspects of sea surface temperature in Monterey Bay. *Progress in Oceanography*, 69(1), 61–89. DOI: 10.1016/j.pocean.2006.02.015. [384]

Breaker, L.C. and Lewis, P.A.W. (1988). A 40–50 day oscillation in sea-surface temperature along the Central California coast. *Estuarine, Coastal and Shelf Science*, 26(4), 395–408. DOI: 10.1016/0272-7714(88)90020-0. [384]

Breidt, F.J. (1996). A threshold autoregressive stochastic volatility model. VI Latin American Congress of Probability and Mathematical Statistics (CLAPEM), Valparaiso, Chile. [80]

Breidt, F.J. and Davis, R.A. (1992). Time-reversibility, identifiability and independence of innovations for stationary time series. *Journal of Time Series Analysis*, 13(5), 377–390. DOI: 10.1111/j.1467-9892.1992.tb00114.x. [333]

Breiman, L. and Friedman, J.H. (1985). Estimating optimal transformations for multiple regression and correlation (with discussion). *Journal of the American Statistical Association*, 80(391), 580–619. DOI: 10.1080/01621459.1985.10478157. [383]

Brillinger, D.R. (1965). An introduction to polyspectra. *Annals Mathematical Statistics*, 36(5), 1351–1374. DOI: 10.1214/aoms/1177699896. [128]

Brillinger, D.R. (1975). *Time Series Data Analysis and Theory*. Holt, Rinehart and Winston, New York. [142]

Brillinger, D.R. and Rosenblatt, M. (1967). Asymptotic theory of kth order spectra. In B. Harris (Ed.) *Spectral Analysis of Time Series*. Wiley, New York, pp. 189–232 (see also pp. 153–188). [149]

Brock, W.A., Hsieh, W.D., and LeBaron, B. (1991). *Nonlinear Dynamics, Chaos, and Instability: Statistical Theory and Economic Evidence*. MIT Press, Cambridge, MA. [282]

Brock, W.A., Dechert, W.D., LeBaron, B., and Scheinkman, J.A. (1996). A test for independence based on the correlation dimension. *Econometric Reviews*, 15(3), 197–235. DOI: 10.1080/07474939608800353. [279]

Brockett, P.L., Hinich, M.J., and Patterson, D.M. (1988). Bispectral-based tests for the detection of Gaussianity and linearity in time-series. *Journal of the American Statistical Association*, 83(403), 657–664. DOI: 10.2307/2289288. [150]

Brockett, R.W. (1976). Volterra series and geometric control theory. *Automatica*, 12(2), 167–176. DOI: 10.1016/0005-1098(76)90080-7. [72]

Brockett, R.W. (1977). Convergence of Volterra series on infinite intervals and bilinear approximations. In V. Lakshmikathan (Ed.) *Nonlinear Systems and Applications*. Academic Press, New York, pp. 39–46. DOI: 10.1016/b978-0-12-434150-0.50009-6. [73]

Brockwell, P.J. (1994). On continuous time threshold ARMA processes. *Journal of Statistical Planning and Inference*, 39(2), 291–304. DOI: 10.1016/0378-3758(94)90210-0. [44]

Brockwell, P.J. and Davis, R.A. (1991). *Time Series: Theory and Methods* (2nd edn.). Springer-Verlag, New York. [1, 3]

Brockwell, P.J., Liu, J., and Tweedie, R.L. (1992). On the existence of stationary threshold autoregressive moving-average processes. *Journal of Time Series Analysis*, 13(2), 95–107. DOI: 10.1111/j.1467-9892.1992.tb00096.x. [100]

Brown, B.W. and Mariano, R.S. (1984). Residual-based procedures for prediction and estimation in a nonlinear simultaneous system. *Econometrica*, 52(2), 321–343. DOI: 10.2307/1911492. [429]

Bryant, P.G. and Cordero–Braña, O.I. (2000). Model selection using the minimum description length principle. *American Statistician*, 54(4), 257–268. DOI: 10.2307/2685777. [249]

Brys, G., Hubert, M., and Struyf, A. (2004). A robustification of the Jarque-Bera test of normality. In J. Antoch (Ed.), *COMPSTAT 2004 Symposium – Proceedings in Computational Statistics*. Physica-Verlag/Springer-Verlag, New York, pp. 753–760. [22]

Buchen, T. and Wohlrabe, K. (2011). Forecasting with many predictors: Is boosting a viable alternative? *Economics Letters*, 113(1), 16–18. DOI: 10.1016/j.econlet.2011.05.040. [381]

Bühlmann, P. (2006). Boosting for high-dimensional linear models. *The Annals of Statistics*, 34(2), 559–583. DOI: 10.1214/009053606000000092. [371]

Bühlmann, P. and Hothorn, T. (2007). Boosting algorithms: Regularization, prediction and model fitting. *Statistical Science*, 22(4), 477–505. DOI: 10.1214/07-sts242. [383]

Bühlmann, P. and Yu, B. (2003). Boosting with the L_2 loss: Regression and classification. *Journal of the American Statistical Association*, 98(462), 324–339. DOI: 10.1198/016214503000125. [371]

Burg, J.P. (1967). Maximum Entropy Spectral Analysis. *Proceeding of the 37th Meeting of the Society of Exploration, Geophysicists, Oklahoma City*. Reprinted in D.G. Childers (Ed.) (1978) *Modern Spectral Analysis*. IEEE Press, New York. [123]

Cai, Y. (2003). Convergence theory of a numerical method for solving the Chapman-Kolmogorov equation. *SIAM Journal on Numerical Analysis*, 40(6), 2337–2351. DOI: 10.1137/s0036142901390366. [428]

Cai, Y. (2005). A forecasting procedure for nonlinear autoregressive time series models. *Journal of Forecasting*, 24(5), 335–351. DOI: 10.1002/for.959. [393]

Cai, Y. and Stander, J. (2008). Quantile self-exciting threshold autoregressive time series models. *Journal of Time Series Analysis*, 29(1), 186–202. DOI: 10.1111/j.1467-9892.2007.00551.x. [79]

Cai, Z., Fan, J., and Li, R. (2000a). Efficient estimation and inferences for varying-coefficient models. *Journal of the American Statistical Association*, 95(451), 888–902. DOI: 10.1080/01621459.2000.10474280. [384]

Cai, Z., Fan, J., and Yao, Q. (2000b). Functional-coefficient regression models for nonlinear time series. *Journal of the American Statistical Association*, 95(451), 941–956. DOI: 10.1080/01621459.2000.10474284. [374, 375]

Cai, Z., Li, Q., and Park, J.Y. (2009). Functional-coefficient models for nonstationary time series data. *Journal of Econometrics*, 148(2), 101–113. DOI: 10.1016/j.jeconom.2008.10.003. [384]

Camacho, M. (2004). Vector smooth transition regression models for US GDP and the composite index of leading indicators *Journal of Forecasting*, 23(3), 173–196. DOI: 10.1002/for.912. [487]

Campbell, S.D. (2007). A review of backtesting and backtesting procedures. *Journal of Risk*, 9(2), 1–18. [430]

Caner, M. (2002). A note on least absolute deviation estimation of a threshold model. *Econometric Theory*, 18(03), 800–814. DOI: 10.1017/s0266466602183113. [248]

Caner, M. and Hansen, B.E. (2001). Threshold autoregression with a unit root. *Econometrica*, 69(6), 1555–1596. DOI: 10.1111/1468-0262.00257. [189]

Casali, K.R, Casali, A.G., Montano, N., Irigoyen, M.C., Macagnan, F., Guzzetti, S., and Porta, A. (2008). Multiple testing strategy for the detection of temporal irreversibility in stationary time series. *Physical Review*, E77(6), 066204-1–066204-7. DOI: 10.1103/physreve.77.066204. [333]

Casdagli, M. and Eubank, S. (Eds.) (1992). *Nonlinear Modeling and Forecasting*, Addison-Wesley, Redwood City. [597]

Chabot–Hallé, D. and Duchesne, P. (2008). Diagnostic checking of multivariate nonlinear time series models with martingale difference errors. *Statistics & Probability Letters*, 78(8), 997–1005. DOI: 10.1016/j.spl.2007.10.003. [472, 473]

Chakraborty, B. (2003). On multivariate quantile regression. *Journal of Statistical Planning and Inference*, 110(1-2), 109–132. DOI: 10.1016/s0378-3758(01)00277-4. [521]

Chan, K.S. (1988). On the existence of the stationary and ergodic NEAR(p) model. *Journal of Time Series Analysis*, 9(4), 319–328. DOI: 10.1111/j.1467-9892.1988.tb00473.x. [74]

Chan, K.S. (1990). Testing for threshold autoregression. *The Annals of Statistics*, 18(4), 1886–1894. DOI: 10.1214/aos/1176347886. [170]

Chan, K.S. (1991). Percentage points of likelihood ratio tests for threshold autoregression. *Journal Royal Statistical Society*, B 53(3), 691–696. [170, 191]

Chan, K.S. (1993). Consistency and limiting distribution of the least squares estimator of a threshold autoregressive model. *The Annals of Statistics*, 21(1), 520–533. DOI: 10.1214/aos/1176349040. [173, 247, 249]

Chan, K.S. (Ed.) (2009). *Exploration of a Nonlinear World: An Appreciation of Howell Tongs Contributions to Statistics*. World Scientific, Singapore. DOI: 10.1142/7076. [597]

Chan, K.S. and Tong, H. (1985). On the use of the deterministic Lyapunov function for the ergodicity of stochastic difference equations. *Advances in Applied Probability*, 17(3), 666–678. DOI: 10.2307/1427125. [111]

Chan, K.S. and Tong, H. (1986). On estimating thresholds in autoregressive models. *Journal of Time Series Analysis*, 7(3), 179–190.
DOI: 10.1111/j.1467-9892.1986.tb00501.x. [73, 74, 189]

Chan, K.S. and Tong, H. (1990). On likelihood ratio tests for threshold autoregression. *Journal Royal Statistical Society*, B 52(3), 469–476. [170, 191]

Chan, K.S. and Tong, H. (2001). *Chaos: A Statistical Perspective*. Springer-Verlag, New York. DOI: 10.1007/978-1-4757-3464-5. [597]

Chan, K.S. and Tong, H. (2010). A note on the invertibility of nonlinear ARMA models. *Journal of Statistical Planning and Inference*, 140(12), 3707–3714.
DOI: 10.1016/j.jspi.2010.04.036. [107]

Chan, K.S. and Tsay, R.S. (1998). Limiting properties of the least squares estimator of a continuous threshold autoregressive model. *Biometrika*, 85(2), 413–426. DOI: 10.1093/biomet/85.2.413. [44, 45]

Chan, K.S., Ho, L.-H., and Tong, H. (2006). A note on time-reversibility of multivariate linear processes. *Biometrika*, 93(1), 221–227. DOI: 10.1093/biomet/93.1.221. [333]

Chan, K.S., Petruccelli, J.D., Tong, H., and Woolford, S.W. (1985). A multiple-threshold AR(1) model. *Journal of Applied Probability*, 22(2), 267–279. DOI: 10.2307/3213771. [100]

Chan, N.H. and Tran, L.T. (1992). Nonparametric tests for serial dependence. *Journal of Time Series Analysis*, 13(1), 19–28. DOI: 10.1111/j.1467-9892.1992.tb00092.x. [271, 272]

Chan, W.S. and Cheung, S.H. (1994). On robust estimation of threshold autoregressions. *Journal of Forecasting*, 13(1), 37–49. DOI: 10.1002/for.3980130106. [248]

Chan, W.S. and Tong, H. (1986). On tests for non-linearity in time series analysis. *Journal of Forecasting*, 5(4), 217–228. DOI: 10.1002/for.3980050403. [129, 147]

Chan, W.S., Wong, A.C.S., and Tong, H. (2004). Some nonlinear threshold autoregressive time series models for actuarial use. *North American Actuarial Journal*, 8(4), 37–61. DOI: 10.1080/10920277.2004.10596170. [486]

Chan, W.S., Cheung, S.H., Chow, W.K., and Zhang, L-X. (2015). A robust test for threshold-type nonlinearity in multivariate time series analysis. *Journal of Forecasting*, 34(6), 441–454. DOI: 10.1002/for.2344. [487]

Chandra, S.A. and Taniguchi, M, (2001). Estimating functions for nonlinear time series models. *Annals Institute of Statistical Mathematics*, 53(1), 125–141. [246, 248]

Chang, C.T. and Blondel, V.D. (2013). An experimental study of approximation algorithms for the joint spectral radius. *Numerical Algorithms*, 64(1), 181–202. DOI: 10.1007/s11075-012-9661-z. [455]

Charemza, W.W., Lifshits, M., and Makarova, S. (2005). Conditional testing for unit-root bilinearity in financial time series: Some theoretical and empirical results. *Journal of Economic Dynamics & Control*, 29(1-2), 63–96. DOI: 10.1016/j.jedc.2003.07.001. [189]

Chatfield, C. (1993). Calculating interval forecasts. *Journal of Business & Economic Statistics*, 11(2), 121–135. DOI: 10.2307/1391361. [425]

Chaudhuri, P. (1992). Multivariate location estimation using extesion of R-estimates through U-statistics type approach. *The Annals of Statistics*, 20(2), 897–916. DOI: 10.1214/aos/1176348662. [522]

Chaudhuri, P. (1996). On a geometric notation of quantiles for multivariate data. *Journal of the American Statistical Association*, 91(434), 862–872. DOI: 10.2307/2291681. [521, 522, 525]

Chen, C.W.S., Gerlach, R., Hwang, B.B.K., and McAleer, M. (2012). Forecasting Value-at-Risk using nonlinear regression quantiles and the intra-day range. *International Journal of Forecasting*, 28(3), 557–574. DOI: 10.1016/j.ijforecast.2011.12.004. [81]

Chen, C.W.S., McCulloch, R.E., and Tsay, R.S. (1997). A unified approach to estimating and modeling linear and nonlinear time series. *Statistica Sinica*, 7(2), 451–472. [249]

Chen, C.W.S., Liu, F.C., and Gerlach, R. (2011a). Bayesian subset selection for threshold autoregressive moving-average models. *Computational Statistics*, 26(1), 1–30. DOI: 10.1007/s00180-010-0198-0. [210, 249]

Chen, C.W.S., So, M.K.P., and Liu, F.C. (2011b). A review of threshold time series models in finance. *Statistics and Its Interface*, 4(2), 167–181. DOI: 10.4310/sii.2011.v4.n2.a12. [73, 111]

Chen, D.Q. and Wang, H.B. (2011). The stationarity and invertibility of a class of nonlinear ARMA models. *Science China, Mathematics*, 54(3), 469–478. DOI: 10.1007/s11425-010-4160-y. [111, 384]

Chen, G., Abraham, B., and Bennett, G.W. (1997). Parametric and non-parametric modelling of time series – An empirical study. *Environmetrics*, 8(1), 63–74. DOI: 10.1002/(sici)1099-095x(199701)8:1%3C63::aid-env238%3E3.0.co;2-b. [381]

Chen, H., Chong, T.T.L., and Bai, J. (2012). Theory and applications of TAR model with two threshold variables. *Econometric Reviews*, 31(2), 142–170. DOI: 10.1080/07474938.2011.607100. [189]

Chen, J. and Huo, X. (2009). A Hessian regularized nonlinear time series model. *Journal of Computational and Graphical Statistics*, 18(3), 694–716. DOI: 10.1198/jcgs.2009.08040. [384]

Chen, M. and Chen, G. (2000). Geometric ergodicity of nonlinear autoregressive models with changing conditional variances. *The Canadian Journal of Statistics*, 28(3), 605–613. DOI: 10.2307/3315968. [111]

Chen, R. (1995). Threshold variable selection in open-loop threshold autoregressive models. *Journal of Time Series Analysis*, 16(5), 461–481. DOI: 10.1111/j.1467-9892.1995.tb00247.x. [249]

Chen, R. (1996). A nonparametric multi-step prediction estimation in Markovian structures. *Statistica Sinica*, 6(3), 603–615. [382]

Chen, R., Liu, J.S., and Tsay, R.S. (1995). Additivity tests for nonlinear autoregression. *Biometrika*, 82(2), 369–383. DOI: 10.1093/biomet/82.2.369. [383]

Chen, R. and Liu, L.-M. (2001). Functional-coefficient autoregressive models: Estimation and tests of hypotheses. *Journal of Time Series Analysis*, 22(2), 151–173. DOI: 10.1111/1467-9892.00217. [383]

Chen, R. and Tsay, R.S. (1991). On the ergodicity of TAR(1) processes, *Annals of Applied Probability*, 1(4), 613–634. DOI: 10.1214/aoap/1177005841. [100]

Chen, R. and Tsay, R.S. (1993a). Nonlinear additive ARX models. *Journal of the American Statistical Association*, 88(423), 955–967. DOI: 10.2307/2290787. [381]

Chen, R. and Tsay, R.S. (1993b). Functional coefficient autoregressive models. *Journal of the American Statistical Association*, 88(421), 298–308. DOI: 10.2307/2290725. [374]

Chen, R., Yang, K., and Hafner, C. (2004). Nonparametric multistep-ahead prediction in time series analysis. *Journal of the Royal Statistical Society*, B 66(3), 669–686. DOI: 10.1111/j.1467-9868.2004.04664.x. [382]

Chen, X., Linton, O., and Robinson, P.M. (2001). The estimation of conditional densities. In M.L. Puri (Ed.) *Asymptotics in Statistics and Probability, Festschrift for George Roussas*. VSP International Science Publishers, The Netherlands, pp. 71–84. Also available as LSE STICERD Paper, No. EM/2001/415 (http://sticerd.lse.ac.uk/dps/em/em415.pdf). [349]

Chen, Y.-T. (2003). Testing serial independence against time irreversibility. *Studies in Nonlinear Dynamics & Econometrics*, 7(3). DOI: 10.2202/1558-3708.1114. [321]

Chen, Y.-T. (2008). A unified approach to standardized-residuals-based correlation tests for GARCH-type models. *Journal of Applied Econometrics*, 23(1), 111–133. DOI: 10.1002/jae.985. [236, 237]

Chen, Y.-T. and Kuan, C.-M. (2002). Time irreversibility and EGARCH effects in US stock index returns. *Journal of Applied Econometrics*, 17(5), 565–578. DOI: 10.1002/jae.692. [321]

Chen, Y.-T., Chou, R.Y., and Kuan, C.-M. (2000). Testing time reversibility without moment restrictions. *Journal of Econometrics*, 95(1), 199–218. DOI: 10.1016/s0304-4076(99)00036-6. [320, 321, 333]

Cheng, B. and Tong, H. (1992). On consistent non-parametric order determination and chaos (with discussion). *Journal of the Royal Statistical Society*, B 54(2), 427–474. DOI: 10.1142/9789812836281_0010. [383]

Cheng, C., Sa-ngasoongsong, A., Beyca, O., Le, T, Yang, H., Kong, Z., and Bukkapatnam, S.T.S. (2015). Time series forecasting for nonlinear and non-stationary processes: A review and comparative study. *IIE Transactions*, 47(10), 1053–1071. DOI: 10.1080/0740817x.2014.999180. [427]

Cheng, Q. (1992). On the unique representation of non-Gaussian linear processes. *The Annals of Statistics*, 20(2), 1143–1145. DOI: 10.1214/aos/1176348677. [333]

Cheng, Q. (1999). On time-reversibility of linear processes. *Biometrika*, 86(2), 483–486. DOI: 10.1093/biomet/86.2.483. [333]

Cheng, Y. and De Gooijer, J.G. (2007). On the uth geometric conditional quantile. *Journal of Statistical Planning and Inference*, 137(6), 1914–1930. DOI: 10.1016/j.jspi.2006.02.014. [521]

Chini, E.Z. (2013). Generalizing smooth transition autoregressions. CREATES research paper 2013-32, Aarhus University. Available at: ftp://ftp.econ.au.dk/creates/rp/13/rp13_32.pdf. Also available at: http://economia.unipv.it/docs/dipeco/quad/ps/RePEc/pav/demwpp/DEMWP0114.pdf. [74]

Christoffersen, P.F. (1998). Evaluating interval forecasts. *International Economic Review*, 39(4), 840–841. DOI: 10.2307/2527341. [419, 420]

Chung, Y.P. and Zhou, Z.G. (1996). The predictability of stock returns – a nonparametric approach. *Econometric Reviews*, 15(3), 299–330. DOI: 10.1080/07474939608800357. [429]

Claeskens, G., Magnus, J.R., Vasnev, A.L., and Wang, W. (2016). The forecast combination puzzle: A simple theoretical explanation. *International Journal of Forecasting*, 32(3), 754–762. DOI: 10.1016/j.ijforecast.2015.12.005. [425]

Clark, T.E. (2007). An overview of recent developments in forecast evaluation. Available at: http://www.bankofcanada.ca/wp-content/uploads/2010/09/clark.pdf. [427]

Clark, T.E. and McCracken, M.W. (2001). Tests of equal forecast accuracy and encompassing for nested models. *Journal of Econometrics*, 105(1), 85–110.
DOI: 10.1016/s0304-4076(01)00071-9. [417, 427]

Clark, T.E. and McCracken, M.W. (2005). Evaluating direct multistep forecasts. *Econometric Reviews*, 24(4), 369–404. DOI: 10.1080/07474930500405683. [427]

Clark, T.E. and West, K.D. (2006). Using out-of-sample mean squared prediction errors to test the martingale difference hypothesis. *Journal of Econometrics*, 135(1-2), 155–186.
DOI: 10.1016/j.jeconom.2005.07.014. [431]

Clark, T.E. and West, K.D. (2007). Approximately normal tests for equal predictive accuracy in nested models. *Journal of Econometrics*, 138(1), 291–311.
DOI: 10.1016/j.jeconom.2006.05.023. [417]

Clements, M.P. (2005). *Evaluating Econometric Forecasts of Economic and Financial Variables*. Palgrave MacMillan, New York. DOI: 10.1057/9780230596146. [412, 422, 430, 431]

Clements, M.P. and Hendry, D.F. (1993). On the limitations of comparing mean squared forecast errors. *Journal of Forecasting*, 12(8), 617–637 (with discussion).
DOI: 10.1002/for.3980120815. [479]

Clements, M.P. and Krolzig, H.-M. (1998). A comparison of the forecast performance of Markov-switching and threshold autoregressive models of US GNP. *Econometrics Journal*, 1(1), C47–C75. DOI: 10.1111/1368-423x.11004. [429]

Clements, M.P. and Smith, J. (1997). The performance of alternative forecasting methods for SETAR models. *International Journal of Forecasting*, 13(4), 463–475.
DOI: 10.1016/s0169-2070(97)00017-4. [407, 429]

Clements, M.P. and Smith, J. (1999). A Monte Carlo study of the forecasting performance of empirical SETAR models. *Journal of Applied Econometrics*, 14(2), 124–141.
DOI: 10.1002/(sici)1099-1255(199903/04)14:2%3C123::aid-jae493%3E3.0.co;2-k. [429]

Clements, M.P. and Smith, J. (2000). Evaluating the forecast densities of linear and non-linear models: Application to output growth and unemployment. *Journal of Forecasting*, 19(4), 255–276.
DOI: 10.1002/1099-131x(200007)19:4%3C255::aid-for773%3E3.0.co;2-g. [430]

Clements, M.P. and Smith, J. (2001). Evaluating forecasts from SETAR models of exchange rates. *Journal of International Money and Finance*, 20(1), 133–148.
DOI: 10.1016/s0261-5606(00)00039-5. [429]

Clements, M.P. and Smith, J. (2002). Evaluating multivariate forecast densities: A comparison of two approaches. *International Journal of Forecasting*, 18(3), 397–407. DOI: 10.1016/s0169-2070(01)00126-1. [480, 492]

Clements, M.P. and Taylor, N. (2003). Evaluating interval forecasts of high frequency financial data. *Journal of Applied Econometrics*, 18(4), 445–456. DOI: 10.1002/jae.703. [430]

Clements, M.P., Franses, P.H., Smith, J., and Van Dijk, D. (2003). On SETAR non-linearity and forecasting. *Journal of Forecasting*, 22(5), 359–375. DOI: 10.1002/for.863. [429]

Cleveland, R.B., Cleveland, W.S., McRae, J.W., and Terpenning, I. (1990). STL: A seasonal-trend decomposition procedure based on loess. *Journal of Official Statistics*, 6(1), 3–73 (with discussion). [386]

Cleveland, W.S. (1979). Robust locally weighted regression and smoothing scatterplots. *Journal of the American Statistical Association*, 74(368), 829–836. DOI: 10.2307/2286407. [353, 385]

Cleveland, W.S. and Devlin, S.J. (1988). Locally weighted regression: An approach to regression analysis by local fitting. *Journal of the American Statistical Association*, 83(403), 596–610. DOI: 10.1080/01621459.1988.10478639. [353]

Cline, D.B.H. (2007a). Stability of nonlinear stochastic recursions with application to nonlinear AR-GARCH models. *Advances in Applied Probability*, 39(2), 462–491. DOI: 10.1239/aap/1183667619. [93, 94]

Cline, D.B.H. (2007b). Regular variation of order 1 nonlinear AR-ARCH models. *Stochastic Processes and their Applications*, 117(7), 840–861. DOI: 10.1016/j.spa.2006.10.009. [92]

Cline, D.B.H. (2007c). Evaluating the Lyapounov exponent and existence of moments for threshold AR-ARCH models. *Journal of Time Series Analysis*, 28(2), 241–260. DOI: 10.1111/j.1467-9892.2006.00508.x. [91, 92, 93]

Cline, D.B.H. and Pu, H.H. (1999a). Geometric ergodicity of nonlinear time series. *Statistica Sinica*, 9(4), 1103–1118. [91]

Cline, D.B.H. and Pu, H.H. (1999b). Stability of nonlinear AR(1) time series with delay. *Stochastic Processes and their Applications*, 82(2), 307–333. DOI: 10.1016/s0304-4149(99)00042-3. [91]

Cline, D.B.H. and Pu, H.H. (2001). Geometric transience of nonlinear time series. *Statistica Sinica*, 11(1), 273–287. [91]

Cline, D.B.H. and Pu, H.H. (2004). Stability and the Lyapounov exponent of threshold AR-ARCH models. *The Annals of Applied Probability*, 14(4), 1920–1949. DOI: 10.1214/105051604000000431. [91]

Coakley, J., Fuertes, A-M., and Pérez, M-T. (2003). Numerical issues in threshold autoregressive modeling of time series. *Journal of Economic Dynamics & Control*, 27(11-12), 2219–2242. DOI: 10.1016/s0165-1889(02)00123-9. [248]

Collomb, G. (1984). Propriétés de convergence presque complète du prédicteur à noyau. *Zeitschrift für Wahrscheinlichkeitstheorie und verwandte Gebiete*, 66(3), 441–460. DOI: 10.1007/bf00533708. [339]

Collomb, G., Härdle, W., and Hassani, S. (1987). A note on prediction via estimation of the conditional mode function. *Journal of Statistical Planning and Inference*, 15 (1986-1987), 227–236. DOI: 10.1016/0378-3758(86)90099-6. [340]

Connor, J.T., Martin, D.R., and Atlas, L.E. (1994). Recurrent neural networks and robust time series prediction. *IEEE Transactions on Neural Networks*, 5(2), 240–254. DOI: 10.1109/72.279188. [75]

Corradi, V. and Swanson, N.R. (2006a). Predictive density evaluation. In G. Elliott et al. (Eds.) *Handbook of Economic Forecasting*, North-Holland, Amsterdam, pp. 197–284. DOI: 10.1016/s1574-0706(05)01005-0. [427]

Corradi, V. and Swanson, N.R. (2006b). Bootstrap conditional distribution tests in the presence of dynamic misspecification. *Journal of Econometrics*, 133(2), 779–806. DOI: 10.1016/j.jeconom.2005.06.013. [427]

Corradi, V. and Swanson, N.R. (2012). A survey of recent advances in forecast accuracy comparison testing, with an extension to stochastic dominance. In X. Chen and N.R. Swanson (Eds.) *Causality, Prediction and Specification Analysis: Recent Advances and Future Directions. Essay in Honour of Halbert L. White Jr.* Springer-Verlag, New York. Available at: http://www2.warwick.ac.uk/fac/soc/economics/staff/academic/corradi/research/corradi_swanson_whitefest_2012_02_09.pdf and http://econweb.rutgers.edu/nswanson/papers/corradi_swanson_whitefest_2012_02_09.pdf. [429]

Corradi, V., Swanson, N.R., and Olivetti, C. (2001). Predictive ability with cointegrated variables. *Journal of Econometrics*, 104(2), 315–358. DOI: 10.1016/s0304-4076(01)00086-0. [429]

Cox, D.R. (1981). Statistical analysis of time series: Some recent developments. *Scandinavian Journal of Statistics*, 8(2), 93–115 (with discussion). [315]

Cox, D.R. (1991). Long-range dependence, non-linearity and time irreversibility. *Journal of Time Series Analysis*, 12(4), 329–335. DOI: 10.1111/j.1467-9892.1991.tb00087.x. [334]

Cressie, N. and Read, T.R.C. (1984). Multinomial goodness-of-fit tests. *Journal of the Royal Statistical Society*, B 46(3), 440–464. [265]

Cryer, J.D. and Chan, K.S. (2008). *Time Series Analysis: With Applications in* R (2nd edn.). Springer-Verlag, New York. DOI: 10.1007/978-0-387-75959-3. [251]

Csiszár, I. (1967). Information-type measures of divergence of probability distributions and indirect observations. *Studia Scientiarum Mathematicarum Hungarica*, 2, 299–318. [264]

Cutler, C.D. (1991). Some results on the behavior and estimation of the fractal dimensions of distributions on attractors. *Journal of Statistical Physics*, 62(3/4), 651–708. DOI: 10.1007/bf01017978. [312]

Cutler, C.D. and Kaplan, D.T. (Eds.) (1996). *Nonlinear Dynamics and Time Series: Building a Bridge between the Natural and Statistical Sciences*. Fields Institute Communications, American Mathematical Society, Providence, Rhode Island. [597]

D'Alessandro, P., Isidori, A., and Ruberti, A. (1974). Realizations and structure theory of bilinear dynamical systems. *SIGMA Journal of Control*, 12(3), 517–535. DOI: 10.1137/0312040. [73]

Dagum, E.B., Bordignon, S., Cappuccio, N., Proietti, T., and Riani, M. (2004). *Linear and Non Linear Dynamics in Time Series*. Pitagora Editrice, Bologna, Italy. [597]

Dai, Y. and Billard, L. (1998). A space-time bilinear model and its identification. *Journal of Time Series Analysis*, 19(6), 657–679. DOI: 10.1111/1467-9892.00115. [485]

Dai, Y. and Billard, L. (2003). Maximum likelihood estimation in space time bilinear models. *Journal of Time Series Analysis*, 24(1), 25–44. DOI: 10.1111/1467-9892.00291. [485]

Dalle Molle, J.W. and Hinich, M.J. (1995). Trispectral analysis of stationary random time series. *Journal of the Acoustical Society of America*, 97(5), 2963–2978. DOI: 10.1121/1.411860. [323]

Daniels, H.E. (1946). Discussion to 'Symposium on autocorrelations in time series'. *Journal of the Royal Statistical Society*, 8 (Suppl.), 29–97. [333]

Darolles, S., Florens, J.-P., and Gouriéroux, C. (2004). Kernel-based nonlinear canonical analysis and time reversibility. *Journal of Econometrics*, 119(2), 323–353. DOI: /10.1016/s0304-4076(03)00199-4. [333]

Davidson, J. (2004). Forecasting Markov-switching dynamic, conditionally heteroscedastic processes. *Statistics & Probability Letters*, 68(2), 137–147. DOI: 10.1016/j.spl.2004.02.004. [75]

Davies, N. and Petruccelli, J.D. (1986). Detecting nonlinearity in time series. *The Statistician*, 35(2), 271–280. DOI: 10.2307/2987532. [193]

Daw, C.S., Finney, C.E.A., and Kennel, M.B. (2000). Symbolic approach for measuring temporal irreversibility. *Physical Review E*, 62(2), 1912–1921. DOI: 10.1103/physreve.62.1912. [333]

De Brabanter, J., Pelckmans, K., Suykens, J.A.K., and Vandewalle, J. (2005). Prediction intervals for NAR model structures using a bootstrap method. *Proceedings of the International Symposium on Nonlinear Theory and its Applications (NOTA 2005)*, Bruges, Belgium, pp. 610–613. Available at: http://www.ieice.org/proceedings/. [429]

De Gooijer, J.G. (1998). On threshold moving average models. *Journal of Time Series Analysis*, 19(1), 1–18. DOI: 10.1111/1467-9892.00074. [248]

De Gooijer, J.G. (2001). Cross-validation criteria for SETAR model selection. *Journal of Time Series Analysis*, 22(3), 267–281. DOI: 10.1111/1467-9892.00223. [235]

De Gooijer, J.G. (2007). Power of the Neyman smooth test for evaluating multivariate forecast densities. *Journal of Applied Statistics*, 34(4), 371–382. DOI: 10.1080/02664760701231526. [487]

De Gooijer, J.G. and Brännäs, K. (1995). Invertibility of non-linear time series models. *Communications in Statistics: Theory and Methods*, 24(11), 2701–2714. DOI: 10.1080/03610929508831644. [105]

References

De Gooijer, J.G. and De Bruin, P.T. (1998). On forecasting SETAR processes. *Statistics & Probability Letters*, 37(1), 7–14. DOI: 10.1016/s0167-7152(97)00092-8. [401, 404, 432]

De Gooijer, J.G. and Gannoun, A. (2000). Nonparametric conditional predictive regions for time series. *Computational Statistics & Data Analysis*, 33(3), 259–275. DOI: 10.1016/s0167-9473(99)00056-0. [384, 413, 415, 429]

De Gooijer, J.G., Gannoun, A., and Zerom, D. (2001). Multi-stage kernel-based conditional quantile prediction in time series. *Communications in Statistics: Theory and Methods*, 30(12), 2499–2515. DOI: 10.1081/sta-100108445. [344, 345, 346]

De Gooijer, J.G., Gannoun, A., and Zerom, D. (2002). Mean squared error properties of the kernel-based multi-stage median predictor for time series. *Statistics & Probability Letters*, 56(1), 51–56. DOI: 10.1016/S0167-7152(01)00169-9. [382]

De Gooijer, J.G., Gannoun, A., and Zerom, D. (2006). A multivariate quantile predictor. *Communications in Statistics: Theory and Methods*, 35(1), 133–147. DOI: 10.1080/03610920500439570. [497, 500, 521]

De Gooijer, J.G. and Kumar, K. (1992). Some recent developments in non-linear time series modelling, testing, and forecasting. *International Journal of Forecasting*, 8(2), 135–156. DOI: 10.1016/0169-2070(92)90115-P. Corrigendum: (1993, p. 145). [190, 428]

De Gooijer, J.G. and Ray, B.K. (2003). Modeling vector nonlinear time series using POLY-MARS. *Computational Statistics & Data Analysis*, 42(1-2), 73–90. DOI: 10.1016/S0167-9473(02)00123-8. [522]

De Gooijer, J.G., Ray, B.K., and Kräger, H. (1998). Forecasting exchange rates using TS-MARS. *Journal of International Money and Finance*, 17(3), 513–534. DOI: 10.1016/S0261-5606(98)00017-5. [381]

De Gooijer, J.G. and Sivarajasingham, S. (2008). Parametric and nonparametric Granger causality testing: Linkages between international stock markets. *Physica*, A 387(11), 2547–2560. DOI: 10.1016/j.physa.2008.01.033. [523]

De Gooijer, J.G. and Vidiella–i–Anguera, A. (2003a). Nonlinear stochastic inflation modelling using SEASETARs. *Insurance Mathematics and Economics*, 32(1), 3–18. DOI: 10.1016/S0167-6687(02)00190-7. [80]

De Gooijer, J.G. and Vidiella–i–Anguera, A. (2003b). Forecasting threshold cointegrated systems. *International Journal of Forecasting*, 20(2), 237–253. DOI: 10.1016/j.ijforecast.2003.09.006. [79, 487]

De Gooijer, J.G. and Vidiella–i–Anguera, A. (2005). Estimating threshold cointegrated systems. *Economics Bulletin*, 3(8), 1–7. [453]

De Gooijer, J.G. and Yuan, A. (2016). Non parametric portmanteau tests for detecting non linearities in high dimensions. *Communications in Statistics: Theory and Methods*, 45(2), 385–399. DOI: 10.1080/03610926.2013.815209. [296]

De Gooijer, J.G. and Zerom, D. (2000). Kernel based multi-step-ahead prediction of the U.S. short-term interest rate. *Journal of Forecasting*, 19(4), 335–353. DOI: 10.1002/1099-131x(200007)19:4%3C335::aid-for777%3E3.3.co;2-v. [381]

De Gooijer, J.G. and Zerom, D. (2003). On conditional density estimation. *Statistica Neerlandica*, 57(2), 159–176. DOI: 10.1111/1467-9574.00226. [348, 351]

Deheuvels, P. (1977). Estimation non paramétrique de la densité par histogramme généralisé. *Revue de Statistique Appliquée*, 25(3), 5–42. [382]

Deheuvels, P. (1981). An asymptotic decomposition for multivariate distribution-free tests of independence. *Journal of Multivariate Analysis*, 11(1), 102–113.
DOI: 10.1016/0047-259x(81)90136-6. [286]

Delgado, M.A. (1996). Testing serial independence using the sample distribution function. *Journal of Time Series Analysis*, 17(3), 271–285.
DOI: 10.1111/j.1467-9892.1996.tb00276.x. [285, 287]

de Lima, P.J.F. (1996). Nuisance parameter free properties of correlation integral based statistics. *Econometric Reviews*, 15(3), 237–259. DOI: 10.1080/07474939608800354. [296]

de Lima, P.J.F. (1997). On the robustness of nonlinearity tests to moment condition failure. *Journal of Econometrics*, 76(1-2), 251–280. DOI: 10.1016/0304-4076(95)01791-7. [190]

Denison, D.G.T., Mallick, B.K., and Smith, A.F.M. (1998). Bayesian MARS. *Statistics and Computing*, 8(4), 337–346. [383]

Denker, M. and Keller, G. (1983). On U-statistics and v. Mises' statistics for weakly dependent processes. *Zeitschrift für Wahrscheinlichkeitstheorie und verwandte Gebiete*, 64(4), 505–522. [310, 518]

Deutsch, M., Granger, C.W.J., and Teräsvirta, T. (1994). The combination of forecasts using changing weights. *International Journal of Forecasting*, 10(1), 47–57.
DOI: 10.1016/0169-2070(94)90049-3. [425]

Dey, S., Krishnamurthy, V., and Salmon–Legagneur, T. (1994). Estimation of Markov-modulated time-series via EM algorithm. *IEEE Signal Processing Letters*, 1(10), 153–155. DOI: 10.1109/97.329841. [250]

Diebold, F.X. (2015). Comparing predictive accuracy, twenty years later: A personal perspective on the use and abuse of Diebold–Mariano tests (with discussion). *Journal of Business & Economic Statistics*, 33(1), DOI: 10.2139/ssrn.2316240. [429]

Diebold, F.X. and Mariano, R.S. (1995). Comparing predictive accuracy. *Journal of Business & Economic Statistics*, 13(3), 253–263. DOI: 10.2307/1392185. [416, 417, 424]

Diebold, F.X., Gunther, T.A., and Tay, A.S. (1998). Evaluating density forecasts with applications to financial risk management. *International Economic Review*, 39(4), 863–883. DOI: 10.2307/2527342. [430]

Diebold, F.X., Hahn, J., and Tay, A.S. (1999). Multivariate density forecast evaluation and calibration in financial risk management: High-frequency returns on foreign exchange. *Review of Economics and Statistics*, 81(4), 661–673. DOI: 10.1162/003465399558526. [430]

Diebold, F.X., Tay, A.S., and Wallis, K.F. (1999). Evaluating density forecasts of inflation: The survey of professional forecasters. In R.F. Engle and H. White (Eds.) *Cointegration, Causality and Forecasting*, Festschrift in Honour of Clive W.J. Granger. Oxford University Press, New York, pp. 76–90. DOI: 10.3386/w6228. [430]

Diks, C. (1999). *Nonlinear Time Series Analysis: Methods and Applications*. World Scientific, Singapore. DOI: 10.1142/3823. [597]

Diks, C. (2009). Nonparametric tests for independence. In R.A. Meyers (Ed.) *Encyclopedia of Complexity and Systems Science*. Springer-Verlag, New York, pp. 6252–6271. DOI: 10.1007/978-0-387-30440-3_369. [262]

Diks, C., Van Houwelingen, J.C., Takens, F., and DeGoede, J. (1995). Reversibility as a criterion for discriminating time series. *Physics Letters*, A 201(2-3), 221–228. DOI: 10.1016/0375-9601(95)00239-y. [321, 327, 328]

Diks, C. and Mudelsee, M. (2000). Redundancies in the Earth's climatological time series. *Physics Letters*, A 275(5-6), 407–414. DOI: 10.1016/s0375-9601(00)00613-7. [24]

Diks, C. and Panchenko, V. (2005). A note on the Hiemstra-Jones test for Granger non-causality. *Studies in Nonlinear Dynamics & Econometrics*, 9(2). DOI: 10.2202/1558-3708.1234. [516]

Diks, C. and Panchenko, V. (2006). A new statistic and practical guidelines for nonparametric Granger causality testing. *Journal of Economic Dynamics & Control*, 30(9-10), 1647–1669. DOI: 10.1016/j.jedc.2005.08.008. [516, 517, 518, 521]

Diks, C. and Panchenko, V. (2007). Nonparametric tests for serial independence based on quadratic forms. *Statistica Sinica*, 17(1), 81–98. [277, 291]

Diks, C. and Wolski, M. (2016). Nonlinear Granger causality: Guidelines for multivariate analysis. *Journal of Applied Econometrics*, 31(7), 1333–1351. DOI: 10.1002/jae.2495. [519, 520]

Diop, A. and Guégan, D. (2004). Tail behavior of a threshold autoregressive stochastic volatility model. *Extremes*, 7(4), 367–375. DOI: 10.1007/s10687-004-3482-y. [80]

Dobrushin, R.L., Sukhov, Yu.M., and Fritz, J. (1988). A.N. Kolmogorov – the founder of the theory of reversible Markov processes. *Russian Mathematical Surveys*, 43, 157–182; translation from *Uspekhi Matematicheskikh Nauk*, 43(6) (1988), 167–188 (Russian). DOI: 10.1070/rm1988v043n06abeh001985. [332]

Donner, R.V. and Barbosa, S.M. (Eds.) (2008). *Nonlinear Time Series Analysis in the Geosciences: Applications in Climatology, Geodynamics and Solar-Terrestrial Physics*. Springer-Verlag, New York. DOI: 10.1007/978-3-540-78938-3. [2, 597]

Doornik, J.A. and Hansen, H. (2008). An omnibus test for univariate and multivariate normality. *Oxford Bulletin of Economics and Statistics*, 70, 927–939. DOI: 10.1111/j.1468-0084.2008.00537.x. [22, 254]

Douc, R., Moulines, E., and Stoffer, D.S. (2014). *Nonlinear Time Series: Theory, Methods, and Applications with R Examples*. Chapman & Hall/CRC Press, London. [250, 597]

Doukhan, P. (1994). *Mixing. Properties and Examples*. Lecture Notes in Statistics 85. Springer-Verlag, New York. [95]

Drunat, J., Dufrenot, G., and Mathieu, L. (1998). Testing for linearity: A frequency domain approach. In C. Dunnis and B. Zhou (Eds.) *Nonlinear Modelling of High Frequency Financial Time Series*. Wiley, New York, pp. 69–86 [151]

Duchesne, P. (2004). On matricial measures of dependence in vector ARCH models with applications to diagnostic checking. *Statistics & Probability Letters*, 68(2), 149–160. DOI: 10.1016/j.spl.2004.02.006. [487]

Dueker, M.J., Psaradakis, Z., Sola, M., and Spagnolo, F. (2011). Multivariate contemporaneous-threshold autoregressive models. *Journal of Econometrics*, 160(2), 311–325. DOI: 10.1016/j.jeconom.2010.09.011. [79, 486]

Dufour, J.-M., Lepage, Y., and Zeidan, H. (1982). Nonparametric testing for time series: A bibliography. *The Canadian Journal of Statistics*, 10(1), 1–38. DOI: 10.2307/3315073. [295]

Dumitrescu, E.L., Hurlin, C., and Madkour, J. (2013). Testing interval forecasts: A GMM-based approach. *Journal of Forecasting*, 32(2), 97–110. DOI: 10.1002/for.1260. [430]

Dunis, C.L. and Zhou, B. (Eds.) (1998). *Nonlinear Modelling of High Frequency Financial Time Series*. Wiley, New York. [597]

Dunn, P.K. and Smyth, G.K. (1996). Randomized quantile residuals. *Journal of Computational and Graphical Statistics*, 5(3), 236–244. DOI: 10.2307/1390802. [240]

Eckmann, J.-P., Amherst, S.O., and Ruelle, D. (1987). Recurrence plots of dynamical systems. *Europhysics Letters*, 4(9), 973–977. DOI: 10.1209/0295-5075/4/9/004. [19]

Eilers, P. and Marx, B. (1996). Flexible smoothing with B-splines and penalties. *Statistical Science*, 11(2), 89–102. DOI: 10.1214/ss/1038425655. [372]

Elman, J.L. (1990). Finding structures in time. *Cognitive Science*, 14(2), 179–211. [74]

El-Shagi, M. (2011). An evolutionary algorithm for the estimation of threshold vector error correction models. *International Economics and Economic Policy*, 8(4), 341–362. DOI: 10.1007/s10368-011-0180-5. [453]

Embrechts, P., Lindskog, F., and McNeil, A.J. (2003). Modelling dependence with copulas and applications to risk management. In S.T. Rachev (Ed.), *Handbook of Heavy Tailed Distributions in Finance*, Elsevier, Chapter 8, pp. 329–384. DOI: 10.1016/b978-044450896-6.50010-8. [306]

Enders, W. and Granger, C.W.J. (1998). Unit-root tests and asymmetry adjustment with an example using the term structure of interest rates. *Journal of Business & Economic Statistics*, 16(3), 304–311. DOI: 10.2307/1392506. [79, 189]

Engen, S. and Lillegård, M. (1997). Stochastic simulations conditioned on sufficient statistics. *Biometrika*, 84(1), 235–240. DOI: 10.1093/biomet/84.1.235. [277]

Engle, R.F. (2002). New frontiers for ARCH models. *Journal of Applied Econometrics*, 17(5), 425–446. DOI: 10.1002/jae.683. [74]

Engle, R.F. and Kozicki, S. (1993). Testing for common features. *Journal of Business & Economic Statistics*, 11(4), 369–380. DOI: 10.2307/1391623. [456]

Engle, R.F. and Manganelli, S. (2004). CAViaR: Conditional autoregressive value-at-risk by regression quantiles. *Journal of Business & Economic Statistics*, 22(4), 367–381. DOI: 10.1198/073500104000000370. [430]

Ephraim, Y. and Merhav, N. (2002). Hidden Markov processes. *IEEE Transactions on Information Theory*, 48(6), 1518–1569. DOI: 10.1109/tit.2002.1003838. [75]

Epps, T.W. (1987). Testing that a stationary time series is Gaussian. *The Annals of Statistics*, 15(4), 1683–1698. DOI: 10.1214/aos/1176350618. [151]

Ertel, J.E. and Fowlkes, E.B. (1976). Some algorithms for linear spline and piecewise multiple linear regression. *Journal of the American Statistical Association*, 71(355), 640–648. DOI: 10.1080/01621459.1976.10481540. [182]

Fan, J. (1992). Design-adaptive nonparametric regression. *Journal of the American Statistical Association*, 87(420), 998–1004. DOI: 10.1080/01621459.1992.10476255. [350]

Fan, J. and Gijbels, I. (1996). *Local Polynomial Modelling and Its Applications*. Chapman & Hall, London. DOI: 10.1007/978-1-4899-3150-4. [349, 382, 409]

Fan, J. and Yao, Q. (2003). *Nonlinear Time Series: Nonparametric and Parametric Methods*. Springer-Verlag, New York. DOI: 10.1007/978-0-387-69395-8_4. [209, 374, 382, 386, 414, 597]

Fan, J., Yao, Q., and Cai, Z. (2003). Adaptive varying-coefficient linear models. *Journal of the Royal Statistical Society*, B 65(1), 57–80. DOI: 10.1111/1467-9868.00372. [374]

Fan, J., Yao, Q., and Tong, H. (1996). Estimation of conditional densities and sensitivity measures in nonlinear dynamic systems. *Biometrika*, 83(1), 189–206. DOI: 10.1093/biomet/83.1.189. [383]

Fan, J. and Yim, T.H. (2004). A cross-validation method for estimating conditional densities. *Biometrika*, 91(4), 819–834. DOI: 10.1093/biomet/91.4.819. [382]

Fassò, A. and Negri, I. (2002). Multi-step forecasting for nonlinear models of high frequency ground ozone data: A Monte Carlo approach. *Environmetrics*, 13(4), 365–378. DOI: 10.1002/env.544. [428]

Feigin, P.D. and Tweedie, R.L. (1985). Random coefficient autoregressive processes: A Markov chain analysis of stationary and finiteness of moments. *Journal of Time Series Analysis*, 6(1), 1–14. DOI: 10.1111/j.1467-9892.1985.tb00394.x. [96, 97]

Feo, T.A. and Resende, M.G.C. (1995). Greedy randomized adaptive search procedures. *Journal of Global Optimization*, 6(2), 109–133. DOI: 10.1007/bf01096763. [74]

Ferguson, T.S., Genest, C., and Hallin, M. (2000). Kendall's tau for serial dependence. *The Canadian Journal of Statistics*, 28(3), 587–604. DOI: 10.2307/3315967. [16]

Fermanian, J.-D., and Scaillet, O. (2003). Nonparametric estimation of copulas for time series. *Jounal of Risk*, 5(4), 25–54. DOI: 10.2139/ssrn.372142. [296]

Fernandes, M. and Néri, B. (2010). Nonparametric entropy-based tests of independence between stochastic processes. *Econometric Reviews*, 29(3), 276–306. DOI: 10.1080/07474930903451557. [269, 271]

Fernández–Rodríguez, F., Sosvilla–Rivero, S., and Andrada–Félix, J. (1997). Combining information in exchange rate forecasting: Evidence from the EMS. *Applied Economics Letters*, 4(7), 441–444. DOI: 10.1080/135048597355221. [522]

Fernández–Rodríguez, F., Sosvilla–Rivero, S., and Andrada–Félix, J. (1999). Exchange-rate forecasts with simultaneous nearest-neighbor methods: Evidence from the EMS. *International Journal of Forecasting*, 15(4), 383–392. DOI: 10.1016/s0169-2070(99)00003-5. [522]

Fernández, V.A., Gamero, M.D.J, and García, J.M. (2008). A test for the two-sample problem based on empirical characteristic functions. *Computational Statistics & Data Analysis*, 52(7), 3730–3748. DOI: 10.1016/j.csda.2007.12.013. [296]

Ferrante, M., Fonseca, G., and Vidoni, P. (2003). Geometric ergodicity, regularity of the invariant distribution and inference for a threshold bilinear Markov process. *Statistica Sinica*, 13(2), 367–384. [111]

Findley, D.F. (1993). The overfitting principles supporting AIC. Statistical Research Division Report: RR-93/04, U.S. Bureau of the Census statistical, Washington, DC. Abstract: http://www.census.gov.edgekey.net/srd/www/abstract/rr93-4.html. [229]

Fiorentini, G, Sentana, E., and Calzolari, G. (2004). On the validity of Jarque–Bera normality test in conditionally heteroskedastic dynamic regression models. *Economics Letters*, 83(3), 307–312. DOI: 10.1016/j.econlet.2003.10.023. [23]

Fitzgerald, W.J., Smith, R.L., Walden, A.T., and Young, P.C. (Eds.) (2000). *Nonlinear and Nonstationary Signal Processing*. Cambridge University Press, Cambridge, UK. [597]

Fong, W.M. (2003). Time reversibility tests of volume-volatility dynamics for stock returns. *Economics Letters*, 81(1), 39–45. DOI: 10.1016/s0165-1765(03)00146-0. [333]

Fonseca, G. (2004). On the stationarity of first-order nonlinear time series models: Some developments. *Studies in Nonlinear Dynamics & Econometrics*, 8(2). DOI: 10.2202/1558-3708.1216. [111]

Fonseca, G. (2005). On the stability of nonlinear ARMA models. Quaderno della Facoltà di Economia, 2005/3, Università dell'Insubria, Varese. Abstract: http://econpapers.repec.org/paper/insquaeco/qf0503.htm. [111]

Forbes, C.S., Kalb, G.R.J., and Kofman, P. (1999). Bayesian arbitrage threshold analysis. *Journal of Business & Economic Statistics*, 17(3), 364–372. DOI: 10.2307/1392294. [488]

Francis, B.B., Mougoué, M., and Panchenko, V. (2010). Is there a *symmetric* nonlinear causal relationship between large and small firms? *Journal of Empirical Finance*, 17(1), 23–38. DOI: 10.1016/j.jempfin.2009.08.003. [523]

Francq, C. and Zakoïan, J.-M. (2005). The L^2-structures of standard and switching regime GARCH models. *Stochastic Processes and their Applications*, 115(9), 1557–1582. DOI: 10.1016/j.spa.2005.04.005. [110]

Francq, C. and Zakoïan, J.-M. (2010). *GARCH Models: Structure, Statistical Inference and Financial Applications*. Wiley, New York. DOI: 10.1002/9780470670057. [25]

Franke, J. (2012). Markov switching time series models. In T. Subba Rao et al. (Eds.) *Time Series Analysis: Methods and Applications*, Handbook of Statistics, Vol. 30. North-Holland, Amsterdam, The Netherlands, pp. 99–122. DOI: 10.1016/b978-0-444-53858-1.00005-3. [75]

Franke, J., Härdle, W., and Martin, D. (1984). *Robust and Nonlinear Time Series Analysis.* Springer-Verlag, New York. [597]

Franke, J., Kreiss, J.-P., and Mammen, E. (2002). Bootstrap of kernel smoothing in nonlinear time series. *Bernoulli*, 8(1), 1–37. Available at http://projecteuclid.org/euclid.bj/1078951087. [382]

Franses, P.H. and Van Dijk, D. (2000). *Nonlinear Time Series Models in Empirical Finance.* Cambridge University Press, Cambridge, UK. DOI: 10.1017/cbo9780511754067. [597]

Friedman, J.H. (1984a). A variable span scatterplot smoother. Laboratory for Computational Statistics, Stanford University Technical Report No. 5. Available at: http://www.slac.stanford.edu/cgi-wrap/getdoc/slac-pub-3477.pdf. [85]

Friedman, J.H. (1984b). SMART user's guide. Technical Report LCS01, Laboratory for Computational Statistics, Stanford University. Available at: https://statistics.stanford.edu/sites/default/files/LCS%2001.pdf. [386]

Friedman, J.H. (1991). Multivariate adaptive regression splines. *The Annals of Statistics*, 19(1), 1–141 (with discussion). DOI: 10.1214/aos/1176347963. [365]

Friedman, J.H. and Stuetzle, W. (1981). Projection pursuit regression. *Journal of the American Statistical Association*, 76(376), 817–823. DOI: 10.1080/01621459.1981.10477729. [364, 386]

Frühwirth–Schatter, S. (2006). *Finite Mixture and Markov Switching Models.* Springer-Verlag, New York. DOI: 10.1007/978-0-387-35768-3. [75]

Fukuchi, J.-I. (1999). Subsampling and model selection in time series analysis. *Biometrika*, 86(3), 591–604. DOI: 10.1093/biomet/86.3.591. [359]

Furstenberg, H. and Kesten, H. (1960). Products of random matrices. *Annals of Mathematical Statistics*, 31(2), 457–469. Available at: https://projecteuclid.org/download/pdf_1/euclid.aoms/1177705909. [88]

Gabr, M.M. (1998). Robust estimation of bilinear time series models. *Communications in Statistics: Theory and Methods*, 27(1), 41–53. DOI: 10.1080/03610929808832649. [248]

Galeano, P. and Peña, D. (2007). Improved model selection criteria for SETAR time series models. *Journal of Statistical Planning and Inference*, 137(9), 2802–2814. DOI: 10.1016/j.jspi.2006.10.014. [235]

Galka, A. (2000). *Topics in Nonlinear Time Series Analysis – With Implications for EEG Analysis.* World Scientific, Singapore. DOI: 10.1142/9789812813237. [2, 597]

Galvão, A.B.C. (2006). Structural break threshold VARs for predicting US recessions using the spread. *Journal of Applied Econometrics*, 21(4), 463–487. DOI: 10.1002/jae.840. [74, 80]

Gannoun, A. (1990). Estimation non paramétrique de la médiane conditionnelle: médianogramme et méthode du noyau. *Publications de l'Institut de statistique de l'Université de Paris*, XXXXV, 11–22. [340]

Gao, J. (2007). *Nonlinear Time Series: Semiparametric and Nonparametric Methods*. Chapman & Hall/CRC, London. DOI: 10.1201/9781420011210. [80, 597]

Gao, J. and Tong, H. (2004). Semiparametric nonlinear time series model selection. *Journal of the Royal Statistical Society*, B 66(2), 321–336.
DOI: 10.1111/j.1369-7412.2004.05303.x. [383]

Gao, J., Tjøstheim, D., and Yin, J. (2013). Estimation in threshold autoregressive models with a stationary and a unit root regime. *Journal of Econometrics*, 172(1), 1–13.
DOI: 10.1016/j.jeconom.2011.12.006. [80]

Gao, W. and Tian, Z. (2009). Learning Granger causality graphs for multivariate nonlinear time series. *Journal of Systems Science and Systems Engineering*, 18(1), 038–052.
DOI: 10.1007/s11518-009-5099-9. [523]

Garth, L.M. and Bresler, Y. (1996). On the use of asymptotics in detection and estimation. *IEEE Transactions on Acoustics, Speech, and Signal Processing*, 44(5), 1304–1307.
DOI: 10.1109/78.502350. [133]

Gasser, T. (1975). Goodness-of-fit for correlated data. *Biometrika*, 62(3), 563–570.
DOI: 10.1093/biomet/62.3.563. [12]

Gaver, D.P. and Lewis, P.A.W. (1980). First order autoregressive gamma sequences and point processes. *Advances in Applied Probability*, 12(3), 727–745.
DOI: 10.2307/1426429. [433]

Gel, Y.R. and Gastwirth, J.L. (2008). A robust modification of the Jarque–Bera test of normality. *Economics Letters*, 99(1), 30–32. DOI: 10.1016/j.econlet.2007.05.022. [22]

Geman, S. and Geman, D. (1984). Stochastic relaxation, Gibbs distribution and the Bayesian restoration of images. *IEEE Transactions on Pattern Analysis and Machine Intelligence*, PAMI 6(6), 721–741. DOI: 10.1109/tpami.1984.4767596. [249]

Genest, C. and Rémillard, B. (2004). Tests of independence and randomness based on the empirical copula process. *Test*, 13(2), 335–369. DOI: 10.1007/bf02595777. [286, 312]

Genest, C. and Zidek, J. (1986). Combining probability distributions: A critique and an annotated bibliography. *Statistical Science*, 1(1), 114–148 (with discussion).
DOI: 10.1214/ss/1177013825. [426]

Genest, C., Ghoudi, K., and Rémillard, B. (2007). Rank-based extensions of the Brock, Dechert, and Scheinkman test. *Journal of the American Statistical Association*, 102(480), 1363–1376. DOI: 10.1198/016214507000001076. [282, 283]

Gerlach, R., Chen, C.W.S., and Chan, N.Y.C. (2011). Bayesian time-varying quantile forecasting for Value-at-Risk in financial markets. *Journal of Business & Economic Statistics*, 29(4), 481–492. DOI: 10.1198/jbes.2010.08203. [81]

Gharavi, R. and Anantharam, V. (2005). An upper bound for the largest Lyapunov exponent of a Markovian product of nonnegative matrices. *Theoretical Computer Science*, 332(1-3), 543–557. DOI: 10.1016/j.tcs.2004.12.025. [111]

Ghoudi, K., Kulperger, R.J., and Rémillard, B. (2001). A nonparametric test of serial independence for time series and residuals. *Journal of Multivariate Analysis*, 79(2), 191–218. DOI: 10.1006/jmva.2000.1967. [286, 287, 288, 312]

Giacomini, R. and White, H. (2006). Tests of conditional predictive ability. *Econometrica*, 74(6), 1545–1578. DOI: 10.1111/j.1468-0262.2006.00718.x. [417, 427]

Giannakis, G.B. and Tsatsanis, K. (1994). Time-domain tests for Gaussianity and time-reversibility. *IEEE Transactions on Signal Processing*, 42(12), 3460–3472. DOI: 10.1109/78.340780. [333]

Giannerini, S., Maasoumi, E., and Dagum, E.B. (2015). Entropy testing for nonlinear serial dependence in time series. *Biometrika*, 102(3), 661–675. DOI: 10.1093/biomet/asv007. [295]

Giordani, P. (2006). A cautionary note on outlier robust estimation of threshold models. *Journal of Forecasting*, 25(1), 37–47. DOI: 10.1002/for.972. [249]

Giordano, F. (2000). The variance of CLS estimators for a simple bilinear model. *Quaderni di Statistica*, 2(2), 147–155. [248, 251]

Giordano, F. and Vitale, C. (2003). CLS asymptotic variance for a particular relevant bilinear time series model. *Statistical Methods & Applications*, 12(2), 169–185. DOI: 10.1007/s10260-003-0061-3. [248, 253]

Gneiting, T. (2011). Making and evaluating point forecasts. *Journal of the American Statistical Association*, 106(494), 746–762. DOI: 10.1198/jasa.2011.r10138. [430]

Godambe, V.P. (1960). An optimum property of regular maximum likelihood equation. *Annals of Mathematical Statistics*, 31(4), 1208–1211. DOI: 10.1214/aoms/1177705693. [248]

Godambe, V.P. (1985). The foundations of finite sample estimation in stochastic processes. *Biometrika*, 72(2), 419–428. DOI: 10.1093/biomet/72.2.419. [248]

Goldsheid, I. Ya. (1991). Lyapunov exponents and asymptotic behaviour of the product of random matrices. In L. Arnold et al. (Eds.) *Lyapunov Exponents*. Lecture Notes in Mathematics, Vol. 1486. Springer-Verlag, New York, pp. 23–37. DOI: 10.1007/bfb0086655. [111]

Gonzalo, J. and Pitarakis, J.-Y. (2002). Estimation and model selection based inference in single and multiple threshold models. *Journal of Econometrics*, 110(2), 319–352. DOI: 10.1016/s0304-4076(02)00098-2. [195, 249]

Gonzalo, J. and Wolf, M. (2005). Subsampling inference in threshold autoregressive models. *Journal of Econometrics*, 127(2), 201–224. DOI: 10.1016/j.jeconom.2004.08.004. [45, 73]

Gouriéroux, C. and Jasiak, J. (2005). Nonlinear innovations and impulse responses with application to VaR sensitivity. *Annales d'Économie et de Statistique*, 78, 1–33. DOI: 10.2139/ssrn.757352. [78]

Grahn, T. (1995). A conditional least squares approach to bilinear time series estimation. *Journal of Time Series Analysis*, 16(5), 509–529. DOI: 10.1111/j.1467-9892.1995.tb00251.x. [217, 218, 220]

Granger, C.W.J. (1969). Investigating causal relations by econometric models and cross-spectral methods. *Econometrica*, 37(3), 424–438. DOI: 10.2307/1912791. [514]

Granger, C.W.J. (1989). Combining forecasts – twenty years later. *Journal of Forecasting*, 8(3), 167–173. DOI: 10.1002/for.3980080303. [425, 430]

Granger, C.W.J. (1993). Strategies for modelling nonlinear time-series relationships. *Economic Record*, 69(3), 233–238. DOI: 10.1111/j.1475-4932.1993.tb02103.x. [246, 429]

Granger, C.W.J. and Andersen, A.P. (1978a). *An Introduction to Bilinear Time Series Models*. Vandenhoeck & Ruprecht, Göttingen. [73, 101, 115, 441, 597]

Granger, C.W.J. and Andersen, A.P. (1978b). On the invertibility of time series models. *Stochastic Processes and their Applications*, 8(1), 87–92. DOI: 10.1016/0304-4149(78)90069-8. [101]

Granger, C.W.J. and Lin, J.-L. (1994). Using the mutual information coefficient to identify lags in nonlinear models. *Journal of Time Series Analysis*, 15(4), 371–384. DOI: 10.1111/j.1467-9892.1994.tb00200.x. [18, 19, 271]

Granger, C.W.J., Maasoumi, E., and Racine, J. (2004). A dependence metric for possibly nonlinear processes. *Journal of Time Series Analysis*, 25(5), 649–669. DOI: 10.1111/j.1467-9892.2004.01866.x. [336]

Granger, C.W.J. and Ramanathan, R. (1984). Improved methods of combining forecasts. *Journal of Forecasting*, 3(2), 197–204. DOI: 10.1002/for.3980030207. [425]

Granger, C.W.J. and Teräsvirta, T. (1992a). *Modelling Nonlinear Economic Relationships*. Oxford University Press, Oxford. [74, 597]

Granger, C.W.J. and Teräsvirta, T. (1992b). Experiments in modeling nonlinear relationships between time series. In M. Casdagli and S. Eubank (Eds.) *Nonlinear Modeling and Forecasting*. Proceedings Volume XII, Santa Fe Institute, New Mexico. Addison-Wesley, Redwood City, pp. 189–197. [383]

Granger, C.W.J., White, H., and Kamstra, M. (1989). Interval forecasting: An analysis based on ARCH-quantile estimators. *Journal of Econometrics*, 40(1), 87–96. DOI: 10.1016/0304-4076(89)90031-6. [426]

Grassberger, P. and Procaccia, I. (1983). Measuring the strangeness of strange attractors. *Physica*, D 9(1-2), 189–208. DOI: 10.1016/0167-2789(83)90298-1. [260]

Grenander, U. and Rosenblatt, M. (1984). *Statistical Analysis of Stationary Time Series*, (2nd edn.). Chelsea Publishing Company, New York. [184]

Gretton, A., Bousquet, O., Smola, A.J., and Schölkopf, B. (2005). Measuring statistical dependence with Hilbert-Schmidt norms. In S. Jain et al. (Eds.) *16th International Conference on Algorithmic Learning Theory*. Springer-Verlag, Berlin, pp. 63–77. DOI: 10.1007/11564089_7. [296]

Grünwald, P.D., Myung, I.J., and Pitt, M.A. (Eds.) (2005). *Advances in Minimum Description Length: Theory and Applications*. MIT Press. [249]

Guay, A. and Scaillet, O. (2003). Indirect inference, nuisance parameter and threshold moving average models. *Journal of Business & Economic Statistics*, 21(1), 122–132. DOI: 10.1198/073500102288618829. [74]

Guégan, D. (1993). On the identification and prediction of nonlinear models. In D.R. Brillinger et al. (Eds.), *New Directions in Time Series Analysis*. Springer-Verlag, New York, pp. 195–210. DOI: 10.1007/978-1-4613-9296-5_11. [428]

Guégan, D. (1994). *Séries Chronologiques Non Linéaires à Temps Discret*. Economica, Paris. [597]

Guégan, D. and Pham, T.D. (1992). Power of the score test against bilinear time series models. *Statistica Sinica*, 2(1), 157–169. [194]

Guégan, D. and Rakotomarolahy, P. (2010). Alternative methods for forecasting GDP. In F. Jawadi and W.A. Barnett (Eds.) *Nonlinear Modeling of Economic and Financial Time-Series*. Emerald Group Publishing Ltd., Bingley, UK, pp. 161–185. DOI: 10.1108/s1571-0386(2010)0000020013. [382]

Guégan, D. and Wandji, J.N. (1996). Power of the Lagrange multiplier test for certain subdiagonal bilinear models. *Statistics & Probability Letters*, 29(3), 201–212. DOI: 10.1016/0167-7152(95)00174-3. [188]

Guo, M. and Petruccelli, J. (1991). On the null recurrence and transience of a first-order SETAR model. *Journal of Applied Probability*, 28(3), 584–592. DOI: 10.2307/3214493. [99]

Guo, M. and Tseng, Y.K. (1997). A comparison between linear and nonlinear forecasts for nonlinear AR models. *Journal of Forecasting*, 16(7), 491–508. DOI: 10.1002/(sici)1099-131x(199712)16:7%3C491::aid-for669%3E3.0.co;2-3. [433]

Guo, M., Bai, Z., and An, H.Z. (1999). Multi-step prediction for nonlinear autoregressive models based on empirical distributions. *Statistica Sinica*, 9(2), 559–570. [400, 401]

Guo, Z.-F. and Shintani, M. (2011). Nonparametric lag selection for nonlinear additive autoregressive models. *Economics Letters*, 111(2), 131–134. DOI: 10.1016/j.econlet.2011.01.014. [383]

Györfi, L., Härdle, W., Sarda, P., and Vieu, P. (1989). *Nonparametric Curve Estimation from Time Series*. Springer-Verlag, New York. DOI: 10.1007/978-1-4612-3686-3. [382]

Haggan, V. and Ozaki, T. (1980). Amplitude-dependent exponential autoregressive model fitting for nonlinear random vibrations. In O.D. Anderson (Ed.) *Time Series*. North-Holland, Amsterdam, pp. 57–71. [73]

Haggan, V. and Ozaki, T. (1981). Modelling nonlinear random vibrations using an amplitude-dependent autoregressive time series model. *Biometrika*, 68(1), 189–196. DOI: 10.1093/biomet/68.1.189. [73]

Haldrup, N., Meitz, M., and Saikkonen, P. (Eds.) (2014). *Essays in Nonlinear Time Series Econometrics*. Oxford University Press, Oxford, UK. DOI: 10.1093/acprof:oso/9780199679959.001.0001. [597]

Hall, P. (1989). On projection pursuit regression. *The Annals of Statistics*, 17(2), 573–588. DOI: 10.1214/aos/1176347126. [383]

Hall, P. and Minotte, M.C. (2002). Higher order data sharpening for density estimation. *Journal of the Royal Statistical Society*, B 64(1), 141–157. DOI: 10.1111/1467-9868.00329. [520]

Hall, P. and Morton, S.C. (1993). On the estimation of entropy. *Annals of the Institute of Statistical Mathematics*, 45(1), 69–88. [525]

Hall, S.G. and Mitchell, J. (2007). Combining density forecasts. *International Journal of Forecasting*, 23(1), 1–13. DOI: 10.1016/j.ijforecast.2006.08.001. [426]

Hallin, M. (1980). Invertibility and generalized invertibility of time series models. *Journal of the Royal Statistical Society*, B 42(2), 210–212. [102, 111]

Hallin, M. and Puri, M.L. (1992). Rank tests for time series analysis: A survey. In D.R. Brillinger et al. (Eds.) *New Directions in Time Series Analysis, Part I*. Springer-Verlag, New York, pp. 111–153. [295]

Hamaker, E.L. (2009). Using information criteria to determine the number of regimes in threshold autoregressive models. *Journal of Mathematical Psychology*, 53(6), 518–529. DOI: 10.1016/j.jmp.2009.07.006. [249]

Hamilton, J.D. (1994). *Time Series Analysis*. Princeton University Press, Princeton, NJ. [68]

Hannan, E.J. (1979). The statistical theory of linear systems. In P.R. Krishnaiah (Ed.) *Developments in Statistics*, Vol. 2. Academic Press, New York, pp. 83–122. [22]

Hannan, E.J. and Deistler, M. (2012). *The Statistical Theory of Linear Systems*. Classics in Applied Mathematics (CL70), SIAM, Philadelphia (Originally published: Wiley, New York, 1988). DOI: 10.1137/1.9781611972191. [22]

Hannan, E.J. and Rissanen, J. (1982). Recursive estimation of a mixed autoregressive - moving average order. *Biometrika*, 69(1), 81–96. DOI: 10.1093/biomet/69.1.81. [164, 211]

Hansen, B.E. (1996). Inference when a nuisance parameter is not identified under the null hypothesis. *Econometrica*, 64(2), 413–430. DOI: 10.2307/2171789. [171]

Hansen, B.E. (1997). Inference in TAR models. *Studies in Nonlinear Dynamics & Econometrics*, 2(1). DOI: 10.2202/1558-3708.1024. [189]

Hansen, B.E. (1999). Testing for linearity. *Journal of Economic Surveys*, 13(5), 551–576. DOI: 10.1111/1467-6419.00098. [172, 249]

Hansen, B.E. (2000). Sample splitting and threshold estimation. *Econometrica*, 68(3), 575–603. DOI: 10.1111/1468-0262.00124. [189]

Hansen, B.E. (2005). Exact mean integrated squared error of higher order kernel estimators. *Econometric Theory*, 21(06), 1031–1057. DOI: 10.1017/s0266466605050528. [305]

Hansen, B.E. (2011). Threshold autoregression in economics. *Statistics and Its Interface*, 4(2), 123–127. DOI: 10.4310/sii.2011.v4.n2.a4. [73]

Hansen, B.E. and Seo, B. (2011). Testing for two-regime threshold cointegration in vector error-correction model. *Journal of Econometrics*, 110(2), 293–318.
DOI: 10.1016/s0304-4076(02)00097-0. [486]

Hansen, L.P. (1982). Large sample properties of generalised method of moments estimation. *Econometrica*, 50(4), 1029–1054. DOI: 10.2307/1912775. [248]

Hansen, M. and Yu, B. (2001). Model selection and the principle of minimum description length. *Journal American Statistical Association*, 96(454), 746–774.
DOI: 10.1198/016214501753168398. [249]

Härdle, W. (1990). *Applied Nonparametric Regression*. Cambridge University Press, Cambridge, UK. [298, 382, 385]

Härdle, W., Lütkepohl, H., and Chen, R. (1997). A review of nonparametric time series analysis. *International Statistical Review*, 65(1), 49–72. DOI: 10.2307/1403432. [382]

Härdle, W. and Marron, J.S. (1985). Optimal bandwidth selection in nonparametric regression function estimation. *The Annals of Statistics*, 13(4), 1465–1481.
DOI: 10.1214/aos/1176349748. [305]

Härdle, W., Tsybakov, A., and Yang, L. (1998). Nonparametric vector autoregression. *Journal of Statistical Planning and Inference*, 68(2), 221–245.
DOI: 10.1016/s0378-3758(97)00143-2. [499]

Härdle, W. and Vieu, P. (1992). Kernel regression smoothing of time series. *Journal of Time Series Analysis*, 13(3), 209–232. DOI: 10.1111/j.1467-9892.1992.tb00103.x. [382]

Harvey, D.I., Leybourne, S.J., and Newbold, P. (1997). Testing the equality of prediction mean squared errors. *International Journal of Forecasting*, 13(2), 281–291.
DOI: 10.1016/s0169-2070(96)00719-4. [418]

Harvey, D.I., Leybourne, S.J., and Newbold, P. (1998). Tests for forecast encompassing. *Journal of Business & Economic Statistics*, 16(2), 254–263. DOI: 10.2307/1392581. [427]

Harvey, D.I., Leybourne, S.J., and Newbold, P. (1999). Forecast evaluation in the presence of ARCH. *Journal of Forecasting*, 18(6), 435–445.
DOI: 10.1002/(sici)1099-131x(199911)18:6%3C435::aid-for762%3E3.0.co;2-b. [429]

Harvill, J.L. and Newton, H.J. (1995). Saddlepoint approximations for the difference of order statistics. *Biometrika*, 82(1), 226–231. DOI: 10.2307/2337643. [133]

Harvill, J.L. and Ray, B.K. (1998). Testing for nonlinearity in a vector time series. Available at: http://citeseerx.ist.psu.edu/viewdoc/download?doi=10.1.1.46.7136&rep=rep1&type=pdf. [486]

Harvill, J.L. and Ray, B.K. (1999). A note on tests for nonlinearity in a vector time series. *Biometrika*, 86(3), 728–734. DOI: 10.1093/biomet/86.3.728. [459, 462]

Harvill, J.L. and Ray, B.K. (2000). An investigation of lag identification tools for vector nonlinear time series. *Communications in Statistics: Theory and Methods*, 29(8), 1677–1702. DOI: 10.1080/03610920008832573. [513, 525]

Harvill, J.L. and Ray, B.K. (2005). A note on multi-step forecasting with functional coefficient autoregressive models. *International Journal of Forecasting*, 21(4), 717–727.
DOI: 10.1016/j.ijforecast.2005.04.012. [506, 522]

Harvill, J.L. and Ray, B.K. (2006). Functional coefficient autoregressive models for vector time series. *Computational Statistics & Data Analysis*, 50(12), 3547–3566.
DOI: 10.1016/j.csda.2005.07.016. [506, 509]

Harvill, J.L., Ravishanker, N., and Ray, B.K. (2013). Bispectral-based methods for clustering time series. *Computational Statistics & Data Analysis*, 64, 113–131.
DOI: 10.1016/j.csda.2013.03.001. [150]

Hastie, T. (1989). Discussion on 'Flexible parsimonious smoothing and additive modeling' (by J. Friedman and B. Silverman). *Technometrics*, 31(1), 23–29.
DOI: 10.2307/1270360. [372]

Hastie, T. and Tibshirani, R. (1990). *Generalized Additive Models*. Chapman & Hall, London. [372, 383]

Hastings, W.K. (1970). Monte Carlo sampling methods using Markov chains and their applications. *Biometrika*, 57(1), 97–109. DOI: 10.1093/biomet/57.1.97. [249]

Haykin, S. (Ed.) (1979). *Nonlinear Methods of Spectral Analysis*. Springer-Verlag, New York. [597]

Heiler, S. (2001). Nonparametric time series analysis: Nonparametric regression, locally weighted regression, autoregression, and quantile regression. In D. Peña et al. (Eds.) *A Course in Time Series Analysis*. Wiley, New York, pp. 308–347.
DOI: 10.1002/9781118032978.ch12. [382]

Hellinger, E. (1909). Neue Begründung der Theorie quadratischer Formen von unendlichvielen Veränderlichen. *Journal für die reine und angewandte Mathematik*, 136, 210–271. [264]

Hendry, D.F. and Clements, M.P. (2004). Pooling of forecasts. *Econometrics Journal*, 7(1), 1–31. DOI: 10.1111/j.1368-423x.2004.00119.x. [426]

Henneke, J.S., Rachev, S.T., Fabozzi, F.J., and Nikolov, M. (2011). MCMC based estimation of Markov switching ARMA-GARCH models. *Applied Economics*, 43(3), 259–297.
DOI: 10.1080/00036840802552379. [75]

Herrndorf, N. (1984). A functional central limit theorem for weakly dependent sequences of random variables. *The Annals of Probability*, 12(1), 141–153.
DOI: 10.1214/aop/1176993379. [96]

Hertz, J., Krogh, A., and Palmer, R.G. (1992). *Introduction to the Theory of Neural Computation*. Addison-Wesley, New York. [74]

Hiemstra, C. and Jones, J.D. (1994). Testing for linear and nonlinear Granger causality in the stock price-volume relation. *The Journal of Finance*, 49(5), 1639–1664.
DOI: 10.2307/2329266. [515, 516]

Hili, O. (1993). Estimateurs du minimum de distance d'Hellinger des modèles EXPARMA (Minimum Hellinger distance estimates from EXPARMA models). *Comptes Rendus de l'Académie des Sciences Paris*, t. 316, Série I, 77–80. [248]

Hili, O. (2001). Hellinger distance estimation of SSAR models. *Statistics & Probability Letters*, 53(3), 305–314. DOI: 10.1016/s0167-7152(01)00086-4. [248]

Hili, O. (2003). Hellinger distance estimation of nonlinear dynamical systems. *Statistics & Probability Letters*, 63(2), 177–184. DOI: 10.1016/s0167-7152(03)00080-4. [248]

Hili, O. (2008a). Hellinger distance estimation of general bilinear time series models. *Statistical Methodology*, 5(2), 119–128. DOI: 10.1016/j.stamet.2007.06.005. [248]

Hili, O. (2008b). Estimation of a multiple-threshold AR(p) model. *Statistical Methodology*, 5(2), 177–186. DOI: 10.1016/j.stamet.2007.08.004. [248]

Hinich, M.J. (1982). Testing for Gaussianity and linearity of stationary time series. *Journal of Time Series Analysis*, 3(3), 169–176.
DOI: 10.1111/j.1467-9892.1982.tb00339.x. [119, 130, 131, 136]

Hinich, M.J. and Patterson, D.M. (1985). Evidence of nonlinearity in daily stock returns. *Journal of Business & Economic Statistics*, 3(1), 69–77. DOI: 10.2307/1391691. [150]

Hinich, M.J. and Rothman, P. (1998). A frequency-domain test of time reversibility. *Macroeconomic Dynamics*, 2(1), 72–88. [322, 323]

Hinich, M.J., Foster, J., and Wild, P. (2006). Structural change in macroeconomic time series: A complex systems perspective. *Journal of Macroeconomics*, 28(1), 136–150.
DOI: 10.1016/j.jmacro.2005.10.009. [319]

Hinich, M.J. and Wolinsky, M.A. (1988). A test for aliasing using bispectral analysis. *Journal of the American Statistical Association*, 83(402), 499–501.
DOI: 10.1080/01621459.1988.10478623. [150]

Hinich, M.J., Mendes, E.M., and Stone, L. (2005). Detecting nonlinearity in time series: Surrogate and bootstrap approaches. *Studies in Nonlinear Dynamics & Econometrics*, 9(4). DOI: 10.2202/1558-3708.1268. [136, 151]

Hjellvik, V. and Tjøstheim, D. (1995). Nonparametric tests of linearity for time series. *Biometrika*, 82(2), 351–368. DOI: 10.2307/2337413. [250]

Hjellvik, V. and Tjøstheim, D. (1996). Nonparametric statistics for testing linearity and serial dependence. *Journal of Nonparametric Statistics*, 6(2-3), 223–251.
DOI: 10.1080/10485259608832673. [250]

Hjellvik, V., Yao, Q., and Tjøstheim, D. (1998). Linearity testing using local polynomial approximation. *Journal of Statistical Planning and Inference*, 68(2), 295–321.
DOI: 10.1016/s0378-3758(97)00146-8. [250]

Hoeffding, W. (1948). A class of statistics with asymptotically normal distribution. *Annals of Mathematical Statistics*, 19(3), 293–325. DOI: 10.1214/aoms/1177730196. [308, 309]

Holst, U., Lindgren, G., Holst, J., and Thuvesholmen, M. (1994). Recursive estimation in switching autoregressions with a Markov regime. *Journal of Time Series Analysis*, 15(5), 489–506. DOI: 10.1111/j.1467-9892.1994.tb00206.x. [110, 250]

Hong, Y. (1998). Testing for pairwise serial independence via the empirical distribution function. *Journal of the Royal Statistical Society*, B 60(2), 429–453. DOI: 10.1111/1467-9868.00134. [272]

Hong, Y. (2000). Generalized spectral tests for serial dependence. *Journal of the Royal Statistical Society*, B 62(3), 557–574. DOI: 10.1111/1467-9868.00250. [274, 275]

Hong, Y. and Lee, T.-H. (2003). Diagnostic checking for the adequacy of nonlinear time series models. *Econometric Theory*, 19(6), 1065–1121. DOI: 10.1017/s0266466603196089. [297]

Hong, Y. and White, H. (2005). Asymptotic distribution theory for nonparametric entropy measures of serial dependence. *Econometrica*, 73(3), 837–901. DOI: 10.1111/j.1468-0262.2005.00597.x. [270, 271, 272, 273]

Hoover, W.G. (1999). *Time Reversibility, Computer Simulation, and Chaos*. World Scientific, Singapore. DOI: 10.1142/9789812815071. [333]

Hornik, K., Stinchcombe, M., and White, H. (1989). Multilayer feedforward networks are universal approximations. *Neural Networks*, 2(5), 359–366. DOI: 10.1016/0893-6080(89)90020-8. [75]

Horová, I., Koláček, J., and Zelinka, J. (2012). *Kernel Smoothing in* MATLAB: *Theory and Practice of Kernel Smoothing*. World Scientific, Singapore. DOI: 10.1142/8468. [385]

Hosking, J.R.M. (1980). The multivariate portmanteau statistic. *Journal of American Statistical Association*, 75(371), 602–607. DOI: 10.1080/01621459.1980.10477520. [473]

Hostinsky, B. and Potocek, J. (1935). Chaînes de Markoff inverses. *Bulletin International de l'Académie de Sciences de Bohème*, 36, 64–67. [332]

Hou, F.Z., Ning, X.B., Zhuang, J.J., Huang, X.L., Fu, M.J., and Bian, C.H. (2011). High-dimensional time irreversibility analysis of human interbeat intervals. *Medical Engineering & Physics*, 33(3), 633–637. DOI: 10.1016/j.medengphy.2011.01.002. [333]

Hristova, D. (2005). Maximum likelihood estimation of a unit root bilinear model with an application to prices. *Studies in Nonlinear Dynamics & Econometrics*, 9(1). DOI: 10.2202/1558-3708.1199. [189]

Hsiao, C, Morimune, K., and Powell, J.L. (Eds.) (2011). *Nonlinear Statistical Modeling*. Cambridge University Press, Cambridge, UK. DOI: 10.1017/cbo9781139175203. [597]

Huang, H. and Lee, T.H. (2010). To combine forecasts or to combine information? *Econometric Reviews*, 29(5-6), 534–570. DOI: 10.1080/07474938.2010.481553. [425]

Huang, J.Z. and Yang, L. (2004). Identification of non-linear additive autoregressive models. *Journal of the Royal Statistical Society*, B 66(2), 463–477. DOI: 10.1111/j.1369-7412.2004.05500.x. [372, 383]

References

Huang, M, Sun, Y, and White, H. (2015). A flexible nonparametric test for conditional independence. *Econometric Theory*. DOI: 10.1017/S0266466615000286. Also available at: http://papers.ssrn.com/sol3/papers.cfm?abstract_id=2277240. [294]

Hubrich, K. and Teräsvirta, T. (2013). Thresholds and smooth transitions in vector autoregressive models. In T.B. Fomby et al. (Eds.) *VAR Models in Econometrics - New Developments and Applications: Essays in Honor of Christopher A. Sims*. Emerald Group Publishing Limited: Bingley, UK, Volume 32, pp. 273–326. Also available as CREATES Research paper 2013-18 at ftp://ftp.econ.au.dk/creates/rp/13/rp13_18.pdf. [74, 80, 486]

Hung, Y. (2012). Order selection in nonlinear time series models with application to the study of cell memory. *The Annals of Applied Statistics*, 6(3), 1256–1279. DOI: 10.1214/12-aoas546. [79]

Hurvich, C.M. and Tsai, C.-L. (1989). Regression and time series model selection in small samples. *Biometrika*, 76(2), 297–307. DOI: 10.1093/biomet/76.2.297. [230]

Hwang, S.Y., Basawa, I.V., and Reeves, J. (1994). The asymptotic distributions of residual autocorrelations and related tests of fit for a class of nonlinear time series models. *Statistica Sinica*, 4(1), 107–125. [250]

Hyndman, R.J. (1995). Highest-density forecast regions for nonlinear and non-normal time series models. *Journal of Forecasting*, 14(5), 431–441. DOI: 10.1002/for.3980140503. [414, 429]

Hyndman, R.J. (1996). Computing and graphing highest density regions. *The American Statistician*, 50(2), 120–126. DOI: 10.2307/2684423. [414, 429]

Hyndman, R.J. and Yao, Q. (2002). Nonparametric estimation and symmetry tests for conditional density functions. *Journal of Nonparametric Statistics*, 14(3), 259–278. DOI: 10.1080/10485250212374. [382]

Ibragimov, R. (2009). Copula-based characterizations for higher order Markov processes. *econometric Theory*, 25(3), 819–846. DOI: 10.1017/S0266466609090720. [296]

Ichimura, H. (1993). Semiparametric least squares (SLS) and weighted SLS estimation in single-index models. *Journal of Econometrics*, 58(1-2), 71–120. DOI: 10.1016/0304-4076(93)90114-k. [378]

Ispány, M. (1997). On stationarity of additive bilinear state-space representation of time series. In I. Csiszár and Gy. Michaletzky (Eds.) *Stochastic Differential and Difference Equations (Progress in Systems and Control Theory)*. Birkhäuser, Boston, pp. 143–155. [110]

Jacobs, P.A. and Lewis, P.A.W. (1977). A mixed autoregressive moving average exponential sequence and point processes (EARMA 1,1). *Advances in Applied Probability*, 9(1), 87–104. DOI: 10.2307/1425818. [74]

Jaditz, T. and Sayers, C.L. (1998). Out-of-sample forecast performance as a test for nonlinearity in time series. *Journal of Business & Economics Statistics*, 16(1), 110–117. DOI: 10.2307/1392021. [387, 429]

Jahan, N. and Harvill, J.L. (2008). Bispectral-based goodness-of-fit tests of Gaussianity and linearity of stationary time series. *Communications in Statistics: Theory and Methods*, 37(20), 3216–3227. DOI: 10.1080/03610920802133319. [134, 147]

Jarque, C.M. and Bera, A.K. (1987). A test for normality of observations and regression residuals. *International Statistical Review*, 55(2), 163–172. DOI: 10.2307/1403192. [10]

Joe, H. (1989). Estimation of entropy and other functionals of a multivariate density. *Annals Institute of Statistical Mathematics*, 41(4), 683–697. DOI: 10.1007/bf00057735. [525]

Joe, H. (1997). *Multivariate Models and Dependence Concepts*. Chapman & Hall, London. DOI: 10.1201/b13150. [305]

Johnson, R.A. and Wichern, D.W. (2002). *Applied Multivariate Statistical Analysis*, (5th edn.). Prentice Hall, New York. [462]

Jones, D.A. (1978). Nonlinear autoregressive processes. *Proceedings of the Royal Society of London*, A 360(1700), 71–95. DOI: 10.1098/rspa.1978.0058. [73, 428]

Jones, D.A. (1978). Linearization of non-linear state equation. *Bulletin of the Polish Academy of Sciences, Technical Sciences*, 54(1), 63–73. [429]

Jose, K.K. and Thomas, M.M. (2012). A product autoregressive model with log-Laplace marginal distribution. *Statistica*, LXXII(3), 317–336. [74]

Kallenberg, W. (2009). Estimating copula densities using model selection techniques. *Insurance: Mathematics and Economics*, 45(2), 209–223.
DOI: 10.1016/j.insmatheco.2009.06.006. [296]

Kalliovirta, L. (2012). Misspecification tests based on quantile residuals. *Econometrics Journal*, 15(2), 358–393. DOI: 10.1111/j.1368-423x.2011.00364.x. [241, 242, 313]

Kalliovirta, L. and Saikkonen, P. (2010). Reliable residuals for multivariate nonlinear time series models. Available at: http://blogs.helsinki.fi/saikkone/research/. [475, 476]

Kalliovirta, L., Meitz, M., and Saikkonen, P. (2015). A Gaussian mixture autoregressive model for univariate time series. *Journal of Time Series Analysis*, 36(2), 247–266. DOI: 10.1111/jtsa.12108. [296]

Kankainen, A. and Ushakov, N.G. (1998). A consistent modification of a test for independence based on the empirical characteristic function. *Journal of Mathematical Sciences*, 89(5), 1582–1589. DOI: 10.1007/bf02362283. [296]

Kantz, H. and Schreiber, T. (2004). *Nonlinear Time Series Analysis* (2nd edn.). Cambridge University Press, Cambridge, UK. DOI: 10.1017/cbo9780511755798. [25, 597]

Kapetanios, G. (2000). Small sample properties of the conditional least squares estimator in SETAR models. *Economics Letters*, 69(3), 267–276.
DOI: 10.1016/s0165-1765(00)00314-1. [246]

Kapetanios, G. (2001). Model selection in threshold models. *Journal of Time Series Analysis*, 22(6), 733–754. DOI: 10.1111/1467-9892.00251. [249]

Kapetanios, G. and Shin, Y. (2006). Unit root tests in three-regime SETAR models. *Econometrics Journal*, 9(2), 252–278. DOI: 10.1111/j.1368-423x.2006.00184.x. [189]

Karlsen, H. and Tjøstheim, D. (1988). Consistent estimates for the NEAR(2) and NLAR(2) time series model. *Journal of the Royal Statistical Society*, B 50(2), 313–320. [74]

Karvanen, J. (2005). A resampling test for the total independence of stationary time series: Application to the performance evaluation of ICA algorithms. *Neural Processing Letters*, 22(3), 311–324. DOI: 10.1007/s11063-005-0956-0. [296]

Keenan, D.M. (1985). A Tukey nonadditivity-type test for time series nonlinearity. *Biometrika*, 72(1), 39–44. DOI: 10.1093/biomet/72.1.39. [179, 180]

Kemperman, J.H.D. (1987). The median of finite measures of Banach space. In Y. Dodge (Ed.) *Data Analysis Based on the L_1-norm and Related Methods*. North-Holland, Amsterdam, pp. 217–230. [496]

Kessler, M. and Sørensen, M. (2005). On time-reversibility and estimating functions for Markov processes. *Statistical Inference for Stochastic Processes*, 8(1), 95–107. DOI: 10.1023/b:sisp.0000049125.31288.fa. [333]

Khan, S., Bandyopadhyay, S., Ganguly, A.R., Saigal, S., Erickson III, D.J., Protopopescu, V., and Ostrouchov, G. (2007). Relative performance of mutual information estimation methods for quantifying the dependence among short and noisy data. *Physical Review*, E 76(2), 026209. DOI: 10.1103/physreve.76.026209. [23]

Kilian, L. (1998). Small sample confidence intervals for impulse response functions. *The Review of Economics and Statistics*, 80(2), 218–230. DOI: 10.1162/003465398557465. [411, 412]

Kilian, L. and Demiroglu, U. (2000). Residual-based tests for normality in autoregressions: Asymptotic theory and simulation evidence. *Journal of Business & Economic Statistics*, 18(1), 40–50. DOI: 10.2307/1392135. [23]

Kiliç, R. (2016). Tests for linearity in STAR models: SupWald and LM-type tests. *Journal of Time Series Analysis*, 37(5), 660–674. DOI: 10.1111/jtsa.12180. [188]

Kim, C.J. and Nelson, C.R. (1999). *State-Space Models with Regime Switching, Classical and Gibbs-Sampling Approaches with Applications*. The MIT Press, Cambridge, MA. [75]

Kim, J.H. (2003). Forecasting autoregressive time series with bias-corrected parameter estimators. *International Journal of Forecasting*, 19(3), 493–502. DOI: 10.1016/s0169-2070(02)00062-6. [411]

Kim, T.S., Yoon, J.H., and Lee, H.K. (2002). Performance of a nonparametric multivariate nearest neighbor model in the prediction of stock index returns. *Asia Pacific Management Review*, 7(1), 107–118. [382]

Kim, W.K. and Billard, L. (1990). Asymptotic properties for the first-order bilinear time series model. *Communications in Statistics: Theory and Methods*, 19(4), 1171–1183. DOI: 10.1080/03610929008830255. [248]

Kim, W.K., Billard, L., and Basawa, I.V. (1990). Estimation for the first-order diagonal bilinear time series model. *Journal of Time Series Analysis*, 11(3), 215–229. DOI: 10.1111/j.1467-9892.1990.tb00053.x. [252, 253]

Kim, Y. and Lee, S. (2002). On the Kolmogorov–Smirnov type test for testing nonlinearity in time series. *Communcations in Statistics: Theory and Methods*, 31(2), 299–309. DOI: 10.1081/sta-120002653. [250]

Klement, E.P. and Mesiar, R. (2006). How non-symmetric can a copula be? *Commentationes Mathematicae Universitatis Carolinae*, 47, 141–148. [325]

Knotters, M. and De Gooijer, J.G. (1999). TARSO modeling of water table depths. *Water Resources Research*, 35(3), 695–705. DOI: 10.1029/1998WR900049. [80, 242, 245, 246, 251]

Ko, S.I.M. and Park, S.Y. (2013). Multivariate density forecast evaluation: A modified approach. *International Journal of Forecasting*, 29(3), 431–441. DOI: 10.1016/j.ijforecast.2012.11.006. [480, 492]

Kočenda, E. (2001). An alternative to the BDS test: Integration across the correlation integral. *Econometric Reviews*, 20(3), 337–351. DOI: 10.1081/etc-100104938. [312]

Kočenda, E. and Briatka, Ľ. (2005). Optimal range for the iid test based on integration across the correlation integral. *Econometric Reviews*, 24(3), 265–296. DOI: 10.1080/07474930500243001. [280]

Kock, A.B. and Teräsvirta, T. (2011). Forecasting with nonlinear time series models. In M.P. Clements and D.F. Hendry (Eds.) *The Oxford Handbook of Economic Forecasting*, Oxford University Press, Oxford, pp. 61–88. DOI: 10.1093/oxfordhb/9780195398649.013.0004. [427]

Koizumi, K., Okamoto, N., and Seo, T. (2009). On Jarque–Bera tests for assessing multivariate normality. *Journal of Statistics: Advances in Theory and Applications*, 1(2), 207–220. Available at: http://www.scientificadvances.co.in/about-this-journal/4. [23]

Kojadinovic, I. and Yan, J. (2010). Modeling multivariate distributions with continuous margins using the copula R package. *Journal of Statistical Software*, 34(9). DOI: 10.18637/jss.v034.i09. [297]

Kojadinovic, I. and Yan, J. (2011). Tests of serial independence for continuous multivariate time series based on a Möbius decomposition of the independence empirical copula process. *Annals of the Institute of Statistical Mathematics*, 63(2), 347–373. DOI: 10.1007/s10463-009-0257-x. Available at: http://www.ism.ac.jp/editsec/aism/pdf/10463_2009_Article_257.pdf. [287, 289]

Kolmogorov, A.N. (1936). Zur theorie der Markoffschenketten. *Mathematische Annalen*, 112, 155–160. [332]

Koop, G. and Potter, S.M. (1999). Dynamic asymmetries in U.S. unemployment. *Journal of Business & Economic Statistics*, 17(3), 298–312. DOI: 10.2307/1392288. [208]

Koop, G. and Potter, S.M. (2001). Are apparent findings of nonlinearity due to structural instability in economic time series? *Econometrics Journal*, 4(1), 37–55. DOI: 10.1111/1368-423x.00055. [249]

Koop, G. and Potter, S.M. (2003). Bayesian analysis of endogenous delay threshold models. *Journal of Business & Economics Statistics*, 21(1), 93–103. DOI: 10.1198/073500102288618801. [79]

Koop, G., Pesaran, M.H., and Potter, S.M. (1996). Impulse response analysis in nonlinear multivariate models. *Journal of Econometrics*, 74(1), 119–147. DOI: 10.1016/0304-4076(95)01753-4. [77, 79, 489]

Kooperberg, C., Bose, S., and Stone, C.J. (1997). Polychotomous regression. *Journal of the American Statistical Association*, 92(437), 117–127. DOI: 10.2307/2291455. [502]

Koul, H.L. and Schick, A. (1997). Efficient estimation in nonlinear autoregressive time series models. *Bernoulli*, 3(3), 247–277. DOI: 10.2307/3318592. [248]

Kreiss, J.-P. and Lahiri, S.N. (2011). Bootstrap methods for time series. In T. Subba Rao et al. (Eds.) *Handbook of Statistics*, Vol. 30. North-Holland, Amsterdam, pp. 3–26. DOI: 10.1016/b978-0-444-53858-1.00001-6. [151]

Krishnamurthy, V. and Yin, G.G. (2002). Recursive algorithms for estimation of hidden Markov models and autoregressive models with Markov regime. *IEEE Transactions on Information Theory*, 48(2), 458–476. DOI: 10.1109/18.979322. [250]

Kristensen, D. (2009). On stationarity and ergodicity of the bilinear model with applications to GARCH models. *Journal of Time Series Analysis*, 30(1), 125–144. DOI: 10.1111/j.1467-9892.2008.00603.x. [110, 116]

Kumar, K. (1986). On the identification of some bilinear time series models. *Journal of Time Series Analysis*, 7(2), 117–122. DOI: 10.1111/j.1467-9892.1986.tb00489.x. [125]

Kumar, K. (1988). Bivariate bilinear models and their specification. In R.R. Mohler (Ed.) *Nonlinear Time Series and Signal Processing*, Lecture Notes in Control and Information Sciences, 106. Springer-Verlag, Berlin, pp. 59–74. [486]

Kunitomo, N. and Sato, S. (2002). Estimation of asymmetrical volatility for asset prices: The simultaneous switching ARIMA approach. *Journal of the Japan Statistical Society*, 32(2), 119–140. DOI: 10.14490/jjss.32.119. [80]

Lai, T.L. and Wei, C.Z. (1982). Least squares estimates in stochastic regression models with applications to identification and control of dynamic systems. *The Annals of Statistics*, 10(1), 154–166. DOI: 10.1214/aos/1176345697. [447]

Lai, T.L and Wong S.P-S. (2001). Stochastic neural networks with applications to nonlinear time series. *Journal of the American Statistical Association*, 96(455), 968–981. DOI: 10.1198/016214501753208636. [75]

Lai, T.L. and Zhu, G. (1991). Adaptive prediction in non-linear autoregressive models and control systems. *Statistica Sinica*, 1(2), 309–334. [429]

Lall, U. and Sharma, A. (1996). A nearest neighbor bootstrap for resampling hydrologic time series. *Water Resources Research*, 32(3), 679–693. DOI: 10.1029/95wr02966. [381, 382]

Lall, U., Sangoyomi, T., and Abarbanel, H. (1996). Nonlinear dynamics of the Great Salt Lake: Nonparametric short-term forecasting. *Water Resources Research*, 32(4), 975–985. DOI: 10.1029/95wr03402. [387]

Lanne, M. and Saikkonen, P. (2002). Threshold autoregression for strongly autocorrelated time series. *Journal of Business & Economic Statistics*, 20(2), 282–289. DOI: 10.1198/073500102317352010. [189]

Lanne, M. and Saikkonen, P. (2003). Modeling the U.S. short-term interest rate by mixture autoregressive processes. *Journal of Financial Econometrics*, 1(1), 96–125. DOI: 10.1093/jjfinec/nbg004. [296]

Lanterman, A.D. (2001). Schwarz, Wallace, and Rissanen: Intertwinning themes in theories of model selection. *International Statistical Review*, 69(2), 185–212. DOI: 10.2307/1403813. [249]

Lapedes, A and Farber, R. (1987). *Nonlinear Signal Processing Using Neural Networks: Prediction and System Modelling*. Technical Report LA-UR-87-2662. Los Alamos National Laboratory, Los Alamos, New Mexico. Available at: http://permalink.lanl.gov/object/tr?what=info:lanl-repo/lareport/LA-UR-87-2662. [75]

Lawrance, A.J. (1991). Directionality and reversibility in time series. *International Statistical Review*, 59(1), 67–79. DOI: 10.2307/1403575. [333]

Lawrance, A.J. and Lewis, P.A.W. (1977). An exponential moving average sequence and point process (EMA1). *Journal of Applied Probability*, 14(1), 98–113. DOI: 10.2307/3213263. [74]

Lawrance, A.J. and Lewis, P.A.W. (1980). The exponential autoregressive-moving average EARMA(p,q) process. *Journal of the Royal Statistical Society*, B 42(2), 150–161. [54]

Lawrance, A.J. and Lewis, P.A.W. (1981). A new autoregressive time series model in exponential variables (NEAR(1)). *Advances in Applied Probability*, 13(4), 826–845. DOI: 10.2307/1426975. [54]

Lawrance, A.J. and Lewis, P.A.W. (1985). Modelling and residual analysis of nonlinear autoregressive time series in exponential variables. *Journal of the Royal Statistical Society*, B 47(2), 165–202 (with discussion). [74]

Le, N.D., Martin, R.D., and Raftery, A.E. (1996). Modeling flat stretches, bursts, and outliers in time series using mixture transition distribution models. *Journal of the American Statistical Association*, 91(436), 1504–1515. DOI: 10.2307/2291576. [296]

Lee, A.J. (1990). *U-statistics: Theory and Practice*. Marcel Dekker, New York. [308]

Lee, O. and Shin, D.W. (2000). On geometric ergodicity of the MTAR process. *Statistics & Probability Letters*, 48(3), 229–237. DOI: 10.1016/s0167-7152(99)00208-4. [100]

Lee, O. and Shin, D.W. (2001). A note on stationarity of the MTAR process on the boundary of the stationarity region. *Economics Letters*, 73(3), 263–268. DOI: 10.1016/s0165-1765(01)00508-0. [99, 100]

Lee, T.-H., White, H., and Granger, C.W.J. (1993). Testing for neglected nonlinearity in time series models: A comparison of neural network methods and alternative tests. *Journal of Econometrics*, 56(3), 269–290. DOI: 10.1016/0304-4076(93)90122-l. [22, 188, 190]

Leistritz, L, Hesse W., Arnold M., and Witte H. (2006). Development of interaction measures based on adaptive non-linear time series analysis of biomedical signals. *Biomedical Engineering*, 51(2), 64–69. DOI: 10.1515/bmt.2006.012. [451]

Lentz, J.-R. and Mélard, G. (1981). Statistical analysis of a non-linear model. In O.D. Anderson and M.R. Perryman (Eds.) *Time Series Analysis*. North-Holland, Amsterdam, pp. 287–293. [73]

León, C.A. and Massé, J.-C. (1992). A counterexample on the existence of the L_1-median. *Statistics & Probability Letters*, 13(2), 117–120. DOI: 10.1016/0167-7152(92)90085-j. [496]

Lewis, P.A.W., McKenzie, E., and Hugus, D.K. (1989). Gamma processes. *Communications in Statistics: Stochastic Models*, 5(1), 1–30. DOI: 10.1080/15326348908807096. [318, 335]

Lewis, P.A.W. and Ray, B.K. (1993). Nonlinear modeling of multivariate and categorical time series using multivariate adaptive regression splines. In H. Tong (Ed.) *Dimension Estimation and Models*. World Scientific, Singapore, pp. 136–169. [362, 384]

Lewis, P.A.W. and Ray, B.K. (1997). Modeling long-range dependence, nonlinearity, and periodic phenomena in sea surface temperatures using TSMARS. *Journal of the American Statistical Association*, 92(439), 881–893. DOI: 10.2307/2965552. [362, 368, 381, 384]

Lewis, P.A.W. and Ray, B.K. (2002). Nonlinear modeling of periodic threshold autoregressions using Tsmars. *Journal of Time Series Analysis*, 23(4), 459–471. DOI: 10.1111/1467-9892.00269. [383]

Lewis, P.A.W. and Stevens, J.G. (1991). Nonlinear modeling of time series using multivariate adaptive regression splines (MARS). *Journal of the American Statistical Association*, 86(416), 864–877. DOI: 10.1080/01621459.1991.10475126. [381]

Li, C.W. and Li, W.K. (1996). On a double-threshold autoregressive heteroscedastic time series model. *Journal of Applied Econometrics*, 11(3), 253–274. DOI: 10.1002/(sici)1099-1255(199605)11:3%3C253::aid-jae393%3E3.0.co;2-8. [80, 225]

Li, D. (2012). A note on moving-average models with feedback. *Journal of Time Series Analysis*, 33(6), 873–879. DOI: 10.1111/j.1467-9892.2012.00802.x. [106, 111]

Li, D. and He, C. (2012a). Testing common nonlinear features in nonlinear vector autoregressive models. Available at: http://ideas.repec.org/p/hhs/oruesi/2012_007.html. [456, 486]

Li, D. and He, C. (2012b). Testing for linear cointegration against smooth-transition cointegration. Available at: http://ideas.repec.org/p/hhs/oruesi/2012_006.html. [487]

Li, D. and He, C. (2013). Forecasting with vector nonlinear time series models. Working papers 2013:8, Dalarna University, Sweden. Available at: http://www.diva-portal.org/smash/get/diva2:606647/FULLTEXT02.pdf. [487, 492]

Li, D., Li, W.K., and Ling, S. (2011). On the least squares estimation of threshold autoregressive and moving-average models. *Statistics and Its Interface*, 4(2), 183–196. DOI: 10.4310/sii.2011.v4.n2.a13. [204, 205, 206, 207, 208]

Li, D. and Ling, S. (2012). On the least squares estimation of multiple-regime threshold AR models. *Journal of Econometrics*, 167(1), 240–253.
DOI: 10.1016/j.jeconom.2011.11.006. [208]

Li, D., Ling, S., and Li, W.K. (2013). Asymptotic theory on the least squares estimation of threshold moving-average models. *Econometric Theory*, 29(03), 482–516.
DOI: 10.1017/S026646661200045X. [248]

Li, D., Ling, S., and Tong, H. (2012). On moving-average models with feedback. *Bernoulli*, 18(2), 735–745. DOI: 10.3150/11-bej352. [106, 111]

Li, D., Ling, S., and Zhang, R. (2016). On a threshold double autoregressive model. *Journal of Business & Economic Statistics*, 34(1), 68–80.
DOI: 10.1080/07350015.2014.1001028. [81]

Li, G. and Li, W.K. (2008). Testing for threshold moving average with conditional heteroscedasticity. *Statistica Sinica*, 18(2), 647–665. [174, 189]

Li, G. and Li, W.K. (2011). Testing a linear time series model against its threshold extension. *Biometrika*, 98(1), 243–250. DOI: 10.1093/biomet/asq074. [174, 175, 176, 190]

Li, J. (2011). Bootstrap prediction intervals for SETAR models. *International Journal of Forecasting*, 27(2), 320–332. DOI: 10.1016/j.ijforecast.2010.01.013. [410, 411, 412]

Li, M.S. and Chan, K.S. (2007). Multivariate reduced-rank nonlinear time series modeling. *Statistica Sinica*, 17(1), 139–159. [80]

Li, Q. and Racine, J.S. (2007). *Nonparametric Econometrics: Theory and Practice*. Princeton University Press, Princeton and Oxford. [298, 485, 597]

Li, W.K. (1992). On the asymptotic standard errors of residual autocorrelations in nonlinear time series modelling. *Biometrika*, 79(2), 435–437.
DOI: 10.1093/biomet/79.2.435. [236, 250]

Li, W.K. (1993). A simple one degree of freedom test for time series nonlinearity. *Statistica Sinica*, 3(1), 245–254. [186]

Li, W.K. (2004). *Diagnostic Checks in Time Series*. Chapman & Hall/CRC, New York. (Freely available at: http://dlia.ir/Scientific/e_book/Science/General/006256.pdf). DOI: 10.1201/9780203485606. [250]

Li, W.K. and Mak, T.K. (1994). On the squared residual autocorrelations in non-linear time series with conditional heteroskedasticity. *Journal of Time Series Analysis*, 15(6), 627–636. DOI: 10.1111/j.1467-9892.1994.tb00217.x. [236]

Liang, R., Niu, C., Xia, Q., and Zhang, Z. (2015). Nonlinearity testing and modeling for threshold moving average models. *Journal of Applied Statistics*, 42(12), 2614–2630.
DOI: 10.1080/02664763.2015.1043872. [189]

Liebscher, E. (2005). Towards a unified approach for proving geometric ergodicity and mixing properties of nonlinear autoregressive processes. *Journal of Time Series Analysis*, 26(5), 669–689. DOI: 10.1111/j.1467-9892.2005.00412.x. [111, 114]

Lientz, B.P. (1970). Results on nonparametric modal intervals. *SIAM Journal of Applied Mathematics*, 19(2), 356–366. DOI: 10.1137/0119034. [429]

Lientz, B.P. (1972). Properties of modal intervals. *SIAM Journal of Applied Mathematics*, 23(1), 1–5. DOI: 10.1137/0123001. [429]

Lii, K.-S. (1996). Nonlinear systems and higher-order statistics with applications. *Signal Processing*, 53(2-3), 165–177. DOI: 10.1016/0165-1684(96)00084-9. [150]

Lii, K.-S. and Masry, E. (1995). On the selection of random sampling schemes for the spectral estimation of continuous time processes. *Journal of Time Series Analysis*, 16(3), 291–311. DOI: 10.1111/j.1467-9892.1995.tb00235.x. [150]

Lim, K.S. (1987). A comparative study of various univariate time series models for Canadian lynx data. *Journal of Time Series Analysis*, 8(2), 161–176. DOI: 10.1111/j.1467-9892.1987.tb00430.x. [293]

Lim, K.S. (1992). On the stability of a threshold AR(1) without intercepts. *Journal of Time Series Analysis*, 13(2), 119–132. DOI: 10.1111/j.1467-9892.1992.tb00098.x. [100]

Lin, C.C. and Mudholkar, G.S. (1980). A simple test for normality against asymmetric alternatives. *Biometrika*, 67(2), 455–461. DOI: 10.2307/2335489. [11]

Lin, T.C. and Pourahmadi, M. (1998). Nonparametric and non-linear models and data mining in time series: A case-study on the Canadian lynx data. *Applied Statistics*, 47(2), 187–201. DOI: 10.1111/1467-9876.00106. [381]

Lindner, A.M. (2009). Stationarity, mixing, distributional properties and moments of GARCH(p,q)-processes. In T.G. Andersen et al. (Eds.) *Handbook of Financial Time Series*. Springer-Verlag, Berlin, pp. 43–69. DOI: 10.1007/978-3-540-71297-8_2. [111]

Lindsay, B.G., Markatau, M., Ray, S., Yang, K., and Chen, S.-C. (2008). Quadratic distances on probabilities: A unified foundation. *The Annals of Statistics*, 36(2), 983–1006. DOI: 10.1214/009053607000000956. [261]

Ling, S. (1999). On the probabilistic properties of a double threshold ARMA conditional heteroskedastic model. *Journal of Applied Probability*, 36(3), 688–705. DOI: 10.1239/jap/1029349972. [100, 102]

Ling, S. and Li, W.K. (1997). Diagnostic checking of nonlinear multivariate time series with multivariate arch errors. *Journal of Time Series Analysis*, 18(5), 447–464. DOI: 10.1111/1467-9892.00061. [487]

Ling, S. and Tong, H. (2005). Testing for a linear MA model against threshold MA models. *The Annals of Statistics*, 33(6), 2529–2552. DOI: 10.1214/009053605000000598. [102, 174, 189]

Ling, S. and Tong, H. (2011). Score based goodness-of-fit tests for time series. *Statistica Sinica*, 21(4), 1807–1829. DOI: 10.5705/ss.2009.090. [250]

Ling, S., Tong, H., and Li, D. (2007). Ergodicity and invertibility of threshold moving-average models. *Bernoulli*, 13(1), 161–168. DOI: 10.3150/07-bej5147. [102, 109]

Ling, S., Peng, L., and Zhu, F. (2015). Inference for a special bilinear time-series model. *Journal of Time Series Analysis*, 36(1), 61–66. DOI: 10.1111/jtsa.12092. [248]

Liu, J. (1989a). A simple condition for the existence of some stationary bilinear time series. *Journal of Time Series Analysis*, 10(1), 33–39. DOI: 10.1111/j.1467-9892.1989.tb00013.x. [111]

Liu, J. (1989b). On the existence of a general-multiple bilinear time series. *Journal of Time Series Analysis*, 10(4), 341–355. DOI: 10.1111/j.1467-9892.1989.tb00033.x. [443]

Liu, J. (1990). A note on causality and invertibility of a general bilinear time series model. *Advances in Applied Probability*, 22(1), 247–250. DOI: 10.2307/1427608. [103]

Liu, J. (1995). On stationarity and asymptotic inference of bilinear time series models. *Statistica Sinica*, 2(2), 479–494. [111]

Liu, J. and Brockwell, P.J. (1988). On the general bilinear time series model. *Journal of Applied Probability*, 25(3), 553–564. DOI: 10.2307/3213984. [111]

Liu, J. and Susko, E. (1992). On strict stationarity and ergodicity of a non-linear ARMA model. *Journal of Applied Probability*, 29(2), 363–373. DOI: 10.2307/3214573. [99, 100]

Liu, S.-I. (1985). Theory of bilinear time series models. *Communications in Statistics: Theory and Methods*, 4(10), 2549–2561. DOI: 10.1080/03610926.1985.10524941. [103]

Liu, S.-I. (2011). Testing for multivariate threshold autoregression. *Studies in Mathematical Sciences*, 2(1), 1–20. DOI: 10.2139/ssrn.1360533. [465, 467]

Liu, W., Ling, S., and Shao, Q.-M. (2011). On non-stationary threshold autoregressive models. *Bernoulli*, 17(3), 969–986. DOI: 10.3150/10-bej306. [247]

Lobato, I.N. and Velasco, C. (2004). A simple test for normality for time series. *Econometric Theory*, 20(04), 671–689. DOI: 10.1017/s0266466604204030. [13]

Lomnicki, Z.A. (1961). Tests for departure from normality in the case of linear stochastic processes. *Metrika*, 4(1), 37–62. DOI: 10.1007/bf02613866. [12, 242]

Lopes, H.F. and Salazar, E. (2006). Bayesian model uncertainty in smooth transition autoregressions. *Journal of Time Series Analysis*, 27(1), 99-117. DOI: 10.1111/j.1467-9892.2005.00455.x. [74]

Lutz, R.W., Kalisch, M., and Bühlmann, P. (2008). Robustified L_2 boosting. *Computational Statistics & Data Analysis*, 52(7), 3331–3341. DOI: 10.1016/j.csda.2007.11.006. [383]

Luukkonen, R., Saikkonen, P., and Teräsvirta, T. (1988a). Testing linearity against smooth transition autoregressive models. *Biometrika*, 75(3), 491–499. DOI: 10.2307/2336599. [159, 165, 181, 188, 193]

Luukkonen, R., Saikkonen, P., and Teräsvirta, T. (1988b). Testing linearity in univariate time series. *Scandinavian Journal of Statistics*, 15(3), 161–175. [180, 188, 193]

Ma, J. and Wohar, M. (Eds.) (2014). *Recent Advances in Estimating Nonlinear Models with Applications in Economics and Finance*. Springer-Verlag, New York. DOI: 10.1007/978-1-4614-8060-0. [430, 597]

MacNeill, I.B. (1971). Limit processes of co-spectral and quadrature spectral distribution function. *The Annals of Statistics*, 42(1), 81–96 DOI: 10.1214/aoms/1177693497. [184]

Mak, T.K. (1993). Solving non-linear estimation equations. *Journal of the Royal Statistical Society*, B 55(4), 945–955. [223]

Mak, T.K., Wong, H., and Li, W.K. (1997). Estimation of nonlinear time series with conditional heteroscedastic variances by iteratively weighted least squares. *Computational Statistics & Data Analysis*, 24(2), 169–178. DOI: 10.1016/s0167-9473(96)00060-6. [223]

Manzan, S. and Zerom, D. (2008). A bootstrap-based non-parametric forecast density. *International Journal of Forecasting*, 24(3), 535–550.
DOI: 10.1016/j.ijforecast.2007.12.004. [356, 357]

Marek, T. (2005). On the invertibility of a random coefficient moving average model. *Kybernetika*, 41(6), 743–756. [102, 103]

Mariano, R.S. and Preve, D. (2012). Statistical tests for multiple forecast comparison. *Journal of Econometrics*, 169(1), 123–130. DOI: 10.1016/j.jeconom.2012.01.014. [429]

Marinazzo, D., Pellicoro, M., and Stramaglia, S. (2008). Kernel method for nonlinear Granger causality. *Physics Review Letters*, A, 100(14). Article 144103.
DOI: 10.1103/physrevlett.100.144103. [523]

Marron, J.S. (1994). Visual understanding of higher order kernels. *Journal of Computational and Graphical Statistics*, 3(4), 447–458. DOI: 10.2307/1390905. [305]

Masani, P. and Wiener, N. (1959). Nonlinear prediction. In U. Grenander (Ed.) *Probability and Statistics: The Harold Cramér Volume*. Wiley, New York, pp. 190–212. [141]

Masry, E. (1996a). Multivariate local polynomial regression for time series: Uniform strong consistency and rates. *Journal of Time Series Analysis*, 17(6), 571–599.
DOI: 10.1111/j.1467-9892.1996.tb00294.x. [382]

Masry, E. (1996b). Multivariate regression estimation: Local polynomial fitting for time series. *Stochastic Processes and their Applications*, 65(1), 81–101.
DOI: 10.1016/s0304-4149(96)00095-6. [382]

Masry, E. and Tjøstheim, D. (1995). Nonparametric estimation and identification of nonlinear ARCH time series: Strong convergence and asymptotic normality. *Econometric Theory*, 11(02), 258–289. DOI: 10.1017/s0266466600009166. [356]

Matilla-Garcia, M. and Ruiz-Marin, M. (2008). A non-parametric independence test using permutation entropy. *Journal of Econometrics*, 144(1), 139–155.
DOI: 10.1016/j.jeconom.2007.12.005. [296]

Matsuda, Y. (1998). A diagnostic statistic for functional-coefficient autoregressive models. *Communications in Statistics: Theory Methods*, 27(9), 2257–2273.
DOI: 10.1080/03610929808832226. [384]

Matsuda, Y. and Huzii, M. (1997). Some statistical properties of linear and nonlinear predictors for stationary time series. Research Report on Mathematical and Computing Sciences, B-325, Tokyo Institute of Technology. Abstract: http://www.is.titech.ac.jp/~natsuko/B/B-325.txt. [145, 437]

Matusita, K. (1955). Decision rules, based on the distance, for problems of fit, two samples, and estimation. *Annals of Mathematical Statistics*, 26(4), 631–641. DOI: 10.1214/aoms/1177728422. [336]

Matzner-Løber, E., Gannoun, A., and De Gooijer, J.G. (1998). Nonparametric forecasting: A comparison of three kernel-based methods. *Communications in Statistics: Theory and Methods*, 27(7), 1593–1617. DOI: 10.1080/03610929808832180. [341, 382]

McAleer, M. and Medeiros, M.C. (2008). A multiple regime smooth transition heterogeneous autoregressive model for long memory and asymmetries. *Journal of Econometrics*, 147(1), 104–119. DOI: 10.1016/j.jeconom.2008.09.032. [76]

McCarthy, M. (2005). The lynx and the snowshoe hare: Which factors cause the cyclical oscillations in the population? Available as a PPT download at: http://www.slideserve.com/angeni. [292]

McCausland, W.J. (2007). Time reversibility of stationary regular finite-state Markov chains. *Journal of Econometrics*, 136(3), 303–318. DOI: 10.1016/j.jeconom.2005.09.001. [332]

McKeague, I.W. and Zhang, M.-J. (1994). Identification of nonlinear time series from first order cumulative characteristics. *The Annals of Statistics*, 22(1), 495–514. DOI: 10.1214/aos/1176325381. [383]

McKenzie, E. (1982). Product autoregression: A time series characterization of the gamma distribution. *Journal of Applied Probability*, 19(2), 463–468. DOI: 10.2307/3213502. [54, 55, 74]

McKenzie, E. (1985). An autoregressive process for beta random variables. *Management Science*, 31(8), 988–997. DOI: 10.1287/mnsc.31.8.988. [318]

McLeod, A.I., Yu, H., and Mahdi, E. (2012). *Time series analysis with* R. In T. Subba Rao et al. (Eds.) *Handbook of Statistics 30: Time Series Analysis: Methods and Applications*. Elsevier, Amsterdam, pp. 661–712. [24]

McQuarrie, A.D.R., Shumway, R., and Tsai, C.-L. (1997). The model selection criterion AICu. *Statistics & Probability Letters*, 34(3), 285–292. DOI: 10.1016/s0167-7152(96)00192-7. [230]

McQuarrie, A.D.R. and Tsai, C.-L. (1998). *Regression and Time Series Model Selection*. World Scientific, Singapore. DOI: 10.1142/9789812385451. [230, 231]

Medeiros, M.C., Teräsvirta, T., and Rech, G. (2006). Building neural network models for time series: A statistical approach. *Journal of Forecasting*, 25(1), 49–75. DOI: 10.1002/for.974. [188]

Medeiros, M.C. and Veiga, A. (2002). A hybrid linear-neural model for time series forecasting. *IEEE Transactions on Neural Networks*, 11(6), 1402–1412. DOI: 10.1109/72.883463. [75]

Medeiros, M.C. and Veiga, A. (2003). Diagnostic checking in a flexible nonlinear time series model. *Journal of Time Series Analysis*, 24(4), 461–482. DOI: 10.1111/1467-9892.00316. [75]

Medeiros, M.C. and Veiga, A. (2005). A flexible coefficient smooth transition time series model. *IEEE Transactions on Neural Networks*, 16(1), 97–113. DOI: 10.1109/tnn.2004.836246. [75, 188, 247]

Medeiros, M.C., Veiga, A., and Resende, M.G.C. (2002). A combinatorial approach to piecewise linear time series analysis. *Journal of Computational and Graphical Statistics*, 11(1), 236–258. DOI: 10.1198/106186002317375712. [73]

Meitz, M. and Saikkonen, P. (2008). Stability of nonlinear AR-GARCH models. *Journal of Time Series Analysis*, 29(3), 453–475. DOI: 10.1111/j.1467-9892.2007.00562.x. [111]

Meitz, M. and Saikkonen, P. (2010). A note on the geometric ergodicity of a nonlinear AR-ARCH model. *Statistics & Probability Letters*, 80(7-8), 631–638. DOI: 10.1016/j.spl.2009.12.020. [91, 111]

Metropolis, N., Rosenbluth, A.W., Rosenbluth, M.N., Teller, A.H., and Teller, E. (1953). Equation of state calculations by fast computing machines. *Journal of Chemical Physics*, 21(6), 1087–1091. DOI: 10.1063/1.1699114. [249]

Meyn, S.P. and Tweedie, R.L. (1993). *Markov Chains and Stochastic Stability*. Springer-Verlag, New York. (Freely available at: http://probability.ca/MT/BOOK.pdf). Second edn. (2009), Cambridge University Press, MA. [111]

Milas, C., Rothman, P.A., Van Dijk, D., and Wildasin, D.E. (Eds.) (2006). *Nonlinear Time Series Analysis of Business Cycles*. Elsevier, Amsterdam, The Netherlands. [597]

Mira, S. and Escribano, A. (2006). Nonlinear time series models: Consistency and asymptotic normality of NLS under new conditions. In W.A. Barnett et al. (Eds.), *Nonlinear Econometric Modeling in Time Series*. Cambridge University Press, Cambridge, UK, pp. 119–164. [247]

Miwakeichi, F., Ramirez–Padron, R. Valdes–Sosa, P.A., and Ozaki, T. (2001). A comparison of non-linear non-parametric models for epilepsy data. *Computers in Biology and Medicine*, 31, 41–57. DOI: 10.1016/s0010-4825(00)00021-4. [23]

Moeanaddin, R. and Tong, H. (1988). A comparison of likelihood ratio test and CUSUM test for threshold autoregression. *The Statistician*, 37(2), 213–225. Addendum & Corrigendum 37(4/5), p. 473. DOI: 10.2307/2348695 and DOI: 10.2307/2348773. [193]

Mohler, R.R. (Ed.) (1987). *Nonlinear Time Series and Signal Processing*. Springer-Verlag, Berlin. [597]

Montgomery, A.L., Zarnowitz, V., Tsay, R.S., and Tiao, G.C. (1998). Forecasting the U.S. unemployment rate. *Journal of the American Statistical Association*, 93(442), 478–493. DOI: 10.1080/01621459.1998.10473696. [23, 276]

Moon, Y-I., Lall, U., and Kwon, H-H. (2008). Non-parametric short-term forecasts of the Great Salt Lake using atmospheric indices. *International Journal of Climatology*, 28(3), 361–370. DOI: 10.1002/joc.1533. [387]

Moran, P.A.P. (1953). The statistical analysis of the Canadian lynx cycle. *Australian Journal of Zoology*, 1(2), 163–173. [293]

Mudholkar, G.S., Marchetti, C.E., and Lin, C.T. (2002). Independence characterizations and testing normality against restricted skewness-kurtosis alternatives. *Journal of Statistical Planning and Inference*, 104(2), 485–501. DOI: 10.1016/s0378-3758(01)00253-1. [23]

Nadaraya, E.A. (1964). On estimating regression. *Theory of Probability and its Applications*, 15(1), 134–137. [302]

Nelsen, R.B. (2006). *An Introduction to Copulas* (2nd edn.). Springer-Verlag, New York. DOI: 10.1007/0-387-28678-0. [305, 306]

Nelsen, R.B. (2007). Extremes of nonexchangeability. *Statistical Papers*, 48(4), 329–336. DOI: 10.1007/s00362-007-0380-9. [325]

Newey, W.K. and West, K.D. (1987). A simple, positive semi-definite, heteroskedasticity and autocorrelation consistent covariance matrix. *Econometrica*, 55(3), 703–708. DOI: 10.2307/1913610. [431, 516, 517]

Nichols, J.M., Olson, C.C., Michalowicz, J.V., and Bucholtz, F. (2009). The bispectrum and bicoherence for quadratically nonlinear systems subject to non-Gaussian inputs. *IEEE Transactions on Signal Processing*, 57(10), 3879–3890. DOI: 10.1109/tsp.2009.2024267. [150]

Nicholls, D.F. and Quinn, B.G. (1981). The estimation of multivariate random coefficient autoregressive models. *Journal of Multivariate Analysis*, 11(4), 544–555. DOI: 10.1016/0047-259x(81)90095-6. [455, 485]

Nicholls, D.F. and Quinn, B.G. (1982). *Random Coefficient Autoregressive Models: An Introduction*. Springer-Verlag, New York. DOI: 10.1007/978-1-4684-6273-9. [73, 90, 455, 485, 597]

Nielsen, H.A. and Madsen, H. (2001). A generalization of some classical time series tools. *Computational Statistics & Data Analysis*, 37(1), 13–31. DOI: 10.1016/s0167-9473(00)00061-x. [23]

Nieto, F. (2005). Modeling bivariate threshold autoregressive processes in the presence of missing data. *Communications in Statistics: Theory and Methods*, 34(4), 905–930. DOI: 10.1081/sta-200054435. [486]

Niglio, M. (2007). Multi-step forecasts from threshold ARMA models using asymmetric loss functions. *Statistical Methods & Applications*, 16(3), 395–410. DOI: 10.1007/s10260-007-0044-x. [429]

Niglio, M. and Vitale, C.D. (2010a). Local unit roots and global stationarity of TARMA models. *Methodology and Computing in Applied Probability*, 14(1), 17–34. DOI: 10.1007/s11009-010-9166-y. [100]

Niglio, M. and Vitale, C.D. (2010b). Generalization of some linear time series property to nonlinear domain. In C. Perna and M. Sibillo (Eds.), *Mathematical and Statistical Methods for Actuarial Sciences and Finance*. Springer-Verlag, New York, pp. 323–331. DOI: 10.1007/978-88-470-2342-0_38. [102]

Niglio, M. and Vitale, C.D. (2013). Vector threshold moving average models: Model specification and invertibility. In N. Torelli et al. (Eds.) *Advances in Theoretical and Applied Statistics*. Springer-Verlag, New York, pp. 87–98. DOI: 10.1007/978-3-642-35588-2_9. [448]

Niglio, M. and Vitale, C.D. (2015). Threshold vector ARMA models. *Communications in Statistics: Theory and Methods*, 44(14), 2911–2923. DOI: 10.1080/03610926.2013.814785. [448]

Nørgaard, M., Ravn, O., Poulsen, N.K., and Hansen, L.K. (2000). *Neural Networks for Modelling and Control of Dynamic Systems*. Springer-Verlag, New York. DOI: 10.1007/978-1-4471-0453-7. [74]

Norman, S. (2008). Systematic small sample bias in two regime SETAR model estimation. *Economics Letters*, 99(1), 134–138. DOI: 10.1016/j.econlet.2007.06.013. [246]

Öhrvik, J. and Schoier, G. (2005). SETAR model selection – A bootstrap approach. *Computational Statistics*, 20(4), 559–573. DOI: 10.1007/bf02741315. [249]

Oja, H. (1983). Descriptive statistics for multivariate distributions. *Statistics & Probability Letters*, 1(6), 327–332. DOI: 10.1016/0167-7152(83)90054-8. [496]

Olteanu, M. (2006). A descriptive method to evaluate the number of regimes in a switching autoregressive model. *Neural Networks*, 19(6-7), 963–972. DOI: 10.1016/j.neunet.2006.05.019. [249]

Ozaki, T. (1982). The statistical analysis of perturbed limit cycle processes using nonlinear time series models. *Journal of Time Series Analysis*, 3(1), 29–41. DOI: 10.1111/j.1467-9892.1982.tb00328.x. [293]

Ozaki, T. and Oda, H. (1978). Non-linear time series models with identification by Akaike's information criterion. In D. Dubuisson (Ed.) *Information and Systems*. Pergamon, Oxford, pp. 83–91. [73]

Pan, L. and Politis, D.N. (2016). Bootstrap prediction intervals for linear, nonlinear and nonparmetric autoregressions (with discussion). *Journal of Statistical Planning and Inference*, 177, 1–27. DOI: 10.1016/j.jspi.2014.10.003. [410]

Paparoditis, E. and Politis, D.N. (2001). A Markovian local resampling scheme for nonparametric estimators in time series analysis. *Econometric Theory*, 17(3), 540–566. DOI: 10.1017/s0266466601173020. [356]

Paparoditis, E. and Politis, D.N. (2002). The local bootstrap for Markov processes. *Journal of Statistical Planning and Inference*, 108(1-2), 301–328. DOI: 10.1016/s0378-3758(02)00315-4. [325, 326, 356]

Parzen, E. (1962). On estimation of a probability function and its mode. *Annals Mathematical Statistics*, 33(3), 1065–1076. DOI: 10.1214/aoms/1177704472. [305]

Patterson, D.M. and Ashley, R.A. (2000). *A Nonlinear Time Series Workshop*. Kluwer Academic Publishers, Norwell, MA. DOI: 10.1007/978-1-4419-8688-7. [150, 151, 597]

Péguin–Feissolle, A., Strikholm, B., and Teräsvirta, T. (2013). Testing the Granger non-causality hypothesis in stationary nonlinear models of unknown functional form. *Communications in Statistics: Simulation and Computation*, 42(5), 1063–1087.
DOI: 10.1080/03610918.2012.661500. [523]

Pemberton, J. (1987). Exact least squares multi-step prediction from non-linear autoregressive models. *Journal of Time Series Analysis*, 8(4), 443–448.
DOI: 10.1111/j.1467-9892.1987.tb00007.x. [393, 428]

Perera, S. (2002). Maximum quasi-likelihood estimation for a simplified NEAR(1) model. *Statistics & Probability Letters*, 58(2), 147–155. DOI: 10.1016/s0167-7152(02)00112-8. [74]

Perera, S. (2004). Maximum quasi-likelihood estimation for the NEAR(2) model. *Journal of Time Series Analysis*, 25(5), 723–732. DOI: 10.1111/j.1467-9892.2004.01886.x. [74]

Pesaran, M.H. and Potter, S.M. (1997). A floor and ceiling model of US output. *Journal of Economic Dynamics & Control*, 21(4-5), 661–695.
DOI: 10.1016/s0165-1889(96)00002-4. [79]

Pesaran, M.H. and Shin, Y. (1998). Generalized impulse response analysis in linear multivariate models. *Economics Letters*, 58(1), 17–29.
DOI: 10.1016/s0165-1765(97)00214-0. [490]

Pesaran, M.H. and Timmermann, A.G. (1992). A simple nonparametric test of predictive performance. *Journal of Business & Economic Statistics*, 10(4), 461–465.
DOI: 10.2307/1391822. [429, 431]

Petruccelli, J.D. (1986). On the consistency of least squares estimators for a threshold AR(1) model. *Journal of Time Series Analysis*, 7(4), 269–278.
DOI: 10.1111/j.1467-9892.1986.tb00494.x. [247]

Petruccelli, J.D. (1990). A comparison of tests for SETAR-type non-linearity in time series. *Journal of Forecasting*, 9(1), 25–36. DOI: 10.1002/for.3980090104. [189, 193]

Petruccelli. J.D. (1992). On the approximation of time series by threshold autoregressive models. *Sankhyā: The Indian Journal of Statistics*, 54, Series B, 106–113. [73]

Petruccelli, J.D. and Davies, N. (1986). A portmanteau test for self-exciting threshold autoregressive-type nonlinearity in time series. *Biometrika*, 73(3), 687–694.
DOI: 10.1093/biomet/73.3.687. [183, 189, 193]

Petruccelli, J.D. and Woolford, S.W. (1984). A threshold AR(1) model. *Journal of Applied Probability*, 21(2), 270–286. DOI: 10.2307/3213639. [100]

Pham, D.T. (1986). The mixing property of bilinear and generalised random coefficient autoregressive models. *Stochastic Processes and their Applications*, 23(2), 291–300.
DOI: 10.1016/0304-4149(86)90042-6. [90, 111]

Pham, D.T., Chan, K.S., and Tong H. (1991). Strong consistency of the least squares estimator for a non-ergodic threshold autoregressive model. *Statistica Sinica*, 1(2), 361–369. [247]

Pham, D.T. and Tran, L.T. (1981). On the first-order bilinear time series model. *Journal of Applied Probability*, 18(3), 617–627. DOI: 10.2307/3213316. [103]

Pham, D.T. and Tran, L.T. (1985). Some mixing properties of time series models. *Stochastic Processes and their Applications*, 19(2), 297–303.
DOI: 10.1016/0304-4149(85)90031-6. [111]

Pinsker, M.S. (1964). *Information and Information Stability of Random Variables and Processes*. Holden-Day, San Francisco. [18]

Pinson, P., McSharry, P., and Madsen, H. (2010). Reliability diagrams for nonparametric density forecasts of continuous variables: Accounting for serial correlation. *Quarterly Journal of the Royal Meteorological Society*, 136(646), 77–90, Part A.
DOI: 10.1002/qj.559. [430]

Pippenger, M.K. and Goering, G.E. (2000). Additional results on the power of unit root and cointegration tests under threshold processes. *Applied Economics Letters*, 7(10), 641–644.
DOI: 10.1080/135048500415932. [486]

Pitarakis, J.-Y. (2006). Model selection uncertainty and detection of threshold effects. *Studies in Nonlinear Dynamics & Econometrics*, 10(1), 1–30.
DOI: 10.2202/1558-3708.1256. [187]

Pitarakis, J.-Y. (2008). Comments on: Threshold autoregression with a unit root. *Econometrica*, 76(5), 1207–1217. DOI: 10.3982/ECTA6979. [189]

Polanski, A. and Stoja, E. (2012). Efficient evaluation of multidimensional time-varying density forecasts, with applications to risk management *International Journal of Forecasting*, 28(2), 343–352. DOI: 10.1016/j.ijforecast.2010.10.007. [487]

Polinik, W. and Yao, Q. (2000). Conditional minimum volume predictive regions for stochastic processes. *Journal of the American Statistical Association*, 95(450), 509–519.
DOI: 10.2307/2669395. [384, 413, 429]

Politis, D.N. (2013). Model-free model-fitting and predictive distributions. *Test*, 22(2), 183–250 (with discussion). DOI: 10.1007/s11749-013-0317-7. [412]

Politis, D.N. (2015). *Model-Free Prediction and Regression*. Springer-Verlag, New York.
DOI: 10.1007/978-3-319-21347-7. [412]

Politis, D.N. and Romano, J.P. (1992). A circular block-resampling procedure for stationary data. In R. LePage and L. Billard (Eds.) *Exploring the Limits of Bootstrap*. Wiley, New York, pp. 263–270. [329]

Politis, D.N. and Romano, J.P. (1994). The stationary bootstrap. *Journal of the American Statistical Association*, 89(428), 1303–1313. DOI: 10.2307/2290993. [321]

Pomeau, Y. (1982). Symétrie des fluctuations dans le renversement du temps. *Journal de Physique*, 43(6), 859–867. DOI: 10.1051/jphys:01982004306085900. [317]

Porcher, R. and Thomas, G. (2003). Order determination in nonlinear time series by penalized least-squares. *Communications in Statistics: Simulation and Computation*, 32(4), 1115–1129. DOI: 10.1081/sac-120023881. [383]

Potter, S.M. (1995). A nonlinear approach to US GNP. *Journal of Applied Econometrics*, 10(2), 109–125. DOI: 10.1002/jae.3950100203. [77]

Potter, S.M. (2000). Nonlinear impulse response functions. *Journal of Economic Dynamics & Control*, 24(10), 1425–1446. DOI: 10.1016/s0165-1889(99)00013-5. [77]

Pourahmadi, M. (1986). On stationarity of the solution of a doubly stochastic model. *Journal of Time Series Analysis*, 7(2), 123–131. DOI: 10.1111/j.1467-9892.1986.tb00490.x. [73]

Pourahmadi, M. (1988). Stationarity of the solution of $X_t = A_t X_{t-1} + \varepsilon_t$ and analysis of non-Gaussian dependent random variables. *Journal of Time Series Analysis*, 9(3), 225–239. DOI: 10.1111/j.1467-9892.1988.tb00467.x. [110]

Priestley, M.B. (1980). State-dependent models: A general approach to non-linear time series analysis. *Journal of Time Series Analysis*, 1(1), 47–71. DOI: 10.1111/j.1467-9892.1980.tb00300.x. [32]

Priestley, M.B. (1981). *Spectral Analysis and Time Series: Vol. 1*. Academic Press, New York. [1]

Priestley, M.B. (1988). *Non-linear and Non-stationary Time Series Analysis*. Academic Press, New York. [73, 149, 597]

Priestley, M.B. and Gabr, M.M. (1993). Bispectral analysis of non-stationary processes. In C.R. Rao (Ed.) *Multivariate Analysis: Future Directions*. North-Holland, Amsterdam, Chapter 16, pp. 295–317. [149, 150]

Psaradakis, Z. (2008). Assessing time-reversibility under minimal assumptions. *Journal of Time Series Analysis*, 29(5), 881–905. DOI: 10.1111/j.1467-9892.2008.00587.x. [329]

Psaradakis, Z., Sola, M., Spagnolo, F., and Spagnola, N. (2009). Selecting nonlinear time series models using information criteria. *Journal of Time Series Analysis*, 30(4), 369–394. DOI: 10.1111/j.1467-9892.2009.00614.x. [249]

Puchstein, R. and Preuß, P. (2016). Testing for stationarity in multivariate locally stationary processes. *Journal of Time Series Analysis*, 37(1), 3–29. DOI: 10.1111/jtsa.12133. [522]

Qi, M. and Zhang, G.P. (2001). An investigation of model selection criteria for neural network time series forecasting. *European Journal of Operational Research*, 132(3), 666–680. DOI: 10.1016/s0377-2217(00)00171-5. [249]

Qian, L. (1998). On maximum likelihood estimators for a threshold autoregression. *Journal of Statistical Planning and Inference*, 75(1), 21–46. DOI: 10.1016/s0378-3758(98)00113-x. [247]

Quade, D. (1967). Rank analysis of covariance. *Journal of the American Statistical Association*, 62(320), 1187–1200. DOI: 10.1080/01621459.1967.10500925. [17]

Quinn, B.G. (1982). Stationarity and invertibility of simple bilinear models. *Stochastic Processes and their Applications*, 12(2), 225–230. DOI: 10.1016/0304-4149(82)90045-x. [103]

Rabemananjara, R. and Zakoïan, J.-M. (1993). Threshold ARCH models and asymmetries in volatility. *Journal of Applied Econometrics*, 8(1), 31–49. DOI: 10.1002/jae.3950080104. [81]

Racine, J.S. and Maasoumi, E. (2007). A versatile and robust metric entropy test of time-reversibility, and other hypotheses. *Journal of Econometrics*, 138(2), 547–567. DOI: 10.1016/j.jeconom.2006.05.009. [333]

Raftery, A.E. (1980). Estimation efficace pour un processus autorégressif exponentiel à densité discontinue. *Publications de l'Institut de statistique de l'Université de Paris*, 25(1), 64–90. [74]

Raftery, A.E. (1982). Generalized non-normal time series models. In O.D. Anderson (Ed.) *Time Series Analysis: Theory and Practice 1*. North-Holland, Amsterdam, pp. 621–640. [74]

Rajagopalan, B. and Lall, U. (1999). A k-nearest-neighbor simulator for daily precipitation and other weather variables. *Water Resources Research*, 35(10), 3089–3101. DOI: 10.1029/1999wr900028. [381]

Ramsey, J.B. (1969). Tests for specification errors in classical linear least squares regression analysis. *Journal of the Royal Statistical Society*, B 31(2), 350–371. [189]

Ramsey, J.B. and Rothman, P. (1996). Time irreversibility and business cycle asymmetry. *Journal of Money, Credit and Banking*, 28(1), 1–21. DOI: 10.2307/2077963. [317, 318]

Rao, C.R. (1973). *Linear Statistical Inference and Its Applications* (2nd edn.). Wiley, New York. DOI: 10.1002/9780470316436. [459]

Rao Jammalamadaka, S., Subba Rao, T., and Terdik, G. (2006). Higher order cumulants of random vectors and applications to statistical inference and time series. *Sankhyā: The Indian Journal of Statistics*, A 68(2), 326–356. Available at: http://eprints.ma.man.ac.uk/188/, and https://www.researchgate.net/publication/266584530_Higher_order_statistics_and_multivariate_vector_Hermite_polynomials_for_nonlinear_analysis_of_multidimensional_time_series. [522]

Rapach, D.E. and Wohar, M.E. (2006). The out-of-sample forecasting performance of non-linear models of real exchange rate behaviour. *International Journal of Forecasting*, 22(2), 341–361. DOI: 10.1016/j.ijforecast.2005.09.006. [430]

Rech, G., Teräsvirta, T., and Tschernig, R. (2001). A simple variable selection technique for nonlinear models. *Communications in Statistics: Theory and Methods*, 30(6), 1227–1241. DOI: 10.1081/sta-100104360. [487, 522]

Rényi, A. (1961). On a measure of entropy and information. *Fourth Berkeley Symposium on Mathematical Statistics and Probability*, Vol. I. University of California Press, Berkeley, pp. 547–561. [228, 264]

Resnick, S. and Van den Berg, E. (2000a). Sample correlation behavior for the heavy tailed general bilinear process. *Communication in Statistics: Stochastic Models*, 16(2), 233–258. DOI: 10.1080/15326340008807586. [23]

Resnick, S. and Van den Berg, E. (2000b). A test for nonlinearity of time series with infinite variance. *Extremes*, 3(2), 145–172. DOI: 10.1023/A:1009996916066. [23]

Rinke, S. and Sibbertsen, P. (2016). Information criteria for nonlinear time series models. *Studies in Nonlinear Dynamics & Econometrics*, 20(3), 325–341. DOI: 10.1515/snde-2015-0026. [249]

Rio, E. (1993). Covariance inequalities for strongly mixing processes. *Annales de l'Institute Henri Poincaré–Probabilité et Statistiques*, Section B, 29(4), 587–597. [96]

Rissanen, J. (1986). Stochastic complexity and modeling. *The Annals of Statistics*, 14(3), 1080–1100. DOI: 10.1214/aos/1176350051. [232]

Robert, C.P. and Casella, G. (2004). *Monte Carlo Statistical Methods* (2nd edn.). Springer-Verlag, New York. DOI: 10.1007/978-1-4757-4145-2. [233, 249]

Robinson, P.M. (1977). The estimation of a nonlinear moving average model. *Stochastic Processes and their Applications*, 5(1), 81–89. DOI: 10.1016/0304-4149(77)90052-7. [73]

Robinson, P.M. (1983). Nonparametric estimators for time series. *Journal of Time Series Analysis*, 4(3), 185–207. DOI: 10.1111/j.1467-9892.1983.tb00368.x. [356]

Robinson, P.M. (1991). Consistent nonparametric entropy-based testing. *Review of Economic Studies*, 58(3), 437–453. DOI: 10.2307/2298005. [271, 333]

Robinzonov, N., Tutz, G., and Hothorn, T. (2012). Boosting techniques for nonlinear time series models. *Advances in Statistical Analysis*, 96(1), 99–122. DOI: 10.1007/s10182-011-0163-4. [381, 383]

Rosenblatt, M. (1952). Remarks on a multivariate transformation. *Annals of Mathematical Statistics*, 23(3), 470–472. DOI: 10.1214/aoms/1177729394. [422, 479]

Rosenblatt, M. (1956). Remarks on some nonparametric estimates of a density function. *Annals of Mathematical Statistics*, 27(3), 832–835. DOI: 10.1214/aoms/1177728190. [305]

Rosenblatt, M. (1969). Conditional probability density and regression estimators. In P.R. Krishnaiah (Ed.) *Multivariate Analysis-II*. Academic Press, New York, pp. 25–31. [382]

Rota, G.C. (1964). On the Foundations of Combinatorial Theory. I. Theory of Möbius Functions. *Zeitschrift für Wahrscheinlichkeitstheorie und verwandte Gebiete*, 2(4), 340–368. DOI: 10.1007/bf00531932. [285]

Rothman, P. (1992). The comparative power of the TR test against simple threshold models. *Journal of Applied Econometrics*, 7(S1), 187–195. DOI: 10.1002/jae.3950070513. [333]

Rothman, P. (1996). FORTRAN programs for running the TR test: A guide and examples. *Studies in Nonlinear Dynamics & Econometrics*, 1(4). DOI: 10.2202/1558-3708.1023. [333]

Rothman, P. (Ed.) (1999). *Nonlinear Time Series Analysis of Economic and Financial Data* Springer Science+Business Media, New York. DOI: 10.1007/978-1-4615-5129-4. [597]

Rusticelli, E., Ashley, R.A., Dagum, E.B., and Patterson, D.G. (2009). A new bispectral test for nonlinear serial dependence. *Econometric Reviews*, 28(1-3), 279–293. DOI: 10.1080/07474930802388090. [136, 147]

Saikkonen, P. (2005). Stability results for nonlinear error correction models. *Journal of Econometrics*, 127(1), 69–81. DOI: 10.1016/j.jeconom.2004.03.001. [455]

References

Saikkonen, P. (2008). Stability of regime switching error correction models under linear cointegration. *Econometric Theory*, 24(01), 294–318. DOI: 10.1017/s0266466608080122. [296, 455]

Saikkonen, P. and Luukkonen, R. (1988). Lagrange multiplier tests for testing non-linearities in time series models. *Scandinavian Journal of Statistics*, 15(1), 55–68. [158, 188, 193]

Saikkonen, P. and Luukkonen, R. (1991). Power properties of a time series linearity test against some simple bilinear alternatives. *Statistica Sinica*, 1(2), 453–464. [193]

Sakaguchi, F. (1991). A relation for 'linearity' of the bispectrum. *Journal of Time Series Analysis*, 12(3), 267–272. DOI: 10.1111/j.1467-9892.1991.tb00082.x. [152]

Sakamoto, W. (2007). MARS: Selecting basis and knots with the empirical Bayes method. *Computational Statistics*, 22(4), 583–597. DOI: 10.1007/s00180-007-0075-7. [383]

Samia, N.I., Chan, K.S., and Stenseth, N.C. (2007). A generalized threshold mixed model for analyzing nonnormal nonlinear time series, with application to plague in Kazakhstan. *Biometrika*, 94(1), 101–118. DOI: 10.1093/biomet/asm006. [79]

Samworth, R.J. and Wand, M.P. (2010). Asymptotics and optimal bandwidth selection for highest density region estimation. *The Annals of Statistics*, 38(3), 1767–1792. DOI: 10.1214/09-aos766. [429]

Sankaran, M. (1959). On the noncentral chi-square distribution. *Biometrika*, 46(1-2), 235–237. DOI: 10.1093/biomet/46.1-2.235. [134]

Schleer–van Gellecom, F. (Ed.) (2014). *Advances in Non-linear Economic Modeling: Theory and Applications*, Springer-Verlag, New York. DOI: 10.1007/978-3-642-42039-9. [597]

Schleer–van Gellecom, F. (2015). Finding starting-values for the estimation of vector STAR models. *Econometrics*, 3(1), 65–90. DOI: 10.3390/econometrics3010065. [486]

Schmid, M. and Hothorn, T. (2008). Boosting additive models using component-wise P-splines as base-learners. *Computational Statistics & Data Analysis*, 53(2), 298–311. DOI: 10.1016/j.csda.2008.09.009. [383]

Schwarz, G. (1978). Estimating the dimension of a model. *The Annals of Statistics*, 6(2), 461–464. DOI: 10.1214/aos/1176344136. [231]

Scott, D.W. (1992). *Multivariate Density Estimation: Theory, Practice, and Visualization*. Wiley, New York (2nd edn., 2015). DOI: 10.1002/9781118575574. [521, 525]

Seo, M.H. (2006). Bootstrap testing for the null of no cointegration in a threshold vector error correction model. *Journal of Econometrics*, 134(1), 129–150. DOI: 10.1016/j.jeconom.2005.06.018. [486]

Seo, M.H. (2008). Unit root test in a threshold autoregression: Asymptotic theory and residual-based bootstrap. *Econometric Theory*, 24(06), 1699–1716. DOI: 10.1017/s0266466608080663. [189]

Serfling, R.J. (1980). *Approximation Theorems of Mathematical Statistics*. Wiley, New York. DOI: 10.1002/9780470316481. [308]

Serfling, R.J. (2002). Quantile functions for multivariate analysis: Approaches and applications. *Statistica Neerlandica*, 56(2), 214–232. DOI: 10.1111/1467-9574.00195. [521]

Serfling, R.J. (2004). Nonparametric multivariate descriptive measures based on spatial quantiles. *Journal of Statistical Planning and Inference*, 123(2), 259–278. DOI: 10.1016/s0378-3758(03)00156-3. [521]

Sesay, S.A.O. and Subba Rao, T. (1992). Frequency-domain estimation of bilinear time series models. *Journal of Time Series Analysis*, 13(6), 521–545. DOI: 10.1111/j.1467-9892.1992.tb00124.x. [486]

Shafik, N. and Tutz, G. (2009). Boosting nonlinear additive autoregressive time series. *Computational Statistics & Data Analysis*, 53(7), 2453–2464. DOI: 10.1016/j.csda.2008.12.006. [381, 383]

Sharifdoost, M., Mahmoodi, S., and Pasha, E. (2009). A statistical test for time reversibility of stationary finite state Markov chains. *Applied Mathematical Sciences*, 3(52), 2563–2574. Available at: http://www.m-hikari.com/ams/ams-password-2009/ams-password49-52-2009/lotfiAMS49-52-2009-4.pdf. [333]

Shorack, G.R. and Wellner, J.A. (1984). *Empirical Processes with Applications in Statistics*. Wiley, New York. DOI: 10.1137/1.9780898719017. [285]

Silverman, B.W. (1986). *Density Estimation for Statistics and Data Analysis*. Chapman & Hall, London. DOI: 10.1007/978-1-4899-3324-9. [268, 270, 305, 358]

Simonoff, J.S. and Tsai, C.-L. (1999). Semiparametric and additive model selection using an improved Akaike information criterion. *Journal of Computational and Graphical Statistics*, 8(1), 22–40. DOI: 10.2307/1390918. [249]

Singh, R.S. and Ullah, A. (1985). Nonparametric time-series estimation of joint DGP, conditional DGP and vector autoregression. *Econometric Theory*, 1(01), 27–52. DOI: 10.1017/s0266466600010987. [356]

Skaug, H.J. and Tjøstheim, D. (1993a). Nonparametric tests of serial dependence. In: T. Subba Rao (Ed.) *Developments in Time Series Analysis*. Chapman & Hall, London, pp. 207–229. [264, 271, 297, 311]

Skaug, H.J. and Tjøstheim, D. (1993b). A nonparametric test of serial independence based on the empirical distribution function. *Biometrika*, 80(3), 591–602. DOI: 10.1093/biomet/80.3.591. [271, 272, 297]

Skaug, H.J. and Tjøstheim, D. (1996). Measures of distance between densities with application to testing for serial independence. In P.M. Robinson, and M. Rosenblatt (Eds.) *Time Series Analysis in Memory of E.J. Hannan*. Springer-Verlag, New York, pp. 363–377. [271, 297]

Sklar, A. (1959). Fonctions de répartition à n dimensions et leur marges. *Publications de l'Institut de statistique de l'Université de Paris*, 8, 229–231. [305, 306]

Small, M. (2005). *Applied Nonlinear Time Series Analysis: Applications in Physics, Physiology and Finance*. World Scientific, Singapore. DOI: 10.1142/5722. [2, 24, 597]

Smith, J. and Wallis, K.F. (2009). A simple explanation of the forecast combination puzzle. *Oxford Bulletin of Economics and Statistics*, 71(3), 331–355. DOI: 10.1111/j.1468-0084.2008.00541.x. [425]

Smith, R.L. (1986). Maximum likelihood estimation for the NEAR(2) model. *Journal of the Royal Statistical Society*, A 48(2), 251–257. [74]

So, M.P., Li, W.K., and Lam, K. (2002). A threshold stochastic volatility model. *Journal of Forecasting*, 21(7), 473–500. DOI: 10.1002/for.840. [81]

Solari, S. and Van Gelder, P.H.A.J.M. (2011). On the use of vector autoregressive (VAR) and regime switching VAR models for the simulation of sea and wind state parameters. In C.G. Soares et al. (Eds.), *Marine Technology and Engineering*, Volume 1. Taylor & Francis Group, London, pp. 217–230. Available at: http://www.tbm.tudelft.nl/fileadmin/Faculteit/CiTG/Over_de_faculteit/Afdelingen/Afdeling_Waterbouwkunde/sectie_waterbouwkunde/people/personal/gelder/publications/papers/doc/solari_015.pdf. [486]

Sorour, A. and Tong, H. (1993). A note on tests for threshold-type non-linearity in open loop systems. *Applied Statistics*, 42(1), 95–104. DOI: 10.2307/2347412. [189]

Stam, C.J. (2005). Nonlinear dynamical analysis of EEG and MEG: Review of an emerging field. *Clinical Neurophysiology*, 116, 2266-2301. [23]

Steinberg, I.Z. (1986). On the time reversal of noise signals. *Biophysical Journal*, 50(1), 171–179. DOI: 10.1016/s0006-3495(86)83449-x. [317]

Stenseth, N.C., Chan, K.S., Tavecchia, G., Coulson, T., Mysterud, A., Clutton-Brock, T., and Grenfell, B. (2004). Modelling non-additive and nonlinear signals from climatic noise in ecological time series: Soay sheep as an example. *Proceedings of The Royal Society London*, B 271(1552), 1985–1993. DOI: 10.1098/rspb.2004.2794. [73]

Stenseth, N.C., Falck, W., Bjørnstad, O.N., and Krebs, C.J. (1997). Population regulation in snowshoe hare and Canadian lynx: Asymmetric food web configurations between hare and lynx. *Proceedings of the National Academy of Sciences USA*, 94(10), 5147–5152. DOI: 10.1073/pnas.94.10.5147. [293]

Stensholt, B.K. and Tjøstheim, D. (1987). Multiple bilinear time series models. *Journal of Time Series Analysis*, 8(2), 221–233. DOI: 10.1111/j.1467-9892.1987.tb00434.x. [441, 442, 443]

Stephens, M.A. (1974). EDF statistics for goodness of fit and some comparisons. *Journal of the American Statistical Association*, 69(347), 730–737. DOI: 10.2307/2286009 and DOI: 10.1080/01621459.1974.10480196. [134]

Stephens, M.A. (1986). Tests based on EDF statistics. In R.B. D'Agostino and M.A. Stephens (Eds.) *Goodness-of-Fit Techniques*. Marcel Dekker, New York, pp. 97–193. [135]

Steuber, T.L., Kiessler, P.C., and Lund, R. (2012). Testing for reversibility in Markov chain data. *Probability in the Engineering and Informational Sciences*, 26(04), 593–611. DOI: 10.1017/s0269964812000228. [333]

Stoica, P., Eykhoff, P., Janssen, P., and Söderström, T. (1986). Model-structure selection by cross-validation. *International Journal of Control*, 43(6), 1841–1878. DOI: 10.1080/00207178608933575. [234]

Stone, C.J. (1977). Consistent nonparametric regression. *The Annals of Statistics*, 5(4), 595–645. DOI: 10.1214/aos/1176343886. [354]

Strikholm, B. and Teräsvirta, T. (2006). A sequential procedure for determining the number of regimes in a threshold autoregressive model. *Econometrics Journal*, 9(3), 472–491. DOI: 10.1111/j.1368-423x.2006.00194.x. [249]

Su, L. and White, H. (2008). A nonparametric Hellinger metric test for conditional independence. *Econometric Theory*, 24(04), 829–864. DOI: 10.1017/s0266466608080341. [294]

Suárez–Fariñas, M., Pedreira, C.E., and Medeiros, M.C. (2004). Local global neural networks: A new approach for nonlinear time series modeling. *Journal of the American Statistical Association*, 99(468), 1092–1107. DOI: 10.1198/016214504000001691. [64, 75, 247]

Subba Rao, T. (1981). On the theory of bilinear time series models. *Journal of the Royal Statistical Society*, B 43(2), 244–255. [103]

Subba Rao, T. (1997). Time-domain and frequency-domain analysis of non-linear astronomical time series. In T. Subba Rao et al. (Eds.) *Applications of Time Series Analysis in Astronomy and Meteorology*. Chapman & Hall, London, pp. 142–157. [150]

Subba Rao, T. and Gabr, M.M. (1980). A test for linearity of stationary time series. *Journal of Time Series Analysis*, 1(2), 145–158. DOI: 10.1111/j.1467-9892.1980.tb00308.x. [119, 126, 128]

Subba Rao, T. and Gabr, M.M. (1984). *An Introduction to Bispectral Analysis and Bilinear Time Series Models*. Springer-Verlag, New York. DOI: 10.1007/978-1-4684-6318-7. [73, 117, 126, 129, 150, 151, 486, 597]

Subba Rao, T. and Terdik, G. (2003). On the theory of discrete and continuous bilinear time series models. In D.N. Shanbhag and C.R. Rao (Eds.) *Stochastic Processes: Modelling and Simulation*, Handbook of Statistics, Vol. 21. North-Holland, Amsterdam, pp. 827–870. DOI: 10.1016/s0169-7161(03)21023-3. [485]

Subba Rao, T. and Wong, W.K. (1998). Tests for Gaussianity and linearity of multivariate stationary time series. *Journal of Statistical Planning and Inference*, 68(2), 373–386. DOI: 10.1016/s0378-3758(97)00150-x. [522]

Subba Rao, T. and Wong, W.K. (1999). Some contributions to multivariate nonlinear time series bilinear models. In S. Gosh (Ed.) *Asymptotics, Nonparametrics and Time Series*. Marcel Dekker, New York, pp. 259–294. [486, 511]

Swanson, N.R. and White, H. (1997a). Forecasting economic time series using flexible versus fixed specification and linear versus nonlinear econometric models. *International Journal of Forecasting*, 13(4), 439–461. DOI: 10.1016/s0169-2070(97)00030-7. [429]

Swanson, N.R. and White, H. (1997b). A model selection approach to real-time macroeconomic forecasting using linear models and artificial neural networks. *The Review of Economics and Statistics*, 79(4), 540–550. DOI: 10.1162/003465397557123. [429]

Székely, G.J., Rizzo, M.L., and Bakirov, N.K. (2007). Measuring and testing dependence by correlation of distances. *The Annals of Statistics*, 35(6), 2760–2794.
DOI: 10.1214/009053607000000505. [296]

Tai, H. and Chan, K.S. (2000). Testing for nonlinearity with partially observed time series. *Biometrika*, 87(4), 805–821. DOI: 10.1093/biomet/87.4.805. [188]

Tai, H. and Chan, K.S. (2002). A note on testing for nonlinearity with partially observed time series. *Biometrika*, 89(1), 245–250. DOI: 10.1093/biomet/89.1.245. [188]

Tay, A.S. and Wallis, K.F. (2000). Density forecasting: A survey. *Journal of Forecasting*, 1(4), 235–254. DOI: 10.1002/1099-131X(200007). Reprinted in M.P. Clements and D.F. Hendry (Eds.), *A Companion to Economic Forecasting*. Blackwells, Oxford (2002), pp. 45–68. [430]

Teles, P. and Wei, W.W.S. (2000). The effects of temporal aggregation on tests of linearity of a time series. *Computational Statistics & Data Analysis*, 34(1), 91–103.
DOI: 10.1016/s0167-9473(99)00072-9. [151]

Teräsvirta, T. (1994). Specification, estimation, and evaluation of smooth transition autoregressive models. *Journal of the American Statistical Association*, 89(425), 208–218.
DOI: 10.2307/2291217. [74, 293]

Teräsvirta, T., Lin, C.-F., and Granger, C.W.J. (1993). Power of the neural network linearity test. *Journal of Time Series Analysis*, 14(2), 209–220.
DOI: 10.1111/j.1467-9892.1993.tb00139.x. [188, 190]

Teräsvirta, T., Tjøstheim, D., and Granger, C.W.J. (2010). *Modelling Nonlinear Economic Time Series*. Oxford University Press, New York.
DOI: 10.1093/acprof:oso/9780199587148.001.0001. [201, 597]

Teräsvirta, T. and Yang, Y. (2014a). Linearity and misspecification tests for vector smooth transition regression models. *CORE Discussion paper* 2014/62. Available at: `http://www.uclouvain.be/cps/ucl/doc/core/documents/coredp2014_62web.pdf`. Also available as CREATES Research Paper 2014-04, Aarhus University. [469, 470]

Teräsvirta, T. and Yang, Y. (2014b). Specification, estimation and evaluation of vector smooth transition autoregressive models with applications. CREATES Research Paper 2014-8. Available at: `ftp://ftp.econ.au.dk/creates/rp/14/rp14_08.pdf`. [487]

Terdik, G. (1999). *Bilinear Stochastic Models and Related Problems of Nonlinear Time Series Analysis*. Lecture Notes in Statistics 14. Springer-Verlag, New York.
DOI: 10.1007/978-1-4612-1552-3. (Freely available at: `http://dragon.unideb.hu/~terdik/PostScr/TerdikGyLNS142.pdf`). [23, 115, 140, 146, 273, 597]

Terdik, G. (1990). Second-order properties for multiple-bilinear models. *Journal of Multivariate Analysis*, 35(2), 295–307. DOI: 10.1016/0047-259x(90)90030-l. [485]

Terdik, G., Gál, Z., Iglói, E., and Molnár, S. (2002). Bispectral analysis of traffic in high-speed networks. *Computers & Mathematics with Applications*, 43(12), 1575–1583.
DOI: 10.1016/s0898-1221(02)00120-7. [146]

Terdik, G. and Máth, J. (1993). Bispectrum based checking of linear predictability for time series. In T. Subba Rao (Ed.) *Developments in Time Series Analysis*. Chapman & Hall, London, pp. 274–282. DOI: 10.1007/978-1-4899-4515-0_19. [141, 146]

Terdik, G. and Máth, J. (1998). A new test of linearity of time series based on the bispectrum. *Journal of Time Series Analysis*, 19(6), 737–753.
DOI: 10.1111/1467-9892.00120. [140, 142, 143, 146, 147]

Thavaneswaran, A. and Abraham, B. (1988). Estimation for non-linear time series models using estimating equations. *Journal of Time Series Analysis*, 9(1), 99–108.
DOI: 10.1111/j.1467-9892.1988.tb00457.x. [248]

Thavaneswaran, A. and Abraham, B. (1991). Estimation of multivariate non-linear time series models. *Journal of Statistical Planning and Inference*, 29(3), 351–363.
DOI: 10.1016/0378-3758(91)90009-4. [485]

Theiler, J., Eubank, S., Longtin, A., Galdrikian, B., and Farmer, J.D. (1992). Testing for nonlinearity in time series: The method of surrogate data. *Physica*, D 58(1-4), 77–94.
DOI: 10.1016/0167-2789(92)90102-s. [150, 188]

Theiler, J. and Prichard, D. (1996). Constrained Monte-Carlo method for hypothesis testing. *Physica*, D 94(4), 221–235. DOI: 10.1016/0167-2789(96)00050-4. [188]

Tiao, G.C. and Tsay, R.S. (1994). Some advances in nonlinear and adaptive modeling in time series analysis. *Journal of Forecasting*, 13(2), 109–131.
DOI: 10.1002/for.3980130206. [46]

Tibshirani, R. (1988). Estimating transformations for regression via additivity and variance stabilization. *Journal of the American Statistical Association*, 83(402), 194–205.
DOI: 10.2307/2288855. [383]

Timmermann, A. (2000). Moments of Markov switching models. *Journal of Econometrics*, 96(1), 75–111. DOI: 10.1016/s0304-4076(99)00051-2. [75]

Timmermann, A. (2006). Forecast combinations. In G. Elliott et al. (Eds.) *Handbook of Economic Forecasting*, North-Holland, Amsterdam, pp. 135–196.
DOI: 10.1016/s1574-0706(05)01004-9. [430]

Tjøstheim, D. (1986a). Some doubly stochastic time series models. *Journal of Time Series Analysis*, 17(1), 51–72. DOI: 10.1111/j.1467-9892.1986.tb00485.x. [39]

Tjøstheim, D. (1986b). Estimation in nonlinear time series models. *Stochastic Processes and their Applications*, 21(2), 251–273. DOI: 10.1016/0304-4149(86)90099-2. [39, 199, 472, 473]

Tjøstheim, D. (1990). Non-linear time series and Markov chains. *Advances in Applied Probability*, 22(3), 587–611. DOI: 10.2307/1427459. [90]

Tjøstheim, D. (1994). Non-linear time series: A selective review. *Scandinavian Journal of Statistics*, 21(2), 97–130. [295]

Tjøstheim, D. (1996). Measures of dependence and tests of independence. *Statistics*, 28(3), 249–284. DOI: 10.1080/02331889708802564. [295]

Tjøstheim, D. and Auestadt, B.H. (1994a). Non-parametric identification of non-linear time series: Projections. *Journal of the American Statistical Association*, 89(428), 1398–1409. DOI: 10.2307/2291002. [355, 358]

Tjøstheim, D. and Auestadt, B.H. (1994b). Nonparametric identification of nonlinear time series: Selecting significant lags. *Journal of the American Statistical Association*, 89(428), 1410–1419. DOI: 10.2307/2291003. [355]

Tong, H. (1977). Discussion of the paper by A.J. Lawrance and N.T. Kottegoda. *Journal of the Royal Statistical Society*, A 140(1), 34–35. DOI: 10.2307/2344516. [73, 78]

Tong, H. (1980). A view on non-linear time series building. In O.D. Anderson (Ed.) *Time Series*. North-Holland, Amsterdam, pp. 41–56. [73]

Tong, H. (1983). *Threshold Models in Non-Linear Time Series Analysis*. Springer-Verlag, New York. DOI: 10.1007/978-1-4684-7888-4. [73, 250, 597]

Tong, H. (1990). *Non-Linear Time Series: A Dynamical System Approach*. Oxford University Press, Oxford. [50, 73, 75, 78, 85, 250, 293, 312, 400, 597]

Tong, H. (2007). Birth of the threshold time series model. *Statistica Sinica*, 17(1), 8–14. [73]

Tong, H. (2011). Threshold models in time series analysis – 30 years on. *Statistics and Its Interface*, 49(2), 107–136 (with discussion). DOI: 10.4310/sii.2011.v4.n2.a1. [73]

Tong, H. (2015). Threshold models in time series analysis – some reflections. *Journal of Econometrics*, 189(2), 485–491. DOI: 10.1016/j.jeconom.2015.03.039. [73]

Tong, H. and Lim, K.S. (1980). Threshold autoregression, limit cycles and cyclical data. *Journal of the Royal Statistical Society*, B 42(3), 245–292 (with discussion). Also published in *Exploration of a Nonlinear World: An Appreciation of Howell Tong's Contributions to Statistics*, K.S. Chan (Ed.), World Scientific, Singapore.
DOI: 10.1142/9789812836281_0002. [41, 73]

Tong, H. and Moeanaddin, R. (1988). On multi-step non-linear least squares prediction. *The Statistician*, 37(2), 101–110. DOI: 10.2307/2348685. [428]

Tong, H. and Yeung, I. (1990). On tests for threshold-type nonlinearity in irregularly spaced time series. *Journal of Statistical Computation and Simulation*, 34(4), 172–194.
DOI: 10.1080/00949659008811226. [189]

Tong, H. and Yeung, I. (1991a). Threshold autoregressive modelling in continuous time. *Statistica Sinica*, 1(2), 411–430. [188]

Tong, H. and Yeung, I. (1991b). On tests for self-exciting threshold autoregressive-type nonlinearity in partially observed time series. *Applied Statistics*, 40(1), 43–62.
DOI: 10.2307/2347904. [189]

Tong, H. and Zhang, Z. (2005). On time-reversibility of multivariate linear processes. *Statistica Sinica*, 15(2), 495–504. [333]

Tong, H., Thanoon, B., and Gudmundson, G.L. (1985). Threshold time series modeling of two Icelandic riverflow systems. In K.W. Hipel (Ed.) *Time Series Analysis in Water Resources*. American Water Research Association, 21, pp. 651–661. [85, 481]

Trapletti, A., Leisch, F., and Hornik, K. (2000). Stationary and integrated autoregressive neural network processes. *Neural Computation*, 12(10), 2427–2450. DOI: 10.1162/089976600300015006. [58]

Tsai, H. and Chan, K.S. (2000). Testing for nonlinearity with partially observed time series. *Biometrika*, 87(4), 805–821. DOI: 10.1093/biomet/87.4.805. [189]

Tsai, H. and Chan, K.S. (2002). A note on testing for nonlinearity with partially observed time series. *Biometrika*, 89(1), 245–250. DOI: 10.1093/biomet/89.1.245. [189]

Tsallis, C. (1998). Generalized entropy-based criterion for consistent testing. *Physical Review*, E 58(2), 1442–1445. DOI: 10.1103/physreve.58.1442. [264]

Tsay, R.S. (1986). Nonlinearity tests for time series. *Biometrika*, 73(2), 461–466. DOI: 10.1093/biomet/73.2.461. [180, 181, 193]

Tsay, R.S. (1989). Testing and modeling threshold autoregressive processes. *Journal of the American Statistical Association*, 84(405), 231–240. DOI: 10.2307/2289868. [182, 184, 185, 193, 293]

Tsay, R.S. (1991). Detecting and modeling nonlinearity in univariate time series analysis. *Statistica Sinica*, 1(2), 431–451. [180, 185, 193]

Tsay, R.S. (1998). Testing and modeling multivariate threshold models. *Journal of the American Statistical Association*, 93(443), 1188–1202. DOI: 10.2307/2669861. [447, 464, 465, 481, 482, 488, 492]

Tsay, R.S. (2010). *Analysis of Financial Time Series* (3rd edn.). Wiley, New York. DOI: 10.1002/0471264105. [488, 501]

Tschernig, R. and Yang, L. (2000). Nonparametric lag selection for time series. *Journal of Time Series Analysis*, 21(4), 457–487. DOI: 10.1111/1467-9892.00193. [358, 359, 522]

Tse, Y.K. and Zuo, X.L. (1998). Testing for conditional heteroskedasticity: Some Monte Carlo results. *Journal of Statistical Computation and Simulation*, 58(3), 237–253. DOI: 10.1080/00949659708811833. [236]

Tsolaki, E.P. (2008). Testing nonstationary time series for Gaussianity and linearity using the evolutionary bispectrum: An application to internet traffic data. *Signal Processing*, 88(6), 1355–1367. DOI: 10.1016/j.sigpro.2007.12.011. [150]

Tukey, J.W. (1949). One degree of freedom for non-additivity. *Biometrics*, 5(3), 232–242. DOI: 10.2307/3001938. [179]

Tutz, G. and Binder, H. (2006). Generalized additive modelling with implicit variable selection by likelihood based boosting. *Biometrics*, 62(4), 961–971. DOI: 10.1111/j.1541-0420.2006.00578.x. [386]

Ubilava, D. (2012). El Niño, La Niña, and world coffee price dynamics. *Agricultural Economics*, 43(1), 17–26. DOI: 10.1111/j.1574-0862.2011.00562.x. [24]

References

Ubilava, D. and Helmers, C.G. (2013). Forecasting ENSO with a smooth transition autoregressive model. *Environmental Modelling & Software*, 40, 181–190. DOI: 10.1016/j.envsoft.2012.09.008. [24, 215, 422]

Ullah, A. (1996). Entropy, divergence and distance measures with econometric applications. *Journal of Statistical Planning and Inference*, 49(1), 137–162. DOI: 10.1016/0378-3758(95)00034-8. [295]

Van Casteren, P.H.F.M. and De Gooijer, J.G. (1997). Model selection by maximum entropy. In T.B. Fomby and R.C. Hill (Eds.), *Advances in Econometrics* (Applying Maximum Entropy to Econometric Problems), Vol. 12. JAI Press, Connecticut, pp. 135–161. DOI: 10.1108/s0731-9053(1997)0000012007. [249]

Van Dijk, D. and Franses, P.H. (2003). Selecting a nonlinear time series model using weighted tests of equal forecast accuracy. *Oxford Bulletin of Economics and Statistics*, 65(s1), 727–744. DOI: 10.1046/j.0305-9049.2003.00091.x. [429]

Van Dijk, D., Teräsvirta, T., and Franses, P.H. (2002). Smooth transition autoregressive models – A survey of recent developments. *Econometric Reviews*, 21(1), 1–47. DOI: 10.1081/etc-120002918. [74]

Van Ness, J.W. (1966). Asymptotic normality of bispectral estimates. *Annals of Mathematical Statistics*, 37(5), 1257–1275. DOI: 10.1214/aoms/1177699269. [149]

Vavra, M. (2013). *Testing for Non-linearity and Asymmetry in Time Series*, Ph.D. thesis, Birbeck college, University of London, UK. Available at: http://bbktheses.da.ulcc.ac.uk/97/1/final%20Marian%20Vavra.pdf. [190]

Ventosa-Santaulària, D. and Mendoza-Velázquez, A. (2005). Non linear moving-average conditional heteroskedasticity. Available at: http://mpra.ub.uni-muenchen.de/58769/. [73]

Vialar, T. (2005). *Dynamiques non linéaires chaotiques en finance et économie*. Economica, Paris. [597]

Vieu, P. (1994). Choice of regressors in nonparametric estimation. *Computational Statistics & Data Analysis*, 17(5), 575–594. DOI: 10.1016/0167-9473(94)90149-x. [383]

Vieu, P. (1995). Order choice in nonlinear autoregressive models. *Statistics*, 26(4), 307–328. DOI: 10.1080/02331889508802499. [383]

Vilar-Fernandez, J.M. and Cao, R. (2007). Nonparametric forecasting in time series – A comparative study. *Communications in Statistics: Simulation and Computation*, 36(2), 311–334. DOI: 10.1080/03610910601158377. [382]

Volterra, V. (1930). *Theory of Functionals and of Integro-differential Equations*. Dover, New York. Abstract: http://www.ams.org/journals/bull/1932-38-09/S0002-9904-1932-05479-9/S0002-9904-1932-05479-9.pdf. [72]

Von Mises, R. (1947). On the asymptotic distribution of differentiable statistical functions. *Annals of Mathematical Statistics*, 18(3), 309–348. DOI: 10.1214/aoms/1177730385. [309]

Wallis, K.F. (2003). Chi-square tests of interval and density forecasts, and the Bank of England's fan charts. *International Journal of Forecasting*, 19(2), 165–175. DOI: 10.1016/s0169-2070(02)00009-2. [430]

Wallis, K.F. (2011). Combining forecasts – forty years later. *Applied Financial Econometrics*, 21(1-2), 33–41. DOI: 10.1080/09603107.2011.523179. [430]

Wand, M.P. and Jones, M.C. (1995). *Kernel Smoothing*. Chapman & Hall, London. DOI: 10.1007/978-1-4899-4493-1. [298, 385]

Wang, H.B. (2008). Nonlinear ARMA models with functional MA coefficients. *Journal of Time Series Analysis*, 29(6), 1032–1056. DOI: 10.1111/j.1467-9892.2008.00594.x. [374]

Watson, G.S. (1964). Smooth regression analysis. *Sankhyā*, A 26, 359–372. [302]

Wecker, W.E. (1981). Asymmetric time series. *Journal of the American Statistical Association*, 76(373), 16–21. Corrigendum: p. 954. DOI: 10.2307/2287034. [74, 116]

Weiss, A.A. (1986). ARCH and bilinear time series models: comparison and combination. *Journal of Business & Economic Statistics*, 4(1), 59–70. DOI: 10.2307/1391387. [188]

Welsh, A.K. and Jernigan. R.W. (1983). A statistic to identify asymmetric time series. *American Statistical Association, Proceedings of the Business and Economic Statistics Section*, pp. 390–395. [194]

West, K.D. (1996). Asymptotic inference about predictive ability. *Econometrica*, 64(5), 1067–1084. DOI: 10.2307/2171956. [427]

West, K.D. (2001). Tests for forecast encompassing when forecasts depend on estimated regression parameter. *Journal of Business & Economic Statistics*, 19(1), 29–33. DOI: 10.1198/07350010152472580. [427]

West, K.D. (2006). Chapter 3: Forecast evaluation. In G. Elliott et al. (Eds.) *Handbook of Economic Forecasting*, Volume 1. North-Holland, Amsterdam, pp. 99–134. DOI: 10.1016/s1574-0706(05)01003-7. [429]

White, H. (1984). *Asymptotic Theory for Econometricians*. Academic Press, Orlando, Florida. [320]

White, H. (1989). An additional hidden unit test for neglected non-linearity in multilayer feedforward networks. In *Proceedings of the International Joint Conference on Neural Networks*, Washington, D.C. (IEEE Press, New York), Vol. I. San Diego, CA: SOS Printing, pp. 451–455. DOI: 10.1109/ijcnn.1989.118281. [188]

White, H. (1992). *Estimation, Inference and Specification Analysis*. Cambridge University Press, New York. DOI: 10.1017/ccol0521252806. [188]

White, H. (2000). A reality check for data snooping. *Econometrica*, 68(5), 1097–1126. DOI: 10.1111/1468-0262.00152. [427, 429]

Wiener, N. (1958). *Non-linear Problems in Random Theory*. Wiley, London. [72, 597]

Wilson, G.T. (1969). Factorization of the covariance generating function of a pure moving average process. *SIAM Journal of Numerical Analysis*, 6(1), 1–7. DOI: 10.1137/0706001. [219]

Wolff, R.C.L. and Robinson, P.M. (1994). Independence in time series: Another look at the BDS test [and Discussion]. *Philosophical Transactions Royal Society London*, A 348(1688), 383–395. DOI: 10.1098/rsta.1994.0098. [296]

Wong, C.-M. and Kohn, R. (1996). A Bayesian approach to estimating and forecasting additive nonparametric autoregression in time series. *Journal of Time Series Analysis*, 17(2), 203–220. DOI: 10.1111/j.1467-9892.1996.tb00273.x. [383]

Wong, C.S. and Li, W.K. (1997). Testing for threshold autoregression with conditional heteroskedasticity. *Biometrika*, 84(2), 407–418. DOI: 10.1093/biomet/84.2.407. [188]

Wong, C.S. and Li, W.K. (1998). A note on the corrected Akaike information criterion for threshold autoregressive models. *Journal of Time Series Analysis*, 19(1), 113–124. DOI: 10.1111/1467-9892.00080. [249]

Wong, C.S. and Li, W.K. (2000a). Testing for double threshold autoregressive conditional heteroskedastic model. *Statistica Sinica*, 10(1), 173–189. [188]

Wong, C.S. and Li, W.K. (2000b). On a mixture autoregressive model. *Journal of the Royal Statistical Society*, B 62(1), 95–15. DOI: 10.1111/1467-9868.00222. [240, 296, 313]

Wong, C.S. and Li, W.K. (2001). On a mixture autoregressive conditional heteroscedastic model. *Journal of the American Statistical Association*, 96(455), 982–995. DOI: 10.1198/016214501753208645. [240, 296]

Wong, W.K. (1997). Frequency domain tests of multivariate Gaussianity and linearity. *Journal of Time Series Analysis*, 18(2), 181–194. DOI: 10.1111/1467-9892.00045. [511, 512]

Wu, E.H.C., Yu, P.L.H., and Li, W.K. (2009). A smoothed bootstrap test for independence based on mutual information. *Computational Statistics & Data Analysis*, 53(7), 2524–2536. DOI: 10.1016/j.csda.2008.11.032. [23, 296]

Wu, T.Z., Yu, K., and Yu, Y. (2010). Single-index quantile regression. *Journal of Multivariate Analysis*, 101(7), 1607–1621. DOI: 10.1016/j.jmva.2010.02.003. [384]

Wu, T.Z., Lin, H., and Yu, Y. (2011). Single-index coefficient models for nonlinear time series. *Journal of Nonparametric Statistics*, 23(1), 37–58. DOI: 10.1080/10485252.2010.497554. [384]

Xia, X. and An, H.Z. (1999). Projection pursuit autoregression in time series. *Journal of Time Series Analysis*, 20(6), 693–714. DOI: 10.1111/1467-9892.00167. [381, 383]

Xia, Y. and Li, W.K. (1999). On single-index coefficient regression models. *Journal of the American Statistical Association*, 94(448), 1275–1285. DOI: 10.2307/2669941. [378]

Xia, Y., Tong, H., and Li, W.K. (1999). On extended partially linear single-index models. *Biometrika*, 86(4), 831–842. DOI: 10.1093/biomet/86.4.831. [378, 379]

Yakowitz, S.J. (1985). Nonparametric density estimation, prediction, and regression for Markov sequences. *Journal of the American Statistical Association*, 80(389), 215–221. DOI: http://dx.doi.org/10.2307/2288075 and DOI: 10.1080/01621459.1985.10477164. [382]

Yakowitz, S.J. (1987). Nearest neighbor methods for time series analysis. *Journal of Time Series Analysis*, 8(2), 235–247. DOI: 10.1111/j.1467-9892.1987.tb00435.x. [353, 382]

Yang, Y. (2012). *Modelling Nonlinear Vector Economic Time Series*, Ph.D. thesis, Aarhus University, Denmark. *CREATES Research Paper 2012-7*. Available at: http://pure.au.dk/portal/files/45638557/Yukai_Yang_PhD_Thesis.pdf. [487, 489]

Yang, K. and Shahabi, C. (2007). An efficient k nearest neighbor search for multivariate time series. *Information and Computation*, 205(1), 65–98. DOI: 10.1016/j.ic.2006.08.004. [522]

Yang, L., Härdle, W., and Nielson, J. (1999). Nonparametric autoregression with multiplicative volatility and additive mean. *Journal of Time Series Analysis*, 20(5), 579–604. DOI: 10.1111/1467-9892.00159. [382]

Yang, Z., Tian, Z., and Zixia, Y. (2007). GSA-based maximum likelihood estimation for threshold vector error correction model. *Computational Statistics & Data Analysis*, 52(1), 109–120. DOI: 10.1016/j.csda.2007.06.003. [486]

Yao, Q. and Tong, H. (1994). On subset selection in non-parametric stochastic regression. *Statistica Sinica*, 4(1), 51–70. [383]

Yao, Q. and Tong, H. (1995). On initial-condition sensitivity and prediction in nonlinear stochastic systems. *Bulletin International Statistical Institute*, IP10.3, 395–412. [408, 413, 429]

Yi, J. and Deng, J. (1994). The ergodicity of vector self excited threshold autoregressive (VSETAR) models. *Applied Mathematics. A Journal of Chinese Universities*, Series A (Chinese Edition), 9(1), 53–59. [486]

Yoshihara, K. (1976). Limiting behavior of U-statistics for stationary, absolutely regular processes. *Zeitschrift für Wahrscheinlichkeitstheorie und verwandte Gebiete*, 35(3), 237–252. DOI: 10.1007/bf00532676. [309]

Young, P.C. (1993). Time variable and state dependent modelling of non-stationary and nonlinear time series. In T. Subba Rao (Ed.), *Developments in Time Series Analysis*. Chapman & Hall, London, pp. 374–413. [384]

Young, P.C. and Beven, K.J. (1994). Data-based mechanistic modelling and the rainfall-flow non-linearity. *Environmetrics*, 5(3), 335–363. DOI: 10.1002/env.3170050311. [384]

Yu, P.L.H., Li, W.K., and Jin, S. (2010). On some models for Value-at-Risk. *Econometric Reviews*, 29(5-6), 622–641. DOI: 10.1080/07474938.2010.481972. [81]

Yuan, J. (2000a). Testing linearity for stationary time series using the sample interquartile range. *Journal of Time Series Analysis*, 21(6), 713–722. DOI: 10.1111/1467-9892.00206. [150]

Yuan, J. (2000b). Testing Gaussianity and linearity for random fields in the frequency domain. *Journal of Time Series Analysis*, 21(6), 723–737. DOI: 10.1111/1467-9892.00207. [150]

Zakoïan, J.-M. (1994). Threshold heteroskedastic models. *Journal of Economic Dynamics & Control*, 18(5), 931–955. DOI: 10.1016/0165-1889(94)90039-6. [81]

Zeevi, A.J., Meir, R., and Adler, R.J. (1999). Non-linear models for time series using mixtures of autoregressive models. Available at: http://citeseerx.ist.psu.edu/viewdoc/summary?doi=10.1.1.44.4549. [296]

Zhang, J. and Stine, R.A. (2001). Autocovariance structure of Markov regime switching models and model selection. *Journal of Time Series Analysis*, 22(1), 107–124. DOI: 10.1111/1467-9892.00214. [75]

Zhang, X., King, M.L., and Hyndman, R.J. (2006). A Bayesian approach to bandwidth selection for multivariate kernel density estimation. *Computational Statistics & Data Analysis*, 50(11), 3009–3031. DOI: 10.1016/j.csda.2005.06.019. [305]

Zhang, X, Wong, H., Li, Y., and Ip, W.-C. (2011). A class of threshold autoregressive conditional heteroscedastic models. *Statistics and Its Interface*, 4(2), 149–157. DOI: 10.4310/sii.2011.v4.n2.a10. [248]

Zhou, Z. and Wu, W.B. (2009). Local linear quantile estimation for nonstationary time series. *The Annals of Statistics*, 37(5B), 2696–2729. DOI: 10.1214/08-aos636. [382]

Zhou, Z. (2012). Measuring nonlinear dependence in time-series, a distance correlation approach. *Journal of Time Series Analysis*, 33(3), 438–457. DOI: 10.1111/j.1467-9892.2011.00780.x. [296]

Zhu, K., Yu, P.L.H., and Li, W.K. (2014). Testing for the buffered autoregressive processes. *Statistica Sinica*, 24(2), 971–984. DOI: 10.5705/ss.2012.311. [81]

Zivot, E. and Wang, J. (2006). *Modeling Financial Time Series with S-Plus* (2nd edn.). Springer-Verlag, New York. DOI: 10.1007/978-0-387-32348-0. Freely available at: http://faculty.washington.edu/ezivot/econ589/manual.pdf. [75]

Zoubir, A.M. (1999). Model selection: A bootstrap approach. In *IEEE International Conference on Acoustics, Speech, and Signal Processing*, Vol. 3, IEEE, Phoenix, AZ, USA, pp. 1377–1380. DOI: 10.1109/icassp.1999.756237. [140]

Zoubir, A.M. and Iskander, D.R. (1999). Bootstrapping bispectra: An application to testing for departure from Gaussianity of stationary signals. *IEEE Transaction on Signal Processing*, 47(3), 880–884. DOI: 10.1109/78.747796. [150]

Books about Nonlinear Time Series Analysis

General
Chan (2009)
Douc et al. (2014)
Franses and Van Dijk (2000)
Granger and Teräsvirta (1992a)
Guégan (1994)
Priestley (1988)
Teräsvirta et al. (2010)
Tong (1990)
Wiener (1958)

Applications
Casdagli and Eubank (1992)
Donner and Barbosa (2008)
Dunis and Zhou (1998)
Galka (2000)
Haldrup et al. (2014)
Ma and Wohar (2014)
Milas et al. (2006)
Patterson and Ashley (2000)
Rothman (1999)
Schleer–van Gellecom (2014)
Small (2005)

Bilinear models
Granger and Andersen (1978a)
Subba Rao and Gabr (1984)
Terdik (1999)

Chaos
Cutler and Kaplan (1996)
Chan and Tong (2001)
Diks (1999)
Kantz and Schreiber (2004)
Vialar (2005)

Proceedings
Barnett et al. (2006)
Casdagli and Eubank (1992)
Dagum et al. (2004)
Fitzgerald et al. (2000)
Franke et al. (1984)
Hsiao et al. (2011)

Semi- and nonparametric
Fan and Yao (2003)
Gao (2007)
Li and Racine (2007)

Spectral and signal analysis
Haykin (1979)
Mohler (1987)

Threshold and RCA models
Tong (1983)
Nicholls and Quinn (1982)

Notations and Abbreviations

The following notation is frequently used throughout the book. The number following the description of a notation marks the page where the notation is first introduced.

Table 1: *List of Symbols.*

Symbol	Description	Page
	General	
$\equiv$	equals, by definition	10
$\perp$	perpendicular, mutually singular (of measures)	457
$\|x\|$	norm of x in L_2 (Euclidean norm)	19
$\|x\|_p$	L_p-norm	112
$!$	factorial	299
$!!$	semifactorial: $(2k-1)!! = 1 \cdot 3 \cdot 5 \cdots (2k-1)$	221
$[x]$	absolute value (integer part) of scalar x (largest integer $\leq x$)	127
$\lfloor x \rfloor$	the largest integer not greater than x	126
$x \wedge y$	$= \min(x,y)$	449
$x \vee y$	$= \max(x,y)$	198
$\log(x)$	natural logarithm of x (with base $e = 2.71828\cdots$)	11
$\log^+(x)$	$= \max\{\log(x), 0\}$	89
B	backward shift (or lag) operator	62
C	$= 0.5772156649\cdots$, Euler's constant	89
δ_{ij}	Kronecker delta, where $\delta_{ij} = 1$ if $i = j$ and $\delta_{ij} = 0$ if $i \neq j$	327
$\exists$	"there exists"	37
$h \equiv h_T$	smoothing parameter or bandwidth	209
h_b	binwidth	270
$K(\cdot), K_h(\cdot)$	kernel function (with bandwidth h)	260
$\forall$	"for all" ("for every")	2
arg min	argument that minimizes a function	58
arg max	argument that maximizes a function	340
exp	exponential	2
inf	infimum (greatest lower bound)	339

min	minimum	44
max	maximum	37
Leb	Lebesgue measure on $\mathbb{R}^m$	98
lim	limit (number); *also* limit (sets)	13
lim inf	inferior limit (number); *also* inferior limit (sets)	91
lim sup	superior limit (number); *also* superior limit (sets)	91
Ran H	range of the function H	306
sign(a)	sign of the real number a	311
sup	supremum (least upper bound)	19
s.t.	"subject to"	93

Sets

$\{\cdot\}$	set designation; *also* sequence, array	2
$\in, \notin$	set membership, does not belong to	2
$\cup$	union	41
$\subset$	subset (strict containment)	198
$\cap$	intersection	41
$\mathcal{F}_t$	σ-algebra (information set)	2
$\emptyset$	empty (null) set	41
$I(\cdot)$	indicator function, i.e. $I(z)=1$ if $z>1$ and $I(z)=0$ if $z \le 0$	16
$\Im(\cdot)$	imaginary part	121
$\mathbb{N}$	$=\{0,1,2,\ldots\}$, i.e. the set of all natural numbers, including zero	10
$\mathbb{R}$	the set of all real numbers	41
$\mathbb{R}^+$	the set all non-negative real numbers	240
$\mathbb{R}^n, \mathbb{R}^{m \times n}$	the set of real $n \times 1$ vectors ($m \times n$ matrices)	37
$\Re(\cdot)$	real part	142
$\mathbb{Z}$	$=\{0,\pm 1,\pm 2,\ldots\}$, i.e. the set of all relative integers	2
$\mathbb{Z}^+$	$=\{1,2,3,\ldots\}$, i.e. the set of all positive integers	19

Special matrices and vectors

$\mathbf{e}$	$=(1,0,\ldots,0)'$, a vector with 1 in the first entry and zeros elsewhere	205
$\mathbf{1}$	$=(1,\ldots,1)'$, a unity row vector	491
$\mathbf{I}_n$	identity matrix of order $n \times n$	42
$\mathbf{O}_{m \times n}$	$m \times n$ null matrix	42
$\mathbf{0}_m, \mathbf{0}_{m \times 1}$	$m \times 1$ null vector	42

Operations on matrix A and vector a

$\mathbf{A}', \mathbf{a}'$	transpose of a matrix or vector	13
$\mathbf{A}^{-1}$	inverse of a matrix	129
$\mathbf{A}^{\#}$	Hankel matrix	219
diag($\mathbf{A}$)	diagonal matrix, containing the diagonal elements of $\mathbf{A}$	262
vec($\mathbf{A}$)	= stacking the elements of $\mathbf{A}$ one underneath the other	441
vech($\mathbf{A}$)	= stacking the elements of $\mathbf{A}$ on and below the main diagonal into one vector	180
$\rho(\mathbf{A})$	maximum absolute eigenvalue of $\mathbf{A}$ (spectral radius)	90
tr($\mathbf{A}$)	trace	229

Notation and Abbreviations

$\|\mathbf{A}\|$, $\det(\mathbf{A})$	determinant of a matrix	232
$\|\mathbf{A}\|$, $\|\mathbf{a}\|$	norm of a matrix or vector	88

Matrix products

$\otimes$	Kronecker product	90
$\odot$	Hadamard product (also known as direct product or tensor product)	524

Statistical symbols

$\mathbb{C}(\cdot)$	copula	267
$\xrightarrow{D}$	convergence in distribution (or weak convergence)	10
$\overset{D}{\sim}$	equivalence in distribution	316
$\mathbb{E}$	expectation	2
$\mathbb{P}$	probability	54
$\mathcal{P}$	probability measure	96
$(\Omega, \mathcal{F}, \mathbb{P})$	probability space	95
i.i.d.	independently and identically distributed	2
Var	variance	16
Cov	covariance	39
Cum	cumulant	510
$\sim$	is distributed as	2
a.s.	almost surely	88
$\mathcal{N}_m(\mathbf{0}, \mathbf{\Sigma})$	m-dimensional normal (or Gaussian) distribution with mean $\mathbf{0}$ and covariance matrix $\mathbf{\Sigma}$	467
t_ν	Student t distribution with ν degrees of freedom	104
χ^2_n	chi-squared distribution with n degrees of freedom	10
$\chi^2_n(\lambda)$	χ^2_n distribution with noncentrality parameter λ	131

"Big O" and "little o"

Suppose $\{x_n\}$ is a scalar non-stochastic sequence of real numbers for integers $n = N, \ldots, \infty$. Then

$x_n = \mathcal{O}(1)$	if $\|x_n\| < c\ \forall n$, and $0 < c < \infty$;	
$x_n = \mathcal{O}(n^m)$	if $n^{-m} = \mathcal{O}(1)$;	130
$x_n = o(n^m)$	if $\lim_{n\to\infty} n^{-m} x_n = 0$.	344

Suppose $\{X_n\}$ is a sequence of random variables for integers $n = N, \ldots, \infty$. Then

$X_n = \mathcal{O}_p(n^m)$	if for any $\epsilon > 0$ there is a constant $c < \infty$ such that $\mathbb{P}(\|n^{-m} X_n\| > c) < \epsilon\ \forall n > N$ (convergence in probability);	351
$X_n = o_p(1)$	if X_n converges in probability to zero as $n \to \infty$.	229

Table 2: *List of abbreviations. The number following the description marks the page where the notation is first introduced. For acronyms given to threshold-type time series models, we refer to Appendix.* 2.B.

Symbol	Description	Page
ACE	alternating conditioning expectations	360
ACF	autocorrelation function	14
ACVF	autocovariance function	12
AD	Anderson–Darling	266
AFPE	asymptotic FPE	358
AIC	Akaike's information criterion	69
AMISE	asymptotic MISE	300
AMSE	asymptotic mean squared error	300
ANN	artificial neural network	56
AO	additive outlier	248
AR(MA)–NN	autoregressive (moving average) neural network	58
asARMA	asymmetric ARMA	47
(G)ARCH	(generalized) autoregressive conditional heteroskedasticity	67
ARMA(X)	autoregressive moving average (exogenous)	1
ASTMA	additive smooth transition moving average	52
AVAS	additive and variance stabilizing	360
BDS	Brock–Dechert–Scheinkman	279
BFI	bootstrap FI	411
BGAR	beta-gamma AR	335
BIC	Bayesian information criterion	69
BL	bilinear	33
BS	bootstrapping	11
cc	conditional coverage	420
CCF	cross-correlation function	292
CPP	compound Poisson process	205
CDF	cumulative distribution function	51
(C)LS	(conditional) least squares	44
CLT	central limit theorem	96
CNF	common nonlinear feature	457
CPI	conditional predictive interval	408
CR	Cressie–Read	265
CUSUM	cumulated sum	183
CV	cross-validation	268
CvM	Cramér–von Mises	261
CVR	coverage rate	412
DE	dynamic estimation	406
DGP	data generating process	4
DM	Diebold–Mariano	416
DP	Diks–Panchenko	291
ECM	error correction model	216

Notation and Abbreviations

EDF	empirical distribution function	134
EEG	electroencephalogram	5
eff	efficiency	300
ELS	empirical least squares	400
ENSO	El Niño–Southern Oscillation	7
ESTAR	exponential STAR	51
ew	equal-weighting	425
ExpARMA	exponential ARMA	37
FC(MA)AR	functional-coefficient (MA) AR	374
FI	forecast interval	408
FPE	final prediction error	358
FR	forecast region	408
FT	Fourier transform	120
GA	genetic algorithm	210
GCI	Granger causality index	451
GCV	generalized cross-validation	367
GFESM	generalized forecast error second moment	479
GIC	generalized information criterion	231
GIRF	generalized impulse response function	36
GJB	generalized JB	12
GMM	generalized method of moments	248
GOF	goodness-of-fit	133
GRASP	greedy randomized adaptive search procedure	74
HDR	highest density region	414
HJ	Hiemstra–Jones	515
HL	Hotelling–Lawley	462
IDR	inter decile range	136
IO	innovational outlier	249
IQR	inter quartile range	132
ISE	integrated squared error	299
IWLS	iteratively weighted least squares	223
JB	Jarque–Bera	10
KL	Kullback–Leibler	18
KS	Kolmogorov–Smirnov	266
LB	Ljung–Box	236
LGNN	local global neural network	62
L^{2}GNN	local linear global neural network	63
LL	local linear	304
LM	Lagrange multiplier	155
LN	linearization	404
LR	likelihood ratio	155
LSTAR	logistic STAR	51
LSTEC	logistic smooth transition error-correction	215
LVSTAR	logistic VSTAR	454

LWR	locally weighted regression	353
MAE	mean absolute error	246
MAFE	mean absolute forecast error	72
MAR	mixture AR	240
MARS	multivariate adaptive regression splines	365
MC	Monte Carlo	11
MCDR	maximum conditional density region	414
MCMC	Markov chain Monte Carlo	305
MDM	modified DM	417
MDL	minimum descriptive length	232
MFD	Markov forecast density	356
MHD	minimum Hellinger distance	248
MISE	mean integrated squared error	300
ML	maximum likelihood	54
MLP	multi-layer perceptron	56
MMSE	minimum mean squared error	391
MS–ARMA	Markov-switching ARMA	67
MSE	mean square error	129
NAIC	normalized AIC	211
NBER	National Bureau of Economic Research	4
NC(S)TAR	neuro-coefficient (S)TAR	65
NEAR	newer exponential AR	53
NFE	normal forecast error	401
NLARMA	nonlinear ARMA	101
NLS	nonlinear least squares	198
NW	Nadaraya–Watson	302
ODP	Ocean Drilling Program	8
PACF	partial autocorrelation function	14
PAR	product AR	54
pdf	probability density function	18
PEE	parameter estimation error	427
PI	plug-in	396
PIT	probability integral transform	305
pmf	probability mass function	326
PPR	projection pursuit regression	363
QML	quasi maximum likelihood	198
RMAFE	relative mean absolute forecast error	403
RCAR(MA)	random coefficient AR(MA)	39
(R)MSFE	(root) mean squared forecast error	72
RNW	re-weighted NW	350
rot	rule-of-thumb	301
SCMI	shortest conditional modal interval	413
SDM	state-dependent model	32
SK	skeleton	399

SRE	stochastic recurrence equation	88
SST	sea surface temperature	7
STAR	smooth transition autoregressive	51
TEAR	transposed exponential AR	54
TFN	transfer function noise	245
TI	traditional impulse	76
TR	time-reversibility	315
TSMARS	time series MARS	365
uc	unconditional coverage	419
VARMA	vector autoregressive moving average	440
VEC	vector error correction	452
VSTAR	vector STAR	453
WN	white noise	1
WS	wind speed	506

List of Pseudocode Algorithms

Page numbers are in parentheses

CHAPTER 3

3.1 Empirical invertibility of an NLARMA(p,q) model (105)

CHAPTER 4

4.1 The Subba Rao–Gabr Gaussianity test (126)
4.2 The Subba Rao–Gabr linearity test (129)
4.3 Goodness-of-fit test statistics (135)
4.4 Bootstrap-based tests (138)
4.5 The MSFE-based linearity test statistic (144)

CHAPTER 5

5.1 $\mathrm{LM}_T^{(3^*)}$ test statistic (161)
5.2 $\mathrm{LM}_T^{(3^{**})}$ test statistic (162)
5.3 $F_T^{(5)}$ test statistic (164)
5.4 $\mathrm{LM}_T^{(7)}$ test statistic (168)
5.5 Bootstrapping p-values of $F_T^{(1,i)}$ test statistic (172)
5.6 Bootstrapping p-values of $\mathrm{LR}_T^{(9)}$ test statistic (176)
5.7 Tukey's nonadditivity-type test statistic (180)
5.8 $F_T^{(\mathrm{O})}$ test statistic (181)
5.9 CUSUM test statistic (183)
5.10 TAR F test statistic (184)
5.11 New F test statistic (185)

CHAPTER 6

6.1 Nonlinear iterative optimization (200)
6.2 A multi-parameter grid search (204)
6.3 The density function of M_- (206)
6.4 Sampling Y_1 from an estimate of $F_1(\cdot|r_0)$ (207)
6.5 k-regime subset SETARMA–CLS estimation (211)
6.6 A simple genetic algorithm (211)
6.7 CLS estimation of the BL model (218)
6.8 Minimum order selection (233)
6.9 Leave-one-out CV order selection (234)
6.10 Selecting a (SS)TARSO model (244)

CHAPTER 7

7.1 Bootstrapped p-values for single-lag tests (276)
7.2 Permutation-based p-values for multiple-lag tests (277)
7.3 Bootstrapping p-values of the BDS test statistic (281)
7.4 Bootstrap-based p-values for multivariate serial independence tests (289)

CHAPTER 8

8.1 The Ramsey–Rothman TR test (319)
8.2 The bispectrum-based TR test (322)
8.3 The trispectrum-based TR test (324)
8.4 Resampling scheme (326)

CHAPTER 9

9.1 Loess/Lowess (353)
9.2 Robust Loess/Lowess (354)
9.3 Resampling scheme for MFDs (357)
9.4 ACE (361)
9.5 Gradient descent boost (371)
9.6 Bootstrap-based LR-type test (376)
9.7 Estimating $\boldsymbol{\theta}$ and h_T for the single-index model (379)

CHAPTER 10

10.1 Bootstrap FI (410)
10.2 Bootstrap bias-corrected FI (411)

CHAPTER 11

11.1 A nonadditivity-type test for nonlinearity (459)
11.2 Tukey's nonadditivity-type test for nonlinearity (460)
11.3 $F_T^{(O)}$ test statistic for nonlinearity (461)
11.4 Multivarite test statistic for VSETAR (464)
11.5 $\mathrm{LM}_{T,p}^{(1)}(m)$ test statistic for LVSTAR (469)
11.6 Bootstrapping the GIRF (489)

CHAPTER 12

12.1 Bootstrap-based p-values for LR_T (509)

List of Examples

Page numbers are in parentheses

CHAPTER 1

1.1 U.S. Unemployment Rate (4)
1.2 EEG Recordings (5)
1.3 Magnetic Field Data (6)
1.4 ENSO Phenomenon (7)
1.5 Climate Change (8)
1.6 Summary Statistics (11)
1.7 Summary Statistics (Cont'd) (14)
1.8 Sample ACF and Kendall's τ (17)
1.9 The Logistic Map (20)
1.10 EEG Recordings (Cont'd) (22)

CHAPTER 2

2.1 A BL Time Series (33)
2.2 Comparing BL Time Series (35)
2.3 Dynamic Effects of a BL Model (36)
2.4 ExpAR Time Series (38)
2.5 Dynamic Effects of an NLMA Model (40)
2.6 Dynamic Effects of a SETAR Model (42)
2.7 A Simulated CSETAR Process (45)
2.8 A Simulated SETAR$(2;1,1)_2$ Model (46)
2.9 Dynamic Effects of an asMA Model (48)
2.10 NEAR(1) Model (53)
2.11 Skeleton of an AR–NN$(2;0,1)$ Model (59)
2.12 Skeleton of an AR–NN$(3;1,1,1)$ Model (60)
2.13 A Simulated L^{2}GNN$(2;1,1)$ Time Series (63)
2.14 A Two-regime Simulated MS–AR(1) Time Series (67)
A.1 Impulse Response Analysis (78)

CHAPTER 3

3.1 Evaluating the Top Lyapunov Exponent (89)
3.2 An Explicit Expression for γ (92)
3.3 Numerical Evaluation of γ (93)
3.4 Geometric Ergodicity of the SRE (97)
3.5 SETAR Geometric Ergodicity (99)
3.6 Invertibility of an RCMA(1) Model (104)
3.7 Invertibility of an ASTMA(1) Model (105)
3.8 Invertibility of a SETMA Model 108)

CHAPTER 4

4.1 Third-order Cumulant and Bispectrum (124)
4.2 Principal Domain of the Subba Rao–Gabr Gaussianity Test (127)

CHAPTER 5

5.1 ENSO Phenomenon (Cont'd) (173)
5.2 U.S. Unemployment Rate (Cont'd) (177)
5.3 Interpretation of the LM_T^* Test Statistic (186)

CHAPTER 6

6.1 NLS Estimation (201)
6.2 U.S. Unemployment Rate (Cont'd) (208)
6.3 U.S. Real GNP (212)
6.4 ENSO Phenomenon (Cont'd) (215)
6.5 CLS-based Estimation of a BL Model (221)
6.6 Daily Hong Kong Hang Seng Index (225)
6.7 U.S. Unemployment Rate (Cont'd) (235)
6.8 Daily Hong Kong Hang Seng Index (Cont'd) (239)

CHAPTER 7

7.1 Some Kernel Functions and their FTs (261)
7.2 An Explicit Expression for $\Delta^Q(\cdot)$
7.3 Magnetic Field Data (Cont'd) (273)
7.4 U.S. Unemployment Rate (Cont'd) (276)
7.5 Dimension of an ExpAR(1) Process (280)
7.6 S&P 500 Daily Stock Price Index (283)
A.1 NW Kernel Regression Estimation (303)
B.1 Gaussian and Student t copulas (307)

CHAPTER 8

8.1 Exploring a Logistic Map for TR (317)
8.2 Exploring a Simulated SETAR Process for TR (321)
8.3 Exploring a Time-delayed Hénon Map for TR (329)

CHAPTER 9

9.1 A Comparison Between Conditional Quantiles (345)
9.2 Old Faithful Geyser (347)
9.3 Hourly River Flow Data (354)
9.4 Canadian Lynx Data (Cont'd) (359)
9.5 Sea Surface Temperatures (362)
9.6 Sea Surface Temperatures (Cont'd) (364)
9.7 Sea Surface Temperatures (Cont'd) (368)
9.8 Quarterly U.S. Unemployment Rate (Cont'd) (372)
9.9 Quarterly U.S. Unemployment Rate (Cont'd) (376)
9.10 A Monte Carlo Simulation Experiment (379)

CHAPTER 10

10.1 Forecast Density (393)
10.2 Comparing LS and PI Forecast Strategies (396)
10.3 Comparing NFE and MC Forecasts (403)
10.4 Forecasts from an ExpAR(1) Model (405)
10.5 Forecasts from a SETAR(2; 1, 1) Model (407)
10.6 FIs for a Simulated SETAR Process (412)
10.7 Hourly River Flow Data (Cont'd) (414)
10.8 ENSO Phenomenon (Cont'd) (422)

CHAPTER 11

11.1 Stationarity and Invertibility of a Bivariate BL Model (445)
11.2 A Two-regime Bivariate VSETAR(2; 1, 1) Model (450)
11.3 An LVSTAR Model with Nonlinear Cointegration (456)
11.4 An LVSTAR Model with a single CNF (458)
11.5 Tree Ring Widths (463)
11.6 Tree Ring Widths (Cont'd) (470)
11.7 Forecasting an LVSTAR(1) Model with CNFs (477)

List of Examples

CHAPTER 12

12.1 A Monte Carlo Experiment (497)
12.2 Daily Returns of Exchange Rates (499)
12.3 Sea Surface Temperatures (Cont'd) (504)
12.4 Sea Surface Temperatures (Cont'd) (509)
12.5 Climate Change (Cont'd) (513)
12.6 Climate Change (Cont'd) (518)

Table 3: *Time series used throughout the book. File names are given in parentheses.*

Series	Example	Exercise	Application (Section number)
U.S. unemployment rate[1] (USunemplmnt_first_dif.dat)	1.1, 1.6, 1.7, 1.8, 5.2, 7.4	4.4, 8.6	4.7, 8.5
U.S. unemployment rate[2] (USunemplmnt_logistic.dat)	6.2, 6.7, 9.8, 9.9	2.10, 6.9	
EEG recordings (eeg.dat)	1.2, 1.6, 1.7, 1.8, 1.10	2.9, 7.5, 8.6	2.11, 4.7, 8.5
Magnetic field (magnetic_field.dat)	1.3, 1.6, 1.7, 1.8, 7.3	8.6	4.7, 8.5
ENSO phenomenon (ENSO.dat)	1.4, 1.6, 1.7, 1.8, 5.1, 6.4, 10.8	8.6	4.7, 8.5
Climate change (deltaC.dat and deltaO.dat) (earthP1.dat - earthP4.dat)	1.5, 1.6, 1.7, 1.8, 7.7, 12.5, 12.6	1.6, 1.8, 8.6, 12.3	4.7, 8.5
Jökulsá Eystri streamflow (jokulsa.dat)		2.11	
Icelandic river flow (ice.dat)		12.4	11.8
West German unemployment (German_unemplmnt.dat)		3.8	
U.S. real GNP (USGNP.dat)	6.3		
Hong Kong Hang Seng Index (HSI_returns)	6.6, 6.8		
Water table depth (WaterT_Precip.dat)			6.4
S&P 500 stock price index (SP500.dat)	7.6		
Canadian lynx (lynx.dat)	9.4	7.6, 7.7	7.5
Old Faithful geyser (geyser_waiting.dat)	9.2	9.4	
Hourly river flow (flow.dat and rain.dat)	9.3, 10.7	9.5, 10.11	
Sea surface temperatures (SST.dat and SSTGranite.dat)	9.5, 9.6, 9.7, 12.3, 12.4		
Great Salt Lake volume (gsl.dat)		9.2	
Intraday transaction (intraday.dat)		11.5	
Tree ring widths (treering.dat)	11.5, 11.6	12.2	
U.S. consumption-income (con_inc.dat)		11.6	
Exchange rates (ExchangeRates.dat)	12.2		

[1] First differences of original data.
[2] Logistic transformation of original data.

Subject index

ACE algorithm, 360, 361
Added variable approach, 179
Akaike's information criterion
    AIC, 69, 208, 216, 228–232, 235, 246
    AIC_c, 230
    AIC_u, 230
    multivariate, 471
    NAIC, 211
Anderson–Darling GOF test, 134
Anosov diffeomorphism, 335
Aperiodic, 66
Arranged autoregression, 182
Artificial neural network (ANN), 56–58
    activation-level, 56
    AR–NN, 58, 59
    ARMA–NN, 61, 62
    back-propagation, 58
    bias, 58
    hidden unit, 56
    L^2GNN, 63, 64
    LGNN, 62, 63
    multi-layer perceptron (MLP), 56
    NCSTAR, 65
    neurons, 56
    shortcut connections, 58
    skip-layer, 57
    training, 57
Asymmetric ARMA (asARMA)
    model, 47
Asymmetry, 4, 10
Asymptotically stationary, 59
Augmented F test, 181
Autocorrelation function (ACF), 14
Autocovariance function (ACVF), 12, 141, 218, 417
AVAS algorithm, 362

Backward shift operator, 62
Bandwidth, 298
    oversmoothing, 328
    plug-in, 301, 341
    rule-of-thumb (rot), 301, 302, 358, 359
    undersmoothing, 328
Bartlett's confidence limits, 15, 451
Base learner, 370
Basis functions, 366
Bayesian information criterion
    (BIC), 69, 231, 243
    multivariate, 471
BDS test statistic, 278
    rank-based, 282
Beta-Gamma transformation, 335
Bilinear model
    multivariate, 441
    super (sub) diagonal, 34
    univariate, 33, 35, 36, 216
Binwidth, 270
Bispectral density function, 121
Bispectrum, *see* Bispectral density function
Boosting, 369
    componentwise, 371
    gradient descent, 370
    greedy, 370
Bootstrapping, 136
    backward (forward), 410
Boundary effects, 269
BRUTO, 372

Calibration, 234, 243, 244
Causality test
    bivariate, 515
        modified, 516
    Hiemstra–Jones (HJ), 515

multivariate, 518
Causally invertible, 30
Cave plot, 9
Chapman–Kolmogorov relationship, 392
Check function, 342, 499
Cholesky decomposition, 478, 490
Cointegration, 452, 456
Common features, 455
    nonlinear (CNF), 457
Commutative, *see* Exchangeable
Companion matrix, 42, 108, 113
Complexity penalty, 232
Compound Poisson process (CPP), 205
Concordant, 15
Conditional least squares (CLS), 44, 198, 202, 210, 214, 217, 218, 221, 234, 244
Conditional mean, 339
Conditional median, 339
Conditional mode, 340
Conditional percentile interval (CPI), 408, 413
Conditional quantile predictor
    multi-stage, 344, 346, 347
    single-stage, 342, 344, 346, 347
Copula, 259, 266
    density, 267, 306
    empirical, 269
    Fréchet–Hoeffding bounds, 307
    Gaussian, 307
    independence, 267
    Student t, 307
Correlation dimension, 280
Covariance matrix, 90
Coverage
    conditional, 420
    unconditional, 419
Coverage rate (CVR), 412
Cramér–von Mises (CvM) GOF test, 134
Cross-correlation function (CCF), 237, 450
Cross-validation (CV), 234
    generalized (GCV), 367, 504
Crossover, 212
Cumulants, 25
    third-order, 120
Cumulative sums (CUSUM) test, 183
Curse of dimensionality, 250, 338
Cut-off threshold, 260

Data generating process (DGP), 4
Data-sharpening, 520

Delay parameter, 42
Dependogram, 290
Descriptive statistics, 10
Design adaptive, 350
Designated frequency, 126, 128
Detailed balance equations, 317
Diagnostic checking, 236, 472
Diebold–Mariano (DM) test, 416, 417, 424
    modified (MDM), 417, 418
Direct method, 123
Directed scatter plot, 21
Disconcordant, *see* Concordant
Distance
    Anderson–Darling (AD), 266
    correlation integral, 260
    Cramér–von Mises (CvM), 265
    Cressie–Read (CR), 265
    Csiszár (C), 264
    functionals, 263
    Hellinger (H), 264
    Kolmogorov (K), 264
    Kolmogorov–Smirnov (KS), 266
    Kullback–Leibler (KL), 18, 227
    quadratic (Q), 260
    Rényi (R), 264
    Tsallis (T), 264
Doubly stochastic, 39
Duration, 421

Embedding dimension, 19
Equilibrium error process, 452
Ergodic, 66, 97
Error correction model (ECM), 216
Essentially linear, 3
Euler's constant, 89, 115
Exchangeable, 317
Exponential AR (EAR) model, 54
Exponential ARMA (ExpARMA) model, 36, 51
Exponential function, 51

Feed-forward network, 56
Feller chain, 97
Final prediction error
    AFPE, 358
    CAFPE, 359
    FPE, 358
Forecast
    interval (FI), 408

SUBJECT INDEX

linear (L), 140
quadratic (Q), 141
region (FR), 408
Forecast combination
   density forecasts, 426
   interval forecasts, 425
   point forecasts, 425
Forecast evaluation
   density forecast, 422
   interval forecast, 419
   point forecast, 415
   vector
      density, 479
      GFESM, 479
      RMSFE, 478
Forecasting
   bootstrap (BS), 399
   combined (C), 396
   dynamic estimation (DE), 406
   empirical least squares (ELS), 400
   encompassing, 427
   exact, 392
   least squares (LS), 395
   linearization (LN), 404
   Monte Carlo (MC), 398
   normal forecasting error (NFE), 401
   plug-in (PI), 396
   recursive, 416, 427
   rolling, 416, 427
   SETARMA, 394
   skeleton (SK), 399
Fourier transform (FT), 120, 126
Frequency bicoherence, 123
Functional-coefficient AR (FCAR) model, 374

Gaussian mixture AR (MAR) model, 313
Generalized impulse response function (GIRF), 36
Generalized information criterion (GIC), 231
Generalized spectrum, 274
Genetic algorithm (GA), 210
   fitness function, 210
Geometric ergodicity, 81, 95, 96
Goodness-of-fit (GOF) test, 133
Gradient vector, 200–202
Granger's causality index (GCI), 451
Grid search, 69

Hénon map, 330
Hamilton filter, 68

Hankel matrix, 219
Hessian matrix, 179, 199, 200
Hidden unit, 58
Highest (conditional) density region (HDR), 414
Hinich's tests, 130, 131, 133, 136
Hotelling–Lawley (HL) trace test, 462
Hyperplane, 46

Impulse response function, *see* Generalized impulse response function (GIRF)
Indirect method, 123
Information matrix, 199, 224, 232, 241
Innovation process, 31, 141
Integrated squared error (ISE), 299
Interdecile range (IDR), 136
Interquartile range (IQR), 132, 136
Intrinsically linear, 55
Invariance, 306
Inversion method, 306
Invertibility, 101, 109
   classical, 101
   empirical, 105
   global, 101
      generalized, 102
      Granger–Andersen, 101
      Pham–Tran, 103
   local, 107
Irreducible, 66
Iteratively weighted least squares (IWLS), 223, 224

Jarque–Bera (JB) test
   generalized (GJB), 12
   independent data, 10
   weakly dependent data, 12
Jensen's inequality, 98, 227
Jittering, 290
Joint entropy, 18

Kendall's (partial) tau, 14, 15, 17
Kernel functions, 298
   biweight, 299
   Cauchy, 261
   Epanechnikov, 299
   Gaussian, 261, 299
   triweight, 299
   uniform, 299
Kolmogorov–Gabor polynomial, 31

Kurtosis, 10

Lag selection, 512
Lag window
    Daniell, 273
    Parzen, 15, 124
    right-pyramidal, 139
    trapezoid, 138
Lagrange multiplier (LM) type tests
    AsMA and SETMA models, 163
    ASTMA model, 165
    bilinear model, 157
    ExpARMA model, 159
    general, 156
    NCTAR and AR-NN models, 166
    STAR model, 159
        augmented first-order, 162
        first-order procedure, 160
        third-order procedure, 161
    VSTAR model, 468
Leakage, 140
Leave-one-out CV, 234, 304
Lebesgue measure, 98, 338, 414
Likelihood ratio (LR) tests
    NeSETAR model, 171
    SETAR model, 168
    SETARMA model, 174
    VSETAR model, 465
Limit cycle, 37
Lin–Mudholkar test, 11
Linear causal, 3
Linear forecast, 140
Linear process, 2
Linear single-index model, 378
Lipschitz continuous, 340
Ljung–Box (LB) statistic, 177, 209, 236
Local linear (LL)
    conditional density
        asymptotic bias, 350
        asymptotic variance, 350
    conditional mean, 375
        asymptotic bias, 409
        asymptotic variance, 409
Logistic function, 51
Logistic map, 20
Logistic smooth transition error correction (LSTEC), 215
Lyapunov exponent, 88
    NLAR–GARCH model, 91

Möbius transformation, 285, 286, 288
Markov chain, 66
    collapsed, 92
    Monte Carlo (MCMC), 210, 249, 305
Markov-switching (MS–ARMA) model, 67
Martingale difference, 2
Maximal test, 136
Mean absolute forecast error (MAFE), 72
Mean integrated squared error (MISE), 300
Mean squared error (MSE), 129, 299
Mean squared forecast error (MSFE), 141, 144
Minimum descriptive length (MDL), 232
Mixing, 95
    α-mixing, 95
    β-mixing, 96
Mixing coefficient, 95
Mixing proportions, 313
Multiple-lag tests, 272
Multivariate adaptive regression splines (MARS), 365
Multivariate quantile, 496
Mutation, 212
Mutual information, 18

Nadaraya–Watson (NW), 302
    conditional density
        asymptotic bias, 349
        asymptotic variance, 349
    conditional mean
        asymptotic bias, 409
        asymptotic variance, 409
    kernel estimator
        re-weighted (RNW), 350
Newer exponential AR (NEAR) model, 53
Newton–Raphson method, 219
Non-anticipative, 89
Nonadditivity-type test
    multivariate
        original F test, 461
        Rao's (R), 459
        Tukey (T), 460
    univariate
        Tukey (T), 179
Nonlinear, 4
Nonlinear ARMA (NLARMA) model, 39, 101
Nonparametric regression
    K-nearest neighbor (k-NN), 352, 501
    local polynomial, 304

SUBJECT INDEX

loess/lowess, 353
projection pursuit regression (PPR), 363, 504
Normality, 10
Normalized bispectrum, 122

Occam's razor, 187

Parameter estimation error (PEE), 427
Parseval's identity, 262
Partial autocorrelation function (PACF), 14
Pearson residuals, 236, 237, 240, 241
Penalty function, 232
Periodic function, 37
Permutation test, 277
Phase space, 19
Pillai's (P) trace test statistic, 462
Poisson equation, 93
PolyMARS (PMARS), 502
Polyspectrum, 121
Portmanteau-type test, 179, 266, 474
Prediction, see Forecasting
Predictive residuals, 412
Principal domain, 121, 126, 128, 131
Probability integral transform (PIT), 241, 422, 479
Product AR (PAR) model, 54, 55
Product kernel, 339

Quantile residuals, 240, 474
Quasi maximum likelihood (QML), 68, 198, 199

Random coefficient AR (RCAR) model, 39
    generalized, 88
Reconstruction errors, 101
Reconstruction vector, 19
Recurrence plot, 19
Recurrent, 61, 62
Recursive partitioning, 366
    backward step, 366
    forward step, 366
Root mean squared forecast error (RMSFE), 72
Roughness, 300

Score vector, see Gradient vector
Selection, 212
Self-exciting, 41
Semi-invariants, see Cumulants
Sensitivity parameter, 358

Shannon entropy, 18, 525
Shortest conditional modal interval (SCMI), 413, 414
Sigma-field, 77
Sign AR model, 311
Single-index coefficient model, 378
Single-lag tests, 270
Skeleton, 59, 61, 202
Skewness, 10, 84
Sklar's theorem, 306
Smooth transition (ST) model, 51
    ASTMA, 52
    cointegration, 456
    ESTAR, 51
    LSTAR, 51
    LVSTAR, 454
    STAR, 51
    VSTAR, 453
Spectral density function, 120
Spectral distribution function, 274
Spectral matrix, 511
Spectral radius, 90, 114, 448, 455
Spectrum, see Spectral density function
Squared tricoherence, 323
State space, 32
State vector, see Reconstruction vector
State-dependent model (SDM)
    multivariate, 440
    univariate, 32
Stochastic permutation, 175
Stochastic recurrence equation (SRE), 88
Subba Rao–Gabr tests, 126
Surrogate data, 188
Switching mechanism, 41
Symmetric-bicovariance function, 318
Szegö condition, 141

Third-order periodogram, 123, 124, 142
Threshold, 41
Threshold model, 41, 45
    TARMA, 41
    CSETAR, 44, 45
    NeSETARMA, 49, 50
    SETARMA, 42
    SSTARSO, 242
    TAR, 78
    TARSO, 50, 242
    VASTAR(X), 502
    VSETAR, 447

VTARMA, 446
Time-irreversibility
    Type I, 318
    Type II, 318
Time-reversible, 6
Tolerance distance, *see* Cut-off threshold
Traditional impulse (TI) response function, 76
Transfer function, 123
Transition function, 51
Transition probability matrix, 66
Transposed EAR (TEAR) model, 54
Triangle inequality, 112
Trispectrum, 323
Truncation point, 124
Tsay's test statistics
    new F test, 185
    original F test, 180
    TAR F test, 184

VSETAR F test, 464

U-statistic, 308
Unit root, 189, 361, 464

V-statistic, 308
Validation, 234, 243
Vector error correction (VEC) model, 452
Vector smooth transition error correction (VSTEC), 455
Volterra, 30, 31, 179, 522

Wald (W) test
    asARMA model, 178
Weak learner, *see* Base learner
White noise (WN)
    conditional, 3
    Gaussian, 3
    strict, 3
    weak, 2